KIRK-OTHMER

ENCYCLOPEDIA OF CHEMICAL TECHNOLOGY

Third Edition

VOLUME 5

**Castor Oil
to
Chlorosulfuric Acid**

EDITORIAL BOARD

HERMAN F. MARK
Polytechnic Institute of New York

DONALD F. OTHMER
Polytechnic Institute of New York

CHARLES G. OVERBERGER
University of Michigan

GLENN T. SEABORG
University of California, Berkeley

EXECUTIVE EDITOR

MARTIN GRAYSON

ASSOCIATE EDITOR

DAVID ECKROTH

KIRK-OTHMER

ENCYCLOPEDIA OF CHEMICAL TECHNOLOGY

THIRD EDITION

VOLUME 5

CASTOR OIL
TO
CHLOROSULFURIC ACID

A WILEY-INTERSCIENCE PUBLICATION

John Wiley & Sons

NEW YORK • CHICHESTER • BRISBANE • TORONTO

Copyright © 1979 by John Wiley & Sons, Inc.

All rights reserved. Published simultaneously in Canada.

Reproduction or translation of any part of this work
beyond that permitted by Sections 107 or 108 of the
1976 United States Copyright Act without the permission
of the copyright owner is unlawful. Requests for
permission or further information should be addressed to
the Permissions Department, John Wiley & Sons, Inc.

Library of Congress Cataloging in Publication Data:

Main entry under title:
    Encyclopedia of chemical technology.

    At head of title: Kirk-Othmer.
    "A Wiley-Interscience publication."
    Includes bibliographies.
    1. Chemistry, Technical—Dictionaries. I. Kirk,
Raymond Eller, 1890–1957. II. Othmer, Donald Frederick,
1904–   III. Grayson, Martin. IV. Eckroth, David.
V. Title: Kirk-Othmer encyclopedia of chemical technology.

TP9.E685   1978   660'.03   77-15820
ISBN 0-471-02041-9

Printed in the United States of America

CONTENTS

Castor oil, 1
Catalysis, 16
Catalysis, phase-transfer, 62
Cellulose, 70
Cellulose acetate and triacetate fibers, 89
Cellulose derivatives, esters, 118
Cellulose derivatives, ethers, 143
Cement, 163
Centrifugal separation, 194
Ceramics, 234
Ceramics as electrical materials, 290
Cerium and cerium compounds, 315
Cesium and cesium compounds, 327
Chelating agents, 339
Chemical grouts, 368
Chemicals from brine, 375
Chemicals in war, 393

Chemiluminescence, 416
Chemotherapeutics, anthelmintic, 451
Chemotherapeutics, antimitotic, 469
Chemotherapeutics, antimycotic and antirickettsial, 489
Chemotherapeutics, antiprotozoal, 513
Chemotherapeutics, antiviral, 542
Chemurgy, 553
Chloramines and bromamines, 565
Chlorine oxygen acids and salts, 580
Chlorocarbons and chlorohydrocarbons, 668
Chlorohydrins, 848
Chlorophenols, 864
Chlorosulfuric acid, 873

EDITORIAL STAFF FOR VOLUME 5

Executive Editor: **Martin Grayson**
Associate Editor: **David Eckroth**
Production Supervisor: **Michalina Bickford**
Editors: **Galen J. Bushey**   **Loretta Campbell**   **Caroline L. Eastman**
         Anna Klingsberg   **Lorraine van Nes**

CONTRIBUTORS TO VOLUME 5

R. C. Ahlstrom, Jr., *Dow Chemical U.S.A., Freeport, Texas,* Methyl chloride under Chlorocarbons and chlorohydrocarbons

T. Anthony, *Dow Chemical U.S.A., Midland, Michigan,* Methylene chloride under Chlorocarbons and chlorohydrocarbons

Wesley L. Archer, *Dow Chemical U.S.A., Midland, Michigan,* Survey; Other chloroethanes both under Chlorocarbons and chlorohydrocarbons

O. A. Battista, *Research Services Corp., Fort Worth, Texas,* Cellulose

R. T. Bogan, *Tennessee Eastman Co., Kingsport, Tennessee,* Cellulose derivatives, esters

Henry I. Bolker, *Pulp and Paper Research Institute of Canada, Pointe Claire, Canada,* Cellulose

H. Kent Bowen, *Massachusetts Institute of Technology, Cambridge, Massachusetts,* Ceramics as electrical materials

R. J. Brewer, *Tennessee Eastman Company, Kingsport, Tennessee,* Cellulose derivatives, esters

H. O. Burrus, *E. I. du Pont de Nemours & Co., Inc., Wilmington, Delaware,* Chlorosulfuric acid

CONTRIBUTORS TO VOLUME 5

T. W. Clapper, Kerr-McGee Corporation, Oklahoma City, Oklahoma, Chloric acid and chlorates under Chlorine oxygen acids and salts

J. Peter Clark, Virginia Polytechnic Institute and State University, Blacksburg, Virginia, Chemurgy

James G. Colson, Hooker Chemicals & Plastics Corp., Grand Island, New York, Benzene hexachloride under Chlorocarbons and chlorohydrocarbons

J. Ross Colvin, National Research Council, Ottawa, Canada, Cellulose

Murray C. Cooperman, NL Industries, Inc., Hightstown, New Jersey, Castor oil

James A. Cusumano, Catalytica Associates, Inc., Palo Alto, California, Catalysis

Robert F. Davis, North Carolina State University, Raleigh, North Carolina, Scope under Ceramics

Aldo DeBenedictis, Shell Chemical Company, Houston, Texas, Allyl chloride under Chlorocarbons and chlorohydrocarbons

Eckehard V. Dehmlow, Technischen Universitat Berlin, Berlin, Germany, Catalysis, phase-transfer

H. D. DeShon, Dow Chemical U.S.A., Midland, Michigan, Carbon tetrachloride; Chloroform both under Chlorocarbons and chlorohydrocarbons

Richard L. Doerr, Olin Corporation, New Haven, Connecticut, Chlorine dioxide, chlorous acid, and chlorites under Chlorine oxygen acids and salts

Hans Dressler, Koppers Company, Monroeville, Pennsylvania, Chlorinated naphthalenes under Chlorocarbons and chlorohydrocarbons

Frank Duneczky, NL Industries, Inc., Hightstown, New Jersey, Castor oil

Donald F. Durso, Johnson & Johnson, New Brunswick, New Jersey, Cellulose

Nathan Eastman, Union Carbide Corporation, Charleston, West Virginia, Cellulose

E. R. Freiter, Dow Chemical U.S.A., Midland, Michigan, Chlorophenols

Samuel Gelfand, Hooker Chemicals & Plastics Corp., Grand Island, New York, Benzyl chloride, benzal chloride, benzotrichloride; Ring-chlorinated toluenes both under Chlorocarbons and chlorohydrocarbons

George E. Gifford, University of Florida, Gainesville, Florida, Chemotherapeutics, antiviral

N. R. Greening, Portland Cement Association, Skokie, Illinois, Cement

G. K. Greminger, Jr., Dow Chemical U.S.A., Midland, Michigan, Cellulose derivatives, ethers

John V. Hamme, North Carolina State University, Raleigh, North Carolina, Raw materials under Ceramics

Thomas M. Hare, North Carolina State University, Raleigh, North Carolina, Thermal treatment; Properties both under Ceramics

B. L. Harris, Edgewood Arsenal, Aberdeen, Maryland, Chemicals in war

Roger E. Hatton, Monsanto Company, St. Louis, Missouri, Chlorinated biphenyls and related compounds under Chlorocarbons and chlorohydrocarbons

Richard A. Helmuth, Portland Cement Association, Skokie, Illinois, Cement

William L. Howard, Dow Chemical U.S.A., Freeport, Texas, Chelating agents

James W. Ingalls, Arnold & Marie Schwartz College of Pharmacy and Health Sciences, Brooklyn, New York, Chemotherapeutics, anthelmintic

Paul R. Johnson, E. I. du Pont de Nemours & Co., Inc., Wilmington, Delaware, Chloroprene under Chlorocarbons and chlorohydrocarbons

Che-I-Kao, Dow Chemical U.S.A., Midland, Michigan, Chlorinated benzenes under Chlorocarbons and chlorohydrocarbons

Reuben H. Karol, Rutgers University, New Brunswick, New Jersey, Chemical grouts

S. L. Keil, Dow Chemical U.S.A., Midland, Michigan, Tetrachloroethylene under Chlorocarbons and chlorohydrocarbons

CONTRIBUTORS TO VOLUME 5

F. W. Keith, Jr., *Pennwalt Corp., Warminster, Pennsylvania,* Centrifugal separation
Theodor N. Kleinert, *Pulp and Paper Research Institute of Canada, Pointe Claire, Canada,* Cellulose
Hans Krassig, *Lenzing, A. G., Oberosterreich, Austria,* Cellulose
Thaddeus Kroplinski, *NL Industries, Hightstown, New Jersey,* Castor oil
C. M. Kuo, *Tennessee Eastman Company, Kingsport, Tennessee,* Cellulose derivatives, esters
A. C. Lavanchy, *Pennwalt Corporation, Warminster, Pennsylvania,* Centrifugal separation
R. St. John Manley, *McGill University, Montreal, Canada,* Cellulose
Edgar J. Martin, *Food and Drug Administration, Rockville, Maryland,* Chemotherapeutics, antiprotozoal
Charles J. Masur, *Lederle Laboratories, American Cyanamid Company, Pearl River, New York,* Chemotherapeutics, antimitotic
C. Glen Mayhall, *Medical College of Virginia, Richmond, Virginia,* Chemotherapeutics, antimycotic and antirickettsial
A. L. McCrary, *Dow Chemical U.S.A., Freeport, Texas,* Chelating agents
W. C. McNeill, Jr., *Dow Chemical U.S.A., Midland, Michigan,* Trichloroethylene under Chlorocarbons and chlorohydrocarbons
F. M. Miller, *Portland Cement Association, Skokie, Illinois,* Cement
G. Alex Mills, *ERDA, Washington, D. C.,* Catalysis
Thomas E. Morris, *Dow Chemical U.S.A., Freeport, Texas,* Ethyl chloride under Chlorocarbons and chlorohydrocarbons
Paul E. Muehlberg, *Dow Chemical U.S.A., Freeport, Texas,* Chemicals from brine
Frank Naughton, *NL Industries, Hightstown, New Jersey,* Castor oil
G. D. Nelson, *Monsanto Industrial Chemicals Co., St. Louis, Missouri,* Chloramines and bromamines
Manfred G. Noack, *Olin Corporation, New Haven, Connecticut,* Chlorine dioxide, chlorous acid, and chlorites under Chlorine oxygen acids and salts
T. R. O'Connor, *Portland Cement Association, Skokie, Illinois,* Cement
William Pearl, *Lederle Laboratories, American Cyanamid Company, Pearl River, New York,* Chemotherapeutics, antimitotic
Noland Poffenberger, *Dow Chemical U.S.A., Midland, Michigan,* Chlorinated benzenes under Chlorocarbons and chlorohydrocarbons
Bing T. Poon, *Walter Reed Army Institute of Research, Washington, D. C.,* Chemotherapeutics, antiprotozoal
Michael M. Rauhut, *American Cyanamid Co., Bound Brook, New Jersey,* Chemiluminescence
R. C. Rhees, *Pacific Engineering and Production Co., Henderson, Nevada,* Perchloric acid and perchlorates under Chlorine oxygen acids and salts
Gregor H. Riesser, *Shell Development Company, Houston, Texas,* Chlorohydrins
J. R. Sanders, *Celanese Fibers Company, Charlotte, North Carolina,* Cellulose acetate and triacetate fibers
B. A. Schenker, *Diamond Shamrock Corp., Painesville, Ohio,* Chlorinated paraffins under Chlorocarbons and chlorohydrocarbons
George A. Serad, *Celanese Fibers Company, Charlotte, North Carolina,* Cellulose acetate and triacetate fibers
Smith Shadomy, *Medical College of Virginia, Richmond, Virginia,* Chemotherapeutics, antimycotic and antirickettsial

CONTRIBUTORS TO VOLUME 5

F. Shanty, *Edgewood Arsenal, Aberdeen, Maryland,* Chemicals in war

Walter L. Silvernail, *Consultant, West Chicago, Illinois,* Cerium and cerium compounds

J. M. Steele, *Dow Chemical U.S.A., Freeport, Texas,* Methyl chloride under Chlorocarbons and chlorohydrocarbons

James E. Stevens, *Hooker Chemicals & Plastics Corp., Grand Island, New York,* Chlorinated derivatives of cyclopentadiene under Chlorocarbons and chlorohydrocarbons

Violete L. Stevens, *Dow Chemical Company, Midland, Michigan,* Dichloroethylenes under Chlorocarbons and chlorohydrocarbons

Robert F. Stoops, *North Carolina State University, Raleigh, North Carolina,* Forming processes under Ceramics

C. Richard Swenson, *NL Industries, Inc., Hightstown, New Jersey,* Castor oil

William Tasto, *Dow Chemical U.S.A., Freeport, Texas,* Ethyl chloride under Chlorocarbons and chlorohydrocarbons

Bruce Tippin, *Great Salt Lake Minerals and Chemicals Corp., Ogden, Utah,* Chemicals from brine

J. R. Tonry, *Portland Cement Association, Skokie, Illinois,* Cement

Albin F. Turbak, *ITT Rayonier Inc., Whippany, New Jersey,* Cellulose

Joseph Welsh, *Hayward Baker Co., Cherry Hill, New Jersey,* Chemical grouts

C. T. Williams, *Tantalum Mining Corp., Bernick Lake, Manitoba, Canada,* Cesium and cesium compounds

W. J. Wiseman, *Edgewood Arsenal, Aberdeen, Maryland,* Chemicals in war

J. A. Wojtowicz, *Olin Corporation, New Haven, Connecticut,* Chlorine monoxide, hypochlorous acid, and hypochlorites under Chlorine oxygen acids and salts

Howard C. Zell, *Food and Drug Administration, Rockville, Maryland,* Chemotherapeutics, antiprotozoal

NOTE ON CHEMICAL ABSTRACTS SERVICE REGISTRY NUMBERS AND NOMENCLATURE

Chemical Abstracts Service (CAS) Registry Numbers are unique numerical identifiers assigned to substances recorded in the CAS Registry System. They appear in brackets in the *Chemical Abstracts* (CA) substance and formula indexes following the names of compounds. A single compound may have many synonyms in the chemical literature. A simple compound like phenethylamine can be named β-phenylethylamine or, as in *Chemical Abstracts,* benzeneethanamine. The usefulness of the Encyclopedia depends on accessibility through the most common correct name of a substance. Because of this diversity in nomenclature careful attention has been given the problem in order to assist the reader as much as possible, especially in locating the systematic CA index name by means of the Registry Number. For this purpose, the reader may refer to the CAS Registry Handbook-Number Section which lists in numerical order the Registry Number with the Chemical Abstracts index name and the molecular formula; eg, **458-88-8,** Piperidine, 2-propyl-, (S)-, $C_8H_{17}N$; in the Encyclopedia this compound would be found under its common name, coniine [*458-88-8*]. The Registry Number is a valuable link for the reader in retrieving additional published information on substances and also as a point of access for such on-line data bases as Chemline, Medline, and Toxline.

In all cases, the CAS Registry Numbers have been given for title compounds in articles and for all compounds in the index. All specific substances indexed in *Chemical Abstracts* since 1965 are included in the CAS Registry System as are a large number of substances derived from a variety of reference works. The CAS Registry System identifies a substance on the basis of an unambiguous computer-language description of its molecular structure including stereochemical detail. The Registry Number is a machine-checkable number (like a Social Security number) assigned in sequential order to each substance as it enters the registry system. The value of the number lies in the fact that it is a concise and unique means of substance identification, which is independent of, and therefore bridges, many systems of chemical nomenclature. For polymers, one Registry Number is used for the entire family; eg, polyoxyethylene (20)sorbitan monolaurate has the same number as all of its polyoxyethylene homologues.

Registry numbers for each substance will be provided in the third edition index (eg, Alkaloids will show the Registry Number of all alkaloids (title compounds) in a table in the article as well, but the intermediates will have their Registry Numbers shown only in the index). Articles such as Absorption, Adsorptive separation, Air conditioning, Air pollution, Air pollution control methods will have no Registry Numbers in the text.

Cross-references have been inserted in the index for many common names and for some systematic names. Trademark names appear in the index. Names that are incorrect, misleading or ambiguous are avoided. Formulas are given very frequently in the text to help in identifying compounds. The spelling and form used, even for industrial names, follow American chemical usage, but not always the usage of *Chemical Abstracts* (eg, *coniine* is used instead of (*S*)-*2-propylpiperidine, aniline* instead of *benzenamine,* and *acrylic acid* instead of *2-propenoic acid*).

There are variations in representation of rings in different disciplines. The dye industry does not designate aromaticity or double bonds in rings. All double bonds and aromaticity will be shown in the *Encyclopedia* as a matter of course. For example, tetralin has an aromatic ring and a saturated ring and its structure will appear in the

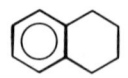

Encyclopedia with its common name, Registry Number enclosed in brackets, and parenthetical CA index name, ie, tetralin, [*119-64-2*] (1,2,3,4-tetrahydronaphthalene). With names and structural formulas, and especially with CAS Registry Numbers, the aim is to help the reader have a concise means of substance identification.

CONVERSION FACTORS, ABBREVIATIONS, AND UNIT SYMBOLS

SI Units (Adopted 1960)

A new system of measurement, the International System of Units (abbreviated SI), is being implemented throughout the world. This system is a modernized version of the MKSA (meter, kilogram, second, ampere) system, and its details are published and controlled by an international treaty organization (The International Bureau of Weights and Measures) (1).

SI units are divided into three classes:

BASE UNITS

length	meter[†] (m)
mass[‡]	kilogram (kg)
time	second (s)
electric current	ampere (A)
thermodynamic temperature[§]	kelvin (K)
amount of substance	mole (mol)
luminous intensity	candela (cd)

[†] The spellings "metre" and "litre" are preferred by ASTM; however "-er" will be used in the Encyclopedia.
[‡] "Weight" is the commonly used term for "mass".
[§] Wide use is made of "Celsius temperature" (t) defined by

$$t = T - T_0$$

where T is the thermodynamic temperature, expressed in kelvins, and $T_0 = 273.15$ K by definition. A temperature interval may be expressed in degrees Celsius as well as in kelvins.

FACTORS, ABBREVIATIONS, AND SYMBOLS

SUPPLEMENTARY UNITS

plane angle	radian (rad)
solid angle	steradian (sr)

DERIVED UNITS AND OTHER ACCEPTABLE UNITS

These units are formed by combining base units, supplementary units, and other derived units (2–4). Those derived units having special names and symbols are marked with an asterisk in the list below:

Quantity	Unit	Symbol	Acceptable equivalent
*absorbed dose	gray	Gy	J/kg
acceleration	meter per second squared	m/s^2	
*activity (of ionizing radiation source)	becquerel	Bq	1/s
area	square kilometer	km^2	
	square hectometer	hm^2	ha (hectare)
	square meter	m^2	
*capacitance	farad	F	C/V
concentration (of amount of substance)	mole per cubic meter	mol/m^3	
*conductance	siemens	S	A/V
current density	ampere per square meter	A/m^2	
density, mass density	kilogram per cubic meter	kg/m^3	g/L; mg/cm^3
dipole moment (quantity)	coulomb meter	C·m	
*electric charge, quantity of electricity	coulomb	C	A·s
electric charge density	coulomb per cubic meter	C/m^3	
electric field strength	volt per meter	V/m	
electric flux density	coulomb per square meter	C/m^2	
*electric potential, potential difference, electromotive force	volt	V	W/A
*electric resistance	ohm	Ω	V/A
*energy, work, quantity of heat	megajoule	MJ	
	kilojoule	kJ	
	joule	J	N·m
	electron volt[†]	eV[†]	
	kilowatt-hour[†]	kW·h[†]	

[†] This non-SI unit is recognized by the CIPM as having to be retained because of practical importance or use in specialized fields (1).

Quantity	Unit	Symbol	Acceptable equivalent
energy density	joule per cubic meter	J/m³	
*force	kilonewton	kN	
	newton	N	kg·m/s²
*frequency	megahertz	MHz	
	hertz	Hz	1/s
heat capacity, entropy	joule per kelvin	J/K	
heat capacity (specific), specific entropy	joule per kilogram kelvin	J/(kg·K)	
heat transfer coefficient	watt per square meter kelvin	W/(m²·K)	
*illuminance	lux	lx	lm/m²
*inductance	henry	H	Wb/A
linear density	kilogram per meter	kg/m	
luminance	candela per square meter	cd/m²	
*luminous flux	lumen	lm	cd·sr
magnetic field strength	ampere per meter	A/m	
*magnetic flux	weber	Wb	V·s
*magnetic flux density	tesla	T	Wb/m²
molar energy	joule per mole	J/mol	
molar entropy, molar heat capacity	joule per mole kelvin	J/(mol·K)	
moment of force, torque	newton meter	N·m	
momentum	kilogram meter per second	kg·m/s	
permeability	henry per meter	H/m	
permittivity	farad per meter	F/m	
*power, heat flow rate, radiant flux	kilowatt	kW	
	watt	W	J/s
power density, heat flux density, irradiance	watt per square meter	W/m²	
*pressure, stress	megapascal	MPa	
	kilopascal	kPa	
	pascal	Pa	N/m²
sound level	decibel	dB	
specific energy	joule per kilgram	J/kg	
specific volume	cubic meter per kilogram	m³/kg	
surface tension	newton per meter	N/m	
thermal conductivity	watt per meter kelvin	W/(m·K)	
velocity	meter per second	m/s	
	kilometer per hour	km/h	
viscosity, dynamic	pascal second	Pa·s	
	millipascal second	mPa·s	
viscosity, kinematic	square meter per second	m²/s	

Quantity	Unit	Symbol	Acceptable equivalent
	square millimeter per second	mm^2/s	
volume	cubic meter	m^3	
	cubic decimeter	dm^3	L(liter) (5)
	cubic centimeter	cm^3	mL
wave number	1 per meter	m^{-1}	
	1 per centimeter	cm^{-1}	

In addition, there are 16 prefixes used to indicate order of magnitude, as follows:

Multiplication factor	Prefix	Symbol	Note
10^{18}	exa	E	
10^{15}	peta	P	
10^{12}	tera	T	
10^9	giga	G	
10^6	mega	M	
10^3	kilo	k	
10^2	hecto	h[a]	[a] Although hecto, deka, deci, and centi are SI prefixes, their use should be avoided except for SI unit-multiples for area and volume and nontechnical use of centimeter, as for body and clothing measurement.
10	deka	da[a]	
10^{-1}	deci	d[a]	
10^{-2}	centi	c[a]	
10^{-3}	milli	m	
10^{-6}	micro	μ	
10^{-9}	nano	n	
10^{-12}	pico	p	
10^{-15}	femto	f	
10^{-18}	atto	a	

For a complete description of SI and its use the reader is referred to ASTM E 380 (4) and the article Units and Conversion Factors which will appear in a later volume of the *Encyclopedia*.

A representative list of conversion factors from non-SI to SI units is presented herewith. Factors are given to four significant figures. Exact relationships are followed by a dagger. A more complete list is given in ASTM E 380-76(4) and ANSI Z210.1-1976 (6).

Conversion Factors to SI Units

To convert from	To	Multiply by
acre	square meter (m^2)	4.047 × 10^3
angstrom	meter (m)	1.0 × 10^{-10}†
are	square meter (m^2)	1.0 × 10^{2}†
astronomical unit	meter (m)	1.496 × 10^{11}
atmosphere	pascal (Pa)	1.013 × 10^5
bar	pascal (Pa)	1.0 × 10^{5}†
barrel (42 U.S. liquid gallons)	cubic meter (m^3)	0.1590
Bohr magneton μ_β	J/T	9.274 × 10^{-24}
Btu (International Table)	joule (J)	1.055 × 10^3

† Exact.

To convert from	To	Multiply by
Btu (mean)	joule (J)	1.056×10^3
Btu (thermochemical)	joule (J)	1.054×10^3
bushel	cubic meter (m^3)	3.524×10^{-2}
calorie (International Table)	joule (J)	4.187
calorie (mean)	joule (J)	4.190
calorie (thermochemical)	joule (J)	4.184†
centipoise	pascal second (Pa·s)	1.0×10^{-3}†
centistoke	square millimeter per second (mm^2/s)	1.0†
cfm (cubic foot per minute)	cubic meter per second (m^3/s)	4.72×10^{-4}
cubic inch	cubic meter (m^3)	1.639×10^{-5}
cubic foot	cubic meter (m^3)	2.832×10^{-2}
cubic yard	cubic meter (m^3)	0.7646
curie	becquerel (Bq)	3.70×10^{10}†
debye	coulomb·meter (C·m)	3.336×10^{-30}
degree (angle)	radian (rad)	1.745×10^{-2}
denier (international)	kilogram per meter (kg/m)	1.111×10^{-7}
	tex‡	0.1111
dram (apothecaries')	kilogram (kg)	3.888×10^{-3}
dram (avoirdupois)	kilogram (kg)	1.772×10^{-3}
dram (U.S. fluid)	cubic meter (m^3)	3.697×10^{-6}
dyne	newton (N)	1.0×10^{-5}†
dyne/cm	newton per meter (N/m)	1.00×10^{-3}†
electron volt	joule (J)	1.602×10^{-19}
erg	joule (J)	1.0×10^{-7}†
fathom	meter (m)	1.829
fluid ounce (U.S.)	cubic meter (m^3)	2.957×10^{-5}
foot	meter (m)	0.3048†
footcandle	lux (lx)	10.76
furlong	meter (m)	2.012×10^{-2}
gal	meter per second squared (m/s^2)	1.0×10^{-2}†
gallon (U.S. dry)	cubic meter (m^3)	4.405×10^{-3}
gallon (U.S. liquid)	cubic meter (m^3)	3.785×10^{-3}
gallon per minute (gpm)	cubic meter per second (m^3/s)	6.308×10^{-5}
	cubic meter per hour (m^3/h)	0.2271
gauss	tesla (T)	1.0×10^{-4}
gilbert	ampere (A)	0.7958
gill (U.S.)	cubic meter (m^3)	1.183×10^{-4}
grad	radian	1.571×10^{-2}
grain	kilogram (kg)	6.480×10^{-5}
gram force per denier	newton per tex (N/tex)	8.826×10^{-2}
hectare	square meter (m^2)	1.0×10^{4}†
horsepower (550 ft·lbf/s)	watt (W)	7.457×10^2
horsepower (boiler)	watt (W)	9.810×10^3
horsepower (electric)	watt (W)	7.46×10^2†
hundredweight (long)	kilogram (kg)	50.80
hundredweight (short)	kilogram (kg)	45.36
inch	meter (m)	2.54×10^{-2}†
inch of mercury (32°F)	pascal (Pa)	3.386×10^3

† Exact.
‡ See footnote on p. xiv.

To convert from	To	Multiply by
inch of water (39.2°F)	pascal (Pa)	2.491×10^2
kilogram force	newton (N)	9.807
kilowatt hour	megajoule (MJ)	3.6[†]
kip	newton (N)	4.48×10^3
knot (international)	meter per second (m/s)	0.5144
lambert	candela per square meter (cd/m^2)	3.183×10^3
league (British nautical)	meter (m)	5.559×10^3
league (statute)	meter (m)	4.828×10^3
light year	meter (m)	9.461×10^{15}
liter (for fluids only)	cubic meter (m^3)	1.0×10^{-3}[†]
maxwell	weber (Wb)	1.0×10^{-8}[†]
micron	meter (m)	1.0×10^{-6}[†]
mil	meter (m)	2.54×10^{-5}[†]
mile (U.S. nautical)	meter (m)	1.852×10^3[†]
mile (statute)	meter (m)	1.609×10^3
mile per hour	meter per second (m/s)	0.4470
millibar	pascal (Pa)	1.0×10^2
millimeter of mercury (0°C)	pascal (Pa)	1.333×10^2[†]
minute (angular)	radian	2.909×10^{-4}
myriagram	kilogram (kg)	10
myriameter	kilometer (km)	10
oersted	ampere per meter (A/m)	79.58
ounce (avoirdupois)	kilogram (kg)	2.835×10^{-2}
ounce (troy)	kilogram (kg)	3.110×10^{-2}
ounce (U.S. fluid)	cubic meter (m^3)	2.957×10^{-5}
ounce-force	newton (N)	0.2780
peck (U.S.)	cubic meter (m^3)	8.810×10^{-3}
pennyweight	kilogram (kg)	1.555×10^{-3}
pint (U.S. dry)	cubic meter (m^3)	5.506×10^{-4}
pint (U.S. liquid)	cubic meter (m^3)	4.732×10^{-4}
poise (absolute viscosity)	pascal second (Pa·s)	0.10[†]
pound (avoirdupois)	kilogram (kg)	0.4536
pound (troy)	kilogram (kg)	0.3732
poundal	newton (N)	0.1383
pound-force	newton (N)	4.448
pound per square inch (psi)	pascal (Pa)	6.895×10^3
quart (U.S. dry)	cubic meter (m^3)	1.101×10^{-3}
quart (U.S. liquid)	cubic meter (m^3)	9.464×10^{-4}
quintal	kilogram (kg)	1.0×10^2[†]
rad	gray (Gy)	1.0×10^{-2}[†]
rod	meter (m)	5.029
roentgen	coulomb per kilogram (C/kg)	2.58×10^{-4}
second (angle)	radian (rad)	4.848×10^{-6}
section	square meter (m^2)	2.590×10^6
slug	kilogram (kg)	14.59

[†] Exact.

To convert from	To	Multiply by
spherical candle power	lumen (lm)	12.57
square inch	square meter (m^2)	6.452×10^{-4}
square foot	square meter (m^2)	9.290×10^{-2}
square mile	square meter (m^2)	2.590×10^6
square yard	square meter (m^2)	0.8361
stere	cubic meter (m^3)	1.0†
stokes (kinematic viscosity)	square meter per second (m^2/s)	1.0×10^{-4}†
tex	kilogram per meter (kg/m)	1.0×10^{-6}†
ton (long, 2240 pounds)	kilogram (kg)	1.016×10^3
ton (metric)	kilogram (kg)	1.0×10^3†
ton (short, 2000 pounds)	kilogram (kg)	9.072×10^2
torr	pascal (Pa)	1.333×10^2
unit pole	weber (Wb)	1.257×10^{-7}
yard	meter (m)	0.9144†

† Exact.

Abbreviations and Unit Symbols

Following is a list of commonly used abbreviations and unit symbols appropriate for use in the *Encyclopedia*. In general they agree with those listed in *American National Standard Abbreviations for Use on Drawings and in Text (ANSI Y1.1)* (6) and *American National Standard Letter Symbols for Units in Science and Technology (ANSI Y10)* (6). Also included is a list of acronyms for a number of private and government organizations as well as common industrial solvents, polymers, and other chemicals.

Rules for Writing Unit Symbols (4):

1. Unit symbols should be printed in upright letters (roman) regardless of the type style used in the surrounding text.

2. Unit symbols are unaltered in the plural.

3. Unit symbols are not followed by a period except when used as the end of a sentence.

4. Letter unit symbols are generally written in lower-case (eg, cd for candela) unless the unit name has been derived from a proper name, in which case the first letter of the symbol is capitalized (W,Pa). Prefix and unit symbols retain their prescribed form regardless of the surrounding typography.

5. In the complete expression for a quantity, a space should be left between the numerical value and the unit symbol. For example, write 2.37 lm, *not* 2.37lm, and 35 mm, *not* 35mm. When the quantity is used in an adjectival sense, a hyphen is often used, for example, 35-mm film. *Exception:* No space is left between the numerical value and the symbols for degree, minute, and second of plane angle, and degree Celsius.

6. No space is used between the prefix and unit symbols (eg, kg).

7. Symbols, not abbreviations, should be used for units. For example, use "A," not "amp," for ampere.

8. When multiplying unit symbols, use a raised dot:

N·m for newton meter

In the case of W·h, the dot may be omitted, thus:

$$Wh$$

An exception to this practice is made for computer printouts, automatic typewriter work, etc, where the raised dot is not possible, and a dot on the line may be used.

9. When dividing unit symbols use one of the following forms:

$$m/s \text{ or } m\cdot s^{-1} \text{ or } \frac{m}{s}$$

In no case should more than one slash be used in the same expression unless parentheses are inserted to avoid ambiguity. For example, write:

$$J/(mol\cdot K) \text{ or } J\cdot mol^{-1} \cdot K^{-1} \text{ or } (J/mol)/K$$

but *not*

$$J/mol/K$$

10. Do not mix symbols and unit names in the same expression. Write:

$$\text{joules per kilogram } or \text{ } J/kg \text{ } or \text{ } J\cdot kg^{-1}$$

but *not*

$$\text{joules/kilogram } nor \text{ joules/kg } nor \text{ joules}\cdot kg^{-1}$$

ABBREVIATIONS AND UNITS

A	ampere	amt	amount
A	anion (eg, H*A*)	amu	atomic mass unit
a	atto (prefix for 10^{-18})	ANSI	American National Standards Institute
AATCC	American Association of Textile Chemists and Colorists	AO	atomic orbital
		APHA	American Public Health Association
ABS	acrylonitrile–butadiene–styrene	API	American Petroleum Institute
abs	absolute	aq	aqueous
ac	alternating current, *n.*	Ar	aryl
a-c	alternating current, *adj.*	ar-	aromatic
ac-	alicyclic	*as*-	asymmetric(al)
ACGIH	American Conference of Governmental Industrial Hygienists	ASHRAE	American Society of Heating, Refrigerating, and Air Conditioning Engineers
ACS	American Chemical Society		
AGA	American Gas Association		
Ah	ampere hour	ASM	American Society for Metals
AIChE	American Institute of Chemical Engineers	ASME	American Society of Mechanical Engineers
AIP	American Institute of Physics	ASTM	American Society for Testing and Materials
alc	alcohol(ic)		
Alk	alkyl	at no.	atomic number
alk	alkaline (not alkali)	at wt	atomic weight

av(g)	average	dp	dew point; degree of polymerization
bbl	barrel		
bcc	body-centered cubic	dstl(d)	distill(ed)
Bé	Baumé	dta	differential thermal analysis
bid	twice daily	(E)-	entgegen; opposed
BOD	biochemical (biological) oxygen demand	ϵ	dielectric constant (unitless number)
bp	boiling point	e	electron
Bq	becquerel	ECU	electrochemical unit
C	coulomb	ed.	edited, edition, editor
°C	degree Celsius	ED	effective dose
C-	denoting attachment to carbon	emf	electromotive force
		emu	electromagnetic unit
c	centi (prefix for 10^{-2})	eng	engineering
ca	circa (approximately)	EPA	Environmental Protection Agency
cd	candela; current density; circular dichroism		
		epr	electron paramagnetic resonance
cgs	centimeter–gram–second		
CI	Color Index	eq.	equation
cis-	isomer in which substituted groups are on same side of double bond between C atoms	esp	especially
		esr	electron-spin resonance
		est(d)	estimate(d)
		estn	estimation
cl	carload	esu	electrostatic unit
cm	centimeter	exp	experiment, experimental
cmil	circular mil	ext(d)	extract(ed)
cmpd	compound	F	farad (capacitance)
COA	coenzyme A	f	femto (prefix for 10^{-15})
COD	chemical oxygen demand	FAO	Food and Agriculture Organization (United Nations)
coml	commercial(ly)		
cp	chemically pure		
CPSC	Consumer Product Safety Commission	fcc	face-centered cubic
		FDA	Food and Drug Administration
D-	denoting configurational relationship	FEA	Federal Energy Administration
d	differential operator	fob	free on board
d-	dextro-, dextrorotatory	FPC	Federal Power Commission
da	deka (prefix for 10^1)	fp	freezing point
dB	decibel	frz	freezing
dc	direct current, n.	G	giga (prefix for 10^9)
d-c	direct current, adj.	g	gram
dec	decompose	(g)	gas, only as in $H_2O(g)$
detd	determined	g	gravitational acceleration
detn	determination	gem-	geminal
dia	diameter	glc	gas-liquid chromatography
dil	dilute	g-mol wt; gmw	gram-molecular weight
dl-; DL-	racemic		
DMF	dimethylformamide	grd	ground
DOE	Department of Energy	Gy	gray
DOT	Department of Transportation		

FACTORS, ABBREVIATIONS, AND SYMBOLS

H	henry	LPG	liquefied petroleum gas
h	hour; hecto (prefix for 10^2)	ltl	less than truckload lots
ha	hectare	lx	lux
HB	Brinell hardness number	M	mega (prefix for 10^6); metal (as in MA)
Hb	hemoglobin		
HK	Knoop hardness number	M	molar
HRC	Rockwell hardness (C scale)	m	meter; milli (prefix for 10^{-3})
HV	Vickers hardness number	m	molal
hyd	hydrated, hydrous	m-	meta
hyg	hygroscopic	max	maximum
Hz	hertz	MCA	Manufacturing Chemists' Association
^{i}(eg, Pr^i)	iso (eg, isopropyl)		
i-	inactive (eg, i-methionine)	MEK	methyl ethyl ketone
IACS	International Annealed Copper Standard	meq	milliequivalent
		mfd	manufactured
ibp	initial boiling point	mfg	manufacturing
ICC	Interstate Commerce Commission	mfr	manufacturer
		MIBC	methylisobutyl carbinol
ICT	International Critical Table	MIBK	methyl isobutyl ketone
ID	inside diameter; infective dose	MIC	minimum inhibiting concentration
IPS	iron pipe size	min	minute; minimum
IPT	Institute of Petroleum Technologists	mL	milliliter
		MLD	minimum lethal dose
ir	infrared	MO	molecular orbital
ISO	International Organization for Standardization	mo	month
		mol	mole
IUPAC	International Union of Pure and Applied Chemistry	mol wt	molecular weight
		mom	momentum
IV	iodine value	mp	melting point
J	joule	MR	molar refraction
K	kelvin	ms	mass spectrum
k	kilo (prefix for 10^3)	mxt	mixture
kg	kilogram	μ	micro (prefix for 10^{-6})
L	denoting configurational relationship	N	newton (force)
		N	normal (concentration)
L	liter (for fluids only) (5)	N-	denoting attachment to nitrogen
l-	levo-, levorotatory		
(l)	liquid, only as in NH$_3$(l)	n (as n_D^{20})	index of refraction (for 20°C and sodium light)
LC$_{50}$	conc lethal to 50% of the animals tested		
		^{n} (as Bu^n),	
LCAO	linear combination of atomic orbitals	n-	normal (straight-chain structure)
lcl	less than carload lots		
LD$_{50}$	dose lethal to 50% of the animals tested	n	nano (prefix for 10^{-9})
		na	not available
liq	liquid	NAS	National Academy of Sciences
lm	lumen		
ln	logarithm (natural)	NASA	National Aeronautics and Space Administration
LNG	liquefied natural gas		
log	logarithm (common)	nat	natural

NBS	National Bureau of Standards	PVC	poly(vinyl chloride)
neg	negative	pwd	powder
NF	*National Formulary*	qv	quod vide (which see)
NIH	National Institutes of Health	R	univalent hydrocarbon radical
NIOSH	National Institute of Occupational Safety and Health	(R)-	rectus (clockwise configuration)
nmr	nuclear magnetic resonance	rad	radian; radius
NND	New and Nonofficial Drugs (AMA)	rds	rate determining step
		ref.	reference
no.	number	rf	radio frequency, *n*.
NOI-(BN)	not otherwise indexed (by name)	r-f	radio frequency, *adj*.
		rh	relative humidity
NOS	not otherwise specified	RI	Ring Index
nqr	nuclear quadrople resonance	RT	room temperature
NRC	Nuclear Regulatory Commission; National Research Council	^{s} (eg, Bu^s); *sec*-	secondary (eg, secondary butyl)
NRI	New Ring Index	S	siemens
NSF	National Science Foundation	(S)-	sinister (counterclockwise configuration)
NTSB	National Transportation Safety Board	S-	denoting attachment to sulfur
O-	denoting attachment to oxygen	s-	symmetric(al)
o-	ortho	s	second
OD	outside diameter	(s)	solid, only as in H_2O(s)
OPEC	Organization of Petroleum Exporting Countries	SAE	Society of Automotive Engineers
OSHA	Occupational Safety and Health Administration	SAN	styrene–acrylonitrile
		sat(d)	saturate(d)
owf	on weight of fiber	satn	saturation
Ω	ohm	SCF	self-consistent field
P	peta (prefix for 10^{15})	Sch	Schultz number
p	pico (prefix for 10^{-12})	SFs	Saybolt Furol seconds
p-	para	SI	Le Système International d'Unités (International System of Units)
p.	page		
Pa	pascal (pressure)		
pd	potential difference	sl sol	slightly soluble
pH	negative logarithm of the effective hydrogen ion concentration	sol	soluble
		soln	solution
		soly	solubility
pmr	proton magnetic resonance	sp	specific; species
pos	positive	sp gr	specific gravity
pp.	pages	sr	steradian
ppb	parts per billion	std	standard
ppm	parts per million	STP	standard temperature and pressure (0°C and 101.3 kPa)
ppt(d)	precipitate(d)		
pptn	precipitation		
Pr (no.)	foreign prototype (number)	SUs	Saybolt Universal seconds
pt	point; part	syn	synthetic

FACTORS, ABBREVIATIONS, AND SYMBOLS

^{t} (eg, Bu^t), *t*-, *tert*-	tertiary (eg, tertiary butyl)	Twad	Twaddell
		UL	Underwriters' Laboratory
		USP	*United States Pharmacopeia*
T	tera (prefix for 10^{12}); tesla (magnetic flux density)	uv	ultraviolet
		V	volt (emf)
t	metric ton (tonne) temperature	var	variable
		vic-	vicinal
TAPPI	Technical Association of the Pulp and Paper Industry	vol	volume (not volatile)
		vs	versus
tex	tex (linear density)	v sol	very soluble
THF	tetrahydrofuran	W	watt
tlc	thin layer chromatography	Wb	Weber
TLV	threshold limit value	Wh	watt hour
trans-	isomer in which substituted groups are on opposite sides of double bond between C atoms	WHO	World Health Organization (United Nations)
		wk	week
		yr	year
		(Z)-	zusammen; together

Non-SI (Unacceptable and Obsolete) Units   *Use*

Å	angstrom	nm
at	atmosphere, technical	Pa
atm	atmosphere, standard	Pa
b	barn	cm^2
bar†	bar	Pa
bhp	brake horsepower	W
Btu	British thermal unit	J
bu	bushel	m^3; L
cal	calorie	J
cfm	cubic foot per minute	m^3/s
Ci	curie	Bq
cSt	centistokes	mm^2/s
c/s	cycle per second	Hz
cu	cubic	exponential form
D	debye	C·m
den	denier	tex
dr	dram	kg
dyn	dyne	N
erg	erg	J
eu	entropy unit	J/K
°F	degree Fahrenheit	°C; K
fc	footcandle	lx
fl	footlambert	lx
fl oz	fluid ounce	m^3; L
ft	foot	m
ft·lbf	foot pound-force	J
gf den	gram-force per denier	N/tex
G	gauss	T
Gal	gal	m/s^2
gal	gallon	m^3; L

† Do not use bar (10^5Pa) or millibar (10^2Pa) because they are not SI units, and are accepted internationally only for a limited time in special fields because of existing usage.

Non-SI. (*Unacceptable and Obsolete*) Units		Use
Gb	gilbert	A
gpm	gallon per minute	(m³/s); (m³/h)
gr	grain	kg
hp	horsepower	W
ihp	indicated horsepower	W
in.	inch	m
in. Hg	inch of mercury	Pa
in. H₂O	inch of water	Pa
in.·lbf	inch pound-force	J
kcal	kilogram-calorie	J
kgf	kilogram-force	N
kilo	for kilogram	kg
L	lambert	lx
lb	pound	kg
lbf	pound-force	N
mho	mho	S
mi	mile	m
MM	million	M
mm Hg	millimeter of mercury	Pa
mμ	millimicron	nm
mph	miles per hour	km/h
μ	micron	μm
Oe	oersted	A/m
oz	ounce	kg
ozf	ounce-force	N
η	poise	Pa·s
P	poise	Pa·s
ph	phot	lx
psi	pounds-force per square inch	Pa
psia	pounds-force per square inch absolute	Pa
psig	pounds-force per square inch gage	Pa
qt	quart	m³; L
°R	degree Rankine	K
rd	rad	Gy
sb	stilb	lx
SCF	standard cubic foot	m³
sq	square	exponential form
thm	therm	J
yd	yard	m

BIBLIOGRAPHY

1. The International Bureau of Weights and Measures, BIPM, (Parc de Saint-Cloud, France) is described on page 22 of Ref. 4. This bureau operates under the exclusive supervision of the International Committee of Weights and Measures (CIPM).
2. *Metric Editorial Guide (ANMC-75-1)*, American National Metric Council, 1625 Massachusetts Ave. N.W., Washington, D.C. 20036, 1975.
3. *SI Units and Recommendations for the Use of Their Multiples and of Certain Other Units (ISO 1000-1973)*, American National Standards Institute, 1430 Broadway, New York, N. Y. 10018, 1973.
4. Based on *ASTM E 380-76 (Standard for Metric Practice)*, American Society for Testing and Materials, 1916 Race Street, Philadelphia, Pa. 19103, 1976.
5. *Fed. Regist.*, Dec. 10, 1976 (41 FR 36414).
6. For ANSI address, see Ref. 3.

R. P. LUKENS
American Society for Testing and Materials

C *continued*

CASTOR OIL

Castor oil [8001-79-4] is derived from the bean of the castor plant, *Ricinus Communis* L., of the family Eurphorbiaceae. The castor plant occurs in practically all tropical and subtropical countries, either wild or cultivated. It is also found widely as both an ornamental and cultivated plant in temperate zones where, because it is frost sensitive, it is grown annually from seed. It is a highly heterogeneous species, with wide variations in the size, form, and color of the plant, as well as the size and color of the seed, and the dehiscence of the capsules. There is relatively little variation in the oil content of fully matured seeds and in the chemical composition of the oil. The seeds are toxic and the ingestion of even one seed can be fatal to humans.

The seeds of the castor plant are produced in racemes, or clusters of capsules. The capsules are usually spiny and each contains three seeds. The hulls surrounding the seeds constitute roughly one third of the weight of the mature capsule. The seeds are mottled to varying extents, most often with shades of dark brown overlaying shades of light brown. Seed size of commercial varieties varies from 250 to 1680 per kilogram. Although laboratory yield of oil from entire (undecorticated) seed of commercial varieties averages about 49%, the factory yield is usually about 45%.

There was considerable production of castor plants in the United States in the 1800s, but by 1900 that production shifted to countries where the necessary manual harvesting and hulling was cheaper. Because of the defense value of castor oil, the United States government sponsored a small domestic castor production during World War I. Starting in World War II and continuing to 1973, there was increasing domestic production. That increase paralleled and was made possible by the development of production technology, including much improvement in varieties, and the mechanization of the harvesting and hulling. For the period of 1957 through 1969, the United

States production averaged over 20,000 metric tons per year. The United States production was competitive with foreign production in all except the years of lowest price. With the discontinuance of meaningful government price supports, the once sizable United States production dropped almost to zero by 1974.

Castor oil is also known as Ricinus Oil, oil of Palma Christi, tangantangan oil, and Neoloid. Typical of most fats, the oil is a triglyceride of fatty acids. What is unique is that the fat contains 87–90% ricinoleic acid, *cis*-12-hydroxyoctadec-9-enoic acid, $CH_3(CH_2)_5CH(OH)CH_2CH=CH(CH_2)_7COOH$, a rare source of an eighteen carbon hydroxylated fatty acid with one double bond. Castor oil, sometimes described as a triglyceride of ricinoleic acid, is one of the few naturally occurring glycerides that approaches being a pure compound.

The oil is pale yellow and viscous, with a slight characteristic odor, and nearly tasteless but familiarly unpleasant through its minor use as a purgative. The industrial uses of castor oil are extensive.

Oil Recovery

The recovery of castor oil from castor beans has followed the historic development of general oil seed processing (see Vegetable oils). The most common method is the use of hydraulic presses or expellers (continuous mechanical screw-presses) followed by solvent extraction. Direct solvent extraction of castor beans with and without the removal of the seed coats (decortication) has also been reported (1).

Cold pressing of the beans yields an oil of high quality, ie, light color and with low fatty acid content, free from toxic proteins. It is suitable for many uses without further treatment.

Hot-pressed oil from hydraulic or continuous mechanical presses requires refining steps to remove toxic proteins, improve the color, and reduce the free fatty acid content. Both hydraulic and expeller pressing removes about 75–85% of the total oil content of the castor bean, leaving behind a press cake with 10–20% oil content.

Solvent extraction in batch or continuous systems is used to recover most of the oil from the press cake, yielding castor pomace with about 1–2% residual oil content. The solvent is usually heptane which is recovered from the oil and the pomace for recycling. The extracted oil is darker and it has a higher free fatty acid content than the pressed oil. Sometimes it is referred to as no. 3 castor oil.

The castor pomace has a protein content of about 35% and it is used mostly as a fertilizer. It is not suitable for animal feeding without further treatment because it is highly toxic (2–3). The toxicity of the castor bean and pomace comes from a very poisonous heat-labile protein called ricin and a toxic alkaloid, ricinine. The ricin, which is present in relatively large amounts, can be inactivated by treating the beans or pomace with steam. The ricinine is present in small amounts and it is not considered to be significantly detrimental (4). In addition, the castor pomace contains about 12.5% of a very potent and heat-resistant allergen (5–8).

Extensive work is reported for detoxification and deallergenation of castor pomace by treatment with ammonia, caustic soda, lime, and heat (9–15).

The hot-pressed oil is processed in several steps:

Degumming. The dissolved or dispersed proteins are removed by settling or centrifuging.

Alkali treatment (optional). The free fatty acid content is reduced by treatment with caustic soda solution. This step is especially difficult with castor oil because of easy, stable emulsion formation. A continuous countercurrent process is used with a stationary contact reactor (16). Treatment in the presence of a solvent is also successful (17).

Decolorization. The color of the oil is improved by treatment with 0.5–1.0% activated carbon and/or clay at about 90°C under vacuum. The additives are removed by filtration.

Deodorization. The odor causing and other volatile components are removed by steam stripping at 160°C under less than 1.5 kPa (11 mm Hg) pressure.

A new process was reported (18) to replace the above alkali treatment procedure. Intensive contact of the oil with steam under reduced pressure in a special reactor removes the fatty acids which are recovered.

Properties

The average fatty acid composition of castor oil is as follows:

ricinoleic acid	89.5%
dihydroxystearic acid	0.7
palmitic acid	1.0
stearic acid	1.0
oleic acid	3.0
linoleic acid	4.2
linolenic acid	0.3
eicosanoic acid	0.3

Ricinoleic acid has an acid value of 180, a saponification value of 186, a Wijs iodine value of 89, and mp 5.5°C. The 12th carbon is asymmetric, resulting in an optical rotation of $[\alpha]_D^{22} + 6.67$. Other properties are Gardner color 5, viscosity at 25°C 4 cm^2/s (= stokes), pour point -12°C, specific gravity (25°C/25°C) 0.940, n_D^{25} 1.4699. Ricinoleic acid cannot be distilled without decomposition unless special precautions are taken to protect the hydroxyl group via derivative formation.

Standards for industrial quality castor oil as specified by the ASTM are given in Table 1.

Other general properties of castor oil are listed in Table 2.

The oil is distinguished from other triglycerides by its high specific gravity, viscosity, and hydroxyl value. Another unique feature is its solubility in alcohol: one volume of castor oil dissolves in two volumes of 95% ethyl alcohol at room temperature, and the oil is miscible in all proportions with absolute ethyl alcohol; also, the oil is typically soluble in polar organic solvents and less soluble in aliphatic hydrocarbon solvents. Its slight solubility in petroleum ether is a characteristic which distinguishes it from other fats.

Chemical Modification

Castor oil serves as an industrial raw material for the manufacture of a number of complex organic derivatives (20–21). The unit processes involved in converting castor oil to chemical compounds include dehydration, sulfonation, alkali fusion, oxidation, pyrolysis, saponification, hydrogenation, and others of a more complex nature.

Table 1. Properties of Castor Oil[a]

Property	Value
acid value, max	2.0
clarity	clear
Gardner color, max	2
hydroxyl value	160–168
loss on heating, max %	0.2
refractive index, 25°C	1.4764–1.4778
saponification value	176–184
solubility in alcohol[b]	complete
specific gravity, 25/25°C	0.957–0.961
unsaponifiable, max %	0.7
viscosity, cm²/s (= stokes)	6.5–8.0
iodine value	84–88

[a] Courtesy of the American Society for Testing and Materials (19).
[b] Soluble 1:2 in 95% ethanol 3A (SDA) at 20°C.

Table 2. General Properties of Castor Oil

Property	Value
Gardner-Holdt viscosity, 25°C	U ± ½
flash point, Tag closed cup, °C	230
Cleveland open cup, °C	285
ignition temperature, °C	449
surface tension, mN/m (= dyn/cm), 20°C	39.0
80°C	35.2
Reichert-Meissl value	<0.5
Polenske value	<0.5
acetyl value	144–150
polarimeter, 200 mm (optical rotation)	+7.5 to +9.0
pour point, °C	−23
coeff. of expansion per °C	0.00066

Dehydration. Catalytic dehydration converts castor oil to an excellent drying oil called dehydrated castor oil (see Drying oils). It is used extensively by the coating industry. Although the reaction was known for a long time, its commercial application became significant during World War II in making a tung oil replacement. In the dehydration reaction the 12-hydroxyl group is removed with a nearby hydrogen atom to form water and a new double bond (21). Further reaction yields both conjugated and nonconjugated linoleic acid isomers. The reaction is first order with an activation energy of about 188 J/mol (43 cal/mol) (22).

Analyses of commercial dehydrated castor oil indicate that using sulfuric acid as the catalyst, about 30% of the linoleic acid double bonds are conjugated in the 9,11 position. Practical dehydration processes are the subject of extensive publications and patents. They deal with improvement of color, reduction of the after-tack after drying and increased conjugation. The most common catalysts are sulfuric acid and its acid salts (23), oxides (24–25), and activated clay (26). In a typical process, castor oil is heated to 230–280°C under vacuum and 3–5% diluted sulfuric acid is added to it at a controlled rate. The viscosity of the product is regulated by the degree of dehydration, reaction time and temperature. Typical properties of two commercial-grade dehydrated castor oils are shown in Table 3. Systems for continuous dehydration have been patented (27–28).

Dehydrated castor oil is well known for its nonyellowing film formation, outstanding color retention, flexibility, and adhesion in protective coatings (29).

The fatty acids of dehydrated castor oil are obtained by hydrolysis or by saponification followed by acidification of the oil. The conjugated acid content of the product remains the same as in the oil itself. Much higher conjugated acid contents, 50% and above, are obtained by dehydration of ricinoleic acid (30). Usually, the crude acids are distilled under vacuum to obtain a light-colored, high quality product. Typical properties of dehydrated castor acids are shown in Table 4. Type I is the distilled grade and Type II is the crude, undistilled acid.

Sulfonation. Sulfonated castor oil, also known as Turkey red oil, represents one of the earliest chemical derivatives of castor oil. The traditional method of preparing Turkey red oil is to add concentrated sulfuric acid to castor oil over a period of several hours while cooling to maintain a temperature of 25–30°C, followed by washing to remove surplus acid and then neutralization with either aqueous sodium hydroxide, potassium hydroxide, ammonia, or an amine such as ethanolamine.

Sulfonation results largely in esterification of the hydroxyl group of ricinoleic acid. Side reactions that can take place include attack of the double bond to produce an ester or the hydroxysulfonic acid (31). Hydrolysis of the esters during the reaction and subsequent treatment forms sulfuric acid and hydroxy acids, which can be further sulfated at the OH group. Despite the many side reactions most commercial products seem to be similar in properties.

Table 3. Specifications for Dehydrated Castor Oil[a]

Specification	Unbodied[b]	Bodied[b]	ASTM Method
viscosity at 25°C	F to I	Z2 to Z4	D 1545, D 445
specific gravity, 25/25°C	0.926–0.937	0.944–0.966	D 1963, D 1475
acid value	6	6	D 1639
saponification value	188–195	188–195	D 1962
iodine value (Wijs)	125–145	100 minimum	D 1959
color no. (Gardner)	6	7	D 1554
gel time at 315°C, min	145	63	D 1955
set-to-touch time, h	2.5	1.4	
refractive index at 25°C	1.4805–1.4825	1.4860–1.4890	

[a] ASTM D-961, 1976.
[b] Bodied refers to polymerized castor oil of higher viscosity; unbodied refers to dehydrated castor oil that has not been additionally heat polymerized.

Table 4. Requirements for Dehydrated Castor Acids[a]

Requirement	Type I	Type II	ASTM Method
acid value	195–200	187–195	D 1980
saponification value	195–200	193–199	D 1962
iodine value	150–156	138–143	D 1959
color, Gardner	1 max	5–8	D 1544
spectrophotometric diene value	28–35	25–32	D 1358

[a] ASTM D-1539-60, 1974.

Commercially sulfated castor oil contains ca 8.0–8.5% combined SO_3, indicating that only one of the reactive sites in the unsaturated, hydroxy-bearing triglyceride has been sulfated. The sulfate group imparts excellent wetting, emulsification, and dispersing characteristics. The product is used in the textile industry for fiber wetting ability and as dye agent to obtain bright, clear colors.

Sulfonation of castor oil with anhydrous SO_3 (2 mol) at temperatures higher than the sulfuric acid treatment produces a product also containing only 8.0–8.5% combined SO_3 which has better hydrolytic stability (32) and contains less inorganic salts and free fatty acids.

Alkali Fusion. Depending on reaction conditions, two different sets of products can be formed (33–34). At 180–200°C, using one mole of sodium or potassium hydroxide, methyl hexyl ketone and 10-hydroxydecanoic acid result. This reaction is favored by the presence of unhindered primary or secondary alcohols, such as 1- or 2-octanol, and 10-hydroxydecanoic acid is also formed in good yield when methyl ricinoleate is used (35–37). Two moles of alkali per mole of ricinoleate at 250–275°C and with a shorter reaction cycle produces capryl alcohol (2-octanol) and sebacic acid. Hydrogen is also formed with excess alkali.

A mechanism has been proposed to describe the reactions (38) which involves a retroaldol fission of a conjugated keto acid to yield the ketone, which can be reduced to capryl alcohol and an aldehyde acid, which can be reduced to 10-hydroxydecanoic acid or oxidized to sebacic acid (Fig. 1).

Sebacic acid is used in the manufacture of nylon-6,10 by reaction with hexamethylenediamine. Nylon-6,10 has better molding properties and resistance to moisture than adipic acid-based (6,6) nylon and is used in moldings, extrusions, filaments, bristles, and some textile fiber applications (see Polyamides). Sebacic acid esters are used as plasticizers, and bis(2-ethylhexyl) sebacate is a high-performance lubricant for jet engines.

Production of Nylon-11. In the production of nylon-11 castor oil is transesterified with methyl alcohol to form methyl ricinoleate and glycerol. The pyrolysis of methyl ricinoleate is accomplished at ca 450–500°C to form methyl 10-undecylenate and heptaldehyde. Hydrolysis of the methyl ester gives 10-undecylenic acid. Hydrogen bromide is added to this molecule in the presence of peroxides to form 11-bromoundecanoic acid which is then converted to 11-aminoundecanoic acid, known as Rilsan monomer (39). Heating the molten monomer gives the polymer by condensation (Fig. 2). Nylon-11 polyamide has a mp range of ca 186–190°C which facilitates high speed, uniform processing. Nylon-11 has the advantage of having a cold impact, no-rupture temperature of −40°C, low moisture sensitivity, and superior zinc chloride resistance. In a 50% solution of zinc chloride, it showed no attack after several thousand hours of testing. Nylon-6 and nylon-6,6 showed attack in ca 20 min and nylon-6,10 was attacked in 20 h. The excellent chemical resistance and stability in contact with all types of fuels, along with vibration and shock resistance, have led to the use of nylon-11 in the automotive industry.

Monofilament fabricated from nylon-11 is used as a covering for braided cable and in industrial fabrics such as filter cloth, bags, netting, flexible fishing lines, and a variety of brush applications where toughness and durability are required. A variety of bearings and rollers in conveyor systems are fabricated by machining. Flat and tubular film is used in packaging demanding superior quality nylon. Glass-reinforced nylon-11 provides an appreciable increase in rigidity and improvement in compressive strength and resistance to cold flow under load.

$$CH_3(CH_2)_5\underset{OH}{\overset{H}{C}}CH_2CH=CH(CH_2)_7COO^-$$

↓ Dehydrogenation

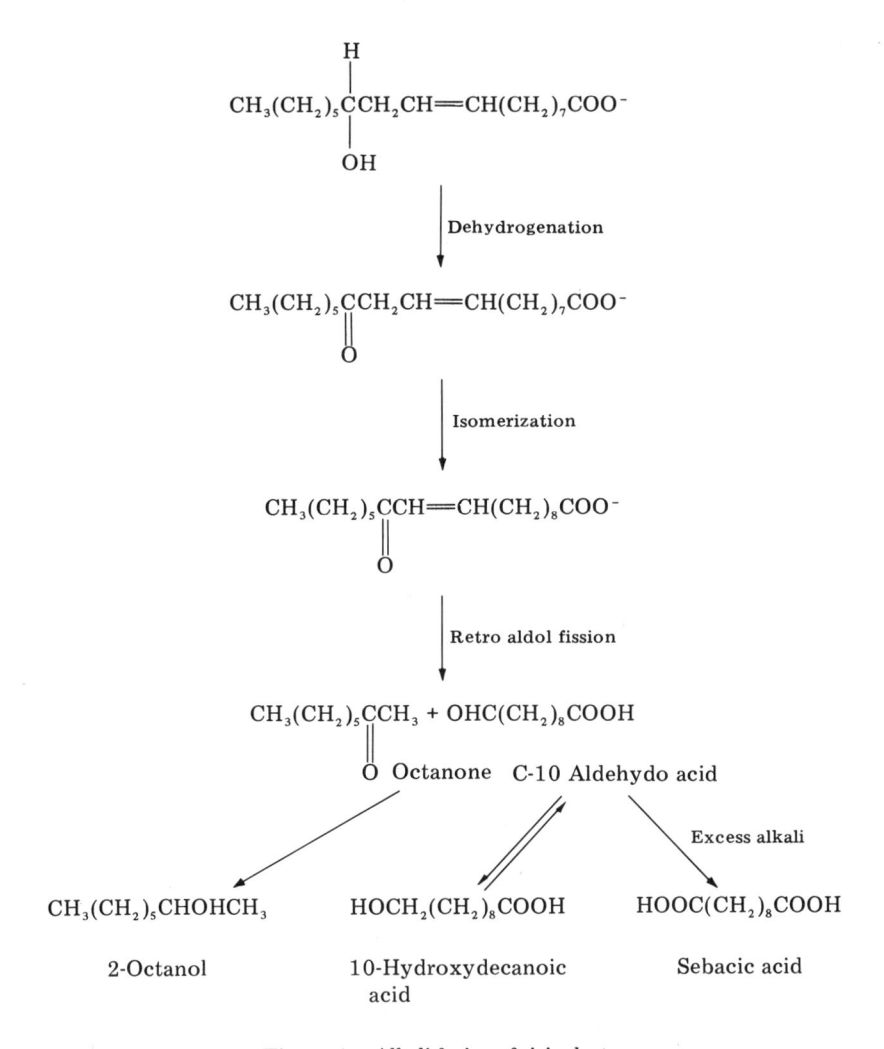

$$CH_3(CH_2)_5\underset{O}{\overset{\|}{C}}CH_2CH=CH(CH_2)_7COO^-$$

↓ Isomerization

$$CH_3(CH_2)_5\underset{O}{\overset{\|}{C}}CH=CH(CH_2)_8COO^-$$

↓ Retro aldol fission

$$CH_3(CH_2)_5\underset{O}{\overset{\|}{C}}CH_3 + OHC(CH_2)_8COOH$$
Octanone   C-10 Aldehydo acid

Excess alkali

$CH_3(CH_2)_5CHOHCH_3$ $HOCH_2(CH_2)_8COOH$ $HOOC(CH_2)_8COOH$

2-Octanol   10-Hydroxydecanoic acid   Sebacic acid

Figure 1. Alkali fusion of ricinoleate.

The use of thermoplastic nylon-11 for powder coatings (qv) or dry coatings that require no solvents has been developed in response to a growing concern for the environment (40). These coatings find use where wear and impact resistance are required.

Hydrogenation. The hydrogenation of castor oil can be performed in a number of ways to produce unique derivatives. The largest use for hydrogenated castor oil is in the manufacture of multipurpose greases.

Simple double bond hydrogenation at 140°C in the presence of Raney nickel catalyst produces glyceryl tris(12-hydroxystearate), mp 86°C (41–42). Care is necessary to avoid decomposition of the sensitive hydroxyl group. This synthetic wax is used commercially in waxes, polishes, cosmetics, and paper coatings. Although it lacks the gloss of some high melting natural waxes, it is preferred in specific applications where its ability to impart grease-proofness is required. Partial hydrogenation results in waxes of modified properties.

Methyl ricinoleate

$$CH_3(CH_2)_5CHOHCH_2CH = CH(CH_2)_7COOCH_3$$

↓ Pyrolysis

$$CH_2 = CH(CH_2)_8COOCH_3 + CH_3(CH_2)_5CHO$$
Methyl undecylenate   Heptaldehyde

↓ Hydrolysis

$CH_2 = CH(CH_2)_8COOH$ (10-Undecylenic acid)

↓ HBr

$BrCH_2(CH_2)_9COOH$ (11-Bromoundecanoic acid)

↓ NH_3

$NH_2CH_2(CH_2)_9COOH$ (11-Aminoundecanoic acid)

↓ Condensation

$H[HN(CH_2)_{10}CO]_nOH$ Nylon-11 polyamide

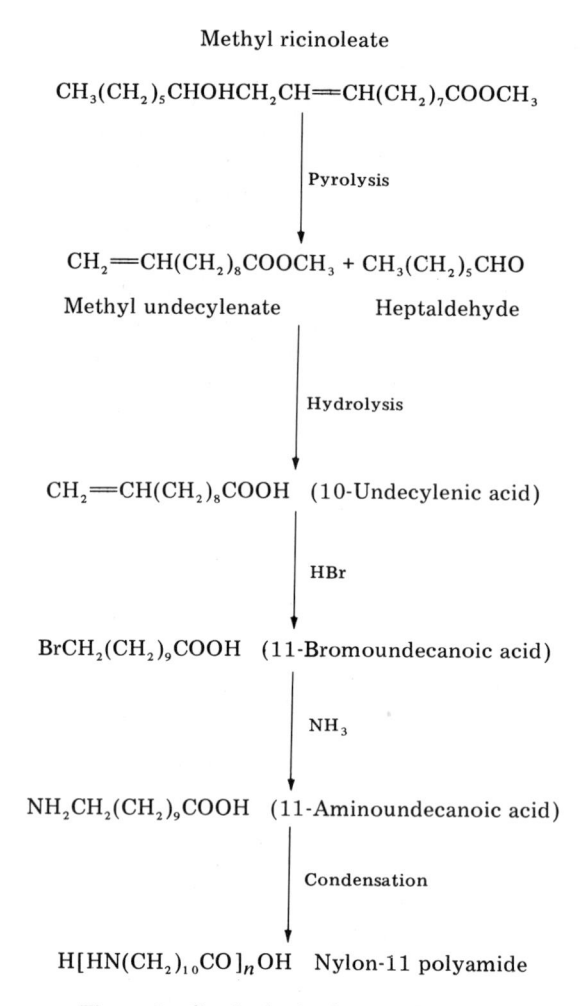

Figure 2. Synthesis of nylon-11 polyamide.

Quantitative reduction of castor oil or the methyl ester to ricinoleyl alcohol, $CH_3(CH_2)_5CH(OH)CH_2CH=CH(CH_2)_7CH_2OH$, is achieved using a secondary reducing alcohol in the presence of sodium suspensions in refluxing xylene. Hydrogenation of ricinoleic acid with a copper–cadmium catalyst at 220°C and 26 MPa (ca 250 atm) yields 70% ricinoleyl alcohol (43).

The preparation of methyl 12-ketostearate from methyl ricinoleate has been accomplished using copper chromite catalyst or by a two-step process using Raney nickel. The first step is a rapid hydrogenation to methyl 12-hydroxystearate, the hydrogen coming from the catalyst, followed by a slower dehydrogenation to the keto acid (44–45).

Stearates can be prepared by dehydrating hydrogenated castor oil or by dehydration of castor oil followed by full hydrogenation of the diene intermediate (46).

The hardness, flexibility, mp, and iodine value of the finished products are controlled by degree of hydrogenation. Esters of castor oil are changed by hydrogenation from fluid products to soft waxes with a mp range of 45–65°C. These products are used in leather coatings requiring oil resistance and water imperviousness, and in roll leaf foils because of their alcohol solubility and excellent wetting adhesion to metallic particles.

Pyrolytic Decomposition. Pyrolysis of castor oil at 340–400°C splits the ricinoleate molecule at the hydroxyl group to form heptaldehyde and undecylenic acids. Heptaldehyde is used in the manufacture of synthetic flavors (qv) and fragrances (see Perfumes). It also is converted to heptanoic acid by various oxidation techniques and to heptyl alcohol by catalytic hydrogenation. When heptaldehyde reacts with benzaldehyde by the aldol reaction, alpha-amyl cinnamic aldehyde, $C_6H_5CH=C(C_5H_{11})CHO$, is produced. 10-Undecylenic acid and its derivatives are used primarily for their fungicidal and bactericidal properties (see Chemotherapeutics, antimycotic). A combination of undecylenic acid and zinc undecylenate is used in the treatment of athlete's foot infections. The copper salt has been compounded into ointments used in treating facial and body infections.

Alkoxylation. Ethylene oxide and propylene oxide react with the hydroxyl groups of castor oil to yield a variety of polyoxyalkylene derivatives. The reaction is carried out at 120–180°C and 0–405 kPa (0–4 atm) using alkaline catalysts such as sodium hydroxide. Free radical acid catalysts are also used at lower temperature and pressure (47).

The ethoxylated derivatives of castor oil and hydrogenated castor oil are nonionic surface active agents with varied degrees of hydrophobic–hydrophilic properties. The low level ethoxylated derivatives are water emulsifiable and are used as defoamers (qv) and deemulsifiers for petroleum emulsions. The highly ethoxylated products are excellent solubilizers for water-insoluble oils in cosmetic compositions (48–50). They are also used as components in detergents, lubricating and cutting oils, hydraulic fluids (51), textile finishing compositions, and as antistatic agents (qv) for nylon carpets and apparel (52–53). The propoxylated derivatives of castor oil are mineral-oil soluble and useful in lubricating oils and hydraulic fluid compositions (see Hydraulic fluids).

Oxidation. Oxidized (blown) castor oils are clear viscous oils resulting from controlled oxidation by intimate mixing (blowing) of castor oil at 80–130°C with air or oxygen, with or without the use of a catalyst. The reaction is a combination of oxidation and polymerization promoted by transition metals like iron, copper, and manganese (54–55). The viscosity, acidity, color, and other properties are highly dependent on the reaction conditions, and a variety of products can be prepared (Table 5).

Oxidized castor oils are excellent nonmigrating, nonvolatile plasticizers for cellulosic resins, poly(vinyl butyral), polyamides, shellac, and natural and synthetic

Table 5. Properties of Oxidized Castor Oils

Property	Value
color, Gardner	2–15 (pale to dark brown)
viscosity, cm^2/s (= stokes), 25°C	11–2000 to gel
sp gr at 25°C	0.970–1.028
acid value	5–25
saponification value	188–230
iodine value	80–50
hydroxyl value	160–125

rubber. The high viscosity products are also used as tackifiers in gasket compounds and as adhesives because of their good oil and solvent resistance. They also serve as excellent pigment grinding media and as a base for lubricating and hydraulic oils (56).

Economic Aspects

The recent United States production, imports, and domestic consumption of castor oil are listed in Table 6.

Principal countries of origin were Brazil (28,500 t in 1975), with smaller amounts from Paraguay (500 t in 1975) and elsewhere in Asia. Estimated world production in 1975 was 349,000 t (USDA updated *World Agricultural Production and Trade Statistical Report*).

The principal grade of castor oil imported into the United States is classified as no. 1 castor oil by the International Castor Oil Association and also as no. 1 technical or imported no. 1.

The 1975 average wholesale price of Brazilian no. 1 technical castor oil in tank cars, fob, New York, was $0.145/kg.

Uses

Castor oil, known primarily for its medicinal use as a cathartic, is now used primarily as an industrial raw material for the preparation of chemical derivatives used in coatings, urethane derivatives, surfactants and dispersants, cosmetics, lubricants, etc.

Urethanes. Castor oil and its many derivatives have been used successfully for at least twenty years as polyols for the preparation of polyurethanes, particularly in the preparation of polyurethane coatings, adhesives, and casting compounds (58–63). Urethanes are prepared by the reaction of polyhydroxy compounds such as castor oil with polyfunctional isocyanates. Urethanes prepared from castor-based polyols are characterized by their excellent hydrolytic stability and shock absorbing and electrical insulating properties (64–73) (see Adhesives; Coatings; Urethane polymers).

Transesterification (alcoholysis) of castor oil with polyols such as trimethylolpropane or propylene glycol results in castor oil polyols with higher or lower hydroxyl

Table 6. United States Castor Oil Supply and Consumption, 1965–1975[a]

Year	Production, t	Imports, t	Consumption, t
1965	12,247	58,730	68,570
1966	9,072	47,230	70,045
1967	9,072	44,060	68,830
1968	6,804	48,090	72,210
1969	18,144	64,440	75,710
1970	8,618	48,290	60,290
1971	3,629	34,690	59,080
1972	1,361	44,240	64,770
1973	1,814	45,960	55,730
1974	2,268	55,660	55,240
1975	454	31,130	30,500

[a] Ref. 57.

functionality (see also Alkyd resins). Polymerization of castor oil, chemical or oxidative, results in higher viscosity or bodied oils which are more useful in urethane coatings than the untreated castor oil (61). Other derivatives used to prepare urethanes are amides prepared by reaction of castor oil with alkanolamines, amides of ricinoleic acid with long chain di- and triamines, and butanediol diricinoleate (74). Other methods of preparing castor-based polyols are by acetylation, dehydration, epoxidation, and hydrogenation.

Castor oil and its polyol derivatives have been found to be very useful in the preparation of rigid, semirigid, and flexible urethane foams. These foams are resistant to moisture, are shock-absorbing, and have good low temperature flexibility. Castor-based urethane foams have been used in the manufacture of foam packaging, in applications requiring high shock absorbing properties, in clothing interliners, and in filters. High density semirigid castor-based urethane foams have also been used for potting electrical components where their excellent electrical and shock-absorbing properties, as well as their excellent hydrolytic stability, are utilized.

Castor oil's resistance to hydrolysis, pigment dispersion ability, and compatibility with polyether polyols has also made it useful as a modifier for polyether-based foam. Foams based on castor oil are semirigid at low densities and become more rigid as the foam density increases. These foams generally possess an open cell structure (75). Castor oil can also be used to formulate commercially acceptable rigid urethane foams for such uses as thermal insulation and structural support (76). Superior rigid urethane foams have been prepared from hydroxymethylated polyol esters of castor acids (77). Recently, considerable emphasis has been placed on developing fire retardant urethane foams. Brominated castor oil has been investigated as a modifier for preparing fire-resistant urethane foams (78–79) (see Foams).

Coatings. Dehydrated castor oil (DCO) fatty acids have been copolymerized with acrylic esters or allowed to react with hexamethoxy methylolated melamine resin and neutralized with ethyl amine to give an aqueous solution reportedly useful as a water-soluble resin for an electrodeposition coating (80) (see Amino resins).

Sulfated castor oil and an emulsifiable polyethylene can be used to make a creamy paste pigment-dispersing agent. This paste can be blended with resins and pigments in high-flash naphtha to develop a white baking enamel with good nonsettling readings after three months (81).

Castor oil condenses with triethylenetetramine to give diamides which, when neutralized with H_3PO_4, act as emulsifiers for nitrocellulose lacquers (82).

The quality of poly(phenylene sulfide) coatings was improved by using polyoxyethylated castor oil as a surfactant (eg, Emulphor EL-620 or Emulphor EL-719) (83).

Castor oil fatty acid reacts with gallic acid using a tetraisopropyl titanate catalyst to give a castor oil fatty acid adduct from which an amide hardener is made using diethylenetetraamine (DETA). This hardener can be mixed with Epikote 1001 epoxide resin coating compound to give an anticorrosive film-forming coating composition (84). The adduct can also be used with commercial epoxy resin EPN 740 with triethylamine in mixed solvents. This will produce a thick anticorrosive film on rusted steel sheet with good salt-water resistance (85).

A vinyl chloride–vinyl propionate copolymer varnish at 40% solids with TiO_2, talc, baryta, dioctyl phthalate (DOP) in ethyl acetate–xylene can be mixed with Castorwax and Gelton 50 (acid treated $CaCO_3$) to yield a stable coating for spray-coating glass plate, giving a 30 μm thick sag-free film (86).

Many other coating applications have been reported (88–91).

Surfactants and Dispersants. A sulfated derivative of castor oil is used as a dispersant for plaster of Paris. It reduces the amount of water needed to form a plastic slurry (92). Sulfated and sulfonated derivatives have also been used in electroplating zinc (93), as softeners in dishwashing compounds (94), and in glass cleaning (95).

A hydrolysis-sensitive biocide was stabilized using an ethoxylated castor oil as the emulsion stabilizer for use in a pesticide (96). Alkylene oxide-modified castor oil has also been used in ink for pigment dispersion (97).

Polyoxyethylated castor oil (42 moles ethylene oxide) has been approved by FDA as an emulsifier in nitrocellulose coatings for paper and paperboard intended for use in contact with fatty foods at a maximum level of 8% of the coating solids (98).

Emulsified noncreosotic pine oil disinfectant compositions were prepared using <7% of an emulsifier mixture and soaps of castor oil and a mixture of soaps of tall oil (88) (see Disinfectants).

A partially saponified castor oil surfactant was prepared by reaction of an ethoxylated castor oil with alkali metal hydroxide. The product had excellent detergency, low foaming characteristics, stability in alkali and acid solutions, complete water solubility, and no cloud point in typical aqueous use solutions (99) (see Surfactants; Dispersants).

Lubricants. A nontoxic lubricating cutting oil for metals at 700–1200°C can be prepared from aluminum powder, graphite, potassium soap, and castor oil (100).

Oils suitable as lubricants for rolling of steels at high temperatures are prepared by mixing castor oil, alkylene oxide-containing castor oil, and C_4–C_{24} fatty acid esters (101).

A metalworking lubricant useful for drawing aluminum tubing can be prepared by mixing sulfur-free wool–wax, aluminum stearate, castor oil wax, and 1,1,1-trichloroethane at 38–49°C (102).

Aviation oils were prepared by esterification of thermally treated or oxidized castor oil with alcohols in the presence of H_2SO_4 or HCl (103).

A highly stable oil was prepared by the reaction of castor oil with an epoxy compound (eg, ethylene oxide, butadiene oxide, styrene oxide). The product was hydrogenated and dispersed in oils or water to yield a highly stable rolling oil (104).

Textiles. Different polyamide fibers with different affinities for anionic dyes are dyed level shades by pretreatment with aqueous acidic solutions containing sulfated castor oil (105).

Readily dispersible antistatic agents (qv) are prepared by reaction of polyoxyalkylated hydrogenated castor oil and a dialkyl oxalate or malonate. A mixture of poly(ethylene terephthalate) is spun with this product to give fibers having static discharge properties (106).

Other uses of sulfated castor oil in textiles have included lubrication (107) and imparting hydrophilicity to polyester tricot (108).

Cosmetics. Ethoxylated castor oils (Cremophor EL) or ethoxylated Castorwaxes (Cremophor RH, a 40/60 ethoxylated-hydrogenated castor oil) are used as solubilizers of hydrophobic substances such as perfumes in cosmetics. Ethoxylated castor oil was the best solubilizer in a multicomponent system, and the hydrogenated oil was more effective than the unsaturated oil. Other ethoxylated triglycerides were not as effective as castor oil. Ethoxylated castor oil was also a good solubilizer for vitamin A palmitate (109).

Castor oil and hydrogenated castor oil impart emollient and lubricant properties to ointments and cosmetic preparations (110).

Miscellaneous Uses. Castor oil and polychlorinated biphenyl (PCB) were principal impregnants used in energy storage capacitors. Castor oil is better than PCB for pulsed applications (111).

A cigarette filter with improved nicotine and tar absorption characteristics is obtained by spraying glyceryl triacetate on crimped acetate fiber tow and subsequently applying castor oil and 75% poly(vinyl alcohol) (112).

Stable protease-containing tablets can be prepared with ethoxylated castor oil (113).

BIBLIOGRAPHY

"Castor Oil" in *ECT* 1st ed., Vol. 3, pp. 237–244, by R. L. Terrill, Spencer Kellogg and Sons, Inc.; "Castor Oil" in *ECT* 2nd ed., Vol. 4, pp. 524–534, by G. J. Hutzler, Spencer Kellogg Division of Textron, Inc.

1. E. L. D'Aquin and co-workers, *J. Am. Oil Chem. Soc.* **37,** 93 (1960).
2. M. Mourgne and co-workers, *Bull. Soc. Chim. Biol.* **40,** 1465 (1958).
3. A. M. Altschul, *Processed Plant Protein Foodstuff,* 1st ed., Academic Press, Inc., New York, 1958, pp. 835–844.
4. M. A. Jones, *Chemurgic Dig.* **15,** 4 (1957).
5. J. R. Spies and co-workers, *J. Am. Chem. Soc.* **65,** 1720 (1943).
6. J. R. Spies and co-workers, *J. Am. Chem. Soc.* **66,** 748 (1944).
7. E. J. Coulson and co-workers, *J. Allergy* **21,** 34 (1950).
8. *Health Aspects of Castor Bean Dust, Public Service Pub. No. 999-AP-36,* U.S. Dept. of Health, Education, and Welfare, Washington, D.C., 1967.
9. H. K. Gardner, Jr., and co-workers, *J. Am. Oil Chem. Soc.* **37,** 142 (1960).
10. A. C. Mottola and co-workers, *J. Am. Oil Chem. Soc.* **49,** 101 (1972).
11. Ref. 10, p. 662.
12. J. W. Hinkson and co-workers, *J. Am. Oil Chem. Soc.* **49,** 196 (1972).
13. A. C. Mottola and co-workers, *J. Am. Oil Chem. Soc.* **48,** 510 (1971).
14. A. C. Mottola and co-workers, *J. Agric. Food Chem.* **16,** 725 (1968).
15. A. C. Mottola and co-workers, *J. Am. Oil Chem. Soc.* **47,** 458 (1970).
16. U.S. Pat. 2,125,544 (Aug. 2, 1938), I. M. Colbeth (to Baker Castor Oil Co.).
17. U.S. Pat. 2,739,164 (Mar. 20, 1956), H. M. Weber (to Sherwin-Williams Co.).
18. U.S. Pat. 3,469,617 (Sept. 30, 1969), E. H. Palmason (to Parkson Industrial Equipment Co.).
19. *1976 Annual Book of ASTM Standards, D-960-73,* Pt. 29, American Society for Testing and Materials, Philadelphia, Pa., pp. 525–526.
20. F. C. Naughton, *J. Am. Oil Chem. Soc.* **51,** 65 (1974).
21. K. T. Achaya, *J. Am. Oil Chem. Soc.* **48,** 759 (1971).
22. Y. Toyama and co-workers, *Fette Seifen Anstrichm.* **71,** 195 (1969).
23. U.S. Pat. 2,140,271 (Dec. 13, 1938), A. Schwarcman (to Spencer Kellogg & Sons, Inc.).
24. U.S. Pat. 2,484,328 (Oct. 11, 1949), M. Agster and R. L. Terrill (to Spencer Kellogg & Sons, Inc.).
25. U.S. Pat. 2,330,181 (Sept. 21, 1943), A. Schwarcman (to Spencer Kellogg & Sons, Inc.).
26. U.S. Pats. 1,979,495 (Nov. 6, 1934); 1,942,778 (Jan. 9, 1934), J. Scheiber.
27. U.S. Pat. 2,392,119 (Jan. 1, 1946), I. M. Colbeth (to Baker Castor Oil Co.).
28. S. B. Radlove, *J. Am. Oil Chem. Soc.* **25,** 268 (1948).
29. F. Frank, *Am. Paint J.* **53**(1), 17 (July 29, 1968).
30. K. T. Achaya, *J. Am. Oil Chem. Soc.* **48,** 759 (1971).
31. D. Burton and G. F. Robert-Shaw, *Sulphated Oils and Allied Products,* Chemical Publishing Co., New York, 1942.
32. F. Kremers, *J. Am. Oil Chem. Soc.* **48,** 314 (1971).
33. U.S. Pats. 2,851,491; 2,851,492 (Sept. 9, 1958), F. C. Naughton and P. C. Daidone (to Baker Castor Oil Co.); 2,851,493 (Sept. 9, 1958), F. C. Naughton (to Baker Castor Oil Co.).
34. A. S. Gupta and J. S. Aggarwal, *J. Sci. Ind. Res.* **13B,** 277 (1954).
35. M. J. Diamond, R. G. Binder, and T. H. Applewhite, *J. Am. Oil Chem. Soc.* **42,** 882 (1965).

36. M. J. Diamond and T. H. Applewhite, *J. Am. Oil Chem. Soc.* **44,** 656 (1967).
37. M. J. Diamond and co-workers, *J. Am. Oil Chem. Soc.* **48,** 678 (1971).
38. R. A. Dyntham and B. C. L. Weedon, *Tetrahedron* **8,** 246 (1960); **9,** 246 (1961).
39. R. Aelion, *Fibres Eng. Chem.* **17,** 78 (1956).
40. J. C. Nesbitt, *Nylon-11 Paper,* Society of Manufacturing Engineers, Dearborn, Mich., Apr. 1973.
41. R. Shripathi, K. S. Chari, and J. S. Aggarwal, *J. Sci. Ind. Res.* **21D,** 89 (1962).
42. R. Shripathi and co-workers, *Indian J. Technol.* **1,** 320 (1963); **2,** 21 (1964).
43. A. J. Pantulu and K. T. Achaya, *J. Am. Oil Chem. Soc.* **41,** 511 (1964).
44. B. Freedman and T. H. Applewhite, *J. Am. Oil Chem. Soc.* **43,** 125, 342 (1961).
45. B. Freedman and co-workers, *J. Am. Oil Chem. Soc.* **42,** 340 (1965).
46. S. N. Modak and J. G. Kane, *J. Am. Oil Chem. Soc.* **42,** 428 (1965).
47. U.S. Pat. 2,542,550 (Feb. 20, 1951), J. P. McDermott (to Standard Oil Development Co.).
48. K. Seig, *Seifen Oele Fette Wachse* **101**(11), 299 (1975).
49. Fr. Pat. 1,554,539 (Jan. 17, 1969), (to Badische Anilin- und Soda-Fabrik A.G.).
50. K. Muller, *Tenside* **3**(2), 37 (1966).
51. U.S. Pat. 3,661,782 (May 9, 1972), M. K. Smith (to Baker Castor Oil Co.).
52. U.S. Pat. 3,475,898 (Nov. 4, 1969), E. E. Magat and W. H. Sharkey (to E. I. du Pont de Nemours & Co., Inc.).
53. U.S. Pat. 2,542,550 (Feb. 20, 1951), J. P. McDermott (to Standard Oil Development Co.).
54. *Fette Seifen Anstrichm.* **62,** 496 (1960).
55. V. Anantakrishnam Sekharipuram and co-workers, *Indian J. Chem.* **9**(11), 1304 (1971).
56. D. Ya. Kolomatskii, *Otkrytiya, Izobret. Prom. Obraztsy. Tovarnye Znaki (USSR)* **52**(11), (1975).
57. *U.S. Dep. Agric. Agric. Res. Serv. Stat. Bull.* (560), (1976).
58. T. C. Patton, H. M. Metz, *Off. Dig.* **32**(421), 202 (1960); T. C. Patton, A. Ehrlich, and M. K. Smith, *Rubber Age* **86,** 639 (1960).
59. T. L. Smith and A. B. Magnusson, *Rubber Chem. Technol.* **35,** 753 (1962).
60. T. C. Patton, *Paint Ind.*, 15 (Sept. 1959).
61. F. M. Frank, *Paint Varn. Prod.* **60**(1), 43 (1970).
62. N. Legue, *J. Paint Technol.* **38**(501), 620 (1966).
63. Jpn. Kokai 75 51,544 (May 8, 1975), R. Nakatsuka and co-workers (to Sumitomo Bakelite Co. Ltd.).
64. U.S. Pat. 3,483,150 (Dec. 9, 1969), A. Ehrlich, J. W. Hayes, and T. C. Patton (to Baker Castor Oil Co.)
65. U.S. Pat. 3,345,311 (Oct. 3, 1967), A. Ehrlich, M. K. Smith, and T. C. Patton (to Baker Castor Oil Co.).
66. U.S. Pat. 3,362,921 (Jan. 9, 1968), A. Ehrlich, T. C. Patton, and M. K. Smith (to Baker Castor Oil Co.).
67. T. F. Kroplinski, "Polyurethane Casting Compounds Based on Castor Oil," *S.P.E. 33rd ANTECH, May 1975,* S.P.E., Greenwich, Conn., pp. 242–245.
68. A. Ehrlich, "Thermal and Hydrolytic Stability of Urethane Resin," *IEEE Proceedings of the 9th Electrical Insulation Conference, Sept. 8–11, 1969,* Institute of Electrical and Electronic Engineers, Piscataway, N.J., pp. 56–60.
69. A. J. Sherburne, *Insulation,* (July 1965).
70. O. K. Olsen and co-workers, *Insulation,* (Dec. 1966).
71. J. J. Mellon, *Insulation,* (June 1968).
72. S. Kaufman and R. Sabia, "Reclamation of Water-Logged Buried PIC Cable," *Wire and Cable Symposium,* U.S. Army Electronics Command, Washington, D.C., Dec. 1972, pp. 64–74.
73. M. Brauer and R. Sabia, "Design Considerations, Chemistry and Performance of a Reenterable Polyurethane Encapsulant," *International Wire and Cable Symposium Proceedings,* U.S. Army Electronics Command, Washington, D.C., Nov. 1975, pp. 104–111.
74. C. K. Lyon and V. H. Garrett, *J. Am. Oil Chem. Soc.* **50**(4), 112 (Apr. 1973).
75. A. Ehrlich and T. C. Patton, *Mod. Plast.* **41,** (Mar. 1964).
76. R. H. Leitheiser and co-workers, *J. Cell. Plast.* **5,** 364 (1969).
77. C. K. Lyon, V. H. Garrett, and E. W. Frankel, *J. Am. Oil Chem. Soc.* **51**(8), 331 (1974).
78. U.S. Pat. 3,704,256 (Nov. 28, 1972), R. T. Shone and T. W. Findley (to Swift & Co.).
79. C. K. Lyon and G. Fuller, *Plast. Aust.* **20**(9), 78 (1969).
80. Jpn. Pat. 76 32,654 (Apr. 14, 1976), (to Dainippon Ink & Chemicals).
81. U.S. Pat. 3,985,568 (Oct. 12, 1976), C. Swenson, F. Naughton, and A. Franco (to NL Industries, Inc.).

82. Pol. Pat. 80,751 (Feb. 20, 1976), S. Ropuszynski and H. Matyschok (to Politechnika Wroclawska).
83. U.S. Pat. 3,968,289 (July 6, 1976), D. E. Higbee (to Phillips Petroleum Co.).
84. Jpn. Kokai 76 68,633 (June 14, 1976), T. Sato (to Kansai Paint Co. Ltd.).
85. Jpn. Kokai 76 68,632 (June 14, 1976), T. Sato (to Kansai Paint Co. Ltd.).
86. Jpn. Kokai 76 65,138 (June 5, 1976), T. Nishimura (to Nippon Paint Co. Ltd.).
87. Jpn. Kokai 76 07,036 (Jan. 21, 1976), K. Saito, H. Terakado, and J. Matsusaka (to Nippon Oil & Fats Co.).
88. Ger. Offen. 2,529,800 (Jan. 22, 1976), J. Mache (to Ugine Kuhlmann).
89. Ger. Offen. 2,523,611 (Dec. 11, 1975), K. J. O'Hara and H. Kilvington (to Coates Bros. & Co. Ltd.).
90. T. A. Amfiteatrova and co-workers, *Lakokras Mater Ikh Primen.* (5), 8 (1975).
91. Pol. Pat. 75,500 (Mar. 25, 1975), S. Ropuszynski and V. Plucinski (to Politechnika Wroclawska).
92. Jpn. Pat. 76 73,527 (June 25, 1976), (to Misubishi Chemical Industries Co.).
93. Ger. Offen. 2,600,216 (July 8, 1976), F. Passal (to M&T Chemical).
94. Jpn. Pat. 76 10,807 (Jan. 28, 1976), A. Mori (to Lion Fat and Oil Co.).
95. U.S. Pat. 3,915,738 (Oct. 28, 1975), J. W. Willard (to CAW Industries).
96. Ger. Offen. 2,503,768 (Aug. 5, 1976), W. Stenger, M. Grossmann, and Z. Oano (to Hoechst A.G.).
97. Jpn. Kokai 75 152,808 (Dec. 9, 1975), T. Funahashi.
98. *Fed. Reg.* **40**(100), 22251 (May 22, 1975).
99. U.S. Pat. 3,663,583 (May 16, 1972), G. M. Haynes (to Whitesone Chemical Corp.).
100. Can. Pat. 974,968 (Sept. 23, 1975), A. V. Kochinashvili and T. A. Kochineshvili (to All-Union Scient. Res. Inst. for Worker Safety (USSR)).
101. Jpn. Kokai 75 146,556 (Nov. 25, 1975), K. Onoda, T. Yahiro, and Y. Hashiguchi (to Miyoshi Oil & Fat Co.).
102. U.S. Pat. 3,692,678 (Sept. 19, 1972), T. F. Stiffler and A. M. Murphy.
103. U.S.S.R. Pat. 75,182 (Mar. 25, 1975), D. Y. Kolomatskii and L. V. Zhirnova.
104. Jpn. Pat. 74 00,965 (Jan. 10, 1974), K. Hirai and co-workers (to Daido Chemical Industrial Co.).
105. Swiss Pat. 571,611 (Jan. 15, 1976), A. Mikula (to Sandoz Ltd.).
106. Ger. Offen. 2,400,906 (July 18, 1974), I. S. Fisher and J. E. McIntyre (to Imperial Chemical Ind. Ltd.).
107. Jpn. Pat. 74 42,876 (Nov. 18, 1974), Y. Kobayashi and T. Kobayashi (to Teijin Ltd.).
108. Jpn. Kokai 75 55,677 (May 15, 1975), M. Ono, T. Ito, and S. Iuchi (to Mitsubishi Rayon Co.).
109. K. Seig, *Seifen Oele Fette Wachse* **101**(11), 299 (1975).
110. M. Windholz, ed., *The Merck Index,* 9th ed., Merck & Co., Inc., Rahway, N.J., 1976, p. 242.
111. G. P. Boicourt, *Report LA-UR-75-2150,* Los Alamos Sci. Lab., Los Alamos, N. Mex., 7 pp.
112. Ger. Offen. 2,140,346 (Feb. 22, 1973), E. Nakatsuka.
113. Jpn. Kokai 75 88,218 (July 15, 1975) T. Morikawa and Y. Sakuma (to Takeda Chem. Industries, Ltd.).

FRANK C. NAUGHTON
FRANK DUNECZKY
C. RICHARD SWENSON
THADDEUS KROPLINSKI
MURRAY C. COOPERMAN
NL Industries, Inc.

CATALYSIS

Catalysis is the key to the greatly expanded modern chemical and petroleum industries. About 90% of chemical manufacturing processes are catalytic, corresponding to 15–20% of all manufactured goods produced in the United States. Sulfuric acid, ammonia, edible oils, aromatic hydrocarbons, butadiene, cyclohexane, vinyl acetate, acetaldehyde, acetic acid, alcohols, acrylonitrile, and synthetic rubber and plastics are now produced almost completely by catalytic processes. Likewise, petroleum refining, providing by far the largest volume of industrial products, now consists almost entirely of a series of catalytic processes, such as cracking, reforming, desulfurization, isomerization, polymerization, and alkylation (see Petroleum).

These immense industrial installations are based on the application of an unusual class of substances known as catalysts, whose technical and scientific bases have been only slowly understood. Ostwald's definition that "A catalyst is a substance that alters the velocity of a chemical reaction without appearing in the end products," was proposed about 1900 and serves equally well today; it is, however, much more than a definition since the concept of change in velocity provides the basis for catalytic action. Changes in reaction rate produced by a catalyst are usually positive, corresponding to acceleration. By increasing the velocity of a desired reaction relative to unwanted reactions, the formation of a desired product can be maximized compared with undesirable by-products. This is one of the most important features of catalysis. Moreover, by increasing the velocity of a normally undetectable, slow reaction to values that can be observed or predominate, a catalyst can, for all practical purposes, actually initiate a heretofore unobserved reaction.

Catalysts are now generally believed to function through an unstable chemical complex formed between catalyst and reactant molecules. This complex reacts to produce new compounds with dissociation of the complex and regeneration of the catalyst. The catalyst can then bring about the transformation of additional reactant molecules (see also Enzymes). This cyclic behavior accounts for the ability of very small amounts of catalysts to convert large amounts of reactants. The complex is considered to obey usual chemical principles in contrast to the concept of a special catalyst force proposed by Berzelius, who coined the term catalyst about 1836.

Chemical compounds may refrain from reacting not because reaction is thermodynamically unfavorable, but because there is a certain inertia. An example is the lack of reaction between hydrogen and oxygen at room temperature. This inertia can be understood by the concept that the bonds in one reactant molecule are sufficiently strong to resist attack by another potential reactant molecule. The catalyst is able to dissociate or break such bonds, making new ones with the catalyst, with formation of a complex that can then react further.

Heterogeneous catalysts, in contrast to homogeneous catalysts, form a separate phase from reactants and products and are usually solids. They have a microporous structure and a very large internal surface area that may reach 1000 m^2/g. Despite these large surface areas these colloidal catalysts are often inactivated by the presence of small amounts of certain chemicals, commonly known as poisons, that combine with an active site on the catalyst surface, rendering it inactive. Active sites usually occupy only a small fraction of the total surface. In contrast, small amounts of certain substances called promoters or cocatalysts may cause striking increases in catalytic activity. Selectivity usually denotes the ratio of the desired reaction to others, whereas activity refers to the overall conversion.

Catalysts are comprised of a very wide range of chemicals represented by both elements and compounds, especially halides, oxides, and sulfides.

In addition to general advances in catalysis, recent developments include: manufacture of stereoregulated polymers (1–8); low-pressure, low-density polyethylene (9); polyurethane foams (10); hydrogen peroxide, acetaldehyde, and vinyl acetate from ethylene (11–16); acrylonitrile (17–18); acetic acid (19); cyclohexane; aromatics by alkylation, hydrodealkylation, isomerization, and disproportionation reactions; hydrodesulfurization of petroleum fractions including residues, gasoline by hydrocracking, use of molecular sieves (qv) in cracking catalysts, and other applications (20–24); multimetallic catalysts in reforming (25–59); and catalytic mufflers (60–81) (see Exhaust control, automotive).

In the future, application of catalysis for fuel synthesis from coal and oil shale (qv) seems certain to offer enormous potential (82–119) (see Fuels, synthetic). Another application is in the growing field of environmental control. Furthermore, certain revolutionary catalytic processes for chemicals manufacture can be expected. Some of these will increasingly utilize $CO + H_2$ as starting materials, whereas others will involve partial oxidation of hydrocarbons. Catalytic fuel cells (120–125) could also be of great future importance (see Batteries and electric cells, primary).

Encouraging theoretical progress has also occurred including a more specific definition of catalyst–reactant complexes (126–137), oxidation catalysis (138–139), the electronic factor (139–142), as well as an understanding of the mechanism as revealed by new techniques such as isotopic tracers, infrared identification of surface species, spectroscopic tools including SIMS, ISS, INS, XPS, UPS, and AES (94,142–180) (see Advanced Techniques below; also Analytical methods).

The most comprehensive information sources are: the seven-volume series on *Catalysis,* the annual *Advances in Catalysis,* the quadrennial *Proceedings of the International Congress of Catalysis,* the *Journal of Catalysis, Catalysis Reviews,* and scores of recent monographs and books (see under General References).

The industrial growth of catalysis has followed an interesting pattern (see Fig. 1). Its earliest applications have been in the production of wine, vinegar, and soap, dating back to antiquity. The first inorganic applications were introduced in the middle of the eighteenth century, and were rapidly expanded. Catalytic production of organic chemicals followed. Finally, the production of liquid hydrocarbon fuels used in energy generation rapidly grew into the largest, by far, of industrial catalytic applications.

CHEMICAL AND PHYSICAL ASPECTS

Catalysts accelerate reactions but do not change equilibria. Therefore, both forward and backward reactions are catalyzed to the same extent. The rate of a chemical reaction is proportional to a rate constant:

$$k = Ae^{-E/RT}$$

Because of the exponential form of the equation, even a small decrease in E, the activation energy, can bring about a large increase in velocity. The energy pattern for a unimolecular reaction pathway is shown in Figure 2. The catalytic pathway is a series of three chemical reactions: formation of the complex (activated adsorption), surface reaction, and desorption. The activation energy for the catalytic reactions, E_c, is much less than that for the thermal reaction, E_t, hence the overall reaction is faster.

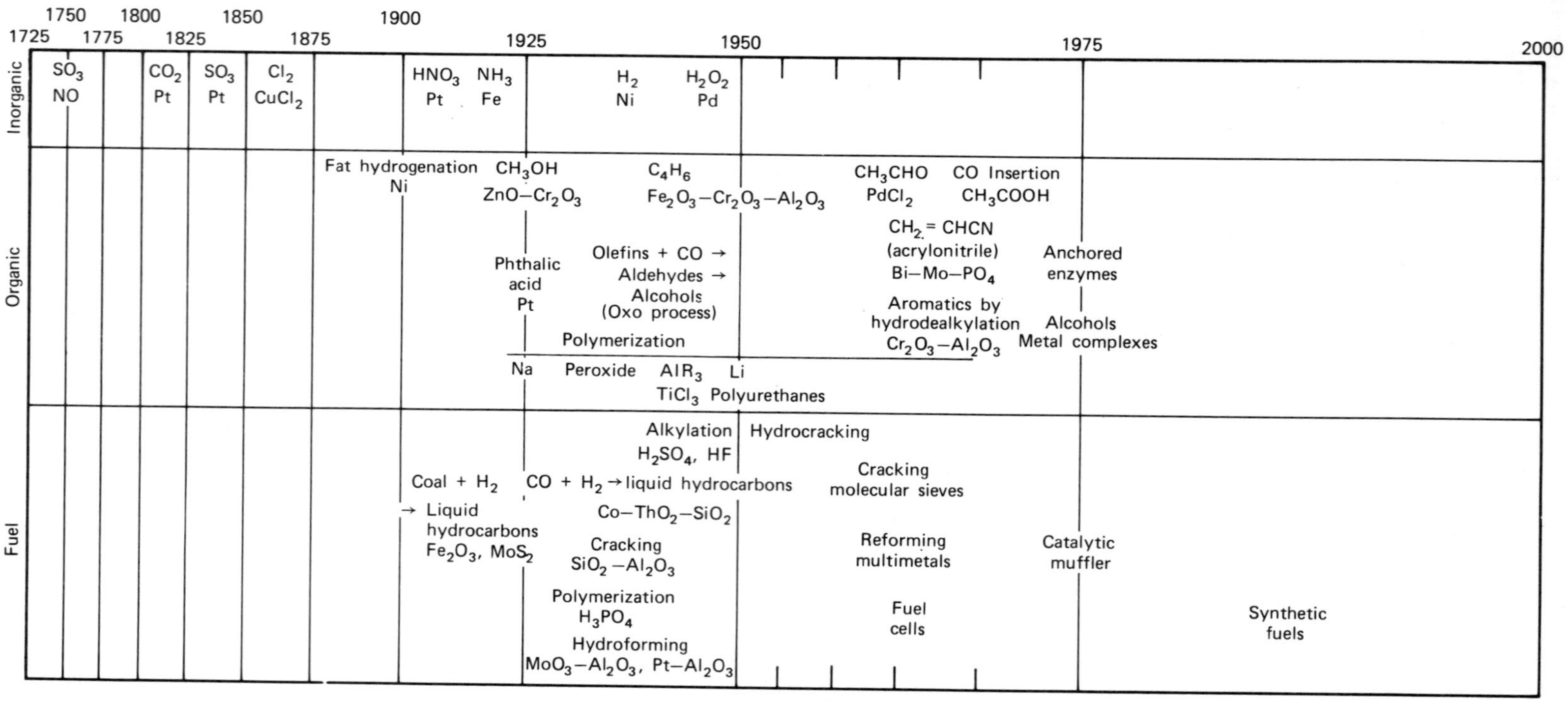

Figure 1. Survey of catalytic development.

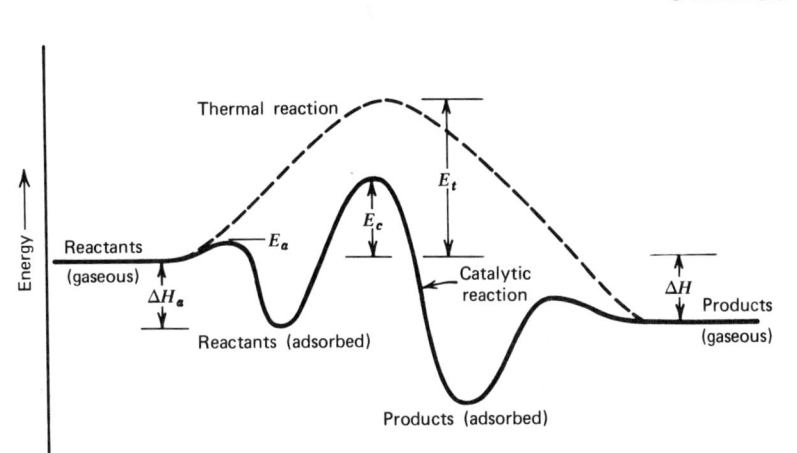

Figure 2. Energy profile for catalytic and thermal reaction. Courtesy of Reinhold Publishing Corp.

Solid surface catalysts invariably adsorb reactant molecules. Chemisorption exerts much stronger effects than physical or van der Waal's adsorption (182). Heats of chemisorption can be large and may lead to dissociation of the adsorbed molecule, eg, hydrogen on tungsten. Chemisorption frequently involves an activation energy (181), and certain atoms, such as those on corners and edges of a solid surface, show a variation in adsorptive characteristics (183). According to this concept, adsorption and chemical reaction would occur predominantly on active centers, identified as certain types of lattice defects.

The geometric factor in catalysis emphasizes the significance of the reactant molecule's spacial arrangement and the catalyst crystalline lattice (184–185). It was proposed that catalytic dehydrogenation of cycloparaffins occurs when a group of surface atoms, appropriately spaced, adsorbs the reactant molecule in a definitely oriented position. This was called the multiplet hypothesis because of the simultaneous action of several surface atoms. However, more fundamental chemical requirements have to be met for catalytic activity (186).

When alumina is dehydrated under severe conditions (600°C), it becomes more and more able to activate molecular hydrogen. This catalytic activity depends upon strain centers created in the surface of the solid structure by loss of OH groups from distances greater than normal for the structure (187), thus it is a type of geometrical catalysis. Such dehydration or other strain centers are possibly responsible for the activity of catalysts induced in solids irradiated with high-energy rays (188).

Diffusion Activity and Selectivity. The activity and selectivity of a catalyst are affected by its pore structure. In a very slow reaction, molecules can diffuse through a pore system to the center of a catalyst particle before they react, and the entire internal surface area will be used. In a fast reaction, diffusion is slower than the conversion and only the outer pore mouths of the catalyst pellet will be used. In the case of slow diffusion, all the important kinetic parameters, such as activity, selectivity, temperature coefficients, and kinetic reaction order, depend on the pore and pellet sizes of the catalyst (189–191).

The fraction of surface available in a given reaction can be calculated. The experimental rate constant, K_{exp}, is compared with

$$K_d = (18/a^2)\rho_B VgD$$

where a = catalyst granule size, ρ_B = bulk density, Vg = porosity, and D = diffusion constant.

The results for first-order reactions are given in Figure 3 and illustrate the great variation in the fraction of internal surface that is available for a number of important reactions. It can be seen that pellet diameter and porosity can affect catalytic activity.

Electronic Factors. Various theories have been applied to catalysis, including quantum mechanical treatment of solids and the electron band, crystal field, resonance valence bond, and molecular orbital ligand field theories. Considerable progress has been made in defining the principles concerning the relation of the electronic levels of reactants and catalytic solids for both metals and semiconductors (139–141).

In the 1950s, attempts were made to interpret catalysis in terms of electronic properties of solids. Although the electronic factor is certainly fundamental to catalysis, the structure of solid surfaces is not sufficiently well known to permit this theory to provide a predictive basis for catalysis.

TECHNIQUES

When choosing a catalyst for a specific reaction, the following factors should be considered: selectivity, activity, stability, physical suitability, regenerability, and cost. Table 1 gives the classification of heterogeneous catalysts and relates conductivity and catalytic functions.

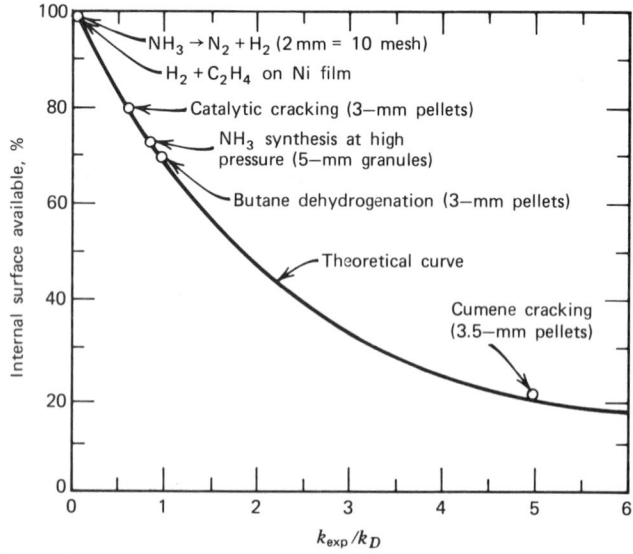

Figure 3. Dependence of fraction of surface available on k_{exp}/k_D (189–190). Note that for olefin hydrogenation, hydrogen transfer, and liquid-phase reactions on 3-mm pills the fraction of surface available is less than 20%.

Table 1. Classification of Heterogeneous Catalysts According to their Principal Functions[a]

Metals	Metal oxides and sulfides		Salts and acids
Conductors	Semiconductors	Insulators	
hydrogenation	oxidation	dehydration	polymerization
dehydrogenation	reduction	isomerization	isomerization
hydrogenolysis	dehydrogenation	(hydrogenation)	cracking
(oxidation)	cyclization		alkylation
(reduction)	(hydrogenation)		hydrogen transfer

[a] Ref. 192.

Selectivity means the efficiency in catalyzing a desired transformation. It is usually expressed by a factor representing the amount of desired product formed, divided by the amount of reactants converted. Low selectivity represents waste of reactants. It often decreases as the percentage of converted charge increases. Thus in industrial practice, a compromise is frequently made between selectivity and conversion. Nevertheless, selectivity often exceeds 90%. Another measure of selectivity is in product quality, eg, the octane number of gasoline or the desirable properties of a plastic polymer.

Activity is frequently expressed as the amount of reactant in contact with the catalyst under a given set of conditions converted to all products. Activity is less desirable than might be thought since construction of a larger catalyst reactor, of say twice the size, is ordinarily inexpensive. However, a catalyst that is active at low temperatures may contribute to selectivity since this permits operation with minimum thermal degradation.

Stability is the ability to retain initial activity and selectivity, over the catalysts' lifetime. Activity is usually lost either by overheating (frequently in the presence of steam) which causes sintering with loss of surface area, or by poisons that deactivate active centers. Loss of selectivity is usually caused by chemical transformation in the catalyst, frequently by addition of poisons which cause unwanted reactions.

Physical Suitability. Catalysts require not only appropriate porosity, which can affect diffusion and thus the activity and regenerability, but also mechanical strength in moving-bed reactors.

Regenerability. A catalyst may lose activity or selectivity because of coke deposition, which in petroleum refining, may include sulfur and nitrogen residues. Even on a small scale, coke can gradually build up to unacceptable levels. It should then be possible to regenerate the catalyst; this is most readily achieved by burning in air. Less frequently, the catalyst is chemically treated to restore activity. In the case of platinum catalysts, the platinum is usually recovered, and its reuse represents complete chemical regeneration.

Cost calculations can be made from the product value and costs of raw material processing, plant amortization, and catalyst. The latter is usually quite small (see under Economic Aspects).

Poisons. A poison, present in small amounts, adversely affects the catalyst performance. Most poisons are strongly sorbed and react quantitatively with active sites to form stable inactive surface compounds. For instance, acidic catalysts are poisoned by basic nitrogen compounds and alkali metal ions, and metallic catalysts by sulfides, arsenic, and lead compounds, and sometimes by carbon monoxide, chlorides, and water.

Some poisons cause loss of selectivity, such as the nickel, vanadium, iron, and copper present in gas oils used in catalytic cracking. These deposit on the catalyst and, because of their dehydrogenation ability, increase gas and coke formation and decrease gasoline production.

In some instances, a poison minimizes an undesired reaction while permitting a desired reaction. Thus in dehydrocyclization of paraffins to aromatics, potassium oxide is added to chromium oxide–aluminum oxide to neutralize acid sites active in cracking.

Reactivation. Decomposition or desorption of poison can sometimes be accomplished by steaming or evacuation at the maximum temperature permissible, or hydrogenation. More often, cokelike deposits are burned off with air. However, sintered catalysts cannot be regenerated. Removal of heavy metals by leaching restores cracking catalysts' selectivity and can sometimes be economically justified.

Maintenance of Catalytic Efficiency. The maintenance of efficient performance during the lifetime of a catalyst can be more important than initial catalytic efficiency. In industrial use, catalysts lose not only activity but sometimes selectivity as well, ie, the ability to produce a desirable distribution of products (quality).

In general, heterogeneous catalysts lose effectiveness because of overheating and contamination. Overheating causes sintering with loss of surface area and consequently of activity. Finely divided oxides and metals sinter at temperatures much lower than their melting points.

Contaminants affecting catalytic efficiency are metals entrained in the feedstock; oxygen, nitrogen, or sulfur compounds; and polynuclear aromatic compounds or coke. The latter can be removed by calcination. Nickel catalysts are notoriously susceptible to sulfur poisoning. However, cracking catalysts are not inactivated by sulfur compounds at the operating temperatures since they promote the formation of hydrogen sulfide which is eliminated.

PREPARATION

A number of important catalysts are comparatively simple definite chemical compounds such as sulfuric acid, aluminum chloride, and certain peroxides. However, most catalysts are complicated, high-area solids whose precise chemical and physical structure determine their effectiveness. Many catalysts consist of a major component (high-area support) with one or more lesser components. Even in high-area catalysts, only a small fraction of the surface is active.

The properties of a catalyst are determined by its preparation (193–195). Care has to be taken to ensure duplication of characteristics of surface area, pore size, crystal structure, and surface hydration and oxidation.

Colloid chemistry plays an important role in the general techniques employed; these are gel preparation, leaching selected components from solids, or decomposition of salts or hydrates with gas evolution. Frequently, the support is manufactured first by one of these techniques, followed by impregnation with additional chemical components. The final product may again require suitable treatment, for example, by precipitation, calcination, reduction, etc.

Gel Formation. Many hydrous oxides are of interest as catalysts or catalyst supports and many of them are amenable to gel formation. A primary problem arising in the preparation of a precipitated catalyst is the occlusion and adsorption of impurities. This can be minimized by the use of ammonia or ammonium salts as precipitants together with nitrates of the desired metals. The resulting precipitate then requires a minimum of washing since any adsorbed material remaining can be removed by calcination of the product. Generally, nitrates are a good source of the desired cations because other ions (chlorides, sulfates) may act as catalyst poisons. With two or more components, the greatest possible degree of homogeneity is desirable and can be achieved by simultaneous precipitation with careful selection and control of pH. Systems containing primarily silica or alumina are particularly amenable to gel formation and are suitable as catalyst constituents. Gel aging conditions are important.

Procedures. Leaching is an important method for high-area catalyst preparation. For example, Raney nickel catalyst is prepared by leaching the aluminum from a nickel–aluminum alloy by means of caustic alkali. Another example is the preparation of cracking catalysts from montmorillonite clays obtained by leaching natural clays with dilute sulfuric acid. Increased surface area is produced by leaching out part of the alumina and other nonsiliceous constituents of the clay.

A similar method is used in the production of high-area, activated carbon and also of silica in the form of kieselguhr.

Decomposition. High-area catalyst components can be prepared by decomposition of salts or hydrates with the evolution of gas, leaving the metal or oxide in a highly dispersed form. Porosity can also be obtained by the chemical reduction of a nonporous oxide; a good example is the iron catalyst used in ammonia manufacture.

Impregnation. Impregnation is a technique in which an active component is placed on a support that may be active or inactive. In its simplest form this method comprises the following steps: (*1*) evacuation of the porous support; (*2*) contacting the support with a solution containing a soluble salt of the component being added; (*3*) drying; and (*4*) calcination and activation. Frequently, a precipitation and washing step is also needed either before or after drying. A peripheral layer or complete penetration can be obtained. Deposition from the vapor phase rather than from a liquid medium is also possible.

Processing. Impurities are removed by washing at controlled pH. The gels are frequently ion exchangers and hence cannot be washed free of counterions. Ammonium ions can be exchanged for sodium ions and the ammonia removed by calcination. Time and temperature schedules of drying and calcination must be carefully controlled in addition to the chemical character of the atmosphere and the partial pressure of the steam.

An activation step may be included, such as a reduction step for hydrogenation catalysts. Nickel, cobalt, iron, or copper catalysts are prepared by reduction of the corresponding oxides in a hydrogen atmosphere at 300–500°C. The activity of these catalysts, as well as those of palladium and platinum, is a function of what extent of reduction and temperature are used. Oxides that do not reduce readily to the metal, such as chromium, molybdenum, and tungsten oxides, are reduced by hydrogen, at least on the surface. After reduction these catalysts should be handled in an inert or reducing atmosphere. Exposure of finely divided metallic catalysts to air frequently destroys their activity.

Promoters. Promoters may be added at the preparation or operational stage. Added in small amounts, a promoter itself has little activity but it improves catalyst activity, stability, or selectivity for the desired reaction. Examples are silica on molybdena and alumina on iron catalysts for the synthesis of ammonia. Certain promoters alter the electronic state and, therefore, the activity of catalysts. Other promoters such as HCl are essential in conjunction with AlCl$_3$ to provide the Brønsted acid nature of the catalyst complex. Some promoters prevent coke formation.

INORGANIC CHEMICALS

Modern industrial catalysis may be said to date from the introduction of the lead-chamber process in 1746. Nitric oxide is added to accelerate the oxidation of sulfur dioxide to sulfur trioxide. Almost seventy years later, Davy discovered that platinum hastens the combustion of mine gases and around 1832 it was found that platinum is also a catalyst for the conversion of sulfur dioxide to sulfur trioxide. The Deacon process for the catalytic oxidation of hydrogen chloride to molecular chlorine was an important milestone in 1860. The platinum-catalyzed oxidation of ammonia for nitric acid production was developed by Ostwald in 1905. The synthesis of ammonia was accomplished by Haber and Mittasch just before World War I. More recently, the catalytic productions of hydrogen and hydrogen peroxide have been notable inorganic developments.

For hydrogen production, a group of catalytic reactions is employed, namely, the reforming reaction where hydrocarbons and steam produce hydrogen and carbon monoxide, the shift reaction where carbon monoxide and steam produce carbon dioxide and hydrogen, and the methanation reaction where traces of carbon monoxide are removed by hydrogenation to methane.

In hydrogen peroxide manufacture, an alkyl anthraquinone is reduced with hydrogen in the presence of a catalyst to the corresponding hydroquinone, followed by an oxidation step where hydrogen peroxide is produced and the anthraquinone is regenerated.

ORGANIC CHEMICALS

Sabatier studied the hydrogenation of unsaturated compounds with the aid of nickel catalysts. The hardening of fat by hydrogenation was accomplished commercially by Normann in 1902. Ipatieff introduced high pressure technique that he applied successfully to catalytic hydrogenation in the early 1900s. The synthesis of methanol from carbon monoxide and hydrogen was achieved in Germany in 1923. Phthalic anhydride was produced by the oxidation of naphthalene in a process that is still used extensively. During World War II, Roelen and co-workers discovered the oxo reaction whereby carbon monoxide and hydrogen are added to olefinic materials, making possible production of aldehydes. These in turn can be catalytically hydrogenated to the corresponding alcohols. During World War II, processes were also developed for the production of butadiene by dehydrogenation of butene or butane (see Butadiene). The most successful process for butane dehydrogenation, devised by Houdry, is still being used on a worldwide basis.

Catalytic reforming of petroleum became the dominant process for aromatics following World War II. More recently, large-scale production of benzene and naphthalene by hydrodealkylation of corresponding methyl homologues has been introduced. Large-scale manufacture of other aromatics include the alkylation of benzene with ethylene and subsequent dehydrogenation to styrene (qv), isomerization of xylenes, and transalkylation and toluene disproportionation (see BTX processes). A new feature is the sophisticated integrated petrochemicals manufacture that uses hydrogen obtained from reforming.

Catalytic polymerization for production of plastics and synthetic rubber has enjoyed great industrial success. Catalysis by sodium polymerization of butadiene was introduced around 1925 by Lebedev in the Union of Soviet Socialist Republics, followed by development of emulsion polymerization techniques in Germany in the mid 1930s employing peroxide catalysts. This was developed further using redox systems (see Elastomers, synthetic). More recently olefins, notably ethylene, have been polymerized at high pressures and even at lower pressures using Ziegler-type or solid catalysts such as molybdena–alumina or chromia–silica–alumina. Of greatest interest is stereospecific polymerization catalyzed by lithium metal or aluminum alkyl and titanium chloride (see Olefin polymers).

The tremendous increase in the production of polyurethane plastics, especially as flexible and rigid foam products, has been accelerated by the discovery of high activity catalyst systems, notably triethylenediamine plus tin compounds that make possible rapid one-shot polymerization. Also important is the direct conversion of nitroaromatics to isocyanates (qv) which are used in polyurethane manufacture (196).

In a new catalytic reaction of olefins, first described in 1964 (197), propylene was disproportionated to ethylene and butene at 94% efficiency at 43% conversion. The mechanism is considered to proceed by a four-center (quasicyclobutane) intermediate involving the four doubly bonded carbon atoms of the two molecules of olefin and the catalyst, alumina–molybdena, although recently a carbene mechanism has also been proposed (198).

The large-scale production of acrylonitrile (qv) from propylene by air oxidation in the presence of ammonia is a good example of a complex catalyzed reaction.

A number of important new processes catalyzed by metallic complexes can be understood in terms of their coordination chemistry. Those operating in solution are known as homogeneous catalysts. These new processes include the oxidation of ethylene to acetaldehyde and vinyl acetate; the insertion of CO in CH_3OH to form acetic acid (qv); and new, more active and selective catalytic complexes in the oxo reaction.

Transition Metal Complexes—Homogeneous Catalysis

In the last two decades, there has been a surge of interest in the chemistry and application of catalysis by certain transition metal complexes (127–131). Many of these are classified as homogeneous catalysts since they can be molecularly dispersed in solution; eg, hydrocobalt tetracarbonyl, $HCo(CO)_4$, and chlorobis(triphenylphosphine)rhodium carbonyl, $RhClCO[P(C_6H_5)_3]_2$. Heterogeneous catalysts are usually metals, metal oxides, or sulfides. This narrow range of chemical types severely limits their chemical and therefore catalytic properties. In contrast, metal complexes can possess a whole range of organic and inorganic ligands with a virtually infinite range of properties.

Theoretical interest in homogeneous catalysts stems from the unique insight provided into the mechanisms of catalytic action. The nature and surfaces of solids are not well known or understood. By contrast, the nature of molecular and ionic species existing in solution have been relatively well established. The coordination bonding properties of the metal are a key feature of the catalytic complexes. The nature of the metal and ligands critically influences the electronic structure of the complex. A further point of interest arises from the relation believed to exist between such catalysts and those important in biocatalysis, the enzymes. The distinction between homogeneous and heterogeneous catalysts has been disappearing as fundamental similarities have been recognized (126).

Organometallic complex catalysts tend to be highly active, specific, and selective, thus minimizing unwanted reactions. Furthermore, their state of molecular dispersion maximizes contact time, and they should be more effective than heterogeneous catalysts. In addition, they may resist poisoning better than heterogeneous systems.

Recent advances in organometallic chemistry such as the synthesis of polynuclear cluster complexes (132) promise to extend the ranges of catalytic properties of organometallic complexes and to give greater control of catalyst selectivity and specificity. Furthermore, the anchoring of complexes and clusters to heterogeneous surfaces has been developed as a mechanism for obviating the separation and corrosion problems that hinder the commercial development of organometallic catalysts.

Some homogeneously catalyzed reactions are summarized in Table 2 (131); the most important industrial reactions are shown in Table 3 (199). The catalysts are generally dissolved in the liquid, and all the reactor designs require the introduction of reactant gas into a liquid phase. Since the reactions are usually exothermic, liberating roughly 127–209 kJ/mol (30–50 kcal/mol) of reactant, the well-mixed liquid medium allows for rapid heat removal from the reactor. Since most of the processes operate with soluble catalysts, it is difficult and expensive to separate the reaction products from the catalysts.

Research on the activation of molecular hydrogen by homogeneous catalysts established the mechanism in terms of homolytic or heterolytic splitting of the hydrogen, including the concept that the catalytic center has two metal sites (133–137).

Organometallic catalysts have a central transition metal atom or ion bonded to atoms known as ligands which form a polyhedron around the metal. Partly filled d-orbitals and the next higher s- and p-orbitals are used to form the metal–ligand bonds. Bonds can be formed by a combination of an empty (Lewis acid) and a filled ligand orbital (Lewis base) or vice versa.

Important reactions of catalyst complexes are ligand exchange, oxidative addition, and the insertion reaction. With the latter, an atom or group is inserted between two atoms initially bound together. The reaction takes place in the coordination sphere of a transition atom or ion. The insertion step (below) can be followed, eg, by reaction with H_2 to form

$$H-\underset{\underset{O}{\|}}{C}-R$$

$$-\underset{|}{\overset{R}{\underset{|}{M}}}-CO \longrightarrow \left[-\underset{|}{M} \cdots \cdots \overset{R}{\underset{}{\cdots}} CO \right] \longrightarrow -\underset{|}{M}\cdots\cdots\underset{\underset{O}{\|}}{C}-R$$

transition state

Table 2. Some Reactions of Olefin Catalyzed By Transition Metal Complexes[a]

Reaction	Ru(II)	Co(II)	Fe	Co(I)	Rh(I)	Ir(I)	Pd(II)	Pt(II)[b]
hydrogenation	RuCl$_6^{4-}$	Co(CN)$_5^{3-}$	Fe(CO)$_5$	CoH(CO)$_4$	RhClL_3^c	IrI(CO)L_2^c		Pt(SnCl$_3$)$_5^{3-\ b}$
hydroformylation	RuCl$_2$L$_4^{b,c}$		FeH(CO)$_4^-$	CoH(CO)$_4$	RhCl(CO)L_2^c	IrCl(CO)L_2^c	PdCl$_4^{2-}$	Pt(SnCl$_3$)$_5^{3-\ b}$
double bond migration				CoH(CO)$_4$	RhCl$_3$(olefin)$_2^{2-}$			
polymerization								
oxidation					RhCl$_2$(C$_2$H$_4$)$_2^-$		PdCl$_4^{2-}$	

[a] Ref. 131.
[b] Active form of catalysts uncertain.
[c] L = P(C$_6$H$_5$)$_3$.

Table 3. Industrial Processes Employing Transition Metal Complex Catalysis[a]

Process	Reaction	Typical catalyst	Approximate temperature, °C	Approximate pressure, MPa[b]
Wacker	C$_2$H$_4$ + 1/2 O$_2$ → CH$_3$CHO	PdCl$_2$–CuCl$_2$, aq solution	110	0.5
vinyl acetate	C$_2$H$_4$ + 1/2 O$_2$ + CH$_3$CO$_2$H → CH$_3$CO$_2$CH=CH$_2$ + H$_2$O	PdCl$_2$–CuCl$_2$, aq solution	130	3
oxo	RCH=CH$_2$ + CO + H$_2$ → RCH$_2$CH$_2$CHO + RCHCHO|CH$_3$	HCo(CO)$_4$, org. solution or RhCl(CO)[P(C$_6$H$_5$)$_3$]$_2$, org. solution	150	25
methanol carbonylation	CH$_3$OH + CO → CH$_3$CO$_2$H	RhCl(CO)[P(C$_6$H$_5$)$_3$]$_2$ with CH$_3$I promoter, org. aq sol	80 175	1.5 1.5
Ziegler-Natta polymerization	C$_2$H$_4$ → $\frac{1}{n}$(C$_2$H$_4$)$_n$	α-TiCl$_3$ (solid) + Al(C$_2$H$_5$)$_2$Cl susp in org. liq	70	0.5
	C$_3$H$_6$ → $\frac{1}{n}$(C$_3$H$_6$)$_n$		100	1

[a] Ref. 199.
[b] To convert MPa to atm, divide by 0.101.

The above equation illustrates a three-center transition state, called cis-insertion since the reactants are bonded adjacent to each other on the metal.

The advantage of complex catalysts including the anchored complexes lies in the predictability of their properties, the ease of interpreting reaction mechanisms, and the consequent potential for extending the homogeneous catalysts to many new applications (200).

The Wacker Process. In the Wacker process, ethylene is oxidized to acetaldehyde (qv) in aqueous solution in about 95% yield over a Pd catalyst (13).

$$CH_2=CH_2 + H_2O + PdCl_2 \rightarrow CH_3CHO + Pd + 2\,HCl$$

The reduced palladium is reoxidized by $CuCl_2$ (11).

$$Pd + 2\,CuCl_2 \rightarrow PdCl_2 + 2\,CuCl$$

The cycle is completed by oxidation of the cuprous chloride.

$$2\,CuCl + 1/2\,O_2 + 2\,HCl \rightarrow CuCl_2 + H_2O$$

All three reactions take place in a single vessel with only small amounts of Pd and Cu salts present. With other olefins, yields vary considerably (12).

In a different process design (14), acetaldehyde formation and reoxidation of metallic Pd take place in one reactor, whereas cuprous ion is reoxidized in a second reactor. Ethylene and the solution containing Pd and Cu salts flow concurrently in a tubular reactor. The solution is then recycled to the second reactor where air reoxidizes the copper. Alternatively, the reaction can be carried out in a single reactor having gas recycle, but not liquid recycle; oxygen replaces air. The concentration of palladium in the reactant solution is 0.02–0.2 molar, and copper concentration may be 100-fold larger. In a fixed-bed process the catalytically active Pd–Cu solution is held within the pores of a fixed bed of particles (201). A solid Wacker catalyst (Pd complexes on V_2O_5) has been described (202).

Vinyl Acetate Synthesis. A similar catalytic system is employed for the manufacture of vinyl acetate for the large-scale production of poly(vinyl acetate) and poly(vinyl alcohol). The reaction of ethylene in anhydrous acetic acid,

$$CH_2=CH_2 + CH_3C(=O)OH + PdCl_2 \rightarrow CH_3C(=O)O-CH=CH_2 + Pd + 2\,HCl$$

is followed by the oxidation of Pd with $CuCl_2$ and of the CuCl by oxygen (15). Combined yields of 95% are reported (16). The reaction mechanism is believed to proceed by π-olefin complex formation, incorporation of a nucleophile into the complex, rearrangement from a π- to a σ-complex, and decomposition of the complex into products (see Vinyl polymers).

The Oxo Reaction. In 1938, Roelen discovered the oxo reaction which led to one of the largest industrial applications of homogeneous catalysis (see Oxo process). The cobalt catalyst employed, $Co_2(CO)_8$, is soluble in hydrocarbons and forms $HCo(CO)_4$ with hydrogen. This hydrocarbonyl reacts with an olefin in a sequence of steps (203–204). In this process, CO and H_2 are added to olefins at 150°C and 20 MPa

(200 atm) to form straight and branched-chain aldehydes (qv), and under certain conditions, alcohols:

$$R-\overset{H}{\underset{|}{C}}=CH_2 + CO + H_2 \longrightarrow RCH_2CH_2CHO$$

The aldehyde can be subsequently hydrogenated to alcohols that are used extensively as solvents and in the manufacture of plasticizers (see Alcohols).

Newer catalysts incorporate phosphine ligands, as in the Shell catalyst, $HCo(CO)_3P(tert\text{-}Bu)_3$, that improve the yield of straight-chain aldehydes. Rhodium complexes such as $Rh(CO)Cl[P(C_6H_5)_3]_2$ show remarkable activity and ability to function at much lower CO partial pressure (205). Thus, the Shell catalyst functions at 10 MPa (100 atm) and the rhodium catalyst at 1.5 MPa (15 atm). The more expensive rhodium catalysts are as much as one thousand times more active than the conventional cobalt catalysts.

Yields of straight-chain aldehydes, usually the desired products, are favored by lower temperature, higher CO pressures, and bulky ligands, such as PR_3. In an interesting reactor design (206), the catalytically active species, $HCo(CO)_4$, is extracted from the organic-phase reaction products by an aqueous Na_2CO_3 solution, forming sodium cobalt tetracarbonyl, $NaCo(CO)_4$. The addition of acid regenerates the volatile $HCo(CO)_4$ which is sparingly soluble in water and is stripped by $H_2 + CO$ being carried overhead where it is dissolved in the olefin reactants.

Methanol Carbonylation. A major advancement in applied, homogeneous catalysis was the development of the production process for acetic acid (qv) from methanol (19).

$$CH_3OH + CO \xrightarrow[RhCl(CO)[P(C_6H_5)_3]_2]{\text{rhodium complex}} CH_3COOH$$

Promotors such as CH_3I or HI are employed commercially conducted at ca 1.5 MPa (15 atm) and 175°C. The reaction mechanism involves the oxidative addition of the CH_3I promotor to the rhodium complex, subsequent bonding of CO cis to the methyl group, insertion of the carbonyl group, followed by a series of steps yielding the product and regenerating the catalyst and the promotor.

Polyhydric Alcohols. In the production of ethylene glycol and other polyhydric alcohols directly from $CO + H_2$ (207), rhodium carbonyl cluster complexes, such as $[Rh_{13}(CO)_{24}H_3]^{2-}$, are believed to be the active catalysts (132,208). Equilibria involving ion-pairing interactions appear to be an essential part of the mechanism (209) (see Glycols; Alcohols, polyhydric).

Stereospecific Polymerization. The catalyzed formations of stereoregular polymers from ethylene and other α-olefins were discovered by Ziegler and Natta (1–4), who were awarded the 1963 Nobel Prize in chemistry (see Ziegler-Natta catalysts).

Ziegler catalysts comprise certain strongly electropositive transition metals of Groups IVA–VIII, eg, Ti, V, Cr, Zr, at a level of oxidation lower than the maximum, together with a compound containing carbon or hydrogen linked to a metal from Groups I–III. The classical example is the precipitate which is formed by the interaction of titanium tetrachloride and triethylaluminum in heptane, where the titanium is reduced to a lower valence state. This extraordinarily active system catalyzes the polymerization of ethylene at ordinary temperature and pressure.

CATALYSIS

In polymers of $CH_2=CHR$, every other carbon atom is asymmetric, exhibiting D and L configurations:

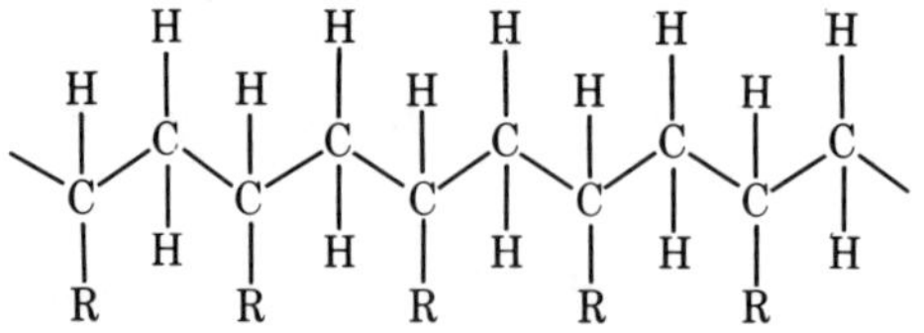

isotactic polymer

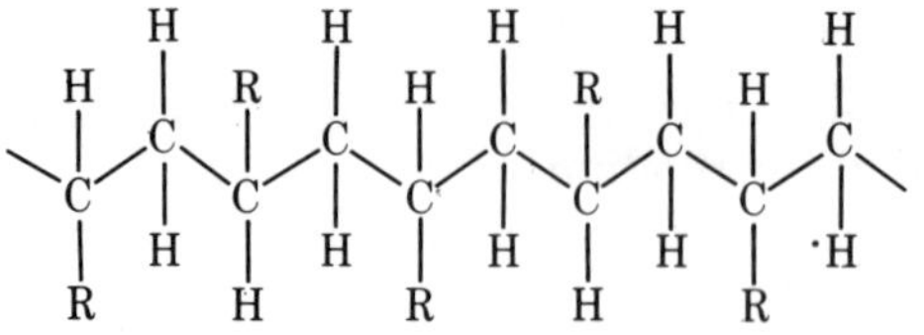

syndiotactic polymer

The Ziegler-type catalyst may be a stable complex formed between aluminum alkyl and reduced transition metal halides. The surface of the catalytic system plays an important role in controlling the stereochemistry of the propagation. The incoming monomer unit is polarized so that it is always inserted in the same head-to-tail fashion between the metal and the attached alkyl or growing chain.

The stereospecificity of the catalyst depends on its crystal structure. The most commonly used catalyst system consists of a slurry of $TiCl_3$ powder in a solution of diethylaluminum chloride, $Al(C_2H_5)_2Cl$, or triethylaluminum chloride, $Al(C_2H_5)_3$, in n-hexane or a similar solvent. Newer industrial catalysts are supported on porous solids prepared, for example, from magnesium hydroxy chloride, $Mg(OH)Cl$ (5). They have better activity and give much higher yields of polymer, up to 100 kg of polymer per gram of catalyst.

The mechanism of stereospecific polymerization (6–7) is believed to be a cis-insertion taking place on $TiCl_3$ surface sites formed by ligand exchange with $Al(C_2H_5)_3$. (The □ represents an electron pair vacancy.)

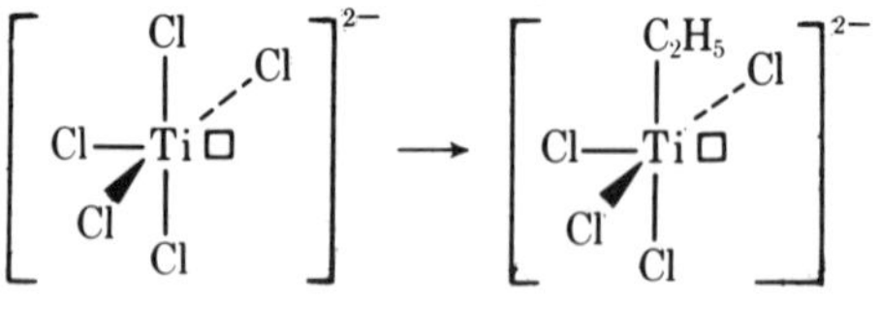

The coordinatively unsaturated, active catalyst probably adds C_2H_4 to form an intermediate leading, with cis-insertion, to an σ-bonded alkyl incorporating the new monomer unit. Addition of hydrogen terminates the chain formation.

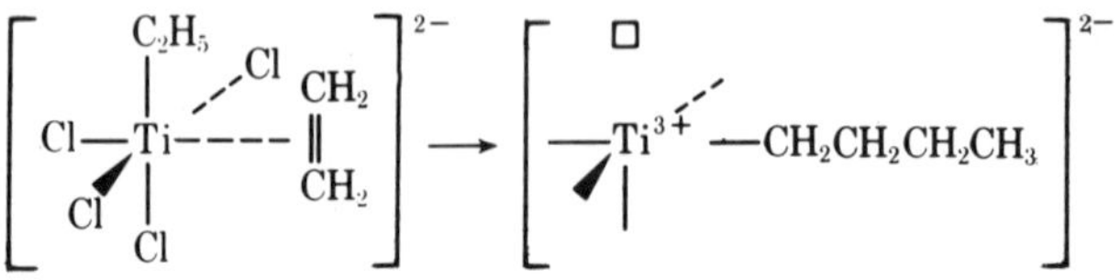

Activated transition of a metal oxide on a support catalyzes stereospecific polymerization; for example, vanadium, molybdenum, or tungsten oxides on alumina or another inert metal oxide are activated by reduction with hydrogen. The activity is increased by sodium or calcium hydride or aluminum alkyl (8). Chromic oxide on silica–alumina is activated by oxidation with air at 500°C. Polymerization is carried out in suspension or hydrocarbon solvents with monomer concentrations of 2–7%, at 100–200°C and 1.5 MPa (15 atm) pressures.

In addition to the coordination catalysts discussed above, stereospecific polymerization can be achieved by cationic, anionic, and free-radical catalysts. Free-radical propagation to form poly(methyl methacrylate) at low temperatures permits syndiotactic growth. Most polymerizations initiated with organometallic compounds are commonly referred to as anionic polymerization. It is now possible to prepare stereoregular polymers from styrene and its derivatives and various acrylate ester monomers by the use of organometallic catalysts, such as Grignard reagents, dialkylmagnesium compounds, and sodium and potassium alkyls; lithium alkyls are the most effective (see Grignard reaction; Organometallics).

Insolubilized Complexes. The difficult recovery of complex catalysts has prevented their general commercial adoption. The chemical anchoring of complexes to a solid, such as polystyrene or silica, shows great promise. These anchored or matrix-bound complexes retain many of the important features of the homogeneous species. Examples are reported for hydroformylations, methanol carbonylation, and the Wacker reaction.

Enzymatic Catalysts. The application of immobilized enzymes (210–211) is a rapidly developing field. The enzymes supported on alumina or cellulose are very selective and active, and therefore, can be used at low temperatures. Enzymatic catalysts derived from living substances may have a great industrial future. The production of dextrose from starch, and glucose from cellulose has been studied, and immobilized enzymes are used on a large scale for conversion of corn starch to a high fructose sugar. The potential production of hydrogen by biophotolysis is still in the research stage (see Enzymes, immobilized).

Oxidation Catalysis

Classical examples of heterogeneous catalysis include oxidation of CO over copper, of NH_3 over Pt, and of SO_2 over Pt or V_2O_5. Heterogenous oxidation catalysis is similar to catalysis by metal complexes, since both utilize a metal of variable valence. Some oxidations are homogeneously catalyzed, eg, the following reaction proceeds with 88% yield in pyridine solvent.

$$2\ C_6H_5\text{—}NH_2 + O_2 \xrightarrow[\text{CuCl}]{\text{pyridine}} C_6H_5\text{—}N\text{=}N\text{—}C_6H_5 + 2\ H_2O$$

Hydrocarbons. Commercially important hydrocarbon oxidation processes include not only the oxidation of benzene to maleic anhydride and toluene to benzoic acid, naphthalene and xylenes to phthalic acids, and ethylene and propylene to their respective epoxides, but also the manufacture of acrylonitrile by propylene ammoxidation (see Acrylonitrile) and of vinyl chloride by oxychlorination of ethylene.

Catalysts employed are V_2O_5, MoO_3, Ag, Cu, Pd, Bi_2O_3, U, Sb, and Fe. They are generally mixed oxides whose surface properties are not well understood. However, it has been established that first the oxygen of the catalyst is added to the hydrocarbon, and then the catalyst is reoxidized by O_2. The oxygen is taken from the oxide and resupplied from the gas phase. As early as 1936 (138), the catalytic activity of oxides for oxidation reactions was correlated with ΔH_{m-o}, the energy necessary to remove oxygen from the oxide surface. Later it was shown that the free energy of oxygen release and the catalytic activity are inversely related. The selectivity of oxide catalysts for partial oxidation of hydrocarbons, in contrast to oxidation to carbon oxides and water, is affected by the thermodynamic function $\partial \Delta H/dx$ where x is the degree of reduction caused by oxygen release. A high gradient favors partial oxidation.

Catalysts having a low oxygen–metal bond strength favor total oxidation of hydrocarbons, whereas catalysts with higher oxygen–metal bond strength catalyze selective oxidation (139) (see also Hydrocarbon oxidation).

Ammoxidation of Propylene. The catalytic ammoxidation of propylene has become the successful process for the manufacture of acrylonitrile.

$$CH_3-CH=CH_2 + NH_3 + 1\tfrac{1}{2}\,O_2 \rightarrow CH_2=CH-CN + 3\,H_2O$$

The catalysts selective for this reaction are oxide combinations containing at least two different metals, one of which is a Group VA element, such as P, As, Sb, and Bi; the last two give the most active catalysts. The second metal is always a transition metal.

Well-known ammoxidation catalysts

$Bi_2O_3 + MoO_3$	$UO_3 + Sb_2O_4$
$Fe_2O_3 + Sb_2O_4$	$SnO_2 + Sb_2O_4$

The catalysts are supported on silica and may also contain small amounts of other elements such as K, P, W, and Te.

The question arises why the acrolein formed is not further oxidized. The concept of site isolation has been proposed (17), ie, that there are islands of active oxygen. Thus, oxidation proceeds only to a limited extent, where the reoxidation of the catalyst is relatively slow (18).

In the most commonly applied ammoxidation process, the Sohio process, a fluid-bed type reactor is fed with a mixture of air, NH_3, propylene, and possibly water, operating at a pressure of about 100–300 kPa (1–3 atm) and at 400–500°C; contact time is about 1–15 s. Selectivities to acrylonitrite of 40–90% have been reported.

Oxychlorination. Ethylene dichloride, the precursor of vinyl chloride, is manufactured by the oxychlorination of ethylene (212) (see Chlorocarbons; Vinyl polymers). Most processes use a supported copper chloride catalyst for the reaction.

$$C_2H_4 + 2\,CuCl_2 \rightarrow C_2H_4Cl_2 + 2\,CuCl$$

$$\frac{2\,CuCl + 2\,HCl + \tfrac{1}{2}\,O_2 \rightarrow 2\,CuCl_2 + H_2O}{C_2H_4 + 2\,HCl + \tfrac{1}{2}\,O_2 \rightarrow C_2H_4Cl_2 + H_2O}$$

The reaction is carried out in small-diameter, cooled tubes, or in a fluidized bed because it is highly exothermic.

Oxychlorination can also be carried out in homogeneous phase by the reaction of ethylene with oxygen in a solution of cuprous and cupric chlorides (213–214).

CATALYSIS APPLIED TO FUELS

Catalysis was first used by Bergius in 1913 for liquefaction of coal by hydrogenation. Subsequently, Fischer and Tropsch synthesized hydrocarbon fuels from carbon monoxide and hydrogen (see Fuels, synthetic). Catalytic cracking of crude petroleum (qv), by far the largest catalytic process, was introduced industrially by Houdry in 1936; at this time catalytic polymerization of olefins to liquid polymers also began, followed shortly by production of C_8-hydrocarbons by alkylation. Hydroforming, ie, the conversion of cycloparaffins (naphthenes) to aromatic compounds, was used commercially in the United States and Germany during World War II. Molybdena–alumina and later platinum–alumina catalysts were employed. Hydroforming has developed into the second largest industrial catalytic process. Hydrodesulfurization is primarily employed to remove objectionable sulfur compounds and has greatly increased in importance.

Acidic and Basic Catalysts

The correlation of catalytic efficiency with the strength of the acid or base is of considerable importance and has been studied by several investigators (215–216). The catalytic coefficients of a large number of acids and their conjugate bases in the mutarotation of glucose (217) constitute the experimental basis for the general theory of acid–base catalysis in which a proton is transferred from the catalyst to the reactant (acid catalysis) or from the reactant to the catalyst (base catalysis). The velocity of the catalyzed change is thought to be determined by a protolytic reaction between the reactant and the catalyst. The molecule, on receiving or giving a proton, is converted into an unstable state which immediately (or very rapidly, compared with the velocity of the protolytic reaction) leads to the reaction under consideration. Thus, acid catalysis of a basic reactant is represented by the general scheme $R + AH^+ \rightarrow RH^+ + A$, whereas $RH^+ + B \rightarrow R + BH^+$ represents basic catalysis of an acidic reactant.

The concepts of acid and basic catalysis, as proposed by Brønsted and Lowry, apply to reactions of industrial importance, particularly hydrocarbon reactions used in large-scale petroleum refining.

Acid catalysts are employed in such diverse areas as Friedel-Crafts reactions (qv) and reactions in the presence of clays. The dimerization of isobutylene, in the presence of a suitable acid catalyst, HX (218), as shown below, illustrates (1) initiation, (2) propagation, and (3) chain-termination reactions.

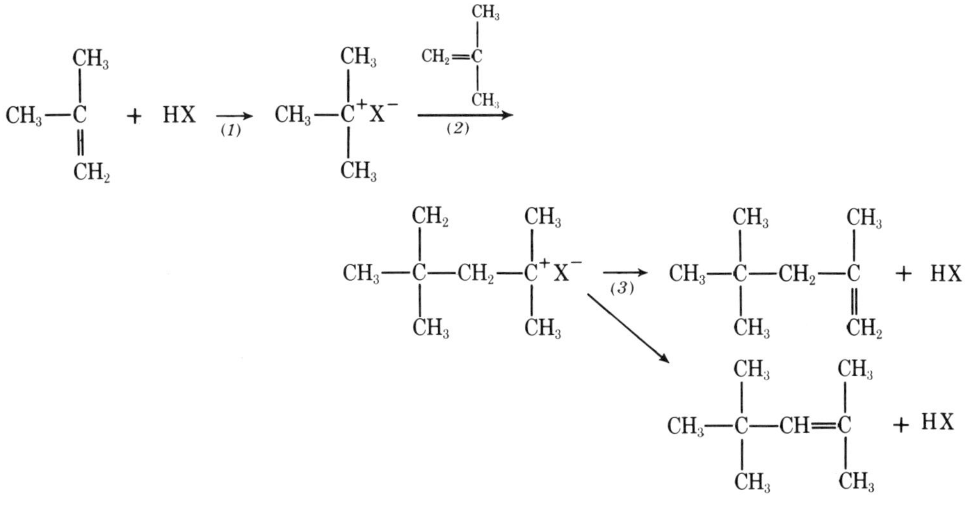

34 CATALYSIS

The concept of transfer or interchange of hydrogen as a charged species was confirmed by the interchange of hydrogen for halogen that occurs when branched paraffins containing a tertiary hydrogen atom react with tertiary alkyl halides in the presence of aluminum halide (219).

The catalytic interchange of hydrogen between two molecules has been observed using isotopic tracers. The mechanism is also understood in terms of the carbonium ion concept plus transfer of hydrogen in the form of a hydride ion (220–221).

The isomerization of n-butane over an aluminum halide–hydrogen halide catalyst takes place only in the presence of olefins (222–223). A chain mechanism was proposed:

$$CH_3-CH_2-CH=CH_2 + HX \longrightarrow CH_3-CH_2-\overset{+}{C}H-CH_3 \ \ X^- \longrightarrow$$

$$CH_3-\underset{X^-}{\overset{CH_3}{\underset{|}{\overset{|}{\underset{+}{C}}}}}-CH_3 \xrightarrow{CH_3CH_2CH_2CH_3} CH_3-CH_2-\overset{+}{C}H-CH_3 \ \ X^- + CH_3-\overset{CH_3}{\underset{|}{C}H}-CH_3$$

Tertiary carbonium ions are more stable than secondary or primary ions and are able to undergo skeletal isomerization rapidly. Paraffins alkylate similarly (224–225).

The complex reactions of catalytic cracking, such as scission, isomerization, cyclization, and hydrogen transfer, can be explained fairly well using the carbonium ion theory (226–227). When a straight-chain paraffin is cracked, large amounts of aromatics and branched paraffins are formed. The isomerization of carbonium ions and hydrogen transfer occur extensively (see also Alkylation).

Basic catalysts are employed less than acid catalysts. They are, however, used industrially for polymerization of conjugated dienes (particularly in the Union of Soviet Socialist Republics), as well as for the isomerization of olefins, dehydrogenation of certain diolefinic materials to aromatics, alkylation of arylalkanes, and polymerization of monoolefins. It is important that base catalysis gives different products than acid catalysis. For instance, ethylene alkylates toluene in the side-chain with a base catalyst whereas acid catalysis gives ring alkylation (228).

$$C_6H_5\overset{H}{\underset{H}{\overset{|}{\underset{|}{C}}}}-H + B^-Na^+ \longrightarrow C_6H_5\overset{H}{\underset{H}{\overset{|}{\underset{|}{C}}}}{}^-Na^+ + BH \xrightarrow{CH_2=CH_2}$$

$$C_6H_5-\overset{H}{\underset{H}{\overset{|}{\underset{|}{C}}}}-CH_2-CH_2{}^-Na^+ \xrightarrow{C_6H_5CH_3} C_6H_5\overset{H}{\underset{H}{\overset{|}{\underset{|}{C}}}}-CH_2-CH_3 + C_6H_5\overset{H}{\underset{H}{\overset{|}{\underset{|}{C}}}}{}^-Na^+$$

Additional substitutions on the α-carbon yield 3-phenylpentane and 3-ethyl-3-phenylpentane. This is a catalytic chain mechanism because the agent that adds to the olefins is regenerated in the last step. Here the negatively charged primary carbanion is more stable than the tertiary carbanion, in contrast to acid systems where the tertiary carbonium ion is the more stable. Basic catalysts include potassium amide, lithium that has reacted with ethylenediamine, and organosodium compounds.

Basic catalysts are extensively employed by the organic chemical industry, eg, in the manufacture of polyurethane foams from isocyanates (11). The ditertiary amine triethylenediamine, 1,4-diazabicyclo[2.2.2]-octane, N(CH$_2$CH$_2$)$_3$N, DABCO, possesses very high activity and is the principal amine catalyst used. The cage structure and sterically exposed nitrogen atoms, no doubt, contribute to its high activity (see Urethane polymers).

Structure of Acid Catalysts. Friedel-Crafts catalysts, mineral acids, such as H$_2$SO$_4$ and HF, and claylike substances, eg, H$_2$O–Al$_2$O$_3$–SiO$_2$, all act in a similar manner because of their essential acidity. In each instance the active catalyst contains a proton; it is a Brønsted acid. In the case of silica–alumina–H$_2$O catalyst, the aluminum ion may be considered as substituting for quadrivalent silicon ion in an oxygen ion lattice with a proton as a counter (ion-exchange) ion.

$$\begin{matrix} O & & O & & O \\ & Si^{4+} & & Al^{3+} & \\ O & & O & & O \\ & & & H^+ & \end{matrix} \quad \text{hypothetical acid}$$

similarly

$$HCl + AlCl_3 \longrightarrow H^+ \begin{bmatrix} Cl & & Cl \\ & Al & \\ Cl & & Cl \end{bmatrix}^-$$

These Brønsted acids are very strong acids indeed, ie, they have a high tendency to donate protons. They do not seem to preexist in the free state but the active complex is formed in the presence of a base such as an olefin:

$$CH_2=CH_2 + HX + AlX_3 \rightarrow C_2H_5^+ + (AlX_4)^-$$

where X is a halide or oxide ion. The shift of the aluminum coordination to the four-coordinated form and back again is an essential feature of the mechanism. The rapid exchange of isotopic oxygen between silica–alumina and water provides conclusive evidence of the highly dynamic structure of such catalysts (229). Alumina shows no Brønsted acidity although after calcination at relatively high temperature, it exhibits Lewis acidity which, after removal of an OH, leaves exposed an aluminum cation capable of accepting an electron pair. Silica shows neither Brønsted nor Lewis acidity, whereas silica–alumina shows both. The Brønsted sites are converted to Lewis sites on calcination and back to Brønsted sites upon addition of water. These results are consistent with the catalytic properties of silica, alumina, and silica–alumina.

The catalytic activity of acid catalysts has been correlated with surface acidity for different reactions (216). In general, Brønsted acidity is necessary for successful catalytic cracking, and the distribution of acid strengths is important. For some reactions, even low acid strength is adequate for catalytic activity.

Zeolite (Molecular Sieves)

Crystalline zeolites are silicates with an open-framework, regular structure, and apertures of molecular dimensions based on a tetrahedron of four oxygen ions surrounding a smaller silicon or aluminum ion (see Molecular sieves). These tetrahedra are arranged to form the porous framework of the zeolite crystal structure, whose pore diameters are determined by the number of oxygen ions arranged in regular rings in the crystal structure. Openings in the oxygen rings are 0.74 nm for 12-membered rings (20). The important zeolite catalysts are known as X, Y (similar to the rare natural mineral, faujasite), erionite, and mordenite (44), and the newer ZSM-5 and ZSM-11 (110,230). The unit cell in faujasite-type zeolites is 25 nm. In some instances, rare earth-exchanged zeolites are commercially employed. Zeolites exhibit a rather narrow distribution of acid-site strengths. It is important that acid sites are fifty times more concentrated in zeolites, that there is increased contact with hydrocarbons because of the highly microporous nature of the zeolite, and that there are high electric fields within the zeolite pore structure.

The first application of molecular-sieve catalysts was based on the concept of shape selectivity (22), ie, only those molecules that can enter the catalyst will react and those excluded on account of shape will not react. Shape-selectivity is the basis of the Selectoforming process for cracking straight-chain paraffins from a naphtha mixture containing aromatics, or branched-chain or cycloparaffins, all of which are excluded from the catalyst pore structure (23). Recently, the ZMS-5 catalyst has been used to dewax high-pour-point gas oils by commercially cracking the long-chain normal and slightly branched paraffins leaving the aromatic, cycloparaffin, and highly branched structures unchanged (231). This type of catalyst is also used successfully in xylene isomerization.

The most significant application of crystalline zeolites has been in catalytic cracking where catalysts containing 10–20% zeolites greatly improves selectivity and activity. Much more gasoline is produced with less coke and light gases (24) as can be seen in Table 4.

Molecular-sieve catalysts revolutionized catalytic cracking and greatly improved process economics. Their very high activity led to the widely adopted riser-cracker, where gas oil cracking occurs in a riser-tube during transport of the fluid catalyst, and thus a conventional reactor is not required.

The recent introduction to cracking catalysts of combustion agents that convert CO to CO_2 permits more complete regeneration of the catalysts (232). Coke deposits can be removed completely, and CO in the flue gas is eliminated by high-temperature regeneration or CO oxidation catalysts. Previously, flue gas from catalytic crackers contained about 10% CO. Now it can be completely converted to CO_2 by adding components to the cracking catalyst that catalyze the CO → CO_2 reaction but do not affect the selectivity. Alternatively, separate addition of a combustion promoter allows more independent control over the operation.

Table 4. Compositions of Gasoline Obtained by Gas Oil Cracking Catalyzed by Silica–Alumina[a] and Zeolite[b,c]

Catalyst	Calif. virgin 5[b]	Calif. virgin 1[a]	Calif. coker 5[b]	Calif. coker 1[a]	Gachsaran 5[b]	Gachsaran 1[a]
paraffins, %	21.0	8.7	21.8	12.0	31.9	21.2
cycloparaffins, %	19.3	10.4	13.4	9.5	14.3	15.7
olefins, %	14.6	43.7	19.0	42.8	16.3	30.2
aromatics, %	45.0	37.3	45.9	35.8	37.4	33.1

[a] Durabead 1 = silica–alumina.
[b] Durabead 5 = early generation zeolite (REHX).
[c] Ref. 24.

Dual Function Catalysts

Reforming (Hydroforming). Catalytic reforming was introduced in 1939 for the purpose of increasing the octane number of naphtha by converting its straight-chain paraffins and cycloparaffins into high-octane number constituents, ie, isoparaffins and aromatics.

Hydroforming, as this process is called, was a radical departure from thermal processes. It employed a molybdena–alumina catalyst requiring periodic regeneration. In 1949, UOP announced the pioneering Platforming process using a fixed bed of platinum-containing catalyst that did not require regeneration and was highly selective. This initiated a remarkable industrial expansion and is today, the second largest industrial catalytic application (see Petroleum).

Reforming represented a significant advance in the utilization of the dual-function catalysts (233–234), eg, platinum on an acidic alumina, where the former is active for hydrogenation–dehydrogenation and the latter for olefin isomerization, as illustrated below for n-pentane.

$$n\text{-pentane} \rightarrow 1\text{- or 2-pentene} + H_2 \text{ (on Pt site)}$$

$$1\text{- or 2-pentene} \rightarrow \text{branched pentylenes (on acid site)}$$

$$\text{branched pentylenes} + H_2 \rightarrow \text{isopentane (on Pt site)}$$

The amount of olefin present in the gas phase is extremely small because of the large excess of hydrogen used in the process and the moderate temperature and high pressure. The rate-determining step is the isomerization of adsorbed olefin while the olefin in the gas phase is in equilibrium with the paraffin and hydrogen. This has been confirmed by the observation that isomerization of the olefin proceeds in the presence of the acid catalyst alone, at the same rate as pentane isomerization, with due consideration being given to olefin concentration.

Another example of dual-functionality is the conversion of C_6-cycloparaffins (cyclohexane and methylcyclopentane) over single- and dual-function catalysts (233). Whereas the dehydrogeneration catalyst can transform cyclohexane and cyclohexene to benzene and the acidic single-function catalyst can isomerize methylcyclopentene to cyclohexene, only the dual-function catalyst can convert methylcyclopentane to benzene. Two types of catalytic sites are required with transferal of an olefinic intermediate between sites (see Fig. 4). Chemical transformations (represented by vertical arrows) occur on the dehydrogenation site, whereas isomerizations take place horizontally. For a diagonal chemical change, a transfer from one to the other type of site is required. In certain cases, dehydrocyclization occurs on metal sites (235–236).

For certain reactions dual-function catalysts can accomplish more than passing the reactants through two reactors in sequence, each filled with a different single-function catalyst. In the reaction

$$A \underset{k_{-1}}{\overset{k_1}{\rightleftharpoons}} B \overset{k_2}{\longrightarrow} C,$$

to convert reactant A, passing it over a bed of one single-function catalyst to give an equilibrium amount of B, followed by passage over a second bed of the other type of single-function catalyst capable of converting B to C, only a small amount of C results if $K = k_1/k_{-1}$ is small. However, when both catalyst functions are present in a single

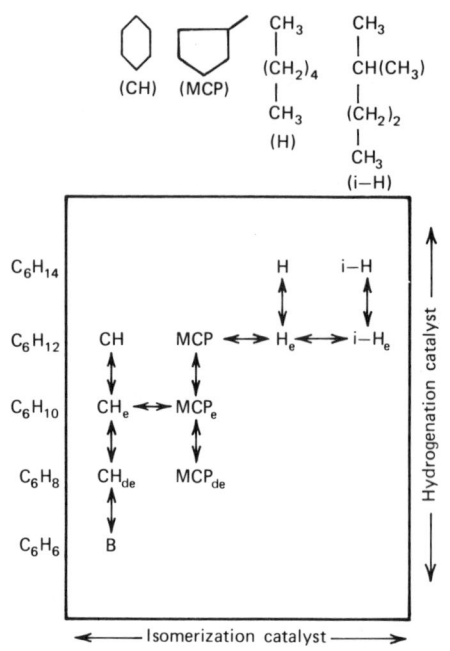

Figure 4. Reforming C$_6$-hydrocarbons with a dual-function catalyst. Courtesy of *Industrial and Engineering Chemistry*.

catalyst particle, it is possible to obtain large yields of C because of removal of small equilibrium amounts of B into C with consequent shift of more A into B and so on. In this way the control of rate by the thermodynamic limitations of A ⇌ B is lifted (234).

The interrelation of functions in a dual-function catalyst is important. It is actually possible to mix powders of two different types of single-function catalysts and obtain excellent performance, eg, for *n*-heptane isomerization (234,237–238). Rapid isomerization occurs when the catalyst particles are small enough, 1.0–10 μm, whereas poor catalytic performance is obtained when catalyst powders are 100–1000 μm. Furthermore, the balance of functions has to be considered. For the Pt–acid alumina system, increase of the platinum content above a certain level does not raise catalyst activity in accordance with the concept that its utility is to provide a small equilibrium amount of olefin. Increase in acidity raises the conversion; however, beyond a certain amount, further increase causes rapid deactivation by a polymerization-type reaction which eventually leads to coke formation. Therefore, it is believed that both a certain physical proximity and a balance of activities are required for optimum performance in dual-function catalysts.

Hydrocracking. In the 1960s, modern hydrocracking processes in petroleum refining were established. From a single demonstration unit of 159 m^3 (1000 bbl) capacity per day, commercial installations in the United States have grown to capacities of 144,000 m^3 (907,000 bbl) per day by 1977. The demand for gasoline and middle distillates and the availability of inexpensive hydrogen, in part from catalytic reforming, have accelerated this growth.

The most significant combination has been the development of superior catalysts of excellent activity which can be maintained at operating conditions far less severe than those used in older processes. In addition, much heavier petroleum fractions including vacuum residues can now be hydrocracked. The dual function catalysts employed have a critical balance of functions. Too high a cracking activity relative to hydrogenation ability would lead to coke formation and inactivation of the catalyst (239).

An interesting new reaction has been termed paring, illustrated by hydrocracking hexamethylbenzene to produce light isoparaffins and tetra- and pentamethylbenzene.

Desulfurization. Sulfur removal from lower-boiling petroleum fractions are necessary for subsequent catalytic refining steps, specifically reforming. Desulfurization of heavy fuels has been given new importance by the Clean Air Act, imposing a limit of 360 g SO_2 per GJ (0.8 lb $SO_2/10^6$ Btu) for boiler fuel. Even stricter regulations have been adopted by some cities and also proposed by the Amendment of 1977.

In 1960, installed hydrotreating capacity in the United States was about 350,000 m^3/d (2.2 million bbl/d). This has now (1977) grown to 885,000 m^3/d (5,567,000 bbl/d). Thus both rapid growth rate and high capacity occurred. Furthermore, hydrotreating has been applied to heavier fractions including residuals. Cobalt or nickel molybdate on alumina catalysts are used, with the nickel-containing catalysts believed to be particularly suited for removal of nitrogen. The Germans used nickel–tungsten sulfide catalysts in World War II. The chemistry of hydrodesulfurization (240) is not as well understood as other petroleum refining processes (see Petroleum).

Multimetallic Catalysis

Much of the early work with multi- and bimetallic catalysts was conducted with conventional metal alloys prepared in a suitable catalytic form. A major area of study was the relation between catalytic activity and the electronic structure of metals (25–27). The approach focused on catalytic activity as a function of alloy composition, the latter determining the electronic properties of the metal. Alloys of a Group VIII and Group IB metal (eg, nickel–copper and palladium–gold) were investigated in catalytic reactions including mainly hydrogenation, dehydrogenation, hydrogenolysis, and isomerization. The d electrons of the metal seem to play an important role in both the electron-band and valence-bond treatment of metals. Other types of alloys include combinations of two metals from Group VIII (eg, nickel–iron and nickel–cobalt) or from Group IB (eg, silver–gold and copper–gold).

Early metallic catalysts had comparatively low dispersion. The term dispersion refers to the fraction of atoms in a metal crystallite that exist on the surface. The degree of metal dispersion increases with decreasing crystallite size, and is a measure of the specific surface area of the metal.

Bimetallic cluster catalysts (28–29) are composed of atoms of two different metals in a state of high dispersion on a carrier. These novel systems offer a new range of catalytic properties (29–33) and are not limited to combinations of metals that form bulk alloys. That is, the degree of dispersion strongly affects the stability of the bimetallic or multimetallic clusters. Virtually any property of a metallic catalyst (ie, activity, selectivity, or stability) is affected by combination with one or more metals in the form of multimetallic clusters.

New bimetallic catalysts with high activity and unprecedented activity maintenance are employed in the reforming of petroleum naphthas to high-octane gasoline

(34–39) and in isomerization (40), hydrocracking (41), and hydrogenation (42). In the petrochemicals field, improved Pd–Au catalysts are utilized for the synthesis of vinyl acetate (43), whereas more selective catalysts (eg, Ag–Au, Cu–Au) are used for the partial oxidation of olefins (44–45). Supported alloys such as Pt–Co (46) increase thermal stability and resistance to sintering. Bimetallic systems optimize product selectivity. For example, dehydrocyclization (47) and aromatic hydrogenation (38) are possible with a minimum of cracking reactions (hydrogenolysis) while maintaining very high hydrogenation activity.

Bimetallic Catalysts. Many commercial catalysts are supported on high-surface-area carriers (48) and are more resistant to surface area loss by sintering. The useful life of a catalyst depends on this factor. Bimetallic catalysts usually have been studied with unsupported systems; supported bimetallic catalysts have only recently been investigated in detail (29). In preparing a supported bimetallic catalyst, possible interaction of the two metallic entities has to be considered, especially for a high-surface-area (100–500 m^2/g) support and a low total metal concentration (ca 1 wt %). In such a case, the metal dispersion, defined as the percentage of metal atoms present in the surface, frequently approaches 100%. Such highly dispersed bimetallic catalysts are prepared like monometallic catalysts. For example, a high-surface-area carrier may be impregnated with an aqueous solution containing the appropriate concentrations of salts of the two metals. The carrier is then dried and subsequently reduced at an elevated temperature in hydrogen. If the two metals do not interact in the highly dispersed state, one would expect additive catalytic behavior. However, if the metals do interact, one might expect to find a different behavior, especially if the individual metals have very different catalytic activities for a given reaction, as observed, eg, with ruthenium–copper and osmium–copper on silica (29). Interaction in the highly dispersed state has been substantiated by chemisorption and x-ray diffraction data, and especially by the marked changes in catalytic properties observed for the highly dispersed bimetallic catalysts (29). These systems are termed bimetallic clusters rather than alloys (28–29,241).

Activity. Bimetallic catalysts show increased activity, eg, a combination of Group IB with Group VIII metals for hydrogenation. Thus, Ni–Cu (30) and Ru–Cu (29) show higher rates for cyclohexane dehydrogenation than do the pure metals. This could be attributed to a modification of the catalytic properties of the metal or a reduced carbon fouling of the bimetallic surface as compared to that of the pure metals (33). Similar effects have been observed with olefin hydrogenations using Pd–Au (42) and Ni–Cu (50–51).

A series of bimetallic catalysts, including Pt–Re (37–38), Pt–Pb (52), Pt_3Cu (53), and a multimetallic proprietary system, KX-130, has been commercialized for the catalytic reforming of petroleum naphthas to high-octane gasoline. The activity maintenance for Pt–Re and KX-130 was found to be far superior to that of a conventional platinum catalyst.

Bimetallic catalysts do require, in some instances, new regeneration procedures. Furthermore, some multimetallic systems are more resistant to sulfur than their constituent metals, whereas others are less so.

Activity Maintenance. In commercial operation a catalyst can be poisoned or fouled. In hydrocarbon reactions, metal sites are usually poisoned by coke accumulation or substances like sulfur. Bimetallic systems can show improved resistance to these effects.

Both Pt–Re and KX-130 show more significant resistance to coke deposition (33–38) than conventional platinum catalysts (34). This may occur for two reasons: first, coke formation may be caused by polymerization of acetylenic residues, the formation of which requires at least three contiguous metal sites (33). Contiguous sites can be disrupted by alloying an active and an inert metal (eg, Pt–Au, Pt–Sn, or Pt–Cu). Therefore, alloy formation can dilute the primary active centers and inhibit coke generation. A second reason is an optimized balance of hydrogenation and hydrogenolysis activity. The latter promotes cracking of surface residues, whereas the former guarantees facile hydrogenation of the cracked species, and therefore inhibits carbon species polymerization.

The balance of hydrogenation to hydrogenolysis activity can be controlled by the appropriate combination of metals and catalyst preparation, such as the addition of Group IB (Cu, Ag, or Au) to Group VIII metals (eg, Ru, Rh, Ir, Ni) with high hydrogenolysis–hydrogenation ratios. This occurs because the Group IB metal selectively inhibits hydrogenolysis but maintains or promotes hydrogenation.

Selectivity. Selectivity is a ratio of activities or rate constants. Perhaps the most widely studied bimetallic system is that of Ni–Cu alloys. The selectivity of Ni is markedly enhanced by alloying with Cu for scission of carbon–hydrogen bonds (55), because of suppression of carbon–carbon bond hydrogenolysis. Similar effects were observed for ethane hydrogenolysis and cyclohexane dehydrogenation (30) over a series of Ni–Cu alloys. The addition of copper to nickel had markedly different effects for the two reactions (see Fig. 5). The data depicted can be rationalized on the basis of the rate-limiting steps for the two reactions.

Selectivity effects have also been observed for supported bimetallic cluster cat-

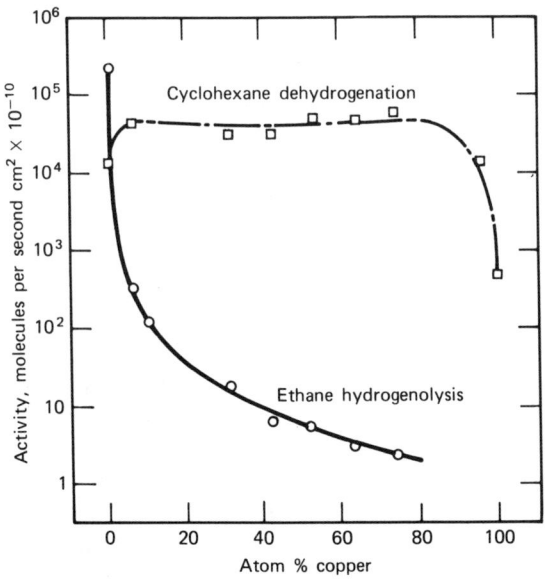

Figure 5. Activities of copper–nickel alloy catalysts for the hydrogenolysis of ethane to methane and the dehydrogenation of cyclohexane to benzene referring to reaction rates at 316°C. Ethane hydrogenolysis activities were obtained at ethane and hydrogen pressures of 3 and 20 kPa (0.030 and 0.20 atm), respectively. Cyclohexane dehydrogenation activities were obtained at cyclohexane and hydrogen pressures of 17 and 83 kPa (0.17 and 0.83 atm), respectively (30).

alysts (29,49). In silica-supported Ru–Cu and Os–Cu catalysts (29), in the highly dispersed state, copper interacts strongly with ruthenium and osmium, and decreases the hydrogenolysis activity of these metals, even though copper is virtually immiscible with these metals in the bulk.

More recent work (56) with unsupported Ru–Cu catalysts has given definite evidence of interaction between the two metals occurring at the surface. The presence of copper decreases the capacity of ruthenium for hydrogen chemisorption, and also suppresses markedly the catalytic activity for hydrogenolysis of ethane to methane. The state of dispersion has a major influence on the effect of the copper, and the atomic ratio of copper to ruthenium required for a given degree of coverage of the surface by copper increases with increasing dispersion.

Other hydrocarbon reactions, eg, cyclization and skeletal isomerization of paraffins, are also enhanced by alloying a catalytic metal such as platinum with an inert metal such as gold or copper (57–58). This enhancement in selectivity can be ascribed to an increase in the number of isolated metal atoms, eg, Pt in an Au matrix. A further effect of dilute active sites on activity maintenance is a decrease in self-poisoning by coke (33).

Synthetic Fuels

Synthetic fuels from coal, oil shale, and biomass are expected (beginning about 1985), to become of enormous industrial importance with catalysis playing a key role. The conversion of these complex solids to more convenient gaseous and liquid forms and the removal of sulfur, nitrogen, and oxygen in addition to inorganic mineral matter, can be carried out in a series of catalytic reactions. Since the hydrogen content of coal is about 5 wt %, whereas for gasoline it is about 14%, the manufacture of gasoline from coal requires the addition of hydrogen to coal.

Recent reviews and conference reports point out that great process improvements can be achieved through catalysis to make synthetic fuels manufacture more efficient and economical (82–97,119) (see Fuels, synthetic).

Liquefaction of Coal. Large-scale coal liquefaction plants were installed in Germany (82) and to a lesser extent in other countries in the 1930s. The two catalytic processes used were hydroliquefaction (Bergius) and synthesis from coal-derived CO + H_2 (Fischer-Tropsch). In Germany during World War II, coal hydroliquefaction provided one-third of the petroleum needs, corresponding to half of automotive gasoline and diesel fuel and 90% of aviation fuel. However, production from several plants amounted to only 16,000 m^3 (100,000 bbl) per day. The Fischer-Tropsch process was used primarily to produce intermediates for subsequent conversion to lube oil, soaps, edible fats, etc, with a production of 2500 m^3 (16,000 bbl) per day. At present, the only commercial operation for synthetic liquid fuels from coal is the South African Coal, Gas, and Oil Company (SASOL) plant, which processes about 6350 metric tons of soft coal per day, producing liquid fuels, chemicals, and industrial gases using the Fischer-Tropsch process (98). A second, much larger plant is under construction.

The four general types of coal liquefaction processes are shown in Figure 6; this includes the catalysts and the five chemical and physical steps required to convert coal to gasoline.

In 1913, Bergius achieved the continuous liquefaction of coal, adding hydrogen and employing iron oxide–alumina for removal of sulfur. Bergius was awarded the

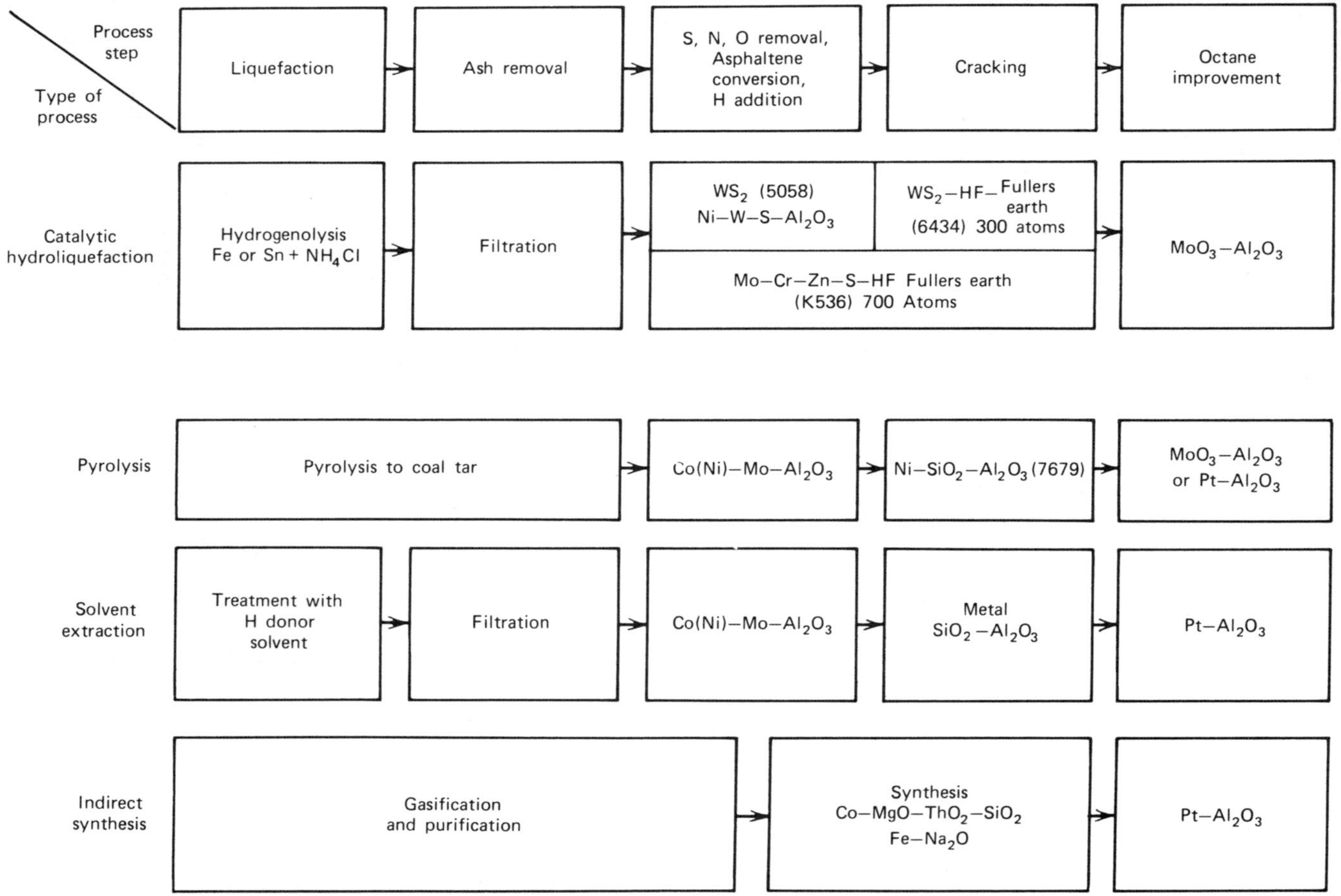

Figure 6. Coal liquefaction processes.

Nobel prize in chemistry in 1931. At BASF, an intensive study of coal liquefaction began in 1924, where, in the next ten years, all the presently used catalysts were developed (99–100). Since sulfur, oxygen, and nitrogen compounds are present in coal, catalysts had to be resistant to H_2S, H_2O, and NH_3, which are decomposition products of coal in the presence of H_2. Low-temperature catalysts such as Pd, Pt, and Ni are easily poisoned in the presence of sulfur compounds, whereas molybdenum and tungsten sulfides, although less active, operate satisfactorily at higher temperature.

Coal liquefaction was carried out in two steps: first, coal was converted to a liquid employing inexpensive catalysts such as molybdenum, tin, and iron. Second, the light-distillate oils obtained were processed in vapor-phase hydrocracking reaction schemes; catalysts were molybdenum or tungsten sulfides promoted by cobalt or nickel sulfides. Furthermore, both cobalt–molybdena and nickel–molybdena catalysts, standard catalysts for present petroleum refining, were developed by the BASF group during this period.

The most widely used catalyst was 1.5–2.0% iron oxide ("red mud" by-product from alumina manufacture) with some added $FeSO_4$ solution. Bituminous coal is more difficult to hydrogenate than brown coal. For bituminous coal, a more active catalyst, stannous oxalate, was employed with the addition of NH_4Cl to obtain acid conditions. However, this did lead to corrosion.

The mechanism of coal liquefaction is believed to consist of a primary thermal pyrolysis of coal, followed by catalytic hydrogenation of the reactive fragments or asphaltenes formed by pyrolysis.

Following a period of low research activity, 1950–1970, a greatly accelerated research effort is now underway to improve coal liquefaction processes. The ERDA-Fossil Energy Reports (101–102) describe the U.S. Federal Program, now conducted by the Department of Energy in cooperation with industry, which includes large pilot plants to test catalytic processes. Other countries have also embarked on massive energy development programs. The German plan (103) to synthesize fuels from coal devotes considerable effort to develop technology to utilize nuclear heat to gasify coal.

Several direct catalytic coal hydroliquefaction processes use active cobalt–molybdenum–alumina catalysts developed for petroleum desulfurization, including the H-Coal process (HRI), Synthoil (Bureau of Mines, now DOE), and CCL (Gulf Oil). In the H-Coal process an ebullated-bed reaction system operates a catalyst system continuously with a combined feed of solid, liquid, and gas. The upward flow of catalyst permits the continuous removal of unconverted coal and ash, and catalyst can be added and removed during operation (105). In the Synthoil process, coal suspended in oil derived from coal is processed through a fixed-bed of $Co–Mo–Al_2O_3$ catalyst under conditions of highly turbulent flow of H_2 to prevent obstruction of flow and to promote catalytic contact (105).

In the SRC process coal is extracted with solvent in the presence of hydrogen. Coal minerals, particularly iron pyrite, show considerable catalytic activity. Hence minerals activity can be increased by building up the desired mineral matter content in the reactor, as has been reported for SRC-II. The two-step donor–solvent process (Exxon) recycles a hydrogenated oil (donor) in the transfer of hydrogen to coal.

Massive amounts of molten halide catalysts, such as tin and especially zinc chloride, have shown remarkable ability to hydrocrack coal directly into hydrocarbons in the gasoline boiling range (106–107). Catalyst recovery is a problem.

A number of other catalytic processes are in the research and development stage, including CO-steam (base catalysis), reductive alkylation, electrochemical coal hydrogenation, and others (86,108).

Indirect Liquefaction—Fischer-Tropsch Process. In contrast to direct coal hydroliquefaction, in indirect liquefaction, coal is gasified to a mixture of CO + H_2, followed by catalytic conversion to fuel products. With nickel, methane is virtually the only product. In 1923, however, Fischer and Tropsch produced higher hydrocarbons (109). Scientists at BASF obtained a variety of organic compounds with cobalt and osmium. The influence of the catalyst is profound, illustrated by the range of product distribution formed from CO + H_2 over various metal catalysts shown in Figure 7. A number of different catalysts have been used commercially. The catalyst developed in Germany is 100 Co:8 MgO:5 ThO_2:200 kieselguhr (diatomaceous earth). Special consideration is required for removal of heat, since the reaction is highly exothermic. In Germany and also in the SASOL process, complicated tube bundles are packed with catalysts for heat exchange. SASOL also uses the fluid or suspended catalyst type with indirect cooling by oil circulation (98). The product distribution obtained varies considerably, as shown in the comparison of SASOL products in Table 5.

Methanol manufacture from CO + H_2 presents an inexpensive alternative for

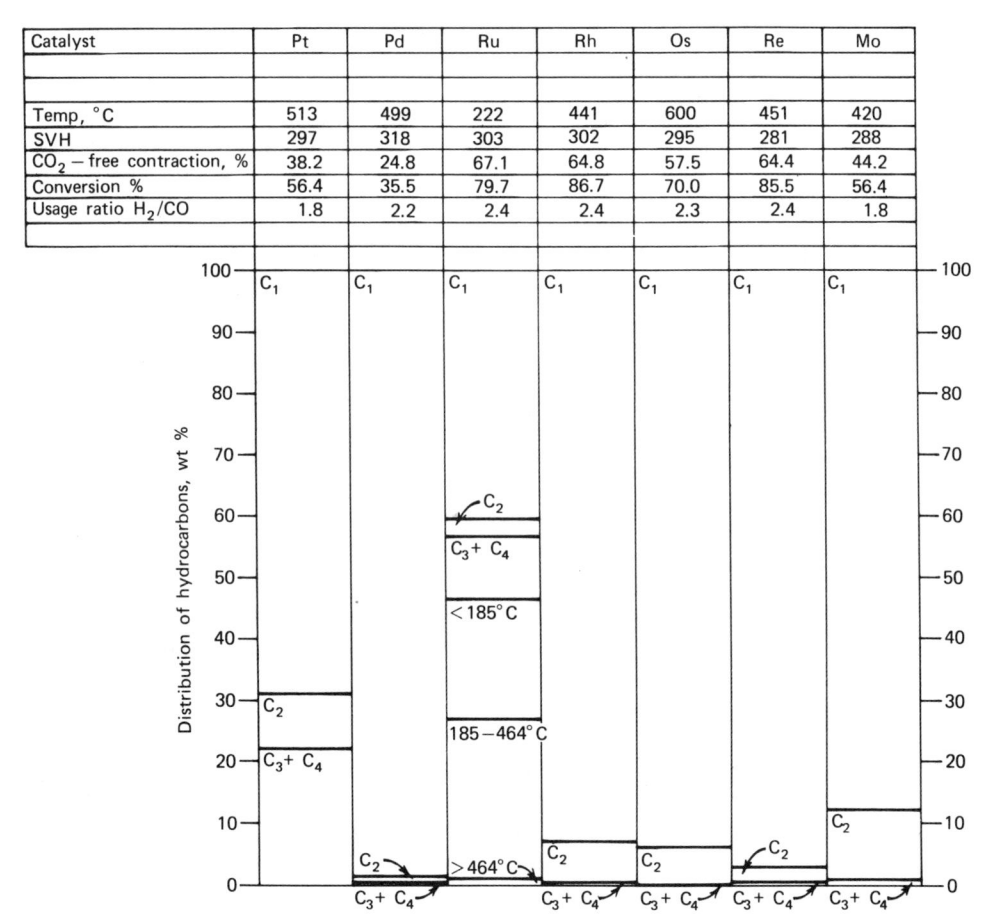

Catalyst	Pt	Pd	Ru	Rh	Os	Re	Mo
Temp, °C	513	499	222	441	600	451	420
SVH	297	318	303	302	295	281	288
CO_2 – free contraction, %	38.2	24.8	67.1	64.8	57.5	64.4	44.2
Conversion %	56.4	35.5	79.7	86.7	70.0	85.5	56.4
Usage ratio H_2/CO	1.8	2.2	2.4	2.4	2.3	2.4	1.8

Figure 7. Comparison of operating conditions and selectivity using 3 H_2 + CO (242). SVH = Gas Hourly Space Velocity.

Table 5. Syngas Process Products[a]

	Process, % of product		
	Sasol-I		
Product	Fixed	Fluid	Mobil
light gas C_1, C_2	7.6	20.0	1.3
LPG C_3, C_4	10.0	23.0	17.8
gasoline C_5—C_{12}	22.5	39.0	80.9
diesel C_{13}—C_{18}	15.0	5.0	0
heavy oil $C_{19}+$	41.0	6.0	0
oxygen compounds	3.9	7.0	0
aromatics, % of gasoline		5	38.6

[a] Ref. 243.

the manufacture of a clean liquid fuel from coal. The catalytic synthesis of methanol from synthesis gas is carried out on a large scale at 300–375°C and 27–35 MPa (270–350 atm). In 1968, Imperial Chemicals Industries announced an improved process operating at 250°C and 5 MPa (50 atm) using a new, highly active copper-containing catalyst instead of the previous zinc oxide catalyst. Methanol is of interest for use in automobiles and for generators of electricity, especially for peak shaving.

Recently, the scientists at the Mobil Company converted methanol almost quantitatively to gasoline and water, using a new type of molecular-sieve catalyst (110). The pores of this catalyst are of such a size as to permit the evolution of molecules only as large as those hydrocarbons found in gasoline. Thus, a much more narrow product distribution range is obtained (see Table 5). Moreover, the gasoline produced is highly aromatic, and consequently, has a high octane number (93–96 research octane number) without lead. Work is in progress to convert $CO + H_2$ directly to high octane gasoline.

Coal Gasification. Coal is gasified by reaction with steam and oxygen, usually at 1000–2000°C giving a mixture that includes hydrogen and carbon oxides. Usually, the product gas is purified and freed of H_2S and CO_2, and the synthesis gas ($H_2 + CO$) is shifted ($CO + H_2O \rightarrow CO_2 + H_2$), to obtain a higher H_2–CO ratio (including complete shift for production of H_2). Subsequently $CO + H_2$ react to form methane or higher hydrocarbons or alcohols (111). Catalysis is effective in all three reactions, ie, gasification, shift, and synthesis (94,112):

ideal	coal + $H_2O \rightarrow CH_4 + CO_2$
gasification	coal + $H_2O \rightarrow CH_4 + CO_2 + H_2 + CO$ (highly endothermic)
shift	coal + $H_2O \rightarrow \underbrace{H_2 + CO}_{\text{synthesis gas}}$
methanation	$CO + 3 H_2 \rightarrow CH_4 + H_2O$ (highly exothermic)

The idealized reaction is balanced with respect to materials and heat, ie, coal can be transformed to methane essentially without the loss of heat. However, the reaction is slow, thus gasification is carried out at 1000°C and above. The coal then reacts with steam but at such high temperatures that methane is no longer stable, and mixtures of $CO + H_2$ are formed. Consequently, in order to manufacture CH_4, pipeline gas, two additional steps, shift and methanation, are needed. Clearly, the first reaction has to be accelerated, and this is effected with certain alkaline catalysts—notably sodium and especially potassium oxide. The alkali oxide is a volatile catalyst, and therefore able to move over the solid coal–char surface and act catalytically. Catalytic coal gasification is being studied at a number of laboratories under DOE support.

In the United States, the synthesis gas is manufactured from gas or oil. Before 1963, the standard shift catalyst was a moderately sulfur-resistant catalyst of 70–85% iron oxide promoted by 5–15% chromia operating at ca 340°C. Because of equilibrium limitations at this relatively high temperature, two catalyst beds were used with interbed cooling and CO_2 removal. Shift conversion was greatly simplified when low-temperature shift catalysts came on the market in 1963; these are copper-based catalysts, sometimes containing chromia which is active at as low as 190°C. They are used in the 190–230°C range where the equilibrium for the shift reaction is so favorable to CO_2 and H_2 formation that the CO content can be reduced to the 0.3–0.5 mol % range without intermediate removal of CO_2. Copper catalysts are more sensitive to poisons, particularly sulfur and chloride, than high-temperature catalysts. Cobalt–molybdena on spinel catalysts are active in the sulfide form, below 300°C (113). At present, shift catalysts are directed primarily to the manufacture of hydrogen. It is expected that they will be used in manufacture of pipeline gas, in the step preceding catalytic methanation.

After their classic report in 1902 on the formation of methane by hydrogenation of CO and CO_2 over nickel (114), Sabatier and Senderens extended their study to other metals. By 1925, all metals active for methanation had been identified; important are Ru, Ni, Co, Fe, and Mo. Further progress over the past 50 years concerned preferred promoters, supports, operating conditions, reactor design, and removal of catalyst poisons, especially sulfur compounds, from the feed. Work was carried out particularly at the British Gas Research Board and the U.S. Bureau of Mines (115–116).

At present, methanation is used commercially to convert relatively small amounts of harmful carbon oxides to CH_4 where, as in ammonia synthesis, CO would especially interfere with catalytic conversion of gas mixtures. However, it now appears that methanation will be used on a large scale in the manufacture of high Btu or pipeline gas, essentially methane. From an industrial viewpoint, high selectivity to methane is not difficult to achieve. Rather, the problems are to prevent catalyst inactivation by carbon deposition or sulfur and catalyst sintering because of the high temperatures reached. With nickel catalysts, sulfur in the gas is limited to less than 1 ppm. Carbon deposition can be avoided by operation with a sufficiently high H_2–CO ratio. Excessive temperatures in methanation are avoided by either limiting CO content of the reactant gases or providing a special heat removal apparatus. Special start-up and shut-down procedures are needed with nickel catalysts to avoid formation of highly poisonous $Ni(CO)_4$ and to maintain temperatures above 260°C until all CO has been purged from the system (117). A comprehensive DOE coal gasification program is underway (101–102), including the testing of a variety of catalytic processes.

In the United States, pipeline gas is manufactured from naphtha and steam over an alkalized nickel reforming catalyst in indirectly heated packed tubes. The CO + H_2 thus formed subsequently reacts to give methane. With heavier naphthas, it is still more important to avoid coke formation.

Chemicals From Coal. Before World War II, the organic chemicals industry relied heavily on coking by-products. Sooner or later a return to coal can be foreseen (118). Chemicals are expected to be derived from the refining of coal hydroliquefaction, especially aromatics, or from gas synthesis. A number of remarkable new and improved catalysts have been discovered to convert CO + H_2 to a wide array of valuable chemicals, which gives promise for an era of petrochemicals from coal (199,207–208).

FUEL CELLS

Catalysts are used at both the anode and cathode in a fuel cell (see Batteries and electric cells, primary). Among numerous fuels, hydrogen is perhaps the most common (120–122). It is oxidized at the anode, and a corresponding catalytic reduction of an oxidant (usually O_2) takes place at the cathode. Typical catalysts are noble metals on electrically conducting supports such as carbon or nickel (see also Hydrogen energy).

A key problem is to minimize the amount of noble metal and decrease the cell overvoltage. In liquid-phase media such as the phosphoric acid electrolyte used in a hydrogen fuel cell, the highly dispersed Pt tends to sinter even at low temperatures, reducing cell performance. A significant contribution to cell overvoltage (low efficiency) comes from the oxygen electrode (cathode). Development of a reversible oxygen electrode would facilitate direct transfer of four electrons to O_2 and minimize two-electron transfer processes that result in H_2O_2 formation. The latter is readily decomposed in a nonelectrical process, thus minimizing the current density at a given cell voltage. Other efforts have been directed at minimizing or replacing the noble metal catalysts with non-noble metals and improving resistance to sulfur; alloy catalysts have been investigated (125).

A small fuel cell was used in the Gemini and Apollo spacecrafts. The first large-scale electric utility demonstration unit (4.8 MW) is planned for 1979 by Consolidated Edison in New York City (123–124). This system was developed by United Technologies Corporation and has an overall thermal efficiency of about 47%.

ENVIRONMENTAL CONTROL

The use of catalysts for the abatement of noxious emissions has increased substantially over the last several years. In the area of automotive emissions control, this has resulted in major growth for the catalyst manufacturing industry (see Exhaust control, automotive). The use of platinum alone has nearly doubled, increasing from 43.5 t (1.4 million troy oz) in 1973 to 84 t (2.7 million troy oz) in 1975 (60–61). Similarly, catalysis is being applied to a number of other environmental problems including nitrogen oxide reduction, odor control, and the development of stationary-source catalytic combustors (see Exhaust control, industrial).

Automotive Emissions. Increasing concern in the United States over automotive emissions led to the 1970 Clean Air Amendment which calls for 90% reduction of CO and unburned hydrocarbons (UHC) emissions between 1970 and 1975, and 90% reduction of NO_x emissions between 1971 and 1978 (see Air pollution). The NO_x requirement has been postponed repeatedly; however, the CO and UHC standards have been enforced since 1975 by employing several mechanical alterations and oxidation catalysts on most American vehicles. Table 6 gives the automobile emissions standards required by U.S. legislation. These levels are determined by the constant volume sampling (CVS) test which incorporates cold and hot starting conditions. The emissions are calculated in grams per kilometer (1.6 grams per mile) from a weighted average of cold and hot starts. In many vehicles, more than 80% of the emissions occur within the first 75 s of the choked, cold start. To meet the above limits, the catalyst must, therefore, begin to operate within 20–30 s of the cold start, and must have a low "light-off" temperature, ie, the temperature at which the conversion of emissions jumps in a step-like fashion from very low to very high conversion levels. The catalyst has

Table 6. United States' Automotive Emission Standards

Year	CO	UHC[a]	NO$_x$
	\multicolumn{3}{c}{Pollutant, g/km (g/mile)}	
1972	24.2 (39.0)	2.1 (3.40)	
1973	17.4 (28.0)	1.9 (3.00)	2.1 (3.1)
1975	2.1 (3.4)	0.25 (0.41)	2.1 (3.1)
1978	2.1 (3.4)	0.25 (0.41)	0.25 (0.4)

[a] UHC = unburned hydrocarbons.

to withstand high operating temperatures (500–1000°C) and should be insensitive to air–fuel ratio, ie, not be affected by the relative concentrations of reactants. It must also operate effectively over a broad range of space velocities (0–150,000 h^{-1}). The expected catalyst life is between 40,000 and 80,000 km of vehicle operation. Thus, it should endure thermal and mechanical shock and accidental engine malfunctions over an extended time period (see Exhaust control, automotive).

To meet these demands, catalysts were developed where the active component and substrate are supported on sturdy ceramic honeycomb structures, called monolithic supports (63–66), composed of small parallel channels of a variety of shapes and diameters. These structures may be in the form of honeycombed ceramics extruded in one piece, oxidized aluminum alloys in rigid cellular configurations, or multilayered ceramic corrugations. In general, low-surface-area ceramics such as mullite (3Al$_2$O$_3$.2SiO$_2$) or cordierite (2MgO.5SiO$_2$.2Al$_2$O$_3$) are used. The refractory monolith is produced with macropores (1–10 μm) and may be coated with thin layers of catalytic materials (5–20 wt %). The two major advantages of monolith supports for catalytic operations are the high superficial or geometrical surface area and low pressure drop during operation (67). Figure 8 shows typical extruded monolith structures.

In addition to the monoliths, highly porous pelleted materials can also be used as the catalyst support. Pellets come in a wide variety of materials, shapes, and sizes, but have the disadvantage of a higher pressure drop through the bed.

The low surface area of the monolith structure, such as cordierite, requires the

Figure 8. Typical extruded monolith structures. Courtesy of Corning Glass Works.

application of a thin coat (10–20 μm) of oxide such as Al_2O_3 or ZrO_2 (66). This wash coat provides a uniform high surface area while still ensuring that the catalytic material is close to the main flow of reactants.

A catalytic component is then deposited on the wash-coated ceramic monolith, either an oxide or a metal (Group VIII, eg, Fe, Co, Ni, Ru, Rh, Pd, Os, Ir, and Pt). Only the noble metals (Ru, Rh, Pd, Os, Ir, Pt) have a possibility of remaining in the metallic state in a high-temperature, oxidizing environment; the others readily form oxides. Platinum and palladium (63–66) are among the most active catalysts for the oxidation of a number of fuels, including methane, methanol, and hydrogen. Their high activity is related to their ability to activate H_2, O_2, and carbon–hydrogen and oxygen–hydrogen bonds. Palladium and platinum are readily prepared in a highly dispersed form on a number of support materials, and only small amounts are necessary for catalytic combustion (0.1–0.5 wt %). Ruthenium metal, another possible candidate for catalytic combustion, forms a volatile oxide (RuO_4) that is rapidly removed from conventional catalyst supports (68). Osmium, costly and limited in supply, is even more volatile and poisonous in an oxidizing environment.

A typical catalyst contains 10–15 wt % Al_2O_3 wash coat on a monolith structure such as cordierite, and about 1–3 wt % Pt–Pd (based on wash coat). These catalysts have been prepared in very stable form using various oxide promoters such as CeO_2 (66).

In addition to CO and UHC emissions, the following reaction also occurs:

$$NO + CO \rightarrow CO_2 + \tfrac{1}{2} N_2$$

The best known catalyst for this reaction is Ru metal which has the disadvantages discussed above. However, ruthenium losses can be minimized by stabilization with basic oxides such as BaO or La_2O_3 (66,68).

Catalysts for NO_x reduction are not yet used on automobiles except to a limited degree in the so-called three-way catalyst. In this system, all three pollutants (NO, CO, UHC) are converted in one reactor using a single catalyst, Pt–Rh on an Al_2O_3-coated monolith. However, this requires working in a very narrow air–fuel ratio (14.7–14.8 wt/wt) not possible with existing carburetors. The three-way catalyst is more appealing and simpler from a systems point of view than having two separate reactors, one for oxidation of CO and UHC, and another for the reduction of NO_x. It may be the trend for future catalytic emissions control. However, eventually the catalytic systems may well be phased out by mechanical improvements in vehicular engines.

Nitric Acid Plant Tail Gas Cleanup. Nitric acid (qv) is manufactured by the catalytic oxidation of ammonia with air to produce nitrogen oxides which are then absorbed in water to give nitric acid (see Nitric acid). Large amounts of unabsorbed nitrogen oxides in the tail gas from the absorption tower pose a pollution problem which has been solved in some cases by the catalytic processing of this tail gas over platinum-group metals to yield useful energy in the form of steam, power, or both (69–70). The basic reactions, using methane as the reducing fuel, are:

$$CH_4 + 4 NO_2 \rightarrow 4 NO + CO_2 + 2 H_2O \text{ (decolorization)}$$

$$CH_4 + O_2 \rightarrow CO_2 + 2 H_2O \text{ (combustion)}$$

$$CH_4 + 4 NO \rightarrow CO_2 + 2 H_2O + 2 N_2 \text{ (abatement)}$$

The tail gas reactor outlet temperature is usually between 650 and 750°C. Support materials for the catalyst include nichrome wire, alumina pellets, and alumina honeycomb materials.

Industrial Odors. The elimination of organic fumes is desirable for air pollution and fire safety reasons. The platinum metals (particularly Pt and Pd) supported on SiO_2, Al_2O_3, and carbon have been used as catalysts for the oxidation of carbon monoxide and a wide range of organic compounds in the presence of air or oxygen (71–72). Most odors are caused by low concentrations of organic sulfur, nitrogen, and oxygen compounds. The catalyst lowers the temperature needed to remove odors and accelerates the process (see Air pollution control methods; also Odor counteractants).

Catalytic Combustors. The use of catalysts to promote hydrocarbon oxidation reactions has a number of advantages for emissions control. For example, flame combustion places a limit on the lower level of NO_x control that can be achieved economically for clean fuels (approximately 25–30 ppm NO_x for small sources), whereas with catalytic combustion much lower levels of NO_x can be reached (perhaps <10 ppm). This is caused primarily by the lower temperatures of catalytic combustion, and therefore the thermodynamically less favorable conditions for fixing nitrogen from the combustion air. In addition, catalytic combustion is soot-free. Overall high combustion efficiency is possible if the excess air can be limited in some way, such as two-stage combustion, flue-gas recirculation, or cooling of the catalyst bed. Furthermore, the presence of a catalyst permits much more efficient and complete combustion at air–fuel ratios that are unacceptable for a homogeneous combustion system (see also Incinerators).

As in the case of automotive catalysts, combustor systems are usually ceramic monolithic substrates coated with a catalytic agent. However, to operate most effectively, temperatures of 1200–1500°C are required, ie, the catalysts must have very high thermal stability, which has been achieved only recently. For example, it is now possible to prepare stabilized ceramic monoliths that can tolerate high temperatures for short periods. These monoliths can be coated with stabilized Al_2O_3 wash coats, and ultimately impregnated with Pt and Pd stabilized by metal-support interactions. An example is 0.3 wt % Pt–Pd on a CeO_2-stabilized alumina coated on a zircon–mullite honeycomb monolith ceramic (73–76). These catalytic materials can produce hot gases to drive gas turbines, or fire stationary boilers in an environmentally acceptable manner.

Hydrogen Sulfide Conversion to Sulfur. Hydrogen sulfide is ubiquitous to many refinery and chemical waste streams, and can be converted to sulfur using the Claus process. There are three major variations of this process, namely partial combustion, split stream, and direct oxidation (78–81). In these processes, part of the H_2S stream is oxidized to SO_2 and S, and the SO_2 then reacts with the remaining H_2S over a bauxite catalyst to give S and H_2O. The process is generally run at atmospheric pressure and GHSV (space velocity) in the range of 1000–5000 h^{-1}. Catalyst life ranges from 1 to 5 years, depending on usage (see Sulfur recovery).

ECONOMIC ASPECTS

Catalyst manufacture has become a major industry with an annual production of over $400 million (244–246). Catalysis contributes directly and indirectly to products accounting for one-sixth of the value of all goods manufactured in the United States; about 60% is used in the petroleum industry (see Table 7) and the remainder for major chemicals and chemical intermediates (see Table 8). An example of the large amounts

Table 7. Catalyst Consumption in the Petroleum Industry, 1972[a]

Process	Type of catalyst	Thousands of metric tons	Millions of dollars
cracking			
fluid bed	zeolite-based alumina–silica	102	56.50
fluid bed	amorphous alumina–silica and clay	6.3	2.30
moving bed	zeolites	12.2	9.50
moving bed	amorphous	3.0	0.77
Total		123.5	69.07[b]
reforming			
replacement	monometallic	0.36	2.80
replacement	bimetallic	1.44	13.60
new	bimetallic	0.45	4.30
new	precious-metal value		9.60
Total		2.25	30.30
hydrotreating			
replacement	cobalt–molybdena, nickel–molybdena, nickel–tungsten	3.6	8.20
new	same	0.9	2.00
Total		4.5	10.20
hydrocracking			
new and replacement	noble metal on amorphous support; non-noble metal and noble metal on molecular sieves		12.80
alkylation	sulfuric acid	1395[c]	39.00[c]
	hydrofluoric acid	10	6.70
Total		1405	45.70
Grand total		1535	168.07

[a] Ref. 244.
[b] By 1977 increased usage of more expensive molecular sieve catalysts increased value of cracking catalysts to ca $140 million.
[c] With the gross totals the net consumption is 12–14% of gross.

of catalyst used in various steps of a major chemical production facility are the data shown in Table 9 for an ammonia synthesis plant. The catalyst industry is expected to grow significantly, not only because of process expansion and the development of new processes, but also because of the need for catalysts in new areas such as environmental control; automotive emissions control is a good example.

For cracking processes employing a clay catalyst or reforming with a platinum catalyst, the catalyst cost is about 0.6¢/m^3 (4¢/bbl) of product which corresponds to about 0.02¢/kg.

The actual prices of catalysts vary widely, ranging from 11–22¢/kg to $22/kg or more for catalysts containing noble metals. Many solid catalysts containing substantial amounts of chromium, cobalt, molybdenum, etc, cost less than $2/kg.

CHARACTERIZATION

Characterization of the catalyst's surface properties is crucial for understanding its operation and for maintaining and improving its performance. Although catalysts are usually complex materials, a number of techniques have been developed to evaluate various physical and chemical properties. Some of these are straightforward and routine and give useful information such as total surface area, pore volume, and

Table 8. Major Products Manufactured by Catalytic Operations, 1972[a]

Operation	Process	Catalyst	Millions of dollars	Thousands of metric tons
hydrogen–ammonia	steam reforming, shift, methanation, etc	supported nickel, iron, etc	10.4	
hydrogenation	edible oil hydrogenation	25% nickel in oil	6–7	2.3–2.7
	inedible oil hydrogenation	25% nickel in oil	2	0.7
	various	Raney nickel	4	
	selective hydrogenation of olefins streams in ethylene plants	nickel–chrome or palladium on alumina	0.25	
	selective hydrogenation pyrolysis gasoline in ethylene plants	noble metals and cobalt–molybdenum	0.1	
dehydrogenation	butadiene from butane	chrome–alumina	1.5–2	0.7–0.9
	styrene from ethyl benzene	promoted iron oxide	1.5	1.4
oxidation	ethylene oxide	silver on alumina	6.3	0.6
	nitric acid from ammonia	platinum–rhodium gauze	9.7	
	contact sulfuric	promoted vanadia	2.5	1.2
	phthalic anhydride	vanadia	1.5–2	
	maleic anhydride	vanadia	0.5–0.6	
ammoxidation	acrylonitrile from propylene	uranium–antimony	2.65	0.45
oxychlorination	vinyl chloride	supported copper chloride	2	0.45
organic synthesis	Friedel-Crafts, etc	aluminum chloride	8.7	26
polymerization	stereopolymers	aluminum alkyls, titanium and vanadium compounds	18–20	
		chrome-based	2–3	
	urethane foams	tertiary amines	9–10	1.35
	cumene, tetramer, etc	phosphoric acid on kieselguhr	1	1.8
Total			90–96	

[a] Ref. 245.

pore-size distribution (142). In this respect, a particularly important parameter for understanding the operation of a catalyst is the specific surface area of the active catalytic function (94).

Specific Surface Area. The specific surface area of a catalyst can be determined and the various components distinguished. In the case of a supported catalyst, the active catalytic material and the support can be identified by x-ray diffraction (143), transmission electron microscopy (144), and gas chemisorption.

Using x-ray diffraction techniques, the width of a line for a particular structure can be related to the crystallite size of aggregates of the compound. From line broadening, one can calculate the average particle size, and therefore the specific surface area. This technique is limited by sensitivity and particle size which cannot be easily detected below 3–4 nm. For larger particles the procedure is routine.

Particle size is determined by electron microscopy. Complications are primarily related to sample preparation and resolution (in the 0.5–1.0 nm range).

A specific surface area is best measured by selective gas adsorption, using gases

54 CATALYSIS

Table 9. Catalyst Charge to 900 Metric Tons Per Day Ammonia Plant[a]

Process	Catalyst	m³	$
feed treatment	impregnated activated carbon	57	42,000
primary reforming	nickel on alumina or calcium aluminate	17	36,000
secondary reforming	nickel on carrier	28	60,000
high-temperature shift	chromium-promoted iron oxide	57	44,000
low-temperature shift	copper–zinc–alumina	57	140,000
methanation	nickel on carrier	23	40,000
ammonia synthesis	promoted iron oxide	57	150,000
Total		296	512,000

[a] Ref. 245.

that chemisorb on only one component of the catalyst. Most of the catalytic metals have been characterized by the selective adsorption of a number of gases (94), but this is not true for oxides. The gases more commonly used are CO and NO. The latter has been used to measure the surface area of oxides of iron (145), chromium (146), copper (147), and nickel (148).

Measurements of specific surface area (94,142) for a catalyst permit calculation of the reaction rate per active site, crucial for determining the intrinsic activity. The importance of such measurements is shown in a study of the methanation activity of a number of transition metals (149). Early activity data, which did not have the advantage of the surface area measurement techniques developed during the last decade, gave the following sequence (150): Ru > Ir > Rh > Ni > Co > Os > Pt > Fe > Pd. By contrast, a different sequence was found when the rate data were corrected to unit surface area of the metal studied: Ru > Fe > Ni > Co > Rh ~ Pd > Pr > Ir. The most significant difference was found for Fe which had been considered to be a poor methanation catalyst. It reflects the difficulty in preparing iron catalysts of high surface area (151).

Advanced Techniques. Detailed information concerning the surfaces of a catalyst can be obtained from surface morphology and composition, as well as the chemical state of surface atoms. These techniques generally require (90) skill in their application to catalysts, and in many instances are still in the developing stages.

X-ray, electron, and ion spectroscopies, all developed over the last decade, have had the most significant impact on catalysis. Extended x-ray absorption fine structure (EXAFS) is beginning to find application (32,247). Among electron spectroscopies, ultraviolet photoelectron spectroscopy (UPS) (156–157), x-ray photoelectron spectroscopy (XPS) (154–155), and Auger electron spectroscopy (AES) (158–172) are used. Ion spectroscopies of interest are secondary-ion mass spectroscopy (SIMS) (173–174), ion-scattering spectroscopy (ISS) (175–178) and ion-neutralization spectroscopy (INS) (175–178) (see Analytical methods).

Reactor Engineering and Catalyst Testing. The laboratory test reactor plays a critical role in catalyst development. It is particularly important to minimize the influence of nonkinetic phenomena such as heat and mass transfer. A number of authors have discussed this subject in detail, in particular Sherwood (246), Peterson (248), Carberry (249–250), Froment (251), Weekman (252–253), Lapidus (254), Difford and Spencer (255), and Cusumano, Dalla Betta, and Levy (256) (see Reactor technology).

BIBLIOGRAPHY

"Catalysis" in *ECT* 1st ed., Vol. 3, pp. 245–272 and Suppl. 1, pp. 144–150, by V. I. Komarewsky, Illinois Institute of Technology, and J. B. Coley, Standard Oil Company of Indiana; "Catalysis" in *ECT* 2nd ed., Vol. 4, pp. 534–586, by G. Alexander Mills, Houdry Process and Chemical Company.

1. K. Ziegler and co-workers, *Angew. Chem.* **67,** 426 (1955).
2. G. Natta, *J. Polym. Sci.* **16,** 143 (1955).
3. G. Natta, *Chem. Ind.* **47,** 1520 (1957).
4. G. Natta and I. Pasquon, *Proceedings of the International Congress on Catalysis,* Editions Technip, Paris, 1960, paper 66.
5. G. Natta and F. Danusso, eds., *Stereoregular Polymers and Stereospecific Polymerizations,* Vols. 1 and 2, Pergamon Press, New York, 1967.
6. P. Cossee, *J. Catal.* **3,** 80 (1964).
7. L. A. M. Rodriguez, H. M. van Looy, and J. A. Gabant, *J. Polym. Sci.* **A-14,** 1905 (1966).
8. E. Field and M. Feller, *Ind. Eng. Chem.* **49,** 1883 (1957).
9. *Chem. Eng. News,* 21 (Dec. 5, 1977).
10. A. Farkas and G. A. Mills in D. D. Eley, H. Pines, and P. B. Weisz, eds., *Advances in Catalysis,* Vol. 13, Academic Press, Inc., New York, 1962.
11. J. Smidt, *Chem. Ind. London* 54, (July 13, 1962).
12. I. I. Moiseev, O. G. Levanda, and M. N. Vargaftek, *J. Am. Chem. Soc.* **96,** 1003 (1974).
13. I. I. Moiseev, *Am. Chem. Soc. Div. Pet. Chem. Prepr.* **14,** B49 (1969).
14. J. Smidt and co-workers, *Angew. Chem.* **71,** 176 (1959).
15. I. I. Moiseev, M. N. Vargaftek, and J. K. Syrkin, *Dokl. Akad. Maml. SSSR* **130,** 801 (1960); **133,** 377 (1960).
16. H. Kekeler and W. Kronig, *Proceedings of the 7th World Petroleum Congress,* Vol. 5, Elsevier Scientific Publishing Co., Inc., New York, 1967, p. 41.
17. J. L. Callahan and R. K. Grasselli, *Am. Inst. Chem. Eng. J.* **6,** 755 (1963).
18. J. Matsuura and G. C. A. Schuit, *J. Catal.* **20,** 19 (1971); **25,** 316 (1972).
19. J. F. Roth and co-workers, *Chem. Technol.,* 600 (1971).
20. D. W. Breck, *Zeolite Molecular Sieves,* John Wiley & Sons, Inc., New York, 1974.
21. J. Turkevich and Y. Ono, *Adv. Chem. Soc.* **102,** 315 (1971).
22. P. B. Weisz and V. J. Frilette, *J. Phys. Chem.* **64,** 382 (1960).
23. N. Y. Chen and co-workers, *Oil Gas J.* **66**(47), 147 (1968).
24. S. C. Eastwood, C. J. Plank, and P. B. Weisz, *Proceedings of the 8th World Petroleum Congress, Moscow,* Applied Sciences, London, 1971.
25. G. M. Schwab, *Discuss. Faraday Soc.* **8,** 166 (1950).
26. D. A. Dowden, *J. Chem. Soc.,* 242 (1950).
27. D. A. Dowden and P. Reynolds, *Discuss. Faraday Soc.* **8,** 184 (1950).
28. J. H. Sinfelt, *Chem. Eng. News,* 18 (July 3, 1972).
29. J. H. Sinfelt, *J. Catal.* **29,** 308 (1973).
30. J. H. Sinfelt, J. L. Carter, and D. J. C. Yates, *J. Catal.* **24,** 283 (1972).
31. M. Boudart, J. A. Cusumano, and R. B. Levy, *New Catalytic Materials for the Liquefaction of Coal,* spons. by the Electric Power Research Institute, Palo Alto, Calif., Report No. RP-415-1, Oct. 30, 1975.
32. J. H. Sinfelt and J. A. Cusumano in *Advanced Materials in Catalysis,* Academic Press, Inc., New York, 1976.
33. J. K. A. Clarke, *Chem. Rev.* **75,** 291 (1975).
34. R. R. Cecil and co-workers, *Advances in Reforming,* 37th Mid-Year Meeting, API, New York, May 1972.
35. E. F. Schwarzenbek, *Adv. Chem. Ser.* **103,** 94 (1971).
36. U.S. Pat. 3,617,520 (Nov. 2, 1971), H. E. Kluksdahl (to Chevron Research Co.).
37. U.S. Pat. 3,558,477 (Jan. 26, 1971), H. E. Kluksdahl (to Chevron Research Co.).
38. C. S. McCoy and P. Munk, *Symposium on Catalytic Reforming,* 68th National Meeting of AIChE, Houston, Tex., Feb. 28–Mar. 4, 1971, paper 42a.
39. U.S. Pat. 3,839,194 (Oct. 1, 1974), J. H. Sinfelt and A. E. Barnett (to Exxon Research & Engineering Co.).

40. U.S. Pat. 3,442,973 (May 6, 1969), J. H. Sinfelt, A. E. Barnett, and G. Dembinski (to Exxon Research & Engineering Co.).
41. U.S. Pat. 3,576,736 (Apr. 7, 1971), J. R. Kittrell (to Chevron Research Co.).
42. S. H. Inami and H. Wise, *J. Catal.* **26**, 92 (1972).
43. E. G. Allison and G. C. Bond, *Catal. Rev.* **7**, 233 (1972).
44. U.S. Pat. 3,844,981 (Oct. 29, 1974), J. A. Cusumano (to Exxon Research and Engineering Co.).
45. W. H. Flank and H. C. Beachell, *J. Catal.* **8**, 316 (1967).
46. U.S. Pat. 2,911,357 (Nov. 3, 1959), J. W. Myers and F. A. Pragne (to Phillips Petroleum Co.).
47. U.S. Pat. 3,617,518 (Nov. 2, 1971), J. H. Sinfelt, A. E. Barnett, and J. L. Carter (to Exxon Research & Engineering Co.).
48. J. H. Sinfelt, *Ann. Rev. Sci.* **2**, 641 (1972).
49. S. C. Robertson, S. C. Kloet, and W. M. H. Sachtler, *J. Catal.* **39**, 234 (1975).
50. W. K. Hall and J. A. Hassell, *J. Phys. Chem.* **67**, 636 (1963).
51. T. Takeuchi and co-workers, *Bull. Chem. Soc. Jpn.* **35**, 1390 (1962).
52. Ger. Pats. 2,123,606 (Nov. 25, 1971); 2,127,348 (Dec. 16, 1971); and 2,141,420 (Mar. 9, 1972), N. Kominami, T. Iwaisako, and K. Ohki (to Asahi Chemical Industrial Co.).
53. Ger. Pat. 2,117,651 (Oct. 28, 1971), E. E. Davies, J. S. Elkins, and R. C. Pitkethly (to British Petroleum Co.).
54. U.S. Pat. 3,567,625 (Mar. 2, 1971), J. H. Sinfelt and A. E. Barnett (to Exxon Research & Engineering Co.).
55. V. Ponec and W. M. H. Sachtler, *J. Catal.* **24**, 250 (1972).
56. J. H. Sinfelt and co-workers, *J. Catal.* **42**, 227 (1976).
57. J. R. H. Van Schaik, R. P. Dessing, and V. Ponec, *J. Catal.*, in press.
58. V. Ponec and W. M. H. Sachtler, *Proceedings of the Fifth International Congress on Catalysis*, North-Holland Publishing Co., Amsterdam, 1973, p. 645.
59. R. B. Moyes and P. B. Wells in D. D. Eley, H. Pines, and P. B. Weisz, eds., *Advances in Catalysis*, Vol. 23, Academic Press, Inc., New York, 1973, p. 121.
60. Southwest Research Institute, *A Literature Search and Analysis of Information Regarding Sources, Uses, Production, Consumption, Reported Medical Cases, and Toxicology of Pt and Pd*, EPA Contract No. 68-02-1274, Apr. 15, 1974.
61. *Platinum Met. Rev.* **18**, 64 (1974).
62. *Consultant Report on an Evaluation of Catalytic Converters for Control of Automobile Exhaust Pollutants*, EPA, NTIS, Sept. 1974.
63. F. G. Dwyer, *Catal. Rev.* **6**(2), 261 (1972).
64. M. Shelef, K. Otto, and H. Gandhi, *J. Catal.* **12**, 361 (1968).
65. M. Shelef and J. T. Kummer, *Sixty-Second Annual AIChE Meeting*, Washington, D.C., Nov. 16–20, 1969, paper 13f.
66. J. E. McEvoy, *Adv. Chem. Ser.* **143**, (1975).
67. H. C. Andersen, P. L. Romeo, and W. J. Green, *Engelhard Ind. Tech. Bull.* **8**, 100 (1966).
68. M. Shelef and H. S. Gandhi, *Platinum Met. Rev.* **18**, 2 (1974).
69. H. C. Andersen, W. J. Green, and D. R. Steele, *Ind. Eng. Chem.* **53**, 199 (1961).
70. M. Yamaguchi, K. Matsushita, and K. Takami, *Hydrocarbon Process.*, 101 (Aug. 1976).
71. G. J. K. Acres, *Platinum Met. Rev.* **15**, 9 (1971).
72. G. J. K. Acres, *Platinum Met. Rev.* **14**, 2 (1970).
73. W. C. Pfefferle and co-workers, *ASME Paper No. 75-WA/FN-1*, ASME Society, New York, Dec. 1975.
74. Ger. Offen. 2,458,122 (June 12, 1975), S. G. Hindin, G. R. Pond, and J. C. Dettling (to Engelhard Minerals and Chemicals Corp.).
75. Ger. Offen. 2,458,221 (June 12, 1975), S. G. Hindin and G. R. Pond (to Engelhard Minerals and Chemicals Corp.).
76. U.S. Pat. 3,928,961 (Dec. 30, 1975), W. C. Pfefferle.
77. V. Haensel and M. J. Sterba, *Energy and Emission Control*, paper presented to EPA, June 25, 1973.
78. B. G. Goar, *Hydrocarbon Process.* **47**(9), 248 (1968).
79. H. Grebel, J. W. Palm, and J. W. Kilmer, *Oil Gas J.* **66**, 88 (1968).
80. P. C. Opekur and B. G. Goar, *Hydrocarbon Process.* **45**(6), 181 (1966).
81. A. R. Valdes, *Hydrocarbon Process. Pet. Refiner.* **43**, 104 (1964).
82. E. E. Donath in H. H. Lowry, ed., *Chemistry of Coal Utilization*, Suppl. Vol., John Wiley & Sons, Inc., New York, 1963, Chapt. 22.

83. S. W. Weller in P. H. Emmett, ed., *Catalysis*, Reinhold Publishing Co., New York, 1956.
84. W. R. K. Wu and H. H. Storch, *U.S. Bureau of Mines, Bull. 633*, Washington, D.C., 1968.
85. J. L. Wiley and H. C. Anderson, *U.S. Bureau of Mines, Bull. 485*, Vol. I, II, and III, Washington, D.C., 1950, pp. 1–2.
86. G. A. Mills, *Ind. Eng. Chem.* **61**(7), 6 (1969).
87. *Liquefaction and Chemical Refining of Coal*, Battelle Energy Program Report, Battelle Memorial Institute, Columbus, Ohio, July 1974.
88. G. A. Mills, *Catal. Rev.* **14**(1), 69 (1976).
89. G. A. Mills and F. W. Steffgen, *Catal. Rev.* **8**(2), 159 (1973).
90. M. Boudart, J. A. Cusumano, and R. B. Levy, *New Catalytic Materials for the Liquefaction of Coal*, EPRI Project 415, Electric Power Research Institute, Palo Alto, Calif., 1975.
91. A. Katzman, *A Research and Development Program for Catalysis in Coal Conversion Processes*, EPRI 207-0-0, May 1974.
92. *Proceedings of the Electric Power Research Institute Conference on Coal Catalysis*, Libby Laboratories, Sept. 1973.
93. S. W. Gouse and E. S. Rubin, *A Program of R. D. & D. for Enhancing Coal Utilization to Meet National Needs*, NSF Workshop, Carnegie-Mellon University, 1973.
94. J. A. Cusumano, R. A. Dalla Betta, and R. B. Levy, *Scientific Resources Relevant to the Catalytic Problems in the Conversion of Coal*, Report No. FE 2017-1, ERDA Contract No. E(49-18)-2017, Catalytica Associates, Inc., Palo Alto, Calif., Oct. 1976.
95. S. W. Weller, J. A. Bergantz, and D. T. Shaw, *Research in Coal Technology: The University's Role*, NTIS, Conference 741, 091, Oct. 1974.
96. G. A. Mills, *Am. Chem. Soc. Div. Fuel Chem. Prepr.*, **16**(2), 107 (1972).
97. J. E. Hightower, *Needs for Fundamental Research in Catalysis as Related to the Energy Problem*, National Science Foundation, Washington, D. C., June 1974.
98. P. E. Rousseau, *World Power Conference*, Tokyo National Committee, Paper 158, No. IV, 1966, p. 912.
99. M. Pier, *Z. Electrochem.* **53**(5) 291 (Oct. 1949).
100. A. G. Oblad, *Catal. Rev.* **14**(1), 83 (1976).
101. *Fossil Energy Program Report 1975–1976*, ERDA 77-70, U.S. Government Printing Office, Washington, D.C.
102. *Fossil Energy Research Program of ERDA*, ERDA 77-33, DOE/ET-0013(78), U.S. Government Printing Office, Washington, D.C., Apr. 1977.
103. *Energy R&D Program of the Government of the Federal Republic of Germany*, Annual Report, German Dept. of Energy, Júlich, FRG, 1975.
104. C. A. Johnson and co-workers, *68th Annual Meeting of AIChE*, Los Angeles, Calif., Nov. 16–20, 1975.
105. P. Yavorsky, S. Aktar, and S. Friedman, *Eng. Prog.* **69**(3), 51 (1973).
106. C. W. Zielke, W. A. Rosenhoover, and E. Gorin, *Adv. Chem. Ser.* (151), (1976).
107. C. W. Zielke and co-workers, *Annual Technical Progress Report for 1976, Including the Quarter Ending January 31, 1977*, Report No. 1743-33, Feb. 20, 1977.
108. *Fossil Energy Program Report*, U.S. Energy Research and Development Administration, ERDA 77-70, U.S. Government Printing Office, Washington, D.C., 1977.
109. H. H. Storch, H. Golumbic, and R. B. Anderson, *The Fischer-Tropsch and Related Synthesis*, John Wiley & Sons, Inc., New York, 1951.
110. S. L. Meisel and co-workers, *Chem. Tech.* **6**, 86 (1976).
111. *Proceedings, Eighth Synthetic Pipeline Gas Symposium*, American Gas Assoc., Arlington, Va., 1976.
112. J. L. Johnson, *Catal. Rev.* **14**(1), 131 (1976).
113. W. Auer, E. Lorenz, and K. H. Grundler, *Paper Presented at 68th National Meeting AIChE*, Houston, Tex., 1971.
114. P. Sabatier and J. B. Senderens, *C. R. Acad. Sci. Paris* **134**, 514 (1902).
115. F. J. Dent and co-workers, *Communication GRB 20*, British Gas Research Board, London, 1945.
116. M. Greyson in P. H. Emmett, ed., *Catalysis*, Vol. 4, Reinhold Publishing Co., New York, 1956, Chapt. 6.
117. D. W. Allen and W. A. Yen, *Chem. Eng. Prog.* **69**(1), 75 (1973).
118. I. Wender, *Catal. Rev.* **14**(1), 97 (1976).
119. *New Catalytic Materials for the Liquefaction of Coal*, Research Project 415, Final Report, Electric Power Research Institute, Palo Alto, Calif., Oct. 1975.

120. D. A. MacInnes, *The Principles of Electrochemistry,* Dover Publications, Inc., New York, 1961.
121. R. F. Gould, ed., *Fuel Cell Systems,* American Chemical Society, Washington, D.C., 1965.
122. R. F. Gould, ed., *Fuel Cell Systems—II,* American Chemical Society, Washington, D.C., 1969.
123. A. Fickett, *Electr. Power Res. Inst. J.,* 17 (Apr. 1976).
124. *Advanced Technology Fuel Cell Program,* Contract RP-114-2, United Technologies Corporation for the Electric Power Research Institute and the Energy Research and Development Administration, 1977.
125. E. J. Cairns and H. Shimotake in B. S. Baker, ed., *Adv. Chem. Ser.* **90,** 392 (1969).
126. H. Heinemann, *Chem. Tech.* **1,** 286 (1971).
127. F. A. Cotton and G. Wilkinson, *Advanced Inorganic Chemistry,* Wiley-Interscience, a division of John Wiley & Sons, Inc., New York, 1972.
128. G. N. Schrauzer, ed., *Transition Metals in Homogeneous Catalysts,* Marcel Dekker, Inc., New York, 1971.
129. D. Forster and J. F. Roth, eds., *Adv. Chem. Ser.* **132,** (1974).
130. J. Chatt and J. Halpern in F. Basolo and R. L. Burwell, eds., *Catalysis, Progress in Research,* Plenum Press, New York, 1973.
131. J. Halpern, *Adv. Chem. Ser.* **70,** 1 (1968).
132. J. R. Norton, *Am. Chem. Soc. Div. Pet. Chem. Prepr.,* (1976).
133. G. A. Mills, S. W. Weller, and A. Wheeler, *J. Phys. Chem.* **63,** 403 (1959).
134. J. Kwiatek, I. L. Modor, and J. K. Seyler, *Adv. Chem. Ser.* (37), (1963).
135. S. W. Weller and G. A. Mills, *Advances in Catalysis,* Vol. 8, Academic Press, Inc., New York, 1956.
136. J. Halpern, *Advances in Catalysis,* Vol. 9, Academic Press, Inc., New York, 1957.
137. R. S. Nyholm, *Proceedings of the 3rd International Congress on Catalysis,* Vol. I, North-Holland Publishing Co., Amsterdam, 1965.
138. R. Schenk, *Angew. Chem.* **49,** (1936); *Z. Anorg. Chem.,* 260 (1949).
139. H. C. Egghart, *Concepts in Heterogenous Catalysts,* Report 2105, AD, Fort Belvoir, Va., 1974.
140. Th. Wolkenstein in ref. 10, Vol. 12, 1960.
141. D. A. Dowden and D. Wells in ref. 4, paper 73.
142. S. J. Gregg and K. S. W. Sing, *Adsorption Surface Area and Porosity,* Academic Press, Inc., New York, 1967.
143. J. R. Anderson, *Structure of Metallic Catalysts,* Academic Press, Inc., New York, 1975, p. 364.
144. *Ibid.,* p. 363.
145. K. Otto and M. Shelef, *J. Catal.* **18,** 184 (1970).
146. K. Otto and M. Shelef, *J. Catal.* **14,** 226 (1969).
147. H. Gandhi and M. Shelef, *J. Catal.* **28,** 1 (1973).
148. H. Gandhi and M. Shelef, *J. Catal.* **24,** 241 (1972).
149. M. A. Vannice, *J. Catal.* **37,** 449 (1975).
150. F. Fischer, M. Tropsch, and P. Dilthey, *Brennst. Chem.* **6,** 265 (1925).
151. M. Boudart and co-workers, *J. Catal.* **37,** 486 (1975).
152. K Siegbahn and co-workers, *Nova Acta Regiae Soc. Sci. Ups.* **20,** (1967).
153. D. C. Frost and co-workers, *Chem. Phys. Lett.* **13,** 391 (1972).
154. L. C. Snyder, *J. Chem. Phys.* **55,** 95 (1971).
155. F. O. Ellison and L. L. Larcom, *Chem. Phys. Lett.* **13,** 399 (1972).
156. D. W. Turner and co-workers, *Molecular Photoelectron Spectroscopy,* John Wiley & Sons, Inc., London, 1970.
157. J. T. Yates, Jr., *Chem. Eng. News,* 19 (Aug. 26, 1974).
158. P. W. Palmberg in D. A. Shirley, ed., *Electron Spectroscopy,* North-Holland Publishing Co., Amsterdam, p. 835.
159. C. C. Chang, *Surf. Sci.* **25,** 53 (1971).
160. C. R. Brundle, *J. Vac. Sci. Technol.* **11,** 212 (1974).
161. C. D. Wagner, *J. Electron, Spectrosc.* **10,** 305 (1977).
162. F. J. Szalkowski and G. A. Somorjai, *J. Chem. Phys.* **56,** 6097 (1972).
163. A. Takeuchi and co-workers, *J. Catal.* **40,** 94 (1975).
164. F. Williams and K. Baron, *J. Catal.* **40,** 108 (1975).
165. F. Williams and M. Boudart, *J. Catal.* **30,** 438 (1973).
166. J. Yates, Jr., T. Madey, and N. Erickson, *Surf. Sci.* **43,** 257 (1974).
167. D. A. Shirley, *Chem. Phys. Lett.* **16,** 220 (1972).
168. D. Eastman and J. Demuth, *Phys. Rev. Lett.* **32,** 1123 (1974).

169. W. E. Spicer and co-workers in J. M. Thompson and M. W. Roberts, eds., *Surface & Defect Properties of Solids,* Vol. 5, Univ. of Bradford, England.
170. C. Helms, K. Yu, and W. Spicer, *Solid State Communications,* in press.
171. P. Ratnasamy, *J. Catal.* **40,** 137 (1975).
172. P. Ratnasamy and co-workers, *J. Catal.* **38,** 19 (1975).
173. S. Rubin, *Nucl. Instrum. Methods* **5,** 177 (1959).
174. H. Werner, *Surf. Sci.* **47,** 301 (1975).
175. H. H. Brogersma, *J. Vac. Sci. Technol.* **11,** 231 (1974).
176. R. F. Groff, *J. Vac. Sci. Technol.* **10,** 355 (1973).
177. E. Taglauer, and W. Heiland, *Surf. Sci.* **47,** 234 (1975).
178. H. Niehus, and E. Bauer, *Surf. Sci.* **47,** 222 (1975).
179. F. W. Lytle in G. R. Mallet, M. Fay, and W. M. Mueller, eds., *Advances in X-Ray Analysis,* Plenum Press, New York, p. 398.
180. D. E. Sayers, F. W. Lytle, and E. A. Stern, *Phys. Rev. Lett.* **27,** 1204 (1971).
181. H. S. Taylor, *J. Am. Chem. Soc.* **53,** 578 (1931); *Chem. Rev.* **9,** 1 (1931).
182. B. M. Trapnell, *Chemisorption,* Academic Press, Inc., New York, 1955.
183. H. S. Taylor, *Proc. R. Soc. London Ser. A* **152,** 445 (1925); *J. Phys. Chem.* **30,** 145 (1926).
184. R. H. Griffith in ref. 10, Vol. 1, 1948, Chapt. 4.
185. N. D. Felinsky, *Ber. Dtsch. Chem. Ges.* **60,** 1723 (1927).
186. O. Beek, *Discuss. Faraday Soc.* **8,** 118 (1950).
187. G. A. Mills and co-workers, *J. Phys. Chem.* **59,** 809 (1955).
188. R. Coekelbergs, A. Crucq, and A. Frennet in ref. 10, Vol. 13, 1962.
189. A. Wheeler in ref. 10, Vol. 3, 1951.
190. P. H. Emmett, *Catalysis,* Vol. 2, Reinhold Publishing Corp., New York, 1955.
191. E. W. Thiele, *Ind. Eng. Chem.* **31,** 916 (1939).
192. G. C. Bond, *Catalysis by Metals,* Academic Press, Inc., New York, 1962.
193. F. G. Ciapetta and C. J. Plank in ref. 190, Vol. 1, 1954, Chapt. 7.
194. F. G. Ciapetta, in C. H. Collier, ed., *Catalysis in Practice,* Reinhold Publishing Corp., New York, 1957.
195. W. B. Innes in ref. 190, Vol. 1, 1954, Chapt. 6.
196. *Chem. Week,* 57 (Nov. 9, 1977).
197. R. L. Banks, and G. C. Bailey, *Ind. Eng. Chem. Prod. Res. Dev.* **3,** 170 (1964)
198. R. L. Banks, *Div. Pet. Chem.* **17**(3), A 21 (1972).
199. B. C. Gates, J. R. Katzer, and G. C. A. Schuit, *Chemistry of Catalytic Processes,* McGraw-Hill Book Co., New York, 1978.
200. M. S. Jarrell and B. C. Gates, *J. Catal.* **40,** 255 (1975).
201. J. Smidt and co-workers, *Angew. Chem.* **71,** 176 (1959).
202. E. B. Evnin and co-workers, *J. Catal.* **30,** 109 (1973).
203. R. F. Heck and D. S. Breslow, *J. Am. Chem. Soc.* **83,** 4023 (1961).
204. M. Orchin and W. Rupilius, *Catal. Rev.* **6,** 85 (1972).
205. J. Falbe, *Carbon Monoxide in Organic Synthesis,* Springer-Verlag, New York, 1970, Chapt. 1; B. Cornils, R. Payer, and K. C. Traencker, *Hydrocarbon Process.* **54**(6), 83 (1975).
206. H. Lemke, *Hydrocarbon Process.* **45**(2), 148 (1966).
207. U.S. Pats. 3,833,634, 3,878,214, 3,878,290, and 3,878,292 (Sept. 3, 1974), R. L. Pruett and W. E. Walker (to Union Carbide Corp.).
208. A. L. Robinson, *Science* **194,** 1150 (1977); E. L. Muetterties, *Science* **196,** 839 (1977).
209. M. Swarz, ed., *Ions and Ion Pairs in Organic Reactions,* Vol. I, Interscience Publishers, Inc., New York, 1922, Chapt. 1.
210. R. A. Messing, *Immobilized Enzymes for Industrial Reactors,* Academic Press, Inc., New York, 1975.
211. K. Mosback, *Immobilized Enzymes,* Vol. 44, *Methods in Enzymology,* Academic Press, Inc., New York, 1976.
212. P. Reich, *Hydrocarbon Process.* **55**(3), 85 (1975).
213. M. L. Spector, H. Heinemann, and K. D. Miller, *Ind. Eng. Chem. Process. Des. Dev.* **6**(3), 327 (1967).
214. L. Friend, L. Wender, and J. C. Yarze, *Adv. Chem.* **70,** 168 (1968).
215. R. P. Bell, *Acid-Base Catalysis,* The Clarendon Press, Oxford, 1941.
216. J. W. Ward, *J. Catal.* **11,** 251 (1968); **13,** 321 (1969); **14,** 365 (1969); **17,** 355 (1970).
217. S. Brønsted and E. A. Groggenhlem, *J. Am. Chem. Soc.* **49** 2554 (1927).

218. A. G. Oblad, G. A. Mills, and H. Heinemann in ref. 190, Vol. 6, 1958, Chapt. 4.
219. P. D. Bartlett, F. E. Condon, and A. Schneider, *J. Am. Chem. Soc.* **66,** 1531 (1944).
220. F. C. Whitmore, *Ind. Eng. Chem.* **26,** 94 (1934).
221. S. G. Hindin, G. A. Mills, and A. G. Oblad, *J. Am. Chem. Soc.* **73,** 278 (1951).
222. F. E. Condon in ref. 190, Vol. 6, 1958, Chapt. 2.
223. H. Pines in ref. 10, Vol. 1, 1948.
224. R. M. Kennedy in ref. 190, Vol. 6, 1958, Chapt. 1.
225. L. Schmerling and J. O. Iverson in K. A. Kobe and J. J. McKetta, Jr., eds., *Advances in Petroleum Chemistry and Refining,* Vol. 1, Interscience Publishers Inc., New York, 1958, Chapt. 7.
226. B. S. Greensfelder, *Chemistry of Petroleum Hydrocarbons,* Vol. 2, Reinhold Publishers Corp., New York, 1955, Chapt. 27.
227. L. B. Ryland, M. W. Tamele, and J. N. Wilson in ref. 190, Vol. 7, 1960, Chapt. 1.
228. H. Pines and L. A. Schaap in ref. 10, Vol. 12, 1960.
229. A. G. Oblad, S. G. Hindin, and G. A. Mills, *J. Am. Chem. Soc.* **75,** 4096 (1953).
230. S. L. Meisel and co-workers, *Recent Advances in the Production of Fuels and Chemicals Over Zeolite Catalysts,* Div. Ind. Eng. Chem., Am. Chem. Soc., Chicago, Ill., Aug. 1977.
231. N. Y. Chen and R. L. Goring, *Oil Gas J.,* (June 6, 1977).
232. L. Rheaume and co-workers, *Oil Gas J.,* 24 (May 17, 1976).
233. G. A. Mills and co-workers, *Ind. Eng. Chem.* **45,** 134 (1953).
234. P. Weisz in ref. 10, Vol. 13, 1962.
235. J. R. Anderson, R. J. MacDonald, and Y. Shimoyama, *J. Catal.* **20,** 147 (1971).
236. J. R. Anderson, *Adv. Catal.* **23,** 1 (1973).
237. P. B. Weisz and E. W. Swegler, *Science* **126,** 31 (1957).
238. S. G. Hindin, S. W. Weller, and G. A. Mills, *J. Phys. Chem.* **62,** 244 (1958).
239. G. E. Langlois and R. F. Sullivan, *Am. Chem. Soc.* **14**(3), D 18 (1969).
240. S. C. Shuman and H. S. Shalit, *Catal. Rev.* **4,** (1971).
241. J. H. Sinfelt, *Catal. Rev. Eng.* **9**(1), 147 (1974).
242. J. F. Shultz, F. S. Karn, and R. B. Anderson, *U.S. Bur. Mines Rep.* **6974,** (1967).
243. G. A. Mills, *Chem. Technol.* **7,** 418 (1977).
244. B. P. Burke, *Chem. Week,* 23 (Nov. 1, 1972).
245. *Ibid.,* 35 (Nov. 8, 1972).
246. C. N. Satterfield and T. K. Sherwood, *The Role of Diffusion in Catalysis,* Addison-Wesley, Reading, Mass., 1963.
247. F. W. Lytle, G. H. Via, and J. H. Sinfelt, *J. Am. Phys.* **67,** 3831 (1977).
248. E. E. Petersen, *Chemical Reaction Analysis,* Prentice-Hall, Englewood Cliffs, N. J., 1965.
249. J. J. Carberry, *Catal. Rev.* **3,** 61 (1969).
250. J. J. Carberry and J. B. Butt, *Catal. Rev.* **10**(2), 221 (1974).
251. G. F. Froment, *Adv. Chem. Ser.* (109), 1 (1972).
252. V. W. Weekman, *Adv. Chem. Ser.* (148), 98 (1975).
253. V. W. Weekman, Jr., *AIChE J.* **120,** 833 (1974).
254. L. Lapidus, *Ann. Rev. Ind. Eng. Chem.* **132,** (1972); J. Seinfeld and L. Lapidus, *Mathematical Methods in Chemical Engineering,* Vol. 3, *Process Modeling, Estimation and Identification,* Prentice-Hall, Englewood Cliffs, N.J., 1973; R. H. Rossen and L. Lapidus, *AIChE J.* **18,** 673 (1972).
255. A. M. R. Difford and M. S. Spencer, *Paper Presented at AIChE Meeting, Pittsburgh, Pa., June 2–5, 1974.*
256. J. A. Cusumano, R. A. Dalla Betta, and R. B. Levy, *Catalysis in Coal Conversion*, Academic Press, Inc., New York, 1978.

General References

D. D. Eley, H. Pines, and P. B. Weisz, eds., *Adv. Cat.* 1–26, (1948–1977).
Proceedings of the International Congress on Catalysis, (a) 1956, **9,** *Adv. Catal.;* (b) 1960, Editions Technip, Paris; (c) 1964, North-Holland Publishing Co., Amsterdam; (d) 1968, J. E. Hightower, Rice Univ. Press, Houston; (e) 1972, North-Holland Publishing Co., Amsterdam; and (f) 1976.
P. Selwood and F. Stone, eds., *J. Catal.* Academic Press, Inc., New York.
H. Heinemann and J. J. Carberry, eds., *Catalysis Reviews,* Marcel Dekker, Inc., New York.
Kinetics and Catalysis, transl. from Russian, Consultants Bureau, New York.
R. B. Anderson, ed., *Experimental Methods in Catalytic Research,* Academic Press, Inc., New York, 1968.

A. A. Balandin and co-workers, *Scientific Selection of Catalysts,* transl. from Russian by A. Aledjene, Israel Program for Scientific Translations, 1968.

R. P. Bell, *Acid-Base Catalysis,* The Clarendon Press, Oxford, 1941.

S. Berkman, J. C. Morrell, and G. Egloff, *Catalysis: Inorganic and Organic,* Reinhold Publishing Corp., New York, 1940.

G. C. Bond, *Catalysis by Metals,* Academic Press, Inc., New York, 1962.

D. P. Burke, *Chem. Week* **85,** 51 (Aug. 17, 1963); 51 (Aug. 24, 1973).

A. Clark, *The Theory of Adsorption and Catalysis,* Academic Press, Inc., New York, 1970.

J. H. deBoer, ed., *The Mechanism of Heterogeneous Catalysis,* Elsevier Publishing Co., Inc., New York, 1960.

P. H. Emmett, *Catalysis Then and Now,* Franklin Publishing Co., Philadelphia, Pa., 1955.

P. H. Emmett, ed., *Catalysis,* Vols. 1–7, Reinhold Publishing Corp., New York, 1954–1960.

B. C. Gates, J. R. Katzer, and G. C. A. Schuit, *Chemistry of Catalytic Processes,* McGraw-Hill Book Co., New York, 1978.

R. H. Griffith and J. D. F. Marsh, *Contact Catalysis,* 3rd ed., Oxford University Press, Inc., New York, 1957.

V. N. Ipatieff, *Catalytic Reactions at High Temperatures and Pressures,* The MacMillan Co., New York, 1937.

O. V. Krylov, *Catalysis by Non Metals,* Academic Press, Inc., New York, 1970.

H. Pines and W. M. Stalick, *Base Catalyzed Reactions of Hydrocarbons and Related Compounds,* Academic Press, Inc., New York, 1977.

M. Prettre, *Catalysis and Catalysts,* Dover Publications, New York, 1963.

E. K. Rideal, *Concepts in Catalysis,* Academic Press, Inc., New York, 1968.

C. N. Satterfield, *Mass Transfer in Heterogeneous Catalysis,* M.I.T. Press, Cambridge, Mass., 1970.

K. Tanabe, *Solid Acids and Bases,* Academic Press, Inc., New York, 1970.

C. L. Thomas, *Catalytic Processes and Proven Catalysts,* Academic Press, Inc., New York, 1970.

J. M. Thomas and W. J. Thomas, *Introduction to the Principles of Heterogeneous Catalysis,* Academic Press, Inc., New York, 1967.

G. ALEX MILLS
U.S. Department of Energy

JAMES A. CUSUMANO
Catalytica Associates, Inc.

CATALYSIS, PHASE-TRANSFER

Phase-transfer catalysis (PTC) is a technique by which reactions between substances located in different phases are brought about or accelerated. Typically, one or more of the reactants are organic liquids or solids dissolved in a nonpolar organic solvent and the coreactants are salts or alkali metal hydroxides in aqueous solution. Without a catalyst such reactions are often slow or do not occur at all; the phase-transfer catalyst, however, makes such conversions fast and efficient. Catalysts used most extensively are quaternary ammonium or phosphonium salts, and crown ethers and cryptates. Although isolated examples of PTC can be found in the early literature, it is only since the middle of the 1960s that the method has developed extensively. Terms partly or fully synonymous with PTC are extractive alkylation, reactions of naked anions, catalytic two-phase reactions, anion activation, and preparative ion pair extraction.

Traditionally, the reactions to be considered here are performed in homogenous medium. In hydroxylic solvents these conversions are slow because of solvation of the anions. Therefore, polar aprotic solvents are often employed. In comparison PTC has the following advantages: no need for expensive aprotic solvents; simpler workup; shorter reaction time and or lower reaction temperature; use of aqueous alkali hydroxides instead of other expensive bases.

Undoubtedly cost factors and environmental considerations (recycling of solvents, less toxic or less hazardous materials) will lead to increasing industrial application of this methodology.

Types of Reactions Catalyzed

There are two distinct classes of reactions performed under PTC conditions: *(1)* reactions without added bases; and *(2)* reactions in the presence of alkali metal hydroxides. Numerous displacement reactions belong to the first class.

$$RX + Y^- \rightarrow RY + X^-$$

The leaving group, X^-, is normally a halide or a sulfonate, and the nucleophile Y^- can be: F^-, Cl^-, Br^-, I^-, CN^-, SCN^-, N_3^-, NO_2^-, NO_3^-, $R-CO_2^-$, SR^-, SH^-, etc.

Oxidation reactions with oxidants such as $KMnO_4$, KO_2, $K_3[Fe(CN)_6]$, $NaOCl$, as well as some reduction reactions are further examples of conversions of class *(1)*.

Aqueous concentrated sodium hydroxide is the most frequently used base for the reactions of class *(2)*. The following types of PTC reactions have been performed: alkylations of weak C—H acidic compounds up to a pK_a limit of 22–25 (depending on the reference acid used); alkylations of ambident anions; alkylations of OH, NH, and SH bonds; isomerizations and H—D-exchange; additions across C=C and C=O bonds; β-eliminations; formation of carbenes by α-eliminations; hydrolysis and saponifications; Darzens reactions; Horner-Wittig reactions; and reactions of sulfonium ylides.

Mechanism of Phase-Transfer Catalysis

The mechanism has been elucidated only for the class (1) reactions. There are indications that at least some of the reactions of class (2) follow a different mechanism. Figure 1(a) shows the mechanistic picture of class (1) reactions developed by C. M. Starks (1–2). The catalyst cation Q^+ extracts the more lipophilic anion Y^- from the aqueous to the nonpolar organic phase where it is present in the form of a poorly solvated ion pair $[Q^+Y^-]$. This then reacts rapidly with RX, and the newly formed ion pair $[Q^+X^-]$ returns to the aqueous phase for another exchange process: $X^- \rightarrow Y^-$. In practice most catalyst cations used are rather lipophilic and do not extract strongly into the aqueous phase so that the anions are exchanged at the phase boundary (Fig. 1(b)).

This general scheme has been supported by a wealth of detailed investigations. Of special importance are the observations that the rate of reaction is proportional to the catalyst concentration and independent of the stirring speed (size of the interphase) above the minimum required for efficient mixing. It can be demonstrated that a certain number of water molecules are coextracted with the anion (3). This phenomenon has a decisive influence on the relative (eg, Cl vs Br vs I) and absolute rate of reaction. In solvents of relative low polarity (eg, CH_2Cl_2, $CHCl_3$, benzene) the ion pair, not the anion, is the dominant nucleophile (4).

For class (2) reactions in the presence of aqueous alkali hydroxide, a gradual transition away from the scheme of Figure 1 can be postulated. For example, relatively strongly acidic compounds (eg, alcohols, β-ketocarbonyl compounds) will be either soluble in aqueous caustic soda or easily deprotonated at the phase boundary. Ion pairs of catalyst cation and deprotonated substrate anion are eventually alkylated by RX in the organic medium. With weaker C—H acids there will be a large amount of unchanged substrate and a small concentration of ion-paired deprotonated substrate in the organic phase in equilibrium with the caustic soda phase. Still weaker acidic compounds will be almost entirely unchanged.

There are three different theories to explain the latter two cases of catalyzed alkylation and other reactions in the presence of concentrated alkali metal hydroxides.

In the first case, the catalyst anion is exchanged at the phase boundary against hydroxide. Then the quaternary ammonium hydroxide functions as base in the organic medium.

Organic phase: $[Q^+OH^-]_{org}$ + H-substrate $\rightleftharpoons$ $[Q^+ \text{substrate}^-]_{org}$ + H_2O

$[Q^+ \text{substrate}^-]_{org}$ + RX $\rightarrow$ R-substrate + $[Q^+X^-]_{org}$

Phase boundary: $[Q^+X^-]_{org}$ + $NaOH_{aq}$ $\rightleftharpoons$ $[Q^+OH^-]_{org}$ + NaX_{aq}

Aqueous phase: $Q^+_{aq}X^-_{aq}$ + $Na^+_{aq}Y^-_{aq}$ $\rightleftharpoons$ $Q^{*}_{aq}Y^-_{aq}$ + $Na^+_{aq}X^-_{aq}$

(a) *Phase boundary*: $Q^+_{aq}Y^-_{aq}$ $\rightleftharpoons$ $[Q^+Y^-]_{org}$

Organic phase: $[Q^+Y^-]_{org}$ + RX $\rightarrow$ $[Q^+X^-]_{org}$ + RY

Phase boundary: $[Q^+X^-]_{org}$ $\rightleftharpoons$ $Q^+X^-_{aq}$

(b) *Phase boundary*: $[Q^+X^-]_{org}$ + NaY_{aq} $\rightleftharpoons$ $[Q^+Y^-]_{org}$ + NaX

Organic phase: $[Q^+Y^-]_{org}$ + RX $\rightarrow$ $[Q^+X^-]_{org}$ + RY

Figure 1. Stark's mechanism for class (1) reactions.

However, hydroxide ion is very hydrophilic and can be extracted in appreciable concentration into an organic medium only if the catalyst cation is very lipophilic and if the competition situation with other anions is favorable. Almost all other common inorganic and organic anions are more lipophilic (see below), and in most cases not only the anion originally brought in with the catalyst, but mainly the one formed in the reaction tend to pair with the catalyst cation. At best the use of a large excess of concentrated aqueous sodium hydroxide or application of molar amounts of catalyst with a very hydrophilic original anion (eg, sulfate) can force this mechanism.

The secord mechanism, developed by Makosza (5), is a modification of the first (Fig. 2). Here substrate molecules are deprotonated at the interface. Without catalyst the anions formed are immobilized because their counterion is in the opposite phase. Catalyst action is seen in detaching substrate anions from the boundary by anion exchange. The competition situation for this is very favorable since most organic weak acids will be much more lipophilic than halide ions except for iodide. Supporting evidence for the importance of interfacial processes is that some closely related alkylations proceed without catalyst in systems with negligible mutual solubility of the phases.

In dihalocarbene generation by phase-transfer catalysis the following steps seem to be involved (Fig. 3) (6): formation of CX_3^- anions dynamically anchored at the boundary; reversible detachment with the help of the catalyst; reversible carbene formation $[Q^+CX_3^-] \rightleftharpoons [Q^+X^-] + CX_2$; addition to olefin.

A free base in the organic phase would ordinarily decompose the haloform irreversibly at room temperature, whether an acceptor is present or not. With PTC, however, only a slow hydrolysis occurs when no olefin is present. The relative rate of hydrolysis–carbene addition and the absolute rate of carbene addition are dependent on the nucleophilicity of the olefin. Contrary to other carbene generating methods the precursor "waits" at the phase boundary until a reactive substrate is at hand (6).

The third mechanistic concept calls for the reactions to occur in a micellar or an inverted micellar phase (7). However, the bulk of recent evidence is contrary to this alternative.

Phase boundary: Substrate-H$_{org}$ + NaOH$_{aq}$ $\rightleftharpoons$ Substrate$^-_{org}$ + H$_2$O + Na$^+_{aq}$

Phase boundary: [Q$^+$X$^-$]$_{org}$ + Substrate$^-_{org}$ $\rightleftharpoons$ [Q$^+$ substrate$^-$]$_{org}$ + X$^-_{aq}$

Organic phase: [Q$^+$ substrate$^-$]$_{org}$ + RX → R-substrate + [Q$^+$X$^-$]$_{org}$

Figure 2. Makosza mechanism for class (2) reactions.

Phase boundary: HCCl$_{3\,org}$ + NaOH$_{aq}$ $\rightleftharpoons$ CCl$^-_{3\,org}$ + H$_2$O + Na$^+_{aq}$

Phase boundary: CCl$^-_{3\,org}$ + [Q$^+$Cl$^-$]$_{org}$ $\rightleftharpoons$ [CCl$_3^-$Q$^+$]$_{org}$ + Cl$^-_{aq}$

Organic phase: [CCl$_3^-$Q$^+$]$_{org}$ $\rightleftharpoons$ [Q$^+$Cl$^-$]$_{org}$ + CCl$_2$

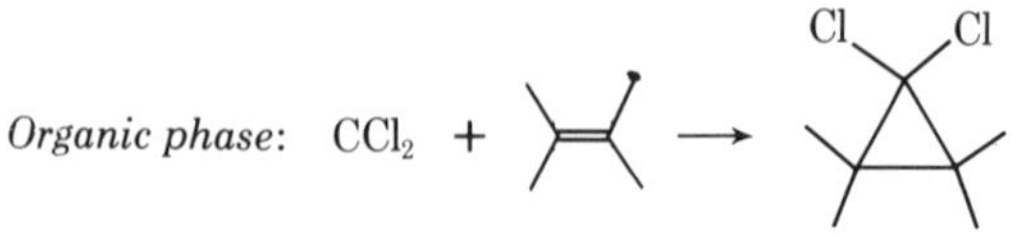

Organic phase: CCl$_2$ + ⋀ → ⋀(Cl)(Cl)

Figure 3. Carbene generation by PTC.

Factors Influencing the Usefulness of Phase-Transfer Catalysis

In the following reaction under phase catalytic conditions

$$RX + KY_{(water)} \xrightarrow{NBu_4X} RY + KX$$

the relative rate for $X = Cl^-, Br^-, I^-$ is determined by two factors: the actual chemical reaction with $I^- > Br^- > Cl^-$; and the relative rates of the competitive extraction of Y^- vs X^-. If the second step is very unfavorable for Y^-, PTC may be impossible. Thus, it is important to have some insight into the order of magnitude of the factors involved. The conditional extraction constant E_{QX} of the salt QX is defined as:

$$E_{QX} = \frac{[QX]_{org}}{[Q^+]_{aq} \cdot [X^-]_{aq}}$$

where the bracketed expressions stand for activities (or concentrations) in organic or aqueous media, respectively.

Solvent. Table 1 gives values for E_{QX} (NBu$_4$Br) between water and different solvents. For PTC the solvents should be nonmiscible with water and nonhydroxylic; CHCl$_3$ and CH$_2$Cl$_2$ are generally best.

Catalyst Cation. It has been found that the logarithms of extraction constants for symmetrical tetra-n-alkylammonium salts (log E_{QX}) rise by ca 0.54 per added C atom. Although absolute numerical values for extraction coefficients are vastly different in various solvents and for various anions this relation holds as a first approximation for most solvent–water combinations tested and for many anions. It is important to note, however, that the lipophilicity of phenyl and benzyl group carrying ammonium salts is much lower than the number of C atoms might suggest. Benzyl is extracted between n-propyl and n-butyl. The extraction constants of tetra-n-butylammonium salts are about 140 times larger than the constants for tetra-n-propylammonium salts of the same anion in the same solvent–water system.

For practical application in mixtures of water–organic solvent only ammonium and phosphonium salts containing 15 or more C atoms are sufficiently lipophilic. In empirical catalyst comparisons crown ethers were often as effective as the best onium salts. Sodium is complexed best by 15-crown-5 and derivatives, potassium by 18-crown-6 and derivatives (1)–(3).

Other complexing agents sometimes advocated are cryptates, especially the

Table 1. Apparent Extraction Constants of 0.1 M Tetra-n-Butylammonium Bromide from Water into Various Solvents[a]

CH$_2$Cl$_2$	35	ClCH$_2$CH$_2$CH$_2$Cl	2.9	(C$_2$H$_5$)$_2$O	<0.1
CHCl$_3$	47	C$_6$H$_5$Cl	<0.1	C$_2$H$_5$COC$_2$H$_5$	1.1
CCl$_4$	<0.1	benzene	very small	n-C$_3$H$_7$COCH$_3$	1.7
CH$_3$CHCl$_2$	0.5	hexane	very small	C$_2$H$_5$COCH$_3$	14
ClCH$_2$CH$_2$Cl	6.1	o-C$_6$H$_4$Cl$_2$	<0.1	n-C$_4$H$_9$OH	69
ClCH$_2$CHCl$_2$	8.6	CH$_2$=CCl$_2$	<0.1	CH$_3$NO$_2$	168
Cl$_2$CHCHCl$_2$	145	trans-/cis- ClCH=CHCl	<0.1 / 33		
Cl$_2$CHCCl$_3$	<0.1	CHCl=CCl$_2$	0.2		
n-C$_3$H$_7$Cl	<0.1	CH$_3$CO$_2$C$_2$H$_5$	0.2		

[a] Ref. 8.

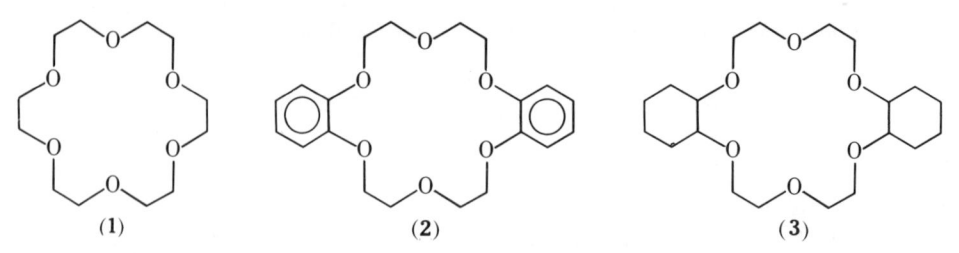

compound dubbed [2.2.2] (Kryptofix 222) (see Chelating agents). Crown ethers were originally advocated for reactions in the presence of solid reagents (liquid–solid PTC). It is now known, however, that onium salts are equally suitable in many cases.

Benzyltriethylammonium chloride is the most widely used catalyst under strongly basic conditions. Choice of catalyst depends on price; methyl trioctylammonium chloride (Aliquat 336, Adogen 464) is probably the least expensive reagent. Other catalysts of high activity and moderate price are tetra-n-butylammonium chloride and hydrogen sulfate, and tetra-n-butyl phosphonium chloride and other phosphonium salts of similar C number. The crown ethers are one to two orders of magnitude more expensive. The stability of the catalysts is another important factor. Generally speaking the phosphonium salts appear to be more stable (up to 200°C) than the ammonium salts (1) in neutral or basic environments. In the presence of concentrated aqueous bases, decomposition including Hofmann elimination, and ylene formation and subsequent further reactions is accelerated. Benzyl-containing ammonium salts can decompose to give benzyl alcohol and benzyl ether.

Competitive Extraction of Anions. The successful extraction of the necessary anion into the organic phase is crucial for PTC. Often three anions compete for the catalyst cation: the one that is to react, the one formed in the reaction, and the one brought in originally with the catalyst. Table 2 lists the widely differing E_{QX} values of tetra-n-butylammonium salts. The big difference in the halide series is evident. Hydroxide is 10^4 times more difficult to extract than chloride (13) and the divalent and trivalent anions SO_4^{2-} and PO_4^{3-} are still more hydrophilic. Thus for practical applications hydrogen sulfate or chloride are recommended as the original anions in the catalysts. Bromide is more lipophilic than the smaller organic anions, and iodide can tie up the catalyst more or less quantitatively. Organic compounds carrying larger groups are

Table 2. Anion Extraction Constants, E_{QX}, of Tetra-n-Butylammonium Ion Pairs Between Water and Chloroform

		Ref.			Ref.
Cl⁻	0.78	9	$CH_3CO_2^-$	7.6×10^{-3}	10
Br⁻	15.9	9	$C_6H_5CH_2CO_2^-$	1.86	10
I⁻	1023	9	$C_6H_5CO_2^-$	2.45	10
ClO_4^-	3020	9	salicylate	263	10
NO_3^-	24.5	9	3-hydroxybenzoate	0.03	10
			phenolate	0.93	10
			picrate	8.1×10^5	10
			p-toluene sulfonate	214	11
			naphthalene-2-sulfonate	2818	11
			anthracene-2-sulfonate	1.3×10^5	12
			trinitrobenzenesulfonate	2.9×10^4	11

naturally more easily extractable than the ones shown. Also noteworthy are the large differences in E_{QX} effected by seemingly slight structural changes (cf acetate–benzoate, salicylate–3-hydroxybenzoate). It must be remembered that, within certain limits, relatively unfavorable extraction constants can be outmaneuvered by using a large excess of the less easily extracted salt. In certain cases it is necessary to apply molar amounts of catalyst, for instance if iodide formed during alkylation with RI halts the course of the reaction. Normally, however, 1–10 mol % catalyst suffices. Regeneration of the catalyst is often possible if deemed necessary. Some authors have advocated systems in which the catalyst is bound to a polymer matrix (triphase-catalysis). Here separation and generation of the catalyst is easy (14–17), but swelling, mixing, and diffusion problems are not always easy to solve (see also Enzymes, immobilized). Commercial anion-exchange resins seem to be rarely suitable for these purposes.

Typical Applications

For particular references to the multitude of PTC reactions in the literature, refer to the survey articles and books mentioned in the bibliography. Only a few typical examples are given here.

Displacement Reactions. The reactions RCl → RCN are carried out by heating a three-fold excess of concentrated aqueous NaCN solution with the alkyl chloride and 2 mol % tributylhexadecylphosphonium bromide for 2 h (1) (see Nitriles).

Formation of Esters. RBr → ROCOCH$_3$ proceeds in 1–2 h by heating a concentrated CH$_3$COOK solution with the halide, with or without solvent, and 1 mole % Aliquat 336 catalyst at 100°C (18). Esterification of higher acids, including sterically hindered ones, is carried out by neutralizing one equivalent of acid and one equivalent of tetrabutylammonium hydrogen sulfate with two moles of aqueous sodium hydroxide and boiling the mixture with the alkylating agent in CH$_2$Cl$_2$ for 15–30 min (19) (see Esterification).

Formation of Ethers. Very high ether yields can be obtained from alcohols and phenols with dialkyl sulfates in CH$_2$Cl$_2$ and concentrated NaOH–tetrabutylammonium chloride at room temperature or slightly elevated temperature within 1–5 h (20–21). Using excess aqueous caustic–NBu$_4$HSO$_4$, unsymmetrical aliphatic ethers can be prepared with alkyl chlorides at 25–70°C in 3–4 h (22) (see Ethers).

Alkylation of Carbanions. Concentrated NaOH–benzyltriethylammonium chloride is the base/catalyst system normally used for this type of process (23). Classes of compounds alkylated in this way include: phenylacetonitriles, benzylketones, simple aliphatic ketones, certain aldehydes, aryl sulfones, β-ketosulfones, β-ketoesters, malonic esters and nitriles, phenylacetic esters, indene and fluorene (see Alkylation).

Horner-Wittig Reactions. Most of these are done by adding a methylene chloride solution of the phosphonium salt or the diethyl phosphonate, with or without added catalyst, to concentrated caustic soda (24–25).

$$(C_2H_5O)_2P(O)CH_2CN + R_2C=O \xrightarrow[\text{NaOH}]{\text{NBu}_4\text{I}} R_2C=CH-CN$$

$$(C_6H_5)_3P^+-CH_2C_6H_5I^- + RCHO \xrightarrow{\text{NaOH}} RCH=CHC_6H_5$$

Oxidations. KMnO₄. In neutral solution with Aliquat 336 and benzene as organic solvent, olefins yield carboxylic acids (1, 26).

NaOCl. Sodium hypochlorite proves to be a useful reagent under PTC conditions for the following transformations (27):

$$ArCH_2OH \rightarrow ArCHO$$
$$R_2CHOH \rightarrow R_2C=O$$
$$RCH_2NH_2 \rightarrow RCN$$

Applications of phase-transfer catalysis with these additional oxidants are known: peracetic acid, air–KOH, $K_3[Fe(CN)_6]$, KO_2, K_2CrO_4, H_2O_2–heavy metal oxide, and H_5IO_6 (28).

Carbene Reactions. The best procedure for preparing dihalocarbene adducts of olefins consists in stirring a haloform–methylene chloride solution with an excess of concentrated aqueous caustic soda in the presence of benzyltriethylammonium chloride. Even sterically hindered and electronically deactivated compounds give excellent yields (29–30). Mixed dihalocarbenes, CXY (X,Y = F, Cl, Br, I), except for CF_2, can be prepared. Insertions into tertiary C—H bonds can be carried out with moderate yields. Among other less common reactions improved by the use of PTC-generated CCl_2 are the carbylamine synthesis ($RNH_2 \rightarrow R—NC$) (31) and diazomethane preparation ($N_2H_4 \rightarrow CH_2N_2$) (32). Alkylidene carbene ($R_2C=C$:) and alkenylidene carbene ($R_2C=C=C$:) adducts have also been prepared (33–36).

$$R_2\overset{\overset{\text{OH}}{|}}{C}-CH_2-\overset{\overset{\text{NO}}{|}}{N}-COCH_3 \xrightarrow[\text{NaOH}]{\text{Aliquat 336}} R_2C=C:$$

$$R_2C=C=CHX \xrightarrow[\text{NaOH}]{\text{catalyst}} R_2C=C=C: \xleftarrow[\text{NaOH}]{\text{catalyst}} R_2\overset{\overset{X}{|}}{C}-C\equiv CH$$

Other reactions assisted by PTC are described in the general references.

BIBLIOGRAPHY

1. C. M. Starks, *J. Am. Chem. Soc.* **93,** 195 (1971).
2. C. M. Starks and R. M. Owens, *J. Am. Chem. Soc.* **95,** 3613 (1973).
3. D. Landini and co-workers, *J. Chem. Soc. Chem. Commun.* 950 (1975).
4. A. Brändström and H. Kolind-Andersen, *Acta Chem. Scan.* **29B,** 201 (1975).
5. M. Makosza and E. Bialecka, *Tetrahedron Lett.* 183 (1977).
6. E. V. Dehmlow, M. Lissel, and J. Heider, *Tetrahedron* **33,** 363 (1977).
7. J. H. Fendler and E. J. Fendler, *Catalysis in Micellar and Macromolecular Systems*, Academic Press, New York, San Francisco, London, 1975.
8. A. Brändström, lecture, *Journees de Chimie Organique d'Orsay, France*, Sept. 17, 1975, and private communication.
9. K. Gustavii, *Acta Pharm. Suec.* **4,** 233 (1967).
10. R. Modin and G. Schill, *Acta Pharm. Suec.* **5,** 311 (1968).
11. *Ibid.*, **4,** 301 (1967).
12. K. O. Borg and D. Westerlund, *Z. Anal. Chem.* **252,** 275 (1970).
13. E. V Dehmlow, M. Slopianka, and J. Heider, *Tetrahedron Lett.* 2361 (1977).
14. S. L. Regen, *J. Am. Chem. Soc.* **97,** 5956 (1975); **98,** 6270 (1976).
15. J. M. Brown and J. A. Jenkins, *J. Chem. Soc. Chem. Commun.* 458 (1976).
16. Ger. Offen. 2,324,390 (Nov. 29, 1973), P. A. Verbrugge and E. W. Uurbanus.

17. M. Cinquini and co-workers, *J. Chem. Soc. Chem. Commun.* 394 (1976).
18. T. Toru and co-workers, *Synthesis*, 867 (1974).
19. A. Brändström, *Preparative Ion Pair Extraction, An Introduction to Theory and Practice*, Apotekarsocieteten/Hässle Läkemedel, Stockholm, Sweden, 1974, p. 155–156.
20. A. Merz, *Angew. Chem. Int. Ed.* **12**, 846 (1973).
21. A. McKillop, J.-C. Fiaud, and R. P. Hug, *Tetrahedron* **30**, 1379 (1974).
22. H. H. Freedman and R. A. Dubois *Tetrahedron Lett.* 3251 (1975).
23. M. Makosza and A. Jonczyk, *Org. Synth.* **55**, 91 (1976).
24. C. Piechucki, *Synthesis*, 869 (1974).
25. G. Märkl and A. Merz, *Synthesis*, 295 (1973).
26. A. W. Herriott and D. Picker, *Tetrahedron Lett.* 1511 (1974).
27. G. A. Lee and H. H. Freedman, *Tetrahedron Lett.* 1641 (1976).
28. U.S. Pat. 3,992,432 (Nov. 16, 1976), D. R. Napier and C. M. Starks.
29. M. Makosza and W. Wawrzyniewicz, *Tetrahedron Lett.* 4659 (1969).
30. E. V. Dehmlow and J. Schönefeld, *Justus Liebigs Ann. Chem.* **744**, 42 (1971); E. V. Dehmlow, H. Klabuhn, and E.-Ch. Hass, *Justus Liebigs Ann. Chem.* 1063 (1973); E. V. Dehmlow and S. S. Dehmlow, *Justus Liebigs Ann. Chem.* 1753 (1973).
31. W. P. Weber, G. W. Gokel, and I. K. Ugi, *Angew. Chem. Int. Ed.* **11**, 530 (1972).
32. D. T. Sepp, K. V. Scherer, and W. P. Weber, *Tetrahedron Lett.* 2983 (1974).
33. M. S. Newman and M. C. Van der Zwan, *J. Org. Chem.* **39**, 761, 1186 (1974).
34. T. B. Patrick, *Tetrahedron Lett.* 1407 (1974).
35. S. Julia, D. Michelot, and G. Linstrumentale, *C.R. Acad. Sci.* **278C**, 1523 (1974).

General References

E. V. Dehmlow, "Phase-Transfer Catalyzed Two-Phase Reactions in Preparative Organic Chemistry," *Angew. Chem. Int. Ed.* **13**, 170 (1974); *Angew. Chem.* **86**, 187 (1974) [German version]; *New Synthetic Methods*, Vol. 1, Verlag Chemie, Weinheim, New York, 1975, p. 1.
E. V. Dehmlow, "Progress in Phase-Transfer Catalysis," *Angew. Chem. Int. Ed.*, **16**, 493 (1977); *Angew. Chem.* **89**, 521 (1977).
E. V. Dehmlow and S. S. Dehmlow, *Phase-Transfer Catalysis*, monograph in press 1978, Verlag Chemie, Weinheim, Germany.
A. Brändström, *Preparative Ion Pair Extraction, An Introduction to Theory and Practice*, Apotekarsocieteten/Hässle Läkemedel, Stockholm, Sweden, 1974.
M. Makosza, "Naked Anions/Phase Transfer," in R. Schefford, ed., *Modern Synthetic Methods 1976*, Schweizerischer Chemiker-Verband, Zürich, Switzerland, 1976, p. 7.
M. Makosza, "Two-Phase Reactions in the Chemistry of Carbanions and Halocarbenes, a Useful Tool in Organic Synthesis," *Pure Appl. Chem.* **43**, 439 (1975).
W. P. Weber and G. W. Gokel, *Phase Transfer Catalysis in Organic Synthesis*, Springer-Verlag, Bohm, Heidelberg, New York, 1977.
A. Brändström, "Principles of Phase-Transfer Catalysis by Quaternary Ammonium Salts," *Adv. Phys. Org. Chem.* **15**, 267 (1978).

ECKEHARD V. DEHMLOW
Technische Universität Berlin

CATECHOL, o-C$_6$H$_4$(OH)$_2$. See Polyhydroxybenzenes.

CATHARTICS. See Gastrointestinal agents.

CAULKING COMPOSITIONS. See Sealants; Chemical grouts.

CAUSTIC SODA. See Alkali and chlorine products.

CELCON. See Acetal resins.

CELLOPHANE. See Film and sheeting materials.

CELLS, ELECTRIC. See Batteries and electric cells.

CELLS, HIGH TEMPERATURE. See Batteries and electric cells, primary.

CELLULOSE

Cellulose [50-98-7] (Fig. 1) is the most abundant organic raw material available in the world today—and it is naturally regenerable. Cellulose, its by-products, and derivatives have always held an important place scientifically and commercially; their value can only increase as other raw materials become depleted (see Chemurgy; Fuels from biomass).

The Raw Material

Biosynthesis. About 5×10^{14} kg of cellulose are biosynthesized in nature (and as much destroyed) yearly on the globe (1–2). A small fraction of this mass is produced by a few bacteria, animals (tunicates), and fungi but by far the greater part is produced by green plants, both marine and terrestrial. In most of the cells of these plants, microfibrils of cellulose, from 10–30 nm in breadth and indefinitely long, strengthen the cell walls (and indirectly the tissues or organs of the plant) as steel rods strengthen reinforced concrete (3).

Most of the information on the chemistry of biosynthesis of cellulose has been obtained from bacteria because they are experimentally convenient, but there is no indication that the process is substantially different in other plants. The following initial series of reactions takes place in cells that produce cellulose:

α-D-glucose + adenosine triphosphate $\rightleftarrows$ α-D-glucose-6-phosphate + adenosine diphosphate
α-D-glucose-6-phosphate $\rightleftarrows$ α-D-glucose-1-phosphate
α-D-glucose-1-phosphate + uridine triphosphate $\rightleftarrows$ uridine diphosphoglucose
 + inorganic pyrophosphate

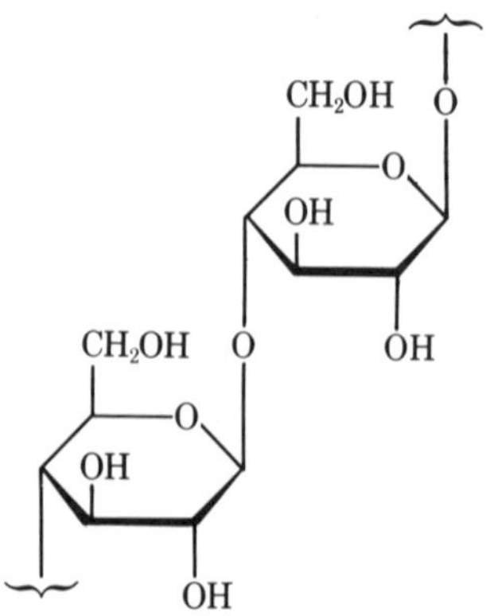

$(C_6H_{10}O_5)_n$

$n = 1500$ to >6000

Figure 1. Repeat unit of cellulose (β-cellobiose residue).

In some plants guanine may be substituted for uridine but the extent of substitution is uncertain.

The subsequent reactions, summarized in Figure 2 take place on or close to the cytoplasmic membrane of the cell. The lipid is not yet fully identified but all evidence is consistent with the hypothesis that it is a polyprenol.

The acceptor in Figure 2 was previously thought to be the growing tip of a microfibril (4) but is now known to be a soluble, hydrated, intermediate polymer of glucose (5–6). The hydrated polymers are transferred to the outside of the cell where they associate longitudinally by London forces to form the nascent microfibril (7). Initially, the polymer chains tend to be separated by water molecules hydrogen-bonded to the hydroxyl groups, but when the water is removed, the chains associate irreversibly via the liberated hydroxyl groups to form the stable, durable, insoluble microfibril. In green plants the membranes of the Golgi bodies are involved in the biosynthesis of the polymer (8–11). The mechanism of formation of the very long chains (high dp) is not yet known.

The nonrandom orientation of microfibrils in plant cells is an important aspect of the biosynthesis of cellulose because the ultimate use of the substance, either biologically or industrially, depends critically upon the relative disposition of the threads in the walls.

The biosynthesis of special forms of cellulose, like that from bacteria, was once important industrially for the manufacture of microbiological filters, nitrocellulose, and glassine-like films. At present, none of these processes are economic.

Microstructure. Extensive investigations with the electron microscope have revealed that the cellulose in cell walls occurs as thread-like structures approximately 3.5 nm in width and of indefinite length. These are the finest structural elements that have been observed. Because of their ubiquity they are generally, but not universally, thought to be the basic structural element and are called elementary fibrils or protofibrils (12–16). The internal molecular constitution of the elementary fibrils remains unclear. On the strength of x-ray and electron diffraction studies, it is generally agreed that the long chain cellulose molecules are arranged with their long axes parallel to the length of the elementary fibrils. Further details of the molecular arrangement are, however, elusive. At present there are basically two schools of thought. According to the one, the elementary fibril is the morphological expression of the individual cellulose

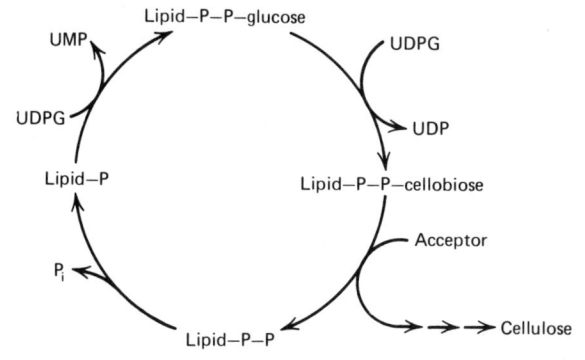

Figure 2. Biosynthesis of cellulose. UMP = uridine monophosphate; UDPG = uridine diphosphate glucose.

chain in the solid state. The elementary fibril contains a single cellulose chain folded in such a way that the short chain segments are parallel to the fibril axis (15). The opposing view is that the elementary fibril is composed of a parallel assembly of about forty cellulose chains (17–19).

In some cell walls, eg, in algae like *Valonia*, the elementary fibrils appear to be laterally associated to form larger fibrils—the so-called microfibrils. In *Valonia* the cross-sectional shape of the microfibrils is rectangular with the longer side of 15–20 nm and the shorter side of 7–10 nm. The longer side lies parallel to the surface of the cell wall (20–23). Two models have been proposed to account for the cross-sectional shape of the microfibrils. The first considers that each microfibril is composed of closely packed elementary fibrils (15,24). The second assumes that the microfibrils consist of a rectangular crystalline core which is embedded in a paracrystalline cortex (25). In higher plants the well-defined microfibrils found in *Valonia* appear to be absent. Instead, the cell wall seems to be a laminated assembly of sheets of laterally associated elementary fibrils.

It is well known that cellulose occurs in several polymorphic modifications (26–27). Of these the best known are cellulose I and cellulose II. The former is characteristic of native celluloses whereas the latter is found in mercerized and regenerated celluloses. In both modifications the chain conformation is that of a 2_1 helix with a 1.03 nm fiber repeat distance. The structures are stabilized by inter- and intramolecular hydrogen bonds (26). According to the most recent crystallographic studies, the unit cell of cellulose I is monoclinic with lattice dimensions that depend upon the source of the cellulose. For ramie and cotton, $a = 0.935$ nm, $b = 1.03$ nm (fiber axis), $c = 0.79$ nm, $\beta = 96°$. There are two chains per unit cell. For bacterial and *Valonia* cellulose there is a super cell containing eight chains with $a = 1.63$ nm, $b = 1.03$ nm (fiber axis), $c = 1.57$ nm, $\beta = 97°$ (28). The cellulose II lattice is also monoclinic with $a = 0.80$ nm, $b = 1.03$ nm (fiber axis), $c = 0.90$ nm, $\gamma = 117°$, and two chains per unit cell (26). Presumably, the unit cell is the unit of pattern in the elementary fibrils. For both modifications, the unit cell (sub cell for cellulose I) contains two cellobiose residues belonging to chain molecules passing through the center and corners of each cell. The older view is that the polarity of the two chains is antiparallel in both modifications, but according to the most recent studies the polarity is parallel in cellulose I (29–30) and antiparallel in cellulose II (31–32). The latter view, however, seems to be at variance with the fact that cellulose I can be converted to cellulose II without disruption of the structural integrity of the elementary fibrils.

X-ray diffraction and other evidence indicate that cellulose is partly crystalline and partly disordered. Within the elementary fibrils there is presumably crystalline regularity in the molecular arrangement in some regions and disorder in others. The relative amounts of disordered and crystalline cellulose have a marked effect on such fiber properties as equilibrium moisture content, strength, flexibility, and reactivity (see also Biopolymers).

Macrostructure. Cellulose is the principal cell-wall material of all higher plants and is found in many of the lower orders as well. The macrostructure of its natural forms shows similarities in all plant cell walls. The discussion here is limited primarily to those cells that are recoverable in the form of industrially useful fibers.

Because the manufacture of paper (qv) and related products represents the largest single use of cellulosic fibers, much attention has been and continues to be focused on the wood cells that constitute raw material for paper manufacture (see Wood). Cells used as textile fibers have many features in common with wood cells, even though they may originate in anatomically different parts of various plants. Thus cellulosic textile fibers can be classified by origin in three main groups (33): (*1*) Bast or stem fibers are fibrous bundles (not single fibers) derived from the inner bark of the stems of certain plants. (*2*) Leaf fibers, also fiber bundles, are recovered from certain monocotyledonous (grassy) plants in whose leaves long, strong fibers run lengthwise. (*3*) Seed and fruit fibers are often single fibers produced by plants whose seeds and fruits bear hair-like processes. Familiar examples of fibers from each of these groups are given in Table 1.

Cotton (qv), which contains the lowest percentage of noncellulosic material (4–12%) of all the commercial sources of cellulose, is not the only example of seed fibers; others include kapok, from the Malaysian and Indonesian tree *Ceiba pentandra*, and coir, the coarse fiber from the husks of coconuts (*Cocos nucifera*). Other sources of considerable scientific if not industrial interest include the nonfibrous walls of most green algae. The most extensively studied is the genus *Valonia*, of which several species, including *V. ventricosa*, *V. utricularis*, and *V. macrophysa*, constitute the main algae of tropical and subtropical seas. The cell wall of *V. ventricosa* is 70% cellulose.

The hemicelluloses (36) are polysaccharides extractable from the plant substance by means of aqueous alkali (see Carbohydrates). They are mainly heteropolymers, often branched, of various sugars and some uronic acids. Certain hemicelluloses, such as the arabinogalactans of larch wood, are water-soluble. The pectins are a group of complex polysaccharides in which D-galacturonic acid is the principal constituent (37).

Lignin (qv) (38–39), the other principal constituent of wood fibers, and minor constituent of the other fibers, is a complex aromatic polymer which is probably a highly cross-linked material with a three-dimensional architecture. Although the precise structures of lignins are unknown, at least three types are recognized by their content of aromatic methoxyl groups: softwood lignin, 15–16% aromatic methoxyl groups; hardwood lignin, 20–22%; monocotyledon (grass) lignin, 14–17%. The distinction arises because softwood lignins are constructed mainly of substituted guaiacylpropane units (**1**), hardwood lignins of both guaiacyl and syringylpropane (**2**) units, and grass lignins of both of these plus *p*-hydroxyphenylpropane (**3**) units. Usually most of the lignin must be removed from wood fibers before the hemicelluloses can be extracted by alkali. Lignin removal on a technological scale is the basic process of the chemical pulping industry (see below).

The fact that cellulose never occurs in absolutely pure form (Table 1) lends support to the view (40) that cell walls are two-phase systems: a crystalline polysaccharide phase of specific composition, usually cellulose, embedded in a matrix (of almost vanishing magnitude in cotton) composed of various polysaccharides and other compounds. In all cell walls the crystalline cellulose-containing fraction, embedded in amorphous lignin and hemicellulose, is seen as fibrils and microfibrils (see section on Microstructure).

Whatever the internal structure of the microfibrils may be, it is now widely accepted that the geometry of their arrangement in fiber walls has a profound effect on the physical properties of the fibers, of which those of wood and cotton are discussed as examples.

The cellular composition of softwoods and hardwoods are somewhat different.

Table 1. Source and Composition of Representative Cellulosic Fibers[a]

Class	Common name	Type or species	Cellulose	Hemicelluloses	Lignin	Pectin	Fats and waxes	Water-soluble, or ash
wood fibers[b]	softwoods	coniferous	42 ± 2	ca 28	25–35			
	hardwoods	deciduous	42 ± 2	ca 38	15–20			
bast fibers	flax[c]	*Linum usitatissimum*	71.2	18.6	2.2	2.0	1.7	4.3
	jute	*Corchorus* sp	71.6	13.3	13.1	0.2	0.6	1.2
	hemp (marijuana)	*Cannabis sativa*	74.4	17.9	3.7	0.9	0.8	2.3
	ramie (China grass)	*Boehmeria nivea*	76.2	14.6	0.7	2.1	0.3	6.1
leaf fibers	sisal	*Agave sisalana*	73.1	13.3	14.0	0.9	0.3	1.4
	abaca (Manila hemp)	*Musa textilis*	70.2	21.8	5.7	0.6	0.2	1.5
seed fibers	cotton, crude	*Gossypium* sp[d]	95.3	0.0	0.0	1.0	0.8	2.9
	cotton, purified[e]		99.9					<0.1

[a] Refs. 34–35.
[b] Analysis after extraction with organic solvents and with water.
[c] Analysis on retted flax, ie, fibers separated by a fermentation process.
[d] Principally, *G. hirsutum*, *G. barbadense*, and *G. arboreum*.
[e] Purification by extraction with solvents and alkali.

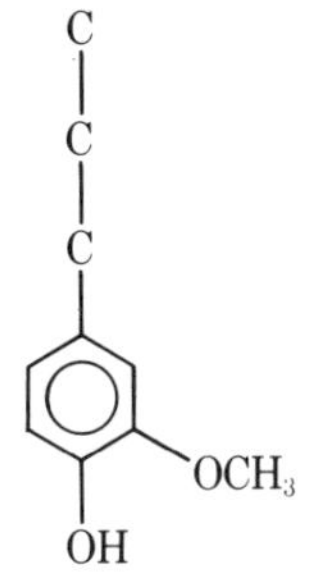

In softwood, 85–95% of the cells are tracheids, each about 3–4 mm long, with tapered ends, and oriented with their long axes parallel to the long axis of the tree trunk. They are characterized by the frequent occurrence of bordered pits along their length. The diameters of tracheids vary, depending on whether they are formed in the spring or in the summer. Thus the diameter of springwood tracheids is about 45 µm including walls about 2 µm thick, and the diameter of summerwood tracheids is about 25 µm including walls about 5 µm thick. The regular radial alternation of these cells is familiar in the annual growth-rings visible in the cross-section of tree trunks. The other 5–15% of cells in softwood are parenchyma cells, which are short, thin-walled, and prismatic in shape.

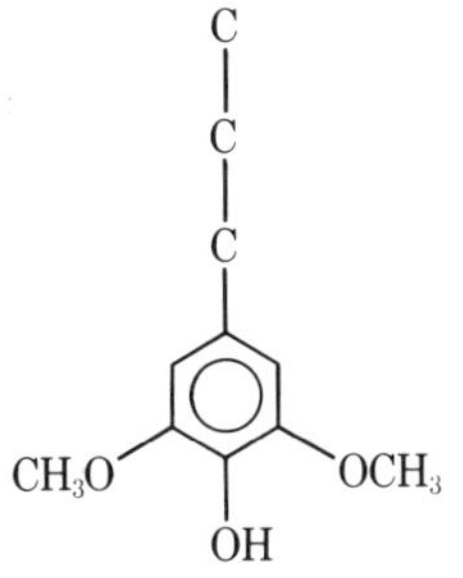

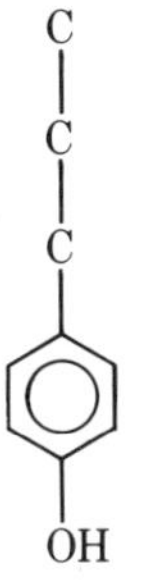

Figure 3 shows the arrangement of tracheids and the orientation of cellulose microfibrils in the wall layers. In cross-section, the tracheids are more or less rectangular, each separated from the other by a compound middle lamella containing lignin and hemicellulose in the proportions of approximately 7 to 3. Lignin and hemicellulose constitute the matrix material of the cell wall. There the proportion of lignin to polysaccharide is about 3 to 7. These ratios were for a long time taken to indicate that the bulk of the lignin in wood was in the middle lamella. However, when the relative

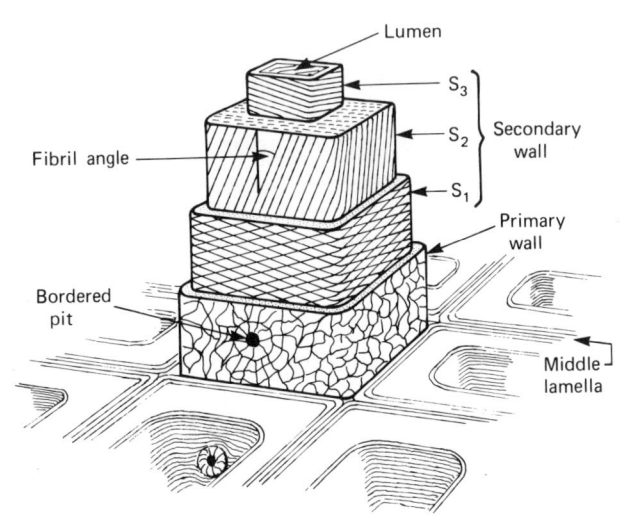

Figure 3. Softwood tracheid set among tracheids in cross-section. The arrangement of the microfibrils in each cell-wall layer is shown.

volumes of the middle lamella and the cell walls are taken into account, about 70% of the total lignin in softwoods (41) is in the cell walls, and about 80% in hardwoods (42).

Within the cell wall, four distinct concentric morphological layers can be distinguished (Fig. 3), each with its own microfibrillar structure. In the primary wall, just inside and contiguous with the middle lamella, the microfibrils are arranged in a loose, apparently random network. The secondary wall contains three layers surrounding the central vacant part of the fiber called the lumen. The microfibrils of the outermost S_1 layer are wound in flat helices. The direction of winding alternates from right-handed to left-handed in successive lamellae. These windings are referred to, respectively, as the Z-pattern (when the fiber is vertical, the helix of the microfibril conforms to the slope of the central part of the letter Z), and the S-pattern (conforming to the slope of the central part of the letter S). The middle S_2 layer is the thickest of the three and constitutes the bulk of the cell wall. In the S_2 layer the microfibrils are arranged in a steep right-handed helix. The pitch of this helix, usually measured as the fibril angle—the angle between the long axis of the fiber and the long axes of the microfibrils—is an important contributor to the ultimate strength of the intact fiber (43). The smaller the fibril angle, the greater the strength. The innermost S_3 layer returns to the pattern of flat helices.

The cell-wall structure observed in tracheids, with the S_2 layer comprising microfibrils wound spirally, has also been demonstrated in sisal and bamboo fibers, and evidence of its presence has been found in other species. It is therefore thought (44) that all elongated plant cells including both hard and soft woods are constructed on this pattern. It should be noted, however, that parenchyma cells and tracheids constitute a much smaller part of hardwood. The tracheids are like those of softwoods, but have thicker walls and a different pattern of bordered pits. Hardwood also contains fiber cells and vessels, which are long, wide, tubular structures. The fiber cells, enclosing small lumina, are described as libriform, or spindle-shaped, and are generally about 0.7–2 mm long. They narrow to points at both ends, but their middle thirds are about 20 µm in diameter, and have sparsely pitted walls about 40 µm thick.

The hardwood fibers are, of course, quite different from the familiar cotton staple fibers which are 15 µm in diameter and about 25 mm long in textile grades, and sometimes as long as 50 mm. The cotton seed also bears short fibers, called linters, which are only 2–3 mm in length and 15–20 µm in diameter. The linters (chemical cotton) are used in industry for the manufacture of cellulose derivatives and regenerated cellulose.

Raw cotton fibers have three layers surrounding a central lumen. The first is an outer noncellulosic cuticle, composed mainly of wax. This encloses the primary wall, 0.1–0.2 µm thick, composed mainly of cellulose but containing also some pectic acids and other compounds. The secondary wall is 1–4 µm thick and almost pure cellulose. Purification removes the cuticle and the pectic acids.

Figure 4 is a composite diagram showing cotton fibers of various lengths, both staple and linters, attached to the seed. There is one lamella, about 100 nm thick, for each day of formation of the fiber. In the outer, primary wall the microfibrils are arranged in a criss-cross structure. In the secondary wall the microfibrils, 15 nm × 5 nm in cross-section, are helically wound at different angles in succeeding lamellae. A special feature of the cotton fibers is that within each lamella, as shown in Figure 4, the helices exhibit frequent reversals of pitch; transitions between S and Z pitches occur about

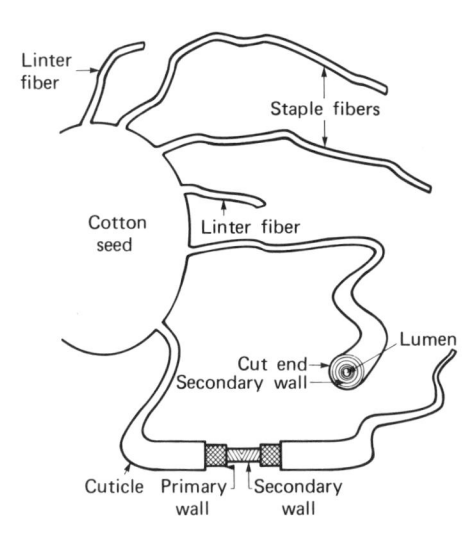

Figure 4. Cotton seed bearing linters and staple fibers. The cut end of a fiber shows the lamellar structure with the small central lumen. The cutaway portion of the bottom fiber shows the reversal of pitch in the microfibrils of the secondary wall.

forty times per centimeter of length. This phenomenon of pitch reversal has not been found in any other naturally occurring cellulosic fiber. It is believed that the 8% extensibility of cotton fibers is a result of the frequency of microfibrillar pitch reversal.

Preparation

Pulping. Pulping consists in separating and recovering more or less delignified cellulosic fibers from fibrous plant materials by mechanical and/or chemical methods. In plant materials, the fibers are biologically intergrown with hemicelluloses (45) and lignin substances (46). Unlike cellulose, lignins soften at relatively mild temperatures (47). These differences in structure and thermal behavior bear upon fiber liberation in mechanical and chemical pulping as well as in combined methods.

In mechanical pulping (48), namely in stone grinding of wood billets, and in mechanical (RMP) and thermomechanical (TMP) wood chip refining, most of the wood substance including the lignin is retained. Thus pulp yield is high and the amount of waste, which causes pollution, relatively small. On the other hand, in order to liberate cellulosic fibers without substantial mechanical defibration, most of the lignin, especially that present in the middle lamella, must be removed. This can be achieved at room temperature by the action of chemicals. For example, acid sodium chlorite destroys the macromolecular lignin structure by oxidation and makes the resulting products water soluble. However, all conventional pulping processes using chemicals operate at elevated temperature, which has a bearing upon delignification. Two processes are commercially used on a large scale, namely the sulfite and the alkaline processes. Recently, a few nonconventional pulping processes have become known, for instance oxygen–alkali pulping. Most chemical pulping processes are designed to produce specific types of paper pulps retaining some of the hemicelluloses and lignin substances. For papermaking pulps the physical properties, in particular bulk, tear, burst, tensile strength, folding endurance, and the optical properties such as brightness and opacity have industrial significance. In modified form, sulfite and alkaline (kraft) pulping are also used for the production of high purity chemical cellulose, so-called dissolving pulps, for use in chemical conversion into cellulose derivatives such as alkali cellulose and various cellulose ethers and esters. The objective in preparing dissolving pulps is not only delignification but also removal of hemicelluloses and controlled chain degradation of cellulose to decrease cellulose viscosity to a desired level. In particular, opening the porous cell-wall system is significant for removal of degraded lignin macromolecules during pulping and pulp washing.

Chemical Pulping. In all chemical wood pulping processes (49–50), size and uniformity of the wood chips are important. To improve penetration of the pulping liquors, the chips should not be thicker than about 5 mm across the grain. Sulfite and sulfate (kraft) pulping, the oldest chemical pulping processes, account for most chemical pulp production. Over the years, although their basic chemistry remained practically unchanged, some process modifications emerged and large-scale equipment was developed to transform the original batch processes into continuous operation. As an example, the widely used continuous Kamyr digester may be mentioned. In all modern chemical pulping operations, chemical recovery and disposal of the organic waste-liquor substances by combustion with adequate heat recovery are economic necessities. The related equipment costs constitute a large part of total mill investment costs. Therefore in new mill design, feasibility, efficacy, and costs of the chemical recovery may be the determining factor in selecting the pulping method to be used.

Sulfite Processes. Wood pulping in hot sulfite solutions is practiced over a range of liquor pH from about 1.2 to about 11. Acid-sulfite pulping is carried out at pH below 2 with free sulfur dioxide present in the liquor. Calcium, magnesium, sodium, and ammonium hydroxides are used as bases. At the low pH of acid-sulfite pulping, considerable hemicellulose hydrolysis takes place concurrently with delignification. For this reason, acid-sulfite pulping is also used in large-scale production of dissolving pulps. Because of the simplicity of chemical recovery, magnesium is the preferred base in acid-sulfite pulping. During waste liquor combustion, magnesium sulfite decomposes into magnesium oxide and sulfur dioxide. Fresh pulping liquor is easily reconstituted by the reaction of the aqueous sulfurous acid solution, recovered from water scrubbing of the combustion gases, with the magnesia ash.

Bisulfite pulping in a pH range of about 2–6 gives an increased yield of easily bleachable pulps from a variety of wood species. Liquor penetration into the wood chips and forced liquor circulation during cooking are important to delignification. Higher unbleached pulp brightness and lower rejects result from sodium-based bisulfite pulping. Two distinct two-stage processes using magnesium-based liquor are also in commercial use. In two-stage magnesium bisulfite pulping, first-stage cooking is carried out at pH 3.5–4 and about 160–166°C to a pulp yield of about 65–60%. In the second stage, cooking is continued in the same liquor at higher pH resulting from injecting magnesium hydroxide slurry into the digester. This cooking sequence increases pulp yield and improves burst and tensile strength.

In two-stage magnesium acid–sulfite pulping, the liquor pH in the first stage is higher (5.5–6.5) producing glucomannan stabilization and further increasing pulp yield. After replacing the liquor by sulfur dioxide and water, second-stage cooking is carried out at low pH and lower temperature (about 140°C). In this pulping sequence, total cooking time to a predetermined level of residual lignin increases with increasing pH of the first-stage cooking liquor.

Yield is still further increased when sodium-based sulfite pulping is carried out at higher pH (7–9) with additional alkali (soda) present to neutralize organic acids, such as acetic acid, split off from the wood during cooking. This applies especially to hardwoods. In this so-called neutral-sulfite pulping, delignification is slow, and hemicellulose hydrolysis is reduced. The process is used mainly for hardwood pulping. The resulting pulps exhibit high stiffness and accordingly are used in the production of corrugating medium.

In the last decade, alkaline sulfite pulping in a pH range of 9–11 (measured at cooking temperature) has been developed (51). The advantages claimed for the process are pulp strength properties similar to those of conventional kraft with elimination of the kraft odor.

Alkaline Processes. The oldest alkaline pulping process is caustic soda pulping which uses dilute aqueous sodium hydroxide solutions as pulping agent. To a limited extent it is still in commercial use. The most significant pulping process is sulfate (kraft) pulping which uses sodium sulfate (salt cake) as source of the alkali. Kraft pulping liquor contains caustic soda and sodium sulfide corresponding to about 25–30% sulfidity (percentage of sodium present as sulfide). Sodium sulfide has a beneficial effect upon delignification and also upon pulp strength properties. In polysulfide pulping excess sulfur is present resulting in some hemicellulose stabilization and yield increase. Kraft liquor alkalinity is usually expressed as gram per liter effective alkali (Na_2O). Effective alkali charge calculated on wood ranges from about 15 to 25% and cooking temperature from 160 to 185°C.

Based upon experiments with low molecular weight lignin model compounds, delignification was suggested to take place by beta-aryl-alkyl ether splitting (52). However, wood lignin is macromolecular and its true structure is still unknown. Remarkably, bulk delignification obeys approximately a pseudo first-order reaction mechanism, with a relatively low activation energy (53–54). This may indicate that distribution of thermal energy in the lignin structure also plays a role in the structure breakdown (55). Kraft pulps have good strength properties, especially high burst and tensile strength as well as folding endurance. However, their strength development in beating is slow compared to that of acid sulfite pulps. Unbleached kraft pulps have low brightness. Their strong coloration requires bleaching in up to six stages. The residual lignin is chemically changed and possibly grafted onto fibrous cellulose portions.

Rapid kraft pulping in the vapor phase at about 182.5–185°C produces bleachable pulp grades in a cooking time of about 20 min. This important reduction in cooking time requires uniformity of chemical distribution throughout the wood chips prior to vapor-phase cooking. First, the deaerated chips are subjected to forced liquor penetration at elevated temperatures (120–140°C) for 15–20 min. The liquor excess is then withdrawn, and the impregnated chips are immediately exposed to saturated steam of pulping temperature. Finally, the cook is quenched and the pulp cooled to about 120°C by immersing it into water. Subsequently, it is washed with hot water, for instance by a diffusion procedure.

Two-stage prehydrolysis kraft cooking is used for the production of dissolving pulps from hardwoods. Prehydrolysis consists of either liquid-phase cooking in acidified water or of high-pressure steam treatment. Acetic acid split from the wood aids in hemicellulose hydrolysis and also in the partial cellulose chain splitting necessary to reduce viscosity for chemical conversion into soluble cellulose derivatives.

Subsequent kraft cooking is carried out with extended cooking time to promote lignin dissolution. Concurrently, the alkali liquor also contributes to chain degradation of the cellulose, as measured by the viscosity drop (56). High purity and uniformity of chain length distribution are requirements for the use of viscose and acetate dissolving pulps. In the washing of dissolving pulps after bleaching and final acidification for removal of residual alkali, deionized water should be used. Carboxyl groups present in chemical cellulose chemisorb calcium ions even from middle-hard tap water.

Drawbacks of commercial kraft pulping are the emission of malodorous gases and the high costs of capital investment. In the decade 1965–1975 these costs rose exponentially (57). Commercial sulfur-free pulping in the oxygen–alkali system is a recent development. The mechanism of delignification has been investigated and the activation energy determined (58). Oxygen–alkali pulping produces good pulp brightness but relatively low tear strength indicating fiber damage.

Of the new chemical pulping processes, organosolv pulping, using a water mixture of an appropriate organic solvent as pulping agent instead of aqueous inorganic chemical solutions, appears to have economic potential and may in the future become of industrial significance. Recently, ethanol–water pulping at isothermal conditions was suggested (59). In a quasi temperature–pressure extraction of the wood, most of the noncellulosic wood portions are dissolved in the liquor, leaving relatively undamaged cellulose in the residue. Distilling alcohol from the used liquor precipitates most of the dissolved lignins, leaving the hemicellulose hydrolysates including various sugars and acetic acid in solution. In this way, lignin substances and hemicellulose hydrolysates can be recovered separately in a single operation. The new pulping and recovery method is still in the experimental stage and needs further development to prove commercial feasibility (see Pulp).

Purification. Originally all steps after pulping were carried out in the bleach plant and the term bleaching referred to the one or two stages of chemical treatment designed to provide whiteness of the necessary level in the final product. In the current broad sense, bleaching refers to all of the stages of pulp manufacturing between the completion of the pulping stage and the beginning of the stock preparation stage.

The pulp producer uses several steps to provide the fibers required by manufacturers of paper or chemical cellulose derivatives. These fibers have well defined physical, chemical, and optical attributes that are specified in the purchase contracts. A complete description of the overall process of pulp production with technical details for all of its many steps is found in Rydholm (60). An especially good coverage of the technical details of bleaching is in a 1963 TAPPI monograph (61).

In those cases where a chemically pure fiber is required, the bleaching process removes the color bodies associated with the lignin and other impurities that survive the pulping operation. At the same time, it is possible to remove hemicellulose and resinous materials. In the ultimate purification process, the steps are also designed to determine the chain length of the cellulose molecule (molecular weight) and the physical length of the fibers (fiber-length distribution). Such a procedure may require 11 or 12 steps to produce pulp for manufacture of cellulose films, sausage casing, textile fibers, and thickening agents (62).

In contrast to this ultimate process, minimum bleaching is designed merely to whiten the fibers to an acceptable level without causing weight loss or other physical effects. The one-step processes for producing newsprint pulp and similar nonpermanent products are minimum bleaching procedures.

Oxidative Bleaching. *Chlorine.* Whether used alone or in combination with other reagents, chlorine (qv) is generally the first chemical treatment for lignin removal. In this step, pulp at 3.0–3.5% consistency mixed with chlorine gas (0.1–5.0 grams per liter of chlorine) at 20–60°C react for 15–30 min. The mixing is usually done in a centrifugal device which then forces the mixture into the bottom of an upflow tower. The tower height serves to provide reaction time and also to prevent loss of gaseous chlorine. The degree of reaction is controlled by the temperature and the amount of chlorine. Addition of chlorine is controlled by oxidation–reduction potential (ORP) cells, visual appearance of the pulp, lignin content, or other arbitrary parameters chosen by the pulp producer.

There is continuing discussion on whether attack on the lignin proceeds via chlorination of the phenolic lignin components, via free-radical oxidation of those impurities, or via another (addition) reaction. In any case, the reaction is accompanied by the liberation of hydrochloric acid as noted by the rapid drop in pH after mixing. In some cases, acid addition enhances the reaction which should always be carried out at pH less than 2 to minimize attack on the cellulose and other carbohydrate components of the pulp. Chlorine as the sole bleaching stage produces pulps with a gray–brown or gray color.

Other Agents. Hypochlorite, chlorine dioxide, hydrogen peroxide, and oxygen are used in pulp bleaching (see Bleaching agents; Pulp). Hypochlorite, the original procedure for pulp bleaching, acts by oxidative degradation of impurities. Chlorine dioxide is a mainstay of bleaching used to produce high brightness levels (63) (see Chlorine oxygen acids and salts). This oxidative treatment attacks impurities but not carbohydrates. Hydrogen peroxide (qv) is used as the final bleaching agent of high brightness, high stability pulps where long life is a prime requirement. The chief value of oxygen is in pollution control (64).

Reductive Bleaching. In the manufacture of newsprint and other papers where only high yield and short life is required, acceptable brightening is obtained via reaction with zinc or sodium hydrosulfite. These chemicals provide the desired level of whiteness without extraction of pulp components; hence, the term brightening implies high yield bleaching. The reaction is carried by mixing groundwood or other mechanical pulp at 4–5% consistency with 0.1–1.0% hydrosulfite at 40–60°C in the presence of polyphosphates or other chelating agents. Reaction time is 10–30 min in upflow or downflow towers.

Other Reactions. To produce pulps for conversion into forms other than the original fibers, specialized purification stages have been developed. These are usually employed on dissolving pulps intended for the manufacture of products such as cellulose acetate film, viscose tire cord, and sausage casing. For these pulps it is necessary to remove as much as possible of all constituents other than cellulose. The usual methods include either or both of these steps: (*1*) Addition of a surfactant (anionic or nonionic) as part of a hot, weakly alkaline extraction step after chlorination. The purpose is to remove resinous materials. (*2*) Incorporation of a separate treatment step in which the pulp reacts with strong NaOH solution (6–10% NaOH) at 10–20°C after most of the oxidative bleaching is completed. The purpose is dissolution of hemicellulose.

In most purification plants, the final stage involves the use of strong acid to reduce the mineral content. This is done by acidifying to pH 2.0–2.5 with sulfuric acid plus optional chelators at 4–5% consistency and 40–50°C for 30–45 min.

82 CELLULOSE

Combined Purification Processes. Depending on the use of the pulp, the manufacturer may employ as many as 12 stages of purification. At present, most newsprint receives a single-step reductive brightening. At the other extreme, acetate-grade dissolving pulp requires 12 purification steps; these include chlorination, hypochlorite bleaching, chlorine dioxide bleaching (61), and concentrated NaOH extraction, with intermediate alkaline extraction–washing after each oxidative stage, and final acid demineralization.

In 1976 the most common sequence was CEDED (C = chlorination; E = alkaline extraction to remove oxidative degradation products; and D = chlorine dioxide) and is used to produce FBK, an industry commodity for many paper products. FBK is derived from some softwood species that has been pulped by the kraft (sulfate) process and then bleached to at least 90 brightness with the CEDED sequence. This 5-stage procedure not only provides the desired whiteness but also accomplishes the necessary physical–chemical fiber and molecular changes to provide the raw material for the high technology paper business.

Woodpulp is the chief source of nontextile fibers in the world. In 1975–1976 the worldwide industry annual production of pulp was of the order of 110 million metric tons (65).

Uses

Fibers. Cellulose in form of refined cotton linters and of high grade wood pulp (α-cellulose content >90%) is used for the manufacture of man-made regenerated cellulose fibers (see Cellulose acetate and triacetate fibers; Fibers, man-made and synthetic; Rayon).

The basis for the development of man-made fibers from cellulose raw materials was the discovery of solvent systems for cellulose, such as cuprammonium by Schweitzer in 1857 (Schweitzer's Reagent), and cellulose xanthate by Cross and Bevan in 1892. Further information can be found in references 66–75.

The viscose fiber processes are used equally well for the manufacture of staple fibers (1975 volume, 2.2 million metric tons) and for filament yarns (1975 volume, 0.92 million metric tons). Viscose rayon staple fibers and filament yarns are highly versatile products available in a wide range of fiber fineness (tex or deniers) and cross-sections (round, flat, serrated, or multilobal), in bright or dull appearance, and as straight or crimped fibers or yarns.

Uses for regular viscose staple fibers and filament yarns include clothing, domestic curtains, carpets, and blankets. Staple fibers are also widely used for nonwovens, disposables, and sanitary products for general, medical or surgical applications. High-wet-modulus or polynosic rayon staple fibers are used mainly in the manufacture of clothing and domestic textiles. Blended with synthetic fibers, they provide sufficient water absorptivity for greater wearing comfort. The high tenacity viscose filament yarns or staple fibers find use as tire cord, especially for radial tires, and for belts, tapes, hoses, coated fabrics, etc (see Rayon).

Films. Cellulose-based films are used extensively in many areas of food packaging, sausage casing, hemodialysis, battery separators, sterilization, shrink seals, and reverse osmosis where select performance characteristics, such as dimensional stability, clarity, wet shrinkage, and gas or solute permeability, make them the material of choice (see Film and sheeting materials).

Cellophane. This flat film is among the oldest packaging materials and is still the best film for clarity, crisp hand, and the dimensional stability vital for good high speed machine printability. Recently, cheaper films based on polyolefins have captured significant food packaging markets, and cellophane production in the United States has dropped from 192,000 t in 1961 to a stable 136,000 t today. This production is about equally divided between (1) snacks and cookies, (2) baked goods and candy, and (3) tobacco and other nonfood uses. The coating, which is always applied to cellophane, gives it almost any desired property. For example, nitrocellulose or Saran coatings are used to reduce moisture or oxygen permeability, respectively; polyethylene or ionomers are used for heat sealing. Coated cellophane has a density of 1.2–1.5 g/cm^3 and thus gives a yield of 267–333 cm^2/(25 μm·g) [18,500–23,000 in.^{2}/(mil·lb)] which translates to about 0.77–1.4¢/1000 cm^2 (5–9¢/1000 in.^{2}).

Attempts to make cellulose films without the viscose process have included low degree of substitution (DS) (0.15) hydroxyethyl cellulose (HEC). This gave excellent dry strength and clarity and shrank less than viscose during casting (ca 35% as compared to ca 50%). However, HEC required more drying and had poorer wet strength owing to the more swollen wet gel state and the residual substitution. More recently, cellulose solutions from dimethyl sulfoxide (DMSO)–paraformaldehyde are being investigated for film and sausage casing production. However, removal of final traces of both DMSO and formaldehyde will present serious technical barriers where such films are to be used with food products.

Sausage Casings. Although cellophane films have declined to $300,000,000 on the United States market, sausage casings have advanced significantly to where they easily represent a $340,000,000 value in world markets. Casings of the pure cellulose (wiener) or the hemp-paper reinforced (large sausage) type each commands about $10/kg and undoubtedly are the most profitable high volume cellulose products sold today. Conservative figures for world casing markets are: wiener, $140,000,000; fibrous, $160,000,000; large diameter cellulose, $40,000,000. These are growing at 8–10%/yr as other nations seek to feed their increasing populations.

Wiener Casings. The production of skinless wieners requires the use of flexible tubular casing to contain the meat emulsion through cooking and smoking, after which the product has sufficient integrity so that the casing is peeled and discarded. Cellulose is the preferred material for all casings since it allows the proper smoke penetration and shrinkage during processing. Wiener casing is normally supplied as shirred sticks in which up to 49 m have been pleated and compressed into 48 cm lengths. Film strength and flex can be controlled by viscose composition, DP, and spinning conditions. Stretch performance, which is the most important parameter for end use is further controlled by tubular blow-up conditions during drying. A typical casing contains 70% cellulose, 12% glycerol, and 18% water. An excellent review of sausage casing technology has recently been published by Karmas (76).

Fibrous Casing. Bolognas, salamis, pepperoni, summer sausage, and liverwurst products require very strong casing since some of these are now stuffed in 1.5 m lengths and processed horizontally. This is achieved by using a very special high-grade hemp paper saturated with viscose. A typical fibrous casing is composed of 23% paper, 46% cellulose, 21% glycerol, and 10% moisture.

Other Casing Uses. Select sizes of wiener casing are used for packaging washed hospital instruments which are subsequently placed in an autoclave and sterilized with steam or ethylene oxide. Cellulose permeability is a major factor in this use since steam can readily permeate to kill the bacteria, but organisms cannot pass through the film, and the sterilized packages can be stored conveniently until needed. Such casings are sold under the trademark of Hos-Pak and are printed with a special red ink that turns green once sterilization conditions have been reached. With only minor modifications casings can be used as battery separators. Some tubular film products are also sold as shrink seals for use on liquor bottles. These seals (Celon seals) are sold in the wet state and then dry and shrink during shipment of the liquor bottles.

Artificial Kidneys. A highly specialized use for tubular cellulose films is in artificial kidneys (see Dialysis). Some 40,000 people annually suffer kidney failure in the United States. Up until 1970 only about 5% could be dialyzed because of a lack of proper dialysis units and facilities. Today government subsidies guarantee that needed treatment can be provided. The main unit of a dialysis machine of the type developed by Kolff costs about $15,000–20,000, and each new dialysis kidney coil costs about $25.00 exclusive of heparin, plasma, manpower, etc. Over the years thin-walled cellulose tubes (12.7–17.8 μm or 0.5–0.7 mil) from either viscose or cuprammonium have been used, but today most units employ the cuprammonium tubular films since they offer faster dewatering and clearance of metabolic by-products from the blood. Recently, hollow fibers from many synthetic polymer sources and from cuprammonium have been investigated and the cuprammonium hollow fibers have found commercial acceptance. Because of the higher surface to volume ratio of fibers, units made from hollow fibers can be made much smaller than those with tubular films (see Hollow-fiber membranes). Flat sheet dialysis units based on the plate–frame filter principle (Kiil type) are also used, especially in Europe. Recently a portable kidney that weighs 4.5 kg and is strapped to the stomach like an accordian was reported by the Kolff Foundation of Cleveland, Ohio.

Reverse Osmosis. Although a wide range of polymers have been investigated for use in reverse osmosis (RO), cellulose acetate, the first material successfully used, is still the best overall polymer for this purpose. RO has been commercially employed for water purification, desalination, and food concentration (maple syrup and cheese whey). RO membrane performance depends on the formation of a very dense, extremely thin surface skin with the rest of the film simply acting as a supporting structure. As this supporting structure compresses under pressure and clogs under use, the flux rates decrease significantly. Hydrolysis of the acetate is a limiting factor for some uses. A great deal of work has already been reported. A spiral design by Gulf General Atomics and a tubular design by Westinghouse, as well as others, are available. A large amount of reference material is available (77–78) (see Reverse osmosis; Membrane technology).

Gums and Thickeners. Cellulose ethers are used in such diverse applications as laundry detergents, water-base paints, oil wells, cement, plaster, lacquers, foods, textiles, and cosmetics. Small amounts of the ethers serve to make the systems more uniform, stronger, more viscous, smoother, more durable, or more workable. Cellulose ethers usually constitute a small portion of various systems, yet they have a disproportionately beneficial effect on the properties of these systems. Cellulose ethers are soluble in water and/or in simple solvents. Two percent solution viscosities range from below five to above 100,000 mPa·s (= cP). The five common ether substituents are sodium carboxymethyl, methyl, ethyl, hydroxyethyl, and hydroxypropyl.

Production starts with the formation of alkali cellulose, either as described for viscose or by the addition of sodium hydroxide to a slurry of cellulose flock in a diluent (see Cellulose derivatives, ethers). Reaction with monochloroacetic acid, methyl chloride, ethyl chloride, ethylene oxide, or propylene oxide (or some combination) follows. Methyl and ethyl chlorides require pressure. The etherifying agent is usually completely consumed. An oxidizing agent may be added for viscosity control (79). Thereafter, the slurry is neutralized, and the product is extracted, dried, and ground. Some ethers are granulated to larger particle sizes in the range of 0.1 to 2.0 mm (80). Glyoxal is added to some ethers prior to drying (81). Surfactants are also added either before or after drying (82). Since the number of producers is limited, production data are not available. Major producers in the United States are Buckeye Cellulose, Dow Chemical, Hercules, Union Carbide, and BASF Wyandotte.

The substituents may be located in any or all of three positions on each anhydroglucose unit, ie, on the C-2, C-3, and C-6 carbon atoms. The average number of sites occupied by the substituent is the degree of substitution (DS), the maximum being three (83). Ethylene and propylene oxides generate new hydroxyl reaction sites which may be further substituted. Molar substitution (MS) indicates the average number of substituents per anhydroglucose unit (83). MS can be measured on hydroxyethyl cellulose (84) and hydroxypropyl cellulose (85), but no chemical method was available for determining the DS of these two substituents through 1977. Many methods have been tested.

Cellulose ethers are white (or nearly white) dry powders. The fiber form remains intact during production, but the fibers become plumper and are broken into short fragments. These fragments may be agglomerated into small lumps in order to aid dispersion when dissolving. Carboxymethyl cellulose and hydroxyethyl cellulose are soluble in both hot and cold water; the more hydrophobic methyl cellulose, ethyl cellulose, and hydroxypropyl cellulose are soluble in cold water but not in hot. The solubility of methyl cellulose and ethyl cellulose in water decreases at high DS levels, but increases in organic solvents. The glyoxal-treated ethers can be dissolved by dispersing in water with mild agitation. The initially thin slurry gradually thickens and becomes clear. The dissolution can be speeded by heating or by increasing the pH. Cooling and mild acidification reverse these effects. The glyoxal functions by internally cross-linking the cellulose chains, thereby preventing rapid swelling and attendant lumping. Ethers made without glyoxal or agglomeration need extremely vigorous agitation to prevent lumping. Surfactant treatment aids dispersion but does not give the thin slurry that glyoxal gives; vigorous agitation is still required. Since the methyl and hydroxypropyl ethers are insoluble in hot water, they can be dispersed in hot water, then easily dissolved on cooling. This thermal dissolving is reversible on warming; either a gel or a flocculated precipitate forms. The solution viscosity changes abruptly at the gel point of the ether.

Solutions of cellulose ethers are clear (or nearly clear), and colorless. The solution viscosity of a given ether rises rapidly with an increase in concentration. Temperature has a strong negative effect on viscosity following the Andrade equation (86). The ethers, except ethyl cellulose, are very soluble in water. Dilute solutions of the very low viscosity ethers are Newtonian; ie, the solution viscosity is independent of the shear rate. Normally, cellulose solutions are pseudoplastic; their measured viscosity decreases as the shear rate is increased. Some solutions exhibit thixotropy; the viscosity depends in part on the previous shear history of the solution. Uses of cellulose ethers are discussed under Cellulose derivatives, ethers.

Foods and Pharmaceuticals. *Foods.* Although pulverized forms of fibrous purified alpha celluloses, such as dissolving pulps, have been used widely as fillers in some foods and pharmaceuticals, their usefulness has been restricted by their highly fibrous nature and resulting poor tongue response. Recently, however, a bread containing 25% cellulose fiber content has gained excellent market acceptance (see Dietary fiber).

The development of colloidal microcrystalline cellulose made available much finer particle forms of highly purified crystalline celluloses, and more importantly aqueous suspensoids of microcrystalline cellulose particles. These have a smooth texture (resembling uncolored butter) and unique pseudoplastic properties, including stable viscosity over a wide temperature range.

Microcrystalline cellulose has enjoyed remarkable success in food and pharmaceutical uses during the 20 years since its commercialization as Avicel by FMC Corporation. Three commercial plants in the United States, Japan, and the Republic of Ireland produce Avicel.

In the food industry, microcrystalline celluloses are important for their heat stability, ability to thicken with favorable mouth feel, and flow control. They serve to extend starches, form sugar gels, stabilize foam, and control formation of ice crystals. A few of the food uses in which colloidal microcrystalline cellulose has been commercially successful are: fillings, meringue (cold process), chocolate cake sauce (frozen), cookie fillings, whipped toppings (both bakery whipped topping, which can be frozen after aeration, and proteinless topping), and imitation ice cream for use as a bakery filling.

Pharmaceuticals. The largest present pharmaceutical use of microcrystalline celluloses, by far, is the tableting of pharmaceuticals, both in the United States and elsewhere. They are used as an excipient to assist in the flow, lubrication, and bonding properties of the ingredients to be tableted, to improve the stability of the drugs in tablet form and especially to give rapid disintegration of the tablet in the stomach.

Cosmetics. A special grade of microcrystalline cellulose known as Avicel-RC581 has been developed for a wide range of cosmetic uses. It has a composition on a dry basis of 89% microcrystalline cellulose and 11% carboxymethyl cellulose (medium viscosity).

A few of the more common uses for this grade are suspensions, shampoos, cream rinses, hair conditioners, hair coloring products, hair bleaches, and toothpaste.

BIBLIOGRAPHY

"Cellulose" in *ECT* 1st ed., Vol. 3, pp. 342–357, by Jack Barsha and Peter VanWyck, Hercules Powder Company; "Cellulose" in *ECT* 2nd ed., Vol. 4, pp. 593–616, by J. Kelvin Hamilton and R. L. Mitchell, Rayonier Inc.

1. K. Hess, *Die Chimie der Zellulose und Ihrer Begleiter,* Akademische Verlagsgesellschaft, M.B.H., Leipzig, Ger., 1928, 5 pp.
2. M. V. Tracey, *Chitin Rev. Pure Appl. Chem.* **7,** 1 (1957).
3. R. D. Preston, *The Physical Biology of Plant Cell Walls,* Chapman & Hall, London, Eng., 1974.
4. J. R. Colvin, *Crit. Rev. Macromol. Sci.* **1,** 47 (1972).
5. J. Kjosbakken and J. R. Colvin, *Can. J. Microbiol.* **21,** 111 (1975).
6. J. R. Colvin, M. Takai, and G. G. Leppard, *Naturwissenschaften* **63,** 297 (1976).
7. G. G. Leppard, L. C. Sowden, and J. R. Colvin, *Science* **189,** 1094 (1975).
8. R. M. Brown and co-workers in F. Loewus, ed., *Biogenesis of Plant Cell Wall Polysaccharides,* Academic Press, Inc., New York, 1973.
9. W. Herth and co-workers, *Cytobiologie Z. Exp. Zellforsch.* **10,** 268 (1975).

10. P. M. Ray, T. L. Shininger, and M. M. Ray, *Proc. Nat. Acad. Sci. U.S.A.* **64,** 605 (1969).
11. G. Shore and G. A. Maclachlan, *J. Cell Biol.* **64,** 557 (1975).
12. K. Mühlethaler, *Z. Schweiz Fortstv.* **30,** 55 (1960).
13. K. Mühlethaler, *Papier (Paris)* **17,** 10a, 546 (1963).
14. A. N. J. Heyn, *J. Cell Biol.* **29,** 181 (1966).
15. R. St. J. Manley, *J. Polym. Sci. A-2* **9,** 1025 (1971).
16. I. Ohad and D. Danon, *J. Cell Biol.* **22,** 302 (1964).
17. R. Muggli, H-G. Elias, and K. Mühlethaler, *Makromol. Chem.* **121,** 290 (1969).
18. R. Muggli, *Cellul. Chem. Technol.* **2,** 549 (1968).
19. R. F. Mark and co-workers, *Science* **164,** 72 (1969).
20. M. C. Probine and R. D. Preston, *J. Exp. Bot.* **12,** 261 (1961).
21. A. J. Hodge and A. B. Wardrop, *Aust. J. Sci. Res.* **38,** 265 (1950).
22. R. D. Preston and A. B. Wardrop, *Discuss. Faraday Soc.* **11,** 165 (1951).
23. T. Goto, H. Harada, and H. Saiki, *Mokuzai Gakkaishi* **19,** 463 (1973).
24. A. Frey-Wyssling, *Science* **119,** 80 (1954).
25. R. D. Preston and J. Cronshaw, *Nature (London)* **181,** 248 (1958).
26. R. H. Marchessault and A. Sarko, *X-ray Structures of Polysaccharides; Advances in Carbohydrate Chemistry,* Vol. 22, Academic Press, Inc., New York, 1967, pp. 421–481.
27. J. Hayashi and co-workers, *J. Polym. Sci. Polym. Lett. Ed.* **13,** 23 (1975).
28. J. Hebert and L. Muller, *J. Appl. Polym. Sci.* **18,** 3373 (19).
29. K. H. Gardner and J. Blackwell, *Biopolymers* **13,** 1975 (1974).
30. A. Sarko and R. Muggli, *Macromolecules* **7,** 486 (1974).
31. F. J. Kolpak and J. Blackwell, *Macromolecules* **9,** 273 (1976).
32. A. Stipanovic and A. Sarko, in press.
33. J. G. Cook, *Handbook of Textile Fibres Natural Fibres,* Merrow Publishing Co. Ltd., Watford, Eng., 1974, chapt. 1.
34. A. J. Turner in A. R. Urquhart and F. O. Howitt, eds., *The Structure of Textile Fibres,* The Textile Institute, Manchester, Eng., 1953, pp. 105–117.
35. H. I. Bolker, *Natural and Synthetic Polymers,* Marcel Dekker, Inc., New York, 1974, p. 52.
36. T. E. Timell, *Adv. Carbohydrate Chem.* **19,** 247 (1964); **20,** 409 (1965).
37. Ref. 35, p. 346.
38. Ref. 35, Chapt. 12.
39. K. V. Sarkanen and C. H. Ludwig, eds., *Lignins, Occurrence, Formation, Structure and Reactions,* Wiley-Interscience, New York, 1971.
40. Ref. 3, p. 163.
41. B. J. Fergus and co-workers, *Wood Sci. Technol.* **3,** 117 (1969).
42. B. J. Fergus and D. A. I. Goring, *Holzforschung* **24,** 113 (1970).
43. F. El-Hosseiny and D. H. Page, *Fibre Sci. Technol.* **8,** 21 (1975).
44. R. D. Preston in J. W. S. Hearle and R. H. Peters, eds., *Fibre Structure,* Butterworths, London, Eng., 1963, p. 250.
45. T. E. Timell, *Adv. Carbohydr. Chem.* **19,** 247 (1964); **20,** 409 (1965).
46. K. V. Sarkanen and C. H. Ludwig, *Lignins,* Wiley-Interscience, New York, 1971.
47. D. A. I. Goring, *Pulp Pap. Mag. Can.* **64**(12), T517 (1963).
48. *1975 International Mechanical Pulping Conference in San Francisco,* Technical Association of the Pulp and Paper Industry, Atlanta, Ga., 1975.
49. S. A. Rydholm, *Pulping-Processes,* Interscience Publishers, a division of John Wiley & Sons, Inc., New York, 1965.
50. R. G. MacDonald and J. N. Franklin, eds., *Pulp and Paper Manufacture,* Vol. I, 2nd ed., McGraw-Hill Book Co., New York.
51. U.S. Pat. 3,630,832 (Dec. 28, 1971), O. V. Ingruber and G. A. Allard (to Canadian International Paper Co.).
52. J. Gierer and I. Kunze, *Acta Chem. Scand.* **15**(4), 803 (1961); J. Gierer and I. Noren, *Acta Chem. Scand.* **16**(7), 1713; 1976 (1962); J. Gierer and L. A. Smedman, *Acta Chem. Scand.* **18**(5), 1244 (1964); J. Gierer, B. Lenz, and N. H. Wallin, *Tappi* **48**(7), 402 (1965).
53. T. N. Kleinert, *Tappi* **49**(2), 53 (1966).
54. T. N. Kleinert, *Papier (Darmstadt)* **21**(10A), 653 (1967).
55. T. N. Kleinert, *Tappi* **49**(3), 126 (1966).
56. T. N. Kleinert, *Tappi* **51**(10), 467 (1968).
57. H. O. Ware, *Tappi* **57**(12), 9 (1974).

58. T. N. Kleinert, *Tappi* **59**(9), 122 (1976).
59. U.S. Pat. 3,585,104 (June 15, 1971), T. N. Kleinert.
60. S. Rydholm, *Pulping Processes,* John Wiley & Sons, Inc., New York, 1965.
61. W. H. Rapson, ed., *The Bleaching of Pulp, TAPPI Monograph Series No. 27,* Technical Association of the Pulp and Paper Industry, Atlanta, Ga., 1963.
62. D. F. Durso, *Forest Prod. J.* **19**(8), 49 (1969).
63. W. H. Rapson, *Tappi* **59**(4), 6 (1976).
64. A. G. Jamieson and L. A. Smedman, *Tappi* **57**(5), 134 (1974); C. J. Myburgh, *Tappi* **57**(5), 131 (1974).
65. L. Haas and J. E. Kalish, *Pulp Pap.* **50**(7), 121 (1976).
66. H. F. Mark, S. M. Atlas, and E. Cernia, *Man-Made Fibers,* Vol. 25, John Wiley & Sons, Inc., New York, 1968.
67. J. L. Hathaway, "Cellulosic Fibers Made from Cuprammonium" in ref. 66, pp. 1–6.
68. P. von Bucher, "Viscose Rayon Textile Fibers" in ref. 66, pp. 1–6.
69. W. J. McGarry and M. H. Priest, "Viscose Rayon Tire Yarn" in ref. 66, pp. 43–80.
70. H. Krassig, "Characterization of Cellulose Fibers" in ref. 66, pp. 121–180.
71. C. B. Chapman, *Fibres,* Butterworths, London, Eng., 1974.
72. H. Siesler and co-workers *Macromol. Chem.* **42**, 139 (1975).
73. H. Krässig, and W. Kitchen, *J. Polym. Sci.* **51**, 123 (1961).
74. H. Krässig and W. Käppner, *Makromol. Chem.* **44–46**, 1 (1961).
75. H. Krassig, *Chemiefasern* **17**, 821 (1967); *Lenzinger Ber.* **24**, 66 (1967); *Das Papier* **21**, 629 (1967); *Lenzinger Ber.* **27**, 5 (1969); *Textilveredlung* **4**, 26 (1969); *Appl. Polym. Symp.* **28**, 777 (1967).
76. E. Karmas, *Sausage Casing Technology,* Noyes Data Corp., Park Ridge, N.J., 1974.
77. A. F. Turbak, ed., *Appl. Polym. Symp.* **13**, (1970).
78. S. Sourirajan, *Reverse Osmosis,* Academic Press, Inc. New York, 1970.
79. U.S. Pat. 2,512,338 (June 20, 1950), E. D. Klug and H. M. Spurlin (to Hercules).
80. U.S. Pat 2,667,482 (Jan. 26, 1954), C. H. Rigby (to Imperial Chemical Industries); U.S. Pat. 2,331,864 (Oct. 12, 1943), R. W. Swinehart and A. T. Maasberg (to The Dow Chemical Co.).
81. U.S. Pats. 2,879,268 (Mar. 24, 1959), I. Jullander (to Mo och Domsjo AB); 3,072,635 (Jan. 8, 1963), J. H. Menkart and R. S. Allen (to Chemical Development Corporation of Canada).
82. U.S. Pats. 2,647,064 (July 28, 1953), 2,720,464 (Oct. 11, 1955), A. W. Anderson and B. V. Moeller (to The Dow Chemical Co.).
83. *ASTM Method D 1695, Definitions,* American Society for Testing and Materials, Philadelphia, Pa.
84. P. W. Morgan, *Ind. Eng. Chem. Anal. Ed.* **18**, 500 (1946); *ASTM Method D 2364, Hydroxyethyl Cellulose,* ASTM, Philadelphia, Pa.
85. R. U. Lemieux and C. B. Purves, *Can. J. Res.* **B-25**, 485 (1947); *ASTM Method D 2363, Hydroxypropyl Methyl Cellulose,* ASTM, Philadelphia, Pa.
86. E. N. da C Andrade, *Nature (London)* **125**, 309, 582 (1930).

ALBIN F. TURBAK (Films)
Co-editor
ITT Rayonier Inc.

DONALD F. DURSO (Purification)
Co-editor
Johnson & Johnson
and
O. A. BATTISTA (Foods and Pharmaceuticals)
Research Services Corp.

HENRY I. BOLKER (Macrostructure)
Pulp and Paper Research Institute of Canada

J. ROSS COLVIN (Biosynthesis)
National Research Council, Ottawa, Canada

NATHAN EASTMAN (Gums and Thickeners)
Union Carbide Corporation

THEODOR N. KLEINERT (Pulping)
Pulp and Paper Research Institute of Canada

HANS KRASSIG (Fibers)
Lenzing, A.G.

R. ST. JOHN MANLEY (Microstucture)
McGill University

CELLULOSE ACETATE AND TRIACETATE FIBERS

Acetate fiber is the generic name for a cellulose acetate [9004-35-7] fiber which is a partially acetylated cellulose, also known as secondary acetate. The desirability and wide textile use of acetate lies in its uniformly high quality, color, and styling versatility, drape, hand, and other favorable esthetic properties (see Economic Aspects). Triacetate is the generic name for cellulose triacetate [9012-09-3] fiber, also known as primary acetate. It is an almost completely acetylated cellulose. Acetate and triacetate differ only moderately in the degree of acetyl substitution on cellulose. Yet they have different chemical and physical properties. Triacetate fiber is hydrophobic; heat treatment develops a high degree of crystallinity which can be used to impart desirable fabric performance characteristics.

Cellulose acetate is the reaction product of cellulose and acetic anhydride (see Cellulose derivatives, esters). The degree of polymerization (DP) of the cellulose used for esterification is generally around 1000–1500, whereas the esterification product DP is around 300. There is evidence that the DP of cellulose acetate is relatively independent of the DP of the cellulose (1). Commercial triacetate is not a precise chemical entity because acetylation is not quite complete.

Secondary cellulose acetate is obtained by acid catalyzed hydrolysis of the triacetate to an average degree of substitution of 2.4 acetyl groups per glucose unit. The primary acetyl groups hydrolyze more readily than the secondary but the exact ratio of primary and secondary substitution depends upon the hydrolysis conditions.

Two separate terms, acetyl value (%) and combined acetic acid (%), are used to specify the degree of acetylation. Since the formula weights are 43 for the acetyl groups (CH_3CO) and 60 for the acid (CH_3COOH), the two terms are always in the ratio of 43:60. The relation of acetyl value and combined acetic acid to the number of hydroxyls acetylated per glucose residue ($C_6H_{10}O_5$) is:

$$\text{combined acetic acid, \%} = \frac{60}{43} \times (\text{acetyl value, \%})$$

$$= \frac{60 \ (\text{acetyls per glucose unit}) \times 100}{162 + 42 \ (\text{acetyls per glucose unit})}$$

Acetyls per glucose	Acetyl value, %	Combined acetic acid, %
2	35.0	48.8
2.5	40.3	56.2
3	44.8	62.5

Commercial cellulose triacetate has a combined acetic acid content of 61.5%, corresponding to 2.92 acetyls per glucose unit. Cellulose acetate with 2.4 acetyl groups per glucose unit has a combined acetic acid content of approximately 55%.

Commercial cellulose acetates contain small amounts of residual free carboxyl and sulfate groups as well as acetyl and hydroxyl groups. The thermal stability of the cellulose acetate depends upon the level and form (free acid or salt) of the sulfate groups. The salt form is ordinarily more stable and commercial acetates are carefully neutralized in the final process steps. Multivalent cations may form cross-links between residual sulfate and carboxyl groups, causing an artificially high solution viscosity. This limits the concentration of cellulose acetate used in solutions for yarn manufacture. The purity of the cellulose used to prepare the cellulose acetate also has an influence on solution rheology (2).

Properties

Structure. Cellulose acetate fibers have a low degree of crystallinity and orientation even after heat treatment. Cellulose triacetate, however, develops considerable crystallinity after heat treatment. This is apparent from x-ray diffraction diagrams which show increased crystalline order in cellulose triacetate that has been heat treated at 240°C for 1 min. Many improved fiber performance characteristics of triacetate over acetate are a result of the increased internal structural order.

Appearance and Color. Fibers of both cellulose acetate and triacetate have a bright, lustrous appearance. A duller, whiter yarn is produced by the addition of pigments (qv) such as titanium dioxide at 1–2%. The bright, nonpigmented fiber reflects light specularly whereas the dull, pigmented fiber reflects diffusely. The application of twist to a fiber bundle reduces the specular reflection and makes it more diffuse. Acetate and triacetate fibers have essentially the same light absorption characteristics in the visible spectrum. Absorption increases slightly in the ultraviolet region.

The exceptionally high level of whiteness of both types of fibers (3) is obtained by careful selection of high purity wood pulp, manufacturing conditions, and pigment inclusion and permits bright, clean, pure colors to be obtained when the fibers are dyed.

Dyeing Characteristics. Disperse dyes (qv) are most frequently used for cellulose acetate and triacetate fibers. They are high melting, crystalline compounds that have low solubility in the dye bath even at high temperatures. They are milled to very small particle size to permit effective dispersion without agglomeration in the dye bath. A small quantity of disperse dye is dissolved in the aqueous medium and then diffuses into the fiber to give a uniform color. The rate-determining step in the dyeing operation is the slow diffusion of the large dye molecules into the fibers. Properties such as dye bath temperature and fiber composition modify the dyeing rate. Triacetate fibers are dyed more slowly than acetate fibers but dye carriers (qv) can be used to accelerate the rate of dyeing. Some typical carriers are based on butyl benzoate, methyl salicylate, diallyl phthalate, and diethyl phthalate (4–14).

Selection of the appropriate azo, anthraquinone, or diphenylamine disperse dye ensures good colorfastness. Fading inhibitors are used to resist the effects of nitrogen oxides and ozone. Triacetate fabrics are heat-treated to raise the safe ironing temperature, drive the dye further into the fiber to gain further gas-fading resistance, and improve colorfastness.

Inherently colored acetate and triacetate yarns are produced by incorporating colored pigments (inorganic or organic) or soluble dyes in the acetate solution prior to spinning. Solution-dyed acetate and triacetate yarns are extremely colorfast to washing, dry cleaning, sunlight, perspiration, sea water, and crocking, and usually surpass the performance of vat dyed yarns. In addition, acetate and triacetate are susceptible to gas- or fume-fading conventionally; such fading is absent with solution dying. Unfortunately, the additional cost of producing solution-dyed acetate fiber has restrained its market acceptance.

Specific Gravity. Fiber cross sections are often irregular and an immersion technique is commonly used to measure specific gravity of fibers. The immersion fluid may interact with the fiber by absorption or surface wetting, and the selection of a suitable fluid is critical for accurate measurements. Values for acetate have been reported (15–17) for different solvents over a range of 1.306 with carbon tetrachloride to 1.415 with n-heptaldehyde. The values of 1.32 for acetate and 1.30 for triacetate are accepted for fibers of combined acetic acid contents of 55 and 61.5%, respectively.

Refractive Index. The refractive index parallel to the fiber axis (ϵ) is 1.478 for acetate and 1.472 for triacetate. The index perpendicular to the axis (ω) is 1.473 for acetate and 1.471 for triacetate. The difference between ϵ and ω (birefringence) is very low for acetate fiber and practically undetectable for triacetate.

Absorption and Swelling Behavior. The absorption of moisture by acetate and triacetate fibers generally depends upon the relative humidity to which the fibers are exposed. However, it varies according to whether equilibrium is approached from the dry or the wet side. This hysteresis effect is noted over the entire range of relative humidities as shown in Figure 1 (18).

Additional moisture regain isotherms for acetate fiber have been reported (19). Heat-treated triacetate fiber has a lower regain than nonheat-treated fiber, and values in the range of 2.5–3.2% have been observed (4–5,20–21).

In the United States, commercial regain is used to calculate commercial weights of yarns or fibers. In effect, commercial moisture regain is added to the weight of bone-dry fiber to account for the moisture normally found in textile fibers. The percentage of commercial regain, taken from ASTM D 1909-68, is 6.5 for acetate fiber and 3.5 for triacetate. In Europe and elsewhere, the Bureau International Pour La Standardisation De La Rayonne Et Des Fibres Synthetiques (BISFA) rules (21) are used for determining the commercial weight of acetate and other man-made fibers. The basic premise differs from that of the ASTM in that the BISFA include a percentage for the normal finish present on the fiber as shipped as well as a figure to account for the moisture normally found in fibers. Thus the commercial weight of acetate according to the BISFA rules includes a figure of 9% in determining the commercial weight of acetate fiber, based on the bone-dry, finish-free weight of the fiber.

Percentage of water imbibition is an important property in ease-of-care and quick-drying fabrics. This value is determined by measuring the moisture remaining in the fiber when equilibrium is established between a fiber and air at 100% rh while the fiber is being centrifuged at forces up to 1000 g. Average recorded values are: acetate, 24%; nonheat-treated triacetate, 16%; heat-treated triacetate, 10%.

92 CELLULOSE ACETATE AND TRIACETATE FIBERS

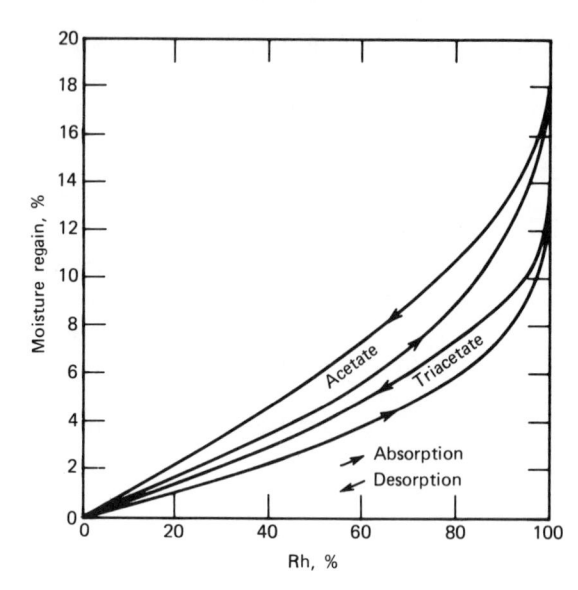

Figure 1. Moisture regains (moisture content on a bone-dry basis) of cellulose acetate and triacetate fibers on absorption and desorption at 22°C (12,19).

Absorption of water by fibers causes swelling roughly proportional to moisture content. Although there is considerable disagreement among investigators in the field, the average value for the increase in length of acetate fibers due to water absorption is about 1%, and the average increase in diameter is approximately 10%. The corresponding values for the swelling of triacetate fiber are lower than those for acetate. Tesi (22) reported a 1.5% cross-sectional area increase for heat-treated triacetate fiber and 4.0% for nonheat-treated fiber.

Thermal Behavior. Acetate, like other thermoplastic fibers, sticks, softens, and even melts when ironed at high temperatures. Sticking and softening temperatures of all fibers depends on such factors as yarn diameter, fabric construction, and general fabric geometry, and varies with different test procedures. Sticking and softening temperatures are not necessarily directly related to the fiber melting point; acetate softens and sticks in the range of 190–205°C but fuses at approximately 260°C. The apparent shining or glazing temperature is usually lower than the sticking temperature and is also influenced by moisture content, fabric construction, and color. (When ironing acetate fabrics the sole-plate temperature of hand irons should not exceed 170–180°C.) The sticking and glazing temperatures of nonheat-treated triacetate fiber are in the same range as those of acetate; but heat-treated triacetate fibers have considerably higher sticking and glazing temperatures. The latter fabrics can be ironed at temperatures as high as 240°C. The melting point of triacetate is approximately 300°C.

Acetate and triacetate, because they are thermoplastic fibers, exhibit changes in mechanical properties as a function of temperature. However, within the range of normal climatic temperatures, the mechanical properties are not altered significantly. As temperature is raised, the modulus of acetate and triacetate fibers is reduced, and the fibers extend more readily under stress (Fig. 2).

Acetate and triacetate are weakened by prolonged exposure to elevated temperatures in air (Fig. 3).

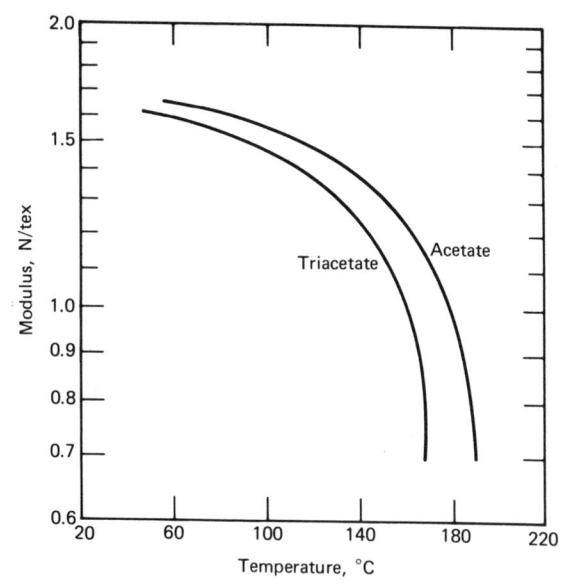

Figure 2. Tensile modulus as a function of temperature. To convert N/tex to gf/den, multiply by 11.33.

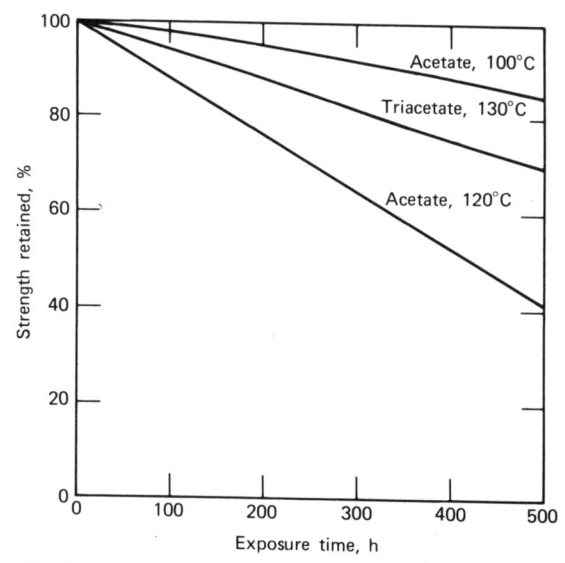

Figure 3. Effect of dry heat exposure on acetate and triacetate (lubricant-free yarns tested at standard conditions of 65% rh and 21°C).

Light Stability. Resistance of textile fibers to sunlight degradation depends upon the wavelength of the incident light, relative humidity, and atmospheric fumes. Although fibers cannot be accurately rated for their resistance to sunlight degradation in terms of Langley units or hours of exposure, certain generalizations appear justified. Acetate and triacetate fibers, when exposed under glass, behave similarly to cotton and rayon, ie, they are somewhat more resistant than unstabilized pigmented nylon and silk and appreciably less resistant than acrylic and polyester fibers. When acetate and triacetate are exposed to direct weathering, their resistance is lowered as compared with exposure under glass. Additional information on actinic degradation is described in refs. 23–29.

Certain pigments, particularly carbon black, offer protection from Fade-Ometer sunlight exposure, as indicated for acetate in Figure 4. Special yarn additives may be used to improve light stability for a specific use such as drapery fabrics (29).

Electrical Behavior. Because acetate and triacetate yarns readily develop static charges, it is sometimes desirable to apply an antistatic finish to aid in textile processing. Both yarns have been used for electrical insulation after lubricants and other finishing agents have been removed by solvent extraction, followed by washing with water. Table 1 contains data from a report of the British Cotton Industry Association (30) and lists the resistivity in MΩ·cm of various fibers over a range of relative humidities (45–95%) (see Antistatic agents).

Cellulose acetate has a high electrical resistivity. Table 2 shows comparative resistivity of scoured acetate and triacetate taffeta fabrics.

Mechanical Properties. The mechanical and esthetic properties of textile fabrics prepared from fibers and yarns depends greatly on the physical form of the fibers themselves and the geometrical construction of the fabric. Some important performance criteria include: hand, drape, wrinkle resistance and recovery, strength, and flexibility. The interactions among fibers and yarns in a fabric array are complex and the most straightforward way of describing the mechanical properties of the fabric is to refer to the inherent mechanical properties of the acetate. Fiber mechanical properties are described by the stress–strain and recovery behavior under conditions of tensile, torsional, bending, and shear loading.

Tensional Properties. Stress–strain curves typical of commercial acetate and triacetate yarns are shown in Figures 5 and 6. The curve is indicative of most commercial acetate and triacetate yarns although some high tenacity triacetate and acetate fibers are being produced. Table 3 shows some mechanical properties characteristic of both acetate and triacetate commercial yarn and fibers.

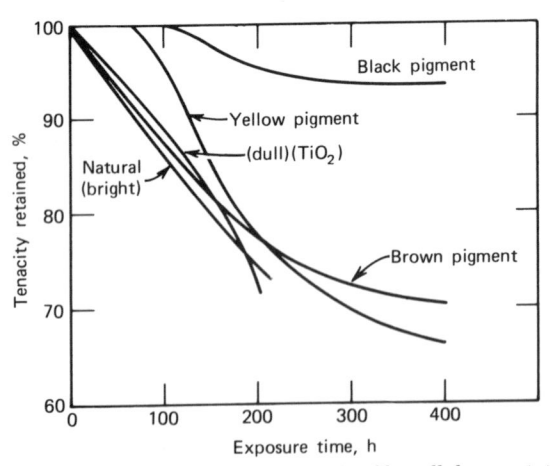

Figure 4. Effect of colored pigments on tenacity retained by cellulose acetate fiber after carbon arc Fade-Ometer exposure. Tenacity measured at standard conditions of 65% rh and 21°C.

Table 1. Resistivity (MΩ·cm) of Various Fibers in Commercial and Purified States[a]

Rh, %	Acetate Commercial	Acetate Purified	Nylon Commercial	Nylon Purified	Cotton Commercial	Cotton Washed[b]	Rayon Commercial	Rayon Purified
45	967,000	81,500,000	813,000	6,430,000	149		543	1,720
50	662,000	21,600,000	585,000	3,200,000	64		235	680
55	424,000	6,040,000	387,000	1,430,000	30		95	266
60	256,000	1,650,000	208,000	525,000	14	530	36	93
65	150,000	448,000	104,000	193,000	6.0	150	13	34
70	74,000	126,000	43,000	70,000	2.4	38	4.6	12
75	28,900	33,200	14,500	20,000	1.02	11.2	1.8	4.4
80	7,200	9,000	4,000	6,000	0.33	3.1	0.63	1.5
85	1,610	2,460	863	1,290	0.106	0.91	0.23	0.52
90	160	370	120	180	0.024	0.23	0.06	0.13
95	11	39	8.5	18.5			0.012	0.02

[a] Ref. 30.
[b] Water-washed only.

Table 2. Resistivity of Acetate and Triacetate Fabrics[a,b]

Fabric	Resistivity, Ω·cm
cellulose acetate	1.27
cellulose triacetate, nonheat-treated	3.81
heat-treated	15.2
specially scoured and heat-treated	1016

[a] Ref. 5.
[b] ASTM Method D 1000-55T.

Table 3. Tenacity and Elongation of Commercial Acetate and Triacetate Yarns and Fibers[a]

Properties	Acetate and triacetate
tenacity, N/tex[b]	
standard conditions[c]	0.10–0.12
wet	0.07–0.09
bone-dry	0.12–0.14
knot, standard conditions[c]	0.09–0.10
loop, standard conditions[c]	0.09–0.10
elongation at break, %	
standard conditions[c]	25–45
wet	35–50

[a] Ref. 5.
[b] To convert N/tex to gf/den, multiply by 11.33.
[c] 65% rh, 21°C.

The ability of a material to resist deformation under an applied tensile stress is measured by the modulus of elasticity (Young's modulus). In viscoelastic materials, the apparent modulus of a textile fiber is defined as the ratio of stress to strain in the initial, linear portion of the stress–strain curve. The apparent modulus at low strain

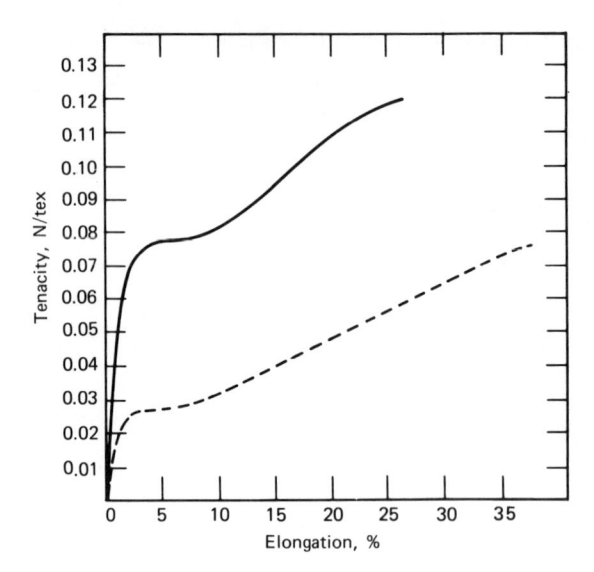

Figure 5. Typical stress–strain curves for bright acetate yarn under standard and wet conditions. Instron tensile tester rate at 60%/min rate of extension, 3.9 cm gage length. ——, 65% rh, 21°C; - - - - -, wet, 21°C. To convert N/tex to gf/den, multiply by 11.33.

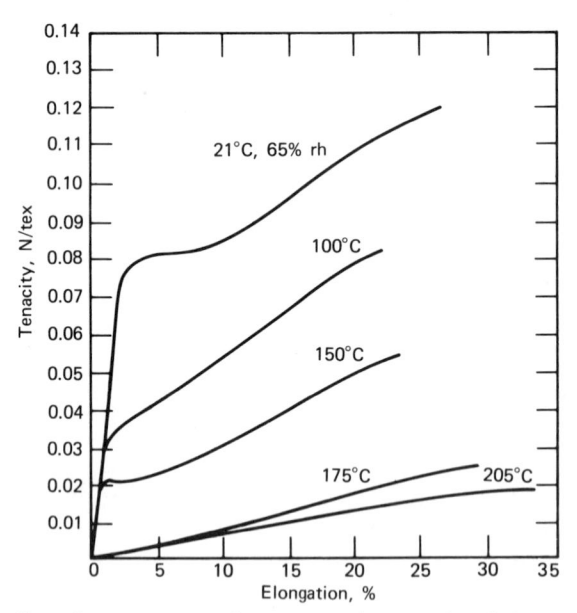

Figure 6. The effect of temperature on the stress–strain properties of triacetate yarn. Instron tensile tester rate at 60%/min rate of extension, 3.9 cm gage length (6). To convert N/tex to gf/den, multiply by 11.33.

levels is directly related to many of the mechanical performance characteristics of textile products. The modulus of elasticity can be affected by drawing (elongating) the fiber and by other manufacturing procedures. Values for commercial acetate and triacetate fibers are generally 2.2–4.0 N/tex (25–45 gf/den).

The wet modulus of fibers at various temperatures is significant for textile applications because it influences the degree of creasing and mussiness caused by laundering. Figure 7 shows the change with temperature of the wet modulus of acetate, triacetate, and a number of other fibers (31).

The ability of a fiber to absorb energy during straining is measured by the area under the stress–strain curve. This property is also known as toughness or work of rupture. Table 4, according to Meredith (32), lists the work of rupture of acetate in comparison with other textile fibers. Work of rupture of triacetate fiber is essentially the same as that of acetate.

A fiber which is strained and allowed to recover will give up a portion of the work absorbed during straining. The ratio of the work recovered to the total work absorbed (measured by the respective areas under the stress–strain and stress–recovery curves) is designated as resilience.

The elongation of a stretched fiber is best described as a combination of instantaneous extension and a time dependent extension or creep. This is called viscoelastic behavior and is common to many textile fibers, including acetate. Conversely, recovery of viscoelastic fibers is typically described as a combination of immediate elastic recovery, delayed recovery, and permanent set or secondary creep. The permanent set

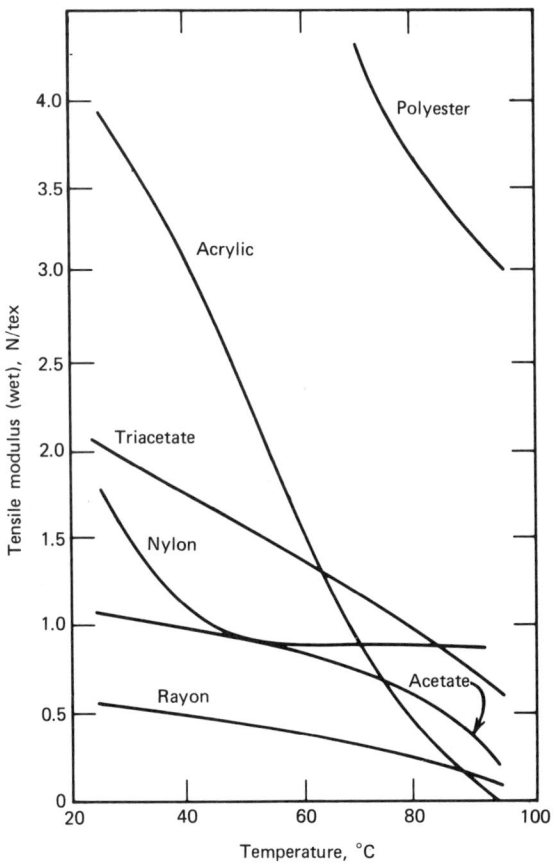

Figure 7. Effect of water temperature on wet modulus of fibers. To convert N/tex to gf/den, multiply by 11.33.

Table 4. Work of Rupture of Acetate and of Some Other Common Fibers[a]

Fibers	Work of rupture (toughness), N/tex[b]
acetate	0.022
cotton	0.010
nylon	0.076
rayon (viscose regular)	0.023
silk	0.072
wool	0.032

[a] Ref. 32.
[b] To convert N/tex to gf/den, multiply by 11.33.

is residual extension which is not recoverable. Susich and Backer (33) have described these three components of recovery for acetate (Table 5).

Temperature and moisture content of the fiber affect viscous behavior and hence modify the stress–strain relationship. Most stress–strain data are reported under standard conditions of 21°C and 65% rh.

Strains of more than 10% are avoided in textile processing so that the dimensional or shape stability of the resultant fabric will be acceptable.

Bending. The bending properties of a fiber generally depend on the viscoelastic behavior of the material. However, in most textile applications the radius of curvature of bending is relatively great, and the imposed strains are of a low order of magnitude. As a first approximation, it is possible to examine the bending properties by the classical methods. The bending stiffness or flexural rigidity of a fiber is the product of the bending modulus and the moment of inertia of the cross section. Thus for fibers of round cross section and constant modulus, the flexural rigidity varies directly with the square of the tex. Table 6 gives data on the flexural rigidity of acetate fibers as a function of fiber tex.

Torsion and Shear. For a discussion of torsional and shear properties of acetate and other fibers, see Kaswell (34).

Chemical Properties. Under slightly acidic or basic conditions at room temperature, acetate and triacetate fibers are very resistant to chlorine bleach at the concentrations normally encountered in laundering.

Triacetate fiber is significantly more resistant than acetate to alkalies encountered in normal textile operations. It is recommended that temperatures no higher than 85°C or a pH above 9.5 be used when dyeing acetate. Under normal scouring and dyeing

Table 5. Elongation Recovery of Acetate Fibers[a]

Fiber	Immediate elastic recovery, %	Delayed recovery, %	Permanent set, %
acetate multifilament			
at 50% of breaking tenacity	74	26	0
at breaking point	14	16	70
acetate staple yarn			
at 50% of breaking tenacity	58	42	0
at breaking point	12	18	70

[a] Ref. 33.

Table 6. Flexural Rigidity of Cellulose Acetate Fibers as a Function of Tex

Tex[a]	Flex rigidity, g·m²
6.7	83
3.3	27
2.2	8.9
1.7	7.4
1.1	2.5
0.56	0.56
0.42	0.35

[a] To convert tex to den, multiply by 9.

conditions, alkalies up to pH 9.5 and temperatures up to 96°C may be used with triacetate with little saponification or delustering. Heat-treated triacetate fiber has even greater alkali resistance, as shown in Figure 8. Strong alkalies and boiling temperatures saponify triacetate as well as acetate fiber.

Acetate and triacetate are essentially unaffected by dilute solutions of weak acids. Strong mineral acids cause serious degradation of both fibers. The results of exposure of nonheat-treated and heat-treated triacetate taffeta fabrics to various chemical reagents have been reported (5).

Acetate and triacetate fibers are not affected by the perchloroethylene dry cleaning solutions normally used in the United States and Canada. Trichloroethylene, employed to a limited extent in Great Britain and Europe, softens triacetate.

Resistance to Microorganisms and Insects. The resistance (based on soil-burial tests) of triacetate to microorganisms is very high, approaching that of polyester, acrylic, and nylon fibers. Figure 9 shows comparative soil-burial test results on acetate, triacetate, and cotton. Neither acetate nor triacetate fiber is readily attacked by moths or carpet beetles; however, there have been a few instances in which test larvae have

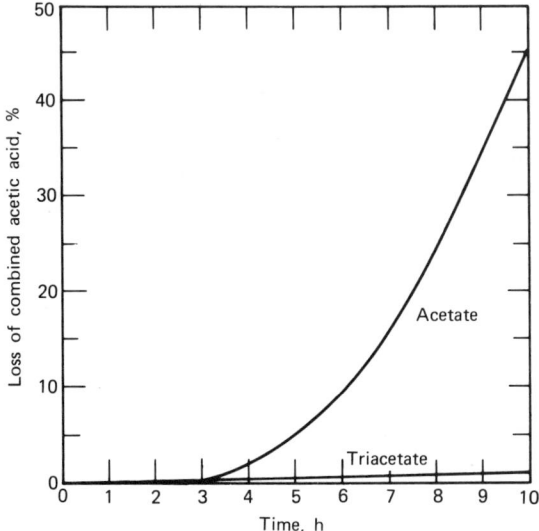

Figure 8. Comparative saponification rates of cellulose acetate and triacetate at scouring conditions of pH 9.5–9.8 and 95°C (23).

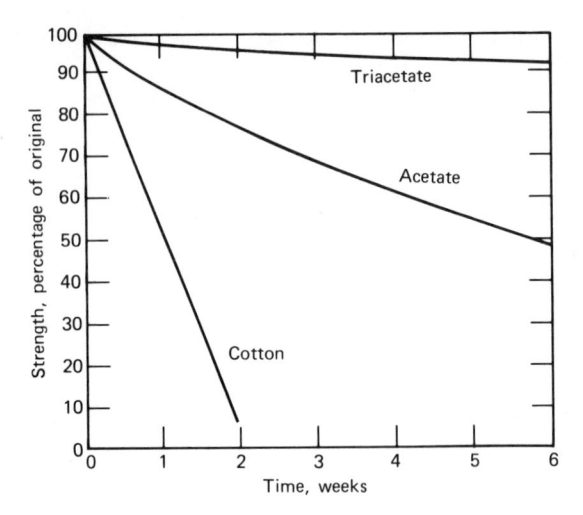

Figure 9. Resistance to biological attack as measured by residual tenacity after various periods of soil burial (12).

cut through acetate to get at wool fibers, or have damaged acetate contaminated with foreign substances containing starches.

Manufacture of Cellulose Acetate and Triacetate Flakes

Acetate and triacetate flakes are prepared by the esterification of high purity chemical cellulose with acetic anhydride (35–37). High purity cellulose is required because the solution properties of the resultant acetate or triacetate flake may be adversely affected by impurities even at low concentrations; extrusion processes require the polymer to flow freely through small diameter capillaries (eg, 30–80 μm). Wood pulp has replaced cotton linters as a source of chemical cellulose except for special plastic-grade acetates where color and clarity are very important.

Wood contains 40–50% cellulose, hemicellulose, lignin, and extractives. The noncellulose impurities must be removed to produce an acetylation-grade wood pulp with a purity of 95–98% α-cellulose. A strong economic incentive to cope with the higher level of hemicellulose impurities in lower purity pulp has led to continued technical interest (38–39). Although several processes have been developed (40–42), none is commercial.

The other raw materials for the manufacture of cellulose acetate, acetic acid, acetic anhydride, and sulfuric acid, are items of commerce and can be obtained in high purity and uniformity.

Secondary Acetate Processes. Three major processes are used to produce cellulose acetate. A solution process is most common (Fig. 10); acetylation is obtained with acetic anhydride using glacial acetic acid as the solvent. The sulfuric acid catalyst increases the solubility of the partially esterified cellulose in acetic acid until sufficient acetyl groups are added to achieve solubility of the cellulose acetate in acetic acid. Two major variants of this process are high catalyst (10–15 wt % sulfuric acid based on cellulose) and low catalyst (<7 wt % sulfuric acid) procedures. The second most common process is the solvent process. Methylene chloride is substituted for all or part of the acetic

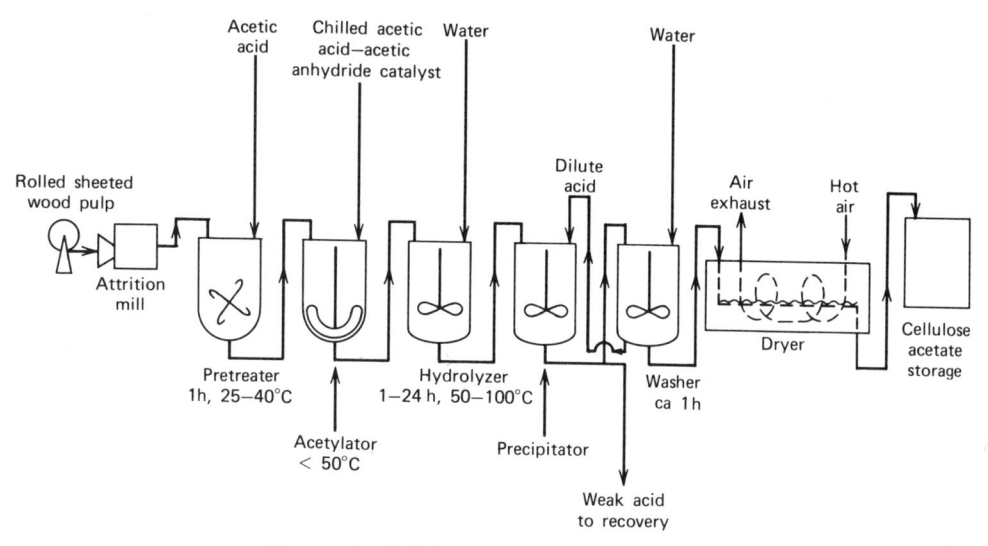

Figure 10. Cellulose acetate flake manufacture.

acid and acts as a solvent for the triacetate as it is formed. Perchloric acid is frequently used as a catalyst as sulfuric acid is not required to form a soluble intermediate. A third method for producing secondary acetate is a heterogeneous process (also called nonsolvent or Schering process) in which an inert organic liquid, such as benzene or ligroin, is used as a nonsolvent to prevent the acetylated cellulose from dissolving as it is formed. The cellulose ester produced is never solubilized, hence its physical form is similar to the original cellulose fiber. More detailed information on each of the above three processes has been presented by Malm and Hiatt (43).

To obtain a secondary acetate which is soluble in acetone, it is necessary to completely acetylate the cellulose during the dissolution step and then hydrolyze it, while still dissolved, to the required acetyl value. The overall process may be separated into four main steps: (1) preparation of cellulose for acetylation, (2) acetylation, (3) hydrolysis, and (4) recovery of cellulose acetate and solvents.

Preparation of Cellulose for Acetylation. Wood pulp is customarily supplied in rolls weighing up to 300 kg but can be obtained in bales of individual sheets. The pulp sheet must be fluffed with a disk refiner to generate a large surface area for acetylation because the rate and completeness of acetylation depends on the accessibility of the cellulose (44–45). The fluffed pulp is treated to further increase its accessibility to the acetylation solution. Several agents can be used for this step (called pretreatment) but an acetic acid–water mixture is most common. Sufficient water for pretreatment is usually present when the moisture content of the pulp is ca 6% of the weight of cellulose. Pretreatment generally involves agitating the pulp–acetic acid mixture for about 1 h at 25–40°C.

The ratio of pulp to pretreat acid depends upon the catalyst level used. An activation stage is added to the low catalyst acetylation procedure. An acetic acid–sulfuric acid mixture is introduced until the sulfuric acid concentration is 1–2% of the pulp weight. The activation period may last for 1–2 h during which the degree of polymerization of the cellulose is reduced. Activation time and temperature are selected so as to achieve the desired degree of polymerization in the acetate flake product. The high catalyst procedure does not usually involve an activation step. The degree of polymerization of the acetate flake is controlled by the conditions selected during acetylation and hydrolysis. The pulp is charged to the acetylation reactor after pretreatment and activation.

Acetylation. The acetylation mixture, consisting of acetic anhydride (esterifying agent), acetic acid (solvent), and sulfuric acid (catalyst), is precooled prior to its addition to the pretreated cellulose. Precooling provides a heat sink for two exothermic reactions that occur in the acetylation process, the esterification of cellulose with acetic anhydride liberates 1.03 kJ/g (246 cal/g) of cellulose, and the reaction of acetic anhydride with water from the pretreat mix generates 3.30 kJ/g (789 cal/g) of H_2O. Because heat is generated quickly and the mixture is viscous, a jacketed vessel does not provide the necessary cooling capacity. A separate vessel called a crystallizer is used to prechill the acetylation mixture and freeze some of the acetic acid. The heat of fusion of the uniformly dispersed acetic acid crystals and the low temperature of the mixture provide readily accessible heat sinks; this is particularly important in the early stages of acetylation where a rapid temperature rise would reduce the degree of polymerization of the cellulose. A cold brine circulates through the jacket of the acetylizer to chill the equipment before esterification. In the methylene chloride process, the required heat transfer is provided by refluxing the solvent. For either process, a 5–15 wt % excess of acetic anhydride assures complete reaction. A series of simultaneous, complex reactions occur during acetylation and a large body of literature is available describing both the solution chemistry and acetylation reaction (35–37, 46–49).

The cellulose chain is partially depolymerized during acetylation by the action of the sulfuric acid catalyst. High temperature and a large amount of catalyst accelerate depolymerization.

Temperatures in excess of 50°C are avoided. At the end point, microscopic examination of the solution should reveal no undissolved residues. The acetylation reaction is terminated by adding water to destroy the excess anhydride and provide a level of 5–10% total water for hydrolysis. A 15–25% cellulose acetate concentration is typical.

Hydrolysis. The number of acetyl groups present in each anhydroglucose unit at the end of acetylation is slightly less than 3.0 and must be reduced to around 2.4 to prepare secondary cellulose acetate. The number of acetyl groups is reduced and the combined sulfate groups minimized by acid hydrolysis under controlled conditions of time, temperature, and acidity. The sulfate groups hydrolyze more easily than the acetyl groups. The sulfuric acid formed increases the acidity of the reaction solution. In the high catalyst acetylation case, a portion of the sulfuric acid is neutralized, eg, by the addition of sodium acetate or magnesium acetate to reduce the free sulfuric acid content and prevent excessive depolymerization of the cellulose acetate. Hydrolysis temperature, normally 50–100°C, is obtained by direct steam injection (50) and reaction time varies from 1 to 24 h.

Flake Recovery. Precipitation, washing, and drying are the final steps in flake preparation. The precipitation procedure varies according to the product desired. In flake preparation, the hydrolyzed cellulose acetate solution is brought to incipient precipitation by mixing it with a stream of dilute (10–15%) acetic acid. Additional dilute acetic acid is rapidly introduced and the solution vigorously agitated. To obtain a powder precipitate, the agitated solution is slowly diluted until precipitation occurs. Another process involves extrusion of the hydrolyzed solution through small holes into a precipitating acid bath; this produces fine strands which are then cut into pellets.

The precipitated cellulose acetate is physically separated from the dilute (25–35%) acetic acid. The flake is sent to a washer which removes the acetic acid and salts remaining from sulfuric acid neutralization. The wet flake is dried to a moisture level of 1–5% in a suitable commercial dryer. The dilute acetic acid that results from the washing and precipitation steps can be used directly in other stages of the process or recovered for reuse. Efficient recovery and recycle of the dilute acetic acid is an economic necessity.

A pressure stabilization step may be required if thermal stability and nonyellowing on heating are critical considerations (eg, in thermoplastic molding application). The acetate flake is heated in deionized water at ≤1.4 MPa (200 psi) for ca 1 h. This treatment removes residual sulfate groups by hydrolysis.

Acetate and triacetate flakes are white amorphous solids produced in granular, flake, powder, or fiber form. They are used as raw materials in the preparation of fibers, films, and plastics. Flake density varies with physical form and ranges from 100–500 kg/m^3 in loose bulk. Acetate flake is shipped by trailer truck or railroad freight car. Smaller shipping quantities are packaged in multiwall paper bags. The bag may include a vapor-barrier layer to protect against a change in moisture content due to atmospheric humidity.

Acid Recovery. Approximately 4.0–4.5 kg of acetic acid per kg of cellulose acetate is used in the homogeneous process. Approximately 0.5 kg is consumed in the product and the remaining 3.5–4.0 kg is recovered. Acetic acid for recovery leaves the main process sequence as an aqueous solution of 25–35% acetic acid. It may also contain dissolved salts from sulfuric acid neutralization, and dissolved and suspended low molecular weight cellulose and hemicellulose acetates. Suspended material is removed in a settling tank or filter. Acetic acid is recovered from the clarified weak acid stream by solvent extraction. Several frequently used organic solvents are benzene, ethyl acetate, and methyl ethyl ketone. The organic extract upper layer from the extractor is sent to a distillation column. The aqueous raffinate phase, containing most of the inorganic salts, is discarded. In the distillation column, the extraction solvent is taken overhead and glacial acetic acid is the bottom product. The energy requirements for acid recovery depend upon the specific solvent used and may be in the range of 4.2–10.5 kJ/g (1–2.5 kcal/g) of acid recovered. A portion of the acetic acid may be subsequently converted to acetic anhydride required for acetylation. This is achieved by catalytic pyrolysis and produces acetic anhydride in good yield and at a low cost (see Acetic acid).

Batch Preparation of Triacetate. The batch triacetate process differs from the preparation of secondary acetate in that there is little hydrolysis and the temperature is in the range of 50–100°C. The triacetate hydrolysis step (or desulfation) removes only the sulfate groups from the polymer by slow addition of a dilute acetic acid solution containing sodium or magnesium acetate (51–52) or triethanolamine (53) to neutralize the liberated sulfuric acid. Meanwhile, the temperature is kept above the peak reaction temperature. The cellulose triacetate product has a combined acetic acid content of 61.5%. Flake recovery is similar to the secondary acetate process.

Sulfuric acid levels as low as 1% can be used for acetylation in the methylene chloride process. Only a minor amount of desulfation is required to reach a combined acetic acid level of 62.0%. If perchloric acid is used as the catalyst, the nearly theoretical value of 62.5% combined acetic acid is obtained. Standard flake recovery methods are used.

The nonsolvent process can also be used to prepare triacetates of nearly theoretically combined acetic acid. The catalyst is normally perchloric acid. However, because the original cellulose was never dissolved, the triacetate retains a fibrous appearance.

Continuous Flake Manufacturing Processes. All processes described thus far have incorporated batch pretreatment, acetylation, and hydrolysis stages. A continuous process offers potential economic advantages owing to reduced labor requirements, lower energy consumption, a more uniform, higher quality product, lower raw materials consumption, and complete automation. Major disadvantages include: more highly specialized equipment, increased maintenance problems with the more sophisticated equipment, and the entire production capability is lost if there are malfunctions in the separate process stages of pretreatment, acetylation, or hydrolysis.

Several continuous processes have been described in the patent literature (54–71). The Societé Rhodiaceta continuous solution process is used commercially to produce both secondary acetate and triacetate. A continuous triacetate process is operated in the United States by Celanese Corporation.

The basic processing stages are identical to those of the batch process but some equipment and materials handling techniques are different. The initial pretreatment segment is actually a batch operation. The cellulose is weighed and combined with a measured amount of 80–90% aq acetic acid. This mixture is slurried 10–15 min to produce a 2–5% suspension of cellulose in the liquid. The slurry is sent to a holding tank which continuously feeds the subsequent processing steps. The water in the pretreatment mixture produces a completely swollen, highly reactive cellulose. It is much more reactive toward acetylation than batch-pretreated cellulose. However, the water must be removed from the pretreated pulp slurry prior to acetylation to reduce the consumption of acetic anhydride. This is accomplished by countercurrent leaching with glacial acetic acid in a multistage, belt-type extractor. In each successive stage the cellulose slurry is contacted with a more concentrated acetic acid solution. The number of stages is selected to reduce the water content to a specified low level at the extractor exit. A reduced pressure is applied at each stage to reduce liquor carryover between stages. The pretreated cellulose mat leaving the extractor is further deliquored prior to entering the acetylizer.

The acetylizer is a specialized, high energy input reactor with a range of materials-handling capability. It receives matted pulp, acetic acid, catalyst, and acetic anhydride, and produces a highly viscous, homogenous solution of cellulose triacetate. The acetylizer must provide: thorough mixing and kneading action, and positive transport of material through the reactor; a plug flow residence time distribution to ensure complete reaction and uniform flake properties; and good heat transfer capability to control the exothermic acetylation reaction. The viscous triacetate solution from the acetylizer is pumped into a high speed blender where water or a dilute acetic acid solution stops the reaction.

The hydrolysis step consists of a series of retention tanks which provide a controlled time–temperature exposure to reach the desired combined acetic acid and residual sulfate level. The remainder of the process (precipitation, washing, and drying) is continuous and similar to the batch processes previously described.

Manufacture of Cellulose Acetate and Triacetate Fibers

Extrusion Processes. Polymer solutions are converted into fiber form by spinning or extrusion. The dry extrusion process is used primarily for acetate and triacetate. A solution of cellulose acetate or triacetate polymer in a volatile solvent is forced through a multihole spinneret into a cabinet of warm air; the fibers are formed by evaporation of the solvent. In wet extrusion, a polymer solution is forced through a spinneret into a nonsolvent liquid which coagulates the filaments and removes the solvent. In melt extrusion, molten polymer is forced into air which cools the strands into filaments.

The dry-extrusion process consists of four main operations: (1) dissolution of the cellulose acetate in a volatile solvent, (2) filtration of the solution to remove insoluble matter, (3) extrusion of the solution to form fibers, and (4) lubrication and take-up of the yarn on a suitable package.

Acetone is the universal solvent for secondary acetate in dry extrusion. The optimum concentration for an acetate spinning solution falls between the highest possible solids concentration and the resulting high solution viscosity. Though higher concentrations of solids produce fibers with better properties, practical limits of viscosity are quickly reached. If water is added to the acetone solvent, the solution viscosity exhibits a minimum at approximately 9:1 acetone to water. Typical solvent composition is about 95% acetone and 5% water, typical solids concentration in the extrusion solution is 20–30% depending on the polymer molecular weight. The viscosity of the solution at room temperature is about 100–300 Pa·s (1000–3000 P).

The solubility of acetates of different combined acetic acid content in an acetone–water and acetic acid–water solution is shown in Figures 11 and 12. Cellulose triacetate is insoluble in acetone and another solvent system must be used for dry extrusion. Triacetate solvents include chlorinated hydrocarbons such as methylene chloride, methyl acetate, acetic acid, dimethylformamide, and dimethyl sulfoxide. Methylene chloride containing 5–15% methanol is often employed.

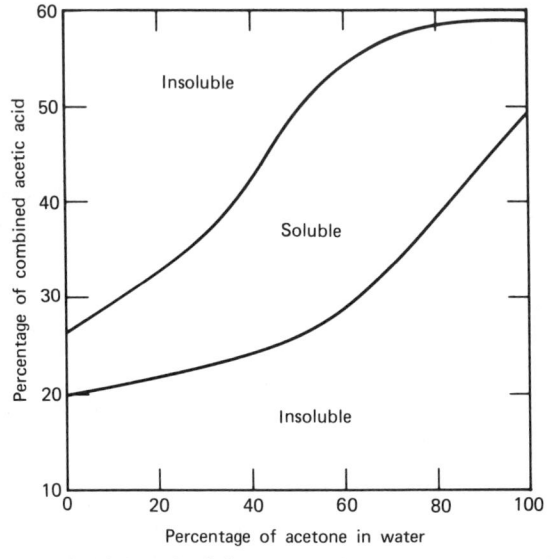

Figure 11. Solubility of cellulose acetate in acetone–water at 25°C (72).

106 CELLULOSE ACETATE AND TRIACETATE FIBERS

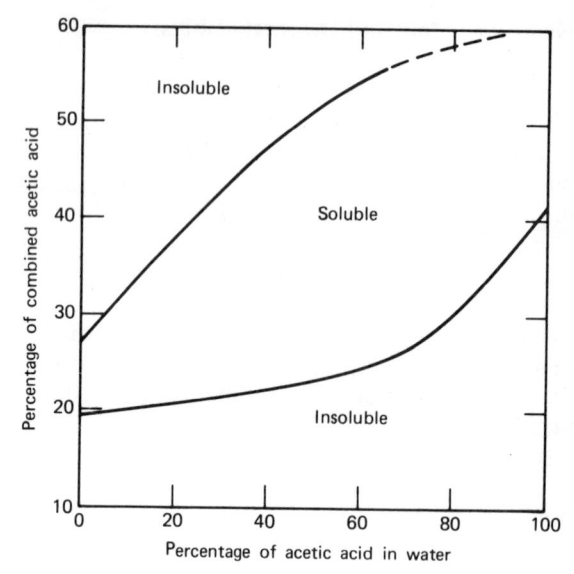

Figure 12. Solubility of cellulose acetate in acetic acid–water at 25°C (72).

Acetate and triacetate dissolution and handling operations are conducted in fully enclosed systems for several reasons. Both acetone and methylene chloride are expensive, acetone is highly flammable in air, and both solvents have strict worker exposure levels administered under the Occupational Safety and Health Act. Ultimately, all solvent vapor must be collected and recycled through the process. This recovery operation represents a significant operating cost.

Acetate or triacetate flake is charged to large, heavy-duty mixers along with solvent and a filter aid such as wood pulp fibers. Concentration, temperature, and mixing uniformity are closely controlled through several mixing stages. If a delustered or dull fiber is desired, 1–2% of a finely ground titanium dioxide pigment may be added in the mixing process or by injection of a pigment slurry after filtration. In the latter method, particular care must be exercised to achieve thorough mixing and maintain strict control of the pigment slurry. The polymer concentration and the composition of the extrusion solvent strongly affect the uniformity and tensile properties of the acetate fiber and must be closely controlled.

For successful extrusion of acetate filaments, the polymer solution should pass through spinnerets whose holes range from 30–80 μm dia. Consequently, the solution must be free of unacetylated cellulose, undissolved gels, and dirt. Multistage filtration, usually consisting of plate-and-frame filter presses with fabric and paper filter media, removes extraneous matter before extrusion. The acetate solution is heated during filtration to reduce viscosity and increase flow rate. The solution may be allowed to degas in holding tanks between each stage of filtration.

Extrusion of the acetate solution is a precisely controlled process. The filtered, preheated solution is delivered to the spinneret at constant volume by very accurate metering pumps. A separate pump for each spinneret position ensures uniform fiber formation. The spinnerets are made of stainless steel or another suitable metal and may contain from thirteen to several hundred holes which are precision-made to close size and shape tolerances. Auxiliary filters are used both prior to the fixture which holds the spinneret and in the spinneret itself as a final precaution against particulate matter in the extrusion solution. A schematic diagram of the extrusion process is shown in Figure 13.

Prior to entering the spinneret the extrusion dope solution is heated to lower the solution viscosity to an appropriate value and to provide some of the heat necessary to flash the solvent from the extruded filament. A thermostatically-controlled heat exchanger may be used to heat the dope or the filter–spinneret assembly may be located inside the heated extrusion cabinet.

The heated polymer solution emerges as a plurality of filaments from the spinneret into a column of warm air. Instantaneous loss of solvent from the surface of the incipient filament causes a solid skin to form over the still liquid interior of the filament. As the filament is heated by the warm air, more solvent evaporates and is carried away. More than 80% of the solvent can be removed during the brief residence time of less than one second in the hot air column. The air column or cabinet height is 2–8 m depending on the extent of drying required and the spinning speed. The air flow may be concurrent or countercurrent to the direction of fiber movement. The fiber properties are contingent upon the solvent removal rate and precise air flow and temperature control is necessary.

A feedroll applies tension to the bundle of acetate fibers, or yarn, to withdraw them from the extrusion cabinet. The product of a spinning position is called a con-

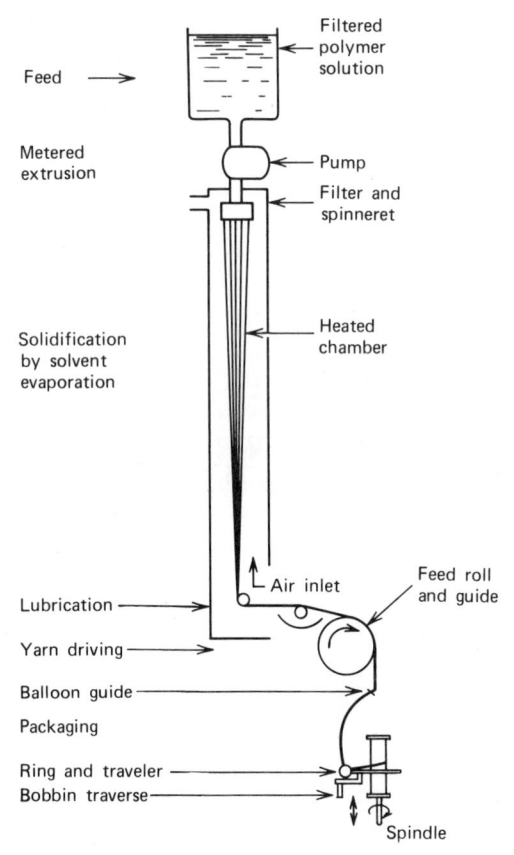

Figure 13. Dry extrusion—dry spinning of cellulose acetate fibers.

tinuous filament yarn (as distinguished from staple). Cellulose acetate yarns are produced from 4 to >100 tex (36 to >900 den). Feedroll speed, metering pump output, and cabinet conditions must be carefully balanced to produce a yarn of specified and uniform denier.

A finish or lubricant gives the extruded yarn the frictional and antistatic properties required for further processing. The finish is applied at levels of 1–5% as the yarn exits from the cabinet. The formulation of a lubricant depends upon the intended use of the yarn; many proprietary types are used. The lubricated yarn, containing only a small amount of residual solvent, is taken up on a ring twister which inserts just enough twist to avoid the handling difficulties of untwisted yarn on a bobbin. Yarn with no twist may be wound on a cylindrical tube. Instead of applying twist to the yarn to aid in handling, the yarn filaments may be compacted or entangled by passing the yarn through a device which intermixes the filaments with air jets (73–75).

The solvent used to form the extrusion dope, which is evaporated during the extrusion process, must be recovered. Adsorption on activated carbon or refrigeration to condense the solvent are the usual processes. Final purification is by distillation. Recovery is about 99% efficient. Approximately 3 kg of acetone must be recovered per kg of acetate yarn produced. Recovery of solvent from triacetate extrusion is similar but about 4 kg of methylene chloride solvent is used per kg of triacetate yarn extruded.

Only a small quantity of triacetate yarn is made by wet extrusion (66) since extrusion speeds are much lower than for dry extrusion and the process is not attractive for producing filament yarns. A solution of the triacetate or a stabilized reaction product in acetic acid is extruded into a nonsolvent bath, such as water, and the filaments form as the triacetate precipitates from solution. A large number of holes in the spinneret (eg, >1000) improves processing economics.

A melt extrusion process is not used to any great extent for fiber formation although a small quantity of triacetate yarn has been produced by this procedure. The residence time of the polymer at the melting point must be minimized to prevent degradation of the acetate.

Types of Yarns and Fibers. Many different acetate and triacetate continuous filament yarns, staples, and tows are manufactured. The variant properties are tex (wt (g) of a 1000 m filament) or denier (wt (g) of a 9000 m filament), cross-sectional shape, and number of filaments. Individual filament deniers (denier per filament or dpf) are usually in the 2–4 dpf range (0.2–0.4 tex per filament). Common continuous filament yarns have total denier of 55, 60, 75, and 150 (6.1, 6.7, 8.3, and 16.7 tex, respectively). However, different fabric properties can be obtained by varying the filament count (dpf) to reach the total denier.

Though the cross-sectional shape of the spinneret hole directly influences the cross-sectional shape of the fiber, the shapes are not identical. Round holes produce filaments with an approximately round cross section, but with crenulated edges; triangular holes produce filaments in the form of a Y. Other cross sections can be achieved by adjusting the extrusion temperatures. Different cross sections are responsible for a variety of esthetics in the finished fabric such as hand, luster, or cover. Some yarn types may also include chemical additives to provide resistance to sunlight degradation and to give fire retardant properties. These additives are usually added to the acetate solution before spinning.

A metier is an array of individual extrusion positions on one common machine.

There are usually 100–200 such positions. The yarn is collected at each position on a package (eg, bobbin, tube, pirn) and removed from the machines at regular intervals to maintain a constant amount of yarn on each package. The package may contain 0.5–7 kg of yarn. As previously indicated, bobbin yarn may contain a low level of twist (about 0.08 turns per centimeter) whereas yarn taken up on tubes may have zero twist. The yarn is transferred from bobbins to different packages for sale. The product may contain twist levels of 0.3–8 turns per centimeter. Compacted yarn is presently more popular than low twist yarn.

Yarn Packages. The principal package types used by the textile industry are tubes, cones, and beams, although specialized packages are available for specific products.

Tubes are wrapped with 1.0–4.0 kg of yarn. The package is built on winders to provide package integrity and easy removal. Some packages are provided with a magazine wrap at the start of winding so that customers can automatically change packages. Zero-twist entangled yarn is packaged on tubes for use in circular knits and on tricot and section beams for warp knits.

Cones contain 0.5–4.0 kg of yarn. The tip of the cone tube must have a smooth finish to prevent damage to the yarn which is drawn over the top. Again, a magazine wrap may be provided for automatic package transfer. Both compacted and twisted yarns are packaged on cones.

Beams (large spools) are usually constructed of an aluminum alloy and vary from 50 to 170 cm in length and from 50 to 90 cm in flange diameter. A beam holds 100–700 kg of yarn. The beam most commonly used by the warp-knitting trade is 107 cm in length and 53–76 cm in dia. Section beams for weaving are usually 137 cm in length and 76 cm in dia. Both types of beams are parallel wound with a large number (up to 2400) of individual yarn ends. The length of yarn on beams varies with yarn denier, beam capacity, and intended use. Lengths are ordinarily in the range of 11,000–78,000 m. When beam winding is complete, the ends are taped in position and the beam is wrapped with a protective cover.

Staple and Tow. The same basic extrusion technology that produces continuous filament yarn also produces staple and tow. The principal difference is that spinnerets with more holes are used, and instead of winding the output of each spinneret on an individual package, the filaments from a number of spinnerets are gathered together into a ribbonlike strand, or tow. A mechanical crimping device uniformly plaits the tow into a carton from which it can be continuously withdrawn without tangling.

Staple is produced by cutting the tow (which may be crimped) into short lengths (usually 4–5 cm) resembling short, natural fibers. Acetate and triacetate staple is shipped in 180–365-kg bales. Conventional staple processing technology used with natural fibers is used to process acetate and triacetate staple into spun yarn.

Economic Aspects

Although cellulose acetate is the second oldest man-made fiber, it continues to be an important factor in the textile industry, with 317,000 metric tons produced worldwide in 1976 (Table 7) (76). Prior to the mid 1950s it had the second largest volume of the man-made fibers but it has subsequently been surpassed by polyester, nylon, and acrylic fibers (see Fibers, man-made and synthetic). It is sold at a relatively low price. Triacetate was introduced to the American market in 1954. The major textile

Table 7. World Production of Cellulose Acetate Textile Fibers (1968–1976)[a,b]

Year	Production, thousands of metric tons		
	Filament	Staple	Total
1968	388	35	423
1969	399	32	431
1970	402	25	427
1971	404	20	424
1972	377	21	398
1973	399	19	418
1974	359	14	373
1975	315	9	324
1976	310	7	317

[a] Textile Economics Bureau.
[b] Ref. 76.

applications of both acetate and triacetate fibers are in women's apparel and home furnishing fabrics (see Textiles). A list of trade names and manufacturers is shown in Table 8.

Although the use of acetate fiber for textile applications has generally declined, the production of cellulose acetate tow for cigarette filters rose from 136,000 to 257,000 metric tons in the 1969–1976 period (Table 9) (76). Because of its superior filtration, impact on cigarette taste, and cost, acetate is projected to supply up to 90% of the growing filter cigarette market.

A list of major acetate and triacetate producers, primary trade names, and production levels is given in ref. 76. The combined annual world acetate production (filament, staple, and tow) has been approximately 575,000 metric tons in the 1968–1976 period. Production in the United States accounts for approximately 51% of the total over that period. Other major acetate producing countries are the United Kingdom, Japan, and the Union of Soviet Socialist Republics.

Health Effects

Both acetate and triacetate remain undigested and cause no harmful reactions when ingested. Toxic effects, skin irritations, or allergic reactions attributable to acetate or triacetate fibers have never been reported.

Uses

The two major markets for cellulose acetate are textiles and cigarette filters. A unique combination of desirable esthetics and low cost accounts for the demand in textiles.

Textiles. Approximately 50% of the acetate and triacetate filament in the United States has been used for tricot knitting, 40% for woven fabrics, 5% for circular knits, and the remaining 5% for other applications (77). This distribution changes according to textile market trends. The principal markets are women's apparel (eg, dresses, blouses, lingerie, robes, housecoats, ribbons) and decorative household applications (eg, draperies, bedspreads, and ensembles). Acetate has been replacing rayon filament in liner fabrics for men's suits and has been evaluated for nonwovens (78–80) (see Nonwoven textiles; Textiles).

Table 8. Trade Names and Manufacturers of Acetate and Triacetate

Trade name[a]	Manufacturer and country
Acele	E. I. du Pont de Nemours & Co., Inc., U.S. (closed in 1977)
Acesil	Montedison Fibre S.p.A., Italy
Albene	Rhodiaseta, Argentina S.A., Argentina
Amcel	Amcel Europe S.A., Belgium
Arnel (ta)	Amcel Europe S.A., Belgium
Arnel (ta)	Celanese Canada Limited, Canada
Arnel (ta)	Celanese Fibers Company, U.S.
Ashi	Asahi Chisso Acetate Co., Ltd., Japan
Avisco	Avtex Fibers Inc., U.S.
Carolan (ta)	Mitsubishi Acetate Co., Ltd., Japan
Celanese	Celanese Fibers Company, U.S.
Celanese	Celanese Venezolana S.A., Venzuela
Celaren	Rayon y Celanese Peruana, S.A., Peru
Celcorta	Celanese Mexicana, S.A., Mexico
Chrysella	Courtaulds Ltd., Australia
Dicel	British Celanese Ltd. (subsidiary of Courtalds), U.K.
Eslon	Diacel Co., Ltd., Japan
Estron	Tennessee Eastman Co., (Division of Eastman Kodak Company), U.S.
Krasil	Ravi Rayon, Ltd., Pakistan
Lanalbene	Montedison Fibre S.p.A., Italy
Lonzona	Lonzana, Gesellschaft für Acetaprodukte m.b.H., FRG
Loteyarn	Teijin, Ltd., Japan
Novaceta	Montedison Fibre S.p.A., Italy
Rhodia	Deutsche Rhodiaceta A.G., FRG
Rhodia	Rhodia Industrias Quimical e Texteis, S.A., Brazil
Rhodia	Société Rhodiaceta, France
Rhodia	Deutsche Rhodiaceta A.G., FRG
Tri-A-Faser (ta)	Deutsche Rhodiaceta A.G., FRG
Tricel (ta)	British Celanese Ltd. (subsidiary of Courtalds), U.K.
Trilan (ta)	Celanese Canada Limited, Canada
Trilbene (ta)	Société Rhodiaceta, France
Velion	Industrias Del Acetato De Cellulosa, S.A., Spain

[a] ta, triacetate

Acetate and triacetate fibers have lower strength and abrasion resistance than most other man-made fibers and are frequently used with nylon or polyester in combination yarns. The latter can be used in markets formerly unavailable to 100% acetate fabrics (eg, men's shirts). Combination yarns can be prepared by twisting or by air entanglement and bulking. Yarns prepared by air-entanglement and bulking have unique characteristics and esthetics that permit their use in casement and upholstery fabric markets. With chemical additives, both acetate and triacetate fibers can pass current United States Government flame retardant fabric legislation (eg, DOC FF 3-71) (see Flame-retardant textiles).

Triacetate can be used in place of secondary acetate in many apparel applications but has the added advantage of ease-of-care properties. A particularly important application of triacetate is in velour and suede-like fabrics for robes and dresses. These fabrics offer superb esthetic qualities at reasonable cost. Triacetate is also desirable for print fabrics as it produces bright, sharp colors.

112 CELLULOSE ACETATE AND TRIACETATE FIBERS

Table 9. World Production of Cellulose Acetate Cigarette Filter Tow (1968–1976)[a], Thousands of Metric Tons

	1968	1970	1972	1974	1976
Europe[b]	25	35	43	48	59
United States	73	84	102	124	134
other American countries[c]	11	15	19	25	27
Asia[d]	19	20	26	30	37
Total	128	154	190	227	257

[a] Textile Economics Bureau.
[b] Belgium, France, FRG, U.K.
[c] Brazil, Canada, Colombia, Mexico, Venezuela.
[d] Japan, South Korea.

Cigarette Tow. Acetate fiber used in the production of cigarette filters is supplied in the form of tow (81). Tow is a continuous band composed of several thousand filaments held loosely together by crimp, a wave configuration set into the band during manufacture (Fig. 14). A tow is formed by combining the output of a large number of spinnerets and crimping the collection of filaments to create an integrated band of continuous fibers. The tow is then dried and baled. The wide range of available acetate filter tow products makes it possible to control selected properties in the finished cigarette filter rod.

Tow Properties. An individual tow item is characterized and identified by the following parameters, which are determined by controlling certain variables in the manufacturing process.

Cross Section. The shape of the filament cross section is related to the shape of the minute orifices in the spinneret used to form the filament. Filament cross sections currently used are shown in Figure 15.

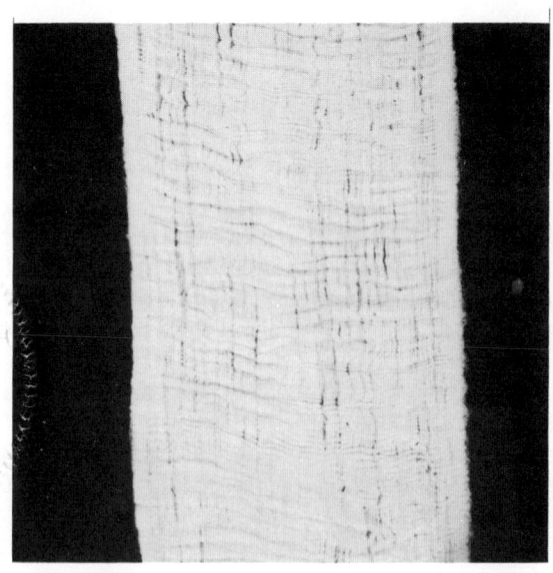

Figure 14. Acetate tow as supplied (81).

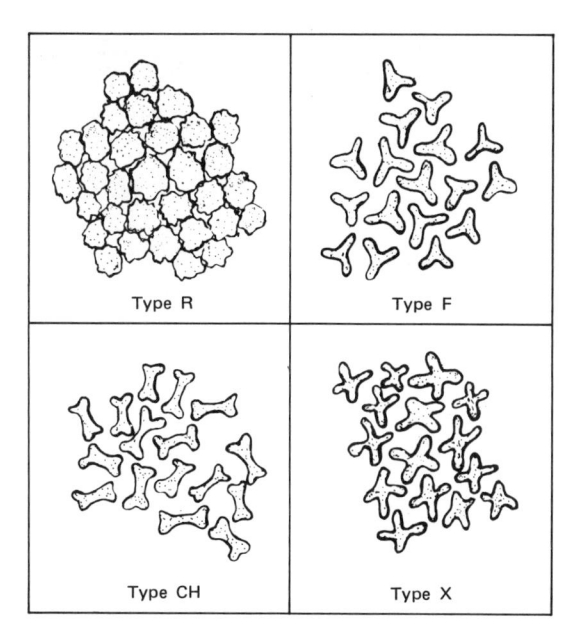

Figure 15. Filament cross sections of acetate filter tows (81).

Denier. Denier (1.111×10^{-7} kg/m or tex = 1.0×10^{-6} kg/m) is a measure of linear density, and is defined as the weight in grams of a 9000 m length of yarn. In filter-tow processing, there are several denier terms to consider: denier per filament (dpf) or 0.1111 tex per filament; total denier (TD) of the uncrimped tow (the product of the dpf multiplied by the number of filaments in the tow band), and crimped total denier (somewhat higher than the total denier).

A tow item described as 8.0 dpf (0.89 tex per filament), 50,000 TD, may therefore be interpreted as: an uncrimped tow band that weighs 50,000 g for each 9000 m of length and which is composed of 6250 individual filaments (50,000 ÷ 8.0) each weighing 8.0 g for each 9000 m in length.

Crimp. The crimp imparted to the tow is normally manifested by a sawtooth or sinusoidal wave shape. Because the filaments are usually crimped as a group, the crimp in parallel fibers is in lateral registry, ie, with the ridges and troughs of the waves aligned, as shown in Figure 16. The presence of crimp in the tow is necessary for two reasons: (*1*) to ensure that the tow can be packaged, processed, and handled easily, and (*2*) to impart bulk to the finished filter. To achieve the latter, it is necessary in production of the filters to open the tow band (Fig. 17) to the desired bulk so that the fibers completely fill the paper wrap without voids and soft spots. Several tow-opening systems are used to reposition the crimp out of lateral registry to create bulk while retaining the crimp in the individual filaments.

Tow Characterization. A linear relationship, constant for any given tow item, exists between the weight of tow in a cigarette filter rod and certain of its properties such as pressure drop and smoke removal efficiency. By preparing filter rods over a range of rod weights, eg, testing the rods for pressure drop, then plotting the results on rectilinear graph paper, a linear curve characteristic of the specific tow item is generated. This characterization provides a means of determining whether the required filter characteristics can be achieved, and if so, the weight of a specific tow item necessary

114   CELLULOSE ACETATE AND TRIACETATE FIBERS

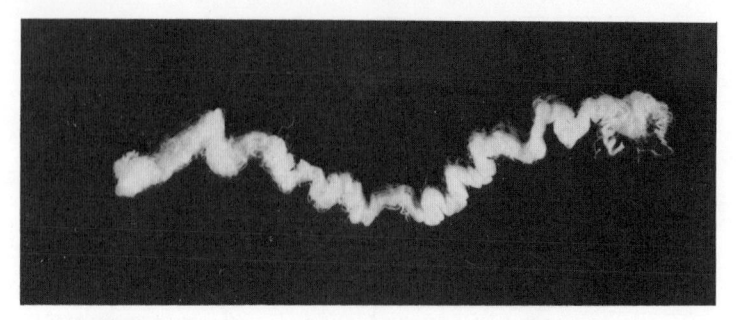

Figure 16. Section of tow showing crimp configuration (81).

Figure 17. Opened acetate tow (81).

to achieve them. By this kind of application, tows can be used to design cigarette filter rods with the desired performance properties.

Effect of Denier and Cross Section. Physical parameters of the tow (eg, denier per filament, total denier, crimp, and cross section) have a marked effect on the physical properties and performance of the finished filter rod.

Other Potential Applications. Additional applications for acetate and triacetate fibers, based on use of their unique properties as well as mastery of their inherent deficiencies, are being explored. Cellulose acetate has been a beneficial membrane material in reverse osmosis (qv) and hollow acetate fibers are now being explored for this application (82–85) (see Hollow-fiber membranes). Water soluble acetate fibers can be produced by selecting the appropriate acetyl value (86). Techniques of improving the antistatic characteristics of acetate fiber (87–93) which might make it competitive with cotton in critical antistatic applications (such as hospital operating rooms) are being developed (see Antistatic agents). Other studies aimed at improved water and solvent resistance (94–97), abrasion resistance (98–101), grafting (102), selective adsorption (103), and timed release of additives (104) offer the potential for developing future applications.

BIBLIOGRAPHY

"Acetate Fibers" treated in *ECT* 1st ed., under "Rayon and Acetate Fibers," Vol. 11, pp. 552–569, by G. W. Seymour and B. S. Sprague, Celanese Corporation of America; "Acetate and Triacetate Fibers" in *ECT* 2nd ed., Vol. 1, pp. 109–138, by L. I. Horner and A. F. Tesi, Celanese Fibers Company, Division of Celanese Corporation of America.

1. *J. Polym. Sci. C* **11**, 161 (1965).
2. C. J. Malm and L. J. Tanghe, *Tappi* **46**(10), 629 (1963).
3. N. F. Getchell and co-workers, *Am. Dyest. Rep.* **45**, 845 (1956).
4. A. Mellor and H. C. Olpin, *J. Soc. Dyers Colour.* **71**, 817 (1955).
5. *Triacetate-General Information and Physical and Chemical Properties, Technical Bulletin TBT30*, Celanese Fibers Marketing Co., Charlotte, N.C., 1974.
6. T. Vickerstaff and E. Water, Jr., *J. Soc. Dyers Colour.* **58**, 116 (1942).
7. C. L. Bird and co-workers, *J. Soc. Dyers Colour.* **70**, 68 (1954).
8. R. K. Fourness, *J. Soc. Dyers Colour.* **72**, 513 (1956).
9. F. Fortess and V. S. Salvin, *Tex. Res. J.* **28**, 1009 (1958).
10. F. Fortess, *Am. Dyest. Rep.* **44**, 524 (1955).
11. *The Physical and Chemical Properties and Dyeability Characteristics of Acetate Filament Yarns and Staple Fiber, Technical Bulletin TBA 8*, Celanese Fibers Marketing Co., Charlotte, N.C., 1974.
12. *Dyeing Triacetate Fabrics, Technical Bulletin TBT 24*, Celanese Fibers Marketing Co., Charlotte, N.C., 1972.
13. R. J. Mann, *J. Soc. Dyers Colour.* **76**, 665 (1960).
14. J. Boulton, *J. Soc. Dyers Colour.* **71**, 451 (1955).
15. F. Fortess, *Text. Res. J.* **19**, 23 (1949).
16. P. M. Heertjes, W. Colthof, and H. I. Waterman, *Rec. Trav. Chim.* **52**, 305 (1933).
17. C. J. Malm, C. R. Fordyce, and H. A. Tanner, *Ind. Eng. Chem.* **34**, 430 (1942).
18. D. K. Beever and L. Valentine, *J. Text. Inst.* **49**, T95 (1958).
19. R. K. Toner, C. F. Bowen, and J. C. Whitwell, *Text. Res. J.* **17**, 14 (1947).
20. *Tricel Technical Service Manual*, British Celanese, Ltd., Coventry, Eng., 1958.
21. BISFA, *Internationally Agreed Methods of Testing Regenerated Cellulose and Acetate Continuous Filament Yarn*, 1970 ed., Bureau International Pour La Standardisation De La Rayonne Et Des Fibres Synthetiques, Basel, Switz., 1971.
22. A. F. Tesi, *Am. Dyest. Rep.* **45**, 512 (1956).
23. L. G. Ray, Jr., *Text. Res. J.* **22**, 144 (1952).
24. H. M. Fletcher, *Am. Dyest. Rep.* **38**, 603 (1949).
25. M. L. Staples and C. J. Brown, *paper presented at the Fourth Canadian Textile Seminar, Queens College, Kingston, Ontario, 1954*, p. 132.
26. G. S. Egerton, *J. Soc. Dyers Colour.* **65**, 765 (1949).
27. L. Hochstaedter, *Text. Res. J.* **28**, 78 (1958).
28. M. Fels, *J. Text. Inst.* **51**, 648 (1960).
29. R. C. Harrington, Jr., and C. A. Jarrett, *Mod. Text.* (4), 67 (1963).
30. *Effect of Conditioning Humidity on the Electrical Resistance of Rayon Yarns*, British Cotton Industry Research Association, London, Eng., 1945.
31. J. C. Guthrie, *J. Text. Inst.* **48**(6), T193 (1957).
32. R. Meredith, *J. Text. Inst.* **37**, T107 (1945).
33. G. Susich and S. Backer, *Text. Res. J.* **21**, 482 (1951).
34. E. R. Kaswell, *Textile Fibers, Yarns, and Fabrics*, Reinhold Publishing Corp., New York, 1953, p. 57.
35. L. Segal, *Cellulose and Cellulose Derivatives* in N. M. Bikales and L. Segal, eds., *High Polymers Series*, Vol. V, Wiley-Interscience, New York, 1971, Chapt. XVII-A.
36. G. D. Hiatt and W. J. Rebel in ref 35, Chapt. VII-B.
37. C. L. Smart and C. N. Zellner in ref. 35, Chapt. XIX-C.
38. P. E. Gardner and M. Y. Chang, *Tappi* **57**(8), 71 (1974).
39. J. D. Wilson and R. S. Tabke, *Tappi* **57**(8), 77 (1974).
40. U.S. Pat. 3,846,403 (Nov. 5, 1974), K. B. Gibney and R. S. Evans (to Canadian Cellulose Company, Ltd.).

41. Can. Pat. 973,174 (Aug. 19, 1975), K. B. Gibney, B. E. Grisack, and R. S. Evans (to Canadian Cellulose Company, Ltd.).
42. Can. Pat. 975,764 (Oct. 7, 1975), K. B. Gibney and R. S. Evans (to Canadian Cellulose Company, Ltd.).
43. C. J. Malm and G. D. Hiatt, *Cellulose and Cellulose Derivatives* in E. Ott, H. M. Spurlin, and M. W. Graffin, eds., *High Polymers Series*, 2nd ed., Vol. V, Pt. II, Wiley-Interscience, New York, 1954.
44. E. Barabash, A. J. Rosenthal, and B. B. White, *Tappi* **38**(12), 745 (1955).
45. E. Dyer and H. D. Williams, *Tappi* **40**(1), 14 (1957).
46. A. Casadevall and co-workers, *Bull. Soc. Chim. Fr. Mem.*, 187 (1974); 196 (1964); 204 (1960); 719 (1970); 1850 (1970); 1856 (1970).
47. S. A. Kadyroba, *Dokl. Akad. Uzb. SSR* **26**(10), 29 (1969).
48. C. J. Clemett, *J. Chem. Soc. B,* 2202 (1971).
49. I. B. Gorovaya, N. A. Kozlov, and O. G. Tarakanov, *Zh. Prikl. Khim. (Leningrad)* **46**(4), 870 (1973).
50. U.S. Pat. 2,539,586 (Jan. 30, 1951), M. E. Martin, T. M. Andrews, and A. R. Franck (to Celanese Corporation).
51. U.S. Pat. 2,259,462 (Oct. 21, 1941), C. L. Fletcher (to Eastman Kodak Co.).
52. Brit. Pat. 566,863 (Feb. 26, 1945), H. Dreyfus.
53. U.S. Pat. 3,525,734 (Aug. 25, 1970), A. Rajon (to Société Rhodiaceta).
54. U.S. Pat. 2,603,634 (July 15, 1952), G. W. Seymour, B. B. White, and M. Plunguian (to Celanese Corporation).
55. U.S. Pat. 2,603,638 (July 15, 1952), G. W. Seymour, B. B. White, and M. Plunguian (to Celanese Corporation).
56. U.S. Pat. 2,731,247 (Jan. 17, 1956), C. Hudry (to Société Rhodiaceta).
57. U.S. Pat. 2,778,820 (Jan. 22, 1957), R. Clevy and J. Robin (to Société Rhodiaceta).
58. U.S. Pat. 2,790,796 (Apr. 30, 1957), R. Clevy and J. Robin (to Société Rhodiaceta).
59. U.S. Pat. 2,801,237 (July 30, 1957), R. Clevy and J. Robin (to Société Rhodiaceta).
60. U.S. Pat. 2,854,445 (Sept. 30, 1957), R. Clevy and J. Robin (to Société Rhodiaceta).
61. U.S. Pat. 2,854,446 (Sept. 30, 1957), R. Clevy and J. Robin (to Société Rhodiaceta).
62. U.S. Pat. 2,966,485 (Dec. 27, 1960), K. C. Laughlin, R. J. Osborne, and J. G. Santangelo (to Celanese Corporation).
63. Can. Pat. 609,900 (Dec. 6, 1960), K. C. Laughlin, R. J. Osborne, and J. G. Santangelo (to Celanese Corporation).
64. U.S. Pat. 3,040,027 (June 19, 1962), H. Bates, F. Hindley, and W. Popiolek (to British Celanese, Ltd.).
65. H. Genevray and J. Robin, *Pure Appl. Chem.* **14**, 489 (1967).
66. J. Corviere, *Faserforsch. Textiltech.* **22**, 71 (1971).
67. U.S. Pat. 3,631,023 (Dec. 28, 1971), C. Horne, Jr., and C. J. Howell, Jr., (to Celanese Corporation).
68. U.S. Pat. 3,755,297 (Aug. 28, 1973), K. C. Campbell and co-workers (to Celanese Corporation).
69. S. A. Kadyroba, *Prom. Arm.* **11**(A), 34 (1969).
70. Brit. Pat. 1,323,200 (July 11, 1973), F. M. Mikhalsky and co-workers.
71. U.S.S.R. Pat. 319,227 (C1.C08b) (Dec. 5, 1975), F. M. Mikhalsky and co-workers; *Otkrytiya Izobret. Prom. Obraztsy Tovarnye Znaki,* **52**(45), 185 (1975).
72. C. J. Malm and co-workers, *Ind. Eng. Chem.* **49**, 79 (1957).
73. U.S. Pat. 2,985,995 (May 30, 1961), W. W. Bunting, Jr., and T. L. Nelson (to E. I. du Pont de Nemours & Co., Inc.).
74. U.S. Pat. 3,110,151 (Nov. 12, 1963), W. W. Bunting, Jr., and T. L. Nelson (to E. I. du Pont de Nemours & Co., Inc.).
75. U.S. Pat. 3,364,537 (Jan. 23, 1968), W. W. Bunting, Jr., and T. L. Nelson (to E. I. du Pont de Nemours & Co., Inc.).
76. *Text. Organon* **48**, 82 (1977).
77. *Knitting Times,* (Mar. 8, 1976).
78. A. A. Lukoshaitis and co-workers, *Fibre Chem.* **6**, 441 (1974); Y. A. Matskevichene and co-workers, *Fibre Chem.* **6**, 446 (1974).
79. R. R. Rhinehart, *International Nonwoven and Disposable Association, Technical Symposium Paper,* 25 (Mar. 1975).
80. Ger. Pat. 2,502,519 (July 31, 1975), C. K. Arisaka and co-workers (to Diacel, Ltd.).
81. *World Smoking Products Technical Bulletin WSP 2.1, Acetate Tow Production and Characterization,* Celanese Fibers Marketing Co., Charlotte, N.C., 1974.

82. G. Rakhmanberdiev and co-workers, *Zh. Prikl. Khim. (Leningrad)* **46**(2), 416 (1973).
83. U.S. Pat. 3,763,299 (Oct. 2, 1973), W. S. Stephen (to FMC Company).
84. G. Rakhmanberdiev and co-workers, *Fibre Chem.* **6,** 219 (1974).
85. Brit. Pat. 1,418,115 (Dec. 17, 1975), R. L. Leonard (to Monsanto).
86. U.S. Pat. 3,482,011 (Dec. 2, 1969), T. C. Bohrer (to Celanese Corporation).
87. O. G. Pikovskaya and Z. G. Serebryakova, *Fibre Chem.* **2,** 378 (1970).
88. F. A. Ismailov and co-workers, *Fibre Chem.* **4,** 584 (1972).
89. P. A. Chakhoya and co-workers, *Fibre Chem.* **5,** 184 (1973).
90. A. V. Kuchmenko, *Fibre Chem.* **5,** 210 (1973).
91. Brit. Pat. 1,381,334 (May 31, 1973), W. Ueno, H. Kawaguchi, and N. Minagawa (to Fuji Photo Film Co., Ltd.).
92. A. P. Fedotov and co-workers, *Fibre Chem.* **6,** 87 (1974).
93. M. N. Skarnlite and Y. Y. Shlyazhas, *Fibre Chem.* **6,** 557 (1974).
94. *Text. World* **123**(9), 37 (1973).
95. U.S. Pat. 3,816,150 (June 11, 1974), K. Ishii and co-workers (to Daicel, Ltd.).
96. U.S. Pat. 3,839,517 (Oct. 1, 1974), A. F. Turbak and J. R. Thelman (to International Telephone and Telegraph Corp.—Rayonier).
97. U.S. Pat. 3,839,528 (Oct. 1, 1974), A. F. Turbak and J. R. Thelman (to International Telephone and Telegraph Corp.—Rayonier).
98. G. F. Kiseleva and co-workers, *Fibre Chem.* **4,** 257 (1972).
99. M. Papikyan and co-workers, *Fibre Chem.* **4,** 441 (1972).
100. M. Mirzaev and co-workers, *Khim. Volokna* (3), 21 (1975).
101. Brit. Pat. 1,414,395 (Nov. 19, 1975), N. V. Mikhailov and co-workers.
102. M. A. Siahkolah and W. K. Walsh, *Text. Res. J.* **44**(11), 895 (1974).
103. Ger. Pat. 2,507,551 (Feb. 28, 1975), B. V. Chandler and R. L. Johnson (to Commonwealth Scientific Org.).
104. U.S. Pat. 3,846,404 (Nov. 5, 1974), L. D. Nichols (to Moleculon Research Corp.).

GEORGE A. SERAD
J. R. SANDERS
Celanese Fibers Company

CELLULOSE DERIVATIVES, ESTERS

ORGANIC ESTERS

Cellulose acetate is by far the most important organic ester of cellulose owing to its broad applications in plastics and fibers (1–3). Industrial production of cellulose acetate began to replace the highly flammable cellulose nitrate as a coating for airplane wing and fuselage fabrics during World War I. Today, cellulose acetate is prepared with varying degrees of substitution, ranging from water soluble monoacetate to triacetate (see Cellulose acetate and triacetate fibers).

Although cellulose acetate remains the most widely used organic ester of cellulose, principally because of its use in fibers, its usefulness is restricted by moisture sensitivity, limited compatibility with plasticizers and other resins, and relatively high flow temperatures. Cellulose esters of higher aliphatic acids circumvent to some degree the shortcomings of cellulose acetate.

Simple triesters such as cellulose formate (4), cellulose propionate (5–6), and cellulose butyrate (7) have been prepared and studied, but are not produced in large quantities. Formate esters prepared by the reaction of cellulose with formic acid have been reported as thermally (8) and hydrolytically (4) unstable. Cellulose propionate and cellulose butyrate triesters can be synthesized by methods similar to those used in the preparation of cellulose acetate. However, mixed esters such as cellulose acetate propionate and cellulose acetate butyrate have desirable properties not possessed by either cellulose acetate or the higher acyl triesters.

There has been little interest in large-scale production of cellulose esters of aromatic acids. They are usually prepared from regenerated cellulose and their physical properties do not differ substantially from those of the more readily available aliphatic esters. In addition the chemicals used in their preparation are too expensive for large-scale production.

Benzoate esters have been prepared from regenerated cellulose with benzoyl chloride with either pyridine–nitrobenzene (9) or benzene (10) as the reaction solvent. The dibenzoate ester prepared from regenerated cellulose is soluble in common solvents such as acetone or chloroform. The benzoate ester as well as the nitro-, chloro-, and methoxy-substituted benzoates have been prepared using the appropriate aromatic acid and chloroacetic anhydride impelling agent with magnesium perchlorate catalyst (11).

Cellulose esters of unsaturated carboxylic acids such as cellulose methacrylate, cellulose acetate methacrylate, and cellulose propionate crotonate have all been prepared, but none have attained commercial importance (12).

Mixed esters such as cellulose acetate phthalate, prepared by the reaction of the free hydroxy group of a hydrolyzed cellulose acetate with phthalic anhydride, have commercially significant characteristics, such as alkaline solubility and excellent film-forming properties. These esters may be prepared by treating a hydrolyzed cellulose acetate with phthalic anhydride in pyridine solution (13–14) or in acetic acid using sodium acetate catalyst (15–17). In general, the solubilities of cellulose acetate phthalate esters in organic solvents approximate those of the simple cellulose acetates having comparable degrees of substitution. Salts of cellulose acetate phthalate are obtained by dispersing the ester in a water–alcohol mixture and adding the proper amounts of ammonium salt, alkali metal hydroxide, or lower molecular weight amine.

Cellulose acetate phthalates are suitable in enteric coatings for medicinal tablets because of their resistance to the acid conditions of the stomach and their solubility in the alkaline environment of the intestinal tract (18–19) (see Pharmaceuticals). The acetate phthalate esters are also used as antihalation backings in photographic film so that the colored backing layer can be removed during treatment in alkaline developer solution (20–21). Other applications such as components for photographic emulsions (22) and water-based paint formulations (23) have been suggested. Cellulose esters of aromatic acids and mixed cellulose esters of these acids with lower aliphatic acids have received considerable attention in the patent literature. Many of these products have unique properties, but for economic reasons, only a very few have broad industrial applications.

Physical and Chemical Properties

The common commercial cellulose acetates are primary (triacetate) and secondary (acetone-soluble). They are usually white, odorless, tasteless, and nontoxic materials. The properties of cellulose acetates are dictated by their combined acetic acid content (acetyl) and molecular weight. For example, the general solubility characteristics of cellulose acetates with various acetyl contents are as follows: (a) acetyl content above 43%—soluble in dichloromethane, insoluble in acetone; (b) acetyl content 37–42%—soluble in acetone, insoluble in dichloromethane; (c) acetyl content 24–32%—soluble in 2-methoxyethanol, insoluble in acetone; (d) acetyl content 13–14%—soluble in water, insoluble in 2-methoxyethanol; (e) acetyl content <13%—insoluble in all the solvents listed above (24).

Moisture regain and vapor permeability rate increase with decreasing acetyl content. The melting point range of secondary acetate is approximately 230–250°C; that of the triacetate is 270–300°C. The bulk density of cellulose acetate varies with the physical form and is 160–481 kg/m^3 (10–30 lb/ft^3). The specific gravity (1.29–1.30), refractive index (1.48), and dielectric strength of most commercial cellulose acetates are very similar.

In fibers, plastics, and film made from cellulose acetate, mechanical properties such as tensile strength, impact strength, elongation, and flexural strength vary with degree of polymerization (DP) and DP distribution. The mechanical properties improve significantly when the DP of cellulose acetate is increased from about 100 to 250.

The thermoplastic characteristics of cellulose acetate are greatly improved as the acetyl content (degree of substitution or DS) is increased from ca 1 to 2.25 max (25). Cellulose acetates are compatible with many liquid plasticizers, as eg, dimethyl, diethyl, and dibutyl phthalates; glyceryl triacetate; and triphenyl phosphate.

Some synthetic resins compatible with cellulose acetates are Acryloid A-10, Bakelite BR-3180, and Vinylite AYAA. The degree of compatibility of cellulose acetate with plasticizers and resins varies according to acetyl content and DP of the cellulose acetate. Detailed descriptions of the properties are given in refs. 26 and 27.

The number of acyl groups per anhydroglucose unit, acyl chain length, and the degree of polymerization, individually and collectively influence the properties of a cellulose ester. A summary of the properties of a number of cellulose triesters is given in Table 1 (28). In this series, with increasing acyl chain length from C_2–C_6, the melting point, tensile strength, mechanical strength, and density generally decrease, and solubility in nonpolar solvents and resistance to moisture increase. Fewer acyl groups per anhydroglucose unit (decreasing degree of substitution) normally increases the solubility of such esters in polar solvents and decreases their moisture resistance. The physical and chemical properties of mixed esters also vary according to acyl–acyl (eg, acetyl–propionyl) ratios of the ester. General trends for the properties of a mixed ester such as cellulose acetate butyrate are illustrated in Figure 1 (29). Increasing butyryl (decreasing acetyl) content increases flexibility, moisture resistance, and solubility, and decreases density and melting point. Gas permeability and crystalline properties are similarly influenced by the type and amount of acyl substitution. Oxygen permeability studies have shown that higher degrees of acyl substitution generally decrease the flow of oxygen through such esters (30). The degree of crystallinity of a cellulose ester is generally increased by increasing the degree of substitution and by decreasing the number of carbons in the aliphatic acid side chains (31).

Manufacture

All commercially important cellulose acetates, except the fibrous triacetate, are manufactured by a solution process using acetic anhydride in the presence of a suitable catalyst and a solvent such as acetic acid. The esterification of cellulose is a heterogeneous topochemical reaction. In order to prepare uniform products of a low degree of substitution (2.2 to 2.6), it is necessary first to prepare the triacetate and then to hydrolyze it in solution to the desired degree of substitution.

Cotton (qv) linters and purified wood pulps are the two major sources of cellulose for the manufacture of cellulose acetate. Acetate from cotton linters is of better color and solution clarity. However, much cellulose acetate is currently produced from wood pulps. Wood pulps suitable for acetylation were made formerly by the sulfite pulping process. However, pollution problems caused by production of sulfite pulps have sparked interest in sulfate or kraft pulps which are now expected to become more important in the production of cellulose acetates (see Cellulose; Pulp).

Catalysts for Acetylation. Sulfuric acid is the most popular esterification catalyst (32–33) and is considered best suited for the commercial manufacture of cellulose acetate (34). Zinc chloride has also been used (35), but the large quantities required make it uneconomical for most producers of cellulose esters. Perchloric acid is the most active catalyst, but its corrosive action on metal acetylation equipment and the danger of explosion during recovery of acetic acid have largely precluded its use. Other compounds such as methanesulfonic, methanedisulfonic, and sulfoacetic acids (34), sulfates containing metallic or ammonium ions (36), aniline sulfate, aniline perchlorate (37), sulfamic acid, ammonium sulfamate, or hydroxylamine (38), and orthotitanic acid (39–40) have also been used as catalysts for the acetylation of cellulose.

Activation. The course of the cellulose acetylation reaction is largely controlled by the rate of diffusion of the reagents into the cellulose fibers. Consequently, the cellulose should be activated before acetylation. The amount of activation needed depends on the source (ie, wood pulp or cotton linters) and the drying history of the cellulose. Cellulose that has never been dried to less than 5% moisture content or subjected to high-temperature drying needs little activation.

Cellulose may be activated with water or aqueous acetic acid. Water is effective

Table 1. Summary of the Properties of Selected Cellulose Triesters[a]

Ester	C atoms	Shrinking point, °C	Melting point, °C	Char point, °C	Water tolerance value[b]	25% rh	50% rh	75% rh	95% rh	Density, g/mL	Tensile strength, MPa[c]
cellulose[d]	0					5.4	10.8	15.5	30.5	1.52	
acetate	2	229	306	315	54.4	0.6	2.0	3.8	7.8	1.28	71.6
propionate	3	178	234	>315	26.9	0.1	0.5	1.5	2.4	1.23	48.0
butyrate	4	178	183	>315	16.1	0.1	0.2	0.7	1.0	1.17	30.4
valerate	5	119	122	>315	10.2	0	0.2	0.3	0.6	1.13	18.6
caproate	6	84	94	>315	5.88	0	0.1	0.2	0.4	1.10	13.7
heptylate	7	82	88	290	3.39	0	0.1	0.2	0.4	1.07	10.8
caprylate	8	82	86	315	1.14	0	0.1	0.1	0.2	1.05	8.8
caprate	10	87	88	310		0	0.1	0.2	0.5	1.02	6.9
laurate	12	89	91	>315		0	0.1	0.1	0.3	1.00	5.9
myristate	14	87	106	315		0	0.1	0.1	0.2	0.99	5.9
palmitate	16	90	105	315		0	0.1	0.1	0.2	0.99	4.9

[a] Ref. 28. Courtesy of the American Chemical Society.
[b] Milliliters of water required to start precipitation of ester from 125 mL of an acetone solution of 0.1% concentration.
[c] To convert MPa to psi, multiply by 145.
[d] Starting cellulose, prepared by deacetylation of commercial, medium-viscosity cellulose acetate (40.4% acetyl content).

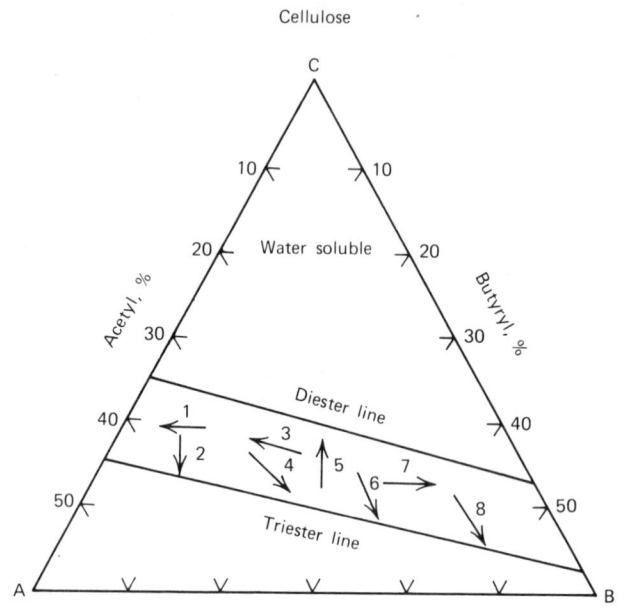

Figure 1. Effects of composition on physical properties. Apices: A = acetyl; B = butyryl; C = cellulose. 1, increased tensile strength, stiffness; 2, decreased moisture sorption; 3, increased melting point; 4, increased plasticizer compatibility; 5, increased solubilities in polar solvents; 6, increased solubilities in nonpolar solvents; 7, increased flexibility; 8, decreased density (29).

because it swells the fiber and partially destroys the hydrogen bonding between chains to increase surface area and accelerate the diffusion of acetylation reagents into the fiber. When cellulose is activated with water or aqueous acid, it must be dehydrated by displacing the water with acetic acid before the addition of the anhydride. The usual commercial practice is to activate cellulose by treatment with acetic acid containing a part of the total amount of catalyst, as the molecular weight of the cellulose must be reduced to obtain a satisfactory product.

The efficiency of activation depends on catalyst and water concentrations as well as on the duration and temperature of the treatment. Increasing water concentration of the wetting liquids has been reported to increase the efficiency of the cellulose activation (41). Significant increases in the acetylation reactivity of cellulose have been obtained by activating cellulose with an amine such as ethylenediamine (42–43), polyhydric alcohols, benzyl alcohol, or acetone (44), and by ultrasonic treatments (45).

Hydrolysis. The primary functions of hydrolysis are removal of some of the acetyl groups and the combined sulfate. Cellulose triacetate is normally hydrolyzed in an aqueous acid solution in the presence of catalyst (46–47). The rate of hydrolysis is controlled by temperature, catalyst concentration, and amount of water in the hydrolysis solution (48) and is determined by a solubility test based on samples of the acetate solution taken at intervals during the hydrolysis. The process is stopped by neutralizing the catalyst with magnesium, calcium, or sodium salts dissolved in aqueous acetic acid. Hydrolyses at temperatures as low as 38°C and as high as 229°C in an autoclave (49–50) and at 129°C in a continuous process (51) have been reported.

For commercial hydrolysis, the water content of the acid dope is usually 5–20%, depending upon the temperature of hydrolysis and the final product desired. An increase in water concentration results in a product of higher primary hydroxy content (52).

Optimum conditions for deacetylation and desulfation of cellulose acetate in aqueous acidic medium are reported in ref. 53.

Precipitation and Purification. Cellulose acetates can be precipitated in either powder or flake form by diluting the acid dope with 5–15% aqueous acetic acid. If a powder precipitate is desired, dilute acid is added to the acid dope with vigorous agitation until precipitation occurs. If a flake precipitate is desired, the acid dope is added to the dilute acetic acid. After precipitation, the acid (25–35%) is drained into the acid recovery system, and the acetate is purified by washing with water. The purified acetate is then stabilized and dried (see below).

Acetic Acid Recovery. Many kilograms of acetic acid are used for each kg of product in the manufacture of cellulose acetate. For economic reasons the acetic acid must be recovered. Extraction or azeotropic distillation is generally used for acid recovery and purification (see Azeotropic and extractive distillation). Methods for purification of acetic acid from aqueous solution are reviewed in ref. 54.

Trends in the Manufacture of Cellulose Acetate. Several continuous processes for the manufacture of cellulose acetate, including continuous cellulose activation, acetylation, hydrolysis, precipitation, washing, and drying, have been used (55–64). Cellulose acetate with improved solubility and filterability has been prepared from partially purified wood pulps by use of a lower alkanoic acid anhydride (which selectively removes hemicelluloses during a preesterification step) (65–66) or by a multistage addition of the esterification mixture (67). A nondegradative acetylation process which involves using acetic anhydride–pyridine reagent containing a catalytic amount of acetyl chloride (68), as well as a process for preparing cellulose acetate from particulate lignocellulosic material (69–70) has also been developed. Other methods for preparing cellulose acetate, such as high-temperature acetylation (71), a dichloromethane process (72), a fibrous esterification process (73–74), and acetylation by ketene have been reported (75–78) (see Cellulose acetate and triacetate fibers).

Manufacture of Other Esters. Cellulose propionate and cellulose butyrate may be prepared by esterification of cellulose with propionic or butyric anhydride in the presence of acid catalysts (79). These anhydrides normally react more slowly with cellulose than does acetic anhydride when sulfuric acid is the catalyst. Therefore, the cellulose must be efficiently activated and the esterification temperature well controlled to avoid substantial loss of molecular weight during the reaction (7). Reaction rates generally decrease with increasing acyl chain length, and cellulose chain degradation becomes more severe in the order: acetic < propionic < butyric < isobutyric anhydride.

Esterification of cellulose with isobutyric anhydride is normally so slow that highly activated cellulose must be used, and the sulfuric acid catalyst must be distributed uniformly throughout the cellulose. Swelling agents containing dissolved catalyst have been used to ensure uniform catalyst distribution for the preparation of isobutyrate esters. The swelling agent is removed and only isobutyric acid and its anhydride are present in the activated cellulose before esterification (80). Other methods of water-activating cellulose in which water is displaced by acetic acid and the latter with isobutyric acid are useful for the preparation of isobutyrate esters that are not excessively degraded (81–82).

Cellulose valerate has been synthesized by conventional methods using valeric anhydride and sulfuric acid catalyst, but higher ester homologues cannot normally be prepared by this method owing to excessive degradation of the cellulose and poor reactivity of the higher acid anhydrides with cellulose.

The mixed esters, cellulose acetate propionate and cellulose acetate butyrate, are produced in large volume by methods similar to those previously described for cellulose acetate (34). These esters are manufactured by combining the desired acyl components in the liquid esterification mixture in the form of acids or acid anhydrides. For example, if acetic anhydride is used in the bath with propionic (or butyric) acid, the cellulose ester product will contain both acyl moieties. It is also possible to esterify cellulose with propionic or butyric anhydride with acetic acid present in the esterification liquids to produce the mixed ester. The ratio of acetyl to higher acyl groups in the product is proportional to the concentration of the components in the esterification solution (Fig. 2) (83). The commercial production of cellulose acetate butyrate has been described (84), and differences in the reactivities of the lower aliphatic acid anhydrides have been investigated in some detail (85). Triesters of lower aliphatic acids, such as cellulose triacetate and cellulose tripropionate, have limited solubility in their esterification mixtures, and diluents such as dichloromethane can be used to prepare these products in solution (86). Dichloromethane has also been used as a diluent for the preparation of cellulose acetate butyrate with an increased degree of polymerization (87). Other mixed esters, including cellulose acetate valerate, propionate valerate, and butyrate valerate, have been prepared by conventional acid anhydride methods and sulfuric acid catalyst (88). The reaction mechanism of the sulfuric acid-catalyzed esterification begins with the formation of cellulose sulfate; transesterification to the appropriate aliphatic ester follows (86). Cellulose acetate isobutyrate (89) and cellulose propionate isobutyrate (90) have been prepared using zinc chloride to catalyze the esterification reaction. Large amounts of the catalyst are required to provide a soluble product, and special methods of anhydride addition are necessary to produce mixed esters containing the acetate moiety. Mixtures of sulfuric and perchloric acids are reportedly useful for the preparation of cellulose acetate propionate in dichloromethane solution at relatively low temperatures (91); such acid mixtures are normally considered too corrosive, however, for large-scale production of cellulose esters.

Cellulose esters are customarily hydrolyzed to eliminate small amounts of com-

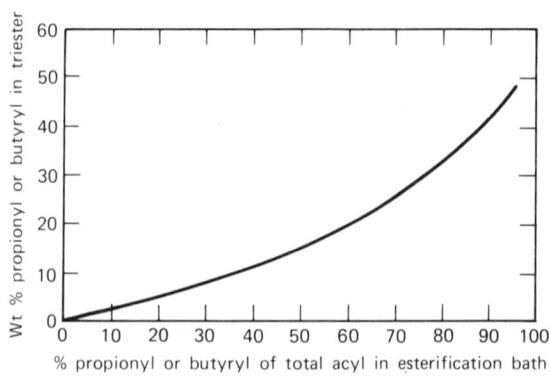

Figure 2. Relationship between composition of esterification bath and percent propionyl or butyryl introduced into the product (83).

bined sulfate and modify the properties of the ester by removing a small number of acyl groups. Sulfuric acid is a preferred hydrolysis catalyst as it is already present in the esterification liquid mixture. Esterification is stopped by the addition of water or a mixture of water and acid such as acetic acid, and the ester is subsequently hydrolyzed for a prescribed period. Partial neutralization of the hydrolysis catalyst may be necessary if large amounts of sulfuric acid are used during esterification. Increasing the amount of water in the hydrolysis solution generally decreases the rates of viscosity reduction and hydrolysis of cellulose propionate and cellulose butyrate esters (92). Several different methods of hydrolyzing esters of higher aliphatic acids have been described (93–94).

The progress of the hydrolysis reaction can be followed by taking appropriate test samples at prescribed intervals until the desired product is produced. Depending upon the degree of hydrolysis, the solution and compatibility properties of the ester can be varied over a wide range. For example, ester hydrolysis generally improves product solubility in polar solvents and reduces tolerance for nonpolar solvents. When a cellulose ester is used for applications that require high purity and/or clarity, the acid solution containing the hydrolyzed ester must be filtered before the ester is precipitated. Mixed esters such as cellulose acetate butyrate are precipitated by methods similar to those used for cellulose acetate. After the precipitated ester is washed free of any residual aliphatic acids, it may be stabilized and then dried.

Stabilization. Instability of commercially prepared cellulose acetate is caused primarily by the presence in the ester of small amounts of free and combined sulfuric acid. Sulfuric acid used in the production of cellulose acetate combines almost quantitatively with the cellulose and is removed during the later stages of acetylation and during hydrolysis (95–97).

Various methods for removing combined sulfuric acid from triacetate have been developed (98). The use of alkali metal salts such as magnesium, calcium, sodium, or aluminum to neutralize the combined and free sulfate at the end of hydrolysis effectively reduces residual sulfate in the hydrolyzed cellulose acetate. There are other stabilization processes, such as the use of boiling water, superheated steam treatment in an autoclave, or treating cellulose acetate with aqueous potassium and cadmium iodide solution (99).

Compounds such as chlorothiophenol (100), neopentyl phenyl phosphite (101), dialkyl esters of 3,3'-thiodipropionic acid (102), and potassium acid oxalate or citrate in combination with a phenol and magnesium stearate (103) are effective in stabilizing cellulose acetate against thermal degradation and discoloration. Improved thermal stability of cellulose acetate can be obtained by subjecting a cellulose acetate solution to an oxidant (eg, potassium permanganate), followed by reduction in solution with a reducing agent (eg, oxalic acid) (104).

Most cellulose esters are stabilized to minimize color development and prevent substantial loss of molecular weight during thermal processing operations such as extrusion and injection molding. It is often difficult to remove the last traces of sulfuric acid catalyst from the cellulose ester. Therefore the tendency of the residual mineral acid to cause degradation and/or discoloration of the ester during processing must be minimized.

Epoxy compounds are used as acid scavengers in end-use applications, especially in cellulose ester formulations subjected to high temperatures during molding or extrusion operations. Phenolic-type antioxidants (qv) are used widely to minimize oxidative degradation and discoloration of cellulose esters (105–108). In addition to antioxidants, some substituted phosphite compounds have been reported to prevent color formation during the thermal processing of cellulose acetate butyrate and cellulose acetate propionate (101,109). Exposure to radiation of less than 400 nm may cause chain scission and ultimate loss of physical properties in a cellulose ester. Consequently, cellulose esters, especially formulations used for outdoor applications, must be stabilized against ultraviolet degradation. Selected resorcinol and benzophenone derivatives such as resorcinol dibenzoate and 2-hydroxy-4-methoxybenzophenone are reported to be effective ultraviolet inhibitors (110). These compounds readily absorb ultraviolet radiation and subsequently dissipate its energy in a harmless manner.

Analytical Methods

   Degree of Substitution. The degree of substitution (DS) or acetyl content of cellulose acetate is determined by saponifying cellulose acetate with a known excess of standard sodium hydroxide solution in the presence of a swelling agent or solvent. The excess sodium hydroxide solution is then back-titrated with a standard solution of hydrochloric acid (111). The acetyl content of cellulose acetate may also be calculated by difference from the hydroxy content which is usually determined by carbanilation and acetylation (112) or near infrared spectroscopy. Methods using pyrolysis gas chromatography (113) and infrared spectroscopy (114) for acetyl determination have also been reported.
   Viscosity. The viscosity of cellulose acetate, a measure of the degree of polymerization (DP), is determined by the falling-ball method. Aluminum or stainless steel balls are usually used. The solvents used depend on the DS of the acetate but the concentration of the solution is normally 20% (111). Methods for the determination of intrinsic viscosity and its conversion to molecular weight have been investigated extensively (115–117).
   Heat Stability. The heat stability of cellulose acetate is determined by heating a weighed sample in a tube at a specified temperature for a specified time. The sample is then dissolved in a given amount of solvent and its viscosity and solution color determined. Solution color is usually determined spectroscopically (111) with the use of platinum–cobalt standards. Alternatively, heat stability may be determined by molding the acetate formulation into a plastic under standard conditions and measuring the color of the plastic or its viscosity when dissolved in a suitable solvent. DTA can also be used to determine the heat stability of cellulose acetate (118–119).
   DP and DS Distributions. DP distribution is commonly determined by gel permeation chromatography (120) or fractional precipitation and extraction (121). The distribution of acetyl groups among C_2, C_3, and C_6 carbon atoms in hydrolyzed cellulose acetate has been studied by an indirect chemical method (122). A direct and more accurate method using nmr spectroscopy has recently been described (123–126).
   ASTM standards for plastics have been adopted for the analysis of cellulose acetate plastic materials (127–128). The relative amounts of acetyl, propionyl, and butyryl in mixed esters may be determined by partition analysis (129). Sodium hydroxide is used to saponify the ester and phosphoric acid is added to liberate the organic acids from their sodium salts. The acids are partitioned with butyl acetate and their mole ratios are determined by comparison with standards. Gas chromatography

has also been used for the determination of acetyl, propionyl, and butyryl groups of cellulose esters by subjecting the esters to quantitative methanolysis and analyzing for the methyl esters of the acids (130).

Uses

Cellulose acetate is used in textile fibers, plastics, film, sheeting, and lacquers. Cellulose acetate film is used in photography and recording tape. The acetate used as photographic film base is now almost exclusively triacetate. Large quantities of cellulose acetate film are used in display packaging because of its excellent clarity (see Film and sheeting material). Cellulose acetate is also used as a plastic in injection molding and extrusion for various applications (131).

Low-viscosity acetate is used in lacquers in protective coatings for paper, metal, and glass. Low molecular weight acetate has been used as an adhesive for photographic films owing to its favorable bonding rate and peel strength (132–133). Heat-sensitive adhesives for textiles have also been prepared from cellulose acetate (134). Films or membranes (135–137) and hollow fibers (see Hollow-fiber membranes) (138) made from cellulose acetate are used as filter media in, for example, reverse osmosis (qv) membranes or dialysis membranes for artificial kidney machines (see Dialysis; Filtration; Membrane technology) (139). Selected physical properties of a typical cellulose acetate plastic are listed in Table 2 (140).

Cellulose acetate propionate and cellulose acetate butyrate have numerous applications such as sheeting compositions, molding plastics, film products, and lacquer formulations. The properties of propionate and acetate propionate esters ordinarily lie between those of cellulose acetates and acetate butyrates. Propionate esters traditionally have covered a narrow composition range. Moreover, through selected variations in acetyl and butyryl contents, a relatively small number (141) of commercially available cellulose acetate butyrate esters have been adopted for broad applications (Table 3). Cellulose acetate propionate and acetate butyrate are thermoplastic; properly formulated, these esters can be processed by methods such as injection molding and extrusion, and can be dissolved and cast from various solvents. The mixed esters possess excellent toughness and clarity, and are easily processed as they are compatible with a wide variety of plasticizers and synthetic resins. For example, acetate butyrate esters are compatible with a variety of polyester, acrylic, and alkyd resin plasticizers, depending on the amount of butyryl substitution and the degree of hydrolysis of the cellulose ester (142).

Cellulose acetate butyrate is extensively used in film-forming applications such as lacquers. The higher butyryl esters generally have greater tolerance for inexpensive lacquer solvents and common diluents (143). Moreover, lower viscosity, higher butyryl esters tolerate significant amounts of alcohol solvent without appreciable increase in solution viscosity. Acetate butyrate esters provide excellent barrier coating properties for molded polystyrene and many other plastics (144). These esters have also been formulated with acrylic polymers to provide weather-resistant automobile coatings which exhibit good color fastness, excellent pigment control, and good flow-out properties (145). Acetate butyrate esters have been used in hot-melt adhesive formulations (146) and have been electrostatically spray-coated on metals to form noncratering fusible films (147).

Certain properties, such as a high tolerance for alcohol solvents, excellent surface

Table 2. Physical Properties of Typical Cellulose Ester Plastics[a]

Processing conditions	Cellulose nitrate	Cellulose acetate Sheeting	Cellulose acetate Molding compd	Cellulose acetate propionate molding compd	Cellulose acetate butyrate Sheeting	Cellulose acetate butyrate Molding compd
melting temp, °C	671	230	230	190	140	140
compression molding temp, °C	85–121		127–216	129–204		129–199
compression molding pressure, MPa[b]	13.8–34.5		0.689–34.5	0.689–34.5		0.689–34.5
injection molding temp, °C			168–255	168–268		168–249
injection molding pressure, MPa[b]			55.2–221	55.2–221		55.2–221
specific gravity (density)	1.35–1.40	1.28–1.32	1.22–1.34	1.17–1.24	1.15–1.22	1.15–1.22
specific volume, cm³/kg	740–714	783–758	819–743	841–808	870–819	870–819
Mechanical properties						
tensile strength, MPa[b]	48.3–55.2	31.0–55.2	13.1–62.1	13.8–53.8	17.9–47.6	17.9–47.6
elongation, %	40–45	20–50	6–70	29–100	50–100	44–88
flexural strength, MPa[b]	62.0–75.8	41.0–68.9	13.8–110.3	20.0–78.6	27.6–62.0	12.4–64.1
impact strength, J/m[c] notched, Izod test	267–374		21.4–278	26.7–614		42.7–336
hardness, Rockwell	R95–R115	R85–R120	R34–R125	R10–R122	R30–R115	R31–R116
flexural modulus, 10² MPa[b]				10.3–23.4		
tensile modulus, 10² MPa[b]	13.1–15.2	20.7–41.4	4.48–27.6	4.14–14.8	13.8–17.2	3.44–13.8
Thermal properties						
deflection temp, °C						
at 1820 kPa[d]	60–71		44–91	44–109		45–94
at 455 kPa[d]			49–98	64–121		54–108
specific heat, J/(g·K)[c]	1.26–1.67	1.26–2.09	1.26–1.76	1.26–1.67	1.26–1.67	1.26–1.67
thermal expansion, 10⁻³ %/°C	8–12	10–15	8–18	11–17	11–17	11–17
Optical properties						
refractive index	1.49–1.51	1.49–1.50	1.46–1.50	1.46–1.49	1.46–1.49	1.46–1.49
transmittance, %		88		88	88	
haze, %		<1		<1	<1	

[a] Ref. 140. Courtesy of McGraw-Hill.
[b] To convert MPa to psi, multiply by 145.
[c] To convert J to cal, divide by 4.184.
[d] To convert kPa to atm, divide by 101.3.

Table 3. Applications of Commercial Grades of Cellulose Acetate Butyrate[a]

Acetyl content, %	Butyryl content, %	Hydroxy content, %	Degree of esterification (no. of groups/glucose residue)			Field of application
			Acetate	Butyrate	Hydroxy	
29.5	17	1	2.1	0.7	0.2	lacquers
20.5	26	2.5	1.4	1.1	0.5	lacquers
13	37	2	0.95	1.65	0.4	plastics, lacquers
6	48	1	0.5	2.3	0.2	melt coatings

[a] Ref. 141.

hardness, and relatively high melting points (of low-viscosity cellulose acetate propionate esters) make them useful in printing ink formulations. They can be used in flexographic and gravure inks without interfering with the repulpability of the printed paper (148). Acetate propionate esters also possess a combination of excellent clarity and high tensile strength which accommodates food-coating applications (see Food additives). Because they are nontoxic, they can be formulated and used in hot-melt dip coatings (149) or dissolved in volatile solvents and applied to food in the form of a lacquer coating (150). Cellulose esters, especially acetate butyrate and acetate propionate, have been used in a wide variety of specialty applications such as nonfogging optical sheeting (151) and several types of semipermeable membranes (152–154).

INORGANIC ESTERS

Cellulose Nitrate. Cellulose nitrate, first prepared in 1832 by Braconnot (155), is the most important and the only commercially available inorganic ester of cellulose (1–3). Schonbein (156) first developed the preferred method of manufacturing cellulose nitrate by using nitric–sulfuric acid mixtures. Cellulose nitrate was used initially for military explosives (gun cotton); later it was combined with camphor to produce the first successful synthetic plastic, Celluloid. Industrial lacquer finishes constitute the largest market for cellulose nitrates; explosives and propellants (qv) are the second largest market. Most recent developments in cellulose nitrate concern continuous nitration processes and newer, lower cost cellulose raw material sources.

No inorganic ester of cellulose other than cellulose nitrate achieved commercial significance. However, considerable interest and effort have been devoted to two other classes of inorganic derivatives: sulfur- and phosphorus-containing esters.

Sulfur-Containing Cellulose Esters. Cellulose dissolves when treated with 70–75% aqueous sulfuric acid and a highly degraded, thermally unstable cellulose sulfate ester can be obtained from a sufficiently aged solution by precipitation. A similar ester with less degradation can be prepared by the reaction of cellulose with chlorosulfonic acid (qv) in pyridine solvent. This ester is also very unstable thermally. Klug (157–158) employed mixtures of sulfuric acid and C_3–C_5 alcohols in an inert solvent (eg, toluene or carbon tetrachloride) to prepare uniform water-soluble cellulose sulfate without severe degradation. The concentration of water, alcohol, and chain length of alcohol used in the sulfation reaction greatly influence the reaction rate and degree of sulfate substitution (159).

Malm (160) successfully prepared cellulose sulfate (as the sodium salt) with excellent stability and without degradation by the reaction of cellulose at low temperature with mixtures of sulfuric acid and aliphatic alcohols (eg, isopropyl alcohol) followed by neutralization with aqueous sodium hydroxide. The water-soluble sodium cellulose sulfate was introduced commercially as a thickening agent and emulsion stabilizer, but was subsequently withdrawn from the market.

Cellulose sulfate with a very high degree of substitution and excellent stability can be prepared by the reaction of cotton linters or wood pulp with a complex of DMF–SO$_3$ in DMF solvent (161–162). Very little cellulose degradation occurs during reaction with the DMF–SO$_3$ complex and high-viscosity cellulose sulfate is obtained.

Cellulose acetate sulfate (163–165), acetate propionate sulfate (166), acetate butyrate sulfate (163–165), and ethyl cellulose sulfate (163–164,167) derivatives have been prepared, but have thus far not achieved commercial significance.

Cellulose sulfate and acyl sulfate derivatives have found limited use in such applications as detergents (168), antistatic coatings for photographic films (169), oil drilling fluids (170), and as thickening agents in food, cosmetics, and pharmaceuticals (171).

Phosphorus-Containing Cellulose Esters. Cellulose phosphate esters (as ammonium salt) can be prepared by the reaction of cellulose (cotton linters or wood pulp) with phosphoric acid in molten urea (172–174). These esters generally have low phosphorus content. Excess urea (174) and short reaction times (ca 15 min) at high temperature (ca 140°C) (173,175) favor higher phosphorus contents with less degradation of the cellulose phosphate esters. Stable, water-soluble cellulose phosphate with a high degree of substitution was prepared in a mixture of phosphoric acid, phosphorus pentoxide, and alcohol diluent (176).

Cellulose phosphites (177–178), phosphinates, and phosphonates (179) have been prepared, but are essentially laboratory curiosities.

Cellulose phosphate esters show considerable flame resistance and are of interest in textiles and paper manufacture. They also have been used to some extent in ion-exchange applications.

Physical Properties

General physical properties of cellulose nitrate are given in Table 4. The most important properties related to its use in lacquer coatings, inks, and adhesives are: solubility; viscosity; compatibility with plasticizers, resins, and pigments; and toughness. The three standard types (RS, AS, and SS) of commercially available cellulose nitrates are described in ref. 180 with viscosity and nitrogen content ranges. RS-type cellulose nitrate (11.8–12.2% nitrogen) is soluble in ketones, esters, and ether–alcohol mixtures and has a high tolerance for aromatic hydrocarbons. The compatibility of RS-type cellulose nitrate with many synthetic resins makes it the most widely used in lacquer-type coatings.

AS-type cellulose nitrate (11.3–11.7% nitrogen) is soluble in the same type of solvents as RS-type cellulose nitrate, but tolerates higher proportions of alcohols in the solvent blend.

Type SS cellulose nitrate (10.9–11.2% nitrogen) is soluble in alcohols (ethyl and isopropyl) and is more thermoplastic than the other types; it is used widely in flexographic inks for paper, foil, and other flexible substrates.

Table 4. Properties of Cellulose Nitrate[a]

odor	none
taste	none
color of film	water-white
clarity of film	excellent
specific gravity of cast film	1.58–1.65
refractive index, principal	1.51
light transmission, lower limit of substantially complete, nm	313
moisture absorption at 21°C in 24 h, 80% rh, %	1.0
water-vapor permeability at 21°C, $(g \cdot cm)/(cm^2 \cdot h) \times 10^{-4}$	2.8
sunlight, effect on discoloration	moderate
effect on embrittlement	moderate
aging, effect of	slight
Electrical properties	
dielectric constant at 25–30°C	
60 Hz	7–7.5
1,000 Hz	7
1,000,000 Hz	6
power factor at 25–30°C, %	
60 Hz	3–5
1,000 Hz	3–6
electric charge on rubbing with silk	negative
Mechanical properties	
tensile strength at 23°C, 50% rh, MPa[b]	62–110.3
elongation at 23°C and 50% rh, %	13–14
flexibility of 0.076–0.102 mm (3–4-mil) film, MIT double folds under 200-g tension	30–500
hardness, Sward, % of glass	90
softening-point range (Parr), °C	155–220
Solubility and compatibility characteristics	
solvents, principal types	esters, ketones, ether–alcohol mixtures
resins, examples of compatible types	almost all
plasticizers, examples of compatible types	almost all, including many vegetable oils
waxes and tars, examples of compatible types	none
compatible cellulose derivatives	ethyl cellulose, cellulose acetate, ethyl hydroxyethyl cellulose
Resistance to various substances	
water, cold	excellent
water, hot	excellent
acids, weak	fair
acids, strong	poor
alkalies, weak	poor
alkalies, strong	poor
alcohols	partly soluble
ketones	soluble
esters	soluble
hydrocarbons, aromatic	good
hydrocarbons, aliphatic	excellent
oils, mineral	excellent
oils, animal	good
oils, vegetable	fair to good

[a] Unplasticized RS* Nitrocellulose (Hercules Powder Co.), from ref. 180.
[b] To convert MPa to psi, multiply by 145.

Purification

The instability of cellulose nitrate is attributed mainly to residual traces of nitrating acids after washing and to unstable combined sulfate ester groups (181) which gradually cleave through hydrolysis. The water slurry of nitrate from either the batch or continuous manufacturing process is subjected to a purification process. The process consists of a series of boiling water and washing treatments at controlled pH designed to remove the last traces of nitrating acid and eliminate by hydrolysis the combined sulfate ester groups. Bennett (182–183) claimed that use of magnesium nitrate as a dehydrating agent in the nitration reaction mixture resulted in a very stable, sulfur-free cellulose nitrate. Magnesium nitrate also prevents gelatinization of the cellulose nitrate during reaction. Netzler (184) has reviewed and evaluated this process. For certain applications stressing low color, cellulose nitrate is bleached by potassium permanganate or sodium hypochlorite by procedures similar to those used in bleaching cellulose.

Manufacture

Prior to World War II, cellulose nitrate was manufactured primarily from cotton linters because the product and yields obtained were superior to those produced from less pure wood pulps and other cellulose sources. However, increased costs and limited availability of cotton linters resulted in the development of methods to prepare uniform and stable cellulose nitrate from wood pulps. Until recently, cellulose nitrate was manufactured only by a batch process (Fig. 3) in which a rather small quantity of cellulose (ca 15 kg wood pulp or cotton linters) is agitated vigorously in a vessel containing ca 800 kg of HNO_3/H_2SO_4 nitrating mixture. The proper $HNO_3:H_2SO_4$ ratio

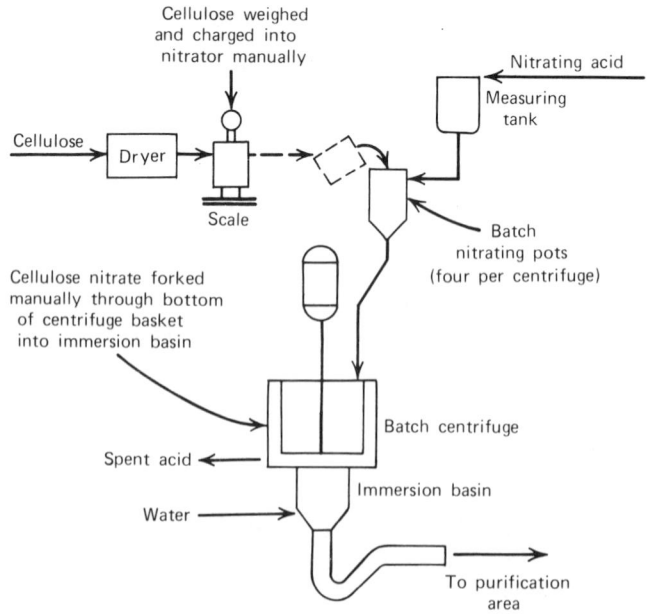

Figure 3. Flow diagram of nitration by batch process (180).

is chosen (Fig. 4 and Table 5) to give the desired degree of nitration and to maintain fibrous structure during the reaction. After 20–30 min of nitration, the mixture is centrifuged and the cellulose nitrate is quickly drowned in a large volume of water. The water slurry is pumped to the purification area, and the spent nitration acids recovered by centrifugation are reconstituted and recycled for further use.

Considerable effort (185–188) has been directed toward the development of a successful continuous process for cellulose nitration (Fig. 5) which results in a more uniform product than that obtained from the batch process. In the continuous process, cellulose and mixed nitrating acids are fed simultaneously into a vessel where nitration occurs. The nitration mixture is continuously withdrawn from the final zone of the reaction vessel and fed to a centrifuge which separates the spent acids and in which the cellulose nitrate is washed in stages with progressively weaker aqueous acid. Finally, a water slurry of cellulose nitrate results and is pumped to the purification area. The spent acids from nitration and the aqueous acid washes are recovered for reuse. Lewis (189) has described a pollution-free continuous, closed-system cellulose nitration process. Ramsey (190) continuously nitrated cellulose strips by drawing them through a bath of nitric acid, phosphoric acid, and water. Balakhnichev (191) also has described a continuous flow apparatus for the cellulose nitration.

The predominant stabilization of cellulose nitrate occurs during the manufacturing processes. However, additional thermal and light stabilization may be required in some uses. Certain organic acids such as citric, tartaric, stearic (193–194), and oxalic (195) acids have been used to stabilize cellulose nitrate against thermal degradation and discoloration during aging. Heat stabilization can be aided by the incorporation

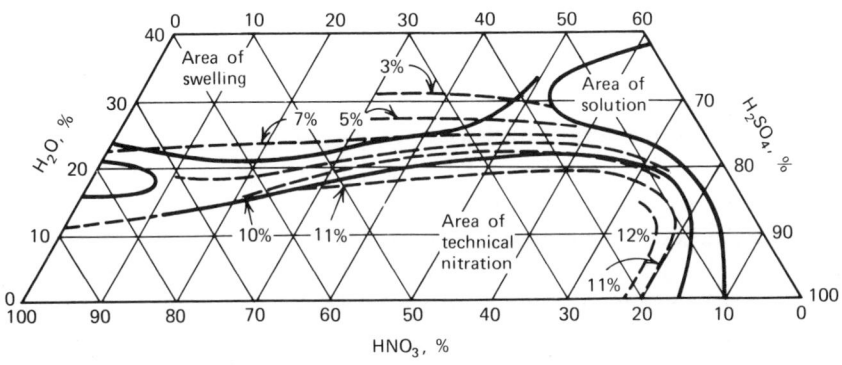

Figure 4. Effect of nitrating acid composition on the nitrogen content of cellulose nitrate (181).

Table 5. Variation of Ester Nitrogen Content With Water Content of Nitrating Bath[a]

Composition of nitrating bath			Cellulose nitrate ester	
Water, %	Sulfuric acid, %	Nitric acid, %	Nitrogen content, %	DS
12.0	66.0	22.0	13.2	2.6
16.0	64.0	20.0	12.5	2.4
21.0	60.0	21.0	10.6	1.9

[a] Ref. 192.

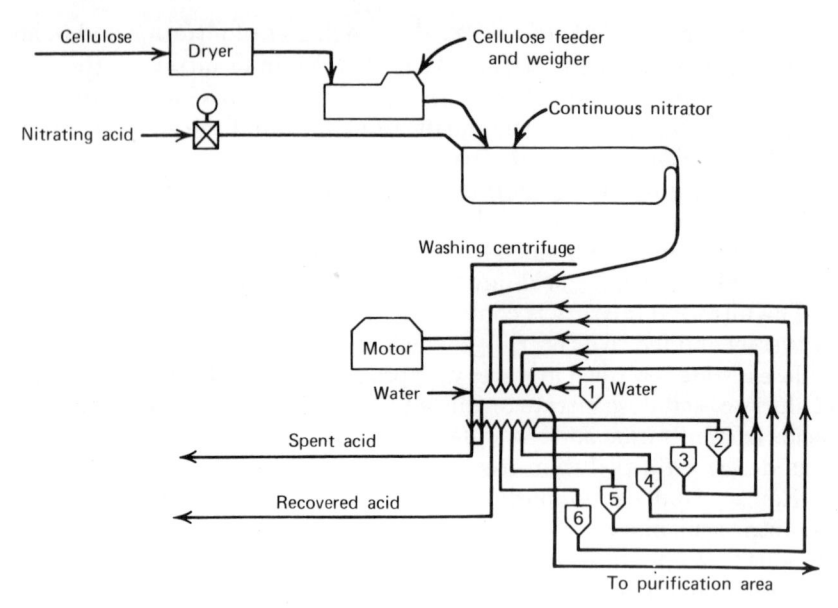

Figure 5. Flow diagram of nitration of cellulose by Hercules continuous process (180).

of stabilizers such as diphenylamine and epoxy compounds (196) (see Heat stabilizers). Discoloration during exposure to light can be reduced by addition of a small amount of dibenzoylmethane (197).

Digestion

Cellulose nitrates are manufactured over viscosities ranging from 0.25 to 5000 s, depending upon the expected use (180). The digestion process is a special treatment given to cellulose nitrate to produce the low-viscosity types used in lacquers. In continuous digestion, cellulose nitrate is heated at 132°C under pressure in water and the heated slurry is pumped through very long coils of piping (ca 1200 m) until the proper viscosity is achieved (Fig. 6). Additional washing is necessary to remove any decomposition products generated during digestion.

Dehydration

Dry cellulose nitrate is extremely flammable; cellulose nitrates with high nitrogen content may explode when subjected to heat or sudden shock. It is, therefore, necessary to ship and handle cellulose nitrate wet with water or, most often, alcohol (ethyl or isopropyl). In the dehydration process, most of the water is removed by pressing, and alcohol is pumped through the compressed cake to displace the remaining water. The alcohol-wet cake is pressed to ca 30–35 wt % alcohol and shredded before packaging and shipment. The alcohol-wet cellulose nitrate fibers are more readily soluble in organic lacquer solvents than dry cellulose nitrate (180).

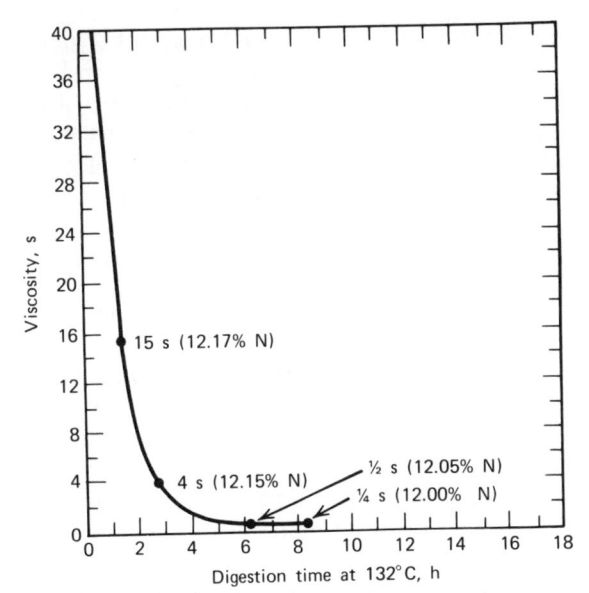

Figure 6. Rate of viscosity reduction of cellulose nitrate on digestion in water at 132°C. The viscosities (falling ball method) were determined in 12.2% solution (180).

Analytical Methods

Specifications and tests for soluble cellulose nitrate have been adopted by ASTM (198). Only brief descriptions of the most important tests are included here.

Flammability increases drastically with dryness, and great care should be exercised when drying cellulose nitrates for analysis. Small quantities of cellulose nitrate (<20 g) for analytical purposes are conveniently dried by spreading the sample in a thin layer on a tray and air drying at room temperature for 12–16 h followed by oven drying at 100–105°C for 1 h. For safety reasons, the door latch of the oven should be removed. Larger samples should be dried by passing a stream of warm compressed air at 60–65°C through the sample held in a container with a screen over one end. Air temperature should not exceed 70°C.

To determine the nitrogen content of cellulose nitrate, a dry sample (1–1.2 g) is decomposed with sulfuric acid over mercury to liberate NO gas, the volume of which is measured and converted to percent nitrogen by means of a calibration curve. The specially designed apparatus (nitrometer) is carefully calibrated with a known quantity of potassium nitrate prior to the cellulose nitrate analysis.

Heat stability is a measure of the relative time required for a specific amount of material held at 134.5°C to decompose and discolor methyl violet test paper. The methyl violet test paper, held 2.5 cm above the top of the sample during heating, should not discolor completely in less than 25 min.

Solution viscosity is determined in solvent blends of toluene, ethyl alcohol, and ethyl acetate. The solvent blend and concentration of cellulose nitrate varies according to the viscosity and nitrogen content of the cellulose nitrate being tested. Viscosities are measured at 25°C by the falling-ball method; the results are ordinarily expressed in terms of seconds required for a 2.4 mm diameter steel ball to fall a distance of 5.08 cm through the designated solution.

Toluene dilution is a measure of the least amount of toluene required to effect separation or precipitation of cellulose nitrate dissolved in butyl acetate solvent. Toluene is added in small increments from a buret to 50 mL of 12.2 wt % solution of cellulose nitrate in butyl acetate. The solution is agitated vigorously after each incremental addition, and the first permanent turbidity is noted. Dilution value is expressed as a percentage by volume of the original solution.

Uses

The largest industrial use of cellulose nitrate is lacquer coatings for decorative and protective purposes. Cellulose nitrates are soluble in a wide variety of organic solvents, yield clear, tough films, and are compatible with many plasticizers (eg, dioctyl phthalate and castor oil) and resins (oil-modified alkyds, acrylics, polyesters, and vinyls). Automobile and wood furniture coatings consume the greatest amount of cellulose nitrate-based lacquers. Rotogravure and flexographic inks consume large quantities of cellulose nitrate because of its toughness and fast drying characteristics. Cellulose nitrate of 10.9–11.2% nitrogen content (SS type) is preferred in flexographic inks owing to its solubility in solvent systems with high alcohol (ethyl or isopropyl) content. A substantial amount of cellulose nitrate is used in coatings for leather finishes (shoes, luggage, upholstery), fabrics (book bindings, artificial leather), nail polishes, and household adhesives.

Cellulose nitrates for use in explosives and propellants are generally higher in viscosity and nitrogen content (12.6–13.4% nitrogen) than are industrial coatings types. Cellulose nitrate has greater energy potential when combined with nitroglycerin (double-base explosives) than alone (single-base). Double-base cellulose nitrate powders (smokeless powders) are used extensively in sporting ammunition and various military explosives and propellants. Very high molecular weight (ca 3000 DP) cellulose nitrate (12.2–12.35% nitrogen) gelled with nitroglycerin is the base for dynamite blasting gelatins used for industrial explosives. An apparatus and method for the semicontinuous manufacture of dynamite-type explosives was described recently (199) (see Explosives and propellants).

Celluloid plastic, prepared by dispersing ca 25% camphor in cellulose nitrate (10.9–11.2% nitrogen) was the earliest successful synthetic plastic (200). Celluloid plastic has some very highly desirable physical properties and performance characteristics in spite of its flammability and relatively high cost of manufacture. Table 2 lists the physical properties of cellulose nitrate plastics and those of other cellulose derivatives for comparison. For the most part, cellulose nitrate plastics have been supplanted by less flammable thermoplastic materials, but celluloid plastic is still used in the manufacture of plastic novelties and as combustible wads and spacers in sporting ammunition cartridges.

Cellulose nitrate was used extensively as a photographic film base for professional, amateur, and x-ray films for many years, but it has been replaced by less flammable cellulose derivatives, eg, cellulose acetate. It is still used as an intercoating on photographic film to improve adhesion of the silver halide emulsion coating to the film base (201).

Very thin films of cellulose nitrate (ca 10-μm thick) supported on polyester base (202–205) are used in fast-neutron detection applications, and are commercially available. After exposure, the cellulose nitrate films are etched (developed) with aqueous sodium hydroxide to reveal the tracks generated by the α-particles.

Recently, Riley (206–207) prepared thin desalination membranes based on combinations of cellulose nitrate and cellulose acetate which showed excellent salt rejection and flow rate properties (see Hollow-fiber membranes).

Economic Aspects

Cellulose ester flake production in the United States showed relatively steady growth from 1960 until 1969, when total flake output reached 466,000 metric tons (Fig. 7), but has been somewhat erratic, showing significant drops in 1971 and 1974. The estimated production of cellulose ester flake in the United States was about 380,000 metric tons in 1975. Cellulose ester flake production for nonfiber applications represents a rather small part of the total U.S. output. Cellulose acetate-based textile fibers and cigarette filter tow consume more than 80% of the total flake produced.

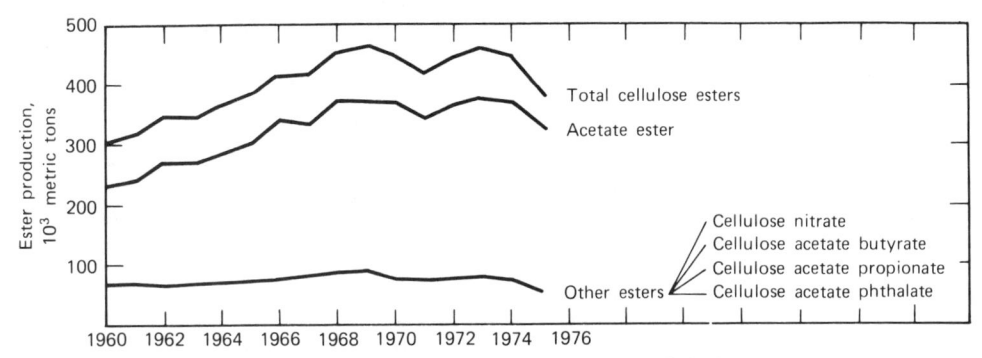

Figure 7. United States production of cellulose ester flake for 1960–1975.

Since 1960, combined annual production of mixed esters and cellulose nitrate in the United States has not exceeded 100,000 metric tons.

The largest current market for cellulose ester plastics is the production of film and sheeting products. Molding applications account for a smaller part of the market. Demand for cellulose nitrate-based plastics has declined sharply, and is expected to decline further. Nevertheless, cellulose nitrate continues as a primary ingredient in lacquer formulations for surface coatings. Cellulose acetate and cellulose acetate butyrate are also used in lacquer formulations both as film formers and as modifiers for other resin systems. The prices of cellulose ester flake have generally increased in line with the general economy of the United States. Resin prices are currently $1.63–3.00/kg, depending on the type of resin. Cellulose acetate and cellulose nitrate are generally lower in cost than the specialty type cellulose resins based on cellulose acetate propionate and cellulose acetate butyrate.

In the United States, plastic-grade cellulose acetate flake is manufactured by E. I. du Pont de Nemours & Company, Inc., and by the Eastman Chemicals Division of Eastman Kodak Company. Celanese Corporation of America markets a fiber-grade cellulose acetate flake that is also suitable for plastics applications. Plastic-grade flakes based on cellulose acetate propionate and cellulose acetate butyrate are produced by the Eastman Chemicals Division. In Western Europe, cellulose acetate, acetate propionate, and acetate butyrate flakes are manufactured by Farbenfabriken Bayer A.G. in the Federal Republic of Germany. Other producers of cellulose acetate flake include Gevaert and Fabelta in Belgium, Rhone-Poulenc S.A. in France, British

138　CELLULOSE DERIVATIVES, ESTERS

Hercules, Ltd. and Courtaulds in Great Britain, Mazzucchelli Celluloide SpA and Montedison in Italy, Daicel and Teijin in Japan, and Celanese in Canada and Mexico. Cellulose nitrate resins are manufactured by Hercules, Inc., and by E. I. du Pont de Nemours & Company, Inc. in the United States and by others in Europe and Japan.

The Eastman Chemicals Division produces cellulose acetate and mixed esters for use in coatings applications.

Table 6 gives an alphabetical list of cellulose esters referred to in text.

Table 6. Alphabetical List of Cellulose Esters Referred to in Text

cellulose acetate	[9004-35-7]	cellulose propionate	[9004-48-2]
cellulose acetate butyrate	[9004-36-8]	cellulose propionate crotonate	[9015-16-4]
cellulose acetate butyrate sulfate	[57485-48-0]	cellulose propionate isobutyrate	[67351-40-6]
cellulose acetate isobutyrate	[67351-38-6]	cellulose propionate valerate	[67351-41-1]
cellulose acetate methacrylate	[9032-34-2]	cellulose sulfate	[9032-43-3]
cellulose acetate phthalate	[9004-38-0]	cellulose sulfate, sodium salt	[9005-22-5]
cellulose acetate propionate	[9004-39-1]	cellulose triacetate	[9012-09-3]
cellulose acetate propionate sulfate	[67351-39-7]	cellulose tributyrate	[9015-12-7]
cellulose acetate sulfate	[51910-28-2]	cellulose tricaprate	[9056-00-2]
cellulose acetate valerate	[55962-79-3]	cellulose tricaproate	[39320-17-7]
cellulose benzoate	[9032-47-7]	cellulose tricaprylate	[67382-71-2]
cellulose butyrate	[9015-12-7]	cellulose triformate	[9036-95-7]
cellulose butyrate valerate	[53568-56-2]	cellulose triheptanoate	[67351-35-3]
cellulose methacrylate	[50823-06-3]	cellulose trilaurate	[67351-36-4]
cellulose nitrate	[9004-70-0]	cellulose trimyristate	[67351-33-1]
cellulose phosphate	[9015-14-9]	cellulose trinitrate	[9046-47-3]
cellulose phosphate, ammonium salt	[9038-38-4]	cellulose tripropionate	[9004-48-2]
		cellulose tripalmitate	[67351-34-2]
cellulose phosphinate	[67351-37-5]	cellulose trivalerate	[39320-21-3]
cellulose phosphite	[37264-91-8]	cellulose valerate	[55962-75-9]
cellulose phosphonate	[37264-91-8]	ethyl cellulose sulfate	[37291-30-8]

BIBLIOGRAPHY

The "Cellulose Derivatives, Esters" are treated in *ECT* 1st ed., under "Cellulose Derivatives, Plastics", Vol. 3, pp. 391–411, by W. O. Bracken, Hercules Powder Company; "Cellulose Derivatives, Esters" in *ECT* 2nd ed., under "Cellulose Derivatives, Plastics," Vol. 4, pp. 653–683 by Ben P. Rouse, Jr., Tennessee Eastman Company.

1. N. M. Bikales, ed., *Encyclopedia of Polymer Science and Technology,* Vol. 3, Interscience Publishers, a division of John Wiley & Sons, Inc., New York, 1965.
2. E. Ott, H. M. Spurlin, and M. W. Grafflin, eds., *Cellulose and Cellulose Derivatives,* Part II of *High Polymers,* Vol. 5, 2nd ed., Interscience Publishers, New York, 1954.
3. N. M. Bikales and L. Segal, eds., *Cellulose and Cellulose Derivatives,* Part V of *High Polymers,* Vol. 5, 2nd ed., Wiley-Interscience, New York, 1971.
4. C. J. Malm and G. D. Hiatt in E. Ott, H. M. Spurlin, and M. W. Grafflin, eds., *Cellulose and Cellulose Derivatives,* Part II of *High Polymers,* Vol. 5, 2nd ed., Interscience Publishers, New York, 1954, p. 766.
5. C. J. Malm, *Sven. Kem. Tidskr.* **73,** 523 (1961).
6. U.S. Pat. 2,600,716 (June 17, 1952), B. B. White, L. J. Rosen, and P. Blackman (to Celanese Corp.).
7. C. J. Malm and co-workers, *Ind. Eng. Chem.* **50,** 1061 (1958).
8. G. Tocco, *G. Chim. Ind. Appl.* **13,** 325 (1931).

9. A. Wohl, *Z. Angew. Chem.* **26,** 437 (1913).
10. K. Atsuki and K. Shimoyama, *Cellul. Ind. Tokyo* **2,** 336 (1926).
11. U.S. Pat. 1,704,283 (March 5, 1929), H. T. Clarke and C. J. Malm (to Eastman Kodak Co.); Brit. Pat. 313,408 (Aug. 27, 1929); Fr. Pat. 653,742 (Dec. 13, 1929).
12. G. D. Hiatt and W. J. Rebel in N. M. Bikales and L. Segal, eds., *Cellulose and Cellulose Derivatives*, Part V of *High Polymers*, Vol. 5, 2nd ed., Wiley-Interscience, New York, 1971, p. 770.
13. U.S. Pat. 2,093,462 (Sept. 21, 1937), C. J. Malm and C. E. Waring (to Eastman Kodak Co.).
14. C. J. Malm and C. R. Fordyce, *Ind. Eng. Chem.* **32,** 405 (1940).
15. C. J. Malm and co-workers, *Ind. Eng. Chem.* **49,** 84 (1957).
16. U.S. Pat. 2,856,400 (Oct. 14, 1958), C. J. Malm and C. L. Crane (to Eastman Kodak Co.).
17. U.S. Pat. 2,856,399 (Oct. 14, 1958), J. W. Mench, B. F. Fulkerson, and G. B. Lapham (to Eastman Kodak Co.).
18. C. J. Malm, J. Emerson, and G. D. Hiatt, *J. Am. Pharm. Assoc. Sci. Ed.* **40,** 520 (1951).
19. U.S. Pat. 2,196,768 (April 9, 1940), G. D. Hiatt (to Eastman Kodak Co.).
20. U.S. Pat. 2,000,587 (May 7, 1935), C. R. Fordyce (to Eastman Kodak Co.).
21. U.S. Pat. 1,954,337 (April 10, 1934), and U.S. Pat. 2,271,234 (Jan. 24, 1942), C. J. Staud (to Eastman Kodak Co.).
22. Brit. Pat. 724,827 (Feb. 23, 1955), R. T. Talbot and T. J. McCleary (to Eastman Kodak Co.).
23. U.S. Pat. 2,338,580 (Jan. 4, 1944), C. R. Fordyce (to Eastman Kodak Co.).
24. B. P. Rouse, Jr., in N. M. Bikales, ed., *Encyclopedia of Polymer Science and Technology*, Vol. 3, Interscience Publishers, a division of John Wiley & Sons, Inc., New York, 1965, p. 344.
25. J. J. Creely, P. Harbrink, and C. C. Conrad, *J. Appl. Polym. Sci.* **19,** 1533 (1965).
26. C. J. Malm and G. D. Hiatt in E. Ott, H. M. Spurlin, and M. W. Grafflin, eds., *Cellulose and Cellulose Derivatives*, Part II of *High Polymers*, Vol. 5, 2nd ed., Interscience Publishers, New York, 1954, pp. 791–805.
27. *Eastman Cellulose Acetate*, Eastman Chemical Product, Inc., Kingsport, Tennessee, 1962.
28. C. J. Malm and G. D. Hiatt in E. Ott, H. M. Spurlin, and M. W. Grafflin, eds., *Cellulose and Cellulose Derivatives*, Part II of *High Polymers*, Vol. 5, 2nd ed., Interscience Publishers, New York, 1954, p. 791, *Ind. Eng. Chem.*, **43,** 688 (1951).
29. C. J. Malm and G. D. Hiatt in E. Ott, H. M. Spurlin, and M. W. Grafflin, eds., *Cellulose and Cellulose Derivatives*, Part II of *High Polymers*, Vol. 5, 2nd ed., Interscience Publishers, New York, 1954, p. 798.
30. T. V. Shakina and co-workers, *Vysokomol. Soedin. Ser. B* **12,** 174 (1970).
31. R. E. Boy, Jr., R. M. Schulken, and J. W. Tamblyn, *J. Appl. Polym. Sci.* **11,** 2453 (1967).
32. C. J. Malm and G. D. Hiatt in E. Ott, H. M. Spurlin, and M. W. Grafflin, eds., *Cellulose and Cellulose Derivatives*, Part II of *High Polymers*, Vol. 5, 2nd ed., Interscience Publishers, New York, 1954.
33. C. J. Malm, *Sven. Papperstidn.* **64,** 740 (1961).
34. C. J. Malm, L. J. Tanghe, and J. T. Schmitt, *Ind. Eng. Chem.* **53,** 363 (1961).
35. B. P. Rouse, Jr., in N. M. Bikales, ed., *Encyclopedia of Polymer Science and Technology*, Vol. 3, Interscience Publishers, a division of John Wiley & Sons, Inc., New York, 1965.
36. A. Takahashi and S. Takahashi, *Kobunshi Kagaku* **27,** 394 (1970).
37. L. M. Eliseeva and co-workers, *Uzb. Khim. Zh.* **12,** 56 (1968); *Chem. Abstr.* **70,** 88962u (1969).
38. A. Takahashi and S. Takahashi, *Kobunshi Kagaku* **26,** 485 (1969).
39. D. P. Mironov and co-workers, *Khim. Drev.* **1,** 9 (1976).
40. K. S. Nikolskii and co-workers, *Khim. Volokna* **15,** 76 (1973).
41. C. J. Malm and G. D. Hiatt in E. Ott, H. M. Spurlin, and M. W. Grafflin, eds., *Cellulose and Cellulose Derivatives*, Part II of *High Polymers*, Vol. 5, 2nd ed., Interscience Publishers, New York, 1954, p. 776.
42. L. I. Makova and N. I. Klenkova, *Zh. Prikl. Khim.* **45,** 140 (1972).
43. L. I. Makova and N. I. Klenkova, *Zh. Prikl. Khim.* **47,** 610 (1974).
44. V. I. Sharkov and M. I. Perminova, *Chem. Abstr.* **83,** 149410e (1975).
45. V. V. Safonova and N. I. Klenkova, *Zh. Prikl. Khim.* **42,** 2636 (1969).
46. C. J. Malm and G. D. Hiatt in E. Ott, H. M. Spurlin, and M. W. Grafflin, eds., *Cellulose and Cellulose Derivatives*, Part II of *High Polymers*, Vol. 5, 2nd ed., Interscience Publishers, New York, 1954, pp. 777–780.
47. U.S. Pat. 2,026,583 (Jan. 7, 1936), C. J. Malm and C. L. Fletcher (to Eastman Kodak Co.).
48. U.S. Pat. 2,013,830 (Sept. 10, 1935), C. J. Malm and C. L. Fletcher (to Eastman Kodak Co.).
49. U.S. Pat. 2,775,529 (Dec. 25, 1956), H. Bates, C. W. Sammons, and J. W. Fisher (to British Hercules).

50. U.S. Pat. 2,836,590 (May 27, 1958), H. W. Turner (to Hercules Powder Co.).
51. U.S. Pat. 2,790,796 (April 30, 1957), J. Robin and R. Clevy (to Société Rhodiaceta).
52. C. J. Malm, L. J. Tanghe, and B. C. Laird, *J. Am. Chem. Soc.* **72,** 2674 (1950).
53. C. J. Malm and L. J. Tanghe, *Ind. Eng. Chem.* **47,** 995 (1955).
54. D. F. Othmer in *Encyclopedia of Chemical Technology,* John Wiley & Sons, Inc., New York, Vol. 2, 2nd ed., 1963, pp. 850–853.
55. G. D. Hiatt and W. J. Rebel in N. M. Bikales and L. Segal, eds., *Cellulose and Cellulose Derivatives,* Part V of *High Polymers,* Vol. 5, 2nd ed., Wiley-Interscience, New York, 1971, p. 765.
56. Brit. Pat 706,522 (March 31, 1954) (to Celanese Corp.).
57. Brit. Pat. 709,430 (May 26, 1954), M. E. Martin and co-workers (to Celanese Corp.).
58. U.S. Pat. 2,484,455 (Oct. 11, 1949), L. E. Herdle and E. L. Perkino (to Eastman Kodak Co.).
59. U.S. Pat. 2,603,634 (July 15, 1952), G. W. Seymour, B. B. White, and M. Plunguian (to Celanese Corp.).
60. U.S. Pat. 2,603,635 (July 15, 1952), G. W. Seymour, B. B. White, and M. Plunguian (to Celanese Corp.).
61. U.S. Pat. 2,603,636 (July 15, 1952), M. E. Martin (to Celanese Corp.).
62. U.S. Pat. 3,631,023 (Dec. 28, 1971), C. G. Horne, Jr., and C. J. Howell, Jr. (to Celanese Corp.).
63. U.S. Pat. 3,755,297 (Aug. 28, 1973), K. C. Campbell, J. M. Davis, G. E. Frye, and R. E. Woods (to Celanese Corp.).
64. U.S. Pat. 3,767,642 (Oct. 23, 1973), K. C. Campbell, J. M. Davis, and R. E. Woods (to Celanese Corp.).
65. U.S. Pat. 3,846,403 (Nov. 5, 1974), K. B. Gibney and R. S. Evans (to Canadian Cellulose Co., Ltd.).
66. U.S. Pat. 3,870,703 (March 11, 1975), K. B. Gibney, J. Howard, and R. S. Evans (to Canadian Cellulose Co., Ltd.).
67. Can. Pat. 973,174 (Aug. 19, 1975), K. B. Gibney, E. Grisack, and R. S. Evans (to Canadian Cellulose Co., Ltd.).
68. D. M. Hall and J. R. Horne, *J. Polym. Sci.* **17,** 3729 (1973).
69. U.S. Pat. 3,479,336 (Nov. 18, 1969), R. P. Taylor and D. Abson (to Columbia Cellulose Co., Ltd.).
70. U.S. Pat. 3,554,775 (Jan. 12, 1971), D. Abson and R. P. Taylor (to Columbia Cellulose Co., Ltd.).
71. U.S. Pat. 2,923,706 (Feb. 2, 1960), N. B. Campbell and L. Berthiaume (to Canadian Celanese Co., Ltd.).
72. U.S. Pat. 2,775,585 (Dec. 25, 1956), H. Bates, J. W. Fisher, and J. R. Smith (to British Celanese, Ltd.).
73. Brit. Pat. 766,293 (Jan. 16, 1957) (to Celanese Corp.).
74. G. D. Hiatt and W. J. Rebel in N. M. Bikales and L. Segal, eds., *Cellulose and Cellulose Derivatives,* Part V of *High Polymers,* Vol. 5, 2nd ed., Wiley-Interscience, New York, 1971, pp. 747–749.
75. C. Hamalainen and J. D. Reid, *Ind. Eng. Chem.* **41,** 1018 (1949).
76. Y. Iwakura, M. Nakajima, and I. Kanda, *Sen'i Gakkaishi* **12,** 861 (1956).
77. *Ibid.,* p. 865.
78. Y. Iwakura and I. Kanda, *Sen'i Gakkaishi* **13,** 17 (1957).
79. U.S. Pat. 2,208,569 (July 23, 1940), L. W. Blanchard, Jr. (to Eastman Kodak Co.).
80. U.S. Pat. 2,790,794 (April 30, 1957), C. J. Malm, L. J. Tanghe, and L. W. Blanchard, Jr. (to Eastman Kodak Co.).
81. U.S. Pat. 2,801,238 (July 30, 1957), C. J. Malm and L. W. Blanchard, Jr. (to Eastman Kodak).
82. U.S. Pat. 2,835,665 (May 20, 1958), C. J. Malm, L. J. Tanghe, and L. W. Blanchard, Jr. (to Eastman Kodak Co.).
83. C. J. Malm and G. D. Hiatt in E. Ott, H. M. Spurlin, and M. W. Grafflin, eds., *Cellulose and Cellulose Derivatives,* Part II of *High Polymers,* Vol. 5, 2nd ed., Interscience Publishers, New York, 1954, p. 788.
84. U.S. Pat. 2,824,098 (Feb. 18, 1958), F. M. Volberg and M. D. Martin (to Eastman Kodak Co.).
85. C. J. Malm and co-workers, *Ind. Eng. Chem.* **43,** 684 (1951).
86. I. Terasaki, *J. Soc. Text. Cellul. Ind. Jpn.* **18,** 331 (1962).
87. E. P. Grishin and co-workers, *Khim. Tekhnol. Proizvod. Tsellyul.* 98 (1971); *Chem. Abstr.* **78,** 31649h (1973).
88. J. W. Mench, B. Fulkerson, and G. D. Hiatt, *Ind. Eng. Chem., Prod. Res. Dev.* **5,** 110 (1966).
89. U.S. Pat. 2,828,303 (March 25, 1958), C. J. Malm and L. W. Blanchard, Jr. (to Eastman Kodak Co.).

90. U.S. Pat. 2,828,304 (March 25, 1958), C. J. Malm and L. W. Blanchard, Jr. (to Eastman Kodak Co.).
91. L. V. Gurkovskoya and co-workers, *Khim. Tekhol. Proizvod. Tsellyul.*, 93 (1971); *Chem. Abstr.* **78**, 112,884h (1973).
92. C. J. Malm and co-workers, *Ind. Eng. Chem. Process Res. Dev.* **5**, 81 (1966).
93. U.S. Pat. 2,801,240 (July 30, 1957), C. J. Malm and L. J. Tanghe (to Eastman Kodak Co.).
94. U.S. Pat. 2,816,106 (Dec. 10, 1957), C. J. Malm, L. J. Tanghe, and H. M. Herzog (to Eastman Kodak Co.).
95. C. J. Malm and co-workers, *Ind. Eng. Chem.* **42**, 2904 (1952).
96. G. A. Richter, L. E. Herdle, and I. L. Gage, *Ind. Eng. Chem.* **45**, 2773 (1953).
97. C. J. Malm, L. J. Tanghe, and B. C. Laird, *Ind. Eng. Chem.* **38**, 77 (1946).
98. U.S. Pat. 3,047,561 (July 31, 1962), C. L. Crane (to Eastman Kodak Co.).
99. U.S.S.R. Pat. 458,560 (Jan. 30, 1975), A. Maciulis and co-workers.
100. U.S. Pat. 2,917,398 (Dec. 15, 1959), H. W. Coover and W. C. Wooten, Jr. (to Eastman Kodak Co.).
101. U.S. Pat. 3,305,378 (Feb. 21, 1967), K. Ritchie and J. W. Addleburg (to Eastman Kodak Co.).
102. U.S. Pat. 3,314,808 (April 18, 1967), G. M. Moulds (to E. I. du Pont de Nemours & Co., Inc.).
103. U.S. Pat. 2,899,315 (Aug. 11, 1959), R. F. Williams, Jr., and C. S. Lowe (to Eastman Kodak Co.).
104. Jpn. Pat. 69 115,674 (July 11, 1969), I. Kato (to Teijin, Ltd.).
105. U.S. Pat. 2,899,316 (Aug. 11, 1959), B. P. Rouse, Jr., and R. O. Hill, Jr. (to Eastman Kodak Co.).
106. U.S. Pat. 2,713,546 (July 19, 1955), R. F. Williams (to Eastman Kodak Co.).
107. E. F. Evans and L. F. McBurney, *Ind. Eng. Chem.* **41**, 1260 (1949).
108. U.S. Pat. 2,617,738 (Nov. 11, 1952), W. B. Horback, E. J. Wickson, and W. J. Myles (to Celanese Corp.).
109. A. M. Kuliev and co-workers, *Issled. Eff. Khim–Dabovak Polim. Mater.* 338 (1969); *Chem. Abstr.* **77**, 6306r (1972).
110. L. M. Malinin and K. F. Yakunina, *Plast. Massy* 47 (1966).
111. *ASTM D-871*, American Society for Testing and Materials, Philadelphia, Pa., 1976.
112. L. J. Tanghe, L. B. Genung, and J. M. Mench in R. Whistler, ed., *Methods in Carbohydrate Chemistry*, Vol. III, Academic Press, New York, 1963, p. 203.
113. E. Isobe and T. Nakajima, *Sen'i Gakkaishi* **31**, T-101 (1975).
114. D. P. Mironov, V. N. Mironova, and V. V. Zharkov, *Zh. Anal. Khim.* **24**, 289 (1969).
115. L. J. Tanghe, L. B. Genung, and J. M. Mench in R. Whistler, ed., *Methods in Carbohydrate Chemistry*, Vol. III, Academic Press, New York, 1963, p. 210.
116. R. N. Shroff, *J. Appl. Polym. Sci.* **9**, 1547 (1965).
117. T. D. Verma and M. Senguta, *J. Appl. Polym. Sci.* **15**, 1599 (1971).
118. A. S. Buntyakov, N. I. Gelkina, and V. M. Averynova, *Plast. Massy* **3**, 71 (1969).
119. Z. B. Komarova and co-workers, *Khim. Volokna* **15**(3), 67 (1973); *Chem. Abstr.* **79**, 93262f (1973).
120. R. J. Brewer and co-workers, *J. Polym. Sci.* [A-1] **6**, 1697 (1968).
121. R. E. Glegg, L. J. Tanghe, and R. J. Brewer in N. M. Bikales and L. Segal, eds., *Cellulose and Cellulose Derivatives*, Part IV of *High Polymers*, Vol. 5, 2nd ed., John Wiley & Sons, Inc., New York, 1971, p. 505.
122. T. S. Gardner and C. B. Purves, *J. Am. Chem. Soc.* **64**, 1539 (1942).
123. V. W. Goodlett, J. T. Dougherty, and H. W. Patton, *J. Polym. Sci.* [A-1] **9**, 155 (1971).
124. D. Horton and J. H. Lauterback, *J. Org. Chem.* **34**, 86 (1969).
125. S. Hirano, *Org. Magn. Reson.* **3**, 353 (1971).
126. H. Friebolin, G. Keilich, and E. Siefert, *Angew. Chem. Int. Ed. Engl.* **8**, 766 (1969).
127. *ASTM Standards in Plastics*, American Society for Testing and Materials, Philadelphia, Pa., 1962.
128. ASTM Standards, Part 21, D817-72, pp. 33–52; D1343-69, 1976, pp. 84–87.
129. *Ibid.*, pp. 37–40.
130. M. Wandel and H. Tengler, *Gummi Asbest Kunst.* **19**, 141 (1966).
131. J. A. Brydson, *Plastic Materials*, D. Van Nostrand Company, Inc., New York, 1966, pp. 369–371.
132. Ger. Offen. 2,104,032 (Aug. 5, 1971), W. Ueno and N. Minagawa (to Fuji Photo Film Co., Ltd.).
133. Brit. Pat. 1,352,605 (May 8, 1974) (to Fuji Photo Film Co., Ltd.).
134. Belg. Pat. 706,838 (April 1, 1968); Brit. Pat. 1,197,570 (July 8, 1970), J. A. Smith, B. W. Cracknell, and H. Bates (to Courtaulds, Ltd.).
135. T. Matsuura, *Yaki Gosei Kagaku Kyokai–Shi* **31**, 717 (1973).
136. S. Yabumoto, *Nippon Kaisui Gakkai–Shi* **27**, 115 (1973).

137. H. D. Saier and H. Strathman, *Angew. Chem.* **87,** 476 (1975).
138. H. I. Mahon and B. J. Lipps in N. M. Bikales and L. Segal, eds., *Cellulose and Cellulose Derivatives,* Part V of *High Polymers,* Vol. 5, 2nd ed., John Wiley & Sons, Inc., New York, 1971, pp. 1261–1263.
139. *Eastman Cellulose Acetate,* Eastman Chemical Products, Inc., 1968.
140. *Modern Plastics Encyclopedia,* Vol. 53(10A), 1976–1977, pp. 458–459.
141. K. J. Saunders, *Organic Polymer Chemistry,* Chapman and Hall, Distributed in U.S.A. by Halsted Press, a division of John Wiley & Sons, Inc., New York, 1973, p. 265.
142. C J. Malm and G. D. Hiatt in E. Ott, H. M. Spurlin, and M. V. Grafflin, eds., *Cellulose and Cellulose Derivatives,* Part II of *High Polymers,* Vol. 5, 2nd ed., Interscience Publishers, New York, 1954, p. 808.
143. *Cellulose Acetate Butyrate Lacquers for Molded and Calendered Plastic Products,* Eastman Chemical Products, Inc., 1972.
144. R. L. Smith, *Paint Varn. Prod. Manager* **59,** 53 (1969).
145. J. W. Lowe and G. S. Teague, Jr., *Paint Varn. Prod. Manager* **55,** 37 (1965).
146. Brit. Pat. 1,077,761 (Aug. 2, 1967), J. W. Lowe and G. S. Teague, Jr. (to Eastman Kodak Co.).
147. Brit. Pat. 1,166,104 (Oct. 8, 1969), J. G. Stranch and E. E. Denison (to Eastman Kodak Co.).
148. C. H. Coney and G. B. Bowen, *Am. Inkmaker* **51,** 20, 24 (1973).
149. U.S. Pat. 3,313,639 (April 11, 1967), F. M. Ball and J. H. Davis (to Eastman Kodak Co.).
150. Ger. Offen. 2,412,426 (Sept. 18, 1975), M. Stemmler and H. Stemmler.
151. Ger. Offen. 2,148,008 (April 5, 1973), K. Landt and P. Neuber (to Winter-Optik G.m.b.H.).
152. T. Wydenen and M. Leban, *J. Appl. Polym. Sci.* **17,** 2277 (1973).
153. U.S. Pat. 3,607,329 (Sept. 21, 1971), S. Manjikian (to U.S. Dept. of the Interior).
154. U.S. Pat. 3,585,126 (June 15, 1971), C. R. Cannon and P. A. Cantor (to Aerojet-General Corp.).
155. H. Braconnot, *Ann.* **1,** 242, 245 (1883).
156. C. F. Schönbein, *Philos. Mag.* **31,** 7 (1847).
157. U.S. Pat. 2,753,337 (July 3, 1956), E. D. Klug (to Hercules Co.).
158. Can. Pat. 542,733 (June 25, 1957), E. D. Klug (to Hercules Co.).
159. G. A. Petropavloskii and M. M. Krunshank, *Kh. Prikl. Khim.* **36,** 2506 (1963).
160. U.S. Pat. 2,539,451 (Jan. 30, 1951), C. J. Malm and C. L. Crane (to Eastman Kodak Co.).
161. U.S. Pat. 3,624,069 (Nov. 30, 1971), R. G. Schweizer (to Kelco Co.).
162. R. G. Schweizer, *Carbohydr. Res.* **21,** 219 (1972).
163. U.S. Pat. 3,075,962 (Jan. 29, 1963), G. D. Hiatt and M. E. Rowley (to Eastman Kodak Co.).
164. U.S. Pat. 3,075,963 (Jan. 29, 1963), G. D. Hiatt and M. E. Rowley (to Eastman Kodak Co.).
165. U.S. Pat. 3,075,964 (Jan. 29, 1963), C. J. Malm and M. E. Rowley (to Eastman Kodak Co.).
166. U.S. Pat. 3,086,007 (April 16, 1963), G. P. Touey and J. E. Kiefer (to Eastman Kodak Co.).
167. W. D. Slowig and M. E. Rowley, *Text. Res. J.* **38,** 879 (1968).
168. U.S. Pat. 3,794,605 (Feb. 26, 1974), F. Diehl (to Proctor and Gamble Co.).
169. Ger. Offen. 2,348,409 (April 11, 1974), G. M. Dodwell (to Leford, Ltd.).
170. U.S. Pat. 3,726,796 (April 10, 1973), R. G. Schweizer (to Kelco Co.).
171. Ger. (East) Offen. 112,456 (April 12, 1975), K. H. Bischoff and H. Dautzenberg.
172. A. C. Nuessle and co-workers, *Text. Res. J.* **26,** 32 (1956).
173. N. I. Irmalenko and I. P. Lyubliner, *Khim. Tekhnol. Proizvod. Tsellyul.* 377 (1971); *Chem. Abstr.* **78,** 138114 (1973).
174. U.S.S.R. Pat. 321,522 (Nov. 19, 1971), S. A. Aleksiev and L. I. Yurchenko.
175. K. Katsuura and N. Inagaki, *Sen'i Gakkaishi* **13,** 24 (1957).
176. U.S. Pat. 2,759,924 (Aug. 21, 1956), G. P. Touey (to Eastman Kodak Co.).
177. D. A. Predvoditelev, E. E. Nifant'ev, and Z. A. Rogovin, *Vysokomol. Soedin.* **8,** 76 (1966).
178. A. Yuldashev and M. Muratova, *Dokl. Akad. Nauk Uzb. SSR* **23,** 42 (1966).
179. U.S.S.R. Pat. 159,524 (1962), A. D. Kiselev and S. N. Danilov.
180. G. N. Bruxelles and V. C. Grassie in N. M. Bikales, ed., *Encyclopedia of Polymer Science and Technology,* Vol. 3, Interscience Publishers, a division of John Wiley & Sons, Inc., New York, 1965, pp. 307–325.
181. J. Barsha in E. Ott, H. M. Spurlin, and M. W. Grafflin, eds., *Cellulose and Cellulose Derivatives,* Part II of *High Polymers,* Vol. 5, 2nd ed., Interscience Publishers, New York, 1954, pp. 713–762.
182. U.S. Pat. 2,776,965 (Jan. 8, 1957), J. L. Bennett and co-workers (to Hercules Co.).
183. Can. Pat. 563,545 (Sept. 23, 1958), J. L. Bennett and co-workers (to Hercules Co.).
184. M. Netzler and T. W. Lott, *Govt. Rep. Announce.* (U.S.) **75,** 50 (1975).
185. Can. Pat. 581,319 (Aug. 11, 1959), W. L. Plunkett (to Hercules Co.).
186. U.S. Pat. 2,950,278 (Aug. 23, 1960), W. L. Plunkett (to Hercules Co.).

187. U.S. Pat. 2,776,944 (Jan. 8, 1957), J. G. McMillan and W. L. Plunkett (to Hercules Co.).
188. U.S. Pat. 2,776,966 (Jan. 8, 1957), J. G. McMillan and W. L. Plunkett (to Hercules Co.).
189. U.S. Pat. 3,714,143 (Jan. 30, 1973), C. W. Lewis and E. O. Haun (to U.S. Dept. of the Army).
190. Can. Pat. 536,919 (Feb. 5, 1957), W. C. Ramsey (to Olin Mathieson Co.).
191. U.S.S.R. Pat. 405,567 (Nov. 5, 1973), I.V. Balakhnichev.
192. J. W. Green in R. L. Whistler and J. N. BeMiller, eds., *Methods of Carbohydrate Chemistry*, Vol. III, Academic Press, New York, 1963, pp. 213–237.
193. U.S. Pat. 2,776,880 (Ján. 8, 1957), D. S. Bruce and H. M. Spurlin (to Hercules Co.).
194. Can. Pat. 563,546 (Sept. 23, 1958), D. S. Bruce and H. M. Spurlin (to Hercules Co.).
195. Ger. Offen. 2,049,379 (April 13, 1972), R. Guenther (to Badische Anilin).
196. S. Oinuma, *Kogyo Kayaku* **37**(1), 2 (1976).
197. U.S. Pat. 3,861,932 (Jan. 21, 1975), B. L. Kabacoff, C. Mohr, and C. M. Fairchild (to Revlon, Inc.).
198. *ASTM Standards,* Part 21, D-1716-62, 7–16, 1976.
199. Norweg. Pat. 127,444 (June 25, 1973), H. Hiorth (to Dyno Industrier A/S).
200. U.S. Pat. 105,388 (July 12, 1870), J. W. Hyatt and I. S. Hyatt.
201. Jpn. Kokai 75 103,324 (Aug. 15, 1975), S. Miyamoto and K. Maki (to Mitsubishi Paper Mills, Ltd.).
202. J. Tripier and co-workers, *Comm. Eur. Communities EUR,* 1975; *Chem. Abstr.* **84,** 171209p (1976).
203. G. M. Hassib and L. Medveczky in ref. 202; *Chem. Abstr.* **84,** 171211h (1976).
204. S. B. Lupica, *Nucl. Sci. Abstr.* **31**(1), 3458 (1975).
205. H. A. Khan, *J. Phys. Soc. Jpn.* **39**(5), 1159 (1975).
206. U.S. Pat. 3,648,845 (March 4, 1972), R. L. Riley (to U.S. Dept. of the Interior).
207. U.S. Pat. 3,676,203 (July 11, 1972), S. B. Sachs and R. L. Riley (to U.S. Dept. of the Interior).

R. T. Bogan
C. M. Kuo
R. J. Brewer
Tennessee Eastman Company

CELLULOSE DERIVATIVES, ETHERS

Cellulose ether polymers have a wide diversity of properties ranging from organic-soluble thermoplastic products to water-soluble food additives. The importance of cellulose ethers has increased in recent years because economic factors have adversely affected the supply and pricing of natural gums (see Gums) and low viscosity products such as starch (qv) derivatives. The rising prices of petroleum-based polymers and chemicals engender interest in cellulose as a renewable resource. This is reflected by the growth of cellulose ethers which reached an estimated United States volume of about 60,000 metric tons in 1976 (U.S. production capacity is estimated at nearly 79,000 metric tons). Current prices of cellulose ethers in the United States are $1.60–7.00/kg. As a class, the water-soluble ethers are essentially minor additives of high value.

The commercial products can be subdivided into two groups, with the water-soluble cellulose ethers having the largest volume of sales:

144 CELLULOSE DERIVATIVES, ETHERS

Water-soluble cellulose ethers	CAS Registry No.
sodium carboxymethyl cellulose	[9004-32-4]
sodium carboxymethyl 2-hydroxyethyl cellulose	[9088-04-4]
2-hydroxyethyl cellulose	[9004-62-0]
methyl cellulose	[9004-67-5]
2-hydroxypropyl methyl cellulose	[9004-65-3]
2-hydroxyethyl methyl cellulose	[9032-42-2]
2-hydroxybutyl methyl cellulose	[9041-56-9]
2-hydroxyethyl ethyl cellulose	[9004-58-4]
2-hydroxypropyl cellulose	[9004-64-2]
Organic-soluble cellulose ethers	
ethyl cellulose	[9004-57-3]
ethyl 2-hydroxyethyl cellulose	[9004-58-4]
2-cyanoethyl cellulose	[9004-41-5]

Cellulose ethers are manufactured from a preformed polymer, cellulose, comprised of linear chains of β-anhydroglucose rings. An idealized structure of a cellulose ether is shown in Figure 1 where the R-groups can be either single representatives or combinations of the above substituent groups. The chemical nature, quantity, and distribution of the substituent groups govern such properties as solubility, surface activity, thermoplasticity, film characteristics, and biodegradation (see Biopolymers; Carbohydrates; Cellulose).

For purposes of process control and specifications, weight percentages of particular substituent groups are sufficient but cumbersome when cellulose ethers containing substituent groups that differ greatly in molecular weight are compared. The degree of substitution (DS), which defines the number of substituted ring sites, is more useful. Since there are three hydroxyl groups on each anhydroglucose ring of the cellulose polymer that are available for reaction, the maximum value of DS (assuming

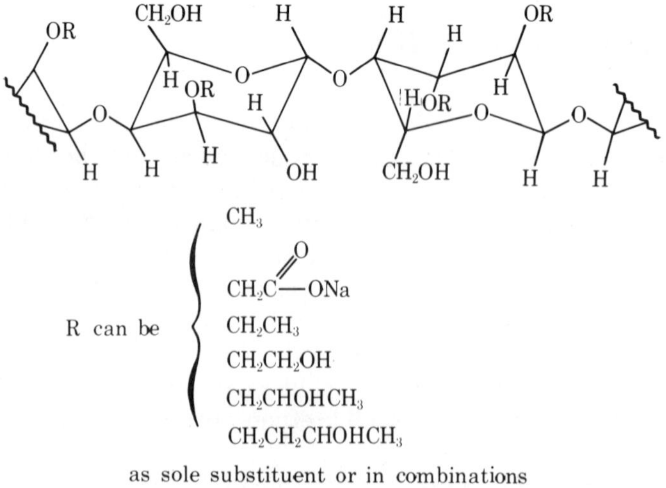

R can be
- CH_3
- $CH_2C(=O)-ONa$
- CH_2CH_3
- CH_2CH_2OH
- $CH_2CHOHCH_3$
- $CH_2CH_2CHOHCH_3$

as sole substituent or in combinations

Figure 1. Idealized structure of a cellulose ether.

a 1:1 relationship between the entering group and the OHs) is 3. In the case of alkylene oxides, which generate a new OH for every glucose–OH reacting, a possibility for side chain formation exists. The indefinite term molar substitution (MS) is used in these circumstances. By dividing MS by DS, the average length of the side chain can be measured.

The manufacture of cellulose derivatives poses some basic problems for the chemist and the engineer. The presence of crystalline and amorphous regions in the cellulose fiber, plus the angular orientation of the cellulose fibrils, requires precise control of the initial treatment of cellulose with sodium hydroxide (used to activate the cellulose). Nonuniformity of sodium hydroxide distribution can cause major problems of product quality and property control. Both wood and cotton-linter celluloses are used, with economics favoring the use of wood pulp. Cotton linters are used in the production of very high molecular weight products. The alkylation and hydroxyalkylation reactions require the use of pressure vessels; the carboxymethylation reaction can be at ambient pressure. Since cellulose is an excellent insulator and these reactions are exothermic, the process design must provide for adequate heat transfer to prevent run-away reactions and hot spots which cause quality deterioration. The capital-intensive processes used for the manufacture of cellulose derivatives require substantial investment for control of gaseous and aqueous effluents. In addition, these processes are also energy intensive with energy conversion values of ca 29–36.2 MJ/kg.

Information on the structure of cellulose and its derivatives is presented in a series of review articles (1). General principles of cellulose etherification have been reviewed (2–3). Additional research is discussed in the *American Chemical Society Symposium Series* (4).

SODIUM CARBOXYMETHYL CELLULOSE

Sodium carboxymethyl cellulose or cellulose gum (CMC, formerly called sodium cellulose glycolate) is an anionic cellulose ether manufactured by the reaction of monochloroacetic acid, as either the acid or the sodium salt, and alkali cellulose. Production of sodium carboxymethyl cellulose began in Germany (5) and has since expanded to fifteen countries. Commercial production in the United States began in 1943. A wide spectrum of uses has resulted in a volume growth from ca 900 metric tons in 1947 to an estimated production capacity of ca 48,000 metric tons in 1977. There are five producers of sodium carboxymethyl cellulose in the United States, but only one U.S. source of purified CMC.

Sodium Carboxymethyl 2-Hydroxyethyl Cellulose. Sodium carboxymethyl cellulose can react with metal ions such as the alkaline earth elements and heavy metal ions to form precipitates. This characteristic of its ionic structure can be modified and suppressed by hydroxyethyl substitution on the CMC structure. This mixed cellulose ether retains some of the useful properties of CMC while gaining lower sensitivity to precipitation by salt solutions and acids. The sensitivity decreases with decreasing carboxyl content. The hydroxyethylation reaction is carried out first in the preparation of this cellulose ether (6).

Properties

Sodium carboxymethyl cellulose is available in purified, technical, and semirefined grades as well as the free acid form. The commercial products span a DS range of 0.38–1.4. The most widely used products have a DS range of 0.65–0.85. Table 1 lists typical properties for purified CMC (7).

Several factors must be considered for their effect on the viscosity of solutions of sodium carboxymethyl cellulose. The polyelectrolyte structure of CMC influences viscosity and stability in solution. Maximum viscosity and best stability occur at 7–9 pH. At ≤4 pH, formation of the less soluble free acid form results in significant increases in viscosity and eventually insolubility at ≤2 pH. Above 10 pH a slight decrease in viscosity occurs. Owing to the polar nature of the carboxyl group, sodium carboxymethyl cellulose is soluble in both hot and cold water. The viscosity of CMC solutions is dependent upon temperature, decreasing as the temperature increases. Under normal conditions, this effect is reversible provided prolonged heating at high temperatures is avoided. The viscosity can also be controlled by choice of equipment (mechanical shear), solvent, or polymer composition. Table 2 illustrates the range of these effects.

The rheological properties of CMC solutions depend upon concentration, rate of shear, and uniformity of substitution. At very low shear rates, solutions approach Newtonian flow. Solutions of medium and high viscosity types with a DS of 0.9–1.2 and other special substitutions show time-independent, shear-thinning pseudoplasticity. The medium and high viscosity types with a DS of 0.4–0.7 generally exhibit time-dependent shear-thinning thixotropy (see Rheological measurements).

Since CMC is a polyelectrolyte, metathesis reactions with other cations can occur. As a general rule, monovalent cations form soluble salts. Divalent cations are borderline depending upon the grade of CMC and the mixing technique used. The trivalent cations form insoluble precipitates.

Table 1. Typical Properties of Sodium Carboxymethyl Cellulose

Property	Value
polymer as shipped	
sodium carboxymethyl cellulose, dry basis, %	99.5
moisture content, maximum %	8.0
browning temperature, °C	227
charring temperature, °C	252
bulk density, g/mL	0.75
biological oxygen demand[a]	
0.7 DS, high viscosity, ppm	11,000
0.7 DS, low viscosity, ppm	17,300
solutions	
sp gr, 2% soln, 25°C	1.0068
refractive index, 2% soln, 25°C	1.3355
pH, 2% soln	7.0
surface tension, 1% soln, mN/m(=dyn/cm) at 25°C	71
bulking value in soln, cm^3/gm	0.544
films	
sp gr, g/mL	1.59
refractive index	1.515

[a] After 5 days of incubation. Under these conditions cornstarch has a BOD of over 800,000 ppm.

Table 2. Factors Affecting Disaggregation of Sodium Carboxymethyl Cellulose[a]

	\multicolumn{3}{c	}{Viscosity in mPa·s(=cP), 25°C}			
	\multicolumn{3}{c	}{Anchor stirrer}	\multicolumn{3}{c}{Waring blender}
DS	Distilled water	NaCl, 4%	Saturated NaCl	Distilled water	NaCl, 4%	Saturated NaCl
0.4	900	11	6	4000	65	16
0.7	1680	140	45	760	1040	2440
0.7	1680	570	165	760	750	1720
0.9	215	160	225	125	95	235
1.2	175	80	180	100	55	140

[a] Ref. 7.

Manufacture

CMC is prepared by treating alkali cellulose with sodium chloroacetate. A side reaction, the conversion of sodium chloroacetate to sodium glycolate, occurs simultaneously.

$$R_{cell}OH + NaOH \rightleftharpoons R_{cell}OH \cdot NaOH$$

$$R_{cell}OH \cdot NaOH + ClCH_2COONa \rightarrow R_{cell}OCH_2COONa + NaCl + H_2O$$

$$NaOH + ClCH_2COONa \rightarrow HOCH_2COONa + NaCl$$

As in the preparation of other cellulose ethers, the degree of substitution is increased and the extent of side reactions is decreased at higher sodium hydroxide concentrations. However, enough water must be present to ensure adequate swelling so that the etherification is uniform. The alkali cellulose can be prepared batchwise by several methods such as steeping and shredder mixing (8–9). Processes for continuous production of alkali cellulose are also used (10). The sodium chloroacetate is added as an aqueous solution, or as free monochloroacetic acid. If the acid is used, sufficient additional sodium hydroxide must be added to neutralize it. Reactions have been carried out in sigma-bladed shredders at 25–100°C. The cellulose retains its fiber structure unless the DS exceeds 0.4, when a dough is obtained. An alternative method of preparation is the slurry process in which an inert organic liquid is used as a diluent (11–12). The product is obtained as a slurry. It is neutralized and, if desired, freed from by-product salts by washing with 70–80% methanol. Viscosity reduction can be achieved by oxidation with air and by use of chemical oxidants (8,13).

Economic Aspects

Statistics on global production of CMC are sparse. Estimated 1977 United States capacity was 48,000 metric tons (14). United States prices (1976) were $1.61–2.32/kg. Markets in the United States are listed in Table 3.

Specifications and Standards

Specifications for CMC are published in *Food Chemicals Codex II*, the *USP XIX*, and in publications of the FAO/WHO (15–17). The European Economic Community (EEC) product number of sodium carboxymethyl cellulose is E461 and the specifi-

Table 3. Major Markets in the United States for Sodium Carboxymethyl Cellulose in 1973[a,b]

Application	Thousand metric tons	Percent of total
detergents	9.9	30
textiles	7.2	22
foods	8.3	25
drilling muds	3.9	12
paper sizing	1.0	3
other	2.6	8
total	32.9	100

[a] Ref. 14.
[b] Estimated.

cation reference is FAS/IV/19. Sodium carboxymethyl cellulose is also classified with "substances that are generally recognized as safe" (GRAS) under Title 21, Part 182.1745 of the Code of Federal Regulations (18). CMC, meeting the above standards, may also be used in standardized foods such as cheese and cheese products, frozen desserts, food dressings, fruit butters, jellies and preserves, and nonalcoholic beverages (see Food additives). The Cosmetic, Toiletry, Fragrance Association's Standard No. 34 also lists specifications for its use.

Analytical and Test Methods

Procedures for the analysis of CMC are described in ASTM Monograph D1439-65 (19). Further information is available in a bulletin from Hercules, Inc. which includes assays in formulations (20). Additional tests are listed in USP XIX and in the U.S. *Food Chemicals Codex II*.

Health and Safety Factors

Dermatological and toxicological studies have been made on cellulose gum by independent laboratories. These studies conclude that "sodium carboxymethyl cellulose shows no evidence of being toxic to white rats, dogs, guinea pigs, or human beings." Feeding, metabolism, and topical use studies show that cellulose gum is physiologically inert. Patch tests on human skin demonstrated that sodium carboxymethyl cellulose was neither a primary irritant nor a sensitizing agent. Detailed reports are on file with the FDA (7).

AID (Acceptable Intake Daily) values of 0–30 mg/kg (taken as the sum of total cellulose derivatives), with higher levels for dietary and calorie control purposes, are listed in a report by the Joint FAO/WHO Food Standards Program Codex Committee on Food Additives (17,21). See Table 1 for data on biological oxygen demand.

Fine dust particles of CMC can form explosive mixtures with air. General precautions are outlined in ref. 16.

Uses

CMC has many uses. Estimated 1973 consumption, by use in the United States, is given in Table 3. In detergents, CMC functions as an antiredeposition agent to inhibit resoiling of fabrics by the soil removed during laundering (see Surfactants). CMC is very effective in washing of cotton clothing. The textile industry uses it for thickening dyestuff printing pastes and for warp-sizing formulations to improve weaving efficiency. The purified CMC products are used in many types of food and pharmaceutical products where thickening, rheology control, emulsion stabilization, and water-loss control are needed. Sodium carboxymethyl cellulose is not metabolized and can be used in low-calorie food products.

Control of water loss to soil formation is important in drilling fluids (qv). They are used in the oil industry to prevent reduction in oil flow from the formation and to inhibit swelling of certain types of clay which reduces the diameter of the drill hole.

Other uses include sizing of paper (to obtain oil and grease resistance), cosmetics, in adhesives (qv) such as wallpaper paste, and as a suspending agent and binder in ceramics and vitreous enamels.

METHYL CELLULOSE AND 2-HYDROXYPROPYL METHYL CELLULOSE

Methyl cellulose was first prepared by Suida in 1905 (22). Commercial production was established in the United States in 1938. As commercial development continued, a need for modified products arose. Alkylene oxides are particularly valuable for modifying the properties of water-soluble methyl cellulose; ethylene, propylene, and butylene oxides are used in commercial processes (23–24). By varying the ratio of alkyl to alkylene oxide substitution, it is possible to generate a family of products with properties directed at specific uses. The type, quantity, and uniformity of substitution on the anhydroglucose ring governs properties such as solubility, surface activity, compatibility with additives in a formulation, biodegradation and enzyme resistance, and thermal gelation. Methyl cellulose and its alkylene oxide modifications are nonionic in structure (see Ethylene oxide; Propylene oxide).

2-Hydroxybutyl Methyl Cellulose. The major use of this modification of methyl cellulose is as a thickener in methylene chloride–alcohol paint removers. The hydroxybutyl substitution is obtained by the use of butylene oxide. The more organophilic hydroxybutyl group increases the solubility in organic solvents.

2-Hydroxyethyl Methyl Cellulose. Small amounts of hydroxyethyl substitution have been used by European manufacturers of methyl cellulose to achieve higher thermal gel points in their products. These products are similar in many properties to methyl cellulose and hydroxypropyl methyl cellulose, and they are used in many of the same nonfood applications.

Properties

Methyl cellulose and its modifications are available in powder and granular forms. Typical physical properties for the powder are listed in Table 4.

Water solutions of these products possess the unusual property of thermal gelation. When the solution is heated to a specific temperature (governed by viscosity type,

150 CELLULOSE DERIVATIVES, ETHERS

Table 4. Typical Properties of Methyl Cellulose[a]

Property	Value
physical appearance	white to slightly off-white, essentially odorless and tasteless powder
apparent density of powder, g/mL	0.25–0.70
browning temperature, °C	190–200
charring temperature, °C	225–230
relative flammability in a furnace at 700°C	90 +
sp gr of film	1.39

[a] Ref. 25.

concentration, and presence of other additives such as salts), a transformation from a sol to a gel structure occurs. This phenomenon is reversible upon cooling. The temperature of gelation and the character and texture of the gel can be controlled by choice of the ratio of substituents on the anhydroglucose ring as shown in Table 5. The decrease in gelation temperature caused by addition of salts is a function of the ions present: the higher the charge on the ion, the greater the decrease will be (25).

Commercial designations of the methyl cellulose products are based on viscosity values determined in water at 20°C and a concentration of 2%. Methods for products produced in the United States are described as ASTM monographs D1347-72 and D2363-72.

The general properties of solutions of methyl cellulose and hydroxypropyl methyl cellulose are given in Table 6.

Rheology of an aqueous solution of Methocel is affected by its molecular weight, concentration, temperature, and the presence of other solutes. In general, at a temperature below the incipient gelation temperature, aqueous solutions of Methocel exhibit pseudoplastic flow. Pseudoplasticity increases with increasing molecular weight or concentration. However, at very low shear rates, all Methocel cellulose ether solutions appear to be Newtonian, and the shear rate below which the solution becomes Newtonian increases with decreasing molecular weight or concentration.

Manufacture

The alkylation reaction with methyl chloride involves the traditional Williamson ether reaction where alkali cellulose reacts with an alkyl halide.

$$R_{cell}OH + NaOH \rightarrow R_{cell}ONa + H_2O$$
$$R_{cell}ONa + R'Cl \rightarrow R_{cell}OR' + NaCl$$

Table 5. Thermal Gel Points of 2% Solutions of Methyl Cellulose and Hydroxypropyl Methyl Cellulose[a]

Property	Gel points					
methoxyl DS	1.8	1.9	1.8	1.4	1.3	2.0
hydroxypropoxyl MS		0.27	0.15	0.86	0.20	
hydroxybutoxyl DS						0.08
gelation temperature, °C	50–55	58–64	62–68	60–70	70–90	45–50
gel texture	firm	semifirm	semifirm	precipitate	mushy	firm

[a] Ref. 25.

Table 6. Properties of Methyl Cellulose and Its Derivatives[a]

Property	Value					
methoxyl DS	1.8	1.9	1.8	1.3	1.4	2.0
hydroxypropoxyl MS		0.27	0.15	0.20	0.86	
hydroxybutoxyl DS						0.08
sp gr, 20/4°C						
1%	1.0012					
5%	1.0117					
10%	1.0245					
refractive index,						
2% soln, n_D^{20}	1.336					
partial specific						
volume, cm^3/g	0.725	0.767	0.734	0.717	0.725	0.774
viscosity, mPa·s (= cP)	4,000	4,000	4,000	4,000	12,000	12,000
freezing point, °C	0	0	0	0	0	0
surface tension						
25°C mN/m(=dyn/cm)[a]	47–53	44–50	44–50	50–56	48–52	49–55
interfacial tension, 25°C						
mN/m(=dyn/cm), paraffin oil	19–23	18–19	19–23	26–28	26–30	20–22
specific heat (10–25% soln)						
20–90°C, J/(g·K)[cal/(g·K)]	3.89 ± 0.21[0.93 ± 0.05]					

[a] For solutions below 500 mPa·s. In many cases, concentrations of 0.001% will give the same values as 1% solutions.

The reaction of alkylene oxides differs from that of alkyl halides in that sites of strong hydrogen bonding are not lost in the product. When an alkylene oxide reacts with cellulose, a new hydroxyl group is generated on the substituent group for every ring hydroxyl substituted. In the case of ethylene oxide, the new hydroxyl is a primary group; with propylene oxide the hydroxyl group is secondary.

Methods for the preparation of the alkali cellulose include the traditional steeping and pressing method of the viscose industry, the use of screw presses (26), and the newer techniques of dipping, spraying, or use of slurries for improving the distribution of the sodium hydroxide. Better heat removal during the reaction has been achieved by the design of continuous reaction systems that use reflux condensers (27) and slurries in excess reagent (28). Finishing comprises washing with hot water, drying, and grinding to obtain the desired physical form.

Specifications and Standards

Specifications for methyl cellulose are given in the *Food Chemicals Codex II*, the *Pharmacopeia of U.S.*, and in publications of the FAO/WHO (15,21). The EEC product number of methyl cellulose is E460, and the specification reference is FAS/IV/17. Methyl cellulose is also classified GRAS under Title 21, Part 182.1480 of the Code of Federal Regulations (18). Methyl cellulose, meeting the above standards, may also be used in the following standardized foods: French dressing, salad dressing, artificially sweetened fruit jelly, fruit preserves, and jams, and soda water. Use as a binder to extend and stabilize meat and vegetable patties, according to USDA regulation 9CFR 318.7, is also permitted.

Specifications for 2-hydroxypropyl methyl cellulose having specified ranges of methoxyl and hydroxypropoxyl substitution, are given in reports of the FAO/WHO

(17,21,29). The EEC product number for hydroxypropyl methyl cellulose is E462 and the specification reference is FAS/IV/16. Food Additive Regulation 172.874 allows the use of hydroxypropyl methyl cellulose is nonstandardized foods as an emulsifier, film former, protective colloid, stabilizer, suspending agent, or thickener. Use in select standardized foods, water ices, fruit and nonfruit sherbets, French dressing and salad dressing is also allowed. Other nonstandardized uses include fruit pie fillings, dietetic items, baked goods, breading batters, fried foods, and whipped toppings.

Both methyl cellulose and hydroxypropyl methyl cellulose are exempt, inert ingredients in pesticide formulations under 40 CFR 180.1001.

Analytical and Test Methods

Procedures for the analysis of methyl cellulose and 2-hydroxypropyl methyl cellulose are recorded in ASTM monographs (19) D 1347-72 and D 2363-72, respectively. Additional tests are listed in refs. 17 and 30. Further references are available (25).

Health and Safety Factors

Detailed reports on the toxicology of methyl cellulose and hydroxypropyl methyl cellulose are on file with the FDA. Two reports in 1973 cover long- and short-term feeding studies and a metabolic study using ^{14}C (31–32).

AID (Acceptable Intake Daily) values of 0–30 mg/kg (taken as the sum of total cellulose derivatives), with higher levels for dietary and calorie control purposes, are listed in a report by the Joint FAO/WHO Food Standards Program Codex Committee on Food Additives (15,21).

Biodegradation of methyl cellulose with activated sludge has been studied (33). Methyl cellulose is slowly biodegradable. In these experiments, 96% of the methyl cellulose was degraded or otherwise removed in 20 days by activated sludge.

As with most fine particulate organic dusts, methyl cellulose can form explosive mixtures with air. With <149 μm (−100 mesh) material, the critical level is reached at 30 g/m^3. See Table 4 for additional data.

Uses

Methyl cellulose and 2-hydroxypropyl methyl cellulose are nonionic cellulose ethers that can also function as polymeric surfactants (qv). Choice of a specific product will depend on the temperature requirements and the properties needed. Applications include adhesives where the thermal gel property provides quick set on heating and control of penetration into the substrate; agricultural chemicals where dispersion, adhesiveness, and low phytotoxicity are necessary; in ceramic shapes that utilize the lubricity, thermal gelation, and binding properties; and construction products such as cement formulations, tile mortars and grouts, dry wall joint cements, and gypsum plasters (see Cement; Chemical grouts). Additional uses are in cosmetics such as shampoos, leather pasting, latex paint, paint removers, reconstituted tobacco sheet, textile printing pastes, and printing inks.

The pharmaceutical industry uses methyl cellulose and its hydroxypropyl modifications as binders and granulating agents for tablets, in bulk laxatives, for film coating of tablets, in suspensions of various types, and for thickening of lotions and gelled products.

The protective colloid properties of methyl cellulose and hydroxypropyl methyl cellulose find extensive use in the suspension polymerization of vinyl chloride to control particle size distribution and porosity of the finished resin (see Vinyl polymers).

HYDROXYETHYL CELLULOSE

2-Hydroxyethyl cellulose (HEC) is a water-soluble, nonionic cellulose ether that differs from methyl cellulose and 2-hydroxypropyl methyl cellulose in that it is soluble in both cold and hot water. The nonthermal gelling property increases the tolerance for ionic salts. Substitution of commercial products covers a range of 1.5 to 3.0 MS. Both untreated products and products that have been cross-linked with glyoxal are offered. The latter requires an alkaline pH for dissolution.

Properties

Typical property data for 2-hydroxyethyl cellulose are given in Table 7. The reaction of ethylene oxide with cellulose can occur in two ways. First, substitution can take place at a ring hydroxyl giving direct attachment to the ring. This reaction generates a primary hydroxyl group which can then further react with ethylene oxide to form a side chain of several ethyleneoxy units. The distribution and ratio of ring to side-chain substitution can affect properties such as rheology, resistance to enzyme attack (35) and compatibility with salts. The quantity of substitution is also important.

Although 2-hydroxyethyl cellulose does not have a thermal gel point, the viscosity of solutions decreases with increase of temperature (36). Solutions of low molecular weight will show a low degree of pseudoplasticity and behave like Newtonian fluids. High mol wt products are highly pseudoplastic.

Manufacture

Since sodium hydroxide is a catalyst in the reaction of ethylene oxide with cellulose, the quantity of sodium hydroxide used in the preparation of HEC is lower than that required for the alkylation of cellulose. A water-miscible diluent such as isopropyl alcohol or *tert*-butyl alcohol is used to improve heat transfer from the exothermic reaction (37). Typical reaction temperatures are 30–35°C with reaction times of 4 hr (38). Purification comprises washing with organic solvents such as acetone–methanol or cross-linking with a dialdehyde such as glyoxal followed by cold water washing (39–40). By-products include ethylene glycol and ethers of ethylene glycol.

Substitution of ethylene oxide results in the formation of a primary hydroxyl group that can react with additional ethylene oxide to form side chains of several units. Several studies have shown that side chain formation can result in lower enzyme resistance (41–42). An alkali mixture of sodium and lithium hydroxides, which achieves increased bioresistance to enzyme hydrolysis, is described in a recent patent (43).

154   CELLULOSE DERIVATIVES, ETHERS

Table 7. Typical Properties of Hydroxyethyl Cellulose [a]

Property	Value
polymer as shipped	
ash content (calculated as Na_2SO_4), %	3.5
effect of heat	
softening range, °C	135–140
browning range, °C	205–210
bulk density, g/mL	0.6
particle size, retained on 40 U.S. mesh (420 μm),%	5
biological oxygen demand [b], ppm	
2.5 MS 1500–2500 mPa·s(=cP), 1%	7,000
2.5 MS 75–150 mPa·s, 5%	18,000
solutions	
sp gr, 2% soln, g/mL	1.0033
refractive index, 2% soln	1.336
pH	7
surface tension, mN/m(=dyn/cm)	
2.5 MS 75–150 mPa·s(=cP), 5% @ 0.1%	66.8
@ 0.001%	67.3
1.8 MS 75–150 mPa·s, 5% @ 0.1%	66.7
@ 0.001%	69.8
interfacial tension vs Fractol, mN/m(=dyn/cm)	
2.5 MS 75–150 mPa·s(=cP), 5% @ 0.001%	25.5
1.8 MS 150–400 mPa·s, 2% @ 0.001%	23.7
bulking value in soln, cm^3/g	.834
films	
refractive index	1.51
sp gr at 50% relative humidity, g/mL	1.34
moisture content (equilibrium), %	
at 23°C, 50% relative humidity	6
at 23°C, 84% relative humidity	29

[a] Ref. 34.
[b] After 5 days of incubation. Under these test conditions, cornstarch has a BOD of over 800,000 ppm.

Economic Aspects

Major production sites of HEC are located in the United States (3 sites), Netherlands, and Belgium. Smaller quantities of this polymer are also produced in Germany and France. Global capacity of major market producers is estimated at about 39,000 metric tons increasing to 46,000 metric tons by 1980. Current prices are $3.09–3.35/kg.

Specifications and Standards

The FDA has included HEC as a permitted ingredient of adhesives and coatings for food packaging and containers in accordance with the following regulations:
  The glyoxal cross-linked products are cleared only under Sections 175.105 and 176.180. Hydroxyethyl cellulose has not been cleared for direct use as a food additive.
  Section 175.300—Resinous and polymeric coatings.
  Section 176.105—Adhesives.

Section 176.170—Components of paper and paperboard in contact with aqueous and fatty foods.
Section 177.1210—Closures with sealing gaskets for food containers.
Section 176.180—Components of paper and paperboard in contact with dry food.
HEC is exempt from the requirements of a residue tolerance under FDA Regulation 182.99 and EPA Regulation 180.1001(c).

Analytical and Test Methods

Procedures for testing HEC are given in ASTM Monograph D-2364-69 (19). Additional information on methods of analysis is also available in the product bulletins of the manufacturers (34,36).

Health and Safety Factors

Although not permitted as a direct food additive, 90-d and 2-yr feeding studies have shown that HEC is of a low order of toxicity. Dosages of 5% of the diet had no adverse effects (34,36). Patch tests showed no evidence of sensitization (34).

Dusts of HEC, like those of other cellulose ethers, can form explosive mixtures with air on the same order of magnitude as that for methyl cellulose.

Uses

The nonionic chemical structure and solubility in both cold and hot water are major factors in the utilization of HEC. Coatings manufacturing and petroleum production are major consumers of HEC. Substantial quantities are used as a protective colloid in vinyl acetate polymerization (see Vinyl polymers), products for the construction industry such as cements, textile processing (see Textiles), and a wide variety of miscellaneous products such as adhesives, agricultural products, cosmetics, paper coatings and sizes, pharmaceutical products such as lotions and jellies, and asphalt emulsions (34,36).

2-HYDROXYETHYL METHYL CELLULOSE

In the section on methyl cellulose, the use of alkylene oxide substitution to modify the properties of methyl cellulose was described. The reverse situation has been disclosed in recent patents (23,44) where methyl substitution has been used to modify the properties of HEC. This provides limitations on side chain formation by capping of the primary hydroxyl end-groups of HEC. It results in increased hydroxyethyl substitution along the cellulose chain with improved enzyme resistance and control of the reaction, plus excellent salt tolerance. Variation of the ratio of methyl to hydroxyethyl substitution permits the design of products for specific uses.

Hydroxyethyl methyl cellulose (HEMC) is similar to HEC in tolerance to most salts. The presence of the methyl group results in surface activity that is intermediate to that of methyl cellulose and hydroxyethyl cellulose. HEMC, like HEC, has no thermal gel characteristics in water at boiling temperatures. Substitution ranges of commercial products are hydroxyethoxyl MS 1.8 to 3.0 and methoxyl DS 0.8 to 1.2. Decomposition begins at 260°C and the bulking value is 0.743 cm^3/g.

A process for HEMC manufacturing has been described (23,44).

Laboratory animal tests and a human panel test showed no contraindications for the use of hydroxyethyl methyl cellulose in cosmetic and dermatological preparations.

HEMC is used in latex paint and formulations where high resistance to enzyme attack is a requisite for long shelf life. The intermediate surface activity properties are also useful in vinyl acetate polymerization.

OTHER WATER-SOLUBLE CELLULOSE ETHERS

2-Hydroxypropyl Cellulose. When propylene oxide reacts with alkali cellulose in a modification of the hydroxyethyl cellulose process, a product is obtained that differs greatly in properties compared to HEC. Hydroxypropyl cellulose has a thermal gel point like methyl cellulose and, in addition, is thermoplastic. The high level of substitution (MS about 4) results in solubility in a number of organic solvents as well as water. Typical properties are given in Table 8.

Hydroxypropyl cellulose can be made in a slurry of hexane (46–47) or propylene oxide (48). Modifications of hydroxypropyl cellulose to obtain higher temperature of gelation have included use of ethylene oxide (49–50) and ethylenimine (51).

Hydroxypropyl cellulose meeting the specifications of Section 172.870 of the FDA may be used as a direct food additive except for the restrictions listed for standardized foods. Specifications for hydroxypropyl cellulose are also listed in ref. 17.

Uses include coatings and glazes for food items, as protective colloid in suspension polymerization, and in alcohol-based cosmetics, micro- and macro-encapsulation, inks, paint removers, and pharmaceuticals. Solubility in alcohols and other polar solvents has permitted uses where a combination of water and organic solvent solubility is needed.

Ethyl 2-Hydroxyethyl Cellulose. Ethyl hydroxyethyl cellulose is a nonionic cellulose ether which, like the methyl cellulose products, is soluble in cold water and insoluble in hot water. Commercial development of water soluble ethyl hydroxyethyl cellulose originated in Sweden in 1945 (52–55). Commercial products are available in two combinations of substitution. Properties of typical products are given in Table 9.

Table 8. Properties of Hydroxypropyl Cellulose[a]

Property	Value
powder	
physical appearance	off-white odorless, tasteless powder
apparent density, g/cm^3	0.5 (can vary with type)
softening temperature, °C	130
burnout temperature, °C	450–500 in N_2 or O_2
solution	
sp gr, 2% at 30°C	1.010
refractive index, 2% soln	1.337
surface tension, mN/m(=dyn/cm)	43.6
interfacial tension, Fractol, mN/m(=dyn/cm)	12.5
bulking value in soln, cm^3/g	0.334

[a] Ref. 45.

Table 9. Properties of Ethyl 2-Hydroxyethyl Cellulose [a]

Property	Value	
ethyl DS	0.9	0.8
hydroxyethyl MS	1.4	0.5
physical appearance	granules; powder	fibrous; granules
bulk density, g/L	400–600; 300–500	100–200; 400–600
sp gr of film	1.33	1.24
refractive index of film	1.49	1.49
tensile strength of film		
65% relative humidity, 20°C, MPa[b]	44.1–53.9	24.5–34.3
elongation of film, 65%		
relative humidity, 20°C, %	25–35	5–15

[a] Ref. 56.
[b] To convert MPa to psi, multiply by 145.

Alkali cellulose is prepared by steeping in sodium hydroxide followed by pressing and shredding. The alkali cellulose is then added to a pressure vessel and allowed to react with a mixture of ethyl chloride and ethylene oxide. Owing to a difference in reactivity, hydroxyethylation precedes ethylation as the temperature is increased. Purification with hot water is similar to that of methyl cellulose.

Uses are similar to those of the methyl cellulose products and include latex paints, wallpaper pastes, mortars, grouts, reconstituted tobacco sheets, and paper coatings.

ETHYL CELLULOSE

Ethyl cellulose and its modification, ethyl hydroxyethyl cellulose, are the major water insoluble cellulose ethers of commerce. Manufacture of these products occurs solely in the United States. Some production of benzyl cellulose [9015-11-6] occurred in Germany in the early 1940s on a pilot plant scale (56) but a low melting point and inferior stability to light and heat precluded extensive commercialization.

Properties

Ethyl cellulose is a white, odorless, granular solid. The ethoxyl substitution values of commercial products range from a DS of 2.2 to 2.6 (44.5–>49.0%). Typical values for a number of properties are shown in Table 10. Commercial products are available in two ranges of DS at ca 2.0–2.3 and ca 2.4–2.6 (57–59).

The ethoxyl content of ethyl cellulose greatly affects the thermal behavior. Figure 2 shows the changes in softening and melting points that occur when the ethoxyl is varied over a range of 44–50% (58).

Ethyl cellulose with a DS of 2.4–2.6 dissolves completely in all organic solvents except naphthas, purely aliphatic hydrocarbons, polyhydric alcohols, and a few ethers. Ethyl cellulose with a lower DS range of 2.0–2.3 yields a clear solution in relatively few single solvents. These include cyclohexene, methyl acetate, dioxane, butyl acetate, and most chlorinated aliphatic hydrocarbons. However, it dissolves readily in mixtures of aromatic hydrocarbons and ethanol or butanol.

Ethyl cellulose is most often used in solvent mixtures consisting of 60–80% aro-

Table 10. Properties of Ethyl Cellulose, 2.4–2.6 DS[a]

Property	Value
powder	
bulking value in soln, cm^3/g	0.826–0.868
film	
tensile strength, MPa[b] (0.076 mm film, dry)	46.8–72.4
tensile strength, % of dry strength (0.076 mm film, wet)	80–85
elongation at rupture (0.76 mm film, 25°C, 50% rH), %	7–30
moisture absorption, %, 24 h at 80% rH	2
hardness index, Sward, 0.076 mm film	52–61
softening point, °C	152–162
gas transmission	
water vapor, g/(m^2·d), 0.076 mm film, ASTM E-96-53T procedure	890
oxygen, ASTM D1434	2000
nitrogen, ASTM D1434	600
carbon dioxide, ASTM D1434	5000
hydrogen, ASTM D1434	7500
electrical properties	
dielectric constant at 25°C, 1 MHz	2.8–3.9
dielectric constant at 25°C, 1 kHz	3.0–4.1
dielectric constant at 25°C, 60 Hz	2.5–4.0
power factor at 25°C, 1 kHz	0.002–0.02
power factor at 25°C, 60 Hz	0.005–0.02
volume resistivity, Ω·cm	10^{12}–10^{14}
dielectric strength, V/0.0254 mm, ASTM step by step	1500

[a] Refs. 58–59.
[b] To convert MPa to psi, multiply by 145.

matic hydrocarbon and 20–40% alcohol. This percentage of alcohol results in the minimum viscosity at any given solids concentration (58). The viscosity is extremely high in toluene alone, decreases very rapidly with small percentages of the alcohol, and reaches a minimum at about 30% alcohol. As the alcohol content increases beyond 30%, the viscosity rises with all alcohols (except methanol) and this increase is greater the higher the molecular weight of the alcohols used. A lower minimum viscosity can be obtained with ethanol or methanol than butanol. Substitution of benzene, xylene, or ethylbenzene for the toluene has little effect on the alcohol content at which the viscosity minimum is obtained and on the actual viscosity at the minimum with any one alcohol. Therefore, the viscosity of an ethyl cellulose solution depends primarily on the kind of alcohol used, and on its percentage in the solvent. The solids content, obtainable at a constant viscosity is also dependent on the alcohol used and on its proportion in the solvent.

The formulation of solvents for 2.21–2.36 ethyl cellulose should be considered as a modification of the formulation procedure used with 2.4–2.6 DS ethyl cellulose. Ethyl cellulose of higher ethoxy content can be used in solvents containing a very low percentage of alcohol when solution viscosity is not very important, but the lower ethoxy-content ethyl cellulose will not produce clear films unless the combined alcohol content is about 20–30%. It yields solutions of minimum viscosity in mixtures of 60–65% aromatic hydrocarbons with 35–45% alcohol. The relative effect of the different alcohols on viscosity is similar to that described above. The particular alcohol to be used is chosen on the same basis as that previously outlined, considering solution flow and the effect of alcohol on resins or plasticizers.

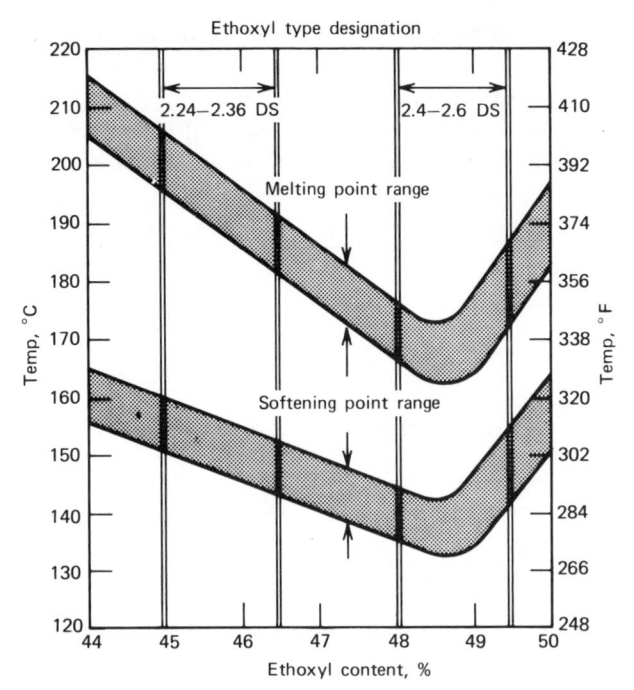

Figure 2. Softening point and melting point temperatures as a function of ethoxyl content of ethyl cellulose (58).

Manufacture

Either chemical grade cotton linters or wood pulp can be used to prepare ethyl cellulose. The sequence of chemical reactions is similar to that for methylation of cellulose. In commercial practice, sodium hydroxide concentrations of 50% or higher are used to prepare the alkali cellulose. Staged additions of solid sodium hydroxide during the reactions can be used to reduce side reactions. Ethyl chloride is added to the alkali cellulose in nickel-clad reactors at 90–150°C and 828 to 965 kPa (120 to 140 psi) for 6–12 h. Diluents such as benzene or toluene can be used.

At the end of the reaction, the volatiles such as ethyl chloride, diethyl ether, ethanol, and diluent are recovered and recycled. The ethyl cellulose in solution is precipitated in the form of granules with further recovery of the carrier solvents. Washing with water completes the processing. Control of metallic impurities is important to achieve stability during storage. Antioxidants can also be incorporated to inhibit loss of viscosity.

Specifications and Standards

Ethyl cellulose resins are used in foods and pharmaceuticals. Specifications applicable to these uses are defined in the NF XIV and ref. 17. Since this is the last edition of the NF, subsequent listings will appear in the USP. The following FDA regulations are also in effect for ethyl cellulose complying with the specifications given in the above compendia (18):

  73.1001—as a diluent in certain color additive mixtures for externally applied drugs.

73.1—as a diluent in inks for marking food supplements in tablet form, gum and confectionary, inks for marking fruits and vegetables, and as a diluent in color additive mixtures for coloring egg shells.

172.868—as a binder and filler in dry vitamin preparations, as a component of protective coatings for vitamin preparations and mineral tablets and as a flavor fixative in flavoring compounds.

573.420—as a binder or filler in dry vitamin preparations to be incorporated into animal food.

Analytical Methods

For methods of analysis refer to ASTM D914-72 (19), NF XIV, and *Food Chemicals Codex II*.

Health and Safety Factors

Commercial ethyl cellulose resins are not soluble in water and thus present no significant ecological hazards. They are nonbiodegradable in aqueous environments and should therefore present no ecological hazards to aquatic life (58). Insolubility in water, the high level of DS, the stability of the ether linkage, and lack of metabolism are factors contributing to its acceptability as a functional additive in sanctioned uses where the ethyl cellulose resin is ingested or in contact with food and pharmaceutical products.

Ethyl cellulose should be stored at 1.7–32°C in a dry area away from all sources of heat. As with any organic material, ethyl cellulose should not be stored next to peroxides or other oxidizing agents. Good housekeeping is required to prevent dusts of ethyl cellulose from reaching explosive levels in air; these levels are approximately 25 g/m^3 for <149-μm (−100 mesh) powder (min explosive limit).

Uses

The early major uses of ethyl cellulose as a molding and extrusion resin have been superseded by a wide spectrum of specialty uses such as:

Lacquers. Chart paper, copy paper, gravure and flexo inks for heat transfer printing on fabrics, screen inks, fabric inks and coatings, electrical coatings, and metal inks.

Binder/Modifier. Publication inks, controlled permeability coatings, compression binders, nitrocellulose lacquers, sandpaper, varnishes, glass frits, and fluorescent tube temporary coatings.

Hot Melts. Strippable protective coatings, transfer and ceramic inks, pressure-sensitive adhesives, paper coating, and decalcomania.

Pharmaceuticals. Binder, filler, protective tablet coatings, delayed-release coatings, microencapsulating polymer, and flavor fixative.

Gel Lacquers. Bowling pins, generator field coils, and glass coatings.

ETHYL 2-HYDROXYETHYL CELLULOSE AND 2-CYANOETHYL CELLULOSE

One of the features of ethyl cellulose is its solubility in a wide range of solvents; however, it is not soluble in aliphatic hydrocarbons. Solubility in mixtures of solvents rich in aliphatics can be achieved by modifying the ethyl cellulose with a small amount (DS ca 0.3) of hydroxyethyl substitution. Since capping of the hydroxyethyl group is necessary, the hydroxyethylation reaction must precede the ethylation of the cellulose (60–61). The addition of small amounts of a cosolvent such as isopropanol will aid in preparing clear solutions which are used in silk-screen printing and rotogravure inks (61).

Ethyl hydroxyethyl cellulose has very low toxicity and has been cleared under the Federal Food, Drug, and Cosmetic Act for specified uses that involve food contact (61).

Cellulose, in the presence of dilute sodium hydroxide (2%), reacts readily with acrylonitrile (qv) to form the cyanoethyl ether (62–63). At higher temperatures and concentrations of NaOH, hydrolysis to the 2-carboxyethyl group can occur. However, concentrations of sodium hydroxide above 10% should be avoided owing to the hazard of highly exothermic runaway polymerization of the acrylonitrile (see Cyanoethylation).

Although cyanoethyl cellulose had some interesting electrical properties (64–65) and cyanoethylation of cotton resulted in improved dyeability and abrasion resistance (66), the introduction of this product did not result in commercial production. Further process and property information can be obtained from a review (67).

BIBLIOGRAPHY

"Cellulose Derivatives" in *ECT* 1st ed., Vol. 3, pp. 357–391, by E. D. Klug, Hercules Power Company; "Cellulose Derivatives" in *ECT* 2nd ed., Vol. 4, pp. 616–652, by E. D. Klug, Hercules Powder Company.

1. N. M. Bikales and L. Segal, eds., *Investigations of the Structure of Cellulose and Its Derivatives in High Polymers*, 2nd ed., Vol. V, Wiley-Interscience, New York, 1971, Part IV, Chapt. XIII, pp. 1–325.
2. A. B. Savage, A. E. Young, and A. T. Maasberg in E. Ott, H. M. Spurlin, and M. W. Graffin, eds., *High Polymers*, 2nd ed., Vol. V, Interscience Publishers, Inc., New York, 1954, pp. 882–958.
3. A. B. Savage, "Derivatives of Cellulose, Ethers," in ref. 2, Part V, Chapt. XVII, pp. 785–810.
4. A. F. Turbak, ed., *Cellulose Technology Research* in *American Chemical Society Symposium Series No. 10*, American Chemical Society, 1975.
5. Ger. Pat. 332,203 (Jan. 10, 1918) (to Deutsche Celluloid Fabrik Eilenberg).
6. U.S. Pat. 2,618,632 (Nov. 18, 1952), E. D. Klug (to Hercules, Inc.).
7. *Hercules Cellulose Gum Chemical and Physical Properties*, Hercules, Inc., Wilmington, Del., 1976.
8. U.S. Pat. 2,523,377 (Sept. 26, 1950), E. D. Klug (to Hercules, Inc.).
9. U.S. Pat. 2,131,733 (Oct. 4, 1938), J. F. Haskins and R. W. Maxwell (to E. I. du Pont de Nemours & Co., Inc.).
10. U.S. Pat. 2,524,024 (Sept. 26, 1950), R. W. Swinehart and S. T. Allen (to The Dow Chemical Co.).
11. U.S. Pat. 2,517,577 (Aug. 8, 1950), E. D. Klug and J. S. Tinsley (to Hercules, Inc.).
12. U.S. Pat. 2,976,278 (Mar. 21, 1961), O. H. Paddisen and R. W. Somer (to E. I. du Pont de Nemours & Co., Inc.).
13. U.S. Pat. 2,512,338 (June 20, 1950), E. D. Klug and H. M. Spurlin (to Hercules, Inc.).
14. *Chem. Mark. Rep.* **212,** 9 (Aug. 22, 1977).
15. *Specifications for the Identity and Purity of Food Additives and Their Toxicological Evaluation: Emulsifiers, Stabilizers, Bleaching and Maturing Agents* in World Health Organization Technical Report Series No. 281, Geneva, 1964.
16. National Fire Protection Association, *Fundamental Principles for the Prevention of Dust Explosions in Industrial Plants*, Bulletin 71063.

162 CELLULOSE DERIVATIVES, ETHERS

17. *U.S. Food Chemicals Codex II*, National Academy of Sciences, 1972.
18. *Code of Federal Regulations, Title 21*, U.S. Gov't Printing Office, Washington, D.C.
19. American Society for Testing Materials, *Monograph D1439-65*, Philadelphia, Pa., 1965.
20. *Analytical Procedures for Assay of CMC and Its Determination in Formulations*, Hercules, Inc., Wilmington, Del., 1971.
21. Joint FAO/WHO Food Standards Programme, *Index of Food Additives CX/FA 72/2*, Codex Committee on Food Additives, Oct. 1971.
22. W. Suida, *Monatsch*, **26,** 413 (1905).
23. U.S. 3,709,876 (Jan. 9, 1973), R. L. Glomski, L. E. Davis, and J. A. Grover (to The Dow Chemical Company).
24. U.S. Pat. 2,835,666 (May 20, 1958), A. B. Savage (to The Dow Chemical Company).
25. *Methocel*, The Dow Chemical Company, Midland, Mich., 1975.
26. U.S. Pat. 3,615,254 (Oct. 26, 1971), F. Eichenseer and H. Kletschke (to Kalle A.G.).
27. U.S. Pat. 3,544,556 (Dec. 1, 1970), F. Eichenseer, S. Janocha, and H. Macholdt (to Kalle A.G.).
28. U.S. Pat. 4,015,067 (Mar. 29, 1977), G. Y. T. Liu and C. P. Strange (to The Dow Chemical Company).
29. *Toxicological Evaluation of Some Antimicrobials, Antioxidants, Emulsifiers, Stabilizers, Flour Treatment Agents, Acids, and Bases,* in *FAO Nutrition Meetings Report Series No. 40 A, B, C, WHO/Food Add. 67.29*, pp. 75.
30. *United States Pharmacopeia*, Vol. XIX, United States Pharmacopeial Convention Inc., 1975.
31. S. B. McCollister, R. J. Kociba, and D. D. McCollister, *Fed. Cosmet. Toxicol.* **11,** 943 (1973).
32. W. H. Braun, J. C. Ramsey, and P. J. Gehring, *Fed. Cosmet. Toxicol.* **12,** 373 (1974).
33. F. A. Blanchard, I. T. Takahaski, and H. C. Alexander, *Appl. Environ. Microbiol.* **32**(4), 557 (Oct. 1976).
34. *Natrosol Hydroxyethylcellulose*, Hercules, Inc., 1974.
35. *Natrosol B Improved Viscosity Stability for Latex Paints*, Hercules, Inc., 1975.
36. *Cellosize Hydroxyethylcellulose*, Union Carbide, 1974.
37. U.S. Pat. 2,572,039 (Oct. 23, 1951), E. D. Klug and H. G. Tennent (to Hercules, Inc.).
38. U.S. Pat. 2,682,535 (June 29, 1954), A. E. Broderick (to Union Carbide).
39. U.S. Pat. 3,347,847 (Oct. 17, 1967), K. Engelskirchen and J. Golinke (to Heubel & Cie.).
40. U.S. Pat. 3,527,751 (Sept. 8, 1970), J. E. Gill (to Imperial Chemical Industries Ltd.).
41. E. T. Reece, R. G. H. Siu, and H. S. Levinson, *J. Bacteriol.* **59**(4), 485 (1950).
41a. E. T. Reece, *Ind. Eng. Chem.* **49,** 89 (1957).
42. M. G. Wirick, *J. Polym. Sci. (A-1)* **6,** 1705 (1968).
43. U.S. Pat 4,009,329 (Feb. 27, 1977), W. C. Arney, C. A. Williams, and J. E. Glass, Jr. (to Union Carbide Corporation).
44. U.S. Pat. 3,903,076 (Sept. 2, 1975), K. L. Krumel, J. A. May, and F. W. Stanley, Sr. (to The Dow Chemical Company).
45. *Klucel Hydroxypropylcellulose, Chemical and Physical Properties*, Hercules, Inc., 1976.
46. U.S. Pat. 3,357,971 (Dec. 12, 1967), E. D. Klug (to Hercules, Inc.).
47. U.S. Pat. 3,351,583 (Nov. 7, 1967), R. G. Bishop (to Hercules, Inc.).
48. U.S. Pat. 3,278,520 (Oct. 11, 1966), E. D. Klug (to Hercules, Inc.).
49. U.S. Pat. 3,278,521 (Oct. 11, 1966), E. D. Klug (to Hercules, Inc.).
50. Jpn. Kokai 50-26888 (July 10, 1973), T. Kitagaki (to Shinetsu Kagaku Kogyo Co.).
51. U.S. Pat. 3,431,254 (Mar. 4, 1969), E. D. Klug (to Hercules, Inc.).
52. S. Sönnerskog, *Su. Papperstidn.* **48,** 413 (1945).
53. S. Sönnerskog, *Su. Papperstidn.* **49,** 409 (1946).
54. S. Sönnerskog, *Tek. Tidskr.* **77,** 133 (1957).
55. S. Sönnerskog, *Acta Polytech. Chem. Met. Ser. 4* **157,** (4), (1954).
56. S. Lindefors and I. Jullander in R. L. Whistler, ed., *Industrial Gums*, 2nd ed., Academic Press, Inc., New York, 1973, pp. 678.
57. *Ethocel Handbook*, The Dow Chemical Company, Midland, Mich., 1940.
58. *Tough Ethocel Ethylcellulose Resins*, The Dow Chemical Co., Midland, Mich., 1974.
59. *Hercules Ethyl Cellulose, Properties Uses*, Hercules, Inc., Wilmington, Del., 1974.
60. U.S. Pat. 2,610,180 (Sept. 9, 1952), E. D. Klug (to Hercules Powder Co.).
61. *Ethylhydroxyethylcellulose(EHEC)*, Hercules, Inc., 1972.
62. U.S. Pat. 2,535,690 (Dec. 26, 1950), H. F. Miller and R. G. Flowers (to General Electric).
63. Can. Pat. 594,306 (Mar. 15, 1960), N. M. Bikales (to American Cyanamid Co.).
64. V. S. Domkin and co-workers, *Vysomokol. Soedin.* **11,** 873 (1969).

65. *Cyanocel, Chemically Modified Cellulose,* American Cyanamid, New York, 1960.
66. J. Compton, *Am. Dyest. Rep.* **43,** 103 (1954).
67. N. M. Bikales in ref. 1, *Ethers from α,β-Unsaturated Compounds,* Part V. Chapt. VII, pp. 811–833.

G. K. GREMINGER, JR.
Dow Chemical, U.S.A.

CELLULOSE DERIVATIVES—PLASTICS. See Plastics technology.

CEMENT

The term cement is used to designate many different kinds of substances that are used as binders or adhesives (qv). The cement produced in the greatest volume and most widely used in concrete for construction is portland cement. Masonry and oil well cements are produced for special purposes. Calcium aluminate cements are extensively used for refractory concretes (see Refractories). Such cements are distinctly different from epoxies and other polymerizable organic materials. Portland cement is a hydraulic cement, ie, it sets, hardens, and does not disintegrate in water. Hence it is suitable for construction of underground, marine, and hydraulic structures whereas gypsum plasters and lime mortars are not. Organic materials, such as latexes and water soluble polymerizable monomers, are sometimes used as additives to impart special properties to concretes or mortars; furthermore, concretes are sometimes impregnated with liquid organic monomers (or liquid sulfur) and polymerized to produce polymer-impregnated concrete. The term cements as used henceforth will be confined to inorganic hydraulic cements, principally portland and related cements. The essential feature of these cements is their ability on hydration to form with water relatively insoluble bonded aggregations of considerable strength and dimensional stability.

Hydraulic cements are manufactured by processing and proportioning suitable raw materials, burning (or clinkering at a suitable temperature), and grinding the resulting hard nodules called clinker to the fineness required for an adequate rate of hardening by reaction with water. Portland cement consists mainly of tricalcium silicate and dicalcium silicate. Usually two types of raw materials are required: one rich in calcium, such as limestone, chalk, marl, or oyster or clam shells; the other rich in the silica, such as clay or shale. The two other major phases in portland cements are tricalcium aluminate and a ferrite phase. A small amount of calcium sulfate in the form of gypsum or anhydrite is also added during grinding to control the setting time and enhance strength development.

The demand for cement was stimulated by the growth of canal systems in the United States during the 19th century. This led to process improvements in the calcination of certain limestones for the manufacture of natural cements, and to its gradual displacement by portland cement. The latter was named by Aspdin in a 1924 patent because of its resemblance to a natural limestone quarried on the Isle of Portland in England. Research conducted in many parts of the world since that time has provided a clear picture of the composition, properties, and fields of stability of the principal systems found in portland cement. These results led to the widely used Bogue calculation of composition based on oxide analysis (1). Recent research is reported in the *International Symposia on the Chemistry of Cements,* and the annual reviews, beginning in 1974, of the American Ceramic Society in *Cements Research Progress* (see under General References).

Clinker Chemistry

The conventional cement chemists' notation uses the following abbreviations for the most common constituents:

CaO = C	MgO = M	K_2O = K
SiO_2 = S	SO_3 = $\bar{S}$	CO_2 = $\bar{C}$
Al_2O_3 = A	Na_2O = N	H_2O = H
Fe_2O_3 = F		

Thus tricalcium silicate, Ca_3SiO_5, is denoted by C_3S.

Portland cement clinker is formed by the reactions of calcium oxide with acidic components to give C_3S, C_2S, C_3A, and a ferrite phase approximating C_4AF.

Phase Equilibria. During burning in the kiln, about 20–30% of liquid forms in the mix at clinkering temperatures. Reactions occur at surfaces of solids and in the liquid. The crystalline silicate phases formed are separated by the interstitial liquid. The interstitial phases formed from the liquid in normal clinkers during cooling are also completely crystalline to x-rays, although they may be so finely subdivided as to appear glassy (optically amorphous) under the microscope.

The high temperature phase equilibria governing the reactions in cement kilns were studied, eg, in the $CaO–Al_2O_3–SiO_2$ system illustrated in Figure 1 (2–3). In such a ternary diagram, the primary-phase fields are plotted, ie, the composition regions in which any one solid is the first to separate when a completely liquid mix is cooled (with negligible supercooling). The primary-phase fields are separated by eutectic points on the sides of the triangle such as that at 1436°C between tridymite and α-CS.

In the relatively small portland cement zone almost all modern cements fall in the high lime portion (about 65% CaO). Cements of lower lime content tend to be slow in hardening and may show trouble from dusting of the clinker by transformation of β- to γ-C_2S, especially if clinker cooling is very slow. The zone is limited on the high lime side by the need to keep the uncombined CaO to low enough values to prevent excessive expansion due to hydration of the free lime. Commercial manufacture at compositions near the $CaO–SiO_2$ axis can present difficulties. If the lime content is high, the burning temperatures may be so high as to be impractical. If the lime content is low, the burning temperatures may even be low, but impurities must be present in the C_2S to prevent dusting. On the high alumina side the zone is limited by excessive liquid-phase formation which prevents proper clinker formation in rotary kilns.

The relations between the compositions of portland cements and some other common hydraulic cements are shown in the $CaO–SiO_2–Al_2O_3$ phase diagram of Figure 2 (4), analogous to Figure 1. In this diagram the Fe_2O_3 has been combined with the Al_2O_3 to yield the Al_3O_3 content used. This is a commonly applied approximation that permits a two-dimensional representation of the real systems.

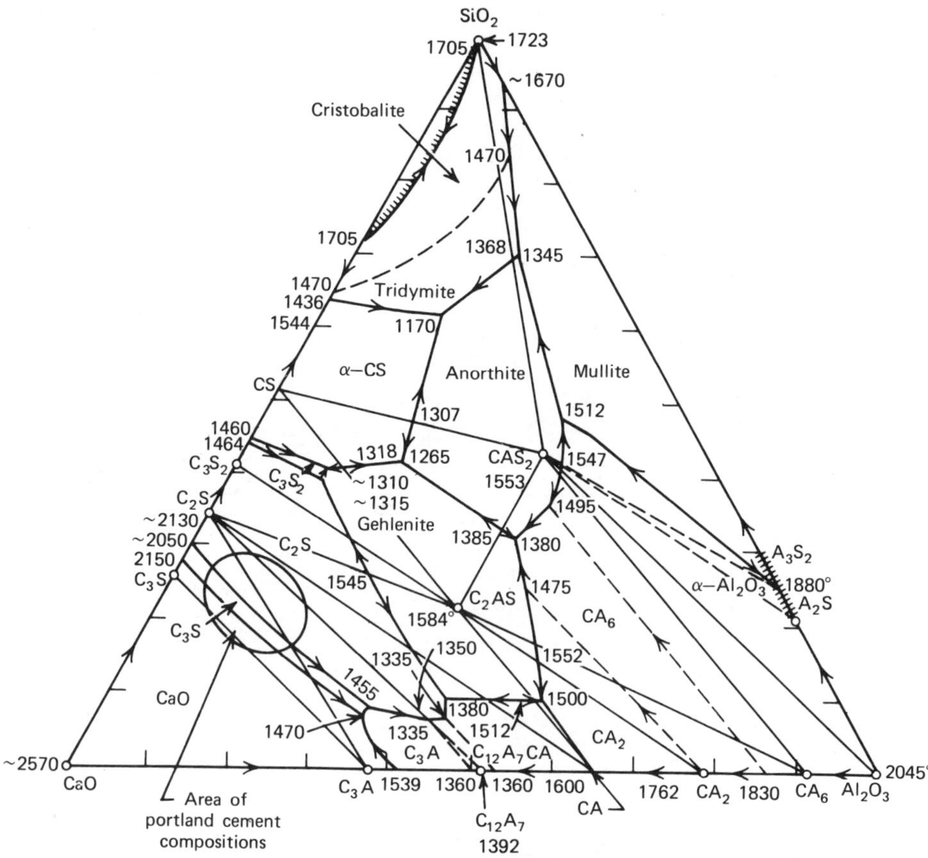

Figure 1. Phase equilibria in the CaO–Al$_2$O$_3$–SiO$_2$ system (2–3), in °C. Shaded areas denote two liquids, compositional index marks on the triangle are indicated at 10% intervals.

Clinker Formation. Portland cements are ordinarily manufactured from raw mixes including components such as calcium carbonate, clay or shale, and sand. As the temperature of the materials increases during their passage through the kiln, the following reactions occur: (1) evaporation of free water; (2) release of combined water from the clay; (3) decomposition of magnesium carbonate; (4) decomposition of calcium carbonate (calcination); and (5) combination of the lime and clay oxides. The course of reactions (5) occurring at the high temperature end of the kiln, just before and in the burning zone, is illustrated graphically in Figure 3 (6).

From the phase diagram of the CaO–SiO$_2$–Al$_2$O$_3$ system, the sequence of crystallization during cooling of the clinker can be derived if the cooling is slow enough to maintain equilibrium. For example, a mix at 1500°C of relatively low lime content, along the C$_3$S–C$_2$S eutectic line in Figure 1, will be composed of solid C$_3$S and C$_2$S and a liquid along the C$_3$S–C$_2$S eutectic at the intersection with the 1500°C isotherm (to the left of the 1470–1455 line). Upon cooling, this liquid deposits more C$_3$S and C$_2$S, moving the liquid composition down to the invariant point at 1455°C, at which C$_3$A also separates until crystallization is complete. Although real cement clinkers contain more components, which alter the system and temperatures somewhat, the behavior is similar.

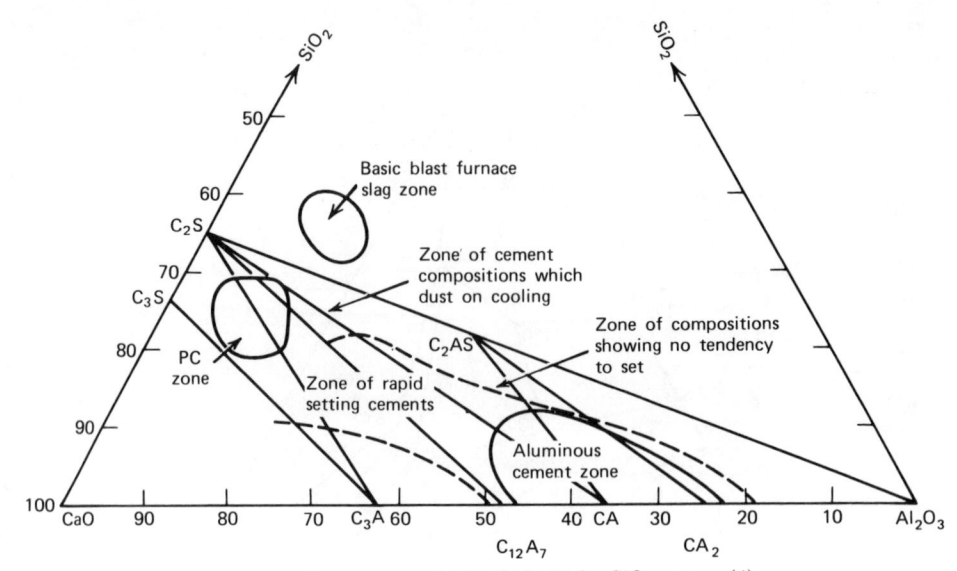

Figure 2. Cement zones in the CaO–Al$_2$O$_3$–SiO$_2$ system (4).

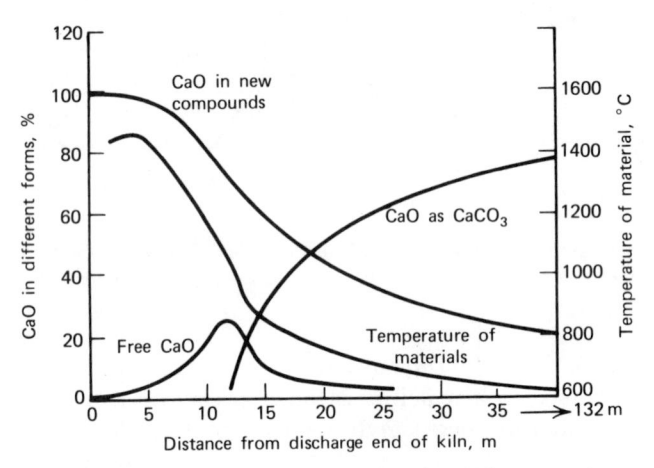

Figure 3. Temperatures and progress of reactions in a 132-meter wet-process kiln.

Cooling is ordinarily too rapid to maintain the phase equilibria. In the above case, the lime-deficient liquid at 1455°C requires that some of the solid C$_3$S redissolve and that more C$_2$S crystallize during crystallization of the C$_3$A. During rapid cooling there may be insufficient time for this reaction and the C$_3$S content will be higher than when equilibrium conditions prevail. In this event crystallization is not completed at 1455°C, but continues along the C$_3$A–C$_2$S boundary until the invariant point at 1335°C is reached. Crystallization of C$_2$S, C$_3$A, and C$_{12}$A$_7$ then occurs to reach complete solidification. Such deviations from equilibrium conditions cause variations in the phase compositions estimated from the Bogue calculation (see below), and cause variations in the amounts of dissolved substances such as MgO, alkalies, and the alumina content of the ferrite phase.

The theoretical energy requirement for the burning of portland cement clinker can be calculated from the heat requirements and energy recovery from the various stages of the process. Knowledge of the specific heats of the various phases, and the heats of decomposition, transformation, and reaction then permits calculation of the net theoretical energy requirement of 1760 kJ (420 kcal) for 1 kg of clinker from 1.55 kg of dry $CaCO_3$ and kaolin (7).

The kinetics of the reactions are strongly influenced by the temperature, mineralogical nature of the raw materials, fineness to which the raw material is ground, percentage of liquid phase formed, and viscosity of the liquid phase. The percentage of liquid formed depends on the alumina and iron oxides (see under Proportioning of Raw Materials). When the sum of these oxides is low, the amount of liquid formed is insufficient to permit rapid combination of the remaining CaO. The viscosity of the liquid at clinkering temperature is reduced by increasing the amounts of oxides such as MnO, Fe_2O_3, MgO, CaO, and Na_2O (8).

The reaction of C_2S with CaO to form C_3S depends upon dissolution of the lime in the clinker liquid. When sufficient liquid is present, the rate of solution is controlled by the size of the CaO particles, which depends in turn on the sizes of the particles of ground limestone. Coarse particles of silica or calcite fail to react completely under commercial burning conditions. The reaction is governed by the rate of solution (9):

$$\log t = \log \frac{D}{A} + 0.43 \frac{E}{RT}$$

t is the time in minutes, D is the particle diameter in mm, A is a constant, T the absolute temperature, and E is the activation energy with a value of 607 kJ/mol (146 kcal/mol). For example, 0.05 mm particles require 59 min for solution at 1340°C but only 2.3 min at 1450°C. A similar relation applies for the rate of solution of quartz grains.

Phases Formed in Portland Cements. Most clinker compounds take up small amounts of other components to form solid solutions. Best known of these phases is the C_3S solid solution called alite. Phases that may occur in portland cement clinker are given in Table 1. In addition, a variety of minor phases may occur in portland cement clinker when certain minor elements are present in quantities above that which can be dissolved in other phases. Under reducing conditions in the kiln, reduced phases, such as FeO and calcium sulfide may be formed.

The major phases all contain impurities. These impurities in fact stabilize the structures formed at high temperatures so that decomposition or transformations do not occur during cooling, as does occur with the pure compounds. For example, pure C_3S exists in (at least) six polymorphic forms each having a sharply defined temperature range of stability, whereas alite exists in three stabilized forms at room temperature depending upon the impurities. Some properties of the more common phases in portland clinkers are given in Table 2.

Structure. Examination of thin sections of clinkers using transmitted light and of polished sections by reflected light reveals details of the structure. A recently developed method (16) employs the polarizing microscope to determine the size and birefringence of alite crystals, and the size and color of the belite to predict later age strength. The clinker phases are conveniently observed by examining polished sections selectively etched with special reagents as shown in Figure 4. The alite appears as clear

168 CEMENT

Table 1. Phases in Portland Cement Clinker[a]

Name of impure form	CAS Registry No.	Chemical name	Cement chemists' notation
free lime	[1305-78-8]	calcium oxide	C
periclase (magnesia)	[1309-48-4] and [1317-74-4]	magnesium oxide	M
alite	[12168-85-3]	tricalcium silicate	C_3S
belite	[10034-77-2]	dicalcium silicate	C_2S
C_3A	[12042-78-3]	tricalcium aluminate	C_3A
ferrite	[12612-16-7]	calcium aluminoferrite[b]	$C_2A_xF_{1-x}$
	[12068-35-8]	tetracalcium aluminoferrite	C_4AF
	[12013-62-6]	dicalcium ferrite[c]	C_2F
mayenite	[12005-57-1]	12-calcium-7-aluminate	$C_{12}A_7$
gehlenite	[1302-56-3]	dicalcium alumino monosilicate	C_2AS
aphthitalite	[12274-74-4] and [17926-93-1]	sodium, potassium sulfate[d]	$N_xK_y\overline{S}$
arcanite	[7778-80-5] and [14293-72-2]	potassium sulfate	$K\overline{S}$
metathenardite	[7757-82-6]	sodium sulfate form I	$N\overline{S}$
calcium langbeinite	[14977-32-8]	potassium calcium sulfate	$2X\overline{S}.K\overline{S}$
anhydrite	[7778-18-9] and [14798-04-0]	calcium sulfate	$C\overline{S}$
calcium sulfoaluminate	[12005-25-3]	tetracalcium trialuminatesulfate	$C_4A_3\overline{S}$
alkali belite	[15669-83-7]	α'- or β-dicalcium (potassium) silicate[e]	$K_xC_{23}S_{12}$
alkali aluminate	[12004-54-3]	8-calcium disodium trialuminate	NC_8A_3
	[65430-58-2]	5-calcium disilicate monosulfate	$2C_2S.C\overline{S}$
spurrite	[1319-44-42]	5-calcium disilicate monocarbonate	$2C_2S.C\overline{C}$
	[12043-73-1]	calcium aluminate chloride	$C_{11}A_7.CaCl_2$
	[12305-57-6]	calcium aluminate fluoride	$C_{11}A_7.CaF_2$[f]

[a] Refs. 10–12.
[b] Solid solution series ($x = A/(A + F)$; $0 < x < 0.7$).
[c] End member of series.
[d] Solid solution series ($1/3 \leq x/y$).
[e] Solid solution series ($x \leq 1$).
[f] Mixed notation.

euhedral crystalline grains, the belite as rounded striated grains, the C_3A as dark interstitial material, and the C_4AF as light interstitial material.

Portland cement clinker structures (17–18) vary considerably with composition, particle size of raw materials, and burning conditions, resulting in variations of clinker porosity, crystallite sizes and forms, and aggregations of crystallites. Alite sizes range up to about 80 μm or even larger, most being 15–40 μm.

Raw Material Proportions. The three main considerations in proportioning raw materials for cement clinker are: the potential compound composition, the percentage of liquid phase at clinkering temperatures, and the burnability of the raw mix, ie, the relative ease, in terms of temperature, time, and fuel requirements, of combining the oxides into good quality clinker. The ratios of the oxides are related to clinker composition and burnability. For example, as the CaO content of the mix is increased, more C_3S can be formed, but certain limits cannot be exceeded under normal burning conditions. The lime saturation factor (LSF) is a measure of the amount of CaO that can be combined (19):

$$\text{LSF} = \frac{\% \text{ CaO}}{2.8\,(\% \text{ SiO}_2) + 1.1\,(\% \text{ Al}_2\text{O}_3) + 0.7\,(\% \text{ Fe}_2\text{O}_3)}$$

Table 2. Properties of the More Common Phases in Portland Cement Clinker[a]

Name	Crystal system	Density	Mohs' hardness
alite	triclinic monoclinic trigonal	3.14–3.25	ca 4
belite	hexagonal	3.04	
	orthorhombic	3.40	
	monoclinic	3.28	>4
	orthorhombic	2.97	
C_3A	cubic	3.04	<6
ferrite	orthorhombic	3.74–3.77	ca 5
free lime	cubic	3.08–3.32	3–4
magnesia	cubic	3.58	5.5–6

[a] Refs. 10–15.

An LSF of 100 would indicate that the clinker can contain only C_3S and the ferrite solid solution. Lime saturation factors of 88–94 are frequently appropriate for reasonable burnability; low LSF indicates insufficient C_3S for acceptable early strengths, and higher values may render the mix very difficult to burn. Several other weight ratios such as the silica modulus and the iron modulus are also important (20).

The potential liquid-phase content at clinkering temperatures range from 18 to 25% and can be estimated from the oxide analysis of the raw mix. For example (21), for 1450°C:

$$\% \text{ liquid phase} = 1.13 \ (\% \ C_3A) + 1.35 \ (\% \ C_4AF) + \% \ MgO + \% \ \text{alkalies}$$

The potential compound composition of a cement or cement clinker can be calculated from the oxide analyses of any given raw materials mixture, or from the oxide analyses of the cement clinker or finished cement. The simplest and most widely used method is the Bogue calculation (22). The ASTM C150 (23) calculation is somewhat modified.

The techniques of determining the proper proportions of raw materials to achieve a mix of good burnability and clinker composition are readily adaptable to a computer program which uses iterative techniques, starting with raw components of known composition. The concept of targets may be utilized, including fixed values of moduli, compound content, and amount of any raw material element in the final clinker. The number of targets that may be set is one less than the number of raw materials. The fuel ash must be considered as one of the raw materials.

Representative chemical analyses of raw materials used in making portland and high alumina cements are given in Table 3, analyses of cements of various types appear in Table 4, and their potential compound compositions in Table 5.

Hydration

Calcium Silicates. Cements are hydrated at elevated temperatures for the commercial manufacture of concrete products. With low pressure steam curing or hydrothermal treatment above 100°C at pressures above atmospheric, the products formed from calcium silicates are often the same as the hydrates formed from their

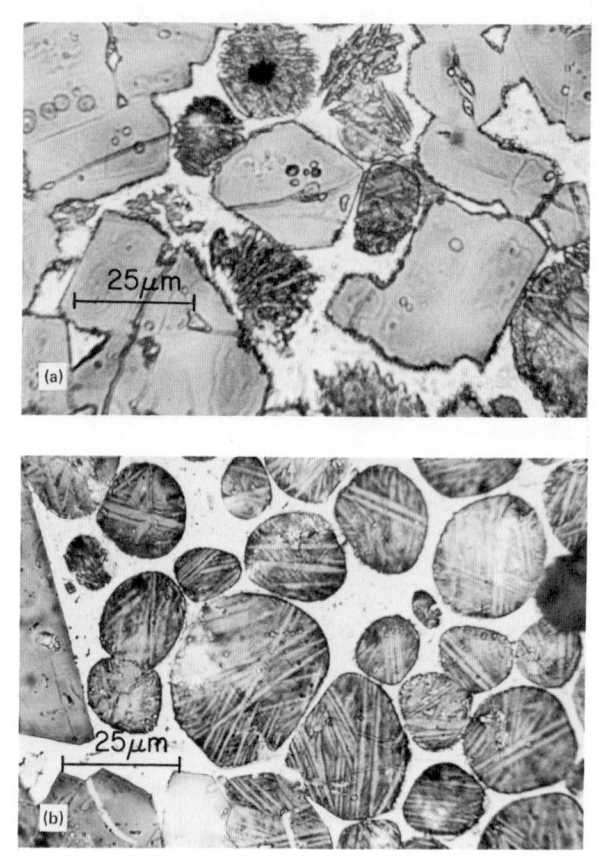

Figure 4. Photomicrograph of polished and etched sections of portland cement clinkers. The C_3A appears as dark interstitial material, the C_4AF as light interstitial material. (**a**) Euhedral and subhedral alite crystals and rounded or ragged belite; (**b**) rounded and striated belite crystals.

oxide constituents. Hence lime and silica are frequently used in various proportions with or without portland cement in the manufacture of calcium silicate hydrate products. Some of these compounds are listed in Table 6.

Although hydration under hydrothermal conditions may be rapid, metastable intermediate phases tend to form, and final equilibria may not be reached for months at 100–200°C, or weeks at even higher temperatures. Hence, the temperatures of formation given in Table 5 indicate the conditions under saturated steam pressure that may be expected to yield appreciable quantities of the compound, although it may not be the most stable phase at the given temperature. The compounds are listed in order of decreasing basicity, or lime/silica ratio. Reaction mixtures with ratios C:S = 1 yield xonotlite at 150–400°C. Intermediate phases of C-S-H (I), C-S-H (II), and crystalline tobermorite are formed in succession. Tobermorite (1.13 nm) appears to persist indefinitely under hydrothermal conditions at 110–140°C; it is a major part of the binder in many autoclaved cement-silica and lime-silica products.

In hydrations at ordinary temperatures (26) pure C_3S and β-C_2S (corresponding to the alite and belite phases in portland cements, respectively) react with water to form calcium hydroxide and a single calcium silicate hydrate (C-S-H), according to the following equations (in cement chemists' notation):

Table 3. Chemical Composition of Raw Materials[a], %

Type	SiO$_2$	Al$_2$O$_3$	Fe$_2$O$_3$	CaO	MgO	Ign. loss
cement rock	13.4	3.5	1.7	42.9	1.0	37.2
limestone	1.2	0.2	0.4	53.4	1.3	43.4
limestone	4.5	0.5	1.6	35.0	14.9	44.0
marl	6.0	0.6	2.3	49.1	0.4	40.4
oyster shells	1.5	0.4	1.2	52.3	0.7	41.8
shale	53.8	18.9	7.7	3.2	2.2	8.2
clay	67.8	14.3	4.5	0.9	1.2	8.0
mill scale			ca 100.0			
sandstone	76.6	5.3	3.1	4.7	1.7	6.6
bauxite	10.6	57.5	2.6			28.4

[a] Courtesy of the American Concrete Institute (24).

Table 4. Chemical Composition of Some Typical Cements, %

	SiO$_2$	Al$_2$O$_3$	Fe$_2$O$_3$	CaO	MgO	SO$_3$	Loss	Insoluble residue
Type I	20.9	5.2	2.3	64.0	2.8	2.9	1.0	0.2
Type II	21.7	4.7	3.6	63.6	2.9	2.4	0.8	0.4
Type III	21.3	5.1	2.3	64.9	3.0	3.1	0.8	0.2
Type IV	24.3	4.3	4.1	62.3	1.8	1.9	0.9	0.2
Type V	25.0	3.4	2.8	64.4	1.9	1.6	0.9	0.2
white	24.5	5.9	0.6	65.0	1.1	1.8	0.9	0.2
alumina	5.3	39.8	14.6	33.5	1.3	0.4	0	4.8

Table 5. Potential Compound Composition of Some Typical Cements[a], %

	C$_3$S	C$_2$S	C$_3$A	C$_4$AF
Type I	55	19	10	7
Type II	51	24	6	11
Type III	56	19	10	7
Type IV	28	49	4	12
Type V	38	43	4	9
white	33	46	14	2

[a] Calculated by the American Society for Testing and Materials C150-76 (23).

$$2 C_3S + 6 H \rightarrow C_3S_2H_3 + 3 CH$$
$$2 C_2S + 4 H \rightarrow C_3S_2H_3 + CH$$

These are the main reactions in portland cements since the two calcium silicates constitute about 75% of the cement. The average lime/silica ratio (C:S) may vary from about 1.4 to about 1.7 or even higher, the average value being about 1.5. The water content varies with the ambient humidity, the three moles of water being estimated from measurements in the dry state and structural considerations. As the lime/silica ratio of the C–S–H increases, the amount of water increases on an equimolar basis, ie, the lime goes into the structure as calcium hydroxide, resulting in less free calcium hydroxide.

Calcium silicate hydrate is not only variable in composition, but is very poorly

Table 6. Calcium Silicate Hydrates Formed at Elevated Temperatures[a]

Name	CAS Registry No	Composition[b]	Temperature of formation, °C	Density, kg/m³
tricalcium silicate hydrate	[54596-90-6]	$C_6S_2H_3$	150–500	2.61
calciochondrodite	[12141-47-8]	C_5S_2H	250–800	2.84
dicalcium silicate hydrates				
α (A)	[15630-58-7]	C_2SH	100–200	2.8
β (B) (hillebrandite)	[18536-02-2]	C_2SH	140–350	2.66
γ (C) (probably a mixture)	[15669-77-9]	($C_5S_2H + C_3S_2H_x$?)	160–300	2.67
δ (D) (dellaite)	[54694-02-9]	C_6S_3H	350–800	2.98
afwillite	[16291-79-5]	$C_3S_2H_3$	100–160[c]	2.63
foshagite	[12173-33-0] and [62520-56-3]	C_4S_3H	300–500	2.7
xonotlite	[12141-77-4]	C_5S_5H	150–400	2.7
C-S-H (II)	[18662-40-3]	$C_{1.5-2.0}SH_x$	<100	2.0–2.2
C-S-H (I)		$C_{0.8-1.5}SH_y$	<130	
1.4 nm tobermorite	[1319-31-9] and [1344-96-3]	$C_5S_6H_9$	60 (?)	2.2
1.13 nm tobermorite	[12028-62-5] and [12323-54-5]	$C_5S_6H_5$	110–140	2.44
0.93 nm tobermorite	[51771-55-2]	C_5S_6H	250–450	2.7
gyrolite	[16225-87-9] [12141-71-8] and [60385-01-5]	$C_2S_3H_2$	120–200	2.39
truscottite	[12425-42-2]	$C_6S_{10}H_3$	200–300	2.36–2.48

[a] Refs. 10–11, and 25.
[b] In cement chemists' notation.
[c] Afwillite can also be formed, and appears to be the thermodynamically stable calcium silicate hydrate in pure systems, at room temperature.

crystallized, and is generally referred to as calcium silicate hydrate gel (or tobermorite gel) because of the colloidal sizes (<0.1 μm) of the gel particles. The calcium silicate hydrates are layer minerals with many similarities to the limited swelling clay minerals found in nature. The layers are bonded together by excess lime and interlayer water to form individual gel particles only 2–3 layers thick. Surface forces, and excess lime on the particle surfaces, tend to bond these particles together into aggregations or stacks of the individual particles to form the porous gel structure.

Significant changes in the structure of the gel continue over very long periods. During the first month of hydration appreciable quantities of the dimer Si_2O_7 are formed, which are reduced by later condensation to higher polysilicates. The amount of the polysilicates and the mean length of the metasilicate chains continues to increase for at least 15 years of moist curing. In one study a mean length of 15.8 silica tetrahedra was found after such prolonged curing (27). These changes appear to have a positive effect on both strength development and reduction of drying shrinkage.

Drying (and other chemical processes) can have significant effects on this structure, there being loss of hydrate water (as well as physically adsorbed water) and collapse of the structure to form more stable aggregations of particles (28–29).

Tricalcium Aluminate and Ferrite. The hydration of the C_3A alone and in the presence of gypsum usually produce well-crystallized reaction products that can be identified by x-ray diffraction and other methods. C_3AH_6 is the cubic calcium aluminate hydrate, C_4AH_{19} and $C_3A C\bar{S}H_{12}$ are hexagonal phases, the latter being commonly referred to as the monosulfate. The highly hydrated trisulfate, ettringite, occurs as needles, rods, or dense columnar aggregations. Its formation on the surfaces of anhydrous grains is responsible for the necessary retardation of hydration of the aluminates in portland cements and the expansion process in expansive cements (30).

The early calcium aluminate hydration reactions in portland cements have been studied in simple mixtures of C_3A, gypsum, calcium hydroxide, and water as shown in Figure 5 (31). The figure shows the progressive reaction of the gypsum, water, and C_3A as ettringite is formed, and then the reaction of the ettringite, calcium hydroxide, and water to form the monosulfate and the solid solution of the monosulfate with C_4AH_{19}. These reactions are important in the portland cements to control the hydration of the C_3A, which otherwise might hydrate so rapidly as to cause flash set, or premature stiffening in fresh concrete.

Other reactions taking place throughout the hardening period are substitution and addition reactions (28). Ferrite and sulfoferrite analogues of calcium monosulfoaluminate and ettringite form solid solutions in which iron oxide substitutes continuously for the alumina. Reactions with the silicate hydrate result in the formation of additional substituted C–S–H gel at the expense of the crystalline aluminate, sulfate, and ferrite hydrate phases

The hydration of the ferrite phase (C_4AF) is of greatest interest in mixtures containing lime and other cement compounds because of the strong tendency to form solid solutions. When the sulfate in solution is very low, solid solutions are formed between the cubic C_3AH_6 and an analogous iron hydrate C_3FH_6. In the presence of water and silica, solid solutions such as $C_3ASH_4 \cdot C_3FSH_4$ may be formed (32). Table 7 lists some of the important phases formed in the hydration of mixtures of pure compounds.

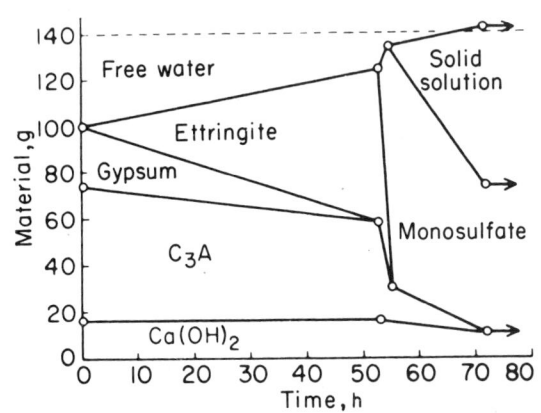

Figure 5. The early hydration reactions of tricalcium aluminate in the presence of gypsum and calcium hydroxide. Initial molar proportions: 1-C_3A; 1-$Ca(OH)_2$; 3/4-$CaSO_4 \cdot 2H_2O$; 0.4 water–solids ratio (31).

Table 7. Cement Phases Hydrated at Normal Temperatures[a]

Name	CAS Registry No.	Approximate composition[b]	Stability range RH (at 25°C)	Stability range Temp, °C	Crystal system	Density, kg/m³
calcium sulfate dihydrate (gypsum)	[10101-41-4] and [13397-24-5]	$C\bar{S}H_2$	100–35	<100	monoclinic	2.32
calcium hydroxide (portlandite)	[1305-62-0]	CH	100–0	<512	trigonal–hexagonal	2.24
magnesium hydroxide (brucite)	[1309-42-8]	MH	100–0	<350	trigonal–hexagonal	2.37
calcium silicate hydrate gel (C–S–H gel)	[12323-54-5]	$C_xS_yH_z$ $1.3 < \frac{x}{y} < 2$ $1 < \frac{z}{y} < 1.5$ (?)	indefinite		indefinite	2.7[c]
tetracalcium aluminate						
19-hydrate	[12042-86-3]	C_4AH_{19}	100–85	<15	trigonal–hexagonal	1.80
13-hydrate	[12042-85-2]	C_4AH_{13}	81–12		trigonal–hexagonal	2.02
7-hydrate	[12511-52-3]	C_4AH_7	2–0	to 120		
tetracalcium aluminate monosulfate						
16-hydrate	[67523-83-5]	$C_4A\bar{S}H_{16}$	aq	<8	trigonal–hexagonal	
14-hydrate	[12421-30-6]	$C_4A\bar{S}H_{14}$	100–95	>9	trigonal–hexagonal	
12-hydrate	[12252-10-7]	$C_4A\bar{S}H_{12}$	95–12	>1	trigonal–hexagonal	1.95
10, 8, x-hydrate	[12252-09-4] and [12445-38-4]	$C_4A\bar{S}H_x$	<12			
ettringite (6-calcium aluminate trisulfate, 32-hydrate)	[12252-15-2]	$C_6A\bar{S}_3H_{32}$	100–4	<60	trigonal–hexagonal	1.73–1.79
	[11070-82-9]	$C_6A\bar{S}_3H_8$	4–2	<110		
garnet-hydrogarnet solid solution series		$C_3(F_{1-x}A_x)(S_{1-y}H_{2y})_3$ $x = \frac{A}{A+F}$ and $y = \frac{2H}{2H+S}$	stable		cubic	
	[12042-80-7]	end member: C_3AH_6	100–0	>15	cubic	2.52

[a] Refs. 10–11, and 33.
[b] In cement chemists notation.
[c] Wet.

Other Phases in Portland and Special Cements. In cements free lime (CaO) and periclase (MgO) hydrate to the hydroxides. The *in situ* reactions of larger particles of these phases can be rather slow and may not occur until the cement has hardened. These reactions then can cause deleterious expansions and even disruption of the concrete and the quantities of free CaO and MgO have to be limited. The soundness of the cement can be tested by the autoclave expansion test of portland cement ASTM C-151 (23).

The expansive component $C_4A_3\bar{S}$ in expansive cements of type K hydrates in the presence of excess sulfate and lime to form ettringite:

$$C_4A_3\bar{S} + 8\ C\bar{S} + 6\ C + 96\ H \rightarrow 3\ C_6A\bar{S}_3H_{32}$$

The reactions in the regulated-set cements containing $C_{11}A_7 \cdot CaF_2$ (mixed notation) as a major phase resemble those in ordinary portland cements. Initial reaction rates are controlled by ettringite formation. Setting occurs with formation of the monosulfate, along with some transitory lower-limed calcium aluminate hydrates that convert to the monosulfate within a few hours.

Pozzolans contain reactive silica which reacts with cement and water by combining with the calcium hydroxide released by the hydration of the calcium silicates to produce additional calcium silicate hydrate. If sufficient silica is added (about 30% of the weight of cement), the calcium hydroxide can be completely combined. Granulated blast-furnace slag is not ordinarily reactive in water, but in the presence of lime reactions occur with the silica framework. This breakdown of the slag releases other components so that a variety of crystalline hydrate phases can also form.

Steam-Curing of Portland Cements. The hydrated silicates formed by portland cements at 100°C are similar to those obtained from lime–silica mixtures or C_3S and C_2S, but with a higher C–S ratio of the C–S–H gel. Some crystalline $C_6S_2H_3$ and $C_5SH(B)$ may also be formed.

Steam curing for 6–12 h at 150–200°C forms C–S–H gel or tobermorite, $C_6S_2H_3$ and $C_2SH(A)$. The formation of the latter is unfavorable for strength development, and silica is often added for its prevention. Small additions result in more $C_2SH(A)$ and no $C_6S_2H_3$. If sufficient silica is added (30–40% of replacement of the cement) formation of 1.1 nm tobermorite is favored, giving optimum strengths. Hydrated calcium aluminates, sulfoaluminates, or hydrogarnets are not usually found (34).

Hydration at Ordinary Temperatures. Portland cement is generally used at temperatures ordinarily encountered in construction (5–40°C). Temperature extremes have to be avoided. The exothermic heat of the hydration reactions can play an important part in maintaining adequate temperatures in cold environments, and must be considered in massive concrete structures to prevent excessive temperature rise and cracking during subsequent cooling.

The initial conditions for the hydration reactions are determined by the concentration of the cement particles (0.2–100 µm) in the mixing water (0.3–0.7 on a weight basis) and the fineness of the cement (2500–5000 cm^2/g). Upon mixing with water, the suspension of particles as shown in Figure 6 (35) is such that the particles are surrounded by films of water with an average thickness of about 1 µm. The anhydrous phases initially react by the formation of surface hydration products on each grain, and by dissolution into the liquid phase. The solution quickly becomes saturated with calcium and sulfate ions, and the concentration of alkali cations increases rapidly. These reactions consume part of the anhydrous grains, but the reaction products tend to fill that space as well as some of the originally water-filled space. The porous gel

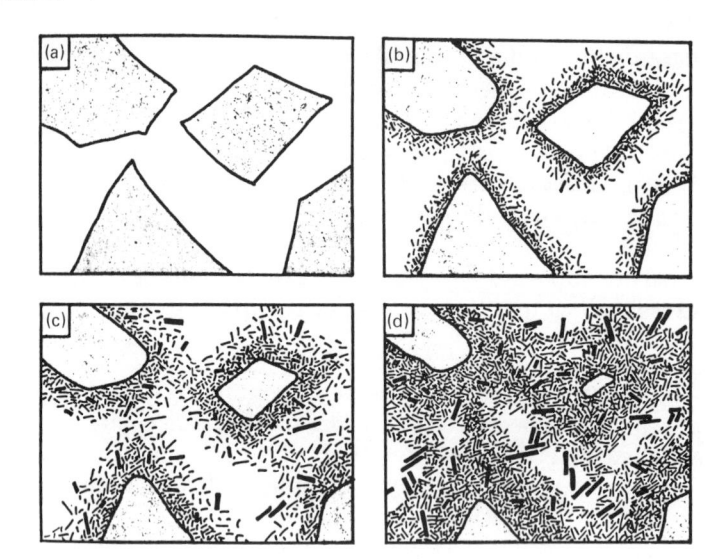

Figure 6. Four stages in the setting and hardening of portland cement: simplified representation of the sequence of changes. (**a**) Dispersion of unreacted clinker grains in water. (**b**) After a few minutes: hydration products eat into and grow out from the surface of each grain; (**c**) After a few hours: the coatings of different clinker grains have begun to join up, the gel thus becoming continuous (setting). (**d**) After a few days: further development of the gel has occurred (hardening). Courtesy of Academic Press Inc. (London) Ltd. (35).

in its most dense configurations occupies about twice the volume of the reacted anhydrous material (36). The hydration products at this stage are mostly colloidal (<0.1 μm) but some larger crystals of calcium aluminate hydrates, sulfoaluminate hydrate, and hydrogarnets form. As the reactions proceed the coatings increase in thickness and eventually form bridges between the original grains. This is the stage of setting. Despite the low solubility and mobility of the silicate anions, growths of the silicate hydrates also form on the crystalline phases formed from the solution and become incorporated into the calcium hydroxide and other phases. With further hydration the water-filled spaces become increasingly filled with reaction products to produce hardening and strength development.

The composition of the liquid phase during the early hydration of portland cements is controlled mainly by the solution of calcium, sulfate, sodium, and potassium ions. Very little alumina, silica, or iron are present in solution. Calcium hydroxide (as calcium oxide) and gypsum (as calcium sulfate) alone have solubilities of about 1.1 and 2.1 g/L at 25°C, respectively. In the presence of alkalies released by the readily soluble alkali sulfate phases in cements (as much as 70–80% may be released in the first 7 min), the composition tends to be governed by the equilibrium:

$$CaSO_4 + 2\ MOH \rightleftharpoons M_2SO_4 + Ca(OH)_2$$

where M represents the alkalies. At advanced stages of hydration of low water–cement ratio pastes, the alkali solution concentration may exceed 0.4 N with a pH above 13. Saturated lime–water has a pH of 12.4 at 25°C.

The exact course of the early hydration reactions depends mainly on the C_3A, ferrite, and soluble alkali contents of the clinker and the amount of gypsum in the cement. Following the rate of reaction by calorimetric measurements, at least two and sometimes three distinct peaks in the rate of heat liberation can be observed (32,37). A large initial peak lasts only a few minutes and may reach 4 J/(g·min) (1 cal/(g·min)) resulting mainly from the solution of soluble constituents and the surface reactions, especially the formation of the sulfoaluminate coating on the highly reactive C_3A phase. After the initial heat peak, the reactions are strongly retarded, producing a 1–2 h delay referred to as the dormant period, during which the cement–water paste remains plastic and the concrete is workable. The C_3A reaction continues slowly to form ettringite. The solution composition remains relatively constant except for a slow increase in supersaturation with respect to calcium hydroxide. Eventually the supersaturation produces nucleation of calcium hydroxide at numerous sites which decreases the calcium concentration in the solution and accelerates the rate of hydration of the alite. This produces the second heat peak which reaches a rate of about 16 J/(g·h) (4 cal/(g·h)), at about six hours of hydration, corresponding to final set. The third peak occurs at a time that depends upon the gypsum and C_3A contents and corresponds to the exhaustion of the solid gypsum, a rapid decrease of sulfate in solution, conversion of the ettringite to monosulfate, and renewed rapid reaction of the remaining C_3A and ferrite phases. At optimum gypsum the third peak ordinarily occurs between 18 and 24 hours with maximum strength and minimum shrinkage.

In these early reactions the reactivities of the individual phases are important in determining the overall reaction rate. However, as the cement particles become more densely coated with reaction products, diffusion of water and ions in solution becomes increasingly impeded. The reactions then become diffusion-controlled at some time depending on various factors such as temperature and water–cement ratio. After about 1 or 2 days (ca 40% of complete reaction) the remaining unhydrated cement phases react more nearly uniformly.

Microscopic examination of sections of hardened cement paste show that the unhydrated cores of the larger cement particles can be distinguished from the hydrated portion or inner product, which is a pseudomorph of the original grain, and the outer product formed in the originally water-filled spaces. Measurements of these cores indicate the depth of penetration of the hydration reactions (38). The overall hydration rate increases with the temperature, the fineness of the cement, and to a lesser extent with the water–cement ratio; measurements of the activation energy indicate that the reaction becomes increasingly diffusion controlled (32). Although more finely ground cements hydrate more rapidly in the first month or more of hydration, these differences gradually disappear at later ages. After one year most portland cements at usual water–cement ratios are more than 90% hydrated if continuously moist-cured. At complete hydration the chemically combined water (the water retained after strong drying) is about 20–25% of the weight of the cement, depending upon its composition. However, a minimum water–cement ratio of about 0.4 is required to provide enough space to permit complete hydration of the cement (36). If moist curing is stopped and the hardened cement is dried sufficiently, say to 80% rh, the hydration process stops.

Cement Paste Structure and Concrete Properties

The properties of both fresh and hardened mortars and concretes depend mainly on the cement–water paste properties. Practical engineering tests are usually made with concrete specimens since their properties also depend on the proportions, size gradation, and properties of the aggregates. Quality control testing and research on cement properties is usually done on cement pastes or cement mortars made with standard sands. The properties of hardened cement pastes, mortars, and concretes are similar functions of the water–cement ratio and degree of hydration of the cement. The properties of fresh concretes that determine the workability, or ease of mixing and placement into forms, also depend strongly on, but are not so simply related to, the cement paste rheological properties (39–40).

The fresh paste even in the dormant period is normally thixotropic, or shear thinning, indicating that the structure is being continuously broken down and reformed during mixing. It is an approximately Bingham plastic body with a finite yield value and plastic viscosity from 5000 to 500 mPa·s (= cP) as the water–cement ratio increases from 0.4 to 0.7 (41). The viscosity and yield values can be greatly reduced by the addition of certain organic water-reducing admixtures especially formulated for this purpose. Workability of concrete is measured by the slump of the concrete determined after removal of a standard slump cone (305 mm high) (42). Workable concretes have slumps of 75 mm or more.

After mixing and casting, sedimentation of the cement particles in the water results in bleeding of water to the top surface and reduction of water–cement ratio in the paste. At high water–cement ratios, some of the very fine particles may be carried with the bleed water to the top resulting in laitance and perhaps the formation of flaws called bleeding channels. In concretes, sedimentation may cause flaws under the larger aggregate particles. If the fresh concrete is not protected from too rapid surface drying, capillary forces cause drying shrinkage which may cause plastic shrinkage cracks. Good construction practices are designed to minimize all of these flaws.

The engineering properties of the concrete, such as strength, elastic moduli, permeability to water and aggresive solutions, and frost resistance, depend strongly on the water–cement ratio and degree of hydration of the cement. A variety of empirical water–cement ratio laws express the strength as functions of water–cement ratio or porosity. The fraction of the original water-filled space which is occupied by hydration products at any stage of hydration, is termed the gel-space ratio X. The compressive strength f_c of hardened cement or mortar then approximately fits the power law:

$$f_c = AX^n$$

where n is about 3.0 and A is the intrinsic strength of the densest ($X = 1$) gel produced by a given cement under normal hydrating conditions. Values of A range upward from ca 100 MPa (15,000 psi), depending on the cement composition (43). Under extreme conditions (hot pressing at very low water–cement ratios) strengths as high as 655 MPa (95,000 psi) have been reported (44). Tensile strengths and elastic moduli are similarly dependent on porosity or gel-space ratio, but the tensile strength is only about one tenth the compressive strength. The Young's modulus of the densest gels produced under normal hydrating conditions is ca 34 GPa (5×10^6 psi) (29).

Under sustained loads hardened cements and concrete creep, or deform continuously with time, in addition to the initial elastic deformation. Under normal working loads this deformation may in time exceed the elastic deformation and must be considered in engineering design. This is especially true in prestressed concrete structural members in which steel tendons under high tensile stress maintain compressive stress in the concrete to prevent tensile cracking during bending. Both creep and drying shrinkage of the concrete may lead to loss of prestress. Some creep in ordinary concrete structures and in the cement paste between the aggregate particles can also be an advantage because it tends to reduce stress concentrations, cracking, and microcracking around aggregate particles.

Drying of hardened cements results in shrinkage of the paste structure and of concrete members. Linear shrinkage of hardened cements is about 0.5% when dried to equilibrium at normal (ca 50%) relative humidities. The cement gel structure is somewhat stabilized during drying so that upon subsequent wetting and drying smaller changes occur. Concretes shrink much less (about 0.05%), depending on the volume fractions of cement paste and aggregates, water–cement ratio and other factors. Drying of concrete structural members proceeds very slowly and results in internal shrinkage stresses because of the moisture gradients during drying. Thick sections continue to dry and shrink for many years. Atmospheric carbon dioxide penetrates the partly dried concrete and reacts with the calcium silicate hydrate gel (as well as with calcium aluminate hydrates) which releases additional water and causes additional shrinkage. The density of the hydration products is increased, however, and the strength is actually increased. This reaction is sometimes used to advantage in the manufacture of precast concrete products to improve their ultimate strength and dimensional stability by precarbonation.

The slowness of drying and the penetration of the hardened cement by carbon dioxide or chemically aggresive solutions (eg, sea water or sulfate ground waters) is a result of the small sizes of the pores. Initially the pores in the fresh paste are the water-filled spaces (capillaries) between cement particles. As these spaces become subdivided by the formation of the hydration products, the originally continuous pore system becomes one of more discrete pores or capillary cavities separated from each other by gel formations in which the remaining pores are very much smaller. These gel pores are so small (ca 3 nm) that most of the water contained in them is strongly affected by the solid surface force fields. These force fields are responsible for a large increase in the viscosity of the water and a decrease in mobility of ionic species in solution. Hence, the permeability of the paste to both water and dissolved substances is greatly reduced as hydration proceeds. This in part accounts for the great durability of concrete, especially when water–cement ratios are kept low and adequate moist curing ensures a high degree of hydration. High water–cement ratios result in large numbers of the capillary spaces (0.1 μm and larger) interconnected through capillaries which are 10 nm or larger. These capillaries not only lower the strength, but also permit the easy penetration of aggressive solutions. Furthermore, these capillary spaces may become filled with water which freezes below 0°C, resulting in destructive expansions and deterioration of the concrete.

To ensure frost resistance, air-entraining admixtures or cements are used to produce a system of small spherical air voids. The amount is adjusted to entrain about 20 vol % of air in the cement paste distributed so that the mean half distance between voids (the void spacing factor) is 0.15–0.20 mm. Such air voids are large enough so that they do not readily fill with water by capillarity. Air entrainment thus ensures the durability of the concrete when exposed to wet and freezing conditions.

Manufacture

PORTLAND CEMENTS

The process of portland cement manufacture consists of (1) quarrying and crushing the rock, (2) grinding the carefully proportioned materials to high fineness, (3) subjecting the raw mix to pyroprocessing in a rotary kiln, and (4) grinding the resulting clinker to a fine powder. A layout of a typical plant is shown in Figure 7 (45), which also illustrates differences between wet process and dry process plants, and newer dry process plants shown in Figure 8 (46). The plants outlined are typical of installations producing approximately 1,000 metric tons per day. Modern installations (47–49) are equipped with innovations such as suspension or grate preheaters, roller mills, or precalciner installations.

Because calcium oxide comprises about 65% of portland cement, these plants are frequently situated near the source of their calcareous material. The requisite silica and alumina may be derived from a clay, shale, or overburden from a limestone quarry. Such materials usually contain some of the required iron oxide, but many plants need to supplement the iron with mill scale, pyrite cinders, or iron ore. Silica may be supplemented by adding sand to the raw mix, whereas alumina can be furnished by bauxites and Al_2O_3-rich flint clays.

Industrial by-products are becoming more widely used as raw materials for cement, eg, slags contain carbonate-free lime, as well as substantial levels of silica and alumina. Fly ash from utility boilers can often be a suitable feed component, since it is already finely dispersed and provides silica and alumina. Even vegetable wastes, such as rice hull ash, provide a source of silica. Probably 50% of all industrial by-products are potential raw materials for portland cement manufacture.

Clinker production requires large quantities of fuel. In the United States, coal and natural gas are the most widely used kiln fuels with coal increasing in importance (47). Residual oil furnishes fuel energy for about 9% of United States clinker production, and petroleum coke is also finding increasing application. It is estimated that by 1990 in the United States 90% of the clinker will be produced in pulverized coal-fired kilns. The feasibility of using supplemental refuse-derived fuel (RDF) together with conventional fuels is also being evaluated (see Fuels from waste).

In addition to the kiln fuel, electrical energy is required to power the equipment. This energy, however, amounts to only about one sixth that of the kiln fuel. The cement industry is carefully considering all measures that can reduce this heavy fuel demand.

Raw Materials Preparation. The bulk of the raw material originates in the plant quarry. Control of the clinker composition starts in the quarries with systematic core drillings and selective quarrying in order to utilize the deposits economically.

A primary jaw or roll crusher is frequently located within the quarry and reduces the quarried limestone or shale to about 100 mm top size. A secondary crusher, usually roll or hammer mills, gives a product of about 10–25 mm top size. Clays may require treatment in a wash mill to separate sand and other high silica material. Combination crusher-dryers utilize exit gases from the kiln or clinker cooler to dry wet material during crushing.

Argillaceous, siliceous, and ferriferous raw mix components are added to the crusher product. At the grinding mills, the constituents are fed into the mill separately, using weigh feeders or volumetric measurements. Ball mills are used for wet and dry processes to grind the material to a fineness such that only 15–30 wt % is retained on

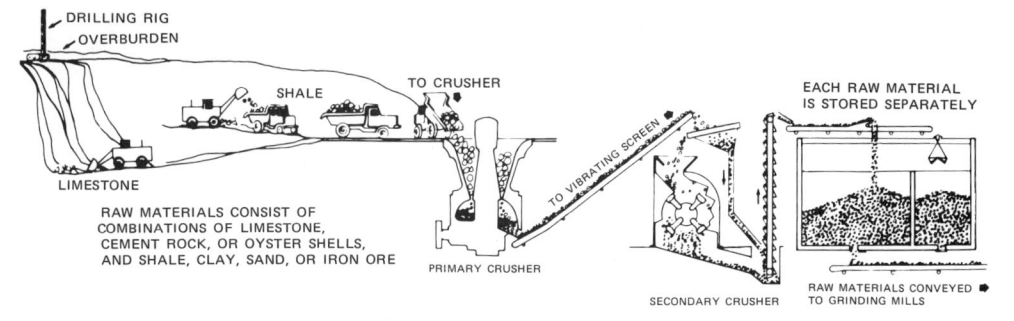

1. Stone is first reduced to 13–cm size, then to 2 cm and stored.

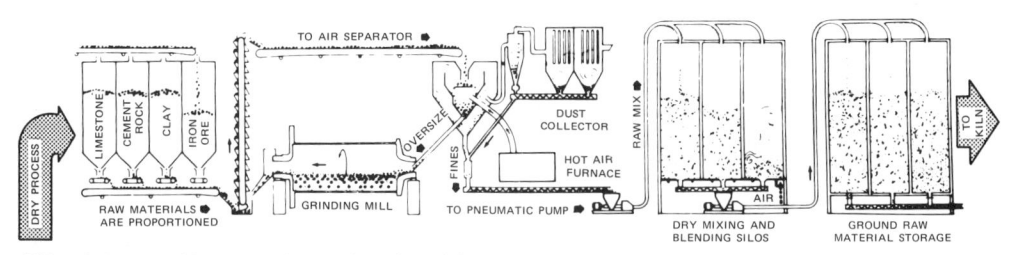

2. Raw materials are ground to powder and blended. (DRY PROCESS)

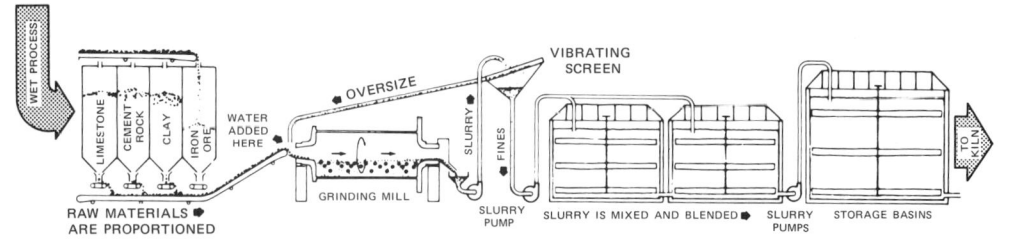

OR

2. Raw materials are ground, mixed with water to form slurry, and blended. (WET PROCESS)

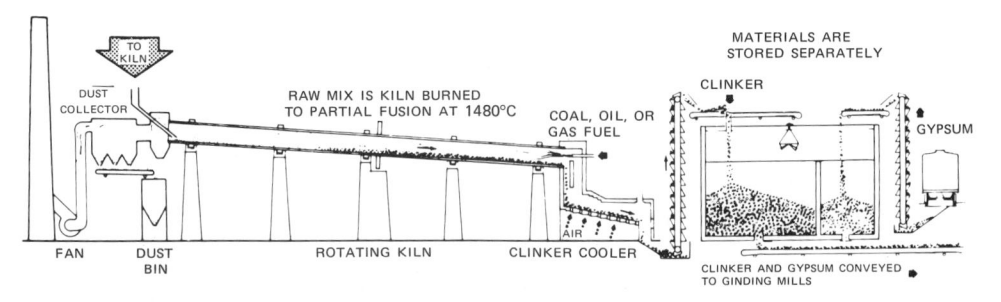

3. Burning changes raw mix chemically into cement clinker.

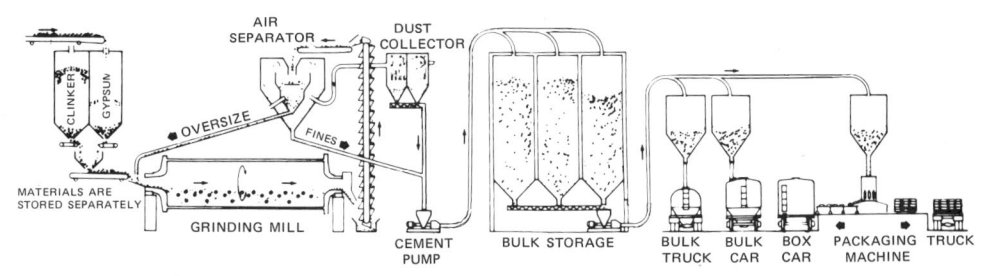

4. Clinker with gypsum is ground into portland cement and shipped.

Figure 7. Steps in the manufacture of portland cement. Courtesy of Portland Cement Association (45).

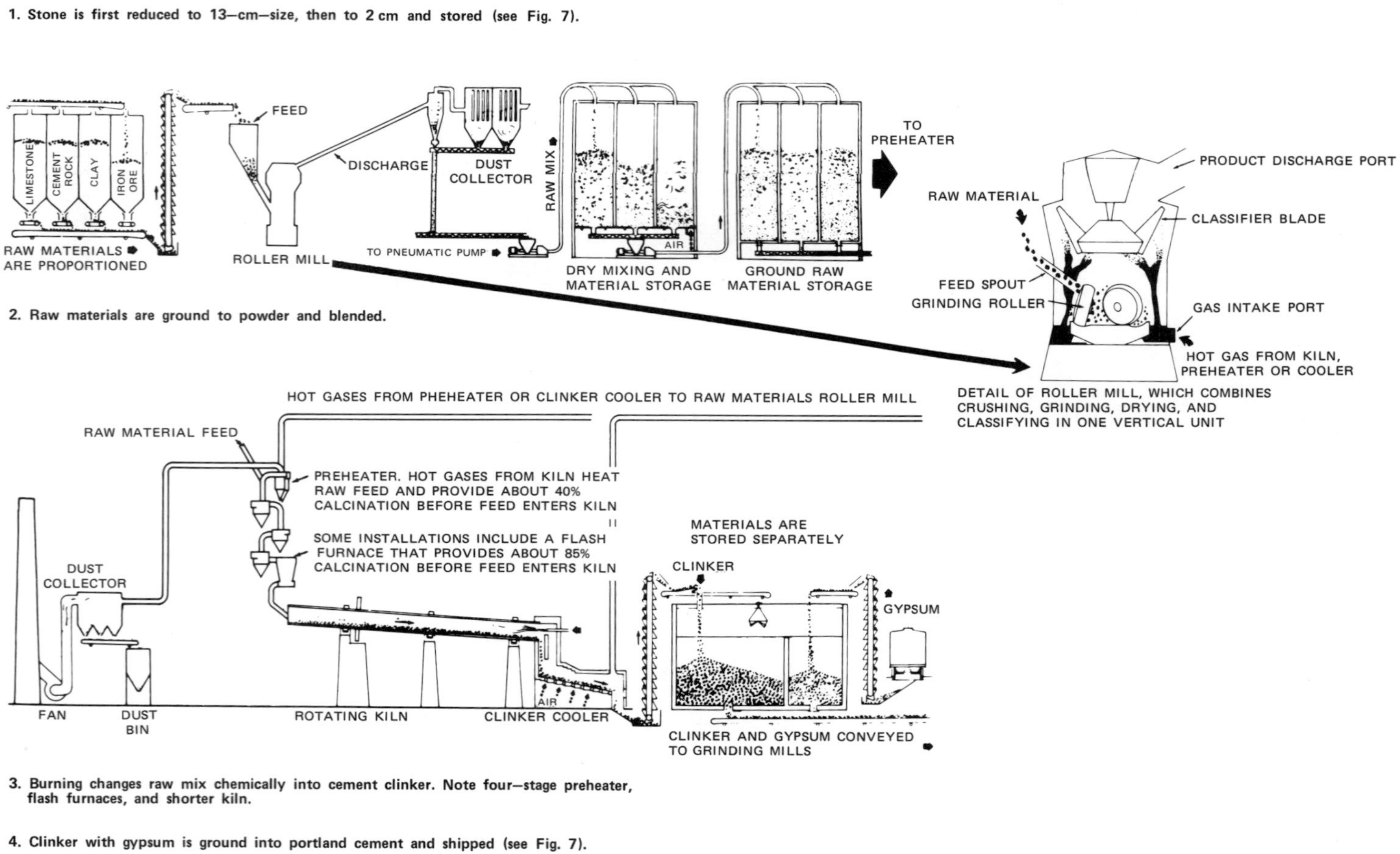

Figure 8. New technology in dry-process cement manufacture. Courtesy of Portland Cement Association (46).

a 74 µm (200 mesh) sieve. In the wet process the raw materials are ground with about 30–40% water, producing a well-homogenized mixture called slurry. Low concentrations of slurry thinners may be added, such as sodium carbonates, silicates, and phosphates, as well as lignosulfonates and modified petrochemicals. Filter presses or other devices remove water from slurries before feeding into the kiln.

Raw material for dry process plants is ground in closed-circuit ball mills with air separators, which may be set for any desired fineness. Drying is usually carried out in separate units, but waste heat can be utilized directly in the mill by coupling the raw mill to the kiln. Autogenous mills, which operate without grinding media are not widely used. For suspension preheater-type kilns, a roller mill utilizes the exit gas from the preheater to dry the material in suspension in the mill.

A blending system provides the kiln with a homogeneous raw feed. In the wet process the mill slurry is blended in a series of continuously agitated tanks in which the composition, usually the CaO content, is adjusted as required. These tanks may also serve as kiln feed tanks, or the slurry after agitation is pumped to large kiln feed basins. Dry-process blending is usually accomplished in a silo with compressed air.

Pyroprocessing. Nearly all cement clinker is produced in large rotary kiln systems. The rotary kiln is a highly refractory-lined cylindrical steel shell (3–8 m dia, 50–230 m long) equipped with an electrical drive to rotate at 1–3 rpm. It is a countercurrent heating device slightly inclined to the horizontal so that material fed into the upper end travels slowly by gravity to be discharged onto the clinker cooler at the discharge end. The burners at the firing end produce a current of hot gases that heats the clinker and the calcined and raw materials in succession as it passes upward toward the feed end (see under Clinker Chemistry). Highly refractory bricks of magnesia, alumina, or chrome–magnesite combinations line the firing end, whereas in the less heat-intensive midsection of the kiln bricks of lower refractoriness and thermal conductivity can be used, changing to abrasion-resistant bricks or monolithic castable lining at the feed end. To prevent excessive thermal stresses and chemical reaction of the kiln refractory lining, it is necessary to form a protective coating of clinker minerals on the hot face of the burning zone brick. This coating also reduces kiln shell heat losses by lowering the effective thermal conductivity of the lining.

It is desirable to cool the clinker rapidly as it leaves the burning zone. This is best achieved by using a short, intense flame as close to the discharge as possible. Heat recovery, preheating of combustion air, and fast clinker cooling are achieved by clinker coolers of the traveling-grate, planetary, rotary, or shaft type. Most commonly used are grate coolers where the clinker is conveyed along the grate and subjected to cooling by ambient air, which passes through the clinker bed in crosscurrent heat exchange. The air is moved by a series of undergrate fans, and becomes preheated to 370–800°C at the hot end of the cooler. It then serves as secondary combustion air in the kiln; the primary air is that portion of the combustion air needed to carry the fuel into the kiln and disperse the fuel.

During the burning process, the high temperatures cause vaporization of alkalies, sulfur, and halides. These materials are carried by the combustion gases into the cooler portions of the kiln system where they condense, or they may be carried out to the kiln dust collector (usually a fabric filter or electrostatic precipitator) together with partially calcined feed and unprocessed raw feed. This kiln dust is reusable. However, ASTM specifications limit the total SO_3 content of the finished cement to 2.3–4.5%, depending upon the cement type and C_3A content. Similarly, an optional ASTM C150 specification limits the total alkali content of the cement to 0.60%, expressed as equivalent

Na$_2$O. Other potential and actual uses of dust include fertilizer supplements, acid mine waste neutralization, boiler SO$_2$ control, and soil stabilization.

Wet-Process Kilns. In a long wet-process kiln, the slurry introduced into the feed end first undergoes simultaneous heating and drying. The refractory lining is alternately heated by the gases when exposed and cooled by the slurry when immersed; thus the lining serves to transfer heat, as do the gases themselves. Because large quantities of water (about 0.8 L/kg of clinker product) must be evaporated, most wet kilns are equipped with chains to maximize heat transfer from the gases to the slurry. Large, dense chain systems have permitted energy savings of up to 1.7 MJ/kg (1.6 × 10^6 Btu/t) clinker produced in exceptionally favorable situations (47). The chain system also serve to break up the slurry into nodules that can flow readily down the kiln without forming mud rings. After most of the moisture has been evaporated, the nodules, which still contain combined water, move down the kiln and are gradually heated to about 550°C where the reactions commence as discussed under Clinker Chemistry. As the charge leaves the burning zone it begins to cool, and tricalcium aluminate and magnesia crystallize from the melt and the liquid phase finally solidifies to produce the ferrite phase. The material drops into the clinker cooler for further cooling by air.

Dry-Process Kilns, Suspension Preheaters, and Precalciners. The dry process utilizes a dry kiln feed rather than a slurry. Early dry process kilns were short, and the substantial quantities of waste heat in the exit gases from such kilns were frequently used in boilers for electric power generation; the power generated was frequently sufficient for all electrical needs of the plant. In one modification, the kiln has been lengthened to nearly the extent of long wet-process kilns, and chains have been added; however, they serve almost exclusively a heat-exchange function. Refractory heat-recuperative devices, such as crosses, lifters, and trefoils, have also been installed. So equipped, the long dry kiln is capable of good energy efficiency. Other than the need for evaporation of water, its operation is similar to that of a long wet kiln.

The second major type of modern dry-process kiln is the suspension preheater system (50). The dry, pulverized feed passes through a series of cyclones where it is separated and preheated several times. The partially calcined feed exits the preheater tower into the kiln at about 800–900°C. The kiln length required for completion of the process is considerably shorter than that of conventional kilns, and heat exchange is very good. Suspension preheater kilns are very energy-efficient (as low as 3.1 MJ/kg or 1334 Btu/lb clinker in large installations).

The intimate mixing of the hot gases and feed in the preheaters promotes condensation of alkalies and sulfur on the feed which sometimes results in objectionably high alkali and sulfur contents in the clinker. To alleviate this problem, some of the kiln exit gases can be bypassed and fewer cyclone stages used in the preheater with some sacrifice of efficiency.

The success of preheater kiln systems, particularly in Europe and Japan where low alkali specifications do not exist, led to precalciner kiln systems. These units utilize a second burner to carry out calcination in a separate vessel attached to the preheater. The flash furnace (51), eg, utilizes preheated combustion air drawn from the clinker cooler and kiln exit gases and is equipped with an oil burner which burns about 60% of the total kiln fuel. The raw material is calcined almost 95%, and the gases continue their upward movement through successive preheater stages in the same manner as in an ordinary preheater.

The precalciner system permits the use of smaller kilns since only actual clinkering is carried out in the rotary kiln. Energy efficiency is comparable to that of a preheater kiln, except that the energy penalty for bypass of kiln exit gases is reduced since only about 40% of the fuel is being burned in the kiln. Precalciner kilns in operation in Japan produce up to 10,000 metric tons of clinker per day; the largest long wet-process kiln, in Clarksville, Missouri, produces only 3270 t/d by comparison.

The burning process and clinker cooling operations for the modern dry-process kiln systems are the same as for long wet kilns.

Finish Grinding. The cooled clinker is conveyed to clinker storage or mixed with 4–6% gypsum and introduced directly into the finish mills. These are large, steel cylinders (2–5 m in dia) containing a charge of steel balls, and rotating at about 15–20 rpm. The clinker and gypsum are ground to a fine, homogeneous powder with a surface area of about 3000–5000 cm^2/g. About 85–96% of the product is in particles less than 44 μm dia. These objectives may be accomplished by two different mill systems. In *open-circuit milling*, the material passes directly through the mill without any separation. A wide particle size distribution range is usually obtained with substantial amounts of very fine and rather coarse particles. In *closed-circuit grinding* the mill product is carried to a cyclonic air separator in which the coarse particles are rejected from the product and returned to the mill for further grinding.

Energy requirements for finish grinding vary from about 33–77 kW·h/t cement, depending also on the nature of the clinker.

Computer Control. Process computer control was introduced to the cement industry in the 1960s and met with varying degrees of success because of complexity of the equipment and control problems. The rotary kiln is the largest and most difficult equipment to operate. Temperature-sensing and gas-analyzing devices present problems. Unless temperature and combustion can be accurately measured, the computer cannot perform its control function. However, progress has been made and a plant of a capacity of 1 million metric tons per year has been built and placed in operation in 1973 with complete computer DDC (direct digital control) process and segmental control (52). Other new plants have been built and some older plants computerized (53). Variables can be measured at intervals of 0.25 s and overall optimum response to operating problems is programmed, not always possible with manual operation.

Quality Control. Beginning at the quarry operation, quality of the end product is maintained by adjustments of composition, burning conditions, and finish grinding. Control checks are made for fineness of materials, chemical composition, and uniformity. Clinker burning is monitored by weighing a portion of sized clinker, known as the liter weight test, a free lime test, or checked by microscopic evaluation of the crystalline structure of the clinker compounds. Samples may be analyzed by x-ray fluorescence, atomic absorption, and flame photometry (see Analytical methods). Wet chemistry is described in ASTM C114 (23). Standard cement samples are available from the National Bureau of Standards. Fineness of the cement is most commonly measured by the air permeability method. Finally, standardized performance tests are conducted on the finished cement (23).

Environmental Pollution Control. With the passage of the *Clean Air Act* and its amendments (54), the cement industry started an intensive program of capital expenditure to install dust collection equipment on kilns and coolers that were not already so equipped. Modern equipment collects dust at 99.8% efficiency. Many smaller dust collectors are installed in new plants, eg, a wet process plant of 430,000 t/yr capacity has 73 collectors connected to points of possible dust emission (55).

The Federal Water Pollution Control Act Amendments of 1972 (56) established limits for cement plant effluents. This includes water run-off from manufacturing facilities, quarrying, raw material storage piles, and waste water. Compliance with these standards has required construction of diversion ditches for surface water, ponds for settling and clarification, dikes and containment structures for possible oil spills, and chemical water treatment in some cases. Since the cement industry obtains most of its raw material by quarrying, the standards for the mineral industry also apply.

SPECIAL PURPOSE AND BLENDED CEMENTS

Special purpose and blended portland cements are manufactured essentially by the same processes as ordinary portland cements but have specific compositional and process differences as noted below.

White cements are made from raw materials of very low iron content. This type is often difficult to burn because almost the entire liquid phase must be furnished by calcium aluminates. As a consequence of the generally lower total liquid-phase content, high burning-zone temperatures may be necessary. Fast cooling and occasionally oil sprays are needed to maintain both quality and color.

Regulated set cements are made with fluorite (CaF_2) additions which also act as fluxing agents, or mineralizers, to reduce burning temperatures. The clinker produced then contains $C_{11}A_7.CaF_2$ (mixed notation) as a major phase.

Expansive cements manufactured in the United States usually depend upon aluminate and sulfate phases that result in more ettringite formation during hydration than in normal portland cements (see under Hydration Chemistry). This can be achieved by three types designated as Type K, Type S, and Type M (57). Type K contains an anhydrous calcium sulfoaluminate, $C_4A_3\bar{S}$, Type S contains a high C_3A content with additional calcium sulfate, and Type M is a mixture of portland cement, calcium aluminate cement, and calcium sulfate. Except for the Type M expansive cement, any of these cements can be made either by integrally burning to produce the desired phase composition, or by intergrinding a special component with ordinary portland cement clinkers and calcium sulfate. Type M can be made by mixing the finished cements in proper proportions, or by intergrinding the clinkers.

Oil well cements are manufactured similarly to ordinary portland cements except that the goal is usually sluggish reactivity. For this reason, levels of C_3A and alkali sulfates are kept low. Hydration-retarding additives are also employed.

The manufacture of blended cements is similar to that of portland cements except for the finish grinding process where the cement clinker is interground with pozzolans, granulated blast-furnace slag, or, in the case of masonry cements, a variety of materials.

Pozzolans include natural materials such as diatomaceous earths, opaline cherts, and shales, tuffs, and volcanic ashes or pumicites, and calcined materials such as some clays and shales. By-products such as fly-ashes and precipitated silica are also employed. In the United States the proportion of pozzolan interground with clinker has varied from 15 to over 30%, whereas in Italy, cements with a 30–40% pozzolan content are produced.

In some European countries portland cement clinker is interground with 10–65%

granulated blast-furnace slag to produce a portland blast-furnace slag cement. The composition of the slag varies considerably but usually falls within the following composition ranges:

CaO	40–50%	MgO	0–8%
SiO_2	30–40%	S (sulfide)	0–2%
Al_2O_3	8–18%	FeO, MnO	0–3%

Most masonry cements are finely interground mixtures with portland cement a major constituent, but also including finely ground limestones, hydrated lime, natural cement, pozzolans, clays, or air-entraining agents. These secondary materials are used to impart the required water retention and plasticity to mortars.

NONPORTLAND CEMENTS

Calcium Aluminate Cements. These cements are manufactured by heating until molten or by sintering a mixture of limestone and bauxite with small amounts of SiO_2, FeO, and TiO_2. In Europe the process is usually carried out in an open-hearth furnace having a long vertical stack into which the mixture of raw materials is charged. The hot gases produced by a blast of pulverized coal and air pass through the charge and carry off the water and carbon dioxide. Fusion occurs when the charge drops from the vertical stack onto the hearth at about 1425–1500°C. The molten liquid runs out continuously into steel plans on an endless belt in which the melt solidifies. Special rotary kilns provided with a tap hole from which the molten liquid is drawn intermittently and electric arc furnaces have also been used.

In a new process called shock sintering (58), finely ground raw materials are pelletized on a disk pelletizer and dried in a drier–preheater. The pellets are heated very rapidly to maximum reaction temperature in the sintering section of a specially designed rotary kiln, and are then ground into cement.

When calcium aluminate cements are made by the fusion process, the solidified melt must be crushed and then ground. The material is very hard to grind and power consumption is high.

Supersulfated Cement. Supersulfated cement contains about 80% slag interground with 15% gypsum or anhydrite and 5% portland cement clinker.

Hydraulic Limes. These materials are produced by heating below sintering temperature a limestone containing considerable clay, during which some combination takes place between the lime and the oxides of the clay to form hydraulic compounds.

Economic Aspects, Production, and Shipment

From the beginning of the United States portland cement industry in 1872, cement consumption grew at an average annual rate of 20% until 1920. As the cement markets matured, the industry grew at an average annual rate of 3% from 1920 to 1975. Annual production and sales tonnages are nearly identical; Table 8 gives United States production figures since 1910, Table 9 the world production.

Since World War II, the cement industry reduced labor and energy costs by increased investment in capital equipment and larger plants to remain competitive with other building materials industries. The average plant size increased more than 65% between 1950 and 1975.

188 CEMENT

Table 8. United States Portland Cement Production[a]

Year	Production, 1000 metric tons	Number of plants	Average capacity per plant, 1000 metric tons
1910	13,056	111	150
1920	17,059	117	213
1930	27,493	163	281
1940	22,209	152	286
1950	38,550	150	304
1960	54,408	176	417
1970	66,378	181	451
1975	60,597	164	511

[a] Refs. 59–60.

Table 9. World Portland Cement Production[a]**, Million Metric Tons**

	1950	1960	1970	1975
Europe	68	168	322	383
France	7	14	29	31
Germany	11	26	37	33
Italy	5	16	33	35
USSR	10	46	95	122
Spain	2	5	16	24
United Kingdom	10	14	18	18
Africa	4	9	18	24
Western Hemisphere	48	77	111	122
United States	38	53	65	59
Exports	0.413	0.032	0.144	0.331
Imports	0.238	0.699	2.356	3.299
Canada	3	5	7	10
Asia	11	57	121	167
Japan	4	22	57	65
Oceania	2	3	6	6
Total	*133*	*314*	*578*	*702*

[a] Refs. 61–62.

The wet process was used in 60% of the plants in the 1960s because it is less labor intensive than the dry process. However, as energy costs escalated in the early 1970s, dry-process manufacturing was preferred because it is generally more energy efficient.

Energy Usage. In the past 25 years, the cement industry has reduced its unit energy usage by 25.2%, from 9.6 MJ/kg (4131 Btu/lb) in 1950 to 7.2 MJ/kg (3098 Btu/lb) in 1975. In the 1950s and 1960s, the industry had shifted from coal to inexpensive, abundant, and clean natural gas, but by 1975 nearly 80% of the industry's capacity had been converted to permit use of coal as the primary kiln fuel; actual coal usage in 1975 was about 50% of all kiln fuels (see Table 10).

Marketing Patterns. Since 1950 the cement industry has reduced its dependence on bag (container) shipments (54.7% in 1950) and turned to the more labor-efficient bulk transport (92.0% in 1975). In addition, the amount of cement shipped by rail transportation declined from 75% of industry shipments in 1950 to less than 13% in 1975. Table 11 gives the shipment distribution by type.

Table 10. United States Portland Cement Industry Energy Consumption

Year	Coal, 1000 metric tons	Oil, 1000 metric tons	Natural gas, million m^3	Power, million kW·h	Energy usage, MJ/t[a]
1950	7206	764	2747	2877	9627
1955	7918	1235	3710	4022	8979
1960	7591	586	4870	5589	8548
1965	8288	649	5635	7485	8222
1970	7227	1455	6003	8717	8008
1975	6866	1065	4531	9315	7201

[a] To convert MJ/t to Btu/lb, divide by 2.32.

Table 11. Portland Cement Shipments by Type[a,b], 1975

Type	Shipments, 1000 metric tons	Average value per metric ton, $
Types I and II	56,987	33.76
Type III	1,911	36.14
oil well	1,016	36.66
white	331	82.51
Type V	314	38.97
portland slag, pozzolan	286	33.54
expansive	83	46.19
miscellaneous	560	41.95
Total	61,488	34.27

[a] Ref. 59.
[b] See Tables 4–5.

In the past 25 years, the ready-mixed concrete industry became the primary customer for cement manufacturers. In 1975 more than 63% of the cement shipped was sold to the ready-mixed concrete industry, compared with 56% in 1960. The other major uses are in building materials, concrete products, and highway construction.

Specifications and Types

Portland cements are manufactured to comply with the specifications established in each country (63). In the United States, several different specifications are used, including those of the American Society for Testing and Materials, American Association of State Highway and Transportation Officials, and various government agencies. The ASTM annually publishes test methods and standards (23) which are established on a consensus basis by its members, including consumers and producers.

In the United States, portland cement is classified in five general types designated by ASTM Specification C150-76 (23) as follows: Type I, when the special properties are not required; Type II, for general use, and especially when moderate sulfate resistance or moderate heat of hydration is desired; Type III, for high early strength. Type IV, for low heat of hydration, and Type V, for high sulfate resistance. Types I, II, and III may also be specified as air-entraining. Chemical compositional, physical, and performance test requirements are specified for each type; optional requirements for particular uses may also be specified.

Other countries have similar types; some, as in Germany and the Union of Soviet Socialist Republics, are based on age-strength levels by standard tests (63). A product made in Italy and France known as Ferrari cement is similar to Type V and is sulfate resistant. Such cements have high iron oxide and low alumina contents, and harden more slowly.

Uses

Hydraulic cements are intermediate products that are used for making concretes, mortars, grouts, asbestos–cement products, and other composite materials. High early strength cements may be required for precast concrete products or in high-rise building frames to permit rapid removal of forms and early load carrying capacity. Cements of low heat of hydration may be required for use in massive structures, such as gravity dams, to prevent excessive temperature rise and thermal contraction and cracking during subsequent cooling. Concretes exposed to seawater, sulfate-containing ground waters, or sewage require cements that are sulfate resistant after hardening.

Air-entraining cements produce concretes that contain a system of closely spaced spherical voids that protect the concrete from frost damage. They are commonly used for concrete pavements subjected to wet and freezing conditions. Cement of low alkali content may be used with certain concrete aggregates containing reactive silica to prevent deleterious expansions.

Expansive, or shrinkage-compensating cements cause slight expansion of the concrete during hardening. The expansion has to be elastically restrained so that compressive stress develops in the concrete (57,64). Subsequent drying and shrinkage reduces the compressive stresses but does not result in tensile stresses large enough to cause cracks.

Regulated-set cement (or jet cement in Japan) is formulated to yield a controlled short setting time (1 h or less) and very early strength (65).

Natural cements (66) may be regarded as intermediate between portland cements and hydraulic limes (see below) in hydraulic activity.

Blended cements. Portland cement clinker is also interground with suitable other materials such as granulated blast furnace slags and natural or artificial pozzolans (see above). These substances also show hydraulic activity when used with cements, and the blended cements (67) bear special designations such as portland blast-furnace slag cement or portland-pozzolan cement. Pozzolans are used in making concrete both as an interground component of the cement and as a direct addition to the concrete mix. It is only when the two materials are interground that the mixture can be referred to as portland–pozzolan cement. Portland-pozzolan cements (68) were developed originally to provide concretes of improved economy and durability in marine, hydraulic, and underground environments; they also prevent deleterious alkali–aggregate reactions. Blast-furnace slag cements (69) also reduce deleterious alkali–aggregate reactions and can be resistant to seawater if the slag and cement compositions are suitably restricted. Both cements hydrate and harden more slowly than portland cement. This can be an advantage in mass concrete structures where the lower rates of heat liberation may prevent excessive temperature rise, but when used at low temperatures the rate of hardening may be excessively slow. Portland blast-furnace slag cements may be used to advantage in steam-cured products which can have strengths as high as obtained with portland cement. Current interest in the use of blended cements is stimulated by energy conservation and solid waste utilization considerations.

Oil well cements (70) are usually made from portland cement clinker and may also be blended cements. They are specially produced for cementing the steel casing of gas and oil wells to the walls of the bore-hole and to seal porous formations (71). Under these high temperature and pressure conditions ordinary portland cements would not flow properly and would set prematurely. Oil well cements are more coarsely ground than normal, and contain special retarding admixtures.

Masonry cements (72) are cements for use in mortars for masonry construction. They are formulated to yield easily workable mortars and contain special additives that reduce the loss of water from the mortar to the porous masonry units.

Calcium aluminate cement (73) develops very high strengths at early ages. It attains nearly its maximum strength in one day, which is much higher than the strength developed by portland cement in that time. At higher temperatures, however, the strength drops off rapidly. Heat is also evolved rapidly on hydration and results in high temperatures; long exposures under moist warm conditions can lead to failure. Resistance to corrosion in sea or sulfate waters, as well as to weak solutions of mineral acids, is outstanding. This cement is attacked rapidly, however, by alkali carbonates. An important use of high alumina cement is in refractory concrete for withstanding temperatures up to 1500°C. White calcium aluminate cements, with a fused aggregate of pure alumina, withstand temperatures up to 1800°C.

Supersulfated cement (74) has a very low heat of hydration and low drying shrinkage. It has been used in Europe for mass concrete construction and especially for structures exposed to sulfate and seawaters.

Trief cements (75), manufactured in Belgium, are produced as a wet slurry of finely ground slag. When activators (such as portland cement, lime, or sodium hydroxide) are added in a concrete mixer, the slurry sets and hardens to produce concretes with good strength and durability.

Hydraulic limes (76) may be used for mortar, stucco, or the scratch coat for plaster. They harden slowly under water, whereas high calcium limes, after slaking with water, harden in air to form the carbonate but not under water at ordinary temperatures. However, at elevated temperatures achieved with steam curing, lime–silica sand mixtures do react to produce durable products such as sand–lime bricks.

Specialty cements. For special architectural applications, white portland cement with a very low iron oxide content can be produced. Colored cements are usually prepared by intergrinding 5–10% of pigment with white cement.

Numerous other specialty cements composed of various magnesium, barium, and strontium compounds as silicates, aluminates, and phosphates, as well as others, are also produced (77).

BIBLIOGRAPHY

"Cement, Structural" in *ECT* 1st ed., Vol. 3, pp. 411–438, by R. H. Bogue, Portland Cement Association Fellowship (Portland Cement); J. L. Miner and F. W. Ashton, Universal Atlas Cement Company (Calcium–Aluminate Cement); and G. J. Fink, Oxychloride Cement Association, Inc. (Magnesia Cement); "Cement" in *ECT* 2nd ed., Vol. 4, pp. 684–710, by Robert H. Bogue, Consultant to the Cement Industry.

1. R. H. Bogue, *The Chemistry of Portland Cement*, 2nd ed., Rheinhold Publishing Corp., New York, 1955.

2. G. A. Rankin and F. E. Wright, *Am. J. Sci.* **39,** 1 (1915).
3. F. M. Lea, *The Chemistry of Cement and Concrete,* 3rd ed., Edward Arnold (Publishers) Ltd., London, Eng., 1971.
4. Ref. 3, p. 88.
5. Ref. 3, p. 122.
6. P. Weber, *Zem. Kalk Gips* Special Issue No. 9, (1963); ref. 3, p. 130.
7. Ref. 3, p. 126.
8. K. Endell and G. Hendrickx, *Zement* **31,** 357, 416 (1942); ref. 3, p. 128.
9. N. Toropov and P. Rumyantsev, *Zh. Prikl. Khim* **38,** 1614, 2115 (1965); ref. 3, p. 135.
10. *Guide to Compounds of Interest in Cement and Concrete Research, Special Report 127,* Highway Research Board, National Academy of Sciences, Washington, D.C., 1972.
11. H. F. W. Taylor, ed., *The Chemistry of Cements,* Vols. 1 and 2, Academic Press Inc. (London) Ltd., London, Eng., 1964, Appendix 1.
12. Ref. 3, p. 121.
13. A. Guinier and M. Regourd in *Proc. of the 5th Int. Symposium on the Chemistry of Cement, Tokyo, 1968,* The Cement Association of Japan, Tokyo, Japan, 1969.
14. G. Yamaguchi and S. Takagi in ref. 13.
15. Ref. 3, p. 42 ff and p. 270.
16. F. A. DeLisle, *Cement Technol.* **7,** 93 (1976); Y. Ono, S. Kamamura, and Y. Soda in ref. 13.
17. F. Gille and co-workers, *Microskopie des Zementklinkers, Bilderatlas,* Association of the German Cement Industry, Beton-Verlag, Dusseldorf, FRG, 1965.
18. L. S. Brown, *Proc. J. Am. Concr. Inst.* **44,** 877 (1948).
19. H. Kuhl, *Zement* **18,** 833 (1929); ref. 3, p. 164.
20. Ref. 3, p. 166.
21. K. E. Peray and J. J. Waddell, *The Rotary Cement Kiln,* Chemical Publishing Co., New York, 1972, p. 65.
22. R. H. Bogue, *Ind. Eng. Chem. Anal. Ed.* **1,** 192 (1929); ref. 1, p. 246.
23. *1976 Annual Book of ASTM Standards,* Part 13, American Society for Testing and Materials, Philadelphia, Pa., 1976, p. 138.
24. F. R. McMillan and W. C. Hansen, *J. Am. Concr. Inst.* **44,** 553, 564, 565 (1948).
25. Ref. 3, p. 188.
26. S. Brunauer and D. K. Kantro in ref. 11.
27. C. W. Lentz in *Special Report 90, Structure of Portland Cement Paste and Concrete,* Highway Research Board, NRC-NAS, Washington, D.C., 1966.
28. L. E. Copeland and G. Verbeck in *The 6th Int. Congress on the Chemistry of Cement, Moscow, 1974, English Preprints,* The Organizing Committee of the U.S.S.R., 1974.
29. G. Verbeck and R. A. Helmuth in ref. 13.
30. W. C. Hansen in E. G. Swenson, ed., *Performance of Concrete,* University of Toronto Press, Toronto, Can., 1968.
31. G. Verbeck, *Research Department Bulletin 189,* Portland Cement Association, Skokie, Ill., 1965.
32. L. E. Copeland and D. L. Kantro in ref. 11.
33. F. M. Lea, *The Chemistry of Cement and Concrete,* 3rd ed., Edward Arnold (Publishers) Ltd., London, Eng., 1971.
34. Ref. 3, p. 203.
35. H. F. W. Taylor in ref. 11, p. 21.
36. T. C. Powers in ref. 11, Chapt. 10.
37. W. Lerch, *ASTM Proc.* **46,** 1251 (1946).
38. Ref. 3, p. 239.
39. G. H. Tattersall, *The Workability of Concrete, Publ. 11.008,* Cement and Concrete Association, Wexham Springs, Eng., 1976.
40. T. C. Powers, *The Properties of Fresh Concrete,* John Wiley & Sons, Inc., New York, 1968.
41. E. M. Petrie, *Ind. Eng. Chem. Prod. Res. Dev.* **15,** 242 (1976).
42. *1976 Annual Book of ASTM Standards,* Part 14, American Society for Testing and Materials, Philadelphia, Pa., 1976.
43. T. C. Powers in *Proc. of the 4th Int Symposium on the Chemistry of Cement, Washington, D.C., 1960,* U.S. Government Printing Office, Washington, D.C., 1962.
44. D. M. Roy and G. R. Gouda, *Cement Concr. Res.* **5,** 153 (1975).
45. *The U.S. Cement Industry: An Economic Report,* Portland Cement Association, Skokie, Ill., Oct. 1974, p. 7.

46. Ref. 45, p. 8.
47. *Energy Conservation in the Cement Industry, Conservation Paper No. 26,* Federal Energy Administration, Washington, D.C., 1975.
48. W. H. Duda, *Cement Data Book,* Bauverlag G.m.b.H., Wiesbaden, FRG, and Berlin, 1976.
49. K. E. Peray and J. J. Waddell, *The Rotary Cement Kiln,* Chemical Publishing Co., New York, 1972.
50. J. R. Tonry, *Report MP-96,* Portland Cement Association, Skokie, Ill., 1961.
51. *Report: 1973 Technical Mission to Japan,* Portland Cement Association, Skokie, Ill., 1973.
52. D. G. Courteney, *Rock Prod.* **78**(5), 75 (1975).
53. D. Grammes in *Mill Session Papers M-195,* Portland Cement Association, Skokie, Ill., 1969.
54. *Clean Air Act,* Public Law 88-206 (1963); Amendments: *Public Law* 89-675 (1966); 91-604 (1970); and 95-95 (1977).
55. W. E. Trauffer, *Pit Quarry* **67**(8), 52 (1975).
56. *Public Law 92-500.*
57. ACI Committee 223, *Proc. Am. Concr. Inst.* **67,** 583 (1970).
58. *Pit Quarry* **66**(8), 104 (1974).
59. *Minerals Yearbook: Cement,* U.S. Bureau of Mines, Washington, D.C., 1975.
60. *Minerals Yearbook: Cement,* U.S. Bureau of Mines, Washington, D.C., 1972.
61. *Statistical Review No. 33, Production-Trade-Consumption 1974–1975,* Cembureau, Paris, Fr., Oct. 1976.
62. *World Cement Market in Figures, Production-Trade-Consumption 1913–1972,* Cembureau, Paris, Fr., 1973.
63. *Cement Standards of the World,* Cembureau, Paris, Fr., 1968.
64. Ref. 23, ASTM C-806.
65. U.S. Pat. 3,628,973 (Dec. 21, 1971), N. R. Greening, L. E. Copeland, and G. J. Verbeck (to Portland Cement Association).
66. Ref. 23, ASTM C-10.
67. Ref. 23, ASTM C-595.
68. Ref. 3, Chapt. 14; R. Turriziani in ref. 11, Chapt. 14.
69. Ref. 3, Chapt. 15; R. W. Nurse in ref. 11, Chapt. 13.
70. *Specifications for Oil-Well Cements and Cement Additives, API Standards 10A,* 19th ed., API, New York, 1974.
71. D. K. Smith, *Cementing,* Society of Petroleum Engineers of AIME, New York, 1976.
72. Ref. 23, ASTM C-91.
73. Ref. 3, Chapt. 16; T. D. Robson in ref. 11, Chapt. 12.
74. Ref. 3, p. 481.
75. Ref. 3, p. 477.
76. Ref. 23, ASTM C-141.
77. L. Cartz in J. F. Young, ed., *Cements Research Progress 1975,* American Ceramic Society, Columbus, Ohio, 1976, Chapt. 11.

General References

J. F. Young, ed., *Cements Research Progress 1976,* American Ceramic Society, Columbus, Ohio, 1977.
Refs. 1, 11, 13, 17, 23, 28, 33, 40, 47–49, and 63 are also good general references.

RICHARD A. HELMUTH
F. M. MILLER
T. R. O'CONNOR
N. R. GREENING
Portland Cement Association

CENTRIFUGAL SEPARATION

Centrifugal separation is a mechanical means of separating the components of a mixture by accelerating the material in a centrifugal field. The centrifugal field acts upon the mixture in the same manner as a gravitational field; however, the former can be varied by changes in rotational speed or dimensions of the equipment, whereas gravity is essentially constant. Commercial centrifugal equipment can reach accelerations of 20,000 times gravity and laboratory equipment can reach up to 360,000 G. Most centrifugation equipment is intended to separate immiscible or insoluble components from a liquid medium. The ultracentrifuge and gas centrifuge represent special cases that establish separation gradients on a molecular scale. The usual gravitational operations, such as sedimentation or flotation of solids in liquids, drainage of liquids from solid particles, and stratification of liquids according to density, are accomplished more effectively in a centrifugal field (see Flotation; Gravity concentration; Size classification).

The slow development of theory for centrifugation equipment is caused by the complex flow patterns in the centrifuge bowls which have defied satisfactory mathematical modelling. The concept of a theoretical capacity factor for sedimentation depends only upon the characteristics of the equipment and is independent of the system (1–2). Furthermore, the theoretical effect of particle size distribution in single and multistage centrifugation has been demonstrated (3). Extensive application of centrifugation to dewatering and thickening of relatively soft solids and hydrogels associated with industrial and municipal waste treatment (qv) has recently resulted in major changes in centrifuge design. A sound theoretical basis for centrifugal drainage of liquids from solids has been only partially developed; many aspects still need amplification.

This article considers centrifugal separation in a liquid medium. For discussion of gaseous separation, see Diffusion separation methods.

Methods of Mechanical Separation

Several fundamentally different methods that overlap in certain areas are used for mechanical separations. For example, the methods of separating immiscible liquid mixtures containing small quantities of solid material may include gravity or centrifugal settling, electrical coalescence, or coalescing filters. Electrical coalescence, requiring media of very high resistivity, is particularly suitable for large volumes of relatively low dollar value, eg, crude petroleum, having flow rates that may range upward from several hundred m³/h and where high clarity is not generally required. Coalescing filters are employed for nearly clean liquid mixtures having interfacial tensions of 10 mN/m (=dyn/cm) or higher and requiring a high degree of final clarity (see Filtration, electrically-augmented). For separation of liquid–liquid–solid mixtures containing appreciable amounts of solids (>0.5%), gravity or centrifugation equipment may accomplish the three-phase separation in one step or may just reduce the solids content sufficiently to complete the operation as a subsequent liquid–liquid separation.

In applications requiring separation of considerable quantities of solids from liquids, alternative or complementary methods may include screening, vacuum, pressure or centrifugal filtration, flotation, gravity or centrifugal thickening or separation, as well as the more specialized methods of electrostatic or magnetic separations (qv). One or more of these methods may overlap. Screening gives relatively incomplete separation since it is based only on particle size; it is, however, particularly valuable for high tonnages where sizes exceed 420 μm (40 mesh) and may operate efficiently down to 74 μm (200 mesh) (see Size measurement). Pressure and vacuum filters have a wide range of applicability, particularly in removal of solids from high concentration slurries, recovery of fine- to medium-sized and fibrous particles, and efficient washing, but final liquid content of the solids is usually not low. Many special filter designs handle particular applications including those requiring the addition of filter aids and flocculating agents (qv); in particular, continuous belt filters using cloth or screen belts in conjunction with high molecular weight coagulating polymers and pressing rolls are frequently applied in waste treatment. Filter presses have also been automated to various degrees for application to waste treatment sludges (see Filtration).

Centrifugal filtration (4–5) is often applied to batch production of fine, slow-draining solids, but it is better suited to handle medium to coarse particles that require fair to good washing and a low residual liquid content. Centrifugal separation by density difference, as opposed to liquid drainage, is adapted to collection and classification of very fine to medium solid particles in concentrations ranging from very low to medium as well as compressible, gelatinous, and amorphous materials that characteristically plug drainage media. The compactness of all centrifugation equipment lends itself to low or medium tonnages where complete clarity of the liquid effluent is not required and to special applications with high or low pressures or temperatures, or with valuable products where equipment volume is important. Gravity settling is most suitable for relatively large volume flows containing low concentrations of solids with settling rates greater than 1×10^{-3} cm/s; such operations are greatly aided by the rapidly developing technologies of flocculation and aggregation by surface-active materials. Froth flotation can clarify fine dispersions of liquids or solid particles, particularly where the density difference between continuous and dispersed phases is small and the flow volume large; separation by selective froth flotation and sink-float systems is common practice in ore dressing. For separation of solutes in liquid media on a molecular scale, the ultracentrifuge is useful only as a laboratory tool; in gas mixtures, separation by gas centrifuge may be competitive with thermal diffusion and selective membrane diffusion methods. The overlap in separation methods is extensive, and final selection can seldom be made without laboratory or plant testing (see Separations synthesis).

Centrifugal separation of a mixture of immiscible components makes use of either density differences between the components or drainage of a liquid phase through a packed bed or cake of solid particles (6).

Separation by Density Difference

A single solid particle or discrete liquid drop settling under the acceleration of gravity in a continuous liquid phase accelerates until a constant terminal velocity is reached. At this point the force resulting from gravitational acceleration and the opposing force resulting from frictional drag of the surrounding medium are equal in magnitude. The terminal velocity largely determines what is commonly known as the settling velocity of the particle or drop under free-fall or unhindered conditions; for a small spherical particle, it is given by Stokes' law (eq. 1):

$$v_g = \frac{\Delta \delta d^2 g}{18 \mu} \quad (1)$$

where v_g = the settling velocity of a particle or drop in a gravity field; $\Delta\delta = \delta_s - \delta_L$ = the difference between the true mass density of the solid particle or liquid drop and that of the surrounding liquid medium; d = the diameter of the solid particle or liquid drop; g = the acceleration of gravity; and μ = the viscosity of the surrounding medium.

Stokes' law can be readily extended to the centrifugal field (eq. 2):

$$v_s = \frac{\Delta\delta d^2 \omega^2 r}{18\mu} = v_g \left(\frac{\omega^2 r}{g}\right) \tag{2}$$

where v_s = the settling velocity of a particle or drop in a centrifugal field; ω = the angular velocity of the particle in the settling zone; and r = the radius at which settling velocity is determined.

Equations analogous to equations 1 and 2 describe the terminal velocity of a light particle or drop rising in a heavy continuous medium.

The settling velocity, v_s, is relative to the continuous liquid phase where the particle or drop is suspended. If the liquid medium exhibits a motion other than the rotational velocity, ω, the vector representing the liquid phase velocity should be combined with the settling velocity vector (eq. 2) to obtain a complete description of the motion of the particle (or drop).

These concepts are used to analyze separations in the bottle centrifuge, the imperforate bowl centrifuge, and the disk centrifuge. Separation by density difference in other types can be analyzed by analogy.

The Bottle Centrifuge. Analysis of the performance of a bottle centrifuge is based on the model shown in Figure 1. A solid or liquid particle is considered in an initial position, X, at a radius, r, from the axis of rotation. If equation 2 is applied to this specific particle, assuming that $v_s = dr/dt$, then (eq. 3):

$$\int_r^{r_c} \frac{dr}{r} = \int_0^t v_g \left(\frac{\omega^2}{g}\right) dt \tag{3}$$

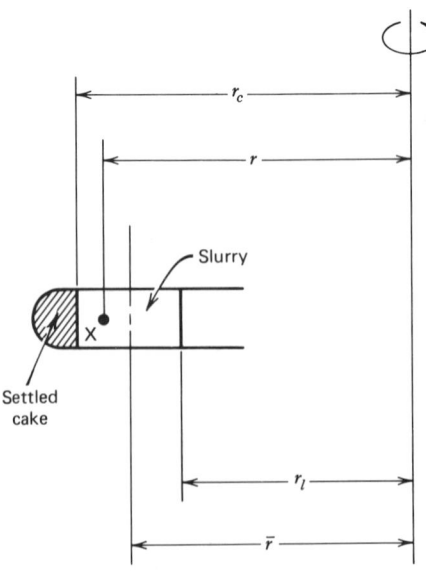

Figure 1. Separation in bottle centrifuge where X = initial position.

where r_C = the radius of the sedimented cake, and t = the time during which the particle is subjected to centrifugal acceleration. Integration of equation 3 leads to (eq. 4):

$$\ln \frac{r_C}{r} = v_g \left(\frac{\omega^2}{g}\right) t \tag{4}$$

A radius, $\bar{r}$, that divides the volume of supernatant into two equal parts can be defined as follows (eq. 5):

$$(\bar{r} - r_l) = (r_C - \bar{r}) \text{ or } \bar{r} = \frac{r_C + r_l}{2} \tag{5}$$

where r_l is the radius of free surface of liquid. Assuming the presence of more than one particle in the feed as well as a uniform initial particle distribution, then each of the two volumes defined by $\bar{r}$ will initially contain the same number of particles. By further assumption that the particles in the suspension are identical, a settling time, t, is chosen so that those particles starting from radius $\bar{r}$ reach the cake r_C after that time t. Under these conditions, one-half of the particles that were in suspension at $t = 0$ are sedimented after the time, t, has elapsed. The other half, initially located above the level defined by the radius $\bar{r}$, remains in suspension. If r is replaced (eq. 4) by $\bar{r}$, a sedimentation condition is established that is referred to as 50% cutoff (see also under Imperforate Bowl Centrifuge). An effective capacity, Q_0, for the bottle centrifuge is determined by the ratio between the volume, V, occupied by the slurry in the bottle, and the spinning time, t, derived from equation 4 (eq. 6):

$$Q_0 = \frac{V}{t} = 2 v_g \left\{ \frac{\omega^2}{g} \frac{V}{2 \ln \frac{2 r_C}{r_C + r_l}} \right\} \tag{6}$$

In equation 6, v_g characterizes the settling behavior of the solid particles or liquid drops in the suspension, whereas the second part of the right-hand side refers to speed and size of the centrifuge and is expressed by the capacity factor Σ_B. For a bottle centrifuge, it takes the form (eq. 7):

$$\Sigma_B = \frac{\omega^2 V}{2g \ln \frac{2 r_C}{r_C + r_l}} \tag{7}$$

This capacity factor has the dimension of an area and represents the area of a gravity settling tank having a separation performance equal to that of the bottle centrifuge handling the same particles. By combining equations 6 and 7, and by eliminating volume V, the following relation is obtained (eq. 8):

$$\frac{Q_0}{\Sigma_B} = \frac{2 g \ln \frac{2 r_C}{r_C + r_l}}{\omega^2 t} \tag{8}$$

Equation 8 provides the basis of comparison for the performance of various bottle centrifuges with the same material, and also, under certain circumstances, of other types of sedimentation centrifuges if geometric dissimilarities are considered.

198   CENTRIFUGAL SEPARATION

The capacity factor, Σ_B, defined by equation 7 is derived from a set of assumptions discussed under The Σ Concept. An additional assumption is specific to the bottle centrifuge: namely, a particle is considered sedimented when it reaches the surface of the cake without contacting the tube wall.

The Imperforate Bowl Centrifuge. In an imperforate bowl centrifuge the flow of continuous liquid phase is effectively axial, except for areas immediately adjacent to the feed inlet and effluent outlet. Tubular solid-bowl basket, and imperforate bowl conveyor-discharge centrifuges satisfy this definition.

The mathematical model chosen for this analysis is that of a cylinder rotating about its axis (see Fig. 2). Suitable end caps are assumed. The liquid phase is introduced continuously at one end so that its angular velocity is identical everywhere with that of the cylinder. The flow is assumed uniform in the axial direction forming a layer bound outwardly by the cylinder and inwardly by a free surface. Initially the continuous liquid phase contains uniformly distributed spherical particles of a given size. The concentration of these particles is sufficiently low so that their interaction can be neglected.

Under these circumstances, the settling motion of the particles and the axial motion of the liquid phase are combined to determine the settling trajectory of these particles.

The trajectory of particles just reaching the bowl wall near the point of liquid discharge defines a minimum particle size that starts from an initial radial location and is separated in the centrifuge. A radius r is chosen to divide the liquid annulus in the bowl into two equal areas initially containing the same number of particles. Half

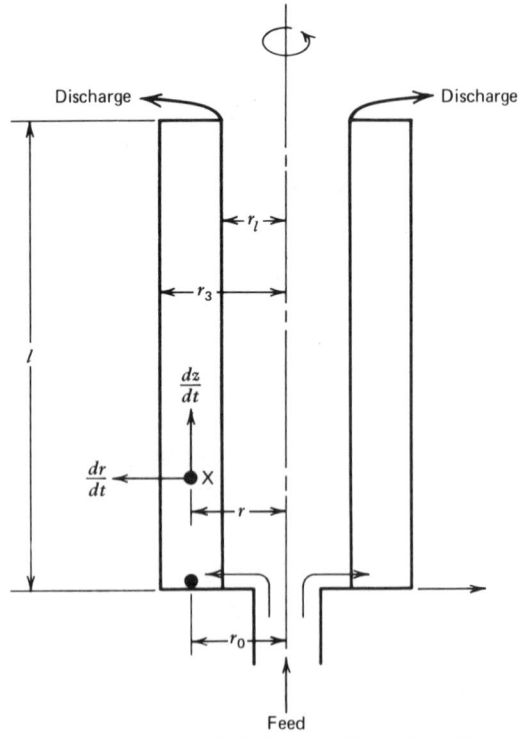

Figure 2.   Separation in basket or tubular centrifuge where X = initial position.

the particles of size d present in the suspension will be separated and the other half will escape. This is referred to as a 50% cutoff.

The feed rate corresponding to this condition is related to the bowl geometry (r_l, r_3, and l), the bowl angular speed (ω), and the Stokes settling velocity (v_g) (2) by equation 9

$$Q_0 = 2 v_g \left(\frac{\omega^2}{g}\right) 2 \pi l \left(\tfrac{3}{4} r_3^2 + \tfrac{1}{4} r_l^2\right) \qquad (9)$$

where v_g = the Stokes settling velocity (see eq. 1); ω = the angular velocity of the centrifuge; g = the gravitational acceleration; l = the length of settling zone; r_3 = the radius of the inside wall of the cylinder; and r_l = the radius of the free surface of the liquid layer in the cylinder. Equation 9 can be rewritten as (eqs. 10–11):

$$Q_0 = 2 v_g \Sigma_T \qquad (10)$$

where

$$\Sigma_T = 2 \pi l \left(\frac{\omega^2}{g}\right) \left(\tfrac{3}{4} r_3^2 + \tfrac{1}{4} r_l^2\right) \qquad (11)$$

Equation 10 estimates the flow or throughput rate above which particles of size d will be less than 50% sedimented and below which over 50% will be mostly collected.

Equations 10 and 11 are applicable to the light particles rising in a heavy phase providing that r_3 and r_l are interchanged in equation 11.

In liquid–liquid separation, the capacity factor relative to one of the phases can also be computed from equation 11 by replacing r_3 and r_l by the limits of the volume occupied by that phase in the bowl.

Equation 11 defines the theoretical capacity factor, Σ_T, which has the dimension of an area and can simply be interpreted as the area of a gravity settling tank that has a separation performance equal to that of the centrifuge, provided that the factor v_g is the same for both. This restriction is required since the particles suspended in the continuous phase can be deaggregated and further dispersed by the vigorous shearing to which the feed is subjected during acceleration in the centrifuge. If this effect is not considered in comparison with the settling tank, centrifugal sedimentation might be less favorable in practice than anticipated. The significance and use of Σ_T is discussed further under The Σ Concept.

Equation 11 can be reduced further to facilitate understanding of its use and application. If, instead of the radii of pond surface r_l and bowl wall r_3, a mean radius r_m is introduced, equation 11 can be rewritten as (eq. 12):

$$\Sigma_T = 2 \pi l \frac{\omega^2}{g} \left(\tfrac{3}{4} r_m^2 + \tfrac{1}{4} r_m^2\right) = (2 \pi r_m l) \left(\frac{\omega^2 r_m}{g}\right) \qquad (12)$$

This equation shows that Σ_T can be expressed as the product of a mean sedimentation area ($2 \pi r_m l$) and the G level ($\omega^2 r_m/g$) and therefore reflects the increased sedimentation rate expected through a defined area using centrifugal acceleration instead of gravity.

The Disk Centrifuge. The separation of particles inside a disk stack is illustrated in Figure 3. The continuous liquid phase, containing solid or liquid particles to be separated, flows from the outside of the disk stack, with radius r_2, to the inside dis-

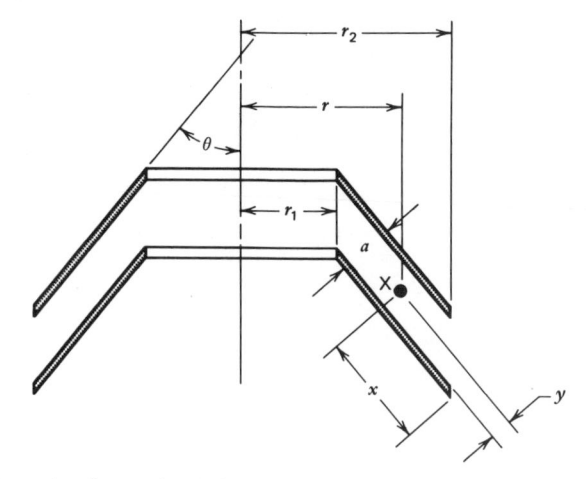

Figure 3. Separation in disk centrifuge where X = initial position.

charge opening, with radius r_1. Assuming that the liquid phase is evenly divided between the spaces formed by the disks, the flow in each disk space is Q_0/n where Q_0 is the total flow through the entire disk stack and n is the number of spaces. The flow of the continuous liquid phase is also assumed to be in a radial plane and parallel to the surfaces of the disks, or as having the same angular velocity as the stack.

Here again, an equation is established (2) to describe the trajectory of a particle under the combined effect of the liquid transport velocity acting in the x direction and the centrifugal settling velocity, in the y direction. This equation determines the minimum particle size which originates from a position on the outer radius, r_2, and the midpoint of the space a between two adjacent disks, and just reaches the upper disk at the inner radius, r_1. Particles of this size initially located above the midpoint of space a are all collected on the underside of the upper disk; those particles initially located below the midpoint escape capture. This condition defines the throughput, Q_0, for which a 50% recovery of the entering particles is achieved (eq. 13).

$$Q_0 = 2 v_g \left\{ \frac{2 \pi n \omega^2}{3 g} \cot\theta \, (r_2^3 - r_1^3) \right\} \tag{13}$$

where v_g = the Stokes settling velocity (see eq. 1); n = the number of spaces between disks in a stack; ω = the angular velocity of the centrifuge; θ = half the included angle of the disks; r_2 = the outer radius of the disks; and r_1 = the inner radius of the disks. If v_g, which describes the settling characteristics of the particles, is separated from parameters relating to the geometry and rotational speed of the disk stack, we obtain (eq. 14):

$$Q_0 = 2 v_g \Sigma_D \tag{14}$$

where

$$\Sigma_D = \frac{2 \pi n \, \omega^2}{3 \, g} \cot\theta \, (r_2^3 - r_1^3) \tag{15}$$

Again, Σ_D has the dimension of a squared length and corresponds to the area of a gravity settling tank capable of the same separation performance as the disk stack defined by the parameters included in equation 15.

The Σ Concept and its Application. The assumptions and conditions set forth as a basis for deriving equations 6, 10, and 14 impose limitations to the application of the Σ concept and fall into two groups:

(1) Concerning the particulate material:

(a) Particles (or drops) are spherically shaped and uniform in size. They should not deaggregate, deflocculate, coalesce, or flocculate during their passage through the zone in which separation occurs.

(b) Initially, particles are evenly distributed in the continuous liquid phase where their concentration is low enough for them to settle as individual particles without interaction.

(c) The settling velocity, v_s, of the particles is such that the Reynolds' number does not exceed 1.0, thus ensuring that the deviation from Stokes' law does not exceed about 10%. The settling velocity, v_g, in a gravity field or v_s, in a centrifugal field, is theoretically, never reached since the accelerating time required for a particle to reach its terminal velocity is infinite. However, a particle of one micrometer in diameter approaches 90% of its terminal velocity in microseconds, and particles up to 100 micrometers approach 90% terminal velocity in milliseconds. Thus, the 0.1 second available for the particles to settle in a centrifuge allows an individual particle continuously to approach its settling velocity despite the variation in G.

(2) Concerning the flow conditions:

(a) Flow is streamlined.

(b) Fresh feed is introduced uniformly into the full space available for its flow. In an imperforate bowl centrifuge, this condition requires that the continuous liquid phase immediately occupies the full liquid layer thickness between the free surface and the inside radius of the cylindrical wall. In a disk centrifuge, the continuous phase is assumed to divide evenly between all the disk spaces.

(c) In the disk stack, flow lines of the continuous phase are directed radially everywhere; in any radial plane, the velocity profile normal to the disk surface is a symmetrical plug flow. In any imperforate bowl centrifuge, the continuous phase rotates everywhere with the same angular velocity as the bowl (ie, not scroll).

(d) The displacement of the flow pattern of the continuous phase by the layer of deposited material is neglected.

(e) Remixing at the interface with the separated materials is negligible.

(f) The detrimental effect resulting from heavy separated material crossing the fresh feed stream outside of the disk stack is also neglected.

Few of these assumptions are fully satisfied in practice. The last three items relate to potential interference between the separated phases. Such interference can occur and may lead to apparently poor sedimentation performance if an excessive volume of the sedimented phase is retained in the centrifuge.

Excessive volume of solids may be retained in the bowl of conveyor centrifuges if: (a) The conveyor volumetric displacement is not sufficient to handle the sedimentation rate of solids. (b) The sedimented solids cannot be successfully conveyed and discharged over the solids port until a sufficient layer has been built up inside the bowl.

In the case of nozzle disk centrifuges, the flow of the solids phase through the discharge nozzles may be so restricted that an excessive layer can accumulate inside the bowl shell. When this layer reaches the zone utilized by the fresh feed stream entering the disk stack, re-entrainment of the sedimented solids by the fresh feed may lead to apparently poor sedimentation performance.

The sedimentation phenomenon that the Σ concept attempts to describe quantitatively is only part of the total task that the centrifuge has to accomplish. Thus, attempts to predict separation performance solely on the basis of the Σ concept have sometimes given disappointing results.

The Σ concept is nevertheless a valuable tool. In theory it allows comparison between geometrically and hydrodynamically similar centrifuges operating on the same feed material. Equations 6, 10, and 14 show that the sedimentation performance of any two similar centrifuges handling the same suspension is the same if the quantity Q_0/Σ is the same for each. In practice, the efficiency factor e is often introduced to extend the use of Σ to compare dissimilar centrifuges. The factor e takes into consideration differences in turbulence, remixing, etc. existing in different types of centrifuges operating on theoretically the same feed material. The flow rate, Q_{02}, of a no. 2 centrifuge can now be compared with a rate, Q_{01}, of a no. 1 centrifuge operating on the same feed. For equal sedimentation performance (eq. 16):

$$Q_{02} = Q_{01} \frac{e_2 \Sigma_2}{e_1 \Sigma_1} \qquad (16)$$

If the two centrifuges are geometrically and hydrodynamically similar, then $e_1 \approx e_2$ and equation 16 can be simplified to equation 17:

$$Q_{02} = Q_{01} \frac{\Sigma_2}{\Sigma_1} \qquad (17)$$

Relation (eq. 17) for similar centrifuges requires identical sedimentation performance characteristics when operating on the same material.

The Σ concept permits scale-up between similar centrifuges solely on the basis of sedimentation performance. Other criteria and limitations, however, should also be considered. Scale-up for a specified solids concentration, for instance, requires knowledge of solids residence time, permissible accumulation of solids in the bowl, G level, and solids conveyability and flowability; limitations of torque and solids loading must be checked as well.

Thus, extrapolation of data from one size centrifuge to another calls for the application of specific scale-up mechanisms for the particular type of centrifuge and performance requirement.

Phase Behavior. Liquid drops, suspended in a continuous liquid medium, separate according to the same laws as solid particles. After reaching a boundary, these drops coalesce to form a second continuous phase separated from the medium by an interface that may be well- or ill-defined. The discharge of these separated layers is controlled by the presence of dams in the flow paths of the phases. The relative radii of these dams can be shown by simple hydrostatic considerations to determine the radius of the interface between the two separated layers. This radius is defined by (eq. 18):

$$r_i^2 - r_h^2 = \frac{\delta_l}{\delta_h}(r_i^2 - r_l^2) \qquad (18)$$

where r_i, r_h, and r_l are the radii of the interface, the liquid surface at the heavy discharge dam, and the liquid surface at the light discharge dam, respectively, and δ_h and δ_l are the densities of the heavy and light phases, respectively. Control of the interface radius, achieved by varying r_h or r_l for the desired range of density ratios, is an important factor in liquid–liquid separation as it determines whether the heavy or light phase is exposed to the greater clarifying effect.

Separation by Drainage

The theory covering drainage in a packed bed of particles is incomplete even for a gravity field, and requires much more development for a centrifugal field. Liquid is held within the bed by various forces and its removal involves several flow mechanisms. In addition, the centrifugal acceleration changes with radius in the bed thus causing changes in packing tendencies of particles and accelerating forces on the residual liquid.

There are three types of liquid content in a packed bed: (a) in a submerged bed, the liquid filling the larger channels, pores, and interstitial spaces; (b) in a drained bed, liquid held by capillary action at points of particle contact, or near contact, as well as a zone saturated with liquid corresponding to a capillary height in the bed at the liquid discharge face of the cake; and (c) essentially undrainable liquid existing within the body of each particle or in fine, deep pores without free access to the surface except perhaps by diffusion.

The last type of liquid can be removed by evaporation or displacement by another liquid but cannot be removed by simple flow in either a gravitational or centrifugal field. There is no sharp distinction between the first two types. The rate and extent of liquid removal from a submerged bed during drainage depends on the physical characteristics of the components, the force of the centrifugal field, and the time of exposure to the field. The residual liquid content of a drained cake consists largely of the capillary and irremovable types at the time of discharge from the separation equipment.

During cake formation and drainage the liquid moves into and through the bed in three different ways: (a) During cake deposition, a continuous head of liquid ranging in composition from feed to clarified supernate may exist over the deposited cake; after feed is stopped, a layer of essentially clarified liquid may still exist over slow-draining cakes; wash liquor may also create a liquid layer over the deposited cake; drainage under these conditions requires continuous flow through the cake with interstitial spaces assumed full; (b) When the free liquid layer no longer exists above the cake, the free liquid surface moves through the cake to an equilibrium position at the capillary height leaving behind the larger voids filled with gas or vapor; and (c) After bulk drainage of the larger voids, liquid still exists in the cake's upper zone in a film covering the surfaces of the solids and in partially filled voids having very restricted outlets; in time, some of this liquid flows as a film to the continuous liquid layer at the capillary height.

If a cake is sufficiently impermeable to permit the build-up of a feed or wash liquid head, flow through the cake approaches steady-state conditions except for changes in compaction or cake thickness as more feed is added, or in the liquid head if the drainage rate differs from the liquid rate addition. An equation for full-pore flow in a centrifugal field has been developed. The hypotheses set forth and, for the most part proved, are as follows: flow radiates out from the rotation axis and the effect of the gravitational field is negligible; voids at all points are filled with liquid that moves in laminar flow through the cake; kinetic energy changes of the liquid in the cake may be neglected; the filter medium is sufficiently permeable so that it does not run full of liquid; and ambient pressure exists at the outer face of the cake. Essentially incompressible solids produce very similar cake permeabilities in vacuum, pressure, and centrifugal filtration, although significant local variations in permeability may occur because of irregularities in the feed and its distribution. With filled pores, the flow rate of the supernatant liquid through the cake is (7–8) (eq. 19):

$$Q = \pi \frac{K\omega^2 h(r_M^2 - r_L^2)}{\mu_L g \ln(r_M/r_C)} \qquad (19)$$

where K is permeability, h is the basket height, r_C and r_M are radii from the axis of rotation to the inner and outer faces of cake, respectively, ω is the angular velocity, and r_L is the radius from the axis to the inner face of the liquid layer. Comparison with the usual term α for cake resistance in pressure filtration shows that $K = \delta_L g/\alpha$. The functions related to cake thickness, $\ln(r_M/r_C)$, liquid layer thickness, $(r_M^2 - r_L^2)$, angular velocity, ω^2, and viscosity, μ_L, have been verified by experiment. The exponent on the angular velocity varies for materials that exhibit cake compression as speed is increased, but compression effects are minor on starch, chalk, and kieselguhr. In practice, the exponent of ω can probably be assumed constant and experimental variations in permeability can be absorbed by changes in the permeability coefficient. At present, the real value of this equation lies in estimating the effect of changes in operating variables for a material whose characteristics are already known from experimental data or plant operation.

Low-permeability cakes draining under the conditions of equation 19 are usually handled in perforate basket centrifuges that have relatively long cycles (20 min to several hours). Following elimination of the supernatant liquid, unsteady-state drainage of the cake may often be neglected. These slow-draining cakes often support such a large capillary height, for example, 90–99% of void volume for chalk (8), that little additional dewatering is obtained after completion of free liquid drainage. Solids of larger particle size or freer drainage characteristics are handled in automatic perforate baskets or continuous screen centrifuges; low final moistures are usually desired. For cakes of high permeability, the period when a liquid layer exists above the cake is either short or nonexistent and the time cycle chiefly depends on the rate of film drainage at the completion of bulk liquid flow. Under these conditions film drainage and permanent residual moisture are most important.

Currently, it is impossible to predict the quantity of undrainable residual moisture (see the third type of liquid content discussed above) without the benefit of experimental data. Equation 20 (9) indicates the important parameters whose exponents were determined with limited experimentation. Introducing the approximation that $s\delta_s$ is proportional to $1/\bar{d}$ (s = specific surface area/weight of solid), the modified equation for undrainable liquid becomes (eq. 20):

$$S_\infty = k \left(\frac{1-\epsilon}{\epsilon}\right)\left(\frac{1}{\bar{d}^2 G}\right)^{1/4} \left(\frac{\sigma \cos\varphi}{\delta_L}\right)^{>1/4} \qquad (20)$$

where S_∞ is the fraction of void volume occupied by liquid after infinite drainage time, k is an experimental coefficient, ϵ is the void fraction, $\bar{d}$ is the mean particle diameter, G is $\omega^2 r/g$, σ is the surface tension, and φ is the wetting angle. Appreciable internal porosity of the particles can badly distort an experimental value of s.

For cakes of high permeability, the capillary drain height may be an insignificant fraction of cake thickness, and film drainage becomes the controlling factor in a centrifugal field (10). Under unsteady-state conditions, equation 21 represents the drainable liquid left in the cake as a function of the centrifugal filtration parameters:

$$S - S_\infty = 3\pi \frac{s'}{\epsilon\omega}\left[\left(\frac{\mu_L}{\delta_L}\right)\left(\frac{h}{2r_M - h}\right)\left(\frac{1}{t-t'}\right)\right]^{1/2} \qquad (21)$$

where S is the fraction of void volume occupied by liquid at time t, s' is surface area/volume of cake, h is cake thickness, and t' is time at which free liquid surface enters cake.

It is difficult to obtain void volume data for a cake under drainage conditions in a centrifuge. Prediction of these values from filter cake data is uncertain since the compressive force in centrifugation increases with radius throughout the cake depth and the effective mass of a particle is proportional to $(\delta_S - \delta_L)$ when the cake is submerged but becomes essentially δ_S after bulk drainage is completed. For engineering purposes, it is simpler to approximate the ratio of volume of liquid to the volume of solid. Assuming that $G \approx \omega^2 (2r_M - h)/g$, and $s' \approx 6(1 - \epsilon)/\bar{d}$ as for spheres, equation 21 becomes (eq. 22):

$$q - q_\infty = \left(\frac{k'}{\bar{d}}\right)\left(\frac{\mu_L}{\delta_L}\right)^{1/2}\left(\frac{h}{Gt_f}\right)^{1/2} \times 100 \qquad (22)$$

where q is the volume of liquid to volume of solid. This measurement is readily obtained from volume percentage of liquid in cake and is also easily converted by a density ratio to a weight ratio of liquid to solid. The value for q_∞ may also be applied when known.

Reasonable experimental agreement was obtained (10) with these exponents for kinematic viscosity, μ_L/δ_L, G, cake thickness, and drain time for relatively slow-draining chalk and kieselguhr. Figure 4 (11) shows the effect of drain time after the disappearance of a free liquid head above the cake. The sharp break probably indicates completion of bulk drainage and start of drainage by film flow only. Drainage before the break is rapid and proportional to $t^{-1.9}$; after the break in the curve, $t_f^{-0.3}$ indicates that film drainage becomes controlling if low residual moisture is required. This ex-

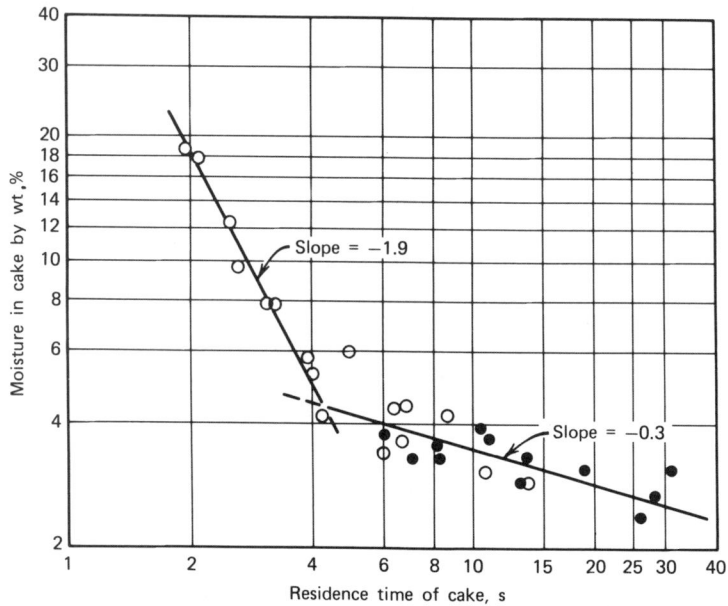

Figure 4. Drainage of salt crystals in cylindrical screen pusher-discharge centrifuge (11). Note that cake thickness = 3.3 cm, centrifugal field = 320 G, crystals = 14% W/W < 250 μm, ● = moisture in discharge cake, and ○ = moisture in cake by material balance with drainage flows.

ponent is appreciably lower than the -0.5 of theory but is close enough to -0.25 previously obtained (12). Considerable data indicate the validity of $G^{-1/2}$ and limited data corroborate the theoretical exponent of 1/2 on kinematic viscosity.

Experimental exponents on cake thickness vary from 0.5 to as much as 3.0, and the theoretical value of 1/2 may be approached only by incompressible cakes of a narrow range of sizes. The proper and characteristic value for the mean particle size, $\bar{d}$, is difficult to ascertain. In practice, the most finely divided particles (10–15 wt % of solids) almost wholly determine the liquid content of a cake regardless of the rest of the size distribution. It seems reasonable to use a $\bar{d}$ closely related to liquid content, for example, the 10% point on a cumulative weight distribution curve.

Power Energy and Noise Considerations

Both power and energy requirements of a centrifuge should be considered in the choice of the motor size for a given application.

Energy Required for Start-up and Speed Cycling. The energy ΔE stored in or liberated from a centrifuge bowl when its angular velocity is changed from ω_1 to ω_2 is (eq. 23):

$$\Delta E = \tfrac{1}{2} I(\omega_2^2 - \omega_1^2) \tag{23}$$

where I is the moment of inertia of the rotating assembly about its axis of rotation. If the centrifuge bowl is started from rest, $\omega_1 = 0$ and equation 23 becomes (eq. 24):

$$E = \tfrac{1}{2} I \omega^2 \tag{24}$$

For high-speed centrifuges, the energy stored in a bowl during start-up is an important factor in the choice of motor size. The energy delivered by the motor as it brings the centrifuge bowl to its full speed is (eq. 25):

$$E_{mot} = \int_0^{\Delta t} m\omega_{mot} dt \tag{25}$$

where Δt is the time required to reach operating speed and m is the torque transmitted to the bowl from the motor rotating at the instantaneous speed, ω_{mot}. The relationship between m and ω_{mot} is a characteristic of the motor, drive inertia and friction, and bowl windage. If these values are known, the time Δt can be determined by integrating equation 25 and equating it to equation 24. Since a motor draws excessive current during start-up, the tolerance period for this condition may be short and should be compared with the value for Δt obtained above. In a high-speed or high-inertia centrifuge, or both, the minimum motor size acceptable is often determined by the time required to bring the machine to full speed rather than by power consumed under normal operating conditions.

The energy required to start the motor is in considerable excess of the value computed by equation 24. Only about 20% of the energy conducted to the motor is transferred to the centrifuge bowl before operating speed is reached; the rest is dissipated in the form of heat in the motor. Even with special measures such as star-delta or depressed torque motors or clutch drives, the frequency of start-ups is limited.

Power Required for Acceleration. The total power required to accelerate the material fed into the centrifuge is the sum of the individual powers dissipated by each individual stream. To compute these powers, the contents of the bowl are assumed to rotate at the same angular velocity as the bowl. Centrifuges are equipped with baffles to ensure that the liquid within the bowl maintains the same angular velocity as the shell.

A fluid stream accelerated from rest to an angular velocity ω and discharged at a radius r consumes a power Pwr (eq. 26):

$$Pwr = Q\delta\omega^2 r^2 \qquad (26)$$

where Q = the flow rate of the discharge stream, and δ = the mass density of the fluid in that phase.

The total power consumed by the feed to the bowl is the sum of the individual powers obtained for each discharge from equation 26.

Equation 26 indicates that, under ideal conditions, a bowl discharging a stream at a radius $r = 0$ requires no power. However, the viscous drag of the fluid in the passages of the bowl must be overcome by an external source since no energy is available from the centrifuge. This energy is supplied in the form of pressure by the pump delivering the feed to the bowl. A seal is placed between the stationary feed tube and the rotating centrifuge bowl to permit transfer of this pressure.

For some disk centrifuges, one of the separated phases is discharged in the form of jets through nozzles at or near the periphery of the bowl. If these jets are directed radially, the power consumed as given by equation 26 may be excessive because of the large radius where these nozzles are usually located. The dissipated power can be mostly recovered by directing the jets tangentially in a direction opposite to the bowl rotation. The following relation is then used to determine the net power requirement (eq. 27):

$$Pwr_N = Q_N \delta_N \omega r_N (\omega r_N - v_N \cos\phi) \qquad (27)$$

where Q_N = the volume rate of the material discharged through the nozzles, δ_N = the mass density of the material discharged through the nozzles, ω = the angular velocity of the centrifuge, r_N = the radius at which the nozzles discharge, v_N = the linear discharge velocity out of the nozzles, and ϕ = the angle measured between the direction of the nozzle and the tangent to the circle of radius r.

The discharge velocity out of the nozzles is given by equation 28:

$$v_N = C\sqrt{\frac{2p_N}{\delta_N}} \qquad (28)$$

where C = a discharge coefficient, and p_N = the hydrostatic pressure at radius r_N resulting from the rotation of the bowl.

The pressure p_N is given by (eq. 29):

$$p_N = \delta \frac{\omega^2(r_N^2 - r^2)}{2} \qquad (29)$$

where r is either the free-surface radius of the light liquid phase in the case of liquid–liquid and liquid–liquid–solids separation, or the free-surface radius of the liquid phase in the case of liquid–solids separation; and δ is intermediate between the mass densities of (a) the heavy and light phases in the case of liquid–liquid separation, or (b) the light and heavy phases and that of the nozzle stream in the case of liquid–liquid–solids separation, or (c) the liquid phase and that of the nozzle stream in the case of liquid–solid separation.

The coefficient C in equation 28 comprises two phenomena taking place in the nozzle. First, the presence of viscous friction accounts for a small loss of energy. Second, energy is lost because of jet contraction near the nozzle inlet and subsequent expansion to fill, or partially fill, the bore of the nozzle. Discharge coefficients range from $C = 0.5$ to 0.85. The coefficient falls in the upper portion of this range when the length of the nozzle is two or three times its diameter, and in the lower end of the range when the nozzle length is more than five times its diameter.

Power Requirements Imposed by Friction and Windage. Bearings, belts, gears, and seals transform small amounts of power into heat. The greatest loss of power derives from the friction of the gas in contact with the surface of the rotating elements of the centrifuge. This friction is called windage. The power lost in windage can exceed that required to accelerate the feed, especially in high pressure operation. For instance, a rotor 30 cm in diameter rotating in air at 10,000 rpm at atmospheric pressures requires about 2 kW to overcome windage. At 1.03 MPa (150 psi), the power requirement increases to 20 kW. Thus, windage power is important in operation of high-speed centrifuges at high pressures.

Noise Level. Centrifuges, as any rotating equipment, create noise and centrifuge manufacturers have made considerable efforts in the last few years to decrease noise level. When the motion of air or gas entrained by a rotating bowl shell is deflected or otherwise disturbed, its energy is transferred to the environment through casing and chutes. This mechanism suggests that the noise level created by the bowl is related to the surface linear speed of the rotor which, in turn, is the product of the angular velocity of the rotating body and its outer radius. Linear speed is, however, important to maintain separating or drainage capacity, so noise level should be reduced without reducing linear speed if possible.

In addition to linear speed, imbalance of the rotating member, surface irregularities, clearance between covers and rotor, resonance in the supporting structure, and type of installation, particularly the drive motor contribute to centrifuge noise. An inadequate supporting platform can amplify centrifuge vibration. Open piping and venting and discharge connections allow noise generated inside the unit to escape. Discharge connections should be tight, yet flexible enough to prevent transfer of vibrational energy to plant piping. Size, spacing, and other properties of the centrifuge room also affect noise level. Sound absorbing materials or enclosures should be provided whenever possible to reduce ambient noise to a minimum.

The electric drive motor is often the single noisiest component of the centrifuge assembly. Most standard motors in the 75–150 kW range develop noise levels in excess of 90 dBA (weighted sound pressure level using filter A as per ANSI standard). Quiet motors can reduce this level by 5 or even 10 dBA and should be used whenever noise is of concern (see Noise pollution).

Equipment

Centrifugation equipment that separates by density difference is available in a variety of sizes and types and can be categorized by capacity range and the theoretical settling velocities of the particles normally handled. Centrifuges that separate by filtration produce drained solids and can be categorized by final moisture, drainage time, G, and physical characteristics of the system, such as particle size and liquid viscosity.

For optimum results a combination of several types of equipment may be used, eg, for oil recovery from a gravity separator sludge at a petroleum refinery. The sludge, an aqueous suspension of 1–5% oil and 5–30% solids below 50 μm, is screened to remove trash and degritted in a cyclone to eliminate the coarse solids that would cause excessive abrasion. An imperforate bowl conveyor-discharge centrifuge then removes 60–70% of the solids in oil-free condition. The resulting oil-in-water emulsion, stabilized by residual fine silt, is screened through 0.25 mm (60 mesh) and sent to a disk centrifuge which discharges an oil stream at 0.5–2% bottom sediment and water and an oil-free peripheral nozzle discharge containing the remaining solids.

Construction Materials and Abrasion Resistance. Stainless steel, especially type 316, is the construction material of choice and can resist a variety of corrosive conditions and temperatures. Carbon steels are occasionally used; however, rusting may cause time-consuming maintenance and can damage mating surfaces which increases the vibration and noise level. Titanium, Hastelloy, or high nickel alloys are used only in special instances and at a considerable increase in capital cost.

Abrasion, a serious problem in some applications, requires the addition of hard-surfacing materials to points exposed to abrasive wear. The severity of wear depends upon the nature, size, hardness, and shape of particles as well as the frequency of contact, the force exerted against the wearing parts, and solids loading as related to feed rate and solids concentration.

A wide range of abrasion resistant materials is available. Nickel–chrome–boron and cobalt–chrome–tungsten hard surfacing alloys have been used for many years. Composite coatings of nickel-base alloys containing crushed tungsten carbide particles, applied by flame spraying and fusing, are a recent improvement. Solid tungsten carbide, pressed to shape and sintered at high temperatures, provides the best protection. Tungsten carbide platelets can be brazed to stainless steel plates which, in turn, can be easily welded to portions of a centrifuge such as conveyor flights. Ceramics have been used successfully where minor impact and abrasive particle pressures are involved (see Abrasives).

Centrifugal Sedimentation. Centrifugal sedimentation equipment is characterized by limiting flow rates and theoretical settling capabilities. Feed rates in industrial applications may be dictated by liquid handling capacities, separating capacities, or physical characteristics of the solids. Sedimentation equipment performance is illustrated in Figure 5 on the basis of nominal clarified effluent flow rates and the applicable Q_0/Σ values; the latter are equivalent to twice the theoretical gravity settling velocities. In liquid–solid separations, the effluent rate represents the clarified stream of the liquid medium and does not include the volume of solids discharged nor the volume of medium discharged with the solids. The effluent rate of liquid–liquid separation refers to the clarified, heavy or light continuous phase that occupies the greater portion of the separating equipment. The flow ranges for a particular piece of equipment never represent its absolute limitations, but the normal flows for good clarification in standard applications. Similarly, large particles can always be sedimented.

As an additional guide, the Q_0/Σ values are correlated with the equivalent spherical particle diameter by Stokes law as in equation 1. A density difference ($\Delta\delta$) of 1.0 g/cm^3 and a viscosity of 1 mPa·s (=cP) are assumed; thus, conversion to other physical characteristics of the system requires that the particle size scale be adjusted to equate a particle of 1.0 μm diameter to its Q_0/Σ in cm/s, according to the relationship

210   CENTRIFUGAL SEPARATION

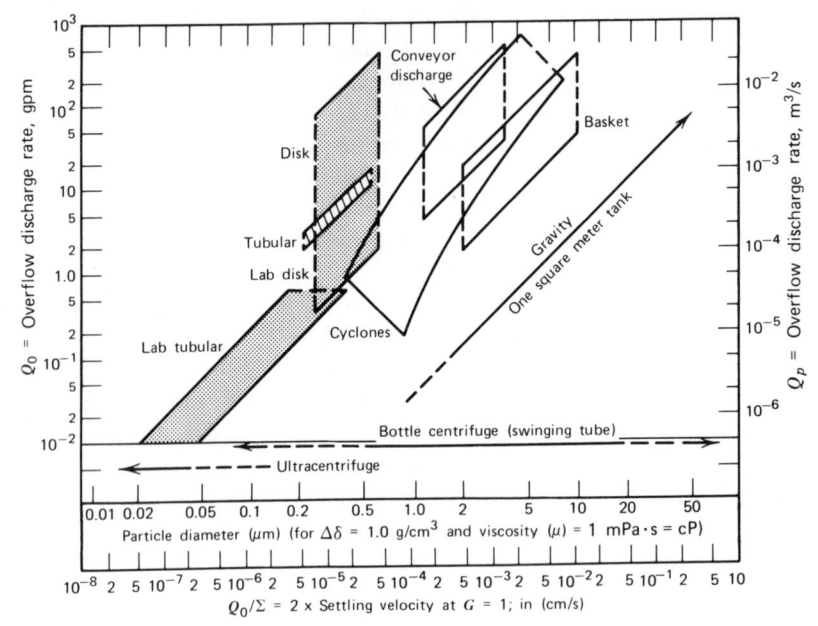

Figure 5. Sedimentation equipment performance.

$Q_0/\Sigma = 10^{-7} \times 1.09 \, \Delta\delta/\mu$ for $\Delta\delta$ in g/cm³ and viscosity (μ) in Pa·s. For interpretation of the particle sizes, two factors should be noted, ie, the scale refers to the 50% cutoff particle size, and under actual centrifugation conditions the value of Σ, determined from Figure 5, must be increased by efficiency factors to give the theoretical value of Σ (see under The Σ Concept).

Figure 5 serves as a guide indicating which types of equipment can handle a given separation; other characteristics narrow further selection. For example, take the separation at $Q_0 = 3.15 \times 10^{-3}$ m³/s (50 gpm) of kaolin clay solids from an aqueous suspension; the particle density is 2.55 g/cm³ and the size range from 0.25 to 30 μm with 55% <2 μm. The 1.0 μm point on the particle size scale would be equivalent to $Q_0/\Sigma = 1.69 \times 10^{-4}$ cm/s as noted above. Assume that a high recovery is desired, a disk centrifuge is required, and recovery of most particles greater than 0.4 μm is satisfactory. The Q_0/Σ equivalent to 0.4 μm on the adjusted scale is about 2.3×10^{-5} cm/s. With a disk machine efficiency of 40% the centrifuge needed has a Σ value of:

$$\Sigma = \frac{3.15 \times 10^{-3} \times 10^6}{0.40 \times 2.3 \times 10^{-5}} = 34.3 \times 10^7 \text{ cm}^2$$

Because clay tends to pack hard, only the continuous nozzle discharge bowl would be satisfactory and intermittently discharging disk centrifuges could not be used.

If classification of solids was desired, several other types of centrifuges could be used as well, assume that only particles over about 2 μm were to be removed from the suspension and that the oversize stream should be highly concentrated for disposal. Figure 5 shows that a conveyor-discharge or basket centrifuge or standard cyclone can, theoretically, be used. However, cyclones cannot concentrate the oversize as much as the centrifuges so they are less satisfactory for this example. If the feed concentration was low, eg, less than a few percent solids, a basket centrifuge would be used with intermittent discharge of solids. However, a conveyor-discharge centrifuge gives almost as good a concentration of oversize as the basket and is a more efficient classifier; it would thus be the better choice and is actually used in the kaolin industry.

With flocculent or soft, aggregated solids, varying degrees of deflocculation of the solid particles may occur during acceleration in the centrifugal field. An additional problem, as noted previously, is the removal of resulting soft cakes under centrifugal force which requires modification of both design and operating conditions for conveyor centrifuges. Specially modified, automated basket centrifuges can handle a broad range of soft sludges, often without polymers; disk centrifuges are particularly well-suited for clarification of streams containing solids such as aluminum hydroxide or waste activated sludge.

The disk centrifuge with high capacity and G level is normally used for separating a liquid–liquid mixture or for clarification of such a mixture containing fine solids. Specially modified conveyor centrifuges have occasionally been used for three-phase separations, but the complexity of control is usually not warranted. Settling velocity is a criterion of selection, but the actual separation of emulsified liquid–liquid mixtures may not strictly follow a settling theory. For an emulsion, a threshold level of centrifugal force may have to be exceeded before the emulsion starts to break; in addition, drainage of liquid from the continuous phase film of the emulsion has a time factor. However, centrifugal acceleration and time cannot be calculated interchangeably as Σ theory would indicate. In centrifugation equipment, coalescence of the dispersed liquid occurs coincidentally with its separation because there is neither time nor space for appreciable interfacial retention of unbroken emulsion.

Batch equipment, such as the bottle centrifuge or ultracentrifuge, does not have a real throughput capacity. By increasing the time of operation, according to equation 8, the smallest particle size of solids usually sedimented in the bottle centrifuge may be 0.1 μm, where Brownian movement becomes significant; in the ultracentrifuge, separation can be achieved down to molecular size, perhaps 0.005 μm, where Brownian movement is controlling. There is clearly no limitation to the larger particles that may be settled in bottle centrifuges so that an arbitrary upper limit is indicated for practical minimum conditions of 1000 rpm and 10 s. Similarly, for the ultracentrifuge the upper limit for Q_0/Σ was estimated for minimum conditions of one hour and 5000 rpm.

Centrifugal Filtration. The important parameters of centrifugal filtration equipment (6–7) are screen area, level of centrifugal acceleration in the final drainage zone, and cake thickness. The latter affects both residence time and volumetric throughput rate. As indicated by equation 22, the particle size of the solids and the kinematic viscosity of the mother liquor also strongly affect the final moisture content. A limited correlation has been developed for performance of perforate basket and conveyor-discharge conical screen bowls, but the range of materials handled and the complexity of the drainage and washing operations do not lend themselves to broad correlations. The variables of correlation may be useful in a particular study, especially if more than one type of centrifuge is involved. A conveniently expressed coordinate is the drainage number, $\overline{d}\sqrt{G/\nu}$ where $\overline{d}$ is the mean particle diameter in micrometers, ν is the kinematic viscosity of the mother liquor in m^2/s (stokes $\times$ 10^{-4}) at the drainage temperature, and G is $\omega^2 r/g$; r is unambiguous for a cylindrical bowl and is the screen radius at the cone's larger end in a conical bowl. $\overline{d}$, at the 15% cumulative weight level of the particle size distribution, is suggested instead of the usual 50% point since the final moisture content of a cake is closely related to the finest 10–15% fraction of the solids and is almost independent of the coarser material.

The function of the other coordinate is $q\sqrt{t}$ where q is the percentage of final moisture on discharged cake, as the volume of mother liquor per unit volume of solid, and t is the drainage time in seconds. A weight ratio may be used for q, but the volume ratio makes the function more universal by eliminating densities. For a conveyor-discharge conical screen bowl, the helical conveyor is assumed to control the residence time of the solids so that time becomes the number of turns of helical flight around the conveyor hub divided by the differential speed between the conveyor and the bowl. For a pusher centrifuge, t is the retention time on the screen as controlled by length and diameter of screen, thickness of cake, and frequency of stroke. In a basket centrifuge, dead and unload times should not be included in the calculation of drain time. With reference to Figure 4, bulk drainage is completed so quickly that film drainage is usually controlling. Thus, drain time t is approximated by the sum of feeding time, spin time prior to rinsing, and spin time after rinsing but prior to unloading according to the theory of cyclical centrifuges (13). Filtration correlations generally show a spread as a function of cake thickness. Conical screen bowls characteristically have short residence times and achieve good drainage by maintaining thin layers of cake. Smaller perforate baskets operate with cakes of 5–10 cm thickness, whereas larger baskets may carry cakes up to 15 cm thick. Pusher centrifuges and high speed peeler baskets may handle cakes ranging in thickness from 5 to 20 cm.

An example of centrifugal filtration is the recovery of salt crystals from a mother liquor of about 3×10^{-6} m^2/s (3 cSt) viscosity; the crystal size is 170 μm at the 15% level and drained cake bulk density about 1.5×10^3 kg/m^3 (95 lb/ft^3). The cone screen is 25.4 cm at the larger diameter and the automatic cycle basket centrifuge is 68.6 cm in diameter; other parameters for final values of $q = 7.6\%$ in both centrifuges are given in Table 1.

EQUIPMENT OPERATING BY DENSITY DIFFERENCE

Bottle Centrifuge. A bottle centrifuge is designed to handle small batches of material for laboratory separations, testing, and control. The basic structure is a motor-driven vertical spindle supporting various heads or rotors; a surrounding cover reduces windage, facilitates temperature control, and provides a safety shield against breakage of the spinning container. Optional accessories include timer, tachometer, and manual or automatic braking. Bench-top bottle centrifuges operate with rheostat control at 500 to 5000 rpm and produce centrifugal fields up to 3000 G in the lower speed range and up to 20,000 rpm with 34,000 G in the high speed units. Larger models operate up to 6000 rpm developing 8000 G with special attachments permitting 40,000 G. These models may also be equipped with automatic temperature control down to $-10°$C.

There are three types of rotors: swinging bucket, fixed-angle head, or small perforate or imperforate baskets for larger quantities of material. In the swinging bucket type, the bottles are vertical at rest but swing to a horizontal radial position during acceleration so that solids are deposited in a pellet at the bottom of the tube. Although sedimenting particles must travel up to the full depth of the liquid layer, which requires appreciable time, the long path of travel and the perpendicularity of the sedimenting boundary to the axis of the tube are distinct advantages in effecting fractional sedimentation. Heads carrying fixed tubes at a 35–50° angle reduce centrifuging time as the maximum distance travelled by a particle is the secant of the tube angle times the

Table 1. Parameter Values in Cone Screen and Automatic Basket Centrifuges Needed to Obtain a Value of $q = 7.6\%$

Properties	Cone screen	Automatic basket
bowl diameter, cm	25.4	68.6
G	1440	865
bowl speed, rpm	3180	1500
$\overline{d}\sqrt{G/\nu}$	3.750×10^6	2.880×10^6
time, s		
feed and spin		12
spin after rinse		27
drain time	0.6^a	39
rinse^b		10
unload and screen rinse		4
cycle time		53
q, %	7.6	7.6
$q\sqrt{t}$, %·$\sqrt{s}$	5.89	47.46
h, cm	.51	3.81
screen area, cm^2	930	7675
approx cake rate, kg/s	1.19 (9450^c)	0.84 (6700^c)

^{a} Differential speed = 60 rpm, ⅝ turn per helical flight.
^{b} Note that the comparison shows rinse only in the basket centrifuge since rinse efficiency in the cone screen is fairly low compared to the basket. The latter would probably be selected for the application if good rinsing was needed and the conveyor-discharge conical screen centrifuge if no rinsing was necessary.
^{c} lb/h.

diameter of the tube; particles strike the wall and slide down the tube to collect near the bottom, but the angle makes it difficult to measure relative volumes of supernatant liquid and sedimented solids.

Rotors carry 2 to 16 metal containers with tubes and bottles of various sizes and shapes. Containers range in capacity from capillaries for microanalysis to a 1-L maximum, limiting the batch capacity of this type of centrifuge to about 4 L. Although glass bottles and tubes are generally used, polyethylene, polycarbonate, aluminum, and stainless steel containers are available for high-speed operation or corrosive liquids. Tubes are usually cylindrical, tapered, or graduated; special shapes for analytical work are available including pear-shaped tubes with capillary tips for measuring small quantities of solids.

The bottle centrifuge is primarily used in the laboratory to separate small quantities of material. It is also used for standard analyses including many ASTM methods and in preliminary testing for scale-up to commercial centrifuges. The Σ_B value for free-swinging tubes is determined by equation 7 and the Q_0/Σ_B value by equation 8; Q_0/Σ_B data can be prepared by bottle centrifuge (14). The bottle centrifuge has been used to study waste treatment sludges for estimation of cake concentrations and feasibility of handling in a conveyor centrifuge (15). Since compaction of solids is largely a function of the centrifugal field force and the exposure time, the bottle centrifuge can be used to study these parameters, although not always in the range applicable to industrial equipment. Similarly, drainage of packed solids can be studied by using tubes with fritted glass or perforated metal bottoms. Closed containers should be used to prevent drying by windage.

Specialty rotors permit ordinary bottle centrifuges to achieve some of the results previously considered possible only in ultracentrifuges. A modified zonal rotor, shown in Figure 6, permits collection of sediment with continuous addition of feed and discharge of centrate. A refrigerated bottle centrifuge using the rotor shown in Figure 7 with a long-path circumferential flow can operate at 6000 rpm as a zonal centrifuge with a density gradient and achieves the equivalent of ultracentrifuge performance.

Preparative Ultracentrifuges. Preparative ultracentrifuges are suitable for a range of applications, such as processing quantities of subcellular particles, viruses, and proteins. Many design variations are available and only the common features are considered here. Preparative ultracentrifuges range in operating speed from 20,000 rpm, generating about 40,000 G, to 75,000 rpm and about 500,000 G. The rotor is surrounded by a high-strength cylindrical casing and underdriven by an electric motor or oil turbine. To avoid overheating of the rotor by air friction at these speeds, the pressure in the casing has to be reduced to about 0.13 Pa (1 μm Hg) for high speeds, and to <130 kPa (<1 mm Hg) for lower speeds. Sensors monitor the temperature and the refrigeration system controls the temperature in the range of -15 to $+30°C$ within $\pm 1°C$. Electronic controls maintain the rotor speed within the required narrow range and may be automatically programmed for sequential changes in speed including control of the acceleration and deceleration (16).

Preparatory ultracentrifuges are guaranteed for several billion revolutions and can be rebuilt with relatively few parts. Among the great number available are batch rotors and those accepting feed and discharging centrate continuously during rotation. Batch rotors include angle and swinging bucket types similar to those in basket centrifuges as well as the newest type whose vertical tubes, parallel to the axis of rotation, present a very short sedimenting distance and time requirement. Swinging bucket rotors are also used for density gradient separations or volume evaluation of the settled cake.

Separation by selective sedimentation on the basis of size and density of the particles may be satisfactory for poly-disperse particle systems; however, the cake contains a range of material depending on its starting position in the container. Selectivity of separation can be improved by introducing the sample near the surface after the container is up to speed; reslurrying and recentrifuging may be necessary to achieve purer fractions. Isopycnic separation improves initial separation efficiency where particles differ in density; if the density of the medium is intermediate to the range of densities of particles, higher density particles will settle, whereas others remain suspended or rise to the surface regardless of size.

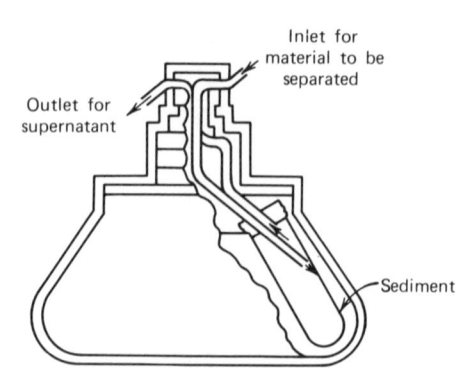

Figure 6. Tube-type continuous flow rotor. Courtesy of Sorvall.

Analytical Ultracentrifuges. The modern analytical ultracentrifuge is a highly sophisticated instrument that allows continuous observation inside the cells of the rotor (14). It may be regarded as a highly specialized form of bottle centrifuge operating on particles so small that diffusion must also be considered.

An analytical ultracentrifuge is driven through gearing from an electric motor, or, less often, by an air or oil turbine. An electro-optical system controls drive speed as follows frequency corresponding to the operating speed is produced by a light chopping disk and compared to a reference signal from the speed selector. Instantaneous speed fluctuation during a run is less than 0.2%. The oval-shaped aluminum or titanium rotor is self-balancing and flexibly suspended from a vertical shaft driven at speeds up to 70,000 rpm to produce fields up to 370,000 G. To enable temperature control at such speeds, the chamber is evacuated to about 0.13 Pa (1 μm Hg) pressure The temperature is held within 0.1°C by a thermal sensor. The cell is located in the rotor at a mean radius from the axis of rotation of 65 mm with a sample radial depth of about 14 mm. Concentration gradients in the cell are observed at full speed through windows in the rotor by Schlieren or Rayleigh fringe interference systems or by absorption of monochromatic or ultraviolet light.

Analytical ultracentrifuges are used to measure sedimentation velocities of micromolecules in solution or to determine the concentration gradient of the molecules after equilibrium between diffusion and sedimentation effects is reached. In the velocity method, the rate of movement of the boundary between clear supernate and solution still containing solute molecules is observed; high rotational speeds minimize diffusion effects and finally the solute molecules are concentrated at the bottom of the cell. In the equilibrium method, relatively low speeds may be used in a concentration gradient established under ideal conditions over the full depth of the cell; higher diffusion and lower sedimentation velocities for the smallest solute molecules concentrate them near the liquid surface, whereas lower diffusion and higher sedimentation rates concentrate the larger molecules near the bottom of the cell. The latter method permits the determination of particle mass or molecular weight independent of particle shape whereas the former directly measures sedimentation coefficients that are related to both size and shape of the sedimenting molecules.

Zonal Centrifuges. The use of density gradients in centrifuge rotors greatly increases the sharpness of separations and the quantities of material that can be handled. In principle, the density gradient is established normal to the axis of rotation of the rotor with the highest density located at the outer radius of the rotor. Low molecular weight solutes such as cesium chloride, sucrose, or potassium citrate, which are compatible with many systems in solution, are frequently used. A natural gradient may be formed by introducing a homogeneous solution and centrifuging for long periods of time. Continuous or step gradients may also be formed by introducing successive layers of solution whose composition varies continuously or stepwise from low to high density with the latter displacing the former toward the center of the rotor (16–17).

In the simpler rotors using batch containers with swinging, angle, or vertical tubes, the gradient is introduced while the rotor is at rest and is then accelerated to speed with the gradient showing relatively little mixing; slowing the rotor gradually at the end of the run allows the gradient to retain its shape and permits the collection of material banded isopycnically as shown in Figure 8. More sophisticated rotors can be loaded with gradient and sample while rotating. When the batch is finished or the

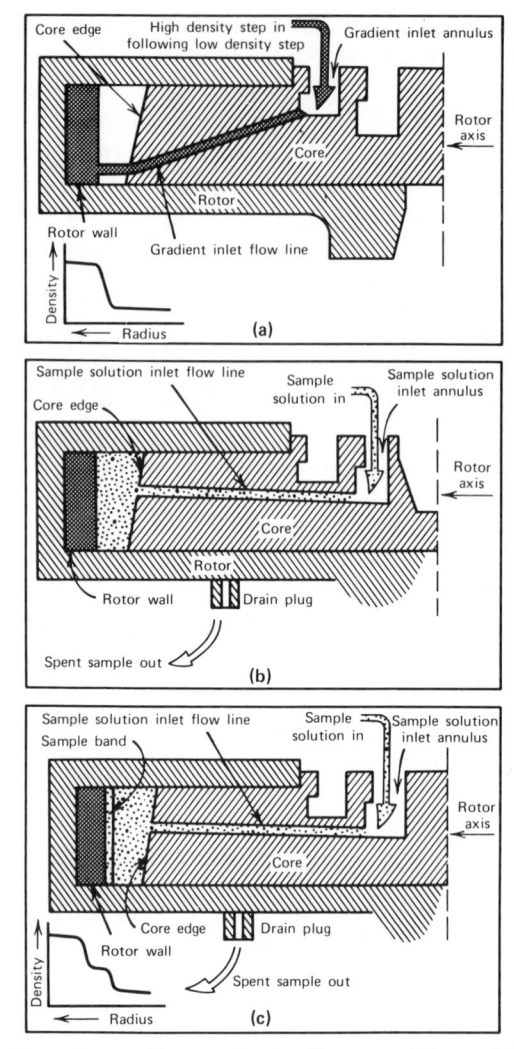

Figure 7. Long path continuous flow zonal rotor. Courtesy of Damon/IEC Division. (a) Loading high density step into spinning rotor to form the density distribution shown at left following loading of low density step in same channel. (b) Sample solution flow through spinning rotor. (c) Increase in concentration of sample band due to particle sedimentation. Note corresponding density distribution in rotor shown at left. (d) CF-6 Planar view diagram (p. 217).

bands are sufficiently loaded with material, the bowl may be stopped slowly and the reoriented layers displaced under static conditions. Rotors may also be designed to establish gradients and isopycnic bands of sample and then be unloaded dynamically by introducing a dense solution near the edge of the rotor as shown in Figure 9.

Particles in the gradient may be separated on the basis of sedimentation rate; a sample introduced at the top of the preformed gradient settles according to density and size of particles but the run is terminated before the heaviest particles reach the bottom of the tube. If the density of all the particles lies within the range of the density limits of the gradient, and the run is not terminated until all particles have reached an equilibrium position in the density field, equilibrium separation takes place. The

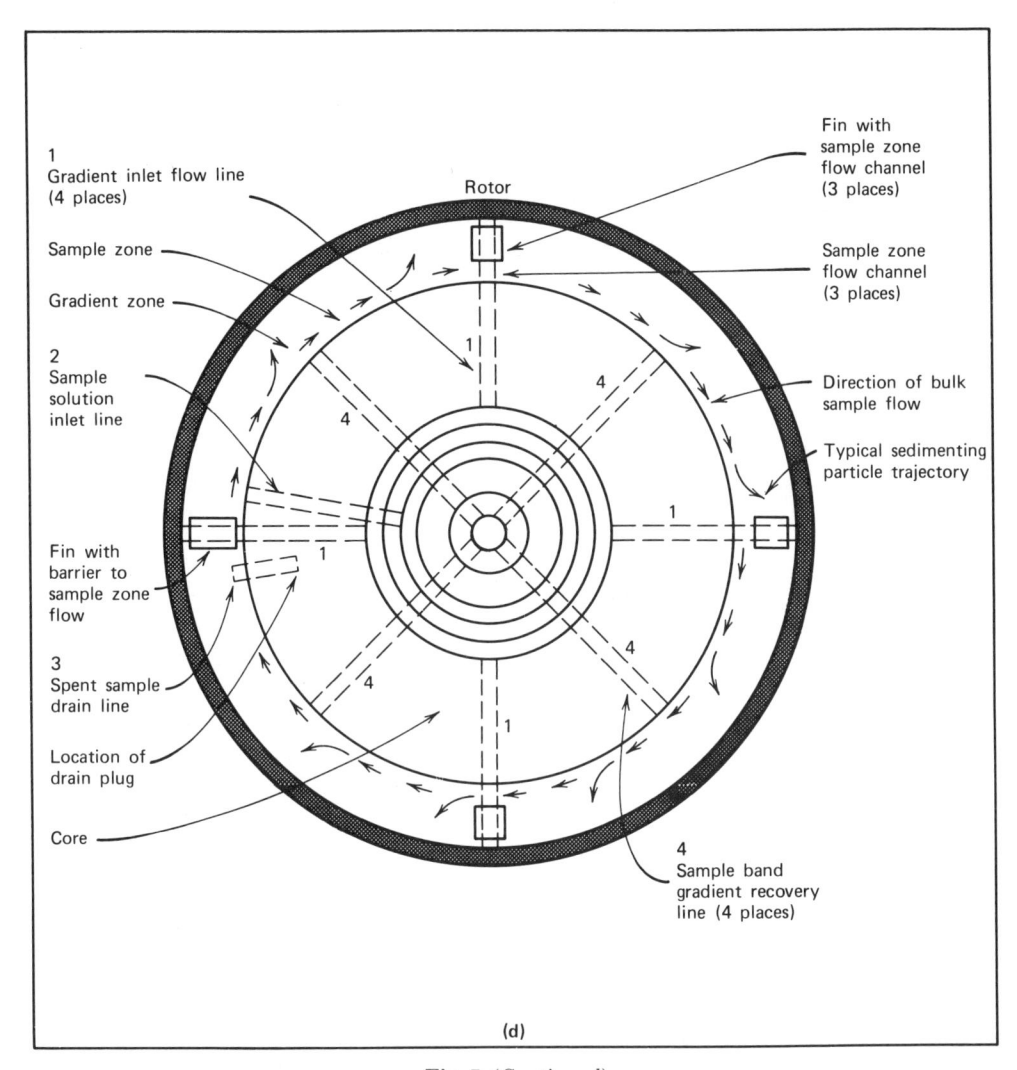

Fig. 7. (*Continued*)

steepness of the gradient can be varied to match the breadth of particle densities in the sample.

Rotors are made of titanium or aluminum and may be cylindrical or bowl-shaped (see Fig. 9) with larger bowls reaching 100,000 G and smaller units 250,000 G. The tubular rotors permit feed rates up to 60 L/h at 150,000 G or 120 L/h in a larger unit at 90,000 G. Such centrifuges may be used to separate relatively large quantities of viral material from larger quantities of cellular and subcellular matter as, for example, in the production of vaccines.

Tubular Centrifuges. Tubular centrifuges (see Fig. 10) separate liquid–liquid mixtures or clarify liquid–solid mixtures with less than 1% solids content and fine particles. Liquid is discharged continuously, whereas solids are removed manually when sufficient bowl cake has accumulated. For industrial use, the cylindrical bowls are 100 to 130 mm in diameter with length-to-diameter ratios ranging from 4 to 8. Bowl

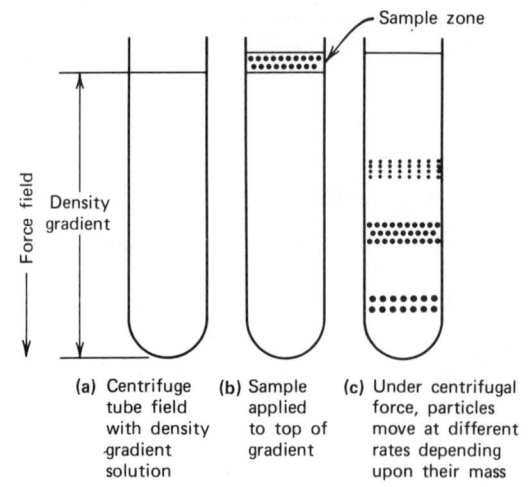

Figure 8. Rate zonal separation in swinging bucket rotor. Courtesy of Spinco Division, Beckman Instruments, Inc.

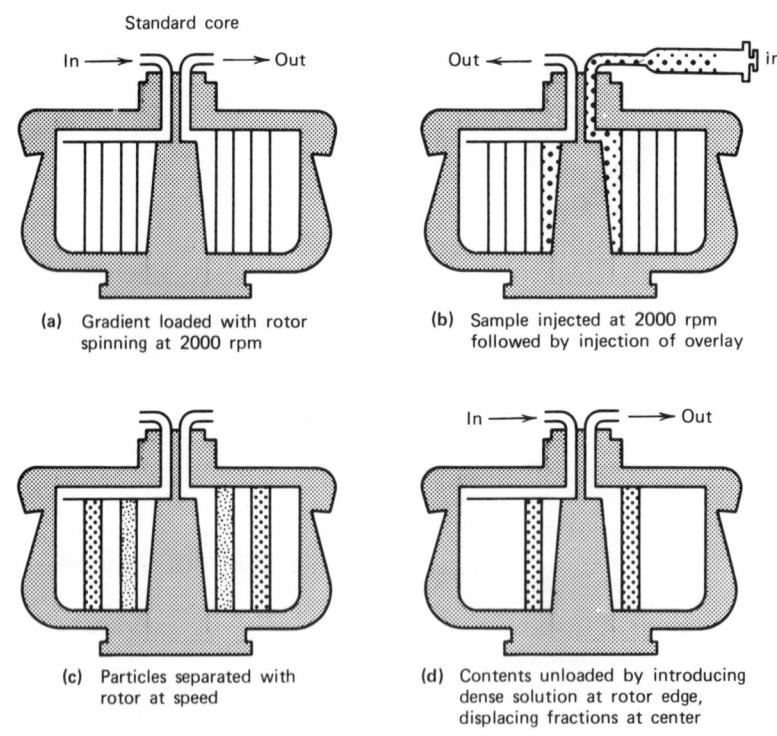

Figure 9. Dynamic loading and unloading of zonal rotor. Courtesy of Spinco Division, Beckman Instruments, Inc.

speeds up to 15,000 rpm generate centrifugal accelerations up to 13,000 G at the bowl wall. Because of the small bowl diameter, however, Σ_T values according to equation 11 are low and flow rates are generally in the range of 0.5–4 m^3/h (2–16 gpm). The tubular centrifuge handles low to medium flows and theoretical particle settling velocities

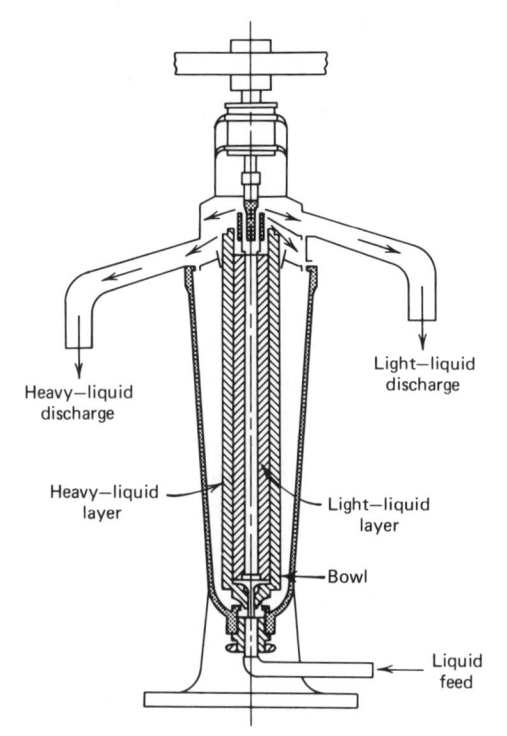

Figure 10. Tubular centrifuge.

in the range of 5×10^{-6} to 5×10^{-5} cm/s. Designed for maximum pressures of less than one kPa (a few inches of water), it can operate up to 200°C driven by motors of 1.5–4 kW.

The laboratory tubular centrifuge is similar to the industrial model. It operates with a motor or turbine drive at speeds of 10,000 to 50,000 rpm, generating 65,000 G at the latter speed in the 4.5 cm diameter bowl. The nominal capacity range is 30 to 2400 cm³/min. This instrument is uniquely capable of separating far finer particles than any other centrifugation equipment except the bottle centrifuge.

A long, hollow, cylindrical bowl is suspended by a flexible spindle and driven from the top as shown in Figure 10. Internal baffles ensure full acceleration of the liquid during its short time in the bowl. Feed is jetted into the bottom of the bowl and clarified liquid overflows at the top leaving deposited solids as compacted cake on the bowl wall. The clarifying performance of the bowl is reduced as the cake decreases the effective outer radius of the bowl in accordance with equation 11. Consequently, cake capacity of the industrial model is limited to 4–6 L. Interface position is determined by selection of ring-dam diameter or of the length of a hollow nozzle-type screw.

The tubular centrifuge is used primarily for the purification of contaminated lubricating oils because of the high centrifugal force developed and the simplicity of its operation. Colloidal carbon and moisture are removed from transformer oils to maintain dielectric strength, carbon and acid sludges from diesel engine lubricating oils, and water and solid contaminants from steam turbine lubricating oil. Polishing operations include the removal of small quantities of solids in the clarification of varnish, cider, fruit juices (qv), and even highly viscous chicle (see Gums). In vegetable oil (qv) refining, oil losses in the semisolid soap stock are kept low by compaction of the phase under high centrifugal force. The laboratory tubular centrifuge is used to recover fine solids in batch preparations too large for bottle centrifuge separation, estimate scale-up rates in larger centrifuges, and analyze particle size distributions involving settling rates too low for feasible gravity sedimentation (18). Bowl diameter is usually 10 cm, temperatures range from 65–120°C, and feed flow rates are between 0.2×10^{-3} and 1.0×10^{-3} m^3/s (3–15 gpm).

Disk Centrifuges. Centrifuges that channel feed through a large number of conical disks to facilitate separation combine high flow rates with high theoretical capacity factors as may be seen in Figure 5; for industrial units, flow rates up to 250 m^3/h are obtained on easy separations and theoretical settling velocities may range from 8×10^{-6} to about 5×10^{-5} cm/s. Both liquid–liquid and liquid–solid separations are performed with solids concentrations below 15% and small particle sizes. As seen from equation 15, the theoretical capacity factor depends on the number of disks (which is limited by the height of the disk stack) and on the cube of the size of the disks. Capacities are maximized with bowls of approximately equal height and diameter. Several of the basic assumptions in the development of Σ_D do not apply in practice (19) and mathematical representation of the actual flow pattern has not been possible. However, as all disk stacks are subject to the same types of deviation, scaling by Σ_D from one speed and size of stack to another is quite accurate.

The outstanding feature of the bowl design is a stack of cones, commonly referred to as disks, so arranged that the mixture to be clarified must pass through the disk stack before discharge. The resulting stratification of the liquid medium greatly reduces the sedimenting distance required before a particle reaches a solid surface and may be considered removed from the process stream. The angle of the cones is great enough so that even solid particles deposited on their surfaces slide either individually or as a concentrated phase according to the difference in their density and that of the medium.

The general flow pattern for a liquid–liquid separator and a recycle clarifier, respectively, is illustrated in Figure 11(**a**) and (**b**). Feed enters near the center of the bowl from either the top or the bottom, depending on the support, and is accelerated by radial vanes to the radius at which it enters the disk stack. When the disk stack is used for one phase, as in clarification or classification, the feed is distributed to the stack through the zone between the outer edges of the disks and the bowl wall. The clarified medium is discharged at a relatively small radius, generally at the top of the bowl. For a liquid–liquid separation, with or without solids, feed is distributed by a number of feed channels. The interface of the two coalesced liquid phases is located by appropriate selection of discharge radii near the feed holes. When handling two liquids, the heavier moves toward the edge of the disks and the lighter moves inward. Separate channels in the bowl and separate cover compartments segregate the discharges.

Solids in either phase are sedimented to the disks and usually slide radially outward along the surfaces because of their higher density. The aggregated solids must move by free settling from the outer edges of the disks to the bowl wall; some may be reentrained and carried into the disk stack which accounts in part for actual performance falling short of theoretical standards.

Disk bowls may be suspended from, or supported by, a flexible spindle that permits the bowl to seek its natural axis of rotation in case of small unbalanced loads. Bowls are usually constructed of corrosion-resistant stainless steel such as type 316,

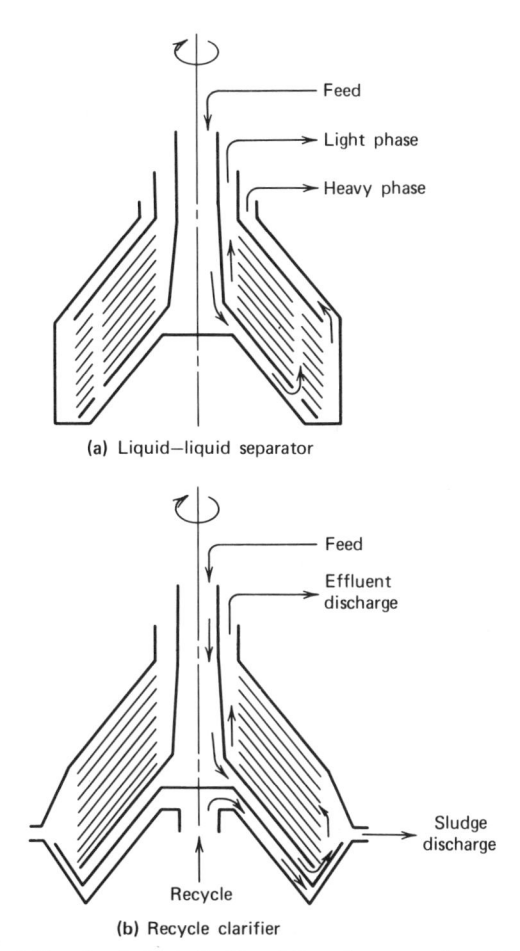

Figure 11. Disk centrifuge bowls. Note that bowl diameters range from 10 to 90 cm; feed flow rates from 0.06×10^{-3} to 38×10^{-3} m³/s (1 to 600 gpm); and operating temperatures from 30 to 90°C.

329, or 431. Disk centrifuges operate at temperatures from −40 to about 200°C and at pressures up to 1.03 MPa (150 psi). Clarified effluents may be discharged through skimmers or centripetal pumps at pressures up to 0.7 MPa (100 psi) to minimize foaming and exposure to the atmosphere.

Small quantities of solids are often permitted to collect in the bowl for periodic, manual removal; for larger quantities or special concentration requirements, intermittent or continuous discharge of the solids during operation is preferred.

To maximize cake capacity, the simplest disk centrifuge bowl (Fig. 11(**a**)) is designed with a nonperforate bowl wall parallel to the axis. Solids should not exceed 0.5%. Bowl diameters of industrial units range from 180 to 600 mm with operating speeds from 8000 to 4500 rpm; the disks—between 30 and 130—are stacked with spacings of 0.4 to 1.3 mm; the half-angle is frequently 35–45° since solids handling is not critical. This type of centrifuge was originally employed for the separation of cream from milk and is still used widely in this field (see Milk products); other uses include purification of fuel and lubricating oils, separation of wash water from fats and vegetable or fish oils, and removal of moisture from jet fuel. Solids moving readily in plastic flow can be continuously discharged as, for example, in the separation of soap stock from oil in vegetable oil refining.

An important variant of the imperforate-bowl disk centrifuge discharges two immiscible liquid phases or one liquid phase and a plastic solids phase through a system of seals around a hollow spindle so that the discharges enter the discharge piping directly and are not sprayed into open covers. The position of the interface in the bowl, and consequently the relative clarification of the two phases, is adjusted externally by applying back pressure to one or both discharge lines instead of establishing a free-surface hydrostatic balance. A change in phase ratio in the feed thus shifts the interface and may require manual or instrumental readjustment of the pressures. The unit can be cleaned by backflushing without opening the bowl. The discharges are under pressure so that an additional pump may sometimes be eliminated. This type of unit is used in the food and vegetable oil industries where exposure of the discharge streams in a finely dispersed spray or the production of foam may harm product quality.

Continuous discharge of solids, as a slurry, is achieved by sloping the inner walls of the bowl toward a peripheral zone containing between 8 and 24 orifices, commonly called nozzles, as shown in Figure 11(b). The nozzles must be spaced closely enough so that the natural angle of repose of the depositing solids does not cause a buildup of cake to reach the disk stack and interfere with clarification. The size of the nozzles is limited because the fluid pressure at the wall is 6.9–13.8 MPa (1000–2000 psi) and can produce large flows as indicated by equation 28. On the other hand, the nozzles must be large enough to prevent obstruction by individual particles; nozzle diameters 2 to 4 times the size of the largest particle are satisfactory. Replaceable wear bushings in the nozzles provide orifices in the range of 0.8 to 2.5 mm. From 5 to 50% of the feed may be discharged with the solids through the nozzles. The upper limit of solids concentration depends on the particle packing characteristics but seldom exceeds 20 times that in the feed. Bowl diameters range from 100 mm in laboratory units to 900 mm in industrial units and speeds from 12,000 to 3000 rpm, respectively; power requirements are between 10 and 100 kW. Centrifugal accelerations to 12,000 G are obtained at the wall in the smaller bowls. Feed rates range from 0.06×10^{-3} to 38×10^{-3} m^3/s (1–600 gpm).

Applications include kaolin clay dewatering, separation of fish oils from press liquor, starch and gluten concentration, clarification of wet-process phosphoric acid, concentration of yeast, bacteria, and fungi from growth media in protein synthesis. A growing application is the thickening of secondary excess activated sludge from biological treatment in industrial and municipal waste water treatment plants. The feed stream must be screened in two stages and cycloned in order to remove fiber and silt and prevent obstruction and wear; thickening of feed from a concentration of 0.5–1.2% to 4.5–5.5% is accomplished at high throughputs and good recoveries without flocculating agents (qv) (4). Thickened sludge concentrations are kept constant despite wide fluctuations in feed conditions by monitoring the underflow with a viscometer controlling the recycle rate (see Wastes, industrial and municipal).

A variant of the continuous-discharge disk centrifuge provides for introduction of a recycle stream as shown in Figure 11(b). Restrictions on number and size of nozzles, discussed above, sometimes prevent concentration of the discharged solids to satisfy further process requirements. To increase this concentration, a portion of the discharged slurry could be returned to the feed, thereby increasing the overall loading of solids in the bowl. A more efficient method is to return some discharged slurry through the recycle system to the region of the nozzles where the higher density recycle stream preferentially joins the nozzle flow.

A further modification of the disk centrifuge provides peripheral ports that are opened only intermittently to discharge sludge. This type is employed for medium quantities of solids (1–4%) for which neither continuous discharge nor batch operation is suitable, and for solids that break down under the shear forces of nozzle discharge and are therefore not suitable for recycle. Intermittent discharge provides longer holding time and better concentration of solids, often at the expense of decreased disk size and reduced bowl throughput. The frequency and duration of opening can be controlled to discharge very high concentrations of sludge with either partial or complete emptying of the bowl. Slots, opened by a sliding annular ring, handle solids considerably coarser and stiffer than those handled by the valved orifices, but in either case the solids must be sufficiently plastic to discharge evenly.

These centrifuges are available in 180-mm bowl laboratory and industrial units (460 to 600 mm bowls). Disks are mostly spaced at 0.6–1.0 mm and the half-angle is 40–45°. In the industrial units, bowls with 60–150 disks operate at maximum speeds of 4400–6500 rpm and require 10–30 kW. Feed flow rates range from 0.6×10^{-3} to 13×10^{-3} m^3/s (10–200 gpm); temperatures from 30–90°C. Theoretical capacity factors are generally lower than in continuous discharge bowls of the same size in order to provide solids-holding space of 8 to 20 L. Applications are limited to free-flowing solids that do not pack, and include recovery of wool grease from wool scouring liquor, orange juice clarification, recovery of soya protein, clarification of animal fats and food extracts, and purification of marine fuels.

Other modifications have special but more limited applications. A centrifuge bowl may contain, instead of disks, several annular baffles that take the liquid through a labyrinth path before discharge. The multiple cylinders increase cake capacity to as much as 70 L for easily sedimented solids. This centrifuge is used for clarification of food syrups and antibiotics, and for recovery of heavy metallic salts and catalysts.

Another type of special bowl recovers solids that are less dense than the liquid medium. The narrow disks leave a larger free volume inward from the disk stack; solids collect in this zone and are discharged by impact against the open end of a stationary skimmer tube under centrifugal force. This type, for example, is used to concentrate cream cheese from an acid whey.

Continuous Conveyor-Discharge Centrifuges. Imperforate bowl conveyor-discharge centrifuges collect solids by sedimentation and continuously discharge both liquid and solid material. These centrifuges have bowl diameters of 150–1400 mm and are essentially tubular shells with a length-to-diameter ratio of 1.5–4.4 as shown in Figure 12. Deposited solids are moved by a helical screw conveyor operating at a differential speed of 5–100 rpm with respect to the bowl. Centrifugal fields are lower than in disk or tubular centrifuges because of the conveyor and associated mechanism; maximum speeds range from 1600 to 8500 rpm and some units operate as low as 500 rpm. Figure 5 shows that particles of intermediate settling velocities, such as 1.5 to about 15×10^{-4} cm/s, are handled at medium to large flow rates. For clarification, this type of centrifuge recovers medium and coarse particles from feeds at high or low solids concentration. Particle sizes less than about 2 µm are normally not collected without the addition of flocculating agents. For classification of solids, the flow rates are higher than for clarification and the overflow usually contains an appreciable portion of the finer solids. Feed flow rates range from 0.2×10^{-3} to 31.5×10^{-3} m^3/s (3–500 gpm).

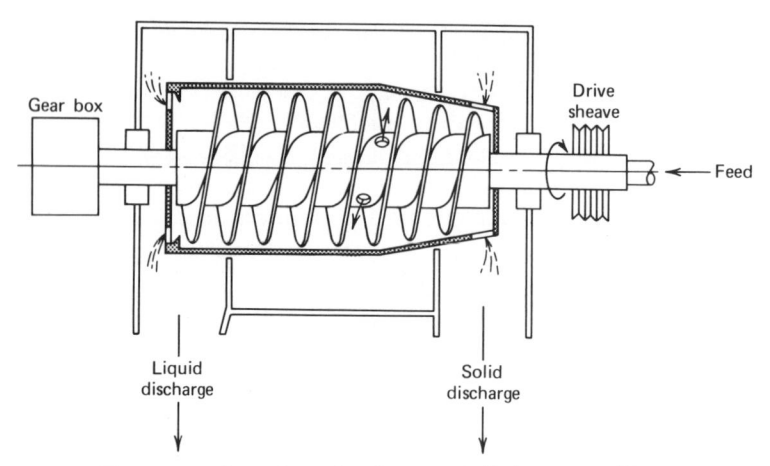

Figure 12. Imperforate bowl conveyor-discharge centrifuge.

Discharged solids are not as dry as those obtained by centrifugal filtration. Coarse crystals may discharge at 2–10% moisture, ground limestone at 15–20%, and kaolin clay in the filler range (1–10 μm) at 30–35%. Amorphous and fibrous materials are higher in liquid content, such as sewage sludges at 70–85% moisture and meat rendering solids at 60–70% moisture plus 6–8% liquid fat. Operating pressures up to 1.03 MPa (150 psi) and temperatures up to 200°C are standard and 300°C is possible.

Feed is introduced through a stationary axial tube. Solids sediment to the bowl wall, are removed from the liquid by a conveyor up a sloping "beach," and discharged at a radius smaller than that of the liquid discharge. Fine and flocculent solids compact under the liquid and show relatively little drainage on the "beach"; coarse crystals and fibers drain to a low residual moisture. The liquid level in the bowl is maintained by ports adjustable to the desired overflow radius. Considerable variation occurs in the bowl shell, flight angle and pitch, beach angle and length, conveyor speed, feed position, and patterns of liquid and solids movement through the bowl in different models. A conical bowl, or a bowl having a long conical section, has lower clarification capacity but supposedly produces drier solids than does a cylindrical shell with a short beach.

Bowl designs include countercurrent or concurrent movement of the phases. In countercurrent flow, feed enters near the conical–cylindrical intersection; liquid flows toward the far end of the bowl to discharge over dams while deposited solids move toward the cake end with the conveyor. In concurrent flow, feed enters at the end away from the cake discharge, liquid and settled solids move in the same direction toward the cake end, axial conduits or a skimming device remove centrate from the pond surface, and solids move up the beach. In either type, solutions of flocculating agents may be introduced in the feed tube, the feed acceleration zone, or the pond after feed acceleration; the last has generally proved most efficient on waste treatment sludges. A wash can be applied to the solids on the beach, but is generally inefficient and equivalent only to direct dilution of retained mother liquor; the wash is not collected separately but is discharged with the mother liquor. A vertical design uses a bowl suspended from a spindle with no bottom bearing; these units are very easy to seal for pressure operation and are well-suited to accommodate bowl expansion at high temperatures.

Although Σ_T (see equation 11) indicates a reduced performance level with increased liquid depth, this occurs only with the coarsest solids. The optimum pond depth varies with feed zone design, tendency of deposited solids to redisperse, conveyor differential speed, and particularly the depth of the cake layer required to produce a given concentration. Deep ponds are generally more effective with soft, slimy solids as conveying problems may reduce performance level and prevent complete clarification even at low flow rates. The capacity of a conveyor centrifuge as a clarifier may be limited by the liquid loading rate that can be related to Σ_T. Its capacity as a thickener may be limited by the compaction and loading rates of the solids or the concentration required in the underflow. As previously noted, the torque developed by the conveyor on some types of solids may also limit capacity; for low-torque, easily settled materials, a large conveyor centrifuge can handle up to 80 t/h of discharged solids although 5–30 t/h is more common.

Currently, the majority of conveyor centrifuges are employed in dewatering or thickening sludges obtained from municipal and industrial waste treatment. Although some of these sludges include particulate solids, most have a broad range of particle types including much flocculent biological material. Dewatering these sludges became feasible due to design modifications including a number of proprietary designs and improved hard surfacing techniques. Operation is helped further by the improved efficiency of current high molecular weight polymeric flocculants. The sludges produce soft cakes with 10–25% solids that are slow to compact in the bowl, drain little on the beach, and are generally difficult to convey (2,20–22). The cake residence time, augmented by the increased centrifugal force represented by G, largely determines the concentration of the soft cakes before discharge from the centrifuge. The differential speed between conveyor and bowl is determined primarily by the gearbox ratio, but control can be extended by driving the pinion of the gearbox at different speeds. Thus the differential speed can be varied during operation downward to zero, although the optimum range seems to be 5–20 rpm differential. A reduced differential is particularly useful on soft sludges and has the advantages of increasing cake residence time and decreasing turbulence in both the liquid and cake layers. Residence time on the beach is increased for cakes that drain, and the erosive effect of abrasive particles is decreased. Thickening of municipal sewage sludges from the usual 1–5% concentration to 6–12% before digestion, barging, or filter pressing is an important application permitting higher feed rates than dewatering (see Wastes, industrial and municipal).

In clarifying operations, the conveyor-discharge centrifuge recovers many types of crystals, meal from fish press liquor, and polymers, such as poly(vinyl chloride) and polyolefins; it is also used to dewater coal (23), and to concentrate solids from flue gas desulfurization sludges. Vertical designs, vapor tight under pressure, are applied to terephthalic acid, polypropylene, and catalyst recovery. Classification includes separation of particles over 2 μm from kaolin coating clay, and of particles over about 5 μm in the mill discharge of ground TiO_2; selective recovery of calcium carbonate from lime-treated waste sludges permits calcining and recycling of the lime without an overwhelming recycle load of inert material. The conveyor centrifuge is frequently used to rough out medium and coarse solids before a second separation stage such as a disk centrifuge handling refinery sludges.

Basket Centrifuges. Imperforate basket centrifuges are suitable for applications where the nature of solids may vary widely with time and for collection of soft or fine solids that are difficult to filter. Solids content of the feed should be low to avoid frequent cleaning. Solid bowl baskets are available in the same range of sizes, speeds, and suspensions as the more extensively used perforate basket centrifuges. The capacity value, Σ_T, is based on equation 11; the efficiency of operation varies widely because of the relatively short clarification length of the bowl and different methods of feeding, particularly if adequate acceleration is not provided for the feed. The approximate sedimentation application zone for imperforate bowl-basket centrifuges is outlined in Figure 5. For this definition, a ten-fold range of flow rates for each size and a full depth of liquid under the lip of the basket are assumed.

Feed is usually introduced near the bottom of the bowl at a rate ranging from 0.75×10^{-3} to 3.8×10^{-3} m³/s (12–60 gpm). Solids collect at the bowl wall under the top flange, and the clarified liquor overflows the flange and is continuously discharged. The quality of clarification is maintained until the cake builds inward from the wall and enters the clarification zone; interference does not usually occur until the cake reaches 60–85% of the depth under the lip ring. At the end of an operating cycle and prior to shutdown, the supernatant liquid may be skimmed off the top of the cake by a skimming tube to produce a drier cake for discharge. Cake may be discharged completely at a reduced rotational speed by a knife.

The fully automated imperforate-basket centrifuge is used to dewater excess biological and alum wastes from municipal sewage and water treatment plants. Recoveries normally run 90–95% without the use of flocculating agents at cycle times of 5–15 min producing cakes of 8–12% solids. The addition of less than 1 kg/t of a polymeric flocculant increases capacity by 10–20% and cake concentration by 2–3% solids.

Liquid Cyclones. Liquid-media cyclones, or hydroclones, are sedimentation-type, imperforate-wall centrifugation equipment (24). Figure 5 indicates an operating range for cyclones of 10–600 mm in diameter. The small units are capable of settling fine particles (down to 0.5 μm) at small flows but are usually grouped with as many as 480 units to a header. Feed flow rates range from 6.3×10^{-5} m³/s (1 gpm) for small units to 66×10^{-3} m³/s (1050 gpm) for larger units. Flow rates in large hydroclones are equivalent to those in disk centrifuges, but the theoretical particle size sedimented is two magnitudes larger. In general, hydroclones can economically handle large volumes of feed to produce a ten-to-twenty-fold concentration of feed solids. The flow ranges in Figure 5 represent normal operating pressure drops of 35–350 kPa (5–50 psi). The underflow is usually 5–10% of the feed for clarification, but for classification operation well outside this range is feasible. The design may include a grit box for intermittent discharge of accumulated underflow solids with no continuous underflow stream. High-efficiency cyclones may give fairly complete recovery of even fine particles at low to medium feed solids concentrations. High-capacity units can increase the concentration of a medium or coarse solids slurry prior to subsequent processing, such as recovery in a centrifuge. Its simple construction adapts the cyclone to high pressure operation and temperatures up to 300°C.

Optimum cyclone performance has been obtained with fairly consistent designs (25–27). In theory, free vortex flow exists throughout the cyclone so that high shearing rates are found in most of the sedimentation zone; consequently, cyclones cannot achieve complete liquid–liquid separation and are inefficient collectors of flocculant and easily deagglomerated solid particles. The inlet flow velocity creates the angular velocity of the liquid mass in the cyclone. Since the pressure drop between inlet and overflow determines the feed rate, the separating capacity factor depends on both the feed rate to the cyclone and the split ratio between overflow and underflow. Hence, it differs from the capacity factor of a centrifuge which is independent of these factors. In addition, increasing the concentration of underflow solids from the cyclone ultimately affects the clarity of the overflow, a condition seldom encountered in centrifuge operation. The adaptability of the cyclone for use with rubber or ceramic liners, the ease of replacing eroded parts, and the ease of operation have resulted in its widespread use in many fields.

Cyclones are used for classification as, for example, in the degritting of kaolin clay where sand is removed from the crude clay suspension before finer classification in a conveyor-discharge centrifuge and final product recovery in a disk centrifuge. They are also widely used in concentrating ores, desliming tailings, and recovering catalysts from catalytic cracker distillation bottoms (see also Air pollution control methods).

EQUIPMENT OPERATING BY FILTRATION

Basket Centrifuges. The simplest and most common form of centrifugal filter is a perforate-wall basket centrifuge consisting of a cylindrical bowl with a diameter ranging from about 100 to 2400 mm and with a diameter-to-height ratio ranging from 1 to 3. The wall is perforated with a large number of holes, more than adequate for the drainage of most liquid loads, and is lined with a filter medium. In the simplest case, the medium is a single layer of fabric or metal cloth or screen; in high-speed basket centrifuges, one or more backup screens of relatively large mesh support a finer mesh filter surface. The method of discharging accumulated solids distinguishes three types of basket centrifuge, ie, those that are stopped for discharge, those that are decelerated for discharge, and those that discharge at full speed.

Basket centrifuges that must be stopped for discharge are available in many sizes. The bowls are usually supported on a vertical spindle. Designs vary from a 300-mm diameter basket of 30 L cake capacity (0.4 kW motor, 2100 rpm, and about 800 G) to a 1500-mm diameter basket with 0.5 m^3 cake capacity (14 kW motor, 600 rpm, and 300 G). Basket cake volumes are always nominal and must be modified according to density. Construction materials include carbon or stainless steel, Monel, Inconel, titanium, and a variety of rubber and plastic coatings. Normal operation includes pressures up to 35 kPa (5 psi) and temperatures up to 180°C. This type of centrifuge is used if a variety of materials must be filtered in small batches, if equipment must be sterilized between batches, or if the production rate is too low to warrant more automation. These centrifuges are also used in removing liquid from crystalline materials, in drying bulk materials such as raw leafy vegetables, and in clarifying process liquids and waste streams. Cycle times vary upward from about 10 min. Large baskets (1300–1500 mm diameter), operating at 700–800 rpm may use an inner perforated container which mounts in the bowl but is removable for bulk loading outside the centrifuge; such units are well suited to laundry and dyeing purposes and for dewatering of textiles, yarn, raw stock, feathers, and hair. Particle sizes range from very fine to 500 μm. Feed flow rates vary from 0.28×10^{-3} to 3.2×10^{-3} m^3/s (4.5–50 gpm). A syphon to control the rate of filtrate removal improves dewatering (28).

Improved control systems and rising labor costs have led to manually or automatically controlled cycles with mechanical unloading at reduced speed. Baskets typically load at low to medium speed, accelerate to 900–1800 rpm for drainage and washing, decelerate to 35–75 rpm for mechanical unloading and then start the cycle again. Cycle times range from 2 to 6 min to 30 min for slow drainage or multiple rinses. Both a top-driven suspended bowl and an underdriven bowl with three-point casing suspension are available. The cake is discharged in 20–120 s through radial and axial motion of a single or multiple plow that leaves a heel of cake on the filter medium. It can be completely removed with the help of a plastic-tipped plow and a perforated protecting plate. Filter media vary from perforated plates with 3-mm holes to 37 μm (400-mesh) Dutch twill. Baskets are always bottom discharge; a valve mechanism may be used to seal the bottom if the basket is fed with the whole charge at one time, so that an appreciable liquid layer develops.

The most fully automated basket or peeler centrifuge discharges at high speed and normally operates with cycle times of less than 3 min at pressures up to 1.03 MPa (150 psi) and temperatures from -70 to $120°C$, and in special cases to $350°C$. It is primarily used for materials draining freely in a high centrifugal field, for medium tonnages, and for multiple rinses where nearly complete segregation of the rinses and mother liquor is desired. Ideally, the feed should have a constant composition and high concentration. For this purpose, a gravity slurry concentrating tank is often installed in front of the centrifuge. The bowl rotates on a horizontal axis with a metal screen as the filter medium and discharges solids by cutting them out with a hydraulically operated knife. Bowls are 300–1200 mm in diameter and handle loads of 28–170 L (1–6 ft^3) of cake at bowl speeds of 1000–2500 rpm producing maximum centrifugal fields of about 1250 G. The solids discharge through a chute which simplifies sealing against pressure. A distributor riding on the cake during loading maintains uniform cake thickness, gives better bowl balance, and improves washing efficiency. Power requirements range from 15 to 150 kW. Materials of construction include carbon or stainless steel and Monel. The high speed during discharge may cause deformation of particles and breakage of crystals, whereas the heel of cake left on the screen may lose permeability through glazing or plugging with fractured fines and is usually conditioned by suitable washes after one or more cycles. Table 2 lists two process cycles.

This type of centrifuge is also used on borax and boric acid, p-xylene, sodium bicarbonate, and sodium chloride from glycerol or electrolytic caustic, in addition to dewatering of various products (eg, potato starch), or dewatering and washing slurries (eg, calcium hypochlorite or affination magma).

Table 2. Basket Centrifuge Operation on Free- and Slow-Draining Particles

Operation	Free-draining[a]	Slow-draining[b]
solids handling rate, t/h	20–24	1.5–2
conditioning rinse, s	1	0
feeding time, s	7	25
wash time, s	5	25
drain time, s	12	30
unloading time, s	2	6
total cycle, s	27	86

[a] For example, ammonium sulfate.

[b] Requiring several rinses; eg, polyolefins.

Continuous Cylindrical Screen Centrifuges. Continuous filtering centrifuges are used for very fast draining solids that do not require extremely dry final products. Rinsing efficiency varies considerably; power requirements are usually low; initial slurry concentrations can be somewhat more variable and not as high as for the high-speed basket. Continuous centrifugal filters are equipped with either a cylindrical or a conical screen. Both types are made without a retaining lip on the solids-discharge end of the bowl and employ various methods to move the solids through the bowl.

The cylindrical screen centrifuge deposits solids at one end of the bowl in a layer of 6–80 mm thick and pushes the annular ring of cake axially through the bowl by means of a reciprocating piston (4,29). Washes are collected in separate sections of the casing but are not as distinctly separated as with basket centrifuge sequential operation. Drained solids at the end of the bowl are thrown into a casing which is sealed from the liquid discharge zones. Bowls rotate on a horizontal axis, range in diameter from 200 to 1200 mm, and have capacities of 1 to 25 tons of solids per hour. Centrifugal fields of 300 to 600 G are common. To reduce the fines loss and to facilitate movement of the cake on the screen, maximum speeds are rarely used. Power requirements range from 4 to 60 kW. The cylindrical screens are generally of the bar screen type with 0.1–0.5 mm spacing. The reciprocating piston operates at 20–100 strokes per minute with stroke lengths up to 80 mm. The thickness of the cake (max about 80 mm) depends on the packing and draining characteristics, the buckling tendency of the layer in front of the pusher, and the frictional resistance of the cake on the screen.

The feed slurry is introduced by gravity flow 20–200 t/h or screw conveyor to an imperforate distributing cone that deposits the slurry at its original concentration immediately in front of the pusher as shown in Figure 13(**b**); as the cake must not buckle at this point, slurries of 40% concentration are generally necessary. In another type of feed system, shown in Figure 13(**a**), the incoming slurry is first deposited at the small end of a conical screen. A dam at the large end restricts the rate of discharge from the cone to the cylindrical screen at the pusher face. Between 50 and 90% of the mother liquor is drained from this conical zone and a firmer cake is deposited in front of the pusher. Thus, initial slurry concentrations as low as 10% solids can be handled. Because fast drainage is required, feed particle size should exceed 150 μm; medium and coarse crystals and granules or fibrous solids can be handled in this type of centrifuge. Crystal breakage is low within the basket but some breakage does occur during discharge.

To handle materials that form a soft or plastic cake or have a high frictional resistance, the cylindrical screen may consist of two to six steps with successively larger diameters as shown in Figure 13(**b**). Alternate steps reciprocate with the piston; thus, the cake is pushed across only a short length of screen before redistribution on the next step at a slightly larger diameter. Drainage and washing efficiency are increased by redistribution of the cake under these conditions. This type of centrifuge is used on sugar where the high viscosity of the mother liquor causes slow drainage, a high degree of plasticity in the partially drained cake, and poor penetration of wash liquor.

Another type of continuous, cylindrical screen centrifuge, used only rarely, discharges the cake by moving it axially through the bowl with a helical conveyor. Crystal breakage through conveyor action is considerably greater than with the pusher-type of mechanism. Applications include the handling of copperas and trisodium phosphate.

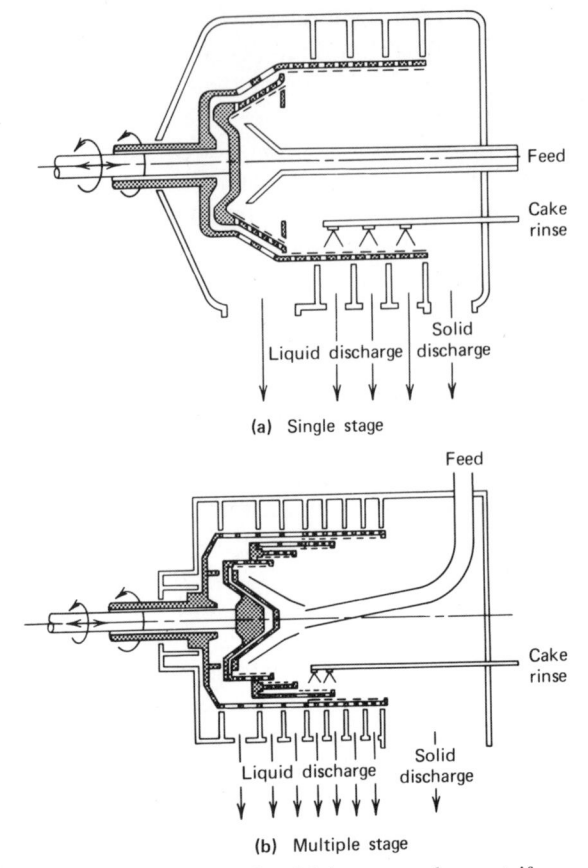

Figure 13. Single and multiple stage pusher centrifuge.

In conical screen centrifuges the angle of the bowl causes or assists the cake to move axially and redistributes it in an increasingly thin layer which improves drainage characteristics. The feed slurry is deposited at the small end of the cone where most drainage occurs. The drained solids are discharged from the large end which has no retaining ring. Screens generally have 0.08 to 1.5 mm slots or perforate plate holes. Screens less than about 0.25 mm thick require a backup screen to extend screen life. There are three types of conical screen centrifuges, those that are self-discharging, those that discharge by means of a helical conveyor, and those that apply an axial vibration or oscillation to the bowl or the bowl and casing.

In its simplest form, the cone angle is slightly larger than the angle of repose of the solids at any stage in their drying cycle. Horizontal or vertical bowls of 500 to 1000 mm large-end diameter operate at speeds up to 2600 rpm producing up to 2500 G; cone angles vary from 20 to 35° and selection of the proper cone angle and suitable screen surface is critical for each application. Temperature control of viscous feeds is necessary to maintain proper distribution and drainage. Feed slurry, usually under gravity flow, is introduced at the small end of the basket where it is accelerated and spread evenly over the periphery of the screen. Viscous sugar massecuites are successfully handled at capacities of 2 to 7 t/h on a 750 mm diameter basket. Loss of fines is greater than on the automatic basket centrifuges, but improved rinse performance increases sugar purity. Rinsing efficiency is not generally high on this type of cone and segregation of rinse liquor is incomplete. This centrifuge is also used for the drying of crystalline materials such as ammonium sulfate and separation or dewatering of fine fiber in wet corn milling.

Conical screens with a helical conveyor turning slightly faster than the bowl can handle a variety of materials. The vertical baskets are underdriven and the larger diameter is downward as shown in Figure 14. Feed is introduced at the small end and the rate at which solids move through the drainage zone is controlled to some degree by the differential speed of the conveyor. Bowls have large-end diameters of 250 to 500 mm and lengths about two thirds of the large diameter; bowl angles range from 10 to 20°, and operating speeds are 2500 to 3800 rpm giving up to 3500 G. Solids capacities range from 1 to 30 t/h with feed capacities of 1 to 15 m^3/h (ca 260–4000 gph). Centrifuge casings may be vapor-tight but are not intended for pressure operation; maximum temperature is about 150°C. Power requirements, low for the tonnages handled, range from 7 to 30 kW.

Applications include dewatering of medium and coarse crystals, deoiling of proteinaceous solids, and removal of solids from fruit and vegetable pulps and other food slurries.

The third type of cone screen centrifuge operates with bowl angles of 13 to 18° and assists solids discharge by a vibratory motion of the bowl or bowl and casing. These units usually have underdriven bowls with the 500–1100 mm larger diameter at the upper end; diameter-to-length ratios range from 1 to 2. Operating speeds are normally 300–500 rpm and solids capacities range from 25 to 150 t/h. Pressurized units are not available and operating temperatures range to 100°C. Power requirements are 15 to 35 kW. Bar screens are frequently used and applications are largely on coal dewatering; particle sizes from about 30 mm down to 0.25 mm (60 mesh) are easily handled. These centrifuges are also used in dewatering of potash and other crystalline solids.

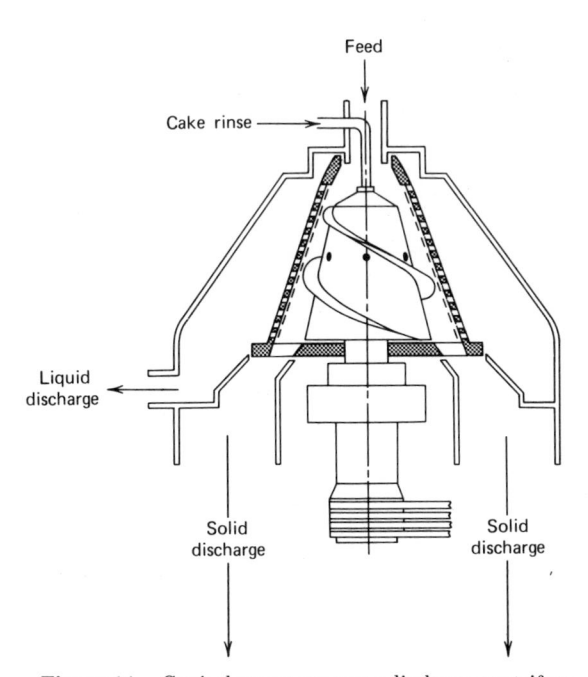

Figure 14. Conical screen conveyor discharge centrifuge.

232   CENTRIFUGAL SEPARATION

Nomenclature

a	= distance; space between adjacent disks
C	= discharge coefficient
d	= diameter of particle; particle size
$\bar{d}$	= mean particle diameter
e	= efficiency factor
E	= energy
g	= acceleration of gravity
G	= ratio of centrifugal acceleration to acceleration gravity
h	= cake thickness; height
I	= moment of inertia of rotating bowl about axis of rotation
k	= experimental coefficient in equation 20
k'	= experimental coefficient in equation 22
K	= cake permeability coefficient, mass/(time)^2
l	= length, particularly from feed zone to centrate discharge
m	= torque
n	= number of disks
Pwr	= power
p	= pressure
q	= volume of undrained liquid (mother liquor)/unit volume of solid, %
q_∞	= volume of undrained liquid (mother liquor)/unit volume of solid at infinite time, %
Q	= flow rate, volumetric
r	= radius
s	= external surface area/weight of solids
s'	= surface area/volume of cake
S	= fraction of void volume occupied by liquid at time t
S_∞	= fraction of void volume occupied by liquid at infinite time
t	= time during which material is exposed to separation effect
t'	= time at which free liquid enters the cake in equation 21
v	= velocity
V	= volume
α	= cake resistance constant in pressure filtration, length/volume
δ	= mass density = mass/unit volume
Δ	= difference, particularly of density
ϵ	= void fraction
μ	= viscosity
ω	= angular velocity of rotating motion
φ	= wetting angle
ϕ	= angle between direction of the nozzle and tangent to circle intersecting nozzle axis at discharge section
σ	= surface tension
Σ	= theoretical capacity factor
θ	= half included angle of disks
ν	= kinematic viscosity

Subscripts

B refers to bottle centrifuge
C refers to cake
D refers to disk centrifuge
f refers to film flow
g refers to settling velocity of particle in gravity field
h refers to heavy phase
i refers to interface
l refers to light phase
L refers to liquid medium
m refers to mean value

mot refers to motor
M refers to filter medium
N refers to nozzle
s refers to settling velocity of particle in centrifugal field
S refers to solid

BIBLIOGRAPHY

"Centrifugal Separation" in *ECT* 1st ed., Vol. 3, pp. 501–521, by M. H. Hebb and F. M. Smith, The Sharples Corporation; "Centrifugal Separation" in *ECT* 2nd ed., Vol. 4, pp. 710–758, by A. C. Lavanchy and F. W. Keith, Jr., The Sharples Company and J. W. Beams (Gas Centrifugal Separation), University of Virginia.

1. C. M. Ambler, *Chem. Eng. Prog.* **48,** 150 (1952).
2. F. W. Keith, Jr. and R. T. Moll in R. A. Young and P. Cheremisinoff, eds., *Wastewater Physical Treatment Processes,* Ann Arbor Science Publ., Ann Arbor, Mich., scheduled 1978.
3. J. Murkes, *Brit. Chem. Eng.* **14,** (12), 636 (1969).
4. D. K. Baumann and D. B. Todd, *Chem. Eng. Prog.* **69,** 62 (Sept. 1973).
5. K. Zeitsch, *Filtr. Sep.* **13,** 223 (May/June 1976).
6. R. Day, *Chem. Eng.* **81,** (May 13, 1974).
7. H. P. Grace, *Chem. Eng. Prog.* **49,** 427 (1953).
8. J. A. Storrow, *AIChE J.* **3,** 528 (1957).
9. W. Batel, *Chem. Ing. Tech.* **33,** 541 (1961).
10. E. Nenninger, Jr. and J. A. Storrow, *AIChE J.* **4,** 305 (1958).
11. Sharples Research Laboratory, *private communication,* 1961.
12. J. O. Maloney, *Ind. Eng. Chem.* **48,** 482 (1956).
13. F. A. Records, *Chem. Proc. Eng.* **52,** 47 (Nov. 1971).
14. C. M. Ambler and F. W. Keith, Jr. in A. Weissberger, ed., *Techniques of Chemistry,* Wiley-Interscience, New York, 1978, 3rd ed. Vol. XII, Chapt. VI.
15. P. A. Vesilind, *Treatment and Disposal of Waste Water Sludges,* Ann Arbor Science Publ., Ann Arbor, Mich., 1974.
16. O. M. Griffith, *Techniques of Preparative, Zonal and Continuous Flow Ultracentrifugation,* Spinco Div. of Beckman Instruments, Inc., Palo Alto, Calif., 1975.
17. G. B. Cline in E. S. Perry and C. F. van Oss, eds., *Progress in Separation and Purification,* Wiley-Interscience, New York, 1971, pp. 299–306.
18. T. Lee and C. W. Weber, *Anal. Chem.* **39,** 620 (1967).
19. C. A. Willus and B. Fitch, *Chem. Eng. Prog.* **69,** 73 (Sept. 1973).
20. D. E. Albertson and E. E. Guidi, Jr., *J. WPCF* **41,** 607 (Apr. 1969).
21. F. A. Records, *Chem. Eng.* **81,** (Jan. 1974).
22. R. J. Woolcock, *Filtr. Sep.* **12,** 174 (Mar./Apr. 1975).
23. J. J. Halloran, *Annual Meeting, #TIS-5039,* Slurry Transport Assoc., Houston, Tex., Aug. 24–25, 1976.
24. R. Day, *Chem. Eng. Prog.* **69,** 67 (Sept. 1973).
25. D. F. Kelsall, *Trans. Inst. Chem. Eng. London* **30,** 87 (1952).
26. D. F. Kelsall, *Chem. Eng. Sci.* **2,** 254 (1953).
27. D. Bradley, *The Hydrocyclone,* Pergamon Press, London, England, 1965.
28. K. Lilley and G. Huhtsch, *Filtr. Sep.* **12,** 70 (Jan./Feb. 1975).
29. P. M. T. Brown, *Chem. Proc. Eng.* **52,** 65 (Nov. 1971).

A. C. LAVANCHY
F. W. KEITH, JR.
Pennwalt Corp.

CEPHOLOSPORINS. See Antibiotics—β-Lactams.

CERAMIC COLORS. See Colorants for ceramics.

CERAMIC COMPOSITE ARMOR. See Cermets.

CERAMICS

Scope, 234
Raw materials, 237
Forming process, 253
Thermal treatment, 260
Properties and applications, 267

SCOPE

"Ceramics comprise all engineering materials or products (or portions thereof) that are chemically inorganic, except metals and alloys, and are usually rendered serviceable through high-temperature processing" (1). Ceramic materials are normally composed of both cationic and anionic species. The primary difference between ceramics and other materials is the nature of their chemical bonding (2–5).

Although there are no distinct boundaries between ceramic and metallic or polymeric materials, it is instructive to compare them in terms of the service requirements in engineering design (3). As a class of materials, ceramics are better electrical and thermal insulators and more stable in chemical and thermal environments than are metals (see Cement). Metals usually have comparable tensile and compressive strengths, whereas ceramics are normally appreciably stronger in compression than in tension. Ceramics exhibit greater rigidity, hardness and temperature stability than polymers; however, polymerization occurs in ceramics, especially in glasses (see Glass; Glass-ceramics).

Modern ceramics encompass a wide variety of materials and products ranging from single crystals and dense polycrystalline materials, through glass-bonded aggregates to insulating foams and wholly vitreous substances (4–7). Such a range of microstructural characteristics allows the considerable versatility evidenced in the range of manufactured industrial products.

On the basis of available statistics, the value of this industrial output, in terms of the value of products shipped during 1975, approximated 25.2 billion dollars. The breakdown by major product classifications is given in Figure 1. Although several of the product areas do not have large dollar values as compared to many industrial commodities, they are nevertheless vital to an industrial economy. Two notable examples are the refractories necessary for the reduction of ores in the metallurgical industries and abrasives (qv) which allow the mass production of machine parts.

The magnitude of the ceramic industry is by no means completely represented by the data in Figure 1. For instance, dielectric and magnetic components in electrical and electronic products, enameled parts of household appliances, refractories in heating systems and fuel materials and other parts of nuclear reactors (qv) are all components of finished goods which should be, but are not currently classified as ceramics (see Enamels; Refractories).

As late as the 1930s, ceramic technology was primarily perceived as applied high-temperature silicate chemistry. Although silicate materials continue to be the inexpensive high-tonnage backbone of the industry, the desire for high performance ceramic materials, particularly those having improved electrical, electronic, piezoelectric, and magnetic and, more recently, electro-optic, pyroelectric, and laser properties has increased steadily in the last ten to twenty years (see Ceramics as

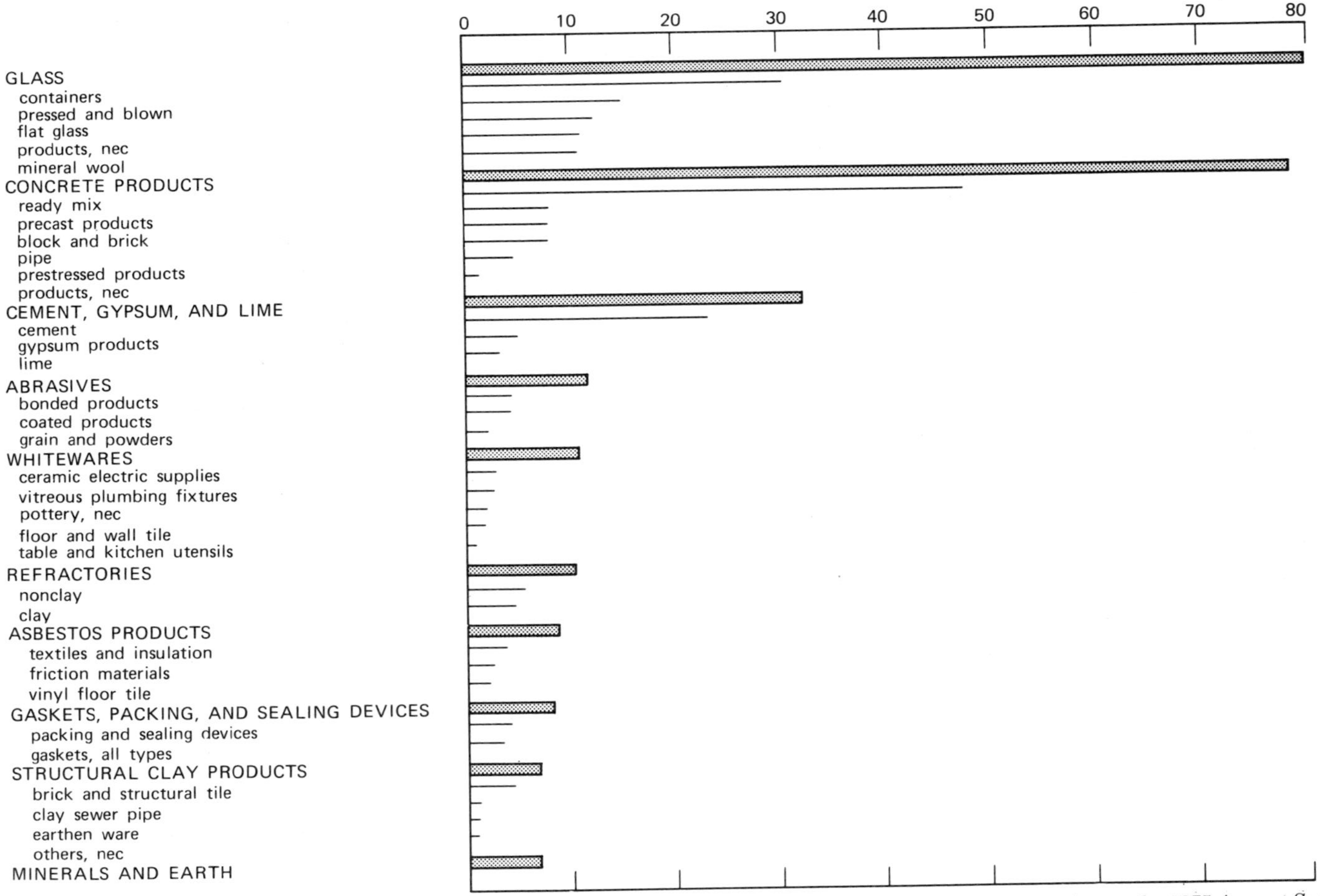

Figure 1. Value of ceramic products shipped in 1975, in billions of dollars; nec = not elsewhere classified. Abstracted from the *1975 Annual Survey of Manufacturers*, U.S. Department of Commerce.

electrical materials). The present urgency to develop materials for energy production, conversion and storage apparatus has stimulated the evolution of solid electrolytes for batteries (qv), refractories for magnetohydrodynamic generators and coal gasification devices (see Coal), strong dense ceramics for high efficiency turbine parts, and new glasses for solar collector panels (see Solar energy). In these newer ceramics, the greatest emphasis has been given to the oxide systems; however, considerable progress has also been made in the synthesis and employment of the borides, carbides, nitrides, and silicides (see Boron compounds, refractory; Carbides; Nitrides; Silicon and silicides).

The development, purification, and utilization of materials often requires the evolution of new processing techniques. Particle preparation, extrusion, dry pressing, slip casting, and sintering remain as important techniques in the ceramic industry, but they have been joined by freeze drying, thermal evaporation or sputtering, and chemical vapor deposition to produce high purity materials or thin films or complex shapes, respectively. Furthermore, the growth of single crystals, improved densification techniques and the glass–ceramic process have led to pore-free crystalline or nearly crystalline ceramics having dramatically improved properties.

Ceramics are frequently termed ionic solids, ie, possessing ionic bonding. In reality the bonding varies as a function of the polarizing power of the cations and the polarizability of the anions and is almost totally ionic in CsF and covalent in SiC. In the layer silicates such as the clays, van der Waals forces also bond the layers together. When these materials are subjected to firing processes, pyrochemical changes result in the formation of new crystalline aggregates dispersed in a vitreous matrix, each having its own ionic-covalent bonding.

Empiricism remains an intrinsic aspect of ceramic technology; however, the latter now involves the cooperative talents of the ceramic scientist, chemist, metallurgist, and solid state physicist in order to effect a more fundamental understanding of these materials. Manufacturing is now highly mechanized with several industries having fully automated, computer-controlled processing (see Instrumentation and control).

The following articles present generalized and unifying characteristics of ceramic technology by means of discussions of raw materials, processing, thermal treatment, and properties.

BIBLIOGRAPHY

"Structural Clay Products" under "Ceramic Industries" in *ECT* 1st ed., Vol. 3, pp. 521–545, by R. M. Campbell, The New York State College of Ceramics; "Whiteware" under "Ceramic Industries" in *ECT* 1st ed., Vol. 3, pp. 545–574, by F. P. Hall, Pass & Seymour, Inc., and Onondaga Pottery Company; "Scope of Ceramics" under "Ceramics" in *ECT* 2nd ed., Vol. 4, pp. 759–762, by W. W. Kriegel, North Carolina State of The University of North Carolina.

1. *Objective Criteria in Ceramic Engineering Education,* American Society for Engineering Education, Urbana, Illinois, 1963.
2. F. H. Norton, *Elements of Ceramics,* 2nd ed., Addison Wesley Publishing Company, Inc., Reading, Mass., 1974.
3. Institute of Ceramics, Textbook Series, published by Maclaren and Sons, Ltd., London: (a) W. E. Worrall, *Raw Materials,* 1964; (b) F. Moore, *Rheology of Ceramic Systems,* 1967; R. W. Ford, *Drying,* 1964; W. F. Ford, *The Effect of Heat on Ceramics,* 1976.

4. W. D. Kingery, H. K. Bowen, and D. R. Uhlmann, *Introduction to Ceramics,* 2nd ed., John Wiley & Sons., Inc., New York, 1976.
5. L. H. Van Vlack, *Physical Ceramics for Engineers,* Addison-Wesley Publishing Company, Inc., Reading, Massachusetts, 1964.
6. J. E. Burke, ed., *Progress in Ceramic Science,* Vols. 1-4, Pergamon Press, Inc., New York, 1962–1966.
7. *Proceedings of the University Conferences on Ceramic Science* (various publishers). This is a series of edited books on cogent topics in the ceramic field beginning in 1964 and continuing to the present (1977 conference at North Carolina State University, Raleigh, in press, Plenum Publishing Co., New York).

ROBERT F. DAVIS
North Carolina State University

RAW MATERIALS

The principal raw materials of the ceramic industry are clay (including shale and mudstone), silica, and feldspar (see Clays). Since clay is used in the production of a large variety of products, such as whitewares, refractories (qv), structural clay products, pottery, stoneware, and fillers, greater tonnages of it are used than all of the other ceramic raw materials combined. Most of the whiteware products including wall tile, floor tile, hard porcelain, electrical porcelain, translucent porcelain, and tableware are produced from combinations of these three ingredients and are known as triaxial ware. Figure 1 is a triaxial (ternary) composition diagram showing areas of commercial whiteware products. Clay, silica, and feldspar are also used in many other products including nontriaxial whiteware, glass (qv), glazes, vitreous enamel (qv), refractories, and fine ceramics.

Other raw materials include a wide variety of rocks, minerals, and synthetic compounds used in the manufacture of abrasives (qv), special refractories, lime, cement (qv), electrical ceramics, magnetic products (see Magnetic materials), and optical ceramics (see Amorphous magnetic alloys; Glassy metals). A discussion of ceramic raw materials classified according to usage is very difficult because most ingredients have more than one use; therefore, for clarity and simplicity, the following discussion is divided into three main groups: clays; nonclay minerals; and special materials. A list of minerals and compounds described in this article including CAS Registry Numbers is provided at the end of the article.

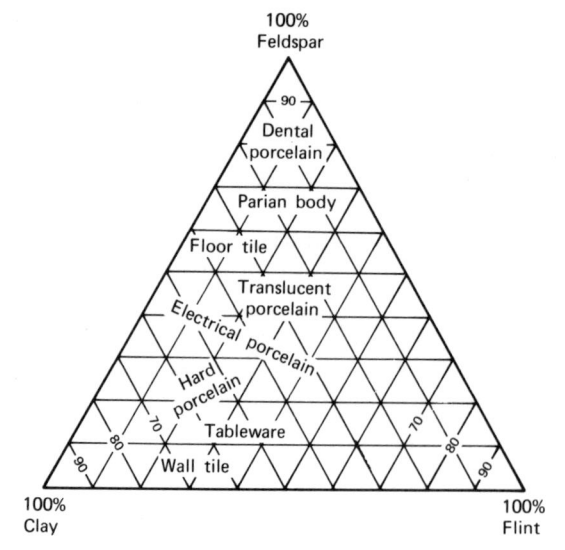

Figure 1. Triaxial diagram showing areas of commercial wares.

Clays

Clay Minerals. The principal clay minerals are kaolinite, $Al_4Si_4O_{10}(OH)_8$ montmorillonite, $X_yAl_2(Al_ySi_{4-y}O_{10})(OH)_2$ where X is usually Na, Mg or Al, and illite, $K_y(AlFeMg_4Mg)(Al_ySi_{8-y})O_{20}(OH)_4$. These are actually three families of minerals since kaolinite has several modifications and since isomorphous substitution occurs in the latter two giving rise to other compositions having different mineral names. Closely associated with the above minerals are gibbsite, $Al(OH)_3$, diaspore, $HAlO_2$, and bauxite (of indefinite composition but usually given as $Al_2O_3 \cdot 2H_2O$ which is an intermediate between the first two). All clays have as the major constituents one or more of the above minerals or minerals of the above families.

The kaolinite group includes kaolinite, halloysite, dickite, and nacrite, all of which are structurally similar and belong to the sheet family of silicates (see Silica; Silicon compounds). Chemically, dacrite and nacrite are identical with kaolinite but differ only in the structural stacking of the sheets or layers. For a more detailed discussion of the structures and properties of the clay minerals and the nonclay minerals, the reader is referred to references 1–4. Since chemical substitutions occur, there are variations in the compositions for each mineral.

The montmorillonite group includes the minerals montmorillonite, nontronite, beidellite, hectorite, and saponite. The latter two are trioctahedral and the other three are dioctahedral. Extensive substitutions occur in the octahedral sites as well as substitutions of aluminum for silicon in the tetrahedral sites. Montmorillonite absorbs water readily with accompanying swelling. It is the principal mineral in bentonite and accounts for its high plasticity and usually very sticky nature. The structure of this group of minerals is like that of talc and is classified with the sheet or layer silicates.

The illite group of clay minerals also belongs to the sheet-type silicates with structures related to that of muscovite. The composition is different from that of muscovite in that these minerals contain less potassium, more silica, and more combined water. Again the wide range of substitution that occurs accounts for the varying compositions of the illite group. The minerals included in this group are illites, the hydromicas, phengite, brammallite, glaucomite, and celadonite. The illites are commonly found in sedimentary clays and are frequently associated with kaolinite or montmorillonite.

In addition to these principal clay minerals which give clays their plastic properties, most clays will also have one or more accessory minerals. Some of the more common of these minerals found in clays include quartz, muscovite, biotite, limonite, hydrous micas, feldspar, and vermiculite. In addition to these accessory minerals, carbonaceous matter is found in sedimentary clays such as the ball clays.

Clay Terminology in the Ceramic Industry. The clays used in the ceramic industry are usually referred to by names that reflect their use. Thus a pottery clay is one used in the making of pottery, a sewer pipe clay is used for sewer pipe and similar products, and so on. Exceptions to this occur, such as the use of the term kaolin to indicate a clay consisting chiefly of kaolinite, or the term ball clay to indicate a sedimentary clay containing some organic matter and having a very high plasticity.

Kaolins. Very pure kaolins, either residual or sedimentary, usually have compositions very nearly that of kaolinite. The oxide formula for kaolinite is $Al_2O_3 \cdot 2SiO_2 \cdot 2H_2O$ which gives a theoretical analysis of 46.5% SiO_2, 39.5% Al_2O_3, and 14.0% H_2O. The analyses of some typical kaolins are given in Table 1.

The residual kaolins usually have much more free silica as quartz than do the sedimentary clays. Most of the kaolins are washed before they are shipped from the mine to the user, thus assuring greater uniformity of the raw material and increased plasticity. Plasticity permits the clay to be worked or shaped readily without rupture and retain its form when the working force is removed. The sedimentary kaolins have much greater plasticity than the residual kaolins and some are practically as plastic as ball clays.

Kaolins are used extensively in whitewares, and are the chief ingredients of most porcelain and china bodies. High purity kaolins fire to a white body, but small amounts of iron will cause the fired body to have a light cream color. A typical porcelain composition may consist of, in wt %: kaolin, 40; feldspar 30; flint, 20; and ball clay, 10. Since kaolins are refractory, the feldspar addition serves as a flux and the flint tends to prevent warping and sagging during firing. The ball clay increases the plasticity and workability of the mixture during the forming process. The ball clay also increases the dried and the fired strength of the body.

Because of the high purity and the composition of the kaolins, they are the most refractory of all the clays, with softening points of 1700–1785°C for kaolinite. Thus

Table 1. Analyses of Typical Kaolins, %

Type of kaolin	SiO_2	Al_2O_3	Fe_2O_3	FeO	CaO	MgO	Alkalies	TiO_2	H_2O
sedimentary (Fla.)	47.0	36.8	0.8		0.2	0.2	0.2	0.2	15.0
crude residual (N.C.)	62.4	26.5	1.4		0.6		1.0		8.8
washed residual (N.C.)	45.8	36.5	0.3	1.1	0.5		0.3		13.4
English china clay	48.3	37.6	0.5		0.1		1.6		12.0
sedimentary (Fla.)	45.0	38.7	0.4				0.5	1.4	13.6
sedimentary (Ga.)	45.0	38.1	0.6					1.7	14.7

kaolins are used for the manufacture of refractories as well as whiteware bodies. The total kaolin sold or used by producers in the United States in 1974 amounted to ca 5,800,000 metric tons valued at about $209,000,000. Georgia was the principal producing state with an output of 4,320,000 t or 75% of the United States production. The paper, rubber, refractories, and face brick industries consumed 66% of the kaolin used in the United States in 1974. The paper industry was by far the largest consumer, accounting for 40% of the total consumption.

Ball Clays. The term ball clay originated in England and designated a very plastic sedimentary clay. The name is currently used for any sedimentary clay of very high plasticity. These clays are composed primarily of kaolinite with varying amounts of impurities and organic matter. Montmorillonite may be present in varying amounts along with other clay minerals. The raw clay may vary in color from dark to very light and may fire to a nearly white color suitable for use in whitewares. The color after firing is dependent upon the impurities present, the most important of which is iron oxide.

The very high plasticity of the ball clays encourages their use whenever increased workability and high dry strength are needed. However, associated with the high plasticity is high shrinkage and the danger of cracking upon drying.

As ball clays have very fine particle sizes (usually less than 1 μm) they tend to remain in suspension and promote the stability of slips used for glazing and enameling. A slip is a slurry or a suspension of a finely divided ceramic material in a liquid. Ball clays also serve as a source of aluminum oxide, Al_2O_3, and silicon dioxide, SiO_2, in glazes and the other bodies in which they might be used. Chemical analyses of several ball clays are given in Table 2 and some physical properties of typical ball clays are given in Table 3.

Tennessee with 61% of the total United States output ranked first in ball clay

Table 2. Chemical Analyses of Typical Ball Clays, %

Clay	SiO_2	Al_2O_3	Fe_2O_3 and FeO	CaO	MgO	Alkalies	TiO_2	Ignition loss
Tennessee ball clay, 1	46.9	33.2	2.0	0.3	0.4	0.7		16.5
Kentucky ball clay	46.9	36.6	1.1	0.5		1.6	0.4	13.2
English ball clay	49.0	32.1	2.3	0.4	0.2	3.3		9.6
New Jersey ball clay	45.6	38.9	1.1		0.1	0.2	1.3	14.1
Tennessee ball clay, 2	54.0	29.3	1.0	0.4	0.3	0.4	1.6	13.0

Table 3. Physical Properties of Some Ball Clays

	Clay			
Property	A	B	C	D
color of raw clay	gray	light gray	grayish white	light pink
water of plasticity, %	40.6	44.7	34.9	34.6
linear drying shrinkage, %	5.6	4.7	4.7	8.8
total shrinkage[a], %	18.1	17.9	15.2	14.1
absorption[a], %	0.7	2.9	4.0	5.8
fired color	grayish white	white	white	white
fusion point, °C	1745	1745	1700	1700

[a] Fired to 1285°C.

production in 1974 with Kentucky second. Total production in the United States in 1974 was ca 741,000 metric tons valued at about $14,200,000. Of this production, approximately 21% was used in pottery and 21% in sanitary ware, 11% in floor and wall tile, 5% in china and tableware, 6% in electrical porcelain, and 36% in other products and uses.

Fire Clays. For a clay to be classed as a fire clay (or a refractory clay), its fusion point must not be below about 1600°C. The fire clays may be divided into the plastic fire clays and the nonplastic fire clays. The plasticity of the plastic fire clays may range from a workability just sufficient to shape the ware to that of a ball clay. The nonplastic fire clays include those refractory clays that have insufficient plasticity to permit shaping. The flint fire clays belong to this latter category.

The composition of fire clays may vary from the kaolinite type to those of higher silicon dioxide and lower aluminum oxide content. In general, the alkalies are low and so are the basic oxides, calcium, magnesium, and iron oxides. The chemical analyses of some fire clays are given in Table 4.

Fire clays are used in the manufacture of fireclay brick, saggers, glass melting pots, crucibles, refractory mortars and cements, and refractories containing large quantities of nonplastic grog. The use of fireclay refractories has declined in recent years in favor of refractories with greater dimensional tolerance control and special properties found in high-alumina refractories, mullite brick, nonclay bonding mortars, nonclay plastic refractories and ramming mixes, and dead-burned magnesia refractories. The total production of fireclay in the United States in 1974 was ca 3,800,000 t valued at $41,100,000.

China Clay. This is a term applied to the white-firing kaolins used in the manufacture of china and similar whitewares.

Slip Clays. The slip clays are usually glacial in origin, very impure, high in alkalies (2–4%), and will form a glass at about 1260°C. They are used for glazing stoneware and electrical porcelain. The better known deposits of slip clay occur around Albany, N.Y.

Bentonite. Bentonite is a highly plastic clay whose composition is chiefly montmorillonite. The particle sizes are largely in the colloidal range which no doubt contributes to its very high plasticity, exceeding that of ball clays. Often it is possible to get the same degree of plasticity as with ball clay by using only one fifth as much

Table 4. Chemical Analyses of Various Fire Clays[a], %

	1	2	3	4	5	6
silica	42.8	43.1	56.8	44.8	76.2	34.6
alumina	39.7	39.4	26.6	35.1	15.8	48.0
iron oxide	1.1	1.4	1.6	1.9	0.5	1.1
titania	2.2	2.0	1.8	2.2	1.7	2.3
lime	0.1	0.1	0.4	0.8	0.3	0.3
magnesia	0.3	0.1	0.2	0.6	0.3	0.6
alkalies	0.2	0.2	0.1	0.7	0.2	0.4
ignition loss	13.8	13.6	12.1	14.0	5.6	13.3
fusion point, °C	1750	1730	1617		1617	1720

[a] 1, Kaolinitic flint clay (Missouri); 2, Kaolinitic flint clay (Pennsylvania); 3, Kaolinitic plastic fire clay (Missouri); 4, Flint fire clay (Kentucky); 5, Siliceous plastic fire clay (New Jersey); 6, Aluminous diaspore fire clay (Missouri).

bentonite. The main use in the ceramic industry is for increasing the plasticity and dried strength of bodies. Its use is limited because of its high drying and firing shrinkage and its high iron oxide content (3–4%). The high alkali content causes fusion to occur at about 1300°C. Wyoming is the principal producing state and accounted for about 65% of the United States production in 1974.

Pottery Clays. Any clay which is used in the manufacture of pottery is called a pottery clay. Porcelain and china are special classes of pottery in which kaolin and ball clays are used; however, the more common usage of pottery clay refers to the crude or less pure clays employed for making stoneware, earthenware, garden pottery, artware, flower pots, etc. These clays must have sufficient plasticity to be used on the potter's wheel or to be suitable for casting slips. Some of the properties required for stoneware clay are the following: good plasticity; free of coarse material; low iron content; forms vitrified body at less than 1200°C; should have vitrification range of 90–100°C; capable of drying and heating at moderate speed; tough, strong body when fired; and free of carbonates, sulfates, or other salts likely to cause blisters or scumming when fired.

Sewer Pipe Clays. The requirements of a clay for sewer pipe manufacture are similar to those of the pottery clays. The clay should have high dry strength and vitrify into a strong dense pipe at about 1100°C. The clay must have a broad vitrification range, and low warpage. A typical analysis of a sewer pipe clay is 60% SiO_2, 21% Al_2O_3, 6% iron oxide, 0.5% CaO, 1.5% MgO, 4% alkalies, and 7% ignition loss.

Brick Clays. Clays used for the manufacture of common brick and face brick have such a range of compositions that no typical compositions will be given. In general, for face brick, the alkalies may range from 2 to 6%, magnesia up to 10%, lime up to 15%, iron oxides up to 8%, alumina 10–25%, and silica 35–65%. For common brick, even wider variations in compositions may be tolerated.

Shales and mudstones, which are sedimentary clay deposits that have been compacted and consolidated, are used extensively for brick making when they may be easily ground to form a plastic extrudable material either alone or when blended with varying amounts of clay. The Triassic mudstones and shales of North Carolina are well known for their brick-making qualities.

Important properties of the clays used for face brick are the plasticity, the fired color, shrinkage, water absorption of the fired brick, and the strength. Considerable attention has been given to controlling the color of face brick in recent years with the result that the fired color may be altered by controlling the firing temperature and gases, and by using various additives in the clay.

Production, Consumption, and Uses of Clay. In 1974 in the United States, shipments of clay refractories amounted to ca $410,000,000 and shipments of the principal construction clay products were valued at $695,000,000. Clay sold or used by producers in the United States in 1974 is given in Table 5 (5).

Nonclay Minerals

Silica. The most common occurrence of silica (qv) is in the form of quartz. Other forms which are found in nature are tridymite, cristobalite, vitreous silica, cryptocrystalline forms (usually as pebbles in chalk), hydrated silica, and diatomite. The principal sources of silica used in the ceramic industry are the sandstones, quartzites, and sands. Quartzites, often called ganister, are firmly consolidated sandstones, whereas sandstones are rather lightly bonded quartz grains or sands.

Silica is the primary ingredient in glass and is usually obtained from high purity sandstones or quartzites by crushing and grinding, or from high-grade sand deposits. The term glass sand may refer to a deposit of sand or, more commonly, it is used to refer to the sand after it has been beneficiated from sandstones, quartzites, or natural sands.

Flint or potter's flint in the strictest usage refers to flint pebbles usually occurring in calcareous or chalk deposits. It is a microcrystalline form of silica with a small amount of combined water. More generally, flint is used to designate either the true flint or quartz.

The principal uses of silica in the ceramic industry are in glass, fused silica, brick, whiteware bodies, whiteware glazes, and enamels (qv). Diatomite (qv), which is over 90% SiO_2, is used for insulating refractories up to a temperature of approximately 980°C. Diatomite is formed from the siliceous skeletons of diatoms and thus is very porous and cellular. When bonded with clay it may be shaped and used for insulation up to a temperature of 1370°C.

Feldspar. The feldspars are used chiefly as fluxes and sources of Al_2O_3, SiO_2, alkalies (K_2O, Na_2O), and CaO. The plagioclase feldspars vary in composition from albite, $NaAlSi_3O_8$, to anorthite, $CaAl_2Si_2O_8$, in a continuous series of solid solutions. Orthoclase and microcline feldspar ($KAlSi_3O_8$) are often referred to as potash feldspar. Anorthoclase, $(Na,K)AlSi_3O_8$, is a combination of albite and potash feldspar.

The principal producing state is North Carolina which, together with Connecticut, Georgia, California, and South Dakota, produced 98% of the United States production in 1974. Feldspar occurs in dikes and in igneous rocks such as granite. Graphic granite is an igneous rock that consists entirely of quartz and feldspar, and it is used as a source of both feldspar and silica. The chemical analyses of some typical feldspars are given in Table 6.

The potash feldspars are used extensively in whiteware bodies and the high-soda feldspars are used in glasses and glazes. All of the feldspars contain some free silica which is not detrimental as long as it is controlled. The iron oxide content of feldspars must be low for most uses to control color. For whiteware bodies, the feldspar is supplied as a finely ground product usually less than 0.074 mm. Ground feldspar sold by merchant mills in the United States in 1974 is given in Table 7 according to use (5).

Nepheline Syenite. Nepheline syenite is a rock that contains a large percentage of the mineral nephelite, $(Na,K)_2Al_2Si_2O_8$, along with some soda and potash feldspars. It is mined in large quantities in Ontario and is used as a substitute for feldspar, particularly in the glass industry. Imports from Canada in 1974 were over 453,600 t of syenite. Canadian production amounted to 522,500 t valued at $7,400,000 in 1974.

Lime. Lime is derived by the calcination of limestone which consists largely of calcium carbonate, $CaCO_3$. The calcined product is calcium oxide, CaO, which is used in manufacturing cement, plaster, and mortars. It has also found limited use as a refractory in the form of fused calcium oxide and more extensive use in lime–silica brick.

Finely ground limestone is used in glazes, enamels, and glasses where the mineral dissociates into CaO and CO_2 which escapes. Whiting is a high purity calcium carbonate which has been chemically precipitated and is used for special bodies of high purity (see Calcium compounds).

Table 5. Clays Sold or Used by Producers in the United States in 1974 by Kind and Use (Metric Tons)

Use	Ball clay	Bentonite	Common clay and shale	Fire clay refractory	Fuller's earth	Kaolin	Undistributed[a]	Total
common brick	[b]	[b]	2,895,290					2,895,975
face brick	11,793	2,651	15,679,350	95,488		385,851	685	16,175,133
cement, portland		416	10,855,880		17,862	121,812		10,995,970
ceramic, hobby						587		587
china/dinnerware	39,273					19,554		58,827
crockery and other earthenware	1,031		6,253	[b]	[b]	62,110	21,473	90,867
electrical porcelain	40,943					14,000		54,943
fiberglass			116			116,105		116,221
firebrick, block, shapes	24,867	[c]		2,234,024		407,551		2,666,442[d]
flower pots			42,970					42,970
flue linings			90,527	51,837				142,364[d]
foundry sand		669,552		208,104		4,873		882,529
glazes, glass, enamels	1,786	190		45,983		4,754		52,713
grogs, and crudes, refractory	[c]			129,504		251,619		381,123[d]
gypsum products		459				4,374		4,833
high alumina (minimum 50% Al_2O_3) refractories	19,975			316,190		54,886		391,051
kiln furniture	8,481			227		2,037		10,745

Use							Total
lightweight aggregate			5,433,406				5,433,406
concrete block			2,691,644				2,691,644
structural concrete			884,956				884,956
highway surfacing			89,747				89,747
other	b			392,594	b	65,530	459,807
mortar and cement, refractory	154,414	279	33,541	18,561	1,683		244,950
pottery	158,486	c			38,155		233,577[d]
sanitary ware					75,091		
sewer pipe, vitrified		91	1,363,053	96,254			1,459,398
tile			394,093				394,093
drain	80,262		129,518				285,149
floor and wall, ceramic	1,036		142,329	59,872	15,497		143,365
quarry		11,910	75,781				87,691
roofing			58,596				58,596
structural			47,196	907			48,103
terra cotta	3,899	45,533	7,943	7,156	36,321		105,647
miscellaneous[e]	4,595	1,788,460	26,648	25,036			
nonceramic-related uses	115,384	9,105		23,313	3,297,810	64,834	6,216,353[d]
undistributed	75,103	474,587	57,140	51,458	884,805		1,617,719
exports							
Total	*741,328*	*3,003,233*	*41,005,977*	*3,756,508*	*5,799,475*	*1,110,973*	*55,417,494*

[a] Total of clays indicated by footnote b.
[b] Included with undistributed to avoid disclosure of confidential company data.
[c] Included with miscellaneous to avoid disclosure of confidential company data.
[d] Incomplete figure, remainder included in miscellaneous.
[e] Includes abrasives, graphite anodes, linoleum, mineral wool and insulation, roofing granules, textiles, unknown, and data indicated by footnote c.
[f] Undistributed total included with the total for each use and the total for each clay.

245

Table 6. Chemical Analyses of Feldspars Found in Various Locations, %

	N.C. 1	N.C. 2	Ontario	Maine	Pa.
SiO_2	69.5	70.0	65.4	65.2	64.6
Al_2O_3	17.5	18.1	18.8	20.1	20.6
Fe_2O_3	0.1	0.1		0.7	
CaO	0.8	1.5			0.1
MgO					2.4
K_2O	8.1	3.5	13.9	11.6	1.9
Na_2O	3.6	6.5	2.0	2.0	10.3
loss	0.3	0.3	0.6	0.4	

Table 7. Ground Feldspar Sold by Merchant Mills in the United States in 1974 According to Use[a]

Use	Amount, t	Value, $
glass	196,920	3,597,000
pottery	246,760	6,852,000
other[b]	112,300	2,609,000
Total	555,980	13,058,000

[a] Ref. 5.
[b] Includes, soaps, abrasives, and other ceramic and miscellaneous uses.

Magnesite. Magnesite is the mineral form of magnesium carbonate, $MgCO_3$. Complete calcination or dead burning of magnesite produces magnesium oxide, MgO, which when fired at 1450–1500°C, sinters into dense grains that are used in refractory brick or as refractory grains in the bottom of open hearth furnaces. Since magnesium oxide fuses at about 2800°C, it has high potential for greater use as a super refractory. Partially calcined magnesite yields a product called caustic magnesia which is used as a cement when combined with a solution of magnesium chloride, $MgCl_2$. Pure grades of magnesite are used in whiteware bodies, glazes, enamels, and in glasses. Magnesite is also used with chrome ore in making basic refractory brick.

Dolomite. Dolomite is a rock consisting of a mixture of calcium and magnesium carbonates. In ceramics it is used as a source of magnesium and calcium oxides in whiteware bodies and in glasses. It is also used alone as dead-burned dolomite for the manufacture of basic refractory brick. Refractory dead-burned dolomite sold or used by producers in 1974 amounted to ca 1,200,000 t valued at about $24,400,000.

Gypsum. Gypsum is a mineral having the formula $CaSO_4 \cdot 2H_2O$. It occurs in several parts of the United States and is best known for its use in the manufacture of plaster of Paris by calcination at 175–200°C:

$$CaSO_4 \cdot 2H_2O \rightarrow CaSO_4 \cdot \tfrac{1}{2}H_2O + 1.5\ H_2O$$

This reaction is reversible at room temperature when water is added. In the process of slip casting, porous molds made of plaster of Paris are used to absorb water or other fluid of a slip; this results in the deposition of the solids on the walls of the mold and thus forms the article in the shape of the mold.

In 1974 ca 10,000,000 t of calcined gypsum valued at about $206,000,000 was produced in the United States.

Chromite. Chromite is a spinel mineral having the formula $FeCr_2O_4$. In the ceramic industry, chromite is used for making refractories in the form of burned bricks and chemically bonded bricks. When pure $FeCr_2O_4$ is burned in the air, the sesquioxides are formed:

$$2\ FeCr_2O_4 + \tfrac{1}{2}\ O_2 \rightarrow Fe_2O_3 + 2\ Cr_2O_3$$

Since some Al_2O_3, MgO, and SiO_2 are present in chrome ores, additional spinels result from the firing operation. Magnesite may be combined with chrome ore to form chrome–magnesite or magnesite–chrome bricks. These are basic refractory bricks and find extensive use in metallurgy. Typical compositions of basic refractory brick are given in Table 8. Imported ore in 1974 came principally, in decreasing order, from Africa, the Union of Soviet Socialist Republics, the Philippines, and Turkey. Approximately 62.3% of the total United States consumption in 1974 was used in the metallurgical industry, 20.3% in the refractory industry, and 17.4% in the chemical industry. The total United States consumption amounted to ca 1,300,000 t of chromite ore containing about 400,000 t of chromium.

Aluminum Silicates. Andalusite, sillimanite, and kyanite have the formula Al_2SiO_5 and may contain iron replacing part of the aluminum. The decomposition ranges are 1350–1450°C for andalusite, 1550–1650°C for sillimanite, and 1100–1480°C for kyanite. They are used as a source of alumina and silica in refractories and in glass batches and for the production of synthetic mullite, $3Al_2O_3 \cdot 2SiO_2$. Since kyanite decomposes into mullite and silica at a lower temperature than the other two, it is used in the calcined form as a grog in high-alumina super refractories. On the other hand, sillimanite and andalusite, which decompose more slowly than kyanite, are preferred for manufacture of spark plugs, pyrometer tubes, and similar whitewares because the volume change upon dissociation is not so severe. Synthetic mullite production in 1974 amounted to ca 25,000 t valued at about $5,000,000. About 90% of all mullite refractories are used in the metallurgical and glass industries.

Pyrophyllite. Pyrophyllite, $Al_2Si_4O_{10}(OH)_2$ is a sheet-type mineral similar to montmorillonite in structure except that there is no substitution of Al^{3+} for Si^{4+}. Pyrophyllite is used extensively in the ceramic industry in whiteware bodies, floor and wall tiles, and some refractories and cements, but it is used more widely as a filler for various products (see Fillers). In 1974 North Carolina produced all of the pyrophyllite reported as mined in the United States. The total sales for 1974 amounted to ca 92,000 t valued at about $1,500,000 of which approximately one fifth was used in the ceramic industry.

Table 8. Typical Chemical Analyses of Basic Refractory Brick, %

	Chrome	Chrome–magnesite	Magnesite–chrome	Magnesite
SiO_2	5.0	5.0	4.0	2.0
TiO_2	0.1	0.1	0.1	0.1
Fe_2O_3	15.0	12.0	6.0	3.0
Al_2O_3	20.0	17.0	5.0	2.0
MnO	0.1	0.1	0.1	0.1
Cr_2O_3	40.0	29.0	7.0	
CaO	2.0	1.0	2.0	2.0
MgO	16.0	38.0	73.0	90.0

Magnesium Silicates. Talc, asbestos, and olivine are the more important magnesium silicates used in the ceramic industry. Talc has a composition, when pure, given by the formula $Mg_3Si_4O_{10}(OH)_2$, with a three-layer structure similar to that of pyrophyllite. The mineral is very soft and greasy to the feel, and it acts as a self-lubricant when it is used in dry-pressing operations. It is used for high-frequency electrical insulators and also for some whiteware bodies such as tile and china. Talc acts as a flux with clays and thus forms bodies that vitrify very readily. It is used also in saggers and kiln furniture where resistance to thermal shock is important. In 1974 the ceramic industry used approximately 202,300 t of talc valued at over $6,850,000. Ceramic bodies made predominately of talc are usually referred to as steatite ware (see Talc).

The asbestos minerals anthophyllite, tremolite, actinolite, and chrysolite are used chiefly for insulating refractories in the low to medium temperature range and in fireproof materials (see Asbestos).

Olivine, $[(Mg,Fe)_2SiO_4]$, may be considered a solid solution of forsterite ($2MgO.SiO_2$) and fayalite ($2FeO.SiO_2$). Since forsterite melts at 1910°C and fayalite at 1204°C, olivines of high forsterite content are very refractory. Properly proportioned magnesite and olivine mixtures when fired yield $MgO.Fe_2O_3$ and forstersite. The resulting product is very refractory. Some olivine has been used for foundry sand, and its use for this purpose has increased in recent years.

Fluxing Minerals. Fluxing minerals are those minerals that lower the vitrification temperature, the melting temperature, or the reaction temperature. In this category are the lithium minerals spodumene, $Li_2Al_2Si_4O_{12}$, lepidolite, $[K_2(Li_3Al_3)(Al_2Si_6O_{20})(OH,F)_4]$, amblygonite, $Li_2F_2Al_2P_2O_8$, and petalite, $LiAlSi_4O_{10}$. Usually the lithium salt is extracted and the pure salt is used in glasses, glazes, and acid-resisting enamels.

The barium minerals used as fluxes are barite, $BaSO_4$, and witherite, $BaCO_3$, used in some glass batches. Fluorspar, CaF_2, after it has been beneficiated to 95% CaF_2 or higher is used for some optical glasses and in enamels. Apatite, $[Ca_5(Cl,F)(PO_4)_3]$, is used in some glass batches and in enamels. Calcined bones are used as a source of phosphate for bone china.

Refractory Minerals. The *zirconium minerals* used as raw materials for refractories are baddeleyite, ZrO_2, zirkite, ZrO_2, and zircon, $ZrSiO_4$. The first two minerals are usually purified from a raw grade of 60–80% ZrO_2 to 90–98% ZrO_2 before use. Most zirconia refractories are stabilized with calcia or magnesia to prevent the destructive crystal transformations that occur at 1000 and 1900°C. The fusion point of pure ZrO_2 is 2700°C. High purity ZrO_2 is used as an opacifier in glazes and enamels.

Zircon, which has a fusion point of 2550°C, is used as a refractory alone and is also the principal source of high purity ZrO_2 obtained by chemical treatment. The consumption of zircon in the United States in 1974 was approximately 150,000 t of which 74% was used in the ceramic and foundry industries. Total zircon, zirconia, and alumina–zirconia–silica-base refractories produced in the United States in 1974 amounted to 49,900 t (see Zirconium and zirconium compounds).

Hydrated alumina minerals were listed with the clay minerals but are included here also since their importance in the ceramic industry is chiefly in the manufacture of refractories. These minerals are diaspore, $HAlO_2$, gibbsite, $Al(OH)_3$, and bauxite (an impure mixture of the first two minerals). These minerals when fired with the andalusite minerals yield mullite, $3Al_2O_3.2SiO_2$, which is a very useful refractory in the 1600–1650°C temperature range. Mullite decomposes at 1830°C.

Titania, TiO_2, which occurs as the minerals rutile, brookite, and anatase, has a fusion point of 1900°C and is used after chemical purification in enamels and glazes as an opacifier and in electrical porcelains (see Titanium compounds).

Thoria, ThO_2, is found in monazite sands and has a melting point of 3050°C. In the ceramic industry it has been used for the manufacture of very refractory crucibles. With the rather large resources in the United States considerable research is needed to develop the nuclear fuel potential for thoria. Estimated industrial demand in the United States in 1974 was 73 t of ThO_2 equivalent. About 36 t was used for producing nuclear fuel and for nuclear research. The remaining 37 t was used in incandescent gas light mantles, refractories, alloys, electronics, and various chemical applications (see Thorium and thorium compounds).

Graphite is one of the most refractory materials available with a decomposition temperature in excess of 3650°C. Its refractoriness and electrical and thermal conductivities are properties that make it suitable for use as electrodes for arc furnaces, resistance heating elements, susceptors for induction furnaces, and crucibles which may be bonded with fire clay. Increasing amounts of graphite are being used as fibers for reinforcing many products (see Carbon and artificial graphite).

The various kinds of nonclay refractories shipped in the United States in 1974 are given in Table 9 (5).

Table 9. Shipments of Nonclay Refractories in the United States in 1974

Product	Unit of quantity	Quantity	Value, $
silica brick and shapes	1000 23-cm equiv	37,626	20,004,000
magnesite and magnesite chrome brick and shapes	1000 23-cm equiv	117,209	178,840,000
chrome and chrome–magnesite brick and shapes	1000 23-cm equiv	20,726	30,536,000
shaped refractories containing natural graphite	t	20,467	23,863,000
other carbon refractories, forsterite, pyrophyllite, dolomite–magnesite, molten cast and other brick and shapes	1000 23-cm equiv	40,418	97,085,000
other mullite, kyanite, sillimanite, or andalusite brick and shapes	1000 23-cm equiv	6,239	14,253,000
other extra high alumina (over 60%) brick and fused bauxite, fused alumina, dense-sintered alumina shapes	1000 23-cm equiv	4,148	16,475,000
silicon carbide brick, shape and kiln furniture	1000 23-cm equiv	5,272	25,644,000
zircon and zirconia brick and shapes	1000 23-cm equiv	2,703	11,380,000
refractory bonding mortars	t	54,658	12,245,000
nonclay refractory castables, hydraulic setting	t	43,006	14,656,000
plastic refractories and ramming mixes	t	212,130	48,737,000
gunning mixes	t	382,926	55,624,000
dead-burned magnesite	t	77,606	10,785,000
other nonclay refractory materials sold in lump or ground form[a]	t	429,104	18,277,000
Total nonclay refractories			578,404,000

[a] Materials for domestic use as finished refractories and all exported refractory material.

Special Materials

Super Refractories. In this category are included the carbides of such elements as silicon, boron, zirconium, hafnium, tantalum, vanadium, molybdenum, tungsten, and niobium (see Carbides). Also included are the borides of the high-melting metals in the fourth, fifth, and sixth periodic groups (see Boron compounds). For high temperature usage (in excess of 2500°C) *in vacuo,* the carbides and borides of these high melting metals are about the only suitable materials available because of their low volatility.

Nitrides (qv) of the transition elements in the third, fourth, and fifth periodic groups plus the nitrides of beryllium, boron, aluminum, silicon, and the elements of the lanthanide and actinide series have high melting points. Usage of the nitrides as refractories has not been extensive since they are very brittle and have very poor oxidation resistance. Nitride-bonded carbides have been used and the additions of nitrides to other refractories have promise. Promising research on Si_3N_4 for high temperature gas turbines has created much interest in this material.

Sulfides as refractory materials are largely in the development process. However, worthy of mention is cerium sulfide, CeS, which was developed in the Manhattan Project during World War II (see Cerium and cerium compounds). Crucibles made of cerium sulfide are capable of melting all metals that have a melting point below 1800°C (with the exception of platinum) with little or no attack upon the crucible. Similar to the cerium sulfide refractory properties are those of thorium sulfide, ThS.

Single oxides which are used as refractories to the greatest extent commercially include silica, alumina, magnesia, and zirconia. These have been discussed earlier. It should be noted here, however, that when these are of high purity, their refractory properties are enhanced. The possible exception to this is zirconia which is usually stabilized by the addition of either calcia or magnesia. These refractory oxides are used in granular form as well as for the fabrication of crucibles, tubes, muffles, and heavy refractory ware such as bricks. The oxides of beryllium and thorium are available commercially, but are used mostly in light refractory ware such as crucibles and tubes. Calcia is a very refractory material of low cost but subject to hydration. Considerable interest exists in overcoming this drawback in the applicability of calcia as a super refractory. To a lesser extent the same problem existed with the use of magnesia, but this has been overcome to a large extent by the dead-burning process.

Complex oxides of commercial importance include mullite, zircon, high alumina firebrick, and the basic refractories discussed earlier. Again it should be noted that the use of high purity materials improves the refractory properties. Magnesia spinel, $MgAl_2O_4$, has been used to a limited extent as a refractory but, in ramming mixes using magnesium oxide plus aluminum oxide, a volume expansion occurs when the spinel forms, and this property has been used to make linings for melting furnaces to retain molten steel.

Nuclear Ceramic Materials. With the advent of nuclear power plants, the need for better nuclear fuels capable of operating at higher temperatures and longer life has given impetus to research in this area. The principal fuel for power generation is uranium oxide, UO_2, which although refractory (melting point above 2800°C), has low thermal conductivity. It has good resistance to irradiation damage, is relatively inert to many coolants such as carbon dioxide, hydrogen, and water, and has long burnup without swelling. Uranium oxide also has good retention of fission products, low fabrication cost, and no crystallographic modifications in inert atmospheres. Its poor thermal conductivity has, however, been its chief disadvantage and consequently considerable research has been directed toward the uranium carbide, nitrides, sulfides, phosphides, and combinations of carbides, nitrides, and oxides (see Nuclear reactors; Uranium).

Electronic Ceramic Materials. Ferrites (qv) with improved magnetic properties have been developed in recent years and have contributed much to the advancement of the electronics field. The ferrites have the general formula $MO\cdot Fe_2O_3$ where M is a divalent metal atom. Thus, some typical ferrites are $ZnO\cdot Fe_2O_3$, $FeO\cdot Fe_2O_3$, and $NiO\cdot Fe_2O_3$. Substitution of nonmagnetic ferrites in magnetic ferrites (6), such as the replacement of part of the nickel monoxide in the nickel ferrite with zinc oxide, improves the magnetic properties of the material (see Magnetic materials). The ferrites are readily fabricated into many shapes by sintering techniques and are used for cores in low-loss coils operating at high frequencies and for strong permanent magnets.

The titanates, principally $BaTiO_3$, have been used in the electronic industry for the manufacture of ceramic capacitors of high capacitance because of the very high dielectric constant. In addition to this usage, $BaTiO_3$ is used for transducers for the conversion of electrical energy into mechanical energy and vice versa. Such transducers are used in ultrasonic cleaners (see Ultrasonics), sonar and depth-sounding apparatus, and accelerometers (see Ceramics as electrical materials).

Other materials too numerous to list here are used in the ceramic industry. *Ceramic Age* (7) has classified over 450 materials used in the manufacture of ceramic products.

Mineral or compound	CAS Registry No.	Mineral or compound	CAS Registry No.
silica	[7631-86-9]	olivine	[1317-71-1]
kaolinite	[1318-74-7]	anthophyllite	[17068-78-9]
montmorillonite	[1318-93-0]	tremolite	[14567-73-8]
illite	[12173-60-3]	actinolite	[13768-00-8]
gibbsite	[14762-49-3]	chrysolite	[25666-97-1]
diaspore	[14457-84-2]	forsterite	[15118-03-3]
bauxite	[1318-16-7]	fayalite	[13918-37-1]
halloysite	[12244-16-5]	spodumene	[1302-37-0]
dickite	[1318-45-2]	lepidolite	[1317-64-2]
nacrite	[12279-65-1]	amblygonite	[1302-58-5]
nontronite	[12174-06-1]	petalite	[1302-66-5]
beidellite	[12172-85-9]	barite	[13462-86-7]
hectorite	[12173-47-6]	witherite	[14941-39-0]
saponite	[1319-41-1]	fluorspar	[14542-23-5]
bentonite	[1302-78-9]	apatite	[1306-05-4]
muscovite	[1318-94-1]	baddeleyite	[12036-23-6]
hydromicas	[12173-56-7]	zirkite	[1314-23-4]
phengite	[12174-17-3]	zircon	[1490-68-2]
brammallite	[12197-36-3]	titania	[13463-67-7]
glaucomite	[1317-57-3]	rutile	[1317-80-2]
celadonite	[12173-00-1]	brookite	[12188-41-9]

252 CERAMICS (RAW MATERIALS)

quartz	[14808-60-7]	anatase	[1317-70-0]
biotite	[1302-27-8]	thoria	[1314-20-1]
limonite	[1317-63-1]	graphite	[7440-44-0]
vermiculite	[1318-00-9]	silicon carbide	[409-21-2]
aluminum oxide	[1344-28-1]	boron carbide	[12069-32-8]
calcium oxide	[1305-78-8]	zirconium carbide	[12070-14-3]
magnesium oxide	[1309-48-4]	hafnium carbide	[12069-85-1]
iron oxide	[1309-37-1]	tantalum carbide	[12070-06-3]
mullite	[1302-93-8]	vanadium carbide	[12070-10-9]
tridymite	[15468-32-3]	molybdenum carbide	[12069-89-5]
cristobalite	[14464-46-1]	tungsten carbide	[12070-12-1]
hydrated silica	[10279-57-9]	niobium carbide	[12069-94-2]
albite	[12244-10-9]	beryllium nitride	[1304-54-7]
nephelite	[1302-72-3]	boron nitride	[10043-11-5]
anorthite	[1302-54-1]	aluminum nitride	[24304-00-5]
magnesite	[13717-00-5]	silicon nitride	[12033-89-5]
magnesium chloride	[7786-30-3]	cerium sulfide	[12014-82-3]
orthoclase	[61076-95-7]	thorium sulfide	[12039-06-4]
calcium carbonate	[13397-26-7]	magnesia spinel	[1302-67-6]
gypsum	[13397-24-5]	uranium oxide	[1344-57-6]
plaster of Paris	[26499-65-0]	carbon dioxide	[124-38-9]
chromite	[1308-31-2]	hydrogen	[1333-74-0]
andalusite	[12183-80-1]	uranium carbides	[12070-09-6]
sillimanite	[12141-45-6]		and [12071-33-9]
kyanite	[1302-76-7]	uranium nitrides	[25658-43-9]
pyrophyllite	[12269-78-2]		and [12033-83-9]
talc	[14807-96-6]	uranium sulfide	[12039-11-1]
asbestos	[1332-21-4]	uranium phosphides	[12037-69-3]
			and [12037-84-2]
		barium titanate	[12047-27-7]

BIBLIOGRAPHY

"Raw Materials" under "Whiteware" under "Ceramic Industries" in *ECT* 1st ed., Vol. 3, pp. 547–551, by F. P. Hall, Pass & Seymour, Inc. and Onondaga Pottery Company; "Ceramic Raw Materials" under "Ceramics" in *ECT* 2nd ed., Vol. 4, pp. 762–776, by John V. Hamme, North Carolina State of The University of North Carolina.

1. W. E. Worrall, *Clays and Ceramic Raw Materials,* Halsted Press, a division of John Wiley & Sons, Inc., New York, 1975.
2. W. A. Deer, R. A. Howie, and J. Zussman, *An Introduction to the Rock-Forming Minerals,* John Wiley & Sons, Inc., New York, 1966.
3. R. W. Grimshaw, *The Chemistry and Physics of Clays,* 4th ed., John Wiley & Sons, Inc., New York, 1971.
4. F. A. Wade and R. B. Mattox, *Elements of Crystallography and Mineralogy,* Harper & Brothers, New York, 1960.
5. *Minerals Yearbook 1974,* Vol. 1, U.S. Bureau of Mines, Dept. of the Interior, Washington, D.C., 1974.
6. H. E. Brown, *Zinc Oxide Rediscovered,* The New Jersey Zinc Company, New York, 1957.
7. *Ceram. Age,* 4 (July 1976).

General References

Editorial Board, American Institute of Mining and Metallurgical Engineers, *Industrial Minerals and Rocks*, 2nd ed., American Institute of Mining and Metallurgical Engineers, New York, 1949.
W. H. Gitzen, *Alumina as a Ceramic Material*, The American Ceramic Society, Columbus, Ohio, 1970.
J. R. Hague and co-workers, *Refractory Ceramics for Aerospace*, The American Ceramic Society, Inc., Columbus, Ohio, 1964.
R. B. Holden, *Ceramic Fuel Elements*, Gordon and Breach, Science Publishers, Inc., New York, 1966.
E. J. Kliff, D. R. Dykes, and J. R. Hickey, *Ceramic Raw Materials*, Charles H. Kline and Co., Fairfield, N.J., 1971.
C. P. McNamara, *Ceramics*, Vol. 2, The Pennsylvania State College, State College, Pa., 1948.
F. H. Norton, *Fine Ceramics*, McGraw-Hill Book Co., Inc., New York, 1970.
F. H. Norton, *Refractories*, 2nd ed., McGraw-Hill Book Company, Inc., New York, 1942.
C. W. Parmelee and C. G. Harman, *Ceramic Glazes*, 3rd ed., Cahners Publishing Company, Inc., Boston, Mass., 1973.

JOHN V. HAMME
North Carolina State University

FORMING PROCESSES

Before ceramics can be used, they must be formed into shapes with known properties. The preparation of raw materials and the forming operation affect the distribution and particle size and shape of each phase present and the amount, location, size, and distribution of pores in the fired ware. All of these are important factors in determining the properties of the ceramic.

Forming usually imparts permanent shape and temporary size and strength to the ceramic ware. Size is often altered by shrinkage, and strength usually increases markedly during drying and firing. The forming methods used in the production of ceramics may be divided into two broad categories: plastic deformation and casting. Extrusion, dry pressing, and all hot forming methods fall under the first category, slip and fusion casting under the latter.

Material Preparation

Particle size and size distribution of raw materials in the ceramic prior to firing are particularly important. Diffusion rates are relatively slow during the firing of ceramics; therefore, distances through which diffusion must occur for reaction between the raw materials should be minimized. This is accomplished by using the smallest practical particle size and thoroughly mixing or blending the ingredients. Small particle size and intimate mixing are particularly important for shapes that are to be sintered in the solid state. Diffusion is the primary mechanism for consolidation in this case. A particle size distribution is usually preferable to a single particle size. This ensures minimum porosity in the green (undried) ware and minimizes the shrinkage required to eliminate porosity during drying and firing. General principles concerning optimum particle size distributions have been developed (1).

Raw materials are usually crushed and ground by conventional techniques to obtain optimum particle size distributions. When more than one raw material is used, it may be necessary to screen the various ingredients and blend the desired size fractions of each. More intimate contact between the ingredients often can be achieved by mixing the ingredients in water in a blunger; this method commonly is used in the whiteware industry. A blunger is a vertical cylindrical tank with a vertical rotating shaft on which are mounted horizontal blades or paddles. The homogeneous mixture of water and solid materials produced in a blunger is called a slip. The slip is stored in agitators which keep it in motion thereby preventing settling of the solids. Filter presses or drum filters are generally used to remove any water prior to soft plastic forming. If sufficient water is to be removed so that the resulting material can be dry pressed, a spray dryer may be used. Dry mixing, where applicable, aids in energy conservation as energy does not have to be expended to remove added water. Slips and dry batches of ingredients are often passed through magnetic separators to remove iron contaminants which have a detrimental effect on the color of many ceramics.

Forming Processes

Cold forming processes, those carried out at ambient temperature, are predominant in the ceramic industry. However, use of processes by which ceramic parts are formed at elevated temperatures is increasing. Such processes include hot pressing, hot extrusion, hot rolling, forging and swaging, hot isostatic pressing, chemical vapor deposition, and injection molding. These processes are discussed in references 2–4. In hot forming, ceramic ware is made very close to its final size. This is a distinct advantage since cold-formed ware must be made oversize so that it will be the desired size after it shrinks during drying and firing. Shrinkage results primarily from the increase in density that occurs as porosity is eliminated from the ware. Precise control of shrinkage is difficult and shrinkage increases the danger of obtaining warped or cracked ware.

Extrusion. The extrusion process for forming ceramics is widely used in the structural clay products industry and is used to a lesser extent in the whiteware and refractories industries (see Refractories). The principal advantage of this method is rapid and economic formation of dense ware. Extrusion can be used to form ware of any shape that has an axis normal to a fixed cross section. The cross section may contain holes produced by pins in the die through which the material is extruded. This method is usually used to produce stiff plastic masses, ie, materials that contain significant amounts of clay and 12–20% water. However, completely nonplastic materials can be extruded with the addition of suitable plasticizers (qv) such as gums, starches, poly(vinyl alcohol), waxes, and wax emulsions. The plasticizers are eliminated from the ware at relatively low temperatures during drying or firing and do not affect the properties of the ceramic.

Extrusion usually comprises three stages: pugging, deairing, and extrusion through a die. In a typical extrusion machine (Fig. 1) the dry material is fed continuously into one end of a long trough, sprayed with water and cut and kneaded (pugged) by rotating knives into a homogeneous plastic mass. The blades are tilted to move the material

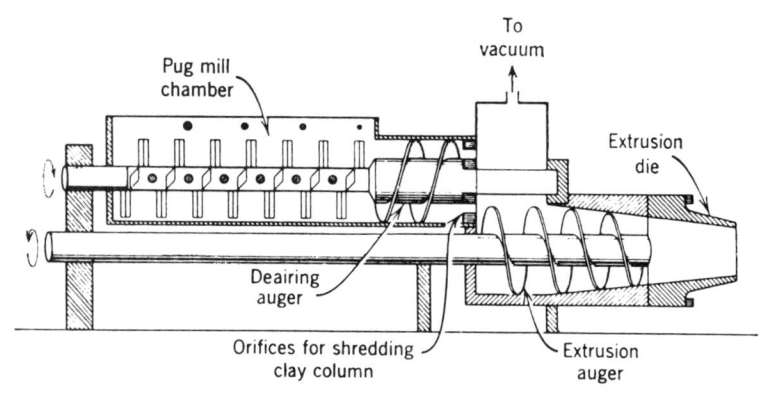

Figure 1. Cross section of an extrusion machine (5).

slowly to the exit end of the pugging chamber. The plastic mass is forced by an auger through a shredding die and into the deairing chamber in which a vacuum is maintained. Removal of air from the material improves extrusion characteristics and permits formation of denser ware. An auger then forces the plastic material through a die having the desired cross section configuration, and the column of material is automatically cut into pieces of proper length. Figure 2A depicts a simplified flow sheet for the forming of bricks by the extrusion process.

Soft Plastic Forming. All materials customarily formed by soft plastic processes contain significant percentages of plastic clays and 20–30% water. Consequently, the materials can be formed with very low pressures. Jiggering, the most common soft plastic forming process, is used principally in the whiteware industries to make chemical porcelain, dinnerware, and electrical insulators. Shapes with symmetrical circular cross sections are formed by this technique. In jiggering, materials prepared by blunging and filter pressing are deaired and extruded, and the extruded column is cut into slugs. A slug of material is placed on a plaster mold with a surface contour the same as either the inside or outside of the ware. The slug is then deformed to contact the mold surface intimately so that one surface of the ware is formed by the plaster mold. The remaining surfaces are formed by a template which contacts the clay as the mold and clay revolve (see Fig. 3). After formation, the ware is dried and separated from the plaster mold. A simplified flow diagram for forming ware by jiggering is shown in Figure 2B.

Ware may be formed by pressing a plastic mass of material in a hydraulic press between two relatively dense master molds. High pressure air is then forced through the pores of the molds to separate the pressed ware from the molds. Advantages of this process are rapid forming rates and low mold cost.

Another soft plastic process is hand molding. A piece of the ceramic batch is thrown into a mold and the excess clay is cut off flush with the top of the mold. This process is employed primarily in making special refractory shapes. Ramming is a modification of the hand-molding technique. The shape is built up gradually by placing successive layers of material in a mold and tamping each layer with pneumatic tools as it is added. Ramming is used to form intricate shapes and ware that is too large to be formed by other methods. Thin sheets of ceramic are sometimes formed by passing the soft plastic ceramic batch materials between two rolls.

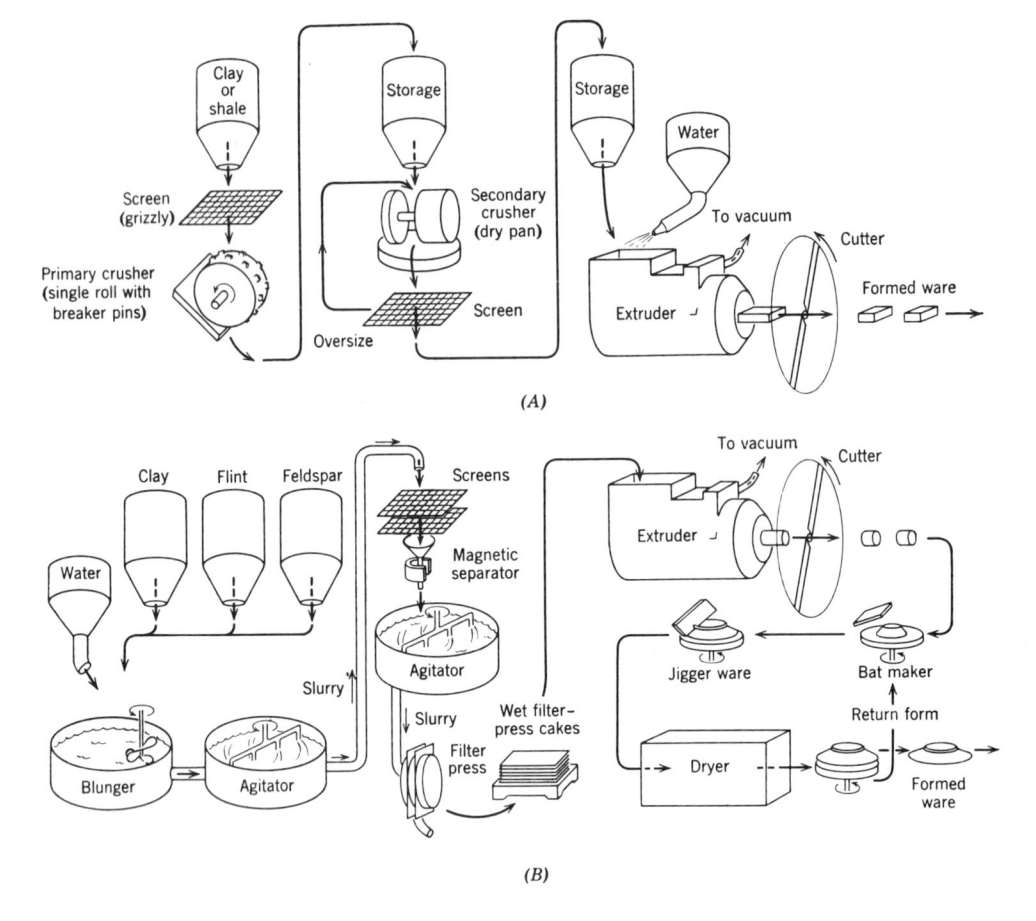

Figure 2. Simplified flow diagrams of forming processes: (*A*) extrusion of structural clay bricks; (*B*) jiggering of dinner plates.

Dry Pressing. Dry pressing is used extensively in the whiteware, refractories, and abrasives industries and in the production of cermets (ceramic–metal composites). This forming method is used when many parts of relatively simple shapes are to be made. The parts must have no undercuts, be fairly uniform in thickness, and their length normally should not exceed twice their diameter. Only when a large number of parts is to be made can the expense of making the necessary die and punches be justified. These are usually made from hardened steel or if the material to be pressed is very abrasive, cemented carbides (see Carbides).

The material to be dry pressed may be dry or it may contain up to 12% water. In either case it is usually in the form of granules such as those obtained by spray drying or other granulation processes. The proper amount of material is placed in the die cavity with the lower punch partially inserted in the die and the granules flow readily and tend to fill the die cavity uniformly. After the top punch is inserted, pressures between 34.5 and 207 MPa (5,000 and 30,000 psi) are applied to both punches to form the part. If the material contains little or no plastic material such as clay, a binder is added to give it enough plasticity to permit it to flow in the die when pressure is applied to the punches. In such cases, 1 or 2% of a wax, starch, poly(vinyl alcohol), acrylate,

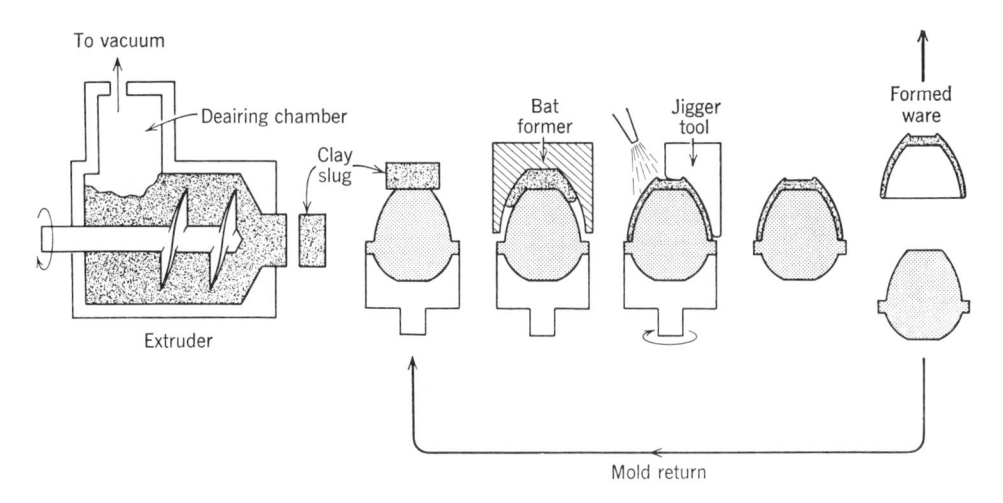

Figure 3. Jiggering operation.

or similar material is mixed as uniformly as possible with the raw materials. The binder also reduces friction with the die walls during pressing and helps to alleviate sticking of the formed part to the die and punches. Modern presses load the die, apply pressure, and eject the part from the die quickly and automatically. The simplified flow chart of Figure 4A illustrates the formation of refractory brick by dry pressing.

In dry pressing, force is applied from opposite directions; because of die wall friction and uneven filling of the die, pressure distribution throughout the formed part varies. The resulting density variations within the parts cause warpage during firing. Several remarkable methods have been developed to eliminate this problem. For example, in isostatic dry pressing the forming pressure is uniform in all directions. This is accomplished by placing the granulated material in a rubber mold and effectively immersing this mold in a liquid under high pressure. Automatic isostatic presses are widely used in making spark plug insulators. Vibratory forming has also been used to overcome the problem of nonuniform density and to permit formation of parts with a length to diameter ratio exceeding 100. Vibrations, used in this process instead of pressure, cause nonplastic particles of material to move into their closest packing configuration. Impact forming is similar to vibratory forming except that the forces are greater. Impacts can actually fragment the particles and produce the optimum particle size distribution for densest packing. Vibratory and impact forming have been used principally in packing ceramic fuel materials into elements for nuclear reactors.

Slip Casting. In slip casting, a slip (suspension of clay and other solid particles in water) is poured into a porous plaster of Paris mold. The interior surfaces of the mold conform to the exterior surface of the desired ware. As the plaster absorbs water from the slip, solid particles are deposited on the interior surface of the mold. The process may be continued until the walls of the ware meet the center, as in solid casting, or the slip may be drained from the mold when the walls reach the desired thickness, as in drain casting. Slips are prepared for casting in several ways. For example, filter cake may be blunged or raw materials may be ball-milled in water as illustrated in the simplified flow diagram, Figure 4B. A deflocculant, such as 0.02% of sodium silicate, is added to keep the solid particles in suspension in the water. Otherwise, the particles

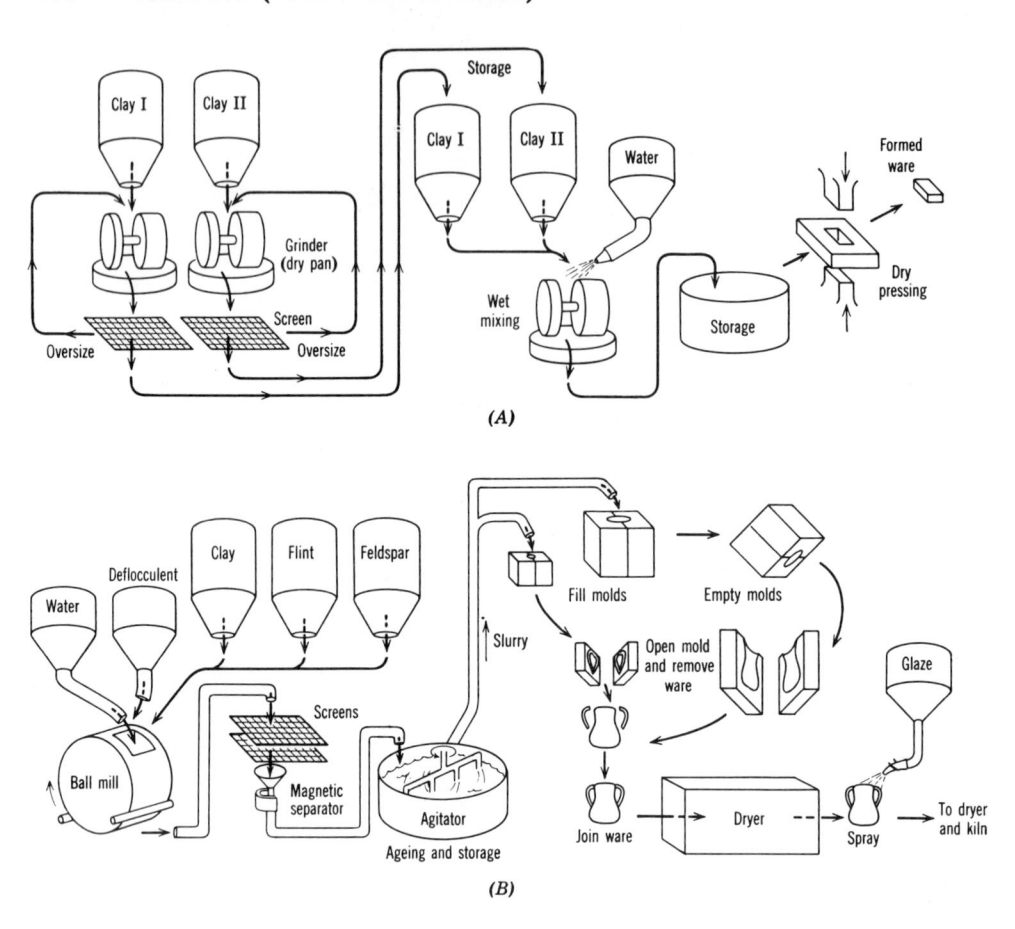

Figure 4. Simplified flow diagrams of forming processes: (*A*) dry pressing of refractories; (*B*) slip casting art ware.

would settle, forming thicker walls at the bottom of the mold. The mold is made in two or more parts to facilitate removal of the ware.

The principal advantage of slip casting is that it permits formation of complex shapes. It is widely used in the whiteware industries to make heavy plumbing ware, art objects, and dinnerware that is not adaptable to jiggering. Slip casting is used to a lesser extent in the refractories industries to make crucibles and other special shapes. Another advantage of slip casting is that the molds are relatively inexpensive and they are reusable. In some cases, pieces of ware are cast separately and joined, using slip as the adhesive, eg, handles for cups and vases. A coating of glaze material is sometimes applied by spraying or dipping before the ware is dried and fired. The glaze material melts during firing to form a glassy coating which improves the strength, cleanliness, and appearance of the ceramic ware. Special techniques can be used to form materials without plasticity, such as refractory oxides and carbides, by slip casting. Thin sheets of ceramics can be made by a casting technique called the doctor-blade process (6).

BIBLIOGRAPHY

"Ceramic Forming Processes" under "Ceramics" in *ECT* 2nd ed., Vol. 4, pp. 776–783, by Robert F. Stoops, North Carolina State of The University of North Carolina.

1. W. D. Kingery, "Pressure Forming of Ceramics" in W. D. Kingery, ed., *Ceramic Fabrication Processes*, The Massachusetts Institute of Technology Press, Cambridge, Mass., and John Wiley & Sons, Inc., New York, 1958, pp. 57–58.
2. J. T. Jones and M. F. Berard, *Ceramics: Industrial Processing and Testing*, The Iowa State University Press, Ames, Iowa, 1972, pp. 46–61.
3. R. W. Rice, "Hot Forming of Ceramics" in J. J. Burke, N. L. Reed, and V. Weiss, eds., *Ultrafine-Grain Ceramics*, Syracuse University Press, Syracuse, N.Y., 1970, pp. 203–250.
4. F. Forbes, M. Crossland, and G. Arthur, "Fabrication of Thin-Wall Alumina Capillary Tubes by Vapour Deposition" in *Proceedings of the British Ceramic Society: Fabrication Science 2*, British Ceramic Society, Stoke-on-Trent, Mar. 1969, pp. 231–243.
5. W. D. Kingery, *Introduction to Ceramics*, 1st ed., John Wiley & Sons, Inc., New York, 1960, p. 38.
6. J. C. Williams, "Doctor-Blade Process" in F. F. Y. Wang, ed., *Treatise on Materials Science and Technology*, Vol. 9, Academic Press, Inc., New York, 1976, pp. 173–198.

General References

W. D. Kingery, H. D. Brown, and D. R. Ullman, *Introduction to Ceramics*, 2nd ed., John Wiley & Sons, Inc., New York, 1976.
"Critical Compilation of Ceramic Forming Methods" in *Air Force Tech. Document Rpt. No. RTD-TDR-63-4069*, Air Force Materials Laboratory, Research and Technology Division, Air Force Systems Command, Wright-Patterson Air Force Base, Ohio, Jan. 1964.
R. Newcomb, Jr., *Ceramic Whitewares*, Pitman Publishing Corp., New York, 1947.
F. H. Norton, *Elements of Ceramics*, 2nd ed., Addison-Wesley Publishing Co., Reading, Mass., 1974.

ROBERT F. STOOPS
North Carolina State University

THERMAL TREATMENT

Thermal treatment is an essential step in the manufacturing of ceramic products. Materials that are stable at room temperature have to be raised to relatively high temperatures for reactions to take place. The temperatures range from 700°C for enamels (qv) to approximately 1650°C for alumina ceramics, and even higher for certain specialty products.

Methods of Thermal Treatment

Temperature and Time Factors. In all thermal treatment of ceramics, control is maintained by adjusting two independent variables: the temperature to which the material is heated and the time at temperature. In some cases the thermal treatment does not consist of a particular temperature held for some time period, but a smoothly varying time–temperature profile. Equivalent energy inputs can be often obtained by firing to a somewhat lower temperature for a longer time or to a higher temperature for a shorter time. For a large class of materials, particularly structural clay products, porcelains, sanitary ware, and other clay-containing products, expendable pyrometric cones are a useful method of determining firing temperature and as a means of quality control.

The degree of slumping of each cone corresponds to a definite temperature when heated at a specific rate and thus can be used to evaluate time and temperature factors. Thermocouples, radiation pyrometers, and optical pyrometers may also be used to sense the temperature (see Temperature measurement).

The microstructures and properties of many specialty ceramics produced using a higher temperature for a shorter time are not exactly equivalent to those fired longer at a lower temperature. Often these ceramics require a very narrow temperature range to develop specific properties: very precise monitoring of temperature and thermal gradients is required with a complex system of adjustable burners and electric heating elements or both to maintain the narrow temperature–time parameters. This is especially true in the manufacture of electronic and magnetic ceramics (see Amorphous magnetic materials; Glassy metals).

Extreme Temperatures. Fuel-fired kilns and furnaces (qv) are usually limited to maximum temperatures on the order of 1800°C unless oxygen enrichment is available. Electrical resistance heating in air using silicon carbide elements is normally limited to about 1500°C and with noble metals (Pt–Rh alloys) to about 1700°C. Modern laboratory furnaces using $MoSi_2$ resistance elements can be used to 1750°C in oxidizing atmospheres. In protective atmospheres (including high vacuum) the refractory metals can be used successfully up to temperatures at which vaporization becomes a limiting factor; for molybdenum the upper limit is 1800–2200°C, for tantalum it is 2500°C, and for tungsten it is 3000°C. Above 3000°C, graphite is about the only practical heating element; its maximum service limit is in the 3500–3600°C range. Still higher temperatures can be attained in plasma devices (up to 15,000°C), and have been used for dynamic thermal processing of ceramic particulates in such specialty operations as plasma flame spraying, spheroidizing, and crystal growth (see Furnaces, electric; Plasma technology).

Control of Furnace Atmosphere. By far the largest tonnages of ceramics are oxides, and a high percentage of them are processed in fuel-fired kilns and furnaces. They are directly or indirectly exposed and may be quite sensitive to the combustion atmosphere during firing. Most oxide ceramics are fired under nominally oxidizing conditions, but neutral or reducing atmospheres must be employed to develop or maintain lower valence states of oxides that may be critical to the stability of the principal phase present or tend to dominate the structure-sensitive properties, eg, color, electrical conductivity, or magnetic properties.

Many oxide ceramics have a volatile component which must be present in the fired ceramic in critical proportions for the attainment of the desired properties. Many magnetic materials contain lead and zinc, both subject to volatilization. Ceramics that are used as ionic conductors usually have a high content of easily volatilized alkali. These problems can be managed with suitable atmospheric control. A common method consists of providing an atmosphere rich in the mobile chemical species, either by using a bed of powder, which provides an enrichment source, or a refractory material containing the volatile component. In extreme cases the ceramics are encapsulated.

For nonoxide ceramics, controlled-atmosphere thermal processing is the rule rather than the exception. Nitrogen and the inert gases are frequently used to exclude oxygen. Cracked ammonia, hydrocarbon gases, hydrogen, carbon monoxide, and carbon provide reducing atmospheres. Thermal processing *in vacuo* has attained considerable importance for specialty oxide and nonoxide ceramics. Vacuum furnaces normally operate in the 133–0.133 mPa (10^{-3} to 10^{-6} torr) range and frequently utilize a refractory metal (W, Mo, Ta) or carbon as the heating element and for thermal (radiation) shielding.

Kilns. Perhaps the most prevalent equipment for thermal treatment is a tunnel kiln. The ware is placed on a flat car with a refractory top; the understructure of the car is sealed from the high temperature gases in the kiln. The car moves progressively along the length of the kiln through zones of increasing temperature. Gas or oil burners or resistance elements located along the middle section of the kiln provide the necessary temperature profile. The movement of the ware through progressively higher temperature zones provides a safe rate of increase to a uniform elevated temperature. The ware then moves through a section where there are no burners and fans provide cooling at a rate which permits removal without undue thermal shock at the kiln exit.

Production tunnel kilns normally vary in length from 23 to 152 m. Cross sections may be as small as 0.2 m wide by 0.3 m high to 3 m wide and 2.4 m high (Fig. 1).

Structural clay products, refractories, tile, porcelain, dinnerware, and electronic

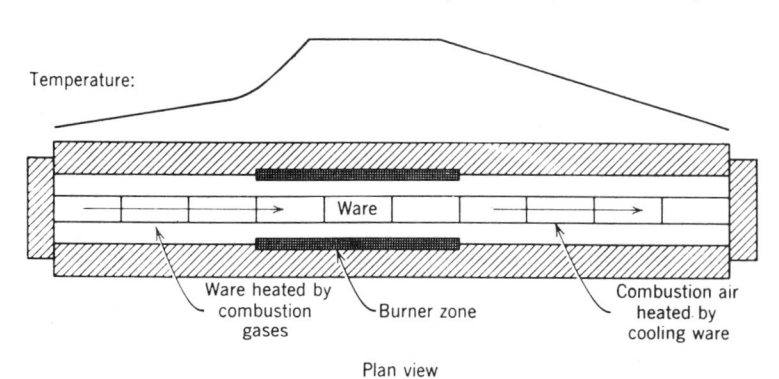

Figure 1. Diagram of a tunnel kiln (1).

262 CERAMICS (THERMAL TREATMENT)

components are normally processed through a tunnel kiln. In many cases a tunnel dryer and kiln are built as one continuous unit. Much thought and research is currently being directed toward energy conservation in tunnel kiln design and operation. Because of its clean burning characteristics, natural gas has been the fuel of choice in the past; however, oil and coal, and in some case wood products, such as sawdust, are being increasingly utilized. The principal areas of concern for energy conservation are: the utilization of waste heat in the cooling cycles for drying and preheating, insulation improvement, the improvement of heat transfer characteristics of the stack of ware to reduce thermal gradients, and the design of the time–temperature profile to speed the firing cycle while maintaining product quality (rate-controlled sintering) (2).

A periodic kiln is used for small-scale production, especially where highly specialized products require adjustments in kiln atmosphere and firing cycle. Periodic kilns cover size ranges required for missile nose cones to very small electronic parts. They can be fired by gas, oil, or electric heating elements. Periodic kilns are primarily used for research and development since they use more fuel and require more frequent refractory replacements than the continuous type.

A rotary kiln is used for the manufacture of products such as cement (qv) and calcined lime or dolomite (Fig. 2). This kiln is a long steel tube with a refractory lining. The material slowly moves from the entrance to the exit, counter to the direction of heat flow, under the influence of the slope and rotation of the kiln. A burner is located at the exit which fires into the kiln, with gases of combustion being exhausted at the entrance so that material progressing through the kiln passes through zones of increasing temperature. After going through the high temperature zone, the product, such as cement clinker, falls from the kiln for further processing. Some lightweight aggregate is made in rotary kilns; other manufacturers use a traveling-grate arrangement.

Arc Melting. A few materials, such as alumina, magnesia, zirconia, and mullite, can be processed by melting in a steel container with an arc struck between graphite electrodes. The pool of molten material which forms in the area of the electrodes also carries the current which is continued until all the material is liquid except for an insulating layer in contact with the container. After the molten mass is cooled, the resultant pig is broken, sorted, crushed, and sized. Many impurities are eliminated from the melt by evaporation. Some refractory shapes are cast from a molten state by the process of fusion casting which gives a refractory shape with much lower porosity than that obtained by tunnel kiln firing. Quite large sections (of the order of 500–600 kg)

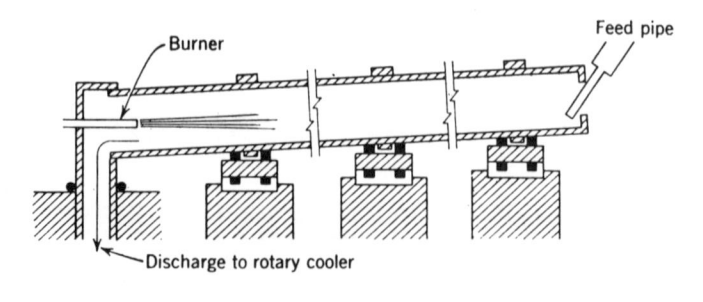

Cross section

Figure 2. Diagram of a rotary kiln (3).

can be fabricated by this method. Refractories for bottoms and sidewalls of glass tanks are made by this method to take advantage of the improved resistance to chemical attack associated with the very dense cast structures.

Hot Pressing. The hot pressing process incorporates forming and thermal treatment in one step. Finely divided particles are placed in a die, usually graphite, which is heated while pressure is being applied to the particles. This method is used to obtain a dense ceramic in chemical systems that cannot be fully densified using conventional firing methods. The use of pressure gives a dense, fine-grained product at a much lower temperature and reduces stringent processing requirements to eliminate flaws. However, the use of hot pressing is limited to very simple shapes and slow rates of production where expensive machining operations are to be performed on materials that are often extremely hard.

Physical and Chemical Changes During Thermal Treatment

The thermal treatment of a ceramic can be separated into two temperature ranges: 0–400°C, where the ware loses physically held water and organic binders which were essential for the forming operation; and higher temperatures (firing), where numerous physical and chemical changes take place to develop the desired properties.

The changes that take place during firing may be classified as dissociation, compound formation, polymorphic transformation, sintering, and vitrification. In each case energy must be supplied to break the attraction holding ions together in a stable structure to allow them to diffuse to more favorable sites, resulting in a product with a structural arrangement having a lower free energy.

Drying and Binder Removal. Because water is required for forming of the majority of common ceramics, the drying operation is an important stage of the thermal treatment, often the key to a successful ceramic process. The water content can be as high as 20 wt % and requires careful removal to produce a defect-free piece. Drying of large pieces too rapidly can cause cracking, and incomplete drying before firing can cause the piece to explode. This is also true of the rate of binder removal on which the integrity of the piece is highly dependent.

Generally, drying is accomplished at or below 100°C, and can take as long as 24 h for a large piece. Temperature–time schedules employed for the removal of organic binder are highly dependent on the chemistry of the binder employed. Although the bulk of organic binders is removed by 200–300°C, some hydrocarbon and carbon residues persist to much higher temperatures.

Dissociation, Compound Formation, and Polymorphic Transformation. Many raw materials lose their original identity after being subjected to a high temperature environment. For some minerals dissociation is the first change observed. For example, many carbonates are constituents of ceramic compositions; thus carbon dioxide may be lost during firing. The clay minerals as well as other hydrous silicates lose chemically combined water at elevated temperatures. At this point the clays have also lost crystallinity; hence, they are in a state where reactions with other materials will proceed more rapidly. Continued heating of clays will result in mullite, $3Al_2O_3 \cdot 2SiO_2$, a product with a lower free energy, with the excess silica changing to the cristobalite modification. However, this process is not one which takes place instantaneously, rather it depends on the rate of diffusion of certain ions. As a general rule, rates of reaction in silicate systems are relatively slow; hence, phase relationships developed at peak firing temperatures deviate extensively from equilibrium conditions unless long soaking times at the peak firing temperature are used to allow the reaction to proceed nearer to a state of equilibrium.

The formation of $MgAl_2O_4$ spinel from its component oxides is another example of compound formation. In this case the reaction takes place in the absence of a liquid phase; hence, volume diffusion through the bulk material determines the reaction rate. In many instances a reaction of this type follows a parabolic law, the thickness of the new phase varying as the square root of the time, since diffusing ions must traverse the increased volume of the reaction product. This reaction rate can be maximized by reduction of the particle size, yielding more contact surface and shorter diffusion distances. That particle size is of utmost concern in the reaction kinetics of ceramic systems is shown in Figure 3.

Some materials are capable of existing in more than one crystallographic arrangement. Silica and zirconia are examples of such materials. As the material is heated, a temperature is reached at which a new structural arrangement is more stable, the higher temperature form usually has higher symmetry. Cracking and fracture of a piece can take place as the material experiences this change. This is a problem in many quartz-bearing materials such as structural clay products and porcelain. Too rapid cooling can cause fracture. However, in the case of zirconium oxide, ZrO_2, additions of alkaline earths result in a cubic solid solution that has no polymorphic forms. This product is known as stabilized zirconia

These changes can often be followed by means of the technique of differential thermal analysis, dta. A small sample of the material is placed in a furnace where the temperature is increased at a steady rate, for example 10°C/min; a thermocouple in the sample, and one in the furnace, are connected, reversed, to a millivoltmeter, which then registers the difference in temperatures between the two. Figure 4 shows a dta of a North Carolina clay, containing illite, which is used for the manufacture of bricks and sewer pipe. The peaks and troughs indicate the magnitude of the various reactions.

Sintering. Often it is desired to fabricate a product from a single material such as aluminum oxide, Al_2O_3, uranium dioxide, UO_2, or titanium dioxide, TiO_2. Crystalline materials with complex chemistries can often be "prereacted" from constituent raw materials to form a single-phase powder material, which can then be fabricated by dry pressing.

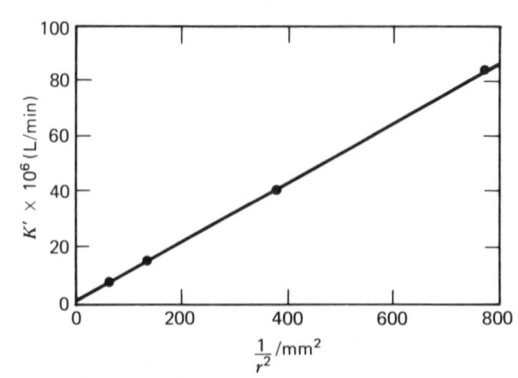

Figure 3. Influence of particle size on reaction rate (4); r is average radius of particles.

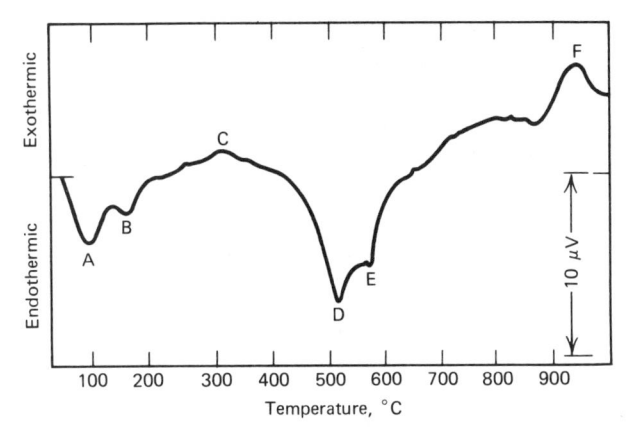

Figure 4. Dta of a triassic shale: A, loss of mechanically held water; B, loss of chemically held water; C, oxidation of organics; D, loss of chemically held water; E, α–β quartz inversion; and F, γ-Al$_2$O$_3$ and mullite formation. Heating rate 10°C/min.

The fabrication of a product from a single phase requiring no intermediate reaction or new phase formation is called sintering. In this case the thermal treatment results in a transformation from a porous compact, whose constituent particles are usually held together by an organic binder, to a strong, dense, coherent product. Although the term sintering originated strictly with the transformation of a single-phase particle compact to a polycrystalline ceramic, a broader meaning implying any densification mechanism is becoming prevalent.

The required reduction of porosity occurs by diffusion of material into the neck region between particles. For diffusional processes to bring about densification, the centers of the particles must move toward each other. The onset of this process is termed the first stage of sintering and is described by equations relating the diffusional characteristics of the individual particles or crystals to the shrinkage or densification rate. This stage often occurs 500–700°C below the highest temperature. The second or intermediate stage covers the bulk of the porosity reduction from about 35 to 5–10%. The rate of sintering is highly dependent on the diffusional characteristics and thus the temperature, the grain size, and the pore sizes and shapes. At some point the pore network becomes sealed off, so that the porosity is said to be closed. During this final stage of sintering, trapped gases must be removed by diffusion through the crystals themselves, rather than out the open channels present in the intermediate stage. In air firings, N_2 is usually a particularly slowly diffusing gas species. Sometimes a H_2 or O_2 atmosphere is employed to provide a rapidly diffusing species during the critical final stage.

The principal concern during final stage sintering is the development of the optimum microstructure by the avoidance of rapid grain growth and the elimination of porosity.

Vitrification. A glass phase is present in appreciable quantities in such products as porcelain, dinnerware, structural clay products, and some electronic components. This glass results either from eutectic melting, eg, with silica and an alkali, or from the melting of one component such as feldspar. In either case the glass serves as a reaction medium through which diffusion can take place at a lower temperature than in a solid. The liquid may react with more refractory particles, taking them into solution. In many cases a crystalline phase may precipitate from a saturated solution. This liquid tends to fill void spaces between particles, resulting in an impervious product. This process is referred to as vitrification. In some compositions, a large quantity of liquid is formed over a small range of temperatures near the eutectic composition. In this case care must be exercised in firing to keep the ware from slumping. However, along with the quantity of liquid phase developed, the viscosity of the liquid phase must be considered. A relatively small quantity of liquid with a low viscosity can be more sensitive to overfiring than a large quantity of a higher viscosity.

BIBLIOGRAPHY

"Kilns" under "Ceramic Industries" in *ECT* 1st ed., Vol. 3, pp. 586–590, by F. P. Hall, Pass and Seymour, Inc., and Onondaga Pottery Company; "Thermal Treatment of Ceramics" under "Ceramics" in *ECT* 2nd ed., Vol. 4, pp. 783–792, by W. C. Hackler, North Carolina State of The University of North Carolina.

1. W. D. Kingery, *Introduction to Ceramics*, 1st ed., John Wiley & Sons, Inc., New York, 1960, p. 60.
2. U.S. Pat. 3,900,542 (Aug. 19, 1975), H. Palmour, III, and M. L. Huckabee (to A. D. Little and Company).
3. Ref. 1, p. 61.
4. Ref. 1, p. 339.

General References

W. D. Kingery, H. D. Brown, and D. R. Ullman, *Introduction to Ceramics*, 2nd ed., John Wiley & Sons, Inc., New York, 1976.
Materials Advisory Board, Division of Engineering, National Research Council, *Ceramic Processing*, Publication 1576, National Academy of Science, Washington, D.C., 1968.
H. Salmang, *Ceramics—Physical and Chemical Fundamentals*, Butterworth and Co. Ltd., London, Eng., 1961.

THOMAS M. HARE
North Carolina State University

PROPERTIES AND APPLICATIONS

Composition and Microstructure

In terms of useful properties and typical applications, there is great diversity among ceramic substances, and indeed between differently processed ceramic items of the same substance. These important differences in materials can, in principle, be described in terms of just two overall parameters, the composition and the microstructure. By composition is meant the chemical and mineralogical makeup of the total substance of the material, including its deliberate and accidental impurities. Microstructure means broadly the crystallographic structure of each component or phase, and in particular, the size, shape, and distribution of each of the phases present, including the unoccupied space, or porosity.

These determinative parameters are fixed by the combined response of the starting materials to all the processing steps. Consequently, microstructural characterization is an important factor in controlling the processing of ceramics. In many applications ceramics are used at service temperatures far below their initial maturing or processing temperature, so that little or no change in composition and microstructure can occur. Thus these two parameters, once fixed during processing, also establish the useful properties of the ceramic product. In service at temperatures near or above the maturing temperature, the parameters are likely to change with temperature, time, environment, etc, so that the useful properties are likely to be transient rather than static.

A ceramic material is a solid and it may be either crystalline or vitreous or of mixed type. It is crystalline if its ions or atoms are in orderly array, row on row, layer upon layer, over long distances compared to one interatomic spacing. Crystalline ceramics include many single oxides, carbides, nitrides, borides, sulfides, etc, as well as their binary and ternary compounds. A material is vitreous if its ions are arranged with some regularity at close range but do not display long-range order. Glass (qv) is the classic vitreous substance. Technically, it can be characterized as an extremely viscous, metastable, supercooled liquid but in practice it behaves in most respects as a solid at room temperature.

A ceramic material may be composed entirely of a single substance and would be described as monophase but it is most likely to be polyphase, containing two or more discrete substances. Monophase crystalline ceramics may occur as monocrystals, having a unique crystallographic orientation throughout, eg, sapphire is the monocrystal form of α-Al_2O_3. Characteristically, most ceramics are polycrystalline masses, with abrupt changes in orientation or composition occurring across each grain or phase boundary. Polyphase materials must, out of necessity, be polycrystalline, and in such materials there may be wide variations in the character of internal boundaries depending on the orientation and composition of the neighboring grain or phase. Boundaries constitute fairly abrupt discontinuities in local atomic arrangements and bonding character, and in properties. Such boundary regions are thermodynamically less stable than the more orderly crystalline material(s) adjoining them.

Another discontinuity exists at the free surface of the ceramic, the final boundary between the ceramic solid and its environment. Many properties of ceramics are surface-sensitive, both with respect to mechanical condition and chemical and atmospheric environment, or both.

In practice, compositional and microstructural parameters are fairly difficult to determine with precision in typically complex ceramic systems. In selected, simple monophase systems, considerable progress has been made in recent years in systematically relating microstructure to processing on the one hand, and in characterizing variations in properties in terms of microstructure on the other. Regardless of whether or not such relationships are amenable to precise measurement, every property of a ceramic is dependent to some degree upon the relative amounts of the phases present, their respective compositions and structures, the spatial disposition of the phases, and the shapes, sizes, and orientations of the grains and pores.

The mechanical behavior of ceramics offers one example of this problem. Whereas it is possible to conceive of the calculation of stress and strain at a particular point within a particular grain of a porous, polyphase polycrystalline ceramic under a particular state of applied stress and strain, it would be almost impossibly difficult to carry out, even if all the boundary conditions were mathematically resolvable. On the other hand, flow or fracture must be related to specific details of stress and strain, in specific directions, on specific planes, within a particular atomic structure if fundamental understanding of the process is to be achieved; similar considerations apply in the case of optical, thermal, and electromagnetic properties. Consequently, there is an important divergence between those properties that must be determined in single crystals so that orientation can be selectively controlled and specifically related to certain discrete atomistic processes, and those properties of polyphase or polycrystalline materials that must of necessity be measured and interpreted in a statistical manner. Experiments of both types are usually required to characterize adequately a particular kind of behavior in ceramic materials.

Many of the elastic, thermal, and optical properties of materials may be quite adequately described by tensor quantities predicated only on an ideal model of the crystal structure. But many of the properties of ceramic materials cannot be so described; mechanical and electromagnetic properties in particular, can be shown to be extremely sensitive to defects existing in the structure. Such properties are termed structure-sensitive; some of the commonly encountered defects in ceramics which account for them are summarized in Table 1.

Chemical Properties of Ceramic Materials

A high degree of chemical stability is characteristic of the oxides, carbides, nitrides, borides, etc, which form the basis for all ceramic materials. In particular, many of the oxides are extremely stable over wide ranges of temperature and environmental conditions, and are resistant to further oxidation and to reduction to suboxides or metals. Some nonoxide ceramics are very stable at extreme temperatures (>2000°C) *in vacuo,* or in neutral or reducing atmospheres, but in general they will not tolerate long-term high temperature exposure to oxidizing environments. The stability of these ceramic compounds is the result of the compactness of the crystal structure, the directed chemical bonding (generally ionic or covalent, or of mixed ionic–covalent character), and of the high field strengths associated with the relatively small, highly charged cations encountered in refractory ceramics. Thermodynamically, the most stable ce-

Table 1. Types of Defects in Crystalline Ceramics

Defect	Cause	Relative scale
impurity	foreign atom or ion introduced by substitution or interstitially; vacancies may be created to balance valence charges	atomic in three dimensions
vacancy	missing positive or negative ion, or atom	atomic in three dimensions
vacancy pair	missing positive ion coupled with missing negative ion	atomic in three dimensions
vacancy cluster	aggregation of vacancies	microscopic (large clusters can be resolved optically)
color center	anion vacancy plus electron or cation vacancy plus electron hole	atomic in two dimensions, microscopic in length
dislocation	structural misfit, linear character	atomic in two dimensions, microscopic in length
subgrain boundary	structural misfit (at small angles), surface character	atomic in one dimension, microscopic in two dimensions
grain boundary	structural misfit (at large angles), surface character	atomic to microscopic in one dimension, microscopic to macroscopic in two dimensions
phase boundary	compositional or crystallographic discontinuity, surface character	atomic to microscopic in one dimension, microscopic to macroscopic in two dimensions

ramic will be that having the greatest negative free energy of formation from the elements; Kingery (1) points out that of the oxides yttria and thoria are the most stable in this respect. Others widely used for their chemical stability are alumina, beryllia, magnesia, and stabilized zirconia (see Refractories).

Although there are exceptions (SnO_2, ZnO, MgO, etc), most refractory ceramics have very low vapor pressures in the most stable valence state, but the suboxides tend to be volatile (SiO, Al_2O, etc). Limiting conditions for more complex oxides are often established by the volatility of a particular component, such as Na in sodium aluminate, etc.

Interactions between dissimilar ceramics, or between ceramics and metals, becomes increasingly likely as operating temperature is increased, or at low pressure. Heating *in vacuo* favors reaction by facilitating removal of vapor phase products, eg:

$$x\,W + ThO_2 \xrightarrow{\Delta} ThO_{2-2x} + x\,WO_2 \uparrow \quad (>2000°C) \qquad (1)$$

$$3\,C + 2\,Al_2O_3 \xrightarrow{\Delta} 4\,Al \uparrow + 3\,CO_2 \uparrow \quad (>1950°C) \qquad (2)$$

Even a system of materials with very high melting points can form fluid liquids at temperatures much below either melting point. For example, Al_2O_3 melts at 2050°C and CaO at 2500°C, but when in contact, they can form a reactive, eutectic liquid at 1450°C. The presence of a few percent impurities can significantly lower the temperature of liquid formation. Johnson (2), Economos and Kingery (3), and Kingery (1) have discussed the stability of oxide refractories in contact with other materials at high temperatures.

Optical Properties

Many ceramic substances, particularly the oxides, are optically transparent in single crystal or vitreous forms; eg, the transparency of glass is perhaps its most useful and most characteristic feature. Optical transmission occurs in dielectric materials such as ceramics through interactions of the oncoming electromagnetic radiation (photons) with the polarized (or polarizable) electron shells surrounding the nuclei of the constituent ions or atoms. Optical properties are closely linked to composition and structure of the ceramic since the degree of polarization is a function of ion size, bounding energy, and crystallographic direction. The index of refraction is a sensitive quantitative measure of these materials parameters.

A material is often characterized by its transparency in certain wavelength regions (% transmission/mm thickness). Many ceramics and most glasses absorb highly in the infrared and ultraviolet regions of the spectrum. MgF_2 and CaF_2 are used for uv transmission. Other specialty materials, such as CsI and KI, are transparent at very long infrared wavelengths.

Isotropic crystals and vitreous glass display isotropic optical properties, but other crystals are optically anisotropic, the index of refraction being highest when light rays traverse the close-packed crystallographic directions. Dispersion is the term employed to describe the variation of index of refraction with change in wavelength. It results from resonance effects between incoming photons in the visible range and the typical ultraviolet frequencies of oscillating electronic states associated with the constituent ions.

Large ions in crystals or in glass compositions are more readily polarized than small ones and produce a high index of refraction. Most glasses are in the range 1.4–2.0, metals are usually >3.0. The index of refraction of some ceramic materials is shown in Table 2.

Optical absorption measures the loss of intensity as light traverses a translucent medium. In addition to the losses attributable to surface reflection and dispersed-phase scattering, a principal cause of absorption is the presence of unfilled electronic energy bands within the material. One source of such donor or acceptor sites is the color center, which can be produced in normally colorless ionic materials by nonstoichiometric cation–anion ratios. Such defects (F-centers, V-centers) are strongly absorbing at some given wavelength (F-centers, usually in or toward the ir, V-centers in the uv), and so characteristic of the particular defects that optical absorption spectra are used to identify them.

Principal and very practical sources of unfilled energy bands in ceramic materials are impurities, or deliberate colorant oxide additives, which can be selected from any one of a number of transition metal oxides having unfilled $3d$, $4f$, or $5f$ electron shells.

Table 2. Refractive Index of Some Ceramic Crystal Phases[a]

MgO	1.74	TiO_2	2.71
Al_2O_3	1.76	SiC	2.68
$MgAl_2O_4$	1.72	Y_2O_3	1.92
SiO_2 (quartz)	1.55	$BaTiO_3$	2.40
$ZrSiO_4$	1.95		
mullite	1.64	LiF	1.392

[a] Ref. 4.

The 3d colorants are especially well known; eg, cobalt blue, chrome green, chrome–alumina pink, ferric iron brick-red, and vanadium yellow are familiar and traditional ceramic colors. The energy band structure responsible for optical effects is quite sensitive to the crystallographic environment of the colorant ion, to the valence state, and a number of other factors. A ceramic (ruby) single crystal, α-Al_2O_3, containing deliberate Cr^{3+} impurities, was the first to be employed as a laser host crystal (other hosts include $CaWO_3$, CaF_2, YlG, YAG, etc) and glasses doped with rare earth and actinide elements are now used in laser devices (see Colorants for ceramics; Lasers).

The microstructure of a polycrystalline ceramic tends to dominate the useful optical properties of the base material. Pores within grains or between boundaries scatter and reflect light and can induce translucency or even opacity in an otherwise transparent medium. Figure 1 shows the decrease in percent transmission of alumina as a function of porosity. Transmission is also reduced with finer pore size, until the pore size approaches the wavelength of light. Materials with extremely small pores <100 nm can be made transparent even at high levels of porosity.

Polyphase polycrystalline materials may develop additional scattering owing to differences in refraction across boundaries. Dispersed phases in particular act as opacifiers, with the degree of opacity depending upon concentration, particle size and

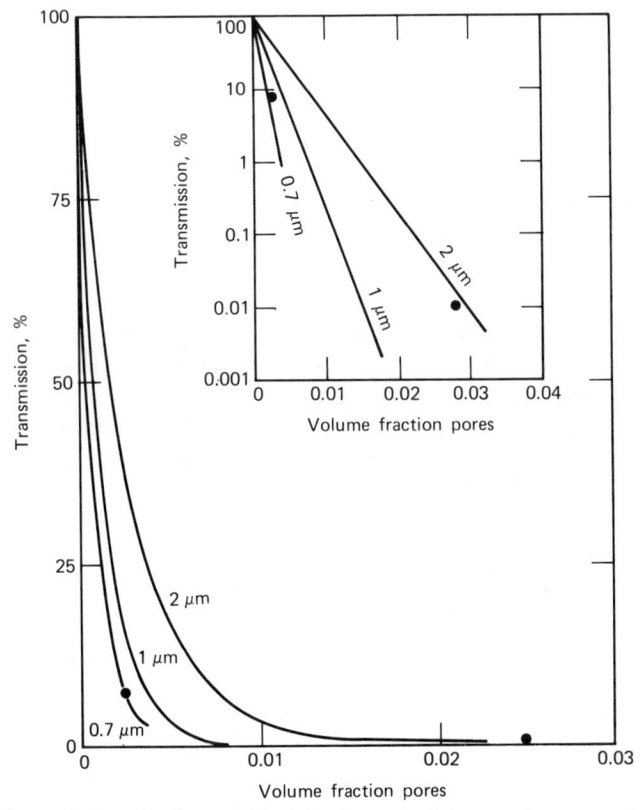

Figure 1. Transmission of polycrystalline alumina containing small amount of residual porosity (equivalent thickness 0.5 mm) (5).

difference in index of refraction of the dispersant phase. Dispersions of SnO_2, $ZrSiO_4$, TiO_2, and/or Sb_2O_3 are commercially employed as opacifiers and whiteners in enamels (qv), glazes, and glasses. They may be inert additions, products of reactions, or nucleated and crystallized from a glass. Dispersed phases are also employed to produce color effects, where the color-absorbing energy bands are not associated directly with the host ceramic matrix but are contained instead in a refractory additive phase called a color stain. In stains, the colorant transition metal oxides have been bound up in a stable crystal structure (the spinel structure is frequently employed) by prior high temperature reaction (calcining), so that it undergoes relatively little change during subsequent thermal processing of the ceramic product. A wide range of calcined, milled color stains for various uses are commercially available.

Just as color and opacity are highly desired in many ceramic applications, the absence of these properties is critically important in others. Highly translucent, almost transparent, pore-free polycrystalline ceramics produced by sintering or the hot-pressing method require careful control of impurities which could jeopardize the desired color-free transparency.

Because of total internal reflection, transparent fibers can transmit light around corners. A bundle of such fibers and its outer sheath can greatly minimize transmission losses over long distances. Proper control of impurities can minimize internal absorption. Other optical devices utilizing important ceramic components are optical wave guides, and electrooptic and acoustioptic materials (4) (see also Fiber optics).

Thermal Properties

The thermal properties of ceramics illustrate the subtle interplay between structure and properties, as well as between properties and applications, which determines the suitability of a particular ceramic material for a given job. The thermal properties of greatest interest include specific heat, thermal conductivity, and thermal expansion. Each is important since ceramics are frequently employed, and are always produced, at some high temperature. In such cases it is useful, or even essential, to know and provide for the quantity of heat stored and/or transferred by the material, and the dimensional changes it undergoes as a result of the added thermal energy. Furthermore, materials with high thermal conductivity and low thermal expansion are most likely to survive severe thermal shock (4).

Thermal energy within a solid takes the form of increased motions of its fundamental particles (electrons, ions, molecules) in a variety of rotational and vibrational modes. The availability of the electronic conduction mode in any oxide materials markedly improves thermal conductivity, eg, in SiC. However, for most ceramics having insulating rather than semiconducting properties, the most important mode involves lattice vibrations; they may be directly correlated with the thermal properties of interest in ceramics. The thermal energy consists of vibrations that are neither random nor independent but take the form of strain waves which move through the lattice with the speed of sound. These strain waves are called phonons. Phonons have quantized energy states and are restricted to the natural vibrational frequencies of the solid, and hence are governed in part by the lattice configuration and in part by the external dimensions of the solid. In essence a phonon is a transient lattice defect and in fact may interact with or be scattered by impurities, grain boundaries, and other defects including other phonons. To avoid such problems in theoretical considerations of the role of phonons in some thermal properties, eg, specific heat, it is convenient to assume an ideal crystal free of defects in which phonons do not interfere with one another.

Specific Heat. At high temperatures, the average vibrational energy of a particle in a solid is $3/2\,kT$ (k = Boltzmann constant), reflecting the three principal orthogonal vibrational modes, and the specific heat of the solid at high temperature becomes a constant, $3\,Nk$ = 24.9 J/(mol·K) [5.96 cal/(mol·°C)] (N = Avogadro's no.). At some lower temperature, however, the vibrational energy decreases and ultimately goes to zero at 0 K. Classical physical concepts are not able to account for this behavior, but the spectral (frequency) distributions of phonons in solids were utilized by Debye in 1912 in arriving at a quantum approach that does correctly predict the variation in thermal properties. His treatment defines a characteristic temperature θ at which the phonons attain their maximum frequency and shortest mean free path in the solid as shown below:

$$\theta = h\nu_{\max}/k \qquad (3)$$

where h = Planck's constant and ν = frequency of phonons. Below θ, the specific heat decreases with decreasing temperatures since only lower frequencies of phonons with longer mean free path are permitted; heat is being conducted through rather than stored in the lattice. Above θ, the three principal modes of lattice vibration are in full effect, and the specific heat remains constant. Kingery (4) notes that most ceramic oxides and carbides have Debye temperatures on the order of 1000°C. At higher temperatures, the thermal energy absorbed in processes other than lattice vibration (disordering, creation of defects, electronic effects) causes the measured heat capacity to increase at a modest rate rather than remain constant at $3\,Nk$.

Heat capacity curves for some ceramic materials are shown in Figure 2. Kingery also points out that although the molar heat capacity is essentially a constant, unaffected by the microstructure of the ceramic, the volume heat capacity frequently reported is a function of the porosity in the structure. Much less heat is required to raise the temperature of a given volume of a porous material than of a dense one, and much less must be removed during cooling. Minimum density insulation is therefore much

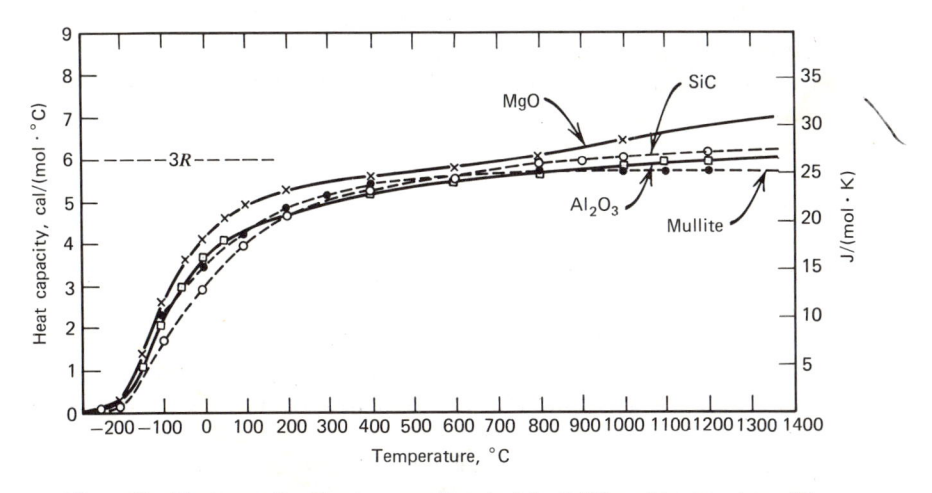

Figure 2. Heat capacity of some ceramic materials at different temperatures (6).

used in the construction of kilns and furnaces which are to be operated in periodic fashion. The selection of minimum density insulation results in energy savings in these kilns.

Modern ZrO_2 fiber insulations for laboratory furnaces have a fractional density of very low thermal conductivity, which allows very light construction and extremely fast heating and cooling rates (10–15 min to 1600°C).

Thermal Conductivity. The mechanism by which thermal energy is transported from a region of higher temperature to one of lower temperature involves a transport of phonons and electrons, or both (7). Electron conduction is of importance in metals, and in semiconductors over some temperature ranges, but in insulators it is phonon transport that is responsible for thermal conductivity:

$$k = pC_p lc \qquad (4)$$

where k = coefficient of thermal conductivity, p = density of solid, C_p = heat capacity, l = mean free path of phonons, and c = velocity of propagation of phonons. The mean free path l reflects the effects of scattering of phonons from permanent structural defects l_s as well as from phonon–phonon scattering l_t as shown below:

$$\frac{1}{l} = \frac{1}{l_s} + \frac{1}{l_t} \qquad (5)$$

At high temperature $1/l_s$ is essentially a constant, but below 100 K it decreases with decreasing temperature. The number of structural defects decreases with decreasing temperature, therefore the mean path between structural defects becomes larger as the temperature is lowered. At about 20 K, the mean path for the structural scattering factor is quite sensitive to defects and microstructure, being lower in single crystals than in polycrystalline ceramics, etc.

The thermal scattering effect, $1/l_t$, is almost proportional to the absolute temperature, increasing as the vibrational frequency increases and the mean path between phonon–phonon collisions shortens. The combined effects of both types of scattering as a function of temperature of α-Al_2O_3 are shown in Figure 3. The curves in Figure 4 illustrate the wide range of thermal conductivities available in ceramic materials.

The low conductivity of porous refractory ceramics combined with resistance to chemical change at very high temperature make them a valuable insulating material for high temperature applications. It should be noted that an increase in heat flow is indicated at temperatures above about 1500°C. This additive effect is attributable to radiation transfer processes.

Thermal conductivity is a tensor property, strongly dependent upon the crystallographic orientation and bonding character of the solid (7). Kingery (4) discusses the decreased thermal conductivity of binary oxide compounds and solid solutions in comparison to their respective end-member oxides. This behavior is also attributable to an increased likelihood of structural scattering of phonons in the more complex structures that result.

Thermal Expansion. It has been noted earlier that, as the temperature of a solid is increased, vibrations about each lattice point (ion, atom, or molecule) are induced, creating elastic waves or phonons. But the oscillations about the equilibrium lattice point encounter different energy fields in the compressive and expansive half cycles. The differences occur because repulsion forces increase much more rapidly with variations in interatomic distance than do attraction forces. This produces a shift

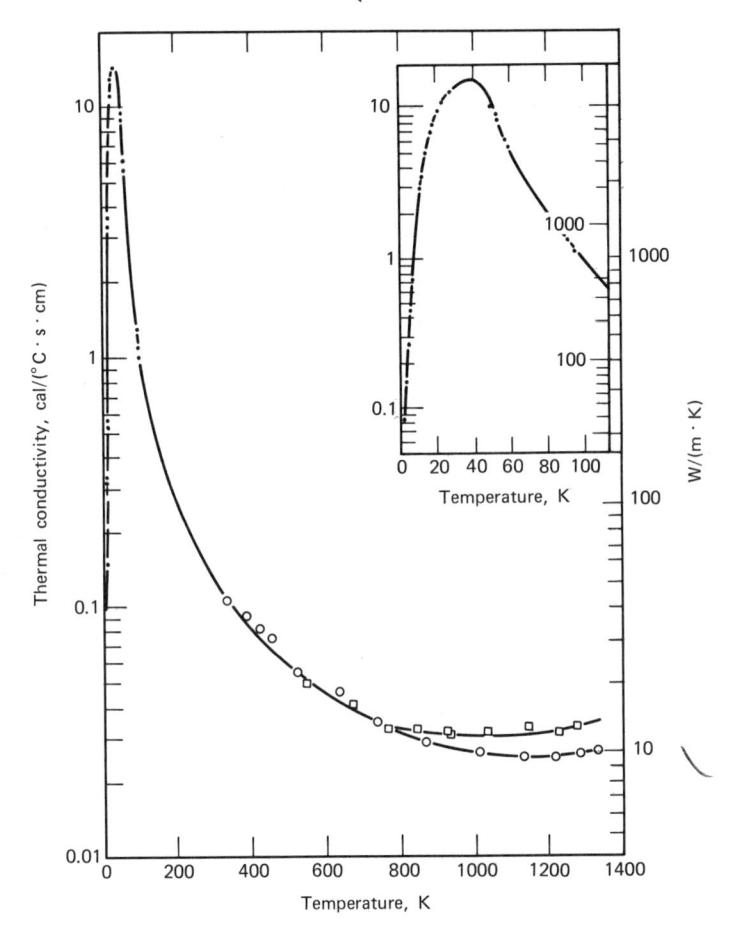

Figure 3. Thermal conductivity of single crystal aluminum oxide (8). ●, Berman (1951); ○, Lee and Kingery (1960) Pt foil interface; □, Lee and Kingery (1960) graphite interface.

toward greater separations of the equilibrium lattice sites as the temperature increases, ie, the crystal expands with increasing temperature (see Table 3).

The expansion for oxide structures with dense packing of oxygen ions is in the range of 6×10^{-6} to $8 \times 10^{-6}/°C$ at room temperature and increases to 10×10^{-6} to $15 \times 10^{-6}/°C$ at high temperatures (4). More open structures can have a much lower expansion in some crystallographic direction. The transverse vibrations of ions provide a vibrational energy component which compacts the structure. In some highly anisotropic materials, the thermal expansion coefficient in one direction may be negative, so that the average value of thermal expansion for a randomized polycrystalline material may be very low. This accounts for the excellent thermal shock resistance of ceramics such as cordierite and the lithium alumino–silicates much used as insulators and coil forms in rapid-heating electrical appliances such as soldering irons and radiant heaters.

In an extreme case, these very low or negative coefficients along certain crystallographic axes can induce such severe grain boundary stresses that the body is subject to internal ruptures during cooling from the maturing temperature. On heating, the cracks recombine; this phenomenon has been analyzed in detail for alumino–titanate

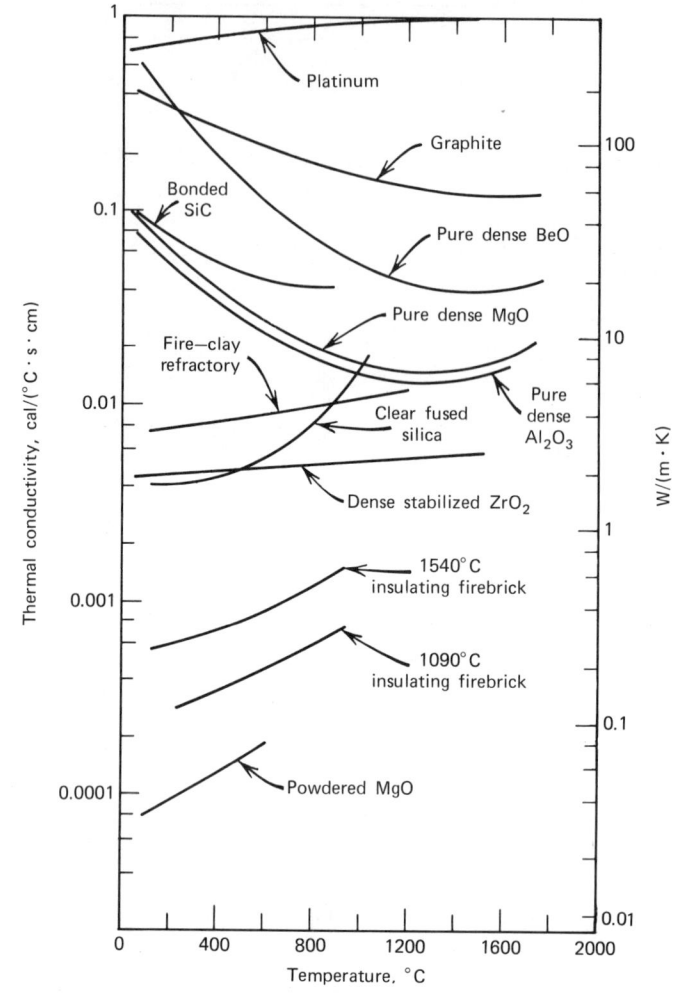

Figure 4. Thermal conductivity as a function of temperature for a variety of ceramic materials (9).

ceramics by Buessem (10). It accounts for the abnormally low strength of such materials at room temperature, and for the typical increases in strength that occur as the temperature is raised. Graphite, some borides, and other highly anisotropic materials also show this anamalous strength–temperature relationship.

Thermal Shock Resistance. The principal properties improving thermal shock resistance are (1) low thermal expansion, (2) high thermal conductivity, and (3) high strength. Ceramic materials in general do not have particularly good thermal shock resistance because of their brittle nature, but are used in many applications where ductile materials could not be used because of other superior high temperature properties. Thermal shock resistance becomes an important property in any application involving rapid changes in temperature or large thermal gradients. Applications such as cookware, spark plug insulators, abrasive wheels, refractory linings, etc, all demand some measure of shock resistance.

Thermal shock resistance is usually measured by quenching from a higher tem-

Table 3. Mean Thermal Expansion Coefficients for Various Ceramic Materials, Range of 0–1000°C

Material	Linear expansion coefficient, $(°C \times 10^6)^{-1}$	Material	Linear expansion coefficient, $(°C \times 10^6)^{-1}$
Al_2O_3	8.8	fused silica glass	0.5
BeO	9.0	soda–lime–silica glass	9.0
MgO	13.5	TiC	7.4
mullite	5.3	porcelain	6.0
spinel	7.6	fire-clay refractory	5.5
ThO_2	9.2	TiC cermet	9.0
zircon	4.2	B_4C	4.5
SiC	4.7	UO_2	10.0
ZrO_2 (stabilized)	10.0	Y_2O_3	9.3

perature and by measuring the strength degradation (as compared to room temperature). Figure 5 shows the effect on strength of quench temperature for aluminum oxide of various grain sizes. Alumina can withstand a quench from 200°C with no effect on strength. From above 200°C, microstructural damage takes place, with a more dramatic effect at finer grain size.

Materials such as fused silica, cordierite, and lithium aluminosilicates have very low thermal expansion and good thermal shock resistance despite being weak when compared to other ceramic materials. Improperly processed low thermal expansion materials such as cordierite can be very weak (owing to thermal expansion anisotropy of individual crystals) and thus have poor thermal shock resistance.

BeO, despite a relatively high thermal expansion, has excellent thermal conductivity and high strength, and thus has good thermal shock resistance. Many nonoxide ceramics also possess high thermal conductivity and can have very high strength. SiC combines high strength with a relatively low thermal expansion and high thermal conductivity. Al_2O_3 is also a good shock resistant material because of its high strength.

A thorough understanding of shock resistance requires a knowledge of microstructural, thermal, and elastic properties, and is discussed in detail by Kingery (4) and Hasselman (12).

Many ceramics contain phases (such as SiO_2 or ZrO_2) that undergo phase transformations accompanied by a volume change. Large particles of quartz, for instance, impart very poor resistance to porcelains in the neighborhood of 573°C. This gives these materials very poor shock resistance. The effect of the transformation can be reduced by additives or with the use of very fine particles.

Elastic Properties

The elastic properties of solids arise from the interaction of mechanical distortion, called strain, with the periodicities of the atomic or ionic structure. Elastic moduli measure the force involved in achieving recoverable, small, unit displacements of atoms or ions from their equilibrium positions. The elastic moduli are consequently strongly dependent upon the type of bonding and upon the electron density configuration of

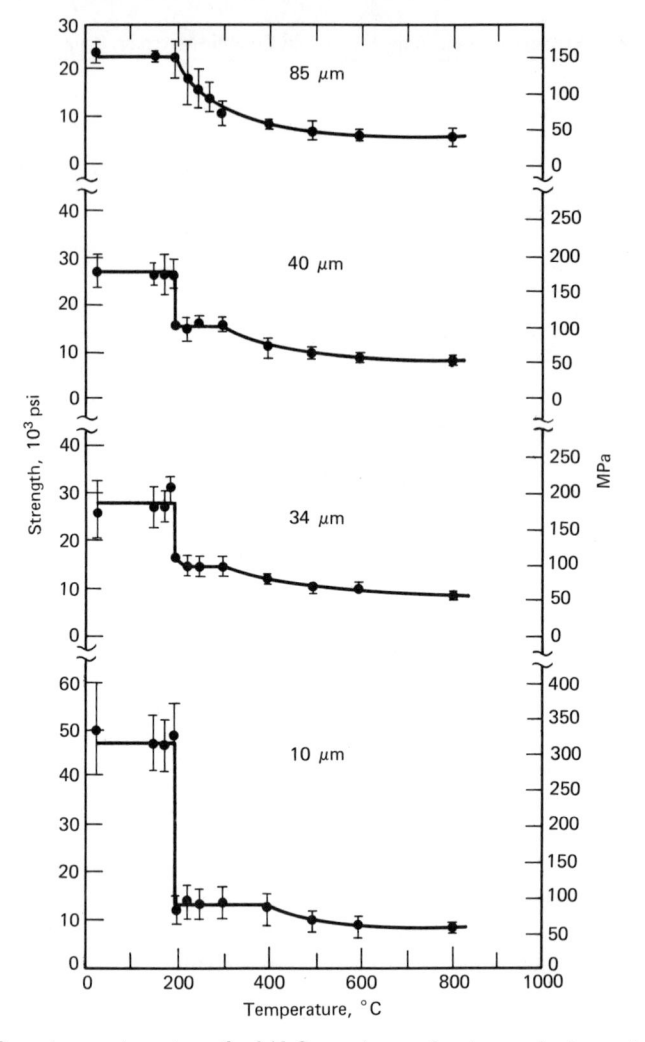

Figure 5. Room temperature strength of Al$_2$O$_3$ specimens of various grain sizes as function of quench temperature (11).

the crystal, just as is the melting point of the materials. This relationship is illustrated in Figure 6.

As in the case of optical (photon) waves and thermally induced strain waves (phonons), the elastic properties are directional, and can be described as tensors with respect to an ideal crystal (7):

$$\sigma_{ij} = C_{ijkl}\epsilon_{kl} \tag{6}$$

$$\epsilon_{ij} = S_{ijkl}\sigma_{kl} \tag{7}$$

where C_{ijkl} = elastic rigidity tensor and S_{ijkl} = elastic compliance tensor.

For isotropic substances such as glass, and for randomized polycrystalline materials having pseudoisotropic behavior, the elastic properties can be expressed adequately in terms of the following three quantities:

$$E = \text{Young's modulus of elasticity} = \frac{\sigma}{\epsilon} \tag{8}$$

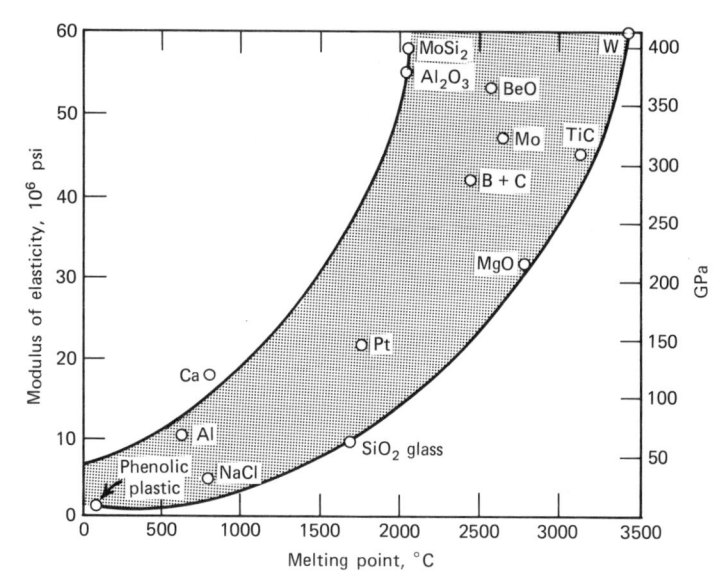

Figure 6. Relationship between elastic modulus and melting point for different materials (4).

$$G = \text{Shear modulus} = \frac{\tau}{\gamma} \quad (9)$$

$$\mu = \text{Poisson's ratio} = \frac{E}{2G} - 1 \quad (10)$$

where σ = normal stress, τ = shear stress, ϵ = normal strain, γ = shear strain. They are strictly valid only in the elastic (ie, recoverable) range of deformation and only when the material is homogeneous and isotropic. Most materials, and certainly most ceramics, are both inhomogeneous and anisotropic, therefore more complete descriptions of elastic behavior are usually required.

The temperature dependence on the elastic moduli for ceramic materials depends in large degree upon the temperature range and upon the types of structural defects present in the material. For monocrystals free from mobile defects within the given temperature span, the temperature dependence over most of the range is almost linear, with the moduli decreasing with increasing temperature. For sapphire (α-Al$_2$O$_3$), Wachtman and co-workers, (13) reported the following over the range 77–850 K:

$$E = E_o - bTe^{-T_o/T} \quad (11)$$

where E = Young's modulus of elasticity; T = temperature, K; and E_o, T_o, b = empirical constants for the curve shown in Figure 7. For polycrystalline ceramics, the temperature dependence of elastic moduli is more complex; this is particularly so over the temperature range where inelastic effects occur, as illustrated in Figure 8. In polyphase polycrystalline ceramics, for cases where Poisson's ratio is equal for both phases and where adequate bonding exists across phase boundaries, the elastic moduli are approximately additive in proportion to their respective volume fractions.

The effect of porosity on elastic properties of ceramics has been treated by a number of authors. Mackenzie (15) has derived a quadratic relationship for closed spherical pores:

$$E = E_o (1 - 1.9 P + 0.9 P^2) \quad (12)$$

CERAMICS (PROPERTIES AND APPLICATIONS)

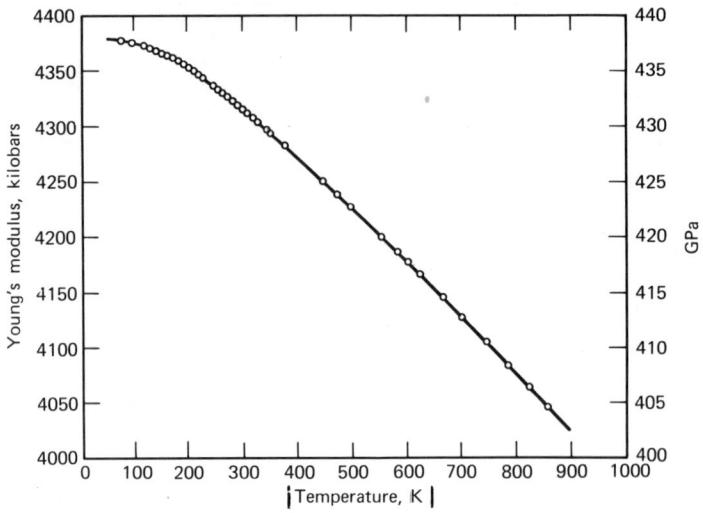

Figure 7. Young's modulus of sapphire as a function of temperature (13).

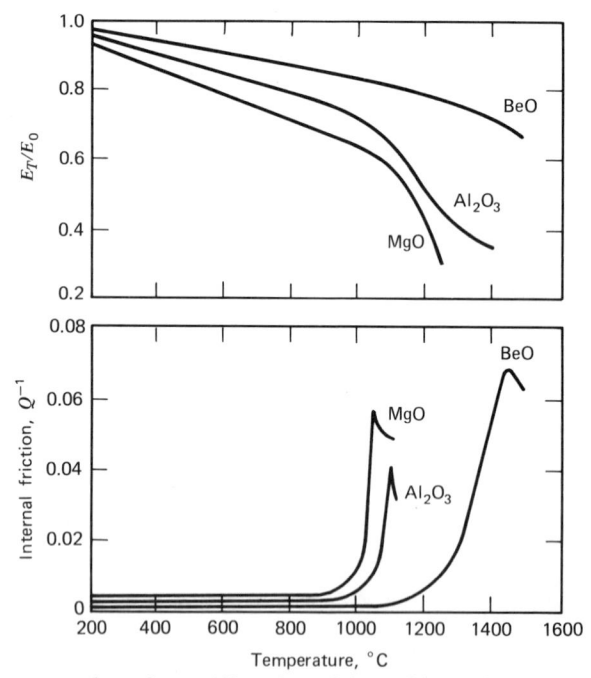

Figure 8. Temperature dependence of Young's modulus and internal friction of polycrystalline BeO, Al_2O_3, and MgO (14).

where E = Young's modulus with porosity, E_o = Young's modulus without porosity, and P = fractional porosity, which reflects the notion that the pore not only fails to contribute its own volume to the load-carrying portion of the structure, but that it shadows an additional, essentially stress-free volume of adjacent good material on either side of the pore along the stress axis. This relationship, and substantiating experimental data for alumina ceramics, are illustrated in Figure 9. Other relationships have been proposed, particularly for low porosity ranges (16).

Inelastic effects are attributable to time-dependent stress–strain relationships. They are particularly evident when the frequency of mechanical oscillation causes resonance of certain structural defects (vacancy pairs, segments of dislocations, grain boundary sliding, etc). At resonance, the defect absorbs energy. Such internal friction peaks are dependent upon the applied frequency, which must be tuned to obtain resonance, and upon the temperature.

Strength

Ceramics are classically considered to be brittle materials. They are strong in compression, but relatively much weaker (frequently by factors of 10 or more) in transverse bending or in tension. In the case of glass at room temperature, it is believed that deformation is entirely elastic up to the point of failure. Brittle fracture always appears to initiate at the specimen surface and can be attributed to the stress-induced catastrophic extension of preexisting defects, the well-known Griffith cracks.

Room Temperature Strengths. The theoretical strengths of ceramic materials, derived from atomic bonding considerations, are far larger than the observed tensile strengths. The presence of flaws, which act as stress concentrators, explain the discrepancy.

The calculation of the stress required to initiate fracture owing to the increased stress in the vicinity of a crack tip was derived by Inglis (17) who derived an expression for stress in the tip vicinity σ_m:

$$\sigma_m = 2\sigma \left(\frac{c}{\rho}\right)^{1/2} \tag{13}$$

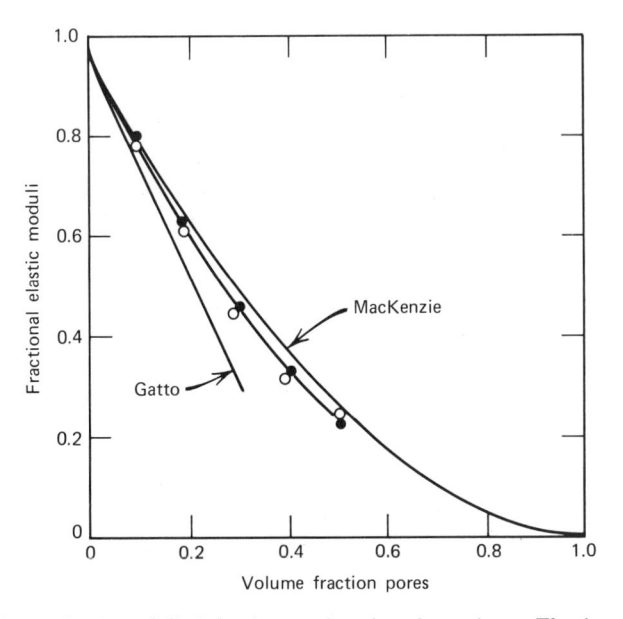

Figure 9. Relative elastic moduli of alumina as a function of porosity. ●, Elastic modulus; ○, rigidity modulus.

where ρ is the crack tip radius, $2c$ is the (elliptical) crack length, and σ is the applied stress.

Using the idea that the crack tip was of atomic spacing dimension and by utilizing the theoretical strength expressed as a function of Young's modulus and surface energy, Orowan (18) expressed the condition for failure as follows:

$$\sigma_f = \left(\frac{E\gamma}{4c}\right)^{1/2} \tag{14}$$

which is similar to the expression derived by Griffith (19) based on elastic and surface energy considerations. The equation assumes that σ_f is smaller than the general plastic yield strength by a factor of two or three. This approach becomes inapplicable when plastic flow becomes possible, which nearly always requires high temperature for ceramic materials.

The crack tip areas, or microflaws from which fracture originates are produced during the fabrication process and can be identified by careful microscopy. An example of such a process-related flaw in alumina is shown in Figure 10.

For many materials the flaw size can be related to the grain size or pore size. It is the largest flaw that affects strength and thus the relative sizes of pores and grains dictates which of these is important (21–22). In fully dense materials there are no large pores and the flaw size is usually related to the grain size. The strength of alumina as a function of grain size is shown in Figure 11. The decreased slope in the smaller region has been attributed to the increased influence of flaws larger than the grain size (21).

High Temperature. At higher temperatures, ceramics can deform plastically by the generation and motion of dislocations. Dislocation nomenclature employs (abc) [abc] or {abc} ⟨abc⟩ Miller indices to denote first the (crystallographic plane) or {family of planes} upon which slip can occur with the dislocations under consideration, and

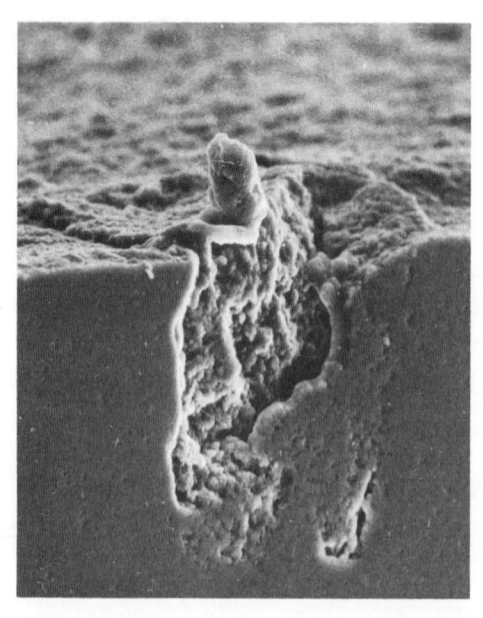

Figure 10. Flaw in well processed Al_2O_3 sintered under rate control (20).

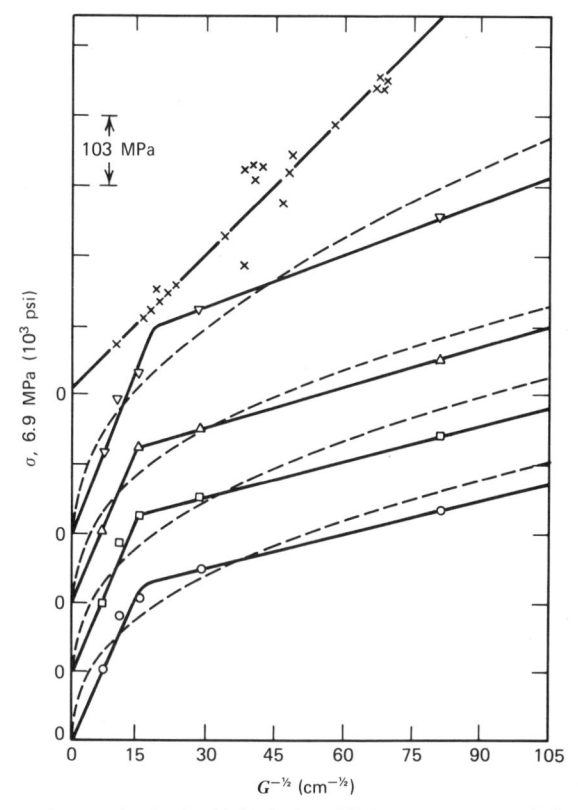

Figure 11. Strength vs grain size in Al$_2$O$_3$ (23). x, High pressure, annealed, machine, annealed; σ, room temperature (20). ▽, High pressure, machine, annealed; σ, room temperature (23). △, High pressure, machine, annealed; σ, 400°C (23). □, High pressure, machine, annealed; σ, 700°C (23). ○, High pressure, machine, annealed; σ, 1000°C (23). To convert MPa to psi, multiply by 145.

second, the [crystallographic direction] or ⟨family of them⟩ of the Burger's vector of the dislocation, which coincides with the slip direction. For further information on the subject of dislocations in ceramics, see ref. 4.

The combined influences of bond character and crystal structure have their greatest effect on the relative mobility of dislocations. When a dislocation moves, bonds must be broken and remade, which is a considerably more difficult process in ionic and covalent-bonded material than in metals having mobile bonding electrons. If the dislocation is a dissociated one, as in alumina or spinel, the partial dislocations and the intervening stacking faults must be moved along together. Only at high temperatures, where ion mobility is high and the several required motions are easy to synchronize, can dislocations be made to move in uniaxially stressed alumina. Even at high temperatures, the stress levels required to induce plastic deformation are high, and are strongly temperature dependent. On the other hand, the simple structure, the predominantly ionic bond, and the relatively simple configuration of the {110} ⟨110⟩ dislocations in magnesium oxide leads to plastic deformation at much lower stress levels, and to a much lower order of temperature dependence (24). Magnesium oxide crystals with properly prepared surfaces may even be deformed in liquid nitrogen, at a temperature where many metals are severely embrittled.

Easy flow requires two particular phenomena; at least five independent slip systems (25) and mobile dislocations. β-SiC has sufficient number of slip systems but does not exhibit ductile behavior until very high temperatures are reached, because of the directionality of covalent bonds. MgO and UO_2 are principally ionic and exhibit flow more readily.

With increases in temperature, each material exhibits first limited and then general plasticity, but the range of behavior is very large. At low temperatures the behavior is as previously discussed: the stresses required for flow are much larger than those required for fracture. As temperature increases, brittle fracture still occurs, but equation 13 no longer strictly applies because flaws are extended by limited plastic processes. At the highest temperature, general plastic flow occurs and the mechanical behavior is similar to metals. Within this framework there are large variations between individual materials.

At high temperatures, ceramics deform in creep, ie, deformation occurs as a function of time under the influence of an applied stress. The general strain rate equation provides an adequate description of such processes:

$$\epsilon = A\sigma n e^{-Q/RT} \qquad (15)$$

where ϵ = strain rate; A = constant; σ = applied stress; n = stress–strain exponent; and Q = activation energy for the flow process. For viscous flow, or grain boundary sliding, the value of the exponent n equals 1: a form of diffusion-controlled deformation, the Nabarro-Herring creep process, also has n equal to 1. With plastic deformation processes, the value of n is much higher, on the order of 4, 4.5, or 5. The creep rate is dependent not only upon the stress level, the stress–strain rate dependence, and the temperature, but also on the average grain size of the material; creep rates are higher for fine-grained material than for coarse-grained. It is important to note that this grain size effect at high temperatures is opposite to the effect of grain size on strength at low temperatures. A ceramic material selected for resistance to creep at high temperatures may be rather weak and prone to brittle fracture at low temperatures, but subject to excessive deformation at high temperatures, creeping even under its own weight. Other microstructural considerations are important to creep behavior; Figure 12 illustrates the marked effect of porosity on creep rates and flow stresses in alumina at high temperature. In many common ceramics a vitreous phase is present and usually is localized at the grain boundary. At elevated temperatures, the strength decreases rapidly, an effect attributable to lowered viscosity and increased lubricity of the intergranular glassy phase.

Engineering with Ceramics. The strength of a polycrystalline ceramic is a very complex quantity, strongly influenced by temperature and environment, the size, shape, and loading geometry of the specific part being tested, its composition, microstructure, and surface condition, as well as its prior thermal and mechanical history.

In structural design with brittle materials, the average breaking strength is of little value. Often a particular sample will break at a considerably lower stress level than the average. Microscopic fracture analysis usually reveals that a large flaw in the form of an impurity crystal, pore or large grain is located at the fracture origin. Failure analyses of ceramic materials, particularly at room temperature, is nearly always concerned with flaw identification.

By subjecting an individual piece to a stress higher than will be seen in service

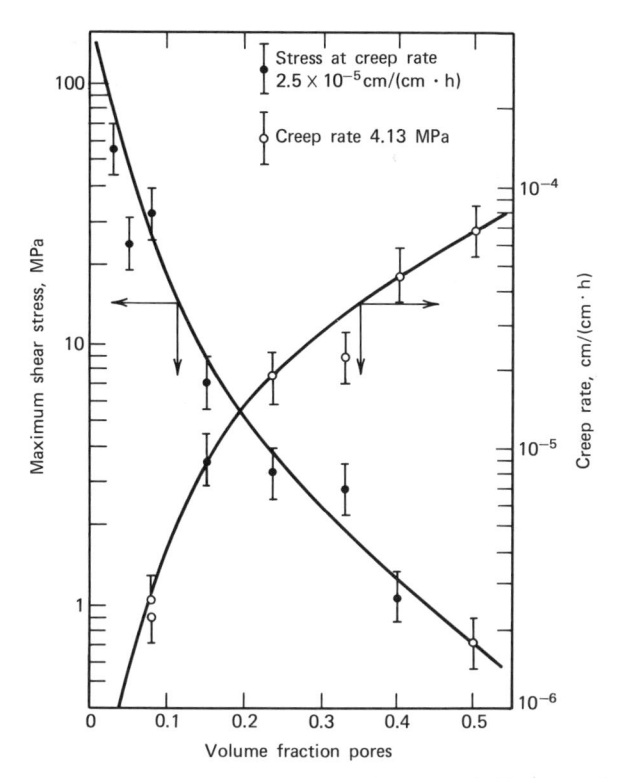

Figure 12. Effect of porosity on torsional creep of Al_2O_3 at 1275°C. To convert MPa to psi, multiply by 145.

(proof testing), the upper limit to the size of flaws which can be present can be calculated. A minimum service life guarantee can be determined from knowledge of crack propagation behavior (11,26).

The contributions of mechanical surface damage to loss of strength and increased likelihood of fracture in ceramics have been described by Westwood as an extreme form of notch sensitivity (27). The action of chemical polishing in reducing notch sensitivity is probably not entirely dependent on complete removal of a crack, but on a blunting of its tip to acceptable proportions. For example, in alkali halides Pulliam (28) found that a 60 s exposure to H_2O caused the radius of curvature of a crack tip to increase from an initial value of 0.2–1 nm to about 200 nm. Chemical polishing and etching operations are essential in detailed studies of plasticity in ceramic materials. Polishing is employed to remove cracks and cold-worked surface layers; etching is used to reveal grain boundaries as grooves and individual dislocations by etch pits at their surface termini.

The value obtained for strength in a test is highly dependent on how the tests are made, particularly with respect to surface preparation. The type of flaw present at the surface can dominate if these flaws are larger than those inherent in the bulk materials.

Strength values of ceramics are also dependent on the time of load, being lower over a longer time as cracks can propagate and cause delayed failure at a load which does not cause catastrophic failure for short loading times. The rate of crack propagation under the influence of stress and chemical environment is very important and is a factor in design criteria. See reference 29 for more information on stress corrosion and time dependence of failure.

The range of transverse bending strengths at room temperature for a variety of ceramic materials is indicated in Table 4. Additional information on strength of ceramic systems can be found in references 4 and 24.

Many ceramics in common usage are not simple, monophase materials, but frequently are quite complex proprietary compositions which have been developed to afford some suitable compromise between a number of conflicting property requirements, together with acceptable workability and firing behavior. The best sources of information on strength and other properties of such ceramic compositions are often to be found in the technical brochures issued by individual ceramic manufacturers.

Electrical and Magnetic Properties

An increasing emphasis on ceramics as electronic and magnetic components in a host of applications has been apparent since World War II, and it continues to intensify (see Ceramics as electrical materials). Ceramic components, ceramic substrates, and ceramic thin films are major constituents in the present development of microcircuits for electronics. Most oxide ceramics in their normal valence states are insulators, and serve very important roles in electronic and electrical devices purely for that reason (see Insulation, electric). Most nonoxide ceramics, and some oxides, are

Table 4. Strength Values for Some Ceramic Materials

Material	Modulus of rupture, MPa[a]
aluminum oxide crystals	345–1034
sintered alumina (ca 5% porosity)	207–345
alumina porcelain (90–95% Al_2O_3)	345
sintered beryllia (ca 5% porosity)	138–276
hot-pressed boron nitride (ca 5% porosity)	48–103
hot-pressed boron carbide (ca 5% porosity)	345
sintered magnesia (ca 5% porosity)	103
sintered molybdenum silicide (ca 5% porosity)	690
sintered spinel (ca 5% porosity)	90
dense silicon carbide (ca 5% porosity)	172
sintered titanium carbide (ca 5% porosity)	1100
sintered stabilized zirconia (ca 5% porosity)	83
silica glass	107
vycor glass	69
pyrex glass	69
mullite porcelain	69
steatite porcelain	138
superduty fire-clay brick	5.2
magnesite brick	27.6
bonded silicon carbide (ca 20% porosity)	13.8
1090°C insulating firebrick (80–85% porosity)	0.28
1430°C insulating firebrick (ca 75% porosity)	1.17
1650°C insulating firebrick (ca 60% porosity)	2.0

[a] To convert MPa to psi, multiply by 145.

semiconductors (qv) with very interesting and very useful properties, eg, SiC, a semiconductor, is much used as high temperature heating element, and as a voltage-dependent resistor to ground (conducting only at high voltages), providing lightning protection to TV antennae, etc.

Those ceramics that contain transition metal ions may display pronounced magnetic effects associated with the spins of unpaired electron orbital states. The magnetic ceramics, notably the ferrites (qv) and the synthetic garnets, constitute a significant segment of technical ceramic endeavor, and are of major importance in computer memory circuits, delay lines, and Faraday rotators in microwave electronic systems, and in a variety of permanent magnet applications which range from TV tube deflection yokes to refrigerator door closures. Magnetic ceramics, in contrast to magnetic metals, do not act as conductors of electricity (see Magnetic materials; Glassy metals; Amorphous magnetic materials; Microwave technology).

In these specialized ceramics, the matter of principal concern is the mobility of electrons, electron holes, and ions under the driving force of alternating electrical and/or magnetic fields. Interactions with these materials occur between nonthermodynamic defects, particularly dislocations, and the electromagnetic properties of ceramics, so that the thermo–mechanical history of the specimens may have a pronounced influence on the observed electrical or magnetic behavior. Many of the interactions of charge carriers with structural defects are analogous to the behavior of photons and phonons with structural defects discussed earlier.

Magnetization. Ceramic materials respond to magnetic fields in a way which is analogous to inelastic and dielectric behavior. The magnetic susceptibility is as follows:

$$\chi_m = M/H = N\alpha_m \qquad (16)$$

where M = magnetic moment; H = magnetizing field; N = number of elementary magnetic dipoles per unit volume; and α_m = magnetizability of the elementary dipoles.

Only those electrons having unpaired spin orbitals can contribute to magnetization; they are associated with conduction electrons, with atoms and molecules having an odd number of electrons, and with atoms and ions having partially filled inner electron shells (mainly the $3d$, $4f$, and $5f$ transition series). Magnetic ceramics rely principally upon the last category, and are usually based on compounds involving one or more of those atoms or ions.

Magnetic materials are generally classified as hard or soft; the hard materials are permanent magnets, soft are materials that can be magnetized and demagnetized readily, as required in electronic applications. The main types are outlined below:

Spinel ferrites have a wide range of composition, based on the spinel structure AB_2O_4, which contains both tetrahedral and octahedral sites for the ions, including those with unpaired spin orbitals. Spinels can be either hard or soft and cover a wide range of magnetic properties.

Rare earth garnets have the general formula $A_3Fe_5O_{12}$, with the iron in two valence states, and are used in magnetic bubble memories (see Magnetic materials).

Hexagonal ferrites are related to the spinel structure with a hexagonally packed structure of the formula $AB_{12}O_{19}$, where A is usually divalent Ba, Sr, or Pd and B is trivalent Al, Ga, Cr, or Fe. Hexagonal ferrites are primarily used for permanent magnets.

Ceramic magnets are of increasing importance in the electronics industry particularly in high frequency devices. For more detail on magnetic properties see reference 30.

Composites and Cermets

An important segment of the ceramic industry is concerned with complex materials or components of which the ceramic phase is only a part; another part is metallic or plastic. The useful properties of the combination are different from those of either phase alone. For example, porcelain enamel, a vitreous or crystal-plus-vitreous ceramic phase, is applied to metal structures to give oxidation and and corrosion resistance, smoothness, color, texture, and other desirable properties. The metal gives shape and strength to the article, and provides for easier forming and generally lighter weight than would be the case if it had been produced entirely as a ceramic (see Enamels).

Composites. As a rule of thumb, composites contain at least one phase which is macroscopic in at least one dimension; thus the minimum qualifying phase would have to have fibrous shape. Ceramic–metal composites, particularly those based on metallic honeycombs or webbings impregnated with a ceramic phase, have been employed for aerospace hardware, including reentry nose cones, rocket nozzles, ram-jet chambers, etc, under extreme conditions of thermal shock, high surface temperature, high-velocity erosion and/or ablation (see Ablative materials). Another important ceramic-based composite is glass–fiber reinforced plastic, which finds a variety of uses, including ablative nose cones in aerospace applications. Ceramic fiber-reinforced metals, metal-impregnated ceramics, and metal–ceramic laminates are receiving considerable research and development attention, particularly for high-temperature, high-strength materials capable of withstanding thermal shock or mechanical impact (see Composite materials).

Cermets. If the ceramic and metallic phases are randomly shaped and intimately dispersed one within the other on a microstructural scale, the resultant material is properly termed a cermet. A number of important cermets have been developed; by far the largest group of these are the metal-bonded hard carbides or cemented carbides (see Carbides).

Uses of Ceramics

Ceramics differ from other engineering materials (metals, plastics, wood products, textiles) in a number of individual properties, but perhaps the most distinctive difference to a designer or potential user of ceramic ware is the particularity of the individual ceramic piece. Ceramics are not readily shaped or worked after firing, except by very costly grinding; consequently, they normally must be used as is. Except for some simple tile, rod, and tube shapes of limited sizes, ceramics cannot be marketed by the foot or by the yard, nor cut to fit on the job.

All the useful properties, including shape, size, etc, must be provided in advance, beginning with the very early stages of ceramic processing, and not as an afterthought. The structural integrity of each piece must be preserved through a variety of thermal and mechanical stress exposures during processing and until it is finally installed and in service. If a ceramic should fail in service as a result of a variety of causes (brittle fracture on impact, thermal shock, dielectric breakdown, abrasion, melting slag corrosion, etc) it is not likely to be repairable, and usually must be replaced *in toto*.

Significant advancements have been made in fundamental understanding and technological control of the properties of ceramics, and of their utilization in many new, demanding, highly technical applications. The industry, in general, and the technical and electronic ceramic portions of it, in particular, have devised production and control techniques for mass producing complex shapes in bodies having carefully controlled electrical, magnetic, and/or mechanical properties, while maintaining dimensional tolerances that are good enough to permit relatively easy assembly with other components.

Many ceramics are produced in volume as standard items; refractory bricks and shapes, crucibles, muffles, furnace tubes, insulators, thermocouple protection tubes, capacitor dielectrics, hermetic seals, fiber boards, etc, are routinely stocked by a number of ceramic producers in a variety of compositions and sizes. It is usually quicker and cheaper to use stock items whenever possible. When stock items will not meet the need, most manufacturers are prepared to produce custom items to specifications. The more stringent the requirements for a given property of the ceramic, or the more restrictive the requirements for specific combinations of properties, sizes, and shapes, the more limited are the acceptable compositional, microstructural, and configurational parameters for the ceramic, and hence the greater the cost and difficulty of manufacture. Most ceramic manufacturers have experienced staff engineers and designers who are well qualified to work with potential customers on details of design of ceramic ware.

BIBLIOGRAPHY

"Properties and Applications of Ceramic Materials" under "Ceramics" in *ECT* 2nd ed., Vol. 4, pp. 793–832, by Hayne Palmour III, North Carolina State of The University of North Carolina.

1. W. D. Kingery, "Oxides for High Temperature Applications," *Proceedings of International Symposium on High Temperature Technology, Stanford Research Institute, Asilomar, Calif., Oct. 6–9, 1959,* McGraw-Hill Book Co., Inc., New York, 1960, pp. 76–89.
2. P. D. Johnson, *J. Am. Ceram. Soc.* **33**(5), 168 (1950).
3. G. Economos and W. D. Kingery, *J. Am. Ceram. Soc.* **36**(12), 403 (1953).
4. W. D. Kingery, H. K. Bowen, and D. R. Uhlmann, *Introduction to Ceramics,* 2nd ed., John Wiley & Sons, Inc., New York, 1976.
5. D. W. Lee and W. D. Kingery, *J. Am. Ceram. Soc.* **43,** 594 (1960); ref. 4, p. 675.
6. Ref. 4, p. 587.
7. G. G. Koerber, *Properties of Solids,* Prentice-Hall, Inc., Englewood Cliffs, N.J., 1962.
8. Ref. 4, p. 616.
9. Ref. 4, p. 643.
10. W. R. Buessem, "Internal Ruptures and Recombinations in Anisotropic Ceramic Materials" in W. W. Kriegel and H. Palmour, III, eds., *Mechanical Properties of Engineering Ceramics,* Interscience Publishers, New York, 1961, pp. 127–148.
11. T. K. Gupta, *J. Am. Ceram. Soc.* **55,** 249 (1972); ref. 4, p. 829.
12. D. P. H. Hasselman, "Thermal Stress Crack Stability and Propagation in Severe Thermal Environments" in W. W. Kriegel and Hayne Palmour, III, eds., *Ceramics in Severe Environments,* Plenum Press, New York, 1971.
13. J. B. Wachtman, Jr., W. E. Tefft, and D. G. Lam, Jr., "Young's Modulus of Single Crystal Corundum from 77 to 850 K" in ref. 10, pp. 221–283.
14. R. Chang, "The Elastic and Anelastic Properties of Refractory Materials for High Temperature Applications" in ref. 10, pp. 209–220.
15. J. K. Mackenzie, *Proc. Phys. Soc. (London) B* **63,** 2 (1950).
16. R. M. Sprigg, *J. Am. Ceram. Soc.* **44**(12), 628 (1961); D. P. H. Hasselman, *J. Am. Ceram. Soc.* **45**(9), 452 (1962).
17. R. Inglis, *Trans. Inst. Nav. Arch.* **55,** 219 (1913).

18. H. Orowan, *Z. Krist.* **A89,** 327 (1934).
19. A. A. Griffith, *Trans. R. Soc.* **A221,** 163 (1920).
20. R. W. Rice in R. K. MacCrone, ed., *Properties and Microstructure,* Academic Press, Inc., New York, 1978.
21. R. W. Rice and co-workers in R. C. Bradt, D. P. H. Hasselman, and F. F. Lange, eds., *Fracture Mechanics of Ceramics,* Vol. 4, Plenum Press, New York, 1978.
22. T. M. Hare and Hayne Palmour, III, in G. Y. Onoda and L. L. Hench, eds., *Processing of Ceramics Before Firing,* John Wiley & Sons, Inc., New York, 1978.
23. S. C. Carniglia, *J. Am. Ceram. Soc.* **55,** 243 (1972).
24. E. R. Parker, "Ductility of Magnesium Oxide" in ref. 10, pp. 65–87.
25. R. Von Mises, *Z. Agnew. Math. Mech.* **8,** 161, 1928.
26. R. C. Bradt, D. P. H. Hasselman, and F. F. Lange, eds., *Fracture Mechanics of Ceramics,* Vols. 3–4, Plenum Press, New York, 1978.
27. A. R. C. Westwood, *Effects of Environment on Fracture Behavior, Technical Report G2-15,* Research Institute for Advanced Study, Baltimore, Md., 1962.
28. G. R. Pulliam, *J. Am. Ceram. Soc.* **42**(10), 477 (1959).
29. S. M. Wiedhorn and co-workers, *J. Am. Ceram. Soc.* **57,** 336 (1974).
30. R. S. Tebble and D. J. Craik, *Magnetic Materials,* Wiley-Interscience, New York, 1969.

THOMAS M. HARE
North Carolina State University

CERAMICS AS ELECTRICAL MATERIALS

There is a great number of applications for ceramic materials for which electrical conduction properties are important. Semiconductor materials are used in many specialized applications such as resistance heating elements; semiconductor devices such as rectifiers, photocells, transistors, thermistors, detectors, and modulators have become an important part of modern electronics. Ceramics are equally important as electrical insulators; porcelains are used for both low- and high-voltage insulation. Consequently, the entire range of electrical conduction properties should be considered (see Photovoltaic cells; Semiconductors; Insulation, electric).

The electrical characteristics of ceramic materials vary greatly. In some cases, oxides, borides, nitrides, and carbides have metallic conductivity; in other instances these same classes of materials behave like semiconductors. Many ceramics are good insulators even at high temperatures. The atomic processes are very different for the various conduction modes and the transport of current may be caused by the motion of electrons or electron holes, or by ions.

This article discusses the background material needed to understand the differences in conduction processes in ceramics and examines the electrical behavior of specific materials. Electrical ceramics are commonly used in special situations where refractoriness or chemical resistance are needed, or other environmental effects are severe. Thus, it is important to understand the effects of temperature, chemical additives, gas-phase equilibration, and interface reactions.

Much of the background for this article can be found in the textbook *Introduction to Ceramics* (1). More detailed reviews on specific classes of ceramic conductors are given for electronic conductors (2) and for ionic conductors (3).

Electrical Conduction Phenomena

Consider a ceramic solid at equilibrium with its environment (temperature, pressure, and composition) and then apply an electrical field, $\epsilon = dV/dx$, in terms of V/m. The ions and electrons reestablish equilibrium by moving in the field which results in a net electric current density, j, defined as C/(m²·s) or A/m². An electron or an electron-hole has a unit charge, $e = 1.601 \times 10^{-19}$ C; an ion has this unit charge times its valence, z, ze. Thus the current density owing to the flow of the particle i is (eq. 1),

$$j_i = n_i z_i e v_i = \left(\frac{\text{particles}}{\text{m}^3}\right)\left(\frac{\text{charge}}{\text{particle}}\right)\left(\frac{\text{C}}{\text{charge}}\right)\left(\frac{\text{m}}{\text{s}}\right) \quad (1)$$

where n_i is the number of i particles per m³ and v_i is their net velocity in the direction of the applied field. This is also called the drift velocity.

In terms of moles of charged particles per m³, N_i, the current density is (eq. 2),

$$j_i = N_i Z_i F v_i = \left(\frac{\text{mol}}{\text{m}^3}\right)\left(\frac{\text{equiv}}{\text{mol}}\right)\left(96{,}500 \frac{\text{C}}{\text{equiv}}\right)\left(\frac{\text{m}}{\text{s}}\right) \quad (2)$$

where Z_i is the valence or number of equivalent charges per mol and F the Faraday constant (96,500 C/equiv).

The electrical conductivity σ is defined as (eq. 3),

$$\sigma = j/\epsilon = \frac{1}{\Omega \cdot \text{m}} = \frac{\text{C/(m}^2\text{·s)}}{\text{V/m}} \quad (3)$$

thus the conductivity owing to the ith particle is (eq. 4),

$$\sigma_i = n_i z_i e \frac{v_i}{\epsilon} = N_i Z_i F \frac{v_i}{\epsilon} \quad (4)$$

The mobility of the particle is defined as the ratio of the velocity and the electric field ($\mu_i = v_i/\epsilon$), but it is best to use the absolute mobility B_i, which is the velocity per unit driving force acting on the particle (eq. 5),

$$B_i = \frac{v_i}{\text{force}} = \frac{v_i}{z_i e \epsilon} \quad (5)$$

The chemical mobility is $B_i' = v_i/Z_i F \epsilon$ (see Table 1). The conductivity σ, of the particle is thus (eq. 6).

$$\sigma_i = n_i z_i^2 e^2 B_i = N_i Z_i^2 F^2 B_i' \quad (6)$$

The diffusion coefficient of the ion, D_i, is related by the Nernst-Einstein relation to the mobility (eq. 7),

$$D_i = \frac{\mu_i k T}{e z_i} = B_i k T = B_i' R T \quad (7)$$

where T is the absolute temperature (K), k is the Boltzmann constant (1.38×10^{-23} J/K), and R the gas constant 8.31 J/(mol·K). The value of these expressions and the unit analysis in Table 1 are important when comparing the factors that control the conduction processes in ceramics, or when comparing electrical and chemical forces ($\partial \overline{G}_i/\partial x = Z_i F \partial V/\partial x$). For example, to determine the flux or conductivity of ions in

Table 1. Dimensional Units for Mobility Where Mobility ≡ Velocity/Unit Force [a,b]

$$B = \frac{m/s}{J/m} = \frac{10^{-5} \text{ m/s}}{\text{erg/cm}} = \frac{10^{-3} \text{ cm/s}}{\text{erg/cm}} = \frac{10^{-3} \text{ cm/s}}{\text{dyn}}$$

$$B_i = \frac{v_i}{(1/N)(\partial \overline{G}_i/\partial x)} = \frac{m^2}{J \cdot s} = \frac{10^{-3} \text{ cm}^2}{\text{erg} \cdot s}$$

v_i = m/s
$\overline{G}_i$ = J/mol = 10^7 erg/mol

= absolute mobility

N = Avogadro's number = atoms/mol
x = m

$$B'_i = \frac{v_i}{\partial \overline{G}_i/\partial x} = \frac{\text{mol} \cdot m^2}{J \cdot s} = \text{chemical mobility}$$

v_i = m/s

$\overline{G}_i$ = J/mol
x = m

$$B''_i = \frac{v_i}{\partial G_i/\partial x} = \frac{m^2}{V \cdot s} = \frac{10^4 \text{ cm}^2}{V \cdot s}$$

v_i = m/s
V = volts

= electrical mobility

x = m

$B_i = NB'_i$
$B''_i = Z_i F B'_i$

Z_i = valence = equiv/mol
F = Faraday constant = 96,500 J/(V·equiv)

$$\frac{\partial \overline{G}_i}{\partial x} = Z_i F \frac{\partial V}{\partial x}$$

[a] $J = C \cdot V = 10^7$ ergs = 0.2389 cal = 6.243 × 10^{18} eV.
[b] $\overline{G}_i$ = chemical potential.

Table 2. Electrical Conductivity of Some Materials at Room Temperature

Materials	Conductivity, $(\Omega \cdot m)^{-1}$
Metals	
copper	5.9 × 10^7
iron	1.0 × 10^7
molybdenum	1.9 × 10^7
tungsten	1.8 × 10^7
ReO$_3$	5.0 × 10^7
CrO$_2$	3.0 × 10^6
Semiconductors	
dense silicon carbide	10
boron carbide	200
germanium, pure	30
Fe$_3$O$_4$	10^4
Insulators	
SiO$_2$ glass	<10^{-12}
steatite porcelain	<10^{-12}
fire-clay brick	10^{-6}
low-voltage porcelain	10^{-10}–10^{-12}

Table 3. Conversion Factors for SI and Esu Units

Quantity	Units Esu	Units SI	To convert from SI to esu units multiply by:
conductivity, σ	s^{-1} [(statampere/statvolt)/cm]	(ohm·m)$^{-1}$ [A/(V·m)]	9×10^9
current density, j	(statcoulomb·s^{-1}/cm^2 [statampere/cm^2]	C/s·m^2 [A/m^2]	3×10^5
electric field, ϵ	statvolt/cm	V/m	$(1/3) \times 10^{-4}$
carrier concentration, n	carriers/cm^3	carriers/m^3	10^{-6}
drift velocity, v	cm/s	m/s	10^2
mobility, μ	(cm/s)/(statvolt/cm)	(m/s)/(V/m)	3×10^6
electronic charge, e	4.803×10^{-10} statcoulomb	1.601×10^{-19} C	3×10^9

a solid electrolyte as compared to electrons in a semiconducting ceramic two terms in equation 6 are of interest: the number of charge carriers and their mobility. The effects of temperature, composition, and structure on each of these terms must be considered. Materials are usually classed according to the specific conductivity mode, such as insulators which have low concentrations and low mobility of carriers. Metallic conductors such as oxides, borides, etc, have a high conductivity value that is not a strong (exponential) function of temperature. Semiconductors are intermediate and have an exponential temperature dependence. Table 2 and Figure 1 give some examples. Note that conductivity is the inverse of resistivity ($\sigma_i = 1/\rho_i$) and the two are used interchangeably. The common units and conversions are given in Table 3. The resistivity in SI units is $\Omega \cdot m$, but much of the literature remains in units of $\Omega \cdot cm$.

The total conductivity is that owing to the net motion of electrons, holes, and each of the ions (eq. 8),

$$\sigma = \sigma_1 + \sigma_2 + \sigma_3 \ldots \sigma_i \qquad (8)$$

The fraction of the total conductivity (at a specific temperature, composition, etc) owing to the particle i is called the transference number of i (eq. 9),

$$t_i = \sigma_i/\sigma \qquad (9)$$

The transference numbers for several compounds are given in Table 4.

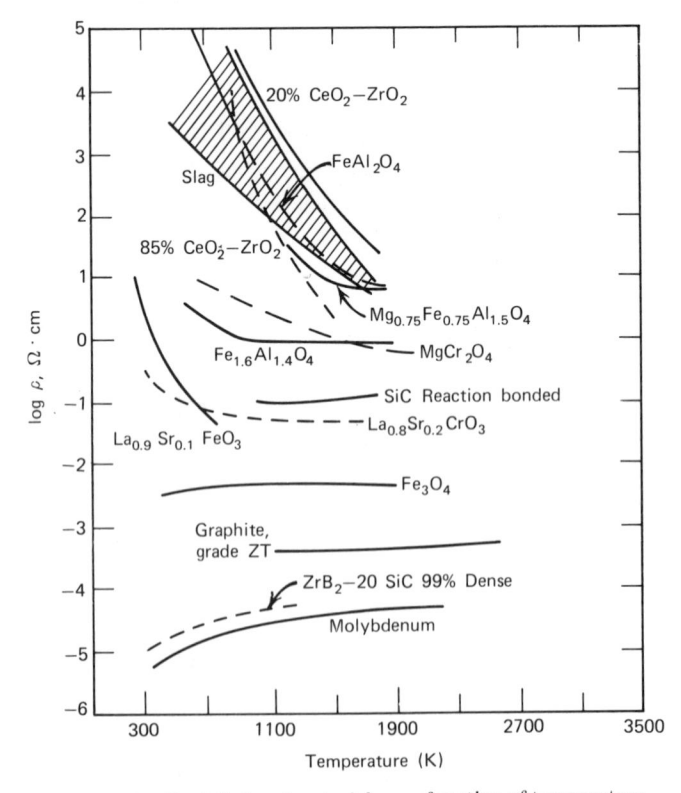

Figure 1. Resistivity of materials as a function of temperature.

Table 4. Transference Numbers of Cations t^+, Anions t^-, and Electrons or Holes $t_{e,h}$ in Several Compounds[a]

Compound	Temperature, °C	t^+	t^-	$t_{e,h}$
NaCl	400	1.00	0.00	
	600	0.95	0.05	
KCl	435	0.96	0.04	
	600	0.88	0.12	
KCl + 0.02% CaCl$_2$	430	0.99	0.01	
	600	0.99	0.01	
AgCl	20–350	1.00		
AgBr	20–300	1.00		
BaF$_2$	500		1.00	
PbF$_2$	200		1.00	
CuCl	20	0.00		1.00
	366	1.00		0.00
ZrO$_2$ + 7% CaO	>700	0	1.00	10^{-4}
Na$_2$O·11Al$_2$O$_3$	<800	1.00 (Na$^+$)		<10^{-6}
FeO	800	10^{-4}		1.00
ZrO$_2$ + 18% CeO$_2$	1500		0.52	0.48
+ 50% CeO$_2$	1500		0.15	0.85
Na$_2$O·CaO·SiO$_2$ glass		1.00 (Na$^+$)		
15% (FeO·Fe$_2$O$_3$)·CaO·SiO$_2$·Al$_2$O$_3$ glass	1500	0.1 (Ca^{+2})		0.9

[a] Ref. 1.

Table 5. Fast Ionic Conductors

Compound	T°C	Conducting ion	σ_{ion} (ohm·m)$^{-1}$
ZrO$_2$ + 12 mol % CaO	1000	O	0.8
NaAl$_{11}$O$_{17}$	300	Na	35
K$_{1.4}$Fe$_{11}$O$_{17}$	300	K	2
Li$_{0.5}$Zr$_{1.5}$Ta$_{0.5}$P$_3$O$_{12}$	200	Li	0.1
Na$_3$Zr$_2$PSi$_2$O$_{12}$	300	Na	20
CeO$_2$ + 12 mol % CaO	700	O	4

Ionic Conduction in Ceramics

Ions such as oxides and halides in crystalline materials constitute an ever present charge carrier that can contribute to electrical conductivity. As illustrated in Table 4, electrical conductivity resulting from ion migration is important in many ceramic materials. Its analysis requires a determination of the concentration and mobility of charge carriers. For an ion to move through the lattice under the driving force of an electric field, it must have sufficient thermal energy to pass over a free energy barrier, ΔG^+, the intermediate position between lattice sites. The result is a mobility that depends on the ion jumping from site to site. By comparison with equation 4, the ion mobility (for a rocksalt-structure crystal) is expressed as (eq. 10),

$$\mu_i = \frac{4\,ez_i a^2 \nu}{kT} e^{-\Delta G^+/kT} \tag{10}$$

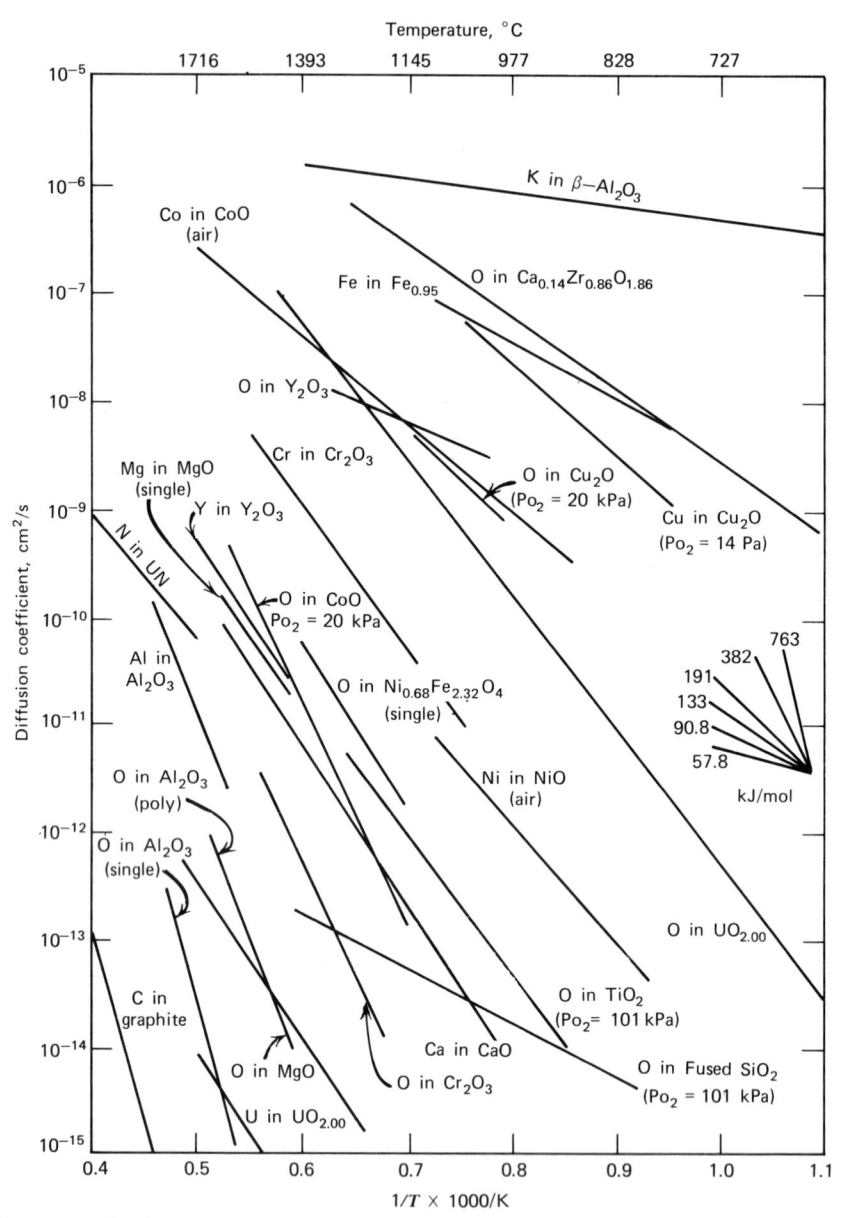

Figure 2. The diffusion coefficient of several single crystal and polycrystalline ceramic materials as a function of temperature. The activation energy is obtained from the slope and the insert; eg, for O diffusion in ZrO_2 + 14 cation % CaO, ΔG^+ ca 121 kJ/mol. To convert J to cal, divide by 4.184. To convert Pa to atm, divide by 101×10^3.

where a is the lattice parameter (ca 3×10^{-8} cm) and ν the vibrational frequency ($\sim 10^{13}$/s). The absolute mobility is (eq. 11),

$$B_i = \frac{4 a^2 \nu}{kT} e^{-\Delta G^+/kT} = \frac{D_i}{kT} \qquad (11)$$

which is again the Nernst-Einstein relationship. For a particular ion having a transference number, t_i, in a crystal (eq. 12),

$$\sigma_i = t_i\sigma = (n_iez_i)(ez_i)\frac{D_i}{kT} = \frac{D_in_iz_i^2e^2}{kT} = \frac{D_iN_iZ_i^2F^2}{RT} \quad (12)$$

If the ion jumps into a normally occupied lattice site, a term for the creation of vacancies must be included. Creation of thermal vacancies by Frenkel (vacancy plus interstial lattice defects) or by Schottky (cation and anion vacancies) mechanisms can be expressed in terms of an equilibrium constant and, therefore, a free energy of formation of defects, ΔG_f. This results in an added exponential term in the conductivity expression in equation 11, and is valid when the concentration of defects caused by impurities is less than that caused by thermal energy (eq. 13),

$$\sigma_i = \frac{n_iz_i^2e^2}{kT}e^{-\Delta G^+/kT}e^{-\Delta G_f/kT} \quad (13)$$

The term $e^{-\Delta G_f/kT}$ is the probability that a site will be vacant. The diffusion coefficients of representative ceramic materials are given in Figure 2.

When aliovalent (different valence) impurities are added to an ionic solid the crystal lattice compensates by forming defects that maintain both electrical neutrality and the anion to cation ratio of the host lattice. For example, addition of a mol of CaO to ZrO_2 requires the formation of a mol of oxygen vacancies. If this concentration is larger than the oxygen vacancies created by the thermal effects, then the conductivity from the motion of doubly charged oxygen ions is directly proportional to the concentration of CaO (eq. 14),

$$\sigma_{O^{2-}} = \frac{[CaO](2)^2e^2}{kT}e^{-G^+/kT} \quad (14)$$

In polycrystalline materials, ion transport within the grain boundary must also be considered. For oxides with close-packed oxygens, the O-ion almost always diffuses much faster in the boundary than in bulk. Other examples are less clear except that second phases at boundaries often promote higher ion diffusivities.

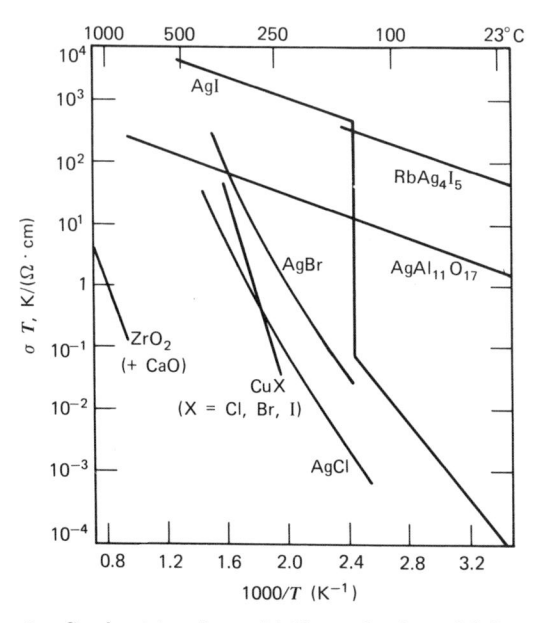

Figure 3. Conductivity of some highly conducting solid electrolytes.

Solid Electrolytes. Several types of compounds show exceptionally high ionic conductivity and have recently become of technological interest (Table 5). Such phases fall into three broad groups: (*1*) halides and chalcogenides of silver and copper in which the metal atom is disordered over several alternative sites; (*2*) oxides with the β-alumina structure in which a monovalent cation is mobile; and (*3*) oxides of the fluorite structure in which large concentrations of defects are caused by either a variable-valence cation or solid solution with a second cation of lower valence; for example, $CaO–ZrO_2$ or $Y_2O_3–ZrO_2$. Figure 3 shows electrical-conductivity data for some representative high conductivity materials. The values are many orders of magnitude larger than normal ionic compounds (compare, for example, K diffusion in β-alumina to Al diffusion in Al_2O_3 in Fig. 2) and are comparable with the conductivity of such liquid electrolytes as dilute solutions of sulfuric acid.

Oxides with Fluorite Structure. The high dopant levels in fluorite type solid solutions lead to large defect concentrations and vacancy ordering. The rapid oxygen migration that occurs in these materials (ZrO_2, UO_2, ThO_2) is presumably caused by the high concentration of vacancies (ca 15%) and correlated ion jumps over distances greater than an interionic separation. The measured oxygen-tracer diffusion coefficients agree with values obtained from conductivity through use of the Nernst-Einstein equation.

Silver and Copper Halides. The silver and copper halides and chalcogenides often have simple arrays of anions. The cations occur in disorder in the interstices among the anions and the number of available sites is larger than the number of cations. In highly conductive phases the energy barrier between neighboring sites is very small. The connectivity of such sites thus provides channels along which the cations are free to move. The potential energy seen by the mobile Ag^+ ions among the body-centered cubic array of I^- ions in α-AgI has been calculated theoretically as a function of position within the unit cell. This calculation not only confirmed the existence of paths with

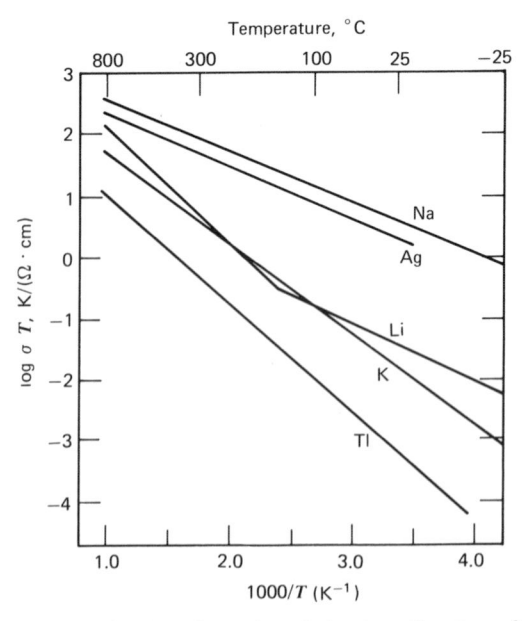

Figure 4. Electrical conductivity for various β-aluminas. Courtesy of R. A. Huggins.

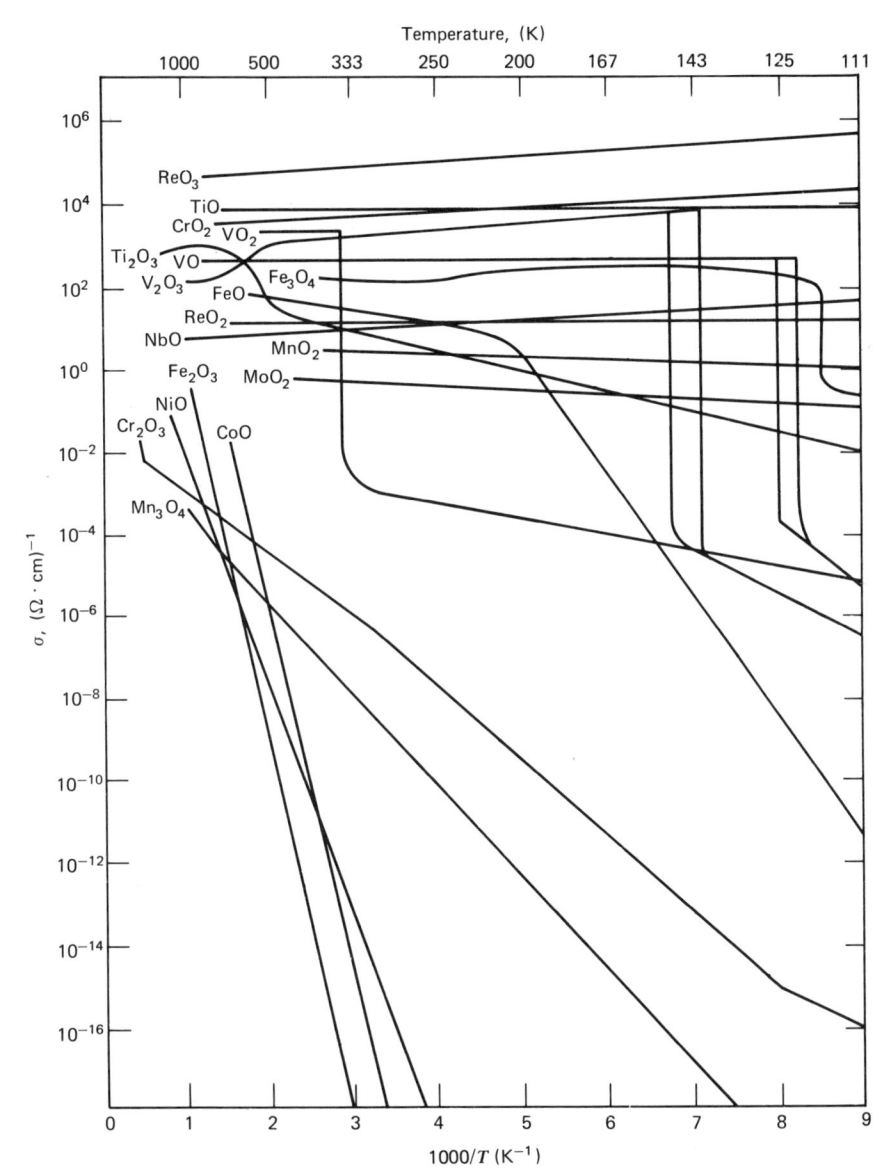

Figure 5. Temperature dependence of the electrical conductivity of several electronically conducting oxides. Courtesy of D. Adler.

very low activation barriers to Ag$^+$ migration, but showed that the height of the barrier increased rapidly with either a small increase or decrease in the size of the migrating ion.

Oxides With the β-Alumina Structure. The β-aluminas are hexagonal structures with approximate composition $AM_{11}O_{17}$. The mobile ion A is a monovalent ion such as Na, K, Rb, Ag, Tl, or Li, and M is a trivalent ion, Al, Fe, or Ga. Related phases also occur with approximate formulas $AM_7O_{11}(\beta')$ and $AM_5O_8(\beta'')$, the latter having extremely high conductivities. The conductivities for several β-aluminas are plotted as a function of temperature in Figure 4.

300   CERAMICS AS ELECTRICAL MATERIALS

Table 6. Value of the Energy Gap at Room Temperature for Intrinsic Semiconduction

Crystal	E_g (eV)	Crystal	E_g (eV)
BaTiO$_3$	2.5–3.2	TiO$_2$	3.05–3.8
C, diamond	5.2–5.6	CaF$_2$	12
Si	1.1	BN	4.8
α-SiC	2.8–3	CdO	2.1
PbS	0.35	LiF	12
PbSe	0.27–0.5	Ga$_2$O$_3$	4.6
PbTe	0.25–0.30	CoO	4
Cu$_2$O	2.1	GaP	2.25
Fe$_2$O$_3$	3.1	Cu$_2$O	2.1
AgI	2.8	CdS	2.42
KCl	7	GaAs	1.4
MgO	>7.8	ZnSe	2.6
Al$_2$O$_3$	>8	CdTe	1.45
		CdGeAs$_2$	0.5

Table 7. Approximate Carrier Mobilities at Room Temperature

Crystal	Electrons	Holes	Crystal	Electrons	Holes
diamond	1,800	1,200	PbS	600	200
Si	1,400	500	PbSe	900	700
Ge	3,900	1,900	PbTe	1,700	930
InSb	10^5	1,700	AgCl	50	
InAs	23,000	200	KBr, 100 K	100	
InP	3,400	650	CdTe	600	
GaP	150	120	GaAs	8,000	3,000
AlN		10	SnO$_2$	160	
FeO			SrTiO$_3$	6	
MnO			Fe$_2$O$_3$	0.1	
CoO		~0.1	TiO$_2$	0.2	
NiO			Fe$_3$O$_4$		0.1
GaSb	2,500–4,000	1,400	CoFe$_2$O$_4$	10^{-4}	10^{-8}
As$_2$S$_3$	1		BaTiO$_3$(+La)	0.5	0.1
CdGeAs$_2$	1,000		AgInSe$_2$	200	

The crystal structure consists of atom planes parallel to the basal plane. Four planes of oxygens in a cubic close-packed sequence comprise a slab wherein aluminum atoms occupy octahedral and tetrahedral sites as in spinel. The spinel blocks are bound together by an open layer of the monovalent ion and oxygen. This loosely bound layer is thought to be disordered and provides a two-dimensional path for atom motion with greater than single jump distances.

As the monovalent conduction ions become larger, their mobility is impeded, σ(Na β-Al$_2$O$_3$) > σ(K β-Al$_2$O$_3$). As the ions become too small, eg, as in Li β-Al$_2$O$_3$, they "rattle around" in the conduction channels and their mobility is again impeded. As expected, the conductivity is extremely anisotropic, $\sigma_{\perp c} \gg \sigma_{\|c}$. However, polycrystalline materials show less than an order of magnitude decrease in conductivity over single crystals measured parallel to the highly conducting basal planes. This may be an indication of high-conductivity paths through grain boundaries. Stable ceramics with completely ionic conductivity ($t_i = 1$) can be used as solid-state electrolytes. Owing to the precise relationship between the voltage and the chemical potential gradient across the electrolyte, it can be used for batteries and fuel cells and as an ion pump or ion-activity probe (see Batteries; Ion-specific electrodes).

The β-alumina fast-ion conductors are used as electrolytes for a sodium–sulfur storage battery. In this case the sodium ion is the conducting ion. Above 300°C the overall chemical reaction occurs between molten sodium metal and sulfur to form sodium polysulfide. The cell voltage is related to the activity of the sodium (a_{Na}) in the sulfide relative to its activity in the metal (eq. 15)

$$V = -\frac{RT}{F} \ln \frac{a_{Na} \text{ (sodium polysulfide/sulfur)}}{a_{Na} \text{ (sodium metal)}} \quad (15)$$

Using excess voltage, ions can be pumped from the low-concentration (activity) side of the electrolyte to the high-activity side, during which the storage battery is charged. In another application, the ion activity on one side, can be fixed at a known value and the activity on the other side determined for various unknown conditions.

Electronic Conduction in Ceramics

The relatively high mobility of conducting electrons or electron holes when present contributes appreciably to electrical conductivity. In some cases, metallic levels of conductivity result; in others, the electronic contribution is extremely small. In all cases, the electrical conductivity can be interpreted in terms of carrier concentrations and carrier mobilities. The range of electronic conductivities in simple transition metal oxides is given in Figure 5. Metallic conduction occurs in a few cases of transition metal oxides such as ReO_3, CrO_2, VO, TiO, and ReO_2; the doped perovskite structures $LaTiO_3$, $CaVO_3$, $CaMoO_3$, $BaTiO_3$, $SrCrO_3$, $SrFeO_3$, $LaNiO_3$, $LaFeO_3$, $LaCoO_3$, and $LaCrO_3$; many of the bronzes, Na_xWO_3, La_xWO_3, and $Na_xNb_2O_5$; and some of the spinels, $Li_{0.5}In_{0.5}Cr_2S_4$. It occurs at specific dopant concentrations. An overlap of electron orbitals within certain temperature ranges results in partially filled d or f bands. These bands, in turn, result in a concentration of 10^{28} to 10^{29} quasi-free electrons per cubic meter and essentially metallic conduction.

Ordinarily, there is an energy gap, E_g, appreciably greater than kT between filled

Table 8. Partial List of Impurity Semiconductors

			n-type		
TiO_2	Nb_2O_5	CdS	Cs_2Se	$BaTiO_3$	Hg_2S
V_2O_8	MoO_2	$CdSe$	BaO	$PbCrO_4$	ZnF_2
U_3O_8	CdO	SnO_2	Ta_2O_5	Fe_3O_4	
ZnO	Ag_2S	Cs_2S	WO_3		
			p-type		
Ag_2O	CoO	Cu_2O	SnS	Bi_2Te_3	MoO_2
Cr_2O_3	SnO	Cu_2S	Sb_2S_3	Te	Hg_2O
MnO	NiO	Pr_2O_3	CuI	Se	
			amphoteric		
Al_2O_3	SiC	$PbTe$	Si	Ti_2S	
Mn_3O_4	PbS	UO_3	Ge		
Co_3O_4	$PbSe$		Sn		

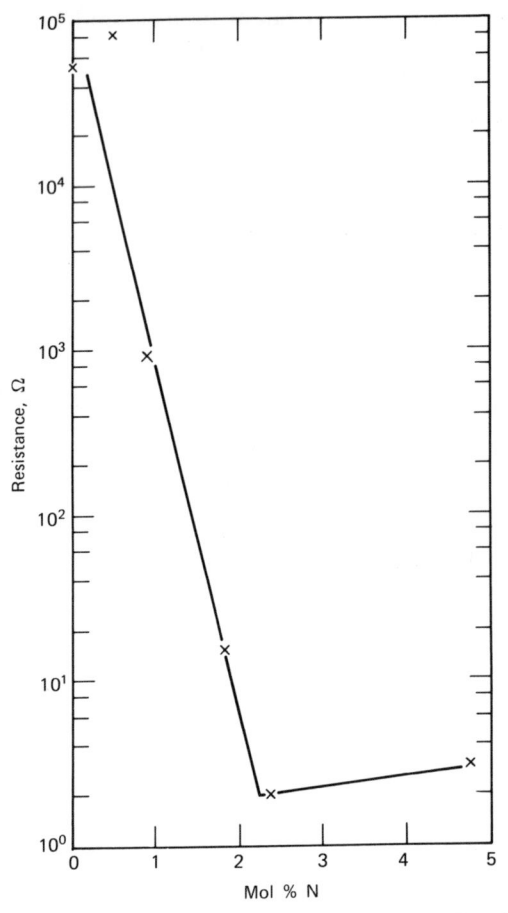

Figure 6. Electrical resistance of SiC hot-pressed with 1.8 mol % B and varying Si$_3$N$_4$ addition (4).

and empty bands. The concentration of conduction electrons, n, in the pure stoichiometric material is equal to the concentration of electron holes, p, and is given by (eq. 16),

$$n = p = 2 \left(\frac{2\pi kT}{h^2}\right)^{2/3} (m_e^* m_h^*)^{3/4} \exp\left(-\frac{E_g}{2kT}\right) \quad (16)$$

where h is Planck's constant and m_e^* and m_h^* are the effective masses of the electron and hole which depend on the strength of interactions between the electron, holes, and lattice and which may be larger or smaller than the rest mass of the electron. The room temperature band gaps for several materials are given in Table 6.

In an ideal covalent semiconductor, electrons in the conduction band and holes in the valence band may be considered as quasi-free particles. The environment of a periodic lattice and its periodic potential may be accounted for by the effective mass of the electron, m_e^*, and hole, m_h^*. The carriers have high drift mobilities in the range of 10 to 10^4 cm^2/(V·s) at room temperature (Table 7). This is the case for both metallic oxides and covalent semiconductors at room temperature (eg, Ge, Se, GaP, GaAs, CdS, CdTe) (see Semiconductors).

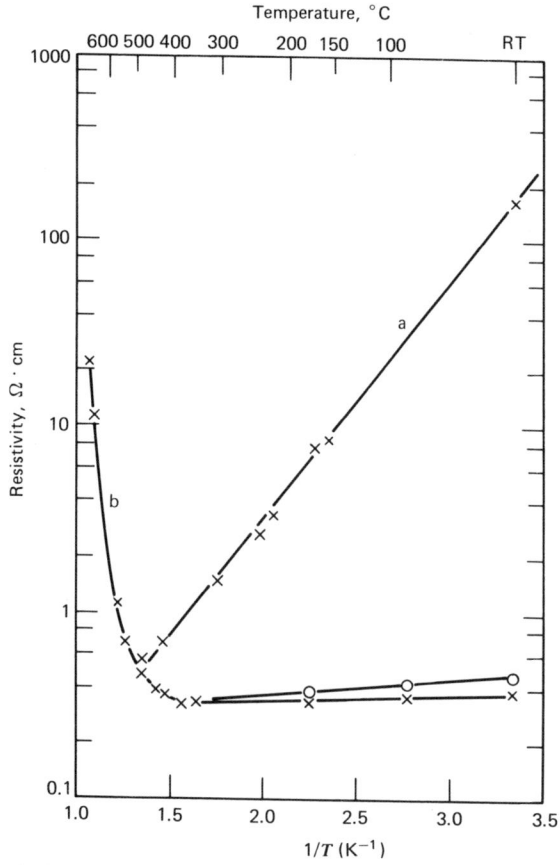

Figure 7. Electrical resistivity of hot-pressed SiC with (a) boron SiC + 0.5% B; and (b) boron and silicon nitride vs temperature SiC + 1% B + 1.8% N (4).

Two types of scattering affect the motion of electrons and holes. Lattice scattering results from thermal vibrations of the lattice and increases with the increasing amplitude of vibrations at higher temperatures, $\mu \propto T^{-3/2}$. A second source of scattering is the presence of impurities which distort the periodicity of the lattice, $\mu \propto T^{+3/2}$. The total mobility is proportional to the sum of these two terms. The temperature dependence of the mobility term for quasi-free electrons and holes is much smaller than that for their concentration (exponential). As a result, the conductivity has a temperature dependence chiefly determined by the concentration term.

In ionic host lattices where there is interaction between orbitals of neighboring ions, a polarization of the lattice associated with the presence of electronic carriers occurs. The associate, consisting of the electronic carrier plus its polarization field, is referred to as a polaron. When the association is weak (large polarons), conductivity similar to quasi-free electrons results with a small effective mass. When the linear dimension of the electronic carrier plus the lattice distortion is smaller than the lattice parameter (small polarons), the mobility is strongly affected by the lattice distortion which must move along with the electronic carrier. This process is often referred to as a hopping mechanism. The mobility of small polarons is usually less than 1 cm²/(V·s) and may be much lower.

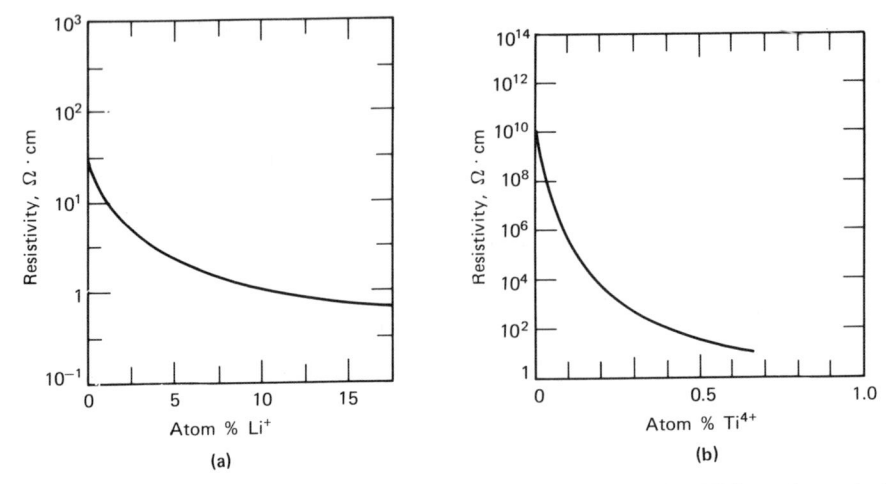

Figure 8. Effect of **(a)** added Li$_2$O on the conductivity of NiO and **(b)** added TiO$_2$ on the conductivity of Fe$_2$O$_3$ (5).

Nonstoichiometric and Solute-Controlled Electronic Ceramics. Most oxide semiconductors are either doped to create extrinsic defects or annealed under conditions in which they become nonstoichiometric. Although these effects have been carefully studied in many oxides, the precise nature of the low mobility value is often difficult to measure. Reported conductivities are often at variance because the effects of variable impurities and past thermal history overwhelm other effects. This section considers several electronically conducting ceramics and important features of their behavior. A list of some impurity semiconductors is given in Table 8. Because the impurity atoms introduce new localized energy levels for electrons, intermediate between the valence and conduction bands, impurities strongly influence the properties of semiconductors. If the new energy levels are unoccupied and lie close to the energy of the top of the valence band, electrons are easily excited out of the filled band into these new acceptor levels. This leaves an electron hole in the valence band that can contribute to electrical conductivity. Positive carrier (p-type) oxide conductors commonly arise as a result of nonstoichiometric composition with a decreased metal content (eg, Cu$_{2-x}$O) and are sometimes called deficit semiconductors. If the impurity additions have filled electron levels close to the energy level of the conduction band, electrons may be excited from impurity atoms into the conduction band; these are called donor levels. The electron excited into the conduction band is able to contribute to conductivity. Negative carrier (n-type) oxide conductors commonly result from a nonstoichiometric composition with an excess metal content (eg, Zn$_{1+x}$O) and are sometimes called excess semiconductors.

Silicon carbide can be doped with boron to provide acceptor levels within the band gap (ca 0.3 eV above the valence band), thus making it a p-type conductor, or nitrogen can be added to provide donor levels and n-type conduction (ca 0.07 eV below the conduction band). Figures 6 and 7 show the data of Prochazka and Smith (4) for polycrystalline SiC.

Strontium titanate is an n-type semiconductor in which two additional electrons are created when oxygen is removed from the material and oxygen vacancies act as donors, $\sigma \propto n \propto P_{O_2}^{-1/6}$. The mobility of the electrons in the conduction band is about

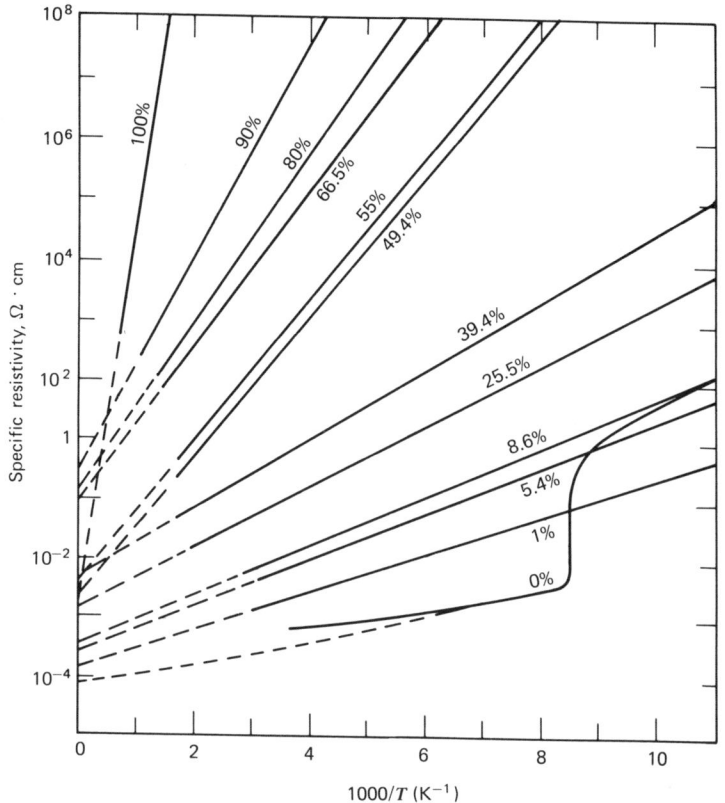

Figure 9. Specific resistivity of solid solutions of Fe_3O_4 and $MgCr_2O_4$. Mol % $MgCr_2O_4$ is indicated on the curves (6).

6 cm²/(V·s) [6 × 10⁻⁴ m²/(V·s)]. On the other hand, when ZnO is reduced, zinc interstitials are formed and these act as donors, $\sigma \propto P_{O_2}^{-1/4}$; each interstitial yielding a free electron.

Copper oxide is an p-type semiconductor, $Cu_{2-x}O$, in which copper vacancies act as acceptors to create electron holes in the valence band. Nickel monoxide forms a deficit semiconductor in which vacancies occur in cation sites similar to those for cuprous oxide. For each cation vacancy two electron holes must be formed; these are normally associated with a lattice cation. Semiconduction results from the transfer of positive charge from cation to cation through the lattice. This charge-transfer process corresponds to a low mobility [0.1 cm²/(V·s)]. Recent results suggest that with very pure stoichiometric crystals the mobility of holes in NiO may be 10–100 cm²/(V·s), and thus lower mobilities may be the result of impurity or vacancy trapping.

If a small amount of lithium oxide is added to the nickel oxide and the mixture is fired in air, a product with much lower resistivity is obtained. The resistivity of pure nickel oxide after firing in air is approximately 10⁶ Ω·m; the addition of 10 atom % lithium gives a product with a resistivity of about 0.01 Ω·m. Appreciable quantities of lithium dissolved in the nickel oxide lattice result in the formation of a Ni^{3+} content equal to the amount of lithium added to the solution. X-ray studies indicate a homogeneous crystal of the same structure as the initial nickel oxide but with a somewhat smaller unit cell. The formation of ions with increased valency (Ni^{3+}) is promoted by

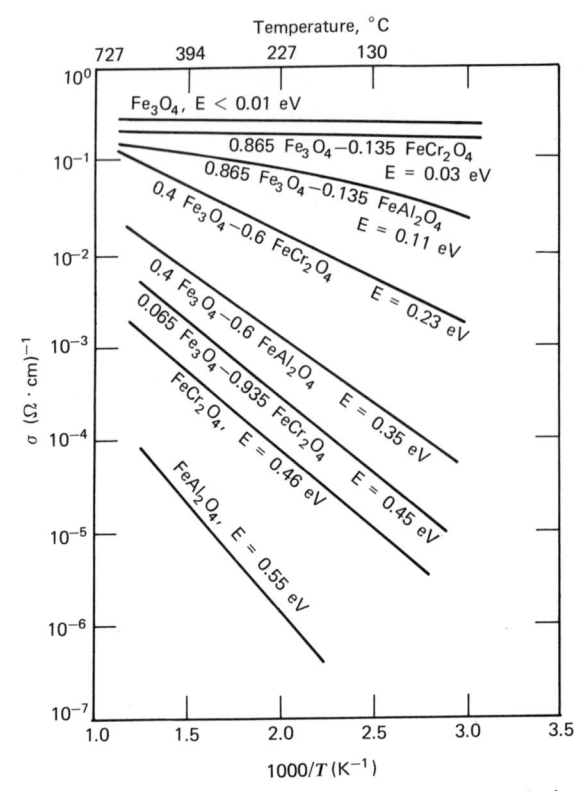

Figure 10. Electrical conductivity of solid solution spinels.

the introduction of lower-valence ions at normal cation sites. The ion must be approximately the same size as the ion being replaced and it must have a fixed valency. The second ionization potential of lithium is more than twice as large as the third ionization potential of nickel, hence this condition is satisfactorily fulfilled in the Li_2O–NiO system.

Similarly, the insulating characteristics of nickel oxide can be improved by the addition of a stable trivalent ion such as Cr^{3+} in solid solution. The addition of trivalent ions to the lattice decreases the fraction of Ni^{3+} ions formed. Since electron transfer between Ni^{2+} and Cr^{3+} does not occur, the overall conductivity is substantially decreased.

The same result can be obtained in many other systems of both n-type and p-type semiconductors. For example, Fe_2O_3 is an n-type semiconductor in which oxygen-ion vacancies are formed along with electrons that tend to be associated with specific cations. This is equivalent to forming a certain fraction of Fe^{2+} ions. If Ti^{4+} is added to Fe_2O_3 in solid solution, an increased fraction of the Fe^{3+} is forced into the Fe^{2+} state—a number equal to the Ti^{4+} additions. As a result, the conductivity of the product is substantially increased; it is determined primarily by the concentration of titanium oxide added and is much less dependent on oxygen pressure and firing conditions than is the pure material. Variations of conductivity with additions of Li_2O in NiO and TiO_2 in Fe_2O_3 are illustrated in Figure 8.

Tin oxide and indium oxide are other important semiconductors that are doped to increase their conductivity. SnO_2, In_2O_3, TiO_2, and $SrTiO_3$, in particular, are also

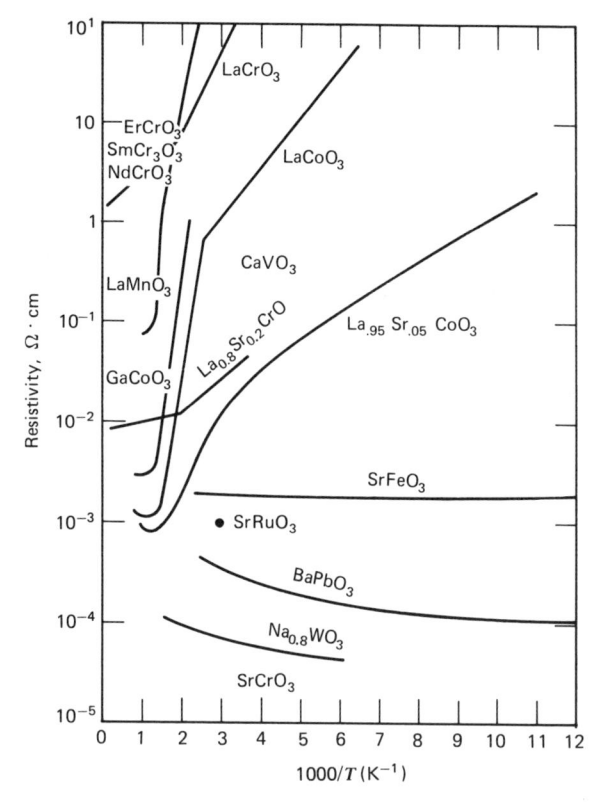

Figure 11. Resistivity of selected perovskite-structure oxides (7).

transparent to visible light and are often used as transparent electrodes on, for example, vidicon tubes.

Another method of obtaining semiconductors with controlled resistivity which avoids difficulties caused by stoichiometry deviations is the formation of solid solutions of two or more compounds with widely different conductivity. Magnetite, Fe_3O_4, is an excellent conductor with a specific resistance of about 10^{-4} $\Omega \cdot m$ compared to values of about 10^8 $\Omega \cdot m$ for most stoichiometric transition element oxides. The good electrical conductivity of magnetite is a function of the random location of Fe^{2+} and Fe^{3+} ions on octahedral sites that allows electron transfer from cation to cation. This is best illustrated by the order–disorder transformation occurring at about 120 K. Below this temperature the Fe^{2+} and Fe^{3+} ions are distributed in an ordered pattern on the octahedral sites; above this temperature Fe^{2+} and Fe^{3+} positions are randomly distributed. This results in a substantial increase in electrical conductivity (see Fig. 5).

In general, a condition for appreciable conductivity in the spinel structure is the presence of ions having multiple valence states as equivalent crystallographic sites. The number of these ions in Fe_3O_4 can be controlled by controlling the solid solutions in which Fe^{2+} or Fe^{3+} are diluted by other ions which do not participate in the electronic exchange. This is illustrated in Figures 9 and 10 for solid solutions of Fe_3O_4 with $MgCr_2O_4$, $FeAl_2O_4$, and $FeCr_2O_4$. The temperature coefficient or activation energy increases with increasing resistivity. Semiconductor materials of this type with ma-

308 CERAMICS AS ELECTRICAL MATERIALS

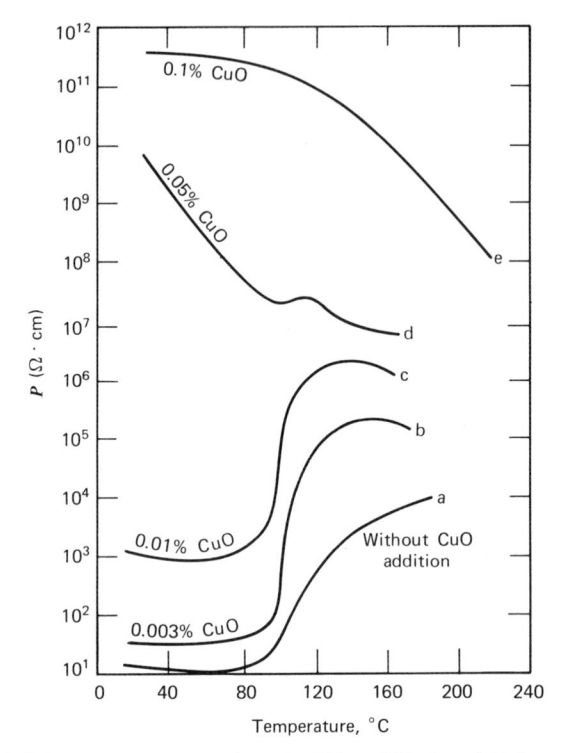

Figure 12. Resistivity vs temperature characteristics of Sb-doped high purity BaTiO$_3$-ceramics containing various amounts of CuO (9).

terials like MgAl$_2$O$_4$, MgCr$_2$O$_4$, and Zn$_2$TiO$_4$ as the nonconducting component can be prepared with a controlled temperature coefficient of resistivity. Semiconductors made in this way are used as thermistors. Since their electrical conductivity has a negative temperature coefficient, these types of materials are used as temperature controllers and monitors and are called NTCs (negative temperature coefficients).

Ceramics With High Electronic Conductivity or With Nonlinear Behavior. As previously indicated, many oxides are metallic conductors. The perovskite- and rutile-structure oxides are the most frequently studied. Figures 5 and 11 show that electronic conductivity can be very high and has little temperature dependence (metallic type). The d- or f-electron states of the transition metal ions overlap causing a partially filled band of ca 10^{28} electrons/m^3 for conduction. The mobility of the electrons is usually less than 1 cm^2/(V·s).

These materials are technologically significant as electrodes and in micro-electronics; the ReO$_2$ and RuO$_2$ compounds are used in organic and inorganic pastes as electrical components and are applied as thick films followed by a firing step.

The material most often studied is doped barium titanate, BaTiO$_3$, which is used as a barrier layer capacitor, intergranular capacitor, and nonlinear positive-temperature-coefficient (PTC) resistor. Controlled doping with lanthanum, copper, or antimony, for example, results in depletion layers near the grain boundaries in polycrystalline material. Figure 12 shows the change in resistivity (7 orders of magnitude) at room temperature caused by small changes in the CuO content. Two recent reviews for this important material are references 8 and 9.

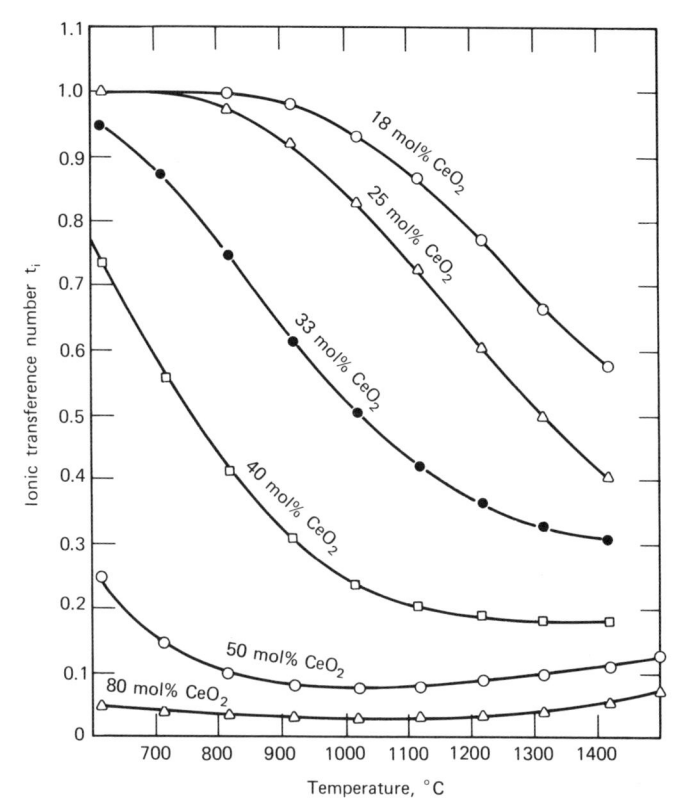

Figure 13. Ionic transference number of CeO_2–ZrO_2 compositions vs temperature (12).

Polycrystalline ZnO, with Bi_2O_3 and other additives, has nonohmic conduction and is used as a voltage-dependent resistor (varistor) to protect electronic equipment against voltage surges (10). The varistor responds very rapidly and allows the overvoltage pulse to be gated (the varistor resistance to ground decreases several orders of magnitude) away from the device circuitry. However, for nominal voltages, the resistance is sufficiently great so that little current leaks from the circuit. The mechanism was first thought to be the result of a Bi-rich grain boundary phase (11), but more recently it is thought to be a depletion layer near the grain boundary, and a resistive second phase in the grain boundary is not considered necessary for nonohmic behavior.

Mixed Conduction in Ceramics

The conductivity of many ceramic materials is significantly influenced by both ionic and electronic defects. Highly resistive ceramics such as MgO and Al_2O_3 fall into this category; however, good electrical conductors may be of a mixed type. The CeO_2–ZrO_2 system (Fig. 13) is an example of mixed conduction (12). The ionic transference number varies from 0.03 to 1.0, depending on temperature and composition.

For ceramics with electrical conductivities below ca 10^{-3} $(\Omega \cdot m)^{-1}$, the concentration-mobility product of the charge carrier is small. As a result, minor variations

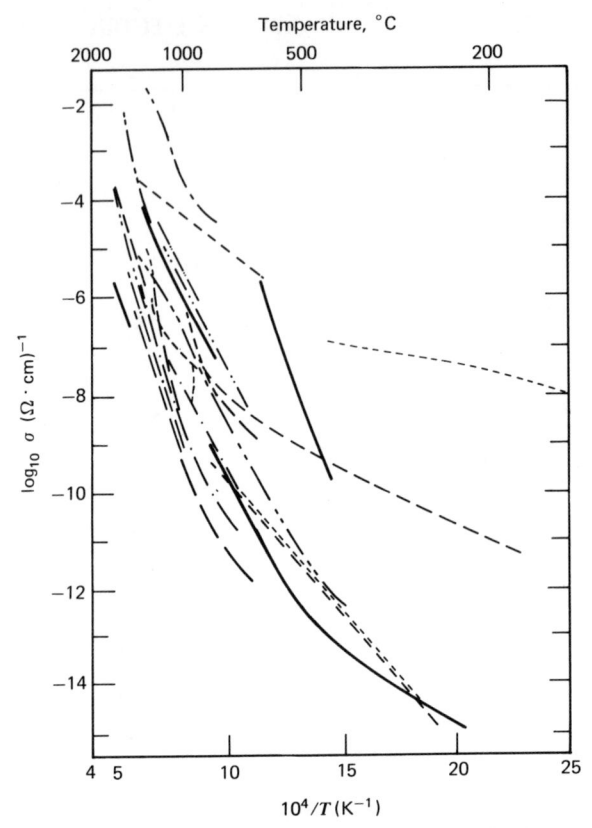

Figure 14. Reported data for the electrical conductivity of aluminum oxide (13).

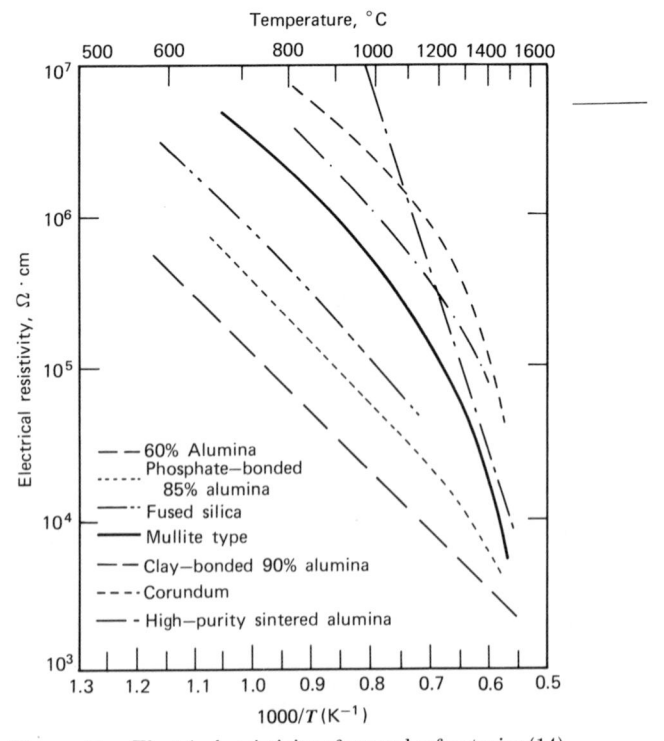

Figure 15. Electrical resistivity of several refractories (14).

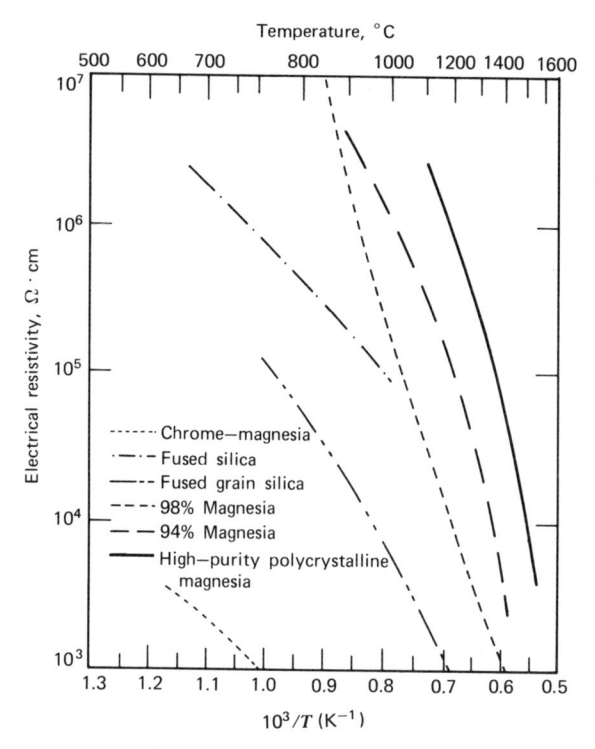

Figure 16. Electrical resistivity of several refractories (14).

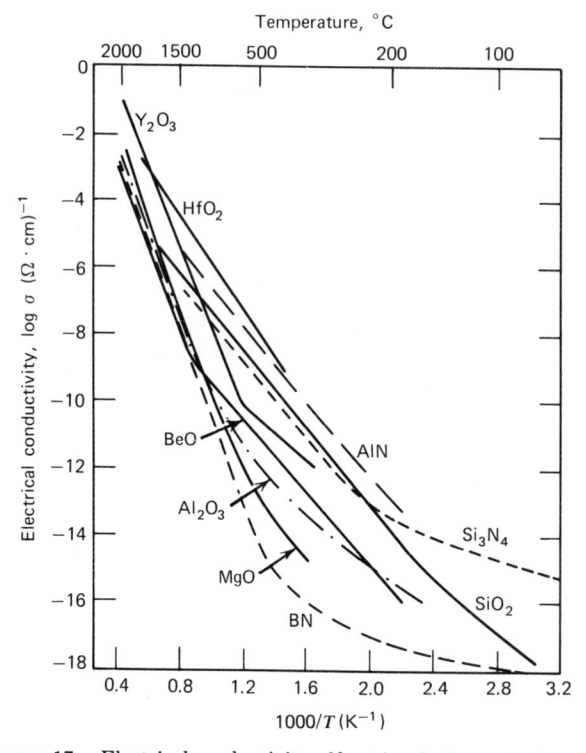

Figure 17. Electrical conductivity of best insulating materials

311

Table 9. List of Compounds Referred to in the Text and Their CAS Registry Numbers

Substance	Name	CAS Registry No.
CaO	calcium oxide	[1305-78-8]
ZrO_2	zirconium oxide	[1314-23-4]
Y_2O_3	yttrium oxide	[1314-36-9]
$CaZr_7O_{15}$	calcium zirconium oxide	[12506-31-9]
AgI	α-silver iodide, cubic	[7783-96-2]
CeO_2	cerium oxide	[1306-38-3]
$CeZr_3O_8$	cerium zirconium oxide	[12525-70-1]
Fe_3O_4	iron oxide, magnetite	[1317-61-9]
$FeAl_2O_4$	hercynite	[1302-61-0]
(LaSr)FeO$_3$	lanthanum strontium ferrite	[12186-37-7]
ZrB_2	zirconium diboride	[12045-64-6]
C	graphite	[7782-42-5]
$MgCr_2O_4$	chromium magnesium oxide, magnesium chrome spinel	[12053-26-8]
$NaAl_{11}O_{17}$	aluminum sodium oxide, β-alumina	[12005-48-0]
$NaAl_5O_8$	β″-alumina	[12005-16-2]
AgBr	silver bromide	[7785-23-1]
$RbAg_4I_5$	rubidium silver iodide	[12267-44-6]
ReO_3	rhenium oxide	[1314-28-9]
TiO	titanium oxide	[12137-20-1]
CrO_2	chromium oxide	[12018-01-8]
VO	vanadium oxide	[12035-98-2]
Ti_2O_3	titanium oxide	[1344-54-3]
V_2O_3	vanadium oxide	[1314-34-7]
FeO	wustite, ferrous oxide	[17125-56-3]
ReO_2	rhenium oxide	[12036-09-8]
NbO	niobium oxide	[12034-57-0]
MnO_2	manganese oxide	[1313-13-9]
MoO_2	molybdenum oxide	[18868-43-4]
Fe_2O_3	hermatite	[1317-60-8]
Mn_3O_4	hausmannite	[1309-53-3]
	manganese oxide	[1317-35-7]
CoO	cobalt oxide	[1307-96-6]
NiO	nickel oxide	[1313-99-1]
Cr_2O_3	chromium oxide	[1308-38-9]
SiC	carborundum; silicon carbide	[12504-67-5]
Si_3N_4	silicon nitride	[12033-89-5]
Li_2O	lithium oxide	[12057-24-8]
$FeCr_2O_4$	chromite, ferrous chromite	[1308-31-2]
$LaCrO_3$	chromium lanthanum oxide; lanthanum chromite	[12017-94-6]
$LaCoO_3$	cobalt lanthanum oxide; lanthanum cobaltite	[12016-86-3]
$CaVO_3$	calcium vanadium oxide	[12138-49-7]
$LaSrCoO_3$	cobalt strontium lanthanum oxide	[58799-80-7]
$SrRuO_3$	ruthenium strontium oxide; strontium ruthenate	[12169-14-1]
$BaPbO_3$	barium lead oxide	[12047-25-5]
WO_2	tungsten oxide	[12036-22-5]
IrO_2	irridium oxide	[12030-49-8]
Al_2O_3	corundum, α-alumina	[12252-63-0]
HfO_2	hafnium oxide	[12055-23-1]
AlN	aluminum nitride	[24304-00-5]
SiO_2	quartz	[14808-60-7]
BN	boron nitride	[10043-11-5]
BeO	beryllium oxide	[1304-56-9]

Table 9 (continued)

Substance	Name	CAS Registry No.
MgO	magnesium oxide	[1309-48-4]
BC	boron carbide	[12069-32-8]
Ge	germanium	[7740-56-4]
Cu	copper	[7440-50-8]
Fe	iron	[7439-89-6]
Mo	molybdenum	[7439-98-7]
W	tungsten	[7440-33-7]
NaCl	sodium chloride	[7647-14-5]
KCl	potassium chloride	[7447-40-7]
AgCl	silver chloride	[7783-90-6]
BaF_2	barium fluoride	[7787-32-8]
PbF_2	lead fluoride	[7783-46-2]
CuCl	copper chloride	[7758-89-6]
$BaTiO_3$	barium titanium oxide; barium titanate	[12047-27-7]
Cu_2O	copper oxide	[1317-39-1]
PbTe	lead telluride	[1314-91-6]
PbSe	lead selenide	[12069-00-0]
PbS	galena; lead sulfide	[12179-39-4]
CaF_2	calcium fluoride	[7789-75-5]
	fluorite	[14542-23-5]
CdO	cadmium oxide	[1306-19-0]
LiF	lithium fluoride	[7789-24-4]
Ga_2O_3	gallium oxide	[12024-21-4]
CdS	cadmium sulfide	[1306-23-6]
ZnSe	zinc selenide	[1315-09-9]
CdTe	cadmium telluride	[1306-25-8]
GaSb	gallium antimonide	[12064-03-8]
As_2S_3	arsenic sulfide	[1303-33-9]
$CoFe_2O_4$	cobalt iron oxide, spinel; cobalt ferrite	[12052-28-7]
$SrTiO_3$	strontium titanium oxide; strontium titanate	[12060-59-2]
SnO_2	tin oxide	[18282-10-5]
UO_2	uranium oxide	[1344-57-6]
$PbCrO_4$	lead chrome oxide	[18454-12-1]
Ag_2O	silver oxide	[20667-12-3]

in composition, impurity content heat treatment, stoichiometry, and other variables significantly affect measured results and experimental measurement techniques are more difficult. The reported data for a single material characteristically covers several orders of magnitude, as illustrated in Figure 14. These results should be interpreted cautiously in the absence of complete information on purity, solute concentration, nonstoichiometry, experimental techniques, possible presence of high-mobility paths such as grain boundaries and dislocations, prior heat treatment, and similar data. At low temperatures, where conductivities are in the range below 10^{-10} $(\Omega \cdot m)^{-1}$, difficulties in measurement caused by surface conductivity, grain boundaries, and other high-conductivity paths greatly impede precise measurements. The resistivity of several commercial refractories is shown in Figures 15 and 16. The best insulating materials (pure, single phase) are given in Figure 17.

For a listing of compounds with their Chemical Abstracts Service Registry Numbers, see Table 9.

BIBLIOGRAPHY

1. W. D. Kingery, H. K. Bowen, and D. R. Uhlmann, *Introduction to Ceramics,* 2nd ed., John Wiley & Sons, Inc., New York, 1976.
2. N. M. Tallen, ed., *Electrical Conductivity in Ceramics and Glasses,* Marcel Dekker, Inc., New York, 1974.
3. R. J. Friauf in J. Hladik, ed., "Basic Theory of Ionic Transport Processes," in *Physics of Electrolytes,* Vol. 1, Academic Press, Inc., New York, 1972.
4. S. Prochazka and R. C. Smith, General Electric Co., Schenectady, N.Y.
5. K. Lark-Horowitz, *Elec. Eng.* **68,** 1087 (1949).
6. E. J. Verwey, P. W. Haagman, and F. C. Romeijn, *J. Chem. Phys.* **15,** 18 (1947).
7. R. W. Vest and J. M. Honig in ref. 2, Chapt. 6.
8. J. Daniels and co-workers, *Philips Research Reports,* Vol. 31, No. 6, 1976, pp. 487–559.
9. W. Heywang, *J. Mater. Sci.* **6,** 1214 (1971).
10. K. Mukae, K. Tsuda, and I. Nagasawa, *Jpn. J. Appl. Phys.* **16,** 1361 (1977).
11. W. G. Morris, *J. Am. Ceram. Soc.* **56,** 360 (1973).
12. B. R. Rossing, L. H. Cadoff, and T. K. Gupta, *The Fabrication and Properties of Electrodes Based on Zirconium Dioxide,* Vol. 2, 6th Intl. Conf. MHD Electrical Power Generation, Washington, D. C., June 1975, pp. 105–117.
13. A. A. Bauer and J. L. Bates, *Battelle Mem. Inst. Rept.,* 1930 (July 31, 1974).
14. R. W. Wallace and E. Ruh, *J. Am. Ceram. Soc.* **50,** 358 (1967).

H. KENT BOWEN
Massachusetts Institute of Technology

CERAMICS, CHEMICAL WARE. See Ceramics.

CERMETS. See High temperature composites; Glassy metals.

CERIUM AND CERIUM COMPOUNDS

Cerium [7440-45-1] is a member of Group IIIA of the periodic table, variously referred to as rare earth metals, lanthanons, or lanthanides. Generally, it is recommended that the name rare earth metals be used for the elements scandium, yttrium, and lanthanum to lutetium, inclusive. The term lanthanum series (lanthanons) should be reserved for the elements from atomic numbers 57 to 71. It is also recommended that the term lanthanides should be further restricted, by the exclusion of lanthanum, to the inner series, atomic numbers 58 to 71 (1). However, this usage is by no means universal and all three terms are frequently used synonymously. In particular, the term rare earth, when used with a specific compound, has come to represent a mixture of lanthanons in their natural ratio of abundance, ie, one in which the cerium content usually is ca 50%; lanthanum, 25%; neodymium, 15%; praseodymium, 5%; and other lanthanons about 5% (see Rare earth elements).

Traditionally, the lanthanons have been classified into the cerium, terbium, and yttrium earth groups. The cerium group includes the elements from lanthanum (at no. 57) through samarium (at no. 62); the terbium group, the elements europium (at no. 63) through dysprosium (at no. 66); and the yttrium group, the elements holmium (at no. 67) through lutetium (at no. 71). The latter group is so named because yttrium (at no. 39), although not a member of the lanthanum series, is always found with these elements in nature. Practically, there is considerable overlapping of the groups and usually only the cerium and yttrium earth groups are differentiated.

Cerium has two common oxidation states: 3+, cerous, Ce(III), and 4+, ceric, Ce(IV). Cerous compounds resemble the other trivalent lanthanons, but ceric compounds are more like those of the elements titanium, zirconium, and thorium. The potential for the reaction $Ce^{3+} \rightleftarrows Ce^{4+} + e$ depends on the nature of the medium. Values of $E°_{298}$ are reported (6–9) between −1.28 and −1.70 V. The electrode potential for the reaction $Ce(s) \rightleftarrows Ce^{3+} + 3e$ is +2.335 V on the hydrogen scale (2). The properties of cerium are listed in Table 1.

Occurrence

The world's rare earth reserves and resources are large, and processing capacity exceeds the demand. It is estimated that the total world resources expressed as contained rare earth oxide is about 35 million metric tons with process reserves being about 7 million metric tons. The lanthanons constitute about 0.008% of the earth's crust and follow the odd-even rule in that the even atomic numbered members are more plentiful than the odd numbered members. Cerium is the most abundant of the group, generally comprising about 50 percent of a natural mixture. Although the lanthanons are widely distributed and there are many minerals containing them, sizable deposits are few in number, and only two ores, bastnasite and monazite, have achieved commercial importance.

In the 1970s bastnasite [12172-82-6], a fluorocarbonate having the general formula $Ln_2F_3(CO_3)_3$, has become the major source for rare earth elements. It is composed chiefly of the light rare earth elements, but contains more lanthanum and fewer heavier lanthanons than monazite. The refined mineral is composed of yellow to reddish-brown, translucent hexagonal crystals (sp gr, 4.9–5.2; Mohs hardness, 4–4.5). The only important commercial occurrence of bastnasite is a massive lode in the Mountain Pass

Table 1. Physical Properties of Cerium

Property	Value[a]
atomic number	58
atomic weight	140.42
isotopes, natural abundance, %	
^{136}Ce	0.193
^{138}Ce	0.250
^{140}Ce	88.48
^{142}Ce	11.07
crystal structure, atomic distance, nm	
α-Ce, fcc	0.485
β-Ce, hex	
γ-Ce, fcc	0.516
δ-Ce, bcc	0.411
melting point, °C	798
boiling point, °C	3257
density, kg/m^3	6773
heat capacity, J/(mol·K)[b]	26.96
thermal expansion per °C	8.5×10^{-6}
thermal conductivity, W/(m·K)	10.9
compressibility, GPa^{-1}[c]	4.18×10^{-2}
Young's modulus, GPa[c]	30
shear modulus, GPa[c]	12
Poisson's ratio	0.248
hardness, Vickers, MPa[d]	235
yield strength, MPa[d]	91.2
ultimate strength, MPa[d]	102
tensile elongation, %	24
thermal neutron cross section, m^2/atom	0.6×10^{-28}

[a] For γ-Ce at 25°C, the predominating modification at room temperature.
[b] To convert J to cal, divide by 4.184.
[c] To convert GPa to psi, multiply by 145,000.
[d] To convert MPa to psi, multiply by 145.

area of California, representing the world's largest single deposit and accounting for about two-thirds of world and about 90% of United States output. Typical ore generally analyzes about 50% calcite, 25% barite, 15% bastnasite, and 10% silica, monazite, galena, etc. The ore, after crushing and grinding, is upgraded by flotation to give a concentrate containing about 60% Ln_2O_3, which may be further upgraded to about 70% Ln_2O_3 by leaching with HCl. A typical analysis of the lanthanon content is: CeO_2, 50%; La_2O_3, 32%; Nd_2O_3, 13%; Pr_6O_{11}, 4%; Sm_2O_3, 0.5%; Gd_2O_3, 0.2%; Eu_2O_3, 0.1%; other Ln_2O_3, 0.2%.

Monazite [1306-41-8], a thorium–rare earth orthophosphate, is found in many countries such as India, Brazil, Australia, Malaysia, the Malagasy Republic, and the United States. It occurs primarily as beach placers and concentrates with other heavy minerals such as zircon, cassiterite, ilmenite, and rutile (See also Thorium and thorium compounds). The heavy mineral content may range up to 50%, with monazite only 1–5% of the heavy mineral content. Consequently deposits of this type are worked for the other heavy minerals, monazite being obtained as a by-product. Separation of such mixtures relies on differences in density, or in electrical or magnetic properties. Monazite is weakly magnetic and may be concentrated using a combination of magnetic cross belts. Material so obtained is a fine, uniform sand, usually less than 177 μm (80 mesh), and containing about 60% equivalent rare earth and thorium oxides. A massive deposit of monazite located in the Van Rhynsdorp region in the Republic of South Africa for many years supplied a high proportion of the world's monazite requirements until it was closed in 1966.

Monazite forms light brown to hyacinth-red monoclinic crystals, (sp gr, 4.9–5.5; Mohs hardness, 5–5.5). A typical analysis of refined ore is: CeO_2, 30%; "Di_2O_3," 32%; ThO_2, 6.5%; P_2O_5, 28%; SiO_2, 1.5%. (Although classically the expression "didymium," "Di," represented a mixture of neodymium and praseodymium, it is now commonly used to designate a mixture containing all the lanthanons other than cerium in their natural ratio of abundance.) The thorium oxide content of monazite varies considerably, ranging from 5–9% for commercial ores. Because of the content of thorium and small amounts of uranium and their radioactive daughter products, its possession and handling in the United States is controlled by the Nuclear Regulatory Commission. Similar requirements exist in all other countries.

Ore Extraction

Direct Chlorination at High Temperatures. Anhydrous rare earth chlorides may be derived directly from bastnasite or monazite by chlorination of an ore–carbon mixture in a special, electrically-heated furnace at 1000–1200°C. Fused anhydrous chlorides are obtained which are suitable for the production of rare earth metals (3).

BASTNASITE

Hydrochloric Acid Treatment. The ore, after being concentrated to about 60% Ln_2O_3 by flotation (qv), is dried and roasted at 480–675°C to drive off CO_2 and oxidize the cerium. The roasted ore is then slurried in water and treated with diluted (3–5%) hydrochloric acid to solubilize the bulk of the trivalent lanthanons, which are then separated from the solids containing the Ce(IV) by filtration. The crude cerium hydrate cake contains about 65–90% equivalent CeO_2, the remainder being other lanthanons and a minor amount of rare earth fluorides which are not solubilized by the hydrochloric acid treatment. The soluble lanthanon chloride solution is used as a source for various products that may be preferentially extracted by suitable organic solvents. The crude cerium hydrate cake is used as a source of cerium for glass batch additives and optical polishes (4–5).

Alternatively, the ore concentrate is treated with 30% HCl at about 100°C. The bulk of the lanthanon values is solubilized as chlorides, leaving an acid insoluble residue of fluorides. The latter is converted to hydroxides by boiling with 50% aqueous sodium hydroxide. After washing, the hydroxides are used to neutralize the excess acid in the initial chloride solution. The final neutral chloride solution may be concentrated by evaporation and cast into a solid form (6).

Sulfuric Acid Treatment. The ore is first roasted to drive off a substantial portion of the carbon dioxide present as carbonate and simultaneously oxidize the cerium to the 4+ state, in which it strongly complexes with the fluoride ion in acid solution. The roasted ore is then dissolved in dilute (6 N) sulfuric acid and the solution of lanthanon sulfate is removed from the gangue by filtration. Subsequent processing depends on the products desired. The entire batch may be precipitated as fluorides by adding sufficient hydrofluoric acid or alkali metal fluoride to the solution. Alternatively, the solution may be treated with a reducing agent such as sulfur dioxide to reduce the Ce(IV) to Ce(III). The liberated fluoride ions precipitate a portion of the lanthanons as fluorides. These are then treated with caustic soda and the hydroxides obtained are used to neutralize the excess acid in the separated sulfate solution, which may then be processed in a variety of ways depending on the products desired (7).

Bastnasite may also be opened by heating the unroasted ore with concentrated sulfuric acid. The fluoride in this case is driven off as hydrogen fluoride and the lanthanons remain as anhydrous sulfates. These are dissolved in water to separate the gangue and the rare earth sulfates are then recovered (8).

Treatment with Aqueous Sodium Hydroxide. Heating unroasted bastnasite with 50% sodium hydroxide solution converts the lanthanons to lanthanon hydroxides, and the carbonates and fluorides to soluble sodium salts. The lanthanon hydroxides are washed free of soluble impurities and dissolved in hydrochloric or nitric acid for further separation (9).

MONAZITE

Sulfuric Acid Treatment. Monazite contains an appreciable amount of thorium that must be removed, whereas bastnasite contains only a very small amount. In commercial practice, a mixture of monazite sand and 98% sulfuric acid is heated to 120–150°C in cast iron pots fitted with cast iron covers and heavy anchor stirrers. The reaction is exothermic and initial heating must be controlled to prevent the reaction from becoming violent. Depending on the quantity of sulfuric acid used, soluble or insoluble thorium compounds are obtained when the reaction mass is added to water. In either case, the lanthanons are converted to water-soluble sulfates. In general, with ores containing a considerable amount of gangue (calcium, silica, etc) both the thorium and lanthanon values are solubilized and the gangue is removed by filtration. The thorium may then be precipitated as crude pyrophosphate.

The lanthanons are precipitated from solution with sodium sulfate as lanthanon sulfate–sodium sulfate double salts (pink salts). The rare earth values may also be recovered as insoluble oxalates by addition of oxalic acid. In the latter case, recovery of the lanthanons is complete, but in the case of the double sulfates, some of the cerium earth elements and most of the yttrium earth elements remain in solution. This in effect gives a partial separation and upgrading of the heavy lanthanons and facilitates their recovery.

Lanthanons recovered as pink salts may be conveniently stored as such or may be converted to hydroxides and subsequently to the chlorides. On drying in air, cerium(III) oxidizes to cerium(IV). Because the precipitation pH of Ce(IV) salts is much below that of the trivalent rare earth elements, the latter may be dissolved by treatment with dilute acid (pH 2–3) leaving behind the hydrated ceric oxide.

Treatment with Aqueous Sodium Hydroxide. By heating monazite with a 65–70% sodium hydroxide solution, the rare earth phosphates in the ore are converted to hydroxides (10). After leaching the sodium phosphate from the reaction mass with hot water, the hydroxides are washed with water or dilute sodium hydroxide solution. The P_2O_5 concentration should be finally less than one percent. Thorium is separated from the lanthanons by a partial dissolution of the hydroxides in hydrochloric acid (pH 2–3).

The lanthanon chloride solution contains Ce(III). Oxidation with sodium hypochlorite gives 90% more of cerium oxide. The pH of the solution is kept at 3–4, in which range the preferential precipitation of hydrous ceric oxide takes place.

Separation and Purification

Regardless of the initial extraction process, separation of cerium from didymium takes advantage of the fact that Ce(IV) is much less basic than the trivalent lanthanons. Accordingly, Ce(III) must be oxidized to Ce(IV), and several methods are available for this purpose. If the Ce(III) is in solution, it may be oxidized anodically or chemically with bromate, permanganate, hypochlorite, or ozone. By maintaining the pH at 3–4, ceric hydroxide precipitates and may be recovered by filtration (11). Oxidation may also be effected by drying a mixture of rare earth hydroxides in air. The Ce(OH)$_3$ then oxidizes to Ce(OH)$_4$, leaving the Ln(OH)$_3$ unchanged. Leaching with dilute hydrochloric or nitric acid at a pH of 2–3 then solubilizes the trivalent lanthanons (didymium) which may be recovered by filtration. If properly carried out, the cerium contains less than 3–4% didymium and the didymium less than 2–3% cerium.

A number of fractionation techniques such as crystallization, precipitation, decomposition of salts, ion exchange, and solvent extraction were used formerly for separating a mixture of lanthanons into its components. Today, however, only ion exchange and solvent extraction have commercial significance. Although Ce(III) may be separated from the other trivalent lanthanons by ion exchange, this method is not economical and is used primarily for separating and purifying the heavy lanthanons (12). On the other hand, solvent extraction takes advantage of the fact that Ce(IV) is much more soluble in organic solvents such as tributyl phosphate (TBP), or di(2-ethylhexyl) phosphoric acid (DEHPA) than the trivalent lanthanons (13). For example, a mixture of Ce(IV) and trivalent lanthanons is dissolved in nitric acid, and the aqueous solution treated with a solution of TBP or DEHPA in kerosene. The Ce(IV) nitrate is much more soluble in the organic solvent than the Ln(III) nitrate and thus a partial separation is effected. The small amount of Ln(III) nitrate that does go into the organic phase is removed by scrubbing with dilute nitric acid or water, after which the purified cerium is recovered by first reducing it with hydrogen peroxide or sodium nitrite and then returning it to the aqueous phase (9). By using a conventional countercurrent stream of organic and aqueous phases, material of quite high purity may be obtained (see Extraction).

Cerium may also be purified and recovered from concentrated nitrate solutions by precipitating orange crystals of ammonium hexanitratocerate (see under Compounds) with an excess of ammonium nitrate. Crystals of 99.5% purity are obtained, which may be purified further by recrystallization. If ceric ammonium sulfate is desired, sulfuric acid may be substituted for nitric acid. However, a product of lower purity is obtained. In either case, considerable excess acid must be used to prevent the precipitation of basic cerium salts. Using this method, Ce(IV) is recovered in the form of a soluble solid salt.

Cerium Metal

Properties. Cerium metal resembles steel in appearance. The freshly cut metal has a bright silvery luster that tarnishes readily in air forming the oxide Ce_2O_3 or, in the presence of water vapor, $Ce(OH)_3$. At higher temperatures further oxidization to CeO_2 or $Ce(OH)_4$ occurs. Cerium dissolves readily in dilute mineral acids and also is attacked by alkaline solutions. Above 200°C vigorous reaction with chlorine, bromine, and iodine takes place to form the corresponding cerium trihalide. At high temperatures cerium reacts directly with carbon, sulfur, nitrogen, and boron to form metalloid compounds. Cerium combines with hydrogen at 345°C to form cerium hydride [15785-09-8], CeH_2. Cerium metal as well as cerium alloys such as mischmetal are strongly pyrophoric when sawed or scratched with a file, accounting for their important use in lighter flints. At elevated temperatures, the metal burns in air forming the stable oxide, CeO_2.

Preparation. Cerium metal and cerium alloys are prepared by metallothermic reduction of the halides using calcium or lithium as the reductant, or electrolytic reduction of the fused chloride.

In the metallothermic method (14), anhydrous cerium trifluoride is prepared by fluorinating cerium oxide with HF gas or NH_4HF_2 at 200–500°C. It is then mixed with granular calcium metal in an excess of about 10% of the stoichiometry required for the reaction:

$$2\,CeF_3 + 3\,Ca \rightarrow 2\,Ce + 3\,CaF_2$$

The charge is placed in a tantalum crucible and heated by induction to about 900°C. The reaction is exothermic and proceeds smoothly to completion in a few minutes. Holding the charge at the reaction temperature for about 15 min allows good separation of the metal and slag, after which the mass is allowed to cool to room temperature and the brittle slag broken away from the metal. The cerium metal may then be vacuum-cast to remove the excess calcium. Anhydrous $CeCl_3$ may be used instead of CeF_3, or lithium may be used as the reductant.

On a commercial scale, cerium metal and mischmetal are prepared by electrolysis of a melt containing the partially dehydrated chlorides mixed with sodium and calcium chloride in a cast iron pot which serves as the cathode. Graphite anodes are used and the pot is heated externally until fusion begins. The metal collects at the bottom of the pot and is easily removed and cast into ingots (15).

In a process developed by the U.S. Bureau of Mines, cerium as well as other lanthanons are prepared by electrolysis of the oxides in a fluoride melt. The electrolyte (63 wt % CeF_3, 16 wt % BaF_2, and 21 wt % LiF) is fused in a graphite crucible fitted with a tubular graphite anode and molybdenum or tungsten cathodes. Cerium oxide is fed to the cell continuously and the metal is reduced using a current of 785 A at 8.5 V. The electrolysis is stopped periodically and the metal tapped into ladles. This method can also be applied to the preparation of cerium alloys (16).

Cerium Alloys

The most important alloys of cerium are those with iron and those with other lanthanon metals (see Iron). Mischmetal [8049-20-5], an alloy in which the cerium comprises about 50 percent of the lanthanon content, is prepared by electrolysis as described above. The specific composition of a given alloy generally depends upon the application.

In particular, the use of mischmetal to increase the ductility of iron (nodular iron) has won acceptance over magnesium because the rare earth metals do not volatilize and thus improve control of the process. The addition usually is as ferro-mischmetal silicide or ferro-cerium silicide. The latter additions also have been used to control the sulfur content in plate and pipeline steels thereby improving the workability and transverse impact values (17–18) (see Iron; Steel).

Mischmetal has been used for many years in lighter flints and other sparking devices. A variety of combinations are used with the mischmetal content generally being about 70% and the remainder chiefly iron to increase the hardness. The use of mischmetal in fire starters in incendiary munitions takes advantage of the pyrophoric nature of mischmetal (19) (see Chemicals in war).

Other alloying uses of cerium have been to impart increased hardness and strength to magnesium and in permanent magnets of the type $LnCo_5$. However, the $CeCo_5$ [12214-13-0] composition is inferior to $SmCo_5$ and the lower cost does not appear to offer sufficient compensation to justify its use.

A joint study (20) by the SAE and the ASTM developed a unified numbering system UNS for over 2,000 metals and alloys in commercial production. The scheme of the (UNS) is an alpha-numeric code and rare earth metals and alloys are included in category E.

Cerium Compounds

Cerous salts, derived from the strong base, cerous hydroxide [15785-09-8], $Ce(OH)_3$, are only slightly hydrolyzed in solution, and like the other trivalent rare earths, the nitrate, chloride, sulfate, acetate, sulfamate, and formate are soluble in water, whereas the carbonate, oxalate, hydroxide, oxide, orthophosphate, fluoride, and many basic salts are insoluble. Cerous salts are colorless and have no absorption bands in the visible region of the spectrum. They do, however, show intense absorption in the ultraviolet region. Cerous hydroxide, when freshly precipitated, is colorless but on standing in air becomes violet and eventually yellow, because of oxidation to hydrated ceric oxide.

Ceric compounds are derived from the weak base ceric oxide, CeO_2, and are more acidic, more highly hydrolyzed, and more susceptible to complex formation than the corresponding cerous compounds. In their solubilities they more closely resemble thorium than the tripositive lanthanons, in keeping with size similarities as well as equality in charge. Salts of the strong acids, nitric and sulfuric, are soluble in concentrated solutions but are readily hydrolyzed to insoluble basic salts when diluted or heated.

Ceric salts are strongly colored. They are strong oxidizing agents and easily reduced by halogen acids, oxalic acid, ferrous salts, sulfurous acid, hydrogen peroxide in acid solution, etc. Likewise, cerous salts are oxidized in acid solutions by ammonium peroxydisulfate with a silver catalyst, sodium bismuthate, lead dioxide, etc, and in basic suspensions by alkali permanganates and hypochlorites. Oxygen and air also oxidize basic suspensions of cerous salts.

Because of the similarity of trivalent cerium to the other trivalent lanthanons, the methods described below usually apply to mixtures and may be used with the mixture representing the natural ratio of abundance, that is, ca 50% cerium.

Cerium Carbonate. The only simple carbonate of cerium is cerous carbonate [537-01-9], $Ce_2(CO_3)_3$. Cerous carbonate precipitates as the pentahydrate when an alkali bicarbonate solution is added to a solution of a cerous salt. Rapid addition gives a gelatinous precipitate that crystallizes on prolonged standing. A granular product may be made directly by slowly mixing two solutions containing equivalent concentrations of alkali carbonate and cerous salt at pH 4.5–6. Boiling hydrolyzes a solution of cerous carbonate to basic cerous carbonate [537-01-9], $CeOHCO_3$. Both the dry, normal, and basic carbonates decompose when calcined at 500°C to form CeO_2, CO, and CO_2. The carbonates are readily converted to other compounds and easily stored.

Cerium Halides (21). Cerous chloride [7790-86-5], $CeCl_3$, is prepared by dissolving cerous carbonate or cerous hydroxide in hydrochloric acid. It also may be prepared by dissolving hydrous ceric oxide in hydrochloric acid, whereupon the cerium is reduced to the cerous state and chlorine is evolved. Cerous chloride heptahydrate [18618-55-8] $CeCl_3.7H_2O$, may be made by saturating a concentrated solution of cerous chloride, with HCl or by evaporating a cerous chloride solution to a syrup and allowing it to cool.

Anhydrous cerous chloride containing a small amount of basic cerium chloride and moisture is made commercially by heating hydrated cerous chloride in cast iron pots under conditions where access to air is restricted. Anhydrous cerous chloride free from basic salts is best prepared by heating the hydrated chlorides in a current of hydrogen chloride gas or by heating a mixture of the hydrated chloride and ammonium chloride in a vacuum. Anhydrous rare earth chlorides are prepared similarly with ammonium chloride. However, the oxide instead of hydrated chloride may be used as starting material.

Ceric chloride exists as chloroceric acid [48015-96-9], H_2CeCl_6. In the presence of excess chloride, Ce(IV) is reduced to Ce(III) with the liberation of chlorine.

Cerous fluoride tetrahydrate [7758-88-5], $CeF_3.4H_2O$, is made by addition of hydrofluoric acid to a soluble cerous salt, or by treating cerous oxalate with hydrofluoric acid. Ceric fluoride [10060-10-3], CeF_4, may be prepared by treating anhydrous cerous fluoride [7758-88-5] with fluorine at 500–650°C.

Cerium Nitrates. *Cerous nitrate hexahydrate* [10294-41-4], $Ce(NO_3)_3.6H_2O$, is prepared by dissolving cerous carbonate in nitric acid or by dissolving ceric hydroxide in nitric acid in the presence of oxalic acid or hydrogen peroxide to reduce the Ce(IV) to Ce(III). The solid cerous nitrate is recovered by crystallization. Ceric nitrate [13093-17-9], made by dissolving ceric hydroxide in nitric acid, exists only in acid solutions. Evaporation leads to precipitation of the basic nitrate. *Ammonium hexanitratocerate* [16774-21-3] (ceric ammonium nitrate), $(NH_4)_2Ce(NO_3)_6$, precipitates when ammonium nitrate is added to a solution of ceric nitrate containing an excess of nitric acid. It also is made by the electrolytic oxidation of cerous nitrate in nitric acid solution. The orange-colored salt is obtained in very pure form (>99.5%; see under Purification) and may be used as a standard of reference in oxidimetry. In the presence of excess concentrated nitric acid, ceric nitrate or ceric ammonium nitrate assumes a cherry-red color, apparently as a result of complex formation.

Cerium Oxalate. Only cerous oxalate nonahydrate [19511-08-1], $Ce_2(C_2O_4)_3.9H_2O$ is stable. It is a sparingly soluble salt formed as a colorless crystalline precipitate when slightly acid cerium solutions are treated with oxalic acid or a soluble oxalate. It is appreciably soluble in dilute acids, but insoluble in excess oxalic acid.

Cerium Oxides and Hydroxides. *Ceric oxide* [1306-38-3] (ceria), CeO_2, is usually made by calcining cerous oxalate or cerous or ceric hydroxide in air. Prolonged calcination of salts of volatile acids also gives ceric oxide. Ceric oxide may range in color from dark brown to pure white. The darker colors are attributed to the presence of other rare earths, notably praseodymium oxide, Pr_6O_{11}. The off-white and light tan colors of pure CeO_2 are ascribed to lattice defects. Pure white CeO_2 can only be prepared by special techniques.

Ceric oxide is refractory and soluble with difficulty in acids. It is more soluble in the presence of a reducing agent such as hydroxylamine.

Cerous oxide [1345-13-1], Ce_2O_3, is formed by the reduction of CeO_2 heated in a stream of hydrogen at 1200–1400°C, or by the reduction of CeO_2 by heating a mixture of CeO_2 and powdered carbon in a static CO atmosphere at 1250°C. The product from the hydrogen reduction is unstable in air, reoxidizing to CeO_2, whereas that obtained by carbothermic reduction is stable (22).

Ceric hydroxide [12014-56-1] (hydrous ceric oxide, cerium hydrate), $CeO_2.xH_2O$ ($x = \frac{1}{2}$ to 2), is a gelatinous precipitate formed when sodium or ammonium hydroxides are added to solutions of ceric salts. Upon drying, a yellow hydrated oxide containing 85–90% CeO_2 is obtained. Granular ceric hydroxide may be made by boiling insoluble cerium salts with concentrated NaOH, followed by washing and drying. In many applications, cerium hydrate may be substituted for cerium oxide.

Cerium Sulfates. *Cerous sulfate* [17193-06-5], $Ce_2(SO_4)_3$, exists in a number of hydrated forms containing 4, 5, 7, 9, and 12 molecules of water. These are prepared under varying conditions of evaporation of cerous sulfate solutions. Anhydrous cerous sulfate is made by heating the hydrated salt at about 350–400°C. Reagent-grade cerous sulfate is made by reducing ceric sulfate in dilute sulfuric acid solution (sulfatoceric acid) with hydrogen peroxide.

Double cerous sulfates of the type $Ce_2(SO_4)_3.M_2SO_4.xH_2O$, where M is an alkali metal or ammonium and x is 2, 4, or 8, are formed by precipitation from cerous sulfate solutions by addition of the alkali sulfate. These salts are crystalline, and many other varieties containing more molecules of alkali sulfate per molecule of cerous sulfate are known. The sodium and potassium salts are sparingly soluble in water, but the ammonium salt is appreciably soluble.

Ceric sulfate [17106-39-7], $Ce(SO_4)_2$, is made by heating ceric oxide or hydrated ceric oxide with concentrated sulfuric acid in which it is insoluble. The excess acid is removed by washing with glacial acetic acid. Orange ceric sulfate is soluble in dilute sulfuric acid where it exists as sulfatoceric acids.

Ammonium trisulfatocerate dihydrate [10378-47-9] (ceric ammonium sulfate), $(NH_4)_2[Ce(SO_4)_3].2H_2O$, is made by methods similar to those described for ammonium hexanitratocerate.

Economic Aspects

Prices, depending on purity and the contained CeO_2 or Ln_2O_3 content, vary from about $1.30/kg for rare earth chloride (46% Ln_2O_3, 22% CeO_2) to about $13/kg for 99.9% CeO_2. Prices for optical-grade cerium oxide range from $4 to $6 per kg. The price of cerium metal (99.9%) is about $100/kg and mischmetal about $7/kg.

In 1974 the estimated United States production expressed as equivalent rare earth oxide was 20,000 metric tons or about 65 percent of total world production. Both of these values represent about 65–70% of production capacity (23); United States consumption was about 16,500 t.

Analysis

Qualitative. Preliminary separation of the lanthanons as a group has to precede the detection and determination (24–25) of cerium. In general, no single reagent can be used exclusively for the detection of lanthanons and some combination of the precipitation of the oxalate, fluoride, or hydroxide is used, depending upon what other elements are present. The oxalate precipitation carried out at pH 1 is the most nearly specific method and generally effects separation from all the elements usually present except thorium and the alkaline earths. Calcium, magnesium, zinc, or aluminum may be eliminated by controlled hydroxide precipitation, whereas precipitation as the fluoride removes iron, zirconium, and titanium. There are no qualitative chemical tests for individual lanthanons in the presence of others. An exception is cerium which can be detected in the absence of iron by the orange precipitate formed with excess ammonium hydroxide and hydrogen peroxide. The test is quite sensitive and indicates quantities as small as 20 µg.

Quantitative Determination. *Gravimetric.* After the necessary preliminary separations have been made, the total lanthanon content of a sample may be determined gravimetrically by calcining the oxalate or hydroxide to the oxide at about 1000°C. Inasmuch as the oxalates or hydroxides are intermediates in the separation procedures, this is a convenient method for quantitative determination unless appreciable quantities of alkaline earths are present and interfere. In this case, the oxide is redissolved and the lanthanon content titrated.

Volumetric methods are used extensively, including titration with complexing agents such as ethylenediaminetetraacetic acid (EDTA) or diethylenetriaminepentaacetic acid (DTPA). Any Ce(IV) is first reduced to Ce(III) with a suitable reducing agent such as ascorbic acid. Such procedures can be performed rapidly and are useful in process control work. Interferences are not usually serious and can be largely avoided by pH adjustment.

The percentage of cerium in the total lanthanon oxide content may be determined by oxidizing the Ce(III) to Ce(IV) with ammonium peroxydisulfate catalyzed by silver ion, followed by titration with standard ferrous ammonium sulfate using ferroin as an indicator. Alternatively, the ratio of Ce(IV) to total cerium (percent oxidation) is measured by comparing the volume ratios of the ferrous ammonium sulfate solutions used before and after the ammonium peroxydisulfate treatment.

Instrumental Methods. Spectrophotometry is used widely to determine both the composition of a crude mixture and pure lanthanons. However, determination of cerium is difficult because of interferences in the ultraviolet region of the spectrum although, in the absence of such interferences, the absorption peak at 253 nm may be used. X-ray fluorescence is a useful and versatile method applicable to concentrations ranging from 100% to about 0.005% and may be used directly on minerals, rare earth concentrate products, or process solutions.

The d-c arc spectrograph is useful for determining small amounts of noncerium lanthanons and some nonlanthanon impurities in a cerium oxide matrix, and also small amounts of cerium in a noncerium lanthanon matrix. Detection levels are usually in the range 10–1000 ppm.

Atomic absorption has been used for lanthanon analyses, but high temperatures are required and the method is better suited to the determination of nonlanthanon elements such as calcium, sodium, etc, in a lanthanon matrix. Other instrumental methods such as mass spectroscopy, neutron activation analysis, etc, are useful in special cases, but are not widely used (see Analytical methods).

Toxicology

Cerium is similar to the other lanthanons in that its compounds have low to moderate toxicity ratings and cause very little change in animals when fed for several months. The low oral toxicity is ascribed to poor intestinal absorption. Local injection or inhalation of lanthanon compounds has been found to induce both skin and lung granulomas (26).

Uses

Cerium compounds are available as (1) purified compounds in which the equivalent cerium oxide content with respect to other lanthanons is greater than 99.5%; (2) commercial-grade compounds in which the cerium oxide content is 80–95% of the total rare earth oxide content; and (3) rare earth mixtures in which the cerium oxide to total rare earth oxide ratio is that found in the ore, that is, approximately one half.

By far the greatest use of cerium is in the form of rare earth mixtures where its dilution is more than offset by the lower cost. Based on equivalent oxide weight, it is estimated that of the total United States consumption of about 16,500 metric tons in 1974, about 44% was used in petroleum-cracking catalysts; about 34% in metallurgical applications; about 17% in ceramics and glass; and the remaining 5% in a variety of uses including arc carbons, research, etc (23).

The largest use of rare earth compounds continues to be in cracking catalysts containing rare earth zeolite as the active ingredient (see Molecular sieves). These catalysts, which have an Ln_2O_3 content of 1–5%, remarkably improve the gasoline output of refineries (27). Other catalytic uses of rare earths have been for the oxidation of carbon monoxide and in air pollution control (28) (see Air pollution control methods; Exhaust control).

Another major use of the rare earth metals is as additives (ca 0.2 wt %) to iron and steel to nodularize the graphite and counteract the influence of various impurities such as sulfur (see Iron; Steel). Cerium is also used in pyrophoric devices such as lighter flints and in military ordnance.

Cerium oxide is used extensively in the glass industry for polishing glass (29) and also as a decolorizer. It is a better and cleaner glass polish than rouge (iron oxide) and finds particular application in the polishing of spectacle lenses, television tube face plates, and mirrors. Hydrous ceric oxide added to conventional glass polishes may enhance polishing activity by as much as 50% (30).

As a decolorizer in glass formulations ceric hydroxide is used in conjunction with sodium nitrate and physical decolorizers such as manganese oxide, didymium carbonate, hydrated rare earth oxides, cobalt oxide, and selenium. The cerium acts as an oxygen carrier and oxidizing agent to oxidize iron so that the resulting glass is stable toward solarization and reheating. Glass decolorized with cerium and didymium (1–2 kg/1000 kg sand) is remarkably clear and does not have a gray appearance. Cerium is also used in glass to impart color and to eliminate the transmission of ultraviolet light. With titanium, cerium in glass gives a yellow color due to the formation of cerium titanate [12185-87-4] . Other uses are in the preparation of signal glasses, ophthalmic lenses, and other optical glasses having special properties (see Glass).

A wide variety of uses of cerium as well as of rare earths in general are described in refs. 31 and 32.

BIBLIOGRAPHY

"Cerium" in *ECT* 1st ed., Vol. 3, pp. 634–647, by H. E. Kremers, Lindsay Light and Chemical Company; "Cerium and Cerium Compounds" in *ECT* 2nd ed., Vol. 4, pp. 840–854, by Walter L. Silvernail and Robert M. Healy, American Potash & Chemical Corporation.

1. Commission on Nomenclature of Inorganic Chemistry, *J. Am. Chem. Soc.* **82,** 5523 (1960).
2. F. H. Spedding and C. F. Miller, *J. Am. Chem. Soc.* **74,** 4195 (1952).
3. W. Brugger and E. Greinacher, *Proc. 6th Rare Earth Conf., May 3–5, 1967, Gatlinburg, Tenn.,* pp. 702–713; *J. Met.* **19**(12), 32 (1967).
4. P. R. Kruesi and N. N. Schiff, "Molybdenum Corporation of America's Mountain Pass Europium Process," *paper presented at TMS–AIME Meeting Feb. 25–29, 1968,* New York.
5. U.S. Pat. 3,812,233 (May 21, 1974), L. K. Duncan (to W. R. Grace and Co.).
6. P. R. Kruesi and G. Duker, *J. Met.* **17,** 847 (1965).
7. U.S. Pat. 2,900,231 (Aug. 18, 1959), H. E. Kremers, D. W. Newman, and F. C. Kautzky (to American Potash & Chemical Corporation).
8. V. E. Shaw, *U.S. Bur. Mines Rep. Invest.,* 5474 (1959).
9. D. J. Bauer and V. E. Shaw, *U.S. Bur. Mines Rep. Invest.,* 6381 (1964).
10. U.S. Pat. 2,783,125 (Feb. 26, 1957), C. de Rhoden and M. Peltier (to Société des Produits Chimiques des Terres Rares, Paris).
11. D. J. Bauer and R. E. Lindstrom in ref. 9, pp. 692–699.
12. W. L. Silvernail and M. M. Woyski in ref. 9, pp. 678–690.
13. J. C. Warf, *J. Am. Chem. Soc.* **71,** 3257 (1949).
14. A. H. Daane in F. H. Spedding and A. H. Daane, eds., *The Rare Earths,* John Wiley & Sons, Inc., New York, 1961, Chapt. 8.
15. I. S. Hirschhorn in ref. 9, pp. 728–739.
16. E. Morrice, E. S. Shedd, and T. A. Henrie in ref. 3, pp. 715–725; *U.S. Bur. Mines Rep. Invest.,* 7146, (1968).
17. K. Reinhardt, *Addition Techniques of Rare Earth Metals for the Treatment of Special Steels with Rare Earths* in ref. 11, pp. 689–696.
18. D. Ghosh and D. A. R. Kay in *Proc. 12th Rare Earth Res. Conf., July 18–22, 1976, Vail, Colorado,* pp. 1083–1092.
19. B. A. Kulp, *Proc. 10th Rare Earth Res. Conf., Apr. 30–May 3, 1973, Carefree, Arizona,* pp. 53–62.
20. K. A. Gschneidner, Jr., *Commercial Rare Earth Metals and Alloys Covered by the Unified Numbering System (UNS),* IS-RIC-8 Rare Earth Information Center, Iowa State University, Ames, Iowa, July 1976.
21. D. Brown, *Halides of the Lanthanides and Actinides,* John Wiley & Sons, Ltd., London, Eng., 1968.
22. E. S. Shedd and T. A. Henrie, *Proc. 3rd Rare Earth Res. Conf., Apr. 21–24, 1963, Clearwater, Florida,* pp. 21–27.
23. J. H. Jolly, *Rare Earth Minerals and Metals,* Mineral Yearbook, Bureau of Mines, U.S. Department of Interior, Washington, D.C., 1974.
24. M. M. Woyski and R. E. Harris in I. M. Kolthoff and Philip J. Elving, eds., *Treatise on Analytical Chemistry,* Part 2, Vol. 8, Interscience Publishers, a division of John Wiley & Sons, Inc., New York, 1963.
25. O. B. Michelsen, ed., *Analysis and Application of Rare Earth Materials, Proc. of the NATO Advanced Study Institute, August 23–29, 1972, Kjeller, Norway.*
26. T. J. Haley, *J. Pharm. Sci.* **54,** 663 (1965).
27. R. L. Koffler, *Proc. 7th Rare Earth Res. Conf., Oct. 28–30, 1968, Coronado, Calif.,* pp. 697–713.
28. M. Steinberg in ref. 25, pp. 263–270.

29. W. L. Silvernail and N. J. Goetzinger, *Proc. 10th Rare Earth Res. Conf., Apr. 30–May 3, 1973, Carefree, Arizona*, Vol. II, pp. 652–661.
30. U.S. Pat. 3,097,083 (July 9, 1963), W. L. Silvernail (to American Potash & Chemical Corporation).
31. R. M. Mandle and H. H. Mandle in L. Eyring, ed., *Progress in the Science and Technology of the Rare Earths*, Vol. 1, Pergamon Press, Inc., New York, 1964, pp. 416–500.
32. *Ibid.* Vol. 2.

General References

T. Moeller, *The Chemistry of the Lanthanides*, Reinhold Publishing Corp., New York, 1963.
R. C. Vickery, *The Chemistry of the Lanthanons*, Academic Press, Inc., London, Eng., 1953.
R. J. Callow, *The Industrial Chemistry of the Lanthanons, Yttrium, Thorium, and Uranium*, Pergamon Press, Inc., New York, 1967.

WALTER L. SILVERNAIL
Consultant

CESIUM AND CESIUM COMPOUNDS

Cesium [7440-46-2], is the most electropositive and the least abundant of the naturally occurring alkali metals. It resembles potassium and rubidium in the metallic state, and its chemistry resembles that of potassium and rubidium much more than that of sodium or lithium. Cesium is used in preference to other alkali metals only when maximum reactivity, maximum vapor pressure, maximum atomic weight, or minimum electronic work function is required.

The element was discovered in 1860 by Bunsen and Kirchoff. Cesium was the first element discovered spectroscopically, and the name, after the Latin *caesius* for sky-blue, refers to the characteristic blue spectral lines of the element. Cesium salts were not successfully reduced to metal until 1881, as electrolysis of the molten chloride did not yield cesium metal under the same conditions that led to reduction of the other alkali metal chlorides.

Setterberg (1) first obtained cesium in the metallic state by electrolysis of a cesium cyanide–barium cyanide melt, and subsequently the now more common thermochemical–reduction techniques were developed by other workers (2–3). Cesium had no significant industrial utility until 1926 when the metal came into use as a getter and an effective agent in reducing the electron work function on coated tungsten filaments in radio tubes (see Vacuum technology). Development of photoelectric cells in the early thirties provided a small, but consistent market for small quantities of cesium; several other applications for cesium in photosensing elements were developed particularly during World War II. The primary source of cesium chemicals has been the rare mineral pollucite [1308-53-8], ideally $Cs_2O.Al_2O_3.4SiO_2$. Known reserves of this rare mineral are sufficient for current and any anticipated future cesium requirements (4).

Many of the cesium chemicals produced since 1958 have been used in research on thermionic power conversion, ion propulsion, and magnetohydrodynamics (MHD); commercial utilization of cesium in any of these fields offers the main potential for the use of cesium chemicals in quantities above the current consumption level of a few metric tons per year of cesium salts and a few hundred kg/yr of cesium metal (5).

Physical Properties

Pure cesium is a silvery-white, soft, ductile metal that melts just above normal room temperature (the color of cesium metal has often been reported as golden, but this is probably due to alteration of the surface by minute traces of oxygen). Of the five naturally occurring alkali metals, cesium has the highest vapor pressure, highest density, lowest boiling point, and the lowest ionization potential. Selected physical properties of cesium are given in Table 1.

Chemical Properties

Cesium is very similar in chemical behavior to potassium and rubidium except that cesium is oxidized more readily than any of the other alkali metals. Thus when cesium is exposed to air, the metal ignites spontaneously and burns vigorously with a reddish-violet flame. It reacts with water to give the hydroxide and hydrogen. If cesium is exposed to both air and water, a hydrogen explosion usually occurs since burning cesium readily ignites the liberated hydrogen gas. Cesium is the most reactive alkali metal in reactions with oxygen and the halogens, but it is the least reactive of the alkali metals with nitrogen, carbon, and hydrogen. The hydroxide absorbs carbon dioxide readily and attacks glass relatively rapidly.

Cesium forms simple alkyl and aryl compounds that are similar to those of the other alkali metals (6). The simple alkyl and aryl compounds are colorless, solid, amorphous, nonvolatile, and insoluble (except by decomposition) in most solvents except diethylzinc. The reactivity of cesium alkyls with halides and nitriles is greater than that of the other alkali metal alkyls and, because of their exceptional reactivity, cesium alkyls should be effective in alkylations wherever other alkaline alkyls or Grignard reagents have failed. Cesium reacts readily with hydrocarbons in which the activity of a C—H link is increased by attachment to the carbon atom of doubly linked or aromatic radicals. A brown, solid addition product is formed when cesium reacts with ethylene, and a very reactive, dark-red powder, triphenylmethylcesium [76-83-5], $(C_6H_5)_3CCs$, is formed by the reaction of cesium amalgam with a solution of triphenylmethyl chloride in anhydrous ether.

Cesium forms neutral complex compounds with two or four moles of salicylaldehyde per mole of cesium, indicating a coordination number of four or six. However, cesium does not form such complexes readily, and the relatively high ionic mobility of the cesium ion in aqueous solutions indicates that any hydration water about the cesium ions is held very loosely if held at all.

Cesium ions can be extracted from aqueous solutions into organic phases by the formation of ion-pair association complexes which are soluble in high dielectric-constant organic solvents.

Cesium salts are generally similar chemically to other alkali metal salts. The

Table 1. Properties of Cesium

Property	Value
atomic number	55
atomic weight (^{12}C = 12.000)	132.905
melting point, °C	28.64 ± 0.17
boiling point, °C	670
density, g/mL	
solid, 18°C	1.892
liquid, 40°C	1.827
viscosity at mp, mPa·s (=cP)	0.686
surface tension at mp, mN/m (=dyn/cm)	39.4
vapor pressure, kPa[a]	
solid, from −23°C to mp	$\log_{10}P = -4120/T + 11.321 - 1.0 \log_{10}T$
liquid, from mp to 377°C	$\log_{10}P = -4042/T + 12.051 - 1.4 \log_{10}T$
heat of fusion, J/g[b]	16.372
entropy of fusion, J/(mol·K)	7.217
heat of vaporization, J/g[b]	611
specific heat, J/(g·K)[b]	
solid, 20°C	0.217
liquid, bp	0.239
vapor, bp	0.156
thermal conductivity, W/(m·K)	
liquid, mp	18.4
vapor, bp	0.0046
ionization potential, eV	3.893
conductivity, solid (mp), S/cm[c]	4.50×10^4
specific magnetic susceptibility	0.22×10^6
neutron absorption cross section, m^2	
thermal	$(29.0 \pm 1.0) \times 10^{-28}$
fast	$(15 \pm 0.2) \times 10^{-31}$
thermal neutron scattering cross section, m^2	$(20.0 \pm 1.0) \times 10^{-28}$

[a] To convert kPa to mm Hg, multiply by 7.5.
[b] To convert J to cal, divide by 4.184.
[c] S/cm = (Ω·cm)$^{-1}$.

differences between cesium and potassium compounds are generally intensifications of the differences between potassium and sodium compounds. The solubilities of alkali-metal salts of simple anions generally increase with the atomic weight of the alkali ion from lithium to sodium to potassium to rubidium to a maximum with cesium. In contrast, the solubility of the alkali metal salts of complex anions generally decreases from lithium to a minimum with cesium. The salts of cesium and simple anions are usually hygroscopic as well as very soluble, but the sparingly soluble salts of cesium and complex anions are seldom hydrated and usually are not hygroscopic.

Occurrence

Cesium is widely distributed in the earth's crust at very low concentrations. Granites contain an average of about 1 ppm (7), sedimentary rocks contain about 4 ppm (8), and sea water contains about 0.2 ppm cesium (9). Higher concentrations of cesium are found in lepidolite [1317-64-2], a lithium mica (0.08–0.72% Cs$_2$O), in carnallite [1318-27-0], a double salt of potassium and magnesium chloride (0.001–0.004% Cs_2O), and beryl, rhodonite, leucite, spodumene, petalite, potash feldspars, and related minerals. Both lepidolite and carnallite have yielded commercial quantities of cesium (10), but by far the most important commercial cesium source is the rare mineral pollucite, a cesium aluminum silicate, $Cs_2O\cdot Al_2O_3\cdot 4SiO_2$.

The cesium content of theoretically pure pollucite is 45 wt % Cs_2O, but natural pollucite usually contains only about 5–32 wt % Cs_2O due to the presence of other gangue minerals intimately associated with the pollucite, and also due to the fact that natural pollucite can contain varying amounts of Rb, K, or Na + H_2O in place of Cs in the pollucite crystal structure. Thus natural pollucite can be considered as an intermediate mineral in a continuous series between analcite, $Na_2O\cdot Al_2O_3\cdot 4SiO_2\cdot 2H_2O$, and theoretical pollucite, $Cs_2O\cdot Al_2O_3\cdot 4SiO_2$ (11–12).

Pollucite is not easily recognized by visual inspection since it is similar in surface luster and transparency to the ordinary quartz of granite pegmatites. Pollucite may occur as translucent or opaque crystals, or the mineral may occur as a dull, gumlike powdery mass. However, when the pollucite is high grade, the mineral has a Mohs hardness of 6.5, sp gr of 2.88–2.90. Pollucite occurs in pegmatites associated with lepidolite, petalite, and other lithium minerals. Pollucite was first found in 1846 in the pegmatites of the Isle of Elba. The major known reserves are at the Tantalum Mining Corporation of Canada Ltd. mine at Bernic Lake, Manitoba, Canada. The pollucite occurs in a pegmatite in sheetlike bodies, the largest of which ranges up to 14 m thick and contains over 272,000 metric tons with an average of 23.8 wt % Cs_2O. A second zone, only partially explored, contains an estimated additional 60,000 t of similar grade, and a probable 100,000 t of 5 wt % Cs_2O.

Cesium-137 is a by-product of nuclear fission reactors, and can be recovered from fuel elements in nuclear fuel reprocessing facilities. This radioactive isotope has particular significance because of the long half life (30 yr), rapid heat-generating rate, and high biological hazard (0.66 MeV γ rays).

Manufacture of Cesium Compounds

The Penn Rare Metals Division of Kawecki Berylco Industries, Inc., Revere, Pa., is the only major producer of cesium compounds in the United States. Since 1960 the production has been based primarily upon pollucite as the raw material (5).

The cesium ore is ground to <149 μm (−100 mesh) in a ball mill, using steel balls with sufficient water to give a 60% solids slurry. The ground ore pulp is diluted to 33% solids in a flotation cell and 2.5–7.5 kg of sulfuric acid per metric ton of ore is added to give a pH of 1.4–2.7. Hydrofluoric acid and then aluminum sulfate are added in quantities of 0.5–1 kg/t ore and 250–500 g/t ore, respectively. After conditioning the pulp and reagents for 10 min, a cationic reagent such as Armac CD (a coco amine acetate) is added (200–500 g/t ore) and conditioned for 3 min. Then the nonpollucite minerals are removed by froth flotation (qv) (13).

Pollucite is decomposed by acid digestion or by melting or sintering with an alkaline flux. Hydrofluoric acid gives the most complete cesium recovery from pollucite but, because of the ease of handling, hydrochloric, hydrobromic, or sulfuric acids are used commercially. The laboratory procedures for the extraction of cesium from pollucite with hydrochloric and sulfuric acids are described in detail (14), and these procedures can be scaled up easily with negligible modification. The purification methods based on $(CsCl)_3.(SbCl_3)_2$ [*14590-08-0*], $Cs_2SO_4.Al_2(SO_4)_3.24H_2O$, and $CsICl_2$ [*15605-42-2*] are also given in detail for the laboratory scale (14).

A process based on sulfuric acid decomposition of pollucite which is suitable for tonnage production of cesium chloride was piloted by the Canadian Mines Branch in Ottawa (15). Pollucite ore, ground to 50% <74 μm (−200 mesh), is leached for 4–6 h in 6.8-kg batches in glass-lined vessels with 35–40% sulfuric acid at 110°C. The ore residue is removed by pressure filtration of the hot leach slurry, and cesium alum is crystallized from the leach filtrate by cooling to 50°C and then to 20°C. The cesium alum is roasted with 4% carbon added to decompose the alum:

$$Cs_2SO_4.Al_2(SO_4)_3.24H_2O + \tfrac{3}{2} O_2 + 3\,C \rightarrow 24\,H_2O + Cs_2SO_4 + 3\,SO_2 + 3\,CO_2 + Al_2O_3$$

The decomposed material is then leached to give a cesium sulfate solution. The cesium sulfate is converted to cesium chloride by ion exchange on Dowex 50 Wx8, and solid cesium chloride is recovered by evaporation and dehydration at 260°C.

Cesium values in pollucite are made soluble by sintering with CaO and $CaCl_2$ at 800–900°C. Leaching of the sinter cake gives an impure, alkaline cesium chloride solution, free of silicon and aluminum, but containing 98% of the cesium values in the ore (10). This solution can then be purified by any of several appropriate methods.

Manufacture of Cesium Metal

The major producers of cesium metal in the United States are The Penn Rare Metals Division of Kawecki Berylco Industries, Inc. and MSA Research Corp., Callery, Pa.

Thermochemical Methods. Cesium halides can be reduced readily with calcium or barium, but not with magnesium. A mixture of CsCl and calcium (50–65% of the weight of the CsCl) is heated to 700–800°C under vacuum, and 90–95% of the cesium metal is distilled from the mixture according to the equation (10,16):

$$2\,CsCl + Ca \rightarrow CaCl_2 + 2\,Cs$$

Magnesium is used to obtain cesium metal from the hydroxide [*21351-79-1*], carbonate [*534-17-8, 29703-01-3*], or aluminate [*20281-00-9*] (or the residue from calcined cesium alum) according to the following equations (10,17):

$$2\,CsOH + 2\,Mg \rightarrow 2\,MgO + H_2 + 2\,Cs$$

$$Cs_2CO_3 + 3\,Mg \rightarrow 3\,MgO + C + 2\,Cs$$

$$Cs_2O.Al_2O_3 + Mg \rightarrow MgO.Al_2O_3 + 2\,Cs$$

Thermochemical–reduction reactions are carried out under vacuum or in an atmosphere of an inert gas (argon or helium), and the metal obtained is then freed of impurities (including oxide) by vacuum distillation at low temperatures (about 300°C).

Cesium for use in vacuum tubes is usually generated within the evacuated tube by ignition (at 850°C) of a pelleted getter charge such as cesium chromate [*15596-54-0*] and zirconium (10).

$$4\,Cs_2CrO_4 + 5\,Zr \rightarrow 2\,Cr_2O_3 + 5\,ZrO_2 + 8\,Cs$$

A nearly quantitative yield of cesium metal is obtained inside the vacuum tube, and the vaporized cesium then serves as a getter for residual gases in the tube or as a coating to reduce the work function of the tungsten filaments or cathodes of the tube.

Calcium metal also can be used to decompose pollucite ore under vacuum at 900°C after first dehydrating the ore at this temperature. Yields of 85% cesium metal have been obtained, but calcium requirements were three parts by weight per unit weight of ore (10,18). Other reducing agents, such as aluminum, silicon, ferrosilicon, etc, can be used to obtain cesium from pollucite if the ore is first decomposed by treatment at 900–950°C with three parts of lime to one part (by wt) of ore. Sodium metal also can be used to decompose pollucite ore at 750–850°C (19). However, the product metal from pollucite reduction is always impure with respect to other alkali metals unless it is subjected to extensive distillation purification after the primary separation.

Thermal Decomposition. Cesium azide [22750-57-8], CsN_3 (prepared by the metathetical reaction between aqueous solutions of cesium sulfate and barium azide), melts at 326°C and decomposes at 390°C (20).

$$2\ CsN_3 \rightarrow 3\ N_2 + 2\ Cs$$

Electrolytic Reduction. Attempts by early investigators to reduce molten cesium chloride, using a graphite anode and an iron cathode, did not yield cesium metal, even though each of the other alkalies (lithium, sodium, potassium, and rubidium) were obtained by such an electrolysis. Setterberg (1) first obtained cesium metal by electrolysis of cesium cyanide–barium cyanide melt containing the two salts in a four-to-one mole ratio. The barium cyanide was used merely to lower the melting point of the electrolyte and barium was not reduced.

The extreme reactivity and relatively high volatility of cesium metal combine to make conventional fused-salt electrolysis of cesium salts impractical as a means for direct production of cesium metal. However, electrolysis of the fused chloride with a molten-lead cathode gives a lead–cesium alloy containing about 8.5% cesium at 700°C (15). Cesium is distilled from the alloy at 600–700°C under vacuum.

Cesium can be reduced from concentrated aqueous solutions by electrolysis with a mercury cathode, and presumably the cesium can be recovered by distillation of the amalgam (21).

Economic Aspects

The price of cesium ore in 1978 was $300–312/t for pollucite ore containing a minimum of 24 wt % Cs_2O. The United States consumes ca 250 t/yr. The prices of the metal are $600/kg (tech grade) and $716/kg (high purity); prices for cesium salts are $64–82/kg (tech grade) and $148–170/kg (high purity).

Specifications, Standards, and Analyses

Cesium metal is marketed in 99, 99.5, 99.9, and 99.9% purities. The nonmetallic impurities, particularly oxygen, critically affect the corrosive properties and hence the utility of cesium metal.

In 1961 the standard specification for technical-grade cesium salts was raised from 97 to 99% min. Flame photometry is the best method for determining alkali-metal impurities in cesium compounds, and an analysis and assay procedure has been reported (22). Special, analytical standard-purity cesium chloride is available from

Jarrell-Ash (div. of Fisher Scientific Co), EM Laboratories, and MC/B Manufacturing Chemists (div. of G. D. Searle Co.). Special grades of cesium chloride for use in the ultra centrifuge separation of macromolecules are available from Penn Rare Metals and Gallard-Schlesinger Corp. Besides cesium chloride special grades of cesium bromide, cesium acetate [3396-11-0], and cesium formate [3495-36-1] are used as density gradient materials in the ultracentrifugation of macromolecules. These and other high purity grades of several cesium compounds are also available from EM Laboratories and from MC/B Manufacturing Chemists.

A special grade of cesium iodide for use as a raw material for growing optical and scintillation crystals is available from EM Laboratories. Sodium doped cesium iodide is used in x-ray image converter tubes. A suitable material is available from EM Laboratories.

The determination of cesium in minerals is accomplished by first decomposing the mineral to obtain the cesium in a soluble form. Either the J. Lawrence Smith method (fusion with calcium carbonate and ammonium chloride) or the Berzelius procedure (decomposition of the ore with sulfuric and hydrofluoric acids) can be used. Quantitative spectrographic or spectroscopic procedures, using comparative standards, can then give reliable quantitative determinations of cesium and other alkalies.

The U.S. Bureau of Mines (23) developed a field test for cesium based on spot testing with phosphomolybdic acid and silicomolybdic acids, and Hosking (24) reported a somewhat simpler though less sensitive field test for cesium (or pollucite) based on the precipitation of cesium bismuth iodide.

Alloys of Cesium

Eutectics melting at about −30, −47, and −40°C are formed in the binary systems, cesium–sodium at about 9 wt % sodium, cesium–potassium at ca 25 wt % potassium, and cesium–rubidium at about 14% rubidium (25). A ternary eutectic with mp ca −72.2°C has the composition 73% (by wt) cesium, 24% potassium, and 3% sodium. Cesium and lithium are essentially completely immiscible in all proportions.

Cesium does not alloy with or appreciably attack tungsten, molybdenum, tantalum, iron, nickel, cobalt, or platinum at temperatures up to 650°C (26). A survey of cesium reactivity and materials compatibility was reported by Lawlor (27).

Cesium Compounds

Cesium Carbonate. Cesium carbonate [29703-01-3], Cs_2CO_3, is a colorless crystalline solid. It is stable to heat up to the melting point of 610°C but, at higher temperatures, the increasing equilibrium partial pressure of carbon dioxide above the molten carbonate leads to some slow decomposition. Cesium carbonate is very hygroscopic, and the anhydrous salt is quite soluble in alcohol. The carbonate is easily prepared from the hydroxide by the addition of carbon dioxide, but it can be prepared also by decomposing the nitrate with excess oxalic acid to give the acid oxalate, and igniting the cesium acid oxalate to give the carbonate.

Cesium Halides

Cesium Bromide. The bromide [7787-69-1], CsBr, is a colorless crystalline solid; sp gr, 4.433; mp, 636°C. This salt is usually prepared from the carbonate or hydroxide by neutralization with HBr (solubility, 123 parts CsBr in 100 parts H_2O, 25°C), but it is also made as the primary product of the patented process involving the opening of pollucite ore with HBr, precipitating CsBr by the addition of isopropyl alcohol, and purifying the CsBr by dissolving it in bromine and evaporating the bromine to leave the purified cesium bromide.

Cesium Chloride. The chloride [7647-17-8], CsCl, crystallizes readily from water in well-defined, colorless, cubic crystals; sp gr, 3.97; mp, 646°C; bp, 1290°C. The chloride can be made from the nitrate by repeated evaporations with excess hydrochloric acid, but it is formed more easily by neutralization of cesium carbonate or hydroxide with hydrochloric acid. Cesium chloride is a primary product of pollucite processing when hydrochloric acid is used to decompose the ore, and the chloride is usually purified by precipitation as a complex double salt with lead or antimony. The double salt is decomposed by hydrolysis or sulfide precipitation, leaving pure cesium chloride in solution. Recrystallization of CsCl from water is readily accomplished since about 270 parts of CsCl dissolve in 100 parts of H_2O at 100°C and only 162 parts in 100 parts of H_2O at 0°C. The oral LD_{50} in mice for CsCl is 2300 mg/kg (28).

Cesium Fluoride. Cesium fluoride [13400-13-0], CsF, is an extremely hygroscopic, colorless, crystalline solid; mp, 682–703°C; bp 1253°C. The fluoride is the most soluble of the cesium salts; 366.5 parts CsF will dissolve in 100 parts H_2O at 18°C. Cesium fluoride is made by exactly neutralizing cesium hydroxide with hydrofluoric acid and evaporating the solution obtained to dryness at 400°C. Excess hydrofluoric acid gives a bifluoride salt that does not decompose at 400°C, and carbonate in the cesium hydroxide starting material gives an alkaline product. The neutral salt can be stored in dilute aqueous solution in glass without etching, but even traces of acid give the typical hydrofluoric acid etching reaction.

It has been used in a variety of catalytic applications (29–31) and as a mild or soft fluorinating agent (32–34). The anhydrous salt is more effective as a drying agent than phosphorus pentoxide (based on equilibrium partial pressures of water above the anhydrous materials). Based on oral LD_{50} values for other fluorides, the oral LD_{50} in rats for CsF is estimated to be 400–700 mg/kg.

Cesium Iodide. Cesium iodide [7789-17-5], CsI, is a colorless, crystalline solid; sp gr, 4.51_4^{25}; mp, 621; solubility, in parts per hundred parts of water, 44 at 0°C and 160 at 61°C. It is made by neutralization of cesium hydroxide or carbonate with hydriodic acid.

Cesium Hydroxide. The hydroxide [21351-79-1], CsOH, is prepared by hydrolysis of cesium amalgam or by the reaction of cesium sulfate with barium hydroxide. Cesium hydroxide is the strongest base known, and hot concentrated cesium hydroxide solutions attack nickel or silver rapidly (whereas these materials are frequently used as container materials for the dehydration of less-reactive hydroxides). Cesium hydroxide solutions can be dehydrated in platinum at 180°C to give the monohydrate, and the anhydrous compound can be obtained by further dehydration at 400°C *in vacuo*. Both the solid and aqueous solutions of cesium hydroxide absorb carbon dioxide readily from the air. Carbon monoxide reacts with solid cesium hydroxide at atmospheric pressure and elevated temperatures, forming cesium formate, oxalate [18365-41-8], and carbonate. The water–cesium hydroxide system has been described by Rollet and co-workers (35). Cesium hydroxide is available commercially, primarily as a 50% aqueous solution packaged in polyethylene containers, but small quantities of solid cesium hydroxide have also been marketed.

Cesium Nitrate. This compound [7789-18-6], $CsNO_3$, crystallizes from aqueous solution in well-defined, glittering, hexagonal prisms; sp gr, 3.68; mp, 414°C. It is soluble in water to the extent of 197 parts to 100 parts of water at 100°C, and 9 parts to 100 parts at 0°C. Cesium nitrate is readily prepared from the chloride by heating with an excess of nitric acid until a negative test for chlorides is obtained; the reverse conversion to chloride with excess hydrochloric acid is also possible. When only limited amounts of cesium are at hand, the nitrate is an ideal salt for obtaining cesium free from the ordinary alkalies, because fractional crystallization from water effectively eliminates traces of lithium, sodium, potassium, and rubidium which remain dissolved in the mother liquors along with only a small proportion of the original cesium.

Cesium Oxides. Cesium forms a series of oxides, ranging from the suboxide [12433-62-4], Cs_7O, to the superoxide [12018-61-0], CsO_2. The suboxides Cs_7O, Cs_4O [12433-60-2], Cs_7O_2 [12433-63-5], and Cs_3O [12433-59-9] are formed by incomplete oxidation of cesium metal or by treating the normal oxide, [20281-00-9], Cs_2O, with cesium metal. Cesium monoxide, Cs_2O, can be formed by direct combination of the elements or by thermal decomposition of the suboxides in the form of polycrystalline, laminated plates, lemon-yellow at −80°C, orange-yellow at room temperature, and cherry-red above 180°C. The crystal structures of Cs_2O and Cs_3O have been determined by single-crystal x-ray studies (36).

Partial oxidation of cesium metal causes the silver-white metal color to change to a golden yellow (37). Continued oxidation with dry oxygen gives a black reaction product, which slowly changes (in the presence of oxygen at 330°C) to a bright yellow solid, CsO_2. Thermal decomposition of cesium superoxide in the temperature range from 280 to 360°C yields the peroxide [12053-70-2], Cs_2O_2, without the formation of an intermediate sesquioxide, and further decomposition yields the monoxide, Cs_2O (37).

Cesium Sulfates. *Cesium sulfate* [10294-54-9], Cs_2SO_4, forms rhombic or hexagonal colorless crystals; sp gr, 4.243, mp, 1010°C; it can be readily obtained from cesium alum by adding a hot solution of barium hydroxide to a boiling solution of the alum until all of the aluminum is precipitated. This equivalence point is very well indicated by spot-testing for alkalinity with bromthymol blue (pH 7.6). Filtration to remove barium sulfate and alumina yields a filtrate solution that gives no test for either aluminum or barium, and pure cesium sulfate is then obtained by concentration and recrystallization from water.

Cesium alum [7784-17-0] (cesium aluminum sulfate), $Cs_2SO_4 \cdot Al_2(SO_4)_3 \cdot 24H_2O$—sp gr, 1.97; mp, 117°C—crystallizes from aqueous solution in well-defined, colorless, octahedral crystals. It has a solubility of about 32 parts in 100 parts of water at 100°C, but at 50°C the solubility is less than 1.5 parts per 100. Being the most insoluble of the alkali alums, it tends to crystallize first from aqueous solution, and can readily be completely separated from lithium, sodium, and potassium alums by fractional crystallization. Owing to the small difference in relative solubility, the rubidium and cesium alums are separated only with great difficulty by repeated fractional crystallization from water. Pure cesium alum is an ideal starting material for the preparation of various salts of cesium. The aluminum content of pollucite is sufficient to permit the formation of cesium alum when sulfuric acid is used to decompose pollucite ore.

Uses of Cesium and Cesium Compounds

The principal use of cesium is in the developmental research on ion propulsion and thermionic, turboelectric, and magnetohydrodynamic (MHD) electric power generation (5,38) (see Coal, MHD; Explosives and propellants; Power generation).

NASA has sponsored development of cesium ion engines. In ion engines, or ion thrusters, cesium is vaporized and then ionized while passing through a heated porous tungsten disk. The cesium ions are accelerated by an electric field and the ions, moving at 483,000 kph (300,000 mph), are neutralized by the injection of electrons and are exhausted from the thruster. The thrust per unit weight in N/kg (lbf/lb) of propellant expelled (the specific impulse) is of the order 7000–30,000 (710–3060), as compared to 500–800 (50–80) for the best chemical propellant systems.

The cesium vapor thermionic converter utilizes cesium to neutralize the space charge above a hot cathode that is emitting electrons toward a cooler anode. Power levels of about 80 kW/m^2 at 17% efficiency have been achieved and power densities of 200 kW/m^2 at 30% efficiency can be expected. Nuclear heating can serve as the source for the high (1900°C) temperatures required.

Cesium has been considered as a possible working fluid for high-temperature, Rankine-cycle, turboelectric generators (39). Another developmental power generation system for which cesium is ideally suited is magnetohydrodynamic (MHD) power generation (see Coal conversion processes, MHD). Although either pure cesium or pure potassium can be used in the seeding of a MHD combustion plasma, the mixed potassium–cesium seeding is preferable to either alone (40). United States funding for MHD research has increased steadily since 1971 and for the fiscal year 1978 is 70 million dollars. The Union of Soviet Socialist Republics program U-25B has constructed and operated a unit at more than 20 MW of power, and recently the United States has shipped a superconducting magnet to the USSR for use in their pilot plant (41).

The principal commercial use of cesium is in the manufacture of vacuum tubes where the cesium acts both as a getter to remove traces of oxygen from the tube and as an agent in inducing electron emission from the cathode. The photoemissive properties of cesium are utilized in photoelectric cells where light energy, falling on the cesium cathode, causes electrons to be emitted. Light-sensitive cathodes of cesium on a conducting base, such as silver, may be constructed for photocell use (42), and numerous alloys of cesium are also photoelectric. However, the photoelectrically active substances in bimetallic photosurfaces are usually intermetallic compounds rather than alloys. The compound SbCs$_3$ *[12018-68-7]* has a particularly high photoelectric sensitivity (43). Cesium is also used in vapor glow lamps (44) and in vapor rectifiers where it acts as an initiator through ionization. It has also been used in atomic beam, microwave frequency standards and in construction of an atomic clock based on the frequency of rotation of cesium electrons (45).

Cesium salts are of value to a limited extent in microanalysis, and in medicine they may be used as antishock reagents following the administration of arsenical drugs (46). However, study of the physiological effects of cesium in comparison with potassium and rubidium on the function of various organs in small animals has shown that cesium has the greatest effect in causing disturbances of heart rhythm (47).

Cesium bromide is used in the manufacture of optical crystals. A cesium–rubidium–lithium bromide brine composition has been patented for use in absorption refrigeration devices (48).

Cesium chloride has been used extensively in biological research where concentrated cesium chloride solutions are used for density-gradient ultracentrifuge separation of DNA, viruses, and other large molecules (49). The chloride is the starting material for cesium metal production by the fused salt electrolysis method.

Cesium fluoride is used as a fluorinating agent in organic syntheses (50) (see Fluorine compounds, organic).

Cesium iodide is used for optical crystals and for the enhancement of absorption of zinc sulfide x-ray screens (51).

The use of cesium hydroxide (70–95% CsOH, molten, 177–343°C) as an agent for the removal of sulfur from heavy oils has been patented (52). The CsOH is regenerated by steam hydrolysis of the Cs_2S formed in the reaction (52). Owing to its strong alkalinity, as well as its high solubility at lower temperatures, cesium hydroxide has been suggested for use, along with rubidium hydroxide, in place of all or part of the sodium hydroxide or potassium hydroxide in alkaline storage batteries for operation at temperatures down to −50°C (53) (see Batteries).

BIBLIOGRAPHY

"Cesium" treated in *ECT* 1st ed., Vol. 1, under "Alkali Metals and Alkali Metal Alloys," pp. 453–458, by Elizabeth H. Burkey, Jean A. Morrow, and Muriel S. Andrew, E. I. du Pont de Nemours & Co., Inc.; "Cesium Compounds" in *ECT* 1st ed., Vol. 3, pp. 648–651, by J. J. Kennedy, Maywood Chemical Works; "Cesium and Cesium Compounds" in *ECT* 1st ed., Suppl. 2, pp. 190–192, by J. N. Hinvard, American Potash & Chemical Corporation; "Cesium and Cesium Compounds" in *ECT* 2nd ed., Vol. 4, pp. 855–868, by Robert E. Davis, American Potash & Chemical Corporation.

1. C. Setterberg, *Ann. Chem.* **211,** 100 (1882).
2. E. Graefe and M. Eckhardt, *Z. Anorg. Allgem. Chem.* **22,** 158 (1900).
3. L. Hackspill, *Compt. Rend.* **141,** 106 (1905).
4. C. E. Berthold, *J. Metals* **14,** 355 (1962).
5. F. W. Wessel, *Minor Metals and Minerals—Cesium and Rubidium,* in *Minerals Yearbook,* Vol. 1, U.S. Dept. of the Interior, Washington, D.C., 1959–1962.
6. N. V. Sidgwick, *Group IA. Alkali Metals* in *The Chemical Elements and Their Compounds,* Vol. 1, Oxford University Press, London, Eng., 1950, pp. 59–102.
7. E. L. Horstman, *Geochim. Cosmochim. Acta* **12,** 1 (1957).
8. A. A. Smales and L. Salmon, *Analyst* **80,** 37 (1955).
9. R. Greenwood, *Min. Eng.* **12,** 482 (1960).
10. V. E. Plyushchev and I. V. Shakhno, *Khim. Nauka Promst.* **1,** 534 (1956).
11. H. J. Nell, *Am. Mineral.* **29,** 443 (1944).
12. R. M. Barrer and L. V. C. Rees, *Trans. Faraday Soc.* **56,** 709 (1961); G. M. Schwartz, *Econ. Geol.* **25**(3), 275 (1930).
13. K. S. Dean and I. L. Nichols, *Bur. Mines Rep. Invest.* **5940** (1962); U.S. Pat. 3,107,215 (Oct. 15, 1963), K. C. Dean (to The United States of America).
14. J. C. Bailor, Jr., *Inorganic Syntheses,* Vol. 4, McGraw-Hill, New York, 1953, pp. 5–11.
15. H. W. Parsons and co-workers, *Mines Branch T.B. 50,* 1963, Dept. Mines, Ottawa, Canada, 1963.
16. L. Hackspill, *Helv. Chim. Acta* **11,** 1003 (1928).
17. H. Erdman and A. E. Menke, *J. Am. Chem. Soc.* **21,** 259 (1899).
18. L. Hackspill and G. Thomas, *Compt. Rend.* **230,** 1119 (1950).
19. Brit. Pat. 935,035 (Aug. 28, 1963), (to National Distillers and Chemical Corp.).
20. Eastman Kodak Co., *Chem. Week* **93,** 42 (1963).

21. R. E. Davis, "Electrowinning of Rubidium and Cesium" in C. A. Hampel, ed., *Encyclopedia of Electrochemistry,* Reinhold Publishing Corp., 1964.
22. M. C. Farquhar and J. A. Hill, *Anal. Chem.* **34,** 222 (1962).
23. K. C. Dean and I. L. Nichols, *U.S. Bur. Mines Rep. Invest.* **5675** (1960).
24. K. F. G. Hosking, *Min. Mag.* **104,** 280 (1961).
25. C. Goria, *Gazz. Chim. Ital.* **65,** 1226 (1935).
26. J. M. Lamberti and N. T. Saunders, *NASA Tech. Note D-1739,* Office of Tech. Svcs., Washington, D.C., 1963, p. 43.
27. P. J. Lawlor, "Cesium Ion Engine Feed System," *Paper presented at the Electric Propulsion Conference, American Rocket Society, Berkeley, Calif., Mar. 14–16, 1962.*
28. *Registry of Toxic Effects of Chemical Substances, 1977* ed., Vol. II, U.S. Department of Health, Education, and Welfare, Washington, D.C., 1977, p. 288.
29. M. Lustig and J. M. Shreeve, *Adv. Fluorine Chem.* **7,** 175 (1973).
30. A. Majid and J. M. Shreeve, *J. Org. Chem.* **38,** 4028 (1973).
31. A. Majid and J. M. Shreeve, *Inorg. Chem.* **13,** 2710 (1974).
32. C. T. Ratcliffe and J. M. Shreeve, *Chem. Comm.,* 674 (1966).
33. S. Yu and J. M. Shreeve, *Inorg. Chem.* **15,** 14 (1976).
34. C. A. Burton and J. M. Shreeve, *Inorg. Chem.* **16,** 1408 (1977).
35. A. P. Rollet, R. Cohen-Adad, and C. Ferlin, *Compt. Rend.* **256**(56), (1963).
36. Khi-Ruey Tsai, P. M. Harris, and E. M. Lassettre, *J. Phys. Chem.* **60,** 338, 345 (1956).
37. G. V. Morris, *The Thermal Decomposition of Cesium Superoxide,* Thesis, University of Rhode Island, Kingston, R.I., 1962.
38. R. Brinsmead, *Precambrian* **33**(8), 17 (1960).
39. T. P. Moffitt and F. W. Klag, *Analytical Investigation of Cycle Characteristics for Advanced Turboelectric Space Power Systems, NASA TN D-472, 1960;* L. Rosenblum, *J. Metals* **15,** 637 (1963).
40. P. D. Bergman and D. Bienstock, "Economics of a Mined Potassium–Cesium Seeding of a MHD Combustion Plasma," U.S. Dept. of the Interior, Bureau of Mines Report of Investigation 7717, 12 p., 1972.
41. J. Melcher, *Min. Eng.* **29**(12), 1977.
42. R. C. Walker, *Photoelectric Cells in Industry,* Sir Isaac Pitman & Sons, London, Eng., 1948.
43. J. W. Mellor, *Comprehensive Treatise on Inorganic and Theoretical Chemistry,* Vol. 2, Suppl. 3, John Wiley & Sons, Inc., New York, 1963.
44. N. C. Beese, *J. Opt. Soc. Am.* **36,** 555 (1946).
45. H. Lyons, *Nat. Bur. Stand. (U.S.) Rep.* **1948** (Aug. 8, 1952).
46. W. R. Barton, "Minor Metals—Cesium and Rubidium" in *Minerals Yearbook,* Vol. 1, U.S. Dept. of the Interior, Washington, D.C., 1956–1958.
47. R. Hazard and co-workers, *Compt. Rend.* **243,** 452 (1956).
48. U.S. Pat. 3,004,919 (Oct. 17, 1961), W. F. Rush and W. G. Walters (to Borg-Warner Corp.).
49. J. Vinograd and J. E. Hearst, "Equilibrium Sedimentation of Micromolecules and Viruses in a Density Gradient" in L. Zechmeister, ed., *Progress in the Chemistry of Organic Natural Products,* Vol. 20, Springer-Verlag, Vienna, Austria, 1962.
50. D. Pilipovich and co-workers, *Inorg. Chem.* **11,** 2189 (1972).
51. U.S. Pat. 2,651,584 (Sept. 8, 1963), R. L. Longini and W. J. Hushley (to Westinghouse Electric Corp.).
52. Brit. Pat. 913,730 (Dec. 28, 1962) (to Esso Research & Engineering Co.).
53. U.S. Pat. 2,683,102 (July 6, 1954), R. S. Coolidge.

C. T. WILLIAMS
Tantalum Mining Corp.

CHELATING AGENTS

A chelating agent is a compound containing donor atoms that can combine by coordinate bonding with a single metal atom to form a cyclic structure called a chelation complex or, simply, a chelate. Because the donor atoms are connected intramolecularly by chains of other atoms, a chelate ring is formed for each donor atom after the first which coordinates with the metal. Each ring gives the appearance of a metal atom being held in a pincer formed by the other atoms. The descriptive word chelate, derived from the chela, or great claw, of the lobster, was proposed for this kind of molecular structure (1).

The technological importance of chelation derives from the almost universal presence of metal ions of one kind or another. They are present either naturally, and often in spite of efforts to eliminate them, or by intentional addition. Chelating agents provide a means of manipulating and controlling metal ions by forming complexes that usually have properties that are markedly different from those of either the original ions or the chelants. These properties may serve to reduce undesirable effects of metal ions as in sequestration, or to create desirable effects as in metal buffering and solubilization (see also Coordination compounds; Dispersants; Water, industrial water treatment).

Chelates and chelation reactions are abundant in nature and hundreds of technological applications are practiced. In nature, chelation phenomena range from delicately balanced life processes, depending on only traces of metal ions, to the extremely stable metal chelates in crude petroleums which are the results of processes on a geological time scale. Photosynthesis, oxygen transport by blood, some kinds of enzyme action, ion transport through membranes, and muscle contraction are life processes involving chelation. In technology, some uses of chelation are scale removal from steam boilers, water softening, ore leaching, textile processing, food preservation, treatment of lead poisoning, chemical analysis, and micronutrient fertilization of agricultural crops (see Mineral nutrients).

Structure and Terminology in Chelate Chemistry

The structural essentials of a chelate are coordinate bonds between a metal atom (M) and two or more atoms in the molecule of the chelating agent, or ligand, as shown in (1), (2), and (3). The coordinating atoms of the ligand are electron donors and the metal atom is the electron acceptor. When coordinate bond formation occurs between the metal and two donor atoms, the atoms of the ligand that connect the donor atoms complete the ring that gives the structure its chelate character. For each additional donor atom of the ligand that coordinates with the metal, another ring is formed. A chelating agent is bidentate, tridentate, tetradentate, and so on, according to whether it contains two, three, four, or more donor atoms capable of complexing with the metal atom. Molecules with only one donor atom, such as dimethylamine, are monodentate and form coordination complexes but not chelates. The principal donor atoms in practical use are N, O, and S, but P, As, and Se also form chelates. Metal atoms are characterized by their coordination numbers which give the maximum number of donor atoms to which the metal atoms can coordinate. The most common coordination numbers are 4 and 6, less often 2 and 8. Some atoms may have more than one coordination number, depending on the valence state. The term is also used to state the actual number of donor atoms bound to the central metal atom in a particular compound.

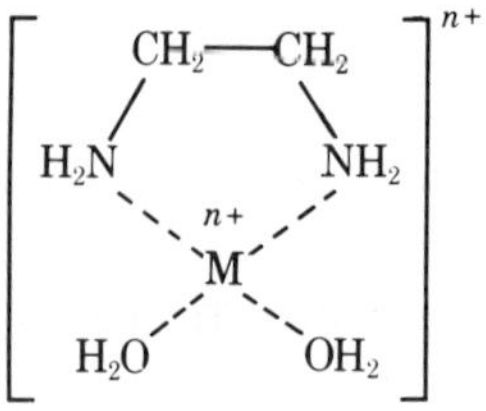

340 CHELATING AGENTS

$$\left[\begin{array}{c} \text{CH}_2\text{—CH}_2 \\ \text{H}_2\text{N} \quad \text{NH}_2 \\ \text{M}^{n+} \\ \text{H}_2\text{O} \quad \text{OH}_2 \end{array} \right]^{n+}$$

(1)

tetracoordinate metal with bidentate
chelant and monodentate water

$$\left[\begin{array}{c} \text{CH}_2 \quad\quad\quad \text{CH}_2 \\ \text{CH}_2\;\text{NH}_2 \quad\;\text{NH}_2\;\text{CH}_2 \\ \text{HN}\text{-----}\text{M}^{n+}\text{-----}\text{NH} \\ \text{CH}_2\;\text{NH}_2 \quad\;\text{NH}_2\;\text{CH}_2 \\ \text{CH}_2 \quad\quad\quad \text{CH}_2 \end{array} \right]^{n+}$$

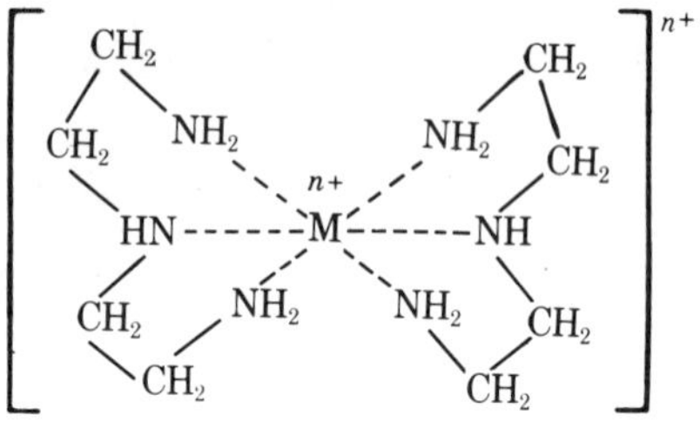

(2)

hexacoordinate metal with tridentate chelant

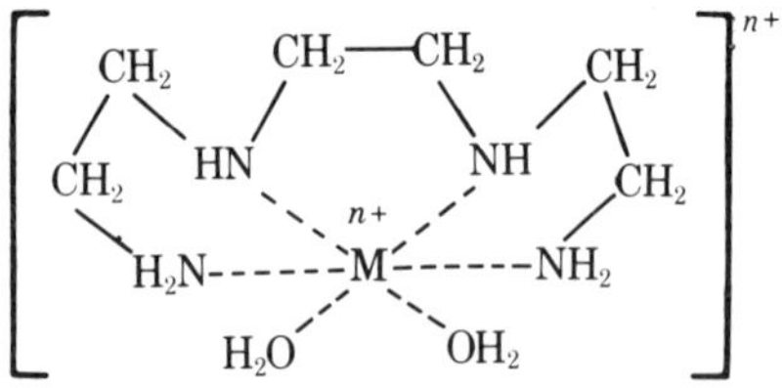

$$\left[\begin{array}{c} \text{CH}_2 \quad \text{CH}_2\text{—CH}_2 \quad \text{CH}_2 \\ \text{CH}_2\;\text{HN} \quad\quad \text{NH}\;\text{CH}_2 \\ \text{H}_2\text{N}\text{-----}\text{M}^{n+}\text{-----}\text{NH}_2 \\ \text{H}_2\text{O} \quad\quad \text{OH}_2 \end{array} \right]^{n+}$$

(3)

hexacoordinate metal with tetradentate
chelant and monodentate water

If the coordination number of a metal atom M is greater than the number of donor atoms in the ligand L, more than one ligand molecule may combine with the metal to form ML_n (2). Different chelating molecules can combine with the same metal atom to form species such as $\text{L}_m\text{ML}'_n$. Remaining unchelated coordination positions of the metal may also bind monodentate molecules, such as water, ammonia, or halogen ions, by coordination as in (1) and (3). Metal ions in solution in water or other monodentate solvents are undoubtedly coordinated with solvent molecules; therefore, formation of chelates should be regarded as displacement of solvent molecules by the donor atoms of the chelating ligand.

Just as a metal ion can coordinate with more than one molecule of chelating agent, a ligand molecule with enough donor atoms in the proper configuration can bind more than one metal atom. The metal atoms bound to such a ligand may be the same or different.

A chelate compound may be either a neutral molecule or a positive or negative complex ion associated with appropriate counterions to produce electroneutrality. The formal charge on the complex is the algebraic sum of the charge developed by ionization of the ligand and the charge of the metal ion chelated, $(M^{2+}L^{4-})^{2-}$. The ligand may be neutral or completely or partially ionized before it chelates the metal ion. Changes in the charge condition of the two reactants, metal atom, and chelating agent, may occur during the chelating reaction as by the displacement of hydrogen ions from donor atoms of the ligand. After the chelate is formed, the charge of the complex can be changed by the ionization of other groups on the ligand that are not involved in the chelate structure, usually as a result of changes in the pH, or by oxidation or reduction of the chelated metal atom.

The donor atoms of most chelating agents are contained in linear or branched chain structures, separated by suitable numbers of other atoms to allow the formation of the chelate rings. However, there are classes of chelating agents whose donor atoms are suitably spaced members of macrocyclic structures. Here the chelated metal atom becomes positioned near the center of the macro ring with the donor atoms coordinated to it. The spacer atoms complete the chelate rings between pairs of coordinated donor atoms, forming a pattern of fused rings centered about the metal. The porphyrins are long-known examples of this type of chelate. Formula (4) represents chelates of porphine [101-60-0] the parent compound of the porphyrins.

(4)

porphine chelate

Recently the complexing properties of another group of macrocyclic ligands have been extensively studied. These are the cyclic polyethers (5) in which the donor atoms are ether oxygen functions separated by two or three carbon atoms. The name crown ethers has been proposed (2) for this class of compounds because of the resemblance of their molecular models to a crown. As in the porphyrins, the chelated metal atom is positioned near the center of the macro ring, coordinated to the donor atoms. But sandwich structures are also known in which the metal atom is coordinated with the oxygen atoms of two crown molecules (see also Antibiotics, polyethers).

Related to the crown ethers are similar compounds (6) which differ by the replacement of one or more of the oxygen atoms by other kinds of donor atoms, eg, N and S. Macrocyclic amine and thioether compounds have been synthesized. Compounds with different kinds of heteroatoms in the large rings are called mixed donor macrocycles. The naturally occurring metabolites nonactin [6833-84-7] and monactin [26446-35-5] have both ether and ester groups incorporated in the macrocyclic structure.

Three dimensional polymacrocyclic chelating agents are formed by joining bridgehead structures with chains that contain properly spaced donor atoms. Bicyclic molecules result from joining nitrogen bridgeheads with chains of ($-OCH_2CH_2-$) groups (7). Such bicyclic structures form a cavity that will hold a metal atom coordinated to the donor atoms in the surrounding chains. The name cryptates has been proposed for these complexes (3). Other groups that are at least trifunctional can serve as bridgeheads, eg, pentaerythritol. The donor atoms of the bridges may all be O, N, or S, or the compounds may be mixed donor macrocycles in which the bridge strands contain combinations of these donor atoms.

N(CH₂COOH)₃
NTA

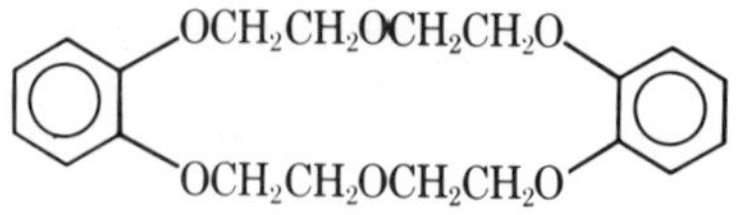

(5)
dibenzo-[18]-crown-6

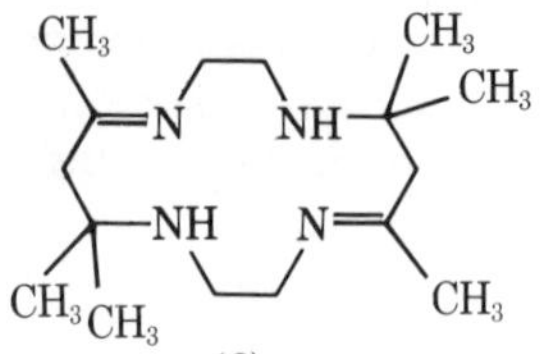

(6)
(CH₃)₆[14]4,11-dieneN₄

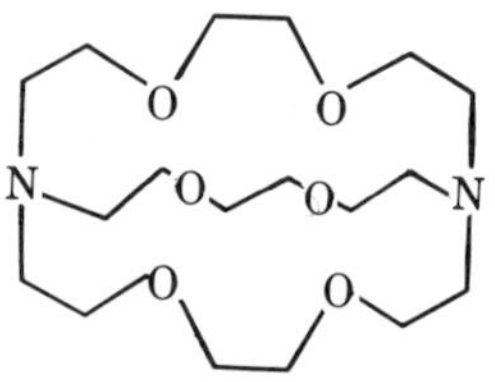

(7)
2.2.2-cryptate

Synthesis, metal binding, and thermodynamic properties of synthetic multidentate macrocyclic complexing agents have been reviewed through 1972 (4). The compounds are compiled according to type of donor atoms together with the metal atoms bound by each. Structural formulas of the parent rings, thermodynamic properties of the complexes, and some kinetic data are given.

Incorporating ligand groups into a cross-linked polymer structure gives the chelate-forming resins which perform ion-exchange functions by chelation. The ligand groups may either be present in the monomer before polymerization, or they may be attached to a preformed polymer by appropriate reactions. Several types are commercially available (see Ion exchange).

Chelating agents may be either organic or inorganic compounds, but the number of inorganic agents is very small. Although many hundreds of organic chelating agents are known, only a few members of a few classes of compounds are extensively used industrially. The best known inorganic chelants are a few polyphosphates whose annual consumption, however, exceeds that of all the organic chelating agents combined. Polyphosphates are less expensive than the organics, but they are hydrolytically unstable at high temperature and pH. An important class of organic chelating agents is the group of phosphonic acids analogous to the amino- and hydroxycarboxylic acids (qv). These phosphonate chelants possess many of the complexing properties of the inorganic polyphosphates, particularly threshold-scale inhibition, but unlike the polyphosphates, they are stable in water at high temperature and pH (see Phosphorus compounds).

Types of Compounds With Chelating Properties

Compounds with chelating properties can be found in almost any class of structures containing two or more donor atoms spatially situated so that they can coordinate with the same metal atom. The chelate rings which form contain four or more members, but for the same donor atoms, the five- or six-membered ring chelates are usually the most stable and most useful. Complexes with chelate rings of more than six members are rare. In the macrocyclic molecules, the stability of the metal complex depends strongly on the relationship between the size of the metal ion and the size of the opening within the crown or the crypt.

Chelating compounds can be classified into more or less arbitrarily selected structural categories based on the kinds of functional groups in which the donor atoms occur. Many compounds can be unambiguously assigned to specific classes, but substances with donor atoms in several kinds of functional groups may be considered members of more than one category. Classifications based on types or extent of use, or other criteria, encounter similar ambiguities.

In Table 1 are listed a number of substances well known for their chelating properties, grouped according to recognized, mainly structural classes. Because systematic nomenclature of chelating agents is frequently cumbersome, they are commonly referred to by trivial names and abbreviations. For the macrocyclic complexing agents, special systems of abbreviated nomenclature have been devised and are widely used. The listing is illustrative rather than comprehensive. Some of the donor atoms involved in chelation and many forms in which they can occur have been reviewed (5).

Nomenclature and Structural Representation

Chelating Agents. Besides the conventional empirical and structural formulas, that are often represented by *type formulas,* ie, formulas which show only certain generalized types of structural features. Ligands are designated by the letter L and their complexes with metal atom M as ML_n with a superscript for ions to show the charge. Chelants with proton acid groups may be shown as H_nA or, if partially dissociated, as H_mA^{n-}. Alcohol or phenol groups that lose protons on chelation are shown as $A(OH)_n$. The letter A is sometimes used to represent an entire multidentate ligand molecule, and sometimes to show only a donor atom as in A-A-A-A for a tetradentate ligand. Examples of some chelates in this system are MA or MA_n, MH_2A, $MAO(OH)_{n-1}$, and

$$\begin{pmatrix} A & A \\ & M \\ A & A \end{pmatrix}$$

with appropriate superscripts to show ionic charge, if any.

For many macrocyclic ligands, simplified names are in common use. The names so obtained are convenient but not always unambiguous. The crown ether nomenclature consists of four parts: (*1*) the number and type of fused rings on the polyether ring, (*2*) in square brackets the number of atoms in the polyether ring, (*3*) the word

Table 1. Some Classes of Chelating Agents

	CAS Registry Number	Abbreviation
polyphosphates		
sodium tripolyphosphate	[7758-29-4]	STPP
hexametaphosphoric acid	[18694-07-0]	
aminocarboxylic acids		
ethylenediaminetetraacetic acid	[60-00-4]	EDTA
hydroxyethylethylenediaminetriacetic acid	[150-39-0]	HEDTA
nitrilotriacetic acid	[139-13-9]	NTA
N-dihydroxyethylglycine	[150-25-4]	2-HxG
ethylenebis(hydroxyphenylglycine)	[1170-02-1]	EHPG
1,3-diketones		
acetylacetone	[123-54-6]	acac
trifluoroacetylacetone	[367-57-7]	tfa
thenoyltrifluoroacetone	[326-91-0]	TTA
hydroxycarboxylic acids		
tartaric acid	[526-83-0]	
citric acid	[77-92-9]	cit
gluconic acid	[133-42-6]	
5-sulfosalicylic acid	[97-05-2]	5-SSA
polyamines		
ethylenediamine	[107-15-3]	en
triethylenetetramine	[112-24-3]	trien
triaminotriethylamine	[4097-89-6]	tren
aminoalcohols		
triethanolamine	[102-71-6]	TEA
N-hydroxyethylethylenediamine	[111-41-1]	hen
aromatic heterocyclic bases		
dipyridyl	[366-18-7]	dipy, bipy
o-phenanthroline	[66-71-7]	phen
phenols		
salicylaldehyde	[90-02-8]	
disulfopyrocatechol	[149-46-2]	Tiron, PDS
chromotropic acid	[148-25-4]	DNS
aminophenols		
oxine, 8-hydroxyquinoline	[148-24-3]	Q, ox
oxinesulfonic acid	[84-88-8]	
oximes		
dimethylglyoxime	[95-45-4]	
salicylaldoxime	[94-67-7]	
Schiff bases		
disalicylaldehyde 1,2-propylenediimine	[94-91-7]	
tetrapyrroles		
tetraphenylporphin	[917-23-7]	
phthalocyanine	[574-93-6]	
sulfur compounds		
toluenedithiol (Dithiol)	[496-74-2]	tdth
dimercaptopropanol	[59-52-9]	BAL
thioglycolic acid	[68-11-1]	
potassium ethyl xanthate	[140-89-6]	
sodium diethyldithiocarbamate	[148-18-5]	

Table 1 (continued)

	CAS Registry Number	Abbreviation
dithizone	[60-10-6]	dz
diethyl dithiophosphoric acid	[298-06-6]	
thiourea	[62-56-6]	
synthetic macrocyclic compounds		
dibenzo[18]crown-6 (5)	[14187-32-7]	
(CH$_3$)$_6$[14]4,11-dieneN$_4$ (6)	[29419-92-9]	
(2.2.2-cryptate) (7)	[23978-09-8]	
polymeric		
polyethylenimine	[151-56-4]	PEI
	[9002-98-6]	
	[25988-99-2]	
	[32167-41-2]	
polymethacryloylacetone	[25120-51-8]	
poly(p-vinylbenzyliminodiacetic acid)	[30395-28-9]	
phosphonic acids		
nitrilotrimethylenephosphonic acid	[6419-19-8]	NTPO, ATMP
ethylenediaminetetra(methylenephosphonic acid)	[1429-50-1]	EDTPO
hydroxyethylidenediphosphonic acid	[2809-21-4]	HEDP

crown, and (4) the number of oxygen atoms in the macro ring (2). The schemes of other systems become apparent on comparing the simplified names with their structures.

Ligand structures are represented by any of the conventional means for depicting structure, as in structures (5–7).

Chelates. Because of their length and complexity, systematic names of chelates are little used except for special purposes, such as indexing or other occasions where unequivocal referencing is essential. Chelates are named in the literature in a variety of other ways, but in context, little or no difficulty in communication is encountered. The name of the ligand in a chelate is usually given a suffix -o or -ato if it is a negative group, and is unchanged if neutral. Prefixes indicate the number of bound ligand molecules. The central atom is the name of the metal, or a derivative name of the metal with the suffix -ate (eg, cuprate, ferrate) if the complex is negatively charged. Oxidation states of the metals are indicated by Roman numerals, as iron(III), and ionic charges are shown as part of the name by Ewens-Bassett numbers, eg, (2+), (1−).

Chelates are often named merely as a complex, eg, cadmium complex with acetylacetone. A common practice in the literature is to give the symbol of the central atom and an abbreviation for the ligand with or without an indication of ionic charges, oxidation states, structure, or counterions, as in the following: Pb-EDTA, Cacit$^-$, Cu(en)$_2$, Co(II)-(phen), [Cu(dipy)$_3$]SO$_4$, [Ru(dipy)$_2$(en)]$^{2+}$, and Na[Co(acac)$_3$].

For the many details of constructing or interpreting systematic names, consult the literature on nomenclature and indexing, such as the Index Guides of Chemical Abstracts (6). Some of the foregoing features of systematic nomenclature are illustrated by the Chemical Abstracts name of the sodium iron(III) EHPG chelate (12): sodium [[N,N'-1,2-ethanediylbis[2-(2-hydroxyphenyl)glycinato]](4-)-N,N',O,O',O^2, O$^{2'}$]-

346   CHELATING AGENTS

ferrate(1−), which has the CAS Registry Number *[16455-61-1]* (also, the free anion *[20250-28-6]* and the potassium salt *[22569-56-8]* (7)).

Several of many ways of representing the structures of chelates are shown in Figure 1, bi- (**8**), (**9**), tetra- (**10**), (**11**), and hexadenticity (**12**), (**13**) are illustrated. Square brackets may be used to emphasize that the structure is a complex ion, but the sum of charges shows that (**12**) is also a charged (1−) complex and (**8**) is neutral. The chelate of a tetracoordinate metal with EDTA is shown in (**11**), and the EDTA complex of a hexacoordinate metal is represented schematically in (**13**), which is also a schematic of (**12**). Form (**13**) is used to show spatial arrangements where the curved lines represent chains of any length connecting the donor atoms. The aromatic rings in (**9**) sterically prevent the sulfo groups from chelating with the metal.

The Chelation Reaction

Chelate Formation Equilibria. In homogeneous solution the equilibrium constant for the formation of the chelate complex from the solvated metal ion and the ligand in its fully dissociated form is called the formation or stability constant. The ligand displaces the solvent molecules coordinated to the metal, but the solvent molecules are usually not shown in equations. When more than one ligand complexes with a metal atom, the reaction usually proceeds stepwise. For a metal with a coordination number of six and a bidentate chelating agent, the following equations represent the equilibria:

$$M + L \rightleftharpoons ML \qquad K_1 = \frac{[ML]}{[M][L]} \tag{1}$$

$$ML + L \rightleftharpoons ML_2 \qquad K_2 = \frac{[ML_2]}{[ML][L]} \tag{2}$$

$$ML_2 + L \rightleftharpoons ML_3 \qquad K_3 = \frac{[ML_3]}{[ML_2][L]} \tag{3}$$

$$overall: M + 3L \rightleftharpoons ML_3 \qquad K = \frac{[ML_3]}{[M][L]^3} \tag{4}$$

The overall stability constant is the product of the step stability constants, $K = K_1 K_2 K_3$, and is often designated by the letter β. Protons displaced from a ligand in the chelation reaction can be shown as follows, with type formulas representing the divalent anion of a hexadentate ligand such as EDTA and the complex formed with a divalent hexacoordinate metal ion:

$$M^{2+} + H_2A^{2-} \rightleftharpoons MA^{2-} + 2H^+ \tag{5}$$

$$k = \frac{[MA^{2-}][H^+]^2}{[M^{2+}][H_2A^{2-}]} \tag{6}$$

Such a reaction is the sum of an acid dissociation reaction

$$H_2A^{2-} \rightleftharpoons A^{4-} + 2H^+ \tag{7}$$

$$K_A = \frac{[A^{4-}][H^+]^2}{[H_2A^{2-}]} \tag{8}$$

and the reaction of chelate formation from the fully dissociated form of the ligand

$$M^{2+} + A^{4-} \rightleftharpoons MA^{2-} \tag{9}$$

$$K = \frac{[MA^{2-}]}{[M^{2+}][A^{4-}]} \tag{10}$$

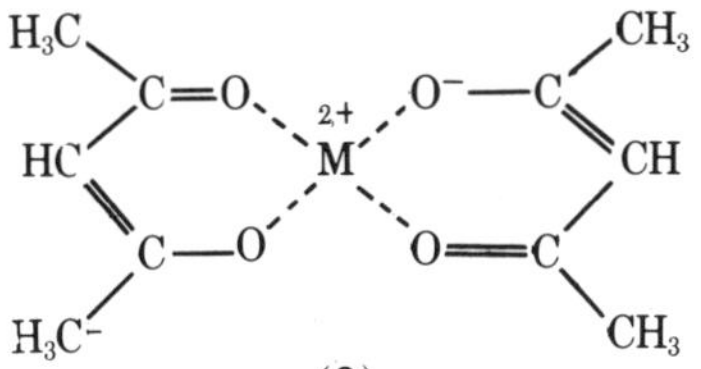

(8)

Metal chelate of acetylacetone, M(acac)$_2$

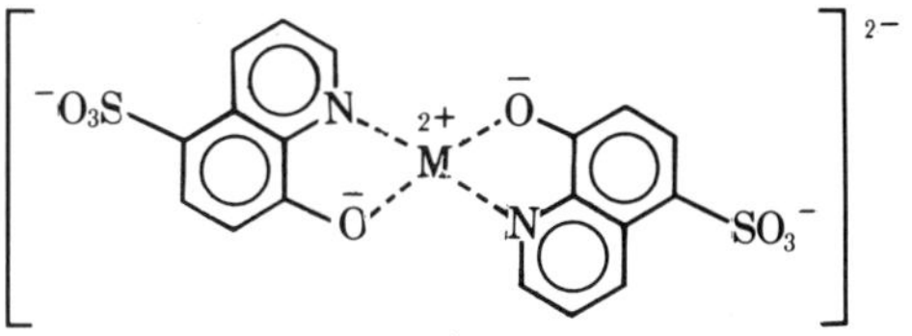

(9)

Metal chelate of 5-sulfo-8-hydroxyquinoline, ML$_2$

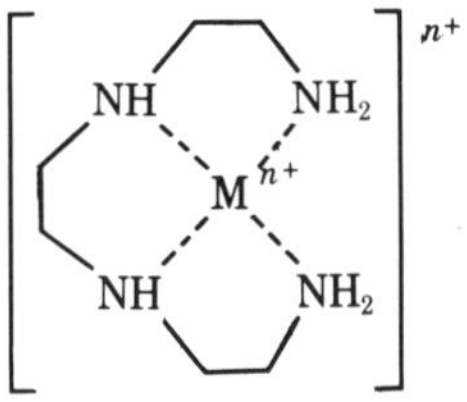

(10)

Metal chelate of triethylene-tetramine, M-Trien

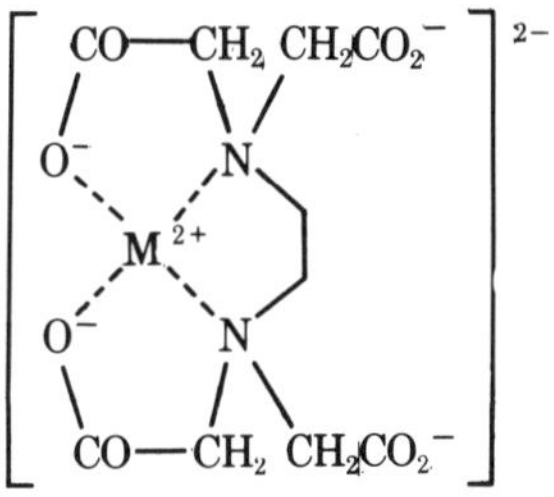

(11)

Chelate of tetracoordinate metal with hexadentate ligand EDTA

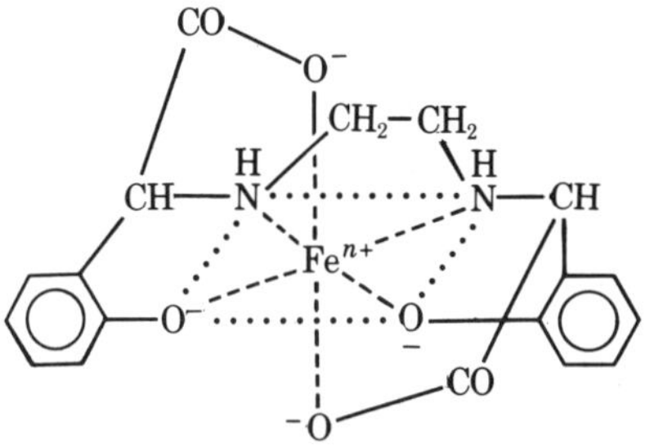

(12)

Fe(III) chelate of EHPG

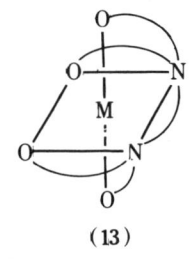

(13)

Spatial arrangement emphasized

Figure 1. Structural representations of chelates.

where K is the chelate formation constant, K_A the overall acid dissociation constant for the two stages, and $k = KK_A$. Similar equations can be written for other protonated ligands such as diphenols, enols of β-diketones, and acid salts of polyamines.

Concentration equilibrium constants are the quantities usually determined experimentally and reported in the literature, with concentrations instead of activities for the species in brackets having been used in the equations. Thermodynamic constants based on activities could be calculated by inserting appropriate activity coefficients in the equations. However, because of the inadequacy of present theory for calculating activity coefficients for the complicated ionic structures involved or the difficulty of determining the ionic activities, the relatively few known thermodynamic constants have usually been obtained by extrapolation of results in dilute solutions to infinite dilution. The constants based on concentration have usually been determined in dilute solution in the presence of excess inert ions to maintain constant ionic strength. Concentration constants are accurate, therefore, only under conditions reasonably close to those used for their determination. Beyond these conditions, concentration constants are useful in estimating probable effects and relative behaviors. Designers of chelation processes need to make allowances for the effects of differences between the conditions of their system and those of the accuracy range of the concentration constants.

Stability constants for a number of industrially important metals with some widely used chelating agents, expressed logarithmically, are given in Table 2. Extensive listings of stability constants are available (8). The entity chelated may be the metal

Table 2. Concentration Formation Constants of Metal Chelates[a]

Metal	STPP	Citric acid	Log K_1 (NTA)	Log K_2 (NTA)	Log K_1K_2 (NTA)	EDTA	EDTPO
V(III)						25.9	
Fe(III)		10.9	15.9	9.9	25.8	25.1	
In(III)			15.0	9.6	24.6	25.0	
Th(IV)			12.4			23.2	
Hg(II)			12.7			21.8	
Cu(II)	8.7	6.1	12.7	3.6	16.3	18.8	23.21
VO(II)						18.8	
Ni(II)	6.7	4.8	11.3	4.5	15.8	18.6	16.38
Y(III)			11.4	9.1	20.5	18.1	
Pb(II)		5.7	11.8			18.0	
Zn(II)	7.6	4.5	10.5			16.5	18.76
Cd(II)		4.2	10.1	4.4	14.5	16.5	
Co(II)	6.9	4.4	10.6			16.3	17.11
Fe(II)	2.5	3.2	8.8			14.3	
Mn(II)	7.2	3.4	7.4			14.0	
V(II)						12.7	
Ca(II)	5.2	3.5	6.4			10.7	9.36
Mg(II)	5.7	2.8	5.4			8.7	8.43
Sr(II)	4.4		5.0			8.6	
Ba(II)	3.0		4.8			7.8	
rare earths			10.4–12.5			15.1–20.0	

[a] STPP = sodium tripolyphosphate; NTA = nitrilotriacetic acid; EDTA = ethylenediaminetetraacetic acid; EDTPO = ethylenediaminetetra(methylenephosphonic acid).

ion itself or a stable oxo cation in the case of some metals, as shown in the table for vanadium. The practical significance of formation constants is that a high value means a large ratio of chelated to unchelated form of the metal when equivalent amounts of metal and ligand are present. From the values shown in the table, it is apparent that the concentration of the free metal ion can be reduced from relatively large to very low values by adding the proper chelant.

Many experimental approaches have been applied to the determination of stability constants. Some of the techniques that have been used are pH titrations, ion exchange, spectrophotometry, measurement of redox potentials, polarimetry, conductometric titrations, solubility determinations, and biological assay. For details of the methods, the reader is referred to books on chelation and coordination chemistry, and to the original literature (9).

Displacement Equilibria. Because various species in solution are in a formation–dissociation equilibrium, displacement reactions of one metal or ligand by another are possible. Thus,

$$ML + M' \rightleftarrows M'L + M \tag{11}$$

or

$$ML + L' \rightleftarrows ML' + L \tag{12}$$

If the stability constants for ML and M'L are K and K', and for the exchange the equilibrium constant is K_x, then for reaction 11

$$K_x = \frac{(M)(M'L)}{(ML)(M')} = \frac{K'}{K} \tag{13}$$

The extent of displacement depends on the relative stabilities of the complexes and the mass action effect of an excess of M'. For equivalent total amounts of M and M', K_x must be of the order of 10^4 for 99% complete displacement. Similar considerations apply for the displacement of L from ML by L'. The situation is quite analogous to the familiar competition of two bases for the hydrogen ion.

If the metals or ligands involved in a displacement reaction form chelates whose type formulas are different, the exchange equilibrium constant will not be the simple ratio of the formation constants of the chelates. Thus, for the reaction

$$2\,ML + M' \overset{K_x}{\rightleftarrows} 2\,M + M'L_2 \tag{14}$$

the equilibrium constant $K_x = K'/K^2$ where K and K' are the formation constants of ML and M'L$_2$. The proper evaluation of K_x can be derived for each particular case.

Metal exchange is the mechanism by which many foods (such as shortenings, shellfish, and dairy products) are stabilized against deleterious effects of traces of metals by the addition of Na$_2$CaEDTA (log K 10.7) (see Food additives). Copper (log K 18.8) and iron (log K 25.1) displace calcium and become sequestered so that the remaining concentration of free iron or copper ions is too low for their catalytic effects to occur at significant rates (10). Ligand exchange occurs when ascorbic acid bound to copper (log K 1.57) is displaced by EDTA, stabilizing this vitamin by disrupting an oxidation mechanism (11) (see Antioxidants). Dyes and bleaches are similarly protected in the textile industry (12) (see Sequestration section).

Corollary to the displacement reaction is the addition of a chelating agent to a solution of two or more metal ions. The order of complexation of the metal ions will be regulated by their displacement equilibrium constants. If the objective is to bind only a particular ion, enough chelant will need to be added to combine with the target ion and all the other ions that are capable of displacing the target ion. For any particular chelating agent under similar solution conditions, a displacement series of metal ions of interest can be assembled by calculating the K_x values from series of stability constants such as those in Table 2. For selective complexation of one metal in the presence of another, a chelating agent with sufficiently different stability constants for the two metals is obviously necessary so that K_x will be large. Its position at the bottom of a displacement series with NTA allows beryllium (qv) to be recovered as the hydroxide by pH adjustment of an ore processing solution, while all of the interfering metals remain sequestered by chelation (13).

Because other metals present will not displace the iron, the chelate of EHPG with iron can be used in highly calcareous soils to supply iron as a trace nutrient in agriculture (14) (see Mineral nutrients).

Selectivity for a single metal of a group is the basis of a solvent extraction process for the recovery of copper (qv) from low concentration ore leach solutions containing high levels of iron and other interfering metals (15) (see also Extractive metallurgy).

Rates of Reaction. The rates of formation and dissociation or displacement reactions are important in the practical applications of chelation. Complexation of many metal ions is almost instantaneous, particularly the divalent ones, but reaction rates of many ions with higher valences are slow enough to measure by ordinary kinetic techniques. Rates with some ions, notably Cr(III) and Co(III), may be very slow. Systems that equilibrate rapidly are termed labile, and those that are slow are inert. Inertness may give the appearance of stability, but a complex that is apparently stable because of kinetic inertness may be unstable in the thermodynamic equilibrium sense. Practical use has been made of both lability and inertness, and of stability and instability. For example, a reaction or handling process may be completed during the period of apparent stability that is actually a result of inertness.

Factors Affecting Stability. A characteristic of chelation which distinguishes it from coordination of metals by monodentate agents is the increased stability of the complex that results from ring formation. For equal numbers of similar coordinated donor atoms, as in ammine complexes compared to chelates of ethylenediamine, eg, $M(RNH_2)_2$ vs M(en) or $M(RNH_2)_4$ vs $M(en)_2$, the stability constants of the chelates are from one to several orders of magnitude greater than those of the monodentate complexes. The greater stability of the chelates is largely the result of an increase in entropy resulting from an increase in the number of free molecules, usually solvent or other monodentate ligand, liberated as the chelate is formed. The extra stabilization produced by the ring formation is called the chelate effect (see Coordination compounds).

Many parameters influence the stability of chelates. Several of the stability factors common to all chelate systems are the size and number of rings, substituents on the rings, and the nature of the metal and donor atoms. In the macrocyclic complexes, the degree to which the size of the metal ion fits the space enclosed by the macro rings is a major factor. In chelation, five- and six-membered rings are more stable than others for much the same reasons as in general organic chemistry. Coordination angles on the metal atoms prohibit the formation of three-membered rings, and ring closure is improbable for rings with more than seven members, with coordination in linear chains being a competing reaction. Formation of each additional ring by the same ligand contributes extra stability from the entropy effect of displacing coordinated solvent molecules, and probably some other effects. Substituents on the ring can produce steric hindrance effects, or they may alter the availability of the donor atom electrons for coordination.

The natures of metal and donor atoms are often complementary. The alkaline and rare earth metals and positive actinide ions generally have greater affinity for $-O^-$ groups. Many transition metals complex preferentially with enolic $-O^-$ and some nitrogen functions. Polarizability of the donor atoms correlates with stability of complexes of the heavier transition metals and the more noble metal ions.

In any series of chelates, the variability of the stability constants is usually influenced by more than one of the parameters that are known to affect chelate stability. The data in Table 3 illustrate some of these relationships. The EDTA homologues show the effect of increasing the size of the ring formed by the coordination of the metal and the two nitrogen atoms. The aminoacetic acid series shows the stability gained by the formation of additional rings with a single ligand molecule. The copper–polyamine series shows combined effects of ring size, number of rings for similar donor atoms, and whether the rings are formed by one or more ligand molecules.

pH Effects. Being Lewis bases, the donor atoms of chelating agents react with Lewis acids among which metal ions and hydrogen ions are of principal interest in chelation chemistry. In the pH range attainable in water solutions, most of the well-known chelating agents exist in both protonated and unprotonated forms which are in equilibrium with each other. Metal ions compete with hydrogen ions for the available donor atoms, and therefore, simultaneous equilibria exist which are treated mathematically by the simultaneous equations for the formation constants of the chelates and the acid dissociation constants of the chelating agents. In aqueous systems, water is a competing ligand, and its dissociation into hydrogen and hydroxyl ions must often be considered in the system of simultaneous equilibria. In nonaqueous solvents, similar treatment is possible with appropriate modifications concerning the concept of acidity in those systems. The pH leveling effect of water affects and limits the acid dissociation behavior of chelating agents in aqueous systems in the same manner as ordinary acids. However, coordination with certain metals can result in loss of a proton in aqueous solution from aliphatic $-OH$ and $-NH_2$ groups.

Consider the equilibria in an aqueous system composed of a bidentate ligand HA (eg, the enol form of acetylacetone) and a tetracoordinate metal M^{2+}, (8) Figure 1. The equations are

$$HA \rightleftharpoons A^- + H^+ \qquad M^{2+} + 2\,A^- \rightleftharpoons MA_2 \qquad (15)$$

$$K_A = \frac{[H^+][A^-]}{[HA]} \qquad K = \frac{[MA_2]}{[M^{2+}][A^-]^2} \qquad (16)$$

which give the relation

$$[M^{2+}] = [H^+]^2 \times \frac{[MA_2]}{[HA]^2} \times \frac{1}{KK_A^2} \qquad (17)$$

This equation shows that an increase in acidity of the solution will increase the concentration of uncomplexed metal which must result from the displacement of M^{2+} from MA_2, causing a simultaneous decrease in the ratio of complexed to free ligand

Table 3. Ring Effects on Stability

Ligand	Type formula	Ring size	Number of rings	Number of coordinated donor atoms		Log K^a	
EDTA, $n = 2^b$	CaL	5				10.5	
homologue, $n = 3$	CaL	6				7.1	
homologue, $n = 4$	CaL	7				5.2	
homologue, $n = 5$	CaL	8				4.6	
					$Cu(II)$	$Ni(II)$	$Co(II)$
H$_2$NCH$_2$COOH	ML	5	1	2	8.6	6.2	5.2
HN(CH$_2$COOH)$_2$	ML	5	2	3	10.6	8.2	7.0
N(CH$_2$COOH)$_3$	ML	5	3	4	12.7	11.3	10.6
					$Log\ K_1$	$Log\ K_2$	$Log\ K = Log\ K_1K_2$
H$_2$NC$_2$H$_4$NH$_2$	CuL	5	1	2	10.72	9.31	20.0
	CuL$_2$	5	2	4			
H$_2$N(CH$_2$)$_3$NH$_2$	CuL	6	1	2	9.77^c	7.1	16.9^c
	CuL$_2$	6	2	4			
HN(C$_2$H$_4$NH$_2$)$_2$	CuL	5	2	3	15.9	5.4	21.3
	CuL$_2$	5	2	4			
(H$_2$NC$_2$H$_4$NHCH$_2$)$_2$	CuL	5	3	4	20.5		

^{a} Ref. 9b.
^{b} (HOOCCH$_2$)N(CH$_2$)$_n$N(CH$_2$COOH)$_2$.
^{c} Ref. 9a.

352

[MA₂]/[HA]. The opposite effects will obviously result on decreasing the acidity. This behavior occurs in the pH range where appreciable amounts of both HA and A⁻ coexist. Outside this range the ligand is present almost entirely as either HA or MA₂ and A⁻, and the system is essentially independent of pH.

Chelating agents that are polybasic acids give two or more hydrogen ions per mole. The four stages of dissociation of EDTA, eg, are represented by the equations

$$H_4A \underset{}{\overset{-H^+}{\rightleftarrows}} H_3A^- \underset{}{\overset{-H^+}{\rightleftarrows}} H_2A^{2-} \underset{}{\overset{-H^+}{\rightleftarrows}} HA^{3-} \underset{}{\overset{-H^+}{\rightleftarrows}} A^{4-} \quad (18)$$

whose equilibrium constants K_1, K_2, K_3, and K_4 are known. The proportion of each species present in aqueous solution as a function of pH is shown in Figure 2, with the

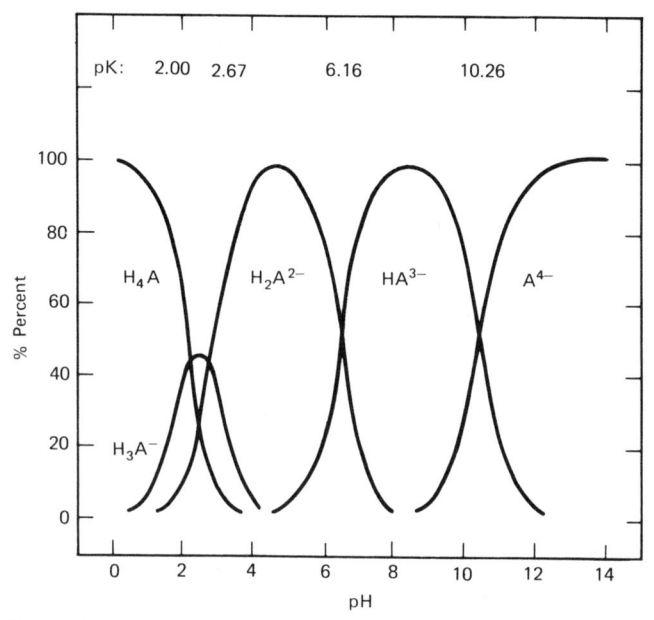

Figure 2. Distribution of ionic species of EDTA as a function of pH. Courtesy of The Dow Chemical Company.

pK values for the four dissociation steps. The reaction of Na₂EDTA (disodium salt) with M²⁺ is represented by equations 5 and 6 which (noting that $KK_3K_4 = k$ of eq. 6) give

$$[M^{2+}] = [H^+]^2 \times \frac{[MA^{2-}]}{[H_2A^{2-}]} \times \frac{1}{KK_3K_4} \quad (19)$$

which is of the same form as equation 16. In general, for the reaction

$$[M^{x+}] + H_nA^{y-} \rightleftarrows [MA^{x-y-n}] + n\,H^+ \quad (20)$$

the equation for the concentration of free metal is

$$[M^{x+}] = [H^+]^n \times \frac{[MA^{x-y-n}]}{[H_nA^{y-}]} \times \frac{1}{KC} \quad (21)$$

where K is the formation constant of MA^{x-y-n} and C is the product of the n stepwise acid dissociation constants involved. Taking the negative logarithm of both sides of the equation and letting $-\log[M^{x+}] = pM$, the equation becomes

$$\text{pM} = n\,\text{pH} + \log\frac{[H_nA^{y-}]}{[MA^{x-y-n}]} + \log KC \tag{22}$$

The corresponding generalized form for equation 16 is

$$\text{pM} = n\,\text{pH} + \log\frac{[HA]^n}{[MA_n{}^{x-n}]} + \log KK_A^n \tag{23}$$

In both cases n is the number of hydrogen ions displaced in the formation of the complex. In solutions with the ratio of free chelating agent to complex, $[H_nA^{y-}]/[MA^{x-y-n}]$, held constant, the slopes of the curves pM vs pH are equal to n in the region where H_nA^{y-} is the principal form of the chelating agent.

The displacement of hydrogen ions by a metal ion from a protonated form of the chelating agent (eq. 20) will generate an autogenous pH that will depend on the base strength of the counterions of the metal salt. The pH of the solution can become quite low if these counterions are those of a strong acid, eg Cl^- or $SO_4{}^{2-}$. However, just as with other types of systems, the pH of chelate solutions can be controlled by the use of compatible buffers. The chelating agent itself can sometimes serve as the pH buffer if one of its acid dissociation stages occurs in the desired pH range.

The variation of pM with pH according to equation 20 is shown by the solid lines in Figure 3 for EDTA with Cu(II) and Mn(II), with three different ratios of free chelant to metal chelate, $[H_nA^{y-}]/[MA^{x-y-n}]$. The pM value at any pH indicates the concentration of the free aquo metal ion in equilibrium with the chelate. From pH 2 to 6, the free form of the EDTA is H_2A^{2-}, and because two protons are displaced by the metal on chelation, the slope of the curves is two. The slope changes to one from pH 6 to 10 where the EDTA is present as HA^{3-} and only one proton is displaced. Above pH 10, the chelates are formed from the fully dissociated chelant, A^{4-}, no hydrogen

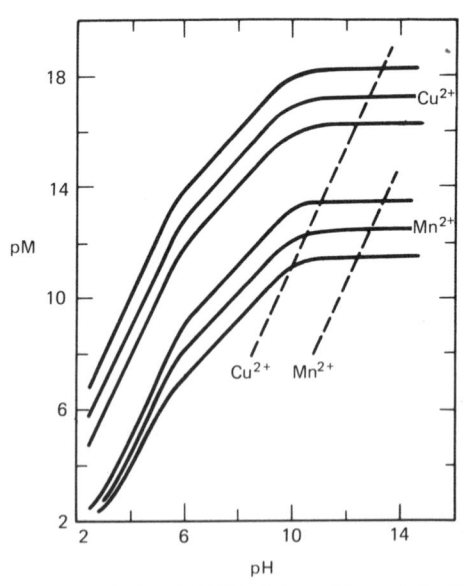

Figure 3. pM vs pH for Cu(II) and Mn(II) EDTA chelates. For each family of curves, the lowest curve represents 1%; the second, 10%; and the top curve, 100% of free ligand species in excess of the amount needed to form the metal chelate. Broken lines represent solid–solution equilibria for corresponding metal hydroxides.

ions are displaced, the slope of the curves is zero, and pM is independent of pH up to the pH of the intersection of the solid lines with the dashed lines for the same metal. Up to pH 10, the rise of pM with pH shows the increasing degree of metal binding as competition by hydrogen ions for the chelant is decreased.

Cu(II) and Mn(II) form insoluble metal hydroxides to which the following equations apply at equilibrium in aqueous solution:

$$M^{2+} + 2\,OH^- \overset{K_{sp}}{\rightleftharpoons} M(OH)_2\,(\text{solid}) \tag{24}$$

$K_{sp} = [M^{2+}][OH^-]^2$, $K_w = [H^+][OH^-]$, and from these

$$pM = 2\,pH + \log(K_w^2/K_{sp}) \tag{25}$$

The dashed lines in Figure 3 are plots of equation 25 for Cu and Mn and indicate the concentration of the aquo metal ions in equilibrium with the solid hydroxides as a function of pH. At any pH where the solid curve is above the dashed curve for the same metal, the EDTA is holding the unchelated metal ion concentration at a value too low for the precipitation of the solid hydroxide. Relatively large quantities of the metal can thus be maintained in solution as the chelate at pH values where otherwise all but trace quantities of the metal would be precipitated, ie, at pH values where pM of the dashed curves is 4 or greater. At the pH of intersection of the solid lines with the dashed lines for the same metal, the free metal ion is in equilibrium with both the solid hydroxide and the chelate, and both can coexist. At higher pH the hydroxyl ion competes more effectively than the chelant for the metal and only a trace of either the chelate or the aquo metal ion can exist in solution. Any excess of the metal will be present as solid hydroxide.

The more stable the chelate, the higher will be the pM that it can maintain, and the higher the pH that will be required to precipitate the metal hydroxide. From equation 25 it can be seen that the smaller the solubility product K_{sp}, ie, the more insoluble the metal hydroxide, the higher is the pM that a chelant must maintain to prevent precipitation. The stability constant of Fe(III)–EHPG, (12) Figure 1, is so large (10^{35}) that iron is not precipitated even in strongly alkaline solutions.

If instead of hydroxyl ion the precipitating agent is the anion of an acid stronger than or comparable in strength to the chelating agent, the solid may be insoluble at low pH where the chelant is protonated but not the precipitant anions, and soluble at higher pH where the complexing form of the ligand is relatively more available. Oxalate with EDTA and Ca(II) shows this behavior; Figure 4 shows the relationships of pM to pH. The solid lines give pM for the Ca–EDTA system and the dashed lines represent the calcium oxalate solubility. As in Figure 3, at pH values where the dashed lines are above the solid lines, the metal is present almost entirely as the insoluble salt. To the right of the intersections of dashed and solid lines the metal is almost entirely chelated and the solid salt phase cannot exist.

Titration Behavior. Protonated chelating agents exhibit titration behavior typical of their acidic groups (eg, carboxyl, phenolic hydroxyl, ammonium, sulfhydryl) if they are titrated with bases whose cations have very weak or no tendency to form chelates. In the presence of a metal ion which coordinates with the donor atom of one of these acidic groups, hydrogen ion will be displaced by the metal, the acid strength of the group will thus appear to be enhanced, and the hydrogen ion concentration developed will be higher than in the absence of the metal. The stronger the coordination tendency, the greater will be the apparent acidity of the group. The driving force of chelate ring

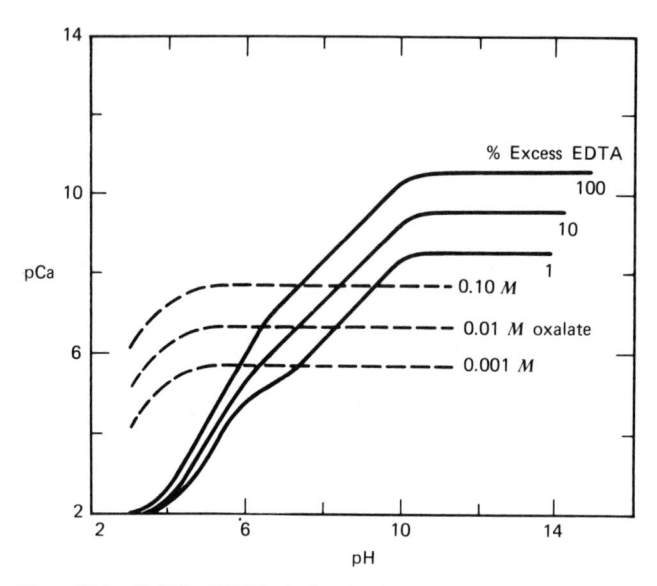

Figure 4. pM vs pH for Ca(II)—EDTA chelate in the presence of excess free ligand. Broken lines indicate solid–solution equilibria of calcium oxalate in the presence of three concentrations of dissolved oxalate.

formation further increases the apparent acidity, often by several orders of magnitude. The cause of this dissociation of hydrogen ion is the familiar tendency for an atom to lose a proton on the development of a nearby positive charge which in chelation is the metal ion.

With A representing donor groups, which may be the same or different, the release of hydrogen ions is shown schematically by the equation

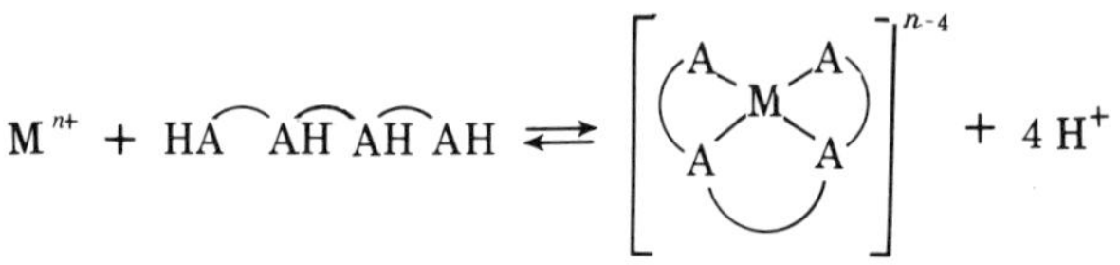

$$M^{n+} + HA\frown AH\frown AH\frown AH \rightleftarrows \left[\begin{pmatrix}A & A \\ & M \\ A & A\end{pmatrix}\right]^{n-4} + 4H^+ \qquad (26)$$

Without coordination, the compound H_4A_4 would titrate as a typical tetrabasic acid with a stepwise titration curve. With strong chelation, all four protons would be displaced and base titration would resemble that of a typical strong acid at four times the equivalent concentration (see Analytical methods). This statement is in agreement with equation 22 which shows that pM can be large (low concentration of free metal) at low pH if K is large (strong chelation). Titration curves for these two types of behavior are shown schematically in Figure 5. The upper curve represents the usual base titration and the lower curve the base titration in the presence of metal ion with strong chelation.

If the chelation is intermediate in strength, the chelation of the groups that coordinate last with the metal may not go to completion at the low pH generated by the hydrogen ions released from the first groups to coordinate. Then, in accordance with equation 22, because K is smaller, the completion of the chelation reaction, as shown by the reduction of the metal ion concentration to a low value (high pM), may not occur until a higher pH is attained where more A^- groups are available to the metal.

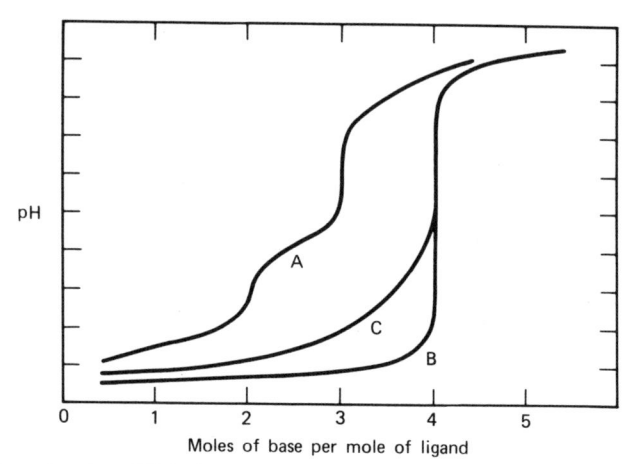

Figure 5. Base titration of H$_4$A chelant; A, free acid without coordination; B, with strongly chelated metal present; C, with less strongly chelated metal.

The curve of pH vs base added will thus rise sooner than the curve for strong chelation, resembling the titration curve of a weak acid and reflecting the lower apparent acid strength resulting from weaker coordination. The intermediate curve of Figure 5 shows this effect of weaker chelation.

Titration of the hydrogen ion liberated from a strong chelating agent is used to determine the concentration of metal ions in solution. The strength of chelation can also be determined from this kind of titration data.

Deprotonation of enols of β-diketones is not considered unusual at moderate pH because of their acidity, although it is facilitated at lower pH by chelate formation. However, chelation can lead to the dissociation of a proton from as weak an acid as an aliphatic amino alcohol in aqueous alkali. Coordination of the O$^-$ atom of triethanolamine with Fe(III) is an example of this effect and results in sequestration of iron in 1 to 18% sodium hydroxide solution (Fig. 6). Even more striking is the loss of a proton from the amino group of a gold chelate of ethylenediamine in aqueous solution (16).

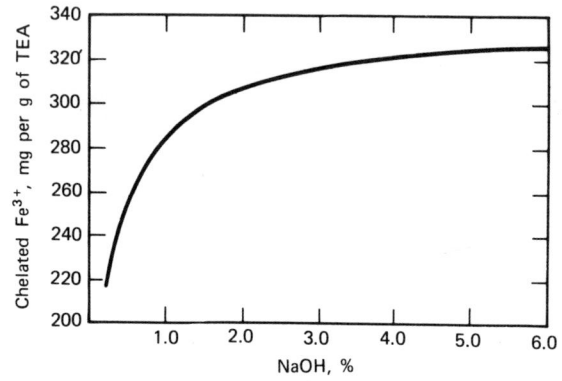

Figure 6. Chelation of iron by triethanolamine (TEA) in aqueous sodium hydroxide. Courtesy of The Dow Chemical Company.

358 CHELATING AGENTS

Another group of chelants that form stable chelates at high pH because of metal–alkoxide coordination are the sugar acids, such as gluconic acid (17). Utility for this group is found in high alkalinity bottle washes and other cleansers (18) (see Hydroxy carboxylic acids).

Metal Buffering. The equation for the formation constant of the reaction

$$M^{n+} + L^{m-} \overset{K}{\rightleftharpoons} ML^{n-m} \tag{27}$$

can be rearranged to give

$$\frac{1}{[M^{n+}]} = \frac{K[L^{m-}]}{[ML^{n-m}]} \tag{28}$$

from which, on taking logarithms and noting that $pM = -\log[M^{n+}]$,

$$pM = \log\frac{[L^{m-}]}{[ML^{n-m}]} + \log K \tag{29}$$

The concentration of the metal ion can be controlled by adjusting the ratio of the concentrations of free ligand and metal chelate, and stabilized at a nearly constant value by the presence of appreciable amounts of these two ligand species so that moderate changes in either will have little effect on their ratio, thus providing buffer capacity. The concentration of the metal ion can thus be buffered in a manner analogous to the buffering of pH by the presence of a weak acid and its anion according to the equation

$$pH = pK_a + \log\frac{[A^-]}{[HA]} \tag{30}$$

In the equation for pM, $\log K$ appears instead of pK because K is a formation constant, the reciprocal of the chelate dissociation constant analogous to the acid dissociation constant K_a.

By choice of chelating agents with the proper stability constants, pM can be regulated over a wide range. Two or more metals may be selectively buffered at different concentrations by a single chelating agent with different stability constants for the metals. Selective buffering of one metal to a low concentration in the presence of other metals is termed masking. The ability to maintain a nearly constant concentration of metal ions at almost any level of concentration is the basis of many of the commercial uses of chelating agents. The buffering capacity may be used to supply metal ions at a definite concentration as in electroplating (qv) and in nutrient media, or to remove or sequester metal ions as in cleaning baths where the fresh stock entering the bath continually introduces additional amounts of the metals to be chelated.

The effect of pH on metal buffering is shown by equations 22 and 23. If a constant pH is imposed on a system by a hydrogen ion buffer, variations in pM are controlled only by variations in the ratio of the free and metal-bound forms of the ligand, and of course by the characteristics (K values) of the ligand. The free form of the ligand is the acid form of its acidic dissociation stage at the imposed pH, ie, HA or H_nA^{y-}. If the acid groups of the chelating agent are fully dissociated at the pH of the buffer, no hydrogen ions are displaced when chelation with the metal occurs, no dissociation constants of the ligand are involved, n in the equations is essentially zero, and pM is independent of pH. Equations 22 and 23 then reduce to the form of equation 29.

Solubilization. The solubility product of a slightly soluble salt determines the concentration of metal ion that can be present in solution with the anion of the salt. For the salt MX the solubility product is

$$K_{sp} = [M^{n+}][X^{n-}] \qquad (31)$$

from which is obtained

$$pM = \log [X^{n-}] - \log K_{sp} \qquad (32)$$

The presence of a sufficiently strong chelating agent (K large in equation 29) will keep the concentration of free metal ion suppressed so that pM will be larger than the saturation pM given by the solubility product relation, equation 32, and no solid phase of MX will form even in the presence of relatively high concentrations of the anion X^{n-}. The metal will be sequestered with respect to precipitation by the anion, such as in the prevention of the formation of insoluble soaps by hardness ions in water.

Deposits of an insoluble salt can be dissolved as a salt of the metal chelate.

$$MX \text{ (solid)} + L^{m-} \underset{}{\overset{K_s}{\rightleftharpoons}} [ML^{n-m}] + X^{n-} \qquad (33)$$

In the presence of the chelating agent and the insoluble salt MX, pM of the solution is subject to both the metal buffering and the solubility equilibria. Equating the right hand sides of equations 29 and 32 and rearranging gives

$$\log [X^{n-}] = \log \frac{[L^{m-}]}{[ML^{n-m}]} + \log KK_{sp} \qquad (34)$$

As the dissolving of the salt progresses $[X^{n-}]$ is approximately equal to $[ML^{n-m}]$, and both represent the amount of MX dissolved. Substituting $[ML^{n-m}]$ for $[X^{n-}]$ in equation 34 gives

$$2 \log [ML^{n-m}] = \log [L^{m-}] + \log KK_{sp} \qquad (35)$$

for the equilibrium concentrations for the process. If $KK_{sp} = 1$, $\log KK_{sp} = 0$, then $[ML^{n-m}] = [L^{m-}]^{1/2}$, and the amount of MX solubilized is equal to the square root of the amount of excess chelating agent required, which is an amount that is in the practical range. An interesting dilution-efficiency effect can be calculated from this relationship. If the amount of MX dissolved gives only a 0.1 M solution of $[ML^{n-m}]$, the excess ligand concentration is 0.01 M and almost 90% of the total amount of ligand is effective in solubilizing salt deposit. However, for 1.0 M $[ML^{n-m}]$ an equal concentration of excess ligand is required and solubilization to 2.0 M requires 4.0 M excess ligand, giving efficiencies of 50 and 33%, respectively. If for economic reasons the chelating agent must be recovered, dilution is a disadvantage, and dilution and efficiency must be compromised. If the stability constant K is large enough, equation 35 shows that only small amounts of the chelating agent in excess of that bound to the metal will be required to dissolve a given amount of the deposit.

The product KK_{sp} is equal to the equilibrium constant K_s for the reaction, equation 33. The rule of thumb, the salt is soluble if $K_s > 1$, that is sometimes encountered means that sequestration or solubilization of moderate amounts of metal ion becomes practical as K_s approaches and exceeds one. For smaller values of K_s, the cost of the required amount of chelating agent may be prohibitive. However, the dilution effect described in the preceding paragraph may allow economical sequestration, or solubilization of small amounts of deposits, at K_s values considerably less than one. In practical applications the parameters affecting ionic activities are difficult to allow for; therefore, calculations based on concentration constants will in general not be exact. Experimental trials based on the calculations as a guide are necessary to determine the actual behavior of particular systems.

The K_s values are shown in Table 4 for some common scale deposits and nitrilotriacetic acid (NTA) which is an effective agent for solubilizing $CaSO_4$ and $CaSiO_3$ ($K_s > 1$). For removal of deposits of $CaCO_3$, a stronger chelating agent for Ca(II) would be required. A large cleaning business which services industry is based on this function of the aminocarboxylic acid chelants.

Electrochemical Potentials. The oxidation potential of a solution containing a metal in two of its valence states, M^{x+} and M^{x+n}, is given by the equation

$$E = E_0 - \frac{RT}{nF} \ln \frac{[M^{x+n}]}{[M^{x+}]} \tag{36}$$

In the presence of a chelating agent, the concentrations of the two forms of the metal are buffered according to the simultaneous equations

$$[M^{x+n}] = \frac{[M_{ox}L]}{K_{ox}[L]} \text{ and } [M^{x+}] = \frac{[M_{red}L]}{K_{red}[L]} \tag{37}$$

where $M_{ox}L$ and $M_{red}L$ are the chelates of the oxidized and reduced forms of the ions and K_{ox} and K_{red} are their formation constants. Substituting these values in the potential equation gives

$$E = E_0 + \frac{RT}{nF} \ln \frac{K_{ox}}{K_{red}} - \frac{RT}{nF} \ln \frac{[M_{ox}L]}{[M_{red}L]} \tag{38}$$

The first two terms of the right member of the equation are sometimes combined and expressed as E'_0, which is called the standard oxidation potential for the chelate system. If the chelation is strong and the ligand is in excess, the metal would be almost entirely in the chelated forms, and $[M_{ox}L]$ and $[M_{red}L]$ would essentially be equal to the total concentrations of the oxidized and reduced forms of the metal. If, as is usual, the oxidized form is the more strongly chelated ($K_{ox} > K_{red}$), the oxidation potential of a system will be increased by the addition of the chelant.

In electrodeposition, the reduced form of the metal is the elemental form M^0, and there is no chelated M^0 in solution. Neglecting activity coefficients, the reversible potential is

$$E = E_0 - \frac{RT}{nF} \ln [M^{x+}] \tag{39}$$

By buffering the metal ion concentration with a chelant, E can be adjusted to and stabilized at values which give desirable properties to the deposit. Selective buffering

Table 4. K_s values for NTA With Calcium Salts (Formation Constant $K = 2.5 \times 10^6$)

Salt	K_{sp}	K_s
$CaSO_4$	6.1×10^{-5}	152
$CaSiO_3$	6.6×10^{-7}	1.65
$CaCO_3$	8.7×10^{-9}	0.022

can sequester the properties of interfering ions, or it can be used to regulate the potentials of two or more ions to approximately the same value in order to effect codeposition.

Applications

Three features of chelation chemistry are fundamental to most of the applications of chelating agents. The first and probably the most extensively used feature is the control of the concentration of the free form (aquo or solvated) of the metal ions by means of the binding-dissociation equilibria. The second is often called the preparative feature in which the special properties of the chelate itself provide the basis of the application. The third feature comprises displacement reactions: metal by other metal or other ions, chelant by chelant, and chelant by other ligands or ions. Overlapping of these features occurs in many uses. An application may be termed defensive if an undesirable property in a process or product is mitigated, or aggressive if a new and beneficial property is induced.

Concentration Control. Sequestration, solubilization, and buffering depend on the concentration control feature of chelation. Traces of metal ions are almost universally present in liquid systems, often arising from the materials of the handling equipment if not introduced with the process materials themselves. Despite their very low concentrations, some of these trace metals produce undesirable effects such as colorations or instability. Sequestration is invaluable in controlling such trace ion effects. Solubilization of insoluble deposits requires both concentration control and the property of solubility and is thus an example of the overlapping of the control and preparative features of chelation chemistry. Buffering may be used at all levels of concentration in applications where an essentially constant metal ion concentration is required as metal is added to or removed from solution. In the concentration control function, the stability constant is a measure of the most important property of the system.

Sequestration. The suppression of certain properties of a metal without removing it from the system or phase is called sequestration. To be practical, the sequestering agent must not cause any undesirable change that would render the system unsuitable for its intended purpose. Chelation produces sequestration mainly by reducing the concentration of free metal ion to a very low value by converting most of the metal to a soluble chelate that does not possess the properties to be suppressed. A sufficiently large stability constant in the medium of the application is required.

The largest single use for sequestration is probably the control of water hardness. Chelating agents are used to prevent the formation of hardness precipitates in a wide variety of aqueous systems such as washing solutions of soaps and detergents, boiler feed water (19), fabric (20), and paper processing solutions (21), cosmetic and pharmaceutical preparations, photographic developing solutions, chemical process water, beverages and foodstuffs (11) (see Surfactants). Oil soluble sequestrants suppress metal catalyzed development of rancidity, gum formation in fuels, and other oxidative degradations (22). In fabric bleaching the catalytic decomposition of the bleaching agent is reduced, and tenderizing of fabrics is minimized by the sequestration of metals that catalyze the reaction of the bleach with the material (23) (see Bleaching agents). In dyeing, metal contaminants that cause spotting, off-color shades, and decomposition of the dyes are sequestered (24). Brightness reversion in paper pulp is diminished (25), and iron is removed in the caustic washing step of the preparation of photographic and other special grades of paper. Discoloration of leather by metal–tannin complexes is prevented. Chelants are used in many metal treating operations such as phosphatizing, alkaline derusting, and etching. Poisonous metal ions are removed from mammals by dimercaptopropanol (BAL). Hundreds of other examples can be found in the literature on sequestration and chelation.

Solubilization. Causing the constituents of a phase that is normally insoluble to dissolve in the medium is termed solubilization. Chelation solubilization depends on the formation of a chelate having solubilizing groups and a stability sufficient to sequester the metal ion to a pM that can exist in the presence of the associated counterions. Usually solubilization into an aqueous phase is thought of in connection with chelation. In this case, the donor atoms involved in the chelation may be sufficiently hydrophilic to produce a soluble species, as in structures (10) and (12), Figure 1. If the organic group is large, more hydrophilic groups may be required for water solubility. Oxine is a well-known precipitant, but its sulfonated derivative, (9) Figure 1, is a solubilizer in water. A chelant whose charged groups neutralize the charge of the metal may solubilize the material into organic media, (8) Figure 1. The macrocyclic chelates derive their dual solubility in aqueous and organic media from both their ionic nature and the largely organic character of the cation.

Dissolving boiler scale, scale in heat exchangers, and hardness scale from pipes is probably the largest industrial example of solubilization (26) (see Dispersants; Water, industrial water treatment). The cleaning of films from dairy equipment and reusable beverage bottles is also a major use (18). Deposits on processing tanks that are unique to the particular industry (paper, textiles, metal treating, photography) are often removed by solubilization. Prevention of the deposition by sequestration is usually preferred where possible because solubilization is sometimes slow and can lead to costly down time. In-service solubilization of boiler scale is a current subject of research. Chelants are used in recovery of metals from ores (27) and in cleaning oxide films from metals in preparation for surface treatments (28) (see Metal surface treatments). Chelation solubilization is especially useful for cleaning up radioactive contamination (see Nuclear reactors).

Macrotetrolides of the valinomycin group of electrically neutral antibiotics form stable 1:1 complexes with alkali metal ions that increase the cation permeability of some biological and artificial lipophilic membranes. This appears to be a solubilization process having implications in membrane transport research (29) (see Antibiotics, macrolides).

An ingenious combination of pH effect and solubilization is the use of a zinc EDTA chelate in achieving permanence of a floor polish to light detergency combined with subsequent strippability (30). The polish is applied as an ammonia solution at whose pH the zinc is chelated. As the pH decreases on evaporation of the ammonia, zinc is released from its chelate and combines with carboxyl groups of the waxes present to form cross-links that impart insolubility and resistance to light washing. When it is desired to remove the polish film, ammonia is added to the washing solution, the zinc recombines with the chelant which is still present, the insolubilizing cross-links are lost, and the film is stripped (see Polishes).

Buffering. If addition or removal of an appreciable amount of a metal ion produces only a relatively small change in the concentration of that ion in a solution, the solution is buffered with respect to the ion. Metal ions are buffered by chelants in a manner exactly analogous to the buffering of hydrogen ions by bases of various strengths. A buffered solution of nearly constant pM can thus be maintained during a process which is withdrawing or supplying the metal ion.

Chelation buffering is particularly useful in supplying micronutrient metal ions to biological growth systems at controlled, very low concentrations (31) (see Mineral nutrients). At the very small subtoxic concentrations required for some metals, the amount present would ordinarily soon be depleted, but with the buffer, a reserve supply of the metal as its chelate is available over long periods with automatic control of the concentration. Examples of this kind of use are found in microorganism cultures in closed, controlled systems and in field use in agriculture in open environments. The metal concentration can be held at optimum values in electroplating (qv), and chelates have supplied the metal in electroless depositions (see Electroless plating).

Control of concentrations enables simultaneous deposition of metals in alloy plating. Increasing attention is being given to the use of chelants to replace cyanide in electroplating baths. Chelants are used to control the activity of redox polymerization catalysts by buffering the metal ions participating in the mechanism. Buffering of the metal is produced by having appreciable amounts of both the chelant and the chelate present simultaneously. The pM is determined mainly by the stability constant of the chelate, and within the region of this major regulation over a small practical range, by the ratio of chelant and chelate concentration.

Special Properties of Chelates. Some of the major applications of the preparative feature of chelation depend on solubility properties, color, or catalytic effects of the chelates. Solubilization of insoluble deposits depends on the formation of a chelate that is soluble in the medium. Other solubility effects are designed to concentrate a metal into a particular phase by extraction, precipitation, or ion exchange onto a chelating resin. Some dyeing processes achieve color or fastness or both by chelation. The phthalocyanine pigments are intensely colored, insoluble chelates. The color of chelates is the basis for many analytical procedures. Catalytic effects may be the property of a chelate, or chelation of the reactant itself may be part of the mechanism of catalysis by a metal. In biological systems the properties of some enzymes and vitamins involve chelation, and the activities of chlorophyll and hemoglobin are associated with their chelate structures.

Catalysis. In catalysis chemistry, the importance of coordination between ligands and metals has long been recognized (see Catalysis). When chelating ligands are used, the special properties of the chelate itself are used for catalytic effects. This is especially evident in asymmetric syntheses catalyzed by chelates of an asymmetric ligand, as in the homogeneous hydrogenation of double bond functions by a cobalt chelate with quinine as the chiral ligand (32) (see Pharmaceuticals, optically-active). In another application, a cobalt chelate is used as an oxygen carrier in the sweetening of gasoline (qv) by oxidation of mercaptans. Chelation is a feature in much research on the development and mechanism of action of catalysts in both industrial and academic disciplines. Enzyme chemistry is aided by the study of reactions of simpler chelates that are models of enzyme reactions (see Enzymes). Certain enzymes, coenzymes, and vitamins (qv) possess chelate structures that must be involved in the mechanism of their action. The activation of many enzymes by metal ions most likely involves chelation, probably bridging the enzyme and substrate through the metal atom. Enzyme inhibition may often result from the formation by the inhibitor of a chelate with a greater stability constant than that of the substrate or the enzyme for a necessary metal ion.

Many reactions catalyzed by the addition of simple metal ions involve chelation of the metal with one of the reactants. The familiar autocatalysis of the oxidation of oxalate by permanganate in analysis results from the chelation of the oxalate with Mn(III) from the permanganate. Oxidation of ascorbic acid is catalyzed by copper (11). The stabilization of preparations containing ascorbic acid by the addition of a chelant appears to be negative catalysis of the oxidation but is due to the sequestration of the copper. Many such inhibitions are the result of sequestration. Catalysis by chelation of metal ions with a reactant is usually accomplished by polarization of the molecule, facilitation of electron transfer by the metal, or orientation of reactants.

Chelation itself is sometimes useful in directing the course of synthesis. The presence of a suitable metal ion facilitates the preparation of the crown ethers, porphyrins, and similar heteroatom macrocyclic compounds. Coordination of the heteroatoms about the metal orients the end groups of the reactants into correct proximity for ring closure, and the product is the chelate from which the metal may be removed by a suitable method. In other catalytic effects of metal chelates reactive centers may be brought into close proximity, charge or bond strain effects may be created, or electron transfers may be made possible.

A feature that has accompanied the introduction of the crown ethers and cryptates is the ability to complex the alkali metals very strongly (33). Applications depend on the appreciable solubility of the compounds in a wide range of solvents and the increase in activity of the co-anion owing to reduced ion-pair association in nonaqueous systems. For example, potassium hydroxide or permanganate can be solubilized in benzene by dicyclohexyl[18]-crown-6 [16069-36-6]. In nonpolar solvents the anions are neither extensively solvated nor strongly paired with the complexed cation, and they behave as naked or bare anions with enhanced activity. Small amounts of the macrocyclic compounds can serve as phase transfer agents and they may be more effective than tetrabutylammonium ion for the purpose (see Catalysis, phase-transfer). These properties have been used extensively in research, but the cost of the macrocyclic agents limits their industrial use.

Precipitation and Extraction. The processes of extraction and precipitation comprise transferring the metal into another phase. If the ligand charges neutralize those of the metal ion, the complex becomes a neutral molecule. As the size of the hydrophobic part of the ligands increases, their neutral chelates become somewhat to much less soluble in water and will precipitate if enough chelate is present to exceed the solubility. Some ligands precipitate certain metals essentially quantitatively, and these properties form the basis for analytical methods (qv) and for recovery of metals from ores or from waste streams (see Recycling, metals). Oxine (8-hydroxyquinoline) is a well-known precipitating agent. Selective and successive precipitations are used in the separation and recovery of the rare earth elements (qv). Passivation of metals by many organic corrosion inhibitors may involve the formation of an insoluble chelate film with the oxide on the surface or with the metal itself (34).

A special kind of transfer of metal ions to a solid phase is found in the chelating resins. These are similar to the ordinary cation-exchange resins except that they have chelating groups in place of salt-forming groups. The behavior of the two kinds of resins is similar except that the special effects of the chelation equilibria must be considered. An important use of chelating resins is for the preconcentration of ions from such media as seawater, body fluids, and geological materials in which their concentration is exceedingly small, so that they may be detected or determined analytically (35).

If a neutral chelate formed from a ligand such as acetylacetone is sufficiently soluble in water not to precipitate, it may still be extracted into an immiscible solvent and thus separated from the other constituents of the water phase. Metal recovery processes, such as from dilute leach dump liquors, and analytical procedures are based on this phase transfer process, as with precipitation. Solvent extraction theory and many separation systems are reviewed by Stary (36).

A cobalt chelate, Fluomine [57693-02-4], absorbs oxygen from air and then releases it pure at a higher temperature. The properties of Fluomine suit it for use in breathing-oxygen systems for high performance aircraft. The precise method of preparing the chelate crystals is crucial to obtaining a product with a sufficiently useful life, ie, high oxygen production per cycle throughout thousands of cycles (37).

Displacement. In many of the applications of chelating agents, the overall effect appears to be a displacement reaction, although the mechanism probably comprises dissociations and recombinations. The basis for many analytical titrations is the displacement of hydrogen ions by a metal, and the displacement of metal by hydrogen ions or other metal ions is a step in metal recovery processes. Some analytical pM indicators function by changing color as one chelant is displaced from its metal by another.

The pH effect in chelation is utilized to liberate metals from their chelates which have participated in another stage of a process, so that the metal or chelant or both can be separately recovered. Hydrogen ion at low pH displaces copper, eg, which is recovered from the acid bath by electrolysis while the hydrogen form of the chelant is recycled (38). Precipitation of the displaced metal by anions such as oxalate as the pH is lowered (Fig. 4) is utilized in separations of rare earths. Metals can also be displaced as insoluble salts or hydroxides in high pH domains where the pM that can be maintained by the chelate is less than that allowed by the insoluble species (Fig. 3).

Rare earths have been separated by elution from ion-exchange resins with chelates of iron, manganese, and cadmium. In another separation, a band of rare earths on an ion-exchange resin was eluted with a chelant and the eluate was passed over another ion-exchange bed loaded with copper. The copper displaced the rare earth metals which deposited on the second bed. In a solution mining process, a leach solution of $Na_2CaEDTA$ and sodium bicarbonate exchanges Ca for Cu, and the copper is then displaced by lime and the leach solution is regenerated (27). In using chelated chromium in leather tanning, the chromium is captured from the chelant by the collagen in the hide.

In the treatment of poisoning by lead and other metals with EDTA, displacement is used to avoid the depletion of calcium in the body fluids that would attend the direct introduction of the chelant (39). By the use of a calcium form of EDTA, the poisonous ion which forms a more stable chelate displaces a small amount of calcium which is tolerated by the body, but no calcium is removed from the fluids. In this manner, an excess of chelant can be used, the more effectively to remove the lead. The calcium form of EDTA is used in many food preparations to stabilize against such deleterious effects as rancidity, loss of ascorbic acid, loss of flavor, development of cloudiness, and discoloration. In these uses, the causative metal ions are sequestered by exchange and possible problems from ingestion of the free chelant are avoided.

When metal ions are involved in a chemical system, chelation and coordination chemistry may be a means either of achieving special effects or of understanding otherwise unusual behavior. The need to control metal ion concentrations has been recognized in many industrial systems, and many of these needs have been met by employing the principles of chelation.

Major Industrial Chelating Agents

Production data for the five principal industrial chelating agents are given in Table 5. The list is dominated by STPP but not all of the amount shown is used in chelation applications (see Phosphoric acids and phosphates). The major use of STPP is in cleaning formulations, mainly as a builder in consumer laundry and cleaning products (see Surfactants). Other polyphosphates can also act as chelating agents, but their volumes are minor compared to STPP.

Citric acid (qv) is another major agent, but like STPP it is difficult to estimate the proportion of the consumption that is used in chelation applications. The primary use of citric acid is in foods and beverages as a pH regulator, flavor enhancer, and antioxidant synergist. Possibly 20% of the citric acid produced is used in chelation and other industrial applications.

The aminopolycarboxylic acids are used principally as chelating agents. The production of EDTA in all its forms (free acid and salts) accounts for about one-third of the total production of this class of chelating agents. Most of the EDTA is sold either as the di- or tetrasodium salt. About 69,500 metric tons of all the nitriloacids and salts was produced in 1975 with an average value of $0.99/kg (40). A large amount of chelating material of this class is used in cleaning formulations, and NTA was a major phosphate replacement until 1970 when a government directive restricted its use in the United States. An estimated 22,700 metric tons of U.S.-produced NTA is exported annually to Canada for use in synthetic detergents.

Gluconic acid and its salts are used primarily in strongly alkaline cleaning preparations, a major market being bottle-washing solutions. Gluconates are also used in rust removers, metal-cleaning and aluminum-etching formulations, and in textile processing to prevent iron contamination during mercerization. Alkaline gluconate solutions will dissolve ferric oxide.

Table 5. United States Production of Major Chelating Agents

Agent	U.S. production, metric tons per year	Major producers[a]	Price, $/kg
STPP	662,000[b]	FMC, Mon, Ol, St	0.49–0.62[h]
citric acid[c]	97,500[d]	Mil, Pf	1.15–1.28[h]
EDTA	22,700[e]	C-G, D, WRG	1.36[e]
organophosphonates	13,600[f]	Mon, Pe	1.10–7.72[f]
gluconic acid[c]	8,160[g]	Pf, St	0.66–1.21[h]

[a] U.S. Department of Commerce, Bureau of the Census, Report M28A(77)-5.
[b] FMC Corporation; Mon—Monsanto Company; Ol—Olin Corporation; St—Stauffer Chemical Company; Pf—Pfizer, Incorporated; Mil—Miles Laboratories, Incorporated; C-G—Ciba-Geigy Corporation; D—The Dow Chemical Company; WRG—W. R. Grace and Company; Pe—Petrolite Corporation.
[c] Includes salt forms.
[d] Chemical Economics Handbook Report No. 636.5021 (1977).
[e] United States International Trade Commission Publication No. 804, Synthetic Organic Chemicals, U.S. Production and Sales, 1975.
[f] Private communication.
[g] Ref. 41.
[h] Ref. 42.

BIBLIOGRAPHY

"Sequestering Agents" in *ECT* 1st ed., Vol. 12, pp. 164–181, by Harry Kroll and Martin Knell, Alrose Chemical Co.; "Complexing Agents" in *ECT* 2nd ed., Vol. 6, pp. 1–24, by Arthur E. Martell, Illinois Institute of Technology.

1. G. T. Morgan and H. D. Drew, *J. Chem. Soc.* **117,** 1456 (1920).
2. C. J. Pedersen, *J. Am. Chem. Soc.* **89,** 7017 (1967).
3. B. Dietrich, J. M. Lehn, and J. P. Sauvage, *Tetrahedron Lett.*, 2889 (1969).
4. J. J. Christensen, D. J. Eatough, and R. M. Izatt, *Chem. Rev.* **74,** 351 (1974).
5. H. Diehl, *Chem. Rev.* **21,** 39 (1937).
6. *Chemical Abstracts Ninth Collective Index,* Index Guide, Appendix IV F., par. 215, pp. 180I–184I (1972–1976).
7. *Chemical Abstracts Eighth Collective Index,* pp. 7004F and 12481S (1967–1971); *Ninth Collective Index,* p. 9668F (1972–1976).
8. J. Bjerrum, G. Schwarzenbach, and L. G. Sillen, *Stability Constants;* Pt. I, *Organic Ligands,* Special Publication No. 6, 1957, and Pt. II, *Inorganic Ligands,* Special Publication No. 7, The Chemical Society, London, 1958; L. G. Sillen and A. E. Martell, eds., *Stability Constants,* 2nd ed., Special Publication No. 17, The Chemical Society, London, Eng., 1964; R. M. Smith and A. E. Martell, *Critical Stability Constants,* Plenum Press, New York, volumes starting from 1974; J. R. Van Wazer and C. F. Callis, *Chem. Rev.* **58,** 1011 (1958); D. T. Sawyer, *Chem. Rev.* **64,** 633 (1964); J. Kragten, *Atlas of Metal Ligand Equilibria in Aqueous Solution,* John Wiley & Sons, Inc., New York, 1978.
9. (a) A. E. Martell and M. Calvin, *Chemistry of Metal Chelate Compounds,* Prentice-Hall, Inc., Englewood Cliffs, N. J., 1952, p. 522; (b) S. Chaberek and A. E. Martell, *Organic Sequestering Agents,* John Wiley & Sons, Inc., New York, 1959, pp. 126–130; (c) F. J. C. Rossotti and H. Rossotti, *The Determination of Stability Constants,* McGraw–Hill Book Co., New York, 1960.
10. T. E. Furia, *Food Technol.* **18,** 1874 (1964).
11. E. Niadas and L. Robert, *Experientia* **14,** 399 (1958).
12. H. W. Zussman, *Am. Dyestuff Reptr.* **38,** *Proc. Am. Assoc. Text. Chem. Color.* P500–4 (1949).
13. K. Vetejska and J. Mazacek, *Czech.* **101,** 864 (Dec. 15, 1961); *Chem. Abst.* 58:11005d.
14. U.S. Pat. 2,921,847 (Jan. 19, 1960), M. Knell and H. Kroll (to Geigy Chemical Corp.).
15. P. J. Bailes, C. Hanson, and M. A. Hughes, *Chem. Eng.* (*N. Y.*) **83**(2), 86 (1976).
16. B. P. Block and J. C. Bailar Jr., *J. Am. Chem. Soc.* **73,** 4722 (1951).
17. D. T. Sawyer, *Chem. Rev.* **64,** 633 (1964).
18. U.S. Pat. 2,584,017 (Jan. 29, 1952), V. Dvorkovitz and T. G. Hawley (to The Diversey Corp.).
19. J. C. Edwards and E. A. Rozas, *Proc. Am. Power Conf.* **23,** 575 (1961).
20. J. H. Wood, *Am. Dyestuff Reptr.* **65**(11), 32 (1976).
21. V. N. Gupta and D. B. Mutton, *Pulp Pap. Mag. Can.,* **70**(1), T174 (1969).
22. U.S. Pat. 2,181,121 (Nov. 28, 1939), F. B. Downing and C. J. Pedersen (to E. I. du Pont de Nemours & Co., Inc.).
23. E. P. Bayha, L. R. Hubbard, and W. H. Martin, *Intern. Dyer* **131,** 529 (1964).
24. H. E. Millson, Jr., *Am. Dyestuff Reptr.* **45,** *Proc. Am. Assoc. Text. Chem. Color.,* P66 (1956).
25. R. H. Dick and D. H. Andrews, *Pulp Pap. Mag. Can.* **66**(3), T201–T208 (1965).
26. J. R. Metcalf, *Ind. Water Eng.* **8**(1), 16 (1971).
27. D. J. Baure and R. E. Lindstrom, *J. Metals* **23**(5), 31 (1971).
28. J. K. Aiken and C. Garnett, *Electroplat. and Met. Finish.* **10**(2), 31 (1957); *Eng. Index,* 667 (1957).
29. W. Simon, W. E. Morf, and P. Ch. Meier, *Struct. Bonding* (*Berlin*) **16,** 113 (1973).
30. W. M. Finn, M. A. Frith, and F. L. McCarthy, *Chem. Spec. Mfrs. Assoc. Proc. Mid-Year Meeting* **52,** 181 (1966).
31. J. J. Mortvedt and co-eds., *Micronutrients in Agriculture,* Soil Science Society of America, Madison, Wisc., 1972.
32. L. Marko and B. Heil, *Catal. Rev.* **8,** 269 (1973).
33. C. J. Pedersen and H. K. Frensdorff, *Angew. Chem. Int. Ed.* **11,** 16 (1972); A. C. Knipe, *J. Chem. Ed.* **53,** 619 (1976).
34. D. C. Zecher, *Mater. Perform.* **15**(4), 33 (1976).
35. G. Schmuckler in N. Bikales, ed., *Encyclopedia of Polymer Science and Technology,* Suppl. Vol. 2, John Wiley & Sons, Inc., New York, 1977, p. 197.
36. J. Stary, *The Solvent Extraction of Metal Chelates,* The Macmillan Company, New York, 1964.
37. A. J. Adduci, *Chemtech* **6,** 575 (1976).
38. M. A. Hughes, *Chem. Ind.* (*London*) (24), 1042 (1975).

39. M. Rubin, and co-workers, *Science* **117**, 659 (1953).
40. *U.S. International Trade Commission Publication 804*, p. 195.
41. *Chemical Statistics Handbook of Manufacturing Chemists Association*, 7th ed., 1971, pp. 158–159.
42. *Chem. Mark. Rep.* (Aug. 15, 1977).

A. L. McCrary
William L. Howard
Dow Chemical U.S.A.

CHEMICAL CLEANING.   See Metal surface treatments.

CHEMICAL GROUTS

Chemical grouting is the practice of injecting chemical grouts into soil, rock, or concrete to alter the physical characteristics of the grouted mass. A chemical grout is defined as a true solution, eg, free of suspended particles, used for grouting purposes. Grouting is normally used when it is desirable to restrict or reroute the flow of water through a formation or to improve formation strength. Grouting is often done in the construction or rehabilitation of dams, buildings, tunnels, shafts, and other structures. The history of grouting can be traced to Genesis, Chapter 11, Verses 2–3, where the use of bitumen for mortar in the construction of the Tower of Babel is discussed. In modern history, Charles Berigny, in 1802, was the first to use grouting to strengthen a masonry wall in Dieppe, France, with a suspension of clay and lime. Portland cement was first utilized as a grouting material in 1838, in the construction of the first Thames tunnel, in England.

In 1886, Jeziorsky patented the first true chemical grouting technique: a two-shot method whereby concentrated sodium silicate was injected through a bore hole into a formation and a solution of calcium chloride was injected through an adjacent bore hole. A precipitate formed when the two liquids came into contact. During the next 70 years all chemical grout materials were variations of the sodium silicate–calcium chloride system, and these materials became synonymous with chemical grouting. In the mid-1950s many other organic chemical products, suitable for use as grouts, were developed. Even so, the current volume of chemicals used for grouting in the United States, is only a small portion of the portland cement grout used (see Cement).

It is estimated that less than 5% of the grouting performed in the United States is done with chemicals; in Europe it is ca 50%. The main reason for this disparity is that European construction is primarily performed on a turn-key basis, where the firm that designs the facility also builds it as well as chooses the best materials for the project. In the United States, very often a consulting engineering firm unfamiliar with chemical grouting designs a facility and a low-bidding construction firm builds the facility at minimum cost. In the United States, chemical grouting has been used after initial construction to rectify unexpected water infiltration problems or to strengthen a distressed foundation, with work being performed by a small number of specialized grouting contractors.

Portland cement was first used as a grout on a large scale in the United States to form a waterproof barrier and strengthen the formations beneath dams built by the various government agencies. These agencies developed practices, theories, and specifications in the first third of this century that became the grouting standards in the United States. However, these practices had little scientific basis and evolved by trial and error from procedures for individual projects. With the recent increase in construction of underground transportation facilities, where grouting is particularly useful, more research is being conducted and more use made of chemical grouts. As a result, it is anticipated that the volume of chemical grouts used in the United States will show a marked increase in future.

An ideal chemical grout would have the following characteristics: permanence; low viscosity, close to that of water; controllable set or gel time; economical qualities, as well as inexpensive handling, storing, and shipping; nontoxic, noncorrosive, nonhazardous, and ecologically safe; and permeability reduction of the substrate or strength increase of that substance by significant amounts, or both.

Silicate Grouts

Silicate-based chemical grouts were the earliest chemical grouts used in the construction industry for both waterproofing and strength improvement of soil formations.

Sodium silicate with a sp gr of 1.40 (41.5° Baumé) and a viscosity of 206 mPa·s (=cP) at 20°C, has been used on the majority of silicate grouting projects in the United States. This material has a weight ratio (SiO_2/Na_2O) of 3.22. The sodium silicate solution gels by reaction with one or more reagents such as acids, polyvalent cations, and certain organic materials. The gelling process involves a decrease in electrical charge of the silicate ions, followed by polymerization of the silicate with the reagent. The gelling time for the grout can be controlled by the quantity of acid or other reagent. The most common silicate systems in the United States are the two-shot or Joosten Process and the one-shot or Siroc System.

The Joosten or Two-Shot Method. Hugo Joosten, a Dutch engineer, was the first to successfully apply Jeziorsky's techniques to large projects (1). He later modified Jeziorsky's methods to allow successive injections of a concentrated solution of sodium silicate and a calcium chloride solution down a single pipe (two-shot method). When injected into sands, stabilized strengths of 2.1 to 4.1 MPa (300 to 600 psi, respectively) result, as determined by unconfined compression tests. In the Joosten method the reaction is almost instantaneous, ground heave is avoided, and the extent of travel of the grout from the injection point is minimal. The main disadvantages of this method are that the high viscosity of the undiluted sodium silicate (50 to 200 mPa·s), and the instantaneous reaction with the calcium chloride require close grout injection points, resulting in a high-cost placing system.

The One-Shot or Siroc System. In 1961, Diamond Alkali Co., now Diamond Shamrock, patented the utilization of formamide as a coagulating agent for sodium silicate (2). The silicate solution in the Joosten process was diluted with water, lowering its viscosity. By the addition of varying amounts of the coagulant and reactants, the *in-situ* gel time could be controlled in a single solution. The concentration of sodium silicate varies between 10–70% giving viscosities, depending on temperature, between 2.5 and 50 mPa·s. The concentration of the formamide varies between 2–30%. Other common reactants are calcium chloride and sodium aluminate and their concentration varies between 2.4–12 g/L (2–10 lb/100 gal) of silicate solution.

For waterproofing purposes, concentrations of sodium silicate less than 30% are adequate; for strength improvement, concentrations normally vary between 35–70% sodium silicate. Grouts containing 35% or more silicate by volume are resistant to deterioration by freezing and thawing and by wetting and drying. The Siroc System can reduce the permeability of sands from 10^{-2} to 10^{-8} cm/s.

Depending upon the sodium silicate concentration, unconfined compressive strengths of sands stabilized with this system can vary between 103 kPa (15 psi) to a maximum of 4130 kPa (600 psi) with the normal practical range between 690–1380 kPa (100–200 psi).

Soils with as much as 20% silt or clay components can be injected with this system. However, with finer soils the rate of injection must be low to avoid fracturing of the formation and forming lenses of chemical grout. All soils have an optimum rate of injection for any specific grout viscosity, depending upon gradation, relative density, void ratio, permeability, depth of overburden, etc.

The pH of the material to be grouted has little effect on the gel time or strength of the grout, except where large amounts of acid or alkali are present in the formation. Sodium aluminate should be used as the reactant when acid soil is being grouted. Alkalinity above 11 pH and acidity below 5 pH may require neutralization before grouting.

Sodium bicarbonate can be used with sodium silicate grouts where low strength, semipermanent grouts are required, eg, in tunnel face stabilizations that need only last until the tunnel rings are in place.

Portland cement can be used as a reactant with sodium silicate grouts, resulting in extremely short gel times. This formulation is very useful for cutting off flowing waters and for grouting cavities to form a seal. Addition of any particulate matter to a grout solution will reduce the penetrability of the resulting suspension, as compared to the original solution. With the addition of portland cement, silicate-based grouts can not be injected into medium sands and finer soils.

The Siroc Grouting System is considered nonhazardous and nonpolluting. Sodium silicate is essentially nontoxic. Formamide is somewhat toxic and corrosive, but does not present any serious hazard if normal safety precautions are followed.

The Siroc chemical grout materials cost 2–5 times as much as portland cement grouts, depending upon the sodium silicate and formamide concentrations. In-place costs are generally much closer to that of cement grouts.

Under certain conditions sodium silicate gels are subject to syneresis, which has been observed only in pure gels or where the skeleton of the sand is coarse. Cement grouting should precede chemical grouting for such conditions. Dissolution (leaching) can also occur in a great excess of water. This can be eliminated by proper proportioning of reagents.

Polyacrylamides

Acrylamide grouts were first introduced in the 1950s; the earliest were mixtures of two organic monomers, acrylamide and methylenebisacrylamide (a cross-linking agent), trade named AM-9, and marketed by the American Cyanamid Company (see Acrylamide; Acrylamide polymers). Cyanamid was the first manufacturer to undertake an extensive research and development program designed to produce a sophisticated technology for field use. Cyanamid discontinued production of AM-9 in July of 1978 because of possible health problems associated with the monomers (3) (see below). Currently, at least two other acrylamide-based grouts are commercially available: Rocagil, from the Rhone Progil Company and Sumisoil, from the Sumitomo Chemical Company Ltd.

Sumisoil is marketed as a powder to be dissolved in water on site. Rocagil is marketed as a concentrated solution, and diluted with water for field use. The dissolved monomers are polymerized with a two-component catalyst system. The catalyst itself can be a per-salt such as ammonium persulfate, a peracid, or a peroxide. The reaction activator is generally an amine such as dimethylaminopropionitrile; however, the EPA has warned that contact with this catalyst should be curtailed because of health risks (3). Acrylamide-based grouts are readily soluble in water, and the 5–20% concentrations used for field work have viscosities of about 1.2 mPa·s, lower than that of any other chemical grout. This permits penetration of finer materials than those that can be treated with other grouts. The other advantage of polyacrylamides is excellent gel time control, coupled with virtually constant viscosity from the time of catalysis to set or gel time.

The gel time in the field can be quite accurately controlled from less than a minute to many hours by varying the concentration of one or both parts of the catalyst system, or by the addition of reaction inhibitors such as potassium ferricyanide. Dissolved salts normally found in groundwater have little effect on gel time and these effects can be completely obviated by using in-site groundwater to dissolve the reagents. Pumping equipment has been developed permitting mechanical or electronic control of gel time of standard stock solutions. Gel times of a few seconds are possible, and have been used to shut off flowing water in construction operations.

Soils stabilized with acrylamide-based grouts generally have unconfined compressive strengths of 345–1380 kPa (50–200 psi, respectively). Strength increases with increasing formation density and decreasing grain size. Below the water table, or in locations approaching 100% humidity, gels are permanent. Exposure to freeze–thaw cycles and wet–dry cycles where the dry periods predominate will cause mechanical deterioration of the gel. Gels in the field are generally 90% water, primarily mechanically bound within the cross-linked polymer lattice. Long exposure to drying will cause the water to migrate out of the gel causing the gel to shrink. In a soil, this will induce forces analogous to capillary tension. Thus, dried stabilized soil samples may show unconfined compressive strengths of 5–20 times the wet strengths. Such high strengths for any dried stabilized soil are generally not representative of actual field conditions. When dried or partially dried gels are reexposed to water or to 100% humidity, the gels will reabsorb the water and swell approximately to the original volume. Shrinkage cracks that may have formed during drying will close, but will not necessarily heal.

Acrylamide gels have a permeability to water of about 10^{-10} cm/s. Unstabilized sands and silts will have permeabilities of between 10^{-1} and 10^{-5} cm/s. Soils whose voids are completely filled with gel will have low permeabilities approaching that of the pure gel.

The cost of acrylamide grouts is many times that of cement, and about twice that of the silicate-based and lignosulfonate grouts. However, the ease of handling and efficiency of acrylamide grout often make it competitive with or less expensive than other materials in place.

Acrylamide grouts are not subject to syneresis, but may, in common with most other chemical grouting materials, be subject to creep and consolidation at high stress points.

The major current use of acrylamide grouts is in stopping infiltration into sewer lines; sophisticated equipment and technology has been developed for this application. Specially designed packers are drawn through the underground lines, and positioned around leaking joints through the use of TV cameras. After treatment, the camera provides immediate judgment of the sealing results.

The largest single usage for acrylamide grout was the construction of the final seal in the grout curtain of the Rocky Reach Dam in Wenatchee, Washington: almost 950 m^3 (250,000 gal). This work has been described in many publications (4). Other large uses include deep shafts and underground mines. Smaller uses include shallow tunnels, and seepage shutoff around foundations, walls, and impounding basins, where inflow or outflow may be coming through soil, rock, or concrete.

Acrylamide is a neurotoxic material, and repeated exposure without normal handling precautions may lead to reversible disturbances of the central nervous system. In at least one instance, acrylamide contamination of a well, resulting in mild acrylamide poisoning of well users, caused a ban on the use of the product in Japan. Gels made from acrylamide are totally nontoxic, except for the minute amounts of unreacted acrylamide they may contain. Organisms normally present in soil consume acrylamide.

In general, the use of acrylamide near potable water sources should be undertaken with caution, and all persons exposed to acrylamide powder or solution should follow the manufacturers recommendation to avoid toxicity problems.

Phenoplasts

Use of these materials began in the 1960s. The prime advantage of the phenoplasts is low viscosity, 1.5–2.5 mPa·s. They can be used for both waterproofing and strength improvement purposes although strengths are generally much lower than with the silicates.

Two currently used chemical grouts in the phenoplast family are Geoseal and Terranier; these have met with limited success in the United States. Terranier, manufactured by ITT Rayonier, contains a polyphenol base.

Geoseal, manufactured by Borden Chemical Company, is prepackaged as a dry powder and is dissolved in the desired amount of water. It includes a mixture of tannin and mimosa extract and, occasionally, a phenolic precondensate which polymerizes in the presence of formaldehyde.

Other characteristics of the phenoplast grouts are that setting time is proportional to the amount of water added, and after setting, the diluted water which is not chemically bound can evaporate and cause the resin to shrink. Unconfined compression strengths of stabilized soils generally range from 345–1380 kPa (50–200 psi).

Lignosulfonate Derivatives

Lignin-based chemical grouts consist of calcium lignosulfonate and a hexavalent chromium salt that react to form a gel. Lignins (qv) are by-products of the wood processing industry and as such cannot be classified as a uniform chemically defined product (see Pulp). Variations in lignosulfate chemical grouts can be anticipated, depending upon the source of material, the time of harvesting, etc.

A precatalyzed dry lignosulfonate grouting material has been sold in the United States under the trade name Terra Firma. Its prime advantages are ease of handling and initial viscosity of 2.5–4 mPa·s. However, the viscosity increases with time until the chemical grout becomes too viscous to pump.

Stabilized samples of lignosulfonate grouts in a water-saturated environment show significant drop in strength over time from initial stabilized strengths in the 345–1390 kPa (50–200 psi) range and wet-dry cycles reduce stabilized strengths with each cycle. Furthermore, lignosulfonates should not be used in conjunction with portland cement grout since the pHs of the materials conflict. The chromium salts generally used are highly toxic. Lignin can cause skin problems and the hexavalent chromium, if not completely incorporated in the polymer, can cause contamination of the environment. These grouts now have limited use in the United States.

Aminoplasts

Urea–formaldehydes, industrially available in large quantities, can be used as chemical grouts (see Amino resins and plastics). Specific formulations (Cyanaloc and Herculox, for example, marketed by American Cyanamid Co. and Halliburton, respectively) will gel with an acid or an acid salt. Gel time control is good, and concentrated solutions will give stabilized soil strengths of ca 1.4 MPa (ca 200 psi). Typically, such solutions will have a viscosity range of 10–20 mPa·s.

One of the reaction's by-products is ammonia. This precludes the use of these grouts in confined areas that are not well ventilated. Also, since the reaction takes place at a low pH, it is inhibited by the presence of cement. Thus, urea–formaldehyde grouts cannot be used in conjunction with or following a cement grouting operation.

The volume of chemical grout used annually in the United States is relatively small at present. However, the acceptability of chemical grout for improving the engineering characteristics of earth and rock masses is increasing. This, in turn, should increase the demand for chemical grout in the future.

BIBLIOGRAPHY

1. U.S. Pat. 1,827,238 (Oct. 13, 1924) and Brit. Pat. 322,182 (May 24, 1928), H. Joosten (to Tiefban und Kalteindustrie A.G.).
2. U.S. Pat. 2,968,572 (Jan. 17, 1961), C. E. Peeler, Jr. (to Diamond Alkali Co.).
3. *Chem. Week,* 18 (May 31, 1978).
4. *MP2-417,* U.S. Army Corps. of Engineers, Mar. 1961.

General References

Siroc Grout Technical Manual, Diamond Shamrock, Cleveland, Ohio, 1964.
M. Polivka, L. R. Witte, and J. P. Gnaedinger, *Field Experience with Chemical Grouting,* Vol. 83, No. SM2, Paper 1204, ASCE Soil Mechanics and Foundations Division, Apr. 1957, pp. 1–31.

R. L. Schiffman and C. R. Wilson, *The Mechanical Behavior of Chemically Treated Granular Soils*, Vol. 58, American Society for Testing and Materials, 1958, p. 1218.

R. H. Karol, *Short Gel Times with Long Pumping Times,* American Cyanamid Co., Princeton, N. J., Apr. 21, 1961.

R. H. Karol and A. M. Swift, *Symposium on Grouting: Grouting in Flowing Water and Stratified Deposits,* Vol. 87, No. SM2, Part 1, Paper 2797, ASCE, Soil Mechanics and Foundations Division, Apr. 1961, pp. 125–145.

Herculox, Technical Data Sheet, Feb. 1961; *Herculox Grout Data Sheet,* Halliburton Co., Duncan, Okla., Nov. 13, 1961.

Cyanaloc 62 Chemical Grout, Explosives and Mining Chemicals Dept., American Cyanamid Co., Wayne, N. J., Oct. 1962.

R. H. Karol, *Gel Extrusion from Grout Holes,* American Cyanamid Co., Princeton, N. J., Feb. 26, 1963.

F. M. Mellinger, *Report of Laboratory Pilot Studies for Rock Bonding with Chemical Grouts,* Technical Report No. 2–31, Ohio River Division Laboratories, U.S. Army Engineer Div., Ohio River, Corps. of Engineers, Cincinnati, Ohio, June 1963.

R. H. Karol, *Eng. Geol.* 1(1), 21 (Jan. 1964).

AM-9 Chemical Grout, Technical Data, American Cyanamid Co., 1965.

G. T. Bator and D. T. Snow, *Grouting,* **61**(2), 128 (Apr. 1966).

Bibliography On Chemical Grouting, Vol. 92, No. SM6, ASCE, Soil Mechanics and Foundations Division, Nov. 1966, pp. 39–66.

R. H. Karol, *Chemical Grouting Technology,* Vol. 94, No. SM1, ASCE, Soil Mechanics and Foundations Division, Jan. 1968, pp. 175–204.

Guide Specifications For Chemical Grouts, Vol. 94, No. SM2, Paper 5830, ASCE, Soil Mechanics and Foundations Division, Mar. 1968, pp. 345–351.

Grouting Methods and Equipment, Department of the Army Technical Manual, TM 5-818-6, Department of the Air Force Manual, AFM 88-32, U.S. Government Printing Office, Washington, D.C., Feb. 1970.

Corps of Engineers, *Engineering Manual, 1110-2-3504, Chemical Grouting,* Department of the Army, Office of the Chief of Engineering, Washington, D.C., May 31, 1973.

Halliburton Co., *List of Patents Pertaining to Grouting Soils for Water Shutoff or Consolidation,* 1974.

R. Bowen, *Grouting in Engineering Practice,* John Wiley & Sons, Inc., New York, 1975.

G. W. Clough, *A Report on the Practice of Chemical Stabilization around Soft Ground Tunnels in England and Europe,* Office of University Research, U.S. Department of Transportation, D. O. T., TST-76-92, Washington, D.C., 1976.

G. W. Clough and co-workers, *Development of Design Procedures for Stabilized Soil Support Systems for Soft Ground Tunneling, Vol. II Preliminary Results,* Office of University Research, U.S. Department of Transportation, D. O. T., TST-77-74, Washington, D.C., Aug. 1977.

J. Herndon and T. Lenahan, *Grouting In Soils, Vol. I, A-State-Of-The-Art Report,* Federal Highway Administration, U.S. D. O. T., FHWA-RD-76-26, Washington, D.C., June 1976.

J. Herndon and T. Lenahan, *Grouting In Soils, Vol. II, Design and Operations Manual,* Federal Highway Administration, U.S. D. O. T., FHWA-RD-76-27, Washington, D.C., June 1976.

G. R. Tallard and C. Caron, *Chemical Grouts For Soils, Vol. I, Available Materials, Vol. II, Engineering Evaluation of Available Materials,* Federal Highway Administration, U.S. D. O. T., FHWA-RD-77-50 and 51, Washington, D.C., June 1977.

REUBEN H. KAROL
Rutgers University

JOSEPH WELSH
Haywood Baker Co.

CHEMICALS FROM BRINE

Various natural brines and artificially produced brines inside salt deposits are important commercial sources of basic industrial chemicals. These chemicals are recovered directly from brines produced by solution mining as well as natural brines (1–2). The brine industry includes the secondary products processed in the same plant as the brine (see p. 388).

The raw material sources include seawater, inland lake as well as subterranean brines, and potash (sylvite) and numerous salt (NaCl) deposits including salt domes and bedded salt. The alternative to these sources are dry-mined evaporite deposits. Raw materials needed for processing are several forms of limestone, dolomite, lime, ammonia, and sulfuric acid, all of which may be produced captively in some cases. The industry is generally energy-intensive to the extent that electrical energy is considered a raw material for several of the large-tonnage products. Electrical power is generated on-site at a number of locations (see Alkali and chlorine products).

Occurrence

Boron Compounds. The saturated liquors pumped from two brine bodies of Searles Lake, California, are the sole commercial domestic brine source of boron compounds yielding almost one-fifth of the boron compounds produced in the United States (3). Possible future commercial brine sources are the geothermal brine wells near the Salton Sea, or even the waters of Great Salt Lake (4) (see Boron compounds).

Bromine. Bromine occurs in the form of bromide ion dissolved in seawater, and to a much lesser extent in underground brines and deposits of marine origin (5). A minute fraction is contained in terrestrial brines (see Bromine).

The earliest commercial production of bromine in the United States, in the mid-1800s, made use of a well-brine at Freeport, Pa. (6). Later, the recovery of bromine from the well-brines of Midland County, Michigan, was developed. Production from natural brines in Michigan, Ohio, and West Virginia supplied the major portion of the United States needs until 1935, and Michigan brines are still one of the principal sources. Minor amounts of bromine are also produced directly from one of the concentrated plant liquors of the KCl recovery process at Searles Lake.

Calcium Chloride. Brines are the sole commercial source of calcium chloride, but are surpassed by limestone as a source of other calcium compounds (see Calcium compounds, calcium chloride).

Natural brines containing a relatively high concentration of calcium ion (>4%) are encountered mainly in Michigan, Ohio, West Virginia, Utah, and California (7), but are, at present, commercially exploited for calcium chloride only in Michigan and California. The California brines occur in near-surface reservoirs resulting from closed-basin drainage. At all other locations mentioned, the brines are of marine origin and occur in much deeper-lying formations.

A secondary, yet commercially important source of calcium chloride is the waste brine resulting from the production of soda ash by the Solvay process that contains 8–9% $CaCl_2$ (8) (see Alkali and chlorine products).

Iodine. Iodine is widely distributed in the lithosphere at low concentrations (about 0.3 ppm) (9), and is present in seawater at a concentration of 0.05 ppm (10). At substantially higher levels, it occurs in specific species of marine plants, the caliche deposits of Chile, rare iodo–silver and iodo–copper minerals, and certain subterranean brines of marine origin. In the latter, the iodine has been concentrated to levels of several hundred ppm (11) by biological activity and geological processes. Underground brines are the sole commercial source in the United States (see Iodine and iodine compounds).

Lithium Compounds. Numerous waters and brines contain lithium in minor concentrations. Commercially valuable natural brines are located at Silver Peak, Nevada (12–13) (400 ppm Li), and at Searles Lake, California (14–15) (40–60 ppm Li). The Great Salt Lake (40 ppm Li) constitutes a major potential source not yet exploited (16). In comparison, seawater contains less than 0.2 ppm Li (see Lithium and lithium compounds).

Magnesium Compounds. Magnesium hydroxide and magnesium chloride are the two commercially important magnesium compounds recovered directly from natural brines. Other magnesium compounds, principally schoenite, $K_2Mg(SO_4)_2 \cdot 6H_2O$; kainite, $KMgClSO_4 \cdot 3H_2O$; and carnallite, $KMgCl_3 \cdot 6H_2O$, are recovered directly in some instances (see p. 387).

The important sources are seawater (ca 1300 ppm Mg), the waters of Great Salt Lake (0.7–1.3%) (17), near-surface and deep-well brines near Wendover, Utah (ca 1%) (18), subterranean brines in Michigan (0.7–2.5%) (19–21), and brine from the Yates formation in the Midland Basin of West Texas (ca 3%) (22). Other subterranean brines containing magnesium at high concentrations (23) are known, but are not commercially exploited (see Magnesium compounds).

Potassium Compounds. Both muriate of potash (KCl) and sulfate of potash salts are produced from brines (18,24–28) in the United States, with three brine potash operations located in Utah, and the fourth in California. All of these plants market the various grades (size) of potash as used by the fertilizer industry. Only a very small amount is sold to other industries (see Fertilizers; Potassium compounds).

Sodium Carbonate. In the United States, natural soda ash (ie, sodium carbonate derived from natural deposits) is produced primarily from nonbrine sources, and secondarily from the brines of Searles Lake. The latter is the sole United States brine source of natural soda ash. The dry mining of immense deposits of trona, $2Na_2CO_3 \cdot NaHCO_3 \cdot 2H_2O$, in the Green River area of southwestern Wyoming, started shortly after World War II, accounted for the rapid gain of the natural soda ash production during the past decade (see Sodium compounds, sodium carbonate).

Sodium Chloride. Common salt (sodium chloride, halite) is obtained by evaporation of brines which may be naturally occurring or obtained by solution mining (see Sodium compounds, sodium chloride).

Sodium Sulfate. Natural brines yielding commercial quantities of Na_2SO_4 are usually, but not always, of terrestrial origin (see Sodium compounds, sodium sulfates). In the United States they include Searles Lake (ca 6.5% Na_2SO_4), the shallow Castile formation underlying Terry and Gaines counties, Texas (ca 11% Na_2SO_4), and the northern arm of Great Salt Lake (ca 2.5% Na_2SO_4). These three sources accounted for slightly more than one-half of the total U.S. sodium sulfate production between 1972 and 1976.

Other natural sodium sulfate brines are found in the dry lakes of southwestern

Saskatchewan, Canada (29); the Laguna del Rey in Coahuila, Mexico; and the Gulf of Kara-Bogaz, U.S.S.R.

Recovery Processes

Boron Compounds. Boron values are recovered from the upper brine structure of Searles Lake by crystallizing borax pentahydrate, $Na_2B_4O_7.5H_2O$, from a plant mother liquor resulting from upstream processes in which sodium and potassium salts are removed (30). In a separate process, boron values are obtained from the liquors from lower-structure brine in a liquid–liquid extraction step (31). Boric acid is then precipitated with diluted sulfuric acid (see Boron compounds, boron oxides).

Bromine. Commercial processes depend upon oxidation of bromide ion to bromine as the initial step (see Bromine). Most of the liberated bromine remains dissolved in the brine, depending on the original bromide concentration and brine temperature. In the second step, the brine is stripped of the dissolved elemental bromine, followed by recovery of bromine from the stripping agent in elemental or combined state, but at a considerably higher concentration than in the original brine. Subsequent purification by distillation may sometimes be the final step.

Direct electrolysis as well as manganese dioxide was used at one time for the oxidation step. Present day processes employ chlorine:

$$Cl_2 + 2\ NaBr \rightarrow 2\ NaCl + Br_2$$

Steam-stripping or steaming-out (Fig. 1), developed in Germany, is the only method currently used in the United States to displace the dissolved elemental bromine from brine. This method is economical for brines containing bromine at concentrations above 1000 ppm (32).

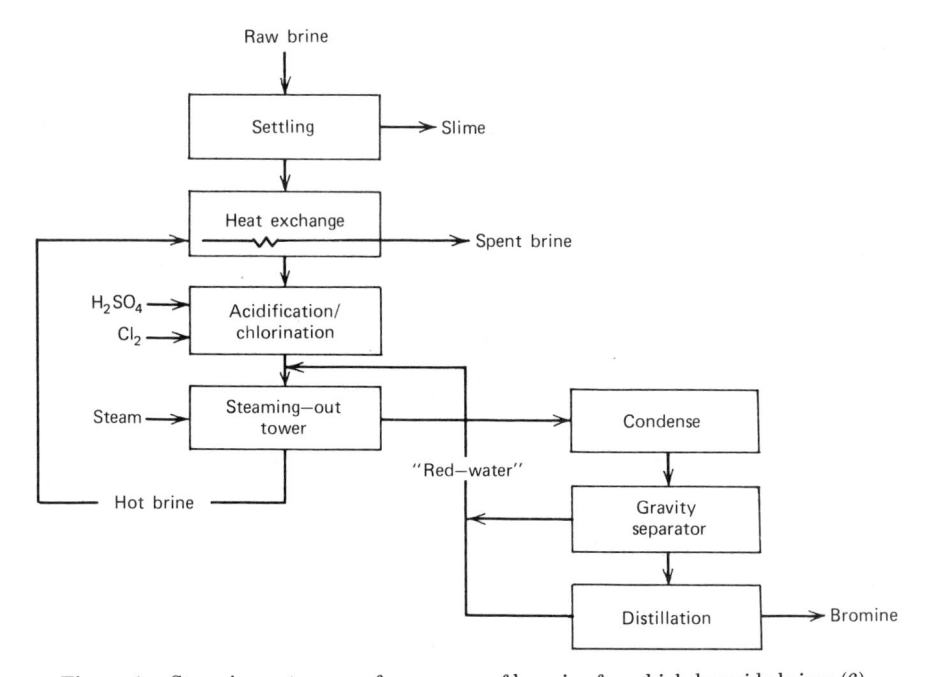

Figure 1. Steaming-out process for recovery of bromine from high-bromide brines (6).

Another method of stripping, usually economical only with brines of low bromine content, is a "blowing-out" process, which uses a stream of air flowing countercurrent to the brine to displace the bromine. The bromine was recovered from the air stream by treatment with wet scrap iron, ammonia, sodium carbonate, or sulfur dioxide, the use of the latter eventually becoming the most useful method (33–35).

In the early 1930s, bromine was extracted from seawater in a plant in North Carolina. After several expansions, operations accounted for nearly 75% of the total U.S. production, and a second seawater plant went on stream in Texas. However, the underground brines found in Arkansas, containing bromine at much higher concentrations, forced the closing of the seawater facilities.

Calcium Chloride. Because of its high solubility compared to that of the other brine constituents, calcium chloride is the final constituent recovered in a multiproduct brine-processing operation. Separation of magnesium ion may be accomplished by precipitation of magnesium hydroxide, or by the crystallization of tachyhydrite [*12194-70-6*], $CaCl_2 \cdot 2MgCl_2 \cdot 12H_2O$ (36–37) or $2CaCl_2 \cdot MgCl_2 \cdot 12H_2O$, both of which yield additional calcium chloride liquor.

Iodine. Because of the relatively low concentrations at which iodine occurs, all recovery processes involve two concentration steps. Domestic production of elemental iodine from natural brines began in 1928 near Shreveport, Louisiana (38). In Long Beach, California, iodine was recovered from oil field brines which remained a source of iodine until 1966. The silver process was used (39).

$$NaI + AgNO_3 \rightarrow AgI\downarrow + NaNO_3$$

$$2\,AgI + Fe \rightarrow FeI_2 + 2\,Ag$$

$$Ag + HNO_3 \rightarrow AgNO_3 + \tfrac{1}{2}\,H_2$$

$$2\,FeI_2 + 3\,Cl_2 \rightarrow 2\,FeCl_3 + 2\,I_2$$

Currently, a blowing-out process (Fig. 2) is employed, similar to that used for recovering bromine from seawater, and ultimately adopted by California producers. Recovery from underground brines at Midland, Michigan, displaced the California operations.

In 1976, Houston Chemicals commenced iodine recovery by a blowing-out process from underground brines from the Anadarko Basin in northwestern Oklahoma in a plant of an annual capacity of about 900 metric tons (40). The brine is produced from a depth of 2200 m from nine wells and the coproduced natural gas is used as fuel in the plant. After neutralization and clarification the spent brine is returned to the producing formation, for repressuring, through five reinjection wells.

In Japan, currently the world's leading iodine producer, most of the recovery plants use a process similar to one formerly used in California. Elemental iodine, previously formed in the brine by treatment with either $NaNO_2$ or Cl_2, is adsorbed on activated carbon, from which it is stripped by NaOH solution. Subsequent acidification yields a slurry of elemental iodine:

$$I_2\,(\text{soln}) \rightarrow I_2\,(\text{adsorbed})$$

$$3\,I_2 + 6\,NaOH \rightarrow 5\,NaI + NaIO_3 + 3\,H_2O$$

$$5\,NaI + NaIO_3 + 3\,H_2SO_4 \rightarrow 3\,Na_2SO_4 + 3\,H_2O + 3\,I_2$$

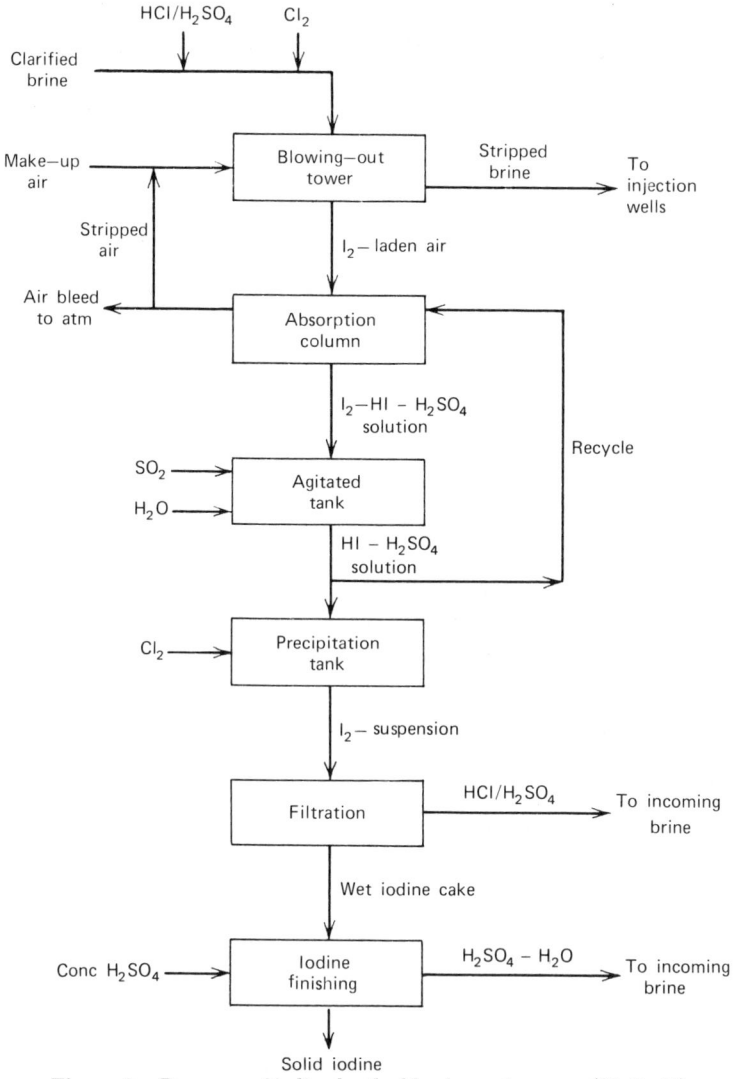

Figure 2. Recovery of iodine by the blowing-out process (38,40–41).

Lithium Compounds. Silver Peak Marsh, Nevada, is the only brine processed solely for its lithium values. The Silver Peak plant has a capacity of about 7,600 metric tons of lithium carbonate per year. The process consists of pumping brines from shallow wells, followed by solar evaporation to concentrate the brine and remove unwanted ions and final processing in the plant to produce a high-purity lithium carbonate [554-13-2]. A major brine field covers about 5 km^2 (2 mi^2) and contains over 40 wells ranging from 100–300 m deep. The well brine, containing about 400 ppm lithium, is pumped into approximately 20 km^2 (5000 acres) of solar evaporation ponds where the lithium concentration is increased to over 5000 ppm and evaporation averages over 125 cm per year. The final pond bittern is pumped to the processing plant where any remaining calcium and magnesium impurities are removed. Soda ash is added to the final brine to precipitate almost pure lithium carbonate which is washed, dried, packaged, and shipped.

In the U.S. the only other producer using brines is Kerr-McGee (14) which makes approximately 227 t per year of lithium carbonate as a by-product from its Trona soda ash plant in California.

Magnesium Compounds. Magnesium hydroxide [1309-42-8] can be recovered in relatively pure form, either from the brine itself or from an intermediate plant liquor, by increasing the alkalinity, for example:

$$MgCl_2 + Ca(OH)_2 \rightarrow CaCl_2 + Mg(OH)_2\downarrow$$

Instead of lime slurry, dolime [39445-23-3] (calcined dolomite) slurries are more frequently used as the source of alkalinity. If a low calcium content is required, NaOH solutions are sometimes employed.

Recovery of magnesium chloride [7786-30-3] as a direct product is usually economically feasible only where other salable products are also recovered; eg, at several locations in Michigan, where NaCl, Br_2, I_2, $CaCl_2$, $Mg(OH)_2$, and KCl are coproducts, and at the Great Salt Lake where NaCl, K_2SO_4, and Na_2SO_4 are also recovered. At one time, $MgCl_2$ was recovered from high-calcium brines as crystals of bischofite $MgCl_2.6H_2O$, by controlled dissolution of tachyhydrite, previously crystallized from intermediate plant liquors. In a current process using one of the Michigan Basin brines, a pure $MgCl_2$ liquor is formed by the reaction of previously precipitated $Mg(OH)_2$ with kiln stack gas and (concentrated) brine:

$$Mg(OH)_2 + CaCl_2 + CO_2 \rightarrow MgCl_2 + H_2O + CaCO_3\downarrow$$

When low-calcium brines are processed, $MgCl_2$, owing to its greater solubility, is usually the principal constituent of the final bitterns remaining after the recovery of NaCl, KCl, and double salts of Mg and K, or of Mg and Na. The double salts subsequently yield additional $MgCl_2$ solution. An example is the $MgCl_2$ liquor obtained in the recovery of KCl from Bonneville brine near Wendover, Utah, resulting, in part from the controlled dissolution of carnallite:

$$KMgCl_3.6H_2O + xH_2O \rightarrow KCl + MgCl_2 \text{ (aq)} + (6 + x) H_2O$$

The exceptionally low Ca:Mg ratio (ca 0.25 by wt) in a brine near Snyder, Texas, enables American Magnesium Company to obtain strong $MgCl_2$ liquor after removal of sodium and calcium ions (22).

Magnesium metal is produced by the electrolysis of molten $MgCl_2$ by two different processes. In the Dow seawater process (Fig. 3), $MgCl_2$ liquor, formed essentially by neutralizing $Mg(OH)_2$ with HCl, is dried to a hydrous cell feed containing about 70% $MgCl_2$. Additional chlorine (or chloride) is required to make up for losses of chlorine values recycled from the electrolysis cells to the neutralization step. The other process recovers $MgCl_2$ instead of $Mg(OH)_2$ from the source brine and forms anhydrous cell feed from which both Mg metal and marketable chlorine are produced (22,42).

Currently, magnesium chloride is recovered directly from underground brines near Snyder, Texas; Ludington, Michigan; and Wendover, Utah; from Great Salt Lake brines at Ogden and Rowley; and from seawater bitterns at Chula Vista, California.

Potassium Compounds. The Texasgulf Cane Creek potash (KCl) operation (44), of Moab, Utah, was originally designed for underground mining, but in 1970, was converted to solution mining (45–48) because of problems with conventional mining of the evaporite deposit. The present process consists of solution mining, solar evaporation, and beneficiation. The ore deposit has a natural dipping slope and is over 1000

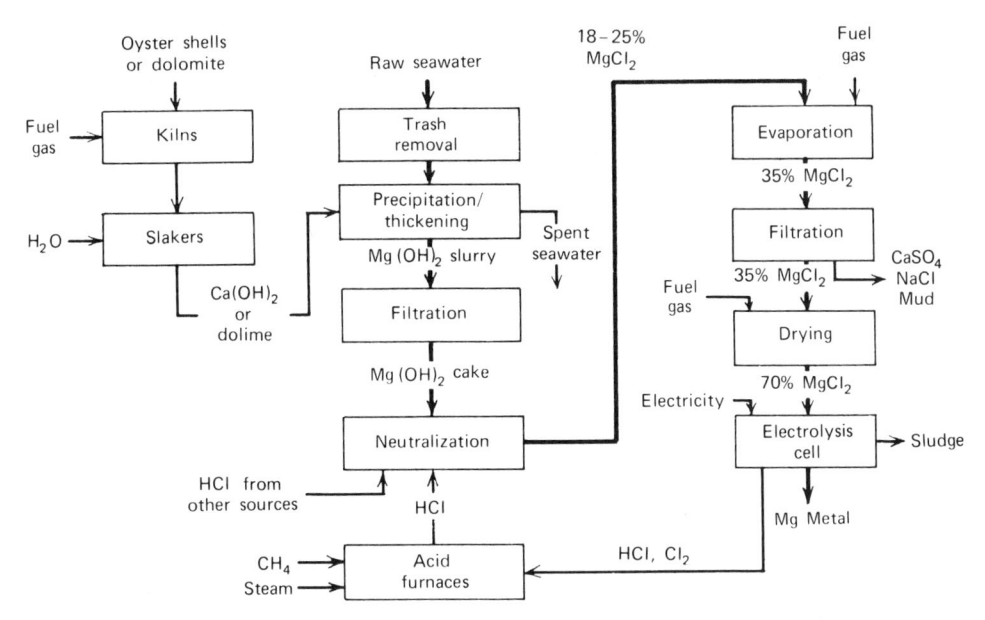

Figure 3. Dow process for recovery of magnesium from seawater (33,43).

m below the surface. Water from the nearby Colorado River is pumped underground through 12 well holes and as the water flows through the mine the brine becomes saturated in potassium and sodium chloride with minor quantities of calcium chloride and insoluble clay slimes. The saturated brine is withdrawn from one large extraction hole at 32–33°C owing to the natural heat in the mine. Pumping takes place 7–12 months per year, and the liquor retention time in the mine is 300–350 d. The water input rate to the mine is controlled by the need for brine in the evaporation pond area. Water is removed from the brine by solar evaporation, and a mixture of KCl and NaCl solids are deposited in ca 1.6 km² (400 acres) of ponds. Production of a solar evaporation pond operation is a function of the evaporation area available, evaporation rate, and brine concentration. Radiation-absorbing dyes are used as brine additives to increase the evaporation rate. To prevent brine leakage, the ponds are sealed with PVC lining protected with a 15-cm salt base. After the water has evaporated, the deposit is harvested with scraper-loaders equipped with lasers to control the depth of cuts. The harvest is transported to a nearby crusher station where the material is slurried with brine and pumped to the nearby plant site.

The slurry from the ponds is processed by standard beneficiation techniques. Potassium chloride is floated away from the waste salt, using an amine collector. After a final water leach to remove the small amount of sodium chloride in the flotation concentrate, the >60%-K_2O [12136-45-7] potash is dried and stored (49). Grades (sizes) include coarse, soluble, standard, and granular. The latter is made in a separate circuit using compaction methods.

Potash has been produced in the Western Utah desert from underground brines since the early 1900s, but large-scale production did not start until the late 1930s. The area (50) borders on the edge of the Great Salt Lake desert and is known as the Bonneville Salt Flats. It encompasses approximately 300 km² (75,000 acres) for brine collection and 40 km² (10,000 acres) for solar evaporation ponding. The operation is highly weather dependent as a result of the quantity and quality of brine available (rainfall) and the evaporation rate. The process (Fig. 4) consists of: (*1*) collection of natural occurring brines; (*2*) concentration of brines by solar evaporation; (*3*) precipitation and harvesting of the evaporite deposit; and (*4*) beneficiation of the deposit to produce potash. Brines from the shallow aquifer are collected by gravity in a network of ditches and subsequently transferred by pumps to a primary evaporation pond as are deep-well brines. The primary evaporation pond is divided into stages where about 3000 t of brine per 4050 m^2 (1 acre) are evaporated annually. About 15 cm of salt buildup is deposited annually, and after eight to ten years the dikes surrounding the pond must be raised or a new pond system built. The brine is pumped through several ponds, and the final brine bitterns contain 28–34% MgCl$_2$ which is sold for well drilling, road applications, and other industrial uses.

The pond deposits are harvested by windrowing with a motor grader. The potash is beneficiated from salt by standard flotation techniques, using an amine–oil collector (see Flotation). After concentration, it is dried and stored for shipment by bulk in railcars. The potash produced at the Bonneville plant, both standard and coarse, is used primarily as fertilizer containing >60% K$_2$O values.

Great Salt Lake Minerals & Chemicals Corporation (51), near Ogden, Utah, processes brine from the Great Salt Lake for potash (K$_2$SO$_4$), salt cake (Na$_2$SO$_4$), and

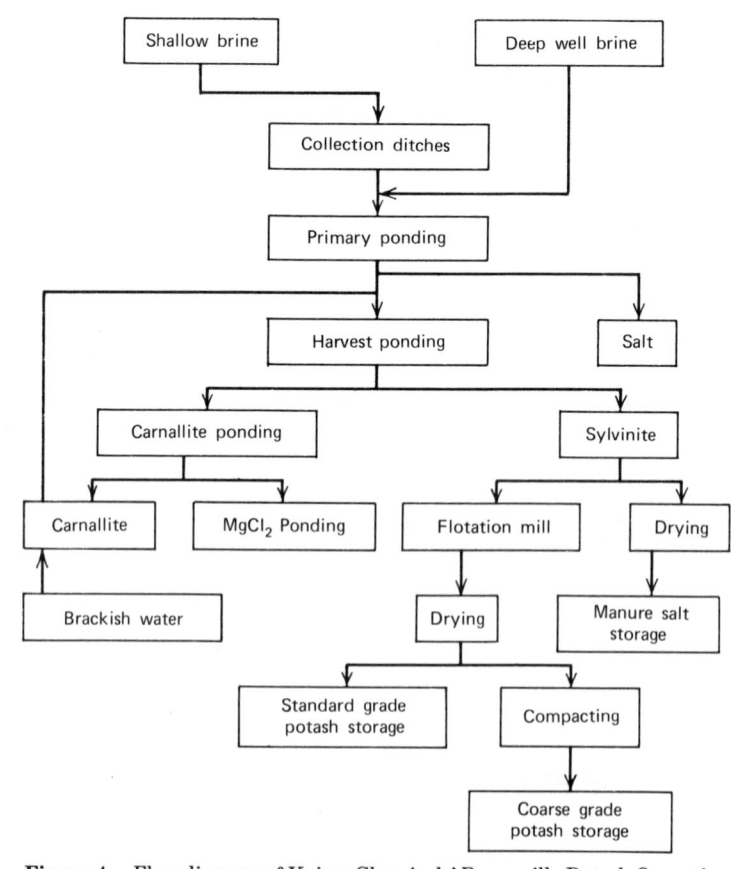

Figure 4. Flow diagram of Kaiser Chemicals' Bonneville Potash Operation.

salt (NaCl). The process (Fig. 5) includes: (*1*) pumping the brine from the Great Salt Lake; (*2*) concentrating the brine by solar evaporation; (*3*) controlled precipitation of evaporite minerals; and (*4*) harvesting and processing the deposits.

The evaporating cycle is from May to October, and production is highly dependent upon weather. In addition to the various potash chemicals produced, the pond end bitterns also contain ca 28–34% $MgCl_2$. During the winter, Glauber's salt is precipitated from the naturally cooled brines to be used for the production of 30,000–50,000 t of Na_2SO_4 per year by the mirabilite process. Potash is a by-product of concentrating Searles Lake brines in the California Trona plant of Kerr-McGee Corporation. Brines processed through the carbonation plant yield boric acid and potassium sulfate in a complex process using an organic extraction step in conjunction with evaporator–crystallizers. In addition, the main plant produces potassium chloride from end liquors of the soda products plant.

Sodium Carbonate. The Solvay process (8), which depends economically on solution-mined salt, takes place in several steps that can be summarized as:

$$CaCO_3 + 2\ NaCl \rightarrow CaCl_2 + Na_2CO_3$$

At Searles Lake, soda ash is recovered from brines of two different compositions. The brine from the lake's lower reservoir is carbonated directly, using kiln stack gas (52).

$$2\ Na^+ + CO_3^{2-} + H_2O + CO_2 \rightarrow 2\ NaHCO_3 \downarrow$$

$$2\ NaHCO_3 \rightarrow Na_2CO_3 + CO_2 \uparrow + H_2O$$

The recovery process from the upper-level brine involves the crystallization of

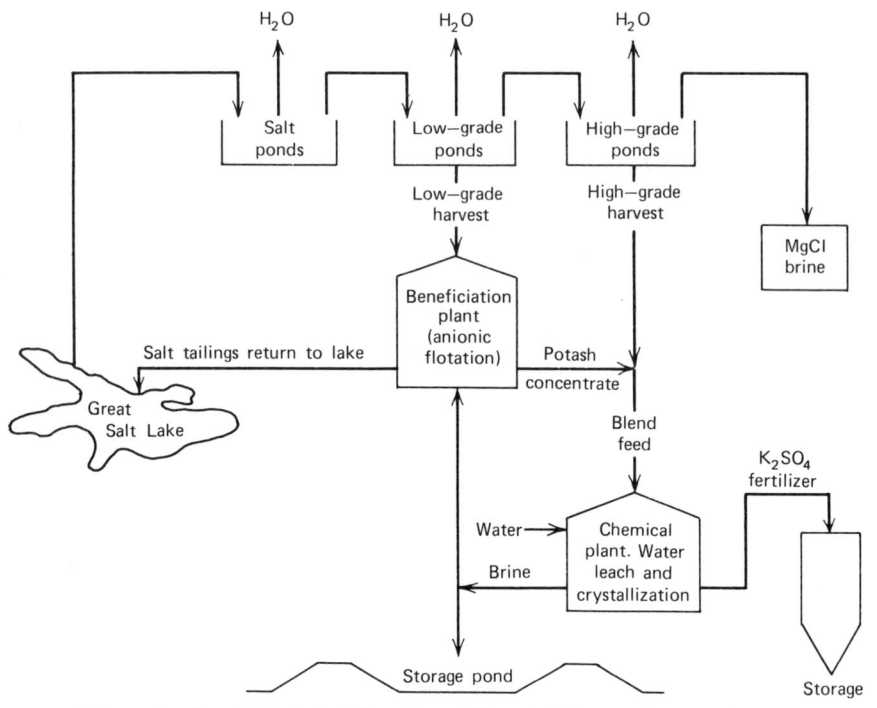

Figure 5. Great Salt Lake Minerals and Chemical Corporation potash complex.

burkeite [12179-88-3], Na$_2$CO$_3$.2Na$_2$SO$_4$, from which sodium carbonate monohydrate [5968-11-6], Na$_2$CO$_3$.H$_2$O, is subsequently separated and calcined to anhydrous soda ash (30).

Sodium Chloride. *Solution Mining.* Solution mining is the recovery of a water-soluble substance from an underground deposit by dissolving the substance *in situ* and forcing the resulting solution to the surface. The process is also known as brining.

Solution mining produced about 23 million metric tons of salt in 1974, representing more than half of the total U.S. salt production (53). The salt brine was derived from bedded salt at more than 18 different locations and from 17 salt domes (54). The bedded salt of the Salina formation is the most widely and intensively exploited by solution mining. Enormous resources of Salina salt are available. The cost of producing saturated salt brine by solution mining, expressed on the basis of a unit of contained NaCl, is less than half the cost of salt produced by dry mining. The factors determining the choice of solution mining vs dry mining are the depth and geological nature of the deposit and the location and nature of the immediate downstream use of the product. The method is particularly well adapted to deep-lying bedded deposits and to salt domes. Compared with dry mining, its economic advantage generally increases with the depth of the deposit.

The essentials of solution mining of a salt dome are shown in Figure 6. This arrangement is also sometimes used in solution mining bedded salt deposits of sufficient thickness. Olfield practice is followed in the drilling operation.

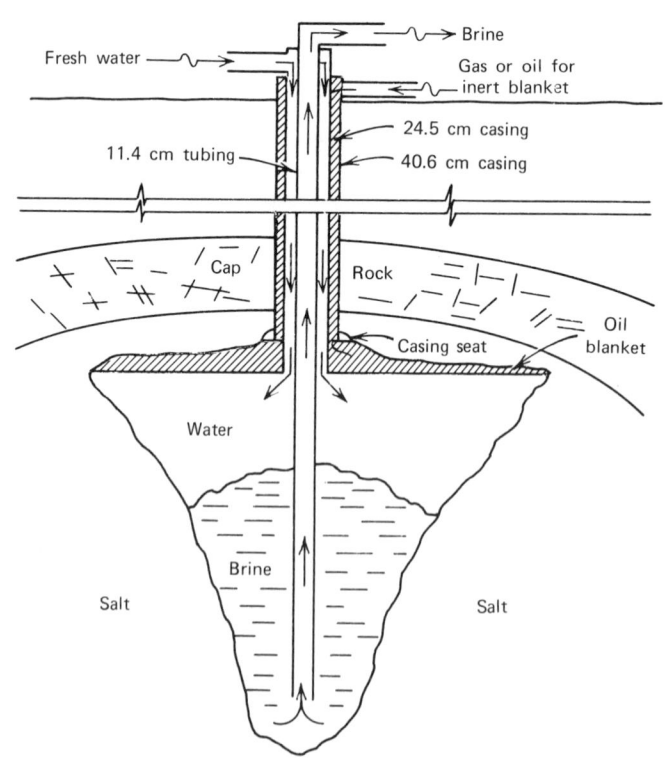

Figure 6. Typical solution mining operation in a salt dome (54).

In order to preserve the structural integrity of the salt plug and maintain conditions of isolation for the cavity, the following three fundamental rules should be observed: (1) creation and maintenance of tight seals between cemented casing and salt; (2) operation at down-hole pressures well below the lithostatic pressure; and (3) maintaining safe distances between the cavity surface and outer surface of the salt plug and between the cavity surface and possible additional cavities.

The production rate of saturated brine is limited at any one time by the overall dissolution rate of the salt. The flow rate of injected water, which directly controls the rate of outflowing brine, is adjusted within the range of values at which the emerging brine will be saturated.

Frequently, in relatively thin salt strata, at least two wells (55) are used communicating by hydraulically formed fracture channels. A hazard affecting operations in both strata and domes is the destruction of the lower portions of the production tubing by massive chunks of salt, anhydrite, or rock which may result as dissolution progresses.

Probably the greatest environmental threat inherent in solution mining is potential surface subsidence. This risk is greater with bedded salt than with salt domes. In fact, no cases of subsidence have been reported over dome cavities dissolved according to industry-approved practice. Solution mining technology has resulted in the widespread use of salt cavities for storage of industrial fluids, chiefly petroleum liquids, and natural gas.

Solar Evaporation. The recovery of salt by solar evaporation (56–58) is favored in a hot, dry climate, but is usually not practical in cold climates or where rainfall exceeds evaporation. Other factors (59–60) include wind, humidity, air temperature, cloud cover, and even the area surrounding the evaporation ponds.

In the design of solar ponds numerous factors must be considered, for example, the type of brine to be evaporated. Seawater evaporates ten times faster than brine saturated with magnesium chloride. Dikes have to protect ponds deeper than 30 cm against erosion. The ponds must be water tight and, therefore, sandy soils are poor bottoms, whereas silt and clay hold the brines well. Weather, seasonal changes, topography and elevation, and the brine concentration desired all have to be considered. Most solar ponds are operated where the sun provides at least 84 kJ/(m^2·min) [20 kcal/(m^2·min)] at noon. To provide the same energy to 4050 m^2 (1 acre) of solar ponds, burning 19 t of coal each day would be required. The sun's energy is more likely to be captured by a colored pond than by a colorless one and therefore a dye, eg, Naphthol Green B or Acid Blue may be added.

Solar ponds are 15–50 cm deep, or deeper in areas of very high evaporation rates. Deeper ponds are easier to control and manage, but produce less salt than shallower ponds. At present, operation of solar ponds is economical only along the coast of California using seawater (61–65); in Utah using the brine from the Great Salt Lake; and in the arid regions of the west using subsurface brines associated with dry lake beds.

Seawater. Salt is produced by solar evaporation of seawater, followed by crystallization. First the seawater is evaporated in a series of concentrating ponds until it is saturated with sodium chloride. At this point, the volume of brine is reduced by 90%, and some of the less soluble salts (eg, CaSO$_4$) have been removed. The saturated brine, or pickle, is transferred to crystallizing ponds, where salt is precipitated upon further evaporation. This bittern may be used for the recovery of other compounds primarily MgCl$_2$. The ratio of the area of concentrating ponds to crystallizing ponds is between 15:1 and 10:1. Adjacent to San Francisco and San Diego bays more than 130 km^2 (33,000 acres) of solar ponds yield about 1.3 million t of salt annually. In order to produce this quantity of salt, approximately 47 million t of water must be evaporated from 49.6 million t of bay water, yielding 1.3 million t of bitterns.

Solar salt should be at least 99% NaCl, but some impurities occur from gypsum in the initial deposition step, and from bittern salts in the final stages of evaporation because of the overlapping of the crystallization phases. The deposit is usually harvested (66) once a year and contains about 97 to 98% sodium chloride.

After harvesting, the salt is cleaned and washed (67–68a) to over 99% NaCl and naturally drained in stockpiles. Most of it is marketed as undried, crude salt after being ground and screened into various sizes. Some salt is rewashed and kiln dried and some is refined by vacuum evaporation to over 99.95% NaCl.

The Great Salt Lake. The Great Salt Lake, located in Northwestern Utah, is the largest lake in the Western hemisphere that does not drain to an ocean. The climate in the valley is characterized by rather large temperature fluctuations with hot, dry summers. Precipitation occurs mainly during winter and spring.

Variations in the lake level are always accompanied by a change in mineral concentration (17) in the brine. However at a given level, the brine concentration (68b) varies considerably at different points and different depths of the lake (see Table 1).

Lake brine is pumped into preconcentrating ponds where, by solar evaporation, saturation with respect to NaCl is reached. The saturated brine contains 8.7% sodium, 15.6% chloride, and about 4% other ions (Table 2). This brine is then transferred to crystallizer ponds (garden ponds) where the crude salt (NaCl) is deposited. As the brine evaporates the concentration of the other minerals increases contaminating the product and resulting in smaller NaCl crystals.

The ponds vary in area from 80,000 to 600,000 m^2 and are 7.5–25 cm deep. The dikes are lined with wood to prevent clay contamination. A deposit of one season is 5–10 cm thick, and beneath it is a base of 30–45 cm used to support the harvesting and hauling equipment. This base salt is deposited by operating the pond for several years without harvesting.

The production season is relatively short and the salt is harvested each year. Actual crystallization of recoverable salt begins sometime in June and lasts until the harvest begins in September. A cleavage plane is made by dragging the ponds to smooth

Table 1. Typical Brine Analysis g/L, of the Great Salt Lake, 1976 [a]

Property and chemical	South arm Shallow	South arm Deep	North arm
density	1082	1175	1221
Na	37.64	84.41	109.93
Mg	4.35	8.31	11.31
K	2.64	5.75	7.64
Cl	66.74	147.75	190.58
SO$_4$	8.75	19.45	23.50

[a] From Morton Salt Co., Hardy Salt Co., American Salt Co., Lake Crystal, and Great Salt Lake Minerals & Chemicals Corporation.

Table 2. Salts from Great Salt Lake Brines

Mineral	CAS Registry No.	Formula
astrakanite (bloedite)	[15083-77-9]	Na$_2$SO$_4$.MgSO$_4$.4H$_2$O
bischofite	[7757-82-6]	MgCl$_2$.6H$_2$O
carnallite	[1318-27-0]	KCl.MgCl$_2$.6H$_2$O
epsomite (bitter salt)	[14457-55-7]	MgSO$_4$.7H$_2$O
glaserite (aphthitalite)	[16349-83-0]	Na$_2$SO$_4$.3K$_2$SO$_4$
Glauber's salt	[7727-73-3]	Na$_2$SO$_4$.10H$_2$O
halite	[14762-51-7]	NaCl
magnesium sulfate hexahydrate	[17830-18-1]	MgSO$_4$.6H$_2$O
kainite	[1318-72-5]	KCl.MgSO$_4$.2.75H$_2$O
kieserite	[14567-64-7]	MgSO$_4$.H$_2$O
langbeinite	[14977-37-8]	2MgSO$_4$.K$_2$SO$_4$
leonite	[15650-69-8]	MgSO$_4$.K$_2$SO$_4$.4H$_2$O
schoenite (picromerite)	[15491-86-8]	MgSO$_4$.K$_2$SO$_4$.6H$_2$O
sylvite (silvine)	[14336-88-0]	KCl

the pond floor and ensure easy separation of the year's crop from the base salt. During the harvest, the salt is brought into the area near the processing plant where it is stockpiled as crude salt and then washed, dried, and cooled, followed by crushing and screening into several sizes (>99.6% pure).

Sodium Sulfate. In the process employed for the Texas brines, Glauber's salt is crystallized directly from the chilled (ca −8°C) brine after the addition of sodium chloride (69). Anhydrous Na$_2$SO$_4$ [7757-82-6] is recovered by submerged combustion evaporation of the liquor formed by remelting the Glauber's salt.

Recovery of the half-dozen chemicals from the two brines pumped from Searles Lake requires a complex sequence of operations (15,52,70). The processing of one of the brines involves the precipitation of a mixture of NaCl and burkeite. After hydraulic separation, the burkeite is redissolved and lithium values are recovered from the liquor which is then cooled to ca 22°C to crystallize Glauber's salt.

In processing brine from the lower level of Searles Lake, a mixture of Glauber's salt and borax is crystallized at ca 5°C from a plant liquor previously freed of most of its carbonate and boron values. The mixed crystals are separated hydraulically.

In processing Great Salt Lake brine, Glauber's salt is crystallized during the winter from a ponded brine resulting, essentially, from the crystals of epsomite (MgSO$_4$.7H$_2$O) rejected by the potash recovery step, and NaCl brine.

Economic Aspects and Uses

In 1975, the chemical brine industry had an estimated total production of about 23 million t (including secondary products), valued at more than $5 billion at 1978 price levels. This production required the daily processing of an estimated 1.2 GL (317 million gal) of seawater, 1.0 GL (264 million gal) of lake brines, 0.3 GL (79 million gal) of underground natural brines, and 0.2 GL (53 million gal) of solution mined brines.

Figure 7 illustrates the processes of the brine industry and the end uses of its products.

A new process is being developed to recover tungsten from the brines of Searles Lake, California. A metal-selective ion-exchange resin is employed to remove tungsten

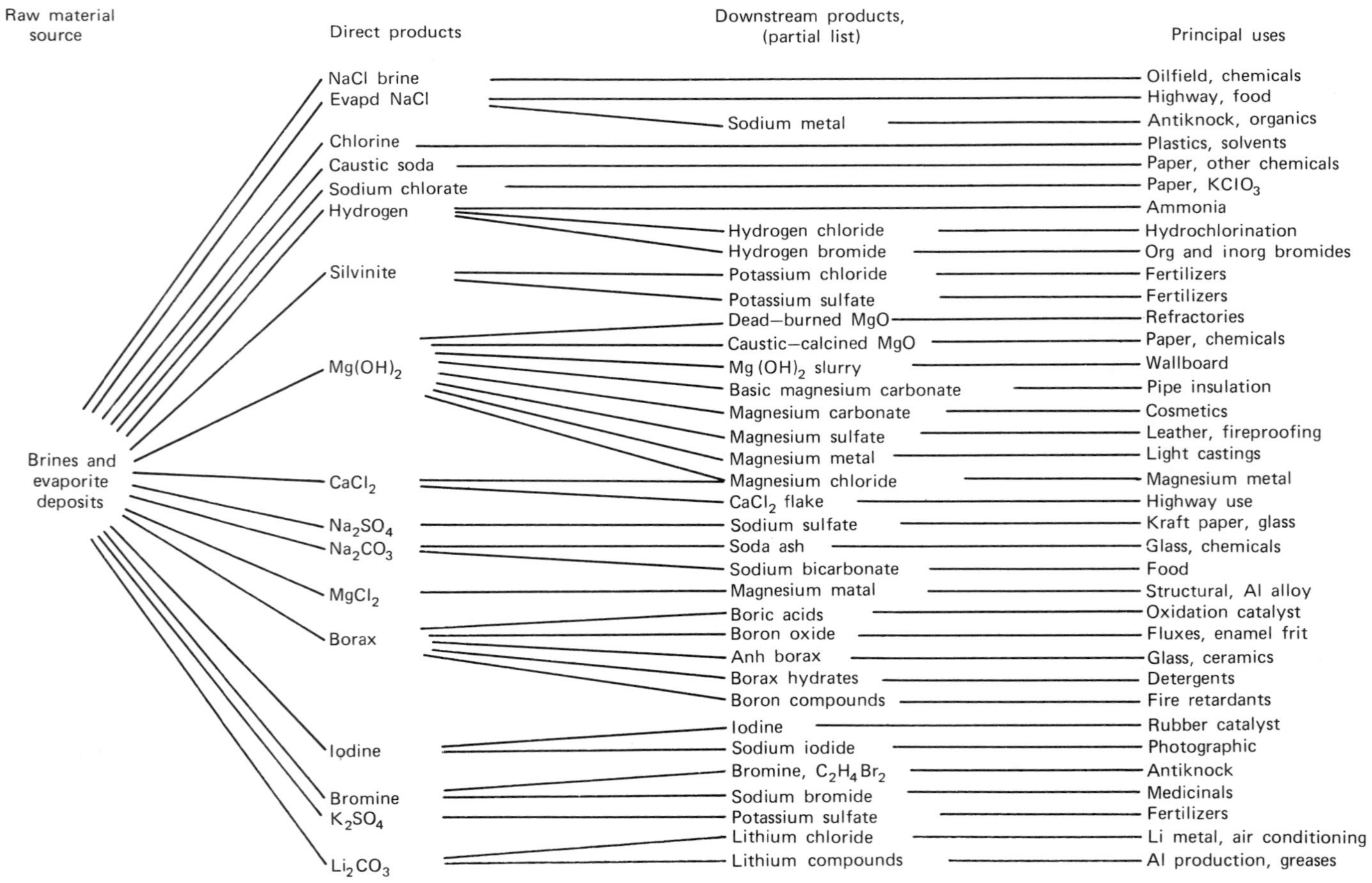

Figure 7. The brine chemical industry and its products.

from the low-grade (70 ppm) brine. In the future this may reduce United States reliance on tungsten imports (71).

Boron Compounds. The principal U.S. producers are U.S. Borax and Chemical Corporation, Kerr-McGee Corporation, and American Borax Corporation. Their combined annual capacity in 1972 was reported to be 600,000 metric tons of equivalent B_2O_3 (72), including the Searles Lake operations of Kerr-McGee at Trona and Westend of about 120,000 t.

During 1975, the total U.S. production was equivalent to 550,000 t of B_2O_3, of which more than 70% was borax (73). About 55% of the production was exported.

Bromine. The 1976 United States capacity for producing elemental bromine was reported to be approximately 272,000 t (74), over two-thirds of which was represented by six plants operating in southeastern Arkansas. Production facilities in Michigan accounted for most of the remainder, and only a minor fraction was represented by the process at Searles Lake.

Traditionally, only between 10 and 15% of the total bromine recovered by the primary producers is marketed in elemental form. About two-thirds is converted to ethylene dibromide at the recovery plants. Other products, including alkali metal bromides, ammonium bromide, and hydrobromic acid, account for about 20% of the total bromine produced. In 1975, the total value of bromine compounds, including elemental bromine, was close to $200 million (75).

Calcium Chloride. The total United States capacity in 1976 was reported to be about 800,000 t per year, anhydrous $CaCl_2$ basis (76). Over 65% of the capacity is found in Michigan, where underground brines are processed at four different locations. Other operations included the underground brine of Bristol Lake, California, and the two artificial brine sources at Tacoma, Washington (77). The distiller waste from a Solvay soda ash plant at Syracuse, New York, accounted for about one-third of the total capacity.

Applications of calcium chloride include 28% dust control, 27% highway deicing, 23% industrial uses, 10% concrete setting control, and 5% oil recovery.

Iodine. Before World War I, the United States depended entirely on foreign sources for its iodine needs. Now two plants, one at Midland, Michigan, of 250 t capacity, and one at Woodward, Oklahoma, of 910 t capacity, furnished about 30% of the total domestic demand estimated for 1977 (78); imports from Japan and Chile supplied the balance of the United States requirement. At least one other company, Ethyl Corporation, has definite plans to recover iodine from Oklahoma brines near Woodward (41).

Several dozen processors convert elemental iodine into downstream products including (78): 22% catalysts (mainly for rubber), 15% stabilizers, 15% animal feed supplements, 14% inks and colorants, 14% pharmaceuticals, 10% sanitary products, and 10% photographic products.

Lithium Compounds. Lithium carbonate recovered from the brines of Silver Peak (Foote Mineral) and Searles Lake (Kerr-McGee) accounted for about one-third of inferred total of about 4500 t of Li equivalent produced in the United States during 1976 (79–80). Brine-derived Li_2CO_3 was valued at about $48 million. In the U.S. which produces and consumes more than one-half of the world supply, the principal applications of lithium chemicals are as a cell-bath additive in primary Al production (ca 38%), in ceramics and glass (ca 27%), lubricants (ca 12%), and, as LiCl and LiBr, in air conditioning (ca 6%). Early commercialization of the developmental lithium–sulfur electric storage cell should materially increase the 3% annual growth forecast for lithium demand.

Magnesium Compounds. Magnesium hydroxide was recovered from brines, including seawater, at an annual rate of about 1,200,000 million t between 1971 and 1976. Only about one-half of this was strictly of brine origin; the remaining half was supplied by the Mg(OH)$_2$ of the dolime usually used in the precipitation. Magnesium hydroxide is recovered from seawater in New Jersey, Mississippi, Texas, and California, and from underground brines in Michigan.

The second-tier products of greatest tonnage were refractory grades of MgO, followed by the reactive (caustic-calcined) forms. Most of the Mg(OH)$_2$ recovered at Freeport, Texas, is used captively to produce magnesium metal, a product whose total annual value ($180–230 million) exceeded that of all other brine-derived magnesium chemicals combined. In other locations (Michigan and California), 48% MgCl$_2$ flake is produced whose major use is in magnesium oxychloride cements. In the Utah locations 30–33% MgCl$_2$ liquor is used in sugar beet processing, drilling muds, and dust-laying.

Potassium Compounds. Brine-derived potassium compounds (KCl and K$_2$SO$_4$) recovered at Moab, Ogden, and Wendover, Utah, and at Searles Lake amounted to about 0.4 million t of K$_2$O equivalent during 1976, about 20% of the total produced in the United States (81–82). Most of the domestic output results from the dry-mining operations of seven companies near Carlsbad, New Mexico. More than 95% of the potash production is used in fertilizers; the remainder is converted to other potassium chemicals.

Domestic potash production currently supplies only one-third the domestic demand; imports from Canada and Europe supply the remainder.

Sodium Carbonate. In 1976, capacity for producing brine-derived soda ash, represented by the recovery operations at Searles Lake and the output of the four Solvay plants then operating, was estimated to be about 3,100,000 metric tons which is about 30% of the total United States capacity (83). The brine-derived product is decreasing in percent because of rapid expansion of dry-mined trona and the decrease in Solvay production (84). A substantial increase in recovery capacity at Searles Lake is currently underway.

The principal uses for soda ash are for glass manufacture (40–50% of total soda ash production), manufacture of other chemicals (24%), pulp and paper production (9%), and manufacture of detergents (5%) (85).

Salt. During 1976 brine industry operations produced about 57% of the nearly 40 million t of salt mined in the United States, including ca 20 million t in solution-mined brine and ca 1.5 million t of solar salt (86). More than 40 companies produce salt brine at about 50 locations, chiefly in Louisiana, Texas, Ohio, New York, and Michigan. Production of chlor–alkali products accounts for about 58% of total salt use (including dry-mined salt), and for almost the entire amount of solution-mined salt. Highway deicing is the second largest single application.

Sodium Sulfate. During 1975, the domestic capacity (87) for producing natural (brine-derived) sodium sulfate was approximately 650,000 t per year (Kerr-McGee at Searles Lake, 60%; Ozark-Mahoning at Brownfield and Seagraves, Texas, 24%; and Great Salt Lake Minerals & Chemicals Corp., 16%). The capacity for producing from nonbrine sources was reported to be approximately 690,000 t per year.

Total United States consumption, recently about 15% greater than total United

States capacity, was principally 70% by the pulp and paper industry, 20% for detergent manufacture, and 7% by the glass industry.

BIBLIOGRAPHY

"Great Salt Lake Chemicals" in *ECT* 2nd ed., Suppl. Vol., pp. 438–467, by G. Flint, Great Salt Lake Minerals & Chemical Corp.

1. P. E. Muehlberg and co-workers in P. W. Spaite and I. A. Jefcoat, eds., *Industrial Process Profile for Environmental Use*, Rept. No. EPA-600/2-77-0230, Environmental Protection Agency, Cincinnati, Ohio, 1977, Chapt. 15.
2. P. W. Spaite and I. A. Jefcoat, eds., *Industrial Process Profile for Environmental Use*, Rept. No. EPA-600/2-77-023a, Environmental Protection Agency, Cincinnati, Ohio, 1977, Chapt. 1, pp. 11.
3. A. Ferguson, D. J. Treskon, and R. E. Davenport, "Boron Minerals and Chemicals" in *Chemical Economics Handbook*, Stanford Research Institute, Menlo Park, 1977, p. 717.1001C.
4. *Ibid.*, p. 717.1002F.
5. V. M. Goldschmidt in A. Muir, ed., *Geochemistry*, Oxford University Press, London, 1954, p. 607.
6. F. Yaron in Z. E. Jolles, ed., *Bromine and its Compounds*, Ernest Benn Ltd., London, 1966, pp. 3–41.
7. A. G. Collins, *Geochemistry of Oilfield Waters*, Elsevier Publishing Co., Inc., New York, 1975, pp. 212–552.
8. T. P. Hou, *Manufacture of Soda*, 2nd ed., *ACS Monograph* 65, Reinhold Publishing Corp., 1942, p. 251.
9. V. M. Goldschmidt, *J. Chem. Soc. London*, 655 (1937).
10. H. U. Sverdrup, M. W. Johnson, and R. H. Fleming, *The Oceans, Their Physics, Chemistry, and General Biology*, Prentice-Hall, Inc., New York, 1942, pp. 176–177.
11. A. G. Collins and G. C. Egleson, *Science* **156**, 934 (1967).
12. I. A. Kunasz in A. H. Coogan, ed., "Lithium Occurrence in the Brines of Clayton Valley, Nevada," *Fourth Symposium on Salt*, Vol. I, Northern Ohio Geological Society, Cleveland, Ohio, 1974, pp. 57–66.
13. K. G. Papke, *Evaporites and Brines of Nevada*, Bull. 87, Nevada Bureau of Mines & Geology, 1976.
14. L. E. Rykken, *Lithium Production from Searles Valley*, U.S. Geological Survey Proceedings Paper, Kerr-McGee, Trona, California, 1976.
15. J. E. Ryan, "Industrial Salts; Production from Searles Lake," *Mining Engineering*, **3**, 447 (1951).
16. E. E. Smith and H. J. Andrews, *Mining Great Salt Lake—A 75 Billion Dollar Reserve of Lithium, Magnesium, Potash and Sodium Salts*, unpublished paper presented at the meeting of the American Institute of Mining Engineers, Los Angeles, Calif., Feb., 1967.
17. D. C. Hahl and A. H. Handy, *Great Salt Lake, Utah: Chemical and Physical Variations of the Brine, 1963–1966*, Water-Resources Bulletin 12, Utah Geological and Mineralogical Survey, Salt Lake City, Utah, 1969, p. 33.
18. P. Hadzeriga, "Some Aspects of the Physical Chemistry of Potash Recovery by Solar Evaporation of Brines," *Transactions American Institute of Mining and Metallurgical Engineers, SME*, **229**, 169–174 (June, 1964).
19. *U.S. Tax Cases, 70-2, 84615,* Commerce Clearing House, Inc., Chicago, 1970.
20. J. J. Christensen and co-workers, *A Feasibility Study on the Utilization of Waste Brines From Desalination Plants, Part I,* OSW Rept. 245, Office of Saline Water, U.S. Dept. of the Interior, Washington, D.C., 1967.
21. D. E. White, J. D. Hem, and G. A. Waring in M. Fleischer, ed., "Chemical Composition of Subsurface Waters," *Data of Geochemistry*, Sixth Ed., Geological Survey Prof. Paper 440-F, U.S. Dept. of the Interior, Washington, D.C., 1964, Chapt. F p. F-2.
22. J. G. Mezoff, American Magnesium Co., *unpublished report*, March, 1977.
23. Ref. 7, p. 410.
24. D. E. Garrett, *Chem. Eng. Prog.* **59**(10), 59 (1963).
25. W. H. Husband, *Application of Solution Mining to the Recovery of Potash*, presented at the Annual Meeting of the AIME, New York, 1971.
26. J. A. Lozano, *Ind. Eng. Chem.* **15**(3), 44 (July 1976).
27. J. A. Fernandez-Lozano in ref. 12, Vol. II, pp. 501–510.
28. W. P. Wilson in J. L. Rau and L. F. Dellwig, eds., "Potassium Chloride Crystallization," *Third Symposium on Salt*, Vol. II, Northern Ohio Geological Society, Cleveland, Ohio, 1970.

29. P. G. Rueffel in ref. 28, Vol. I, pp. 429–451.
30. G. H. Bixler and D. L. Sawyer, *Ind. Eng. Chem.* **49**(3), 322 (March 1957).
31. C. R. Havighorst, *Chem. Eng.* **70,** 229 (Mar. 11, 1963).
32. Ref. 6, p. 23.
33. C. M. Shigley, *J. Met.* **3**(1), 25 (1951).
34. U.S. Pat. 2,143,223 (Jan. 10, 1939), S. B. Heath (to The Dow Chemical Co.).
35. U.S. Pat. 2,143,224 (Jan. 10, 1939), G. W. Hooker (to The Dow Chemical Co.).
36. U.S. Pat. 1,627,068 (May 3, 1927), A. K. Smith and C. F. Prutton (to The Dow Chemical Co.).
37. U.S. Pat. 1,843,761 (Feb. 2, 1932), W. R. Collins (to The Dow Chemical Co.).
38. F. G. Sawyer, M. F. Ohman, and F. E. Lusk, *Ind. Eng. Chem.* **41,** 1547 (1949).
39. U.S. Pat. 1,837,777 (Dec. 22, 1931), C. W. Jones (to The Dow Chemical Co.).
40. *Chem. Mark. Rep.* **211**(10), 7 (1977).
41. *Chem. Wk.* **119**(22), 23 (Dec. 1, 1976).
42. E. F. Emley, *Principles of Magnesium Technology,* Pergamon Press, London, 1966, pp. 34–37.
43. W. F. McIlhenny in J. P. Riley and G. Skirrow, eds., *Chemical Oceanography,* 2nd ed., Vol. 4, Academic Press, London, 1975, pp. 155–218.
44. "Potash is Where You Find It" in *Agi-Chemical Age* (May, 1974).
45. "Combining Two Production Methods Proves Successful at Texasgulf" in *Fertilizer Solutions* (Sept.–Oct., 1973).
46. R. L. Curfman, *Min. Congr. J.,* **60**(3), 32 (Mar., 1974).
47. D. Jackson, Jr., *Eng. Min. J.,* **174**(7), 57 (July, 1973).
48. J. Walden, A. P. McCue, and H. P. Chen in ref. 28, Vol. I, pp. 371–382.
49. J. P. Tailor in ref. 28, pp. 10–12.
50. M. W. Lallman and G. D. Wadsworth, *Kaiser Chemicals—Bonneville Potash Operation,* presented at AIME Fall Meeting, Denver, Colorado, Sept., 1976.
51. *Miner. Process.* (Aug. 1971).
52. J. V. Hightower, *Chem. Eng.* **58**(8), 162 (1951).
53. C. L. Klingman, preprint from "Salt" in *Bureau of Mines Mineral Yearbook—1974,* U.S. Dept. of the Interior, Washington, D.C., 1974.
54. M. T. Halbouty, *Salt Domes, Gulf Region, United States and Mexico,* Gulf Publishing Company, Houston, Tx., 1967, pp. 125–149.
55. K. Henderson in ref. 12, Vol. II, pp. 211–218.
56. A. Delyannis and E. Delyannis, *Chem. Eng.,* **77,** 136 (Oct., 1970).
57. F. C. Standiford in ref. 12, Vol. II, pp. 481–488.
58. E. C. Pendery in ref. 28, pp. 85–95.
59. R. N. Jacobsen and F. Ore in ref. 12, Vol. II, pp. 359–368.
60. S. Pancharatnam, *Ind. Eng. Chem. Process. Des. Develop.,* **11,** 287 (1972).
61. W. F. McIlhenny and M. A. Zeitoun, *Chem. Eng.* **76** (Nov. 3, 1969), p. 81, and (Nov. 17, 1969), p. 251.
62. G. Baseggio in ref. 12, pp. 351–358.
63. P. deFlers, A. Caillaud and P. Charuit in ref. 28, pp. 51–62.
64. D. P. Brice in ref. 28, pp. 96–102.
65. W. E. VerPlank and R. F. Heiser, *Salt in California,* Bulletin 175, California Department of Natural Resources, Mar. 1958.
66. E. H. Rivera in ref. 12, pp. 411–413.
67. E. Chemtob and J. C. Brooks in ref. 12, pp. 393–398.
68. (a) P. A. Stoffel in ref. 28, pp. 3–19, (b) A. H. Handy and D. C. Hahl, *Great Salt Lake; Chemistry of the Water, Guidebook to the Geology of Utah,* No. 20, Utah Geological Society, 1966.
69. W. I. Weisman and R. C. Anderson, *Min. Eng.* **5**(7), 711 (1953).
70. C. H. Chilton, *Chem. Eng.* **65**(16), 116 (1958).
71. *Chem. Eng.,* **77,** 71 (Feb. 27, 1978).
72. K. P. Wang in A. E. Shreck, *Minerals Yearbook—1972,* Vol. 1, Bureau of Mines, U.S. Dept. of the Interior, Washington, D.C., 1974, pp. 217–218.
73. J. W. Pressler, U.S. Bureau of Mines, *personal communication,* 1977.
74. *Chem. Mark. Rep.* **209**(14), 9 (Apr. 5, 1976).
75. C. L. Klingman, preprint from "Bromine" in *Bureau of Mines Mineral Yearbook—1975,* Vol. 1, U.S. Bureau Mines, U.S. Dept. of the Interior, Washington, D.C., 1975.
76. *Chem. Mark. Rep.* **209**(1), 9 (Jan. 5, 1976).

77. A. H. Reed, preprint from "Calcium and Calcium Compounds," *Bureau of Mines Yearbook—1975*, Vol. 2, U.S. Bureau of Mines, U.S. Dept. of the Interior, Washington, D.C., 1975.
78. J. W. Pressler in *Commodity Data Summaries-1977,* U.S. Bureau of Mines, U.S. Dept. of the Interior, Washington, D.C., 1977, pp. 80–81.
79. R. H. Singleton in ref. 78, pp. 94–95.
80. R. H. Singleton and H. B. Wood, *Mineral Facts and Problems—1975, U.S. Bureau of Mines Bulletin 667,* U.S. Department of the Interior, Washington, D.C., pp. 619–635.
81. R. H. Singleton in ref. 78, pp. 128–129.
82. W. F. Keyes in ref. 80, pp. 855–870.
83. *Directory of Chemical Producers—USA,* Stanford Research Institute, Menlo Park, 1976, p. 816.
84. R. J. Foster in ref. 78, pp. 156–157.
85. C. L. Klingman in ref. 80, p. 1017.
86. R. J. Foster in ref. 78, pp. 144–145.
87. Ref. 83, p. 884.

General References

The Symposia on Salt, Northern Ohio Geological Society, Cleveland, Ohio: A. C. Bersticker, K. E. Hoekstra, and J. F. Hall, eds., *First Symposium,* 1963, 661 pp.; J. L. Rau, Ed., *Second Symposium,* Vol. I, 443 pp. and Vol. II, 422 pp., 1966; J. L. Rau and L. F. Dellwig, eds., *Third Symposium,* Vol. I. 474 pp. and Vol. II, 486 pp., 1970; A. H. Coogan, ed., *Fourth Symposium,* Vol. I, 530 pp. and Vol. II, 517 pp., 1973.

R. BRUCE TIPPIN
Great Salt Lake Minerals and Chemicals Corp.

PAUL E. MUEHLBERG
Dow Chemical Company

CHEMICALS IN WAR

Throughout military history there are references to the four major types of chemicals discussed in this article—flame agents, incendiaries, smokes, and toxic chemical agents. In war, toxic chemicals are used to produce casualties in personnel and to inflict harassment. Flame and incendiaries also produce casualties in personnel and are employed to destroy structures and material. Smokes, employed for screening, signaling, and target-marking, have both offensive and defensive applications. A fifth type of chemical is included here, the riot control agent, CS, although it has a limited defensive role in war (see Irritants). An alphabetical list of compounds mentioned in the text with their CAS Registry Numbers is given on page 413.

Although chemical warfare, so-called, has not been used on a large scale since World War I, the world is seeing a revival of interest in this area of military technology. The United States is concerned about chemical warfare because of the stress placed on it by other major powers, notably the USSR and the Warsaw Pact countries. Contrary to popular belief, chemical warfare has not been outlawed internationally. It remains as a potential weapon in some future conflict (1).

The use of chemical agents and weapons in war is commonly thought to be a modern military technique. However, Rothschild (2) documents several such uses that are reported as having taken place before the birth of Christ. These primitive actions included the contamination of drinking water and an attack against ships with earthen pots filled with live serpents. The modern history of the use of toxic chemicals in war began in World War I. Since that time the use of mustard in the Italian-Abyssinian war of 1935–1936 represents the only major appearance of toxic chemicals in battle.

The use of toxic chemicals in World War I and the resultant propaganda have colored all discussions of such weapons since that time. Kelly (3) presents an excellent survey of the many factors affecting the use of toxic chemicals in war, including historical background, legal aspects, attempts at international agreement, and reasons why toxic chemicals were not employed in World War II or Korea.

Toxic chemical agents may be defined as chemical substances in gaseous, liquid, or solid state intended to produce casualty effects ranging from harassment through varying degrees of incapacitation to death. A few such agents are true gases but most are solids or liquids that, in actual use, are converted into a gaseous state or disseminated as aerosols. For contamination of terrain or material the agent can also be disseminated in bulk form.

Survey. The modern history of the military use of toxic chemical agents (2,4–6) dates from the first full-scale (chlorine) gas attack on April 22, 1915, near Ypres, Belgium. There were a few reports of the limited use of toxic chemicals after World War I. The Italians employed mustard (a blister agent) during the Ethiopian war in 1935 and 1936, and the Japanese used toxic chemicals in a number of small-scale engagements in the early years of their war with China.

Since World War I many thousands of chemical compounds have been studied in various countries, but only the Germans succeeded in finding a new and potent class of toxic agents—the nerve agents Tabun, Sarin, and Soman (6–8). In the United States these are the G-agents. A less volatile group is the V-agents.

In the period since World War II many of the world's major powers have continued vigorous research and development programs in the field of toxic chemical agents. Toxic chemicals did not play a significant role during or after World War II. However, they are still considered threats in possible future conflicts. The new classes of compounds are a great deal more toxic and are capable of being disseminated by munition systems that include bombs, artillery rounds, rockets, grenades, missiles, and aerial spray (1,7).

Characteristics. Toxic chemical munitions have unique characteristics in contrast to other weapons systems. Their nature is such that they seek out the enemy whether he is widely dispersed or concentrated in fortifications. Gases and aerosols can turn corners and penetrate crevices, thereby reaching personnel physically protected from high explosives. In addition, toxic chemicals are minimum-destruction weapons as regards material. They are directly effective against personnel, but they leave intact the cities and industrial facilities which are destroyed by high explosives and nuclear weapons (6).

Toxic chemical agents produce a variety of physiological effects depending on the nature of the agent. These effects range from death to the mild incapacitation used for riot control (7). There is a whole spectrum of casualty effects, between these two extremes, applicable to almost any military situation. These include blistering, choking, blood poisoning, lacrimation, nerve poisoning, laxation, and various forms of mental and physical disorganization.

Some of the criteria used in the selection of a suitable agent are (a) effectiveness in extremely small concentrations, (b) time to onset of symptoms and duration of action, (c) effectiveness through various routes of entry into the body, such as the respiratory tract, eyes, and skin, (d) stability to long-term storage, and (e) ease of dissemination in feasible munitions.

Lethal Agents. Research on chemical agents after World War I led to the elimination of all but a handful of chemicals as being of practical battlefield significance. At the time of World War II, for example, the only chemicals considered to be of practical significance to the United States and its allies included the mustard gases (sulfur- as well as nitrogen-mustards) and phosgene.

The discovery of the nerve agents in Germany led to the availability of a class of compounds at least one order of magnitude more lethal than previously known chemical agents. Death may occur in a matter of minutes as contrasted to hours required by the previous agents. With such an increase in potency it is possible to consider chemical agents for application to other than local tactical situations, ie, delivery by aircraft or missiles at long range.

At present, mustard and the nerve agents are among the most important lethal agents available for military application. Although mustard is not as lethal as the nerve agents, its unique characteristic of producing casualties at low concentration renders it an important agent.

Mustard and Related Vesicants. Mustard, bis(2-chloroethyl) sulfide (Chemical Agent Symbol: HD), $ClCH_2CH_2SCH_2CH_2Cl$, is a colorless, oily liquid when pure; most samples have a characteristic garlic-like odor. It is primarily a vesicant, blisters being formed by either liquid or vapor contact. Mustard also attacks the eyes and lungs and is a systemic poison, so that protection of the entire body must be provided. It is insidious in its action; there is no pain at the time of exposure. The symptoms usually do not appear until several hours after exposure.

Its primary military application is to restrict the use of terrain or lower the mobility of an enemy force in a contaminated area. It is volatile enough to be effective as a vapor in warm weather. In this case relatively moderate expenditures of munitions yield severely incapacitating vapor dosages within less than an hour, although the time of onset of effects is several hours after the brief exposure period (7).

In the period following World War I and during World War II, a wide variety of sulfur analogues of mustard were investigated and many potent vesicants were discovered. These compounds resembled mustard to the extent that they had two 2-chloroethyl groups attached to a sulfur atom. Examples of such compounds are 1,2-bis(2-chloroethylthio)ethane (Chemical Agent Symbol: (Q), $ClCH_2CH_2SCH_2CH_2SCH_2CH_2Cl$, and bis(2-chloroethylthioethyl) ether: (T), $ClCH_2CH_2SCH_2CH_2OCH_2CH_2SCH_2CH_2Cl$. It is noteworthy that each 2-chloroethyl group is associated with its own sulfur atom and not with the same one as with mustard. Both Q and T are more vesicant than mustard, but neither is as volatile and for that reason is not as effective by the vapor route.

One disadvantage of sulfur mustard is that it melts at ca 10°C. Solidification in airplane spray tanks also presents a serious problem.

The procedure by which mustard is manufactured can be modified to yield either a mixture of mustard and Q (Chemical Agent Symbol: HQ) or a mixture of mustard and T (HT). These mixtures have several advantages over mustard alone (unless the agent is to be used only for vapor effects). HQ and HT are both more toxic, more vesicant, have lower melting points, and are more persistent than mustard alone.

In 1935 nitrogen analogues of sulfur mustards were first synthesized and found to have marked vesicant action (9). These are tertiary amines containing at least two 2-chloroethyl groups, $RN(CH_2CH_2Cl)_2$. A large number of nitrogen mustards were synthesized and tested during World War II. The most important of these compounds were tris(2-chloroethyl)amine (Chemical Agent Symbol: HN3), $N(CH_2CH_2Cl)_3$; N-methyl-2,2'-dichlorodiethylamine: (HN2), $CH_3N(CH_2CH_2Cl)_2$; and 2,2'-dichloro-triethylamine: (HN1), $CH_3CH_2N(CH_2CH_2Cl)_2$. The nitrogen mustards are colorless liquids when pure, but turn yellow to amber in storage. They have faint odors varying from fishy or soft soaplike to practically odorless. They act on the body in a manner similar to sulfur mustard.

Physical and Chemical Properties. The physical properties of the mustards are summarized in Table 1. The sulfur mustards are only slightly soluble in water. The nitrogen mustards are slightly soluble at neutral pH but form water soluble salts under acid conditions. Both sulfur and nitrogen mustards are extremely soluble in most organic solvents.

Reactions. Although the sulfur and nitrogen mustards have limited solubility in water at neutral pH, the small quantity which dissolves is extremely reactive. The reaction proceeds via a cyclic sulfonium or imonium intermediate which is illustrated in the following equation for sulfur mustard:

$$ClCH_2CH_2SCH_2CH_2Cl \rightleftharpoons ClCH_2CH_2-\overset{+}{S}\begin{pmatrix} CH_2 \\ | \\ CH_2 \end{pmatrix} Cl^-$$

This intermediate will attack compounds containing a variety of functional groups, such as primary, secondary, and tertiary amino nitrogen atoms, carboxyl groups, and sulfhydryl groups (10).

With nitrogen mustards, the imonium ion, which apparently forms even in the absence of any solvent, readily attacks another molecule to form a dimer. For this reason, the nitrogen mustards are less stable than sulfur-mustards in long term storage.

An excellent reagent for the detection and quantitative estimation of the mustards is p-nitrobenzylpyridine (11). On treatment of the reaction product with alkali, a blue color appears which will detect as little as 0.1 µg of mustard.

The mustards readily alkylate inorganic thiosulfates to form Bunte salts (12–13). With sulfur mustard, the product is $ClCH_2CH_2SCH_2CH_2SSO_3^-$. They alkylate phosphates in a similar manner and the product isolated from sulfur mustard is $ClCH_2CH_2SCH_2CH_2OPO_3^{2-}$.

Table 1. Physical and Chemical Properties of Mustards

Property	HD	Q	T	HN1	HN2	HN3
mol wt	159.08	219.08	263.25	170.08	156.07	204.54
boiling pt^a, °C^{kPa}	80$^{0.67}$		120$^{0.003}$	85$^{1.3}$	87$^{2.4}$	144$^{2.0}$
melting pt, °C	14.5	56	10	−34	−60	−4
d^{25}	1.2682		1.24	1.086	1.118	1.2347
volatility at 25°C, mg/m^3	925	0.4	2.8	2.29	3.581	0.120

^{a} To convert kPa to mm Hg, multiply by 7.5.

Sulfur mustard reacts rapidly with chlorine or with bleach, and this reaction has been found suitable as a means of decontamination. Nitrogen mustards chlorinate extremely slowly and, for that reason, chlorination is not suitable as a means of decontamination. The formation of water soluble salts of nitrogen mustards, such as by neutralization with sodium bisulfate, is usually the method of choice for removal from contaminated surfaces. These salts are much less vesicant than the corresponding free bases.

The mustards can be oxidized with such oxidizing agents as hydrogen peroxide or potassium bichromate in sulfuric acid. Oxidation occurs at the sulfur atom of sulfur mustard or at the nitrogen atom of nitrogen mustard. The product formed on strong oxidation of sulfur mustard, bis(2-chlorethyl)sulfone, $(ClCH_2CH_2)_2SO_2$, exhibits vesicant properties.

Physiological Effects (5). The sulfur and nitrogen mustards act first as cell irritants and finally as a cell poison on all tissue surfaces contacted. The first symptoms usually appear in 4–6 h. The higher the concentration, the shorter the interval of time between the exposure to the agent and the first symptoms. The local action of the mustards results in conjunctivitis (inflammation of the eyes); erythema (redness of the skin), which may be followed by blistering or ulceration; and inflammatory reaction of the nose, throat, trachea, bronchi, and lung tissue. Injuries produced by mustard heal much more slowly and are more liable to infection than burns of similar intensity produced by physical means or by other chemicals.

The rate of detoxification is slow. Even very small repeated exposures are cumulative in their effects or more than cumulative owing to sensitization.

Uses and Applications. The nitrogen mustards are used clinically in the treatment of certain neoplasms (14). They have been used in treatment of Hodgkin's disease, lymphosarcoma, and leukemia (see Chemotherapeutics, antimitotic).

Nerve Agents. The term nerve agents refers to two groups of highly toxic chemical compounds that generally are organic esters of substituted phosphoric acid (6–8). The nerve agents inhibit cholinesterase (qv) enzymes and thus come within the category of anticholinesterase agents. The three most active compounds investigated by the German discoverers of G-agents during World War II are:

Tabun, ethyl phosphorodimethylamidocyanidate (Chemical Agent Symbol: GA), $((CH_3)_2N)P(O)(CN)OC_2H_5$.

Sarin, isopropyl methylphosphonofluoridate (GB), $CH_3P(O)(F)OCH(CH_3)_2$.

Soman, pinacolyl methylphosphonofluoridate: (GD), $CH_3P(O)(F)OCH(CH_3)C(CH_3)_3$.

Although the G-agents are liquids under ordinary atmospheric conditions, their relatively high volatility permits them to be disseminated in vapor form. They are generally colorless, odorless, or nearly so, and readily absorbable through not only the lungs and eyes but also the skin and intestinal tract without producing any irritation or other sensation on the part of the exposed individual. They are sufficiently potent so that even a brief exposure may be fatal. Death may occur in 1–10 min, or be delayed for 1–2 h, depending on the concentration of the agent.

Another class of nerve agents, discovered after World War II, is the V-agents. These are generally colorless and odorless liquids which do not evaporate rapidly at normal temperatures. In liquid or aerosol form, V-agents affect the body in a manner similar to the G-agents. Their advantage lies in the fact that they produce casualties when absorbed through the skin in concentrations much lower than those required by the G-agents. Aerosolized V-agents are also quite lethal by inhalation.

Physical and Chemical Properties. Some physical properties of nerve agents are given in Table 2. The G-agents are miscible in both polar and nonpolar solvents. They hydrolyze slowly in water at neutral or slightly acidic pH and more rapidly under strong acid or alkaline conditions. The hydrolysis products are considerably less toxic than the original agent.

Reactions. GB is unstable in the presence of water. Maximum stability in aqueous solutions occurs in the pH range 4.0–6.5 with the hydrolysis rate increasing rapidly as the pH increases. The half-life in distilled water at 25°C is ca 36 h, but hydrolysis is accelerated by the presence of acids or bases. Since bases are far more effective in this respect than acids, caustic solutions are useful for decontamination.

GB decomposes thermally to form a variety of phosphorus-containing products as well as propylene. The rate of decomposition increases with increase in temperature, and in the presence of acids. At the boiling point of GB, under atmospheric conditions, decomposition is fairly rapid.

GB and the other G-agents react with perhydroxyl ions at pH 8–10 to form a perphosphonate ion, $CH_3P(O)(OC_3H_7)OO^-$, which has a sufficiently high redox potential to oxidize indole or o-dianisidine to produce colored products. This reaction is thus useful as a method of detection, and less than 1 μg of GB can be detected in this manner (15).

Another useful reagent for detection and estimation of G-agents is diisonitrosoacetone (16). A magenta color is produced with 1 μg of GB at pH 8.5. Coupling agents, such as p-phenylenediamine increase the reaction rate.

Physiological Effects. Inhalation of G-agent vapor at realizable field concentrations is immediately incapacitating. The symptoms in normal order of appearance are running nose; tightness of chest; dimness of vision and pinpointing of the eye pupils (myosis); difficulty in breathing; drooling and excessive sweating; nausea, vomiting; cramps and involuntary defecation or urination; twitching, jerking, and staggering; and headache, confusion, drowsiness, coma, and convulsion. These symptoms are followed by cessation of breathing and death.

Although GB is also effective by penetration through the skin, the dose required to produce toxic effects by this route is very high, so that a masked person is well-protected.

VX (6–7) is estimated to be about three times as potent a respiratory agent as sarin (GB), but about a hundred times as potent as a percutaneous agent.

$$CH_3-\underset{\underset{OCH_2CH_3}{|}}{\overset{\overset{O}{\|}}{P}}-SCH_2CH_2N(Pr^i)_2$$

VX

Table 2. Physical and Chemical Properties of Nerve Agents

Property	GA	GB	GD	VX
formula wt	162.13	140.10	182.18	267.38
boiling pt, °C	246	147	167	298
freezing pt, °C	−50	−56	unknown	below −51
$d^{25°C}$	1.073	1.0887	1.0222	1.0083
volatility at 25°C, mg/m^3	610	21,900	3,060	10.5

Binary Munitions. In the United States the modern development of the binary munitions concept would greatly enhance the safety in manufacturing, storing, and handling of nerve agent. The binary munition concept consists of two nonlethal components in one casing, separated by a membrane. They remain separated and harmless until the round is fired. When the round is fired, the membrane ruptures and the two components mix to form a standard nerve agent. In addition to the reduced storage and handling hazards, the binary system would enable the manufacture of the nonlethal components in ordinary civilian chemical plants which are not necessarily equipped with the stringent safety measures required for producing nerve agent munitions. The binary system would also overcome the restricting problem of toxic chemicals that are too unstable to be stored for any length of time.

Other Lethal Agents. The limit of toxic effectiveness has by no means been reached with the nerve agents. There are a number of substances, many found in nature, which are known to be more effective (6). Examples of these toxic natural products include *shellfish poison,* isolated from toxic clams; *puffer fish poison,* isolated from the viscera of the puffer fish; the active principle of *curare;* "heart poisons" of the digitalis type; the active principle of the *sea cucumber;* active principles of *snake venom;* and the protein *ricin,* obtained from castor beans.

Incapacitating Agents. Incapacitating agents, or incapacitants, are just what the name implies. In wartime, soldiers and civilians must be physiologically, physically and mentally able to perform their jobs. Thus, any agent which renders an individual incapable of performing his job may be classified as an incapacitating agent (6–7).

Such agents are most suitable for consideration in limited-warfare situations, eg, when enemy troops are intermingled with a friendly population, or a city that is a key military objective. The purpose is to capture the enemy without killing the civilians. For such applications, the incapacitating agent should produce no permanent aftereffects and allow for complete recovery.

Agent BZ (3-quinuclidinyl benzilate) is a typical incapacitating agent. BZ is one of a group of substances, many of them glycolate esters, sometimes known as atropinemimetics, their action on the central and peripheral nervous systems resembling that of atropine. The effects of BZ are those of an anticholinergic psychotomimetic drug (6). They follow about a half-hour after exposure to BZ aerosol and reach peak effects in 4–8 h. They may then take up to 4 d to pass. Effects include disorientation, with visual and auditory hallucinations. The agent disturbs the higher integrative functions of memory, problem-solving, attention, and comprehension. There is a gradual return to normal.

BZ

By U.S. Army criteria, incapacitating agents *do not* include the following (7).

1. Lethal agents that are incapacitating at sublethal doses, such as the nerve agents.

2. Substances that cause permanent or long-lasting injury, such as blister agents, choking agents, and those causing eye injury.

3. Medical drugs that exert marked effects on the central nervous system, such as barbiturates, belladonna alkaloids, tranquilizers, and many of the hallucinogens. These drugs are logistically infeasible for large-scale use because of the high doses required (see Alkaloids; Neurochemical agents; Psychopharmacological agents).

4. Agents of temporary effectiveness that produce reflex responses interfering with performance of duty. These include skin and eye irritants causing pain or itching (vesicants or urticants), vomiting- or cough-producing compounds (sternutators), and tear compounds (lacrimators).

5. Agents that disrupt basic life-sustaining systems of the body and thus prevent physical activity. These include agents that lower blood pressure, paralyzing agents such as curare, fever-producing agents, respiratory depressants, and blood poisons. Such agents almost invariably have a narrow margin of safety between the effective doses and possibly lethal doses. (The basic purpose of an incapacitating agent is to reduce military effectiveness without endangering life.)

Use. A detailed operational analysis of a toxic chemical weapons system would show that lethal chemical agents should be employed in close military engagements where their rapidity of action, area-search, and multiple-casualty capabilities can influence the immediate tactical situation. However, rapidity of action and inherent nondestructiveness and, in some cases, persistency, could make toxic chemicals weapons of choice for deep tactical targets. Owing to inherent physical characteristics, chemical agents can be adapted to a variety of munitions, including grenades, mines, artillery shells, bombs, bomblets, spray tanks, rockets, and missiles. Variety is required because of the size of targets, distance to targets, and weapons available to conduct a chemical attack at any given time.

A commander is provided a degree of flexibility in applying available combat power, because the physiological effects of toxic chemical agents range from temporary incapacitation through death. Tactically, chemical agents have defensive and offensive capabilities, under nuclear or nonnuclear conditions, and in limited or general wars. Toxic chemical agents may be used in conjunction with other types of weapons.

Chemical weapons are not intended to replace the integrated weapons systems, but to complement them. Such chemical weapons systems increase the flexibility of the commander's choice of weapons. They do not destroy materials, but allow physical preservation of industrial complexes and other facilities that may be useful to friendly forces. Incapacitating agents also may be used to preserve life and avoid permanent injury.

In situations where lethal effects are not required or desired, an incapacitating agent would be the weapon of choice. Typical situations of this type would arise during international police actions or limited wars where friendly forces are in close contact with enemy troops or where it is necessary to capture enemy personnel for interrogation. In cases where enemy troops are intermingled among the friendly or neutral population of a city (that is a key military objective) employment of an incapacitating agent could make it possible to capture the city without causing death or permanent injury to the civilians.

Irritants

Irritant compounds such as the lacrimators and sternutators used in World War I are traditional examples of harassing agents whose effects are reversible and briefly incapacitating. Riot control agent CS is a modern irritant compound (6–7).

CS(o-chlorobenzylidenemalononitrile) (ClC$_6$H$_4$CHC(CN)$_2$) causes physiological effects that include extreme burning of the eyes, accompanied by a copious flow of tears; coughing; difficulty in breathing; chest tightness; involuntary closing of the eyes; stinging sensation of moist skin; runny nose; and dizziness or "swimming" of the head. Heavy concentrations will also cause nausea and vomiting.

The effects are immediate, even in extremely low concentrations. The median effective concentration for respiratory effects is 12–20 mg/m^3; for eye effects it is 1–5 mg/m^3. The onset of maximum effects is 20–60 s, and the duration of effects is 5–10 min after the affected individual has been removed to fresh air (7).

A water-soluble white crystalline solid, CS is disseminated as a spray, as a cloud of dust or powder, or as an aerosol generated thermally from pyrotechnic compositions. The formulation designated CS1 is CS powder mixed with an anti-agglomerant; when dusted on the ground it may remain active for as long as five days. CS2, formulated from CS1 treated with a silicone water repellent, may persist for as long as 45 days (6).

The principal uses of CS are in riot control and training; it has a limited tactical use in (17) defensive military modes to save lives.

Flame

In the modern weapons arsenal flame agents are defined as various hydrocarbons, blends of hydrocarbons, and other readily-flammable liquids, usually thickened with additives, that are ignited easily and can be projected or delivered to military targets (7). Although flame agents may be employed against buildings and other flammable targets, their primary role is in employment against personnel in hardened structures or emplacements. In the United States the major application of flame agents is now in flame throwers and flame projectors, including flame rockets. The fire bomb is rapidly becoming obsolete. The replacement for the portable flame thrower is the multi-shot flame rocket which delivers its encapsulated flame warhead directly to its target from a significantly greater and safer range. The large-caliber flame rounds could become the all-weather, highly-aimable replacements for the obsolete air-deliverable fire bombs and the mechanized flame throwers.

Flame Throwers and Projectors. Flame thrower improvements since World War II were primarily in more effective and efficient fuel and in weight reduction of the portable units. Mechanized flame throwers were also improved in fuel quality, and one advance consisted of the development of a mechanized flame thrower kit adaptable to a variety of armored vehicles other than the main battle tank. A major advance was the fielding of the multi-shot, lightweight, shoulder-fired, four-tube launcher capable of firing one to four flame rounds semiautomatically. This flame system is replacing the portable flame thrower. The mechanized flame thrower will probably be replaced by the family of large-caliber flame rounds.

The U.S. Army's M202/M74 Flame System is a shoulder-fired, four-tube launcher equipped with front and rear hinged protective covers. A folding sight and trigger handle assembly provide compact carrying and storage capabilities. An adjustable sling is used to carry the launcher over the shoulder. The rocket system is aimed and fired from the right shoulder from the standing, kneeling, sitting, or prone position. Ammunition for the launcher is provided in rocket clips preloaded with four rockets that slip-fit into the four launcher tubes. The user can fire from one to four rockets semiautomatically at a rate of one per second; the launcher can be reloaded with a new rocket clip repeatedly. The flame agent payload is thickened pyrophoric agent, TPA, a polyisobutylene-thickened metal alkyl formulation (18).

Fire Bombs. After World War II the fire bomb was refined into a standard item of military equipment. The fire bomb was a cigar-shaped, thin-cased tank similar in appearance to the aircraft fuel tank from which it evolved. The typical fire bomb consisted of the basic tank, two igniters, two fuses, an arming wire, and the flame agent payload. Most fighter and fighter-bomber aircraft carried two fire bombs, one under each wing. Ideal delivery was at low-altitude almost-level flight, with the fire bomb impacting along the aircraft flight path and producing a rolling wall of flame covering an area of ca 35 m wide and up to 100 m in length. The initial fireball, a direct function of flame agent quality, burned for about 10 s with intense heat. The dispersed particles could burn for up to 10 min but at greatly reduced intensity.

Fire bombs were employed against armored vehicles and personnel carriers, light and heavy weapons positions and emplacements, supply and ammunition concentrations, command posts and observation posts, airfields and aircraft, and other military targets, especially manned targets. Fire bombs were effective against personnel because of the psychological effects such attacks produced.

With the advent of high-performance jet-powered aircraft, the fire bomb, developed for low- and medium-speed, low-altitude lay-down delivery, became obsolete. When delivered at speeds approaching Mach one (345 m/s), the design characteristics were often grossly exceeded and many units broke up or functioned while still on the aircraft. Delivery from high altitudes created major problems when the fire bombs dug craters and deposited most of the flame agent payload in these craters. Aimability was lost, and when fins were added to improve aimability the craters were even bigger. Attempts to incorporate proximity fusing into the fire bombs did not materially improve performance since the mass of the fire bomb payload still continued along its trajectory and continued to create craters filled with flame agent.

Studies of fire bomb modifications, and flame agent payloads for them, were terminated when the controlled fireball damage mechanism was developed and demonstrated by the Army. In field firings it was shown that large-caliber flame rounds could produce effects equivalent to those produced by fire bombs in the old low-velocity, low-altitude delivery mode. Once again the rolling wall of reacting flame agent was generated, this time with a highly-aimable, all-weather flame system capable of sustained fire in the troop support role.

Flame Agents. For some time after World War II major effort was expended in improvement of napalm-thickened hydrocarbons as the standard flame agents for all applications. The major problem remained—the breakdown of the flame agent in the presence of traces of water and the need for peptizers in cold-weather applications. For almost the last decade, thickened pyrophoric flame agents (7) have been deployed in the field as field expedients and as the payload for the U.S. Army's flame rocket system. Advantages of this flame agent include (a) the ability to prepackage warheads and other containers and store them for indefinite periods with no deterioration and (b) the fact that these pyrophoric flame agents do not require an ignition system to function on target.

Computer-aided research by Army personnel has resulted in development of the controlled-fireball damage mechanism for efficient and effective coupling of heat energy with military targets to effect maximum thermal damage coupled with the ability to permit the tailoring of these flame agents, optimizing them for each candidate flame system. Optimization of several low-viscosity metal alkyl formulations has also removed the problem of temperature by making these flame agents relatively independent of temperature variations.

Incendiaries

Incendiary agents are compositions of chemical substances designed for use in the planned destruction of buildings, property, and material by fire (7). Incendiaries burn with an intense, localized heat. They are very difficult to extinguish and are capable of setting fire to materials that normally do not ignite and burn readily. Although there are tactical applications for incendiary agents and munitions, they have played primarily a strategic role in modern warfare.

Incendiary Requirements. The mechanics of starting fires with incendiary agents may be considered to involve: (a) a source of heat acting as a "match" to initiate combustion in a larger mass, (b) combustible material which serves as "kindling," and (c) "fuel." The first two requirements are supplied by the incendiary munition, and the target is the fuel. All incendiary munitions, except those containing materials that are spontaneously combustible, must contain an initiator of some sort, such as a fuse or an ignition cup. The second element of the incendiary munition, the "kindling," is the important factor, and both the amount and nature of the combustible material in the munition have been the subject of most of the research and development effort in this field.

The maximum total heat output of an incendiary agent can be calculated readily, and obviously it is desirable to use a filling that has a high heat evolution. The rate of heat release will vary with the nature of the agent and depends on: (a) flash, fire, or decomposition temperature; (b) particle size of the agent after ejection from the munition which controls the surface-to-volume ratio of the agent; and (c) oxidizing agents blended with the combustible material to increase the rate of heat evolution. The incendiary agent must be capable of heating the target, the "fuel," until its ignition temperature is reached. The temperature varies from 200–440°C and, to be really effective as an intensive incendiary, an agent must generate at least four times the amount of heat necessary to raise the temperature to this point.

Metal Incendiaries. Metal incendiaries include those consisting of magnesium in various forms and powdered or granular aluminum mixed with powdered iron oxide. Magnesium is a soft metal which, when raised to its ignition temperature, burns vigorously in air. In either solid or powdered form it is used as an incendiary filling. In alloyed form it is used as the casing for small incendiary bombs.

Magnesium has an ignition temperature of 623°C and a burning temperature of 1982°C. (The burning temperature is variable, as it depends upon the rate of heat dissipation, rate of burning, and other factors.) Magnesium, burning with a blinding white flame, melts as it burns and the burning liquid metal drops to lower levels, igniting all combustible materials in its path. Burning stops if oxygen is prevented from reaching the metal, or if the metal is cooled to a point below its ignition temperature. Magnesium does not have the highest heat of combustion of the metals, but none of the other metals have been successfully used as air-combustible incendiaries. Certain other metals may be alloyed with magnesium without affecting its ignitability. The alloyed metal has strength to withstand distortion, whereas pure magnesium does not.

In massive form, magnesium is difficult to ignite. This problem is overcome by packing a hollow core in the bomb with thermite, an easily-ignited mixture that supplies its own oxygen and burns at a very high temperature.

Thermite is essentially a mixture of powdered iron(III) oxide (Fe_2O_3) and powdered or granular aluminum. The aluminum has a higher affinity for oxygen than iron, and if a mixture of iron oxide and aluminum powder is raised to the combustion temperature of aluminum, an intense reaction occurs: $Fe_2O_3 + 2\,Al \rightarrow Al_2O_3 + Fe$, $\Delta H = -3.35$ kJ/g (-800 cal/g). Under favorable conditions, the thermite reaction produces temperatures of about 2200°C. This is high enough to turn the newly-formed metallic iron into a white-hot liquid that acts as a heat reservoir to prolong and spread the heat (igniting action).

Thermite is composed of approximately 73% ferric oxide and 27% fine granular aluminum. The thermate mixture, composed of thermite with various additives, is used as a component in igniter compositions for magnesium bombs. A number of such compositions have been developed. Three of these were Therm-8, Thermate-TH2 (formerly Therm-8-2), and Thermate-TH3 (formerly Therm-64-C). Therm-8 was the precursor of later, improved igniting formulations. TH2 differed from Therm-8 in that it contained no sulfur and slightly less thermite. TH3 was found to be superior to the others and was adopted for use in the incendiary magnesium bomb. The composition by weight of TH3 is:

Ingredient	Percent
thermite	68.7
barium nitrate	29.0
sulfur	2.0
oil (binder)	0.3

The TH3 core is ignited by the primer, and the burning core then melts and ignites the magnesium alloy body of the bomb. The incendiary action on the target is localized since there is little scattering of the incendiary material.

Oil and Metal Incendiary Mixtures. PT1 is a complex mixture based on a paste composed of magnesium dust, magnesium oxide, and carbon, with an adequate amount of petroleum distillate and asphalt to form the paste (7). U.S. developers adopted the following formula:

Ingredient	Percent
type C paste ("goop")	49.0 ± 1.0
IM polymer AE	3.0 ± 1.0
coarse magnesium	10.0 ± 1.0
petroleum oil extract	3.0 ± 0.2
gasoline	30.0 ± 3.0
sodium nitrate	5.0 ± 0.5

PTV is an improved oil and metal incendiary mixture with the following composition:

Ingredient	Percent
polybutadiene	5.0 ± 0.1
gasoline	60.0 ± 1.0
magnesium	28.0 ± 1.0
sodium nitrate	6.0 ± 0.1
p-aminophenol	0.1

Incendiary bombs containing PT1 or PTV mixtures are easily ignited by nose and tail fuses since they contain many combustible ingredients. As these formulations contain both metal and an oxidizer (sodium nitrate), condensed phase reaction products are obtained and the resulting heat flux conducted to the target is increased, resulting in greater potential target damage.

Smokes

Military smokes are aerosols of gaseous, liquid, or particulate matter that are tactically employed to defeat enemy surveillance, target acquisition, and weapons guidance devices. The traditional battlefield applications of smoke are screening and marking. Screening smokes are normally employed to obscure a military objective from enemy observation by creating an aerial blanket, vertical curtain, or ground haze. Signalling smokes are pyrotechnic mixtures that incorporate organic dyes for target marking or for transmitting messages by prearranged color code (see Pyrotechnics).

Battlefield smoke in recent years has taken on great significance in the countermeasure role. Threat weapons no longer rely simply on optical devices which operate in the visible spectrum. Modern lethal weapons are augumented with sophisticated surveillance, target acquisition, and guidance devices which operate in the visible, infrared, and microwave (radar) regions of the electromagnetic spectrum. Accordingly, current research and development is concentrated on multispectral screening systems which are designed to defeat modern weaponry such as the antitank guided missile (ATGM), laser-guided munitions, and heat-seeking missiles.

Screening Smokes. Military smoke screens are produced by dispersing either finely-divided solids or minute liquid droplets in air, and many substances have been investigated for applicability to screening requirements. To be useful, a smoke screen must be sufficiently opaque to provide the desired screening power and long-lasting enough to achieve effective military results. In designing a screen for the modern battlefield, it is necessary to defeat sophisticated surveillance technologies that operate in the visible and near-, mid-, and far-infrared regions. Other factors considered in the evaluation of potential screening agents include cost, ease of dispersion, and efficiency of dispersion. Agents used for screening friendly areas must be as nonirritating as possible.

Both the opacity and persistency of smoke screens are largely dependent on the nature of the individual smoke particles. All present U.S. smoke agents produce aerosols whose principal mechanism of obscuration is the *scattering effect*. To optimize its effectiveness, the aerosol's particle diameter should be approximately in the same size range as the wavelength of the light to be screened. Present U.S. smoke agents produce aerosols whose particle-size distribution is log-normal with a number mean diameter of about 0.6 μm. This allows them to have maximum effectiveness in the visible and near-infrared wavelength regions.

The obscuring action of screening smokes is largely caused by reflection and refraction of light rays by the individual suspended solid or liquid particles of which the smoke is composed (7). This obscuring action occurs to the greatest extent in the absence of light-absorbing particles such as carbon; white smokes, therefore, have the greatest screening action.

The life persistency of a smoke cloud is determined chiefly by wind and convection currents in the air. Ambient temperature also plays a part in the continuance or disappearance of fog oil smoke. Water vapor in the air has an important role in the formation of most smokes. For this reason, high relative humidity improves the effectiveness of most smokes. The water vapor not only exerts its effects through hydrolysis but also it assists the growth to effective size of hygroscopic (deliquescent) smoke particles by a process of hydration.

Smoke may be generated by mechanical or thermal means, or by a combination of the two processes (7).

Types of Screening Smokes. The generation of oil smoke is based on the production of minute oil particles by purely physical means. The most desirable droplet size of these particles is 0.5–1.0 μm. The tiny droplets of oil scatter light rays and produce a smoke that appears to be white. Actually, an individual droplet would be transparent under magnification. These droplets are produced as the vaporized oil passes through the nozzle of a generator and is cooled by the surrounding air. The air cools the oil vapor so quickly that only very small droplets are able to form. The process depends upon a high oil temperature followed by quick cooling. The final smoke cloud is stable, and the life of the cloud is determined almost entirely by meteorological conditions. The smoke generator uses a low-viscosity petroleum oil (U.S. Army designation: SGF No. 2 smoke generator fog). SGF corresponds somewhat to an SAE 10 (light) motor oil in viscosity. Below 0°C, a mixture of SGF No. 2 and paraffin-free kerosene is used.

Another type of smoke mixture, a volatile hygroscopic chloride for thermal generation, has the U.S. Army designation *HC, type C*. It is made up of a mixture of grained aluminum, zinc oxide (ZnO), and hexachloroethane (C_2Cl_6). Percentages by weight are:

Ingredient	Percent (approx)
grained aluminum	6.7
zinc oxide	46.7
hexachloroethane	46.7

The ratio of zinc oxide to hexachloroethane is held between 1.04 and 1.00, but the aluminum may be varied slightly to regulate the burning time. Because the mixture is composed of solids, it can be compressed to provide high payloads in small volumes, and it is used as a filling for smoke grenades, smoke pots, and artillery shells. The initial heat needed to start the burning of the HC smoke mixture is provided by a starter mixture, typically silicon, potassium nitrate, charcoal, iron oxide, grained aluminum, cellulose nitrate, and acetone. A burning HC mixture produces zinc chloride which, in turn, is volatilized by reaction heat and condenses to form a zinc chloride–water aerosol smoke.

A third screening smoke type is *white phosphorus* (WP). This material will react spontaneously with air and water vapor to produce a dense cloud of phosphorus pentoxide. An effective screen is obtained as the P_2O_5 hydrolyzes to form droplets of dilute phosphoric acid aerosol. WP produces smoke in great quantity, but it has certain disadvantages that limit its applications and effectiveness. Because WP has such a high heat of combustion, the smoke it produces has a tendency to rise in a pillarlike mass. This behavior too often nullifies the screening effect, particularly in still air. Also, WP is very brittle and the exploding munitions in which it is used break it into very small particles that burn rapidly. These two major disadvantages have been overcome to some degree by the development of *plasticized white phosphorus* (PWP). PWP is produced by melting WP and stirring it into cold water, which results in a slurry of WP granules ca 0.5 mm dia. The slurry is mixed with a viscous solution of synthetic rubber so that the granules are coated with a film of rubber and thus separated from each other. When PWP is dispersed by an exploding munition, it does not break into such small particles, the burning rate is slowed, and the tendency of the smoke to pillar is reduced. WP and PWP are used as fill for grenades, artillery shells, mortar shells, bombs, and rockets.

Recent developments in several countries have resulted in *red phosphorus* (RP) as a screening smoke agent with less performance problems. RP is an allotropic form of elemental phosphorus which is made by heating white phosphorus at high temperatures in the absence of air. RP is less reactive than WP and thus lends itself to the manufacture of pre-sized subunits that can be packaged in artillery and mortar shells. These subunits, which are dispersed as multiple smoke-producing sources, enhance target effectiveness by the rapid formation of a large homogeneous and persistent screen. The pillaring phenomenon (of WP) is minimized. Another advantage of RP is that it does not undergo a change of state at operational temperatures (as does WP) thereby precluding the munition instability problems that sometimes occur with WP which liquefies above 43°C.

A solution of *sulfur trioxide* (SO_3) dissolved in chlorosulfonic acid ($ClSO_3H$) has been used as a smoke agent (U.S. designation: FS). It is not currently a U.S. standard agent. When FS is atomized in air, the sulfur trioxide evaporates rapidly from the small droplets and reacts with atmospheric moisture to form sulfuric acid vapor. This vapor condenses into minute droplets that form a dense white cloud. FS produces its smokeless effect almost instantaneously upon mechanical atomization into the atmosphere, except at very low temperatures. At low temperatures, the small amount of moisture normally present in the atmosphere requires that FS be thermally generated with the addition of steam to be effective. FS can be used as a fill for artillery and mortar shell and bombs, and can be effectively dispersed from low-performance aircraft spray tanks. FS is both corrosive and toxic in the presence of moisture, which imposes limitations on its storage, handling, and use.

Signaling Smokes. Screening smokes also can be adapted for signalling purposes if necessary. For example, phosphorus-filled artillery and mortar rounds can be used to mark targets and determine range corrections. However, a good signaling smoke must be clearly distinguishable from the smoke incident to battle. The standard signaling smokes (Table 3) afford good visibility and unmistakable identity. But all colors become gray and indistinguishable at great distances, and even excellent signaling smokes have a maximum effective visual range. With signaling munitions in use by the U.S. Army today, the visual range varies from approximately 1000–3000 m, depending upon the color intensity of the smoke and the type and size of the munition.

The known method of producing colored signaling smokes is that of volatilizing and condensing a mixture containing an organic dye. Of the dyes tested by the U.S. Army, the most satisfactory ones are the general types of azo (qv), anthraquinone (qv), azine (qv), or diphenylmethane dyes (7). The filling for a colored smoke munition is essentially a pyrotechnic mixture of fuel and a dye, with a cooling agent sometimes added to prevent excessive decomposition of the dye. The heat produced by the fuel

Table 3. Colored Smoke Fillings

Type mixture	Ingredient	Percentage (approximate)
red smoke	dye: 1-N-methylaminoanthraquinone 85%; dextrin 15%	42
	sodium bicarbonate	19
	potassium chlorate	28
	sulfur	11
green smoke	1,4-di-p-toluidinoanthraquinone 70%; indanthrene golden yellow 10.5%; benzanthrone 19.5%	40.0
	sodium bicarbonate	22.6
	potassium chlorate	27.0
	sulfur	10.4
yellow smoke	benzanthrone 65%; indanthrene golden yellow 35%	38.5
	sodium bicarbonate	33.0
	potassium chlorate	20.0
	sulfur	8.5
violet smoke	1,4-diamino-2,3-dihydroanthraquinone 80%; 1-N-methylaminoanthraquinone 20%	42.0
	sodium bicarbonate	18.0
	potassium chlorate	28.8
	sulfur	11.2
starter	potassium nitrate	37.8
	silicon	28.0
	charcoal	4.2
	cellulose nitrate	1.2
	acetone	28.8

volatilizes the dye, and the dye condenses outside the munition to form the colored smoke. In U.S. munitions, the fuel is made up of a mixture of an oxidizing agent, such as potassium chlorate ($KClO_3$), and a combustible material such as sulfur or sugar. The burning time can be regulated by adjusting the proportions of oxidant and combustible material, and by use of coolants such as baking soda. A typical starter mixture is composed of potassium nitrate, silicon, charcoal, cellulose nitrate, and acetone. Colored smoke mixtures are used in hand and rifle grenades and in canisters for use with larger projectiles. Table 3 gives composition ranges of basic ingredients for colored smoke mixtures.

Defense Against Toxic Agents

Defensive measures against toxic chemical agents may be divided into three major categories: agent detection and identification, individual and collective protection, and decontamination. To these three a fourth may be added which perhaps is the most important of all: a high degree of training in defensive measures and discipline in using them.

Detection. Many modern toxic chemicals, particularly the nerve agents, are colorless and odorless; therefore, chemical and physical means must be employed in their detection and identification (see General references). A number of chemical kits have been developed, applicable to a variety of detection problems. Devices that provide automatic detection and warning of the presence of chemical agents in the air have also been developed.

A recent innovation is the self-contained sampler which is greatly simplified compared to older kits. All the reagents are contained in ampules on a plastic card, and no sampling pump is used. All that is required is to break the ampules, expose the sampler to the air, and observe a change in color.

An automatic nerve agent detector with electric-powered air-sampling pump has been developed based on an electrochemical principle. It weights approximately 7 kg complete with battery for 12-h operation. The detector contains a small horn and can be connected with field wire to a remote alarm equipped with loud horn and warning light 400 m away. The detector contains a recirculating reagent solution and small scrubber–cell. Outside air to be sampled is drawn through a dust–interference filter. The device is rugged enough to work under field conditions.

Other principles of operation that have been utilized in the development of warning devices are infrared absorption, the Raman effect, conductivity, and enzyme inhibition.

Protection. The primary item of individual protection against toxic chemicals is the protective ("gas") mask. The current U.S. Army standard mask is the M17A1 (Figure 1), which has as its basic component a molded rubber facepiece with large cheek pouches that hold filter pads. The filter pads consist of six sheets of core layer laminated between two sheets of backing layer. The backing layers are composed of viscose rayon, vinyon, and glass fibers. The core layers are composed of the same materials as the backing layers plus 75 vol % Whetlerite. Whetlerite is a finely ground activated carbon which has been impregnated by immersion in an ammoniacal solution of silver, copper, chromium, and carbon dioxide, and then dried at temperatures sufficient to expel substantially all ammonia from the granules. Inhalation valves are attached to the filter pad units through the cheek-pouch portions of the mask faceblank. A speech diaphragm, incorporating an outlet valve, is attached to the faceblank at the mouth position. Large eyepieces in the faceblank provide a wide field of vision. The M17A1 mask provides complete respiratory protection against all known military toxic chemical agents, but it does not afford protection against some industrial toxics such as ammonia and carbon monoxide.

A new and refined protective mask is being developed by the Army. A prototype of a mask for civilian use, based on the M17A1 Army mask, has been designed. The design for this mask also eliminates the canister and places the filter pads in cheek pouches. The mask is made of vinyl plastic and provides adequate respiratory protection against military toxic chemicals. Other special purpose items for individual respiratory protection also have been designed, including an infant protector, a mechanical respirator operable from any compressed air source and usable in contaminated and uncontaminated atmospheres, a mask for persons with head wounds, and items to protect other types of hospital cases.

Obviously, the face mask alone will not provide complete protection against the nerve and blister agents, which act not only via the respiratory tract but also on contact with exposed skin. Protection against these types of agents requires that the entire body be covered.

Airtight, impermeable clothing was developed for personnel who must enter heavily contaminated areas. This clothing is made of butyl rubber or a coated fabric and provides complete protection against liquid agents. Such clothing is cumbersome

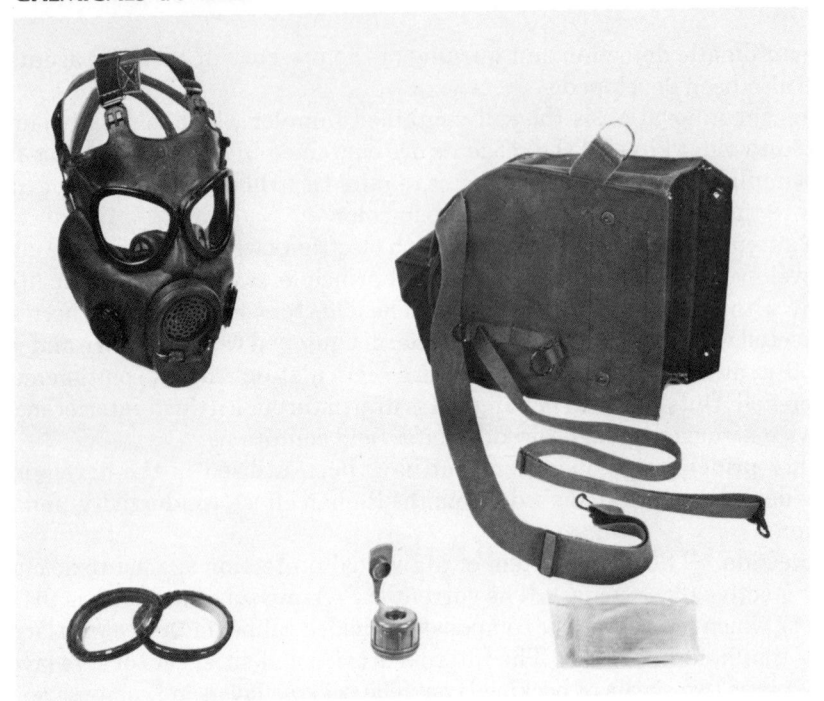

Figure 1. The M17A1 protective mask has replaced the M17 mask as Army standard. Major improvements are the added capabilities of a masked individual to drink liquids and to give respiratory resuscitation to a casualty (mask-to-mouth resuscitation). Courtesy U.S. Army. 1, ABC field protective mask, M17A1; 2, carrier, field protective mask, M15; 3, "outserts" to reduce or eliminate lens fogging and also to prevent abrasion of the eyepieces; 4, canteen cap for use with the mask drinking device; 5, M1 waterproofing bag.

and enervating because it retards the release of body heat and moisture, and personnel efficiency is lowered considerably when it is worn for protracted periods. Although resistant to liquid chemical agents, impermeable protective clothing may be penetrated after a few hours of exposure to heavy concentrations of agent. Consequently, liquid contamination on the clothing must be neutralized or removed as soon as possible.

Chemically-treated clothing was developed for issue to combat troops. Chemicals impregnated in the cloth will neutralize all of the known blister agents, but the porosity of the cloth still allows the release of body heat and moisture. Clothing of this type is effective against the vapors and spray-size droplets of the blister agents, but it does not provide protection against large drops or splashes of such agents.

There are two versions of the U.S Chemical Protective Suit, one impregnated with modified XXCC3 treatment, and the other treated with repellents to repel liquid agents (outer layer, Figure 2) and an inner layer of a charcoal impregnated foam/nylon tricot laminate which absorbs chemical agents.

The British have a two piece suit, trousers and a middie blouse with an integral hood, that is made from a nonwoven charcoal-impregnated material.

Over the past decade many nations have developed and/or fielded permeable protective garments that incorporate the use of charcoal. Among these are: U.S. Chemical Protective Suit; U.K. NBC Mk III Chemical Protective Suit; French Over-

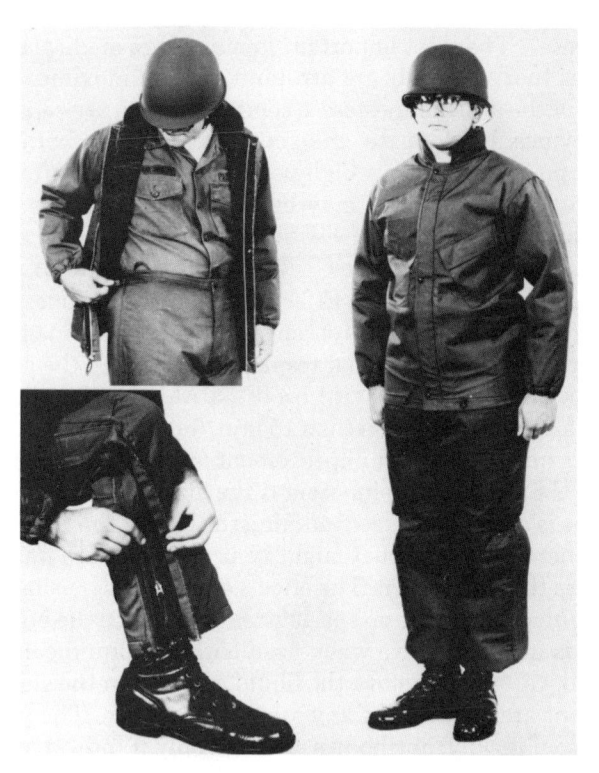

Figure 2. The overgarment is issued to combat troops operating in forward areas (19). It was developed for troops that do not have access to any type of decontamination procedures. The overgarment is designed to protect the environmental clothing from contamination. When contaminated, the overgarment is discarded and replaced. In addition it is used in the chemical laboratory and the salvage section and ammunition supply depot where chemical ammunition is handled (20).

garment; Belgian Overgarment; Federal Republic of Germany Overgarment; Canadian Chemical Battle Dress.

Collective protection involves primarily the use of shelters where personnel can work or rest. Such shelters must be airtight to prevent the inward seepage of chemical agents and, thus, require a means of providing uncontaminated air. Such an air supply can be obtained by two methods.

In small shelters a filter material called diffusion board can be incorporated into the construction or used as a liner in existing structures. The board looks somewhat like ordinary fiberboard but is composed of special fibers and activated impregnated charcoal material. The board not only effectively filters chemical agents out of inflowing air, but it also allows the carbon dioxide and moisture created by the occupants to diffuse outward at the same time. Approximately one square meter of this board will provide enough breathable air for one man for an extended period.

For large shelters, a mechanical collective protector can be used. Essentially, such a mechanical protector is nothing more than a greatly enlarged version of a mask canister through which air is forced by a fan. These collective protectors can furnish from 8.5–142 m^3 (300–5000 ft^3) of fresh, purified air per minute, depending upon the size of the unit.

Medical Defense. The most important items of U.S. medical defense against the organophosphorus (nerve) agents are atropine and pralidoxime. These agents neutralize the effects of the anticholinesterase compounds and are capable of reactivating the inhibited enzymes. If adequate emergency treatment is immediately available, it is theoretically possible to save a high percentage of those affected by the agent. Death from nerve agent poisoning may be caused by paralysis of the respiratory muscles, airway obstruction by excessive salivary and bronchial secretions, and perhaps constriction of the bronchial tree (see Cholinesterase inhibitors).

Immediate treatment is essential. Effective treatment includes the administration of atropine, (2 mg, im (intramuscularly) repeated every 3–8 min until signs of atropinization occur); institution of artificial respiration, clearing the airway if indicated; and the intravenous administration of 1.5 g of 2-PAM chloride (pralidoxime, Protopam) slowly over a period of not less than 15 min, followed by doses of 1.0 g repeated at 30-min intervals until significant improvement is seen in muscle strength (21). The current standard U.S. Army atropine item is the automatic injector (Atropen). The automatic injector is designed for self-administration by the individual in the field.

Other toxic chemical agents that might be used in modern war are the vesicants, of which mustard is the best known. The effects of mustard are similar to those of any other vesicant agent, and there is no special treatment once the burns have occurred. Copious washing is quite effective when used early for liquid contamination of the eyes, and soap and water will remove the liquid agent from the skin. Burns resulting from mustard agent are treated like any other severe burn. The pulmonary injuries are treated symptomatically; antibodies are used only if indicated for the control of infection.

Decontamination. Only persistent and thickened chemical agents normally require decontamination procedures. If contaminated areas or material do not have to be used immediately, natural aeration is the best decontaminant. All effective chemical agents, including the blister and V agents, are volatile to a certain degree. Wind accelerates their evaporation and hastens their dissipation. Rain and dew cause a more or less rapid hydrolysis of chemical agents. Sunlight causes catalytic decomposition of chlorinated compounds, such as mustard; heat from the sun also promotes rapid evaporation of these compounds. Actinic rays (chemically active radiations) from the sun cause chlorine to split from chlorinated organic compounds in the form of hydrogen chloride. As a result, the chemical nature of such compounds is completely altered. Thus, natural processes eventually will destroy all chemical agents.

If decontamination cannot be left to natural processes, chemical neutralizers or removal must be used. In general, these neutralizers mix with the chemical agent and release as a free vapor the chlorine ingredient, thus rendering the agents inert. Some of these neutralizers have been developed especially for the decontamination of chemical agents.

One such specially developed decontaminant is supertropical bleach (STB). STB is a white powder containing about 30% available chlorine which, because of a very low moisture content, is somewhat more stable in storage than ordinary bleaches. STB can be used either as a dry mix or a slurry to decontaminate exterior surfaces and ground that is contaminated with chemical agents. The dry mix is prepared with two parts bleach to three parts earth by volume. A typical slurry consists of 850 L (225 gal) of water, 3 kg of citric acid, and 590 kg of STB, which makes 1500 L (400 gal) of slurry and will treat about 1100 m^2 of smooth surface (see Bleaching Agents).

STB is an effective decontaminant for mustard, lewisite, and VX, but apparently it is only about 50% effective against nerve agents other than VX. As long as STB retains its chlorine content, it also serves to seal in vapors or neutralize them as they rise to the surface of ground that has been heavily contaminated with liquid agents.

Decontaminating agent DS2 is a general-purpose decontaminant. It consists of 70% diethylenetriamine, 28% ethylene glycol monomethyl ether and 2% sodium hydroxide. DS2 reacts with both the nerve agents and blister agents to effectively reduce their hazards within 5 min after application. Important limitations in the use of DS2 are: (a) personnel must remain masked because of the vapor; (b) rubber gloves must be worn to protect the hands; (c) it is a combustible liquid, therefore it must not be allowed to get on hot metal surfaces such as running engines or exhaust pipes. It cannot be used near open flames, and it will also cause fires if contact is made with STB; (d)

Table 4. Alphabetical List of Chemicals Referred to in the Text

Compound	CAS Registry No.	Compound	CAS Registry No.
aluminum	[7429-90-5]	(HN3), tris(2-chloroethyl)-amine	[555-77-1]
p-aminophenol	[123-30-8]		
atropine	[51-55-8]	indanthrene	[81-77-6]
barium nitrate	[10022-31-8]	indanthrene golden yellow	[128-66-5]
benzanthrone	[82-05-3]	indole	[120-72-9]
bis(2-chloroethyl)sulfone	[471-03-4]	Lewisite	[541-25-3]
(BZ), 3-quinuclidinyl benzilate	[6581-06-2]	magnesium	[7439-95-4]
cellulose nitrate	[9004-70-0]	1-N-(methylamino)-anthraquinone	[82-38-2]
charcoal	[7440-44-0]		
chloramine B	[127-52-6]	p-nitrobenzyl pyridine	[1083-48-3]
chloramine T	[127-65-1]	p-phenylenediamine	[106-50-3]
chlorine	[7782-50-5]	phosgene	[75-44-5]
chlorosulfonic acid	[7790-94-5]	phosphorus	[7723-14-0]
(CS), O-chlorobenzylidene malonitrile	[2698-41-1]	phosphorus pentoxide	[1314-56-3]
		polybutadiene	[9003-17-2]
dextrin	[9004-53-0]	potassium chlorate	[3811-04-9]
dianisidine	[119-90-4]	pralidoxime, Protopam, 2-PAM chloride	[51-15-0]
diethylenetriamine	[111-40-0]		
diisonitrosoacetone	[41886-31-1]	(Q), 1,2-bis(2-chloroethylthio)-ethane	[3563-36-8]
O-ethyl S-[2-diisopropylamino]-ethyl methylphosphonothiate	[50782-69-9]		
		silicon	[7449-21-3]
ethylene glycol monomethyl ether	[109-86-4]	sodium bicarbonate	[144-55-8]
		sodium nitrate	[7631-99-4]
ferric oxide	[1309-37-1]	sulfur	[7704-34-9]
(GA), Tabun, ethyl phosphoro-dimethylamidocyanidate	[77-81-6]	sulfur trioxide	[7446-11-9]
		(T), bis(2-chloroethylthioethyl) ether	[63918-89-8]
(GB), Sabrin, isopropyl methylphosphonofluoridate	[107-44-8]		
		thermite	[8049-32-9]
(GD), Soman, pinacolyl methylphosphonofluoridate	[96-64-0]	1,4-di-p-toluidinoanthraquinone	[128-80-3]
		VX	[50782-69-9]
(HD), mustard, bis(2-chloroethyl)sulfide	[505-60-2]	whetlerite	[7440-44-0]
		zinc oxide	[1314-13-2]
hexachloroethane	[67-72-1]		
(HN1),2,2-dichloro-triethylamine	[13426-57-8]		
(HN2), N-methyl-2,2'-dichlorodiethylamine	[51-75-2]		

DS2 is not corrosive to most metal but it may cause corrosion to aluminum, cadmium, tin, and zinc after prolonged contact; and (e) it also softens and removes paint.

The U.S. Army's M13 Decontaminating and Reimpregnating Kit is used by the individual soldier for decontamination of personal equipment and clothing, and for reimpregnating the Clothing Outfit Chemical Protective (Liner System). The M13 kit contains a pad of Fuller's earth powders for decontaminating the inside of the protective mask and, in the absence of the M258 Skin Decontaminating Kit (see below), for exposed skin; two bags of chloramide powder, each containing a dye capsule, for decontaminating clothing and reimpregnating the Clothing Outfit Chemical Protective (Liner System); and a cutter for cutting away heavily contaminated areas of clothing.

The M258 Skin Decontaminating Kit is used for exposed skin areas that have been contaminated with liquid or thickened chemical agents. The M258 kit contains two scraper sticks for removing thickened chemical agents, two plastic capsules containing decontaminants, four gauze pads, and a waterproof plastic case with a piercing tool incorporated in the lid. Decontaminating solution contained in capsule No. I is made by weight of 72% ethanol, 10% phenol, 5% sodium hydroxide, 0.5% ammonia, and the remainder water. Capsule No. II contains a solution of 45% ethanol, 5% zinc chloride, and the remainder water; also contained in this capsule is a glass ampoule containing 17.5 g of dry chloramine B. In the use of the M258 kit, precautions must be taken to avoid getting the solution in the eyes.

Besides these specially-designed decontaminants and decontaminating kits there are a number of commercially available materials that will decontaminate chemical agents to various degrees. Such materials and procedures are outlined in U.S. Army Technical Manual 3-220 (22).

The decontamination of chemical agents, and especially thickened agent, can be greatly enhanced by the thorough mixing of the decontaminant with the contaminant by use of agitation.

Table 4 is an alphabetical list of chemicals referred to in the text.

BIBLIOGRAPHY

"Chemicals in War" is treated in *ECT* 1st ed. under "Gas Warfare Agents," Vol. 7, pp. 117–145, by Rudolph Macy and co–workers, Chemical Corps., Chemical and Radiological Laboratories, Army Chemical Center, and H. A. Charipper, New York University and "Chemical Warfare" *ECT* 2nd ed. pp. 869–907, by Toivo E. Puro, United States Army, Edgewood Arsenal.

1. *Commander's Call, Chemical Warfare*, Department of the Army Pamphlet 360–831, Jan.–Feb. 1977.
2. J. H. Rothschild, *Tomorrow's Weapons*, McGraw-Hill Book Co., New York, 1964.
3. J. B. Kelly, *Gas Warfare in International Law*, Master's thesis, Georgetown University, Washington, D.C., June 1960.
4. B. H. Liddell-Hart, *A History of the World War, 1914–1918*, Little, Brown & Co., Boston, Mass., 1948.
5. *Chemistry*, in W. A. Noyes, Jr., ed., *Science in World War II*, Little, Brown & Co., Boston, Mass., 1948.
6. Stockholm International Peace Research Institute (SIPRI), *The Problem of Chemical and Biological Warfare, Volume LL: CB Weapons Today*, Humanities Press, New York, 1973.
7. U.S. Army Field Manual 3-9/U.S. Air Force Field Manual 355-7, *Military Chemistry and Chemical Compounds*, U.S. Government Printing Office, Washington, D.C., Oct. 1975.
8. R. N. Sterlin, V. I. Yemel'yanov, and V. I. Zimin, "Chemical Weapons and Defense Against Them," *Khim. Oruzhiye i Zashchita ot Nego*, 1975.
9. K. Ward, Jr., *J. Am. Chem. Soc.* **57,** 914 (1935).
10. S. B. Davis and W. F. Ross, *J. Am. Chem. Soc.* **69,** 1177 (1947).

11. B. Gehauf, *CWS Field Lab Memo 1-2-8*, National Technical Information Service, U.S. Department of Commerce, Washington, D.C., April 1943.
12. W. H. Stein and co-workers, *J. Org. Chem.* **11,** 664 (1946).
13. C. Golumbic and co-workers, *J. Org. Chem.* **11,** 518 (1946).
14. A. Gilman and F. S. Philips, *Science* **103,** 409 (1946).
15. B. Gehauf and co-workers, *Anal. Chem.* **29,** 278 (1957).
16. S. Sass and co-workers, *Anal. Chem.* **29,** 1346 (1957).
17. Executive Order 11850, *Renunciation of Certain Uses in War of Chemical Herbicides and Riot Control Agents,* April 8, 1975.
18. U.S. Army Technical Manual 3-1055-456-12, *Launcher, Rocket: 66MM,* 4-Tube, M202A1(NSN 1055-00-021-3909), U.S. Government Printing Office, Washington, D.C., Oct. 1969.
19. *Items of Combat Clothing and Equipment,* U.S. Army Pamphlet 385-3 (DA PAM 385-3), May 1976.
20. *Protective Clothing and Equipment,* Department of the Army Pamphlet 385-3 (DA PAM 385-3), May 1976.
21. V. M. Sim, *Diagnosis and Therapy for Anticholinesterase Poisoning and Other Chemicals Used in Warfare, Drill's Pharmacology in Medicine,* 4th ed., McGraw-Hill Book Co., New York, 1971.
22. U.S. Army Technical Manual 3-200, *Chemical, Biological, and Radiological (CBR) Decontamination,* U.S. Government Printing Office, Washington, D.C., 1967.

General References

A. Goodman and H. Martens, *Studies on the Use of Electric Eel Acetylcholinesterase for Anticholinesterase Agent Detection,* Edgewood Arsenal Report No. Ed-TR-74096, Feb. 1975.*
A. Silvestri and co-workers, *Development of a Kit for Detection of Hazardous Material Spills into Waterways,* Edgewood Arsenal Special Publication No. ED-SP-76023, Aug. 1976.*
J. S. Parsons and S. Mitzner, "Gas Chromatographic Method for Concentration and Analysis of Traces of Industrial Organic Pollutant in Environmental Air and Stacks," *Environ. Sci. Technol.* **9,** 1053 (Nov. 1975).
F. W. Karasek, Detection Limits in Instrumental Analysis, Research/Development, July, 1975.
J. E. Estes, *Remote Sensing Techniques for Environmental Analysis,* Hamilton Publishing Company, 1974.
G. P. Wright, *Designing Water Pollution Detection Systems,* Ballinger Publishing Company, Cambridge, Mass., 1974.
W. S. C. Chang, *Lasers and Applications,* Ohio University, 1963.
R. M. Gamson, D. W. Robinson, and A. Goodman, *Environ. Sci. Technol.* **7,** 1137 (1973).
L. H. Goodson and W. B. Jacobs, *Real Time Monitor, Immobilized Enzyme Alarm and Spare Parts,* Edgewood Arsenal Report No. ED-CR-77015, Feb. 1977.*
J. P. Mieure and M. W. Dietrich, "Determination of Trace Organics in Air and Water," *J. Chrom. Sci.* **13,** 559 (Nov. 1973).
B. J. Ehrlich and S. F. Spencer, *Development of an Automated Mustard Stack Monitor,* Edgewood Arsenal Report No. ED-CR-76084, Tracor, Inc., June 1976.*
L. Schwartz and co-workers, *Evaluation of M15/M18 Enzyme Detector Ticket System with Low Concentration of GB,* Edgewood Arsenal Report No. ED-TR-74018, June 1974.*
L. H. Goodson, *Feasibility Studies on Enzyme System for Detector Kits,* Edgewood Arsenal Report No. ED-CR-77019, Dec. 1976.*
H. W. Levin and E. S. Erenrich, *Enzyme Immobilization Alternatives for the Enzyme Alarm,* Edgewood Arsenal Report No. ED-CR-76005, Aug. 1975.*
R. M. Gamson and co-workers, *Detection of GB, VX and Parathion in Water,* Edgewood Arsenal Report No. ED-TR-74015, June 1974.*
D. P. Soule, *Agent Concentrator Feasibility Studies,* Edgewood Arsenal Report No. ED-CR-76075.*
J. W. Scales, *Air Quality Instrumentation,* Vols. I and II, Instrument Society of America 1974.
H. Tannenbaum, *Laser Applications in Remote Sensing, Proceedings of Society of Photo-Optical Instrumentation Engineers* **49,** 1975, *Impact of Lasers in Spectroscopy.*

---

* Available from National Technical Information Service, U.S. Department of Commerce, Springfield, Virginia 22151.

T. Hirschfeld and co-workers, "Remote Spectroscopic Analysis of Parts-Per-Million-Level Air Pollutants by Raman Spectroscopy," *Appl. Phys. Lett.* **22**(1), (Jan. 1973).
H. A. Walter, Jr., and D. F. Flanigan, "Detection of Atmosphere Pollutants: A Correlation Technique," *Appl. Opt.* **14**, (June 1975).

B. L. HARRIS
F. SHANTY
W. J. WISEMAN
Chemical Systems Laboratory, U.S. Department of Defense

CHEMILUMINESCENCE

Chemiluminescence is the emission of light from chemical reactions at ordinary temperatures (1). Chemiluminescent reactions produce a reaction intermediate or product in an electronically excited state, and radiative decay of the excited state is the source of the light (2). When the excited state is a singlet, the radiative process is identical to fluorescence; when the excited state is a triplet, phosphorescent emission results (3). Electronically excited states can emit ultraviolet or infrared radiation as well as visible light, and the definition of chemiluminescence is no longer restricted to visible light emission. Moreover, the formation of electronically excited reaction products can be detected by their photochemical reactions, even when radiation is negligible. Thus, chemiluminescence is a special case of the more general process of chemiexcitation. It is observed in liquid-, gas-, and solid-phase reactions.

Chemiexcitation is uncommon because most chemical reactions follow a ground-state potential energy surface and release chemical energy as vibrational excitation of ground-state products, which is observed as heat (4). Chemiluminescence is even less common because most electronically excited states decay to ground states by nonradiative processes (3). Nevertheless, a significant number of chemiluminescent reactions are known and a few, such as firefly, *Cypridina*, or bacterial bioluminescence, and peroxyoxalate chemiluminescence combine high chemiexcitation efficiency with high fluorescence yield to provide substantial light production.

Chemiluminescence has been studied extensively (2) for several reasons: (*1*) chemiexcitation relates to fundamental molecular interactions and transformations and its study provides access to basic elements of reaction mechanisms and molecular properties; (*2*) efficient chemiluminescence can provide an emergency or portable light source; (*3*) chemiluminescence provides means to detect and measure trace elements and pollutants for environmental control, or metabolites for disease detection; (*4*) classification of the bioluminescent relationship between different organisms defines their biological relationship and pattern of evolution; (*5*) since bioluminescence involves enzymatic catalysis, the effect of enzyme or substrate structural modification on the reaction is easily followed by measuring the emitted light, which facilitates study of enzyme mechanisms.

Mechanism

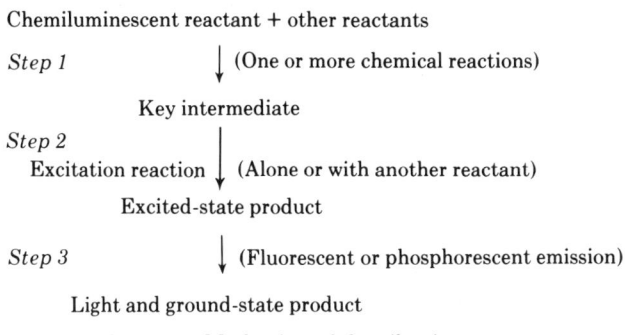

The mechanism of chemiluminescence is still being studied and most mechanistic interpretations should be regarded as tentative. Nevertheless, most chemiluminescent reactions can be classified into: (1) peroxide decomposition, including bioluminescence and peroxyoxalate chemiluminescence; (2) singlet oxygen chemiluminescence; and (3) ion radical or electron transfer chemiluminescence, which includes electrochemiluminescence.

In principle, one molecule of a chemiluminescent reactant can react to form one electronically excited molecule, which in turn can emit one photon of light. Thus one mole of reactant can generate Avogadro's number of photons defined as one einstein(ein). Light yields can therefore be defined in the same terms as chemical product yields, in units of einsteins of light emitted per mole of chemiluminescent reactant. This is the chemiluminescence quantum yield Qc which can be as high as 1 ein/mol or 100%.

The theoretical yield is approached by the firefly reaction, discussed later, which is reported to have a Qc of 88% (5). In practice, however, most chemiluminescent reactions are inefficient, with Qc values on the order of 1% or much less. The factors influencing yields can be discussed in terms of the generalized chemiluminescent mechanism shown in Figure 1(6).

As in any multistep process the overall yield Qc is the product of the yields of the separate steps. In most reactions, step 1 is subject to competitive side reactions which not only reduce the yield but can also obscure the nature of the chemiluminescent reaction itself. Step 2 requires a special chemistry which is discussed below. Step 3 is reasonably well understood from fluorescence and phosphorescence studies (7). Most molecules have low or negligible fluorescence quantum yields, Q_f, and phosphorescence is always inefficient in liquids. Thus, efficient chemiluminescence requires a selective reaction producing a key intermediate and efficient conversion of the key intermediate to the singlet excited state of a highly fluorescent product. The yield of the second step is called the excitation yield and the product of the yields of the first two steps is the yield of excited state. Interest focuses on the chemiexcitation, step 2. In general, efficient excitation requires a large energy release in a single-reaction step (8) and a reaction pathway that promotes crossing of the ground-state potential energy surface to an electronically excited potential energy surface (4).

Chemiluminescent reactant + other reactants

Step 1 ↓ (One or more chemical reactions)

Key intermediate

Step 2
Excitation reaction ↓ (Alone or with another reactant)

Excited-state product

Step 3 ↓ (Fluorescent or phosphorescent emission)

Light and ground-state product

Figure 1. Mechanism of chemiluminescence.

Energy Requirement. Visible light has an energy content of 167 kJ/ein (40 kcal/ein) (red) to 293 kJ/ein (70 kcal/ein) (blue), and an excited state radiating visible light must have that same energy with respect to its ground state. The excitation energy requirement is met by the sum of reaction enthalpy and activation energy. Few chemical reactions are sufficiently energetic. Moreover, release of the entire chemical energy requirement must occur in a single reaction step because excitation must be essentially instantaneous (8). Energetic two-step reactions where a part of the energy is released in each step cannot be chemiluminescent in solution because the energy released in the first step will be lost as vibrational energy to the solvent before the second step can raise the energy level to the excitation requirement.

The Chemiluminescent Pathway. Theory regarding the crossing of ground- to excited-state potential energy surfaces is still developing, with several potential criteria related to efficient chemiexcitation being considered. First, since the energy released by a reaction can evolve as either vibrational or electronic excitation energy, small or rigid product molecules, which have relatively few vibrational degrees of freedom, should favor electronic excitation (6,9). Most likely, the conversion of substantial chemical energy to low energy vibrational excited states is a "forbidden" process analogous to the low probability of the transfer of excitation energy to vibrational energy when the energy gap between available electronic and vibrational quantum states is large. Thus a large energy release combined with a paucity of vibrational modes should favor electronic excitation and chemiluminescence.

Second, excited-state molecular geometry is often different from ground-state geometry. A reaction producing a bent carbonyl group, eg, may favor chemiluminescence because the carbonyl excited state configuration is unfavorable compared to the planar ground state. Electronic excitation would then be preferred because it requires less molecular motion in the transition state (10–11). Orbital symmetry conservation and spin-orbit coupling may also be factors (12–13) (see under 1,2-Dioxetanes).

Third, singlet excitation, which is required for efficient chemiluminescence, may be favored over triplet excitation when the developing excited state is $\pi \rightarrow \pi^*$ rather than $n \rightarrow \pi^*$ (14). Finally, electron transfer between an anion radical–cation radical pair can produce a neutral excited state–ground state product pair, as discussed later, and it has been suggested that reactions of certain peroxides with electron rich fluorescers can produce an ion radical pair comprising the fluorescer cation radical and a carbonyl anion radical derived from the peroxide. Electron transfer within the solvent cage then provides the electronically excited fluorescer (15). Alternatively, it has been suggested that electron-rich fluorescers form charge transfer complexes with such peroxides, and that reversal of charge during peroxide decomposition is related to fluorescer excitation (6,9).

Liquid-Phase Chemiluminescence

PEROXIDE DECOMPOSITION

In many chemiluminescent reactions of peroxides, two carbonyl groups are formed simultaneously by decomposition of an intermediate such as compound (1):

9,10-diphenyl-9,10-dicarboxy-9,10-dihydroanthracene anhydride + H_2O_2 → (1) → (2) + 2 CO_2 + H_2O (1)

In such reactions the substantial heat of the simultaneous (concerted) formation of the carbonyl groups produced meets the energy requirement (8,16). In equation 1 (8) the product is the highly fluorescent excited state of 9,10-diphenylanthracene. Note that it is not necessary for the new carbonyl groups to be a part of the structure of the excited product, only that the excited state be formed synchronously with two carbonyl groups.

1,2-Dioxetanes. Simple dioxetanes (3) decompose thermally near or below room temperature to generate excited states of carbonyl products (17).

$$R_2C=O + [R_2C=O]^*$$
(3)

Excitation appears to be general for this reaction but yields of excited products vary substantially with the substituent, R. The highest yield reported is from tetramethyl-1,2-dioxetane (TMD) where the yield of triplet acetone is 50% of total acetone formed (18–19). Probably only one carbonyl of the two produced can be excited by the thermal decomposition and TMD provides 100% of the possible yield of triplet acetone. Singlet excited acetone is also formed, but at the low yield of 0.1–0.3% (17–21). Other tetraalkyldioxetanes behave similarly to TMD (22).

Since the fluorescence and phosphorescence radiative yields from acetone are very low, chemiluminescence is weak, with a quantum yield of 1×10^{-6} ein/mol (21). Light emission increases significantly when a fluorescent acceptor, such as 9,10-diphenylanthracene, is added to trap the singlet excited state by energy transfer and provide an efficient singlet emitter, or when a triplet acceptor, such as 9,10-dibromoanthracene, is added to trap some of the triplet product (17,19,20,23–24). Neither fluorescer changes the decomposition rate. In the latter case the heavy bromine substituent permits moderately efficient energy transfer from triplet acetone to moderately fluorescent singlet dibromoanthracene by weakening the spin conservation rule through spin–orbit coupling. Even with the addition of fluorescent acceptors, however, chemiluminescence efficiencies remain low because of the inefficient transfer processes at attainable fluorescer concentrations. The highest chemiluminescence quantum yield from TMD in the presence of diphenylanthracene is about 0.1% (20).

Most other dioxetanes investigated provide lower triplet yields, but some provide higher yields of excited singlets. Tetramethoxy-1,2-dioxetane gives only one third the triplet yield of TMD but gives a 1% yield of excited singlet dimethyl carbonate (20). The higher singlet yield of the methoxy derivative may relate to its higher heat of decomposition, −502 kJ/mol (−120 kcal/mol) vs −372 kJ/mol (−89 kcal/mol) for TMD. Despite its higher heat of decomposition, the methoxy derivative is more stable than TMD with a half-life of 14 h at 80°C compared to 14 min for TMD (20). Since the activation energies for both are nearly the same, about 117 kJ/mol (28 kcal/mol), the methoxy derivative clearly has a low activation entropy which may also relate to the higher singlet yield. The following reaction was reported to give an excited singlet yield of 0.9%. The singlet yield is increased substantially by silica gel, which also increases the reaction rate (25).

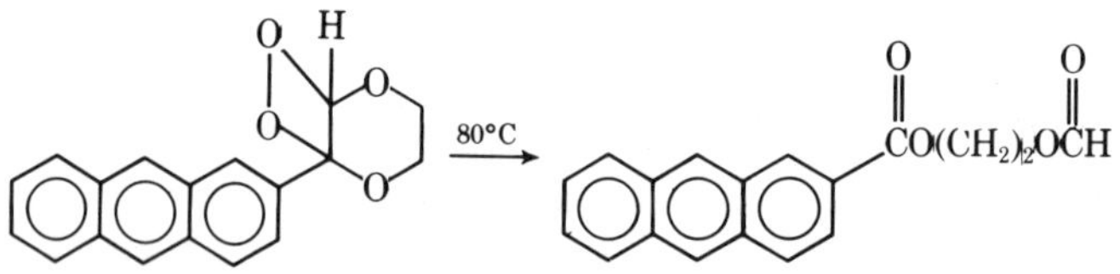

Dioxetanes that can decompose to N-methylacridone have been reported to give excited singlet yields as high as 25% (25).

Formation of the carbonyl group does not appear to be concerted with O–O bond cleavage in 1,2-dioxetane decomposition, since replacement of methyl in the 3-position with phenyl, which would conjugate with the forming carbonyl group and stabilize it, does not change the activation energy (26). Singlet excitation, however, may involve concerted decomposition.

Yields of excited states from 1,2-dioxetane decomposition have been determined by two methods. Using a photochemical method (17–18) excited acetone from TMD is trapped with trans-1,2-dicyanoethylene (DCE). Triplet acetone gives cis-1,2-dicyanoethylene with DCE, whereas singlet acetone gives 2,2-dimethyl-3,4-dicyanooxetane. By measuring the yields of these two products the yields of the two acetone excited states could be determined. The yields of triplet ketone (6) from dioxetanes are determined with a similar technique.

In principle, the excitation energy would be expected to be distributed between ketones (5) and (6) in a ratio dependent on the substituent R, and the distribution would be expected to favor the ketone having the lowest triplet excitation energy.

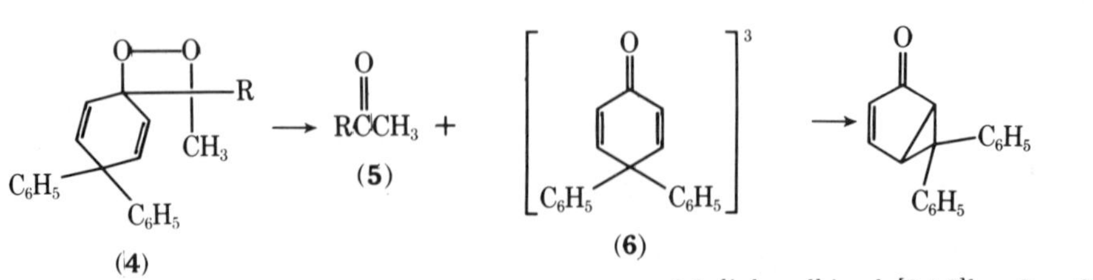

6,6-diphenylbicyclo[3.1.0]hex-3-en-2-one
4,4-diphenyl-2,5-cyclohexadien-1-one

However, contrary to expectation, triplet (**6**) was formed in 17% yield when its triplet energy was less than that of (**3**) (R = phenyl), and triplet (**6**) was still formed in 12% yield when ketone (**5**) (R = β-naphthyl) had the lower triplet energy. The $n \to \pi^*$ triplets, such as (**6**) might be favored over $\pi \to \pi^*$ triplets, such as (**5**) (R = β-naphthyl), even when energy considerations would indicate the opposite (26).

Excited state yields from several dioxetanes have been determined (19–20,24) by the chemiluminescent method (6). Singlet yields were determined by measuring the increasing chemiluminescence yields obtained on adding increasing concentrations of a singlet acceptor, such as 9,10-diphenylanthracene, and extrapolating the chemiluminescence yield to infinite fluorescer concentration where all of the singlet product would be trapped. Triplet yields were determined similarly by adding the triplet acceptor 9,10-dibromoanthracene.

1,2-Dioxetanes were isolated in 1969 (27). Previously, they had been expected to be thermally unstable in view of their high decomposition energies of 270–500 kJ/mol (65–120 kcal/mol). Many are kinetically stable at room temperature with activation energies near 96 kJ (23 kcal), but most decompose below 80°C, and careful synthesis and storage are necessary. An exception is adamantylideneadamantane-1,2-dioxetane which decomposes at about 140°C to give a 2% yield of singlet and 15% yield of triplet adamantanone (28).

In addition to ready thermal decomposition, 1,2-dioxetanes are also rapidly decomposed by transition metals (29–30), amines, and electron-donor olefins (10). However, these catalytic reactions are not chemiluminescent as determined by the temperature-drop kinetic method.

Ultraviolet light also initiates decomposition. Photolysis tends to give a higher yield of singlet products than thermolysis, with the singlet yield increasing with the energy of the exciting light. On the other hand, triplet sensitized photolysis gives exclusively triplet excited products (17). The triplet ketone from thermal decomposition also sensitizes dioxetane decomposition, except in the presence of a triplet quencher such as oxygen (19,24,31).

1,2-Dioxetanes are obtained from an α-halohydroperoxide by treatment with base (eq. 4)(32–33), or reaction of singlet oxygen with an electron-rich olefin such as tetraethoxyethylene (eq. 5) or (**7**) (eq. 6) (25,34).

A number of chemiluminescent reactions may proceed through unstable dioxetane intermediates (12,35). For example, the classical chemiluminescent reactions of lophine (eq. 7), lucigenin (eq. 8), and transannular peroxide decomposition (eq. 10).

Classical chemiluminescence from lophine (**9**) is derived from its reaction with oxygen in aqueous alkaline dimethyl sulfoxide or by reaction with hydrogen peroxide and a cooxidant such as sodium hypochlorite or potassium ferricyanide (36). The hydroperoxide (**10**) has been isolated and independently emits light in basic ethanol (37).

Classical chemiluminescence from lucigenin (**11**) is obtained from its reaction with hydrogen peroxide in water at a pH of about 10; Qc is reported to be about 0.5% based on lucigenin, but 1.6% based on the product N-methylacridone which is formed in low yield (38). Lucigenin dioxetane (**8**) has been prepared by the reaction of equation 9, ie, singlet oxygen addition to an electron-rich olefin at low temperature (39). Thermal decomposition of (**8**) gives a Qc of 1.6% (39). Several workers have provided evidence for the dioxetane intermediate (**12**) of equation 10 (40).

Tetrakis(dimethylamino)ethylene (**13**) (TMAE) reacts spontaneously with

422 CHEMILUMINESCENCE

$$(CH_3)_2C=C(CH_3)_2 \;+\; \text{1,3-dibromo-5,5-dimethylhydantoin} \;+\; H_2O_2 \longrightarrow$$

$$(CH_3)_2\overset{Br}{\underset{|}{C}}-\overset{OOH}{\underset{|}{C}}(CH_3)_2 \xrightarrow[CH_3OH,\,0°C]{NaOH} (CH_3)_2\overset{O-O}{\underset{|\quad\;\;|}{C}}-C(CH_3)_2 + HBr \quad (4)$$

$$(C_2H_5O)_2C=C(OC_2H_5)_2 \xrightarrow{O_2} (C_2H_5O)_2\overset{O-O}{\underset{|\quad\;\;|}{C}}-C(OC_2H_5)_2 \quad (5)$$

$$(7) \xrightarrow[CH_2Cl_2\,(-78°C)]{O_2\,+\,\text{sensitizer}} (8) \quad (6)$$

(7) 10,10′-dimethyl-9,9′-biacridan

(8) 10,10″-dimethyldispiro[acridine-9(10H), 3′ [1,2]dioxetane-4,9″(10″H)-acridine

oxygen to generate light (eq. 11) (41–45) with (13) itself being the emitting excited state. This is in agreement with an energy transfer process from an excited state produced by dioxetane decomposition to a second molecule of (13) (43). Although the reaction rate is first order in (13) and oxygen, the chemiluminescence intensity and Qc are second order in (13) (41), indicating that the second TMAE molecule is involved in a fast, nonrate-determining step. An alternative mechanism involving a chemical dimerization and regeneration of excited (13) through decomposition of the dimer has also been suggested to account for these results (41). Chemiluminescence quantum yields ranging from 10^{-5} to 10^{-3} ein/mol have been reported under varying conditions (41–43). The reaction products are primarily tetramethylurea and tetramethyloxamide (44), which increasingly quench chemiluminescence as the reaction proceeds (42).

Although Qc for reaction (11) is low, a significant yield can be maintained at high TMAE concentrations, and moderate brightness and lifetime can be achieved (see under Chemical Light Applications). Several syntheses of (13) and analogues have been described (45).

Dioxetane decomposition has also been proposed to account for chemiluminescence from other reactions (35), including gas-phase reactions of singlet oxygen with ethylene and vinyl ethers (46).

α-Peroxylactones (1,2-Dioxetanones). Alkyl substituted 1,2-dioxetanones (15) are prepared using low-temperature techniques by the reactions indicated in equations 12 (47) and 15 (48).

The α-hydroperoxy acids (14) can be prepared in high yield and cyclized to the dioxetanone (15) with dicyclohexylcarbodiimide in carbon tetrachloride at low temperatures.

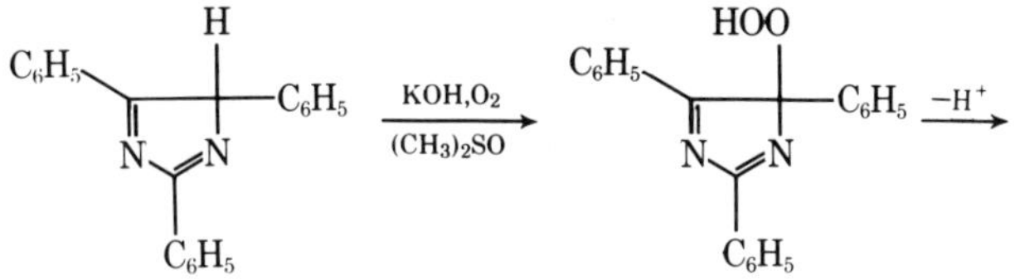

$$(7) + (O_2)' \longrightarrow (8) \qquad (9)$$

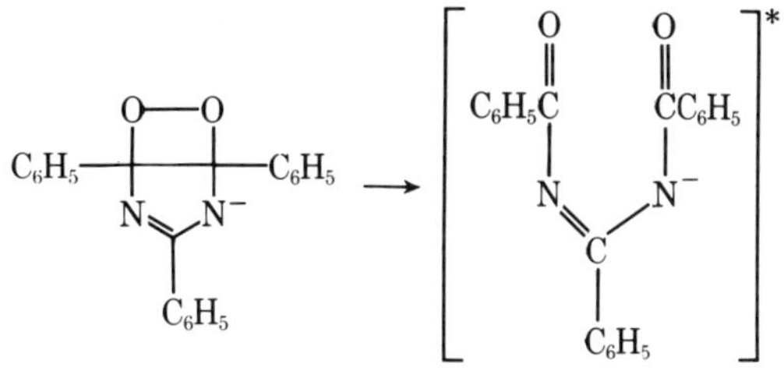

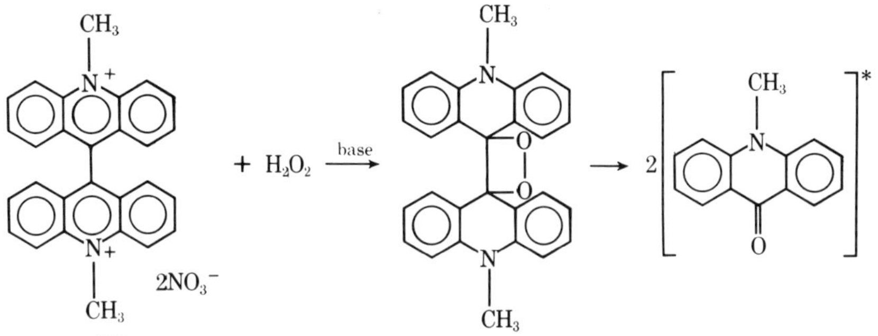

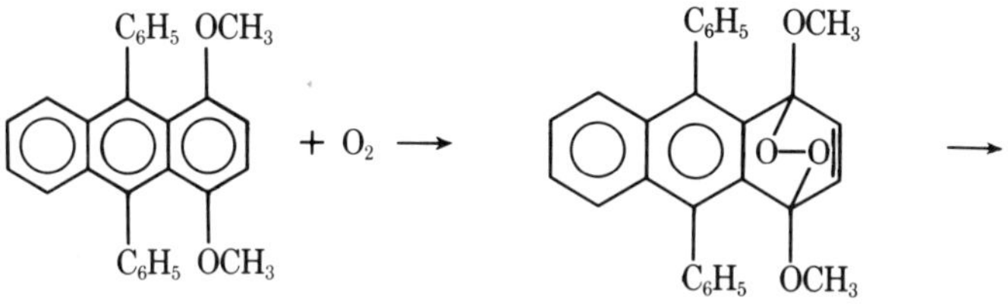

1,4-dihydro-1,4-dimethoxy-9,10-diphenyl-1,4-epidoxyanthracene

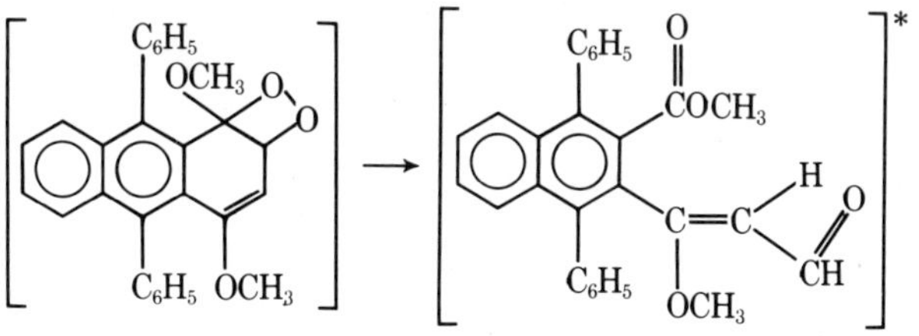

(12)

3-(2-formyl-1-methoxyvinyl)-1,4-diphenyl-2-napthoic acid, methyl ester

$$[(CH_3)_2N]_2C{=}C[N(CH_3)_2]_2 + O_2 \longrightarrow [(CH_3)_2N]_2C\underset{O-O}{-}C[N(CH_3)_2] \longrightarrow$$

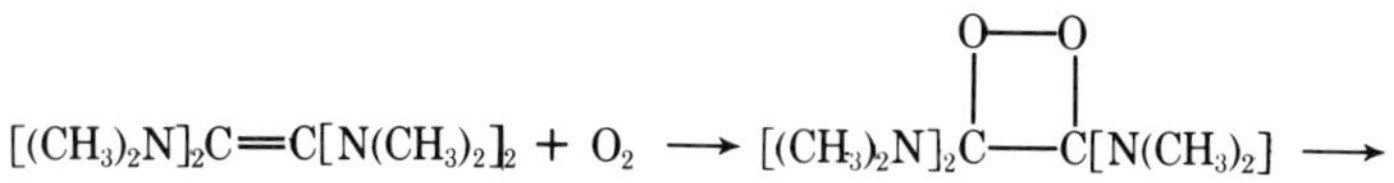

(13)

$$2\,([(CH_3)_2N]_2C{=}O)^* + (13) \longrightarrow [(CH_3)_2N]_2C{=}O + (10)^* \qquad (11)$$

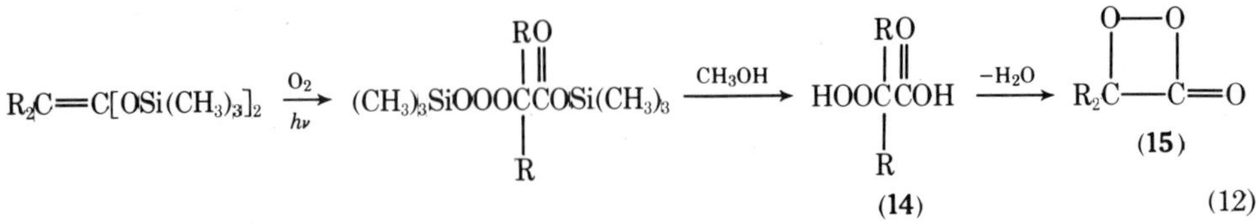

424 CHEMILUMINESCENCE

$$R_2C{=}C[OSi(CH_3)_3]_2 \xrightarrow[h\nu]{O_2} (CH_3)_3SiOOOC(R)(OR)COSi(CH_3)_3 \xrightarrow{CH_3OH} HOOC(R)(OR)COH \xrightarrow{-H_2O} R_2C\text{—}C(O\text{—}O){=}O$$

(14) (12) (15)

Dioxetanones decompose near or below room temperature to aldehydes or ketones (49). The decomposition reactions are weakly chemiluminescent (Qc ca 10^{-7} ein/mol) because the products are poorly fluorescent. However, addition of 10^{-3} M rubrene provides a Qc ca 10^{-3} ein/mol, and a Qc on the order of 3–7% was calculated at rubrene concentrations above 10^{-2} M after correcting for yield loss factors (50a). The decomposition rates are first order in (15) and are independent of added fluorescer at concentrations below 10^{-3} M, where Qc is about 10^{-3} ein/mol. At higher fluorescer concentrations, where Qc increases strongly, rubrene substantially increases the decomposition rate (50a) suggesting catalysis (see under 1,2-Dioxetanediones below). More recently, catalysis by fluorescence has been confirmed and an electron transfer mechanism proposed (50b).

Long before 1,2-dioxetanones were isolated, they were proposed as key intermediates in bioluminescence (51–53). This idea led to the discovery of a number of new chemiluminescent reactions. For example:

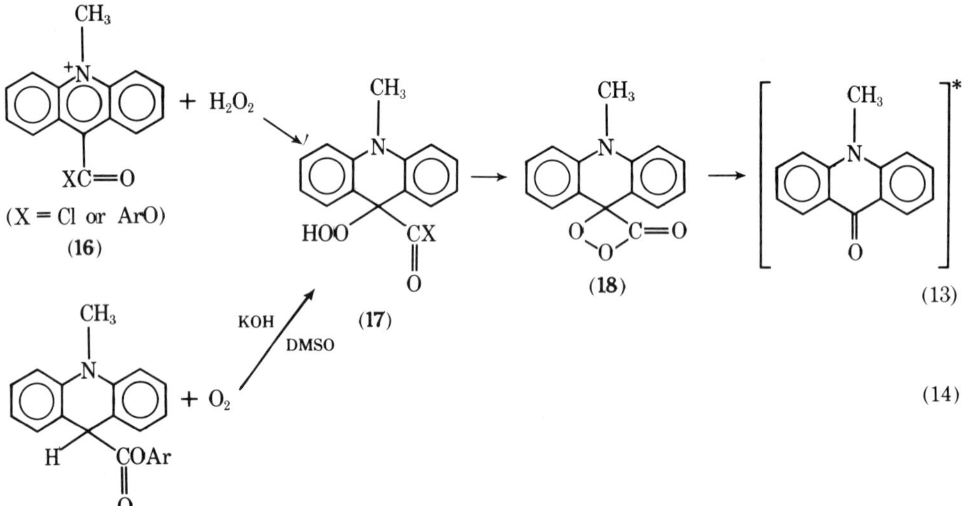

(16) + H$_2$O$_2$ → (17) → (18) → (13) (14)

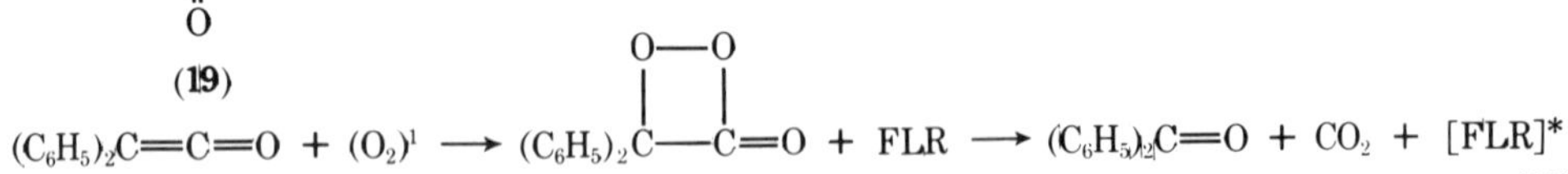

(X = Cl or ArO)
(16)

(19) + O$_2$ (KOH/DMSO) →

$$(C_6H_5)_2C{=}C{=}O + (O_2)^1 \rightarrow (C_6H_5)_2C\text{—}C{=}O + FLR \rightarrow (C_6H_5)_2C{=}O + CO_2 + [FLR]^*$$

(15)

(O$_2$, KOBu^t / DMSO) → → → (16) + CO$_2$

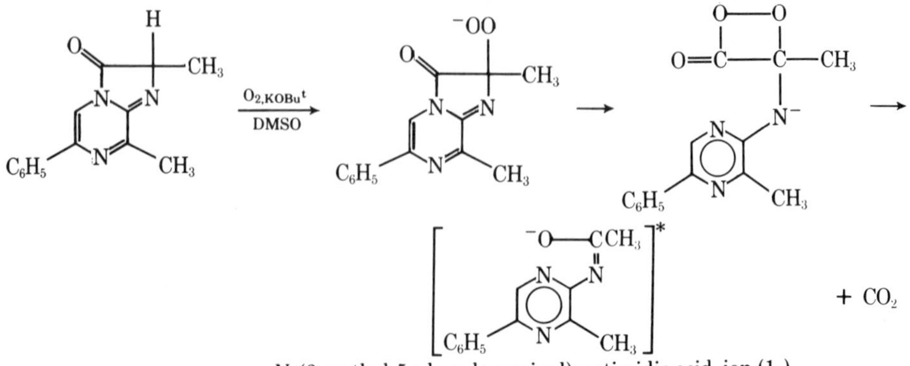

N-(3-methyl-5-phenylpyrazinyl)acetimidic acid, ion (1-)

Reactions 13 and 14 have the common key intermediate (18) (35). The hydroperoxide (17) has been isolated and is independently chemiluminescent under basic conditions (35). Both reactions are efficient with a Qc of 3% reported for (16) (X = Cl) (6) and a Qc of 10% reported for (19) (Ar = phenyl) (54). Discovery of (16) (X = Cl) was based independently on the dioxetanone (55) and concerted peroxide decomposition (6,8,56) theories (see under Mechanism). More recent work, discussed under Bioluminescence, indicates that reactions such as (14) may not involve dioxetanone intermediates in every case.

For reaction 15, it was proposed that singlet oxygen would add to ketenes to produce chemiluminescence in the presence of a fluorescer (FLR) via corresponding dioxetanones. Light was indeed observed using the triphenyl phosphite ozonide complex as the singlet oxygen source in the presence of the fluorescer 9,10-bis(phenylethynyl)anthracene(57). More recently, dioxetanones have been isolated from this reaction, which appears to be a relatively simple and general synthetic method (48) for these compounds.

Reaction 16 (58) and several analogues (59) were designed as a chemiluminescent model of *Cypridina* bioluminescence. Other possible examples of dioxetanones in bioluminescence are discussed below.

Peroxyoxalate (1,2-Dioxetanedione). Peroxyoxalate chemiluminescence (6) is illustrated by the reaction of an oxalic ester, such as bis(2,4,6-trichlorophenyl) oxalate, with hydrogen peroxide and a fluorescer, such as 9,10-diphenylanthracene or rubrene, in benzene or dimethyl phthalate (eq. 17).

$$\underset{\text{ROC—COR}}{\overset{\text{O O}}{\| \ \|}} + H_2O_2 \longrightarrow \underset{\text{ROC—COOH}}{\overset{\text{O O}}{\| \ \|}} + ROH$$

$$\downarrow$$

$$\underset{(20)}{\overset{\overset{\text{O O}}{\| \ \|}}{\underset{\text{O—O}}{\text{C—C}}}} \xrightarrow{\text{FLR}} 2\ CO_2 + [\text{FLR}]^* \qquad (17)$$

The mechanism indicated in equation 17 is tentative, and the proposed key intermediate 1,2-dioxetanedione(20) has not been detected. However, a considerable amount of experimental evidence is in agreement with the mechanistic proposal (6,9). A key aspect of the mechanism is the proposed catalytic decomposition of (20) by the fluorescer in the excitation step.

Peroxyoxalate chemiluminescence is the most efficient nonenzymatic chemiluminescent reaction known, with quantum yields as high as 22–27% having been reported for oxalate esters prepared from 2,4,6-trichlorophenol, 2,4-dinitrophenol, and 3-trifluoromethyl-4-nitrophenol (9,60–61) with the fluorescers rubrene (62–63) or 5,12-bis(phenylethynyl)naphthacene (63). The reaction is efficient because decomposition of a dioxetanedione–fluorescer complex generates the singlet excited state of the fluorescer with an excitation yield as high as 60% (6) and because high yield fluorescers can be used.

Moreover, the key intermediate is metastable toward unimolecular decomposition, and nonluminescent intermolecular side reactions can be minimized by careful selection of reaction conditions. Thus, the yield of excited fluorescer can be high at fluorescer concentrations as low as $10^{-3}\ M$. Dioxetanedione would seem to be an optimum chemiluminescence key intermediate (6) since reactions which simultaneously produce two carbonyl groups in small product molecules should be efficient sources of electronically excited reaction products (see under Mechanism).

High efficiency peroxyoxalate chemiluminescence requires an oxalic acid derivative with a good leaving group to facilitate ring closure to the dioxetanedione (6,9,60,64). Efficient chemiluminescent oxalates include electronegatively substituted aromatic and aliphatic esters (9,60–65), amides (66–68), sulfonamides (66,69), O-oxalyl hydroxylamine derivatives (70), mixed oxalic-carboxylic anhydrides (71), and oxalyl chloride (72). Among the amides, bis[1-(1H)-2-pyridonyl]glyoxal is particularly efficient with a quantum yield of 17% reported (67).

Since the fluorescer is independent of the key intermediate, a variety of fluorescers can be used to provide emission spectra encompassing the visible (62–63,73–74a) and near infrared regions (75). Excitation yields generally decrease as the excitation energy of the fluorescer increases, and Qc for higher energy (blue) fluorescers tends to be relatively low (6).

Most peroxyoxalate chemiluminescent reactions are catalyzed by bases and the reaction rate, chemiluminescent intensity, and chemiluminescent lifetime can be varied by selection of the base and its concentration. Weak bases such as sodium salicylate are generally preferred (76). Alternatively, weak acids and certain salts have been found to extend the lifetimes of inherently rapid reactions which occur with highly reactive esters, such as bis(2,4-dinitrophenyl)oxalate (77). A chemiluminescent demonstration based on the oxalic ester reaction has been described (78), and the reaction has been developed into a practical lighting system (see under Applications).

A number of other chemiluminescent reactions appear to be related to peroxyoxalate chemiluminescence although their mechanistic details may vary. For example, various chlorinated esters and ethers react with H_2O_2 and a fluorescer to emit light (79–82). Other examples are given in equations 18 and 19 (8,83–84).

Peroxyoxalate chemistry has been used to carry out photochemical reactions (74b) but does not appear to produce triplet excited states (74a).

Luminol (Phthalhydrazide). Chemiluminescence from luminol (3-aminophthalhydrazide) (21) and analogues, illustrated in equation 20 has been studied extensively (85).

Reaction takes place in aqueous solution with hydrogen peroxide and a supplemental oxidant including ferricyanide, hypochlorite, persulfate, or the hydroxyl radical generated from hydrogen peroxide, and a metal derivative such as hemin (86). Chemiluminescent reaction also takes place with oxygen and a strong base in a dipolar aprotic solvent such as dimethyl sulfoxide (87). Under both conditions Qc is about 1% (88–89).

The mechanism appears to follow equation 20 (88,90). Dianion (24) has been shown to be the emitting fluorescer (91), and reaction of luminol (21), with oxygen-18 in KOH–dimethylsulfoxide produced one labeled oxygen in each carboxylate group (92). A kinetic study of the reaction of (21) with aqueous alkaline persulfate indicated a one-step, two-electron oxidation of the mono anion of structure (21) to the azoquinone (22) (88), and Omote and co-workers (93) have demonstrated the presence of structure (22) during the chemiluminescent reaction. Compound (22) and several analogues have been synthesized (94–96) and have been shown to be chemiluminescent under luminol conditions. A charge transfer mechanism has been proposed (15).

A substantial effort has been applied to increasing Qc by structural modification (97). Recent work (98–99) has produced derivatives such as the phthalazine-1,4-diones (25) and (26) which have chemiluminescence quantum yields substantially higher than luminol. The fluorescence quantum yield of the dicarboxylate product from (26) is 14%, and the yield of singlet excited state is calculated to be 50% (99).

Organometallics. Arylmagnesium halides, especially bromides, react with oxygen in ether to generate light (100). p-Chlorophenylmagnesium bromide is the most efficient with Qc ca 10^{-4} ein/mol (101), and ArOOMgX is probably an intermediate (102); free radicals may be involved (103). The emitting species are brominated biphenyls (103) and, since these are only weakly fluorescent, the excitation yield must be high. Alkyl Grignard reagents are weakly chemiluminescent in reaction with oxygen (104). Weak chemiluminescence is also seen in reactions of aryl Grignard reagents with benzoyl peroxide (105) and nitro compounds (106).

Lithium diphenylphosphide and related organophosphides are chemiluminescent in reaction with oxygen (107). Chemiluminescence is observed from the solid phosphides.

Autooxidation. Liquid-phase oxidation of hydrocarbons, alcohols, and aldehydes by oxygen produces chemiluminescence in quantum yields of 10^{-8} to 10^{-10} ein/mol (108–110). Although the efficiency is low, the chemiluminescent reaction is important because it provides an easy tool for study of the kinetics and properties of autooxidation

(25) (3%)
2,3-dihydrobenzo[f]phthalazine-1,4-dione

(26) (7%)
2,3-dihydroperylo[1,12-fgh]phthalazine-1,4-dione

reactions including industrially important processes (108,111). The light is derived from combination of peroxyl radicals (112), which are primarily responsible for the propagation and termination of the autooxidation chain reaction. The chemiluminescent termination step for secondary peroxy radicals is indicated in equation 21.

$$2\ R_2CHOO\cdot \longrightarrow (27) \longrightarrow [R_2CO]^* + O_2 + HOCHR_2 \qquad (21)$$

Since ground-state oxygen is a triplet, spin conservation during the decomposition of transition state (27) must lead to an excited triplet-state ketone or (excited) singlet oxygen. The emitting species has been found to be the triplet excited state of the carbonyl product (112), although singlet oxygen has also been detected.

Since the chemiluminescence intensity can be used to monitor the concentration of peroxyl radicals, factors that influence the rate of autooxidation can easily be measured. Included are the rate and activation energy of initiation, rates of chain transfer in cooxidations, the activities of catalysts such as cobalt salts, and the activities of inhibitors (108).

Tertiary peroxyl radicals also produce chemiluminescence although with lower efficiencies. For example, the intensity from cumene autooxidation, where the peroxyl radical is tertiary, is a factor of 10 less than that from ethylbenzene (112). The chemiluminescent mechanism for cumene may be the same as for secondary hydrocarbons, however, since methylperoxy radical combination is involved in the termination step.

$$C_6H_5-C(CH_3)_2-CO\cdot \longrightarrow C_6H_5COCH_3 + CH_3\cdot + O_2 \longrightarrow CH_3OO\cdot$$

The primary methylperoxyl radical terminates according to chemiluminescent reaction (21).

Addition of fluorescent energy acceptors such as 9,10-dibromoanthracene substantially increases chemiluminescence intensity by transferring excitation energy (112–113), as is the case with dioxetanes.

$$\text{(structure)} + H_2O_2 \xrightarrow[\text{THF}]{R_3N} [(28)] \longrightarrow [(29)]^* + 2\ CO_2 \qquad (22)$$

Other Oxidation Reactions. Dihydrophthaloyl cyclic peroxides such as (28) appear to be efficient chemiluminescent key intermediates as indicated by an investigation of reaction 22 (114). Relatively weak chemiluminescence obtained from the excited singlet state of p-terphenyl (29) is strongly intensified by addition of an energy accepting fluorescer, such as 9,10-diphenylanthracene or perylene, which changes the emission spectrum to that of the added fluorescer. By determining Qc as a function of perylene concentration, the yield of excited singlet (29) was estimated to be about 1.5%. The triplet yield was estimated by a similar technique to be 35%. After correction for side reactions, the yield of excited (29) from (28) was estimated to be about 60%. It was pointed out that concerted decomposition of (28) by a sterically favored but electronically forbidden $2s + 2a + 2s$ path is predicted by orbital symmetry theory to produce excited (29) rather than the normally expected ground state.

Direct formation of an excited aromatic hydrocarbon has also been demonstrated in reaction 1 (8). Moderate chemiluminescence emission from singlet excited (2) was observed, and key intermediate (1) was inferred. In contrast to reaction 22 it is unlikely that reaction 1 can produce a cyclic peroxide as a key intermediate.

Decomposition of diphenoylperoxide (30) in the presence of a fluorescer such as perylene in methylene chloride at 24°C produces chemiluminescence matching the fluorescence spectrum of the fluorescer; Qc with perylene was reported to be $10 \pm 5\%$ (115). The reaction follows pseudo first-order kinetics with the observed rate constant increasing with fluorescer concentration according to $k_{obs} = k_1 + k_2 [\text{FLR}]$. Thus the fluorescer acts as a catalyst for peroxide decomposition, as in the case of peroxyoxalate chemiluminescence, with catalytic decomposition competing with spontaneous thermal decomposition. An electron transfer mechanism has been proposed (115).

$$\text{(30)} + \text{FLR} \longrightarrow \text{product} + CO_2 + [\text{FLR}]^* \qquad (23)$$

(30)

Weak to moderate chemiluminescence has been reported from a large number of other liquid-phase oxidation reactions (1,108,116). The list includes reactions of carbenes with oxygen (117), phenanthrene quinone with oxygen in alkaline ethanol (118), coumarin derivatives with hydrogen peroxide in acetic acid (119), nitriles with alkaline hydrogen peroxide (120), and reactions that produce electron-accepting radicals such as HO· in the presence of carbonate ions (121). In the latter, exemplified by the reaction of iron(II) with H_2O_2 and $KHCO_3$, the carbonate radical anion is probably a key intermediate and may account for many observations of weak chemiluminescence in oxidation reactions.

SINGLET OXYGEN

The electronically excited singlet state of oxygen (122) can be produced by passing ground-state (triplet) oxygen through a microwave discharge (123–124), by reaction of hydrogen peroxide with hypochlorite ion (125), by energy transfer from triplet excited states formed by irradiation to ground state oxygen (126), and by low-temperature thermal decomposition of the triphenyl phosphite–ozone complex (127). Two singlet states are produced: (a) $^1\Delta g$, a relatively long-lived, low-energy (92 kJ; 22 kcal) state with all its electrons paired, and (b) $^1\Sigma g^+$, a short-lived, higher-energy (159 kJ; 38 kcal) state with two singly-occupied orbitals (122). Both singlet states are chemiluminescent in the gas phase (128). Emission from $^1\Delta g$ occurs principally at 1269, 634, and 703 nm with the latter two bands derived from simultaneous decay of two $^1\Delta g$ molecules. Infrared emission from $^1\Sigma g^+$ is observed at 762 nm. Green emission is observed at 478 nm from simultaneous decay of a $^1\Delta^- g\ ^1\Sigma g^+$ pair. The red band at 634 nm is prominently visible in the aqueous hydrogen peroxide–hypochlorite reaction where the $^1\Delta g$ oxygen pair emits from the bubbles of product oxygen (128). The intensity of the red band is proportional to the square of the Δg oxygen concentration as would be expected for a double-molecule decomposition (128–129).

Chemiluminescence from $^1\Delta g$ oxygen can be strong at high concentrations, but addition of $5 \times 10^{-4}\ M$ violanthrone increases the intensity 100-fold (130). The emission spectrum, centered at 630 nm, is identical to violanthrone fluorescence and the intensity remains proportional to the square of the $^1\Delta g$ concentration. Energy pooling of two $^1\Delta g$ molecules is clearly also required for violanthrone excitation, but it appears that a different process is involved than in singlet oxygen emission itself. Neither singlet oxygen emission nor violanthrone fluorescence are quenched by oxygen, but the $^1\Delta g$-violanthrone chemiluminescence is strongly oxygen-quenched. It was suggested that one $^1\Delta g$ molecule excites violanthrone to its triplet state and that the triplet is excited further to the singlet by a second $^1\Delta g$ molecule in a subsequent step (123,130). Chemiluminescence from singlet oxygen and rubrene (131) probably involves the same mechanism.

Most likely singlet oxygen is also responsible for the red chemiluminescence observed in the reaction of pyrogallol with formaldehyde and hydrogen peroxide in aqueous alkali (132). It is also involved in chemiluminescence from the decomposition of secondary dialkyl peroxides and hydroperoxides (133), although triplet carbonyl products appear to be the emitting species (112).

ELECTRON TRANSFER CHEMILUMINESCENCE

Electron transfer reactions appear to be inherently capable of producing excited products when sufficient energy is released (134–137). This ability may be related to the speed of electron transfer, which is fast relative to atomic motion, so that vibrational excitation is inhibited (138).

Examples include luminescence from anthracene crystals subjected to alternating electric current (139), luminescence from electron recombination with the carbazole free radical produced by photolysis of potassium carbazole in a frozen glass matrix (140), reactions of free radicals with solvated electrons (135), and reduction of ruthenium(III)tris(bipyridyl) with the hydrated electron (141). Other examples include the oxidation of aromatic radical anions with such oxidants as chlorine or benzoyl peroxide (142–143), and the reduction of 9,10-dichloro-9,10-diphenyl-9,10-dihydroanthracene with the 9,10-diphenylanthracene radical anion (142,144). Many other examples of electron transfer chemiluminescence have been reported (136,145).

Stable anion radicals are easily prepared from aromatic hydrocarbons, eg, 9,10-diphenylanthracene, by electrochemical reduction in acetonitrile or dimethylformamide-containing electrolytes such as tetrabutylammonium perchlorate. Reversal of electrode polarity generates cation radicals from the hydrocarbon, and their reaction with the anion radical reservoir generates chemiluminescence (134–135,146). More simply, an alternating current may be used so that cation and anion radicals are continuously formed and annihilated to produce light and regenerate the original hydrocarbon (147). When hydrocarbons with stable ion radicals are used and impurities reactive with ion radicals are rigorously excluded, long-lasting electrochemiluminescence can be achieved. The oxidation-reduction potentials of the hydrocarbon can be determined by cyclic voltammetry and the stabilities of the ion radicals assessed (146).

Electrochemiluminescence is somewhat complicated in that three processes can produce light, depending on the energy released by the electron transfer process and the excitation energy of the aromatic hydrocarbon (134–135,146). In each case a charge transfer complex between the oppositely charged radicals is probably formed. If sufficient energy is available, the complex can dissociate to one ground-state molecule and one excited singlet molecule, and luminescence is relatively efficient. If only enough energy is available for triplet excitation, a triplet excited state results that can produce excited singlets by triplet–triplet annihilation. If insufficient energy is released even for triplet excitation, luminescence can still be produced by excimer (excited dimer) emission from the complex itself (148). In the first two cases, the luminescence spectrum matches the normal fluorescence spectrum of the hydrocarbon, whereas in the latter case typical, red-shifted, broad-band excimer emission results. Excitation energy transfer from an excimer produced by electrochemiluminescence to a europium chelate has been reported to produce narrow-band europium emission (149).

Electron transfer reactions producing triplet excited states can be diagnosed by a substantial increase in luminescence intensity produced by a magnetic field (150). The intensity increases because the magnetic field reduces quenching of the triplet by radical ions (137).

Under optimum conditions electron transfer can produce excited states efficiently. Triplet fluoranthrene was reported to be formed in nearly quantitative yield from reaction of fluoranthrene radical anion with the 10-phenylphenothiazine radical cation (151), and an 80% triplet yield was indicated for electrochemiluminescence of fluoranthrene by measuring triplet sensitized isomerization of *trans-* to *cis-*stilbene (152).

Electrochemiluminescence quantum yields of 8–10% from 9,10-diphenylanthracene and 14–20% from the 9,10-diphenylanthracene anion–thianthrene cation combination have been reported using the rotating-ring disk electrode technique (137,153).

Gas-Phase Chemiluminescence

Gas-phase chemiluminescence is illustrated by the classic sodium chlorine cool flame (154):

$$Na + Cl_2 \rightarrow NaCl + Cl\cdot \qquad (24)$$

$$Cl\cdot + Na_2 \rightarrow NaCl + [Na]^* \qquad (25)$$

Intense sodium D-line emission results from excited sodium atoms produced in a highly exothermic step (eq. 25) (155). Many gas-phase reactions of the alkali metals are chemiluminescent, in part because their low ionization potentials favor electron transfer to produce intermediate charge transfer complexes such as $[Cl^- \cdot Na_2^+]$ (156).

There appears to be an analogy with solution-phase electron transfer chemiluminescence in such reactions.

Excitation energy can be provided by kinetic (translational) and vibrational energies as well as from reaction enthalpy as demonstrated by molecular-beam experiments. An alkali metal vapor can be accelerated to energies above 20 eV and passed through a chamber containing a substrate gas at low temperature. The cross section for reaction can then be determined, as a measure of the probability for an excitation interaction, as a function of the beam energy and the substrate structure (156). Such reactions between an alkali metal M and substrate S (eq. 26) involve no net chemical change and simply convert kinetic energy to light via excited alkali metal atoms.

$$M + S \rightleftarrows [M^+ \cdot S^-] \rightarrow [M]^* + S \qquad (26)$$

The importance of electron transfer was indicated by the much lower threshold translational energy required to excite substrates with high electron affinities than, eg, required for noble gas substrates (157).

Investigation of excitation probability as a function of substrate structure revealed that reaction 26 was efficient for such gases as N_2, NO, CO, O_2, SO_2, and olefinic and aromatic hydrocarbons (156). Such molecules contain a vacant, weakly antibonding orbital which can accept electron transfer. The bond length in the charge transfer complex intermediate is short and a substantially higher energy (ca 3 eV) is required to produce the free ions $M^+ + S^-$ than to excite the alkali metal (1.6 eV for potassium and 2.1 eV for sodium). Thus, strong chemiluminescence is produced at kinetic energies between 1.6 and 3 eV in reaction between potassium and SO_2; however, the chemiluminescence weakens as ions are formed above 3 eV. A second set of substrates, including H_2, HCl, Cl_2, and saturated hydrocarbons is inefficient in exciting alkali metals through reaction 26 because the lowest-lying vacant orbitals are strongly antibonding, σ bond lengths in the intermediates are long, and the energy required for free-ion formation is only slightly above the energy required for excitation (156,158).

A study of interaction of sodium atoms with vibrationally excited nitrogen molecules at the intersection of two gas beams shows that conversion of vibrational energy to electronic excitation is substantially more efficient than conversion of translational energy (159). This has also been indicated in other reactions (160).

The use of molecular and atomic beams is especially useful in studying chemiluminescence because the results of single molecular interactions can be observed without the complications that arise from preceding or subsequent energy transfer collisions. Such techniques permit determination of active vibrational states in reactants, the population distributions of electronic, vibrational and rotational excited products, energy thresholds, reaction probabilities, and scattering angles of the products (161).

A number of chemiluminescent reactions have been studied by producing key reactants through pulsed electric discharge, by microwave dissociation, or by observing the reactions of atoms and free radicals produced in the inner cone of a laminar flame as they diffuse into the flame's cool outer cone (162–163). These are either combination reactions or atom transfer reactions involving transfer of chlorine (164) or oxygen atoms (161,165–167), the latter giving excited oxides.

The rates and chemiluminescent intensities of atom transfer reactions are proportional to the concentrations of the reactants, but the intensity is inversely proportional to the concentration of inert gas present. The latter quenches the excited state through collision with an efficiency dependent on the structure of the inert gas.

Chemiluminescence Q_c increases with temperature, indicating that excitation has a higher actuation energy than the ground state reaction (163). Such reactions generally provide banded, but very broad, emission spectra.

Electronic excitation from atom transfer reactions appears to be relatively uncommon, with most such reactions producing chemiluminescence from vibrationally excited ground states (168,170–172). Examples include reactions of oxygen atoms with carbon disulfide (171), acetylene (172), or methylene (172), all of which produce emission from vibrationally excited carbon monoxide. When such reactions are carried out at very low pressure (13 mPa or 10^{-4} torr), energy transfer is diminished, as with molecular beam experiments, so that the distribution of vibrational and rotational energies in the products can be discerned (170). Laser emission at 5 μm has been obtained from the reaction of methylene and oxygen initiated by flash photolysis of a mixture of SO_2, C_2H_2, and SF_6 (172).

Combination chemiluminescence is illustrated by equations 27–30 (162–163,173–174):

$$N\cdot + N\cdot \rightarrow N_2^* \text{ (nitrogen afterglow)} \quad (27)$$

$$O\cdot + NO \rightarrow NO_2^* \text{ (air afterglow)} \quad (28)$$

$$H\cdot + NO \rightarrow HNO^* \quad (29)$$

$$O + CO \rightarrow CO_2^{|*} \quad (30)$$

These reactions are often called afterglows in recognition of the luminescence which persists following the dissociation of molecules in an electric discharge. Such reactions are also often produced by microwave discharge dissociation.

Recombination reactions are highly exothermic and are inefficient at low pressures because the molecule, as initially formed, contains all of the vibrational energy required for redissociation (eq. 31). Addition of an inert gas increases chemiluminescence by removing excess vibrational energy by collision (eq. 32) (175–176). Thus, in the nitrogen afterglow chemiluminescence efficiency increases proportionally with nitrogen pressure at low pressures up to about 33 Pa (0.25 torr) (177). However, inert gas also quenches the excited product (eq. 33) and above about 66 Pa (0.5 torr) the two effects offset each other, so that chemiluminescence intensity becomes independent of pressure (175,178). ($\mp$ indicates vibrational excitation).

$$N\cdot + N\cdot \leftrightarrows [N_2]_1^* \quad (31)$$

$$[N_2]_1^* + N_2 \rightarrow [N_2]_2^* + N_2^\mp \quad (32)$$

$$[N_2]_2^* + N_2 \rightarrow N_2^\mp + N_2 \quad (33)$$

$$[N_2]_1^* \rightarrow N_2 + h\nu \quad (34)$$

$$[N_2]_2^* \rightarrow N_2 + h\nu \quad (35)$$

White Phosphorus Oxidation. Emission of green light from the oxidation of elemental white phosphorus in moist air is one of the oldest recorded examples of chemiluminescence. Although the chemiluminescence is normally observed from solid phosphorus, the reaction actually occurs primarily just above the surface with gas-phase phosphorus vapor. The reaction mechanism is not known, but careful spectral analyses of the reaction with water and deuterium oxide vapors indicates that the primary emitting species in the visible spectrum are excited states of $(PO)_2$ and HPO or DPO. Ultraviolet emission from excited PO is also detected (179).

Solid-Phase Chemiluminescence

Siloxene. Siloxene, obtained from reaction of calcium silicide with hydrochloric acid (184), is a yellow polymer with the basic formula $(Si_6H_6O_3)_n$. The silicon atoms are arranged in hexagonal rings joined in a laminar polymeric structure by oxygen atoms (185). The basic structure appears to be substituted randomly by hydroxyl and chlorine groups that affect the chemiluminescence spectrum and efficiency.

Siloxene is fluorescent and red chemiluminescence results from oxidation with ceric sulfate, chromic acid, potassium permanganate, nitric acid, and several other strong oxidants. The chemiluminescence spectrum peaks at 600 nm and has been reported (186) to give a maximum brightness of 3.43 cd/m² (1 footlambert).

Solid lithium organophosphides are chemiluminescent in reaction with oxygen (107).

Bioluminescence

Bioluminescence is characteristic of numerous marine and a few land organisms (180), extending from single cell microorganisms such as bacteria and dinoflagellates, to marine vertebrates, such as the hatchetfish. Certain fish, such as the flashlight fish which has a light organ under its eyes, use photobacteria symbiotically to generate light (181). Marine bioluminescence includes sponges, worms, crustaceans, corals, snails, squids, clams, shrimp, and jellyfish. Bioluminescent land species include fungi, centipedes, millipedes, worms, beetles, and fireflies.

Bioluminescence functions in mating (fireflies, the Bahama fireworm), in the search for prey (angler fish, Photinus fireflies), camouflage (hatchetfish, squid), and no doubt to aid deep water fish to see in the dark ocean depths.

The chemistry of bioluminescence is complex and not yet well understood. In general the reactions involve oxygen, a luciferin, and a luciferase enzyme (59,82). Other reactants are sometimes also essential. The reactants can be isolated from the organisms, and bioluminescence can be demonstrated and studied *in vitro*. Microchemical techniques are required since, eg, one ton of sea pansy must be dredged to obtain 1 mg of its luciferin (183). Different families use different luciferins and luciferases but relationships among different genera have been established (14). Most studies have been carried out with the American firefly (*Photinus pyralis*), the crustacean *Cypridina hilgendorfi*, the coelenterates *Renilla reformis* (sea pansy) and *Aequorea*, and the genus *Photobacterum*.

Firefly. Firefly bioluminescence is the most efficient bioluminescent reaction known, with Qc reported to be 88% (5), and λ_{max} at 562 nm (49). The reaction occurs according to equation 36 where luciferin (**31**) reacts with adenosine triphosphate (ATP) and the enzyme to give complex (**32**) of the adenylate ester (AMP = adenosine monophosphate). Reaction of the complex with oxygen produces the excited state of the highly fluorescent carbonyl compound (**33**) and carbon dioxide. In common with the specificity of other enzyme reactions, only the D(−) enantiomorph of luciferin produces light (182,187). Luciferin (**31**) has been synthesized (188) and its structure established. A number of firefly species appear to use the same luciferin (189), but the color of the emitted light can differ because of variations in enzyme structure (190).

The carbonyl compound (**33**) has also been synthesized, and its fluorescence spectrum has been shown to match the bioluminescence spectrum under equivalent

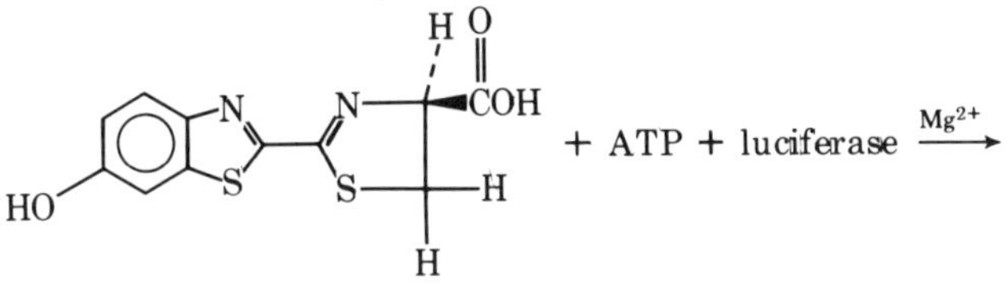

(31)
(R)-4,5-dihydro-2-hydroxy-
2-benzothiozolyl)-4-thiazole
carboxylic acid

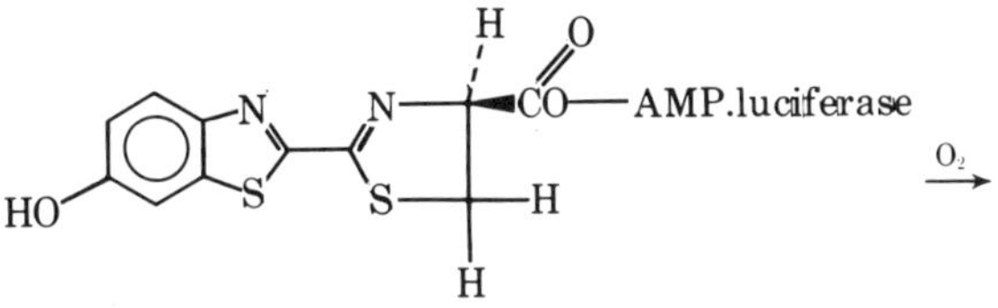

(32)

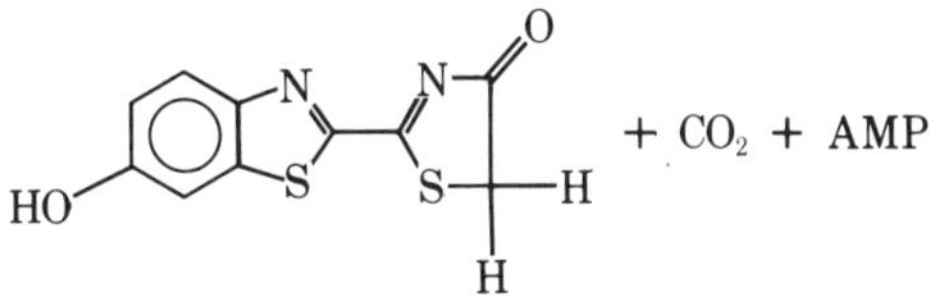

(33)

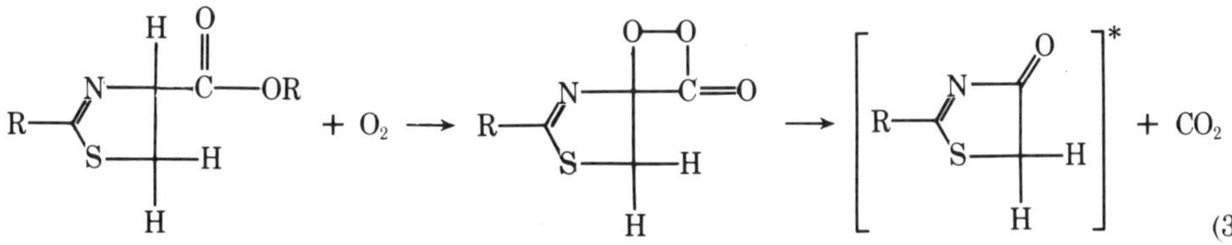

(37)

conditions (191). The details of the excitation step are unclear. The dioxetanone mechanism (eq. 37) (51–52,58) may apply to the reaction, but tracer studies with oxygen-18 indicate that the label does not appear in the CO_2 as required by this mechanism. Instead the oxygen introduced into CO_2 produced by the reaction is derived from water (192–193). The oxygen tracer results, however, have been questioned (194).

Chemiluminescence is also obtained by anionic autooxidation of (31) with oxygen in alkaline dimethyl sulfoxide (DMSO) (54). Qc has been reported to be 10% and ketone (33) and CO_2 are obtained. As in bioluminescence, the oxygen introduced into the CO_2 appears to be derived from water rather than oxygen (192–193). However, several analogues of luciferin have been prepared that are also chemiluminescent when they react with oxygen in alkaline DMSO (54), and in one case substantial oxygen-18 incorporation into the carbon dioxide product was observed (194).

Cypridina. In *Cypridina* the luciferin (34) reacts with oxygen and its luciferase enzyme to generate the excited state of amide (36) and by-products (37) and (38) (195). The amide is formed in 86% yield (195), Qc is 30% (196) and λ_{max} is at 465 nm (49). Luciferin (34) also produces chemiluminescence by reaction with oxygen in alkaline diglyme; however, Qc is only about 3% (197a).

When oxygen-18 is used, 80% of the label is incorporated into carbon dioxide in the aqueous, enzyme-catalyzed reaction, and 62% of the label is incorporated during

436 CHEMILUMINESCENCE

$$(34) + O_2 \xrightarrow{\text{luciferase}} (35) \rightarrow$$

$$CO_2 + (36) + (37) + R'''C-COH \quad (38)$$

$$\left(R' = \text{[indole]}, \ R'' = CH_2CH_2CH_2NHCNH_2, \ R''' = CH(CH_3)CH_2CH_3 \right)$$

chemiluminescence in DMSO with KOBu^t (197b). This result is in accord with the dioxetanone intermediate (35). Both luciferin (34) (198), and the amide (36) (199) have been synthesized. A number of analogues of structure (34) (58,197,200) have also been synthesized and shown to be chemiluminescent in organic solvents with base and oxygen.

Coelenterate. Coelenterates *Renilla reformis* (sea pansy), *Aequorea*, and others produce bioluminescence by similar processes (201). The basic *Renilla* luciferin structure is (39) (202) and excited amide (40) is the emitter. The structural relationship to *Cypridina* is evident (183). Although the substituent R has not been identified, a structural analogue where R = CH$_3$ is active in bioluminescence (183). The quantum yield is about 4% (201), with λ_{max} at 509 nm (49). Unlike *Cypridina*, carbon dioxide oxygen is reported to come from water rather than oxygen (192,203).

$$(39) + O_2 \xrightarrow[-CO_2]{\text{luciferase}} [(40)]^* \quad (39)$$

The jellyfish *Aequorea* can be extracted to yield a photoprotein, a complex of luciferin and enzyme that generates light on treatment with calcium ions (204). A structural unit related to *Renilla* luciferin has been isolated from this photoprotein (183).

Bacteria. Photobacteria enzymatically oxidize reduced flavin mononucleotide and a long-chain aldehyde to flavin mononucleotide and the carboxylic acid corresponding to the aldehyde (205–209). The quantum yield has been reported as 20% (209), but it is only half as much based on the reduced flavin as on the aldehyde (208), indicating that two reduced flavin molecules are involved in the stoichiometry (210). The quantum yield increases with aldehyde chain length up to about eight carbon atoms (206). The emitting fluorescer is the protonated flavin mononucleotide reaction product (211).

Latia. The fresh water snail *Latia* has been reported to provide bioluminescence by reaction 40 (212).

$$\text{(cyclohexene structure)}-(CH_2)_2C=CHO-CH + O_2 \longrightarrow \text{(cyclohexene structure)}-(CH_2)_2CCH_3 \quad (40)$$

4-(2,6,6-trimethyl-2-cyclohexene-1-yl) 2-butanone

Applications of Chemical Light

Marking and Illumination. Chemical light is well suited for lighting applications where distributed electric power is unavailable or hazardous (213). As discussed below, a substantial amount of light can be generated from a small, lightweight package easily transported and stored, and the cold light of chemiluminescence can be used safely in situations where a thermal light could cause fire or explosion. The many uses of chemical light include: as an emergency light for power failures in homes, office buildings, theaters, and factories, for disabled vehicles and aircraft, for lifeboats and lifejackets; as a marker light for pedestrians and bicycles on dark streets; and as a portable light for hikers, campers, and military units.

The product requirements for such applications include adequate brightness and lifetime, convenient utilization, low toxicity, high flash point, long shelf-life, and low cost (213).

The brightness and lifetime limits of a formulation are determined by its light capacity, L_{cap} in (lm·h)/L, which is defined as the integral of luminous intensity I, in lumens, with respect to reaction time t in hours for one liter of formulation (213):

$$L_{cap} = \int_0^\infty I \, dt / \text{vol} \qquad L_{cap} = 4.07 \times 10^4 \, C \cdot Qc \cdot P$$

C is the concentration of limiting reactant in mol/L, Qc is the chemiluminescence quantum yield in ein/mol, and P is a photopic factor that is determined by the sensitivity of the human eye to the spectral distribution of the light. Since the human eye is most responsive to yellow light, where the photopic factor for a yellow fluorescer such as fluorescein can be as high as 0.85, blue or red formulations have inherently lower light capacities.

As indicated earlier, high quantum yields are uncommon and, moreover, the quantum yield almost always decreases as the concentration of chemiluminescent reactant approaches practical levels. Thus, even reactions with high inherent quantum yields at low concentrations, such as the firefly reaction, do not necessarily provide high light capacities (213).

The theoretical limit of light capacity has been estimated for an ideal reaction that provides yellow light with a photopic factor of 0.85 in a quantum yield of one at 5 M concentration as 173,000 (lm·h)/L, equivalent to the light output of a 40 W bulb burning continuously for 2 weeks (213). The most efficient formulation now available, based on oxalic ester chemiluminescence and discussed below, produces about 0.5% of that limit, with a light capacity of 880 (lm·h)/L (213).

Since chemiluminescent brightness decreases as the reactants are consumed, not all of the light capacity is actually useful. Some light at the onset of reaction is emitted at intensities brighter than required by an application, and some light near the end of reaction is emitted below a threshold brightness requirement. A term called decay curve efficiency is used to quantify the fraction of total light capacity that is effective in practical use. The decay curve efficiency is the ratio of the area of largest rectangle that can be drawn under the intensity–time graph of an emission to the total area under the curve (214). Useful brightness and lifetime increase with this ratio for any given light capacity. Maximizing decay curve efficiency is an important goal in applied chemiluminescence. Efficiencies as high as 70% have been reported (214).

Several chemiluminescent reactions have been considered for illumination and marking. Luminol formulations have been proposed for air–sea rescue signaling (215) and a one-package, solid-state formulation comprising luminol, potassium persulfate, potassium perborate, potassium carbonate, and sodium fluorescein has been patented as a chemical light tablet which generates yellow light when added to water (216). The maximum light capacity of current luminol formulations, however, is only about 1.3 (lm·h)/L, because of low inherent Qc and the decrease in Qc at practical luminol concentrations (217).

Chemical light formulations based on air oxidation of tetrakis(dimethylamino)-ethylene (TMAE) have been used on aircraft emergency escape slides (218). The reaction is convenient to use, since the components only have to be exposed to air for activation. Maintenance of an oxygen-free environment during storage, however, can be difficult unless the application permits the use of sealed metal or glass containers (218).

The quantum yield for TMAE chemiluminescence is only about 0.1% (41–43) but, unlike other chemiluminescent reactions, light emission can be obtained at high TMAE concentrations (218). Light capacities of about 60 (lm·h)/L for practical high-concentration formulations have been estimated. Chemiluminescence of TMAE is quenched by its reaction products and is affected by moisture. A number of formulations have been devised to minimize these problems. In particular, solvents have been recommended (219) that dissolve TMAE but not the quenching tetramethyl-urea and tetramethyloxamide reaction products. Formulations that include alkali or alkaline earth group metal salts, such as lithium chloride, give improved performance in humid environments (220). The use of dehydrating agents (eg, calcium chloride) in combination with thickeners, such as finely divided silica, also improves high humidity performance (221). A formulation containing sulfolane is described as providing superior sub-zero temperature performance (222). A wax formulation (223), a polyethylene sponge impregnated with TMAE and lithium chloride (224), and an aerosol TMAE device have been patented (225).

The only chemical light reaction with significant use at present for illumination is the American Cyanamid Company's Cyalume lightstick (213). The Cyalume chemical light product is based on oxalic ester chemiluminescence (60):

This reaction has four special features which promote its utility: (1) the quantum yield is high, about 14% in practical formulations, and this high yield can be maintained at ester concentrations as high as 0.2M with certain fluorescers such as 1-chloro-9,10-bis(phenylethynyl)anthracene and certain esters, such as (41). Light capacities as high as 880 (lm·h)/L have been reported (73). A number of high light capacity esters and fluorescers have been described (64–65,73).

(2) The fluorescer is independent of the oxalic ester and can be selected separately to provide a desired color and to accommodate other practical requirements such as solubility, stability under the oxidizing conditions, excitation efficiency, fluorescence quantum yield, and shelf-life. Moreover, since the fluorescer acts as a catalyst and is recycled, it can be formulated in low concentrations (below 10^{-2} M) to avoid efficiency loss through fluorescence self-quenching, while permitting use of the high ester concentrations required for high light capacity. A number of fluorescers have been disclosed, including 9,10-diphenylanthracene (blue) (63), 9,10-bis(phenylethynyl)anthracene (green) (63–64,76), 1-chloro-9,10-bis(phenylethynyl)anthracene (yellow) (73), rubrene (orange) (62), and 5,12-bis(phenylethynyl)tetracene (red) (63).

(3) Since the reaction is accelerated by weak bases, such as sodium salicylate, the brightness–lifetime performance of the reaction can be varied to meet the needs of an application by varying the catalyst or its concentration (76).

(4) The reaction proceeds efficiently in a number of solvents facilitating selection in regard to safety requirements. Formulations using dimethyl or dibutyl phthalates are preferred because these solvents are essentially nontoxic and have high flash points (64,73,213).

The Cyalume lightstick is a 15-cm long, translucent, flexible plastic tube (226) containing a thin-walled glass ampoule (227), which floats in a solution of oxalate ester and fluorescer. The ampoule contains a dilute solution of hydrogen peroxide and catalyst. The plastic tube is easily bent to break the ampoule and mix the reactants. Light emission is immediate and, depending on the formulation, can last up to 12 hours or longer. The Cyalume green lightstick provides an intensity above 1 lumen up to 1 hour after activation and an intensity above 0.1 lumen up to 3 hours. The shelf-life is well in excess of two years. Water causes the formulation to deteriorate and the lightstick must be packaged in a hermetic aluminum foil plastic-laminate wrapper. As in any chemical reaction, the rate of oxalic ester chemiluminescence is temperature dependent, so that the lightstick is brighter warm than cold. An activated lightstick can be deactivated and preserved for many days by placing it in a freezer.

A number of patents disclose practical refinements to the oxalic ester system related to increasing curve-shape efficiency and light output (76–77,228), brightness-lifetime control (76–77) and lengthening of shelf-life (229). Variations in lightstick design (230), use of lightsticks in emergency lighting devices (231), and other means of packaging (232), displaying (233), or dispensing (234) chemical light formulations have been described.

A modified oxalic ester reaction that is activated by air rather than hydrogen peroxide has been provided by combining a 9,10-dihydroxyanthracene or benzoin with the ester and fluorescer (235). Oxygen from air is converted to hydrogen peroxide by the dihydroxyanthracene.

Analytical Applications. Chemiluminescence is useful in analytical chemistry for several reasons: (1) Modern low noise phototubes when properly instrumented can detect light fluxes as weak as 100 photons/s or 1.7×10^{-22} ein/s. Thus those chemiluminescent reactions in which intensity depends on the concentration of a reactant of analytical interest can be used to determine as little as 10^{-14} g of compound (236).

This is especially useful for biochemical, trace metal, and pollution control analyses (236–237) (see Trace and residue analysis). (2) Light measurement is easily automated for routine measurements as, for example, in clinical analysis. (3) Certain bioluminescent reactions are specific for their reactants, and such reactants are easily detected.

Direct Metal Analyses. Calcium ion can be detected to a lower limit of 10^{-7} M by *Aequorea* bioluminescence. Strontium interferes to a minor extent (238,239).

Divalent copper, cobalt, nickel and vanadyl ions promote chemiluminescence from the luminol–hydrogen peroxide reaction, which can be used to determine these metals to concentrations of 1–10 ppb (240–241). The light intensity is generally linear with metal concentration of 10^{-5} to 10^{-9} M range (240). Manganese(II) can also be determined when an amine is added to increase its reduction potential by stabilizing Mn(III) (240). Since all of these ions are active, ion exchange must be used for determination of a particular metal in mixtures (242).

Chromium(III) can be analyzed to a lower limit of 5×10^{-10} M by luminol–hydrogen peroxide without separating from other metals. Ethylenediaminetetraacetic acid (EDTA) is added to deactivate most interferences. Chromium(III) itself is deactivated slowly by complexation with EDTA; measurement of the sample after Cr(III) deactivation is complete provides a blank which can be subtracted to eliminate interference from such ions as iron(II), iron (III), and cobalt(II), which are not sufficiently deactivated by EDTA (243).

Iron(II) can be analyzed by a luminol–air reaction in the absence of hydrogen peroxide (244). Iron in the aqueous sample is reduced to iron(II) by sulfite; other metals which might interfere are also reduced to valence states that are inactive in the absence of hydrogen peroxide. The detection limit is 10^{-10} M.

Titration Indicators. Concentrations of arsenic(III) as low as 2×10^{-7} M can be measured (240) by titration with iodine, using the chemiluminescent iodine oxidation of luminol to indicate the end point. Oxidation reactions have been titrated using siloxene; the appearance of chemiluminescence indicates excess oxidant. Examples include titration of thallium (245) and lead (246) with dichromate and analysis of iron(II) by titration with cerium(IV) (247).

Hydrogen Peroxide Analysis. Luminol has been used for hydrogen peroxide analysis at concentrations as low as 10^{-8} M using the cobalt(III)triethanolamine complex (248) or ferricyanide (249) as promoter. With the latter chemiluminescence is linear with peroxide concentration from 10^{-8} to 10^{-5} M. Other luminol methods have also been described (250).

Hydrogen peroxide has also been analyzed by its chemiluminescent reaction with bis(2,4,6-trichlorphenyl) oxalate and perylene (see under Peroxyoxalate) in a buffered (pH 4–10) aqueous ethyl acetate-methanol solution (251). Using a flow system, intensity was linear from the detection limit of 7×10^{-8} M to at least 10^{-3} M.

Clinical Analyses. Glucose in urine has been determined by converting it to hydrogen peroxide with immobilized glucose oxidase and determining the hydrogen peroxide by either the luminol (249) or oxalic ester (251) methods. Uric acid interferes with the luminol analysis but not with the oxalic ester procedure.

Lactate dehydrogenase activity in serum has been determined by measuring its catalysis of the reduction of nicotinamide-adenine dinucleotide (NAD) by alkaline lactate using the oxalic ester method (251) which provides a light intensity that is proportional to lactate dehydrogenase activity (252). The detection limit for reduced

NAD is 2×10^{-7} M, and response is linear to 10^{-4} M.

Bacteria in urine can be detected in concentrations as low as 10^4 cells/mL, well below the clinically significant concentration of 10^5 cells/mL, using luminol (253) or firefly (254) reaction methods (see below).

Bacteria and Biomass Determination. Firefly bioluminescence specifically requires adenosine triphosphate (ATP) and has been used to determine ATP by its reaction with firefly luciferin, luciferase, and magnesium(II) in buffered aqueous oxygenated solutions (254–259). Since ATP is specific to living organisms, its analysis provides a rapid method for determining biomass concentrations. The emitted light is proportional to ATP concentration which in turn is dependent on biomass. The method has been suggested as a means for detecting life on other planets, although not yet used for that purpose (see Planetary exploration).

As little as 10^{-14} g of ATP can be detected with carefully purified luciferase. Commercial luciferase contains enough residual ATP to cause background emission and increase the detection limit to 10^{-12} g (259). The method has been used to determine bacterial concentrations in water. As few as 10^4 cells/mL of *Escherichia coli*, which contains as little as 10^{-16} g of ATP per cell, can be detected (259). More than 19 species of bacteria have been studied using this technique (254,256,259–260).

Other applications of firefly bioluminescence include measurement of the activity of bacteria in secondary sewage treatment activated sludge (261–262), detection of bacteria in clean rooms (see Sterile techniques) and operating rooms, measurement of bacteria in bottled foods and pharmaceuticals (254), determination of the antimicrobial activity of potential drugs (257), determination of the viability of seeds (258), and measuring marine biomass concentrations as a function of ocean depth or geographical location (263). The firefly method has also been used to determine inorganic phosphate in amounts as low as 10^{-11} mol (264).

Bacterial concentrations have also been determined by using the enzyme-catalyzed chemiluminescent reaction of reduced flavin mononucleotide (FMN) with oxygen and aldehydes. The detection limit was reported to be 10^5 cells of *E. coli*, which contains 7×10^{-17} g of FMN per cell (265).

Luminol chemiluminescence has also been recommended for measuring bacteria populations (266,267). The luminol–hydrogen peroxide reaction is catalyzed by the iron porphyrins contained in bacteria, and the light intensity is proportional to the bacterial concentration. The method is rapid, especially compared to the 2-day period required by the microbiological plate-count method, and it correlates well with the latter when used to determine bacteria in cooling tower water, paper pulp, activated sludge in secondary water treatment plants, drinking water and air. The limit of detectability depends somewhat on the water source, varying from 5×10^4 cells/mL for cooling water to 5×10^5 cells/mL in paper pulp. Using a concentration technique, bacterial counts as low as 6×10^3 cells/mL could be determined in distilled water (266).

Oxidation Analyses. Polymers, petroleum fuels, lubricating oils, foods, and other materials are degraded by air oxidation. Lifetimes depend on the conditions of use and on the inherent oxidizability of the material. The latter is conventionally determined by accelerated, high temperature techniques, since failure under use conditions may require many years of use. However, the autooxidation mechanism can change with temperature, so that the results of accelerated high temperature tests do not always correlate well with practical performance. A method based on a computer controlled, photon-counting apparatus measures the oxidizability of liquids and solids under use conditions by determining the minute chemiluminescence which accompanies autooxidation (268) (see under Autooxidation). Chemiluminescence from oxidation reactions as slow as 10^{-9} mol/yr in solids and 10^{-14} mol/yr in liquids can be detected. Since the chemiluminescence intensity is a function of oxidation rate, the latter can be determined to estimate the time to failure of the material. The method is also useful in determining the activities of antioxidants (qv).

Air Pollution Analyses. *Ozone.* Air pollution (qv) levels are commonly estimated by determining ozone through its chemiluminescent reaction with ethylene. A relatively simple photoelectric device is used for rapid routine measurements. The device is calibrated with ozone from an ozone generator, which in turn is calibrated by the reaction of ozone with potassium iodide (269,237). Detection limits were found to be 6–9 ppb with commercially available instrumentation (270).

Nitrogen oxides are major pollutants produced by cars, trucks and fossil fuel power plants. Nitric oxide (NO) can be determined by its chemiluminescent reaction with ozone using a photomultiplier detector. Total nitrogen oxides can be determined by thermally dissociating nitrogen dioxide (NO_2) to NO before measurement. Ammonia, which interferes with the latter determination, can be removed from the gas stream by oxidation with dichromate. The method thus permits continuous monitoring of NO, NO_2, and NH_3. Nitrogen oxide concentrations as low as 0.02 ppm are easily detected, and response is linear with concentration up to several percent (237,271); detectable limits are 22 μg m^{-3} (272).

Sulfur dioxide concentrations as low as 40 mg/m^3 in air have been determined by passing air samples through an aqueous solution of tetrachloromercurate, which converts SO_2 to the dichlorosulfitomercurate complex. Oxidation of the complex by potassium permanganate is chemiluminescent and the intensity, as measured by a photomultiplier, is proportional to sulfur dioxide concentration (273).

Total sulfur in air, most of which is sulfur dioxide, can be measured by burning the sample in a hydrogen-rich flame and measuring the blue chemiluminescent emission from sulfur atom combination to excited S_2 (237,274). Concentrations of about 0.01 ppm can be detected.

Hydrocarbons. The gas-phase chemiluminescent reaction of oxygen atoms with hydrocarbons has been proposed as a method for measuring hydrocarbon concentrations in automobile exhaust (275). Olefinic, saturated, and aromatic hydrocarbons are detected, although the method is most sensitive to olefins. The detection limit is 0.2 ppm ethylene equivalent and response is linear with concentration to >1000 ppm. Contaminants such as CO, CO_2, SO_2, and NO_2 do not interfere nor do methane or acetylene, which are thought not to be important contributers to photochemical smog. The method involves combining atomic oxygen, generated by passing an oxygen-helium mixture through a microwave discharge, with the hydrocarbon sample in a reaction zone monitored by a photomultiplier tube through 308.9 and 312.3 nm interference filters. Emission from olefins is detected at 308.9 nm, emission from acetylene is detected at both wavelengths. The 312.2 nm signal is subtracted from the 308.9 nm signal to eliminate the contribution at 308.9 nm from acetylene. Emission at both wavelengths arises from electronically excited hydroxyl radicals, but the vibrational excitation differs for the two reactions, causing a spectral difference. The concentration of atomic oxygen for calibration was also measured by chemiluminescence, using the reaction of atomic oxygen with excess NO, where intensity is proportional to oxygen concentration.

Nickel Carbonyl. The extremely toxic gas nickel carbonyl can be detected at 0.01 ppb by measuring its chemiluminescent reaction with ozone in the presence of carbon monoxide. The reaction produces excited nickel(II) oxide by a chain process which generates many photons from each pollutant molecule to permit high sensitivity (276).

Notice

The information and statements herein are believed to be reliable but are not to be construed as a warranty or representation for which we assume legal responsibility. Users should undertake sufficient verification and testing to determine the suitability for their own particular purpose of any information or products referred to herein.

No warranty of fitness for a particular purpose is made.

Nothing herein is to be taken as permission, inducement, or recommendation to practice any patented invention without a license.

BIBLIOGRAPHY

1. E. N. Harvey, *A History of Luminescence,* American Philosophical Society, Philadelphia, Pa., 1957.
2. E. J. Bowen, *Pure Appl. Chem.* **9,** 473 (1964); K. D. Gunderman, *Top. Current Chem.* **46,** 61 (1974); E. H. White and co-workers, *Angew. Chem. Intern. Ed. Engl.* **13,** 229 (1974); M. J. Cormier, D. M. Hercules, and J. Lee, eds., *Chemiluminescence and Bioluminescence,* Plenum Press, New York, 1973; F. McCapra in W. Carruthers and J. K. Sutherland, eds., *Progress in Organic Chemistry,* John Wiley & Sons, Inc., New York, 1973, p. 231.
3. R. M. Hochstrasser and G. B. Porter, *Quart. Rev. (London)* **14,** 146 (1960); C. A. Parker, *Photoluminescence of Solutions,* Elsevier, New York, 1968; G. G. Guilbault, *Practical Fluorescence; Theory, Method and Techniques,* Marcel Dekker, New York, 1973.
4. M. G. Evans, H. Eyring, and J. F. Kincaid, *J. Chem. Phys.* **6,** 349 (1938).
5. H. H. Seliger and W. D. McElroy, *Arch. Biochem. Biophys.* **88,** 136 (1960).
6. M. M. Rauhut, *Acc. Chem. Res.* **2,** 80 (1969).
7. C. A. Parker, *Advan. Photochem.* **2,** 305 (1964).
8. M. M. Rauhut and co-workers, *Photochem. Photobiol.* **4,** 1097 (1965); U.S. Pats. 3,470,103 (Sept. 30, 1969); 3,637,784 (Jan. 25, 1972); 3,914,255 (Oct. 21, 1975), D. Sheehan (to American Cyanamid).
9. M. M. Rauhut and co-workers, *J. Am. Chem. Soc.* **89,** 6515 (1967).
10. D. C. Lee and T. Wilson in M. J. Cormier, D. M. Hercules, and J. Lee, eds., *Chemiluminescence and Bioluminescence,* Plenum Press, New York, 1973, p. 265.
11. R. W. Dixon, *Diss. Faraday Soc.* **35,** 105 (1963).
12. F. McCapra, *Chem. Commun.,* 155 (1968).
13. D. R. Kearns, *J. Am. Chem. Soc.* **91,** 6554 (1969); W. H. Richardson and H. E. O'Neal, *J. Am. Chem. Soc.* **94,** 8665 (1972); D. R. Kearns, *Chem. Rev.* **71,** 395 (1971); M. J. S. Dewar and S. Kirshner, *J. Am. Chem. Soc.* **96,** 7578 (1974); C. Eaker and J. Hinze, *Theor. Chem. Acta* **40,** 113 (1975); D. R. Roberts, *Chem. Commun.,* 683 (1974); N. J. Turro and A. Devaquet, *J. Am. Chem. Soc.* **97,** 3859 (1973).
14. J. W. Hastings and T. Wilson, *Photochem. Photobiol.* **23,** 461 (1976).
15. J. Koo and G. B. Schuster; *J. Am. Chem. Soc.* **99,** 6107 (1977); **100,** 4496 (1978); K. A. Zalika, A. L. Thayer, and A. P. Schaap, *J. Am Chem. Soc.* **100,** 4916 (1978); F. McCapra, *Chem. Commun.,* 946 (1977); J. Michl, *Photochem. Photobiol.* **25,** 141 (1977).
16. F. McCapra, and D. G. Richardson, *Tetrahedron Lett.* 3167 (1964).
17. N. J. Turro and P. Lechtken, *Pure Appl. Chem.* **33,** 363 (1973); N. J. Turro and co-workers, *Acc. Chem. Res.* **7,** 97 (1974); W. Adams, *Adv. Heterocyclic Chem.* **21,** 438 (1977).
18. N. J. Turro and P. Lechtken, *J. Am. Chem. Soc.* **94,** 2886 (1972); P. Lechten, A. Yekta, and N. J. Turro, *J. Am. Chem. Soc.* **95,** 3027 (1973).
19. N. J. Turro and co-workers, *J. Am. Chem. Soc.* **96,** 1623 (1974).
20. T. Wilson and co-workers, *J. Am. Chem. Soc.* **98,** 1086 (1976).
21. W. Adam, *J. Am. Chem. Soc.* **97,** 5464 (1975).
22. E. J. Becharct, A. L. Baumstark, and T. Wilson, *J. Am. Chem. Soc.* **98,** 4648 (1976).
23. N. J. Turro and H. Steinmetzer, *J. Am. Chem. Soc.* **96,** 4679 (1974).

24. T. Wilson and A. P. Schaap, *J. Am. Chem. Soc.* **93**, 4126 (1971).
25. A. P. Schaap, P. A. Burns, and K. A. Zaklika, *J. Am. Chem. Soc.* **99**, 1270 (1977); *ibid.*, **100**, 318 (1978); F. McCapra and co-workers, *Chem. Commun.*, 944 (1977).
26. W. H. Richardson, M. B. Yelvington, and H. E. O'Neal, *J. Am. Chem. Soc.* **94**, 1619 (1972); H. E. Zimmerman and G. E. Keck, *J. Am. Chem. Soc.* **97**, 3527 (1975).
27. K. R. Kopecky and C. Mumford, *Can. J. Chem.* **47**, 709 (1969).
28. G. B. Schuster and co-workers, *J. Am. Chem. Soc.* **97**, 7110 (1975); J. H. Wieringa and co-workers, *Tetrahedron Lett.* 169 (1972).
29. T. Wilson and co-workers, *J. Am. Chem. Soc.* **95**, 4765 (1973).
30. P. D. Bartlett, A. L. Baumstark, and M. E. Landis, *J. Am. Chem. Soc.* **96**, 5557 (1974).
31. N. J. Turro and P. Lechtken, *J. Am. Chem. Soc.* **94**, 2888 (1972); P. Lechtken, A. Yekta, and N. J. Turro, *J. Am. Chem. Soc.* **95**, 3027 (1973).
32. K. R. Kopecky and co-workers, *Can. J. Chem.* **53**, 1103 (1975).
33. *Ibid*, **51**, 468 (1973).
34. (a) P. D. Bartlett and A. P. Schaap, *J. Am. Chem. Soc.* **92**, 3223, 6055 (1970); (b) S. Mazur and C. S. Foote, *J. Am. Chem. Soc.* **92**, 3225 (1970); (c) W. H. Richardson, M. B. Yelvington, and H. E. O'Neal, *J. Am. Chem. Soc.* **94**, 1619 (1972).
35. F. McCapra in W. Carruthers and J. K. Sutherland, eds., *Progress in Organic Chemistry*, Vol. 8, John Wiley & Sons, Inc., New York, 1973, p. 231; *Pure Appl. Chem.* **24**, 611 (1970).
36. E. W. Evans, *J. Chem. Ed.* **14**, 236 (1937); G. E. Philbrook and M. A. Maxwell, *Tetrahedron Lett.*, 1111 (1964); I. Nicholson and P. Poretz, *J. Chem. Soc.* 3067 (1965); T. Hayashi and K. Maeda, *Bull. Chem. Soc. Jpn.* **36**, 1052 (1963); *Bull. Chem. Soc. Jpn.* **35**, 2057 (1962); *Bull. Chem. Soc. Jpn.* **33**, 565 (1960).
37. E. H. White and M. J. Harding, *J. Am. Chem. Soc.* **86**, 5686 (1964); J. Sonnenberg and D. M. White, *J. Am. Chem. Soc.* **86**, 5685 (1964).
38. J. R. Totter, *Photochem. Photobiol.* **3**, 231 (1964).
39. K. Lee, L. A. Singer, and K. D. Legg, *J. Org. Chem.* **41**, 2685 (1976).
40. T. Wilson, *Photochem. Photobiol.* **10**, 441 (1970); G. W. Lundeen and A. H. Adelman, *J. Am. Chem. Soc.* **92**, 3914 (1970).
41. A. N. Fletcher and C. A. Heller, *J. Phys. Chem.* **71**, 1507 (1967); *J. Catal.* **6**, 263 (1966).
42. H. E. Winberg, J. R. Downing, and D. D. Coffman, *J. Am. Chem. Soc.* **87**, 2054 (1965); J. P. Paris, *Photochem. Photobiol.* **4**, 1059 (1965).
43. W. H. Urry and J. Sheeto, *Photochem. Photobiol.* **4**, 1067 (1965).
44. N. Wiberg and J. W. Buckler, *Z. Naturforsch.* **19b**, 5 (1964).
45. R. L. Pruett and co-workers, *J. Am. Chem. Soc.* **72**, 3646 (1950); N. Wiberg and J. W. Bucher, *Naturwissenschaften* **19b**, 953 (1964); H. E. Winberg and co-workers, *J. Am. Chem. Soc.* **87**, 2055 (1965); U.S. Pat. 3,239,519 (Mar. 8, 1966), H. E. Winberg (to E. I. du Pont de Nemours & Co., Inc.).
46. D. J. Bogan, R. S. Shienson, and F. W. Williams, *J. Am. Chem. Soc.* **98**, 1034 (1976); D. J. Bogan and co-workers, *J. Am. Chem. Soc.* **97**, 2560 (1975).
47. W. Adam and J. C. Liu, *J. Am. Chem. Soc.* **94**, 2894 (1972); W. Adam and H. C. Steinmetzer, *Angew. Chem. Intern. Ed. Engl.* **11**, 540 (1972); W. Adam and co-workers, *J. Am. Chem. Soc.* **99**, 5768 (1977).
48. N. J. Turro and co-workers, *J. Am. Chem. Soc.* **99**, 5836 (1977); W. Adam and co-workers, *J. Am. Chem. Soc.* **99**, 5768 (1977).
49. W. Adam, *J. Chem. Ed.* **52**, 138 (1975).
50. (a) W. Adam, C. A. Simpson, and F. Yang, *J. Phys. Chem.* **78**, 2559 (1974); (b) G. B. Schuster and S. P. Schmidt, *J. Am. Chem. Soc.* **100**, 1966, 5559 (1978); W. Adam, O. Cueto, and F. Yang, *J. Am. Chem. Soc.* **100**, 2587 (1978).
51. T. A. Hopkins and co-workers, *J. Am. Chem. Soc.* **89**, 7148 (1967).
52. F. McCapra, T. C. Chang, and V. P. Francois, *Chem. Commun.*, 22 (1968).
53. E. H. White and co-workers, *J. Am. Chem. Soc.* **91**, 2178 (1969).
54. F. McCapra, *Acc. Chem. Res.* **9**, 201 (1976).
55. F. McCapra, D. G. Richardson, and Y. C. Chang, *Photochem. Photobiol.* **4**, 1111 (1965).
56. M. M. Rauhut and co-workers, *J. Org. Chem.* **30**, 3587 (1965); U.S. Pat. 3,352,791 (Nov. 14, 1967), D. Sheehan, R. A. Clarke, and M. M. Rauhut (to American Cyanamid); U.S. Pat. 3,539,574 (Nov. 11, 1970) (to American Cyanamid).
57. L. Bollyky, *J. Am. Chem. Soc.* **92**, 3230 (1970).
58. F. McCapra and Y. C. Chang, *Chem. Commun.*, 1011 (1967).

59. T. Goto, *Pure Appl. Chem.* **17,** 421 (1968).
60. U.S. Pat. 3,597,362 (Aug. 3, 1971), L. J. Bollyky and M. M. Rauhut (to American Cyanamid).
61. U.S. Pat. 3,704,231 (Nov. 28, 1972), L. J. Bollyky (to American Cyanamid).
62. U.S. Pat. 3,701,738 (Oct. 11, 1972), B. G. Roberts and M. M. Rauhut (to American Cyanamid).
63. U.S. Pats. 3,729,426 (Apr. 24, 1973); 3,557,233 (Jan. 19, 1971), A. Zweig and D. R. Maulding (to American Cyanamid).
64. U.S. Pat. 3,749,679 (July 31, 1973), M. M. Rauhut (to American Cyanamid).
65. U.S. Pat. 3,781,329 (Dec. 25, 1973), L. J. Bollyky (to American Cyanamid); 3,816,326 (June 11, 1974), L. J. Bollyky (to American Cyanamid).
66. D. R. Maulding and co-workers, *J. Org. Chem.* **33,** 250 (1968).
67. L. J. Bollyky and co-workers, *J. Org. Chem.* **34,** 836 (1969); U.S. Pat. 3,843,549 (Oct. 22, 1974), L. J. Bollyky (to American Cyanamid).
68. U.S. Pat. 3,442,815 (May 6, 1969), M. M. Rauhut and L. J. Bollyky (to American Cyanamid).
69. U.S. Pat. 3,400,080 (Sept. 3, 1968), D. R. Maulding (to American Cyanamid).
70. L. J. Bollyky, R. H. Whitman, and B. G. Roberts, *J. Org. Chem.* **33,** 4266 (1968); U.S. Pat. 3,909,440 (Sept. 30, 1975), L. J. Bollyky and R. H. Whitman (to American Cyanamid); U.S. Pat. 3,978,079 (Aug. 31, 1976), L. J. Bollyky and R. H. Whitman (to American Cyanamid).
71. L. J. Bollyky and co-workers, *J. Am. Chem. Soc.* **89,** 5623 (1967); U. S. Pats. 3,399,137 (Aug. 27, 1968), 3,804,891 (Apr. 16, 1974), M. M. Rauhut and L. J. Bollyky (to American Cyanamid).
72. E. A. Chandross, *Tetrahedron Lett.* 761 (1963); M. M. Rauhut, B. G. Roberts, and A. M. Semsel, *J. Am. Chem. Soc.* **88,** 3604 (1966); U.S. Pat. 3,325,417 (June 13, 1967), M. M. Rauhut (to American Cyanamid); U.S. Pats. 3,442,813 (May 6, 1969), 3,644,517 (Feb. 22, 1972), L. J. Bollyky (to American Cyanamid).
73. U.S. Pat. 3,888,786 (June 10, 1975), D. R. Maulding (to American Cyanamid).
74. (a) P. Lechtken and N. J. Turro, *Mol. Photochem.* **6,** 95 (1974); (b) H. Guston and E. E. Ullman, *Chem. Commun.*, 28 (1977)
75. M. M. Rauhut and co-workers, *J. Org. Chem.* **40,** 330 (1975); U.S. Pat. 3,630,941 (Dec. 28, 1971), W. R. Bergmark (to American Cyanamid).
76. U.S. Pats. 3,775,336 (Nov. 27, 1973); 3,704,231 (Nov. 18, 1972), L. J. Bollyky (to American Cyanamid).
77. U.S. Pat. 3,691,085 (Sept. 12, 1972), B. G. Roberts and M. M. Rauhut (to American Cyanamid).
78. A. G. Mohan and N. J. Turro, *J. Chem. Ed.* **51,** 528 (1974).
79. D. R. Maulding and B. G. Roberts, *J. Org. Chem.* **37,** 1458 (1972).
80. U.S. Pat. 3,677,957 (July 18, 1972), D. R. Maulding (to American Cyanamid).
81. U.S. Pats. 3,697,432 (Oct. 10, 1972), 3,894,050 (July 8, 1975), D. R. Maulding (to American Cyanamid).
82. U.S. Pat. 3,749,677 (July 31, 1973), D. R. Maulding (to American Cyanamid).
83. U.S. Pat. 3,425,949 (Feb. 4, 1969), M. M. Rauhut and G. W. Kennerly (to American Cyanamid).
84. U.S. Pat. 3,329,621 (July 4, 1967), M. M. Rauhut and A. M. Semsel (to American Cyanamid).
85. For phthalhydrazide chemiluminescence reviews see references 86, 88, 94–95.
86. W. R. Vaughan, *Chem. Rev.* **43,** 496 (1948); A. Bernanose, T. Bremer, and P. Goldfinger, *Bull. Soc. Chem. Belges.* **56,** 269 (1947); *Dis. Faraday. Soc.* **15**(2), 221 (1947); *Bull. Soc. Chim. Fr.* **15,** 946 (1948).
87. E. H. White, *J. Chem. Ed.* **34,** 275 (1957).
88. M. M. Rauhut, A. M. Semsel, and B. G. Roberts, *J. Org. Chem.* **31,** 2431 (1966).
89. J. Lee and H. H. Seliger, *Photochem. Photobiol.* **4,** 1015 (1965).
90. H. D. Drew, *Trans. Faraday Soc.* **35,** 207 (1939).
91. E. H. White and M. M. Bursey, *J. Am. Chem. Soc.* **86,** 941 (1964).
92. E. H. White and co-workers, *J. Am. Chem. Soc.* **86,** 940 (1964).
93. Y. Omote, T. Miyake, and N. Sugiyama, *Bull. Chem. Soc. Jpn.* **40,** 2446 (1967).
94. K. D. Gundermann in ref 10, p. 209.
95. E. H. White and D. F. Roswell, *Acc. Chem. Res.* **3,** 54 (1970); E. H. White and R. B. Brundrett in ref. 10, p. 231.
96. K. D. Gundermann, H. Fiege, and G. Klockenbring, *Ann.* **738,** 140 (1970); **734,** 200 (1971).
97. K. D. Gundermann and M. Drawert, *Chem. Ber.* **95,** 2018 (1962); H. D. K. Drew and F. H. Pearman, *J. Chem. Soc.,* 586 (1937); H. D. K. Drew and R. E. Garwood, *J. Chem. Soc.*, 836 (1939); B. E. Cross and H. D. Drew, *J. Chem. Soc.*, 1532 (1949).
98. K. D. Gundermann, W. Horstman, and G. Bergman, *Ann.* **684,** 127 (1965).
99. C. C. Wei and E. H. White, *Tetrahedron Lett.*, 3559 (1971).

100. R. T. Dufford, S. Calvert, and D. Nightingale, *J. Am. Chem. Soc.* **45,** 2058 (1923); *J. Opt. Soc. Am.* **9,** 405 (1924); W. V. Evans and E. M. Diepenhorst, *J. Am. Chem. Soc.* **48,** 715 (1926).
101. R. L. Bardsley and D. M. Hercules, *J. Am. Chem. Soc.* **90,** 4545 (1968).
102. C. Walling and S. A. Buckler, *J. Am. Chem. Soc.* **77,** 6032 (1955).
103. P. H. Bolton and D. R. Kearns, *J. Am. Chem. Soc.* **96,** 4651 (1974).
104. R. T. Dufford, *J. Am. Chem. Soc.* **50,** 1822 (1928).
105. H. Gilman and C. E. Adams, *J. Am. Chem. Soc.* **47,** 2816 (1925).
106. H. Gilman, J. McGlumphy, and R. E. Fothergill, *Rec. Trav. Chem.* **49,** 526, 726 (1930).
107. K. Issleib and A. Tzschach, *Chem. Ber.* **92,** 1118 (1959); R. A. Strecker, J. L. Snead, and G. P. Sallot, *J. Am. Chem. Soc.* **95,** 210 (1973).
108. V. Ya. Shlyapintokh and co-workers, *Chemiluminescence Techniques in Chemical Reactions*, Consultants Bureau, New York, 1968.
109. V. Ya. Shlyapintokh and co-workers, *J. Chim. Phys. (Russ.)* **57,** 1113 (1960).
110. R. E. Kellogg, *J. Am. Chem. Soc.* **91,** 5433 (1969).
111. A. A. Vichutinskii, *J. Phys. Chim. (Russ)* **38,** 1242 (1964); O. N. Karpukhin, V. Ya. Shlyapintokh, and I. V. Mikhailov, *J. Phys. Chim. (Russ)* **38,** 81 (1964).
112. V. A. Belyakov and R. F. Vassil'ev, *Photochem. Photobiol.* **11,** 179 (1970); R. F. Vassil'ev, *Optics Spect.* **18,** 131,254 (1965); R. F. Vassil'ev and A. A. Vichutinskii, *Nature* **194,** 1276 (1962).
113. R. F. Vassilev, *Nature* **196,** 668 (1962); **200,** 773 (1963). G. Lundeen and R. Livingston, *Photochem. Photobiol.* **4,** 1085 (1965); V. A. Belyakov and R. F. Vassil'ev, *Photochem. Photobiol.* **6,** 35 (1967); S. R. Abbott, S. Ness, and D. M. Hercules, *J. Am. Chem. Soc.* **92,** 1128 (1970); J. H. Helberger and D. B. Hever, *Chem. Ber.* **72B,** 11 (1939); P. Rothemund, *J. Am. Chem. Soc.* **60,** 2005 (1938); G. D. Dorough, J. R. Muller, and F. F. Huennekens, *J. Am. Chem. Soc.* **73,** 4315 (1951); H. Linschitz and E. W. Abrahamson, *Nature* **72,** 909 (1953).
114. G. B. Schuster, *J. Am. Chem. Soc.* **99,** 651 (1977).
115. J. Koo and G. B. Schuster, *J. Am. Chem. Soc.* **99,** 6107 (1977); J. P. Smith and G. B. Schuster, *J. Am. Chem. Soc.* **100,** 2564, 4496 (1978).
116. K. D. Gundermann, *Top. Current Chem.* **46,** 61 (1974); *Naturwissenschaften* **56** 62 (1969); *Chemilumineszenz Organisher Verbindungen*, Springer Verlag, New York, 1968; *Angew. Chem. Int. Ed. Engl.* **11,** 566 (1965).
117. E. Wasserman, L. Barash, and W. A. Yager, *J. Am. Chem. Soc.* **87,** 4974 (1965).
118. B. Lachowicze, *Chem. Ber.* **16,** 332 (1883).
119. W. Dilthey, and W. Hoschen, *J. Prakt. Chem.* **38**(27), 42 (1933).
120. E. McKeown and W. A. Waters, *Nature* **203,** 1063 (1964).
121. J. Stauff, V. Sander and W. Jaeschke in ref. 10, p. 131.
122. C. S. Foote, *Acc. Chem. Res.* **1,** 104 (1968); D. R. Kearns, *Chem. Rev.*, **71,** 395 (1971).
123. E. A. Ogryzlo and A. E. Pearson, *J. Phys. Chem.* **72,** 2913 (1968).
124. E. J. Corey and W. C Taylor, *J. Am. Chem. Soc.* **86,** 3881 (1964); R. P. Wayne, *Adv. Photochem.* **7,** 311 (1969).
125. R. J. Browne and E. A. Ogryzlo, *Can. J. Chem.* **43,** 2915 (1965); C. S. Foote and S. Wexler, *J. Am. Chem. Soc.* **86,** 3879 (1964).
126. C. S. Foote, S. Wexler, and W. Ando, *Tetrahedron Lett.* **4111** (1965).
127. E. Wasserman and co-workers, *J. Am. Chem. Soc.* **90,** 4160 (1968); R. W. Murray and M. L. Kaplan, *J. Am. Chem. Soc.* **90,** 4161 (1968); **91,** 5385 (1969).
128. S. J. Arnold, E. A. Ogryzlo, and H. Witzke, *J. Chem. Phys.* **40,** 1769 (1964); R. J. Browne and E. A. Ogryzlo, *Proc. Chem. Soc. (London)* 117 (1964); A. V. Khan and M. Kasha, *J. Am. Chem. Soc.* **92,** 3293 (1970).
129. S. J. Arnold and E. A. Ogryzlo, *Can. J. Phys.* **45,** 2053 (1967).
130. R. J. Browne and E. A. Ogryzlo, *Can. J. Chem.* **43,** 2915 (1965).
131. T. Wilson, *J. Am. Chem. Soc.* **91,** 2387 (1969).
132. E. J. Bowen and R. A. Lloyd, *Proc. Chem. Soc. (London)* 305 (1963).
133. I. Stauff, H. Schmidkunz, and G. Hartman, *Nature* **198,** 281 (1963); E. McKeown and W. A. Waters, *Nature* **203,** 1063 (1964); E. J. Bowen, *Nature* **201,** 180 (1964); R. A. Lloyd, *Trans. Faraday Soc.* **61,** 2173 (1965); R. F. Vasilev, *Prog. Reaction Kinetics*, **4,** 304 (1967); J. A. Howard and K. U. Ingold, *J. Am. Chem. Soc.* **90,** 1057 (1968); S. R. Abbott, S. Ness, and D. M. Hercules, *J. Am. Chem. Soc.* **92,** 1128 (1970).
134. D. M. Hercules, *Acc. Chem. Res.* **2,** 301 (1969); A. Zweig, *Adv. Photochem.* **6,** 425 (1968).
135. E. A. Chandross, *Trans, New York Acad. Sci. Ser II* **31,** 571 (1969).
136. A. Weller and K. Zachariasse in ref. 10, p. 181, p. 193.

137. A. J. Bard and co-workers in ref. 10, p. 193.
138. E. A. Chandross and F. I. Sonntag, *J. Am. Chem. Soc.* **88**, 1089 (1966); D. M. Hercules, R. C. Lansbury, and D. K. Roe, *J. Am. Chem. Soc.* **88**, 4578 (1966); R. A. Marcus, *J. Chem. Phys.* **43**, 2654 (1965).
139. D. C. Northrop and O. Simpson, *Proc. Roy. Soc. (London)* **234**, 124 (1956); H. P. Kallmann and M. Pope, *J. Chem. Phys.* **36**, 2482 (1962); W. Helfrich and W. G. Schneider, *J. Chem. Phys.* **44**, 2902 (1966).
140. G. N. Lewis and J. Bigeleisen, *J. Am. Chem. Soc.* **65**, 2424 (1943); H. Linschitz, M. G. Berry, and D. Schweitzer, *J. Am. Chem. Soc.* **76**, 5833 (1954).
141. J. E. Martin and co-workers, *J. Am. Chem. Soc.* **94**, 9238 (1972).
142. E. A. Chandross and F. F. Sonntag, *J. Am. Chem. Soc.* **86**, 3179 (1964); **88**, 1089 (1966).
143. U.S. Pat. 3,391,069 (July 2, 1968), M. M. Rauhut and G. W. Kennerly (to American Cyanamid); T. D. Santa Cruz, D. L. Akins, and R. L. Birke, *J. Am. Chem. Soc.* **98**, 1677 (1976).
144. U.S. Pat. 3,391,068 (July 21, 1968), M. M. Rauhut (to American Cyanamid); C. P. Keszthelyi and A. J. Bard, *J. Org. Chem.* **39**, 2936 (1974).
145. A. Weller and K. Zachariasse, *J. Chem. Phys.* **46**, 4984 (1967); *Chem. Phys. Lett.* **10**, 424 (1971).
146. D. M. Hercules, *Science* **145**, 808 (1964); D. M. Hercules and F. F. Lytle, *J. Am. Chem. Soc.* **88**, 4745 (1966); J. Chang, D. M. Hercules, and D. K. Roe, *Electrochem. Acta* **13**, 1197 (1968); R. E. Visco and E. A. Chandross, *J. Am. Chem. Soc.* **86**, 5350 (1964); *Electrochem. Acta* **13**, 1187 (1968). S. V. Santhanum and A. J. Bard, *J. Am. Chem. Soc.* **87**, 3259 (1965); A. Zweig and co-workers, *J. Am. Chem. Soc.* **88**, 2864 (1966); **89**, 4091 (1967); D. L. Maricle and A. H. Maurer, *J. Am. Chem. Soc.* **89**, 188 (1967). A. Zweig, A. H. Maurer, and B. G. Roberts, *J. Org. Chem.* **32**, 1322 (1967); A. Zweig and co-workers, *J. Am. Chem. Soc.* **89**, 473 (L967); A. Zweig and co-workers, *Chem. Commun.* 106 (1967).
147. U.S. Pat. 3,319,132 (May 9, 1967), E. A. Chandross and R. E. Visco (to Bell Telephone Laboratories); U.S. Pat. 3,654,525 (Apr. 4, 1972), D. L. Maricle and M. M. Rauhut (to American Cyanamid); U.S. Pat. 3,816,795 (June 11, 1974); D. L. Maricle and M. M. Rauhut (to American Cyanamid); U.S. Pat. 3,900,418 (Apr. 19, 1975), A. J. Bard and N. E. Takvoryan (to Bell-Northern Research, Ltd.).
148. E. A. Chandross, J. W. Longworth, and R. E. Visco, *J. Am. Chem. Soc.* **87**, 3259 (1965); A. Weller and K. Zachariasse, *J. Chem. Phys.* **46**, 4984 (1967); S. M. Park and A. J. Bard, *J. Am. Chem. Soc.* **97**, 2978 (1975).
149. R. E. Hemingway, S. M. Park, and A. J. Bard, *J. Am. Chem. Soc.* **97**, 200 (1975).
150. L. R. Faulkner and A. J. Bard, *J. Am. Chem. Soc.* **91**, 209, 6495, 6497 (1969). L. R. Faulkner, H. Tachikawa, and A. J. Bard, *J. Am. Chem. Soc.* **94**, 691 (1972); *Chem. Phys. Lett.* **26**, 246, 568 (1974); P. W. Atkins and G. T. Evans, *Mol. Phys.* **29**, 921 (1975).
151. D. J. Freed and L. R. Faulkner, *J. Am. Chem. Soc.* **93**, 2097, 3565 (1971); **94**, 6324 (1972).
152. D. J. Freed and L. R. Faulkner, *J. Am. Chem. Soc.* **94**, 4790 (1972).
153. C. P. Keszthelyi, N. E. Tokel-Takvoryan, and A. J. Bard, *Anal. Chem.* **47**, 249 (1975).
154. M. G. Evans and M. Polanyi, *Trans Faraday Soc.* **35**, 178,192,195 (1939).
155. W. S. Strive, T. Kitagawa, and D. R. Herschback, *J. Chem. Phys.* **54**, 2959 (1971).
156. D. R. Herschbach in ref. 10, p. 29.
157. R. W. Anderson, V. Aquilante, and D. R. Herschback, *Chem. Phys. Lett.* **4**, 5 (1969); V. Kempter and co-workers, *Chem. Phys. Lett.* **6**, 97 (1970); **11**, 353 (1971).
158. K. Lacmann and D. R. Herschback, *Chem. Phys. Lett.* **6**, 106 (1970).
159. J. E. Mentall and co-workers, *Diss. Faraday Soc.* **44**, 157 (1967); *J. Chem. Phys.* **56**, 4593 (1972).
160. W. Braun and M. J. Kurylo, *J. Chem. Phys.* **61**, 461 (1974).
161. C. Ottinger and R. N. Zare, *Chem. Phys. Lett.* **5**, 243 (1970); D. M. Manos and J. M. Parsons, *J. Chem. Phys.* **63**, 3575 (1975); F. Engelke, R. K. Sandar, and R. N. Zare, *J. Chem. Phys.* **65**, 1146 (1976); L. C. Loh and R. R. Herm, *Chem. Phys. Lett.* **38**, 263 (1976).
162. M. F. Golde and B. A. Thrush, *Adv. At. Mol. Phys.* **11**, 361 (1975).
163. B. A. Thrush, *Chem. Br.* **2**, 287 (1966).
164. D. O. Ham, *Diss. Faraday Soc.* **55**, 313 (1973); *J. Chem. Phys.*, **60**, 1802 (1974).
165. J. C. Greaves and D. Garvin, *J. Chem. Phys.* **30**, 348 (1959); M. A. A. Clyne, B. A. Thrush, and R. P. Wayne, *Trans. Faraday Soc.* **60**, 359 (1964); P. N. Clough and B. A. Thrush, *Trans. Faraday Soc.* **63**, 915 (1967).
166. H. L. Welsh, C. Cumming, and E. J. Stansburg, *J. Opt. Soc. Am.* **41**, 712 (1951).
167. C. J. Halstead and B. A. Thrush, *Nature* **204**, 992 (1964); *Photochem. Photobiol.* **4**, 1007 (1965); B. A. Thrush, *Annual Rev. Phys. Chem.* **19**, 371 (1968).
168. I. W. M. Smith in ref. 10, p. 43.
169. W. D. McElroy and H. H. Seliger, *Sci. Am.*, 76 (Dec. 1962); F. H. Johnson, ed., *The Luminescence*

of Biological Systems, AAAS Press, Washington, D.C., 1961; F. H. Johnson and Y. Haneda, eds., *Bioluminescence in Progress*, Princeton University Press, Princeton, N. J., 1966; H. H. Seliger and W. D. McElroy, *Light, Physical and Biological Action*, Academic Press, New York, 1965; W. D. McElroy and B. Glass, eds., *Light and Life*, Johns Hopkins Press, Baltimore, Md., 1961.
170. A. G. Anlauf and co-workers, *J. Chem. Phys.* **53,** 4091 (1970); P. E. Charters and J. C. Polanyi, *Diss. Faraday Soc.* **33,** 107 (1962); I. W. M. Smith, *Acc. Chem. Res.* **9,** 161 (1976).
171. G. Hancock and I. W. M. Smith, *Chem. Phys. Lett.* **3,** 469 (1969); *Trans. Faraday Soc.* **67,** 2856 (1971); *Chem. Phys. Lett.* **8,** 41 (1971); *Appl. Optics* **10,** 1827 (1971).
172. M. C. Lin in ref. 10, p. 61.
173. M. A. A. Clyne and B. A. Thrush, *Trans. Faraday Soc.* **57,** 1305 (1961).
174. B. H. Mahon and R. B. Solo, *J. Chem. Phys.* **37,** 2669 (1962); M. A. A. Clyne and B. A. Thrush, *Proc. Roy Soc. Ser. A* **269,** 404 (1962); T. G. Slanger, B. J. Wood, and G. Black, *J. Chem. Phys.* **57,** 233 (1972).
175. F. Kaufman in ref. 10, p. 83.
176. B. A. Thrush and M. F. Golde in ref. 10, p. 73.
177. M. Jeunehomme and A. B. F. Duncan, *J. Chem. Phys.* **41,** 1692 (1964); K. H. Becker, W. Groth, and D. Thran, *Chem. Phys. Lett.* **6,** 583 (1970); **15,** 215 (1972).
178. R. A. Young and R. L. Sharpless, *Chem. Phys. Lett.* **39,** 1071 (1963); I. M. Campbell and B. A. Thrush, *Chem. Commun.,* 250 (1965).
179. R. J. VanZee and A. V. Khan, *J. Am. Chem. Soc.* **96,** 6805 (1974).
180. References 10, 14, 49, 54, 59, 169, and 182 provide reviews.
181. J. G. Morin and co-workers, *Science* **190,** 74 (1975); J. E. McCosker, *Sci. Am.* **236,** 106 (Mar. 1977).
182. T. Goto and Y. Kishi, *Angew. Chem. Int. Ed. Engl.* **7,** 407 (1968).
183. K. Hori and M. J. Cormier in ref. 10, p. 361.
184. F. Kenny and R. B. Kurtz, *Anal. Chem.* **22,** 693 (1950).
185. H. Kautsky and E. Gaubatz, *J. Anorg. Chem.* **191,** 384 (1930).
186. J. L. Dyer and W. Lusk, *U.S. National Technical Information Service, AD 631458,* Washington, D.C.
187. W. D. McElroy and M. DeLuca in ref. 10, p. 285.
188. E. H. White, F. McCapra, and G. F. Field, *J. Am. Chem. Soc.* **83,** 2402 (1961); **85,** 337 (1963).
189. H. H. Seliger and W. D. McElroy, *Proc. Nat. Acad. Sci. U.S.A.* **52,** 75 (1964).
190. O. Shimomura, F. H. Johnson, and T. Masugi, *Science* **164,** 1299 (1969).
191. E. H. White and co-workers, *Bioorg. Chem.* **1,** 92 (1971); N. Suzuki and T. Goto, *Tetrahedron* **28,** 4075 (1972); *Tetrahedron Lett.* **2021** (1971).
192. M. DeLuca and M. E. Dempsey in ref. 10, p. 345.
193. M. Deluca and M. E. Dempsey, *Biochem. Biophys. Res. Commun.* **40,** 117 (1970); M. Deluca and co-workers, *Biophys. Res. Commun.* **69,** 262 (1976).
194. E. H. White, J. D. Miano, and M. Umbreit, *J. Am. Chem. Soc.* **97,** 198 (1975).
195. O. Shimomura and F. H. Johnson in ref. 10, p. 337.
196. O. Shimomura and F. H. Johnson, *Photochem. Photobiol.* **12,** 291 (1970).
197. (a) T. Goto, S. Inoue and S. Sugiura, *Tetrahedron Lett.* 3873 (1968); (b) O. Shimomura and F. H. Johnson, *Biochem. Biphys. Res. Commun.* **44,** 340 (1971); **51,** 558 (1973).
198. Y. Kishi and co-workers, *Tetrahedron Lett.,* 3427 (1966).
199. T. Goto and co-workers, *Tetrahedron Lett.,* 4035 (1968); T. Goto, S. Inoue, and S. Sugiura, *Chem. Commun.,* 3873 (1968); T. Goto and co-workers, *Chem. Commun.,* 4035 (1968); J. G. Morin and J. W. Hastings, *J. Cell. Physiol.* **77,** 305 (1971).
200. F. McCapra and co-workers in ref. 10, p. 313.
201. M. J. Cormier, J. E. Wampler, and K. Hori, *Fortschr. Chem. Org. Naturst.* **30,** 1 (1973); M. J. Cormier, K. Hori, and J. M. Anderson, *Biochem. Biophys. Acta* **346** 137 (1974); O. Shimomura and F. H. Johnson, *Nature* **256,** 236 (1975).
202. K. Hori and M. J. Cormier, *Proc. Nat. Acad. Sci. U.S.A.* **70,** 120 (1973).
203. M. DeLuca and co-workers, *Proc. Nat. Acad. Sci. U.S.A.* **68,** 1658 (1971).
204. O. Shimomura, F. H. Johnson, and Y. Saiga, *J. Cell. Comp. Physiol.* **59,** 223 (1962); **62** 19 (1963); O. Shimomura and F. H. Johnson, *Biochemistry* **8,** 3991 (1969).
205. J. W. Hastings and co-workers in ref. 10, p. 369.
206. J. Lee and C. L. Murphy in ref. 10, p. 381.
207. J. W. Hastings and co-workers, *Proc. Nat. Acad. Sci. U.S.A.* **70,** 3468 (1973); J. W. Hastings and C. J. Balny, *J. Biol. Chem.* **250,** 7288 (1975).
208. J. Lee, *Biochemistry* **11,** 3350 (1972).

209. O. Shimomura, F. H. Johnson, and Y. Kohama, *Proc. Nat. Acad. Sci. U.S.A.* **69**, 2086 (1972); J. E. Becvar and J. W. Hastings, *Proc. Nat. Acad. Sci. U.S.A.* **72**, 3374 (1975).
210. J. Lee and C. L. Murphy, *Biochemistry* **14**, 2259 (1975).
211. M. Eley and co-workers, *Biochemistry* **9**, 2902 (1970).
212. O. Shimomura and F. H. Johnson, *Biochemistry* **7**, 1734 (1968).
213. M. M. Rauhut in ref. 10, p. 451.
214. U.S. Pat. 3,775,336 (Nov. 27, 1973), L. J. Bollyky (to American Cyanamid).
215. U.S. Pat. 2,420,286 (May 6, 1947), H. T. Lacey, H. E. Millson, and F. H. Heiss (to American Cyanamid); U.S. Pat. 2,453,578 (Nov. 9, 1948), H. T. Lacey and R. E. Brouillard (to American Cyanamid).
216. U.S. Pat. 3,366,572 (Jan. 30, 1968), J. M. W. Scott and R. F. Phillips (to American Cyanamid).
217. M. M. Rauhut and A. M. Semsel, unpublished work.
218. U.S. Pat. 3,239,406 (Mar. 8, 1960), D. D. Coffman and H. E. Winberg (to E. I. du Pont de Nemours & Co., Inc.).
219. U.S. Pat. 3,264,221 (Aug. 2, 1966), H. E. Winberg (to E. I. du Pont de Nemours & Co., Inc.).
220. U.S. Pat. 3,728,271 (April 17, 1973), W. S. McEwan, H. B. Jonassen, and C. H. Morley (to U.S. Government); U.S. Pat. 3,728,270 (1973), E. M. Bens and C. H. Morley (to U.S. Government); U.S. Pat. 3,558,502 (Jan. 26, 1971), A. F. Tatyrek and B. Werbel (to U.S. Government).
221. U.S. Pat. 3,311,564 (Mar. 28, 1967), E. T. Cline (to E. I. du Pont de Nemours & Co., Inc.).
222. U.S. Pat. 3,726,802 (Apr. 10, 1973), C. H. Morley and E. M. Bens (to U.S. Government).
223. U.S. Pat. 3,392,123 (July 9, 1968), H. E. Winberg (to E. I. du Pont de Nemours & Co., Inc.).
224. U.S. Pat. 3,729,425 (Apr. 24, 1973), C. A. Heller, H. P. Richter, and W. S. McEwan (to U.S. Government).
225. U.S. Pat. 3,697,434 (Oct. 10, 1972), S. Shafler (to U.S. Government).
226. U.S. Pat. 3,539,794 (Nov. 10, 1970), M. M. Rauhut and G. W. Kennerly (to American Cyanamid); U.S. Pat. 3,576,987 (May 1, 1971), H. K. Voight and R. L. Myers (to American Cyanamid).
227. U.S. Pat. 3,752,406 (Aug. 14, 1973), P. A. McDermott and A. M. Semsel (to American Cyanamid).
228. U.S. Pat. 3,969,263 (July 13, 1976), H. P. Richter, C. A. Heller, and R. E. Tedrick (to U.S. Government); U.S. Pat. 3,994,820 (Nov. 30, 1976), D. R. Maulding and M. M. Rauhut (to American Cyanamid).
229. U.S. Pat. 3,718,599 (Feb. 27, 1973), M. M. Rauhut (to American Cyanamid); U.S. Pat. 3,974,086 (Aug. 10, 1976), M. M. Rauhut (to American Cyanamid); U.S. Pat. 3,948,797 (Apr. 6, 1976), M. L. Vega (to American Cyanamid).
230. U.S. Pat. 3,813,534 (Sept. 2, 1974), C. W. Gilliam (to U.S. Government); U.S. Pat. 3,819,925 (June 25, 1974), H. P. Richter and R. E. Tedrick (to U.S. Government); U.S. Pat. 3,764,796 (Oct. 9, 1973), C. W. Gilliam and T. N. Hall (to U.S. Government).
231. U.S. Pat. 3,829,678 (Aug. 13, 1974), G. B. Holcombe; U.S. Pat. 3,940,604 (Feb. 24, 1976), M. M. Rauhut (to American Cyanamid).
232. U.S. Pat. 3,816,325 (June 11, 1974), M. M. Rauhut and A. M. Semsel (to American Cyanamid); U.S. Pat. 3,800,132 (June 28, 1974), R. H. Postal (to American Cyanamid); U.S. Pat. 3,500,033 (Mar. 10, 1970), W. T. Cole and B. K. Daubenspek (to Remington Arms); U.S. Pat. 3,578,962 (May 18, 1971), R. L. Gerber (to Remington Arms); U.S. Pat. 3,671,450 (June 20, 1972) M. M. Rauhut and A. M. Semsel (to American Cyanamid).
233. U.S. Pat. 3,808,414 (May 30, 1974), B. G. Roberts (to American Cyanamid); U.S. Pat. 3,893,938 (July 8, 1975) M. M. Rauhut (to American Cyanamid); U.S. Pat. 3,875,602 (Apr. 8, 1975), R. Miron (to American Cyanamid); U.S. Pat. 3,934,539 (Jan. 27, 1976), S. M. Little and co-workers (to U.S. Government); U.S. Pat. 3,933,118 (Jan. 27, 1976), J. H. Lyons, S. M. Little, and V. J. Esposito (to U.S. Government); U.S. Pat. 3,895,455 (July 22, 1975), C. J. Johnston.
234. U.S. Pat. 3,511,612 (May 12, 1970), G. W. Kennerly and M. M. Rauhut (to American Cyanamid); U.S. Pat. 3,584,211 (June 8, 1971), M. M. Rauhut (to American Cyanamid).
235. U.S. Pat. 3,850,836 (Nov. 26, 1974), H. P. Richter and co-workers (to U.S. Government).
236. H. H. Seliger in ref. 10, p. 461.
237. A. Fontijn, D. Golomb, and J. H. Hodgeson in ref. 10, p. 393.
238. O. Shinomura, F. H. Johnson, and Y. Saiga, *Science* **140**, 1339 (1963); E. B. Ridgway and C. C. Ashley, *Biochem. Biophys. Res. Commun.* **29**, 229 (1967).
239. O. Shinomura and F. H. Johnson in E. W. Chappelle and G. L. Picciolo, eds., *Analytical Applications of Biochemiluminescence and Chemiluminescence, NASA-SP-388,* NASA, Washington, D.C., 1975, p. 89.
240. W. R. Seitz and D. M. Hercules in ref. 10, p. 427.
241. A. K. Babko and N. M. Lukovskaya, *Zh. Anal. Khim.* **17**, 50 (1962); *Zavod. Lab.* **29**, 404 (1963); A.

K. Babko and L. I. Dubovenko, *Z. Anal. Chem.* **200,** 428 (1964); A. K. Babko and I. E. Kalinichenko, *Ukr. Khim. Zh.* **31,** 1316 (1965).
242. K. A. Krause in J. H. Yoe and H. J. Koch, eds., *Trace Analysis,* John Wiley & Sons, New York, 1957, pp. 34–101.
243. W. R. Seitz, W. W. Suydam, and D. M. Hercules, *Anal. Chem.* **44,** 957 (1972).
244. W. R. Seitz and D. M. Hercules, *Anal. Chem.* **44,** 2143 (1972).
245. I. Buyas and L. Erdey, *Talanta* **10,** 467 (1963).
246. F. Kenny and R. B. Kurtz, *Anal. Chem.* **25,** 1550 (1953).
247. F. Kenny and R. B. Kurtz, *Anal. Chem.* **23,** 382 (1951).
248. V. Patrovsky, *Talanta* **23,** 553 (1976).
249. D. T. Bostick and D. M. Hercules, *Anal. Chem.* **47,** 447 (1975); J. P. Auses, S. L. Cooks and J. T. Maloy, *Anal Chem.* **47,** 244 (1975); D. C. Williams, G. F. Huff, and W. R. Seitz, *Clin. Chem.* **22,** 372 (1976).
250. O. Ojima and R. Juaki, *Nippon Kagaku Zasshi* **78,** 1632 (1957); J. Kubal, *Chem. Listy* **62,** 1478 (1968); **64,** 113 (1970); V. K. Zinchukh and A. I. Kampler, *Zh. Analit. Khem.* **28,** 616 (1973).
251. D. C. Williams, G. F. Huff, and W. R. Seitz, *Anal. Chem.* **48,** 1003 (1976).
252. D. C. Williams and W. R. Seitz, *Anal. Chem.* **48,** 1478 (1976).
253. L. Ewetz and K. Strangert, *Acta Path. Microbiol. Scand.* **B82,** 375 (1974).
254. G. L. Picciolo and co-workers in ref. 239, p. 1.
255. E. A. Knust, E. W. Chappelle and G. L. Picciolo in ref. 239, p. 27.
256. V. N. Bush, G. L. Picciolo and E. W. Chappelle in ref. 239, p. 35.
257. H. Vellend and co-workers in ref. 239, p. 43.
258. T. M. Cheng in ref. 239, p. 49.
259. E. W. Chapelle and G. V. Levin, *Biochem. Med.* **2,** 41 (1968).
260. R. D. Hamilton and O. Holm-Hansen, *Limnol. Oceanog.* **12,** 319 (1967).
261. J. W. Patterson, P. L. Brezonik, and H. D. Putnam, *Environ. Sci. Technol.* **4,** 569 (1970).
262. G. V. Levin, J. R. Schrot, and W. C. Hess, *Environ. Sci. Technol.* **9,** 961 (1975).
263. O. Holm-Hansen, *Limnol. Oceanogr.* **14,** 740 (1969); *Plant Cell Physiol.* **11,** 689 (1970).
264. P. E. Stanley and S. G. Williams, *Anal. Biochem.* **29,** 381 (1969).
265. E. W. Chapelle, G. L. Picciolo, and R. H. Altland, *Biochem. Med.* **1,** 252 (1967).
266. N. D. Searle in ref. 239, p. 95.
267. U.S. Pat. 3,959,081 (1976) S. Witz, and W. H. Hartung (to Apzonia Corp.).
268. W. Worthy, *Chem. Eng. News,* 30 (Nov. 24, 1975); R. A. Nathan and co-workers, *Ind. Res.,* 62 (Dec. 1975); G. D. Mendenhall, *Angew. Chem. Int. Ed. Engl.* **16,** 225, (1977).
269. *Fed. Reg.* **36,** Appendix D, 84 (Apr. 30, 1971).
270. H. C. McKee, *J. Air Pollut. Control Assoc.* **26,** 124 (1976).
271. J. E. Sigsby and co-workers, *Environ. Sci. Technol.* **7,** 51 (1973).
272. P. A. Constant, M. C. Sharp, and G. M. Scheil, *Report EPA-650/4-75-013* U.S., National Technical Information Service, Springfield, Va., PB-246 843, Feb. 1975.
273. J. Stauff and W. Jalschke, *Atmos. Enviro.* **9,** 1038 (1975).
274. R. K. Stevens, A. E. O'Keeffe, and G. C. Ortman, *Environ. Sci. Technol.* **3,** 652 (1969).
275. A. Fontijn and R. Ellison, *Environ. Sci. Technol.* **9,** 1157 (1975).
276. D. H. Stedman and D. A. Tommuro, *Anal. Lett.* **9,** 81 (1976).

MICHAEL M. RAUHUT
American Cyanamid Co.

CHEMOTHERAPEUTICS, ANTHELMINTIC

Anthelmintic drugs are used to relieve disease caused by parasitic flatworms or roundworms. Flatworms that infect humans are of two major types: flukes, which have the shape of a simple lanceolate leaf, and tapeworms, which are ribbonlike, made up of serially repeated sections behind a neck and an attachment organ called the scolex. All roundworms have a basic cylindrical shape, with major variations in proportions, in size, and in structure. There is a group of drugs for treatment of blood flukes, another for flukes that live in the lungs, the liver, or the intestine, a third for tapeworms, another for intestinal roundworms, and one for roundworms of blood and tissues. However, this classification of drug use is not rigidly exclusive, and a drug that has its greatest effectiveness against one class of worm may also have effects against some species of another class; also a drug that is very effective against one species may be fairly effective against another species of the same class.

The drugs considered in this article, and the worms against which they are effective, are presented in Table 1. The term drug of choice is avoided because the true drug of choice is sure to change with time. Drugs discussed are specific chemotherapeutic agents. Anthelmintics are used to injure or destroy worm parasites, or to facilitate their removal from the body, or to interfere in some way with their protective mechanisms against the natural defenses of the host.

Frequently, helminthic disease will require medication in addition to the chemotherapeutic agent itself. Allergic conditions are common when there is tissue invasion by worms, so that antihistamines and corticosteroids may be necessary adjuncts to therapy. In fact, some allergic responses can be seriously aggravated by successful anthelmintic therapy. Anemia, indigestion, and secondary bacterial infections may be associated with the presence of worms, and under these conditions hematinics and suitable antibiotics are needed. Some practitioners use cathartics to help remove certain helminths from the digestive tract, after the worms have been temporarily incapacitated by a specific drug (see Gastrointestinal agents).

Drugs that have been found safe and effective in other parts of the world may not be available in the United States by prescription because the market is too small for a manufacturer to undertake the cost of approval through FDA procedures (24–25). However, physicians can obtain such drugs (3), (6), (7), and (12) for their patients, and information about their use, from the Parasitic Diseases Drug Service, Center for Disease Control. U.S. Public Health Service (USPHS), Atlanta, Ga. In an emergency the Service can be reached by telephone (404-633-3311). In Table 1 these drugs are referred to as available from the USPHS.

The danger of infection must be balanced against the hazards of drug use. The situation is not as serious as it was in the 1940s when the anthelmintic armamentarium included some of the most dangerous drugs then available.

Treatment of Blood Fluke Disease (Schistosomiasis)

Antimony potassium tartrate (1) (Fig. 1) is used against *Schistosoma japonicum*, stibophen (2) against *S. haematobium*, and stibocaptate (3) against *S. mansoni*. Each

Table 1. Anthelmintics: Uses and Properties

Drug and structure no.	CAS Registry No.	Trade names	Physical properties	Solubility	Disease (organism)	Refs.
antimony potassium tartrate USP (1)	[28300-74-5]	Tartar emetic	colorless crystals or white powder, mp 100°C– $\frac{1}{2}$ H$_2$O and 210°C – 1 H$_2$O	1 g in 12 mL water; insoluble in ethanol	oriental blood fluke disease (*Schistosoma japonicum*)	1–2
bephenium hydroxynaphthoate USP (15)	[3818-50-6]	Alcopar(a), Befenio, Lecibis, Nemex	crystals or yellow powder, mp ca 170°C	insoluble in water; soluble in hot ethanol	hookworm disease (*Ancylostoma duodenale*)	3–4
bithionol (7)[a]	[97-18-7]	Actamer, Bithin, Lorothidol	crystals or gray–white powder, mp 188°C	insoluble in water; soluble in acetone and 4% NaOH	lung fluke disease (*Paragonimus westermani*)	5–6
chloroquine USP (10)	[54-05-7]	Aralen, Artrichin, Bemophate, Bipiquin, Nivaquine B, Résoquine, Reumachlor, Sanoquin, Tankan	slightly yellowish crystalline powder, mp 87–92°C	very slightly soluble in water; soluble in dilute acids, in CHCl$_3$ and in ether	Chinese liver fluke disease (*Clonorchis sinensis*)	7
dichlorophen BAN (14)	[97-23-4]	Antiphen, Dicestal, Didroxane, Di-phentahane-70, Hyosan, Parabis, Plath-Lyse, Preventol G-D, Teniathane, Teniatol	cream-colored powder with slight phenolic odor, mp 174–178°C	almost insoluble in water; freely soluble in 95% ethanol and in ether	dwarf tapeworm disease (*Hymenolepis nana*)	8
diethylcarbamazine citrate USP (21)	[1642-54-2]	Dirocide, Filazine, Franocide, Longicid, Loxuran	white crystalline powder, mp 141–143°C	soluble in water and hot ethanol; insoluble in acetone	elephantiasis and other filariases (*Wuchereria bancrofti, Loa loa,* and other filaria)	9
hexylresorcinol NF (8)	[136-77-6]	Caprocol, Crystoids, Gelovermin, Sucrets	yellow–white, needle-shaped crystals turning brownish-pink on exposure, mp ca 68°C	1 g in 2000 mL water; freely soluble in ethanol and acetone	roundworm disease (mixed intestinal nematodes)	10

hycanthone mesylate (5) (hycanthone base)	[23255-93-8] [3105-97-3]	Etrenol	hycanthone base is a yellow–orange crystalline powder, mp ca 68°C	base is soluble in water	intestinal blood fluke disease, urinary tract blood fluke disease (*Schistosoma mansoni* and *Schistosoma haematobium*)
lucanthone hydrochloride (4)	[548-57-2]	Miracil D, Miracol, Nilodin, Tixantone	yellow crystals or yellow–orange powder, mp ca 195°C, water solution is orange and stains the skin	1 g in 110 mL water; 1 g in 85 mL ethanol	urinary tract blood fluke disease (*Schistosoma haematobium*)
mebendazole (19)	[31431-39-7]	Pantelmin, Telmin, Vermox	crystals or yellow amorphous powder, mp ca 289°C	very slightly soluble in water and most organic solvents	whipworm disease (*Trichuris trichiura*)
niclosamide (12)[a]	[50-65-7]	Bayer 2353, Cestocid, Fenesal, Lintex, Mansonil, Nasemo, Sulqui, Tredemine, Vermitid, Yomesan	pale yellow crystals, or yellow–white powder, mp 225–230°C	insoluble in water; sparingly soluble in ethanol	tapeworm disease (*T. saginata*, *T. solium*, *D. latum*, and *H. nana*)
niridazole (6)[a]	[61-57-4]	Ambilhar	yellow crystalline powder, mp ca 261°C	sparingly soluble in water and most organic solvents	urinary tract blood fluke disease, guineaworm disease (*S. haematobium*, and *D. medenensis*)
piperazine citrate USP (16)	[144-29-6]	Antepar, Arpezine, Exelmin, Exopin, Helmazine, Multifuge, Nematidal, Oxucide, Oxyzin, Parazine, Pinozan, Pinrou, Piperaverm, Pipazan Citrate, Pipracid, Piptelate, Rhomex, Uvilon	white crystalline powder, mp ca 185°C (dec)	freely soluble in water; insoluble in ethanol	large roundworm disease (*Ascaris lumbricoides*)
pyrantel pamoate (17)	[22204-24-6]	Antiminth, Cobrantil, Combrantin, Helmix, Piranver	white or yellow or tan crystalline powder, melts and decomposes, mp ca 250°C	insoluble in water and nearly so in ethanol	pinworm disease and large roundworm disease (*Enterobius vermicularis*, and *Ascaris lumbricoides*)

Table 1 (Continued)

Drug and structure no.	CAS Registry No.	Trade names	Physical properties	Solubility	Disease (organism)	Refs.
pyrvinium pamoate USP (18)	[3546-41-6]	Alnoxin, Altolat, Molevac, Neo-Oxypaat, Pamovin, Povan, Povanyl, Pyrcon, Tolapin, Tru, Vanquil, Vanquin	orange, red, or darker crystalline powder stains skin, textiles and stool red, softens ca 190°C, melts ca 210°C	insoluble in water; very slightly soluble in ethanol	pinworm disease and threadworm disease (*Enterobius vermicularis*, and *Strongyloides stercoralis*)	18
quinacrine hydrochloride (13)	[69-05-6]	Acrichine, Acriquine, Atabrin(e), Chinacrin, Erion, Italchin, Metoquin, Palacrin	yellow crystals, mp ca 250°C	1 g in 35 mL water; slightly soluble in ethanol	beef tapeworm and broad fish tapeworm (*Taenia saginata* and *Diphyllobothrium latum*)	19
stibocaptate (3)[a]	[3064-61-7]	Astiban	white to yellow-green crystalline powder	soluble in water and unstable in solution at RT	intestinal blood fluke disease (*Schistosoma mansoni*)	20
stibophen (2)	[15489-16-4]	Corystibin, Fantorin, Fouadin, Fuadin, Neoantimosan, Pyrostib, Repodral, Sodium Antimosan, Trimon	white to slightly yellow or pink crystals, mp >300°C, chars 300°C	soluble in water; solns oxidize and turn yellow; insoluble in ethanol	blood fluke disease (*Schistosoma mansoni*, and *S. haematobium*)	21
tetrachloroethylene USP (9)	[127-18-4]	Ankilostin, Didakene, Nema, Perclene, Tetracap, Tetropil	colorless, nonflammable fluid with ethereal odor, deteriorates in warm climates, bp 121°C	soluble 1:10,000 in water; miscible with ethanol	hookworm disease (*Necator americanus*)	22
thiabendazole USP (20)	[148-79-8]	Bovisole, Eprofil, Equizole, Lombristop, Mertect, Mintezol, Minzolum, Nemapan, Omnizole, Polival, Thiaben, Thiabenzole	white crystals, mp ca 300°C	almost insoluble in water, but readily soluble in dilute acids or alkalis; slightly soluble in ethanol	threadworm disease, larva migrans (*Strongyloides stercoralis*, *Toxacariss A. braziliensis*)	23

[a] Available from USPHS.

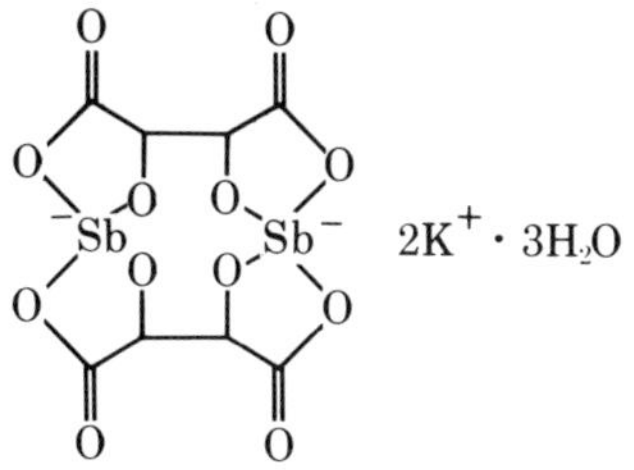

Figure 1. Drugs effective against schistosomes.

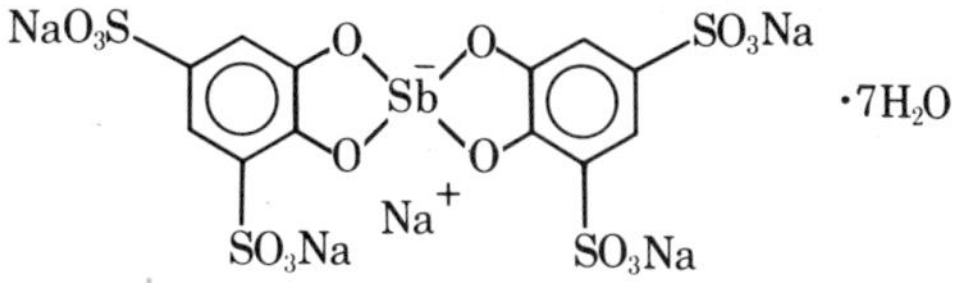

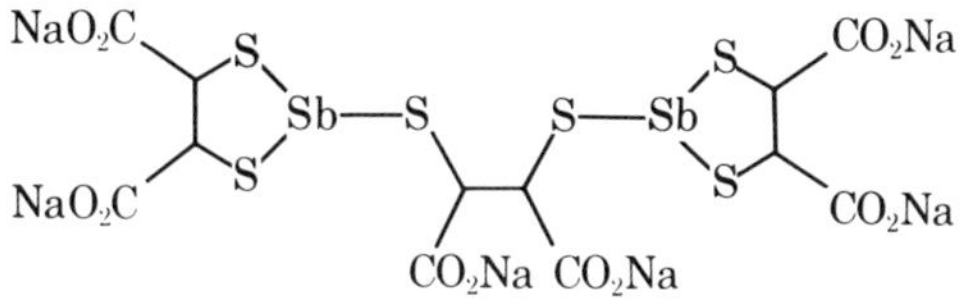

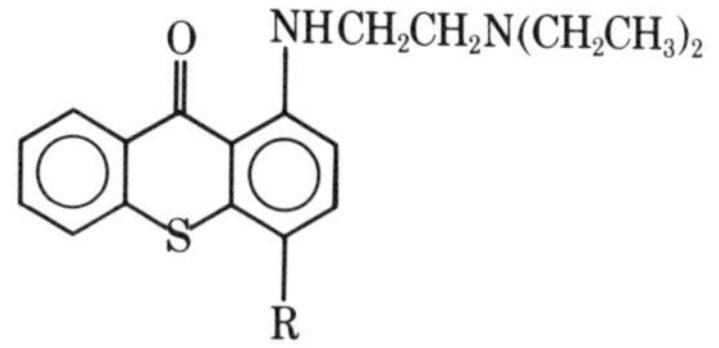

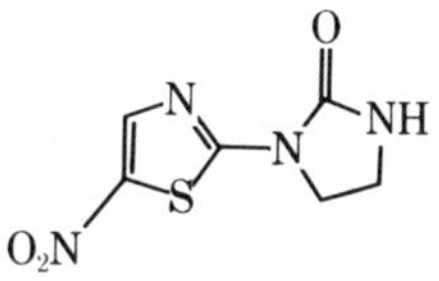

of these antimonial anthelmintics can be employed to treat any one of the three infections, but it is most successful against the species mentioned.

These compounds traditionally have been called trivalent antimonials: antimony potassium tartrate (1) (tartar emetic) was used first with a regimen for administration of many intravenous injections over a period of a month or more. Stibophen (2) can be given in fewer doses by intramuscular administration over a shorter period and it is stable in solution.

Stibocaptate (3) is the newest of these compounds. The antimony in this drug is bound to sulfur atoms rather than to oxygen atoms as with the others, so that stibocaptate has a higher degree of stability (26). The antimony is less easily dissociated because it is bound tightly to dimercaptosuccinate, which resembles BAL (British Anti-Lewisite, dimercaprol, $HOCH_2CH(SH)CH_2SH$) (27). Stibocaptate is given intramuscularly; fewer doses are needed than (1) or (2). These three drugs are essentially specialties for human schistosomiasis, although (1) and (2) are used in veterinary medicine; and (1) has unrelated uses as mordant in the textile and leather industries.

All antischistosomal antimonials are presumed to have the same mechanism of action. They inhibit phosphofructokinase, an enzyme involved in the anaerobic metabolism of glucose (28). The drugs arrest the conversion of fructose-6-phosphate to fructose-1,6-diphosphate by a reversible inhibition that reduces the rate of glycolysis; and this causes the worms to lose their energy supply. Schistosomes normally hold their position in the blood vessel by use of suckers, but when treatment with antimonials is started they lose their hold and are swept away to be destroyed, in time, through the action of white cells.

Because the enzyme in schistosomes is much more sensitive than is the corresponding enzyme of human tissues, the Embden-Meyerhof-Parnas scheme of glycolysis (by which glucose is metabolized to lactic acid) is disrupted in schistosomes, but not in man. This selectively toxic effect on schistosome phosphofructokinases is accomplished at low levels of concentration of the drug. In higher concentrations, antimonials can inhibit the activity of glutamic–pyruvic transaminase, and as drug concentration is increased further, many enzyme systems of both parasite and man are affected, probably by union of the antimony with sulfhydryl groups. Therefore, as drug concentration rises in the host, the important effects cease to be therapeutic and become toxic.

Adverse effects from these drugs are legion, thus they are not satisfactory chemotherapeutic agents. Tartar emetic is the most toxic, but all tend to cause the same problems: abdominal pain and diarrhea, nausea and vomiting; fall in blood pressure and a tendency toward fainting; coughing and difficulty in breathing; electrocardiographic changes; headache and joint pain; liver damage; kidney damage; hemolytic anemia; drug rash; exacerbation of concurrent diseases; and other effects. Because of these toxic effects, the search has continued for different, nonmetallic antischistosomal agents.

Lucanthone (4) and hycanthone (5) (thioxanthones) represented a new direction in the development of drugs for schistosomiasis. Lucanthone was introduced first, but hycanthone, which is an hydroxymethyl analogue, was recognized as the active metabolic product in humans (29). Lucanthone, which had been synthesized as one of a series of drugs starting from quinacrine (see below), is important because it was the first oral drug for a schistosomal disease, and the first one that did not contain a metal. Hycanthone worked by intramuscular injection in a single large dose and was used in mass population treatment before it was withdrawn because of hepatotoxicity. The initial effect of this drug on the worms was to cause a decline in egg production. The mechanism of action may have been through shifts in the tissue distribution of 5-hydroxytryptamine (serotonin), which acts as a neurotransmitter in schistosomes (see Neuroregulators).

A more successful line of drug development produced the nitrothiazole derivative niridazole (6). This is given orally twice a day for a week to hospitalized patients infected with *S. haematobium* or *S. mansoni*; and is also effective in treating the guinea worm (*Dracunculus medinensis*). It is active against amoebae but too toxic for that use as compared to other drugs.

Niridazole (6) arrests egg shell formation (30). In the female worm the ovary degenerates; and in the male, spermatogenesis is arrested. Schistosomes feed on blood sugar, and on red blood cells, from which they digest the globin. Normally they absorb glucose through their body surface as well as by way of their intestines, but (6) inhibits uptake of glucose, accelerates the breakdown of glycogen in the parasites, and is suspected of interference with the activity of hexokinase and of phosphorylase phosphatase. Selective toxicity is related to drug concentrations which are three to four times as high in the portal blood, where the worms are found, as compared to the concentration in the general circulation (because the drug is largely metabolized on first pass through the liver).

Among the adverse effects niridazole may cause are abdominal cramps, vomiting, dizziness, agitation, confusion, hallucinations, and convulsions. Liver disease predisposes to toxicity. Persons with glucose-6-phosphate dehydrogenase deficiency may suffer hemolytic anemia. In various laboratory models the drug has been carcinogenic, mutagenic, immunosuppressive, and inhibited cell-mediated hypersensitivity. It may have adverse effects on T lymphocytes.

The efficacy and toxicity of drugs used in the treatment of schistosomiases leaves something to be desired. A number of other compounds are known to have antischistosomal effects, and there has been extensive clinical experience with some of these in limited geographical areas (31).

The influence of antischistosomal drugs on the disease depends upon the decrease or arrest of egg production, since it is the movement of eggs through tissues that produces the pathology. Adult worms about 2 cm long occur as mating pairs within venules, but at intervals the female travels as far as possible toward the capillary bed to lay eggs that break out of the blood vessel and do extensive damage. The number of pairs of worms in a patient varies from a few to hundreds, and in untreated patients the worms may survive for 5 yr, or sometimes for decades.

Treatment of Fluke (Trematode) Infections in the Lungs, Intestines, and Liver

Bithionol (7) (Fig. 2) is used to treat persons infected with *Paragonimus* (lung fluke) or with *Fasciola* (sheep liver fluke). It is a phenolic compound similar to hexachlorophene (see Disinfectants). Both have been used as veterinary anthelmintics (see Veterinary drugs). In the past, bithionol was incorporated for its antibacterial effects in at least 20 medicated skin cleansers manufactured in the United States but such use had to be discontinued when the compound was discovered to be the cause of skin irritations.

Bithionol (7) interferes with the neuromuscular physiology of helminths, impairs egg formation, and may cause the protective cuticle covering the worm to become defective. At the biochemical level, oxidative phosphorylation is inhibited, and the bithionol molecule can chelate iron so that it may inactivate iron-containing enzyme systems.

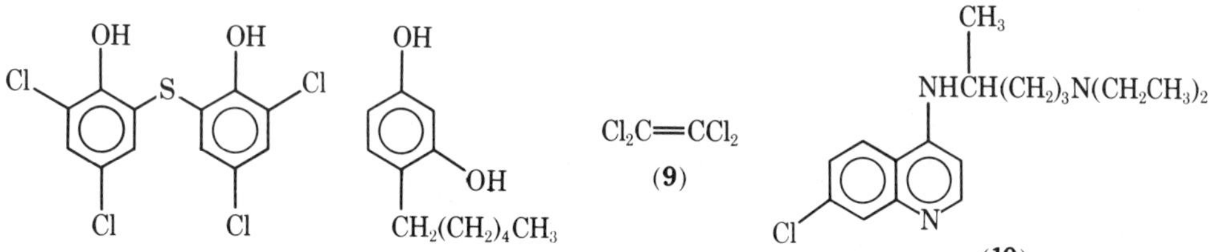

Figure 2. Drugs effective against trematode infections.

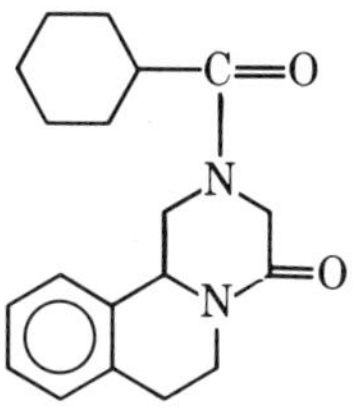

Paragonimus about 1 cm long are normally in the lungs where a population of 20–50 worms will cause chest pain and shortness of breath; but worms may occur in viscera or the brain causing tumor-like symptoms. *Fasciola* about 3 cm long, after extensive migration, invade the bile ducts where heavy infections result in changes of the bile duct wall and surrounding tissue, and cirrhosis; or as a result of incomplete migrations the worms may arrive at other locations and cause intense tissue reactions. A 10-d oral regimen of bithionol may resolve the pathology of lung disease in about 3 mo, but improvement in cerebral paragonimiasis and in sheep liver fluke infections is variable. Untreated adult worms of both species live more than 5 yr. Adverse effects in patients taking oral bithionol commonly include abdominal pain, diarrhea, vomiting, rashes and photosensitivity.

Hexylresorcinol (8) is another phenolic anthelmintic. It is effective in a single oral dose against trematodes in the intestinal tract, and is used in persons infected with the giant intestinal fluke *Fasciolopsis buski*. It was an important anthelmintic in the 1940s through the 1960s when it was used against intestinal nematodes and the dwarf tapeworm. The drug, which was originally introduced as a stainless antiseptic (ST 37), continues to be used that way, and as an anthelmintic in veterinary medicine.

Hexylresorcinol (8) produces blisters and cuts in the surface of the parasites, presumably because it is a phenolic compound (32). It burns and kills superficial tissues of the body wall, precipitates superficial cell protein, and alters permeability as phosphorus compounds leak from the parasites. Perhaps as a consequence of these effects, it paralyzes the muscles of intestinal nematodes.

The selective toxicity of (8) presumably derives from the natural coatings of the human gastrointestinal tract, but the worms are unprotected. When (8) is administered by mouth as in crystalline form enclosed in hard gelatin-coated pills, it is not readily absorbed by the human intestine (ie, there is little or no systemic toxicity). However, the drug can burn unprotected mucous membranes of the mouth and upper esophagus if it comes into contact with these surfaces by eructation.

Fasciolopsis buski grows to 8 cm long and lives attached to the wall of the small intestine where a population of thousands may cause inflammatory ulceration and toxemia with danger of death in children. Treatment is oral medication with hexylresorcinol or tetrachloroethylene.

Tetrachloroethylene (9) also has a broad anthelmintic spectrum, and it can be used to eradicate either of the two tiny intestinal flukes *Heterophyes heterophyes* and *Metagonimus yokogawai*. It is effective against the hookworm *Necator* (see below in the section on intestinal nematodes). The compound's other uses are as a solvent for dry cleaning (qv) and for degreasing metals (see Chlorocarbons).

As an anthelmintic, (9) replaced carbon tetrachloride, which if absorbed, could destroy the liver. The mechanism by which (9) works is not established but some workers believe the halogenated hydrocarbon dissolves in the lipid of cells of the helmintic neuromuscular system and interferes with their function. Investigators do not agree on the anthelmintic action of tetrachlorethylene or the basic pharmacology of the drug.

For use against either flukes or the hookworm *Necator,* a single oral dose is given on an empty stomach after a day of low-fat diet, because fat in the gastrointestinal tract may increase absorption of the drug. This drug is popular in veterinary medicine (in the United States only the veterinary-type of drug preparation can be obtained) (33).

Adverse effects can be a burning sensation in the stomach, nausea and vomiting, and headache, giddiness, dizziness, inebriation, and loss of consciousness. For these reasons patients are kept quiet and restricted, perhaps in bed, for 4 h after treatment, and are not permitted alcoholic beverages. In those exposed to the liquid and its vapors a defatting action on the skin can lead to dermatitis.

The flukes *Heterophyes* and *Metagonimus* are 2.5 mm or less in length, on or in the wall of the small intestine, and the resulting condition is frequently symptomless; rarely, eggs may enter the circulatory system to be carried to heart and brain and damage those areas. Niclosamide (see structure (12)) used mainly for tapeworms, is also effective against these small flukes.

Chloroquine (10) will reduce the egg output of the Chinese liver fluke (*Clonorchis sinensis*). The primary uses of chloroquine are for malaria and amebiasis, in which situations it is essentially nontoxic (see Chemotherapeutics, antiprotozoal). It has also been used in collagen diseases, such as lupus erythematosus, and for rheumatoid arthritis, but is toxic in large doses for long times as used in those conditions. Chloroquine is a 4-aminoquinoline that bears the same alkyl side chain as the acridine dye quinacrine (see Dyes). Chloroquine, like quinacrine, intercalates with deoxyribonucleic acid (DNA) and this upsets the role of DNA as a template in replication and transcription. It inhibits DNA- and ribonucleic acid, RNA-polymerase activity. Chloroquine accumulates in the liver cells, in red cells parasitized with malaria, and binds to melanin (eg, in the retina). It is more toxic to some tissues than others and one of the factors that makes tissues sensitive to the drug may be cell reproduction and growth, as in the helmintic egg production process. Short term administration of chloroquine is relatively harmless. Patients may have nausea, vomiting and loss of appetite, skin rashes and itch, headaches, and visual difficulty. Long term high dose administration as formerly used for collagen diseases led to blindness and other serious toxicities. The drug may be teratogenetic.

Clonorchis adults grow up to 2 cm long and live in the biliary tree where they cause inflammation. The presence of 20–200 individuals is common but the number can be over 20,000; infection is the consequence of a diet of raw fish. Untreated worms can live for 25 yr. Treatment is chloroquine diphosphate by mouth daily for 3 wk to a year; but results are unsatisfactory because, although egg count decreases, most adult worms are not killed (34).

Treatment of Tapeworm (Cestode) Infections

Niclosamide (12) (Fig. 3) is a convenient, effective drug for treatment of persons who have tapeworm infections. More than 80% of the patients are cured by this orally administered drug, and adjunctive therapy is not required when it is used against the beef tapeworm (*Taenia saginata*), the dwarf tapeworm (*Hymenolepis nana*) or the broad fish tapeworm of man (*Diphyllobothrium latum*). In connection with therapy for the pork tapeworm (*Taenia solium*), a laxative may be used afterward. Other helminthic infections have also been successfully treated with this drug, including small intestinal flukes and the whipworm. Additionally, niclosamide ethanolamine salt is used as a snail control agent against the intermediate host of *Schistosoma mansoni*.

Modes of action for niclosamide (12) are interference with respiration and blocking

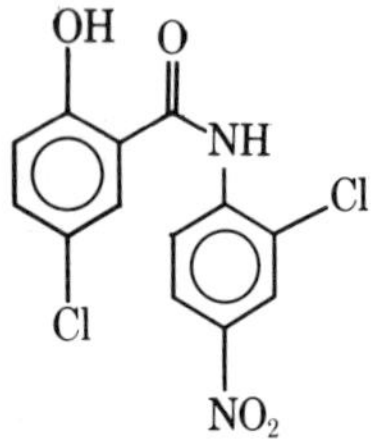

Figure 3. Drugs effective against cestode infections.

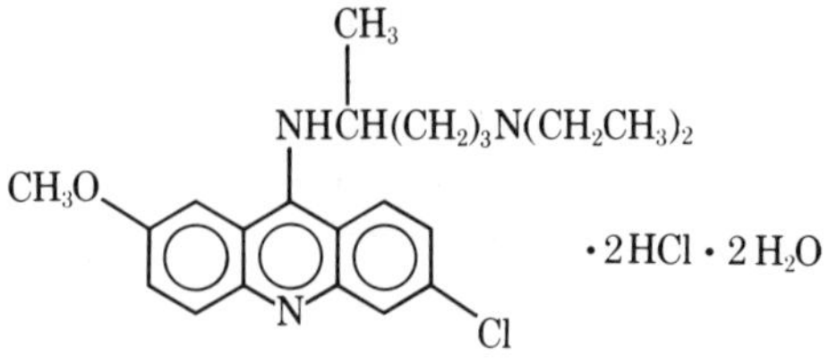

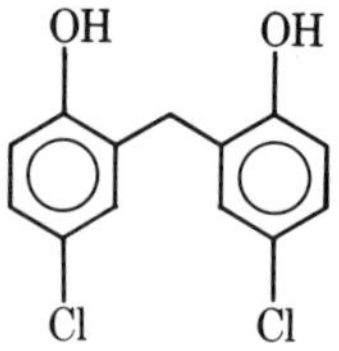

of glucose uptake. This drug uncouples oxidative phosphorylation in both mammalian and taenioid mitochondria (35–36) inhibiting the anaerobic incorporation of inorganic phosphate into ATP. Tapeworms are very sensitive to (12) because they depend on anaerobic metabolism of carbohydrate as their major source of energy. The selective toxicity for the parasites as compared to the host is that very little (12) is absorbed from the gastrointestinal tract (37). Adverse effects are nausea, abdominal pain and malaise in 10% of the patients; a few become lightheaded.

A single dose of niclosamide is used for the large tapeworms. Niclosamide causes the tapeworm head to disengage from the intestinal wall of the host, and the body wall of the parasite disintegrates. Some investigators believe the drug sensitizes the wall of the worm to the action of proteolytic enzymes. When infection by a dwarf tapeworm is treated, the drug taken the first day disposes of the worms that are in the intestinal cavity; however, there are also developing stages in the tissue of the intestinal wall, and these are unharmed. Therefore, dosage must be repeated for 5–7 d to destroy these worms as they finish their development and reenter the intestinal lumen.

There have never been any drugs that are effective against tissue inhabiting stages of cestodes in main; however praziquantel [55268-74-1] (11) (Droncit) is being used against intermediate tissue stages of tapeworms in veterinary medicine.

Quinacrine hydrochloride (13) is an older drug that is effective against tapeworms (38). It is a second choice drug for infections with the beef tapeworm (*T. saginata*) and the large fish tapeworm (*D. latum*), but it is less successful against the dwarf tapeworm (*H. nana*) and some workers advise against its use to treat infection with the pork tapeworm (*T. solium*). It concentrates in the scolex and causes the muscles needed for holding to relax. Worms are stained yellow and pass from the body, still alive. Quinacrine can intercalate with DNA and inhibit nucleic acid synthesis. Quinacrine creates fluorescent bands in deoxyadenylate–deoxythymidylate-rich regions of DNA and has been used as a stain in the study of human genetics.

The selective toxicity of this drug is unexplained except for accumulation in the head of the tapeworm. The drug is readily absorbed from the gastrointestinal tract and may persist in human tissues for 2 mo after administration. Prolonged treatment with the large dose used in helminthiases may cause yellow discoloration of skin. An anthelminthic dose may commonly cause vomiting in one out of four patients and possible regurgitation of *T. solium* eggs into the stomach, with the subsequent possibility of producing an autoinfection of incurable cysticercosis. Quinacrine (13) is more toxic than niclosamide (12) without being more efficacious and is troublesome to administer.

Dichlorophen (14) was used for many years in veterinary medicine before it was introduced for human disease. This is an oral drug available in Europe but not in the United States, and is used against intestinal tapeworms. The drug is still employed in veterinary medicine, and as a germicide in soaps and shampoos, and as a fungicide in agriculture (39) (see Fungicides). Some authors regard dichlorophen as structurally similar to niclosamide (12) and bithionol (7), considering them as halogenated derivatives of diphenylmethane, and related substances (40). The tapeworm detaches under the influence of dichlorophen, dies, and is digested, and does not come out of the host in a recognizable state. Adverse effects are gastrointestinal disturbance, lassitude and depression, and hives. The drug itself is laxative.

Paromomycin [7542-37-2] (Humatin) is a broad-spectrum antibiotic that has been employed for treatment of *H. nana* and *T. saginata*, and for amoebiasis, as well as for its antibacterial effects in diarrheas and dysenteries (see Antibiotics, aminoglycosides).

The broad fish tapeworm grows to a length of 10 m with 4000 proglottides that tend to remain attached and discharge eggs which appear with the feces. Besides absorbing host foodstuffs, this worm avidly sequesters vitamin B_{12} and folic acid. The nearer to the stomach that it is attached in the small intestine, the more vitamin it takes up. This may so deplete the supplies available to the host that pernicious anemia can develop with associated neurological symptoms. There is usually only one worm in a patient and it can survive for 10 yr.

The beef tapeworm is of worldwide distribution in people who eat rare beef. Worms are often 5–10 m long with about 1000 proglottides, and usually a patient carries only one worm. Cases are frequently asymptomatic; however, detached sections of worm 0.5 cm × 2 cm may creep out of the end of the digestive tract.

Both the adult and the larval bladderworm of the pork tapeworm are able to live in man, the parasite being sporadic in occurrence wherever uncooked pork is eaten. The adult is 5 m long and untreated adult worms may survive for 25 yr.

The dwarf tapeworm is only 2–4 cm long, and is of universal distribution in mice and man in temperature zones, where children, especially those in institutions, are the group most frequently infected. Often symptomless, this tapeworm can cause abdominal discomfort and diarrhea if infection is heavy.

Treatment of Intestinal Roundworm (Nematode) Infections

Bephenium hydroxynaphthoate (15) (Fig. 4) is effective against the hookworm *Ancylostoma duodenale* and a one-day oral treatment cures up to 100% of infections (41). The drug can also be used in patients with *Necator* and *Ascaris* but is less effective and may fail against resistant strains of *Necator*. It is used in veterinary medicine.

Bephenium [7181-73-9] is a quaternary ammonium compound with a structure similar to that of acetylcholine, and is more potent than acetylcholine in causing contraction of isolated roundworm muscle strips, an action that can be blocked by *d*-tubocurarine (see Alkaloids). Hookworms become discoordinated under treatment with bephenium, lose their hold on the intestinal mucosa and are carried along in a contracted state with the fecal stream. The selective toxicity for intestinal roundworms may be dependent in part on the fact that it goes essentially unabsorbed through the digestive tract. This is not a dangerous drug but the taste is so bitter that administration can cause nausea and vomiting.

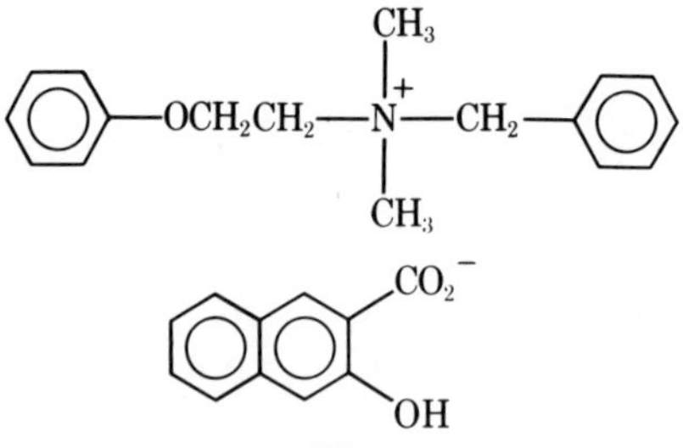

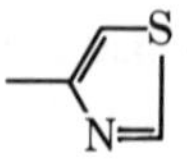

Figure 4. Drugs effective against nematode infections.

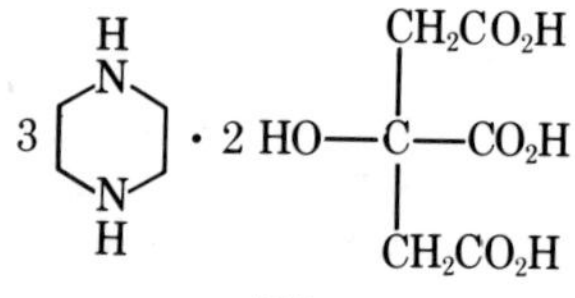

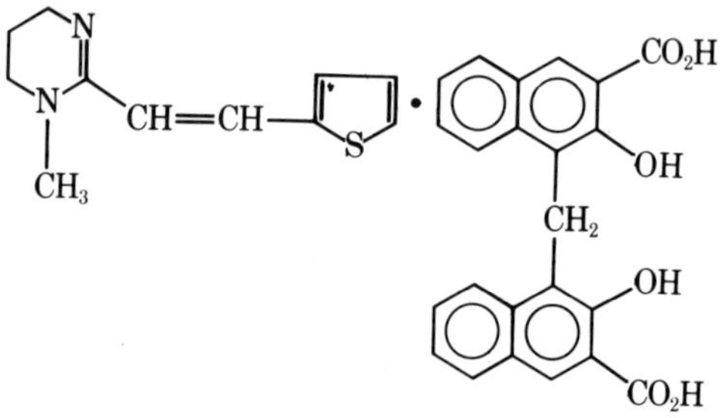

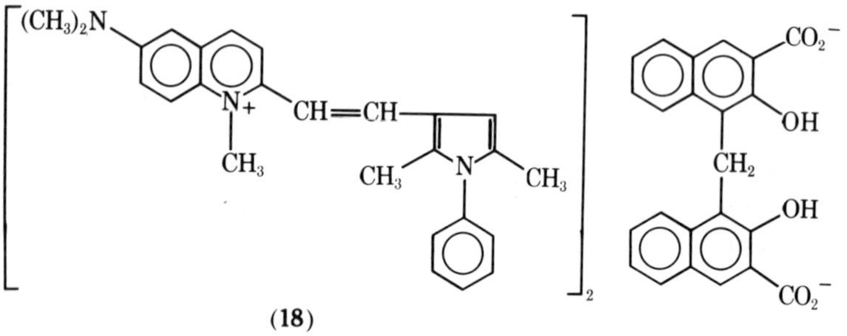

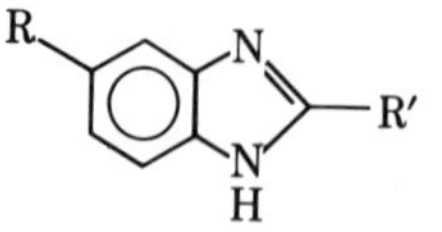

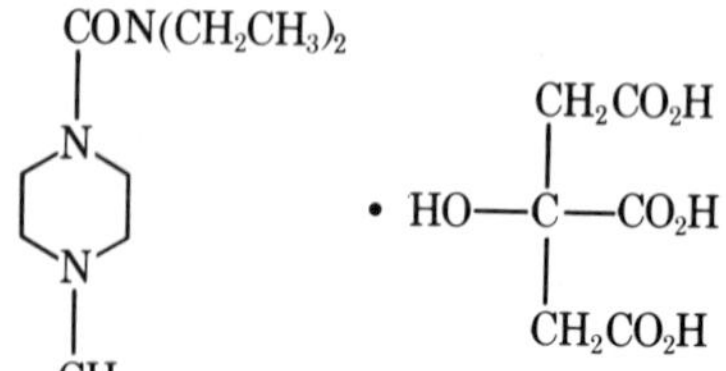

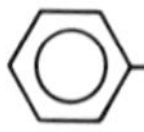

A. duodenale adults are about 1 cm long and live throughout the small intestine holding onto a mouthful of mucosa and draining blood from the capillary circulation. A patient may harbor 1000 worms, and one worm may pump out 2/3 mL blood/d; thus hookworm anemia is caused by blood loss and the severity is related to the number of worms. Untreated worms may survive in the intestine 1–15 yr.

Piperazine (16) (see Amines, cyclic) is taken orally by nonfasting subjects, one dose a day on two consecutive days for large roundworms (*Ascaris*), and reports continue to indicate that it will cure about 90% of cases treated (42–44). This drug can be used for the human pinworm (*Enterobius*) with a 7-d course of treatment and it is used in veterinary medicine.

Piperazine causes flaccid paralysis of *Ascaris* by blocking the ability of the worm to respond to acetylcholine and they lose their position in the digestive tract, and are carried out, still alive. In the helminth, acetylcholine may be a modulating neurohormone rather than a chemotransmitter. The drug is well absorbed from the human intestine and the reason for its selective toxicity is uncertain.

Occasional side effects from piperazine include abdominal cramps, nausea, vomiting, and diarrhea. Rarely patients may have headache, vertigo, and tremors, or they may feel weak and have difficulty focusing. Red patches can appear on the skin, sometimes with the flat, elevated, itching welts of urticaria.

Female worms of *Ascaris lumbricoides* are about 30 cm $\times$ 0.5 cm, and as adults they live and feed in the small intestine where they hold their position by taking the shape of a simple spring. Sometimes they cause abdominal discomfort or pain; but frequently light infections are symptomless. Adult worms can migrate in response to disease or surgery when they may penetrate suture lines, perforate the intestinal wall, or travel up ducts. Untreated adult worms may live for a year and a half before they disappear.

Pyrantel pamoate (17) in a single oral dose cures pinworm infections, and is close to 100% effective against *Ascaris* (45–46). It may be repeated in a month if the worms return. This is also a principal drug for treating hookworm infections of both species, thus it is useful in patients with mixed infections. For hookworms, treatment is continued 3 d. In roundworms, muscle is persistently activated resulting in spastic paralysis. The drug was introduced in veterinary medicine and then applied to clinical medicine.

Since about 90% of the administered dose passes out in the feces of the patient, the selective toxicity of the drug can be attributed primarily to poor absorption. Abdominal pain, nausea, vomiting, diarrhea, and lack of appetite may occur. Patients may have headache and feel dizzy or drowsy. Skin rashes can develop. In the blood chemistry, SGOT (Serum Glutamic Oxaloacetic Transaminase) levels may be elevated.

The pinworm (*Enterobius vermicularis*) is of worldwide distribution in temperate zones, including cities, where it is frequently a disease of households or institutions. Female worms are about 1 cm long. While growing to maturity, they live in the gut lumen holding on by their mouths to the intestinal mucosa. A symptom is an itchy perianal region, because the mature female migrates out the anus and discharges eggs with itchy secretions. A usual intestinal population is less than 100 worms; however, more than 5000 have been recovered from a single patient. Eggs also get into the household dust and are resistant to drying and may even be inhaled and swallowed, leading to very light infections with very few worms in adult relatives of infected children.

Pyrvinium pamoate (18) is an oral drug used in a single dose against pinworm and repeated in a week. When the drug is used for treatment of threadworm (*Strongyloides*), one dose a day is given for seven days.

Pyrvinium (18) is the salt of an asymmetrical cyanine dye with a resonating amidinium system in the molecule (47). In anaerobic worms, such as pinworm and threadworm, it irreversibly interferes with the absorption of glucose. This causes relative failure of muscular activity, then reduced motility. As adenosine triphosphate (ATP) levels steadily fall within the worms, they eventually die. (In some other aerobic parasitic worms that are not parasites of man, pyrvinium causes respiratory inhibition, thus decreasing oxygen uptake.)

Tetrachloroethylene (9) can be used for hookworm infections that have been diagnosed as caused by *Necator americanus*, in which a single oral dose will cure more than half of the cases treated, and reduce worm burdens and egg production in the rest. (The other hookworm, *Ancylostoma duodenale*, is less responsive to this medication.) Under the influence of tetrachloroethylene the worms detach from the mucosa and do not reattach. They move along with the fecal stream to appear in the stool alive and motile; this effect has been interpreted as reversible paralysis of the worms.

Mebendazole (19), by oral administration repeated 3 or 4 d in a row, is most effective against whipworm (*Trichuris trichiura*). In addition, a single dose usually cures infection of the pinworm, and the drug is effective against hookworms and *Ascaris*, and partially effective against threadworms (*Strongyloides*) and taenioid tapeworms. This benzimidazole derivative is a broad-spectrum anthelmintic (48), and is used in veterinary medicine as well as in clinical practice. It belongs to the same series of drugs as thiabendazole (20), and they are closely related to fungicides (qv) such as benomyl [17804-35-2].

Mebendazole (19) interferes with the glucose metabolism of helminths, irreversibly inhibiting glucose uptake. Thereafter glycogen stores in the worms are depleted, and then ATP supplies fail. The worms are slowly immobilized, die, and are lost from the body over a period of three days. In developing *Trichuris* eggs, larvae fail to form normally. Part of the selective toxicity may be that mebendazole has an antimicrotubule action on the intestinal cells of nematodes but not mammals. Furthermore, relatively little is absorbed from the gastrointestinal tract. Adverse effects that may occur in patients under treatment are abdominal pain and diarrhea.

The adult female whipworm (*Trichuris trichiura*) is 5 cm long, and these worms lie with the thin anterior whip ends buried in the mucosa of the intestine, where they feed on tissue juice and small amounts of blood. Infections of several hundred worms may cause irritation and inflammation of the mucosa with abdominal pain, diarrhea, and gas. Untreated adult worms live for years.

Treatment of Tissue Roundworm (Nematode) Infections

Thiabendazole (20) is an effective oral drug against many roundworms that are affected by other anthelmintics, and also works against some nematode tissue parasites that are not successfully treated with other agents. The threadworm (*Strongyloides stercoralis*) that lives embedded in the intestinal wall was a therapeutic problem until the effectiveness of thiabendazole was discovered; and it was the first drug reported to have a beneficial influence on trichinosis, the disease in which *Trichinella spiralis* larvae migrate through skeletal muscle. This drug is prophylactic against trichinosis in persons known to have just eaten infected pork. It can be used in persons with guineaworm (*Dracunculus medinensis*), or for treating cutaneous larva migrans (*Ancylostoma braziliense*) and visceral larva migrans (*Toxocara*). Among the intestinal nematodes, it works against pinworm (*Enterobius*); and somewhat against hookworms, roundworm, and whipworm; the drug can be given to persons with multiple infection of the intestine (47). Extensively used in veterinary medicine, thiabendazole was the first broad-spectrum benzimidazole anthelmintic.

The mechanism by which nematodes are killed is unknown; however, one of the biochemical effects of thiabendazole is to inhibit fumarate reductase, an enzyme that generates ATP as it converts fumarate to succinate in the mitochondria of helminths. The selective toxicity may be caused by differences between enzymes of parasite and host, since the drug is rapidly absorbed, with 90% of a dose passed in the urine in the first day (mostly as metabolic products, one of which may give urine and sweat a distinctive odor).

One ppm of thiabendazole (20) will prevent *Ascaris* eggs from maturing and, in therapeutic topical concentrations, it is used to prevent development of dog and cat hookworm larvae that have invaded human skin to cause cutaneous larva migrans. In laboratory animals thiabendazole is antipyretic, analgesic, and antiinflammatory, as well as anthelmintic. Probably this is also true in humans and may account for the symptomatic improvement in persons with early systemic trichinosis.

Among the adverse effects are digestive disturbances, neurological symptoms, and manifestations of allergic responses. As many as half the patients are incapacitated for several hours after taking the drug. Overall, effects are dose related and transient.

Niridazole (6), already discussed with drugs used for schistosomaisis, is employed in the treatment of infection with the guineaworm (*Dracunculus medinensis*), as is metronidazole [443-48-1] (Flagyl), the broad-spectrum antiprotozoal agent (see Chemotherapeutics, antiprotozoal). Stringlike worms about 1 m long but less than 2 mm dia can be seen and felt below the skin surface, with the head of the worm usually exposed in an ulcer on the foot. The far end of the worm is hooked, so that although it is possible to grasp the front end and pull, the worm does not slide out easily and is apt to break with infectious, allergic, and toxic consequences as the remainder under the skin dies and deteriorates. The treatment is to give niridazole (6) orally two or three times a day for 7 d. With the effect of the drug, either the worm comes out or it can be pulled easily without danger. If left untreated, within a month the worm may come out naturally, or withdraw from the opening and be resorbed.

Diethylcarbamazine (21) has been successfully employed for decades by oral administration to cure infection with the species that causes elephantiasis, and to treat the other filariases (50). It kills microfilariae of the filarial worms: *Wuchereria bancrofti, Brugia malayi, Loa loa, Mansonella ozzardi, Dipetalonema streptocerca, D. perstans*, and *Onchocerca volvulus* (however, onchocerciasis recurs after therapy). There is a disease condition called eosinophilic lung, or tropical eosinophilia, the symptoms of which can be relieved by diethylcarbamazine. This drug is one of the anthelmintics that may have pharmacological effects that enhance immunological response. Another such anthelmintic drug is levamisole [14769-73-4, 53631-68-8] (22). Diethylcarbamazine is used in veterinary medicine.

(22)

This derivative of piperazine affects microfilariae so that they become susceptible to phagocytosis by fixed macrophages of the reticulo–endothelial system. Adult *Wuchereria* and *Loa* are killed. Why this drug is selectively toxic is not known. It is well absorbed and widely distributed, and does not cumulate in parasites, neither microfilariae nor adults. Adults of *Wuchereria bancrofti,* the filarial worm that causes elephantiasis, are coiled in the lymph system where females attain a length of 10 cm. Over the years, tissue reactions result in obstruction to lymph return while lymph nodes, lymph vessels, and the spleen may enlarge. Elephantiasis is a late and unusual complication when dependent parts of the body become edematous, enlarge, and then in time become firm and covered with a rough nodular skin. Untreated adult worms live for 5–10 yr.

Loa loa females are about 6 cm long and migrate constantly through connective tissues. The disease caused is Calabar swellings, which are local responses to worms and/or their products that last two or three days and then regress. A worm may get into the anterior chamber of the eye and the patient may be able to see it, or sometimes one eye is puffed shut when a worm is in the vicinity; because of these effects *Loa* is called the eye worm. The untreated worms will live as long as 10 yr. The seriousness of other filarial infections varies all the way from symptomless with *Mansonella* to cosmetic disfiguration and blindness with *Onchocerca*. The regimen for diethylcarbamazine varies with the condition being treated.

Adverse effects of nausea, fever, headache, and dizziness may occur during diethylcarbamazine therapy and simultaneous administration of corticosteroids and antihistamines is intended to reduce the probability of rash, itch, edema, and other more severe allergic responses to the disintegration of worms. Care must be exercised in the management of patients infected with *Onchocerca*. They have a violent reaction within the first day, and toxicity and fever can last as long as a week. Supplementary treatment with Suramin [129-46-4] (an antitrypanosomal drug) is required to rid patients of adult *Onchocerca* (51).

BIBLIOGRAPHY

"Parasitic Infections, Chemotherapy" in *ECT* 2nd ed., Vol. 14, pp. 532–551, by Helmut Mrozik, Merck & Co., Inc.

1. U.S. Pat. 2,335,585 (Nov. 30, 1943), N. A. Davies (to American Cream Tartar).
2. U.S. Pat. 2,391,297 (Dec. 18, 1945), N. A. Davies (to Stauffer Chemical Co.).
3. U.S. Pat. 2,918,401 (Sept. 9, 1959), F. C. Copp (to Burroughs Wellcome).
4. Ger. Pat. 1,117,600 (Nov. 23, 1961), F. C. Copp (to Wellcome Foundation).
5. Ger. Pat. 583,055 (Aug. 28, 1933), F. Muth (to I. G. Farbenind A.G.).
6. U.S. Pat. 2,849,494 (Aug. 26, 1958), R. H. Cooper and K. Goldberg (to Monsanto).
7. U.S. Pat. 2,232,970 (Mar. 4, 1941), H. Ardersag, S. Breitner, and H. Jung (to Winthrop Chemical Co.).
8. U.S. Pat. 2,334,408 (Nov. 16, 1944), W. S. Gump and M. Luthy (to Burton T. Bush).
9. U.S. Pat. 2,467,893; 2,467,894; 2,467,895 (Apr. 19, 1949), S. Kushner and L. Brancone (to American Cyanamid).
10. U.S. Pat. 1,717,101 (June 11, 1929), H. Hirzel (to Sharp & Dohme).
11. U.S. Pat. 3,294,598 (Apr. 4, 1967), G. P. Rosi and D. Peruzzotti (to Sterling Drug).
12. W. Kikuth and R. Gönnert, *Ann. Trop. Med. Parisitol.* **42,** 256 (1948).
13. U.S. Pat. 3,657,267 (Feb. 18, 1971), J. C. H. Van Gelder (to Janssen).
14. U.S. Pat. 3,079,297 (Feb. 26, 1963), E. Schraufstraller and R. Gonnert (to Bayer).
15. Belg. Pat. 632,989 (Nov. 29, 1963) (to Ciba Ltd.).
16. U.S. Pat. 2,901,482 (Aug. 25, 1959), G. F. MacKenzie and K. L. Turbin (to The Dow Chemical Co.).

17. S. Afr. Pat. 68 00516 (June 27, 1968), R. V. Kasubrick and J. W. McFarland (to Chas. Pfizer and Co.).
18. U.S. Pat. 2,952,419 (Feb. 6, 1960), E. F. Elslager and D. F. Worth (to Parke, Davis).
19. U.S. Pat. 2,113,357 (Apr. 15, 1939), F. Mietzsch and H. Mauss (to Winthrop Chemical Co.).
20. U.S. Pat. 2,880,222 (Mar. 31, 1959), E. A. H. Friedheim.
21. U.S. Pat. 1,549,154 (Aug. 11, 1925); 1,873,668 (Aug. 23, 1932), H. Schmidt (to Winthrop Chemical Co.).
22. U.S. Pat. 3,040,109 (June 9, 1959), R. E. Feathers and R. H. Rogerson (to Pittsburgh Plate Glass).
23. U.S. Pat. 3,017,415 (June 16, 1962), L. Sarett (to Merck & Co.).
24. H. Most, *N. Engl. J. Med.* **287,** 495 (1972).
25. *Ibid.,* p. 698.
26. *The Roche Vademecum 1976,* F. Hoffmann-LaRoche & Co. Limited, Basle, Switz., 1976, pp. 31–33.
27. *Informational Material for Physicians Sodium Antimony Dimercaptosuccinate (Astiban),* Parasitic Diseases Branch, Center for Disease Control, Atlanta, Ga., 1968, pp. 1–15.
28. H. J. Saz and E. Bueding, *Pharmacol. Rev.* **18,** 871 (1966).
29. S. Archer and A. Yarinsky, *Progr. Drug Res.* **16,** 11 (1972).
30. *Informational Material for Physicians Niridazole (Ambilhar),* Parasitic Diseases Branch, Center for Disease Control, Atlanta, Ga., 1976, pp. 1–11.
31. N. Katz, *Adv. Pharmacol. Chemother.* **14,** 1 (1977).
32. G. J. Frayha, *Leban. Med. J.* **25,** 507 (1972).
33. U. K. Sheth, *Prog. Drug Res.* **19,** 147 (1975).
34. M. Katz, *Drugs* **13,** 124 (1977).
35. J. Putter, *Z. Parasitenk.* **34,** 23 (1970).
36. J. Putter and P. Andrews, *Conference of the German Society for Parasitology, Wuppertal, Apr. 9–11, 1970.*
37. *Informational Material Yomesan (Niclosamide),* Parasitic Diseases Branch, Center for Disease Control, Atlanta, Ga., 1976, pp. 1–7.
38. J. S. Swartzwelder, *J. La. State Med. Soc.* **111,** 394 (1959).
39. R. B. Burrows, *Progr. Drug Res.* **17,** 108 (1973).
40. G. Gras, *Med. Afr. Noire* **21,** 11 (1974).
41. D. A. Ogunmekan, *West Afr. Med. J.* **22,** 47 (1973).
42. H. S. Pond and co-workers, *South. Med. J.* **63,** 599 (1970).
43. M. A. Haleem, F. A. Lari, and R. J. Rahimtoola, *J. Pak. Med. Assoc.* **22,** 276 (1972).
44. W. Hatchuel, M. Isaacson, and D. J. de Villiers, *South Afr. Med. J.* **47,** 91 (1973).
45. W. J. Bell and S. Nassif, *Am. J. Trop. Med. Hyg.* **20,** 584 (1971).
46. T. Ishizaki and M. Yokogawa, *Yonsei Rep. Trop. Med.* **4,** 159 (1973).
47. E. Bueding and C. Swartzwelder, *Pharmacol. Rev.* **9,** 329 (1957).
48. J. W. McFarland, *Prog. Drug Res.* **16,** 157 (1972).
49. E. Barrett-Connor, *Am. J. Gastroenterol.* **63,** 105 (1975).
50. J. F. Maldonado and co-workers, *P. R. J. Public Health and Trop. Med.* **25,** 291 (1950).
51. *Informational Material for Physicians Suramin (Bayer 205),* Parasitic Diseases Branch, Center for Disease Control, Atlanta, Ga., 1966, pp. 1–11.

General References

American Medical Association Department of Drugs, "Anthelmintics" in *AMA Drug Evaluations,* 3rd ed., Publishing Sciences Group, Inc., Littleton, Mass., 1977, pp. 862–876.
C. L. Bailey and J. D. Shoft, eds., *APhA Drug Names,* American Pharmaceutical Association, Washington, D.C., 1976.
R. Berkow, ed., "Diseases Caused by Worms" in *The Merck Manual,* 13th ed., Merck Sharp & Dohme Laboratories, Rahway, N. J., 1977, pp. 166–177.
D. R. Botero, "Chemotherapy of Human Intestinal Parasitic Diseases" in *Annu. Rev. Pharmacol. Toxicol.* **18,** 1 (1978).
L. A. Bulla, Jr., and T. C. Cheng, eds., "Pathobiology of Invertebrate Vectors of Disease" in *Ann. N.Y. Acad. Sci.* **266,** 332 (1975).
W. C. Campbell and H. Mrozik, "Antiparasitic Agents" in *Annu. Rep. Med. Chem.* **9,** 115 (1974).
L. L. Corrigan, ed., *Evaluations of Drug Interactions,* 2nd ed., American Pharmaceutical Association, Washington, D.C., 1976, 476 pp.

E. C. Faust, P. F. Russell, and R. C. Jung, "Helminths and Helminthic Infections" in *Craig and Faust's Clinical Parasitology,* 8th ed., Lea & Febiger, Philadelphia, Pa., 1970, pp. 251–570.

F. C. Goble and B. G. Maegraith, eds., "The Pharmacological and Chemotherapeutic Properties of Niridazole and other Antischistosomal Compounds" in *Ann. N.Y. Acad. Sci.* **160,** 423 (1969).

A. Goth, "Anthelmintic Drugs" in *Medical Pharmacology* 8th ed., C.V. Mosby, St. Louis, Mo., 1976, pp. 625–635.

G. W. Hunter, III, J. C. Swartzwelder, and D. F. Clyde, "Helminthic Diseases" in *Tropical Medicine,* 5th ed., W. B. Saunders, Philadelphia, Pa., 1976, pp. 451–621.

B. H. Kean and D. W. Hoskins, "Drugs for Intestinal Parasitism" in W. Modell, ed., *Drugs of Choice 1976–77,* C.V. Mosby, St. Louis, Inc., 1976, pp. 356–369.

B. H. Kean, K. E. Mott, and A. J. Russell, *Tropical Medicine and Parasitology, Classic Investigations,* Vols. I and II, Cornell University Press, Ithaca, N. Y. 1978, 678 pp.

A. Korolkovas and J. H. Burckhalter, "Anthelmintic Agents" in *Essentials of Medicinal Chemistry,* John Wiley & Sons., Inc., New York, 1976, pp. 387–402.

K. MacDonald, "Hookworm Disease" in F. H. Top, Sr., and P. F. Wehrel, eds., *Communicable and Infectious Diseases,* 8th ed., C.V. Mosby, St. Louis, Mo., 1976, pp. 359–361.

K. MacDonald, "Larva Migrans, Visceral and Cutaneous" in F. H. Top, Sr., and P. F. Wehrle, eds., *Communicable and Infectious Diseases,* 8th ed., C.V. Mosby, St. Louis, Mo., 1976, pp. 379–383.

E. K. Markell and M. Voge, *Medical Parasitology,* 14th ed., W.B. Saunders, Philadelphia, Pa., 1976, pp. 167–304.

P. D. Marsden and K. S. Warren, "Helminthic Disease" in P. B. Beeson and W. McDermott, eds., *Textbook of Medicine,* 14th ed., W.B. Saunders, Philadelphia, Pa., 1975, pp. 506–539.

E. J. Martin, "Antiparasitic Agents" in *Annu. Rep. Med. Chem.* **11,** 121 (1976).

E. J. Martin, "Antiparasitic Agents" in *Annu. Rep. Med. Chem.* **10,** 154 (1975).

A. W. Mathies, Jr., and K. MacDonald, *Trichinosis* in F. H. Top, Sr., and P. F. Wehrle, eds., *Communicable and Infectious Diseases,* 8th ed., C.V. Mosby, St. Louis, Mo., 1976, pp. 719–723.

J. J. Plorde, I. L. Bennett, Jr., and R. G. Petersdorf, "Diseases Caused by Worms" in W. M. Wintrobe and co-eds., *Harrison's Principles of Internal Medicine* 7th ed., Mack Publishing Co., Easton, Pa., 1974, pp. 1035–1058.

W. B. Pratt, "Chemotherapy of Helminthic Diseases" in *Chemotherapy of Infection,* Oxford University Press, New York, 1977, pp. 373–407.

I. M. Rollo, "Drugs Used in the Chemotherapy of Helminthiasis" in L. S. Goodman and A. Gilman, eds., *The Pharmacological Basis of Therapeutics,* 5th ed., MacMillan, New York, 1975, pp. 1018–1044.

I. S. Rossoff, *Handbook of Veterinary Drugs,* Springer Publishing Company, New York, 1974, 730 pp.

E. A. Swinyard, "Parasiticides Anthelmintics" in A. Osol and J. W. Hoover, eds., *Remington's Pharmaceutical Sciences,* 15th ed., Mack Publishing Company, Easton, Pa., 1975, pp. 1172–1178.

J. H. Thompson, "Drugs Used in the Treatment of Helminthiasis" in J. A. Bevan, ed., *Essentials of Pharmacology,* 2nd ed., Harper & Row, Hagerstown, Md., 1976, pp. 512–518.

A. Wade, ed., *Martindale, The Extra Pharmacopeia,* 27th ed., The Pharmaceutical Press, London, Eng., 1977, pp. 98–115, 1370–1379.

C. C. Wang and M. H. Fisher, "Antiparasitic Agents" in *Annu. Rep. Med. Chem.* **12,** 140 (1977).

C. Wilcocks and P. E. C. Manson-Bahr, "Diseases Caused by Helminths" in *Manson's Tropical Diseases,* 17th ed., Williams & Wilkins Co., Baltimore, Md., 1972, pp. 192–354.

M. Windholz, ed., *The Merck Index,* 9th ed., Merck & Co., Inc., Rahway, N.J., 1976, 9856 pp.

JAMES W. INGALLS
Arnold & Marie Schwartz College of Pharmacy and Health Sciences

CHEMOTHERAPEUTICS, ANTIMITOTIC

Cancer is still one of the scourges of man, but encouraging results with various types of treatment offer hope for continuing advances in the fight against this disease. The most frequently used treatments at present are surgery, radiation (including radioactive isotopes), and chemotherapy, with future possibilities for immunotherapy.

The earliest reference to the chemical treatment of cancer describes the use of arsenic paste in ancient Egypt; later, caustic pastes were used by Hippocrates (1). Little progress was made until the 19th century, when potassium arsenite [10124-50-2], the first systemically effective antitumor agent, was for a time the treatment of choice for leukemia (2), and toxins were used to treat malignancies (3). Subsequently, a great variety of local and systemic agents were used for antineoplastic therapy with varying degrees of success. Hormones (qv) represented the next chemotherapeutic advance, with the first reports of the use of sex hormones in breast cancer in 1896 (4) and in prostatic cancer in 1941 (5). The modern era in cancer chemotherapy, however, began with the introduction of the polyfunctional alkylating agents, such as nitrogen mustard [555-77-1] in malignant lymphoma, in the early 1940s (6–8). Soon after, the folic acid antagonist aminopterin [54-62-6] was found to induce remission in acute lymphocytic leukemia in children (9). By 1950 useful agents had begun to appear in rapid succession (10–12); since then, whole classes of antitumor drugs have become available. With the establishment of a national development program in 1955, the chemical treatment of cancer has joined surgery and radiation therapy as a standard approach.

Although most current antineoplastic drugs were discovered empirically, considerable insight has been gained into the mechanisms by which many of these compounds affect cell growth, and this has allowed a more rational therapeutic application of these agents. This insight has also led to the development of new drugs designed to kill the cancer cell either directly or by depleting its essential growth elements. In addition to the discovery of new experimental compounds, chemotherapy research has been directed toward alteration in dosimetry and employment of various drug combinations to enhance tumoricidal effects and decrease toxicity. Palliative benefits are definitely attainable and include prolongation of useful survival as well as subjective and objective remission of physical and emotional disability. Decreases in tumor mass and metastatic involvement may be obtained with chemotherapy alone. Moreover, it has proved to be a valuable adjunct to surgical and irradiation procedures. Today, several neoplastic diseases (Burkitt's lymphoma, choriocarcinoma, acute leukemia, etc) can be associated with a normal life expectancy after drug treatment, alone or in combination with other types of therapy.

Drug Classification

Many compounds have been investigated in experimental animals, and a few have proven sufficiently useful in the clinical treatment of human neoplasms, at acceptable levels of toxicity, to deserve the designation of chemotherapeutic agents. The drugs currently used in the chemotherapy of malignant diseases may be divided into several classes according to their general pharmacologic activity: alkylating agents; antimetabolites; antibiotics; plant alkaloids; miscellaneous agents; and hormones.

Agents that do not fit any of the other categories are discussed as a separate miscellaneous group. This classification serves as a convenient framework for describing the various types of agents. In addition, functional classification becomes increasingly important as investigators attempt to use this information to design rational chemotherapeutic regimens.

The compounds discussed here all fall under the jurisdiction of the FDA and are, for the most part, those with long and successful clinical use, although a few have been included either because they illustrate special circumstances or because they represent newer developments. Excluded are several compounds whose structural variations offer no particular advantage over existing drugs or that need additional investigation.

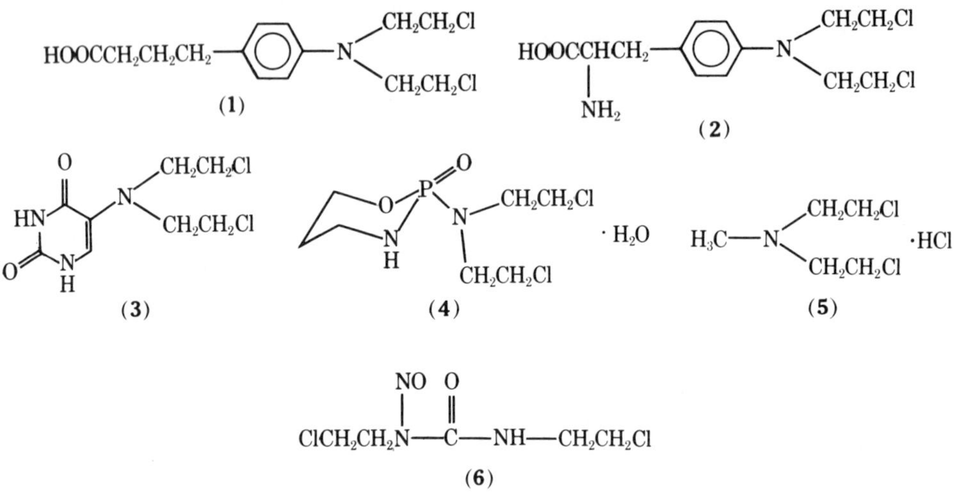

Figure 1. Alkylating agents.

Table 1. Alkylating Agents

Drug, trade name	CAS Registry No.	mp, °C	Dosage form and dosage
(1) chlorambucil USP, Leukeran	[305-03-3]	64–66	tablet: 2 mg oral initial: 0.1–0.2 mg/(kg·d) for 3–6 wk maintenance: 2–4 mg/d
(2) melphalan USP, Alkeran	[148-82-3]	182–183 (dec)	tablet: 2 mg oral initial: 6 mg once a d for 2–3 wk maintenance: 2–4 mg once a d
(3) uracil mustard NF, Uracil Mustard	[66-75-1]	206 (dec)	capsule: 1 mg oral: 1–2 mg/d for 3 wk, repeat after 1 wk interval or 3–5 mg/d for 7 d, then 1 mg/d for 3 wk

Also excluded are hormones, which are not specific oncolytic agents but act by altering the hormonal environment of endocrine-dependent cancers.

Dosage schedules (which depend on such factors as the clinical indication, drug toxicity, and the patient's nutritional and functional status, and vary more widely than indicated in this survey), available sizes, toxicity (not a comprehensive review), and the clinical indications for use (not an exhaustive listing), are shown under the individual drugs. The proprietary names of the drug are given wherever possible.

Alkylating Agents. These agents (Fig. 1, Table 1) include a diverse group of chemicals (nitrogen mustards (1–6), nitrosoureas (7–8), triazenes (10), etc) that have

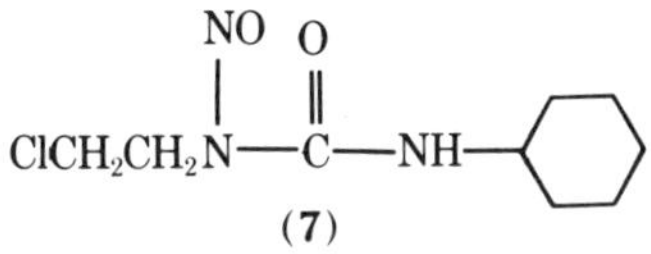

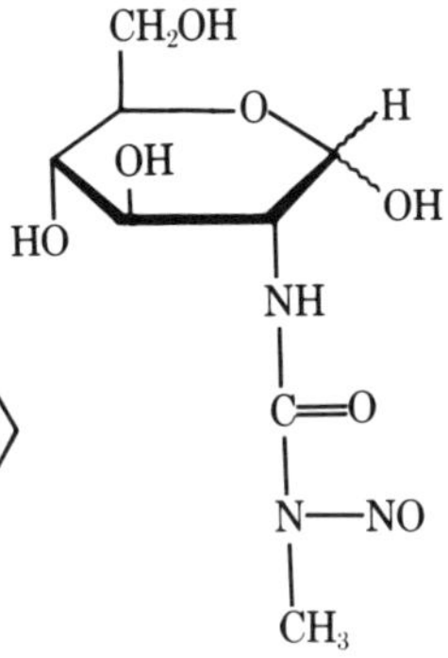

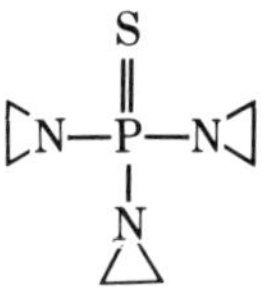

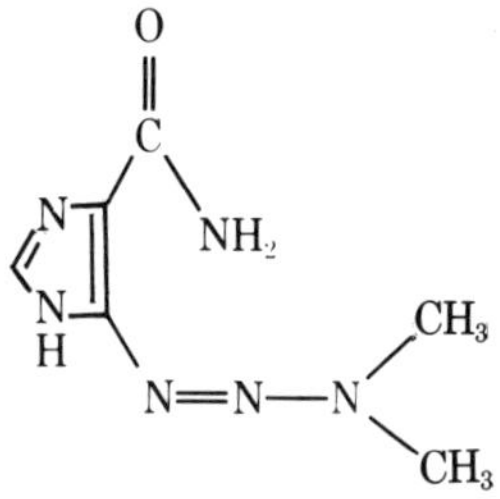

Fig. 1. (*continued*)

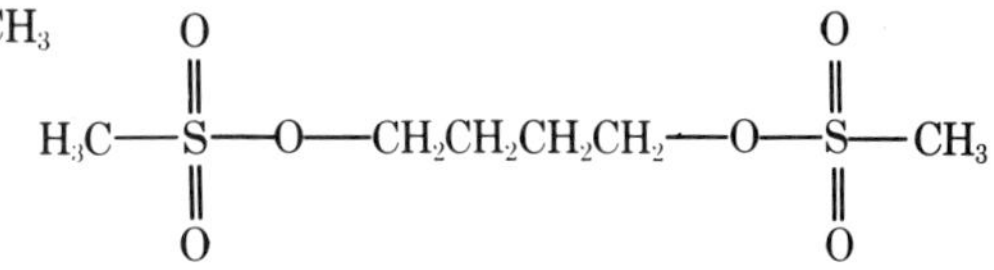

Disease	Toxic effects	Manufacturer
chronic lymphocytic leukemia; cancer of ovary, breast, testis; Hodgkin's disease; non-Hodgkin's lymphomas	bone-marrow depression; nausea; vomiting	Burroughs Wellcome
multiple myeloma; plasmacytic myeloma; cancer of breast and ovary	bone-marrow depression; nausea; vomiting; anorexia	Burroughs Wellcome
Hodgkin's disease; non-Hodgkin's lymphomas; cancer of ovary; chronic lymphocytic leukemia; primary thrombocytosis	bone-marrow depression; dermatitis; diarrhea; nausea; vomiting	Upjohn

Table 1. (continued)

Drug, trade name	CAS Registry No.	mp, °C	Dosage form and dosage
(4) Cyclophosphamide USP, Cytoxan	[6055-19-2]	41–45	vials: 100, 200, 500 mg tablets: 25, 50 mg intravenous (iv), initial: 10–20 mg/kg once a d for 2–5 d maintenance: 10–15 mg/kg every 7–10 d or 3–5 mg/kg two times a wk oral: 1–5 mg/kg once a d
(5) mechlorethamine hydrochloride USP, Mustargen hydrochloride	[55-86-7]	108–111	vial: 10 mg iv: 0.4 mg/kg as a single dose or 0.1 mg/kg once a d for 4 d intraperitoneal (ip) or intrapleural: 0.2–0.4 mg/kg as a single dose
(6) carmustine, BCNU	[154-93-8]	30–32	vial: 100 mg iv: 100–200 mg/m^2 once a d for 1–2 d; do not repeat for at least 6 wk
(7) lomustine, CCNU	[13010-47-4]	89	capsules: 10, 40, 100 mg oral: 100–130 mg/m^2 as a single dose every 6 wk or 75 mg/m^2 as a single dose every 3 wk
(8) streptozocin, investigational drug	[18883-66-4]	115 (dec)	investigational drug iv, intraarterial: 1 g/m^2 once a wk for 4 wk
(9) thiotepa NF, Thiotepa	[52-24-4]	52–57	vial: 15 mg iv, intramuscular (im): 0.2 mg/(kg·d) or 10–30 mg once a wk intracavitary: 45–60 mg/wk or 0.6–0.8 mg/kg once a wk every 1–4 wk bladder instillation: 60 mg once a wk for 4 wk
(10) dacarbazine, DTIC	[4342-03-4]	204–207	vials: 100, 200 mg iv: 2.0–4.5 mg/(kg·d) for 10 d, repeat every 28 d or 250 mg/(m^2·d) for 5 d, repeat every 21 d
(11) busulfan USP, Myleran	[55-98-1]	114–118	tablet: 2 mg oral initial: 2–8 mg once a d for 2–3 wk maintenance: 1–3 mg once a d

Disease	Toxic effects	Manufacturer
acute and chronic lymphocytic leukemia; lung cancer; rhabdomyosarcoma; neuroblastoma; ovarian and mammary carcinoma; multiple myeloma; lymphosarcoma; Burkitt's lymphoma; Hodgkin's disease; retinoblastoma; mycosis fungoides	bone-marrow depression; hepatic toxicity; cystitis; alopecia; nausea; vomiting	Mead Johnson (division of Bristol-Meyers)
Hodgkin's disease; non-Hodgkin's lymphomas; lymphosarcoma; cancer of breast, ovary, lung; neoplastic effusion	bone-marrow depression; nausea; vomiting; anorexia; diarrhea; local irritation	Merck, Sharp, and Dohme
Hodgkin's disease; non-Hodgkin's lymphomas; meningeal leukemia; brain tumor; malignant melanoma; GI cancer; renal cell cancer; breast cancer; lung cancer	bone-marrow depression; hepatic toxicity; nausea; vomiting	Bristol Labs (division of Bristol-Meyers)
malignant brain tumors; Hodgkin's disease; non-Hodgkin's lymphomas; malignant melanoma; multiple myeloma; cancer of lung, GI tract, breast, renal cell	bone-marrow depression; hepatic toxicity	Bristol Labs (division of Bristol-Meyers)
malignant pancreatic islet-cell tumors; malignant carcinoid	bone-marrow depression; renal and hepatic toxicity; nausea; vomiting	
cancer of breast, ovary, lung, bladder; Hodgkin's disease; non-Hodgkin's lymphomas; retinoblastoma; neoplastic effusion	bone-marrow depression; amenorrhea; anorexia; nausea; vomiting	Lederle (division of American Cyanamid)
malignant melanoma; Hodgkin's disease; soft-tissue sarcomas	bone-marrow depression; flu-like syndrome; alopecia; nausea; vomiting; anorexia	Dome (division of Miles Laboratories)
chronic granulocytic leukemia; other myeloproliferative disorders	bone-marrow depression; hyperuricemia; gynecomastia; amenorrhea; skin pigmentation; nausea; vomiting; diarrhea	Burroughs Wellcome

in common the ability to form covalent linkages with various substances, including such important moieties as phosphate, amino, sulfhydryl, hydroxyl, carboxyl, and imidazole groups, in biologically vital macromolecules (see Biopolymers). The key biological compound affected is the purine base, guanine, in the nucleic acids of deoxyribonucleic acid (DNA), in which an alkyl group is substituted for the hydrogen on the N-7 (alkylation). However, less extensive alkylation of the other DNA bases, or of an amino group, or of a sulfhydryl radical of a cell protein, may also occur. These reactions usually lead to gene miscoding, serious damage to the DNA molecule, or major disruption in nucleic acid function, any of which could explain both the mutagenic and cytotoxic effects of alkylating agents. In addition, inhibition of a wide variety of other cell functions, such as glycolysis and respiration, can lead to equally harmful effects on cell viability. The reactions result in the rapid disruption or destruction of

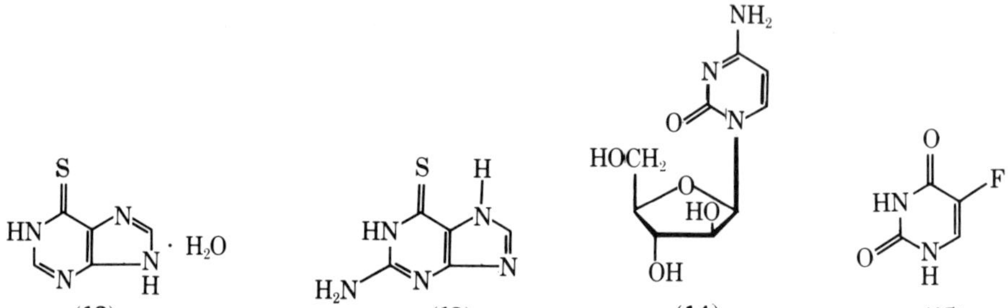

Figure 2. Antimetabolites.

Table 2. Antimetabolites

Drug, trade name	CAS Registry No.	mp, °C	Dosage form and dosage
(12) mercaptopurine USP, Purinethol	[6112-76-1]	308 (dec)	tablet: 50 mg oral: 2.5–5.0 mg/kg once a d
(13) thioguanine USP, Thioguanine	[154-42-7]	>360	tablet: 40 mg oral: 2 mg/kg once a d for 4 wk; if no improvement, increase to 3 mg/kg once a d
(14) cytarabine USP, Cytosar	[147-94-4]	212–213	vials: 100, 500 mg initial iv: 2 mg/kg once a d for 10 d, then 4 mg/kg once a d iv infusion: 0.5–1.0 mg/kg once a d over a period of 1–24 h for 10 d, then 2 mg/kg once a d maintenance sc: 1 mg/kg once or twice a wk

the fundamental mechanisms concerned with cell division, growth, differentiation, and function. Tumor shrinkage can occur in 1 or 2 days with intravenous drug administration. With few drug exceptions, if one alkylator is ineffective in an individual patient, so are others. All of these agents are potentially mutagenic, teratogenic, and carcinogenic themselves. The toxicity of the alkylating agents for cell functions related and unrelated to cell proliferation explains the antineoplastic activity of these agents throughout the mitotic cycle (cell-cycle independent).

Antimetabolites. This group of antineoplastic drugs (Fig. 2, Table 2) is antagonistic to normal metabolites essential for the synthesis of DNA. These compounds compete with and displace the substrate of specific enzymes involved in DNA synthesis. The reaction between the antimetabolite and the enzyme interferes with the synthesis of nucleic acid for DNA production and therefore inhibits cell reproduction. Antimetabolites can be classified according to their specific inhibitory action (antagonists of purine, pyrimidine, glutamine, etc). These drugs act considerably slower than the alkylating agents, with tumor shrinkage observed only after 4–8 wk of treatment; their cytotoxicity is cell-cycle dependent.

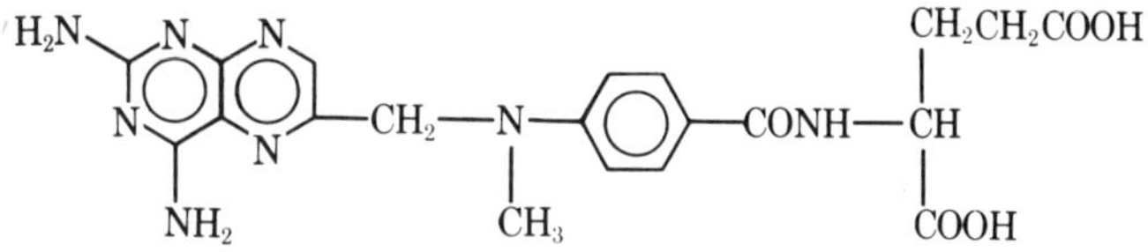

(16)

Fig. 2. (*continued*)

Disease	Toxic effects	Manufacturer
acute leukemia (more effective in children than in adults); chronic granulocytic leukemia	bone-marrow depression; hepatic toxicity; anemia; gastrointestinal (GI) ulceration; nausea; vomiting	Burroughs Wellcome
acute leukemia; chronic granulocytic leukemia	bone-marrow depression; stomatitis; anorexia; nausea; vomiting	Burroughs Wellcome
acute granulocytic leukemia (adults); acute lymphocytic leukemia (children); Hodgkin's disease	bone-marrow depression; hepatic toxicity; megaloblastosis; nausea; vomiting; diarrhea	Upjohn

476   CHEMOTHERAPEUTICS, ANTIMITOTIC

Table 2. (*continued*)

Drug, trade name	CAS Registry No.	mp, °C	Dosage form and dosage
(15) fluorouracil USP, Fluorouracil	[51-21-8]	282–283 (dec)	vial: 500 mg iv initial: 12 mg/kg once a d for 4 d, then 6 mg/kg every other d for 4 doses maintenance: repeat initial dose once a month or 10–15 mg/kg, not exceeding 1 g, once a wk as a single dose
(16) methotrexate USP, Methotrexate	[59-05-2]	185–204 (dec)	tablet: 2.5 mg vials: 5, 50 mg oral: 2.5–30.0 mg/d iv, im: 2.5–30.0 mg/d intraarterial: 50 mg/d plus leucovorin im 6–9 mg/4–6 h intrathecal: 0.2–0.5 mg/kg every 2–5 d

Antibiotics. These compounds (Fig. 3, Table 3), which are chemical substances produced by certain microorganisms (actinomycetes, fungi, bacteria), suppress the growth of or destroy other microorganisms, and are also being used as antagonists of tumor cells. The mechanisms of their antineoplastic, bactericidal, and bacteriostatic actions are similar (see Antibiotics, peptides). In general, these agents bind to DNA, thus inhibiting DNA-dependent ribonucleic acid (RNA) synthesis and consequently the synthesis of proteins required by the cell. The apparent mechanism of action of one antibiotic, bleomycin (**21**), is unique. Its cytotoxicity seems most likely to be related to its ability to cause chain scission, nicking, and fragmentation of DNA molecules.

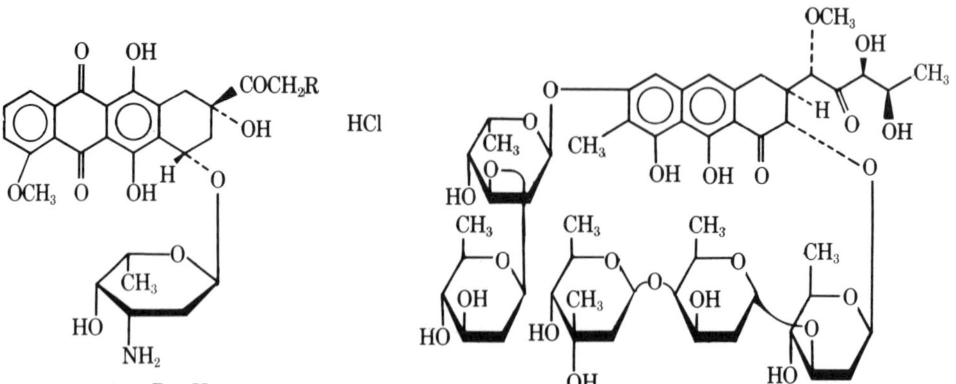

(17) R = H
(18) R = OH

(19)

Figure 3. Antibiotics.

Disease	Toxic effects	Manufacturer
cancer of breast, colon, stomach, pancreas, liver, ovary, prostate, esophagus, bladder, rectum	bone-marrow depression; dermatitis; alopecia; nausea; vomiting; diarrhea; stomatitis; anorexia; GI ulcers; skin pigmentation	Roche
acute lymphocytic leukemia; meningeal leukemia; choriocarcinoma; chorioadenoma destruens; lymphosarcoma; osteogenic sarcoma; cancer of lung, neck, head, cervix; mycosis fungoides; hydatidiform mole	bone-marrow depression; renal and hepatic toxicity; enteritis; stomatitis; alopecia; abdominal distress; erythematous rash; oral and GI ulceration; diarrhea; nausea; vomiting	Lederle (division of American Cyanamid)

In addition, repair of scission is inhibited by this antibiotic, leading to progressive fragmentation of the DNA chain. The bleomycin now used clinically is a mixture consisting mainly of the A$_2$ [*11116-32-8*] and B$_2$ [*9060-10-0*] forms. Daunorubicin (**17**) and doxorubicin (**18**) are representatives of an extensive series of anthracycline antibiotics being investigated in antimitotic chemotherapy. Antibiotics are not as frequently used as are alkylators and antimetabolites. Their cytotoxicity is cell-cycle dependent.

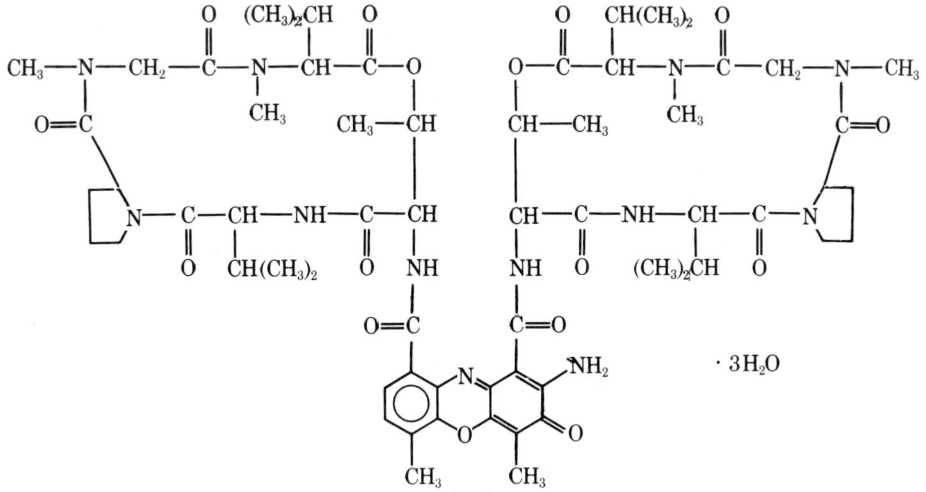

(**20**)

Fig. 3. (*continued*)

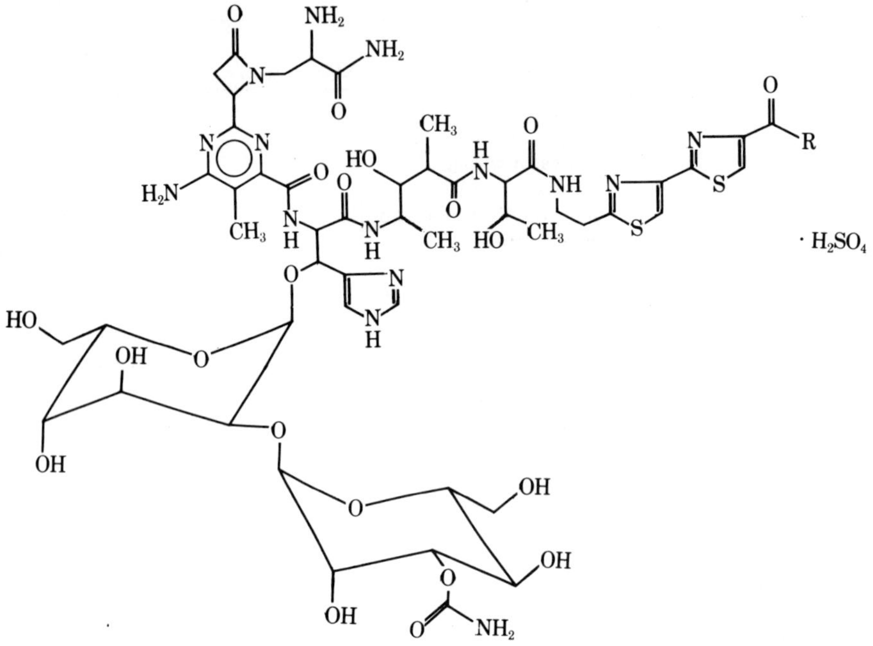

Fig. 3. (*continued*)

Table 3. Antibiotics

Drug, trade name	CAS Registry No.	mp, °C	Dosage form and dosage
(17) daunorubicin, investigational drug	[20830-81-3]	188–190 (dec)	vial: 20 mg iv: 30–60 mg/(m^2·d) for 3 d or once a wk or 0.8–1.0 mg/(kg·d) for 3–6 d
(18) doxorubicin, Adriamycin	[23214-92-8]	204–205	vials: 10, 50 mg iv: 60–75 mg/m^2 as a single dose every 21 d or 20–30 mg/m^2 d for 3 d, repeated every 28 d
(19) mithramycin USP, Mithracin	[18378-89-7]	180–183	vial: 2.5 mg iv infusion: 0.025–0.030 mg/kg over a period of 4–6 h once a day for 8–10 d
(20) dactinomycin USP, Cosmegen	[50-76-0]	241–243 (dec)	vial: 0.5 mg iv: 0.5 mg once a day for 5 d or 0.015 mg/(kg·d) for 5 d; if tolerated, repeat at 2–4 wk intervals

Bleomycin A$_2$, R = NHCH$_2$CH$_2$CH$_2$—$\overset{+}{S}$(CH$_3$)(CH$_3$)

Bleomycin B$_2$, R = NHCH$_2$CH$_2$CH$_2$CH$_2$NHC(=NH)(NH$_2$)

(21)

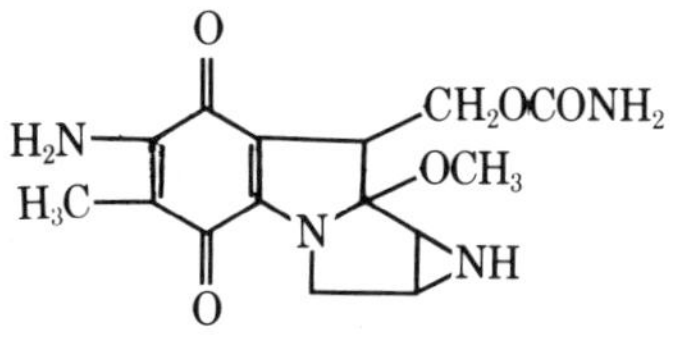

(22)

Fig. 3 (*continued*)

Disease	Toxic effects	Manufacturer
acute lymphocytic and granulocytic leukemia; lymphomas; solid tumors (children)	bone-marrow depression; cardiac toxicity; alopecia; stomatitis; GI disturbance	
soft-tissue and osteogenic sarcomas; Hodgkin's disease; non-Hodgkin's lymphomas; acute leukemia; cancer of thyroid, breast, lung, genitourinary (GU) tract; Wilm's tumor; neuroblastoma	bone-marrow depression; cardiac toxicity; alopecia; stomatitis; GI disturbance	Adria
testicular tumors; hypercalcemia and hypercalciuria associated with advanced malignancies	bone-marrow depression; hepatic and renal toxicity; hypocalcemia; hemorrhage; stomatitis; nausea; vomiting; anorexia; diarrhea	Pfizer
Wilm's tumor; Ewing's tumor; choriocarcinoma; testicular carcinoma; rhabdomyosarcoma; neuroblastoma; melanoma; soft-tissue and osteogenic sarcomas	bone-marrow depression; renal and hepatic toxicity; alopecia; mental depression; stomatitis; nausea; vomiting; diarrhea; anorexia; local irritation	Merck, Sharp, and Dohme

480 CHEMOTHERAPEUTICS, ANTIMITOTIC

Table 3. (continued)

Drug, trade name	CAS Registry No.	mp, °C	Dosage form and dosage
(21) bleomycin sulfate, Blenoxane	[11056-06-7]	196 (dec)	vial: 15 units iv, im, subcutaneous (sc): 0.25–0.50 units/kg (10–20 units/m^2) weekly or twice weekly; total doses exceeding 400 units should be given with great caution
(22) mitomycin, Mutamycin	[50-07-7]	>360	vial: 5 mg iv infusion: 10–20 mg/m^2 as a single dose iv: 2 mg/(m^2·d) for 5 d, 2 d rest, then repeat dose for 5 d; not recommended for use as a single agent for primary therapy; use only in combination with other drugs

Plant Alkaloids. The two clinically useful alkaloids (23–24) (Fig. 4, Table 4) are derived from the periwinkle plant, *Vinca rosea*, a species of myrtle. These substances interfere with mitosis by aborting it in the metaphase portion of the cycle (metaphase arrest), binding with the protein (tubulin) associated with the formation of the mitotic structure (spindle) used by the dividing cell. These alkaloids may also affect other

Table 4. Plant Alkaloids

Drug, trade name	CAS Registry No.	mp, °C	Dosage form and dosage
(23) vinblastine sulfate USP, Velban	[143-67-9]	284–285	vial: 10 mg iv: 3–10 mg/m^2 once a wk or every 2 wk or 0.1–0.5 mg/kg in weekly increments of 0.05 mg/kg once a wk or every 2 wk
(24) vincristine sulfate USP, Oncovin	[2068-78-2]	273–281	vials: 1, 5 mg iv: 1–2 mg/m^2 once a wk or 0.02–0.15 mg/kg once a wk

Disease	Toxic effects	Manufacturer
squamous cell carcinoma of head, neck, esophagus, skin, GU tract; testicular tumor; Hodgkin's disease; non-Hodgkin's lymphomas	pulmonary fibrosis; skin reactions; alopecia; nausea; vomiting; anorexia; fever; stomatitis	Bristol Labs (division of Bristol-Meyers)
chronic myelogenous leukemia; reticulum cell sarcoma; Hodgkin's disease; non-Hodgkin's lymphomas; cancer of stomach, pancreas, lung; epithelial tumors	bone-marrow depression; renal toxicity; alopecia; stomatitis; anorexia; nausea; vomiting	Bristol Labs (division of Bristol-Meyers)

cellular functions associated with tubulin, such as cellular movement and phagocytosis. Other cytologic effects involve aberrations of the cell nucleus (abnormal cleavage, condensation, etc). The cytotoxic effects, which are similar to that of older antimitotic agents such as colchicine [64-86-8] and podophyllotoxin [518-28-5], result in tumor cell death during replication, so that these agents are cell-cycle dependent (see Al-

Disease	Toxic effects	Manufacturer
Hodgkin's disease; lymphosarcoma, reticulum-cell sarcoma; neuroblastoma; choriocarcinoma; carcinoma of breast, lung, oral cavity, testis, bladder; acute and chronic leukemia; histiocytosis; mycosis fungoides	leukopenia; neurological toxicity (paresthesias, mental depression, loss of deep tendon reflexes, etc); dysfunction of autonomic nervous system (ileus, constipation, urinary retention, etc); alopecia; stomatitis; anorexia; diarrhea; nausea; vomiting; local irritation	Lilly
acute leukemia in children; lymphocytic leukemia; Hodgkin's disease; non-Hodgkin's lymphomas; Wilm's tumor; neuroblastoma; rhabdomyosarcoma; reticulum-cell sarcoma	neurological toxicity (paresthesias, foot drop, double vision, etc); constipation; ileus; alopecia; leukopenia (occasional); hyponatremia	Lilly

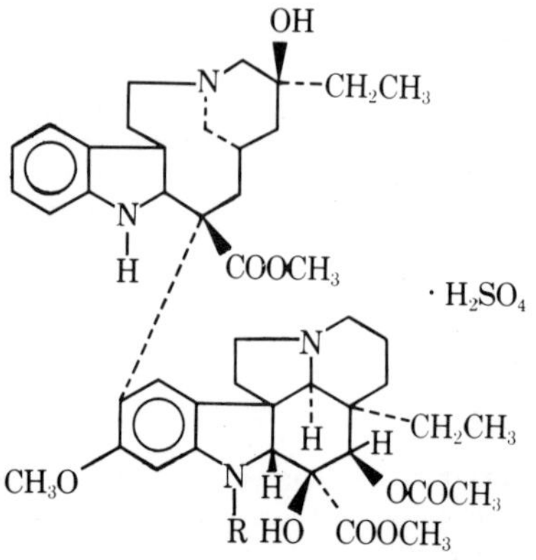

482 CHEMOTHERAPEUTICS, ANTIMITOTIC

(23) R = CH₃
(24) R = CHO

Figure 4. Plant alkaloids.

kaloids). The two alkaloids differ notably in their potency, clinical use, and toxicity, and this may be caused by variation in their ability to enter specific types of cells. The relatively low toxicity of vincristine (24) for normal cells makes it unusual among antineoplastic drugs and useful in the presence of impaired bone-marrow function.

Miscellaneous Agents. The following agents, which do not fit any of the preceding categories, are shown in Figure 5 and listed in Table 5.

Table 5. Miscellaneous Agents

Drug, trade name	CAS Registry No.	mp, °C	Dosage form and dosage
(25) asparaginase, Elspar	[9015-68-3]		vial: 10,000 IU iv: 200 IU/(kg·d) for 28 d or 1000 IU/(kg·d) for no more than 10 d
(26) hydroxyurea USP, Hydrea	[127-07-1]	133–136	capsule: 500 mg oral: 80 mg/kg as a single dose every 3 d or 20–30 mg/kg once a d
(27) Mitotane USP, Lysodren	[53-19-0]	75–81	tablet: 500 mg oral: 8–10 g/d in 3 or 4 divided doses or 3 g three or four times a d
(28) procarbazine hydrochloride USP, Matulane	[366-70-1]	223–226	capsule: 50 mg oral initial: 100–200 mg once a d for 1 wk, then 300 mg once a d until maximum response obtained maintenance: 50–100 mg once a d

mol wt = (133 ± 5) × 10³
an enzyme
(25)

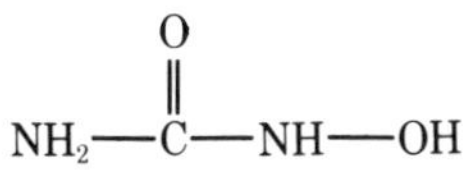

$$NH_2-\underset{\underset{O}{\|}}{C}-NH-OH$$
(26)

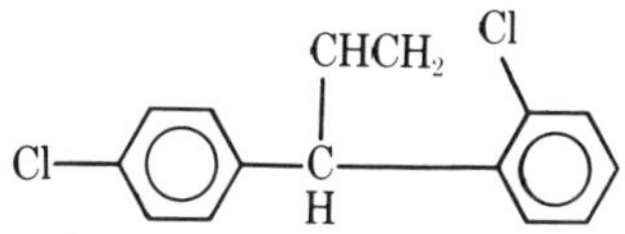

(27)

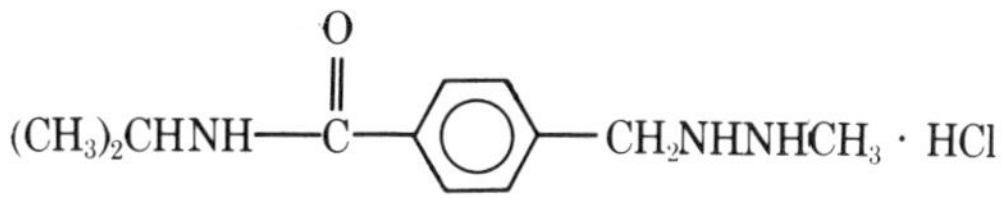

$$(CH_3)_2CHNH-\underset{\underset{O}{\|}}{C}-\underset{}{\bigcirc}-CH_2NHNHCH_3 \cdot HCl$$
(28)

Figure 5. Miscellaneous agents.

Enzymes as antitumor agents are at present represented by only one compound, L-asparaginase (25). The amino acid L-asparagine is an essential growth factor for certain malignant cells, including the leukemic cell, since it cannot be synthesized by these cells and must be supplied to them. Most normal cells, including lymphocytes, synthesize their own asparagine. The enzyme, by catalyzing the hydrolysis of asparagine to L-aspartate and ammonia, deprives the malignant cell of an essential amino acid used in protein synthesis, thus producing tumor cell death without similarly damaging normal tissues. This compound was originally thought to represent the first antineoplastic agent to utilize a qualitative biochemical difference between normal cells and certain tumor cells. However, it is now known that several functions of some

Disease	Toxic effects	Manufacturer
acute lymphocytic leukemia	hepatic, renal, and pancreatic toxicity; neurological effects; hypersensitivity reactions; clotting abnormalities; nausea; vomiting; anorexia; fever	Merck, Sharp, and Dohme
chronic granulocytic leukemia; melanoma; cancer of ovary, head, neck	bone-marrow depression; anorexia; nausea; vomiting; diarrhea	Squibb
only in palliative treatment of inoperable adrenal cortical carcinoma	skin toxicity; dizziness; vertigo; lethargy; somnolence; anorexia; nausea; vomiting; diarrhea	Calbio
Hodgkin's disease; non-Hodgkin's lymphomas; lung cancer	bone-marrow depression; neurological and dermatological toxicity; nausea; vomiting	Roche

normal cells, such as the synthesis of specific proteins (plasma albumin, insulin, etc), are also sensitive to L-asparaginase and may be inhibited by this enzyme. Among the complications encountered with its use are rapid drug resistance as leukemic cells utilize alternative metabolic pathways with glutamine instead of asparagine, poor tissue permeability owing to the large molecular weight, and severe hypersensitivity reactions, including anaphylaxis, since it is a large, foreign protein and therefore antigenic. Unfortunately, tumor remissions with L-asparaginase appear to be transient.

Hydroxyurea (26) is representative of a group of compounds (substituted urea, guanazole, thiosemicarbazones, etc) that have as their mechanism of antitumor action the inhibition of the enzyme ribonucleoside diphosphate reductase. This enzyme catalyzes the conversion of ribonucleotides to deoxyribonucleotides, a critical step in the biosynthesis of DNA. Enzyme inhibition presumably occurs through chelation or complexing of the nonheme iron component of the enzyme.

Mitotane (27), a compound chemically related to the insecticides DDT and DDD, is an adrenocortical suppressant which acts selectively on cells of the adrenal cortex, normal and neoplastic. Administration of mitotane causes a rapid reduction in blood and urinary levels of adrenocorticosteroids and their metabolites as the drug suppresses the adrenal tumor.

Antineoplastic effects have been reported with several methylhydrazine derivatives, including procarbazine (28), through inhibition of DNA, RNA, and protein synthesis. The conversion of hydrogen peroxide (formed by auto-oxidation of the drug) to hydroxyl radicals may be responsible for the degradation of DNA. Also, production of formaldehyde and its derivatives may play an important role in cytotoxicity.

Hormones. These agents, including estrogens, androgens, progestins, and adrenocorticosteroids, are not specific oncolytic agents but are employed to manipulate the hormonal environment of endocrine-dependent cancers such as those of the breast, ovary, and prostate. By changing this environment and thus depriving the tumors of the required hormonal growth factor, it is possible to alter, to some degree, the neoplastic process (see Hormones).

Progestational agents, which have been found useful in the management of endometrial carcinoma previously treated with surgery and radiotherapy, were tried initially because of the concept that endometrial carcinoma results from the prolonged, unopposed overstimulation by estrogen. Progesterone [57-83-0], it was thought, would correct this situation because of its physiological effect in producing maturation and secretory activity of the normal endometrium. Apparently, a proportion of neoplastic cells arising from this tissue is still influenced by normal hormonal controls. Because of their lympholytic effects and their ability to suppress mitosis in lymphocytes, leading to atrophy of lymphoid tissue and reduction in lymphocytes, the greatest value of the adrenocorticosteroids (usually prednisone [53-03-2]) is in the treatment of acute lymphocytic leukemia in children, and malignant lymphomas. In acute leukemia of childhood, these steroids may produce prompt clinical improvement and objective hematological remissions of variable duration (2 wk to 9 mo), unfortunately with invariable relapse and eventual drug resistance. Corticosteroids are therefore usually employed in conjunction with other agents. Estrogens and androgens are of value in the treatment of certain neoplastic diseases because the organs that are often the site of malignant growth, notably the prostate and the mammary gland, are dependent upon hormones for their continuing viability. Carcinomas arising from these organs often retain some of the normal hormonal requirements for varying periods of time. Thus, androgen administration in breast cancer represents an attempt to block estrogen stimulation of the tumor cells. Similarly, androgen control by estrogens produces clinical improvement in prostatic carcinoma, even though relapse eventually occurs. Although chemical cytotoxic agents are associated with untoward and damaging side effects, the anabolic effects of many steroidal agents may be of benefit to the patient (see Steroids).

Combination Therapy

The current philosophy of cancer treatment emphasizes the more aggressive use of combined treatment methods. So far, the combined method that has shown the most effectiveness is adjuvant chemotherapy. Chemotherapy as an adjunct to radiation therapy and/or surgery has improved the curability of two childhood solid tumors, Wilm's tumor and Ewing's sarcoma. The concomitant use of drugs, mainly methotrexate (16) or bleomycin (21), with other treatment has also been reported to improve survival in head and neck cancer, and fluorouracil (15) is useful in conjunction with surgery in colorectal cancer. It is possible that chemotherapy may be most effective when small numbers of tumor cells remain following surgery or radiation. Some drugs that are apparently ineffective against the presurgical tumor cells may cure metastatic disease if given shortly after surgery. As with drug combinations, the use of drugs along with another type of treatment may minimize additive host toxicity and drug resistance.

Multidrug Treatment

The entire population of neoplastic cells must be destroyed in order to obtain optimum results with chemotherapy or with any other treatment. To achieve this, several chemicals, acting as a synergistic combination, are used simultaneously or sequentially. Because different classes of chemotherapeutic agents have different mechanisms of action, they are effective in different phases of the cell cycle. Also, since tumors may differ in biochemical, cytokinetic, or chromosomal characteristics, multitargeted treatment is rational and further explains why combination drug therapies may work when single agents fail. Thus certain tumors, usually slow-growing with only a small fraction of dividing cells, will respond to initial treatment with drugs such as the alkylating agents that can kill cells at any stage of their cycle, even if they are not engaged in DNA synthesis (cycle-independent agents). The cells that survive are the rapidly dividing ones more vulnerable to subsequent attack by cycle-dependent agents (antimetabolites, etc). These agents, representing many of the most cytotoxic drugs, act at specific phases of the cell cycle by inhibiting DNA synthesis and therefore are active only against dividing cells. Accordingly, the human malignancies that are currently most susceptible to single-agent therapy are those with a high percentage of tumor cells in the process of division. In addition, combination chemotherapy using drugs with different modes of action may suppress or delay the emergence of drug-resistant cell lines. There is always the possibility, of course, that one antitumor agent may interfere with the activity of another because of a conflict in their mechanisms of action. Finally, if equally effective drugs with different mechanisms of toxicity are combined (cyclophosphamide (4) with bone-marrow toxicity plus vincristine (24) with neurotoxicity), additive cell-kill may be obtained without additive toxicity, and the patient can tolerate anticancer doses that would be fatal with either drug alone. Multiple-drug programs, therefore, rather than dependence on one agent, are the rule and have been developed for most responsive human tumors. Examples are cyclophosphamide and doxorubicin (18) in breast cancer, cytarabine (14) and thioguanine (13) in acute myelocytic leukemia, and cyclophosphamide, vincristine, methotrexate (16), and cytarabine in diffuse histiocytic lymphoma.

Immunology

The immunological system may be involved in the treatment of cancer, either as the target for the action of a drug or as the focus for its adverse effects.

Immunologic techniques for the treatment of malignant disease (immunotherapy) are undergoing extensive evaluation, all having the fundamental aim of manipulating the patient's defense (immune) mechanisms to combat the neoplastic process. Immunotherapy involves the administration of bacteria, bacterial adjuvants, or other materials, as antigens to amplify cellular or humoral immune responses and activate macrophages. The aim is to enhance the body's natural ability to suppress tumor growth, produce regression of nodules, and prevent metastases. The ultimate aim is to prolong remission and reduce recurrence of the neoplastic process. The most widely tested therapeutic materials to intensify immune responses and increase resistance include BCG (Bacillus-Calmette-Guérin), a live attenuated strain of mycobacteria commonly used for vaccination against tuberculosis, BCG derivatives such as MER (methanol extraction residue) and Ribi vaccine (BCG wall skeleton on oil droplets), the bacillus *Corynebacterium parvum*, tumor cells or antigens from patients or from tissue culture, lymphocytes, antitumor antiserum, purified immunoglobulin derived from antiserum, and levamisole [14769-73-4] (an imidazole compound). From a theoretical point of view and from experience so far, immunotherapy, like chemotherapy, may be most effective when small numbers of tumor cells remain following surgery or radiation which has removed or killed the bulk of cells. The cells most likely to respond to immune effects are those in operable solid tumors that don't metastasize; disseminated cells that cause cancer recurrence rarely respond to immunologic antitumor therapy. Among the neoplasms thus far reported to benefit most from the use of immunological agents, alone or combined with other treatment, are acute myelogenous leukemia, malignant melanoma, soft-tissue sarcomas, colon cancer, lung cancer, and breast cancer. As with any antigenic materials, severe or even fatal adverse hypersensitivity reactions, including asthma and anaphylaxis, may occur. Antitumor immunotherapy cannot yet be recommended as routine cancer therapy.

Many antineoplastic agents have a destructive effect on lymphocytes and thus have the potential to produce profound suppression of the immunological system (immunosuppression), including inhibition of such immune responses as antibody synthesis, delayed hypersensitivity, and graft rejection. It is this very property of immunosuppression that is the basis for the use of some of these agents (cyclophosphamide (4), etc) to block rejection of surgical transplants and to treat autoimmune diseases. However, in cancer chemotherapy inhibition of the body's defense mechanisms is considered an undesirable side effect since some immune responses are thought to play an important role in the natural host resistance to malignant tumors. Immunosuppression may be additionally deleterious in being among the factors responsible for the increased susceptibility of chemically-treated cancer patients to infections. Increased infection may also be a threat when corticosteroids are used as immunosuppressants to treat leukemia and lymphoma.

Since cytotoxic agents can selectively suppress or enhance the immune responses, depending upon dosages or administration schedules, antineoplastic chemotherapy may cause marked alterations of the delicate balance between the patient and his tumor or microbial population. It is essential then, that chemotherapy be carefully designed and monitored to consider the subtle interactions between these drugs and the immunological defenses so as not to critically compromise the patient's ability to withstand either infection by microorganisms or recurrence of the neoplastic process. Examples of such considerations are the observations that small, repeated doses of drugs are far more immunosuppressive than larger, intermittent ones, and that estrogens and androgens appear to stimulate the host defense mechanism.

Drug Toxicity

Anticancer drugs are potent substances and may be expected to have severe adverse effects. For example, normal tissues that proliferate rapidly (bone-marrow, gastrointestinal epithelium, hair follicles) are often subject to damage by some of the cycle-dependent antineoplastic drugs, since these agents do not confine their effects to tumor cells. This toxicity, sometimes severe as with many other potent drugs which have only moderate selectivity, often limits drug utility. Therapeutic usefulness is obtained by careful balancing of dosage so that maximum cytotoxicity is obtained against the proliferating cancer cells with minimum effect against dividing normal ones. Wherever possible, intermittent courses of chemotherapy are used to allow restoration of any normal cells whose numbers may have been reduced by treatment. Certain drugs are toxic because they reduce body levels of essential metabolites, but their adverse effects can be decreased by concomitant administration of the normal metabolite. For example, leucovorin [58-05-9] (citrovorum factor, folinic acid) can ameliorate the toxicity of methotrexate (16) by protecting normal cells against the lethal effects of the drug and also hastening their recovery (leucovorin rescue). The toxicity of anticancer drugs may be markedly increased if liver, kidney, or bone-marrow function has been impaired by previous treatment or disease.

Various aspects of the problem of selective cytotoxicity are being analyzed through the study of cell kinetics (biokinetics of cell growth) and pharmacokinetics (drug metabolism, distribution, and uptake by the cell) (see Pharmacodynamics). In addition, research into cell kinetics is helping to clarify such concepts of carcinogenesis as the role of viruses and immunology, and the biochemical development of the cancer cell. It is now believed, eg, that there is no single cause of cancer. Exposure to a carcinogenic agent such as a chemical or virus is deemed insufficient to induce cancer without other predisposing factors or determinants, such as the immune responses of the host to certain viruses.

Radiation Therapy

Radiation therapy is the primary treatment for some types of malignancy, such as Hodgkin's disease, and can also be useful, in combination with surgery and drugs, against a wide variety of other malignancies. Radiation (α, β, γ) from radioactive isotopes (^{32}P, ^{131}I, ^{198}Au) or electromagnetic sources (x-ray) interacts with atoms and molecules and leads to their excitation and ionization. In irradiated tissue, a variety of chemical radicals are formed from water; these can further interact with altered irradiated proteins and nucleic acids and lead to cell damage, often first expressed at cell division. The value of radiation rests on its capacity to destroy certain types of malignant growth *in situ* without simultaneously destroying the normal tissue in which the tumor is growing. Growths so destroyed must be relatively sensitive to this treatment. Radiation is thus a selectively cytotoxic agent and is not a refined form of cautery. This selectivity is the basis for the following concepts formulated to explain the successful empirical treatment of patients: the lethal dose of radiation to a tumor, the tolerance of normal tissue to radiation, and the ratio between the two (therapeutic ratio) (see Radioactive drugs; Radioisotopes; X-ray techniques).

BIBLIOGRAPHY

"Cancer Chemotherapy" in *ECT* 2nd ed., Suppl. Vol., pp. 81–90, by Charles J. Masur, Lederle Laboratories.

1. S. Perry, "Cancer Chemotherapy: A Broad Overview" in *Proceedings of the Seventh National Cancer Conference,* J. B. Lippincott Co., Philadelphia, Pa., 1973, p. 103.
2. M. B. Shimkin, "Cancer Research" in *Cancer, Diagnosis, Treatment and Prognosis,* 4th ed., C. V. Mosby Co., St. Louis, Mo., 1970, p. 28.
3. W. B. Coley, *Am. J. Med. Sci.* **105,** 487 (1893).
4. G. T. Beatson, *Lancet ii,* 104, 162 (1896).
5. W. P. Herbst, *Trans. Am. Assoc. Genito-Urin. Surg.* **34,** 195 (1941).
6. L. S. Goodman and co-workers, *J. Am. Med. Assoc.* **132,** 126 (1946).
7. A. Gilman, *Am. J. Surg.* **105,** 574 (1963).
8. C. P. Rhoads, *J. Am. Med. Assoc.* **131,** 656 (1946).
9. S. Farber and co-workers, *N. Eng. J. Med.* **238,** 787 (1948).
10. O. H. Pearson and co-workers, *Cancer* **2,** 943 (1949).
11. J. H. Burchenal and co-workers, *Blood* **8,** 965 (1953).
12. A. Haddon and G. M. Timmis, *Lancet i,* 207 (1953).

General References

J. R. Bertino, "Recent Developments in Chemotherapy of Malignancy," *Can. J. Otolaryngol.* **4,** 12 (1975).
J. F. Holland and E. Frei, III, eds., *Cancer Medicine,* Lea and Febiger, Philadelphia, Pa., 1973.
I. Brodsky, S. B. Kahn, and J. H. Moyer, eds., *Cancer Chemotherapy II,* Grune and Stratton Inc., New York, 1972.
A. C. Sartorelli and D. G. Johns, eds., *Antineoplastic and Immunosuppressive Agents,* Part II, Springer Verlag, Berlin, 1975.
B. A. Stoll, ed., *Endocrine Therapy in Malignant Disease,* W. B. Saunders Co., Philadelphia, Pa., 1972.
S. K. Carter and M. Slavik, "Chemotherapy of Cancer," *Annu. Rev. Pharmacol.* **14,** 157 (1974).
E. S. Greenwald, *Cancer Chemotherapy,* Medical Examination Publishing Co., Inc., Flushing, New York, 1973.
R. B. Livingston and S. K. Carter, *Single Agents in Chemotherapy,* IFI/Plenum, New York, 1970.
P. Calabresi and R. E. Parks, Jr., "Chemotherapy of Neoplastic Diseases" in L. S. Goodman and A. Gilman, eds., *The Pharmacological Basis of Therapeutics,* 5th ed., Macmillan Publishing Co., Inc., New York, 1975, pp. 1254–1307.

V. T. DeVita, R. C. Young, and G. P. Canellos, "Combination Versus Single Agent Chemotherapy: A Review of the Basis for Selection of Drug Treatment of Cancer," *Cancer* **35**, 98 (1975).
J. Q. Matthias, "Advances in Oncology," *Practitioner* **211**, 465 (1973).
C. G. Zubrod, "Present Status of Cancer Chemotherapy," *Life Sci.* **14**, 809 (1974).
L. H. Einhorn, "Cancer Chemotherapy," *J. Indiana State Med. Assoc.* **66**, 235 (1973).

CHARLES J. MASUR
WILLIAM PEARL
Lederle Laboratories
American Cyanamid Company

CHEMOTHERAPEUTICS, ANTIMYCOTIC AND ANTIRICKETTSIAL

MYCOTIC INFECTIONS

Fungi adversely affect the health and well being of mankind in numerous ways. The most direct of these include a variety of disease processes discussed in this article. Others, perhaps less direct or apparent but in some ways more deleterious, include the adverse economic and societal effects of plant diseases; occasional incidents of severe illness and even deaths associated with ingestion of toxic mushrooms; mycotoxicoses in animals and humans including potential adverse genetic effects resulting from ingestion of fungal toxins such as ergot in contaminated foodstuffs; and atopic allergic manifestations such as hay fever, asthma and rhinitis from repeated exposure to fungal spores and metabolites (see Fungicides).

Fungi are distinct from other microbial pathogens. Like plants and animals, they are eukaryotic organisms with organized nuclei within a nuclear membrane and possess cytoplasmic structures and biosynthetic pathways similar to mammalian cells. They lack the photosynthetic pigments of plants and, as a result, are heterotrophic organisms requiring preformed energy sources.

Morphologically, fungi can exist in a variety of different forms of varying complexity. Yeast cells represent the least complex form. These are several μm in diameter and reproduce either by asexual budding (formation of blastospores) or by a sexual process leading to the formation of ascospores (see Yeasts). Molds are filamentous fungi that form elongated tubelike structures called hyphae. These structures are 2–10 μm dia and usually are separated by crosswalls or septa into individual cellular elements. One class of filamentous fungi, the *Phycomycetes*, lack such crosswalls and are characterized by common, multinucleated cytoplasmic volumes.

The medically important fungi include some 50 species out of a total of 40,000–50,000 different fungi. The infections produced by these fungi can be grouped into 2 major categories: superficial mycoses and systemic or generalized mycoses. In terms of absolute numbers of infections, the superficial mycoses, which involve infections of skin, hair, and nails, are the most important. The systemic mycoses are the most important in terms of pathology or severity of disease, including mortality.

Superficial Mycoses

Dermatophytic infections of the superficial mycoses (Table 1) are infections of keratinized tissues including skin, hair, and nails. They are the most common of the human mycoses and the only fungal infections capable of direct host-to-host transmission. They include six different clinical manifestations depending upon the anatomical site involved. The three genera of the *Deuteromycetes* causing ringworm infections differ in terms of the types of tissues invaded. *Trichophyton* (21 species) can invade hair, skin and nails; *Microsporum* (15 species) usually invade only hair and skin and *Epidermophyton floccosum* invades only skin and nails.

An important distinction between the different dermatophytic fungi involves sources in infection. Certain species are transmitted primarily between humans. These are called the anthropophilic species and are capable of causing epidemics of ringworm infection, particularly in institutionalized children. Anthropophilic species are associated with mild, chronic infections. Other dermatophytes are acquired either from infected animals (zoophilic species) or contaminated soil (geophilic species). Infections in humans caused by these latter organisms generally are more acute than those caused by the anthropophilic species.

Table 1. Superficial and Cutaneous Mycoses

Infection	Etiologic agents	Principal pathology or symptoms
tinea (pityriasis) versicolor	*Pityrosporum orbiculare*	superficial asymptomatic infection of smooth skin only; characterized by brownish, discolored and elevated scaly patches
piedras (black and white piedra)	*Piedraia hortai* (black) *Trichosporon beigelii* (white)	infection of hair only, characterized by hard nodules along hair shafts
tinea nigra palmaris	*Cladosporium werneckii*	asymptomatic fungus infection of the palmar surfaces of the hand characterized by brown to black patches
dermatophytoses	*Epidermophyton floccosum* *Microsporum* sp. *Trichophyton* sp.	ringworm infections of skin, nails, and hair; infected skin may become macerated (tinea pedis) or scaly (tinea corporis); scalp infections may become inflamed (tinea capitis) with broken hairs; infected nails become brittle and discolored (tinea unguium)
candidiasis (see Table 2)	*Candida albicans* and other *Candida sp.*	mucocutaneous infections of mouth, vagina, etc, characterized by white, adherent patches; skin infections characterized by red, weeping lesions with vesicles

Other superficial fungal infections include tinea versicolor, the piedras and tinea nigra palmaris (Table 1). These are totally asymptomatic infections and are important only for cosmetic reasons. In tinea versicolor, characteristic, irregular, diffuse patches develop on the trunk, arms, neck, and face. These patches often are associated with failure of the skin to tan evenly which may be the chief complaint. The piedras can be best described as nuisance infections in that they are limited to hair shafts of the scalp and beard and are characterized by hard concretions which are gritty feeling when palpated. Tinea nigra palmaris, a totally asymptomatic infection of the palmar surface of the hand, is characterized by superficial, brownish to black patches which may be either discrete or confluent.

Correct diagnosis of a superficial fungus infection, and particularly of a dermatophytic infection, is essential for effective chemotherapy. Such diagnosis can be made on the basis of microscopic examination of infected skin, nails, or hair as well as on the basis of cultural studies. Microscopic examination is usually done with the aid of a 10 or 20% KOH soln which acts as a digestant for cellular debris. In skin and nail specimens, infecting dermatophytic fungi appear as branching hyphae. In hairs, sheaths or rows of spores are seen either along the shaft (ectothrix) or within it (endothrix). Hairs infected by certain zoophilic or geophilic species of *Microsporum* will fluoresce when examined under ultraviolet light.

Systemic and Generalized Mycoses

The deep mycoses include two distinct groups of life-threatening fungal infections of humans (Table 2). First, there are the systemic fungal infections caused by pathogenic fungi in normal hosts. Second, there are the opportunistic infections caused by fungi of low virulence in individuals with compromised resistance factors.

The systemic fungal pathogens share certain common features. Being true pathogens they are capable of causing infections in normal hosts. However, in most instances such infections are usually recognized only by x-ray examination or immunological testing. Some of these organisms, such as *Coccidioides immitis*, are found in nature in highly specific geographic regions, some of which overlap and most of which occur in the Americas. Infections caused by some of these organisms such as *Coccidioides immitis*, *Cryptococcus neoformans*, and *Histoplasma capsulatum* show highly specific patterns of organ involvement, as well as sex, age and racial predilections. Most of the systemic mycoses primarily infect the lungs by inhalation of infective spores. Other infections are acquired by traumatic implantation of contaminated plant materials. Most of the truly pathogenic fungi have two distinct forms depending upon such conditions as temperature, carbon dioxide pressure, presence or absence of sulfhydryl groups and local oxidation–reduction potential.

The opportunistic infections also share common features. Aspergillosis and cryptococcosis occur primarily in individuals with altered resistance owing to leukemia or lymphomas, immunosuppression and corticosteroid therapy. Phycomycosis occurs in diabetics, burn victims and patients with leukemia, lymphomas, or other chronic illnesses. Candidiasis occurs in the same situations as well as with indwelling catheters and broad-spectrum antibiotic treatment. In nearly all instances, these opportunistic invaders are either normal flora of the host or from the environment.

Diagnosis of the deep mycoses is based upon one or more of four factors: epidemiologic fact-finding; demonstration of fungal pathogens in tissue or other specimens;

Table 2. Systemic and Generalized Mycoses

Infection	Etiologic agent	Principal pathology or symptoms
Systemic		
blastomycosis	*Blastomyces dermatitidis*	chronic, granulomatous disease either limited to skin and lungs or widely disseminated; often fatal
chromomycosis	*Phialophora* sp.; *Cladosporium* sp.	chronic, granulomatous infection of skin and lymphatics developing slowly over months or years; rarely fatal
cladosporiosis	*Cladosporium trichoides*; *Phialophora* sp.	(1) brain abscess; usually fatal (2) small, asymptomatic subcutaneous cysts, rarely fatal
coccidioidomycosis	*Coccidioides immitis*	(1) acute, but benign, self-limiting primary, pulmonary infection (2) progressive, multiorgan, disseminated infection involving skin, viscera and bones (less than 1%)
cryptococcosis	*Cryptococcus neoformans*	slowly developing chronic meningitis; often with remissions and relapses; universally fatal if untreated
histoplasmosis	*Histoplasma capsulatum*	(1) asymptomatic pulmonary infection with granulomatous foci (2) clinical pneumonia with protracted illness but self-limiting (3) disseminated infection with multiple organ involvement, particularly of the reticuloendothelial system with enlarged spleen and liver and focal areas of necrosis and granulomas in multiple organs; often fatal
lobomycosis	*Loboa loboi*	chronic, cutaneous infection with formation of fibrous tumorlike or keloidlike nodules; no inflammatory reaction and nonfatal
maduromycosis (mycetoma)	*Allescheria boydii*; *Madurella* sp.	slowly progressive and highly destructive infection of subcutaneous tissues of foot or hand; characterized by deep-seated abscesses and chronic, draining sinuses; infection involves both muscle and bone causing malformities and loss of function; rarely fatal

Table 2. (continued)

Infection	Etiologic agent	Principal pathology or symptoms
paracoccidioidomycosis	*Paracoccidioides brasiliensis*	chronic, granulomatous disease of skin, mucous membranes, lymph nodes, and internal organs; often fatal
sporotrichosis	*Sporothrix schenckii*	local abscess or ulcer at site of implantation of contaminated thorn or splinter followed by multiple subcutaneous abscesses involving the local lymphatics; may disseminate to produce secondary infections of bones and joints
Opportunistic		
aspergillosis	*Aspergillus fumigatus* *Aspergillus* sp.	(1) necrotizing bronchopneumonia with obstruction of blood vessels and production of local infarcts of lung tissue (2) granuloma (aspergilloma) (3) obstructive but noninvasive fungus ball
candidiasis (see Table 1)	*Candida albicans* and other *Candida* sp.	systemic infection of lungs, urinary tract, heart, etc, with no specific symptoms; often fatal
mucormycosis (phycomycosis)	*Mucor* sp. *Absidia* sp. *Rhizopus* sp.	(1) acute infection of paranasal sinuses characterized by inflammation, vascular obstruction and local necrosis; may ultimately invade brain with subsequent infarction; frequently fatal (2) subcutaneous infection of tissues, thorax, abdomen and limbs

isolation, recovery, and identification of the responsible pathogen; and serologic techniques. Epidemiologic fact-finding involves development of a history to include travel in areas endemic for fungal infections; predisposing health factors; and possible occupational exposure to certain fungal pathogens. Demonstration of fungal pathogens in tissues or other pathologic materials includes direct examination of body fluids, staining of pathologic materials, and histological preparations of infected tissues. In some instances, such as cryptococcal meningitis, demonstration of a fungus in pathologic materials may be the only justification required for initiating specific antifungal chemotherapy. Isolation and identification of the responsible pathogen is difficult in some instances as certain of the pathogenic fungi are slow to grow under laboratory conditions. In other situations, such as candidiasis, the significance of the presence of the responsible fungus is a matter of clinical judgement since these organisms are frequently encountered in specimens such as sputa or urines. A variety of serologic techniques now are available for diagnosis of fungal infections. Some have only diagnostic value, others provide both diagnostic and prognostic information. These techniques include agglutination tests, immunodiffusion tests for either antibody or antigen, complement fixation tests, and, most recently, counterimmunoelectrophoresis tests for either antibody or antigen.

Antifungal Agents

Antifungal chemotherapy is limited both in the number of available agents and in therapeutic applications. They can be divided into two categories: antibiotics (secondary metabolites) with antifungal properties, and synthetic compounds. Unlike the current trend in antibacterial chemotherapy (see Antibiotics), there is little emphasis upon development of semisynthetic modifications of antifungal antibiotics. Therapeutic applications include intravenous medication in treatment of the progressive or disseminated cases of systemic mycoses and oral and topical medication in treatment of dermatophytic infections and certain yeast infections. Effective modes of therapy are lacking for most cases of pulmonary aspergillosis, phycomycosis, chromomycosis and maduromycosis. An alphabetical list of compounds mentioned in the text and their CAS Registry Numbers is provided at the end of the article.

The majority of antifungal antibiotics are polyenes produced by different species of *Streptomyces,* a genus of aerobic actinomycetic bacteria (1). In spite of the large number of such antibiotics which have been discovered and characterized, only a few are used in clinical medicine and even fewer have important chemotherapeutic roles (2).

The Polyene Antibiotics. The most important group of antifungal antibiotics are the polyenes (see Antibiotics, polyenes). These are high molecular weight, polyhydroxy compounds belonging to the macrolides but differing from them in possessing conjugated chains of chromophoric double bonds in the macrolactone ring (see Antibiotics, macrolides). Properties of the polyenes include chemical instability; poor solubility in water; strong, characteristic uv absorption spectra; and poor stability to light and temperature. Biologically, the polyenes have potent antifungal activity, some antiprotozoal activity but little or no antibacterial activity. The antifungal mode of action of the polyenes is related to their ability to produce profound changes in cell membrane permeability. This results from binding of the compounds to sterols located in the cell membranes.

Both toxicity and efficacy of the polyenes is determined by the degree of binding and by the specificity and avidity of certain polyenes for specific sterols. For example, the principal sterol in membranes of fungi is ergosterol to which amphotericin B, one clinically useful polyene, binds specifically. A similar sterol, cholesterol, is found in membranes of mammalian cells (see Hormones; Steroids). Amphotericin B also binds to this sterol but to a significantly lesser degree. However, sufficient binding does occur to account for some of the clinical toxicity associated with this compound.

Nystatin. Nystatin (1) (Mycostatin) was the first of the polyene antifungals to become clinically important. Discovered in 1949, nystatin had a major impact in clinical medicine in that it provided the first antibiotic useful in the topical treatment of superficial infections caused by *Candida albicans* (3).

Nystatin is an amphoteric tetraene soluble in dimethyl sulfoxide (DMSO), dimethyl formamide (DMF) and short chain alcohols. It is inactivated by heat, light, and acid or alkali. Its absorption maxima are 235, 291, 304, and 318 nm. It is produced by *Streptomyces noursei.*

Biologically, (**1**) is active *in vitro* at concentrations of 1–10 µg/mL against most pathogenic fungi including yeasts, filamentous pathogens and dermatophytic fungi. However, its activity against the latter group of organisms is not sufficient to be clinically useful. It has no antibacterial or antiviral properties. Nystatin is highly toxic when given to experimental animals via the intravenous or intraperitoneal routes, LD_{50} values are 3 and 45 mg/kg, respectively. Fortunately, it is not absorbed in significant amounts when given orally [LD_{50} in mice per os (orally), >8000 mg/kg] nor is it absorbed percutaneously.

(**1**)

The parenteral toxicity and lack of oral absorption of nystatin limited the clinical usefulness of this drug to topical applications. Other medicinal uses of nystatin include veterinary medicine where it is used in treatment of yeast infection in poultry, swine, and cattle. Laboratory applications include use in tissue culture media to suppress yeast overgrowth and in the clinical laboratory for isolation of bacteria from specimens heavily contaminated with fungi.

Amphotericin B. Just as nystatin had a major impact on treatment of superficial *Candida* infections, amphotericin B (**2**) had its impact on the treatment of systemic fungal infections. It was the first effective agent to be used in the treatment of such life-threatening infections as coccidioidomycosis and cryptococcosis.

(**2**) R = H
(**3**) R = CH_3
(**4**) tentative formula: $C_{63}H_{85}N_2O_{19}$

Amphotericin B is an amphoteric heptaene with a sugar (mycosamine) moiety. It is insoluble in water except at extreme pH (0.1 mg/mL at pH 2 or 11) and partially soluble in DMF (2–4 mg/mL) and DMSO (30–40 mg/mL). The intravenous preparation contains sodium desoxycholate to provide a dispersible colloidal suspension. Its absorption maxima are 345, 363, 382, and 406 nm. Amphotericin B is part of a complex produced by *Streptomyces nodosus* consisting of amphotericins A and B (1).

Biologically, (2) is active *in vitro* against pathogenic yeasts and dimorphic fungi at concentrations of 0.01–2 µg/mL (4). Its activity against opportunistic, subcutaneous and dermatophytic pathogens is less, 3.12–30 µg/mL. It is even less active against the etiological agents of chromomycosis and maduromycoses (5). Amphotericin B is less toxic than nystatin for laboratory animals. The intravenous LD_{50} is 4–6 mg/kg and the oral LD_{50} is >8000 mg/kg. Unlike (1), (2) is absorbed from the gut both in animals and in humans and a dose of 5 mg/kg given orally to mice is protective against some pathogenic fungi. Unfortunately, the gastrointestinal absorption of the drug in humans is not predictable and oral use of amphotericin B in treatment of systemic fungal infections has not been fully successful (6). Therefore, the clinical application of the drug is limited to the intravenous and topical routes, although oral amphotericin B is available in some countries. Use of intravenous amphotericin B is indicated in the systemic mycoses, opportunistic infections and subcutaneous infections caused by susceptible organisms (7). Some cases of invasive pulmonary aspergillosis, disseminated sporotrichosis and most cases of mycetoma and chromoblastomycosis are clinically resistant to the drug (2).

Amphotericin B normally is administered intravenously at dosages of 1.0–1.5 mg/kg after starting at an initial dose of 0.25 mg/kg (7). It is administered in intravenous glucose over a 2–6-h period. The intravenous preparation consists of 50 mg of active drug, 41 mg of sodium desoxycholate and 25.2 mg of sodium phosphate buffer (Amphotericin B, USP, Fungizone). When prepared for infusion in 5% dextrose, the final concentration is 0.1 mg/mL. Serum levels of the drug range from ca 0.2–4 µg/mL but corresponding spinal fluid levels rarely exceed 0.1 µg/mL. The drug is bound to serum lypoproteins and is slowly excreted in urine over a prolonged period (8).

Amphotericin B is a highly toxic substance when given parenterally. The lesser side effects include fever, headache, nausea, vomiting, and malaise. The most important reaction is nephrotoxicity. Some impairment of renal function is seen in all patients regardless of dosage given. This is evidenced by increases in blood urea (BUN) and creatinine levels. Not infrequently, treatment will be stopped until these parameters return to acceptable levels. Efforts to reduce the nephrotoxicity include concurrent infusion of mannitol, alternate-day treatment, and use of serum drug levels as therapeutic guides. Other toxic reactions associated with amphotericin B include anemia, thrombocytopenia, liver function abnormalities, and thrombophlebitis.

Amphotericin B also has been used with varying degrees of success in certain other clinical indications using other routes of administration. Oral (2) (Fungillin) is used in some countries for treatment of superficial and intestinal *Candida* infections. Serious meningeal fungal infections such as cryptococcal and coccidioidal meningitis may respond only to intraventricular or intrathecal therapy.

Amphotericin B, methyl ester. The water-soluble methyl ester of amphotericin B (3) has *in vitro* antifungal properties comparable to the parent compound, but reduced acute *in vivo* toxicity (9). Although (3) is less nephrotoxic in animals, it also appears to be less active *in vivo* than the parent compound (10). *In vitro*, (3) is nearly as active as (2) against pathogenic fungi and also appears to be active against certain viruses such as herpes simplex, vaccinia, simbis, and vesicular stomatitis virus (11–12). *In vivo*, (3) is effective in treatment of experimental infections in mice caused by a variety of fungal pathogens such as *C. neoformans*, *B. dermatitidis*, and *C. immitis* but to a lesser degree than (2) (10,13).

Limited clinical and pharmacologic data are available regarding use of (3) in humans. In one study, alternate day intravenous dosages of 5.0 mg/kg using an ascorbate preparation resulted in peak serum levels of 15–30 µg/mL and trough levels of 0.5–2.5 µg/mL (14). Adverse reactions were said to be qualitatively similar to those of (2) but less severe (14). The exact clinical role for (3) has not yet been defined.

Candicidin. Candicidin (4) is a heptaene complex produced by *Streptomyces griseus* (15). Its chemical structure is not fully characterized. Two forms exist: Complex A with maxima at 340, 360, 380, and 403 nm, and Complex B with maxima at 340, 362, 381, and 404 nm. Candicidin is more active *in vitro* than (1) or (2) against pathogenic yeasts. However, its *in vivo* toxicity is greater than either (2) or (1): oral LD_{50} 98–400 mg/kg, intraperitoneal 2–7 mg/kg. The medical indications for (4) are restricted to topical treatment of vaginal and mucocutaneous candidiasis. It is available in the form of ointments (0.06%), vaginal tablets (0.3%) or capsules (Candicidin, NF; Cadeptin). These are used topically twice daily for 2 wk.

Pimaricin. Pimaricin (5) (Natamycin) a tetraene polyene, is produced by *Streptomyces natalensis* (16). It is soluble in propylene glycol (2%). It is thermostable. Its absorption maxima are 279, 290, 303, and 318 nm (17). Pimaricin is inhibitory for fungi at concentrations of from 1–15 µg/mL (18). It is toxic with intraperitoneal and intravenous LD_{50} values of 250 and 5–10 mg/kg, respectively; the oral LD_{50} is 1500 mg/kg. Therapeutic use of (5) is limited to topical treatment of fungus infections with a 5% ointment. The principal value of (5) appears to be in treatment of mycotic keratitis such as the very destructive ocular infections caused by *Fusarium* and *Cephalosporium* species. It is not an approved drug in the United States [Pimafuncin (Brit. Pharm)]. There appears to be no toxicity associated with topical application (19).

(5)

Nonpolyene Antifungal Antibiotics. The number of nonpolyene antifungal antibiotics is limited and only five are discussed here. These include griseofulvin (6), the only one produced by a fungus; sinefungin (7) (20) and cycloheximide (8), both produced by *Streptomyces* species; and pyrrolnitrin (9) which is produced by a bacterium. The fifth compound, ambruticin (10), which is produced by a myxobacterium, has been only recently described and is of unknown clinical potential (21–22).

Griseofulvin. Griseofulvin (6) is a phenolic, benzofuran cyclohexane produced by several species of *Penicillium* including *Penicillium griseofulvum*, *Penicillium patulum*, and *Penicillium janczewskii*. It is insoluble in water and petroleum ether and only slightly soluble in such solvents as DMF, DMSO, ethanol, benzene, and chloroform. It is thermostable and uv absorption maxima occur at 324, 291, 252, and

236 nm. Its antifungal activity is restricted to the dermatophytic species; MIC (min inhibiting conc) values range from 0.14–2.5 µg/mL (18). Griseofulvin has no activity against other filamentous fungi or against yeastlike organisms or bacteria.

The initial uses of (6) were in protection of plants against fungal infections when it was shown in 1951 that (6) introduced into soil was taken up by the roots and transported to the leaves. It was not until 1958 that (6) was shown to be orally effective in guinea pigs experimentally infected with the dermatophytes *Microsporum canis* and *Trichophyton mentagrophytes* (23–24). This revelation was immediately followed by clinical demonstrations of the efficacy of the drug in treatment of human dermatophytic infection (25–26).

The pharmacology and mode of action of griseofulvin are unique. When administered orally in animals, the drug is widely distributed in body fluids and also is deposited in an active form in keratinized tissues (27). Absorption from the gastrointestinal tract generally is poor with as much as 20% remaining unabsorbed (28). Poor absorption, and excretion in the urine in the form of inactive metabolites, accounts for low serum levels. Other factors affecting serum levels include particle size, fat intake, dosage schedule, and dissolution rate. In attempts to obtain higher serum and skin levels micronized formulations of reduced particle size have been used (28).

The mode of action of (6) in fungi is not clearly established. Early studies showed the drug to have a profound effect on cell wall formation (29), mitosis, inhibition of nucleic acid synthesis and alterations in cytoplasmic microtubules (30). All of these suggest that in fungi (6) acts by specific inhibition of mitotic spindle formation.

Although early *in vivo* studies with griseofulvin showed it to be active topically in experimental infections against *Trichophyton mentagrophytes*, similar topical activity rarely has been seen in human infections. Thus, administration is limited to the oral route. A variety of oral preparations are available (Fulvicin, Grifulvin, Grisactin, Gris-Peg, etc). The majority of these are so-called microsize preparations which have replaced the older particulate form of the drug. The average daily dosage of the microsize preparation in adults is 0.5 g/d although 1.0 g daily dosages may be required in certain chronic or severe infections. Generally, the time required for treatment of ringworm infections with griseofulvin is dependent upon the site of infection and the time required for total replacement of infected tissues by normal tissues. In vivo, the drug is fungistatic only and thus infections will relapse if all infected tissues are not replaced. This means that 2–4 wk of medication will be required in most infections. Treatment of infections involving the soles of the feet or nails will require more time, at least 1 yr in the case of toenail infections.

Griseofulvin is relatively free of toxicity, even in patients receiving large dosages over prolonged periods (31). Occasionally, a patient will complain of gastrointestinal upset, and hypersensitivity reactions as well as photosensitivity rashes have been reported. The drug is contraindicated in patients with porphyria or liver failure. An etiological diagnosis should be obtained before using the drug.

Cycloheximide. This antibiotic (**8**), produced by *Streptomyces griseus*, was used in treatment of cryptococcal infections before the development of (**2**) (32). Its current use is limited to laboratory and agricultural applications. Cycloheximide (**8**) (Acti-dione) is a white powder slightly soluble in chloroform, alcohols and other organic solvents. Its activity is mainly against nonpathogenic fungi. It is used in laboratory media as a selective agent to permit isolation of pathogenic fungi from specimens such as sputa and feces which may be contaminated by other fungi. One important exception is that it is active against *C. neoformans* (MIC <0.2 µg/mL). Agricultural use is limited because of phytotoxicity to topical control of downy mildew of onions and shoot blight of larch (33). The mode of action of (**8**) is at the level of polypeptide formation and nucleic acid synthesis (34).

Other antifungal agents. Pyrrolnitrin (**9**) is a potent antifungal metabolite produced from tryptophan by certain species of *Pseudomonas* (35). Chemically, it is slightly soluble in water and petroleum ether and soluble in ethanol and methanol. Pyrrolnitrin is unstable in presence of acid and is bound by serum. The in vitro spectrum of (**9**) includes *T. mentagrophytes* (MIC of 0.78–6.25 µg/mL), *Trichophyton rubrum* (MIC of 1.56–3.12 µg/mL), *Candida albicans* (MIC of 25–50 µg/mL), and systemic fungal pathogens such as *C. immitis*, *C. neoformans*, and *Sporothrix schenckii* (MIC 0.78 µg/mL). Topically, (**9**) is effective in guinea pigs experimentally infected with *T. mentagrophytes* when applied in 1–2% formulations. The medical use of (**9**) is limited to topical treatment of dermatophytic infections since it is inactive orally, toxic if given by parenteral routes, and inactivated by serum. It is not an approved drug in the U.S. but is marketed elsewhere.

Sinefugin (**7**), A9145, is produced by *Streptomyces griseolus* (36,37). Sinefungin is active against *Candida albicans* both in vitro (MIC of 0.06–1.0 µg/mL) and in vivo (10 mg/(kg·d)). Peak blood levels of 64 µg/mL are obtained in mice with subcutaneous dosages of 100 mg/kg (38). The LD_{50} for mice is 185 mg/kg (37). The mode of action of (**7**) involves inhibition of S-adenosylmethionine-mediated transmethylation, particularly histamine N-methyltransferase (39). The clinical role for (**7**) has not yet been defined.

Ambruticin (**10**), W7783, is of a new class of cyclopropyl–pyran antibiotic. Produced by the myxobacterium *Polyangium cellulosum* var. *fulvum* (40). It is active in vitro at concentrations as low as 0.025 µg/mL against systemic and dermatophytic fungal pathogens (41–42). In vivo, (**10**) is active topically and orally against *T. mentagrophytes* and orally against *H. capsulatum* and *C. immitis* (43–45). It is both protective and curative in mice infected with *C. immitis* at oral dosages of 25 and 50 mg/kg. Ambruticin is orally absorbed in laboratory animals with peak serum levels of 40 µg/mL attainable one h following a single 50 mg/kg dose (41); it is well distributed throughout the body in a biologically active form (41). Although the exact clinical role of (10) is yet to be defined, both its spectrum as well as its topical and oral activity suggest a role in management of both systemic and superficial mycoses.

Synthetic Antifungal Agents. *Nonspecific Topical Medications.* Use of specific, synthetic antifungal agents is only a recent development in antifungal chemotherapy. This is particularly true for chemotherapy of the systemic mycoses. Earlier forms of therapy were concerned primarily with the superficial mycoses and depended extensively upon the use of nonspecific topical fungicides, drying agents, and keratolytics. Benzoic and salicylic acids combined, 6 and 12%, respectively, in petrolatum and known as Whitfield's ointment is the best example of such preparations. Salicylic acid acts as a keratolytic or desquamating agent, softening and loosening the infected cornified epithelium, and the benzoic acid acts as a fungistat in preventing reinfection of newly formed epithelium. Whitfield's ointment was used in treatment of smooth skin infections only, and because of the fungistatic nature of the ointment, prolonged therapy of several weeks to months was required (46).

Treatment of tinea unguium or fungal nail infections often was accomplished by either close filing of the nail or surgical removal followed by daily soaks in solutions of potassium permanganate (1:4000). After soaking, the infected nails were painted with a fungicide such as a 10% solution of salicylic acid. Potassium permanganate also was used topically (1:5000) and vaginally (1:1500) for treatment of cutaneous and mucocutaneous candidiasis (47).

Gentian or crystal violet (11) (hexamethyl-*p*-rosaniline) is both bacteriostatic and fungicidal and, at one time, was used extensively in topical treatment of superficial fungus infections at concentrations of 0.02–1.0%. Gentian violet solutions also were used in treatment of oral and vaginal candidiasis and in local treatment of ringworm infections of the groin; disadvantages included staining of skin and clothing (48).

Undecylenic acid (12) (10-undecenoic acid) is a fungistatic fatty acid. It and various salts, such as zinc undecylenate, have been used for many years in topical treatment of smooth skin ringworm infections. One popular preparation, Desenex, containing 5% of the acid and 20% of the zinc salt, is available in several forms. The zinc salt acts both as a fungistatic agent and as an astringent which also aids in reduction of inflammation. Preparations of (12) are of value primarily in treatment of ringworm infections of the foot and have no role in treatment of hair or nail infections (49).

Other fatty acids or their derivations also have been used in treatment of superficial fungal infections. They include calcium propionate [4075-81-4] (Sopronol) and sodium propionate [137-40-6] (see Carboxylic acids).

Sulfur-containing compounds that once played an important role in treatment of superficial fungal infections include: sulfur-salicylic ointment; sodium thiosulfate [7772-98-7] (25%) and salicylic acid (1.0%) in isopropyl alcohol and propylene glycol (Tinver) is still highly effective for treatment of tinea versicolor; and selenium sulfide [7488-56-4] for treatment of dandruff, seborrheic dermatitis and tinea versicolor. Both preparations are used once or twice weekly as a shampoo or lotion. Selenium sulfide is less toxic than water soluble selenium salts; its oral LD_{50} in rats in 138 mg/kg (50).

Iodochlorhydroxyquin (13) (5-chloro-7-iodo-8-quinolinol, Vioform) is used topically in treatment of the dermatophytic infections and *Candida* vaginal infections. Since (13) is water insoluble and poorly absorbed from the gastrointestinal tract, it is used topically in the form of a 3% suppository or cream in treatment of vaginal trichomoniasis or candidiasis. Side effects are rare but include pruritis and mild iodine reactions (51). Iodochlorhydroxyquin is available in a variety of preparations including topical creams and ointments with and without hydrocortisone and antibiotics, as well as suppositories and oral tablets.

Acrisorcin (14) (9-aminoacridine-4-hexylresorcinolate) is a broad-spectrum nonspecific acridine dye active against bacteria, fungi, and protozoa. It is "cidal" for most such organisms at concentrations of 1–78 µg/mL. Used topically as a 2% cream (Akrinol), (14) is effective only in treatment of tinea versicolor; relapses following treatment are not uncommon. Toxicity includes hives, blisters, erythematous vesicles and photoinduced pruritis (50).

Nonspecific Systemic Medications. The first systemic antifungal chemotherapeutics were the iodides. Local, cutaneous sporotrichosis is readily treated orally with a saturated solution of potassium iodide given three times daily with water or milk for as long as 6 wk. Side reactions include indigestion, rashes, lacrimation, and cardiac problems. The mode of action is unknown; potassium iodide is not inhibitory for *Sporothrix schenckii in vitro*. Oral iodides are still the drug of choice in nondisseminated cases of sporotrichosis (52).

Various sulfonamides have been employed in treatment of the human mycoses. In cases of blastomycosis, some success was obtained with sulfadiazine (15) but only after prolonged treatment (53). Cases of South American blastomycosis, a disease distinct from that seen in the United States and caused by *Paracoccidioides brasiliensis*, responded dramatically to sulfonamides but, again, effective therapy required long periods of treatment (54). Sulfonamide therapy was not effective in cases of coccidioidomycosis. Today the principal antifungal indication for the sulfonamides is in treatment of paracoccidioidomycosis, or South American blastomycosis, where (15) and other sulfonamides have, at best, only a suppressive effect bringing about clinical improvement but not a true cure (55).

The use of aromatic diamines in the treatment of blastomycosis was first reported in 1950 (56). The first compound so used was stilbamidine (16) which proved effective in patients with disseminated disease when given intravenously at dosages of up to 200 mg/d for up to 30 d. However, this drug also proved to be excessively toxic and was associated with a high rate of relapsing infections. Toxic reactions included breathlessness, tachycardia, dizzyness, and depressed blood pressure. The most serious side effect seen with stilbamidine was a peripheral neuropathy involving facial nerves. Stilbamidine was replaced by 2-hydroxystilbamidine isethionate (17) which also proved partially effective but which was not associated with neuropathic toxicity. 2-Hydroxystilbamidine (18) did produce some toxic reactions including malaise, nausea and headache as well as changes in liver function. Hydroxystilbamidine isethionate (17) was administered intravenously in saline at dosages of 225 mg/d with a maximum dose of 8 g (57). The drug was light-sensitive and infusions had to be shielded. Apart from being concentrated in the liver, little is known about the pharmacology of the drug. With the advent of amphotericin B (2), the use of (18) has been limited to treatment of less severe, primary pulmonary infections (58).

H_2N—SO_2NH—(pyrimidine)

(15)

(16) R = H
(17) R = OH · 2HOCH$_2$CH$_2$SO$_3$H
(18) R = OH

5-Fluorocytosine. 5-Fluorocytosine (19) (Flucytosine, 5-FC) was originally developed in 1957 as an antimetabolite for use in treatment of leukemia but, although devoid of anticancer activity (59), it was found to have antifungal activity *in vivo* (60). It is a white powder decomposing at 295°C and partially soluble in water (15 gm/L). It is nontoxic to experimental animals (LD$_{50}$ for mice, >2000 mg/kg, either orally or subcutaneously) and is readily absorbed from the gastrointestinal tract of humans and animals, thus permitting oral medication. In humans, tissue and fluid penetration is good with fungistatic serum and cerebrospinal fluid concentrations of 10–40 µg/mL persisting for 6–10 h after ingestion of a single 2 g dose (61).

The *in vitro* spectrum of (19) includes most isolates of *Cryptococcus neoformans*, many but not all isolates of *Candida albicans* and other species of pathogenic *Candida* strains as well as some isolates of *Aspergillus* species and of dematiaceous fungi such as *Cladosporium trichoides*. Most susceptible organisms are inhibited and killed at concentrations of 0.2–12.5 µg/mL (62); (19) appears to be only fungistatic and not fungicidal for filamentous fungi (63). There is evidence that the hematopoietic toxicity associated with (19) in patients with impaired renal function may be due to 5-fluorouracil (20) (5-FU) (64).

Currently, the therapeutic use of (19) (Ancobon) is limited to treatment of yeast infections caused by organisms of known susceptibity (65).

The mode of action of (19) involves, in part, conversion (deamination) to (20) and subsequent incorporation of (20) into RNA leading to miscoded protein synthesis (66). There is some evidence that inhibition of DNA synthesis by the metabolite 5-fluoro-2′-deoxyuridine-5′-monophosphate (21) also may be involved (67–68). At least four enzymatic actions are involved: permeation, deamination, uptake of (20) as 5-fluorouridine (22) (5-FUR) and incorporation of (22) into RNA. Resistance through mutational changes in enzyme function can result. Mutants with altered permeases are partially susceptible to (19) and mutation at any other site confers total resistance (69).

5-Fluorocytosine has been used with varying degrees of success in a variety of mycotic infections. It has provided a less toxic alternative to amphotericin B (2) in treatment of cryptococcal meningitis, pulmonary cryptococcosis and in pulmonary, urinary and disseminated candidiasis. Unfortunately, the emergence of resistant organisms during therapy has been a major problem in all such infections, particularly in inadequately treated patients with cryptococcosis (70).

Resistance to (19) most commonly involves changes in permease activity. Organisms with altered permeases are still susceptible to the drug if the permeability barrier can be overcome, as with combination therapy of (2) and (19). Some clinicians feel that this is the therapy of choice in patients with cryptococcal meningitis (71). In such patients, the dosage of (2) is reduced to 0.3–0.5 mg/(kg·d) and (19) is given at its full oral dosage of 150 mg/(kg·d). This regimen has proven effective in providing effective cures, reducing toxicity of (2) and preventing the development of (19)-resistant organisms (71).

Toxic reactions observed with (19) include hypersensitivity skin rashes, gastrointestinal disturbances, anemia and liver enlargement (72). The most serious side reactions involve hematopoietic toxicity. These include bone marrow depression and fatal aplastic anemia as well as anemia, neutropenia and thrombocytopenia. Many of these reactions have occurred in patients either previously treated with (2) or receiving combined treatment with the two drugs. Thus it is recommended that all patients receiving (19) be monitored in terms of both renal and marrow functions (73).

(19)

(20)

(21) R = H, R′ = PO$_3$H$_2$
(22) R = OH, R′ = H

Imidazole Compounds. The imidazoles represent the most versatile, and perhaps, most valuable source of antifungal compounds. As a group, they have a uniquely broad spectrum of activity which includes bacteria, fungi as well as protozoa, helminths and nematodes (74). Specific activity and spectra of individual compounds is highly dependent upon structure. The first of the antifungal imidazoles was 1-chlorobenzyl-2-methylbenzimidazole (23) (Myco-Polycid) which has some topical activity against *Candida* and dermatophytes (75). Subsequently, thiabendazole (24) (Mintezol) was introduced as an oral antihelminthic for medical and veterinary use in treatment of roundworm infections. Thiabendazole also has antifungal properties (76), but they have not been exploited extensively.

Clotrimazole (25) was first described in 1970 as a broad-spectrum orally-active imidazole (77), and active *in vitro* against a variety of fungal pathogens at concentrations as low as 0.1 µg/mL (74,78). Some reports showed (25) is active *in vivo* against both pathogenic yeasts and dermatophytic pathogens and others revealed hepatic enzymatic inactivation both in animals and humans (79–80) which appears to have restricted applications to topical use.

Most isolates of *Candida albicans* are inhibited by concentrations of 0.1–0.5 µg/mL; dermatophytic organisms are inhibited by 0.5–2 µg/mL. *In vivo*, (25) has low toxicity in animals when given by the oral route (LD$_{50}$ of 500–2000 mg/kg) and produces peak serum levels of 10–20 µg/mL (77). Oral dosages of (25) at 100 mg/kg are protective in acute animal experiments (77) but hepatic enzymatic inactivation of the drug nullifies the antifungal action of the drug in tests with chronic experiments (79,81). Thus the principal medical use of (25) appears to be in topical treatment of *Candida* and dermatophytic infections. Vaginal *Candida* infections, including those clinically resistant to (1), as well as *Candida* skin infections respond well to topical treatment with either a 1% clotrimazole cream or ointment (Lotrimin) as well as a vaginal tablet preparation (Gyne-Lotrimin). Topical (25) appears to be as active as topical (1) or Whitfield's ointment for treating cutaneous *Candida* infections (82). Gastrointestinal intolerance was observed in many patients receiving the oral medication.

Miconazole (26) also has important broad spectrum antifungal activity. Available both as a base (R 18,134) and as the nitrate (27) (R 14,889), (26) is a white, crystalline substance insoluble in water (0.03%) but soluble in organic solvents such as DMSO (83). *In vitro*, miconazole is fungistatic toward many pathogenic fungi at concentrations of ≤ 5 µg/mL (74,84). The *in vitro* activity of (26) is antagonized by complex culture media (85). There appears to be no difference in the *in vitro* activities of (26) and (27) (84).

Studies in animals revealed oral (26) to be relatively nontoxic with LD$_{50}$ values ranging from >160 mg/kg in dogs to >640 mg/kg in rats. There was no systemic absorption following topical application (86).

In humans, a 522 mg dose of intravenous (26) gives peak plasma levels of 2–9 µg/mL in individuals with normal renal function; the half-life was 0.38 h (87). Penetration of (26) from serum into cerebrospinal fluid is negligible (88–89). Miconazole is active both orally and topically in experimental animals infected with dermatophytic fungi, yeasts, and systemic pathogens such as *Coccidioides immitis* (83,90).

Topical miconazole cream (2% miconazole nitrate, Monistat) is highly effective in treatment of vaginal candidiasis with cure rates of 80–99% being reported (91). These results include patients who failed on therapy with other agents such as (1), (2), and (4). The relapse rate in such patients is approximately 5%. The usual course of therapy with miconazole cream for vaginal candidiasis is 5 to 6 g once daily for 14 d.

Topical miconazole ointments (2% miconazole nitrate, Micatin) are effective in treatment of fungal skin infections including both dermatophytic and *Candida* infections (91). Cure rates ranged from 63% in cases of tinea corporis to 100% in cases of tinea pedis and tinea cruris. Cure rates of 80–90% also have been reported for a (26) varnish in treatment of nail infections (91).

The role of oral, intravenous, and intrathecal miconazole in systemic fungal infections is difficult to assess. Responses of 35–67%, in cases of coccidioidomycosis, depending upon the site of infection, have been reported (92). The most favorable responses were in patients with chronic pulmonary disease; the least favorable in patients with meningitis. The relapse rate was high: 36% in cases of pulmonary disease, and 63% in patients with meningitis. Similar results were obtained in patients with paracoccidioidomycosis. In another study a recovery rate of 59% in 54 patients with proven systemic fungal infection was observed; cultural examinations of 16 of the 35 patients in this study during posttreatment were still mycologically positive for the infecting organism (93).

Topical miconazole preparations are essentially devoid of toxicity although burning, itching, and irritation have been reported (94). Intravenous medication has been associated with a variety of adverse reactions including vision changes, itching, hyperlipidemia, erythrocyte aggregation, and thrombophlebitis. Some of these reactions have been attributed to the vehicle used for the intravenous preparation (95).

At present, the approved therapeutic uses of miconazole in the United States are limited to topical treatment of cutaneous and mucocutaneous infections by dermatophytes and *C. albicans*. Miconazole is available elsewhere as an oral preparation which has been reported valuable for treatment of oral and gastrointestinal candidiasis (96).

Tolnaftate. Tolnaftate (28) was first described in 1962 as a result of a series of studies on the relationship between chemical structures and selective toxicity of aryl thiocarbamoylthiocarbonates. It is soluble in organic solvents but not water. One of three naphthiocarbamates with antifungal activity (MIC of 0.0125 to 0.025 µg/mL for *Trichophyton* sp.) (28) has low toxicity for animals and is curative in experimental infections with *Trichophyton mentagrophytes*. It is inactive toward bacteria and *C. albicans*. Topical 1% tolnaftate is as effective as oral (6) (10 mg per animal) in treatment of experimental infections. It is highly effective as a topical agent for treating human dermatophytic infections and it is virtually nontoxic and nonsensitizing. Tolnaftate is used topically against smooth skin dermatophyte infections caused by species of *Epidermophyton*, *Microsporum*, and *Trichophyton*. It is not effective for infections of the scalp or nails. Preparations include 1% cream, ointment and powder (86).

Haloprogin. Haloprogin (**29**) is an analogue of the antibiotic lenamycin (**30**). It is a white powder soluble in alcohols and slightly soluble in water. The *in vitro* activity of haloprogin includes pathogenic yeasts, gram-positive cocci, *Mycobacterium tuberculosis* and many species of dermatophytic fungi (MIC values of 0.25 µg/mL or less). It is orally nontoxic for animals (LD$_{50}$ for mice = 1000 mg/kg) and sparingly toxic parenterally (LD$_{50}$ for mice intraperitoneally = 510 mg/kg). Haloprogin is used topically in treatment of superficial fungus infections caused by dermatophytes or *C. Albicans*. It is not effective in treatment of scalp or nail infections. Topical (**29**), as a 1% cream or solution, is not without adverse reactions. These include local irritation, burning sensation and vesicle formation with increased pruritis and exacerbation of preexisting lesions (97).

Agricultural Use of Antifungal Agents

In addition to their use in human and veterinary medicine, antifungal agents also have an important role in control of fungal plant pathogens (phytopathogens), particularly in Japan where government regulations preclude the use of organic fungicides and pesticides containing mercury (see Fungicides).

Griseofulvin (**6**) is fungistatic for most phytopathogenic fungi and is readily absorbed by roots and transported systemically within treated plants. Its agricultural use includes treatment for powdery mildews, *Fusarium* wilt of melon and apple blossom blight (33). However, the high cost of the drug precludes large-scale applications. Pimaricin (**5**) is used as a dip to prevent fungal growth in skins of harvested apples. Cycloheximide (**8**) has a limited application in treatment of downy mildew of onions and shoot blight of Japanese larch.

Four antifungal antibiotics not used in medicine are extensively used for agricultural purposes. These include kasugamycin, blasticidin S, polyoxin D, and validamycin A (98) (see Antibiotics, aminoglycosides and nucleosides).

Economic Aspects of Antifungal Agents

Antifungal agents represent a significant component of the antiinfective drug market, especially the sales of vaginal and topical preparations. In 1975 sales of antifungal products in the United States amounted to 87.6 million dollars including 23 million dollars for vaginal preparations, 64 million dollars for antidermatophytic drugs, including both topical preparations and griseofulvum, and 0.6 million dollars for antisystemic antifungal agents. The international sales figure for the same year was approximately 300 million dollars. The insignificant contribution of antisystemic agents to the above figures is reflected in invoice price data for 1976 for (**2**) and (**19**); they were $490,000 and $89,000, respectively.

Agricultural use of antifungal agents represents a lucrative market particularly in the Orient (33,98). For example, in 1968 6.8% of the pesticides produced in Japan were antibiotics (33). These consisted primarily of kasugamycin, 17,565 metric tons; blasticidin-S, 7030 t; and polyoxin D, 3470 t. In 1975, sales for these same antibiotics in Japan totaled 9800 t and the net sales of all agricultural antibiotics was ca 17 million dollars.

RICKETTSIAL INFECTIONS

Infections caused by *Rickettsia* occur worldwide. Although the causative organisms have characteristics of both viruses and bacteria, they are more similar to the latter. Like bacteria, they have cell walls and contain muramic acid, they have metabolic enzymes and are capable of independent respiratory activity, and their growth is inhibited by antibiotics. The only characteristic they share with viruses is their inability to propagate (except for *Rickettsia quintana*) outside of living host cells. The *Rickettsia* were named in honor of H. T. Ricketts who died of typhus fever in 1910 while studying the etiology of rickettsial infections. All rickettsial infections except Q fever are transmitted to humans by bloodsucking arthropod vectors including lice, fleas, ticks and mites. Rickettsial infections are ordinarily classified into five categories: (*1*) typhus group, (*2*) spotted fever group, (*3*) scrub typhus, (*4*) Q fever, and (*5*) trench fever. Those that occur in the United States are Rocky Mountain spotted fever, murine typhus, Brill-Zinsser disease, rickettsialpox, and Q fever. All rickettsial infections are characterized clinically by fever, headache, and (except Q fever) rash. A summary of the epidemiologic characteristics of Rickettsial diseases and their prophylaxes are shown in Table 3.

Treatment of Rickettsial Infections

Miticidal chemicals and the antibiotics used in treatment of rickettsial infections are shown in structures (**31–37**).

All rickettsial infections respond to treatment with chloramphenicol (**35**) or tetracycline (**36**). The latter is usually the drug of choice. In rare cases, (**35**) may cause aplastic anemia due to destruction of bone marrow cells; this untoward reaction has a high mortality. *Para*-aminobenzoic acid was the first antimicrobial agent available for therapy of rickettsial infections. It has been used to treat epidemic typhus (99) and Rocky Mountain spotted fever (100). However, it is less effective than tetracycline and chloramphenicol and has greater toxicity, including suppression of the white blood cell count and formation of crystals in the urine with resultant kidney damage.

m-CH$_3$C$_6$H$_4$—CN(C$_2$H$_5$)$_2$

N,N-diethyl-m-toluamide

(**31**)

o-C$_6$H$_4$(CO$_2$R)$_2$
(**32**) dimethyl phthalate R = CH$_3$
(**33**) dibutyl phthalate R = n-Bu

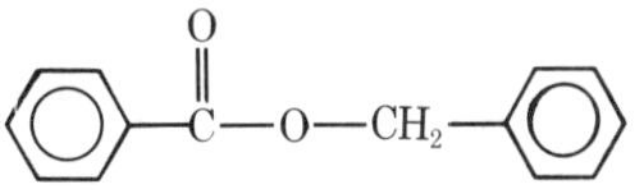

benzyl benzoate

(**34**)

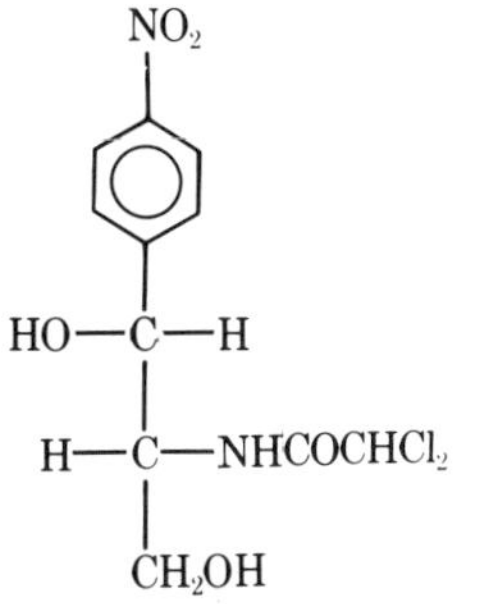

(**35**)

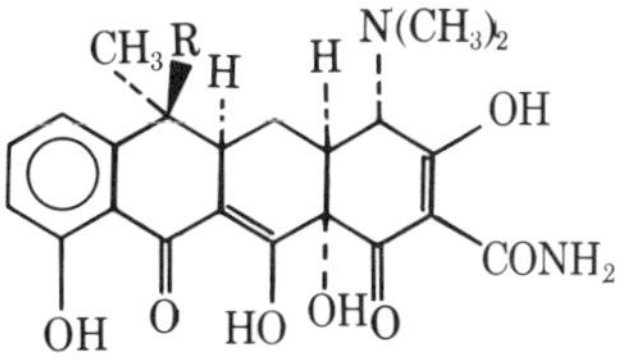

(**36**) R = OH
(**37**) R = H

Table 3. Characterics of Rickettsial Diseases

Disease	Geographic occurrence	Organism	Vector	Reservoir	Control and prophylaxis	Usual incubation period, d
Typhus group						
epidemic typhus	worldwide	*Rickettsia prowazekii*	human body louse	humans	appropriate insecticides[a]	10–14
Brill-Zinsser disease[b]	worldwide	*R. prowazekii*	body louse	humans	appropriate insecticides[a]	7–11
endemic typhus[c]	worldwide	*R. mooseri*	rat flea	rats and mice	appropriate insecticides[a]	6–4
Spotted fever group						
Rocky Mountain spotted fever	eastern and northwestern U.S.	*R. rickettsii*	wood tick, dog tick	wild rodents	vaccination of humans[d], application of (31) and (32) or appropriate insecticides[a]	2–6
tick-borne rickettsioses of eastern hemisphere	Africa; North Asia; Australia	*R. conorii; R. sibirica; R. australis*		wild rodents		5–7
rickettsialpox	U.S. and U.S.S.R., especially urban areas thereof	*R. akari*	mouse mite	house mice	appropriate insecticides[a]	7–10
Scrub typhus	Asiatic–Pacific area	*R. tsutsugamushi*	chigger	mites, rodents	appropriate insecticides[a]; application of (32–34) to clothes; and (35)	6–21
Q fever	worldwide	*Coxiella burnetii*	ticks between animals; to humans via dust or aerosols	mammals	strict hygienic measures	14–28
trench fever	western Europe	*R. quintana*	body louse	humans	appropriate insecticides[a]	10–30

[a] See Insect control technology.
[b] Refs. 99 and 100.
[c] Ref. 101.
[d] Vaccine Technology and ref. 102.

Table 4. Alphabetical List of Compounds Referred to in the Text

Compound	CAS Registry No.
acrisorcin (14)	[7527-91-5]
ambruticin (10)	[58857-02-6]
amphotericin B (2)	[1397-89-3]
amphotericin B, methyl ester (3)	[36148-89-7]
benzyl benzoate (34)	[120-51-4]
blasticidin S	[2079-00-7]
candicidin (4)	[1403-17-4]
chloramphenicol (35)	[56-75-7]
1-chlorobenzyl-2-methylbenzimidazole (23)	[3689-76-7]
clotrimazole (25)	[23953-75-1]
cycloheximide (8)	[66-81-9]
dibutyl phthalate (33)	[84-74-2]
N,N-diethyl-m-toluamide (31)	[134-62-3]
dimethyl phthalate (32)	[131-11-3]
doxycycline (37)	[564-25-0]
5-fluoro-2'-deoxyuridine-5'-monophosphate (21)	[134-46-3]
5-fluorocytosine (19)	[2022-85-7]
5-fluorouracil (20)	[51-21-8]
5-fluorouridine (22)	[316-46-1]
gentian violet (11)	[548-62-9]
griseofulvin (6)	[126-07-8]
haloprogin (29)	[777-11-7]
2-hydroxystilbamidine (18)	[495-99-5]
2-hydroxystilbamidine isethionate (17)	[533-22-2]
iodochlorhydroxyquin (13)	[130-26-7]
kasugamycin	[6980-18-3]
lenamycin (30)	[543-21-5]
miconazole (26)	[22916-47-8]
miconazole nitrate (27)	[22832-87-7]
nystatin (1)	[1400-61-9]
pimaricin (5)	[7681-93-8]
polyoxin D	[22976-86-9]
pyrrolnitrin (9)	[1018-71-9]
sinefungin (7)	[58944-73-3]
stilbamidine (16)	[122-06-5]
sulfadiazine (15)	[68-35-9]
tetracycline (36)	[60-54-8]
thiabendazole (24)	[148-79-8]
tolnaftate (28)	[2398-96-1]
10-undecenoic acid (12)	[112-38-9]
validamycin A	[37248-47-8]

Although (36) is usually the drug of first choice for rickettsial infections, it may not be the first choice for treatment of children less than eight years of age. In individuals of this age group (36) may cause a brownish discoloration of the permanent teeth. Both (35) and (36) are well absorbed orally and should be given by that route unless the patient has nausea and vomiting or is unconscious. In patients too ill to take the medications orally, both may be given by the intravenous route. Treatment should be initiated as early as possible in the illness. Tetracycline (36) is given in a dose of 500 mg orally every six h for adults or in children at a dose of 25 mg/kg body wt per day divided into equal portions given at six hourly intervals. Chloramphenicol (35) should be given orally in a dose of 50 mg/kg body wt per day divided into four equal doses. Intravenous therapy with (36) is started with 1 gm followed by 500 mg every 6 h. Therapy with (35) is begun with 1 gm and then given in a dose of 500 mg every 4–6 h. When the patient has improved to the point where intravenous therapy is no longer needed, treatment should be switched to the oral route (see Antibiotics, chloramphenical and tetracyclines).

Treatment should be continued until the patient has been without fever for two or three days. Therapy needs to be given for a few days past the termination of the illness to prevent relapse. This is due to the fact that both (35) and (36) are rickettsiostatic and not rickettsiocidal. The antibiotics hold the growth of rickettsiae in check until the body's immune system can erradicate them. Another effective tetracycline antibiotic in the treatment of rickettsial infections is doxycycline (37). Several investigators have found that it is effective in the treatment of epidemic typhus when given in a single oral dose of 100–200 mg (101–103). The penicillin-antibiotics and streptomycin are inactive against rickettsiae. The combination of trimethoprim and sulfamethoxazole (co-trimoxazole) has been found to be ineffective when given to patients with rickettsial infections and are contraindicated (see Antibacterial agents, sulfonamides).

Of course, prevention is attained by elimination of lice, mites, ticks, rodents, and use of the appropriate insecticides (see Insect control technology).

An alphabetical list of compounds referred to in the text is shown in Table 4.

BIBLIOGRAPHY

"Bacterial, Rickettsial, and Mycotic Infections, Chemotherapy" in *ECT* 2nd ed., Vol. 3, pp. 1–36, by Morris Solotorovsky, Rutgers University, and Gordon Kemp, American Cyanamid Company.

1. W. Mechlinski in A. I. Laskin and H. A. Lechevalier, eds., *Handbook of Microbiology, Vol. III*, CRC Press, Cleveland, Ohio, 1973, pp. 93–107.
2. C. W. Emmons and co-workers, *Medical Mycology*, Lea and Febiger, Philadelphia, Pa., 1977, p. 74.
3. E. L. Hazen and R. Brown, *Proc. Soc. Exp. Biol. Med.* **76,** 93 (1951).
4. W. Mechlinski in ref. 1, p. 104.
5. S. Shadomy and A. Espinel-Ingroff in E. H. Lennette, E. H. Spaulding, and J. P. Truant, eds., *Manual of Clinical Microbiology*, 2nd ed., American Society for Microbiology, Washington, D.C., 1974, p. 571.
6. C. Halde and co-workers, *J. Invest. Dermatol.* **28,** 217 (1957).
7. A. Kucers and N. M. Bennett, *The Use of Antibiotics*, 2nd ed., Wm. Heinemann Medical Books, London, Eng., 1975, p. 561.
8. J. E. Bennett. *N. Engl. J. Med.* **290,** 30, 320 (1974).
9. W. Mechlinski and C. P. Schaffner, *J. Antibiot.* **25,** 256 (1972).
10. H. H. Gadebusch and co-workers, *J. Infect. Dis.* **134,** 423 (1976).
11. N. M. Stevens and co-workers, *Arch. Virol.* **48,** 391 (1975).
12. G. W. Jordan and E. C. Seet, *Antimicrob. Agents Chemother.* **13,** 199 (1978).
13. D. P. Bonner and co-workers, *Antimicrob. Agents Chemother.* **7,** 724 (1975).
14. P. D. Hoeprich, L. K. Henth, and R. M. Lawrence, *Abstr. 16th Intersci. Conf. Antimicrob. Agents Chemother.* abst. **306** (1976).
15. H. Lechevalier and co-workers, *Antibiot. Chemother.* **11,** 640 (1961).
16. E. Drouhet in ref. 1, p. 693.
17. W. Mechlinski in ref. 1, p. 101.
18. W. Mechlinski in ref. 1, p. 104.
19. Ref. 7, p. 604.
20. D. R. Berry and B. J. Abbott, *J. Antibiot.* **31,** 185 (1978).
21. S. M. Ringel and co-workers, *J. Antibiot.* **30,** 371 (1977).
22. H. B. Levine, S. M. Ringel, and J. M. Cobb, *Chest* **73,** 202 (1978).

23. J. C. Gentles, *Nature (London)* **183**, 476 (1958).
24. A. R. Martin, *Vet. Rec.* **70**, 1232 (1958).
25. H. Blank and F. J. Roth, Jr., *AMA Arch. Dermatol.* **79**, 259 (1959).
26. D. I. Williams, R. H. Martin, and I. Sarkany, *Lancet ii*, 1212 (1958).
27. G. Hildick-Smith, H. Blank, and I. Sarkany, *Fungus Diseases and Their Treatment*, Little, Brown and Co., Boston, Mass., 1964, p. 438.
28. C. Lin and S. Symchowicz, *Drug Metab. Rev.* **4**, 79 (1975).
29. P. W. Brian, *Ann. Bot. (London)* **13**, 59 (1949).
30. G. Evans and N. H. White, *J. Exp. Bot.* **18**, 465 (1967).
31. Ref. 7, p. 598.
32. A. J. Whiffen, *J. Bacteriol.* **56**, 283 (1948).
33. J. Dekker, *World Rev. Pest Control* **10**, 9 (1971).
34. M. R. Siegel and H. D. Sisler, *Biochim. Biophys. Acta* **87**, 70 (1964).
35. R. S. Gordee and T. R. Mathews, *Antimicrob Agents Chemotherapy-1967*, American Society for Microbiology, Ann Arbor, Mich., 1968, p. 378.
36. L. D. Boeck and co-workers, *Antimicrob. Agents Chemother.* **11**, 49 (1973).
37. R. Hamill and M. M. Hoehn, *J. Antibiot. (Tokyo)* **26**, 463 (1973).
38. R. S. Gordee and T. F. Butler, *J. Antibiot. (Tokyo)* **26**, 466 (1973).
39. J. R. Turner and co-workers, *Abst. 17th Intersci. Conf. Antimicrob. Agents Chemother.* abst **49**, (1977).
40. S. M. Ringel and co-workers, *J. Antibiot. (Tokyo)* **30**, 371 (1977).
41. S. M. Ringel, *Antimicrob. Agents Chemother.* **13**, 762 (1978).
42. S. Shadomy and co-workers, *Antimicrob. Agents Chemother.* **14**, 99 (1978).
43. H. B. Levine and S. M. Ringel in L. Ajello, ed., *Coccidioidomycosis: Current Clinical Diagnostic Status*, Symposia Specialists Publishers, Miami, Fl., 1977, p. 319.
44. H. B. Levine, S. M. Ringel, and J. M. Cobb, *Chest* **73**, 202 (1978).
45. S. Shadomy, C. J. Utz, and S. White, *Antimicrob. Agents Chemother.* **14**, 95 (1978).
46. N. F. Conant and co-workers, *Manual of Medical Mycology*, 2nd ed., W. B. Saunders Co., Philadelphia, Pa., 1954, p. 298.
47. *Ibid.*, pp. 191 and 300.
48. *Ibid.*, pp. 302 and 401.
49. *Ibid.*, pp. 298 and 428.
50. Ref. 27, p. 10.
51. *Ibid.*, p. 163.
52. Ref. 2, p. 409.
53. M. S. Silva, *Rev. Brasil Med.* **2**, 918 (1945).
54. Ref. 46, p. 91.
55. J. W. Rippon, *Medical Mycology, The Pathogenic Fungi and the Pathogenic Actinomycetes*, W. B. Saunders Co., Philadelphia, Pa., 1974, p. 398.
56. A. C. Curtis and E. R. Harrell, *AMA Arch. Dermatol. Syphilol.* **66**, 676 (1952).
57. Ref. 2, p. 78.
58. J. F. Busey in H. A. Buechner, ed., *Management of Fungus Diseases of the Lungs*, Charles C Thomas, Springfield, Ill., 1971, p. 47.
59. R. Duschinsky, E. Pleven, and C. Heidelburg, *J. Am. Chem. Soc.* **79**, 4559 (1957).
60. E. Grunberg, E. Titsworth, and M. Bennett, *Antimicrob. Agents Chemother.* **3**, 566 (1963).
61. B. A. Koechlin and co-workers, *Biochem. Pharmacol.* **15**, 435 (1966).
62. D. C. E. Speller and M. G. Davis, *J. Med. Microbiol.* **6**, 315 (1973).
63. G. E. Wagner and S. Shadomy, *Antimicrob. Agents Chemother.* **11**, 299 (1977).
64. P. D. Hoeprich, *Ann Rev. Pharmacol. Toxicol.* **18**, 205 (1978).
65. Ref. 55, p. 541.
66. A. Polak and H. J. Scholer, *Chemotherapy* **21**, 113 (1975).
67. G. E. Wagner and S. Shadomy, *Chemotherapy*, **24**, (1978).
68. R. B. Diasio, J. E. Bennett, and C. E. Myers, *Biochem. Pharmacol.* **27**, 703 (1978).
69. G. Vandervelde, A. A. Mauceri, and J. E. Johnson, III, *Ann. Int. Med.* **77**, 43 (1972).
70. S. Shadomy, *Appl. Microbiol.* **17**, 871 (1969).
71. J. P. Utz and co-workers, *J. Inf. Dis.* **132**, 368 (1975).
72. H. J. Scholer, *Chemotherapy* **22**, 103 (1976).
73. J. Schönebeck and co-workers, *Chemotherapy* **18**, 321 (1973).
74. R. J. Holt, *Infection* **2**, 95 (1974).

75. H. P. R. Seeliger, *Mykosen* **1,** 162 (1958).
76. L. J. Sorensen and R. J. Robinson in J. C. Sylvester, ed., *Antimicrobial Agents and Chemotherapy-1964,* American Society for Microbiology, Ann Arbor, Mich., 1965, p. 742.
77. M. Plempel and co-workers, *Deut. Medizinische Wochenschr.* **94,** 1356 (1969).
78. S. Shadomy, *Infect. Immun.* **4,** 143 (1971).
79. S. Shadomy in G. L. Hobby, ed., *Antimicrobial Agents and Chemotherapy-1970,* American Society for Microbiology, Bethesda, Md., 1971, p. 169.
80. J. A. Waitz, E. L. Moss, and M. J. Weinstein, *Appl. Microbiol.* **22,** 891 (1971).
81. G. K. Crompton and L. J. R. Milne, *Br. J. Dis. Chest* **67,** 301 (1973).
82. Y. M. Clayton and B. L. Connor, *Br. J. Dermatol.* **89,** 297 (1973).
83. J. M. Van Cutsem and D. Thienpont, *Chemotherapy* **17,** 392 (1972).
84. S. Shadomy and co-workers, *J. Antimicrob. Chemother.* **3,** 147 (1977).
85. H. Van den Bossche, G. Willemsens, and J. M. Van Cutsem, *Sabouraudia* **13,** 63 (1975).
86. R. J. Holt, *J. Cutaneous Pathol.* **3,** 45 (1976).
87. P. J. Lewi and co-workers, *Eur. J. Clin. Pharmacol.* **10,** 49 (1976).
88. P. D. Hoeprich and E. Goldstein, *J. Am. Med. Assoc.* **230,** 1153 (1974).
89. J. F. Fisher and co-workers, *Antimicrob. Agents Chemother.* **13,** 965 (1978).
90. H. B. Levine and co-workers, *J. Inf. Dis.* **132,** 407 (1975).
91. P. R. Sawyer and co-workers, *Med. Prog.,* 25 (May, 1975).
92. D. A. Stevens, *Am. Rev. Resp. Dis.* **116,** 801 (1977).
93. J. Symoens, *Proc. R. Soc. Med.* **70**(Suppl. 1), 4 (1977).
94. J. E. Davis, J. H. Frudenfeld, and J. L. Goddard, *Obstet. Gynecol.* **44,** 403 (1974).
95. H. B. Nield, *N. Engl. J. Med.* **296,** 1479 (1977).
96. H. Brincker, *Proc. R. Soc. Med.* **70**(Suppl 1), 29 (1977).
97. American Medical Association, *AMA Drug Evaluations,* 3rd ed., Publishing Sciences Group, Inc., Acton, Mass., 1977, pp. 821–838.
98. T. Misato in J. R. Plimmer ed., *Pesticide Chemistry in the 20th Century,* American Chemical Society, Washington, D.C., 1977, pp. 170–192.
99. J. C. Snyder and co-workers, *Ann. Intern. Med.* **27,** 1 (1947).
100. T. E. Woodward and W. R. Raley, *South. Med. J.* **41,** 997 (1948).
101. J. Huys and co-workers, *Trans. R. Soc. Trop. Med. Hyg.* **67,** 718 (1973).
102. P. L. Perine and co-workers, *Lancet ii,* 742 (1974).
103. D. W. Krause and co-workers, *East Afr. Med. J.* **52,** 421 (1975).

General References

Human Mycoses

E. S. Beneke, *Scope® Monograph on the Human Mycoses,* The Upjohn Co., Kalamazoo, Mich., 1974, 1–48.
G. S. Kobayashi and G. Medoff, *Ann. Rev. Microbiol.* **31,** 291 (1977).
J. F. Martin, *Ann. Rev. Microbiol.* **31,** 13 (1977).
K. Iwata, ed., *Recent Advances in Medical and Veterinary Mycology,* University of Tokyo Press, Tokyo, Jpn., 1977, 316 pp.
G. E. W. Wolstenholme and R. Porter, eds., *Systemic Mycoses, A Ciba Foundation Symposium,* Little, Brown and Co., Boston, Mass., 1967, 287 pp.
Pan American Health Organization, *Paracoccidioidomycosis, Proceedings of the First Pan American Symposium,* Sc. Publ. 254, WHO, Washington, D.C., 1972, 319 pp.
E. W. Chick, A. Balows and M. L. Furcolow, *Opportunistic Fungal Infection,* Charles C Thomas, Springfield, Ill., 1975, 359 pp.
L. Ajello, ed., *Coccidioidomycosis, Current Clinical and Diagnostic Status,* Symposia Specialists, Miami, Fl., 1977, 475 pp.
N. F. Conant and co-workers, *Manual of Clinical Mycology,* W. B. Saunders, Philadelphia, Pa., 1971, 255 pp.

Rickettsial Infections

F. L. Horsfall, Jr., and I. Tamm, eds., *Viral and Rickettsial Infections of Man*, J. B. Lippincott Co., Philadelphia, Pa., 1965.
P. B. Beeson and W. McDermott, eds., *Textbook of Medicine*, W. B. Saunders Co., Philadelphia, Pa., 1975.
G. W. Thorn and co-eds., *Harrison's Principles of Internal Medicine*, McGraw Hill Book Co., New York, 1977.
P. D. Hoeprich, ed., *Infectious Diseases, A Modern Treatise of Infectious Processes*, Harper and Row, Publishers, Inc., New York, 1977.
J. C. Snyder in F. R. Moulton, ed., *Rickettsial Diseases of Man*, American Association for the Advancement of Science, Washington, D.C., 1948.
M. W. Rytel and J. D. Coonrod, *Wis. Med. J.* **70,** 116 (1971).
M. J. Snyder and T. E. Woodward, *Med. Clin. N. Am.* **54,** 1187 (1970).
H. L. Dupont and co-workers, *N. Engl. J. Med.* **282,** 53 (1970).
M. A. W. Hattwick, R. J. O'Brien, and B. F. Hanson, *Ann. Intern. Med.* **84,** 732 (1976).
W. Burgdorfer, *Acta Tropica* **34,** 103 (1977).

SMITH SHADOMY
C. GLEN MAYHALL
Medical College of Virginia

CHEMOTHERAPEUTICS, ANTIPROTOZOAL

Protozoa are single-cell organisms that constitute the most primitive group of the animal realm. About 45,000 species of protozoa have been described. Some of them parasitize higher animals and man. Antiprotozoal chemotherapy was developed in response to health problems of individuals and to economic needs (1–4). Knowledge of the history of antiprotozoal drugs is useful for future development in this area (5). In this article the term chemotherapy is used for protozoocidal and protozoostatic compounds. The protozoocidal drugs are curative in that they eradicate the parasite in all its stages within the host. Protozoostatic drugs suppress the parasite's clinically relevant developmental stages but not certain latent stages. The surviving, suppressed protozoa may cause either clinical relapse of the disease or stimulate immune defenses which in turn may either cure or control the latent infection.

Many of the antiprotozoal drugs have a narrow safety margin, eg, 1:2. The safety margin is determined on a statistical basis, but as the susceptibility to a drug often varies unpredictably from one patient to another, medication always involves some risk. The narrower the safety margin, the greater the risk. The practitioner must weigh the potentialities of toxicity against the seriousness of the protozoal infection and thus evaluate the risk-to-benefit ratio of the medication.

An adequate prediction of a chemical agent's effectiveness against a certain protozoan cannot be made from the present range of knowledge. For some drugs, eg, antifolates, the site of action is well established, yet the reason for better action in some protozoon species than in others, is unknown. *In vitro* studies have shown many compounds to affect one or more biochemical systems, but quite often, *in vitro* activities are not predictive for clinical efficacy and *in vivo* screening is still important in antiprotozoal drug development.

This article describes those drugs that have reached commercial and clinical significance and a few selected experimental drugs that appear promising. Both human and veterinary drugs are described since many of the principal parasitic diseases discussed infect animals of considerable value in the economy (see also Veterinary drugs).

Coccidiosis

The taxonomic classification of *Coccidia* species is still in a state of flux (6–8). *Coccidia* species are cosmopolitan and widespread in the animal kingdom, although most of them are limited to a narrow range of host species and organ systems. Some *Coccidia*, eg, *Toxoplasma,* may pass from one host species to another changing their target organs from host to host. The life cycle of *Coccidia* includes cysts, sporozoites, schizonts, merozoites, and gametocytes. *Eimeria* and *Toxoplasma* are economically and medically important *coccidia*.

Eimeria infect many vertebrates, including fish, poultry, farm animals, dogs, and cats. The parasite first invades the epithelial lining of the digestive tract and may cause diarrhea, sloughing, and ulceration of the intestinal lining, and hemorrhage, leading to possible metastatic spread, malabsorption of nutrients and vitamins, metabolic imbalance, and anemia and bacteremia. Depending on severity and duration, the consequences may range from growth retardation to death.

Eradication of *Eimeria* species with presently available drugs would be impractical because of toxicological and economic considerations; anticoccidials are currently used in animal husbandry as coccidiostats. Mixed with feeds, the coccidiostats serve to minimize the intensity of infection or to reduce clinical symptoms to a nonfatal course. In this way, time is gained for the host to build up immune defenses. *Eimeria* species develop drug resistance quite rapidly so that development of new anticoccidials is economically vital for animal farming (8–9) (Fig. 1). Furthermore, the efficacy of any given anticoccidial against *Eimeria* and the toxicity to the host can vary from one species of either to another. The situation can become even more complicated as mixed infections by several *Eimeria* species may occur, especially in cattle.

Reports on drug efficacy should be evaluated carefully by an analysis of the assay protocol; uncontrolled studies are meaningless. The course of *Eimeria* infection is often self-limited in individual subjects and in epidemics of population groups. If treatment is initiated at the peak of a clinical case or an epidemic, the kind of drug used is irrelevant as the disease would subside in any case. A system known as floor pen is a good standard for drug assays in chicken coccidiosis (10).

For each of the few drugs described below, a large number of congeners are available for both experimental and commercial use. Some of the anticoccidials belong to drug categories with well known biochemical effector systems. The biochemical rationale for others is still uncertain, a failing that does not diminish their practical value. The safety range of anticoccidials varies widely. Contributing variables of toxicity are dosage and duration of the drug administration, species and metabolic condition of the host, environmental conditions, and nutrition. The anticoccidials with defined biochemical targets are thiamine competitors, antifolates, antibiotics, nitrobenzamides, and nitrofurans.

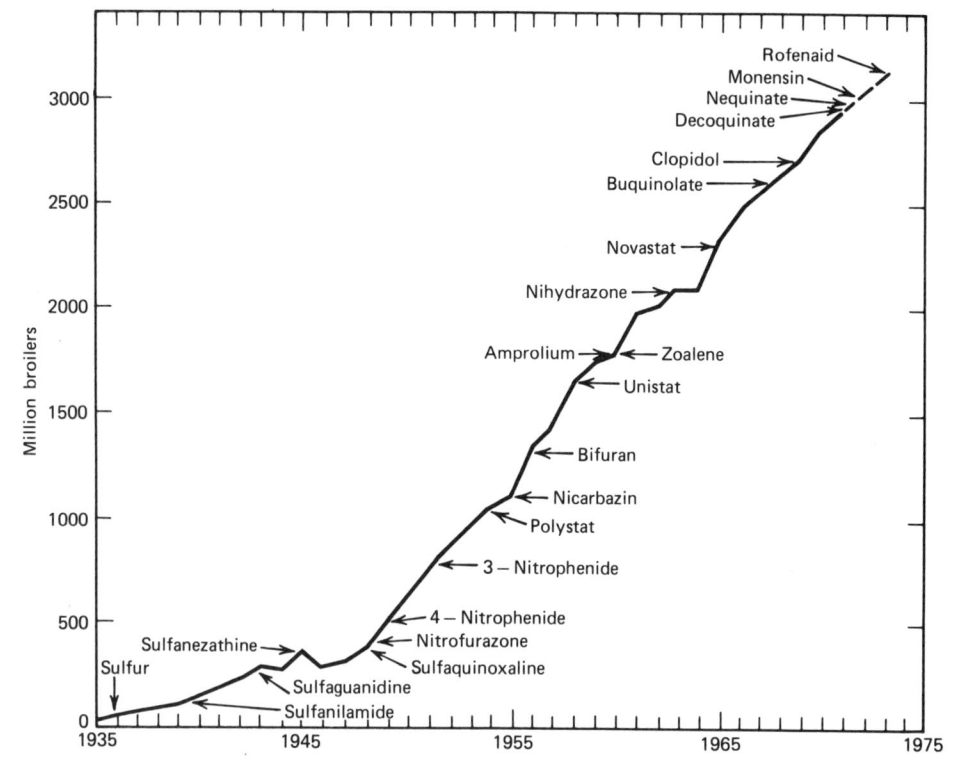

Figure 1. Correlation between growth of the broiler industry and the introduction of coccidiostats (9).

Thiamine Competitors. These are thiamine (1) derivatives in which either the side chains are modified, or the thiazolium ring is substituted or opened, or both (11). Examples are amprolium (2) and aminoalkenyl sulfides, eg, (3).

Amprolium is prepared by reaction of 2-propyl-4-amino-5-chloromethylpyrimidine dihydrochloride with 2-methylpyridine in acetonitrile (12). The synthesis of (3) is described in a Japanese patent: a thiamine-type compound is treated with an alkali and the resulting ring-opened thiol is oxidized with potassium iodide to give the disulfide (13).

Toxicity of the thiamine competitors manifests itself as polyneuritis. Administration of thiamine can prevent this condition but will not cure the adverse effects if nerves and their dependent organs have suffered permanent damage (14).

Antifolates. Sulfonamides, pyrimethamine, and p-aminobenzoic acid competitors such as Ethopabate (4) are in this category (see Antibacterial agents).

Ethopabate is synthesized from the potassium salt of 4-acetamido-2-hydroxybenzoic acid methyl ester and ethyl 4-toluenesulfonate (15).

Antibiotics. Among the various antibiotics, monensin (5) has practical relevance (see Antibiotics, polyethers). It is isolated from *Streptomyces cinnamonensis* and is also produced by a fermentation process (16).

Like other monocarboxylic acid antibiotics, monensin increases the monovalent cation permeability of biological membranes. The loss of these cations impairs the enzyme systems that they regulate (17). Information in reports or patents is inadequate

516 CHEMOTHERAPEUTICS, ANTIPROTOZOAL

on the anticoccidial activity of several newer antibiotics. The reported efficacy of oxytetracycline and similar antibiotics in coccidial infections may be explained by their control of secondary infections (see Antibiotics, tetracyclines).

Nitrobenzamides and Nitrofurans. Efforts to develop anticoccidials in these classes seem justified since compounds with a 5-nitro group interfere with the cytochrome system of other protozoa (18) (see Antibacterial agents); however, no satisfactory anticoccidial has as yet been marketed in this class. The available analogues require high dosages for efficacy and have a very narrow safety margin. Information on the mutagenicity of this class will also be needed.

Reports which dealt with relevant biochemical targets of the following drugs could not be located: Robenidine (6), bisthiosemicarbazones (7), Clopidol (8), quinolones (9,10), and quinazolinones (11).

Robenidine (6) is a very effective compound. However, birds whose diet has contained Robenidine, produce tainted eggs, and their flesh reportedly has an unpleasant taste. Compound (6) is 1,3-bis[(4-chlorobenzylidene)amino]guanidine and is obtained by the reaction of 4-chlorobenzaldehyde with 1,3-diaminoguanidine nitrate (19).

Bitipazone (7) is an example of a bisthiosemicarbazone (20). In this class a basic side chain seems to be necessary for activity, and asymmetric compounds are more efficacious than symmetrical ones. The safety margin in this class is very narrow.

Clopidol (8) is thus far the only simple pyridone derivative with good anticoccidial activity. It has been synthesized by chlorination of 2,6-dimethyl-4-pyridinol in aqueous hydrochloric acid (21).

Among the quinolones, the most active is nequinate (9). Its two-step preparation starts with 4-butyl-3-benzyloxyaniline and diethyl ethoxymethylenemalonate (22).

The quinolones are virtually water-insoluble. For buquinolate (10), the micronized formulation of 1.8 μm particle size was more active than the 6.1-μm milled formulation. Its synthesis starts with catechol diisobutyl ether which is nitrated, catalytically reduced, and the ring closure of an anil diester is accomplished by heating in Dowtherm under reflux (23). Safety studies have not yielded evidence of serious toxicity. But the potential of quinolones for producing drug resistance is very great. Some of the quinazolines are highly active but have a low safety margin. They are compounds with halogen substitutents at positions 6 to 8 derived from the antimalarial febrifugine (11). Febrifugine was isolated from *Dichroa febrifuga* and from the common hydrangea. It was synthesized via the key intermediates 2-methoxy-5-carbobenzoxyaminovaleric acid and 1-carbethoxy-3-methoxy-2-piperidineacetic acid (24).

Drug mixtures including members of the above classes are also successful. Patents have been granted for numerous analogues of each of the above classes but the drugs have not been offered for commercial use. Some new antimicrobials and anthelmintics have anticoccidial potential but often low efficacy or prohibitive toxicity.

Toxoplasmosis. Toxoplasma infections are cosmopolitan and quite common among man and animals (25). Species differentiation between the toxoplasmas of man and other mammals has not been definitely established. According to some estimates one-third or more of human populations may carry a quiescent infection. Though clinical manifestations of toxoplasmosis are relatively rare, they are very serious when they do occur. They are more prevalent in children than in adults and their incidence is increased in immunodeficient individuals (26). In humans, contamination is either transplacental or by consumption of infected meat or animal material. Antifolates are effective in toxoplasmosis chemotherapy. Sulfonamides and pyrimethamine act individually on different steps of the folic acid synthesis and, when combined, they potentiate each other as in Fansidar [*37338-39-9*] (27).

Sulfonamides that dissolve readily in intracellular fluid are effective against toxoplasmosis, but those distributed mainly extracellularly are not effective (28). The use of antifolates during certain stages of pregnancy can have serious teratogenetic effects. In addition, sulfonamides displace bilirubin from its protein binding which causes the free bilirubin plasma to increase to levels that are toxic to the newborn. Hence the administration of sulfonamides at the end of pregnancy can lead to kernicterus in the newborn.

The antibiotic spiramycin (**12**) has also been reported to be useful in the treatment of toxoplasmosis (28–31) (see Antibiotics, macrolides).

(**12**)
R = H, COCH$_3$, COC$_2$H$_5$

Anaplasmosis

It has not been established whether the Anaplasma species be classed among protozoa or rickettsia (32). They parasitize the red blood cells of cattle and can cause heavy mortality. Tetracyclines have been found effective by both oral and parenteral administration. Certain dithiosemicarbazones, eg, gloxazone (**13**), are effective. Its synthesis was accomplished by heating a mixture of crotonaldehyde and ethanol under reflux in the presence of selenium dioxide. The ketoaldehyde obtained was converted to the bisthiosemicarbazone (33). The carbanilide derivative, Imidocarb (**14**), was found effective (34). It has a very narrow safety margin and anticholinesterase activity in the host, and it can cause adverse neuromuscular and cardiovascular effects.

Babesiasis

Members of the genus *Babesia*, of the order *Haemosporidia*, are transmitted by ticks and parasitize the red blood cells of mammals and birds (3,32). Mortality rates can be higher than 10% in cattle depending on the *Babesia* subspecies. Rare cases of human infection have also been reported (35).

Diamidine and related compounds have proven to be effective in the treatment

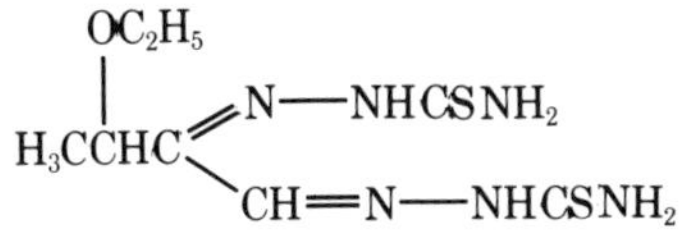

(13) 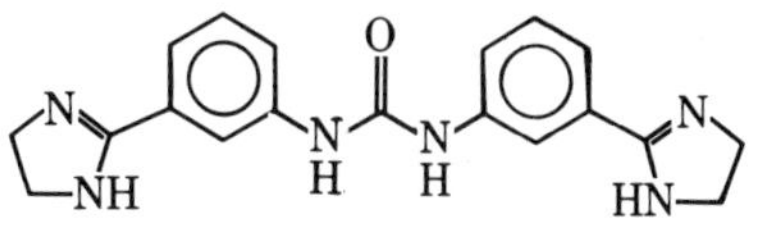   (14)

of babesiasis, eg, (15), (16), and (17).

Pentamidine (15) may be prepared by saturating an anhydrous, alcoholic solution of 1,5-di(4'-cyanophenoxy)pentane with dry hydrogen chloride, and treating the iminoether obtained from the hydrochloride with ammonium β-hydroxyethanesulfonate (36). Amicarbalide (16) is prepared as (15) by the Pinner reaction of the corresponding dinitrile (37). Diminazene aceturate (17), Berenil, is formed via a seven-step synthesis proceeding from 4-toluenesulfonamide through the 4-nitrobenzonitrile, the imidate, amidine and 4-aminobenzamidine with diazotization to the product (38).

Quaternary compounds of the quinuronium class are effective but have a narrow safety range. They are toxic to the parasympathetic nervous system and are potent cholinesterase inhibitors (see Choline; Cholinesterase inhibitors).

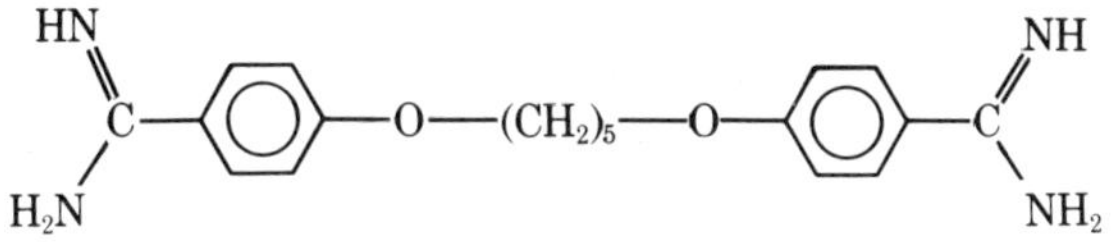

(15)

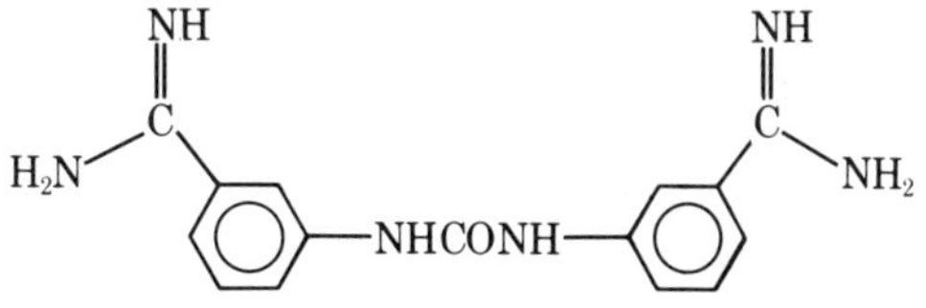

(16)

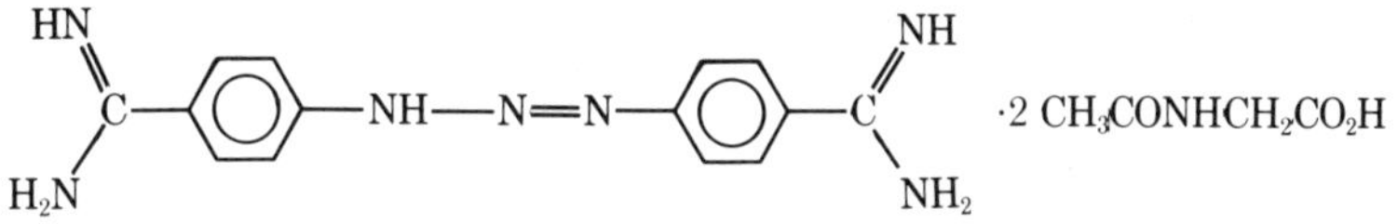

(17)

Theileriasis

Theileria species are transmitted by ticks to cattle and other ungulates (39). Various developmental stages of the parasite are in lymphocytes, histiocytes, and finally erythrocytes. In east and south Africa the species *T. parva* causes considerable damage to herds of farm animals with mortality reaching 90%. Attempts at chemotherapy with almost all known antiparasitic drugs have failed. However, good chemoprophylaxis is obtained when tetracyclines are administered during the period of incubation. This reduces the severity of infection and allows development of immune defenses which keep the infection under clinical control. Another species, *T. annulata*, has a wide geographic distribution. The natural recovery rates from this infection are higher than from *T. parva*. A large number of uncontrolled studies with a variety of drug combinations has been reported.

CHEMOTHERAPEUTICS, ANTIPROTOZOAL

Trypanosomiasis

Species of *Trypanosoma brucei* cause a variety of diseases, many with serious economic consequences, in wild and domestic animals and in man in tropical and subtropical areas (4,40). *T. cruzi* species cause Chagas' disease which is prevalent in Central and South America.

The trypanosomes are transmitted by insect vectors. They have distinct developmental cycles in the host and in most vectors and the epidemiologic control is difficult since certain wild and domestic animals are disease reservoirs. The available drugs are quite toxic and frequently inadequate for some of the diseases (40c). Retrospective studies *in vitro* showed interference with several enzyme systems or with DNA metabolism of the trypanosomes, or both, although the actual basis for their clinical efficacy has not been established (41).

African Trypanosomiasis. The African trypanosomes of the brucei type are dreaded scourges for man and animals (42). In man, *T. br. Gambiense* and *T. br. rhodesiense* cause the African sleeping sickness, and various *T. brucei* types cause damage to livestock in endemic areas. Unless treated successfully, the African trypanosomiases are invariably fatal in man. In the mammalian host the trypanosomes progress from blood to lymphatic tissues and, finally, to the central nervous system. Chemotherapy early in the infection is more promising than at the later stages. The available drugs have a narrow safety margin and side effects of some can be fatal. Drug resistance develops readily in African trypanosomiasis (43).

So far two drugs, suramin sodium (18) and pentamidine (15), have been found promising for safe chemoprophylaxis.

Suramin sodium is prepared by condensing 1-naphthylamine-4,6,8-trisulfonic acid with 3-nitro-4-methylbenzoyl chloride, reducing the product, condensing with 3-nitrobenzoyl chloride, reducing the product, condensing with 3-nitrobenzoyl chloride and reducing again. The product is treated with phosgene and the mixture neutralized with sodium hydroxide (44).

These drugs are firmly bound to tissue proteins from which they are released at a slow rate. Among the hemoflagellates they are effective prophylactics only for African

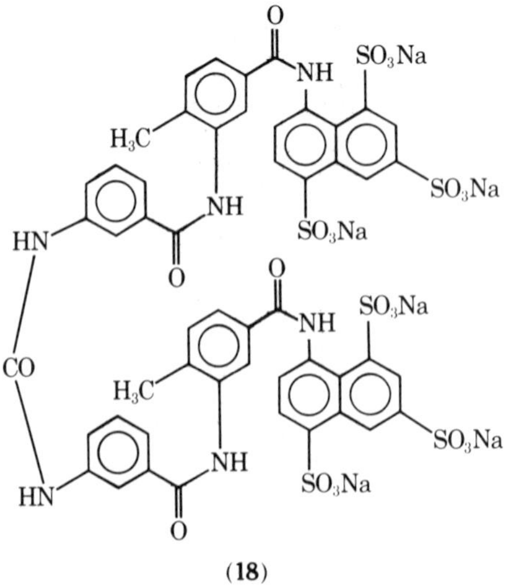

(18)

trypanosomiasis. The prophylactic efficacy of (15) exceeds that of (18). It appears to give protection against *T. br. gambiense* for ca 6 mo but its protective efficacy against *T. br. rhodesiense* is not generally accepted. Suramin (18) and pentamidine (15) are quite toxic. Pentamidine in particular is nephrotoxic. Furthermore (18) is teratogenetic in mice and causes abortions in rats (45). Medical field workers have not reported such effects in humans, but the reliability of reporting and patient follow-up in the concerned geographic areas remain to be evaluated. Thus the risk–benefit ratio must be weighed carefully for these drugs.

Pentamidine and suramin are useful in the early stages of African sleeping sickness but they fail to penetrate the blood–brain barrier. At the advanced stages, where parasites have invaded the central nervous system, drugs that can penetrate the blood–brain barrier are used. They are organic arsenicals such as tryparsamide (19), and derivatives of Melarsen oxide (20) such as melarsoprol (21) and melarsonyl potassium (see Arsenic compounds).

Tryparsamide (19) is formed by heating a solution of arsanilic acid in aqueous sodium hydroxide, sodium carbonate, and chloracetamide (46). Melarsen (20) may be obtained by the reaction of sodium arsanilate and cyanuric chloride to obtain *p*-(2,4-dichloro-*s*-triazinyl-6)aminophenylarsonic acid. This acid is treated with ammonia under pressure and the sodium salt of the resulting diamine is Melarsen (47). Melarsoprol (21) is prepared from 1-hydroxypropane-2,3-dithiol, *p*-aminophenylarsenic dioxide, and 2-chloro-4,6-diamino-1,3,5-triazine (48). Melarsonyl potassium is a water soluble derivative of melarsoprol.

Other useful drugs in the early stage are the aromatic diamidine, Berenil, which is, however, too toxic for humans; the nitrofuran, nifurtimox (22) (see Antibacterial agents, nitrofurans); the phenanthridines such as homidium bromide (23); quinapyramine (24); and some of their derivatives and antimonials such as ethylstibamine (25) (see Antimony compounds).

Nifurtimox (Lampit) is manufactured by the reaction of 5-nitrofurfural and 4-amino-3-methyltetrahydro-1,4-thiazine-1,1-dioxide (49).

Homidium bromide may be obtained by the reaction of 2,7-bis(carbethoxyamino)-9-phenylphenanthridine and the sulfonic acid ester of ethanol. The resulting quaternary salt is hydrolyzed with sulfuric acid, the pH adjusted to 7–7.5 and ammonium bromide added to precipitate the crude quaternary bromide (50). The synthesis of quinapyramine is described in a United States patent (51). Ethylstibamine (Stibosamine, Neostibosan) is a complex of p-aminobenzenestibonic acid (present largely as a tetramer), p-acetylaminobenzenestibonic acid (present largely as a dimer), antimonic acid, and diethylamine in the approximate molar ratio of 1:2:1:3. The manufacture requires special care since the compound is both a colloid and a molecular addition compound, and slight variations greatly affect its toxicity (52).

Structure-activity relationships have been studied under laboratory conditions for the nitroheterocycles (26) (R′ = H [30579-13-6]; R″ = R‴ = H [55330-02-4]) (53).

Similar studies were made on nitrothiazoles (27) (54).

Chagas' Disease. The chemotherapeutic armamentarium against *T. cruzi* is very unsatisfactory (55). This situation has been practically unchanged for many years (56). A large number of compounds, which had been promising when tested on infected laboratory mice, proved inadequate in man. Either they were too toxic or they failed to remove the amastigotes from the tissues. Studies are now under way on 8-aminoquinolines. The nitrofuran (22) gave good results in some but not all clinical trials. Its practicality is questionable since treatment of 120 d is required and side effects can be severe.

A nitroimidazole, Ro-7-1051 (28), showed encouraging results in acute clinical cases but has not been adequately studied in chronic infections. As with other nitro compounds there is concern whether it has mutagenic and carcinogenic potentials. It is prepared by the reaction of the sodium derivative of 2-nitroimidazole with methyl chloroacetate and treatment of the resulting ester with benzylamine (57).

Great circumspection is recommended in appraising reports on the efficacy of

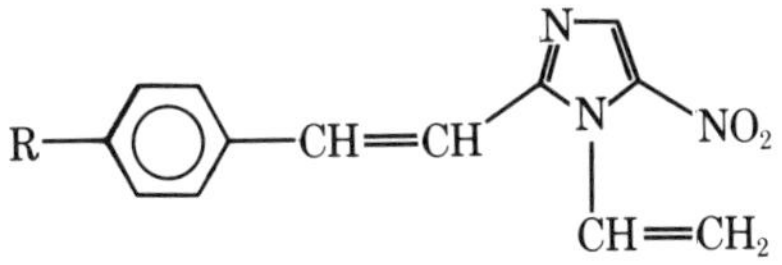

$$(26) \quad R = CH_2R', \quad C\begin{smallmatrix}NHR''\\NR'''\end{smallmatrix}$$

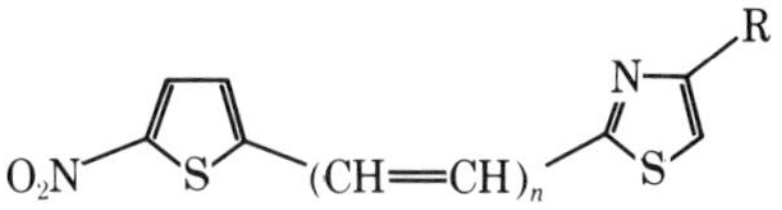

(27)

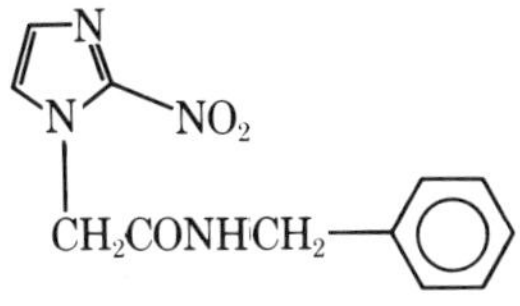

(28)

drugs in Chagas' disease. *T. cruzi* is not a single taxon but a collective term comprising regional varieties of physiologically distinct organisms (58). Hence differences in the clinical course and drug response may occur in different geographic areas.

Leishmaniasis

Members of the genus *Leishmania* have a broad geographic distribution (4,40). They are transmitted by insect vectors, and wild and domestic animals are disease reservoirs. They cause a variety of important diseases in man. *L. donovani*, *L. tropica*, and *L. brasiliensis* cause visceral cutaneous, and mucocutaneous leishmaniases, respectively. Some species cause chronic progressive diseases, others cause self-limiting conditions. The leishmania amastigote stage presents a particularly serious problem. The parasites invade the reticulo-endothelial system and their amastigotes penetrate and survive in the host's macrophages, thus reducing the host's immune defense capabilities. This sheltered location also interferes with direct drug contact and few drugs, if any, are selective against leishmania. One of the oldest antileishmanials is tartar emetic [28300-74-5], antimony potassium tartrate, which is highly toxic.

A number of less toxic, organic antimonials have been developed. These are linear polymers containing pentavalent antimony. Examples are sodium antimonyl gluconate, Pentostam (29), which is prepared from antimony pentoxide, gluconic acid and sodium hydroxide (59); the antimonate of *N*-methylglucamine, Glucantime (30) (60); and ethylstibamine (25). Various investigators differ in their opinions of the efficacy and toxicity of any of these drugs.

Other compounds reportedly effective are amphotericin B [1397-89-3] and pentamidine isethionate (15). They may cause severe adverse effects such as thrombophlebitis, abscesses at the respective injection sites, and nephrotoxicity. Hydroxystilbamidine isethionate (31) is also effective. It is prepared via a five-step synthesis starting with 4-methyl-2-nitrobenzonitrile and 4-cyanobenzaldehyde. The subsequent steps involve reduction of the nitro group of 2-nitro-4,4'-dicyanostilbene and diazotization to obtain 2-hydroxy-4,4'-dicyanostilbene. This product is converted to the amidine (31) (61).

Some 8-aminoquinolines seemed promising in tests on the golden hamster but are too toxic for human use. Cycloguanil embonate (32), which may cause necrosis at the injection site, was also reported effective, as were salts of Berberine (33). Cyclo-

524 CHEMOTHERAPEUTICS, ANTIPROTOZOAL

guanil hydrochloride is produced by condensation of p-chloraniline, dicyandiamide, and concentrated hydrochloric acid in acetone. The embonate salt is made from the hydrochloride (62). Berberine is an alkaloid isolated from *Hydrastis canadensis* L., *Berberedaceae*. The total synthesis of berberine iodide was achieved through a very involved route starting with 3,4-methylenedioxyphenethylamine (63) (see Alkaloids).

Leishmania amastigotes can be maintained in tissue culture cells. This system is suitable for rapid and inexpensive mass screening of drugs (64). However, encouraging findings of *in vitro* laboratory methods are not always predictive for the drug effects under clinical conditions.

Pneumocystosis

The infective agent that causes this condition is *Pneumocystis carinii* (65–66). Although this organism has features of both fungi and protozoa, it is treated here as a protozoon (67–68). *P. carinii* is cosmopolitan among man and other vertebrates (69). It has not been established whether the organisms detected in man and various vertebrates are identical. Epidemiologic observations suggest that the parasite is present but clinically silent in most individuals and is transmitted directly. It is an opportunistic pathogen that proliferates and becomes clinically manifest only in individuals whose immune defenses are impaired, eg, cancer patients who have immunosuppressive medications or infants whose immunodefenses are not yet fully developed and are further weakened by malnutrition. Pneumocystosis manifests itself by a particular pneumonitis which, if untreated, has a high mortality. The laboratory drug-testing system uses immunodepressed rats (70). An *in vitro* culture method for *P. carinii* has been recently developed (71). The available drugs are parenteral pentamidine (15) and the oral antifolate combination of a pyrimidine with a sulfonamide such as in Fansidar [37338-39-9] (72) and Bactrim [8064-90-2] (73). They have similar levels of curative efficiency. In the groups of high risk subjects chemoprophylaxis would be important. Fansidar and Bactrim have been reported as being useful for this purpose (72,74). In contrast, (15) as a prophylactic would be risky because of its high incidence of nephrotoxicity and tissue injury at the injection site.

Trichomoniasis

A variety of trichomonad infections occur in vertebrate hosts (2). They are extracellular, flagellated parasites transmitted by direct contact and reside in the superficial layers of the infected organs. *Trichomonad* species that are supposedly innocuous have been found in various cavities of their hosts. But some species are important pathogens of the genitals. In humans, *Trichomonas vaginalis* may cause significant disease in both sexes. The severity of the clinical manifestations may depend on strain differences of the parasite. Symptomless carriers do occur. Various nitroimidazoles are effective against *Trichomonad* species (75), eg, metronidazole (34) and tinidazole (35).

Metronidazole is prepared by heating 2-methyl-5-nitroimidazole with excess 2-chloroethanol to obtain the crude product which is purified by extraction with chloroform and recrystallization from ethyl acetate (76). Tinidazole was obtained by heating a mixture of $4\text{-}CH_3C_6H_4SO_3CH_2CH_2SO_2C_2H_5$ and 2-methyl-5-nitroimidazole under nitrogen at 145–150°C for 4 h (77). The mutagenic and carcinogenic potentials present problems in nitro group-containing drugs. Assay systems for *T. vaginalis* include *in vitro* culture and inoculation of mice.

T. foetus is another important species. It causes abortion in cattle. It is self-limiting in females. Despite the economic importance of the infection there is so far no satisfactory chemotherapy. Nitroimidazoles were used but resistant strains were a problem (75). Bulls can be treated by topical application of trypaflavine (36) and surfen (37) on the penis, a procedure difficult to apply. A species of economic relevance is *T. equi* which can invade the digestive tract and be fatal to horses. No specific chemotherapy is known.

Trypaflavine, acriflavine hydrochloride, is a mixture of 3,6-diamino-10-methylacridinium chloride hydrochloride and 3,6-diaminoacridine dihydrochloride. The 3,6-diaminoacridine is prepared by the condensation of *m*-phenylenediamine and formic acid (78). The synthesis of surfen is described in a German patent (79).

The species *T. gallinae* parasitizes a variety of birds and is of economic significance in domestic fowl. It is transmitted either by drinking water or by direct contact. It invades the upper digestive tract. The virulence of the infection varies with the strain of the protozoon and can be fatal. Enheptin, 2-amino-5-nitrothiazole (38), is therapeutically effective. It is synthesized by deacetylation of 2-acetamido-5-nitrothiazole (80). Screening tests for drugs against *T. gallinae* include culture methods.

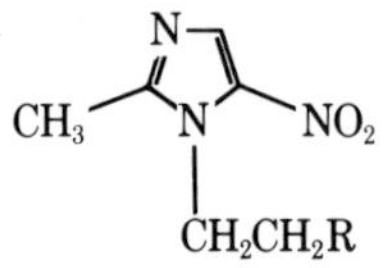

(34) R = OH
(35) R = SO₂C₂H₅

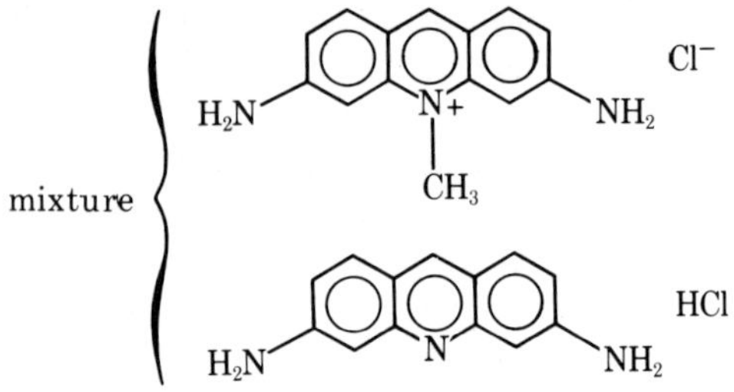

(36)

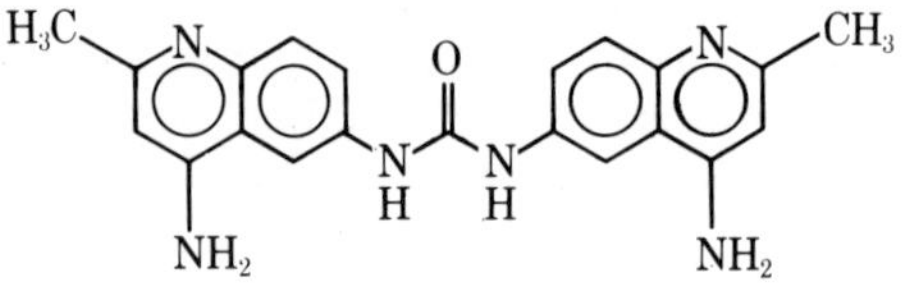

(37)

Hexamitosis

The agent of this condition is the flagellate *Hexamita meleagridis* (2,4). It is transmitted by contaminated food and has been reported in Great Britain and the Americas. The parasite lives in the duodenum and small intestines of birds. It is of economic importance in that it may cause up to 80% mortality in young birds. No effective drug is known against this parasite.

Balantidial Dysentery

The agent of this condition, *Balantidium coli*, occurs in man, primates, and several other vertebrates (2,4). It is cosmopolitan and transmitted by contaminated food and beverages. In man the cysts of *B. coli* pass through the small intestines and the trophozoites are found in the large intestines. There they form deep ulcers and can be carried into the mesenteric lymphatic glands. Effective drugs are metronidazole (34), tetracycline, and paromomycin (39).

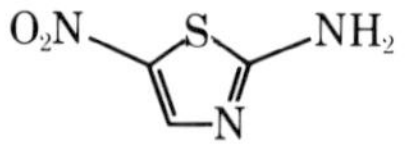

(38)

Paromomycin is an oligosaccharide antibiotic that was isolated from various *Streptomyces*. It is prepared by a fermentation process and isolated from the broth by resin exchange (81) (see Antibiotics, aminoglycosides).

Giardiasis

The agent of this condition is the flagellate *Giardia lamblia* (82). This parasite lives on the epithelial lining of the human duodenum and jejunum and can cause a variety of digestive disturbances, including malabsorption, though symptom-free carriers are common. The parasite is cosmopolitan and is transmitted by contaminated food and beverages. Drug screening methods include inoculation of mice (83). An *in vitro* culture method has been developed but has not yet been used extensively for drug screening (84). There is no specific chemotherapeutic agent for *G. lamblia*. Several antimalarials such as quinacrine (40) are effective, as are the antiamebic drugs metronidazole (34), tinidazole (35), and furazolidone (41).

Quinacrine is obtained by the condensation of 1-diethylamino-4-aminopentane with 3,9-dichloro-7-methoxyacridine (85). The synthesis of furazolidone is accomplished by the reaction of 2-benzylideneamino-2-oxazolidinone and 5-nitro-2-furaldehyde (86).

Amebiases

The agents of these conditions are amebae. They are cosmopolitan and transmitted by ingestion of contaminated material.

Intestinal Amebiasis. The common pathogenic species, *Entamoeba histolytica*, occurs in many vertebrate hosts. It causes amebic dysentery in man although symptomless carriers of the parasite occur. Man becomes infected by ingesting the amebic cysts that descend in the intestinal tract. Amebic trophozoites then develop and reside in the caecum, colon, and sigmoid. They can produce the next generation of infective cysts which are eliminated with the feces. When the trophozoites invade the intestinal wall they cause ulcerations and clinical intestinal symptoms of varying severity. Metastatic amebic lesions may develop in various organs, particularly the liver.

For screening antiamebic drugs, both axenic culture methods and laboratory animals are suitable (87–88). Amebicides have been developed empirically. The nitroimidazoles, metronidazole (34), tinidazole (35), and the more toxic nitro heterocycle

528 CHEMOTHERAPEUTICS, ANTIPROTOZOAL

niridazole [61-57-4], are effective amebicides at all sites. Parenteral emetine (42) and dehydroemetine (43) are not effective against cysts, but are effective against amebiasis in the bowel wall and liver.

Emetine (42) is the principal alkaloid of ipecac, the ground roots of *Uragoga ipecacuanha* (see Alkaloids). Its total synthesis has been described (89) and a stereospecific synthesis of dehydroemetine has been published (90). Chloroquine (44) has been reported effective against liver amebiasis. But amebic abscesses of, eg, liver or brain require surgical treatment in addition to specific chemotherapy. Paromomycin (39) which is nonabsorbable from the bowel (41); iodoquinoline analogues, eg, iodochlorhydroxyquin (45) (now in disrepute because of toxicity); and various oral arsenical and bismuth preparations such as carbarsone (46), Milibis (47), and emetine-bismuth iodide [8001-15-8] are effective in the lumen of the bowel against either tropnozoites or cysts, or both.

Chloroquine is prepared by the condensation of 4,7-dichloroquinoline with 1-diethylamino-4-aminopentane (91). The synthesis of (45), Entero-Vioform, is accomplished by treating an aqueous solution of an alkali salt of 5-chloro-8-hydroxyquinoline with potassium iodide and a hypochlorite (92). Carbarsone (46) is obtained by reaction of the sodium salt of arsanilic acid with potassium cyanate or cyanogen bromide (93), and glycobiarsol is isolated from the reaction of bismuth nitrate and sodium N-glycoloylarsanilate (94).

Mebinol (48) and diloxanide furoate (49) have been reported to be effective against cysts (95).

Among methods of preparing chlorophenoxamide (Mebinol), is the reaction of 4-(4'-nitrophenoxy)benzaldehyde with 2-aminoethanol, reduction of the resulting Schiff base and acetylation of the amine with dichloroacetyl chloride (96). Diloxanide 2-furoic acid ester is prepared from 4-hydroxy-N-methylaniline and dichloroacetyl chloride followed by esterification with 2-furoic acid (97).

The antibiotics Fumagillin [23110-15-8] (see Chemotherapeutics, antimitotic)

(48) (49)

and tetracyclines (see Antibiotics, tetracyclines) have been reported in conjunction with antiamebic treatment. They act on bacterial associations of the amebic lesions and thus may be adjuvants to antiamebic therapy.

Primary Amebic Meningo-Encephalitis. This condition is caused by amebae of the genera *Naegleria* and *Hartmanella*. The parasites occur in contaminated mud of ponds and puddles. The incidence of the infection in man is rare, but the disease is invariably fatal. No chemotherapy is available. *Naegleria* can be cultivated *in vitro* (98).

The Malarias

The agents of these infections are intracellular parasites of the genus *Plasmodium* (4,99). They infect a wide variety of vertebrate hosts and are specific for either a host species or a taxonomically narrow category of hosts. Medically and economically important are *P. falciparum, P. vivax, P. ovale,* and *P. malariae,* which cause in man the historically termed malignant, benign, oval tertian, and quartan malaria, respectively. The normal transmission of these plasmodia between host individuals occurs through mosquito vectors. When feeding on an individual with malaria, the mosquito ingurgitates blood containing male and female gametocytes of the plasmodium. In the mosquito they undergo a developmental cycle yielding the infective sporozoites which the mosquito then transmits to a new host. A suitably warm environment is needed for the development of the sporozoites in the mosquito and, thereby, for the endemic prevalence of malaria. In man the malaria parasite develops through sequential stages. Sporozoites find their way into parenchyma cells of the liver where they develop the primary tissue form, the exoerythrocytic, pre-erythrocytic schizonts. Within the schizonts nuclear divisions yield a large number of nucleated parasites. These are eventually released from the limiting membranes as individual merozoites. In *P. falciparum* infection the merozoites get into the blood circulation and penetrate the red cells. In some of the red cells the merozoites develop into the sexual stage, a single gametocyte per red cell, thus closing the chain of the plasmodium life cycle. However, in other red cells the merozoites start asexual blood forms. Erythrocytic schizonts develop, within which nuclear division produces new generations of merozoites. These finally break out and infect new red cells. This erythrocytic circuit of the developmental cycle of *P. falciparum* is repeated at intervals and causes the clinical manifestations of the disease. If adequate host defenses develop, they can stop this cycle. In contrast, *P. vivax* and *P. ovale,* which cause relapsing malaria, have a somewhat more complicated development. While some of the exoerythrocytic schizonts in liver cells release merozoites within a few days (as do those of *P. falciparum*), others remain dormant for extended time periods. When these release their merozoites, clinical relapses can occur even after years of latency. As to *P. malariae* infection, clinically inactive parasitemia may become clinically activated even after years.

Antimalarial drugs, given at their usual dosages, can affect one or more developmental stages of plasmodia species. Only those of clinical relevance to man are considered and they are classified here according to their target stages although some of them act on more than one stage.

Drugs Acting on Asexual Blood Forms. Drug categories acting primarily on the asexual blood forms of susceptible strains of plasmodia include quinine salts, eg, quinine monohydrochloride (**50**); acridines, eg, (**40**); biguanides, chlorguanide (**51**), dihydrotriazines, eg, cycloguanil (**32**); pyrimidines, eg, pyrimethamine (**52**); sulfones and sulfonamides, eg, dapsone (**53**) (see Antibacterial agents, sulfonamides); antibiotics, eg, tetracycline, clindamycin (**54**) (see Antibiotics, lincosaminides); and 4-aminoquinolines, such as chloroquine (**44**), amodiaquin (**55**); and the promising experimental drugs from the current United States Army antimalarial drug development program, mefloquine HCl (**56**) and WR 030,090HCl (**57**), both quinolinemethanols, and WR 033,063HCl (**58**), a phenanthrenemethanol (100). Clindamycin (**54**) is disreputed since it has caused cases of colitis, with some of them being fatal.

Quinine monohydrochloride (**50**) is a salt of one of the cinchona bark alkaloids. The base is obtained by extraction. Synthesis was achieved by two different routes (101). However, the natural product is preferred because it is less expensive. Chlorguanide (**51**) is the product of the reaction of 4-chlorophenyldicyandiamine and isopropylamine (102). Pyrimethamine (**52**) is prepared by the reaction of guanidine nitrate with 3-isobutoxy-2-(4-chlorophenyl)pent-2-enonitrile in ethanol in the presence of sodium ethoxide (103). Dapsone (**53**) is obtained by reaction of 4-acetamidobenzenesulfinic acid sodium salt with 4-chloronitrobenzene followed by reduction with stannous chloride in concentrated hydrochloric acid (104). Clindamycin (**54**) is manufactured by modification of lincomycin (105), which is produced by *Streptomyces lincolnensis* var. *lincolnensis*, and is prepared by a fermentation process (see Antibiotics, lincosaminides). The pH of the broth is adjusted and the product is isolated by extraction with butanol (106). Amodiaquin (**55**) is prepared from 4,7-dichloroquinoline and 4-acetamido-6-diethylamino-*o*-cresol (107). One route to the quinolinemethanols related to (**58**) utilizes the appropriate cinchophen and involves the reactions of diazomethylation of the acid chloride to the diazomethyl ketone, hydrobromination, aluminum isopropoxide reduction and condensation with the appropriate amine to obtain the amino alcohol (108). The phenanthrenemethanol may be made by this scheme or by a variation as in the original preparation (109). The latter involves the bromomethyl ketone as the common intermediate that reacts with diheptylamine, followed by reduction of the aminomethyl ketone. An improvement, devoid of diazomethane usage and suitable for scale-up, is described in a United States patent (110). Applicable to a wide variety of acids, the process gives the bromomethyl ketone by way of the acid chloride, the acylmalonate, acylbromomalonate, and the unstable acylbromomalonic acid which undergoes double decarboxylation. Mefloquine (**56**) is prepared by the condensation of the 2,8-bis(trifluoromethyl)-4-quinolinecarboxylic acid ester with 2-pyridyllithium and reduction of the pyridyl quinolyl ketone (111).

The drugs that affect the asexual blood form alone can eradicate susceptible strains of *P. falciparum* and can be curative in these infections. However, they can only suppress the relapsing type of infections since they do not affect their exoerythrocytic tissue forms.

(50) (51) (52) (53) (54) (55) (56) (57) (58)

Drugs Affecting Tissue Forms. Drugs affecting the tissue forms are 8-aminoquinolines, eg, pamaquine (**59**) and primaquine (**60**). Their efficiency is enhanced when given in combination with 4-aminoquinolines. Pamaquine (**59**), plasmoquine, is obtained by reductive condensation of 5-diethylamino-2-pentanone with 6-methoxy-8-aminoquinoline (**112**). The synthesis of primaquine involves the condensation of 6-methoxy-8-aminoquinoline with 2-bromo-5-(phthalimido)pentane (**113**).

Drugs Acting on Gametocytes. Quinacrine (**40**) and 4-aminoquinolines affect the gametocytes of *P. vivax* and *P. malariae* only, chlorguanide (**51**) and 8-aminoquinolines affect those of *P. vivax* and *P. falciparum*.

An important aspect of antimalarial drugs is their use in chemoprophylaxis. Drugs that suppress the merozoites and the asexual blood forms of the plasmodia are commonly used to prevent clinical patency and permit time for immunodefenses to de-

532   CHEMOTHERAPEUTICS, ANTIPROTOZOAL

(59) R = C$_2$H$_5$
(60) R = H

velop. Their dose and frequency of administration are far below those used in the treatment of clinical attacks. However, as they are often used for extended periods the only practical ones are those whose chronic administration is least likely to cause adverse effects. Endemic prophylaxis is also obtained with the drugs that antagonize the sexual forms of the plasmodia. An interesting but undeveloped approach would be long-acting drugs against sporozoites (see Pharmaceuticals, sustained-release).

Resistance of plasmodium strains against given drugs poses a serious problem. It appears regionally, can be unique or cross between drugs of apparently different chemical categories, and the proportion of resistant to susceptible strains varies geographically. Owing to these variables, the antimalarial of choice will vary between geographic areas and may change with time in any given area. Certain drugs such as chlorguanide (51), pyrimethamine (52), and sulfonamides cause higher incidences of resistance than do quinine (50), pamaquine (59), or quinacrine (40). Drug combinations and sequentials seem to reduce the resistance problem (114).

The oldest antimalarial, quinine (50), is a herbal product discovered by serendipitous methods. Most of the other antimalarials were developed empirically by screening methods. Information is available on some of their biochemical effects but their mode of therapeutic action is not known. The only categories whose mode of therapeutic actions are well defined are the antifolates and the antibiotics, both spinoffs of antibacterial drug research. In developing antimalarial drugs the initial tests are often done in rodents and birds that are infected with their specific plasmodium species. Information obtained from such tests is often not applicable to human plasmodia. The owl monkey, *Aotus trivirgatus*, is susceptible to *P. falciparum* and *P. vivax*, and the squirrel monkey, *Saimiri sciureus*, to *P. vivax* infection; they are good models for the final drug testing. *In vitro* culture methods, for both plasmodium sporozoites and asexual blood forms, are under development (115–118).

Action Spectra of Antiprotozoal Drugs

Some of the antiprotozoal drugs are effective against more than one protozoal species or nonprotozoal conditions, in fact, some of these drugs, eg, antimetabolites and antibiotics, have been developed primarily against other classes of infective agents and, also, were later found active against protozoa. Within its action spectrum, the efficacy of a drug can vary greatly between classes, species, and strains of the infective agent. Table 1 gives an overview of the action spectra.

Table 1. Clinical Action Spectra of Antiprotozoal Drugs

Class	Group	Drugs	Protozoal infections	Nonprotozoal conditions
Antibiotics		amphotericin B	leish	fung
		clindamycin	mal	bact
		fumagillin	ameb	
		monensin	cocci	
		paromomycin	ameb, giard	bact, helminth
		spiramycin	toxo	bact
		tetracyclines	mal, cocci, ameb, anapl, theileriasis, balant	bact, trachoma, fung, mycoplasma
Antimetabolites	sulfonamides	various congeners	mal, toxo, cocci, pneumocyst	bact, trachoma, fung, lymphogranuloma venerum, dermatitis herpetiforme
	sulfones	dapsone	mal, cocci	bact, leprosy, dermatitis herpetiforme
	pyrimidines	pyrimethamine	mal, cocci, leish, pneumocyst,[b] toxo	bact[b]
		trimethoprim	mal,[b] cocci, leish, pneumocyst[b]	bact[b]
		amprolium, thiamine analogue	cocci	
	biguanide	chlorguanide	mal	
	triazine	cycloguanil	mal, leish	
	guanide	robenidine	cocci	
	thioguanide	bitipazone	cocci	
	PABA analogue	ethopabate	cocci	
Organometallics	As	tryparsamid	Afr tryp	
		melarsoprol (Mel B)	Afr tryp	
		melarsen	Afr tryp	
		carbarsone	ameb, trich, balant	
	Bi	bismuth subgallate	ameb	
		emetine Bi iodide	ameb	
	As + Bi	glycobiarsol	ameb, trich	
	Sb, trivalent	tartar emetic	Afr tryp, leish	filariasis, schistosomiasis, granuloma inguinale, mycosis fungoides,
	Sb, pentavalent	pentostam	leish	
		glucantime	leish, Kala azar	helminth
		neostibosan	leish	helminth, granuloma inguinale

Table 1. (*continued*)

			Conditions of drug use[a]	
Class	Group	Drugs	Protozoal infections	Nonprotozoal conditions
Benzamidines		hydroxystilbamidine	Afr tryp	fung
		pentamidine	Afr tryp, pneumocyst, bab, leish	
		berenil	Afr tryp, anapl, bab	bact
Anilides	ureas	imidocarb	anapl, bab	
		suramin	Afr tryp	onchocerciasis
	amidine	amicarbalide	bab	
	benzamidine	diloxanide furoate	ameb, giard	
		berenil	Afr tryp, anapl, bab	bact
Halogenated hydrocarbons		clopidol	cocci	
		robenidine	cocci	
	phenanthryls	homidium bromide	Afr tryp, bab	
		WR 033,063	mal	
	quinoline derivatives	mefloquine	mal	
		WR 030,090	mal	
		diiodo–oxy-quinoline	ameb, trich, balant	
	dithiosemi-carbazone	contrapar	anapl	
Quinoline derivatives	alkaloid	quinine	mal	parturition, muscle cramps
	4-aminoquino-lines	chloroquine	mal, ameb, giard, bab	helminth, autoimmune conditions
		amodiaquine	mal, giard	leprosy, autoimmune cond
		quinapyramine	Afr tryp	
	8-aminoquino-lines	pamaquine	mal	
		primaquine	mal	
	ureas	quinuronium sulfate	bab	
		surfen	trich fetus	
	quinazoline	febrifugine	mal, cocci	
	hydroxyquinoline	buquinolate	cocci	
	quinolone	nequinate	cocci	
	halogenated derivatives	mefloquine	mal	
		WR 030,090	mal	
		diiodo–oxy-quinoline	ameb, trich, balant	
	acridinyls	quinacrine	mal, giard	helminth, radiosensitizer
		trypaflavin	trich fetus	bact
Nitro compounds	furans	nitrofurazone	cocci, Afr tryp	bact
		furazolidone	ameb, giard, trich, cocci	bact (incl cholera), fung

Table 1. (continued)

Class	Group	Drugs	Protozoal infections	Nonprotozoal conditions
	imidazoles	nifurtimox	Chagas disease	
		metronidazole	trich, giard, ameb, balant	bact, ulcerative gingivitis radiosensitizer
		tinidazole	trich, giard, ameb	
		RO 7-1051	Chagas disease, leish	
	thiazoles	Enheptin	trich (fowl), histomonas	
		niridazole	ameb	schistosomiasis
	hydrocarbon	Mebinol	ameb	
Alkaloids		emetine	ameb	helminth
		berberine salts	leish	

Conditions of drug use[a]

[a] Abbreviations within the table are: Afr tryp = African trypanosomiasis, African sleeping sickness; ameb = amebiasis; anapl = anaplasmosis; bab = babesiasis; bact = bacterial infections; balant = balantidiasis; cocci = coccidiosis; fung = fungal (mycotic) infections; giard = giardiasis; helminth = helminthiasis; leish = leishmaniasis; mal = malaria; pneumocyst = pneumocystosis; toxo = toxoplasmosis; trich = trichomoniasis.

[b] In these conditions, combination drugs of a pyrimidine with a sulfonamide, eg, Fansidar or Bactrim, are more effective than the single ingredients.

Economic Considerations

References 2, 3, 4, and 7 show the broad geographic distribution of protozoal endemic areas. The *Animal Health Yearbook* gives the annual incidence of protozoal diseases of animals and of those common to animals and man (121). It does not specify, for restricted localities, the density of the incidence and the clinical severity of the diseases. This information may be found after some delay in specialized periodicals (122). The *World Epidemiology Review* gives a monthly report on public health events, some of which are retrospective (123).

For estimating market possibilities of antiprotozoal drugs, the figures of overt disease cases are not the only guide. Equally important are the unknown numbers of individuals with latent nonapparent infections, individuals in endemic areas who take chemoprophylaxis, and individuals who are exposed to infection but take no precautions. As examples, the global incidence of malaria cases in 1974 was estimated to be ca 120 million whereas about 25% of the world population lived in endemic areas (124–125). The global incidence for Chagas' disease was estimated in 1977 at about 8 million cases (125), but for Argentina alone, in 1972, the number of infected individuals was estimated at 2.3 million whereas 12 million people, or about 50% of the nation's population, lived in endemic areas (126). There is a considerable need for new drugs in this market (56). The biggest potential market for antiprotozoal drugs is the developing countries of the tropical and subtropical regions. In many cases, however, reported disease figures are unreliable. The greatest drug needs in the developed countries are for coccidiosis in animal husbandry. However, the use of single anticoccidials can fluctuate widely and rapidly owing to the spreading of drug-resistant protozoal strains and the use of drug combinations (8).

Table 2. Selected Generic and Trade Names with Sources of Antiprotozoal Drugs

WHO proposed name, generic	CAS Registry no.	Structure no.	Trade name	Sources
acriflavine	[8063-24-9]	(36)	Euflavin	Bayer; Imperial Chemical Ind.
			Gonacrine	
			Panflavin	Specia
			Trypaflavine	Hoechst
amicarbalide	[3459-96-9]	(16)	Diampron	May & Baker
aminoalkenyldisulfide	[31482-85-6]	(3)		
aminochinuridum	[3811-56-1]	(37)	Surfen	Hoechst
aminonitrothiazolum	[121-66-4]	(38)	Enheptin T	Lederle
			Nitramin	Ferrosan
amodiaquine	[86-42-0]	(55)	Camoquin Hydrochloride	Parke, Davis
			Flavoquine	Roussel
amprolium	[121-25-5]	(2)	Amprol	Merck
berberine	[2086-83-1]	(33)	Canadine	Merck
bitipazone	[13456-08-1]	(7)	Bitipazone	Burroughs Wellcome
buquinolate	[5486-03-3]	(10)	Bonaid	Eaton
carbarsone	[121-59-5]	(46)	Carbarsone	Lilly
chloroquine salts	[54-05-7]	(44)	Aralen	Sterling-Winthrop-Ross
			Avloclor	Imperial Chemical Ind.
			Nivaquine	Specia; May & Baker
			Resochin	Bayer
clefamide, chlorphenoxamide	[3576-64-5]	(48)	Mebinol	Erba
clindamycin	[18323-44-9]	(54)	Cleocin	Upjohn, USA;
			Sobelin	Upjohn, Germany
clioquinol, iodochlorhydroxyquin	[130-26-7]	(45)	Colicid	Chemedica
			Enteritan	Grossman
			Entero-Vioform	Ciba
clopidol	[2971-90-6]	(8)	Coyden	Dow
cycloguanil	[516-21-2]	(32)	Camolar	Parke, Davis; May & Baker; Hess and Clark
decoquinate	[18507-89-6]			
dehydroemetine	[4914-30-1]	(43)	Dametin	Merck
			Dehydroemetine	Roche
			Mebadine	Glaxo
diloxanide furoate	[579-38-4]	(49)	Furamide	Boots; Clin-Comar
diminazene aceturate	[908-54-3]	(17)	Berenil	Hoechst
diphenyl sulfone,	[80-08-0]	(53)	Alvosulfon	Ayerst; Imperial Chemical Ind.
			Dapsone	
emetine	[483-18-1]	(42)	Emetine	Lilly
ethopabate	[59-06-3]	(4)	Ester Amidobenzoate	Merck
febrifugine	[24159-07-7]	(11)	Febrifugine	Roussel
furazolidone	[67-45-8]	(41)	Enterotoxon	Bieff
			Furazon	Daiko
			Furoxone	Eaton; Norwich; Boehringer
gloxazone	[2507-91-7]	(13)	Contrapar	Burroughs Wellcome

Table 2. (continued)

WHO proposed name, generic	CAS Registry no.	Structure no.	Trade name	Sources
glycobiarsol	[116-49-4]	(47)	Amoebicon	Consolidated Midland Corporation
			Broxolin	Breon
			Milibis	Sterling-Winthrop-Ross
homidium bromide	[1239-45-8]	(23)	Ethidium	Boots
hydroxystilb-amidine	[533-22-2]	(31)	Hydroxystilb-amide	May & Baker
imidocarb	[5318-76-3]	(14)	Imizol	Burroughs Wellcome
lincomycin	[154-21-2]		Cillimycin	Hoechst
			Lincocin	Upjohn
			Mycivin	Boots
mefloquine	[51773-92-3]	(56)	Mefloquine	U.S. Army Surgeon General
melarsen	[3599-28-8]	(20)	Melarsen	Hoffmann-La Roche
melarsoprol	[494-79-1]	(21)	Arsobal Mel B	Specia
mepacrine	[69-05-6]	(40)	Atabrine Hydrochloride	Sterling-Winthrop-Ross
			Atebrin	Bayer
			Quinacrine	Specia
			Tenicridine	Norgan
methylglucamine, antimonate	[133-51-7]	(30)	Glucantime	Specia
metronidazole	[443-48-1]	(34)	Clont	Bayer
			Efloran	Kirka
			Elyzol	Dumex
			Flagyl	Specia; May & Baker; Searle
			Metronidal	Kisser
monensin	[17090-79-8]	(5)	Coban; Monelan	Lilly
nequinate	[13997-19-8]	(9)	Statyl	Imperial Chemical Ind.; Ayerst
nifurtimox	[23256-30-6]	(22)	Lampit	Bayer
pamaquine	[491-92-9]	(59)	Plasmochin	Bayer
			Praequine	May & Baker
paromomycin	[7542-37-2]	(39)	Humagel	Parke, Davis
			Pargonyl	Roussel
paromomycin sulfate	[1263-89-4]		Farmiglucina	Farmitalia
			Humatin	Parke, Davis
pentamidine	[140-64-7]	(15)	M&B 800	May & Baker
			Lomidine	Specia
primaquine	[90-34-6]	(60)	Primaquine	Bayer; Sterling-Winthrop-Ross; Imperial Chemical Ind.
proguanil	[500-92-5]	(51)	Chlorguanide	Abbott; Merck
			Guanatol	Lilly
			Paludrine	Ayerst; Imperial Chemical Ind.
pyrimethamine	[58-14-0]	(52)	Daraprim	Burroughs Wellcome
			Malocide	Specia
quina-	[20493-41-8]	(24)	Antrycid	Imperial Chemical Ind.

538   CHEMOTHERAPEUTICS, ANTIPROTOZOAL

Table 2. (continued)

WHO proposed name, generic	CAS Registry no.	Structure no.	Trade name	Sources
pyramine quinine	[130-89-2]	(50)	various quinine salts	Amsterdamsche Chininefabriek
Ro 5-9754	[8076-37-7]		Ormetotrim; Rofenaid	Roche
Ro 7-1051	[22994-85-0]	(28)		
robenidine	[25875-51-8]	(6)	Robenz	Merck
spiramycin	[8025-81-8]	(12)	Rovamycin	Specia; May & Baker
			Selectomycin	Chemie Grunenthal
			Calactin	Leo, Helsingborg
			Suanovil	Biokema, Switzerland
stiboglucon- ate sodium	[16037-91-5]	(29)	Pentostam	Burroughs Wellcome
			Solustibosan	Bayer
stibosamine, ethylstib- amine	[1338-98-3]	(25)	Neostibosan	Bayer
suramine	[129-46-4]	(18)	Antrypol	Bayer; Imperial Chemical Ind.; Sterling-Winthrop-Ross
			Bayer 205, Germanin	Bayer
			Moranyl	Specia
			Naphuride	Sterling-Winthrop-Ross
thiamine	[67-03-8]	(1)	Anevryl	Stella, Belgium
			Benerva	Roche
tinidazole	[19387-91-8]	(35)	Fasigyn, Simplotan	Pfizer
trypars- amide	[554-72-3]	(19)	Tryparsamide	Wallau
WR 030,090	[56162-51-7]	(57)	WR 030,090	U.S. Army Surgeon General
WR 033,063	[58523-33-4]	(58)	WR 033,063	U.S. Army Surgeon General

Sources of Antiprotozoal Drugs

For information on the current status of a drug, the supplier should be consulted. Reference 127 may be useful for locating generic names, trade names, chemical name, and the structural and empirical formula of a drug. The *Index Nominum* gives similar information plus references to monographs, and in many instances names of sources (128). For drugs and sources in the United States consult the *National Drug Code Directory* (129). Table 2 gives an abridged list of generic names, trade names, and sources of antiprotozoal drugs. The listing or omission of a trade name or a source is in no way a recommendation or the reverse.

Notice

The views of the authors in this article do not necessarily represent those of their institutions.

BIBLIOGRAPHY

"Therapeutic Agents, Protozoal Infections" in *ECT* 2nd ed., Vol. 20, pp. 70–99, by Edward F. Elslager, Medical Research and Scientific Affairs Div., Parke, Davis, & Company.

1. E. A. Steck, *Chemotherapy of Protozoon Diseases*, United States Government Printing Office, Washington, D.C., 1972.
2. N. D. Levine, *Protozoon Parasites of Domestic Animals and of Man*, Burgess Publishing Company, Minneapolis, Minn., 1973.
3. L. Hussel and co-workers, *Die Protozoaren Blutparasitosen der Haustiere in Warmen Landern*, S. Hirzel Verlag, Leipzig, GDR, 1966.
4. C. Wilcocks and P. E. C. Manson-Bahr, *Manson's Tropical Diseases*, The Williams and Wilkins Company, Baltimore, Md., 1972.
5. F. Hawking, *Exp. Chemother.* **1,** 1 (1963).
6. L. P. Pellerdy, *Coccidia and Coccidiosis*, Paul Parey Publishing Co., Berlin, 1974.
7. D. M. Hammond and P. L. Long, *The Coccidia*, University Park Press Publishing Co., Baltimore, Md., 1973.
8. J. F. Ryley and M. J. Betts, *Adv. Pharmacol. Chemother.* **11,** 221 (1973).
9. W. M. Reid, *Georgia Agri. Res.* **16,** 4 (1974).
10. J. H. Collins, *Ann. N.Y. Acad. Sci.* **52,** 515 (1949).
11. E. F. Rogers, *Ann. N.Y. Acad. Sci.* **98,** 412 (1962).
12. U.S. Pat. 3,020,277 (Feb. 6, 1962), E. F. Rogers and L. H. Sarett (to Merck and Co., Inc.).
13. Jpn. Pat. 70,34585 (Nov. 6, 1970), I. Uchimi, T. Watanabe, and K. Hayashi (to Tanabe Seiyaku Co., Ltd.).
14. D. A. Roe, *Drug-Induced Nutritional Deficiencies*, The Avi Publishing Company, Inc. Westport, Conn., 1976.
15. U.S. Pat. 3,211,610 (Oct. 12, 1965), E. F. Rogers and R. L. Clark (to Merck and Co., Inc.).
16. M. E. Haney, Jr. and M. M. Hoehn, *Antimicrob. Agents Chemother.* **349,** (1967); W. M. Stark, "Monensin a New Biologically Active Compound Produced by a Fermentation Process" in D. Perlman, ed., *Fermentation Advances*, Academic Press, Inc., New York, 1969, pp. 517–540.
17. D. T. Wong and co-workers, *Biochem. Pharmacol.* **20,** 3169 (1971).
18. D. G. Lindmark and M. Muller, *Antimicrob. Agents Chemother.* **10,** 476 (1976).
19. Ger. Pat. 1,933,112 (Jan. 8, 1970), A. S. Tomcufcik (to American Cyanamid Co.).
20. Fr. Pat. 2,024,194 (Oct. 2, 1970), A. J. S. Evans (to Farbwerke Hoechst A.G.).
21. Neth. Pat. 6,409,766 (Mar. 26, 1965), (to The Dow Chemical Co.).
22. Neth. Pat. 6,602,994 (Sept. 12, 1966), (to Imperial Chemical Industries Ltd.).
23. Belg. Pat. 659,237 (Aug, 3, 1965), E. J. Watson, Jr., (to Norwich Pharmacal Co.).
24. U.S. Pat. 2,651,632 (Sept. 8, 1953), B. R. Baker and M. V. Querry (to American Cyanamid Co.).
25. J. K. Frenkel, *Curr. Top. Pathol.* **54,** 27 (1971).
26. J. K. Frenkel, *Human Pathol.* **6,** 97 (1975).
27. S. R. M. Bushby and G. H. Hitchings, *Br. J. Pharmacol. Chemother.* **33,** 72 (1968).
28. D. E. Eyles, *Exp. Chemother.* **1,** 641 (1963).
29. A. Meyer, *Nouv. Press Med.* **3,** 1383 (1974).
30. J. de Vries and L. E. Francis, *J. Can. Dent. Assoc.* **41,** 101 (1975).
31. J. G. Williams, *J. R. Nav. Med. Serv.* **61,** 44 (1955).
32. L. P. Joyner and D. W. Brocklesby, *Adv. Pharmacol. Chemother.* **11,** 321 (1973).
33. B. D. Tiffany and co-workers, *J. Am. Chem. Soc.* **79,** 1682 (1957).
34. G. Schmidt, R. Hirt, and R. Fischer, *Res. Vet. Sci.* **10,** 530 (1969).
35. A. E. Anderson, P. B. Cassaday, and G. R. Healy, *Am. J. Clin. Pathol.* **62,** 612 (1974).
36. U.S. Pat. 2,410,796 (Nov. 5, 1946), G. Newbery and A. P. T. Easson (to May & Baker Ltd.).
37. J. N. Ashley, S. S. Berg, and J. M. S. Lucas, *Nature* **185,** 461 (1960); S. S. Berg, *J. Chem. Soc.,* 5097 (1961).
38. U.S. Pat. 2,838,485 (June 10, 1958), R. Brodersen, H. Loewe, and H. Ott (to Hoechst A.G.).
39. F. Hawking, *Exp. Chemother.* **1,** 625 (1963).
40. (a) B. A. Newton, *Trypanosomiasis and Leishmaniasis, Ciba Foundation Symposium*, Vol. 20, Associated Scientific Publishers, Amsterdam, 1974, p. 285; (b) E. A. Steck, *Prog. Drug Res.* **18,** 289 (1974); (c) L. G. Goodwin, *Trypanosomiasis and Leishmaniasis, Ciba Foundation Symposium*, Vol. 20, Associated Scientific Publishers, Amsterdam, 1974, p. 303.
41. WHO Technical Report Series No. 411, Geneva, Switz., 1969.

42. J. Williamson, *Trop. Dis. Bull.* **73,** 531 (1976).
43. W. Peters in ref. 40(a), p. 309.
44. E. Fourneau and co-workers, *Compt. Rend.* **178,** 675 (1924).
45. L. Mercier-Parot and H. Tuchmann-Duplessis, *C.R. Soc. Biol.,* **167,** 1518 (1973).
46. W. A. Jacobs and M. Heidelberger, *J. Am. Chem. Soc.* **41,** 1587 (1919); W. A. Jacobs and M. Heildeberger in R. Adams, ed., *Organic Synthesis,* Vol. 8, John Wiley & Sons, Inc., New York, 1928, p. 100.
47. E. A. H. Friedheim, *J. Am. Chem. Soc.* **66,** 1775 (1944).
48. U.S. Pat. 2,659,723 (Nov. 17, 1953), E. A. H. Friedheim.
49. Ger. Pat. 1,170,957 (May 27, 1964), H. Herlinger and co-workers (to Farbenfabriken Bayer A.G.).
50. T. I. Watkins, *J. Chem. Soc.,* 3059 (1952).
51. U.S. Pat. 2,585,917 (Feb. 19, 1952), F. H. S. Curd (to Imperial Chemical Industries, Ltd.).
52. U.S. Pat. 1,988,632 (Jan. 22, 1935), H. Schmidt (to Winthrop Chemical Company, Inc.).
53. W. J. Ross and W. B. Jamieson, *J. Med. Chem.* **18,** 430 (1975).
54. J. P. Verge and P. Roffey, *J. Med. Chem.* **18,** 794 (1975).
55. W. E. Gutteridge, *Trop. Dis. Bull.* **73,** 699 (1976).
56. WHO Technical Report Series No. 202, 1960, p. 14.
57. Brit. Pat. 1,138,529 (Jan. 1, 1969), (to Hoffmann-La Roche A.G.).
58. W. Peters in ref 40(a), p. 311.
59. S. Datta and T. N. Ghosh, *Sci. Cult.* **11,** 699 (1946).
60. P. Karrer, and E. Herkenrath, *Helv. Chem. Acta* **20,** 83 (1937).
61. U.S. Pat. 2,510,047 (May 30, 1950), A. J. Ewins (to May & Baker, Ltd.).
62. E. J. Modest and co-workers, *J. Am. Chem. Soc.* **74,** 855 (1952); U.S. Pat. 2,900,385 (Aug. 18, 1959), E. J. Modest (to Children's Cancer Research Foundation, Inc.).
63. T. Kametani and co-workers *J. Chem. Soc.,* 2036 (1969).
64. N. M. Mattock and W. Peters, *Am. Trop. Med. Parasitol.* **69,** 449 (1975).
65. J. Vanek and O. Jirovec, *Zentralbl. Bakteriol.* **158,** 120 (1952).
66. W. Dutz, *Pathol. Ann.* **5,** 309 (1970).
67. J. Vavra and K. Kucera, *J. Protozool.* **17,** 463 (1970).
68. W. C. Campbell, *Arch. Pathol.* **93,** 312, (1972).
69. F. G. Poelma, *Z. Parasitenk.* **46,** 61 (1975).
70. J. K. Frenkel, J. T. Good, and J. A. Shultz, *Lab. Invest.* **15,** 1559 (1966).
71. L. L. Pifer, W. T. Hughes, and M. J. Murphy, Jr., *Pediat. Res.* **11,** 305 (1977).
72. C. Post and co-workers, *Curr. Ther. Res. Clin. Exp.* **13,** 273 (1971).
73. W. T. Hughes, S. Feldman, and S. K. Sanyal, *Can. Med. Assoc. J.* **112,** 47S (1975).
74. W. T. Hughes and co-workers, *Pediat. Res.* **11,** (776) 501 (1977).
75. C. Rufer and co-workers, *Chim. Ther.* **8,** 567 (1973); D. K. McLoughlin, *J. Parasit.* **53,** 646 (1967).
76. U.S. Pat. 2,944,061 (July 5, 1960), R. M. Jacob, G. L. Regnier, and C. Crisan (to Société Usines Chimiques Rhone-Poulenc).
77. S. Afr. Pat. 66,07,466 (Apr. 25, 1968), K. Butler (to Chas. Pfizer and Co., Inc.).
78. A. Albert, *J. Chem. Soc.* **121,** 484 (1941).
79. Ger. Pat. 591,480 (Jan. 22, 1934), H. Jensch (to I. G. Farbenindustrie A.G.).
80. U.S. Pat. 2,573,641 (Oct. 30, 1951), H. L. Hubbard (to Monsanto Chemical Company); U.S. Pat. 2,573,656 (Oct. 30, 1951), G. W. Steahly (to Monsanto Chemical Company).
81. U.S. Pat. 2,916,485 (Dec. 8, 1959), R. P. Frohardt and co-workers (to Parke, Davis and Company).
82. M. S. Wolfe, *J. Am. Med. Assoc.* **233,** 1362 (1975).
83. C. Rufer, H. J. Kessler, and E. Schroder, *J. Med. Chem.* **18,** 253 (1975).
84. E. A. Meyer, *Exp. Parasitol.* **39,** 101 (1976).
85. U.S. Pat. 2,113,357 (Apr. 5, 1938), F. Mietzsch and H. Mauss (to Winthrop Chemical Company, Inc.).
86. U.S. Pat. 2,759,931 (Aug. 21, 1956) G. D. Drake, G. Gever, and K. J. Hayes (to The Norwich Pharmacal Company).
87. L. S. Diamond, *J. Parasitol.* **54,** 1047 (1968).
88. C. F. T. Mattern and T. B. Keister, *Am. J. Trop. Med. Hyg.* **26,** 393 (1977).
89. E. E. van Tamelen and co-workers, *J. Am. Chem. Soc*: **91,** 7359 (1969); C. Szantay and co-workers, *J. Org. Chem.* **31,** 1447 (1966).
90. D. E. Clark and co-workers, *J. Chem. Soc.,* 2479 (1962).
91. A. R. Surrey and H. F. Hammer, *J. Am. Chem. Soc.* **68,** 113 (1946).

92. U.S. Pat. 641,491 (Jan. 16, 1900), A. Bischler (to Basle Chemical Works).
93. R. W. E. Stickings, *J. Chem. Soc.,* 3131 (1928).
94. B. Reichert, ed., *Hagers Handbook Pharmacological Praxis,* Band I, Suppl. 2, Springer-Verlag, Berlin, 1958, p. 759.
95. G. Woolfe, *Progr. Drug Res.* **8,** 43 (1965).
96. U.S. Pat. 2,824,894 (Feb. 25, 1958), W. Lagemann and L. Almirante (to Carlo Erba).
97. Brit. Pat. 767,148 (Jan. 30, 1957), P. Oxley and co-workers (to Boots Pure Drug Co. Ltd.).
98. L. Cerva, V. Ziman, and K. Novak, *Science* **163,** 575 (1969); L. Cerva and K. Novak, *Science* **160,** 92 (1968).
99. P. E. Thompson and L. Werbel, *Antimalarial Agents and Chemistry and Pharmacology,* Academic Press, Inc., New York, 1972.
100. C. J. Canfield and co-workers, *Antimicrob. Agents Chemother.* **3,** 224 (1973); D. F. Clyde and co-workers, *Antimicrob. Agents Chemotherapy* **9,** 384 (1976).
101. J. Gutzwiller and M. R. Uskokovic, *Helv. Chim. Acta.* **56,** 1494 (1973).
102. F. H. S. Curd and F. L. Rose, *J. Chem. Soc.,* 729 (1946).
103. U.S. Pat. 2,576,939 (Dec. 4, 1951), G. H. Hitchings, P. B. Russell, and E. A. Falco (to Burroughs Wellcome & Co., Inc.).
104. C. W. Ferry, J. S. Buck, and R. Baltzly in L. F. Smith, ed., *Organic Synthesis,* Vol. 22, John Wiley & Sons, Inc., New York, 1942, p. 31.
105. B. J. Magerlein, R. D. Birkenmeyer, and F. Kagan, *Antimicrob. Agents Chemother.* 727 (1966); R. D. Birkenmeyer and F. Kagan, *J. Med. Chem.* **13,** 616 (1970).
106. U.S. Pat. 3,155,580 (Nov. 3, 1964), M. E. Borgy, R. R. Herr, and D. J. Mason (to the Upjohn Company).
107. J. H. Burckhalter and co-workers, *J. Am. Chem. Soc.* **70,** 1363 (1948).
108. R. E. Lutz and co-workers, *J. Am. Chem. Soc.* **68,** 1813 (1946).
109. E. L. May and E. Mosettig, *J. Org. Chem.* **11,** 627 (1946).
110. U.S. Pat. 3,714,168 (Jan. 30, 1973), R. E. Olsen (to United States of America as represented by the Secretary of the Army).
111. C. J. Ohnmacht, A. R. Patel, and R. E. Lutz, *J. Med. Chem.* **14,** 926 (1971).
112. R. C. Elderfield and co-workers, *J. Am. Chem. Soc.* **70,** 40 (1948).
113. R. C. Elderfield and co-workers, *J. Am. Chem. Soc.* **77,** 4816 (1955).
114. R. M. Pinder in A. Burger, ed., *Medicinal Chemistry,* 3rd ed., Vol. 1, Wiley-Interscience, New York, 1970, p. 515; *Chemotherapy of Malaria and Resistance to Antimalarials,* No. 529, WHO Technical Report Series, 1973.
115. H. R. Wolfensberger, *Far East Med. J.* **6,** 48 (1970); A. P. Hall and co-workers, *Br. Med. J.* **1,** 1626 (1977).
116. W. Traeger and J. B. Jensen, *Science* **193,** 673 (1976).
117. R. L. Beaudoin and co-workers, *Exp. Parasitol.* **39,** 438 (1976).
118. J. P. Haynes and co-workers, *Nature* **263,** 767 (1976).
119. M. Windholz, ed., *The Merck Index,* 9th ed., Merck & Co., Inc. Rahway, N. J., 1976.
120. L. S. Goodman and A. Gilman, *The Pharmacological Basis of Therapeutics,* 5th ed., Macmillan Publishing Co., New York, 1975; Martindale, *The Extra Pharmacopeia,* 27th ed., The Pharmaceutical Press, London, Eng., 1977; *United States Dispensatory,* 27th ed., J. B. Lippincott, Co., Philadelphia, Pa., 1973; P. B. Beeson and W. McDermott, *Textbook of Medicine,* 14th ed., W. B. Saunders Co., Philadelphia, Pa., 1975; G. T. Strickland, *Chemotherapy of Parasitic Diseases, CRC Handbook of Clinical Laboratory Science,* Sec. E, Vol. II, Chemical Rubber Company Press, Cleveland, Ohio, 1977.
121. *Animal Health Yearbook FAO-WHO-OIE,* Food and Agricultural Organization of the United Nations, Rome, Italy.
122. *Tropical Diseases Bulletin,* Bureau of Hygiene and Tropical Diseases, London, Eng., *Bulletin of the World Health Organization,* Geneva, Switz.
123. *World Epidemiology Review,* Joint Publications Research Service Publ., Arlington, Va.
124. A. W. A. Brown, J. Haworth, and A. R. Zahar, *J. Med. Entomol.* **13,** 1 (1976).
125. M. G. Schultz, *N. Engl. J. Med.* **297,** 1259 (1977).
126. P. Garaguso, *International Symposium on Chagas' Disease,* Secretario de Estado de Salud Publica, Buenos Aires, Argentina, 1972.
127. M. Negwer, *Organisch-Chemische Arzneimittel u. Ihre Synonyma,* Akademie-Verlag, Berlin, 1971.

128. *Index Nominum,* Société Suisse de Pharmacie, Zurich, Switz., 1975.
129. *National Drug Code Directory 1976,* Vol. 1 and 2, U.S. Department of Health, Education, and Welfare, Public Health Service, Food and Drug Administration, U.S. Government Printing Office, Washington, D.C.

General References

Guidelines: Manufacturing and Controls for IND's and NDA's, FDA Papers, June, 1971, FDA 72-3013, U.S. Government Printing Office, Washington, D.C., 1972-482-082/5.
The United States Pharmacopeia, 20th rev., U.S.P. Convention, Inc., 1975, pp. 707–709.
Remington's Pharmaceutical Sciences, 15th ed., 1975, pp. 1419–1428.
R. M. Bushby, *Exp. Chemother.* **1,** 25 (1963).
B. Basil, *Exp. Chemother.* **1,** 55 (1963).
J. W. Drake and co-workers, *Science* **187,** 503 (1975); R. P. Batzinger and co-workers, *Cancer Res.* **38,** 608 (1978).
J. M. Sontag and co-workers, *Guidelines for Carcinogen Bioassay in Small Rodents, Carcinogenesis Report Series,* No. 1, Nat. Cancer Inst., NCI-CG-TR-1., U.S. Dept. HEW,PHS, 1976.
I. M. Rollo, *Myler's Side Effects of Drugs,* Vol. 8, Excerpta Medica, Amsterdam, 1975, p. 659.

EDGAR J. MARTIN
HOWARD C. ZELL
Food and Drug Administration

BING T. POON
Walter Reed Army Institute of Research

CHEMOTHERAPEUTICS, ANTIVIRAL

The search for antiviral agents began at the time when antimicrobial substances were first shown to be effective as prophylactic and therapeutic agents in bacterial diseases (see Antibiotics, nucleosides). Many hundreds of compounds with antiviral activity have been synthesized or isolated from nature over the last three decades, but those available for clinical trials are few. Only three, idoxuridine, adenine arabinoside (ara-A), and amantadine, have been approved by the FDA for use but they are effective in only a few types of viral infections. Idoxuridine and adenine arabinoside are licensed for use only in topical ophthalmic preparations. These limited successes have stimulated the search for other antiviral substances and an increasing number with potential human use are in various stages of conceptual design, development, or evaluation.

The difficulty in designing and developing effective antiviral substances is caused largely by the very nature of viruses. Viruses are a diverse group of infectious agents that differ greatly in size, shape, chemical composition, host range, and effects on hosts. The uniqueness of viruses includes the following characteristics: (*1*) they consist of a genome which is either ribonucleic acid (RNA) or deoxyribonucleic acid (DNA); (*2*) the genome is surrounded by a protein shell (capsid) which protects the genome and provides receptors for recognition of susceptible cells (these capsids are constructed from repeating polypeptide subunits called protomers) (see Biopolymers); (3) viruses replicate only within cells and are entirely dependent upon the host cells' synthetic and energy-yielding apparatuses; (4) replication initially requires the separation of the genome from the capsid so that the genetic material can communicate with the cell; and (5) the parts of the virus replicate individually and then assemble into the finished product, thus they do not grow as do other living things. At times, virions (the name for the mature virus particle) contain an additional external structure referred to as an envelope or peplos. The envelope is derived partly from the cell membrane and contains host lipid with virus induced proteins and glycoproteins. In addition, many viruses contain enzymes that are indispensible in their replication.

Inhibitors of cellular processes will often prevent viral replication but are also toxic for the host. Most of the antiviral drugs that have been discovered cannot be prescribed because of toxicity. Thus the clinical usefulness of idoxuridine is limited to topical application in ophthalmic solution for treatment of herpetic keratitis.

However, many of the inhibitors have been useful in research laboratories and have served as powerful probes for studying the pathways of virus replication. As more is learned about these pathways, and unique processes for virus replication are discovered, it should be possible to design inhibitors that have specific effects upon virus-induced function, eg, RNA viruses either contain or induce a RNA-dependent RNA polymerase (replicase). This enzyme is unique to viruses since cells make their RNA from a DNA template using a DNA-dependent RNA polymerase (transcriptase). Since the viral enzyme is unique, it should be possible to find a chemotherapeutic agent which specifically inhibits its function. Another group of RNA viruses, the retroviruses, contain their own unique enzyme which is a RNA-dependent DNA polymerase (reverse-transcriptase) and should also be amenable to control by chemical substances. Indeed, some rifamycin derivatives inhibit reverse-transcriptase and prevent transformation of cells by retroviruses (1) (see Antibiotics, ansamacrolides).

Methods

Each step in virus replication should be amenable to control. The major systems for screening potential antiviral agents are the use of tissue or cell cultures, the chick embryo, and animals. Cell cultures are very convenient for screening compounds for antiviral activity since viruses readily grow in such cultures and usually demonstrate a cytopathic effect. Compounds are initially selected that prevent the cytopathic effect without inhibiting cell proliferation, and thus have a permissible chemotherapeutic index. The chemotherapeutic index of a compound can be defined as the ratio between the lowest effective antiviral concentration and the highest nontoxic concentration. The spectrum of activity of the compound against various classes of viruses can also be determined in a similar way.

The disease process in animals is much more complex than that in tissue culture and, for practical purposes, a compound should not be considered as a chemotherapeutic agent until efficacy in an animal has been measured. Furthermore, for various reasons, toxicity not apparent in cell cultures may be pronounced in the whole animal. Most animal studies and preliminary studies in humans, after FDA approval, are carried out with challenge infection. In these studies a virus inoculum is introduced into the test animal or human volunteer and the effect of the drug is compared with similarly infected but placebo-treated control groups. Once efficacy is established by such tests, the drug is then evaluated for the prevention or treatment of naturally occurring infections.

Areas of Application

Viral vaccines have been highly successful in controlling many of the more serious viral diseases such as smallpox, yellow fever, rabies, poliomyelitis, and measles. The dramatic success in controlling these diseases by vaccines can partly be attributed to the fact that they are caused by viruses that remain reasonably constant in their chemical composition. In other diseases, however, vaccines have not been as successful for various reasons and control by chemotherapy may be the method of choice. Two major reasons for the inability of vaccines to control some viral diseases are antigenic changes of the virus which occur in nature and the plethora of viral serotypes that may cause some diseases (see also Vaccine technology).

All viral surfaces contain specific antigens which induce the host to make specific antibodies to neutralize the infectivity of the virus. A single antibody molecule is usually sufficient to neutralize one virion; a very effective mechanism for the elimination of extracellular virus. Antibody production either occurs after natural infection or can be induced by vaccines consisting of inactivated or attenuated viruses. With specific antibody, the host will remain resistant to infection by the same virus unless the virus undergoes a change in antigenic structure (antigenic drift). Influenza A viruses undergo a continuous drift and for this reason vaccines have not been very effective. Major antigenic changes (antigenic shift) occur periodically, as in 1957 with the advent of the A2 Asian-type influenza and again in 1968 with the A2 Hong Kong influenza. The efficacy of antiviral compounds is not influenced by these antigenic changes since they do not usually react with the virus surface. Therefore, chemotherapy offers a method of coping with such diseases until a vaccine effective against the new variant is available.

Another difficulty in vaccine production is exemplified by the rhinoviruses which are a major cause of the common cold. These viruses, unlike the influenza viruses, are antigenically stable but over a hundred different antigenic types are known. Thus, to be effective, a vaccine against rhinovirus infection would have to contain each antigenic type and this is not very practical. Laboratory studies show that certain compounds can have activity against all rhinovirus strains. In this case, the use of such compounds would be the method of choice regardless of the strain causing the infection.

A third case for chemotherapy is use against a virus such as herpes simplex which causes disease even in the presence of circulating antibody. These diseases are not amenable to control with vaccines that only stimulate humoral antibody. Of course, cell-mediated immunity is important in some diseases and vaccines may be developed that stimulate this component of the immune system.

Antiviral chemotherapy and prophylaxis would also be useful in those individuals with immunodeficiency disorders, such as hypogammaglobulinemia and Wiskott-Aldrich syndrome, and for those individuals who are immunologically compromised by immunosuppressants following transplantation.

In all instances antiviral drugs have another advantage over vaccines in their immediate action. Vaccines take several days or weeks to induce immunity. Chemical agents can be useful even in cases where very effective vaccines are available. For example, now that smallpox has been virtually eliminated from the face of the earth, vaccination against this disease is no longer required or recommended in the United States. If this virus should emerge again, devastating epidemics could result before the vaccine could be distributed and immunity reestablished in the population. The antiviral drug methisazone could be immediately employed to protect individuals who come in contact with active cases of smallpox.

In addition to the specific immune system, the interferon system is another natural defense mechanism against viral infection. Administered interferon should be an ideal antiviral agent since it is a natural product, relatively nontoxic, and acts against a broad spectrum of viruses. Furthermore, it is effective against replicating viruses that ordinarily evade human defense. To date, it is difficult to purify and concentrate and thus not amenable for general use. Another approach is to induce the body to produce its own interferon: several chemotherapeutic agents owe their antiviral activity to their ability to induce interferon in the recipient. Antiviral antibiotics, such as statalon and helenine, and chemical agents, such as tilorone, and synthetic double-stranded polynucleotides, such as poly (rI:rC) (rI = riboinosinic acid; rC = ribocytidylic acid), are examples. Unfortunately, poly (rI:rC) and others are toxic, causing adverse effects on hematopoiesis and liver function. Thus there is a need to synthesize new interferon inducers with lower toxicity.

Antiviral Agents Effective in Humans

Only the few compounds extensively tested and reported to be effective are described here. A few of the structures are shown in Figure 1 and the major antiviral agents are listed in Table 1. Interferon, the naturally occurring antiviral protein is included since it is currently being extensively studied and shows promise as a prophylactic and therapeutic agent. Many review articles on virus chemotherapy and interferon are available and some are listed in the bibliography.

Thiosemicarbazones. The first report of antiviral activity for this series of compounds was in 1950, when the antituberculosis agent, p-aminobenzaldehyde thiosemicarbazone [7420-39-5], provided protection to chick embryos and mice infected with vaccinia virus (2). A program of analogue synthesis followed, and in 1955 Bauer (3) reported that indole-2,3-dione-3-thiosemicarbazone [487-16-1] (isatin-3-thiosemicarbazone) provided almost complete protection for mice infected with high doses of vaccinia virus. The compound appears to interact with some early product of infection, blocking the translation of late vaccinia messenger-RNA and thereby preventing the formation of some of the virus structural components. The result is that mature virions are not formed. Probably the most effective compound in the series is the methyl derivative methisazone (N-methylisatin 3-thiosemicarbazone, Marboran) (4–5). Methisazone has been shown to exert an effect against variola and vaccinia viruses in both tissue cultures and experimental animals.

Methisazone was shown to have a prophylactic effect against variola in humans in four separate trials (Madras, India, 1963 and 1965–1967; Brazil, 1964–1965; and Pakistan, 1964–1970) (5–8) with a protective effect of 47–83%. It had no therapeutic benefit in the treatment of established clinical disease. In the United States it has been used in the therapy of vaccinia reactions such as eczema vaccinatum and vaccinia gangrenosum but has not been proven effective. The drug is not licensed for use in the United States but is available as Marboran for experimental use.

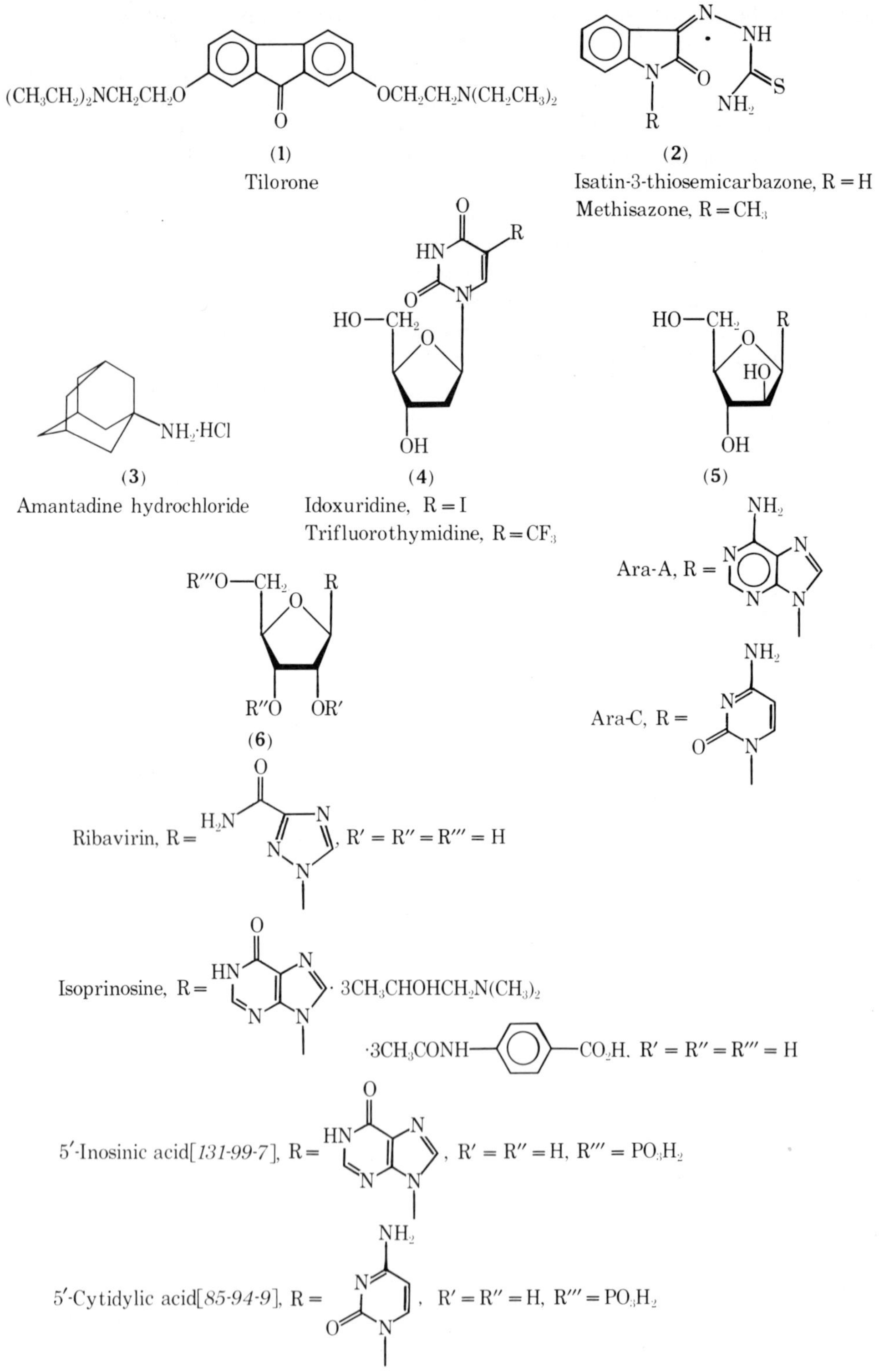

Figure 1. Antiviral agents.

Table 1. Antiviral Agents

Drug	CAS Registry Number	Structure number	mp, °C	Manufacturer
amantadine hydrochloride	[665-66-7]	(3)	360 (dec)	DuPont
ara-A	[5536-17-4]	(5)	257–257.5	Parke Davis Div., Warner Lambert
ara-C	[147-94-4]	(5)	212–213	Upjohn
helenine	[1407-14-3]			Merck, Sharp, and Dohme
idoxuridine	[54-42-2]	(4)	240 (dec)	Smith Kline and French
interferon	[9008-11-1]			Calbio; HEM Research, Inc.
isoprinosine	[36703-88-5]	(6)		Newport Pharmaceutical
levamisole	[14769-73-4, 53631-68-8]		60–61.5	Lederle Div., American Cyanamid
methisazone	[1910-68-5]	(2)	245 (dec)	Burroughs Wellcome; Aldrich
poly(rI:rC)	[24939-03-5]	(6) (monomers)		Miles
ribivarin	[36791-04-5]	(6)	174–176	ICN Pharmaceuticals
statolon	[11006-77-2]			Lilly
tilorone	[27591-97-5]	(1)	235–237	Richardson-Merrell
trifluorothymidine	[70-00-8]	(4)	169–172	Heinrich Mack

More recently, methisazone has been shown to have antitumor activity in experimental animals. Transformation of chicken cells by Rous sarcoma virus (RSV) and inhibition of its reverse transcriptase have been reported. Three other derivatives, 2-formylpyridine thiosemicarbazone [3608-75-1], 1-formylisoquinoline thiosemicarbazone [2365-26-6], and diphenyl ketone thiosemicarbazone [7341-60-8], also inhibit transformation by the virus (9). Since the need of these compounds for poxviruses has essentially been eliminated owing to the eradication of smallpox, the use of this series of agents may have a future as antitumor drugs.

Amantadine Hydrochloride. In 1963 it was shown that amantadine hydrochloride (1-adamantanamine hydrochloride, aminoadamantane hydrochloride, Symmetrel) influenced the course of influenza A2 infections in humans (10). In 1964 amantadine hydrochloride was reported to be effective against infections with influenza A, A1, and A2 in tissue culture, in ovo, and in mice (11). In addition to its protective effect against influenza A infections of humans, amantadine hydrochloride has also shown an antiviral effect against equine influenza in horses and avian influenza A strains in quail and turkeys. There is also evidence for in vitro activity against a number of other viruses such as rubella, parainfluenza, and vaccinia. More recently, the replication of the arenaviruses in cell cultures have been reduced in the presence of amantadine (12). The compound seems to act at an early stage in the influenza replication cycle by blocking or slowing the penetration of virus into the host cell. In the case of arenaviruses, the drug also seems to affect late viral synthesis and release of progeny virus from the cell (12).

In 1966 amantadine hydrochloride was licensed by the FDA for use in the prevention of respiratory illness owing to influenza A viruses by prophylactic treatment of contacts of patients and when index cases appear in the area (13–14).

The results of representative double-blind studies designed to assess the role of amantadine against influenza A2 in humans under natural conditions indicated a protective efficacy of approximately 66%. Amantadine may be more effective in the prevention of clinical disease when a significant preexisting antibody is present (15–19). Amantadine has also been shown to have a therapeutic effect if given early to persons with clinical disease caused by influenza A2 virus (20–21).

Side effects include nervousness, insomnia, depression, confusion, hallucinations, and drowsiness, although some studies have reported no adverse side reactions.

5-Iodo-2′-deoxyuridine. 5-Iodo-2′-deoxyuridine (idoxuridine, IDU, IUDR, Stoxil) is a halogenated pyrimidine originally synthesized in 1958 as an experimental drug for the inhibition of tumors. In 1961 it was shown that IDU inhibited plaque formation because of herpes simplex virus in tissue culture (22). It is also effective against varicella-zoster virus, cytomegalovirus, and vaccinia in tissue culture. In 1962 the successful treatment with IDU of herpes simplex corneal infections (herpetic keratitis) in humans was reported (23). It has now been used in thousands of patients against acute herpetic keratitis with favorable results but with less significant results in chronic and recurrent herpetic keratitis. Therapeutic success rates with idoxuridine have ranged 47–92% in seven studies (24). An overall success of ca 71% resulted in these studies as compared to 24% for placebo-control groups. Treatment of cutaneous herpetic lesions with topical IDU has yielded conflicting results probably owing to the insolubility of the drug. Addition of dimethylsulfoxide (DMSO) increases the solubility of IDU and the combination has been reported to be effective against cutaneous herpes simplex and herpes zoster. DMSO is not licensed for use and cannot be given in any form to humans in the United States. IDU is licensed for the topical treatment of herpetic keratitis and is available as an ophthalmic solution as Stoxil.

In cells, idoxuridine is phosphorylated and converted to the triphosphate. It acts as a thymidine analogue and is incorporated into both host and viral DNA in place of thymidine. Since both herpes and vaccinia viruses induce a thymidine kinase, more of this analogue may be incorporated into infected cells than in uninfected cells. The resultant DNA is thus altered and can neither be replicated with fidelity nor transcribed into functional mRNA. Defective virions and partially assembled components result. Host cell DNA is also adversely affected, especially rapidly proliferating cells such as in the bone marrow and gastrointestinal tract. Leukopenia, thrombocytopenia, stomatitis, and alopecia result from intravenous administration. Consequently, use of the drug is limited to topical application. Heroic attempts have been made to use the drug in herpetic encephalitis, a serious disease with high mortality. However, it has been clearly shown that idoxuridine is both ineffective and overly toxic in the treatment of herpes encephalitis (25).

Trifluorothymidine. A related thymidine analogue, trifluorothymidine (5′-trifluoromethyl-2′-deoxyuridine) has been reported to be a significantly superior alternative to iododeoxyuridine for the treatment of human ocular infections with herpes simplex virus (26). It is more soluble than idoxuridine and the resultant higher concentration and greater potency results in more rapid healing of the lesion.

Arabinosylcytosine. This pyrimidine nucleoside analogue (1-β-D-arabinofuranosylcytosine, ara-C) has arabinose in place of 2-deoxyribose. It has a similar, but not identical, antiviral spectrum to idoxuridine and is equally effective in treating herpetic and vaccinial infections of the eye. However, it is more toxic to the epithelium and is rapidly deaminated to an inactive form *in vivo*.

Arabinosyladenine. This purine nucleoside analogue (9-β-D-arabinofuranosyladenine, ara-A, Vidarabine) is virtually identical in potency and activity to idoxuridine and cytosine arabinoside (27). It is not as readily deaminated as arabinosylcytosine. It is also less toxic to host cells than either ara-C or idoxuridine. The lesser toxicity is caused by not being incorporated into cellular DNA but it seems to inhibit viral DNA polymerase activity. In systemic administration in therapeutically active doses, it does not suppress the hematopoietic system nor does it have immunosuppressive effect on antibody formation or on cellular immunity and produces only minimal systemic symptoms. At high doses it causes vomiting, nausea, weight loss, weakness, and megaloblastosis in the erythroid series. Its primary drawback is its low solubility; large volumes of fluid must be administered. Adenine arabinoside is slowly deaminated *in vivo* to the hypoxanthine derivative which is less effective but does retain some antiviral activity. New derivatives are being developed that are more soluble and not so readily deaminated.

Adenine arabinoside has been licensed by the FDA for topical use in ophthalmic ointments. Topical treatment of primary or recurrent herpetic genital infection with 3% ara-A failed to influence the course of the disease (28). However, treatment of herpes zoster in immunosuppressed patients by intravenous administration has shown some success without serious side effects (29). More recently, it was evaluated for treatment of herpes simplex encephalitis in a placebo-controlled study (30). In 28 cases proved by isolation of type 1 herpes simplex virus from brain biopsy, treatment reduced mortality from 70–28% and over 50% of treated survivors had no or only moderately debilitating neurological sequelae. Thus this drug seems to be the first antiviral drug for systemic treatment of herpetic infections.

Ribavirin. Ribavirin (1-β-D-ribofuranosyl-1,2,4-triazole-3-carboxamide, Virazole) is a ribose containing synthetic purine nucleoside which is active *in vitro* and *in vivo* against a number of RNA and DNA viruses including influenza, viral hepatitis, and herpes viruses (31).

The prophylactic effectiveness of ribavirin against challenge with types A and B influenza viruses was evaluated in double-blind clinical trials (32–33). Ribavirin did not affect the development of type A influenza illness and showed only minimal effectiveness in suppressing type B influenza illness in humans. A more recent study, however, using a different strain of influenza A and different treatment regimen indicates a significant reduction in clinical manifestations and virus in ribavirin-treated young individuals (34). Caution should be used in giving the drug to women of child bearing age since congenitial anomalies of limbs, ribs, eyes, and central nervous system, as well as fetal deaths, have occurred when the drug was given to pregnant hamsters (35).

Interferon. Interferon was discovered in 1956 by Isaacs and Lindenmann (36). It is a glycoprotein produced by cells in response to most viral infections and plays an important role in host defense during viral disease. Interferon has a broad spectrum of antiviral activity and can prevent the replication of most viruses in pretreated cells. More recently, it has been shown that interferon synthesis is also induced by a variety of other substances ranging from bacteria and their products to synthetic polymers.

The expression of the antiviral activity of interferon is a two-step process; interferon is not active directly but induces the antiviral state in treated cells by the synthesis of a short-lived antiviral protein. A great deal of evidence suggests that the antiviral protein interferes with the translation of viral messenger-RNAs. Being a natural product it has the advantage over chemical agents in that it is relatively nontoxic to normal cells, although when used *in vivo* at high doses it may be immunosuppressive. Prolonged interferon treatment is well tolerated. Low-grade fever and mild malaise may occur and a transient depression in white cell counts as well as platelets and reticulocyte counts occur during therapy with interferon (37). All hematologic measurements promptly reverted to normal when interferon therapy was stopped. It is not clear whether the hematopoietic suppression is an intrinsic property of interferon or caused by a contaminating protein.

Two approaches have been utilized to investigate the efficacy of interferon use in humans: (1) the use of inducers to cause the host to make its own interferon and antiviral protein, and (2) the direct application of interferon. Synthetic double-stranded RNA polymers, such as polyriboinosinic acid-polyribocytidylic acid complexes [poly(rI:rC)], have been used with some success (38). However, many of these inducers are toxic and studies in humans are usually limited to topical application. Thus topical application of poly(rI:rC) has been shown to prevent illness from rhinovirus challenge (39). It has also been shown to have a definite but slight therapeutic effect in herpetic keratitis (40).

Initially the results from the direct use of human interferon in human disease was discouraging probably due to the low activity of the preparations. Recently, high-titer human interferon (10^7 units/mL) has become available and encouraging results are being obtained (37,41–42).

Immunopotentiating Agents. Another more recent approach is to use chemicals that augment or modulate immune responses of the host. Inosiplex (Isoprinosine) the *p*-acetamidobenzoic acid salt of inosine *N,N*-dimethylaminoisopropanol, has shown antiviral activities in tissue culture, animal models, and human studies but the results have been contradictory. Controlled human-challenge studies with rhinovirus and influenza virus using the drug in a prophylactic fashion have been disappointing. Clinical studies have suggested that the drug may be more effective when used therapeutically instead of prophylactically and it has been suggested that the antiviral effect is caused by its immunopotentiating action (43–45). Levamisole, the levoisomer of tetramisole, may act in a similar manner (see Chemotherapeutics, antihelmintic) (44).

Future of Antiviral Chemotherapy

After some three decades of study, the limited number of antiviral agents available for the treatment and prevention of human disease appears to be discouraging. Furthermore, because of their toxicity, many of them are limited to topical application. However, as more information becomes available about the mechanisms of virus replication, it may be possible to synthesize substances that specifically interfere with unique viral function. Indeed, some drugs seem to act in this way and therefore should be less toxic to the host. The success with vidarabine in systemic treatment of herpes encephalitis gives encouragement for the future. The success with interferon trials also indicate that virus-specific functions can be specifically inhibited without being detrimental to the host. Its wide-spectrum activity would also indicate that drugs might also be developed that will have similar activity.

BIBLIOGRAPHY

"Viral Infections, Chemotherapy" in *ECT* 2nd ed., Vol. 21, pp. 452–460, by Conrad E. Hoffmann, E. I. du Pont de Nemours & Co., Inc.

1. T. E. O'Connor, C. D. Aldrich, and V. S. Sethi, *Ann. N.Y. Acad. Sci.* **284,** 544 (1977).
2. D. Hamre, J. Bernstein, and R. Donovick, *Proc. Soc. Exp. Biol. Med.* **73,** 275 (1950).
3. D. J. Bauer, *Br. J. Exp. Pathol.* **36,** 105 (1955).
4. D. J. Bauer and P. W. Sadler, *Br. J. Pharmacol.* **15,** 101 (1960).
5. D. J. Bauer and co-workers, *Lancet ii,* 494 (1963).
6. A. R. Rao and co-workers, *Indian J. Med. Res.* **57,** 477 (1969).
7. L. A. R. DoValle and co-workers, *Lancet ii,* 976 (1965).
8. G. G. Heiner and co-workers, *Am. J. Epidemiol.* **94,** 435 (1971).
9. W. Levinson and co-workers, *Ann. N.Y. Acad. Sci.* **284,** 525 (1977).
10. G. G. Jackson, R. L. Muldoon, and L. W. Akers, *Antimicrob. Agents Chemother.,* 703 (1963).
11. W. L. Davies and co-workers, *Science* **144,** 862 (1964).
12. R. Welsh and co-workers, *Virology* **45,** 679 (1971).
13. *J. Am. Med. Assoc.* **201,** 374 (1967).
14. *Med. Lett.* **9**(2), 5 (1967).
15. J. J. Quilligan, M. Hirayama, and H. D. Baerstein, *J. Pediatr.* **69,** 572 (1966).
16. A. W. Galbraith and co-workers, *Lancet ii,* 1026 (1969).
17. N. Oker-Blum, *Br. Med. J.* **3,** 676 (1970).
18. A. W. Galbraith and co-workers, *Bull. W.H.O.* **41,** 677 (1969).
19. A. A. Smorodintsev, *Ann. N.Y. Acad. Sci.* **173,** 44 (1970).
20. R. B. Hornick and co-workers, *Bull. W.H.O.* **41,** 671 (1969).
21. A. W. Galbraith and co-workers, *Lancet ii,* 113 (1971).
22. E. C. Herrmann, Jr., *Proc. Soc. Exp. Biol. Med.* **107,** 142 (1961).
23. H. E. Kaufman, A. B. Nesburn, and E. D. Maloney, *Arch. Ophthalmol.* **67,** 583 (1962).
24. R. Jawetz and co-workers, *Ann. N.Y. Acad. Sci.* **173,** 282 (1970).
25. Boston interhospital virus study group and the NIAID-Sponsored Cooperative Antiviral Clinical Study Group, *N. Engl. J. Med.* **292,** 599 (1975).
26. P. C. Wellings and co-workers, *Am. J. Ophthalmol.* **73,** 932 (1972).
27. H. E. Kaufman, E. D. Ellison, and W. M. Townsend, *Arch. Ophthalmol.* **84,** 783 (1970).
28. H. G. Adams and co-workers, *J. Infect. Dis.* **133**(Suppl.), A151 (1976).
29. L. T. Ch'ien and co-workers, *J. Infect. Dis.* **133**(Suppl.), A184 (1976).
30. R. J. Whitley and co-workers, *N. Engl. J. Med.* **297,** 289 (1977).
31. R. W. Sidwell and co-workers, *Science* **177,** 705 (1972).
32. A. Cohen and co-workers, *J. Infect. Dis.* **133**(Suppl.), A114 (1976).
33. Y. Togo and E. A. McCracken, *J. Infect. Dis.* **133**(Suppl.), A109 (1976).
34. F. Salido-Rengell, H. Nassen-Quinones, and B. Briseno-Garcie, *Ann. N.Y. Acad. Sci.* **284,** 272 (1977).
35. L. Kilham and V. H. Ferm, *Science* **195,** 413 (1977).
36. A. Isaacs and J. Lindenmann, *Proc. R. Soc. London Ser. B* **147,** 258 (1957).
37. H. B. Greenberg and co-workers, *N. Engl. J. Med.* **295,** 517 (1976).
38. G. P. Lampson and co-workers, *Proc. Nat. Acad. Sci. U.S.A.* **58,** 782 (1967).
39. C. Panusarn and co-workers, *N. Engl. J. Med.* **291,** 57 (1974).
40. H. E. Kaufman, E. D. Ellison, and S. R. Waltman, *Am. J. Ophthalmol.* **68,** 486 (1969).
41. B. R. Jones and co-workers, *J. Infect. Dis.* **133**(Suppl.), A169 (1976).
42. B. R. Jones and co-workers, *Lancet ii,* 128 (1976).
43. T. Ginsberg and A. J. Glasky, *Ann. N.Y. Acad. Sci.* **284,** 128 (1977).
44. J. W. Hadden and co-workers, *Ann. N.Y. Acad. Sci.* **284,** 139 (1977).
45. R. H. Waldman and R. Ganguly, *Ann. N.Y. Acad. Sci.* **284,** 153 (1977).

General References

Review Articles

W. H. Prusoff and B. Goz, "Potential Mechanisms of Action Antiviral Agents," *Fed. Proc.* **32,** 1679 (1973).
R. A. Bucknall, "The Continuing Search for Antiviral Drugs," *Adv. Pharmacol. Chemother.* **11,** 295 (1973).
D. Parkes, *Adv. Drug. Res.* **8,** 11 (1974).
J. P. Luby, M. T. Johnson, and S. R. Jones, "Antiviral Chemotherapy," *Ann. Rev. Med.* **25,** 251 (1974).
D. Pavan-Langston, R. A. Buchanan, and C. A. Elford, Jr., eds., *Adenine Arabinoside: An Antiviral Agent,* Raven Press, New York, 1975.
J. P. Luby, M. T. Johnson, and S. R. Jones, "Antiviral Chemotherapy," *Ann. Rev. Med.* **25,** 251 (1974).
M. Ho and J. A. Armstrong, "Interferon," *Ann. Rev. Microbiol.* **29,** 131 (1975).
N. B. Finter, *Interferon and Interferon Inducers,* American Elsevier Publishing Co., Inc., New York, 1973, pp. 598.
S. Baron and F. Dianzani, eds., "The Interferon System: A Current Review," *Texas Rep. Biol. Med.* **35,** (Oct. 1977).
C. E. Hoffman, "Virus Chemotherapy," *Chemtech.* **8,** 726 (1978).

Symposia

Symposium on Clinical Use of Interferon, Ninth International Immunobiological Symposium, Zagreb, Yugoslavia, Izdavacki zavod, Jugoslovenske academe, Zagreb, Yugoslavia, 1975, 262 pp.
T. Merigan, ed., "Antivirals with Clinical Potential, A Symposium at Stanford University, Aug. 1975," *J. Infect. Dis.* **133**(Suppl.), A1 (1976).
E. C. Hermann, Jr., ed., "Third Conference on Antiviral Substances," *Ann. N.Y. Acad. Sci.* **284,** (1977).

GEORGE E. GIFFORD
University of Florida

CHEMURGY

Chemurgy is defined as that branch of applied chemistry devoted to industrial utilization of organic raw materials, especially from farm products. A more modern and general definition might be the use of renewable resources for materials and energy. By this more modern definition practically one half of the topics in the *Encyclopedia of Chemical Technology* could be considered to be under the category of chemurgy. Accordingly, this article is a review and general discussion of the topic; specific details will be found in the various articles on more narrow topics (see Fuels from biomass; Fuels from waste).

It is appropriate to consider briefly some social history before discussing technology. Chemurgy was really a social movement during the 1920s and 1930s, a time when there were large surpluses of agricultural materials and severe economic problems in the farm areas. The idea of using farm commodities as chemical or industrial raw materials was seen as a major contribution to solution of these economic problems. Many great men were associated with the movement, including Henry Ford, George Washington Carver, Henry Wallace, Leo Baekelund, and others. References 1–5 are a few of the tracts written about the movement; in particular, reference 5 describes some of the early accomplishments of the chemurgy movement. Probably one of the movement's major accomplishments was the inspiration for the founding of the regional laboratories of the U.S. Department of Agriculture. These laboratories still exist as major centers for research in the application of agricultural materials. At the time of their founding, they were considered chemurgic laboratories (see also ref. 6).

Some of the early work in chemurgy was disrupted and became somewhat irrelevant owing to World War II. Interestingly, many of these early ideas are now being revived with the modern energy and materials crises. Thus it is appropriate to consider briefly some of these early efforts. For example, one of the early successes was by Herty, who found a way to make strong paper from southern pine. This led to the foundation of the southern pulp and paper industry which today is the largest segment of that business in the United States (see Pulp; Paper). Prior to his work, the paper industry had been largely centered in the north and northeast where slower growing species were used.

Some of the early work on the manipulation of proteins was based on the objective of forming synthetic fibers from natural sources, eg, peanut meal or soy meal (see Nuts; Soybeans and other seed proteins). Although these fibers turned out to be deficient as textiles, the same technology has found applications today in synthetic meat production (see Proteins). A few other technologies that are so large in their own right that it is often forgotten that they are chemurgic processes include such things as the cellulosic fibers, rayon (qv) and cellulose acetate (qv), and cellophane (see Cellulose; Cellulose derivatives). These were among the original synthetic fibers and films; they are mature industries today. Due to the rapid recent increase in the cost of fossil fuel raw materials, there is renewed interest in the cellulosic polymers.

The naval stores industry is another example that is often overlooked as a chemurgic industry. Although the traditional sources of naval stores, namely gum collection and stump distillation, are declining in importance, the recovery of turpentine and rosin from paper and pulping processes is increasing in importance (see Tall oil; Terpenoids). These unusual hydrocarbons have many applications and more may be expected to be discovered in the future because of their renewable base. Still another example which is an old and renewable-based industry is the oils and fatty acids business (see Fats and fatty oils; Vegetable oils; Carboxylic acids). These materials have applications in a wide variety of chemical products including soaps (qv) and detergents (see Surfactants), cosmetics (qv), and coatings (qv). Fatty acids, of course, are derived from vegetable or animal oils or fats, although they can also be produced from hydrocarbons.

Industrial Materials from Renewable Resources

One distinction that can be made in the very broad area of chemurgy is between the use of natural products that are grown solely for industrial purposes as compared to those that are grown primarily for food. In the latter, industrial materials are either by-products from food production or, as the chemurgists intended, substitutes for food uses when the commodity is in surplus.

Trees are by far the largest commodity grown solely for industrial use. About two thirds of the trees harvested are used for construction or structural uses and about one third are used for pulp and paper making. About one third of the pulp is made from residues of other timber conversion, such as trimmings and sawdust. Small quantities are used for fuel.

Other purely industrial crops include cotton (qv), grown for its fiber, and flax grown for its linseed oil (see Drying oils). A large quantity of soybean oil is epoxidized to make an important plasticizer. In addition to linseed, there are some other purely industrial oils but the quantities are very small, eg, tung oil and castor oil (qv). Crambe is a new industrial oil seed which is just being introduced in the United States. Other new crops are constantly under consideration. An example is jojoba whose oil bears some resemblance to sperm whale oil. Jojoba is a wild crop and has not yet been domesticated but it is currently being studied (7).

The industrial use of food materials has generally begun with the detailed analysis of the chemicals present, followed by the purification of the chemicals to obtain reliable and consistent materials, and very often, the chemical modification of these purified components. For example, starch (qv) from cereal grains is used by itself; it is also hydrolyzed to yield glucose and sweeteners (qv) and it can be chemically modified to yield plasticizers (qv) and polymers of unique properties. Likewise, proteins from oilseed meals such as soy, peanuts, and cottonseed, have been incorporated into glues (see Glue) and adhesives (qv) and modified for various industrial purposes. Starch has also been refined from potatoes and sweet potatoes, as well as the cereal grains, and protein has been isolated from the cereal grains and other crops as well as the oilseed meals (see Wheat and other cereal grains).

By far the largest volume of natural products for industrial use aside from the forest products are wastes or by-products of food processing. For example, soybean oil foots or soap stock is a material removed from soybean oil. This is normally used only for industrial purposes. Animal tallow may be edible or inedible depending on the circumstances of its manufacture. The inedible animal tallow is used solely for industrial purposes in the United States although some is exported and refined for human food. Probably the largest use of various food processing wastes is as animal feeds where they are not modified very much, perhaps dried or sterilized. They are not, strictly speaking, chemical or industrial products but they are used rather than disposed of and so they are considered to be chemurgic products. To the extent that wastes are used for animal feeds where potentially human food might have been used, such a use frees food for human consumption and extends our food supply.

There are major wastes from the pulp and paper industry which are finding increasing applications. These include lignin (qv) and tall oil, as well as sugars in the form of sulfite waste liquor. Lignin is used in largest amounts as a fuel in the pulping process but a wide variety of products can be made from isolated lignin. Lignin is composed of aromatics and phenolic compounds which can be separated and purified. These occur in large variety and only relatively small quantities of any one substance can be isolated. In this respect, lignin is more like coal tars from coking than like fractions from crude oil. These are used as dispersants (qv), adhesives, additives to drilling muds (see Drilling fluids), and fillers (qv). Tall oil has been referred to as the largest and fastest growing source of extractives such as turpentine and resin. It can be refined to give tall oil fatty acids (see Carboxylic acids) and tall oil pitch as well as resins. These fatty acids compete with fatty acids from vegetable sources for many of the same industrial markets. Sulfite waste liquor has a large amount of hydrolyzed sugars. Some sulfite waste liquors are now being fermented to give ethanol (qv) and single cell protein in the form of yeast (see Foods, nonconventional; Yeast). All of these applications amount to upgrading to a saleable product materials that otherwise represent a disposal problem.

Some chemurgic processing and research may be seen as defensive in the sense of retaining markets for renewable materials that compete with materials made from nonrenewable sources. For example, cotton and wool (qv) have large markets as textiles but are increasingly threatened by synthetic fibers. A great deal of research has been directed to overcoming the perceived disadvantages of cotton and wool, such as wrinkling, shrinking, and felting, in order to enable these textiles to retain their traditional markets or expand them. These have been relatively successful. For example, cotton is still about one half of the textile fiber market. Wool has declined but, because sheep are relatively efficient meat producers, it is possible that the supply of wool may increase and that its price may decline, provided that it offers a competitive fiber.

The fermentation industry deserves separate mention because it is based almost exclusively on renewable materials in the form of molasses, starch and other materials (see Fermentation). Most products of the fermentation industry are of very high value and relatively low volume such as antibiotics (qv). However, there are some high volume chemicals. Many years ago, chemicals that are now made from hydrocarbons, such as acetone and butanol, were made by fermentation. These are not made by fermentation today because petrochemical sources are less expensive, but if petrochemicals continue to increase in price faster than cereal grains or other fermentable substrates, it is possible that these solvent sources may again become important (8).

Ethanol is the oldest chemical made by fermentation. Industrial alcohol may be made by fermentation or by hydration of ethylene. Beverage alcohol must be made by fermentation (see Beverage spirits; Beer; Wine). Since 1973 the increase in the price of ethylene owing to the increase in price of crude oil has made industrial alcohol by fermentation again competitive at ca $0.25/L ($1/gal). Fermentation plants that had been shut down for many years have been reopened and are likely to gain an increasing portion of this business. However, the efficiency of conversion of glucose to ethanol is less than 50% because so much carbon dioxide is also made. Other products, such as acetic acid (qv) or other fatty acids, can also be made by fermentation and these are articles of commerce in large volume.

When there were large grain surpluses, the fermentation of these grains to ethanol for use as a motor fuel was advocated as a solution to the farmers' economic problems as well as a way of becoming independent of foreign energy sources. This rather prescient proposal was never adopted although it was quite seriously discussed at the time. Recently, the Nebraska state government has revived this same idea and passed legislation providing economic incentives for the use of fermentation-based ethanol ("Gasohol") as a motor fuel. Brazil has also developed an ethanol fuel program. References 9–14 describe the Nebraska program and provide up-to-date information on the fermentation process (see Gasoline and other motor fuels).

Grain usable by humans or animals is an expensive substrate for this fermentation process. A cheaper substrate might be some source of cellulose such as wood or agricultural waste. This requires hydrolysis of cellulose to yield glucose (see Cellulose). Such a process was used in Germany during the second World War to produce yeast as a protein substitute. Another process for the hydrolysis of wood was developed by the U.S. Forest Products Laboratory in Madison, Wisconsin. These processes use mineral acid as a catalyst. The hydrolysis industry is very large in the Union of Soviet Socialist Republics but it is not commercial elsewhere at this time. There are two journals in the Union of Soviet Socialist Republics devoted to this subject. More recently, interest has developed in the use of enzymes to catalyze the hydrolysis of cellulose. Research on cellulase hydrolysis has been conducted at the U.S. Army Natick Laboratories, the University of California at Berkeley, and Virginia Polytechnic Institute and State University, among others (see below). Briefly, the concept is to take domestic or forest product wastes and hydrolyze them to glucose which would then be used as a fermentation substrate. The literature on alcohol fermentation, whether from cereal grains, molasses, or wood hydrolysis is very large but the only commercial practice of this technology is for the industrial alcohol and beverage alcohol industries.

An area that seems to have been a common thread through the history of chemurgy is that of generating energy from biomass in some form. Ethanol fuel as mentioned earlier, is one of these concepts, but the direct combustion of plant materials or the fermentation of plant materials to yield methane have both been additional topics for research. Wood, of course, has been burned as a fuel for many years and in 1974 generated more energy in the United States than did nuclear sources. The bulk of this use was in forest products industries where wood wastes were burned on-site for energy generation. Small-scale anaerobic fermentations of agricultural wastes to yield methane (qv) are practiced but large-scale processes have not yet been built, except in sewage treatment plants to dispose of sludge (see Wastes; Water, sewage).

Specific Processes. Specific chemurgic processes can be organized under several categories according to the various raw materials that are used: those grown specifically for industry; those grown primarily for food; and wastes.

Processes Based on Industrially-Known Raw Materials. The most important of those materials grown for industrial use are trees. Trees are used for structural purposes and also for pulp and paper. The process to convert trees to structural purposes may be as simple as trimming to make telegraph poles or rough sawing for railroad ties. Structural lumber and furniture stock require additional finishing. A more complicated process is the manufacture of plywood by peeling large strips from logs and then laminating these with adhesives (see Laminated and reinforced wood). Still more complicated wood products are made from chips or slivers pressed together with a binder which may be a phenolic resin or other polymer.

Cotton is an industrial product grown primarily for its fiber although the protein and oil contained in the seed also contribute to farm income from cotton. The manufacture of cotton is simply a mechanical separation of the fiber from the seeds followed by cleaning and mechanical processing into thread and ultimately textiles. These textiles may be treated to give them special properties such as water and wrinkle resistance.

Flax is grown primarily for its seed which yields linseed oil, a drying oil that is used in coatings. The fiber from flax grown in the United States is short and is used in certain types of fine papers such as cigarette papers. Longer flax fibers grown in other parts of the world are manufactured into the fine textile, linen. Linseed oil's use in industry is declining in this country. There are relatively few other crops that are grown solely for industrial purposes. Among these is tung, a tree nut that can be used as a drying oil in coatings. Of larger interest are several new crops, none of which are yet commercial but which hold some promise of providing new domestic sources for important industrial materials.

First is jojoba, a desert crop which gives a small bean containing about 50% of a wax, a fatty acid ester with a fatty alcohol. This is an unusual chemical to be found in nature. The only other large source is sperm whale oil which has traditionally been used in fine lubricants. Because the sperm whale is an endangered species, relatively little sperm whale oil is available and there is a large market for a substitute. Jojoba oil has been found to be usable for most of the applications of sperm whale oil. It is obtained simply by pressing the nut followed by conventional refining.

The problems with jojoba as a commercial crop are the usual ones of domestication and cultivation. It is a slow growing plant available only in the wild at present and therefore has very wide genetic variability. Efforts are underway to select the most promising variants and cultivate these as a crop in the southwestern United States deserts (7).

Guayule is a crop that is potentially a source of natural rubber. This is an unusual new crop in that it had been an article of commerce in the past. It grows wild in northern Mexico and the southwestern United States. When the leaves are milled in water, a latex is released which coagulates into natural rubber "worms" which can be easily collected and relatively easily refined to give a product that is almost identical with natural rubber from southeast Asia. During World War II there were several thousand acres of guayule planted in California and a small plant established to extract the rubber for military use. After the war, however, this effort was shut down. Since it is believed that natural rubber trees may have reached their genetic potential and their sources are in politically vulnerable areas such as southeast Asia or in areas where the trees are vulnerable to natural pests as in South America, there is considerable concern about sources of natural rubber and it is very likely that guayule will become an important crop in the future (15).

Crambe is a crop that has been recently introduced to the United States. It is an oilseed whose oil is very high in erucic acid (13-docosenoic acid), which can be used to provide industrial lubricants, especially those needed for the basic oxygen furnace process for making steel. Crambe is grown in relatively small volume in the midwestern United States.

Kenaf is a grass fiber crop which has been proposed as an alternative papermaking source. It is an annual crop and can be grown in many parts of the country. There have been agronomic problems with kenaf, primarily its vulnerability to nematode pests. There is also a problem with kenaf in that it is harvested once a year and must then be stored, usually as silage. There is some loss of the fiber in storage.

Lesquerella is an oil seed crop that has been mentioned as a potential industrial crop but is not yet grown in significant quantities. Castor has been a crop in the United States and provides a well-known lubricating oil; however, it is grown only in small quantities today. Most of the castor used in the United States is imported from Brazil (16).

The economic disposal of the oil seed meal is a problem with almost every new oil seed. The large-volume oil seeds, such as soy and peanut, are particularly valuable because their oil seed meals, which contain large amounts of protein, can be used almost directly for animal or even human food. It is important in developing processes for new oil seeds that a profitable use for the meal be found. Many oil seeds contain toxic substances or undesirable materials which complicate their use. For example, the crambe seed contains toxic and allergenic materials which must be removed before they can be fed to monogastric animals. A large part of the research on new crops and new oil seeds is devoted to upgrading the meal once the oil has been obtained by conventional pressing and solvent extraction methods. The seeds may be cooked first to release the oil. This cooking is often by direct contact with steam.

Processes Based on Food Crops. Crops that are grown primarily for food can be used for industrial purposes when the crops are in surplus or are found to be unfit for human consumption or for their intended purpose. Historically, when agricultural surpluses were high, the distinction between industrial and food use was not significant. With modern demands on food production capacity, it is likely that food value will usually exceed industrial value for food materials. Nonetheless, there are large industries based on food commodities; eg, the corn and wheat starch separation processes to give starch used in paper sizing and textiles. The starch is separated primarily by gravity in water slurries. Other starchy cereals could as easily be used, eg, milo or grain sorghum, rice, oats, and barley, although rice, oats, and barley are usually too valuable to be diverted to starch production. Potatoes and sweet potatoes also can be fractionated to give a very good starch product. Usually only the culls from fresh or processed potato manufacture are used for this purpose, but this is an important economic use because it consumes materials that would otherwise be wasted. Potato starch manufacture has suffered recently because the effluent from the process is a severe pollutant. Research has been directed to the utilization of this material which contains very high quality protein (17).

Oil seeds grown primarily for use in salad dressings, margarines, and cooking oils also produce an important by-product in the form of high protein meal. Historically, this meal was fed to animals but increasingly it is refined and fractionated for human consumption. Soy, in particular, is used to produce a large variety of specialized protein concentrates and isolates by various refining processes usually involving caustic solution and precipitation. Some other oil seeds that are important in the United States include rapeseed, sunflower seed, and safflower (see Vegetable oils).

By-products from meat animals are also commercially significant (see Meat products). The important meat products in the United States are beef, poultry, pork and lamb. Each of these produces (in addition to edible meat) bones, trimmings, fat, hides, and in the case of sheep, wool. The hides are turned into leather (qv) by tanning or are converted to commercial gelatin or other animal glue products (see Glue and gelatin). The fats may be edible or inedible in the United States depending on the conditions under which they are isolated. Their chemical composition is essentially identical. Inedible animal tallow may be exported and refined in foreign countries for edible use; some is added to animal feeds. A large amount is used in soaps and detergents or converted to fatty acids which may be refined or used as a mixture in soaps. Bristles from pork are used to a small extent in brushes. Animal hair is a waste product which creates disposal problems. Uses are still being sought for this material.

Processes Based on Wastes. The trimmings and slash from forest operations historically have been left in the forest but increasingly are being used in much the same way as higher quality timber, primarily for pulping or chipping. Agriculture produces large amounts of wastes in the form of animal manures, branches, stems, stalks, and straws. These also have historically been returned to the land, but as in the case of animal manures, changing patterns of production have made it less convenient to use these for their fertilizer value and aggravate the pollution problem they create. It is possible to recycle animal wastes as animal feed because a large amount of nutritive value is retained in the waste. However, the enormous volume of animal waste is being considered as a potential energy source, probably by anaerobic fermentation to produce methane. Other agricultural wastes, such as straw, have been considered for pulp and paper making and for digestion to produce fuel. They are also candidates for hydrolysis of their cellulose content to produce glucose.

The processing of agricultural products to make foods and feeds yields additional wastes that have traditionally been considered valueless and were disposed of by dumping. With modern constraints on what may be dumped, these must now be considered as raw materials. A good bit of research to discover uses for these materials has thus been motivated primarily by environmental considerations. For example, Kraft black liquor, sulfite waste liquor, and other dilute streams from pulp and paper making are potential fermentation substrates because of the dissolved sugar, the very material that is the most serious pollutant. Edible-oil refining produces soap stock or foots, a combination of free fatty acids with caustic. These can be isolated and sold for soap manufacturing. Cereal grain milling produces a variety of fractions that have found historic uses in animal feeds. Research has been conducted on isolation of protein concentrates from these materials for upgrading as human foods. The processes for isolation usually involve caustic extraction followed by neutralization and precipitation of the protein.

One relatively simple way to upgrade cellulosic materials, such as farm wastes, for animal feed is simply to contact it with ammonia. This seems to swell the cellulose and make it more digestible, and at the same time, leaves behind some ammonia nitrogen which is a nutrient for ruminants (see Pet and other livestock feeds).

Some of the forest wastes and pulp and paper processing wastes contain dilute concentrations of important acids such as acetic and solvents such as methanol (qv). These are formed from the natural constituents of wood in the high temperature digestion of pulping. These streams have been discarded as too dilute for commercial recovery. It is increasingly attractive to use such processes as liquid–liquid extraction (see Extraction) or adsorption on charcoal followed by steam stripping to remove and concentrate these organic chemicals from such wastes (see Adsorptive separation).

Applications of Industrial Materials from Renewable Resources. These are very broad but can be illustrated by a few examples, for instance, clothing, the outlet for cotton and wool fibers which are increasingly being displaced by synthetic fibers. The synthetic fibers are cheaper and offer easier care features. However, they are generally not as comfortable as natural fibers. Inevitably, economics will determine the distribution of textile fibers. The competitive position of natural fibers can be improved by such modifications as wrinkle-proofing and waterproofing.

Transportation is another large market for materials such as fuels, lubricants, and coatings. Here again, there are synthetic substitutes for materials that could be made from renewable resources. We have discussed the interest in ethanol as a motor fuel. At the present time, this does not seem competitive even with crude oil at 7–8¢/L ($12–13/barrel). However, the key point is that it could be produced by simple fermentation of cereal grains and used in existing automobiles with little change. Lubricants today are subject to higher stress because of antipollution devices and smaller engines. Historically, castor oil (qv) has been an important automobile lubricant; its application in that use is now small, but in principle, vegetable oil could be used to replace petroleum-based lubricants. Modern paints and lacquers are based on synthetic chemicals, such as acrylics, but could be made from natural products such as the drying oils, resins, or acrylates from lactic acid. Housing is a large market for natural products such as sawn timber. Additional components of the housing market include coatings and insulation (qv). Wood appears to be the cheapest material of construction generally available and it has the advantage of being a good thermal insulation material. In the area of health, drugs are largely based on fermentation technology which in turn uses farm products as a substrate. In addition to the use as foods, large numbers of food additives can be made from agricultural products. These include modified starches, cellulose gums, flavors and aromas from turpentine, and other less obvious applications (see Flavors; Gums; Perfumes; Terpenoids; Tall oil). Also, rosin can be used to produce resins which compete directly with resins from petroleum (see Hydrocarbon resins; Resins, natural).

Processing Methods. By far the largest class of processes applied to farm commodities are separations. These usually are rather simple, based on some physical property such as density or particle size. For example, the milling process for cereal grains involves size reduction followed by screening to yield products that have varied concentrations of starch, fiber, and protein. Size and density differences can be correlated with chemical differences. Milling of water slurries is practiced to obtain finer separation of starch, fiber, protein, and oil. Another class of separations is based on solubility. A good example of this is the refining of oil seed meals to give protein isolates and concentrates. Proteins are highly soluble in basic solutions and the processes of isolation and concentration involve repeated dissolution, physical separation by centrifuging or filtration, and then precipitation of the protein. Solubility in solvents is used to extract oil from oil seed meals. The usual solvents are light hydrocarbons such as hexane. There is some separation on the basis of volatility such as the distillation of essential oils, the recovery of solvent from edible oils, and the distillation of esters of fatty acids. In general, the separation processes used in chemurgy are relatively simple.

In contrast, the organic chemical reactions that are possible with chemurgic materials can be very sophisticated. The chemistry involved is often reductive because so many of the chemurgic materials, such as cellulose and starch, are carbohydrates whose functionality is based on their hydroxyl groups. Also important are the carboxylic acids and the amine groups of protein. Reactions such as the esterification of sugar (qv) and the epoxidation of soy oil are of commercial significance. Also important are the acetylation of cellulose to make it soluble for subsequent spinning into fibers, and the solubilization of starch and cellulose as xanthates.

Increasingly, biochemical transformations are used to modify renewable resources into useful materials. Fermentation to ethanol is the oldest of such conversions. Another example is the cell-free enzyme catalyzed isomerization of glucose to fructose for use as a sweetener (qv). The enzymatic hydrolysis of cellulose is a biochemical competitor for the acid catalyzed reaction. Fermentation to produce acetone and butanol is another example. Still another would be the production of fatty acids, such as acetic acid, by fermentation. In general, chemurgic processes embrace the range of chemical technology from sophisticated synthetic organic chemistry to crude mechanical screening and gravity separation.

Potential for Renewable Resources

In 1976 the National Academy of Science conducted a study (18) on renewable resources for industrial materials. This study emphasized forest products because, as previously mentioned, these are the largest volume of renewable resources used today. But it also considered such things as special crops, new fats, oils, and fibers, animal by-products, and marine by-products. This study concluded that competition between a material made from a renewable or a nonrenewable resource would be resolved by economics. However, the report concluded that there were significant energy economies in the use of nonrenewable resources for a great many materials. One exception was the cellulosic fibers (acetate and rayon) which consume relatively large amounts of energy. However, a comparison, eg, of lumber with plastics or aluminum for housing construction, paper with polyolefin film, and several other competitive situations, indicated that the chemurgic product has an economic advantage or would likely have one so long as fossil fuels continued to be expensive. This study was summarized by Sarkanen (19). Goldstein has considered the potential for converting wood to plastics (20). Since plastics and polymers are among the largest volume industrial materials, the conversion of renewable resources to polymers and plastics is clearly important. With few exceptions, modern polymers and plastics based on petrochemicals are derived primarily from ethylene, propylene, and benzene. Each of these important building blocks could in theory be obtained from trees or agricultural wastes by an integrated series of processes. The hydrolysis of cellulose using mineral acids or enzymes has been mentioned. Enzyme-catalyzed hydrolysis is a subject of intense research but is not yet a commercial venture. In principle, using one of these techniques, the cellulose component of wood or agricultural waste can be converted to glucose, which can then be converted to ethanol. Ethanol in turn can be dehydrated to ethylene (qv). It can also be used to make butadiene (qv). This gives the olefin functionality that is necessary for a large class of polymers, including the polyolefins and styrene (qv), which is made by alkylation of benzene with ethylene followed by dehydrogenation (see Olefin polymers; Styrene plastics).

Hydrogenolysis of glucose, sucrose, starch, or cellulose can lead to glycerol (qv) and propylene glycol (see Glycols). These are major uses of propylene (qv). Presumably, they could be dehydrated to give propylene if necessary. The yields at present are unattractive, but at least in theory there is a route to propylene from renewable materials.

Lignin can be hydrogenated or hydrolyzed to yield large fractions of aromatics and phenolics. Unfortunately, the distribution of aromatics and phenolics from lignin is rather diverse with no dominant chemical present, but in principle these could be used in much the same way that petrochemical aromatics are now used. Thus the entire petrochemical base of polymers could be replaced by lignocellulosic materials. It is likely that ethanol for industrial use could be made competitively from wood hydrolysis in the near future. It has been contended that phenols from lignin could also be made competitively so long as the price of petrochemical phenol (qv) stays high. The inevitable by-products of such a chemical plant include extractives and the products of acid-catalyzed degradation of glucose. These latter products include hydroxymethylfurfural, levulinic acid, and, from the degradation of pentoses, furfural, which can be recovered and used as chemicals (see Furan derivatives).

There are many problems that need to be solved before this concept is useful. Among these problems are the recovery, purification, and fractionation of the diverse materials. Some of the by-products are toxic and would have to be removed or microorganisms found that are more tolerant towards these substances. However, none of these problems are insurmountable. More serious concerns are: the supply of raw material, the relative costs of competitive materials, and competition with other uses for the raw materials. Competition is particularly significant because the materials that would be sought for hydrolysis could easily be used in many cases for pulping and possibly for even higher value products, such as structural timber.

A rather impressive list of materials and products are made today from renewable resources. For example, the forest industry is truly enormous. The cellulosic fibers, rayon and cellulose acetate, are among the oldest and still relatively popular textile fibers and plastics. Soy and other oil seeds including the cereals are refined into important commodities such as starch, protein, oil, and their derivatives. The naval stores, turpentine, pine oil, and resin, are still important although their sources are changing from the traditional gum and pine stumps to tall oil recovered from pulping. These are all chemurgic industries. Other chemurgic processes, such as the manufacture of ethanol for motor fuel or the anaerobic fermentation of wastes for gaseous fuel, are on the verge of becoming important. The replacement of petrochemicals by chemurgic-based processes is a more questionable matter. This may occur for certain commodities but it is likely that enough crude oil and natural gas will always be available for the important petrochemicals. Nonetheless, it is significant that chemurgy could provide the olefins and aromatics that are the basis of the petrochemical industry (21).

The future of chemurgy is interconnected with the questions of energy, environment, and food. One can envision several scenarios concerning future sources of energy and materials. For example, nuclear energy may provide most of our power, leaving coal and oil available for chemicals and transportation fuels. In this case, chemurgic or renewable materials would probably be used much as they are today, ie, where they have some unique advantage or where conversion of a waste to some product is preferable to simple disposal. A second possibility, certainly more desirable in some respects, would be the use of solar energy (qv) in some fashion, releasing coal and oil for materials. Solar energy could be used directly as heat or it could be converted to electricity through semiconductor devices or it might be collected by photosynthesis in the form of biomass (see Fuels from biomass). Depending on how the biomass was subsequently converted to energy, there might then develop a large supply of unique raw materials such as extractives from woods, foliage (22) or other by-products that could be converted by chemurgic processes. Finally, in the unlikely event that coal and oil are best used for energy, then materials must come from other sources. This would place great demands on chemurgic processes.

Most likely the competition of chemurgic materials with synthetic materials will be settled in the market place. Some examples of the type of competition involved include the use of wood or paper as containers vs the use of plastics. When polyolefins were cheap, they were able to displace paper and cellulose film in wrapping and grocery bags because the polyolefins had more consistent properties and a higher strength-to-weight ratio. However, this advantage is lost when their cost exceeds a certain point. Thus paper bags, paper meat trays, and so forth may obtain a larger share of the market in the future. Similarly, natural and cellulosic fibers may find renewed popularity compared with synthetic fibers when the price advantage of the synthetic fibers no longer exists. Another substitution taking place is the use of vegetable protein for meat protein. This does not directly affect any nonrenewable resource but it has consequences for the increased availability of by-products of vegetable protein processing. We have also previously discussed the competition between fermentation alcohol and synthetic alcohol from ethylene. In 1974 fermentation alcohol for industrial use was competitive owing to the increased price and lack of availability of ethylene. Because ethylene is used in a large number of products, such as ethylene oxide (qv) and polyethylene (see Olefin polymers), and because alcohol is one of its less valuable uses, it is likely that fermentation alcohol will become a permanent fixture in the market place.

A rather complex competition exists among the various vegetable oils, animal tallow, and hydrocarbons. Most of the vegetable oils are essentially interchangeable although the processes needed to make them useful vary depending on the type of contamination they may contain. An example of this is the removal of color from palm nut oil. Tallow can be fractionated to give materials equivalent to vegetable oils as well as higher melting materials. Triglycerides in fats, oils, and tallows can be used for many of the same purposes as hydrocarbons such as lubrication or, as in the past, for illumination, or as a fuel. These applications do not apply today but merely illustrate their interchangeability. Fats and tallow can be converted to fatty acids by saponification. Hydrocarbons can be oxidized to make fatty acids. Each of these materials is thus competing for essentially the same place in the soap and detergent market. As meat production declines or stabilizes, the relative amount of tallow available will also decline in comparison with the quantities of vegetable oils that will become available. Their attractiveness compared with hydrocarbons for similar purposes should increase.

The prospects for hydrolyzing cellulose to give glucose have been mentioned. Starch is more easily hydrolyzed to give glucose and this is the traditional source. Both are equivalent materials for fermentation substrates. If cellulose hydrolysis were practiced on a large scale, it would probably displace starches as raw materials for glucose preparation. This may make starch available for modified products such as starch aldehydes, starch xanthates, etc.

The balance must be considered between the cost involved for recovery, reuse and modification of a waste material or by-product as compared with the cost for its disposal. It is probable that in future there will be increasing practice of recycle, recovery, modification, and upgrading of wastes of all sorts, and a reduction in disposal by incineration, biochemical oxidation, or discharge to the environment (see Recycling).

The problems of chemurgic processes may be summarized as those: (1) caused by variable and complex raw materials, (2) involving natural, ie, degradable, raw materials, (3) involving reductive rather than oxidative organic chemistry because of the composition of the raw materials, and (4) involving relatively simple and unsophisticated unit operations. Relatively little recent research has been done on these unit operations in comparison with the large volume industrial unit operations such as distillation and heterogeneous catalysis. As a result, there is a need for increased research and development in chemurgic processes involved with solid–liquid separations, solid–solid separations, and liquid–liquid extraction.

BIBLIOGRAPHY

1. W. McMillen, *New Riches from the Soil,* Van Nostrand Co., New York, 1946.
2. W. J. Hale, *The Farm Chemurgic,* The Stratford Co., Boston, Mass., 1934.
3. W. J. Hale, *Farmer Victorious,* Coward-McCann, Inc., New York, 1949.
4. W. J. Hale, *Farmward March,* Coward-McCann, Inc., New York, 1939.
5. C. Borth, *Pioneers of Plenty,* Bobbs-Merrill Co., New York, 1942.
6. U.S. Dept. of Agriculture, *Crops in Peace and War,* Government Printing Office, Washington, D.C., 1950.
7. National Research Council, *Jojoba: Feasibility for Cultivation on Indian Reservation in the Sonoran Desert Region,* National Academy of Sciences, Washington, D.C., 1977.
8. D. Perlman, *Chemtech* **4,** 210 (1974).
9. W. A. Scheller and B. J. Mohr, "Net Energy Analysis of Ethanol Production," *paper presented at the ACS National Meeting, New York, 1976.*
10. W. A. Scheller, "Agricultural Alcohol in Automation Fuel—Nebraska Gasoline," *paper presented at the 8th National Conference on Wheat Utilization Research, 1973.*
11. G. S. Santini and W. G. Vaux, *AIChE Symp. Ser.* **72**(158), 99 (1976).
12. W. A. Scheller and B. J. Mohr, *Grain Alcohol—Process, Price and Economic Information,* University of Nebraska, Lincoln, Nebraska, 1975.
13. W. A. Scheller and B. J. Mohr, "Production of Ethanol and Vegetable Protein by Grain Fermentation," *paper presented at the ACS National Meeting, Philadelphia, Pa., 1975.*
14. B. J. Mohr and W. A. Scheller, *State of the Art Report and Literature Survey of the Enzymatic Decomposition of Cellulose,* University of Nebraska, Lincoln, 1976.
15. National Research Council, *Guayule: An Alternative Source of Natural Rubber,* National Academy of Sciences, Washington, D.C., 1977.
16. G. A. White and co-workers, *Econ. Bot.* **25**(1), 22 (1971).
17. E. O. Strolle, J. Cording, and N. C. Aceto, *J. Agric. Food Chem.* **21**(6), 974 (1973).
18. National Research Council, *Renewable Resources for Industrial Materials,* National Academy of Sciences, Washington, D.C., 1976.
19. K. V. Sarkanen, *Science* **191**(4228), 773 (1976).
20. I. S. Goldstein, *Science* **189**(4206), 847 (1975).
21. E. S. Lipinsky, *Sugar J.,* (Aug. 27, 1976).
22. J. L. Keays and G. M. Barton, *Recent Advances in Foliage Utilization Information Report VP-X-137,* Western Forest Products Laboratory, Vancouver, British Columbia, Can., 1975.

General References

U.S. Dept. of Agriculture, *Crops in Peace and War,* Government Printing Office, Washington, D.C., 1950, still valuable compendium of chemurgic accomplishments.
S. M. Barnett, J. P. Clark, and J. M. Nystrom, eds., "Biochemical Engineering Energy, Renewable Resources and New Foods," *AIChE Symp. Ser.* **72**(158), (1976).

J. Peter Clark
Virginia Polytechnic Institute and State University

CHICLE. See Gums and mucilages.

CHLORAL. See Hypnotics, sedatives, anticonvulsants.

CHLORAMINES AND BROMAMINES

The term chloramine specifically denotes NH_2Cl, $NHCl_2$, and NCl_3 where NH_2Cl is the monochlor-, $NHCl_2$ the dichlor-, and NCl_3 the trichloramine. Historically, the term chloramine includes over one thousand compounds containing one or more chlorine atoms attached to a nitrogen atom, that is, chloramines, chlorimines, chloramides, and chlorimides.

A description of the manufacture and properties of cyanuric acid and the chlorinated s-triazinetrione products, as well as their uses, are given elsewhere in the Encyclopedia (see Cyanuric and isocyanuric acids).

Properties

The N–Cl bond is covalent but has properties different from those of the covalent C–Cl bond. For many purposes it is convenient to regard chlorine when bonded to nitrogen as positive chlorine. In organic chloramines the nitrogen to which the chlorine is attached also holds one or two organic radicals. Since these radicals may also contain chlorine, organic chloramines are better termed N-chloramines to indicate attachment of a chlorine atom to a nitrogen so that the chlorine is positive. When the bond is broken, the chlorine may be replaced by a hydrogen that is considered positive (eqs. 1–2):

$$RR'NCl + OH^- \rightarrow RR'NH + ClO^- \qquad (1)$$

$$RR'NCl + H_2O \rightarrow RR'NH + HClO \qquad (2)$$

The chlorine in the hypochlorite ion or hypochlorous acid may also be considered positive and, hence, chloramines are similar in some respects to this acid and its salts. The N–Cl bond is formed by treating an amine, imine, amide, or imide with hypochlorous acid. This reaction is the reverse of equation 2.

All compounds containing the N–Cl bond liberate iodine from an acidified iodide solution, although strong acid may be required, and the formation of iodine may be slow. This reaction (eq. 3) is commonly regarded as a test for N-chloro compounds (1).

$$RR'NCl + 2\,HI \rightarrow RR'NH + I_2 + HCl \qquad (3)$$

Also, this reaction is similar to the reaction of free chlorine. However, one atom of chlorine in a chloramine liberates as much iodine as two atoms of elemental chlorine (eq. 4):

$$Cl_2 + 2\,HI \rightarrow I_2 + 2\,HCl \qquad (4)$$

The available-chlorine content of a hypochlorite source or a chloramine is expressed in terms of the amount of elemental chlorine that would have the same oxidizing power. Available-chlorine content is measured by the ability to liberate iodine from an acidified iodine solution and is usually expressed as volume percent (grams of available-chlorine per 100 mL of solution). It is a quantitative measure of the oxidizing power and not a measure of the reactivity of the compound or of the oxidation potential it develops in solution, which is an intensity factor.

The term active chlorine was used until about 1925, and is still being used in some fields such as pharmacy; this unfortunately leads to considerable confusion. The active chlorine content of a compound is the content of chlorine bonded to nitrogen, or of positive chlorine. Elemental chlorine can be defined as having zero valence and as being reduced to +1 valence, whereas positive chlorine oxidizes from 1+ to 1− and requires two valence changes. Thus commercial sodium dichloro-s-triazinetrione has been described as containing 31% active chlorine, although it has an available-chlorine content of 62% and contains no other chlorine than the positive or active chlorine.

In the sanitation field, a differentiation is made between free available chlorine and combined available chlorine. Free available chlorine refers to chlorine present as hypochlorous acid or hypochlorites, or in the form of a more active chloramine; it rarely refers to dissolved elemental chlorine. Combined available chlorine is present as chloramines, usually having much less potency than hypochlorites. This difference is important since the time required for a given content of combined available chlorine to destroy a specific organism may be from ten to twenty times that required by the same amount of free available chlorine. In water sanitation, the term combined available chlorine generally refers to monochloramine, NH_2Cl. However, commercial chloramines are usually much more effective than NH_2Cl, and some approach the effectiveness of hypochlorites in their bactericidal properties under some conditions. When albuminous substances are present, any chlorine bound to them is in a form that, although still available, is nevertheless very inactive. This chlorine is usually referred to as albuminoid chlorine.

Hypochlorous Acid. The oxidizing and bactericidal properties of chloramines and hypochlorites are related to the hypochlorous acid content in their solutions (see Chlorine oxygen acids). This content depends on the pH and the concentration and hydrolysis constant of the chloramine. Chlorine dissolved in water gives hypochlorous acid and hydrogen and chloride ions (eq. 5):

$$Cl_2 + H_2O \rightarrow HOCl + H^+ + Cl^- \tag{5}$$

Published values of the hydrolysis constant vary greatly; some are listed in Table 1.

Table 2 gives the total chlorine present as HOCl in a water solution of chlorine at 25°C. It can be seen that strong solutions at low pH may have an appreciable amount of the total chlorine present as molecular chlorine.

With such solutions, chlorination reactions may be obtained. For chlorhydrination or oxidation by hypochlorous acid, solutions should be chosen where 99% or more of the available chlorine is present as HOCl.

Above 5 pH, hypochlorite ion concentration increases and that of molecular chlorine is negligible. The ionization constant for hypochlorous acid is 3.2×10^{-8} at 25°C (falling to 2.0×10^{-8} at 0°C) (3), giving a pK of 7.50.

Table 1. Hydrolysis Constants of HOCl at Various Temperatures[a]

Temperature, °C	$K \times 10^{-4}$	Temperature, °C	$K \times 10^{-4}$
0	1.46	35	5.10
15	2.81	45	6.05
25	3.94		

[a] Ref. 2.

Table 2. Total Available Chlorine Present as HOCl, % [a]

pH	10 ppm	100 ppm	1000 ppm	5000 ppm
1.0	96.5	73.7	21.8	(6.5)[a]
1.5	99.0	89.9	46.9	(18.2)[b]
2.0	99.6	96.5	73.7	41.2
2.5	99.9	99.0	89.9	69.0
3.0	99.96	99.6	96.5	87.5
3.5	99.99	99.9	99.0	95.7
4.0	99.99	99.96	99.6	98.5
4.5	99.99	99.99	99.96	99.5

[a] In an acid, aqueous chlorine solution at 25°C.
[b] Values for solutions with greater than 101 kPa (1 atm) pressure of chlorine.

At 7.5 pH at 25°C, the amount of available chlorine present as hypochlorous acid and as hypochlorite ion is equal. Furthermore, the fraction present as hypochlorous acid is independent of the total concentration of chlorine at values above about 5 pH. This fraction may be calculated against values from 5 to 12 pH as shown in Table 3.

At high and low pH values not all the bactericidal property resides in the hypochlorous acid alone. Strong alkali (>10 pH) has its own bactericidal action by dissolving the organism or attacking the cell walls. Acid concentrations below 4.5 pH inhibit or destroy bacteria, especially at 3.0 pH or lower.

Hydrolysis. Hydrolysis constants of chloramines may be expressed as follows (eq. 6):

$$\frac{[HOCl][RR'NH]}{[RR'NCl]} = K \qquad (6)$$

With hydrolysis constants as high as 10^{-4}, the amount present as hypochlorous acid is large in dilute solutions, and the action is comparable to that of hypochlorites of the same available chlorine content. Chloramines with constants over 10^{-6} yield bactericidal concentrations of hypochlorous acid, although they may or may not be less effective than the same content of available chlorine in hypochlorites. Chloramines with constants as low as 10^{-10} may have useful sanitizing value, although they may be bacteriostatic rather than bactericidal (4–5).

Equation 6 illustrates the major difference between hypochlorite and chloramines

Table 3. Percent of Total Chlorine in Hypochlorite Solutions Present as Hypochlorous Acid at Various pH Levels at 25°C

pH	HOCl, %	pH	HOCl, %
5.0	99.7	9.0	3.1
5.5	99.1	9.5	0.99
6.0	96.9	10.0	0.31
6.5	91.0	10.5	0.10
7.0	76.0	11.0	0.03
7.5	50.0	11.5	0.01
8.0	24.0	12.0	0.003
8.5	9.1		

for sanitation. As a hypochlorite solution is used up by a reducing process, it changes to chloride; the remaining available chlorine is distributed into hypochlorous acid and hypochlorite ion in a ratio dependent only on pH. The pH usually falls appreciably as hypochlorite oxidizes a substance. With chloramines, on the other hand, the free amine formed as the available chlorine is consumed tends to repress further hydrolysis so that the amount of available chlorine present as hypochlorous acid decreases. Furthermore, some chloramines have more than one positive chlorine atom. Thus, 1,3-dichloro-5,5-dimethylhydantoin loses the first positive chlorine with a hydrolysis constant of 2.54×10^{-4}, whereas the second splits at a constant of 1.14×10^{-4}. These constants seem to indicate that both positive chlorine atoms, on this molecule, would titrate as free available chlorine. However, the second chlorine does not become available until the solution is heated to about 70°C. Trichloroisocyanuric acid has three positive chlorines. The first chlorine ionizes with a fairly high constant, 1.6×10^{-4}, whereas the second and third ionize with a constant of 4×10^{-10}. The constant of the monochloroisocyanurate is 4.9×10^{-6}, being repressed further by the concentration of cyanuric acid which may be formed by consumption of the chloro compound or added as a conditioner.

These ionization reactions are believed to be extremely rapid (on the order of microseconds) since chlorine derived from chlorinated-s-triazinetriones are as fast acting as hypochlorite solutions of the same available chlorine content and pH in most applications (6).

It may be seen from the hydrolysis constant (eq. 6), that the amount of available chlorine present as hypochlorous acid tends to be low in strong solutions, but rises in weak solutions. Thus, a saturated solution of potassium dichloroisocyanurate does not contain sufficient hypochlorous acid to damage cloth seriously, although sodium hypochlorite solutions of weaker concentrations may cause damage at the same pH. However, weak solutions are used in laundry bleaching and the portion of the available chlorine present as hypochlorous acid and hypochlorite ion becomes much higher (in the range 10–50%), so that the same degree of bleaching takes place when compared to sodium or calcium hypochlorite. At lower hydrolysis constant values, such as 10^{-6}, the portion present as hypochlorous acid and hypochlorite ion becomes quite low. Thus, chloramine-T with a constant of 4.9×10^{-8}, is not used for bleaching and is a slow-acting bactericide at concentrations where hypochlorite is very fast.

In alkaline solutions (slightly below 10 pH), chloramines may react with hydroxide ions to form hypochlorite. Thus the constant for the reaction, $NH_2Cl + OH^- \rightarrow NH_3 + OCl^-$, is 1.6×10^{-3} (7). Values at or above pH 12 are required for measurable formation of hypochlorite ion. However, a number of chloramines may, in high concentrations, react with strong alkali to yield hypochlorites and an amine that can be separated from the mixture, thus giving a solution of hypochlorite that may be further processed. In swimming pools and drinking water with a pH close to 7.5, the ionization and hydroxide ion reactions can be neglected and the sanitizing action related to the hydrolysis constant. Several hydrolysis constants are listed in Table 4 (see Water, swimming pool treatment; Bleaching agents).

Table 5 lists dissociation constants of cyanuric acid and chlorinated s-triazinetriones obtained from acid–base titrations with an automatic titrator (8–9,13–15).

A computer program has been developed to calculate individual species as a function of reservoir chlorine, total cyanuric acid, and pH from individual acidity and hydrolysis constants. Although chlorinated s-triazinetriones serve as a reservoir for

Table 4. Hydrolysis Constants[a] of Various Chloramines

Chloramine	CAS Registry no.	K	References
trichloroisocyanuric acid	[88-90-1]	1.6×10^{-4} [b]	8–9
dichloroisocyanuric acid	[2782-57-2]	4×10^{-10}	2–3, 9
1,3-dichloro-5,5-dimethylhydantoin	[118-52-5]	2.54×10^{-4} [b]	10
monochloroisocyanuric acid	[13057-78-8]	4.9×10^{-6}	8–9
3-chloro-5,5-dimethylhydantoin	[34979-51-6]	1.14×10^{-4}	10
N-chlorosuccinimide	[128-09-06]	6.6×10^{-5}	10
N-chloropropionanilide	[67097-66-9]	7.5×10^{-7}	11
N-chlorobutyranilide	[67097-67-0]	7.0×10^{-7}	11
dichloramine-T	[473-34-7]	8×10^{-7}	12
N-chloroacetanilide	[539-03-7]	6.70×10^{-7}	12
N-chloroformanilide	[2596-93-2]	1.26×10^{-7}	12
chloramine-T	[127-65-1]	4.9×10^{-8}	12
monochloramine	[10599-90-3]	2.8×10^{-10}	7
albuminoid chloramines		up to 10^{-2} [c]	

[a] $[(HOCl)(RR'NH)]/(RR'NCl)$.
[b] First chlorine.
[c] Estimated.

free chlorine, antibacterial efficacy is closely related to the free chlorine present at equilibrium (16).

Reactions. The N-chloro derivatives of aniline rearrange (24–25) as follows (eq. 7):

$$\text{RNCl-C}_6\text{H}_5 \longrightarrow \text{4-Cl-C}_6\text{H}_4\text{-RNH} \tag{7}$$

where R may be a hydrogen, acetyl, benzyl, or other radical. The amine or amide formed, with the chlorine attached to the carbon atom in the ring and thus rendered unreactive, may be treated further with hypochlorous acid and a new N-chloramine formed; the new positive chlorine then migrates to the ring, preferring the ortho position. This process can be repeated until four chlorine atoms have been introduced into the ring (one para, two ortho, and one meta); the rate of rearrangement decreases as substitution in the ring proceeds. With suitable ring substitutions the shift may be largely blocked and a somewhat stable N-chloramine produced.

Other reactions may involve two chloramines and an amine or other molecule. Diazo and other chromophore groups can be formed; thus unstable chloramines sometimes give strongly colored compounds as decomposition proceeds.

Most chloramines can only be synthesized in acid solutions. Many chloramines decompose in alkaline solution, since hypochlorites tend to break the C–N bond adding an ROH or RO$^-$ and adding positive chlorine to the nitrogen. This reaction varies according to the electronic structure of the molecule, and becomes increasingly serious as the pH increases from 9 to 11. Few chloramines can withstand the action of strongly

Table 5. Dissociation Constants[a] of Cyanuric Acid, the Chlorinated s-Triazinetriones, and Related Products

NaOH Plus reacting acid	Range, pH	Product	% at 7.5 pH	Dissociation Constant, K[b] Measured	Dissociation Constant, K[b] Reported	References
(cyanuric)						
H_3CY	5–9	NaH_2CY	78	1.16×10^{-7}	1×10^{-7}	8–9
NaH_2CY	9–11	Na_2HCY[c]	0	1×10^{-11}	<0.8 or 2.5×10^{-11}	2–3, 9
Na_2HCY	>11	Na_3CY	0		< Ca 1×10^{-13}	8–9
(dichlorocyanuric)						
HCl_2CY	2–7	$NaCl_2CY$	100	1.13×10^{-4}	1×10^{-4}	8–9
(trichlorocyanuric)						
Cl_3CY	3.5–8	HCl_2CY[d]	100	1.58×10^{-4}	$<5 \times 10^{-1}$	8
$NaCl_2CY$	8.5–12	NaH_2CY[d]				
HCl_2CY[d]	5.5–9	$NaCl_2CY$	100	4×10^{-8}	none	
(cyanuric + dichloro- cyanuric)[e]						
H_2ClCY	3–8	$NaHClCY$	99	4.9×10^{-6}	2.7×10^{-6}	8
$NaHClCY$		Na_2ClCY		0.815×10^{-10}	4×10^{-10}	8
(tri or dichloro- cyanuric or Na or K dichloro- cyanurate)						
NaH_2CY[d]	9–13	Na_2HCY	0	4.5×10^{-12}	none	
$HOCl$	6–8	OCl^-	50	$3–4 \times 10^{-8}$	3.5×10^{-8}	17
$HOBr$	6–11	$NaOBr$	6	3.33×10^{-9}	2.8×10^{-9}	18–19
HOI	8–13	$NaOI$	0		2.3×10^{-11}	20
NH_2Cl	7–11	$NaOCl$	0.8	3.3×10^{-10}	$2.85 \times 10^-$	21–22
NH_2Br	7–11	$NaOBr$	0	6.3×10^{-10}		
NH_4OH			0	1.87×10^{-5} [f]	1.87×10^{-5}	23

[a] Determined by acid–base titrations at 25°C of 0.01 molar solutions with 0.1N base, using Sargent Model D Automatic Titrometer; $C_3N_3O_3$ = CY.
[b] Calculated from $K_{diss} = 10 - pK_1$ and $pK_1 = pH + \log(\text{ionized–unionized}) = pK - \log \gamma_1/\gamma_2$ where log γ_1/γ_2 is assumed to equal 0 for solutions <0.01 molar, and where γ_1 and γ_2 are the activity coefficients for the ionized and unionized species present respectively. It is further assumed that ionized–unionized species where ½ an equivalent of base has been added.
[c] H^+ liberated by addition of Ag^+.
[d] Excess NaOCl present.
[e] Mixture.
[f] $K_b = K_w/K_a = 1 \times 10^{-14}/5.37 \times 10^{-10}$ (13).

alkaline hypochlorite and, therefore, it is usually necessary to work in an acid solution. If an alkaline preparation is to be studied, however, small amounts should be used since explosive nitrogen trichloride may form from many amides.

Inorganic Chloramines

Inorganic chloramines are formed by the action of hypochlorites or hypochlorous acid on ammonia, or on other nitrogenous materials, such as urea (eqs. 8–10):

$$NH_3 + HOCl \rightarrow NH_2Cl + H_2O \quad (8)$$
$$NH_3 + 2\,HOCl \rightarrow NHCl_2 + 2\,H_2O \quad (9)$$
$$NH_3 + 3\,HOCl \rightarrow NCl_3 + 3\,H_2O \quad (10)$$

or (eq. 11):

$$NH_3 + 3\,Cl_2 \rightarrow NCl_3 + 3\,HCl \tag{11}$$

The product obtained from chlorine or hypochlorite and ammonia depends on the pH of the solution. If the ammonia or ammonium ion is in excess, only monochloramine is formed above pH 9.5, whereas nitrogen trichloride is only formed at low pH values (below pH 4.5). However, with an excess of hypochlorite, nitrogen trichloride is formed in small amounts at all pH values. It decomposes faster in the alkaline solutions than in neutral or acid ones (15,26–27).

In both concentrated and dilute solutions, hypochlorite in excess reacts with ammonia to form nitrogen gas and chloride ion. Ammonium chloride is sometimes formed as an intermediate, and this is later converted to mono- or dichloramine. With the excess of hypochlorite, some dichloramine is formed. This reaction is important in the chlorination of water supplies (see Water).

Monochloramine. Monochloramine, NH_2Cl, is an important reagent in organic syntheses (28). It is a colorless liquid with a strong odor, soluble in water, and may be prepared and handled in an ether solution free of water. This is the preferred procedure for some organic syntheses, especially for phosphorus compounds. The pure compound has a freezing point of $-66°C$. At room temperature it may explode, and is generally handled in solution.

Dichloramine. Dichloramine [3400-09-0], $NHCl_2$, is mostly of theoretical interest. Although monochloramine is stable in a water solution over a long period of time, dichloramine decomposes rapidly.

Nitrogen Trichloride. Nitrogen trichloride [10025-85-1], NCl_3, is a bright-yellow liquid with a powerful, irritating odor and limited solubility in water (mp below $-40°C$, and bp $70°C$). It is explosive and exceedingly dangerous except in very low concentrations (below about 2000 ppm) in water solution (15). The fumes are lacrimatory. Small concentrations (<7 ppm) give a distinctive and objectionable taste to drinking water.

Other Inorganic Chloramines. Other inorganic chloramines include N-chlorosulfamic acid [17172-27-9] (**1**), N,N-dichlorosulfamic acid [17085-87-9] (29), sodium N-chloroimidodisulfonate [67700-32-7] (**2**), and the trichlorimidometaphosphates (**3**) (30).

The N-chloro compounds of sulfamic acid are prepared in aqueous solution. In very dilute solution, as in swimming pools, monochlorosulfamate is quite stable, although dichlorosulfamate is not. In stronger solutions (1% available chlorine content) the decomposition of the N-chlorosulfamates is fairly rapid. The formation of the monochlorosulfamate is slow, taking hours or even days to reach final equilibrium. The dichloro compound may be formed first, and then react with the excess sulfamate to form the mono compound.

(**3**) trichlorometaphosphimic acid, monosodium salt [67651-15-14]

N-Chloroimidodisulfonate decomposes fairly rapidly in neutral or alkaline solutions. Polychlorimidometaphosphates are used only in dilute acid solutions and have enjoyed some commercial use.

Organic Chloramines

The main classes of organic chloramines are (a) chloroisocyanurates (see Cyanuric and isocyanuric acids); (b) heterocyclic chloramines with the chlorine attached to the nitrogen in the ring; (c) N-chloroamine condensation products from cyanamide derivatives; (d) N-chloroanilides (see Amines, aromatic (aniline)]; and (e) N-chlorosulfonamides.

HETEROCYCLIC COMPOUNDS

Glycolurils. Glycolurils were developed after World War II (31–32) for use in impregnating clothing (33–35) and in ointments (36–40) for protection against mustard gas and other chemical warfare agents.

Glycoluril, tetrahydroimidazo[4,5-d]imidazole-2,5-(1H,3H)-dione (4) has an eight-membered ring which is stabilized by a bridge that may include a methylene group (structures (4) and (5), respectively). Like the hydantoins (see below) the glycolurils are sparingly soluble in water.

The most important glycoluril chloramine is the unsubstituted tetrachloro compound, tetrachloroglycoluril [776-19-2], 2,4,6,8-tetrachloro-2,4,6,8-tetrazobicyclo(3.3.0)octa-3,7-dione (5) (40–41).

Tetrachloroglycoluril was marketed at one time (Diamond Alkali Company, now Diamond Shamrock), as a bleaching and sanitizing agent (41), and in briquet form (Dacsan) as a swimming pool disinfectant (42). The use of these compounds was limited by their low solubility and the fact that only a portion of the chlorine titrates as free chlorine at pool water pH. Specifications of Diamond Alkali tetrachloroglycoluril (DAC-559) are shown below (43a).

Property	Value
available chlorine, %	
free	ca 10–20
total, theoretical	101.6
solubility at 25°C, ppm	77
pH of saturated solution	4.6
acute oral toxicity to rats, mg/kg	1780

Diamond Shamrock currently markets tetrachloroglycoluril for wastewater treatment (43b).

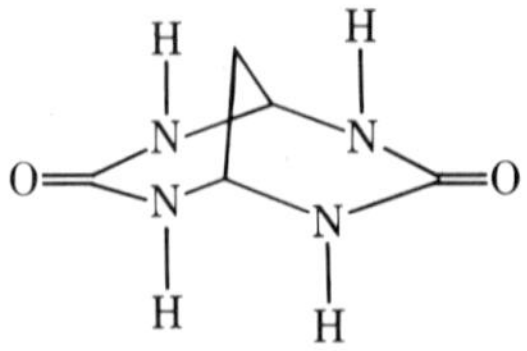

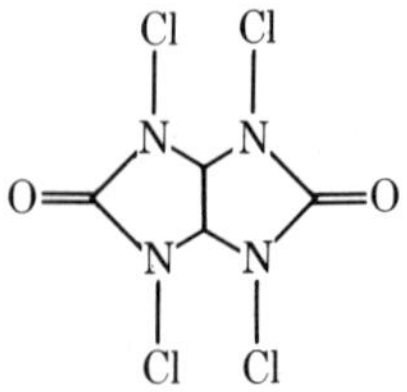

Hydantoins. Chlorinated hydantoins were introduced in the 1930s. However, because of low solubility in water and a very slow rate of solution, they did not find a large market. During World War II, 1,3-dichloro-5,5-dimethylhydantoin [*118-52-5*] (**6**) in tetrachloroethane solution was used as a decontaminating agent (44).

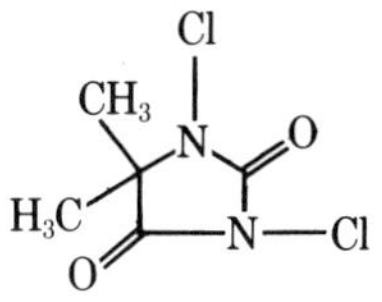

(6)

A mixture with surfactants increased the rate of solution, and blends were offered both as commercial and household bleaches. Dichlorohydantoin is no longer used in home laundering because of the changing laundering needs of synthetic fabrics. Halane (BASF Wyandotte, mol wt 197, available chlorine = 72.08%) is used in commercial bleaching where temperatures of 71°C can be used. It is also a hypochlorite source for use in cleanser products. Dichlorodimethylhydantoin is obtained in 64% yield by chlorination of 5,5-dimethylhydantoin (45). The monochlorinated compounds can be made from the dichloro and the parent compound in several solvents (46).

The 5,5-dimethyl-1,3-dichlorohydantoin has a solubility of 2100 ppm in water, whereas the 5-methyl-5-isobutyl chloroderivative dissolves to the extent of only 260 ppm. Other chloro derivatives have a lower solubility in water. They are, however, highly soluble in some organic solvents and may be used to impregnate clothing and in ointments as protection against allergies.

1,3,5-Trichloro-2,4-Dioxohexahydrotriazine. 1,3,5-Trichloro-2,4-dioxohexahydrotriazine [*67700-33-8*] is a related compound (**7**) that has been reported to be useful as a bactericide (47–48).

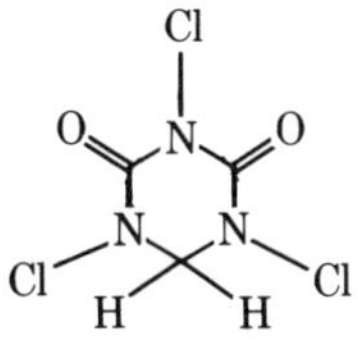

(7)

It may be prepared by the self-condensation and chlorination of methylene diurea, $CH_2(NHCONH_2)_2$.

Succinchlorimide. *N*-Chlorosuccinimide [*128-09-6*] (**8**) is a white crystalline powder, with an odor of chlorine (50–54% available chlorine). It is slightly soluble in water (1.4%), chloroform, and carbon tetrachloride. It is reported to be nontoxic and was used to some extent for water treatment in isolated areas and as a source of hypochlorite chlorine in bleach and sanitizing products.

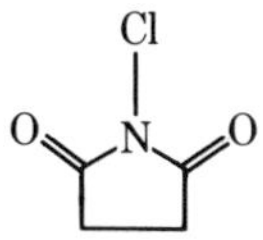

(8)

AMINE CONDENSATION COMPOUNDS

Chlorinated amine condensation compounds were once used as antiseptics and sterilizing agents.

Dichloroazodicarbonamide. Chloroazodin [502-98-7], $NH_2C(=NCl)N=NC(=NCl)NH_2$, occurs as light-yellow needles or flakes with a faint chlorine-like odor and burning taste. It decomposes explosively at 155°C, but is stable at room temperature. Decomposition is accelerated by metals. The compound is only slightly soluble in water (0.3%), chloroform, glycerol, and alcohol. It is supplied as a weak solution in glyceryl triacetate and was once used as a surgical antiseptic in the treatment of wounds (see Disinfectants and antiseptics). It is used in vulcanization of rubber (49) (see Rubber compounding).

The compound may be prepared by treating guanidine nitrate in cold, aqueous acetic acid solution with sodium hypochlorite solution; other methods are also available (50–51).

Melamines. Like many other chloramines, melamines (see Cyanamides) are most effective in acid solution. The hexachloro compound (52) is less stable than N^2, N^4, N^6-trichloromelamine [12379-38-3] (9) as the oxidation potential and activity are proportional to chlorine content (about 70% available chlorine). It used to be employed as a sterilizing agent, but has been more or less replaced by chlorinated-s-triazinetriones in the marketplace.

(9)

N-CHLOROSULFONAMIDES AND RELATED COMPOUNDS

Aromatic N-chlorosulfonamides have been marketed since World War I. In these compounds, the sulfonyl group prevents migration of the positive chlorine atom from the nitrogen atom to the ring. The monochloramines and their salts may be represented as follows (eq. 12):

$$R-C_6H_4-\underset{O}{\overset{O}{S}}-N(H)-Cl + NaOH \rightarrow R-C_6H_4-\underset{O}{\overset{O}{S}} \rightarrow \bar{N}-Cl \quad Na^+ \qquad (12)$$

Only the aromatic derivatives have been commercialized although aliphatic derivatives have been studied (53).

Chloramine-T. Chloramine-T (10) occurs as white crystals. The solid may be dried without decomposition at 90–100°C, whereas solutions decompose slowly in air and under the influence of light. The available chlorine content is 24–25%; it is soluble in water and insoluble in organic solvents. Chloramine-T does not liberate chlorine from acid solutions, and does not chlorinate substances that are attacked by hypochlorite.

$$\left[CH_3-\underset{}{\bigcirc}-SO_2N(Cl)Na \right] \cdot 3H_2O$$
(10)

It was introduced in 1916 as a germicide for the treatment of wounds, and was the first chloramine to meet market acceptance. It is toxic when introduced into the bloodstream but harmless if dilute solutions, such as mouthwash, are accidentally ingested. Chloramine-T may be blended with other bactericides for better action (54).

Chloramine-T is obtained by treating a water solution of p-toluenesulfonamide with sodium hypochlorite. Further chlorination yields dichloramine-T. This compound gradually decomposes in air, yielding chlorine. Treatment with acids also liberates chlorine. Dichloramine-T forms pale, greenish-yellow crystals, mp 71–75°C, dec 160°C. It is almost insoluble in water, but soluble in chloroform, paraffins, etc, and has been used as a topical dressing (55a) and as antivesicant ointment (55b).

Chloramine-B. Chloramine-B [127-52-6], $C_6H_5SO_2NClNa$, forms white crystals that decompose on heating to about 170°C. It is soluble in water and alcohol, stable in dry air, and may be blended with acid and neutral (56) salts to give stable mixtures that have been used to disinfect dairies. The available chlorine content is 29.5%. Further chlorination yields dichloramine-B, $C_6H_5SO_2NCl_2$, which has been used to deodorize and bleach certain oils.

Halazone. p-(N,N-Dichlorosulfamoyl)benzoic acid [80-13-7], $HOOCC_6H_4SO_2NCl_2$, is a white crystalline powder with a characteristic odor that decomposes at about 195°C; it is slightly soluble in water and chloroform. The available-chlorine content is 48–52%. It was issued for emergency sterilization of water during World War II.

Halazone is prepared from p-toluenesulfonamide by oxidizing the methyl group to a carboxyl and treating the resulting compound with hypochlorite.

N-Chloro-N-Methyl-p-Toluenesulfonamide. N-Chloro-N-methyl-p-toluenesulfonamide [2350-10-9] melts at 82°C; similar substitution products have been studied (57). 3-(Dichlorosulfamoyl)phthalic acid [67700-34-9], $(HOOC)_2C_6H_3SO_2NCl_2$, melts at 107–109°C and forms a trihydrated sodium salt. It has been reported to have useful disinfectant properties (58), but is currently not of commercial significance.

Related compounds may be prepared from N-alkyl or N-aryl sulfamates. Thus, sodium N-chloro-N-cyclohexylsulfamate [67700-35-0] has been recommended as a sanitizing agent (59).

Other Organic Chloramines

N-Chloroimines, prepared from the oxidant of certain redox systems, are stable (51). Thus, quinone dichlorimide [637-70-7] (N,N'-dichloroquinonediimine), $ClN=C_6H_4=NCl$, may be dissolved in concentrated sulfuric acid or cold fuming nitric acid and reprecipitated by dilution. It is not hydrolyzed by strong alkali. With potassium iodide and starch, hydrochloric acid slowly gives a blue color. Other compounds resist strong reagents; ethyl N-chlorobenzimidate [1006-93-5] ($C_6H_5C=NClO(C_2H_5)$) resists strong alkali but liberates chlorine slowly in strong hydrochloric acid. N-Chlorodibenzylamine [42393-65-7] is likewise unaffected by boiling alkali. Such apparently inactive compounds are more reactive in nonaqueous solvents.

Monoalkylamines form dichloramines that are oily and fairly stable at room temperature, but unstable or explosive on heating. Representatives are *N,N*-dichloroethylamine [24948-83-2], bp 88°C; *N,N*-dichloromethylamine [7651-91-4], bp 59°C; and *N,N*-dichloropropylamine [10218-84-5], bp 117°C. Dialkylamines give unstable chloramines with strong oxidizing power. Diarylamines give unstable chloramines.

Several hundred chloramines have been tested for usefulness in preparing available-chlorine solutions. Useful compounds are derived from amines with only one hydrogen on the nitrogen atom which may be replaced by positive chlorine. For example, *N*-acetyl-*p*-toluenesulfonamide, *N*-acetylbenzenesulfonamide, *N*-benzoylbenzenesulfonamide, as well as several substituted *p*-nitroacetanilides (60).

Many potentially useful compounds yield a weak solution and liberate an amine which decomposes the solution (Hofmann degradation). For example, chloramines derived from acetamide, phthalimide, benzamide, ethylenediamine, and several of the urethanes (61).

Chain-type molecules, based on bisbiguanidines and known as *N*-chlorophenyldiguanidino compounds (62–73), were tried out for laundry use in the 1950s. *N,N*-Dichloro-1,8-diformamido-*p*-methane [67700-31-6] (64) was recommended as a reagent for preparing synthetic resins. Chlorinated dicyandiamide preparations have been made (65). Several *N*-chloro-2-substituted imidazolines have been prepared for impregnating clothing for protection against mustard gas, etc (66). *N*-Chloro-*tert*-alkyl cyanamides are oils that may be used in ointments and for a few other uses, but not in dry package blends (67). 2-*N*-Chloro-4-thiazolines (68a) have been reported as useful reagents in certain organic syntheses. 1,3-Dichlorotetrahydroquinazoline-2,4-dione [23767-45-5] has been reported to be a useful laundry bleach.

Urea and its monosubstituted derivatives are destroyed by reaction with alkaline hypochlorites which form chloramines of urea and hydrolyze the N–C bond (see Urea and derivatives). If hydrolysis takes place before complete replacement of the hydrogens on the nitrogen with positive chlorine, mono- and dichloramines are formed. Substitution of one hydrogen on both nitrogens improves stability (eqs 13–15) (35, 68b).

$$NH_2-CO-NH_2 + 2\ HOCl \rightarrow NHCl-CO-NHCl + 2\ H_2O \qquad (13)$$

$$NH_2-CO-NH_2 + 2\ HOCl \rightarrow 2\ NH_2Cl + H_2CO_3 \qquad (14)$$

$$NHCl-CO-NHCl + 2\ HOCl \rightarrow NCl_2-CO-NCl_2 + 2\ H_2O \qquad (15)$$

If hydrolysis takes place after formation of 1,1,3,3-tetrachlorourea [67700-36-1], then nitrogen trichloride will be formed (eq. 16):

$$NCl_2-CO-NCl_2 + 2\ HOCl \rightarrow 2\ NCl_3 + H_2CO_3 \rightarrow 2\ NCl_3 + CO_2 + H_2O \qquad (16)$$

Mixed Chlor- and Bromamines. Active halogen compounds, similar to the products described above but containing bromine or both bromine and chlorine, have been prepared. For example, *N*-bromo-*N*-chloro-5,5-dimethylhydantoin [6079-88-2], and *N*-bromo-*N*-chloro-5,5-diphenylhydantoin, sodium *N*-bromo-*N*-chlorocyanurate [20367-88-8] (active halogen 43.7%), and *N*-bromo-*N*-chloro-*p*-toluene sulfonamide [27824-67-5] (active halogen 40.6%), and the bromochlorodihydrochloride of triethylenediamine, (38.8% active chlorine) (70). (Note that available halogen has twice the value of active halogen.)

The following glycolurils were also prepared: *N,N*-dibromo-*N*-monochloro-3a,6a-dimethylglycoluril [67700-37-2], *N,N*-dibromo-*N,N*-dichloro-3a,6a-dimethylglycoluril [67700-38-3], N-bromo-N,N-dichloro-3a,6a-diphenyl-glycoluril [67700-39-4], and N-tribromo-N-monochloro-3a,6a-substituted glycolurils.

Bromochlorohydantoins and bromochloroglycolurils at one time were used as swimming pool disinfectants (72–74). Formation of active halogen from these compounds was complete in less than 15 seconds, in contrast to formation from compounds containing only chlorine which took 240–480 s (74). Sodium dichlorocyanurate [2893-78-9] was converted with potassium bromide to potassium dibromocyanurate [15114-46-2] which had an available-bromine content of 94% (98% theor). Tribromocyanurate [17497-85-7] and sodium dibromocyanurate [15114-34-8] have also been reported.

Analysis

Identification of compounds containing available chlorine in dilute solutions has become more difficult in recent years because of the large number of products. Chloroisocyanuric acid derivatives are attacked by strong hypochlorite in alkaline solution, liberating NCl_3 easily detected by odor. This test is seldom satisfactory for chlorinated hydantoins, because of their slow rate of solution (75). Hydrogen peroxide in alkaline solution attacks hypochlorites vigorously, evolving oxygen. Chlorinated s-triazinetriones and cyanuric acid can be identified in solution since they form a melamine salt at pH 6 which can be used for quantitative determination (15).

Glycoluril and its chloro derivatives can be determined in water systems by color development with potassium ferricyanide in alkaline solutions (36). With the chlorinated s-triazinetriones the attack is slower, and the liberated cyanuric acid forms a white precipitate which is soluble in caustic soda. Frequently, preparations that contain chloroisocyanurates, which had been kept in a closed bottle, have a chloramine odor. After ignition with a small amount of sodium carbonate, N-chlorosulfonamides give a test for sulfates.

All chloramines oxidize iodide in acid solution and chlorine is liberated. With very inactive compounds, such as albuminous chloramines, the release may be very slow. Such chloramines are apt to interfere in tests distinguishing chlorine dioxide from other forms of available chlorine.

A series of tests (77) can be used to distinguish between some chloramines and hypochlorites. The simplest of these is the manganous chloride test. Hypochlorites in solution very rapidly turn manganous chloride solution brown, whereas the chloramines do not. If chloramines, including the chlorinated s-triazinetriones are present in an alkaline mixture, a white manganous precipitate may form that dissolves on addition of dilute acetic acid. This test is quite generally applied.

Similarly, the DPD (dimethyl-p-phenylenediamine) test distinguishes among hypochlorite and the various forms of chloramine that occur in water and wastewater (4).

All of the chlorine contained in the chlorinated s-triazinetriones is free chlorine when measured amperometrically (78), or with the DPD method (79). Test kits based on this method are available and can be used at poolsides or in the field to measure NCl_3, free chlorine, and mono-, as well as dichloramines in potable, pool, or waste water (80).

Economic Aspects and Uses

Chlorinated s-triazinetriones are used as swimming pool disinfectants, bleaches, sanitizers, and in dishwasher detergents and cleanser products. Annual worldwide production is estimated at 40,500 metric tons of which an estimated 22,000 t is used for swimming pools (see Water). The 1977 price was $2.40/kg (in carloads, fob factory). These products were sold under the Monsanto ACL and FMC Corporation CDB registered trademarks and include trichloro-s-triazinetrione [87-90-1], sodium dichloro-s-triazinetrione anhydrous [2893-78-9], sodium dichloro-s-triazinetrione dihydrate [51580-86-0], and potassium dichloro-s-triazinetrione [2244-21-5]. Products that make pesticidal claims require labels registered by the Environmental Protection Agency (81).

Chlorinated hydantoins are mostly used for bleaching; annual worldwide production is estimated at 1000 t; the 1977 price (carloads, fob factory) was $1.72/kg.

BIBLIOGRAPHY

"Chloramines and Chloroamines" in *ECT* 1st ed., Vol. 3, pp. 664–676, by H. L. Robson, Mathieson Chemical Corporation; "Chloramines and Chloroamines" in *ECT* 2nd ed., Vol. 4, pp. 908–928, by H. L. Robson, Olin Mathieson Chemical Corporation.

1. H. D. Dakin and co-workers, *Proc. R. Soc. London Ser. B*, 232 (1916).
2. R. E. Connick and Y. T. Chia, *J. Am. Chem. Soc.* **81,** 1280 (1959).
3. J. C. Morris, *J. Phys. Chem.* **70,** 3798 (1966).
4. E. Wattie and C. T. Butterfield, *Public Health Rep. U.S.* **59,** 1661 (1944).
5. Ibid., **61,** 157 (1946).
6. J. Gardner, *Water Research*, Vol. 7, Pergamon Press, Great Britain, 1973, pp. 823–833.
7. R. E. Corbett, W. S. Metcalf, and F. G. Soper, *J. Chem. Soc.*, 1927 (1953).
8. A. P. Brady, K. M. Sancier, and G. Sirine, *J. Am. Chem. Soc.* **85,** 3101 (1963).
9. J. P. Busscher and co-workers, *Chim. Anal. Paris* **54**(2), 69 (1972).
10. B. M. Israel, *Hydrolysis of Some Organohalogenating Agents*, Univ. Wisconsin, 1962.
11. F. G. Soper and G. F. Smith, *J. Chem. Soc.*, 138 (1928).
12. F. G. Soper, *J. Chem. Soc.* **125,** 1899 (1924); **127,** 98 (1925).
13. S. Glasstone, *Textbook of Physical Chemistry*, 2nd ed., D. Von Nostrand Co., New York, 1961.
14. E. J. Cohn, *J. Am. Chem. Soc.* **49,** 173 (1922).
15. G. D. Nelson, *Swimming Pool Disinfection with Chlorinated Cyanurates*, Special Report 6862, Monsanto Company St. Louis, Mo., Mar. 8, 1969.
16. J. E. O'Brien, J. C. Morris, and J. N. Butler in A. Rubin, ed., *J. Ann Arbor Sci.*, 333 (1974).
17. J. W. Ingham and J. Morrison, *JCS London*, 1200 (1933).
18. L. Farkas and M. Lewis, *J. Am. Chem. Soc.* **72,** 5766 (1950).
19. M. Kiese and A. B. Hastings, *J. Am. Chem. Soc.* **61,** 1291 (1939).
20. Y. Chia, *USAEC Report UCRL*, 1958, pp. 8311.
21. R. E. Corbett, W. S. Metcalf, and F. G. Soper, *JCS London* (II), 1927 (1953).
22. M. Chapin, *J. Am. Chem. Soc.* **51,** 2112 (1929).
23. J. S. Sconce *Chlorine, Its Manufacture, Properties and Uses*, ACS Monograph 154, Reinhold Publishing Co., New York, 1962.
24. F. D. Chattaway, *J. Chem. Soc.* **87,** 145 (1905).
25. K. J. P. Orton, F. G. Soper, and G. Williams, *J. Chem. Soc.*, 998 (1928).
26. A. T. Palin, *Water Water Eng.* **54**(10), 151; (11), 189; (12) 248 (1950).
27. E. W. Moore, *Water Sewage Works* **98,** 130 (1951).
28. H. H. Sisler and co-workers, *J. Am. Chem. Soc.* **81,** 2982 (1959).
29. J. A. McCarthy, *J. New Engl. Water Works Assoc.* **74,** 166 (1960).
30. U.S. Pats. 2,796,321 and 2,796,322 (June 18, 1957), M. C. Taylor.
31. U.S. Pat. 2,654,763 (Oct. 1953), H. B. Adkins (to United States represented by Secretary of the Navy).

32. U.S. Pat. 2,777,856 (Jan. 15, 1957), A. J. Stokes (to United States under Title 35, Section 266).
33. U.S. Pat. 2,628,174 (Feb. 1963), A. J. Stokes and H. W. Carhart (to United States granted under Title 35, Section 266).
34. U.S. Pat. 2,649,389 (Aug. 1963), J. W. Williams (to United States granted under Title 35, Section 266).
35. U.S. Pat. 3,003,971 (Oct. 1961), W. W. Prichard (to E. I. du Pont de Nemours & Co., Inc.).
36. U.S. Pat. 2,638,434 (May 12, 1953), H. B. Adkins (to United States as represented by the Secretary of the Navy).
37. U.S. Pat. 2,725,335 (Nov. 1955), W. A. Lazier, W. J. Peppel, and P. L. Salzbert (to United States as represented by the Secretary of War).
38. U.S. Pat. 2,885,305 (May 1959), J. C. Speck, Jr. (to United States granted under Title 35, Section 266).
39. U.S. Pat. 3,003,971 (Oct. 1961), W. W. Prichard (to E. I. du Pont de Nemours & Co., Inc.).
40. U.S. Pat. 3,002,975 (Oct. 1961), F. B. Slezak (to Diamond Alkali Co.).
41. F. B. Slezak, A. Hirsch, and I. Rosen, *J. Org. Chem.* **25,** 660 (1960).
42. *Bozeclor 70,* Technical Bulletin Farbwerke Hoechst, June 1970.
43. (a) *DAC 559 Tetrachloroglycoluril,* Technical Bulletin, Diamond Alkali Co.; (b) Sanuril,® *Wastewater Chlorinator,* Technical Bulletin Diamond Shamrock, 1977.
44. C. H. Greenwalt, *Chem. Corps J.,* 9 (1948).
45. O. O. Orazi and O. A. Orio *An. Asoc. Quin Argentina* **41,** 153 (1953).
46. U.S. Pat. 2,430,233 (Nov. 1947), P. L. Magill (to E. I. du Pont de Nemours & Co., Inc.).
47. U.S. Pat. 3,040,044 (June 1962), A. Hirsch and F. B. Slezak (to Diamond Alkali).
48. U.S. Pat. 3,035,055 (May 15, 1962), F. B. Slezak and H. A. McElravy (to Diamond Alkali).
49. U.S. Pat. 2,171,901 (Sept. 5, 1940), N. R. Wilson and A. J. Lang (to Rare Metals Products Co.).
50. F. C. Schmelkes and H. C. Marks, *J. Am. Chem. Soc.* **56,** 1610 (1934).
51. U.S. Pats. 1,958,370 and 1,958,371 (May 1934), F. C. Schmelkes (to Wallace and Tiernan Products).
52. U.S. Pats. 2,184,883 and 2,184,886, and 2,184,888 (Dec. 1939), I.E. Muskat and A. G. Chenicek (to Pittsburgh Plate and Glass Co.).
53. A. Sturzenegger, *Ueber Einige Aliphatische Sulfochloroamide* (thesis), Juris-Verlag, Zürich, 1948.
54. U.S. Pat. 2,898,264 (Aug. 1949), F. J. C. Weber (to Colgate-Palmolive Co.).
55. (a) G. L. Jenkins and W. H. Hartung, *The Chemistry of Organic Medicinal Compounds,* John Wiley & Sons, Inc., New York, 1943, p. 417; (b) U.S. Pat. 2,618,584 (Nov. 1952), R. L. Evans and E. G. McDonough (to R. L. Evans Assoc.).
56. U.S. Pat. 2,393,716 (Jan. 1946), E. W. Smith (to Solvay Process Co.).
57. H. C. Marks, O. Wyss, and F. B. Stranskov, *J. Bacteriol.* **49,** 299 (1945).
58. O. K. Kononenko, *J. Appl. Chem. USSR Engl. Transl.* **19,** 411 (1946).
59. U.S. Pat. 2,288,976 (July 1942), M. Sveda (to E. I. du Pont de Nemours & Co., Inc.).
60. U.S. Pat. 1,716,014 (June 1929), M. C. Taylor (to Mathieson Alkali Works).
61. G. R. Elliot, *J. Chem. Soc.* **121,** 202 (1922).
62. U.S. Pat. 2,684,924 (July 1954), R. L. Rose and G. Swain (to Imperial Chemical Industries, Ltd.).
63. Ger. Pat. 1,001,254 (Sept. 24, 1959), (to Henkel et Cie).
64. U.S. Pat. 2,653,169 (Sept. 1953), M. D. Hurwitz and R. W. Auten (to Rohm & Haas Co.).
65. U.S. Pat. 2,841,474 (July 1, 1958), R. R. Dorsett (to Mangels, Herold Co., Inc.).
66. U.S. Pat. 2,678,930 (May 18, 1954), H. A. Weldon (to United States as represented by the Secretary of the Army).
67. U.S. Pat. 2,686,203 (Aug. 1954), I. Hechenbleikner (to American Cyanamid Co.).
68. (a) U.S. Pat. 2,626,950 (Jan. 1953), J. T. Gregory (to B. F. Goodrich Co.); (b) P. C. de la Saulniere, *Ann. Chim.* **17,** 353 (1942).
69. U.S. Pat. 3,007,876 (Nov. 1961), J. R. Schaeffer (to Procter and Gamble Co.).
70. Can. Pat. 730,743 (Mar. 1966), L. O. Paterson (to Drug Research).
71. U.S. Pat. 3,071,591 (Jan. 1963), L. O. Paterson (to Drug Research).
72. U.S. Pat. 2,779,764 (Jan. 29, 1957), L. O. Paterson (to Drug Research).
73. U.S. Pat. 2,868,787 (Jan. 13, 1959), L. O. Paterson (to Drug Research).
74. Can. Pat. 776,623 (Jan. 1964), L. O. Paterson (to Drug Research).
75. R. E. Connick, *J. Am. Chem. Soc.* **69,** 1509 (1947).
76. R. S. DePablo, *J. Am. Water Works Assoc.* **58**(3), 379 (1966).
77. H. W. Van Urk, *Chem. Week Blad* **26,** 9 (1929).
78. *Standard Methods of Test for Residual Chlorine in Water D1253-68* in *1973 Annual Book of ASTM Standards,* American Society for Testing and Materials, Phila., Pa., 1973.

79. *Residual Chlorine—DPD Colorimetric Method 409F* in *Standard Methods for the Examination of Water and Wastewater*, 14th ed. American Public Health Association, 1976.
80. W. A. Taylor Co., Baltimore Maryland; Hach Chemical Co., Ames, Iowa; and CVB LaMotte Co. Chestertown, Md.
81. *Code of Federal Regulations Part II EPA Pesticides Programs*, Data Requirements, Feb. 2, 1976, pp. 7218–7376.

General Reference

J. A. Szilard *Bleaching Agents and Techniques*, Noyes Data Corp. Park Ridge, N.J., 1973.

G. D. NELSON
Monsanto Industrial Chemicals Co.

CHLORAMPHENICOL. See Antibiotics—Chloramphenicol.

CHLORINATED BIPHENYLS. See Diphenyl and terphenyls.

CHLORINE. See Alkali and chlorine products.

CHLORINE OXYGEN ACIDS AND SALTS

Chlorine monoxide, hypochlorous acid, and hypochlorites, 580
Chlorous acid, chlorites, and chlorine dioxide, 612
Chloric acid and chlorates, 633
Perchloric acid and perchlorates, 646

CHLORINE MONOXIDE, HYPOCHLOROUS ACID, AND HYPOCHLORITES

Oxidation States

Chlorine has positive oxidation states in oxychlorine compounds since its appreciable electronegativity (2.83) on the Allred-Rochow scale is exceeded by that of oxygen (1). All oxychlorine compounds are strong oxidants because the transfer of electrons to the orbitals of electronegative chlorine is favored in reactions with compounds of less electronegative elements. Chlorine oxides and oxo-acids exhibit the lack of stability expected of compounds having bonds between two strongly electronegative elements. The decomposition reactions of these compounds are, therefore, always energetic and violent in many cases (Table 1). The chemical properties of chlorine oxides and oxo-acids indicate a trend toward greater thermodynamic and

Table 1. Chlorine Oxides and Oxo-Acids

Formula	CAS Registry No.	Oxidation state	Stability
			Oxides
Cl$_2$O	[7791-21-1]	+1	brownish-yellow gas at 25°C; explodes when heated or sparked
Cl$_2$O$_3$	[17496-59-2]	+3	explodes below 0°C
ClO$_2$	[10049-04-4]	+4	yellowish gas; explodes at >6.7 kPa[a]
Cl$_2$O$_4$	[27218-16-2]	+1 and +7	pale yellow liquid; decomposes to Cl$_2$, O$_2$, ClO$_2$, and Cl$_2$O$_6$
Cl$_2$O$_6$	[12442-63-6]	+6	red liquid; decomposes to ClO$_2$ + O$_2$
Cl$_2$O$_7$	[12015-53-1]	+7	oily liquid; can be distilled under reduced pressure
			Acids
HClO (or HOCl)	[7790-92-3]	+1	very weak acid, $K_a = 2.9 \times 10^{-8}$; cannot be concentrated
HClO$_2$	[13898-47-0]	+3	decomposes rapidly at 25°C; K_a ca 10^{-2}
HClO$_3$	[7790-93-4]	+5	decomposes slowly at conc ca 40% and 25°C
HClO$_4$	[7601-90-3]	+7	can be isolated

[a] To convert kPa to mm Hg, multiply by 7.5.

kinetic stability with increasing oxidation state. It is possible to isolate concentrated perchloric acid and its anhydride, Cl$_2$O$_7$, whereas concentrated hypochlorous, chlorous, and chloric acids have not been obtained. The reduction potentials of the oxo-acids exhibit a similar trend in that the strongest oxidants have chlorine in its lower states of oxidation. Compounds of chlorine having intermediate oxidation states exhibit a strong tendency to disproportionate.

The oxo-anions of chlorine are weaker oxidants than the corresponding acids. Since they are also more stable, it is not too difficult to isolate certain salts of those acids that can be obtained only in dilute solutions. Hypochlorites and chlorites are hydrolyzed in aqueous solution since HOCl and HClO$_2$ have acid dissociation constants of ca 10^{-8} and 10^{-2}, respectively; however, chloric and perchloric acids are fully ionized in aqueous solutions.

Chlorine monoxide, hypochlorous acid, and ionic hypochlorites are compounds in which chlorine is in the +1 oxidation state. In addition to Cl$_2$O and HOCl, other covalent compounds where univalent chlorine is bonded to oxygen are the alkyl, aryl, and acyl hypochlorites and other positive chlorine compounds such as ClOSF$_5$, ClOSO$_2$F, ClONO$_2$, and ClOClO$_3$.

CHLORINE MONOXIDE (DICHLORINE OXIDE)

Chlorine monoxide is the anhydride of hypochlorous acid; the two compounds are readily interconvertible via the equilibrium: Cl$_2$O + H$_2$O $\rightleftharpoons$ 2 HOCl. It has an endothermic heat of formation and is thus thermodynamically unstable with respect to decomposition into chlorine and oxygen. Chlorine monoxide typifies the chlorine oxides as a highly reactive and explosive compound with strong oxidizing properties. Nevertheless, it can be handled safely and is generated on a commercial scale and converted to hypochlorous acid for the production of calcium hypochlorite.

Physical Properties

At ordinary temperatures chlorine monoxide is a brownish-yellow gas resembling bromine. It is condensed to a red–brown liquid whose vapor pressure over the range of 173–288 K is given by the equation: $\log p$ (kPa) = $6.995 - 1373/T$ [$\log p$ (mm Hg) = $7.87 - 1373/T$] (2). Its bp, 2.0°C, and ΔH_{vap}, 25.9 kJ/mol (6.2 kcal/mol), were calculated from vapor pressure data. The mp of Cl_2O is -120.6°C (3). It readily dissolves in water to give a solution of hypochlorous acid containing a small equilibrium amount of Cl_2O. At -9.4°C the saturation solubility is 143.6 g Cl_2O/100 g of H_2O. Henry's constant (using molarity in place of mole fraction) for the vapor–liquid equilibrium at 3.46°C is $K = [Cl_2O]_{(g)}/[Cl_2O]_{(aq)}$ = 14.23 kPa/(M) [106.7 mm Hg/(M)] (4). Theromodynamic properties of Cl_2O are: $\Delta H°$ of solution, 36.6 kJ/mol (8.74 kcal/mol), $\Delta H_f°$, 80.3 kJ/mol (19.2 kcal/mol), $\Delta G_f°$, 97.9 kJ/mol (23.4 kcal/mol), $S°$, 265.9 J/(mol·K) [63.60 cal/(mol·K)], and C_p, 45.40 J/(mol·K) [10.85 cal/(mol·K)] (5). Microwave spectroscopic measurements have established that Cl_2O is a nonlinear molecule of C_{2v} symmetry (6). The uv spectrum shows a maximum near 260 nm with a molar absorptivity of ca 610 cm^{-1} (7).

Chemical Properties

Explosion of gaseous chlorine monoxide can be initiated by spark or heat. It is shock sensitive in the liquid phase. The minimum explosive concentration of gaseous Cl_2O in oxygen at 23°C and 101 kPa (1 atm) in faint daylight is 23.5 mol % (8). Explosions are mild in the 25–30% range but become progressively more violent at higher concentrations. The effect of various diluent gases on the explosion limit at various total pressures has been determined in the complete absence of light. The extrapolated explosion limit in oxygen is ca 33 mol % (9). The threshold for spark initiated decomposition of pure Cl_2O is 0.53 kPa (4.0 mm Hg).

Chlorine monoxide decomposes thermally and photochemically into Cl_2 and O_2. The photolytic decomposition, initiated by cleavage into ClO, O, and Cl free radicals, is sensitized by Cl_2 (10). Significant concentrations of ClO_2 are generated from Cl_2O by controlled thermal decomposition or irradiation of CCl_4 solutions. Photolysis of matrix-isolated Cl_2O yields ClClO, (ClO)$_2$, and ClO (11). Gaseous Cl_2O decomposes thermally in 12–24 h at 60–100°C, but at 150°C the reaction is complete in only a few minutes; above 110°C the reactions terminate in explosion (12). The decomposition is preceded by an induction period inversely proportional to the starting Cl_2O concentration. Several mechanisms, some involving chains, have been proposed for this heterogeneous reaction (13–14).

Chlorine monoxide reacts with a variety of inorganic substances, eg, its reaction with N_2O_5 is a convenient route to $ClNO_3$ (15). The latter is also formed from other nitrogen oxides. The stoichiometry depends on the reaction phase, and the key step appears to be: $ClO + NO_2 \rightarrow ClNO_3$ (16). Chlorine dioxide is an intermediate in certain reactions of Cl_2O as in the preparation of chloryl fluoride and in the formation of ionic complexes from SO_3 and AsF_5 (17–18). The transformation of metal halides into oxy-

Table 2. Reactions of Cl$_2$O With Inorganic Compounds

Reaction	Reference
Cl$_2$O + 3 F$_2$ $\xrightarrow{\text{CsF, }-78°\text{C}}$ ClF$_3$O + ClF$_3$ >80%	20
2 Cl$_2$O + AgF$_2$ $\xrightarrow{65-70°\text{C}}$ ClO$_2$F + AgF + 3/2 Cl$_2$	21
5 Cl$_2$O + 3 AsF$_5$ $\xrightarrow{-78 \text{ to } -50°\text{C}}$ 2 ClO$_2$AsF$_6$ + AsOF$_3$ + 4 Cl$_2$	17
Cl$_2$O + N$_2$O$_5$ $\xrightarrow{0°\text{C}}$ 2 ClNO$_3$ 90%	15
4 Cl$_2$O + 3 SO$_3$ $\xrightarrow{\text{CFCl}_3, -27°\text{C}}$ (ClO)(ClO$_2$)[S$_3$O$_{10}$] + 3 Cl$_2$	18
Cl$_2$O + TiBr$_4$ → TiOBr$_2$ + Br$_2$ + Cl$_2$[a]	22
3 Cl$_2$O + 2 BCl$_3$ $\xrightarrow{-78°\text{C}}$ B$_2$O$_3$ + 6 Cl$_2$[b]	23
2 Cl$_2$O + P(NCl$_2$)$_3$ → PO$_2$Cl + 3 NCl$_3$	24

[a] Similar reactions occur with TiCl$_4$, SnBr$_4$, SnCl$_4$, and PbCl$_4$ (19,25).
[b] AlBr$_3$ similarly gives Al$_2$O$_3$ but with AlCl$_3$ there is no reaction.

halides by Cl$_2$O apparently involves hypochlorite intermediates (19). Some reactions of Cl$_2$O are summarized in Table 2.

The organic chemistry of Cl$_2$O has not been extensively explored. Some representative reactions are shown in Table 3. The liquid-phase reaction of Cl$_2$O with alkanes exhibits the following overall stoichiometry: 2 RH + Cl$_2$O → 2 RCl + H$_2$O. The photoinduced reaction proceeds by a free radical chain mechanism, giving a mixture of chlorination products (27). In contrast, the dark reaction results in high selectivity for tertiary chlorination (28). The proposed mechanism for the reaction of Cl$_2$O with cyclohexene involves direct addition to the olefin, forming the intermediate *trans*-2-chlorocyclohexyl hypochlorite, and a simultaneous molecule-induced homolysis of Cl$_2$O to produce a number of products arising from radical and ionic pathways (29). Chlorine monoxide reacts with COF$_2$ in the presence of CsF to give CF$_3$OCl (26). In contrast, F$_2$O produces CF$_3$OOOCF$_3$ via the intermediate CF$_3$OOF and illustrates the reversal of polarity of the halogen–oxygen bond in the case of the more electronegative fluorine (35).

Chlorine monoxide may be used for the preparation of trichloroisocyanuric acid from cyanuric acid (36) and sodium dichloroisocyanurate dihydrate from sodium cyanurate monohydrate (37) in the absence of a solvent (see Cyanuric and isocyanuric acids).

Preparation

Gaseous chlorine monoxide is conveniently generated by reaction of chlorine gas with mercuric oxide in a packed tubular reactor:

$$2 \text{ Cl}_2 + n \text{ HgO} \rightarrow \text{HgCl}_2 \cdot (n-1)\text{HgO} + \text{Cl}_2\text{O}$$

Table 3. Reaction of Cl₂O With Organic Compounds

Reaction	Reference
$F_2CO + Cl_2O + CsF \xrightarrow{-20°C} CF_3OCl + CsOCl$ >95%	26
$C_3H_7Cl \xrightarrow[CCl_4, h\nu]{Cl_2O} CH_3CH_2CHCl_2 + CH_3CHClCH_2Cl + ClCH_2CH_2CH_2Cl$ 43% 42% 15%	27
$C_2H_5Cl \xrightarrow[40°C, 67 h]{Cl_2O, CCl_4} CH_3CHCl_2 + CH_3CCl_3 + ClCH_2CH_2Cl + ClCH_2CHCl_2$ 16% 81% 2% 1%	28
$CCl_2=CHCl \xrightarrow[CCl_4]{Cl_2O} CCl_3CHCl_2 + CCl_3CHO + (CCl_3CHCl)_2O$	29
cyclohexene $\xrightarrow{Cl_2O}$ chlorocyclohexene + chlorohydrin (Cl, OH)[a]	30
$C_6H_5OH \xrightarrow[CCl_4]{Cl_2O} o\text{-}ClC_6H_4OH + p\text{-}ClC_6H_4OH$ 22% 65%	31
$t\text{-BuOOH} + Cl_2O \xrightleftharpoons[<-30°C]{CCl_3F} t\text{-BuOOCl} + HOCl \xrightarrow{t\text{-BuOOH}} t\text{-BuOOO-}t\text{-Bu} + 2 HOCl$	32
$C_2H_5NH_2 + Cl_2O \rightarrow C_2H_5NHCl + HOCl \rightarrow C_2H_5NCl_2 + H_2O$	33
cyanuric acid + 3 Cl₂O → trichloro derivative + NCl₃ + 2 H₂O	34

[a] Main products; numerous other minor products are also formed.

High chlorine conversion and Cl₂O yield are obtained (38–40). An exothermic, competing side reaction results in formation of mercuric chloride and oxygen. High yields are favored by cooling the reactor, using pure chlorine diluted with an inert gas (air or N₂) and admixing the HgO with an inert material, eg, crushed glass tubing, sand, pumice, crushed brick, or kieselguhr. The HgO must be thoroughly dry. The spent HgO can be regenerated by treatment with aqueous caustic, filtering, washing with water, and drying at 120–130°C. If liquid Cl₂O is desired, the exit gases can be passed through a trap cooled with dry ice and chloroform. In the static reaction of Cl₂ with HgO at −78°C, high yields of Cl₂O are obtained, and with excess Cl₂ the reaction is (41):

$$2\ Cl_2 + HgO \xrightarrow{Cl_2} Cl_2O + HgCl_2$$

A commercial method for producing Cl₂O involves reaction of Cl₂ with moist sodium carbonate in either a tower or a rotating tubular reactor (42).

$$2\ Cl_2 + 2\ Na_2CO_3 + H_2O \rightarrow Cl_2O + 2\ NaHCO_3 + 2\ NaCl$$

$$2\ Cl_2 + 2\ NaHCO_3 \rightarrow Cl_2O + 2\ CO_2 + 2\ NaCl + H_2O \rightleftharpoons 2\ HOCl + 2\ CO_2 + 2\ NaCl$$

$$Na_2CO_3 + CO_2 + H_2O \rightarrow 2\ NaHCO_3$$

Chlorine monoxide has been prepared in high yields by the reaction of Cl_2, diluted with moist air, with activated soda ash (43).

A solution of Cl_2O can be prepared by addition of HgO to a solution of Cl_2 in CCl_4 followed by filtration to remove the basic mercuric chloride. A dilute solution containing a few percent Cl_2O is obtained in 90% yield and is stable when stored in the dark under refrigeration. Moist soda ash can be used in place of HgO but the yield is reduced to about 40% (38).

Chlorine monoxide can also be prepared from concentrated HOCl solutions by vacuum distillation (44), stripping with air (45–46), or treatment with anhydrous $Ca(NO_3)_2$ (47).

Analysis

Chlorine monoxide can be quantitatively determined alone or in admixture with Cl_2 by iodometry; the acid consumed is a direct measure of Cl_2O (48). The reactions involved are:

$$Cl_2O + 4\,I^- + 2\,H^+ \rightarrow 2\,I_2 + 2\,Cl^- + H_2O$$

$$Cl_2 + 2\,I^- \rightarrow I_2 + 2\,Cl^-$$

Gas chromatography has also been employed for analysis of Cl_2O (41).

Uses

Chlorine monoxide is an intermediate in the manufacture of calcium hypochlorite. It has been used in sterilization for space applications (49) (see Sterile techniques). Its use in the preparation of chlorinated solvents (50) and chloroisocyanurates has been described. Chlorine monoxide has been shown to be effective in bleaching of pulp (qv) and textiles (51–52).

HYPOCHLOROUS ACID

Hypochlorous acid is a highly reactive, relatively unstable compound that is known primarily in aqueous solution. Although a solid dihydrate exists and HOCl has been observed in the vapor phase, the pure compound has not been isolated. It is the most stable and the strongest of the hypohalous acids and is one of the most powerful oxidants among the chlorine oxyacids. It is an intermediate in the manufacture of hypochlorites. Generated *in situ* via chlorine hydrolysis, it is an intermediate in the production of chlorohydrins and chloroisocyanurates. It is generally believed to be the active species that kills bacteria and other microorganisms in municipal water treatment or in swimming pool sanitation when Cl_2, hypochlorites or chloroisocyanurates are used (53) (see Water; Bleaching agents).

Physical Properties

Hypochlorous acid is a weak acid with a dissociation constant of 2.90×10^{-8} at 25°C. The temperature dependence is given by $pK_a = 0.0253\,T + 3000/T - 10.0686$ (54). It is about ten times weaker than carbonic acid. The dissociation energy of HOCl from the latter study is 16.3 kJ (3.9 kcal) which is in reasonable agreement with the value of 15.1 kJ (3.6 kcal) based on the heat of neutralization (55). Hypochlorous acid solutions contain small equilibrium amounts of Cl$_2$O which probably imparts their yellow color. The equilibrium constant at 0°C is 3.55 × 10^{-3}.

$$2\ HOCl_{(aq)} \rightleftharpoons Cl_2O_{(aq)} + H_2O_{(l)}$$

At 0°C a 5% solution contains <0.05% Cl$_2$O and a 25% solution contains <1% Cl$_2$O.

The temperature–concentration diagram is shown in Figure 1. Below the eutectic point (A, 11.7 mol % Cl$_2$O) the solid phase in equilibrium with the liquid phase is ice and at higher concentrations the solid phase is HOCl.2H$_2$O. Liquid chlorine monoxide and aqueous HOCl are only partially miscible. Mixing stoichiometric amounts of Cl$_2$O and water does not give pure HOCl but results, instead, in separation of two liquid phases which, on freezing, give the solid compound HOCl.2H$_2$O.

Hypochlorous acid and chlorine monoxide coexist in the vapor phase (56–59). Vapor pressure measurements of aqueous HOCl solutions show that HOCl is the main chlorine species in the vapor phase over ≤1% solutions (60–61), whereas at higher concentrations, Cl$_2$O becomes dominant (62).

Carbon tetrachloride extracts chlorine monoxide (but not HOCl) from concentrated HOCl solutions. For the equilibrium, Cl$_2$O$_{(aq)}$ ⇌ Cl$_2$O$_{(CCl_4)}$, the partition coefficient at 0°C is 2.22 (44,63).

The thermodynamic properties of aqueous HOCl are: $\Delta H_f° = -120.9$ kJ/mol (-18.9 kcal/mol), $\Delta G_f° = -79.9$ kJ/mol (-19.1 kcal/mol), and $S° = 142.3$ J/(mol·K) [34 cal/(mol·K)] (5). The structure of HOCl in the vapor phase, determined by infrared and microwave spectroscopy, shows $d_{O-H} = 0.097$ nm, $d_{Cl-O} = 0.169$ nm, and $\angle_{H-O-Cl} = 104.8°$ (57–58). It closely resembles water in bond angle (104.7°) and O–H bond length (0.096 nm), and Cl$_2$O in Cl–O bond length (0.170 nm). The uv absorption

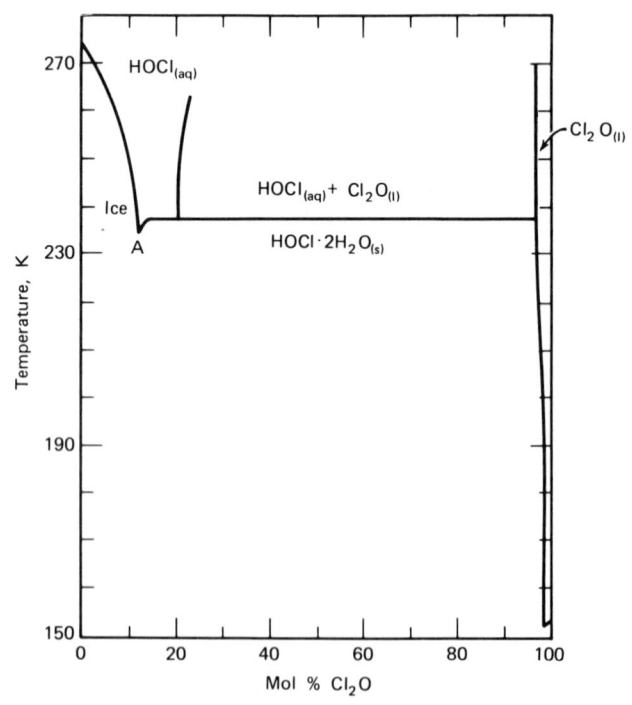

Figure 1. Temperature–concentration diagram for the system chlorine monoxide–water (3).

spectrum of aqueous HOCl shows a maximum at 235 nm with a molar absorptivity of 100 cm^{-1} (54). The densities of chloride-free solutions of HOCl at 10°C as a function of molarity are (M, d$_4^{10}$ g/cm^3): 1, 1.02; 2, 1.05; 4, 1.09; and 6, 1.14.

The standard electrode potentials for the reduction of HOCl in acid solution are given below (64).

$$HOCl + H^+ + e^- \rightleftharpoons \tfrac{1}{2} Cl_2 + H_2O \quad E° = 1.63 \text{ V}$$

$$HOCl + H^+ + 2e^- \rightleftharpoons Cl^- + H_2O \quad E° = 1.49 \text{ V}$$

Chemical Properties

Dilute hypochlorous acid solutions are quite stable if pure, especially if kept cool and in the dark. For example, at 0°C the decomposition rate of a $1M$ solution is only about 0.3%/d. At 20°C the decomposition rate is about tenfold higher. Decomposition occurs in two ways:

$$2\ HOCl \rightarrow 2\ HCl + O_2 \xrightleftharpoons{2\ HOCl} 2\ Cl_2 + O_2 + 2\ H_2O$$

$$\begin{cases} 2\ HOCl \rightarrow [HClO_2] + HCl \\ 2\ HOCl + [HClO_2] \rightarrow HClO_3 + HCl \xrightleftharpoons{HOCl} HClO_3 + Cl_2 + H_2O \end{cases}$$

Chlorous acid is an intermediate in the formation of HClO$_3$. Kinetic studies have shown that both decomposition pathways increase with concentration, temperature (65), and exposure to light (66–69) and are pH-dependent (70). The first reaction is also accelerated by catalysts, and the second reaction is favored by the presence of other electrolytes, notably chloride ion (71).

Hypochlorite solutions oxidize numerous inorganic substrates. However, since HOCl and ClO$^-$ coexist over a wide pH range, kinetic studies are necessary to establish their respective roles; both species are seldom active in the same reaction. The kinetic parameters for various reactions are summarized in Table 4. There is isotopic evidence for complete oxygen atom transfer to the reducing agent in the oxidation of NO$_2^-$, but with SO$_3^{2-}$, some sulfate is apparently formed via ClSO$_3^-$ (79–80). The oxidation of CN$^-$ is an important reaction in the treatment of wastewater and proceeds via the intermediate ClCN (81):

$$ClO^- + CN^- + H_2O \rightarrow ClCN + 2\ HO^- \rightarrow NCO^- + Cl^- + H_2O$$

Cyanate can be further oxidized by HOCl to nitrogen and bicarbonate along with small amounts of N$_2$O and NCl$_3$. Hypochlorous acid reacts with peroxide with evolution of oxygen via the postulated intermediate formation of peroxyhypochlorous acid:

$$HOCl + H_2O_2 \rightarrow H_2O + [HOOCl] \rightarrow H_2O + HCl + O_2$$

Addition of HOCl to ammonia results in stepwise formation of chloramines:

$$NH_3 + HOCl \rightarrow NH_2Cl + H_2O \xrightarrow{HOCl} NHCl_2 + 2\ H_2O \xrightarrow{HOCl} NCl_3 + 3\ H_2O$$

Dichloramine decomposes with regeneration of some HOCl by the overall reaction (82):

$$2\ NHCl_2 + H_2O \rightarrow N_2 + HOCl + 3\ HCl$$

Table 4. Kinetic Parameters for Reactions of Hypochlorous Acid and Hypochlorite

Oxidizing[a] agent	Reducing agent	Oxidation[b] product	Log k[c], 25°C	ΔE, kJ[d]	Log A	$S^{\ddagger}$, J/K[d]	Reference
OCl⁻	IO₃⁻	IO₄⁻	−5.04	109.2	14.1	16.8	72
OCl⁻	OCl⁻	ClO₂⁻	−7.63	103.8	10.6	−52.3	73
OCl⁻	ClO₂⁻	ClO₃⁻	−5.48	87.0	9.8	−66.9	73
OCl⁻	SO₃²⁻	SO₄²⁻	3.93	31.4	9.4	−73.2	72
HOCl	NO₂⁻	NO₃⁻	0.82	27.2	5.6	−146.4	72
HOCl	HCOO⁻	H₂CO₃	−1.38	28.0	3.5	−186.2	75
HOCl	Br⁻	BrO⁻	3.47	18.8	6.8	−123.4	76
HOCl	OCN⁻	HCO₃⁻, N₂	−0.55	63.2	10.5	−52.3	77
HOCl	HC₂O₄⁻	CO₂	1.20	62.8	12.2	−20.9	78
HOCl	I⁻	IO⁻	8.52	3.8	9.2	−77.4	74

[a] In rate controlling step.
[b] HOCl and ClO⁻ are reduced to Cl⁻ in these reactions.
[c] k is a second-order rate constant [mol/(L·s)]⁻¹. A is the pre-exponential factor, and ΔE is the activation energy in the Arrhenius equation $k = Ae^{-\Delta E/RT}$, $\Delta S^{\ddagger}$ is the entropy of activation.
[d] To convert J to cal, divide by 4.184.

A small amount of NO₃⁻ is also formed. This reaction forms the basis of breakpoint chlorination which is important in disinfection of municipal water supplies, in swimming pool sanitation, and in wastewater treatment (see Chloramines).

Hypochlorous acid undergoes a variety of reactions with organic substances including C- and N-chlorination, oxidation, addition, and ester formation. Its most important industrial reaction is with olefins to form chlorohydrins via aqueous chlorination.

$$Cl_2 + H_2O \rightleftharpoons HCl + HOCl \xrightarrow{\text{C=C}} \text{C(Cl)—C(OH)} + HCl$$

Dichlorides and ethers are the main by-products in this reaction. Pure HOCl reacts very slowly with olefins; the reaction is accelerated by the presence of acid and hypochlorite ion (83). Kinetic evidence indicates the participation of Cl₂O in the addition of HOCl to olefins (84). Addition of HOCl to acetylenic compounds produces dichloro ketones (85) (see Acetylene-derived chemicals).

$$RC\equiv CH + 2\,HOCl \longrightarrow RC(O)CHCl_2 + H_2O$$

Hypochlorous acid is also formed *in situ* in the preparation of trichloroisocyanuric acid from chlorine and trisodium cyanurate (86).

$$(NaOCN)_3 + 3\,Cl_2 \xrightarrow{H_2O} (ClNCO)_3 + 3\,NaCl$$

Chloroisocyanurates can also be prepared from cyanuric acid and preformed HOCl (87–89).

Preparation

Chloride-containing solutions of HOCl are readily prepared by the reaction of chlorine with aqueous base. The reaction involves the initial rapid hydrolysis of Cl_2:

$$Cl_2 + H_2O \rightleftharpoons HOCl + HCl$$

A saturated solution of Cl_2 (0.091 M) at 101 kPa (1 atm) and 25°C will be 33% hydrolyzed to HOCl. When Cl_2 is added to strong bases such as caustic or lime the reaction occurs stepwise:

$$Cl_2 + 2\,HO^- \rightarrow ClO^- + Cl^- + H_2O \xrightarrow{Cl_2} 2\,HOCl + 2\,Cl^-$$

With weak bases such as $NaHCO_3$ or $CaCO_3$, hypochlorite is not formed as an intermediate product. Although the presence of chloride increases the decomposition rate of HOCl, the solutions can be utilized for synthetic and manufacturing purposes, especially if used promptly and kept cold and relatively diluted. Buffering of the solutions with bicarbonate reportedly improves stability.

Chloride-free solutions of HOCl can be obtained by several routes. A commonly employed method is from gaseous Cl_2O generated from Cl_2 and either HgO or soda ash. The exit gas is bubbled through cold water using good agitation until the desired strength is obtained. Dissolved Cl_2 or CO_2 can be sparged with air. Alternatively a Cl_2O solution in CCl_4 can be extracted with H_2O to give Cl^-- and Cl_2-free HOCl solutions of up to 5 M (38). More concentrated solutions can be obtained by use of liquid Cl_2O. Chlorination of an aqueous slurry of bismuth oxide reportedly gives insoluble bismuth oxychloride. However, the reaction is slow and will not produce high concentrations of HOCl (90).

$$Bi_2O_3 + 2\,Cl_2 + H_2O \rightarrow 2\,BiOCl \downarrow + 2\,HOCl$$

In contrast, chlorination of aqueous HgO slurries followed by filtration of the basic mercuric chloride, gives HOCl solutions contaminated with significant amounts of $HgCl_2$.

Dilute (1–3%), chloride-containing solutions of either HOCl, hypochlorite, or aqueous base can be stripped in a column against a current of Cl_2, steam, and air at 95–100°C and the vapors condensed giving virtually Cl^--free HOCl solutions of higher concentration in yields as high as 90% (91–93). Distillation of more concentrated solutions requires reduced pressure, lower temperature, and shorter residence times to offset the increased decomposition rates.

Preparation of aqueous HOCl substantially free of chloride ion from either aqueous Cl_2 or HOCl-salt solutions has been accomplished by electrodialysis (qv) using semipermeable membranes (94).

Organic solutions of HOCl can be prepared in near quantitative yield (98–99%) by extraction of chloride-containing aqueous solutions of HOCl with polar solvents such as ketones, nitriles, and esters (95). These organic solutions of HOCl have been used to prepare chlorohydrins (96) and are especially useful for preparation of water insoluble chlorohydrins. Hypochlorous acid in methyl ethyl ketone has also been used to prepare $Ca(OCl)_2$ by reaction with CaO or $Ca(OH)_2$ (97).

Analysis

The analysis of HOCl is usually carried out using acidic KI as described under Cl$_2$O. The reaction is:

$$HOCl + 2\,I^- + H^+ \rightarrow I_2 + Cl^- + H_2O$$

Any chlorine present liberates iodine without consumption of acid. Small concentrations of HOCl can be determined spectrophotometrically via formation of I$_3^-$ which has an absorption maximum at 353 nm (98). Hypochlorous acid in the vapor phase has been observed chromatographically.

Uses

Hypochlorous acid, preformed or generated *in situ* from chlorine and water, is employed in the manufacture of chlorohydrins from olefins, en route to epoxides, and in the production of chloramines (qv), especially chloroisocyanurates from cyanuric acid (see Cyanuric and isocyanuric acids).

METAL HYPOCHLORITES

Some hypochlorites, either as solutions or solids, are much more stable than hypochlorous acid, and because of their high oxidation potential and ready hydrolysis to the parent acid, find wide use in bleaching and sanitizing applications. One of the novel uses of hypochlorites was for disinfection of Apollo Eleven on its return from the moon (99).

The only known stable solid hypochlorites are those of lithium [*13840-33-0*], calcium [*7778-54-3*], strontium [*14674-76-1*], and barium [*13477-10-6*]. Sodium hypochlorite [*7681-52-9*] does not have good stability. Potassium hypochlorite [*7778-66-7*] exists only in solution and attempts to isolate the solid have resulted in decomposition (100). Magnesium hypochlorite [*10233-03-1*] has not been isolated (101) but it forms two stable basic hypochlorites. Impure silver (102) and zinc hypochlorite (103) compositions have been prepared.

Calcium hypochlorite is the major commercial solid hypochlorite; it is produced on a large scale and marketed as a 65–70% product containing sodium chloride and water as the main diluents. It is also manufactured to a much smaller extent in the form of bleaching powder, primarily in less-developed nations. Lithium hypochlorite is produced on a small scale and is sold as a 35% product for specialty applications. Small amounts of sodium hypochlorite are employed in the manufacture of crystalline chlorinated trisodium phosphate.

Physical Properties

The solubilities of Li, Na, and Ca hypochlorites in H$_2$O at 25°C are 40, 45, and 21.4%, respectively. Solubility isotherms in water at 10°C have been determined for the following systems: Ca(OCl)$_2$–CaCl$_2$, NaOCl–NaCl, and Ca(OCl)$_2$–NaOCl (104). The densities of approximately equimolar solutions of NaOCl and NaCl are given in several product bulletins (105–106). The uv absorption spectrum of hypochlorite ion shows a maximum at 292 nm with a molar absorptivity of 350 cm^{-1} (54). Heats of formation, $\Delta H°$ kJ/mol (kcal/mol), of alkali and alkaline earth hypochlorites are:

LiOCl$_{(aq)}$ −383.3 (−91.6), NaOCl$_{(c)}$ −346.0 (−82.7), NaOCl$_{(aq)}$ −304.0 (−72.65), Ca(OCl)$_{2(aq)}$ −753.1 (−180.0), Sr(OCl)$_{2(aq)}$ −738.5 (−176.5), and Ba(OCl)$_{2(aq)}$ −811.7 (−194.0) (107). Thermodynamic properties of the hypochlorite ion are: $\Delta H_f°$ −107.1 kJ/mol (−25.6 kcal/mol), $\Delta G_f°$ −36.8 kJ/mol (−8.8 kcal/mol), and $S°$ 41.8 J/(mol·K) (10 cal/(mol·K)) (5). The reduction potential of ClO$^-$ in basic solution is: ClO$^-$ + H$_2$O + 2 e ⇌ Cl$^-$ + 2 OH$^-$, $E°$ = 0.89 V (64).

Chemical Properties

Hypochlorites yield HOCl when treated with stoichiometric amounts of acid and are converted to Cl$_2$ when excess HCl is used. They react quantitatively with iodide in acid media liberating iodine and with hydrogen peroxide liberating oxygen. These two reactions are employed in the analysis of hypochlorites. The oxidation of various inorganic anions by hypochlorite has been studied kinetically. It is a strong oxidant capable of oxidizing MnO$_4^{2-}$ to MnO$_4^-$, IO$_3^-$ to IO$_4^-$, and Fe^{3+} to FeO$_4^{2-}$. Its reaction with ammonia to form chloramine is the basis of the manufacture of hydrazine (qv).

$$NH_3 + NaOCl \rightarrow NH_2Cl + NaOH \xrightarrow{NH_3} N_2H_4 + NaCl$$

When ammonia, hydrazine, or amido compounds such as urea are treated with excess NaOCl they are converted to N$_2$.

Anhydrous hypochlorites are oxidized to chlorates by Cl$_2$O (108).

$$M(OCl)_n + 2n\ Cl_2O \rightarrow M(ClO_3)_n + 2n\ Cl_2$$

The rate of chlorate formation decreases in the order Na > Ba > Sr > Li > Ca. In the presence of gaseous chlorine, dry hypochlorites decompose in two ways:

$$M(OCl)_n + n\ Cl_2 \rightarrow MCl_n + n\ Cl_2O$$

$$2\ M(OCl)_n + 2n\ Cl_2 \rightarrow 2\ MCl_n + 2n\ Cl_2O \xrightarrow{M(OCl)_n} 2\ MCl_n + M(ClO_3)_n$$

Strontium, lithium, and calcium hypochlorites react primarily by the first path and sodium hypochlorite mainly by the second. In the presence of moisture, chlorate formation is the predominate reaction in all cases.

Although hypochlorite solutions are much more stable than HOCl, they are subject to decomposition which is influenced by concentration, ionic strength, pH, temperature, light, and impurities. Decomposition occurs in two ways:

$$2\ NaOCl \rightarrow 2\ NaCl + O_2$$

$$2\ NaOCl \xrightarrow{k_1} NaCl + NaClO_2 \xrightarrow[NaOCl]{k_2} 2\ NaCl + NaClO_3$$

The rate controlling step to chlorate is the bimolecular formation of chlorite which reacts rapidly with hypochlorite. The temperature dependence of the rate constants is expressed by the equations: $k_1 = 2.1 \times 10^{12} \cdot e^{-103.8/RT}$ and $k_2 = 3.2 \times 10^{11} \cdot e^{-87.0/RT}$ L/(mol·s) (109). The uncatalyzed decomposition to oxygen is bimolecular with an activation energy of 111.3 kJ/mol (26.6 kcal/mol). Although it is much slower than chlorate formation, it is susceptible to catalysis by trace metal impurities (110–113). The most powerful catalysts for accelerating the decomposition to oxygen are Co, Ni, and Cu; Fe and Mn are much less effective. Although these metals do not catalyze chlorate formation, iridium apparently does (114). Certain oxides promote the catalytic activity of other metallic oxides.

The stability of solid calcium hypochlorite is a function of its moisture, lime, and impurity content as well as the temperature and humidity at which it is stored. Anhydrous $Ca(OCl)_2$ containing about 1% H_2O loses 1–3% available chlorine per year, whereas partially hydrated material will generally lose somewhat more. In addition to chlorate formation, which is the main mode of decomposition, release of some oxygen and chlorine also occurs.

When calcium hypochlorite is heated in a stream of N_2 during dta it undergoes dehydration at ca 65–70°C; further heating results in an exothermic decomposition (Fig. 2) at ca 200–210°C forming $CaCl_2$, O_2, and a small amount of $Ca(ClO_3)_2$. Heating $Ca(OCl)_2$ under conditions that do not allow complete dehydration will result in decomposition at a lower temperature giving a higher proportion of $Ca(ClO_3)_2$.

The available chlorine, in a hypochlorite, is a measure of the oxidizing power of its active chlorine expressed in terms of elemental chlorine; one hypochlorite ion is equivalent to one Cl_2 molecule. Thus, pure calcium hypochlorite has an available chlorine content of 2·mol wt Cl_2/mol wt $Ca(OCl)_2$ or $(2·70.9/143.0) \times 100\% = 99.2\%$. A 5% solution of $Ca(OCl)_2$ contains 4.96% or ca 53 g/L available chlorine (see Chloramines).

Hypochlorite ion acts as a chlorinating and oxidizing agent toward organic compounds. In addition to its use in the preparation of carboxylic acids by the haloform oxidation and amines by the Hoffmann rearrangement it has numerous interesting and useful synthetic applications (115). Aromatically bound methylene groups in acetyl substituted aromatics are oxidized by NaOCl to carboxylic acids (116). Acetylenic protons are displaced to give chloroacetylenes (117–119). Cyclopentadiene and indene are readily chlorinated by hypochlorite to perchlorocyclopentadiene (120) and 1,1,3-trichloroindene (121), respectively. Aliphatic oximes (122) and primary and secondary nitro compounds (123) are converted to geminal chloro nitro alkanes. Symmetrical dialkyl hydrazines and methylenediamine sulfate are oxidized to azo compounds (124) and diaziridine (125), respectively. o-Nitroanilines are oxidized in good yields by alkaline hypochlorite to benzofurazan oxides. 2,4-Dinitroaniline, on treatment with NaOCl in alkaline methanol, is converted to 5-chloro-4-methoxybenzofurazan-1-oxide via a haloalkoxy substitution reaction (126); the haloalkoxy reaction has been applied to additional heterocycles, eg, 6-nitroanthroxanic acid (127). Unsaturated aldehydes, ketones, and nitriles are epoxidized in one step in high yield

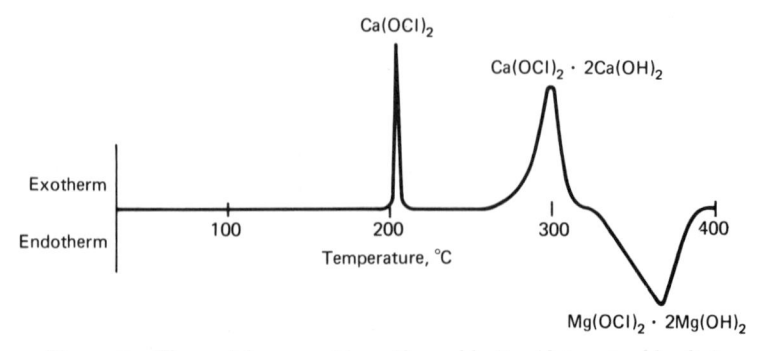

Figure 2. Thermal decomposition of hypochlorites (determined by dta).

via nucleophilic attack by hypochlorite ion (128–131) (see Epoxidation). Hypochlorite readily chlorinates phenols to mono-, di-, and tri-substituted compounds (132). Degradation of chlorophenols to aliphatic acids by excess hypochlorite is employed in wastewater treatment since as little as 0.1 ppb of chlorophenols can impart a definite off-flavor to the water supply (133).

Preparation

Hypochlorite solutions are prepared in near quantitative yield by chlorination of caustic or a lime slurry.

$$2\ NaOH + Cl_2 \rightarrow NaOCl + NaCl + H_2O$$

$$2\ Ca(OH)_2 + 2\ Cl_2 \rightarrow Ca(OCl)_2 + CaCl_2 + 2\ H_2O$$

These solutions are employed in concentrations of 3–15% in various bleaching and sanitizing applications. In the preparation of a solid hypochlorite, particularly $Ca(OCl)_2$, the presence of the coproduct $CaCl_2$ is objectionable because it prevents formation of large easily filterable crystals of $Ca(OCl)_2$, and owing to its hygroscopic nature, it impedes drying and has a deleterious effect on product stability. Industrial processes, therefore, are designed to eliminate or minimize $CaCl_2$ during processing and in the product (see Calcium compounds). This is accomplished in one process via formation of a triple salt:

$$Ca(OH)_2 \xrightarrow[(b)\ -15°C]{(a)\ NaOCl,\ Cl_2(H_2O)} Ca(OCl)_2 \cdot NaOCl \cdot NaCl \cdot 12H_2O$$

and the recovered triple salt is treated with chlorinated lime slurry:

$$2\ Ca(OCl)_2 \cdot NaOCl \cdot NaCl \cdot 12H_2O + Ca(OCl)_2 + CaCl_2 \rightarrow 4\ Ca(OCl)_2 + 4\ NaCl + 24\ H_2O$$

Since the salt remains in solution much of it is removed in filtration of the final $Ca(OCl)_2$ paste; the dried product contains about 70% available chlorine.

In another process, a $Ca(OCl)_2$ slurry prepared by chlorination of a suspension of hydrated lime and dibasic calcium hypochlorite is filtered. The $CaCl_2$ content of the filter cake may be reduced by addition of NaOCl solution.

$$CaCl_2 + 2\ NaOCl \rightarrow Ca(OCl)_2 + 2\ NaCl$$

The available chlorine value in the filtrate is recovered by precipitation of a dibasic salt.

$$Ca(OCl)_2 + 2\ Ca(OH)_2 \rightarrow Ca(OCl)_2 \cdot 2Ca(OH)_2$$

The problem of $CaCl_2$ formation is circumvented by use of hypochlorous acid essentially free of chloride and chlorine and treating it with hydrated lime.

$$Ca(OH)_2 + 2\ HOCl \rightarrow Ca(OCl)_2 + 2\ H_2O$$

Calcium hypochlorite can also be prepared by reaction of solid lime or $CaCl_2$ with Cl_2O; chlorate formation is a competing side reaction (134–135).

$$Ca(OH)_2 + Cl_2O \rightarrow Ca(OCl)_2 \cdot H_2O$$

$$CaCl_2 + 2\ Cl_2O \rightarrow Ca(OCl)_2 + 2\ Cl_2$$

Several modifications of the preparation of neutral $Ca(OCl)_2 \cdot 2H_2O$ do not involve intermediates. In a batch process, NaOH is chlorinated in the presence of recycled neutral Ca(OCl)$_2$ mother liquor. After separation of salt, lime slurry is added and chlorinated (136). The precipitated Ca(OCl)$_2$.2H$_2$O is recovered by filtration. In another version, classification of the Ca(OCl)$_2$ slurry gives a Ca(OCl)$_2$-rich fraction which is filtered and the filtrate is recycled along with the NaCl-rich fraction to the first chlorinator (137). Also, 50% caustic and solid slaked lime are used in the second chlorination.

An earlier patent (138) of a continuous cocrystallization process also utilizes classification. Lime slurry containing caustic and Ca(OCl)$_2$ mother liquor is chlorinated under reduced pressure (to remove the heat of reaction), and the resulting slurry is separated in a classifier into Ca(OCl)$_2$- and NaCl-rich regions from which slurry is withdrawn to obtain Ca(OCl)$_2$ filter cake and solid salt, respectively. A batch modification of this process was patented by Nippon Soda (139).

In another continuous process, lime slurry is chlorinated in the presence of NaOH, NaOCl, and Ca(OCl)$_2$ mother liquor (140). After concentration, the resulting slurry is filtered and the cake is dried. A portion of the filtrate is treated with caustic, the recovered lime is recycled, and the mother liquor is used to prepare the required NaOCl solution in an evaporator–chlorinator which, after separation of salt, is sent to the main reactor. In a slightly modified version, a lime purification step is added (141).

Two solvent processes for preparation of Ca(OCl)$_2$ have been described. In one, a CCl$_4$ solution of t-BuOCl is allowed to react with a thin lime slurry and the aqueous phase, a solution of Ca(OCl)$_2$, is evaporated to a product with a purity of greater than 95% (142–143). In the other, a solution of HOCl in methyl ethyl ketone reacts with either CaO or Ca(OH)$_2$ (97). Following filtration, the residual solvent in the product is removed under vacuum.

Materials of Construction. Because of the corrosive nature of moist chlorine and hypochlorite solutions, chemically resistant materials are necessary to prevent metallic contamination of the products and to ensure proper functioning of equipment. Chlorination vessels and reactors can be constructed from FRP (fiberglass reinforced polyester) or carbon steel with a suitable resistant coating or liner made of rubber, saran, PVC, or poly(vinylidene fluoride) (Kynar) (see Vinyl polymers; Vinylidene polymers). Although dry chlorine can be conveyed in carbon steel, moist chlorine and hypochlorite solutions require the use of solid plastic pipe or plastic-lined steel pipe. Chlorination coils or sparge tubes have been fabricated from silver, lead, glass, or PVC. However, all of these have drawbacks such as high cost, fragility, or deterioration and have been supplanted by solid Kynar pipe. Valves with corrosion resistant fittings such as Teflon-lined ball valves or Kynar-lined diaphragm valves are widely employed. Hypochlorite solutions are pumped by centrifugal pumps with resistant linings such as polypropylene. Heat exchangers or cooling coils are typically made of titanium.

Germicidal Activity

The germicidal activity of aqueous chlorine is caused primarily by hypochlorous acid. Although the detailed mechanism by which HOCl kills bacteria and other microorganisms has not been established, sufficient experimental evidence has been obtained to strongly suggest that the mode of action involves penetration of the cell wall followed by reaction with the enzymatic system. The efficiency of destruction is affected by the temperature, time of contact, pH, and type and concentration of organisms (144).

Although the hypochlorite ion itself is a relatively poor disinfectant (145) in comparison to hypochlorous acid, it serves as a reservoir of the latter by hydrolysis:

$$ClO^- + H_2O \rightleftharpoons HOCl + HO^-$$

The relative concentrations of ClO$^-$ and HOCl are a function of pH: at 25°C they are equal at 7.54 pH. This is within the usual operating pH range of 7.2–7.6 for swimming-pool water. For adequate germicidal activity, the free available chlorine (HOCl + ClO$^-$) should be maintained at 1.0 ppm. The hypochlorite ion absorbs uv light in the range of 250–350 nm and therefore can be decomposed to O_2, Cl^-, and ClO_3^- by sunlight. Addition of cyanuric acid decreases this decomposition by formation of a chloroisocyanurate that reduces the concentrations of HOCl and ClO$^-$ but hydrolyzes rapidly to maintain the equilibria:

[structural equation: chlorinated cyanurate + H$_2$O ⇌ cyanurate + HOCl]

$$HOCl + HO^- \rightleftharpoons H_2O + ClO^-$$

The equilibrium constant for the first reaction is 2.40 × 10^{-6} (146). When cyanuric acid is used in conjunction with a hypochlorite for sanitizing swimming pool water, the free available chlorine is usually kept within 1.0–1.5 ppm.

HYPOCHLORITE SOLUTIONS

Sodium Hypochlorite (Liquid Bleach). Commercial strength liquid bleach, used by industries, laundries, and in swimming pool sanitation, contains 12–15% available chlorine and is sold in 3.8- and 7.6-L polyethylene bottles and 23–57-L carboys, 205-L drums, and tank trucks of about 3-kL capacity and greater. Household bleach contains about 5% available chlorine and is sold in 1–5.7-L polyethylene containers. Shipping is limited within a short radius of the plant because of transportation costs. Liquid bleach for use in pulp or textile bleaching is usually prepared on site at concentrations of 30–40 g/L of available chlorine.

Calcium Hypochlorite (Bleach Liquor). Bleach liquor is a solution of calcium hypochlorite and calcium chloride containing some dissolved lime. The available chlorine content can vary but is typically about 30–35 g/L. It is used primarily in pulp bleaching. Although it has some disadvantages in comparison to sodium hypochlorite, it has considerable use because of lower cost. No production statistics are available since it is invariably prepared on site and consumed captively.

Manufacture

Sodium Hypochlorite. Sodium hypochlorite solution is usually prepared by chlorination of NaOH. Cooling is necessary since excessive temperatures can result in chlorate formation, representing loss of yield, and may contribute to lower product stability. Since chlorate formation is also a function of concentration, the upper limit of 30°C usually employed with <6% NaOCl is reduced to about 20°C for 10–15%

concentrations. The chlorination of caustic is exothermic and liberates 103.1 kJ/mol (24.65 kcal/mol) when gaseous chlorine is used. Liquid chlorine feed is preferred since advantage can be taken of its heat of vaporization, 16.5 kJ/mol (3.94 kcal/mol), to reduce the cooling requirement. Since 50% caustic is normally employed, and diluted to the desired strength, its heat of dilution, 20.3 kJ/mol (4.86 kcal/mol), is an additional cooling load. The dilution is controlled by conductivity or density. For preparation of dilute NaOCl solutions, cooling of the caustic is sufficient to absorb the heat of reaction, but for higher concentrations cooling of the hypochlorite solution is also necessary. In batch preparation of dilute hypochlorite, ice can be added directly to the caustic to provide the necessary precooling. Although manual titration is probably still employed in some small-scale operations for determining the extent of chlorination, monitoring of the oxidation potential is not only more convenient but mandatory in continuous operation. The oxidation potential of the solution is dependent on the half cell reaction $Cl^- + 2\ HO^- \rightleftharpoons ClO^- + H_2O + 2\ e^-$, and its magnitude is given by the equation $E = E_o - 0.0296 \log_{10} [[ClO^-]/[HO^-]^2 [Cl^-]]$ where E_o (−0.89 V) is the standard electrode potential at 25°C when concentrations are at unit activity. Although the oxidation potential is essentially a function of pH, its measurement is preferred to pH measurement because of its faster response and lower maintenance requirements (147). The oxidation potential is measured with a potentiometer using an ORP (oxidation–reduction potential) cell consisting of a calomel or silver reference electrode and a platinum indicating electrode. Agitation of the hypochlorite solution is necessary to minimize local overchlorination and also to reduce any time lag in response of the ORP cell. Air has been used in batch operation for this purpose; however, sufficient agitation can be obtained from the turbulence of the Cl_2 feed itself or by recirculating through an external loop. Complete absorption of the Cl_2 is ensured by maintaining an adequate depth of solution in the chlorinator. In many continuous systems the Cl_2 and caustic are allowed to react in a mixing zone, which may be a packed tubular reactor, with the hypochlorite solution going either to a surge tank or to storage. The conversion of NaOH is usually limited to about 92–94% for ease of control and to avoid overchlorination. A small excess of NaOH is necessary for adequate storage stability. In certain applications, such as textile-bleaching, little or no free caustic alkalinity is desirable. In bleaching of pulp, a pH adjustment may be made to obtain the necessary degree of activity. Prior to bottling, any suspended solids are allowed to settle or they may be removed by filtration.

Sodium hypochlorite is also prepared electrolytically using small diaphragmless or membrane cells, with a capacity of 1–150 kg/d of equivalent Cl_2, which produce a dilute hypochlorite solution of 1–3 and 5–6 g/L from seawater and brine, respectively (see Chemicals from brine). They are employed in sewage and wastewater treatment and in commercial laundries, large swimming pools, and aboard ships. These electrolyzers are produced or distributed by a number of companies.

Calcium Hypochlorite. The mechanics of the preparation of bleach liquor by chlorination of lime slurry is only slightly different than that of sodium hypochlorite and the heat liberated per mol of chlorine is approximately the same. Continuous systems are preferred because of their lower maintenance requirements. On a sufficiently large scale a lime slaker can be justified. In general, good quality, slaked lime is employed to make a thick slurry that is subsequently diluted to the desired concentration by means of a density controller. The lime slurry is conveyed to a suitable reactor (such as a tower) where it is contacted with chlorine under the control of an

ORP cell. The finished bleach liquor can go to a settling tank (for sludge separation) or be put through a centrifugal cleaner before going to storage.

Economic Aspects

In the United States, ca 227–272 t/d of chlorine is processed into sodium hypochlorite solution for household bleach alone. In addition, there are markets for swimming pool sanitation, commercial laundries, paper and pulp manufacture, and textile and food processing. Total United States production is probably >180,000 t/yr on a dry basis (148).

Uses

Sodium hypochlorite is used in the bleaching of wood pulp and textiles where it is largely produced on site for captive consumption. It is also used as a commercial laundry and household bleach, as a sanitizer for swimming pools, and as a disinfectant for municipal water and sewage. In food processing, sodium hypochlorite is employed as a disinfectant and sanitizer. In the production of oil it is added to water used for flooding to prevent the formation of fungi which could cause plugging. In oil refineries it is employed as a sweetening agent. Large quantities of sodium hypochlorite are used in the chemical industry, primarily for the manufacture of hydrazine (qv) as well as in the synthesis of organic chemicals. Sodium hypochlorite is also used in the manufacture of chlorinated trisodium phosphate (see Bleaching agents).

SOLID HYPOCHLORITES

Sodium Hypochlorite–Trisodium Phosphate Complex. Commercial crystalline trisodium phosphate (TSP) is a complex of the type $(Na_3PO_4.x\,H_2O)_n\,NaY$ where n = 4 to 7, x = 11 or 12, and Y is a monovalent anion (see Phosphoric acids and phosphates). Chlorinated trisodium phosphate is also a complex of this type with the formula $(Na_3PO_4.11H_2O)_4.NaOCl$ (149). The crystalline, efflorescent product is trigonal with refractive indexes ω = 1.450 and ϵ = 1.455 and it is uniaxial-positive. Its melting point of 62°C is really a transition to $Na_3PO_4.8H_2O$. A high purity product can be obtained by crystallization from a liquor having the proper $Na_2O-P_2O_5$ ratio containing an excess of NaOCl. Although the calculated available chlorine is 4.66%, commercial material usually contains about 3.65% owing to the presence of salt as well as chlorate from decomposition of NaOCl. The loss in storage is about 0.05% available chlorine per month. The initial process via crystallization–filtration (150) was simplified by addition of NaOCl solution to a hot, concentrated Na_3PO_4 solution containing some Na_2HPO_4 followed by cooling and granulation of the solidified mass (151). The product is shipped in 45-kg polyethylene-lined paper bags, 23-kg Fiberpak drums, and 57- and 136-kg Leverpak drums.

Dibasic Magnesium Hypochlorite. The magnesium analogue of dibasic calcium hypochlorite has been synthesized by addition of either a sodium or calcium hypochlorite solution to an excess of aqueous $MgCl_2$ or $Mg(NO_3)_2$ (152). The reaction probably proceeds as follows:

$$2\,ClO^- + 2\,Cl^- + 2\,H_2O \rightleftharpoons 2\,Cl_2 + 4\,HO^-$$

$$3\,Mg^{2+} + 2\,ClO^- + 4\,HO^- \rightarrow Mg(OCl)_2.2Mg(OH)_2$$

overall:

$$3\,Mg^{2+} + 4\,ClO^- + 2\,Cl^- + 2\,H_2O \rightarrow Mg(OCl)_2\cdot 2Mg(OH)_2 + 2\,Cl_2$$

A competing side reaction is chlorate formation ($3\,ClO^- \rightarrow 2\,Cl^- + ClO_3^-$) which decreases the yields of dibasic magnesium hypochlorite [11073-21-5] and of evolved chlorine. Highest yields are about 35% based on hypochlorite. The product is a white, fine, powdery solid with available chlorine in the range of 52–58%.

Dibasic magnesium hypochlorite not only possesses a higher thermal stability than either its calcium analogue or calcium hypochlorite but it decomposes endothermically rather than exothermically (Fig. 2).

$$Mg(OCl)_2\cdot 2Mg(OH)_2 \xrightarrow{325°C} 3\,MgO + Cl_2 + \tfrac{1}{2}O_2 + 2\,H_2O$$

In contrast, decomposition of dibasic calcium hypochlorite begins at ca 265°C to give $Ca(OH)_2$, $CaCl_2$, and O_2. Dibasic magnesium hypochlorite exhibits a high degree of stability to moisture as shown by the following relative available chlorine losses at 24°C and 80% rh for 60 d: 2% $Mg(OCl)_2\cdot 2Mg(OH)_2$, 6% $Ca(OCl)_2\cdot 2Ca(OH)_2$, and 100% $Ca(OCl)_2$.

Calcium Hypochlorite. High assay calcium hypochlorite was first commercialized in the United States in 1928 by Mathieson Alkali Works, Inc. (now Olin Corp.) under the tradename HTH. It is now produced by two additional manufacturers in the United States (Table 5). Historically, it usually contained about 1% water and 70–74% available chlorine but in 1970, a hydrated product was introduced (153). It is similar in composition to anhydrous $Ca(OCl)_2$ except for its higher water level of about 6–12% and a slightly lower available chlorine content. This product has improved resistance to accidental initiation of self-sustained decomposition by a lit match, a lit cigarette, or a small amount of organic contamination. Current United States production consists primarily of partially hydrated $Ca(OCl)_2$ which is sold as a 65% product mainly for swimming pool use. Calcium hypochlorite is also sold as a 50% product as a sanitizer used by dairy and food industries and in the home, and as a 32% product for mildew control.

Calcium hypochlorite (65%) contains salt and water as the main diluents along with small amounts of lime, $CaCl_2$, $Ca(ClO_3)_2$, and $CaCO_3$. It is shipped as a granular or tabletted product in polyethylene containers of various sizes from 1 to 16 kg capacity. It is also shipped in 23-, 34-, and 45-kg fiber drums with a polyethylene-coated aluminum lining and a galvanized steel cover. The product is also packaged in a polyethylene bag within a fiber drum. An epoxy-coated steel drum is also used, especially for exported materials.

A high purity, completely hydrated lime of high reactivity is employed in the manufacture of calcium hypochlorite. It should contain low levels of impurities such as silica, MgO, $CaCO_3$, $CaSO_4$, Al_2O_3, Fe_2O_3, and trace metals such as Co, Ni, Cu, and Mn, since some of these can cause process difficulties and can also affect the quality and stability of the final product.

Table 5. Calcium Hypochlorite Producers and Capacities as of 1977

Producer	Plant location	Capacity, t/yr
Olin Corporation	Niagara Falls, New York	20,000
Olin Corporation	Charleston, Tennessee	29,000
Pennwalt Corporation	Wyandotte, Michigan	11,800
PPG Industries, Inc.	Barberton, Ohio	6,500
	Total	67,300

Bleaching Powder. In the preparation of calcium hypochlorite, various double salts are formed, eg, bleaching powder, as shown in Table 6; a knowledge of their relationships is essential to the understanding of the various methods of manufacturing calcium hypochlorite. This material, known since 1798, is made by chlorination of slightly moist hydrated lime. Its composition, long a subject of controversy, was established by phase studies, microscopy, and x-ray diffraction techniques (154). The initial chlorination products are a basic chloride and dibasic hypochlorite:

$$5\ Ca(OH)_2 + 2\ Cl_2 \rightarrow CaCl_2.Ca(OH)_2.H_2O + Ca(OCl)_2.2Ca(OH)_2 + H_2O$$

On further chlorination, the dibasic compound is converted to a mixed crystal consisting largely of $Ca(OCl)_2$. Ordinary bleaching powder is a mixture of this substance and the basic chloride. Prolonged chlorination partially converts the basic chloride to hydrated $CaCl_2$ but the mixed crystal persists with gradually changing properties. The nonhygroscopic nature of ordinary bleaching powder is attributed to the absence of free $CaCl_2$. Poor storage stability results from the presence of water, and heating above 40°C causes the material to lose its free-flowing and nonhygroscopic characteristics. The addition of quicklime to react with water improves stability; this gave rise to a so-called stabilized bleaching powder suitable for use in tropical areas.

Manufacture

Sodium Hypochlorite–Trisodium Phosphate Complex. Chlorinated TSP is made batchwise by addition of a 15% NaOCl solution containing some NaOH to a hot (75–80°C) concentrated liquor consisting of di- and trisodium phosphates, in a mole ratio of about 1:10, in a suitable reactor, eg, a pan mixer. The mixture is allowed to cool slowly under constant agitation until crystallization occurs (62°C). When crystallization is complete, cooling is continued to about 45°C and the slightly moist crystals are air dried. Overdrying can result in decreased stability.

Calcium Hypochlorite. Calcium hypochlorite is made by drying a filter cake of neutral calcium hypochlorite dihydrate that is usually prepared from hydrated lime, caustic, and chlorine. Filter cakes prepared by vacuum filtration of badly twinned dihydrate crystals may contain 50–55% water, whereas cakes obtained from well-formed crystals contain 40–45% water. However, the crystal form is unimportant when using pressure filtration of ca 30 MPa (ca 300 atm), and the water content can be reduced to 30%. The cake can be air or vacuum dried. During air drying some reaction with CO_2 occurs resulting in formation of $CaCO_3$. Some loss of available chlorine also occurs leading to formation of chlorate, chloride, oxygen, chlorine, and possibly HOCl and Cl_2O. Instead of air or vacuum drying the filter cake, a slurry containing about 45% solids, can be sprayed into a fluidized bed of granular $Ca(OCl)_2$ (160). The intermediate product containing about 20% water can be dried further by conventional means.

Commercial Processes. In the *Olin process* (previously known as the Olin Mathieson process), a slurry of hydrated lime in sodium hypochlorite solution is chlorinated and then cooled to $-15°C$; about 80% of the available chlorine content crystallizes in the form of a triple salt, $Ca(OCl)_2.NaOCl.NaCl.12H_2O$ (161). Above 16°C, the difficultly filterable $Ca(OCl)_2.2H_2O$ crystallizes, but below this temperature the stable phase is the triple salt that crystallizes as large hexagonal prisms. The coarse triple salt is easily separated from the mother liquor allowing much of the insolubles in the lime to pass through the filter. The filter cake is added to a chlorinated lime slurry

Table 6. Calcium Hypochlorite and Related Compounds

Compound	CAS Registry No.	Mol formula	Crystalline form	Refractive index[a] α	β	γ
neutral anhydrous calcium hypochlorite	[7778-54-3]	Ca(OCl)$_2$	irregular plates and masses	1.545	1.555	1.69
				1.53	1.55	1.65
neutral calcium hypochlorite dihydrate[b]	[22464-76-2]	Ca(OCl)$_2$.2H$_2$O	thin tetragonal plates	1.535		1.63
hemibasic calcium hypochlorite[c]	[62974-42-9]	Ca(OCl)$_2$.½Ca(OH)$_2$	pointed laths, strong basal cleavage	1.52	1.56	1.61
dibasic calcium hypochlorite	[12394-14-8]	Ca(OCl)$_2$.2Ca(OH)$_2$	thin hexagonal plates, thick hemihedral plates	1.512		1.585
tetrabasic calcium chlorohypochlorite[d]	[64175-93-5]	Ca(OCl)$_2$.CaCl$_2$.4Ca(OH)$_2$.24H$_2$O	needles up to several inches in length			
triple salt	[64147-46-2]	Ca(OCl)$_2$.NaCl.12H$_2$O	hexagonal prisms			
bleaching powder	[64175-94-6]	Ca(OCl)$_2$.CaCl$_2$.Ca(OH)$_2$.2H$_2$O[f]				
monobasic calcium chloride[e]	[14031-58-4]	CaCl$_2$.Ca(OH)$_2$	hemimorphic hexagonal	1.634		1.637
calcium hydroxide	[1305-62-0]	Ca(OH)$_2$	hexagonal plates	1.543		1.571

[a] Ref. 154.
[b] Occluded liquor is responsible for frequent reference to this compound as a 2.5 hydrate (155) and a trihydrate (156).
[c] The fine crystals obtained under 30°C are difficult to filter because they have not been adequately separated from the mother liquor and have given rise to names of monobasic and two-thirds basic in the literature prior to 1940 (157–158).
[d] This is an objectionable compound that can form if the Cl$_2$ supply is interrupted while chlorinating lime slurry at below 22°C, it can cause a reaction slurry to "freeze."
[e] This compound was described by Millikan (159) as a monohydrate, although his published data do not prove the presence of water in the compound.
[f] Empirical formula.

in such proportion that the calcium chloride in the chlorinated lime equals the sodium hypochlorite content of the triple salt crystals on a mole basis (162). This mixture is warmed to precipitate crystals of neutral calcium hypochlorite dihydrate that are filtered and dried (163). Air drying was used from 1928 to 1938, then vacuum drying, and air drying again in 1960.

In the *Pennwalt process* (previously known as the Pennsalt process) a suspension of hydrated lime and dibasic calcium hypochlorite is chlorinated producing a slurry of $Ca(OCl)_2 \cdot 2H_2O$ which is filtered. The filter cake may be treated with a concentrated NaOCl solution to reduce its $CaCl_2$ content and is then dried. The filtrate, saturated with $Ca(OCl)_2$, is treated with hydrated lime and chlorinated at about 40°C forming a slurry of the dibasic salt which is filtered and sent to the first chlorinator. Addition of NaCl decreases the solubility of $Ca(OCl)_2$ in equilibrium with the dibasic salt and lime. A lime impurities-removal step may be employed in this process (155,164–165).

The *PPG process* is based on hypochlorous acid (166). Carbon dioxide, containing about 10% chlorine (about an eightfold excess) and saturated with water vapor at 20–30°C, is passed through a rotating tubular reactor countercurrent to a flow of $Na_2CO_3 \cdot H_2O$. The exit gas containing 1–2% Cl_2O (and HOCl) is scrubbed in water forming an HOCl solution of about 10–15%. The excess Cl_2, along with most of the CO_2, is recycled; some CO_2 is purged from the system. The hypochlorous acid reacts with lime slurry at 10.0–10.5 pH producing a $Ca(OCl)_2$ solution of ca 15–20%. Sodium hypochlorite solution may also be added to reduce the $CaCl_2$ concentration (167). This solution is spray dried to a product with an intermediate water level. The product can be mixed with salt, compacted, granulated, and dried further.

The following European processes have been employed for the manufacture of $Ca(OCl)_2$.

The *Imperial Chemical Industries process* starts with crystals of dibasic calcium hypochlorite, which are separated and slurried in water. The crystals are chlorinated to obtain a slurry of neutral calcium hypochlorite dihydrate in a mother liquor of reduced calcium chloride content which is then filtered and air dried (168–169).

In the *Potasse et Produits Chimiques process*, heavy lime slurry is chlorinated at 40–45°C forming large crystals of hemibasic calcium hypochlorite. The isolated hemibasic crystals are suspended in a thin chlorinated-lime slurry and then chlorinated, producing laminar crystals of neutral calcium hypochlorite dihydrate which are filtered. Mother liquors are treated with a lime slurry to recover the dibasic material which is then suspended in a liquor of lower calcium chloride content and chlorinated to form the neutral salt. Crystallization operations are heavily seeded and much of the mother liquor may be returned to dilute the slurries. The volumes of liquor and slurries handled in this process are larger than in other processes because of the thinner slurries employed (158,170–171).

Other European processes use the dibasic salt as an intermediate. Dibasic calcium hypochlorite can be prepared from filtrates (from chlorinated lime slurries) in various ways. In the *Thann process*, the filtrate is returned to the slurry being chlorinated to keep it thin. This is designed to improve crystal growth. The dibasic crystals, together with water, are added to the slurry during chlorination and some dibasic salt is prepared by chlorination in addition to the dibasic salt made from filtrates (158).

Before World War II, the calcium hypochlorite produced in Japan was prepared mainly from bleaching powder and bleach liquor, and the product was characterized by eroded dibasic crystals and the impurities found in bleaching powder.

For bleaching powder, numerous improvements over the original chamber process were developed to deal primarily with heat and moisture removal, and mechanical apparatus for batch or continuous manufacture of bleaching powder of improved quality and stability (172). For example, in a two-stage process (173), hydrated lime is chlorinated in a jacketed reactor equipped with an agitator at 20–25°C and 5.3–12.0 kPa (40–90 mm Hg) and then at 50–55°C and 4.0–6.7 kPa (30–50 mm Hg) to give a product containing 36.8% Ca(OCl)$_2$, 31.7% CaCl$_2$, 26.2% Ca(OH)$_2$, 3.0% CaCO$_3$, 0.2% Ca(ClO$_3$)$_2$, and 0.9% H$_2$O. This material, with hardly any chlorine odor, was much more stable and much less corrosive than the product (ca 11% H$_2$O) made by the older chamber process and was suitable for use in tropical climates. In less developed or remote areas, bleaching powder may still be prepared on a small scale by generating Cl$_2$ from MnO$_2$, NaCl, and H$_2$SO$_4$, and treating it with hydrated lime placed on shelves in an absorption chamber (174).

Economic Aspects

Lithium Hypochlorite. Lithium hypochlorite is produced by Lithium Corporation of America at its plant in Bessemer, North Carolina, which has a capacity of about 4500 metric tons per year. Its total demand is low owing to its relatively high price of about $1.14/kg. Estimated sales for swimming pool sanitation were 726 t in 1969.

Sodium Hypochlorite–Trisodium Phosphate Complex. Chlorinated TSP was commercialized in 1930. It is manufactured by Olin and Stauffer. The demand in 1973 was 80,812 metric tons (148). The recent, impressive growth rate of about 20%/yr has been stimulated in part by the strong growth associated with automatic dishwasher compounds for household consumption.

Calcium Hypochlorite. Demand statistics for the period 1970–1975 are given in Table 7 (175). The market has been growing at an average rate of about 10%/yr. Calcium hypochlorite is used mainly for swimming pool sanitation (Table 8). Although the chloroisocyanurates are the fastest growing segment (ca 10–12%/yr), the hypochlorites and chlorine have the major share (81%) of the market (176). Exports amount to 15% of annual United States production. Complete worldwide statistics are not available; however, there are plants in Japan, France, India, Mexico, South Africa, and Chile. Apparently, several other European countries also produce sufficient material for domestic requirements. A plant in the Union of Soviet Socialist Republics reportedly began production in 1977. In Japan, the demand for Ca(OCl)$_2$ was 22,919 and 26,121 metric tons for the fiscal years 1974 and 1975, respectively; about 60–65% was exported (177). Japan's Nippon Soda Co., Ltd. is probably the world's second

Table 7. United States Calcium Hypochlorite Demand

Year	Demand, metric tons
1970	42,184
1971	45,178
1972	48,081
1973	51,710
1974	55,338
1975	65,517

Table 8. Swimming Pool Sanitizer Use in the United States (Based on Chlorine Demand)

Compound	%
calcium hypochlorite	34
sodium hypochlorite (solution)	24
chlorine gas	23
chloroisocyanurates	15
other[a]	14

[a] Chlorine dioxide, hydantoin compounds, iodine, lithium hypochlorite, ozone, and silver.

largest producer. Two other Japanese producers are Nissin Denka Co., Ltd., and Nankai Chemical Works.

Bleaching Powder. The production of bleaching powder has been steadily declining. Peak United States production was 133,400 metric tons in 1923; it decreased to 23,600 t in 1955 and has not been reported since. It is undoubtedly still manufactured, especially in less developed countries.

Uses

Lithium Hypochlorite. Lithium hypochlorite, first introduced in 1964, has limited use in swimming pool sanitation and dry laundry bleaches.

Sodium Hypochlorite–Trisodium Phosphate Complex. The major uses are in acid metal cleaners for dairy equipment, automatic dishwasher detergents and scouring cleansers. About 48% is used in the consumer market and the remainder for industrial and institutional uses.

Calcium Hypochlorite. Calcium hypochlorite is used for disinfection in swimming pools and drinking water supplies. It is also used for treatment of industrial cooling water for slime control (of bacterial, algal, and fungal origin) and for disinfection, odor control, and BOD reduction in sewage and wastewater effluents. Calcium hypochlorite is employed as a sanitizer in households, schools, hospitals, and public buildings. It is also used for microbial control in restaurants and other public eating places. Calcium hypochlorite is used for bacterial control, odor control, and general sanitation in dairies, wineries, breweries, canneries, food processing plants, and beverage bottling plants. It is employed as a mildewcide in and around the home, boats, campers, and trailers (see Disinfectants and antiseptics; Industrial antimicrobial agents).

Bleaching Powder. Bleaching powder can be employed for general sanitation and may also be used to disinfect sea water, reservoirs, and drainage ditches where the volume of insolubles is not important. It can be used as a decontaminating agent for areas sprayed with chemical warfare agents such as mustard gas (see Chemicals in war). The high insoluble (lime) content is an undesirable feature of bleaching powder owing to the loss of hypochlorite in the sludge that is formed in aqueous slurries.

Safety and Handling

Calcium hypochlorite is heated to about 90°C during drying and thus is stable at normal temperatures encountered in transportation, storage, and use. It decomposes exothermically when heated to ca 175°C releasing oxygen which will intensify a fire reaching containers of calcium hypochlorite. It should be stored in a cool, dry, and well ventilated area. Since calcium hypochlorite can react vigorously and sometimes explosively with certain organic and inorganic materials, they should be kept away from it during shipment, storage, use, or disposal. In the event of decomposition or contamination, the container should not be resealed but should be isolated if possible and flooded with sufficient water to completely dissolve and wash away the contents.

ORGANIC AND NONMETAL HYPOCHLORITES

Alkyl hypochlorites (esters of hypochlorous acid) are volatile liquids with irritating odors and can be extremely lachrimatory. Primary and secondary hypochlorites are very unstable but tertiary hypochlorites exhibit good stability. t-Butyl hypochlorite has been the alkyl hypochlorite of choice for experimental work owing to its stability and ease of preparation in high yield and purity. It is readily prepared by chlorination of an alkaline solution of the alcohol. It is a convenient source of positive chlorine and is soluble in most organic solvents. It is used in organic synthesis because of its high selectivity as an oxidant and for C- and N-chlorinations. There is a potential fire hazard associated with the use of t-BuOCl owing to its flash point $<-10°C$, but this risk is reduced by use of inert solvents.

During the past decade numerous fluorinated alkyl hypochlorites have been synthesized and characterized. These compounds are much more thermally stable than the corresponding parent compounds and can be prepared by reaction of ClF with the appropriate carbonyl compound or alcohol.

Although no acyl hypochlorites [RCO$_2$Cl] have been isolated in pure form, they have been characterized in solution and employed as reactants via *in situ* generation from Cl$_2$O and carboxylic acids, or from Cl$_2$ and silver salts of carboxylic acids (178). Perfluoroacyl hypochlorites have recently been prepared for the first time (179).

The inorganic nonmetal oxychlorine compounds include ClONO$_2$, ClOClO$_3$, and the hypochlorites derived from monobasic fluorine-containing oxyacids of the group VIA elements S, Se, and Te, eg, HOSO$_2$F, HOSO$_2$CF$_3$, HOSF$_5$, HOSeF$_5$, and HOTeF$_5$. In addition, two members of a new class of positive chlorine compounds derived from the hydroperoxides CF$_3$OOH and SF$_5$OOH have been prepared (180).

Physical Properties

There is a paucity of data on physical properties of organic hypochlorites. Some boiling points and densities of alkyl hypochlorites are given in Table 9. Additional data are available on viscosity (181), uv spectra (7) and partition coefficients between CCl$_4$ and water (182). The liquid-phase equilibria for the system t-BuOH-t-BuOCl-H$_2$O have been determined (183). Only a few boiling points have been determined for the fluoroalkyl hypochlorites. The volatilities of the fluoroalkyl hypochlorites are somewhat lower than those of the parent alcohols. Some vapor pressure data are also given as well as ir and nmr spectral data.

Physical data on the inorganic nonmetal hypochlorites and the peroxy hypochlorites are tabulated in Table 10.

Table 9. Physical Properties of Alkyl Hypochlorites (ROCl)

R	CAS Registry No.	bp, °C (kPa)[a]	Density, g/cm³, °C
CH₃	[593-78-2]	12 (96.8)	
C₂H₅	[624-85-1]	36 (100.3)	1.013, −6
t-C₄H₉[b]	[507-40-4]	79.6 (100.0), 31–33 (9.1–9.3)	0.9599, 20
t-C₅H₁₁	[24251-12-5]	76[c] (100.3)	0.8547, 25
CF₃	[22082-78-6]	−47	
C₂F₅	[22675-67-8]	−10	
i-C₃F₇	[22675-68-9]	22	

[a] Pressures not given are approximately atmospheric. To convert kPa to mm Hg, multiply by 7.5.
[b] n_D^{20}, 1.40354.
[c] With some decomposition.

Chemical Properties

The primary and secondary alkyl hypochlorites decompose vigorously on warming and explosively when exposed to light. The major initial thermal decomposition products are aldehydes and ketones.

$$RCH_2OCl \rightarrow RCHO + HCl$$

$$R_2CHOCl \rightarrow R_2C=O + HCl$$

Secondary reactions can occur leading to chlorination products, carboxylic acids, and their esters. In contrast, the tertiary hypochlorites are significantly more stable; t-butyl hypochlorite, eg, can be distilled and stored for months at ambient temperature with little or no decomposition. However, in bright sunlight it will slowly decompose giving chiefly acetone and methyl chloride.

$$(CH_3)_3COCl \rightarrow (CH_3)_2C=O + CH_3Cl$$

The kinetics of formation and hydrolysis of t-BuOCl has been investigated (193).

The chemistry of alkyl hypochlorites (t-BuOCl in particular) has been extensively explored (178). t-Butyl hypochlorite reacts with a variety of olefins via a photoinduced radical chain process to give good yields of allylic chlorides (194). Steroid alcohols can be oxidized and chlorinated with t-BuOCl to give good yields of ketosteroids and chloroketosteroids (195) (see Steroids). t-Butyl hypochlorite is a more satisfactory reagent than HOCl for N-chlorination of amines (196). Sulfides are oxidized in excellent yields to sulfoxides without concomitant formation of sulfones (197). 2-Amino-1,4-quinones are rapidly chlorinated at room temperature; chlorination occurs specifically at the position adjacent to the amino group (198). Anhydropenicillin is converted almost quantitatively to its 6-methoxy derivative by t-BuOCl in methanol (199).

In contrast to the alkyl hypochlorites, the fluoroalkyl hypochlorites are extremely susceptible to hydrolysis but are much more thermally stable. Trifluoromethyl hypochlorite, eg, showed no decomposition when heated for several days at 100°C. When decomposition does occur, several products are formed; C_2F_5OCl gives COF_2, CF_3Cl, CF_3COF, and ClF, whereas $(CF_3)_3COCl$ gives $(CF_3)_2CO$, Cl_2, CF_3Cl, and C_2F_6 (26).

Table 10. Physical Properties of Inorganic Oxychlorine Compounds

Compound	CAS Registry No.	bp, °C	mp, °C	Density, g/cm³, °C	Vapor pressure[a] A	Vapor pressure[a] B	References
ClONO₂	[14545-72-3]	18	−107				15
ClOClO₃	[27218-16-2]	44.5	−117	1.75, 21.2	6.9406	−1568.0	184
ClOSO₂F	[13637-84-8]	45.1	−84.3	1.711, 20	7.2685	−1674.5	185
ClOSO₂CF₃	[23313-87-8]						186
ClOSF₅	[22675-70-3]	8.9					26, 187
ClOSeF₅	[39961-91-6]	31.5	−115		6.70104	−1324.37	188
ClOTeF₅	[41524-13-4]	38.5	−121				189
ClOOSF₅	[58249-49-3]	26.4	−130				190
ClOOCF₃	[32755-26-6]	−22	−132		6.867	−1221	191-192

[a] Log p (kPa) = A-B/T. To convert kPa to mm Hg, multiply by 7.5.

The fluoroalkyl hypochlorites readily react with CO and SO_2 to form the corresponding chloroformates and chlorosulfates in near quantitative yields (200). They add to olefins giving α-chloroethers (201). Borate esters are obtained by reaction of perfluoroalkyl hypochlorites with BCl_3 (202).

$$3 R_fOCl + BCl_3 \rightarrow (R_fO)_3B + 3 Cl_2$$

R_f = perfluoro alkyl group

Although the perfluoroacyl hypochlorites are thermally unstable and explosive, CF_3CO_2Cl and $C_3F_7CO_2Cl$ are easily handled and are well characterized.

Uses

t-Butyl hypochlorite has been found useful in upgrading vegetable oils (qv) (203). It is useful in the preparation of α-substituted acrylic acid esters (204) and in the preparation of esters of isoprene halohydrins (205). Numerous patents describe its use in cross-linking of polymers (qv) (206), in surface treatment of rubber (qv) (207) and in odor control of polymer latexes (208). A recent patent describes its use in the preparation of propylene oxide (qv) in high yield with little or no by-products (209). Fluoroalkyl hypochlorites are useful as insecticides, initiators for polymerizations, and bleaching and chlorinating agents (210).

BIBLIOGRAPHY

"Hypochlorites" under "Chlorine Compounds, Inorganic" in *ECT* 1st ed., Vol. 3, pp. 681–696, by H. L. Robson, Mathieson Chemical Corp.; "Chlorine Monoxide, Hypochlorous Acid, and Hypochlorites" under "Chlorine Oxygen Acids and Salts" in *ECT* 2nd ed., Vol. 5, pp. 7–27, by H. L. Robson, Olin Mathieson Chemical Corporation.

1. F. A. Cotton and G. Wilkinson, *Advanced Inorganic Chemistry*, 3rd ed., John Wiley & Sons, Inc., New York, 1972, p. 114.
2. C. F. Goodeve, *J. Chem. Soc.*, 2733 (1930).
3. C. H. Secoy and G. H. Cady, *J. Am. Chem. Soc.* **62**, 1036 (1940).
4. W. A. Roth, *Z. Phys. Chem. A* **191**, 248 (1942).
5. D. D. Wagman and co-workers, *Natl. Bur. Stand. Tech. Note* **270-3**, (Jan. 1968).
6. G. E. Heberich, R. H. Jackson, and D. J. Millen, *J. Chem. Soc. A.*, 336 (1966).
7. M. Anbar and I. Dostrovsky, *J. Chem. Soc.*, 1105 (1954).
8. G. H. Cady and R. E. Brown, *J. Am. Chem. Soc.* **67**, 1614 (1945).
9. G. Pannetier and M. B. Caid, *Bull. Soc. Chim.*, 1628 (1961).
10. H. J. Schumacher and R. V. Townend, *Z. Phys. Chem. B* **20**, 375 (1933).
11. W. G. Alcock and G. C. Pimentel, *J. Chem. Phys.* **48**, 2373 (1968).
12. J. J. Beaver and G. Stieger, *Z. Phys. Chem. B* **12**, 93 (1931).
13. E. A. Moelwyn-Hughes and C. N. Hinshelwood, *Proc. R. Soc. London* **131**, 177 (1931).
14. N. N. Semenov, *Phys. Z. Sowjetunion* **1**, 561 (1932).
15. M. Schmeisser in S. Y. Tyree, Jr., ed., *Inorganic Synthesis*, Vol. 9, McGraw-Hill Book Co., New York, 1967, p. 127.
16. H. Martin and W. Meise, *Z. Elektrochem.* **63**, 162 (1959).
17. C. J. Shack and D. Pilipovitch, *Inorg. Chem.* **9**, 387 (1970).
18. F. Wenisch, *Dissertation*, Technische Hochschule, Aachen, GDR, 1961.
19. K. Dehnicke, *Z. Anorg. Allg. Chem.* **309**, 266 (1961).
20. C. J. Schack and co-workers, *Inorg. Chem.* **11**, 2201 (1972).
21. Y. Macheteau and L. Gillardeau, *Bull. Soc. Chim.*, 4075 (1967).
22. K. Dehnicke, *Angew. Chem.* **73**, 763 (1961).
23. M. Schmeisser and K. Brändle in H. J. Emeleus and A. G. Sharpe, eds., *Advances In Inorganic and Radiochemistry*, Vol. 5, Academic Press Inc., New York, 1963, p. 50.

24. K. Dehnicke, *Chem. Ber.* **97,** 3358 (1964).
25. K. Dehnicke, *Angew. Chem.* **73,** 535 (1961).
26. D. E. Gould and co-workers, *J. Am. Chem. Soc.* **91,** 1310 (1969).
27. D. D. Tanner and N. Nychka, *J. Am. Chem. Soc.* **89,** 121 (1967).
28. U.S. Pat. 3,872,176 (Mar. 18, 1975), G. L. Kochanny and T. A. Chamberlain (to The Dow Chemical Co.).
29. D. D. Tanner, N. Nychka, and T. Ochai, *Can. J. Chem.* **52,** 2573 (1974).
30. S. Goldschmidt and H. Schussler, *Ber.* **58B,** 566 (1925).
31. M. Fujii, *Ph.D. Thesis,* McGill Univ., Montreal, Can., 1969.
32. J. van Ham, A. Schors, and E. C. Kooyman, *Recl. Trav. Chim. Pays-Bas* **92,** 393 (1973).
33. S. Kolbe, *Dissertation,* Technische Hochschule, Aachen, Ger., 1960.
34. U.S. Pat. 4,055,719 (Oct. 25, 1977), J. A. Wojtowicz (to Olin Corp.).
35. L. R. Anderson and W. B. Fox, *J. Am. Chem. Soc.* **89,** 4313 (1967).
36. U.S. Pat. 3,993,649 (Nov. 23, 1976), D. L. Sawhill and H. W. Schiessl (to Olin Corp.).
37. U.S. Pat. 3,988,336 (Oct. 26, 1976), J. A. Wojtowicz (to Olin Corp.).
38. G. H. Cady in T. Moeller, ed., *Inorganic Synthesis,* Vol. 5, McGraw-Hill Book Co., New York, 1957, p. 156.
39. V. M. Gallak, N. I. Bellinskaya, and T. A. Pavlova, *Zh. Prikl. Khim. Leningrad* **38,** 1225 (1965); *J. Appl. Chem.* **38,** 1212 (1965).
40. J. J. Renard and H. I. Bolker, *Chem. Rev.* **76,** 487 (1976).
41. C. J. Schack and C. B. Lindahl, *Inorg. Nucl. Chem. Lett.* **3,** 387 (1967).
42. U.S. Pats. 2,157,524 and 2,157,525 (May 9, 1939), G. H. Cady (to Pittsburgh Plate Glass Co.).
43. Ger. Offen. 2,326,601 (Dec. 12, 1974), Brit. Pat. 1,358,839 (July 3, 1974), and Can. Pat. 952,258 (Aug. 6, 1975), W. A. Mueller (to Pulp and Paper Research Institute of Canada).
44. S. Goldschmidt, *Ber.* **52B,** 753 (1919).
45. W. A. Noyes and T. A. Wilson, *J. Am. Chem. Soc.* **44,** 1630 (1922).
46. W. A. Noyes and T. A. Wilson, *Recl. Trav. Chim. Pays-Bas.* **41,** 557 (1922).
47. A. J. Balard, *Ann. Chim. Phys., Ser. 2* **57,** 225 (1834).
48. J. W. T. Spinks, *J. Am. Chem. Soc.* **53,** 3015 (1931).
49. O. M. Lidwell and J. E. Lovelock, *Med. Res. Counc. Spec. Rep. Ser.* **262,** 68 (1948).
50. U.S.S.R. Pat. 181,068 (Apr. 15, 1966), V. M. Gallak and N. I. Bellinskaya.
51. H. I. Bolker and N. Liebergott, *Pulp. Pap. Mag. Can.* **73,** 7332 (1972).
52. U.S. Pat. 3,619,349 (Nov. 9, 1971), N. Liebergott and H. I. Bolker (to Pulp and Paper Institute of Canada).
53. G. E. White, *Handbook of Chlorination,* Van Nostrand Reinhold Co., New York, 1972.
54. J. C. Morris, *J. Phys. Chem.* **70,** 3798 (1966).
55. B. Neumann and G. Müller, *Z. Anorg. Chem.* **182,** 235 (1929); **185,** 428 (1930).
56. W. C. Furguson, L. Slotin, and D. W. G. Style, *Trans. Faraday Soc.* **32,** 956 (1936).
57. R. A. Ashby, *J. Mol. Spec.* **23,** 439 (1967).
58. D. C. Lindsay, D. G. Lister, and D. J. Millen, *Chem. Commun.,* 950 (1969).
59. H. Imagawa, *J. Electrochem. Soc. Jpn.* **18,** 382 (1950); **19,** 271 (1951).
60. B. M. Israel, *Hydrolysis of Some Organohalogenating Agents, Ph.D. Thesis,* University of Wisconsin, 1962.
61. J. Ourisson and M. Kastner, *Bull. Soc. Chim.* **6,** 1307 (1939).
62. C. H. Secoy and G. H. Cady, *J. Am. Chem. Soc.* **63,** 2504 (1941).
63. W. A. Roth, *Z. Phys. Chem.* **145A,** 289 (1929).
64. J. A. Dean, ed., *Langes Handbook of Chemistry,* 11th ed., McGraw-Hill Book Co., New York, 1973.
65. M. W. Lister, *Can. J. Chem.* **30,** 879 (1952).
66. A. J. Allmand, P. W. Cunliffe, and R. E. W. Maddison, *J. Chem. Soc.,* 822 (1925); 655 (1927).
67. A. J. Allmand and W. W. Webb, *Z. Phys. Chem.* **131,** 189 (1928).
68. L. Bonnet, *Rev. Gén. Mater. Color.* **39,** 29 (1935).
69. K. W. Young and A. J. Allmand, *Can. J. Res.* **27B,** 381 (1949).
70. J. I. Aznárez and J. V. Vinadé, *An. R. Soc. Esp. Fis. Quim. B* **48,** 653 (1952); **49,** 341 (1953).
71. *Ibid.* **48,** p. 673.
72. M. W. Lister and P. Rosenblum, *Can. J. Chem.* **39,** 1645 (1961).
73. M. W. Lister, *Can. J. Chem.* **34,** 465 (1956).
74. M. W. Lister and P. Rosenblum, *Can. J. Chem.* **41,** 3013 (1963).
75. *Ibid.,* p. 2727.

76. L. Farkas, M. Lewin, and R. Bloch, *J. Am. Chem. Soc.* **71,** 1988 (1949).
77. M. W. Lister, *Can. J. Chem.* **34,** 489 (1956).
78. R. O. Griffith and A. McKeown, *Trans. Faraday Soc.* **28,** 518 (1932).
79. M. Anbar and H. Taube, *J. Am. Chem. Soc.* **80,** 1073 (1958).
80. J. Halperin and H. Taube, *J. Am. Chem. Soc.* **74,** 375 (1952).
81. N. S. Chamberlain and H. B. Snyder, *Technology of Treating Plating Wastes, Tenth Annual Wastes Conference,* Purdue Univ., Lafayette, Ind., May 1955.
82. I. W. Wei and J. Carrell Morris in A. J. Rubin, ed., *Chemistry of Water Supply, Treatment and Distribution,* Ann Arbor Science Publ., Inc., Ann Arbor, Mich., 1975.
83. E. A. Shilov, N. P. Kanyaev, and A. P. Otmennikova, *J. Phys. Chem. USSR* **8,** 909 (1936).
84. E. A. Shilov, G. V. Kupinskaya, and A. A. Yasnikov, *Dokl. Akad. Nauk SSSR* **81,** 435 (1951).
85. F. Strauss and co-workers, *Ber.* **63,** 1868 (1930).
86. U.S. Pat. 2,964,525 (Dec. 13, 1960), W. L. Robinson (to Monsanto Chemical Co.).
87. U.S. Pat. 3,712,891 (Jan. 23, 1973), S. Berkowitz and R. N. Mesiah (to FMC Corp.).
88. U.S. Pat. 3,835,134 (Sept. 10, 1974), H. W. Schiessl, D. L. Sawhill, and S. K. Bhutani (to Olin Corp.).
89. U.S. Pat. 4,024,140 (May 17, 1977), J. A. Wojtowicz (to Olin Corp.).
90. Fr. Pat. 851,659 (Jan. 12, 1940), J. Ourisson and M. Kastner (to Potasse et Produits Chimiques).
91. U.S. Pat. 2,347,151 (Apr. 18, 1944), C. C. Crawford and T. W. Evans (to Shell Dev. Co.).
92. Brit. Pat. 543,944 (Mar. 20, 1942), C. Carter and E. R. B. Jackson (to Imperial Chemical Industries).
93. U.S. Pat. 1,510,790 (Oct. 7, 1924), K. P. McElroy (to Carbide and Carbon Chemicals Corp.).
94. U.S. Pat. 3,616,385 (Oct. 26, 1971), R. K. Kloss and G. W. Claybaugh (to Proctor and Gamble).
95. U.S. Pats. 3,578,400 (May 11, 1971), and 3,718,598 (Feb. 27, 1973), J. A. Wojtowicz, M. Lapkin, and M. S. Puar (to Olin Corp.).
96. U.S. Pat. 3,845,145 (Oct. 29, 1974), J. A. Wojtowicz, M. Lapkin, and M. S. Puar (to Olin Corp.).
97. U.S. Pat. 3,578,393 (May 11, 1971), J. A. Wojtowicz, M. Lapkin, and M. S. Puar (to Olin Corp.).
98. A. W. Awtry and R. E. Connick, *J. Am. Chem. Soc.* **73,** 1842 (1951).
99. H. A. Podoliak, *A Review of The Literature On the Use of Calcium Hypochlorite In Fisheries,* NTIS, U.S. Dept. Commerce, Mar. 1974.
100. *Gmelins Handbuch der Anorganischen Chemie,* 8 Aufl., System Nr. 22, Verlag Chemie, Weinheim, Ger., 1937, p 471.
101. *Ibid.,* System Nr. 27B, 1939, p. 150.
102. P. Pierron, *Bull. Soc. Chim.* **8,** 660 (1941).
103. J. A. W. Luck, *J. Am. Pharm. Assoc.* **13,** 710 (1924).
104. S. Z. Makarov and E. F. Shcharkova, *Russ. J. Inorg. Chem.* **14,** 1632 (1969).
105. *A Practical Guide to Chlorine Bleach Making, Technical and Engineering Service Bulletin 72-19,* Allied Chem. Corp., Morristown, N.J., 1974.
106. *Soda Bleach Solutions,* Diamond Alkali Co., 1952.
107. *Selected Values of Chemical Thermodynamic Properties,* Circ. 500, National Bureau of Standards, Washington, D.C., 1952.
108. P. Pierron, *Bull. Soc. Chim.* **8,** 664 (1941).
109. M. W. Lister, *Can. J. Chem.* **34,** 465 (1956); **40,** 729 (1962).
110. *Ibid.,* **34,** 479.
111. E. Chirnoaga, *J. Chem. Soc.,* 1693 (1926).
112. O. R. Howell, *Proc. R. Soc. London Ser. A* **104,** 134 (1923).
113. J. R. Lewis, *J. Phys. Chem.* **32,** 243, 1808 (1928); **35,** 915 (1931); and **37,** 917 (1933).
114. G. H. Ayers and M. H. Booth, *J. Am. Chem. Soc.* **77,** 825 (1955).
115. R. B. Wagner and H. D. Zook, *Synthetic Organic Chemistry,* John Wiley & Sons, Inc., New York, 1953, pp. 422, 674.
116. D. D. Neiswender, Jr., W. B. Moniz, and J. R. Dixon, *J. Am. Chem. Soc.* **82,** 2876 (1960).
117. F. Strauss and co-workers, *Ber.* **63,** 1873 (1930).
118. R. A. Jacobson and W. H. Carothers, *J. Am. Chem. Soc.* **55,** 4667 (1933).
119. J. Cologne and L. Cumet, *Bull. Soc. Chim.* **14,** 842 (1947).
120. R. Riemschneider, *Chim. Ind.* **34,** 266 (1952).
121. F. Strauss and co-workers, *Ber.* **63,** 1885 (1930).
122. E. M. Cherkasova and co-workers, *Z. Obsc. Chim.* **19,** 321 (1949).
123. F. Asinger, *Chemie Und Technologie Der Paraffin-Kohlenwasserstoffe,* Akademie Verlag, Berlin, Ger., 1956, p. 300.

124. R. Ohme, H. Preuschof, and H. U. Heyne, *Org. Synth.* **52,** 11 (1972).
125. R. Ohme and E. Schmitz, *Ber.* **97,** 297 (1964).
126. F. B. Mallory, C. S. Wood, and B. M. Hurwitz, *J. Org. Chem.* **29,** 2605 (1964).
127. D. R. Eckroth, T. G. Cochran, and E. C. Taylor, *J. Org. Chem.* **31,** 1303 (1966).
128. S. Marmor, *J. Org. Chem.* **28,** 250 (1963).
129. U.S. Pat. 2,887,498 (May 19, 1959), G. W. Hearne, D. S. LaFrance, and H. de V. Finch (to Shell Dev. Co.).
130. C. Schaer, *Helv. Chim. Acta.* **41,** 560, 614 (1958).
131. D. H. Rosenblatt and G. H. Broome, *J. Org. Chem.* **28,** 1290 (1963).
132. B. F. Clark, *Chem. News* **143,** 265 (1931).
133. H. R. Eisenhauer, *J. Water Pollut. Control Fed.* **36,** 1116 (1964).
134. U.S. Pat. 2,157,559 (May 9, 1939), I. E. Muskat and G. H. Cady (to Pittsburgh Plate Glass Co.).
135. U.S. Pat. 2,225,923 (Dec. 24, 1940), I. E. Muskat and G. H. Cady (to Pittsburgh Plate Glass Co.).
136. U.S. Pat. 3,572,989 (Mar. 30, 1971), S. Tatara and co-workers (to Nippon Soda Co., Ltd.).
137. U.S. Pat. 3,767,775 (Oct. 23, 1973), S. Tatara and co-workers (to Nippon Soda Co., Ltd.).
138. U.S. Pat. 3,251,647 (May 17, 1966), B. H. Nicolaisen (to Olin Mathieson Chemical Corp.).
139. U.S. Pat. 3,950,499 (Apr. 13, 1976), N. Miyashin and co-workers (to Nippon Soda Co., Ltd.).
140. U.S. Pat. 3,895,099 (July 15, 1975), W. J. Sakowski (to Olin Corp.).
141. U.S. Pat. 3,954,948 (May 4, 1976), W. J. Sakowski (to Olin Corp.).
142. U.S. Pat. 1,481,040 (Jan. 15, 1924), M. C. Taylor, R. B. MacMullin, and R. E. Gegenheimer (to Mathieson Alkali Works, Inc.).
143. U.S. Pats. 1,623,483, 1,623,485 (June 14, 1927), R. B. MacMullin (to Mathieson Alkali Works, Inc.).
144. O. Wyss, *Water Sewer Works* **109,** 12155 (1962).
145. G. M. Fair, J. C. Morris, and S. L. Chang, *J. New Engl. Waterworks Assoc.* **61,** 285 (1947).
146. J. E. O'Brien, *Hydrolytic and Ionization Equilibria of Chlorinated Isocyanurate in Water*, Ph.D. Thesis, Harvard Univ., 1972.
147. D. J. Pye, *J. Electrochem. Soc.* **97,** 245 (1950).
148. Yu-R. Chin, private communication, Stanford Research Institute.
149. R. N. Bell, *Ind. Eng. Chem.* **41,** 2901 (1949).
150. U.S. Pat. 1,555,474 (Sept. 29, 1925), L. D. Mathias (to Victor Chemical Works).
151. U.S. Pat. 1,965,304 (July 3, 1934), H. Adler (to Victor Chemical Works).
152. U.S. Pat. 3,582,265 (June 1, 1971), J. J. Bishop and S. I. Trotz (to Olin Corp.).
153. U.S. Pat. 3,544,267 (Dec. 1, 1970), G. R. Dychdala (to Pennwalt Corp.).
154. C. W. Bunn, L. M. Clark, and I. L. Clifford, *Proc. R. Soc. London Ser. A* **151,** 141 (1935).
155. U.S. Pats. 2,441,337 (May 11, 1948), 2,469,901 (May 10, 1949), and 2,587,071 (Feb. 26, 1952), J. W. Sprauer (to Pennsylvania Salt Corp.).
156. J. Ourisson, *Bull. Soc. Ind. Mulhouse* **103,** 217 (1938).
157. J. Ourisson, *Atti Congr. Intern. Chim. 10th Congr. Rome* **4,** 40 (1939).
158. Fr. Pat. 825,903 (Mar. 17, 1938), and Brit. Pat. 487,009 (June 14, 1938), J. Ourisson.
159. J. Millikan, *Z. Phys. Chem. Leipzig* **92,** 59, 496 (1916).
160. U.S. Pat. 3,969,546 (July 13, 1976), W. C. Saeman (to Olin Corp.).
161. U.S. Pat. 1,787,048 (Dec. 30, 1930), R. B. MacMullin and M. C. Taylor (to Mathieson Alkali Works, Inc.).
162. U.S. Pat. 1,713,650 (May 21, 1929), A. George and R. B. MacMullin (to Mathieson Alkali Works, Inc.).
163. U.S. Pats. 2,195,755 and 2,195,757 (Apr. 2, 1940), H. L. Robson and G. A. Petroe (to Mathieson Alkali Works, Inc.).
164. U.S. Pat. 3,030,177 (Apr. 17, 1962), J. C. Mohan, Jr. (to Pennsalt Chem. Corp.).
165. U.S. Pat. 3,094,380 (June 18, 1963), E. A. Bruce (to Pennsalt Chem. Corp.).
166. U.S. Pat. 2,240,344 (Apr. 29, 1941), I. E. Muskat and G. H. Cady (to Pittsburgh Plate Glass Co.).
167. U.S. Pat. 3,134,641 (May 26, 1964), R. D. Gleichert (to Pittsburgh Plate Glass Co.).
168. Brit. Pat. 378,847 (Aug. 16, 1932), F. N. Kitchen (to Imperial Chemical Industries, Ltd.).
169. Brit. Pat. 404,627 (Jan. 15, 1934), F. T. Meehan and F. N. Kitchen (to Imperial Chemical Industries, Ltd.).
170. Fr. Pats. 858,057 (Nov. 16, 1940), 862,483 (Mar. 7, 1941), J. Ourisson, P. Camescasse, and M. Kastner (to Potasse et Produits Chimiques).
171. Fr. Pat. 1,019,027 (Jan. 15, 1953), A. J. Vorburger (to Potasse et Produits Chimiques).
172. Can. Pat. 194,216 (Nov. 25, 1919), A. Rudge.

173. Ger. Pat. 656,413 (Feb. 4, 1938), E. Renschler and A. Remelé (to I. G. Farbenindustrie A.G.).
174. R. Stone, *J. Am. Water Works Assoc.* **42,** 283 (1950).
175. *Chem. Mark. Rep.,* (Oct. 1, 1976).
176. *Chem. Week,* 34 (Mar. 2, 1977).
177. *Jpn. Chem. Ann.,* 38 (1976).
178. M. Anbar and D. Ginsburg, *Chem. Rev.* **54,** 925 (1954).
179. D. D. DesMarteau, *174th ACS Meeting, Chicago, Ill., Aug. 1977.*
180. F. Aubke and D. D. DesMarteau in P. Tarrant, ed., *Fluorine Chemistry Reviews,* Vol. 8, Marcel Dekker, Inc., New York, 73 (1977).
181. M. Anbar, *Ph.D. Thesis,* Hebrew Univ., Jerusalem, Isr. 1953.
182. M. C. Taylor, R. B. McMullin, and C. A. Gammal, *J. Am. Chem. Soc.* **47,** 395 (1925).
183. J. W. Westwater and L. F. Audrieth, *Ind. Eng. Chem.* **46,** 1281 (1954).
184. C. J. Schack and D. Pilipovich, *J. Am. Chem. Soc.* **9,** 1387 (1970).
185. W. P. Gilbreath and G. H. Cady, *Inorg. Chem.* **2,** 496 (1963).
186. D. D. DesMarteau, *175th ACS Meeting, Anaheim, Calif., Mar. 1978.*
187. C. J. Schack and co-workers, *J. Am. Chem. Soc.* **91,** 2907 (1969).
188. K. Sepplet, *Chem. Ber.* **106,** 1571 (1973).
189. K. Seppelt and D. Nothe, *Inorg. Chem.* **12,** 2727 (1973).
190. D. D. DesMarteau and R. M. Hammaker, *Isr. J. Chem.* **17,** 103 (1978).
191. N. Walker and D. D. DesMarteau, *J. Am. Chem. Soc.* **97,** 13 (1975).
192. C. T. Ratcliffe and co-workers, *J. Am. Chem. Soc.* **93,** 3886 (1971).
193. M. Anbar and I. Dostrovsky, *J. Chem. Soc.,* 1094 (1954).
194. C. Walling and W. Thaler, *J. Am. Chem. Soc.* **83,** 3877 (1961).
195. J. J. Beereboom and co-workers, *J. Am. Chem. Soc.* **75,** 3500 (1953).
196. W. E. Bachmann, M. P. Cava, and A. S. Dreiding, *J. Am. Chem. Soc.* **76,** 5554 (1954).
197. C. R. Johnson and J. J. Rigau, *J. Am. Chem. Soc.* **91,** 5398 (1969).
198. H. W. Moore and G. Cojipe, *Synthesis,* 49 (1973).
199. J. E. Baldwin and co-workers, *J. Am. Chem. Soc.* **95,** 2401 (1973).
200. D. E. Young and co-workers, *J. Am. Chem. Soc.* **92,** 2313 (1970).
201. L. R. Anderson and co-workers *J. Org. Chem.* **35,** 3730 (1970).
202. D. E. Young, L. R. Anderson, and W. B. Fox, *Inorg. Chem.* **10,** 2810 (1971).
203. H. M. Teeter and co-workers, *Ind. Eng. Chem.* **41,** 849 (1949).
204. U.S. Pat. 2,514,672 (July 11, 1950), D. D. Reynolds and W. O. Kenyon (to Eastman Kodak Co.).
205. U.S. Pat. 2,511,870 (June 20, 1950), W. Oroshnik (to Ortho Pharmaceutical Co.).
206. U.S. Pat. 3,960,822 (June 1, 1976), P. Davis, H. C. Vogt, and C. F. Deck (to BASF Wyandotte Corp.).
207. U.S. Pat. 3,940,548 (Feb. 24, 1976), and Brit. Pat. 1,396,090 (May 29, 1975), Y. Todani and T. Ohkawa (to Nippon Zeon Co., Ltd.).
208. Ger. Offen. 2,541,526 (Apr. 1, 1976), A. P. Gelbein and J. T. Kwon (to Lummus Co.).
209. U.S. Pat. 3,756,976 (Sept. 4, 1973), C. A. Uranek and J. E. Burleigh (to Phillips Petroleum Co.).
210. U.S. Pat. 3,689,563 (Sept. 5, 1972), D. E. Young, L. R. Anderson, and W. B. Fox (to Allied Chem. Corp.).

J. A. WOJTOWICZ
Olin Corporation

CHLORINE DIOXIDE, CHLOROUS ACID, AND CHLORITES

Chlorous acid [13898-47-0], HClO$_2$, and chlorine dioxide [10049-04-4], ClO$_2$, have insufficient stability as articles of commerce, but technical sodium chlorite [7758-19-2], containing approximately 80 wt % NaClO$_2$, is manufactured and distributed in commercial quantities. Addition of acids to dilute solutions of sodium chlorite yields chlorous acid which decomposes producing variable quantities of chlorine dioxide, chlorate, and chloride. However, chlorine dioxide can be made more conveniently, and in near-quantitative yields, by oxidizing chlorites under suitable conditions.

Chlorous acid has no great significance in technology, except that it is thought to be the first in a series of short-lived active species in the bleaching of textiles and in reactions of acidified solutions of chlorites. However, considerable quantities of chlorine dioxide are produced at points of use for immediate consumption. Its production from sodium chlorite is the preferred approach for comparatively small requirements, up to ca 100 kg/d, but economics favor sodium chlorate as the raw material if requirements are much greater, as in the bleaching of pulp. The manufacture of sodium chlorite requires sodium chlorate as the source of chlorine dioxide (see Chloric acid and chlorates).

In 1977 sodium chlorite was produced commercially by Degussa, Farbwerke Hoechst, Pechiney Ugine Kuhlmann, and Interox in Europe and by several companies in Japan. Olin Corporation is the only manufacturer in the United States.

CHLORINE DIOXIDE

Physical Properties

Chlorine dioxide, ClO$_2$, mol wt 67.46, has a mp of $-59°$C and a bp of $11°$C. The vapor pressure at the freezing point is 1.3 kPa (10 mm Hg). Liquid chlorine dioxide has a deep-red color and is explosive at temperatures above $-40°$C. Density of the liquid is 1.765 g/cm^3 at $-56°$C and 1.62 g/cm^3 at $11°$C with intermediate points falling on a straight line. Chlorine dioxide exists entirely or almost entirely as the monomeric free radical (1–4). Infrared and Raman spectra of crystalline ClO$_2$ indicate that dimerization does not occur even in the solid state (5). The chlorine–oxygen bonds have predominantly double bond character (6) and the O–Cl–O angle is about 117.5° with a chlorine–oxygen bond length of 0.147 nm (4,7–12). The compound has a dipole moment of 5.64 × 10^{-30} C·m (1.69 debye units) (13). The enthalpy and free energy of formation for ClO$_2$ gas at 298.15 K in kJ/mol (kcal/mol) are given as $\Delta H° = 102.6$ (24.5) and $\Delta F° = 120.6$ (28.8), and elsewhere (14) as $\Delta H° = 95.0$ (22.69) and $\Delta F° = 119.1$ (28.44) (15). Corresponding data for aqueous ClO$_2$ from gaseous Cl$_2$ and O$_2$ are $\Delta H° = 75.0$ (17.9) and $\Delta F° = 117.7$ (28.1).

The electronic absorption spectrum of chlorine dioxide shows a broad absorption band with characteristic fine structure and a maximum at ca 360 nm and a molar extinction coefficient of ca 1150 (M·cm)$^{-1}$ (16–22). The extinction coefficient varies only slightly with the acidity of the medium and remains unchanged in the presence of chloride ion, chlorate ion, and chlorine at 25–50°C (21,23). The electronic spectrum is essentially the same in the gas phase and in organic and aqueous solutions (16). The infrared (4,10), electron diffraction (8), epr (24–26), microwave (12), and Raman spectra (11) have been studied as well as the electronic fine structure in the region of 400 to 700 nm (27).

Chlorine dioxide vapor resembles chlorine in appearance and odor with a somewhat deeper shade of green. Cigarette smokers can readily detect low concentrations of chlorine dioxide vapor by the characteristic sweet taste produced. Decomposition of vapor at a partial pressure of about 10.7 kPa (80 mm Hg) may produce mild explosions or "puffs". The velocity of the decomposition wave at 17.3 kPa (130 mm Hg) and 2.75°C is 1 m/s and 2 m/s at 27.6 kPa (207 mm Hg) (28). Chlorine dioxide may detonate above 40 kPa (300 mm Hg). The gas is soluble in water where it forms a yellow solution that is quite stable if kept cool and away from light. Various solid polyhydrates have been described including a hexahydrate (29), an octahydrate (1), and a decahydrate (30). The partition coefficient of chlorine dioxide between water and the gaseous state is 70.0 ± 0.7 at 0°C and 21.5 ± 0.8 at 35°C; therefore, at 25°C chlorine dioxide is about 23 times more concentrated in the aqueous phase than it is in the gas phase with which it is in equilibrium (31). Solubility of chlorine dioxide in water at partial pressures up to 21.3 kPa (160 mm Hg) is shown in Figure 1 (28).

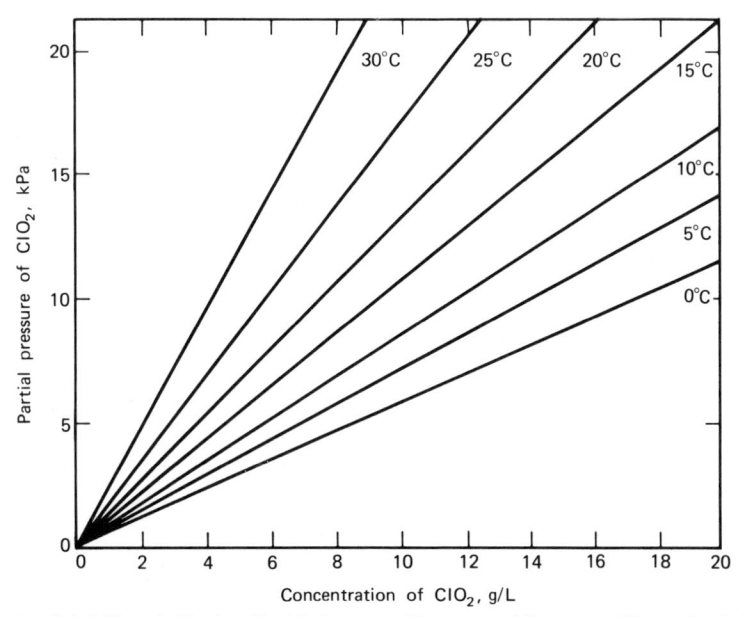

Figure 1. Solubility of chlorine dioxide in water. To convert kPa to mm Hg, multiply by 7.5.

Chemical Properties

The oxidation potential of chlorine dioxide in aqueous solution is 0.95 V at 4–7 pH. Reduction of chlorine dioxide in basic solution by hydrogen peroxide is shown in equation 1:

$$2\,ClO_2 + H_2O_2 \rightarrow 2\,HClO_2 + O_2 \qquad (1)$$

Other reducing agents that convert chlorine dioxide to chlorite at neutral pH or above include potassium iodide (32), sodium sulfite (33), sodium arsenite (34), and plumbous oxide (35). Agents that reduce chlorine dioxide completely to chloride ion include borohydride (36), iodide at 1 pH, and sulfurous acid in acidic solution.

Oxidation of chlorine dioxide by chlorine or hypochlorous acid is (eqs. 2–3):

$$2\ ClO_2 + Cl_2 + 2\ H_2O \rightleftharpoons 2\ ClO_3^- + 2\ Cl^- + 4\ H^+ \qquad (2)$$

$$2\ ClO_2 + HOCl + H_2O \rightleftharpoons 2\ ClO_3^- + Cl^- + 3\ H^+ \qquad (3)$$

Chlorine dioxide is thus oxidized quite readily to chlorate in neutral solutions (eq. 3) but not in acidic solutions where the equilibrium (eq. 2) is shifted to the left. The equilibrium constant for equation 3 is 3.39×10^{-9} at 25°C (37). This is consistent with the observation that the concentration of sulfuric acid, in manufacture of chlorine dioxide from chlorate and chloride, should be greater than 8 N (38). Chlorine dioxide is oxidized to Cl_2O_6 by ozone (39).

The disproportionation reaction of aqueous chlorine dioxide solutions is very slow except in alkaline solution (eq. 4) (16,31).

$$2\ ClO_2 + 2\ OH^- \rightarrow ClO_3^- + ClO_2^- + H_2O \qquad (4)$$

In the dark, at 0°C, an aqueous solution containing 0.155 M ClO_2 and 0.011 M H^+ with Cl^- present as a catalyst loses only 1% of the initial ClO_2 in 7 wk. There is no appreciable exchange of ^{38}Cl between chlorate ion and chlorine dioxide in one hour (40). In sunlight, dry chlorine dioxide gas decomposes into chlorine and oxygen radicals (41). Subsequent chain reactions yield chlorine trioxide in addition to chlorine. Some evidence of chlorine dioxide decomposing in the dark to produce chlorine and oxygen was found by Bray (1) who observed much oxygen at 100°C, but none at 60°C. The chlorine formed reacts with chlorine dioxide in the presence of moisture to produce chloride and chlorate ions. Formation of mist containing chloric and perchloric acids is readily observed with concentrated solutions of chlorine dioxide.

In thermal decomposition of aqueous chlorine dioxide solutions (eq. 5), hydrogen and chloride ions, but not chlorate ion, are reported to accelerate the reaction; however, the rate is only appreciable at high levels of acidity and temperature (1).

$$6\ ClO_2 + 3\ H_2O \rightarrow 5\ HClO_3 + HCl \qquad (5)$$

A highly colored complex is formed when chlorine dioxide is dissolved in an aqueous solution of barium chlorite (eq. 6) (7).

$$ClO_2 + ClO_2^- \rightleftharpoons Cl_2O_4^- \qquad (6)$$

The dark mahogany-colored complex gradually decomposes with evolution of a gas that is believed to be a mixture of chlorine and oxygen (7,42). Postulated structures for the complex are represented (1) and (2). There is indirect paramagnetic resonance evidence supporting existence of the complex in the form of structure (2) (43).

Organic Chemistry. Chlorine dioxide exhibits a pattern of reactivity toward organic compounds considerably different from the behavior of other oxidants commonly employed in the laboratory. For example, olefins react more rapidly with permanganate than with chlorine dioxide (44–45) whereas triethylamine is nearly 10^4 times more reactive with chlorine dioxide than with permanganate (46–47). The organic compounds most reactive with ClO_2 are aliphatic tertiary amines and phenols; aromatic amines may also belong to this category (48).

Alcohols, carbonyl compounds, and carbohydrates are reported to react rather slowly with chlorine dioxide (49–52). Carbon–carbon cleavage may yield a variety of oxygen-containing products.

$$\left[\begin{array}{c} O\diagdown \diagup O \\ Cl-Cl \\ O\diagup \diagdown O \end{array}\right]^{-}$$

(1)

$$\left[Cl\diagdown \begin{array}{c} O-O \\ O-O \end{array}\diagup Cl\right]^{-}$$

(2)

Studies of chlorine dioxide reactions with sulfur compounds include thiamine (53), thiothiamine (54), mercaptans, organic disulfides, and thiourea (55–56).

Phenol reacts rapidly with chlorine dioxide and the reaction is first order with respect to each reactant. Products identified from relatively high reactant concentrations include: 1,4-benzoquinone, 2-chloro-1,4-benzoquinone, 2,5-dichloro-1,4-benzoquinone, 2,6-dichloro-1,4-benzoquinone, 2-chlorophenol, oxalic acid, and maleic acid (57–62) (see also Lignin).

Hydroquinones are readily oxidized to the corresponding quinones without ring chlorination (63–64). 4-Nitrophenol is converted in low yield to 2,6-dichloro-1,4-benzoquinone (65). By contrast, both 2,4-dinitrophenol and 2,4,6-trinitrophenol do not react with chlorine dioxide. 4-Hydroxybenzaldehyde is rapidly oxidized to 1,4-benzoquinone by chlorine dioxide at pH 6 and below, but the yield is only 21–27%. Reaction of chlorine dioxide with thiophenols appears to be confined to the sulfur atom with formation of disulfide, disulfoxide, and sulfonic acid. The disulfide can react further to form disulfoxide and sulfonic acid (66).

In dilute aqueous solution, there is no reaction of chlorine dioxide with ethanol or 2,3-butanediol at pH 7 and 80°C, but reactions take place at pH 1 (67). The carbonyl compounds acetaldehyde, butyraldehyde, acetoin, and biacetal react more rapidly than the alcohols at pH 1, and they are even more reactive at pH 7. Principal oxidation products of biacetal and 2,3-butanediol at pH 1 are acetic acid and carbon dioxide.

A study of the chlorine dioxide oxidations of cellulose (qv) oligosaccharides at pH 3 and 56°C (49) indicated some oxidation of both the aldehyde group (before and after hydrolysis of the glycosidic linkage) and the 2,3-glycol function; in all cases much starting material was recovered. Small amounts of chlorine dioxide inhibit chlorine oxidation of a number of typical low molecular weight carbohydrates (qv), probably by scavenging the free radicals of a chlorination chain-reaction mechanism (68).

Manufacture

Because of its explosive character, chlorine dioxide must be manufactured where it is used. Requirements for the chemical dictate the method of manufacture: large consumption, eg, the bleaching of wood pulp calls for reduction of sodium chlorate [from Cl(V) to Cl(IV)] by means of rather complex processes; in smaller applications, the less complicated oxidation of chlorite is favored. The reduction of sodium chlorate to chlorine dioxide is usually carried out in strongly acidic solution which converts chlorate to chloric acid. Suitable reducing agents include sodium chloride, hydrochloric acid, sulfur dioxide, and methanol. Oxalic acid, nitrogen dioxide, ethanol, and sugar are often used in the laboratory. The three commercial processes based on chlorate use sulfuric acid, hydrochloric acid, or sulfur dioxide; the reducing agents are NaCl, HCl, and SO_2, respectively. Hydrochloric acid is always the actual reducing agent, although little or none of its oxidation product, chlorine, appears as product when sufficient SO_2 is fed to the process to reduce Cl_2 to Cl^- as fast as it is formed. The sulfur dioxide, or Mathieson process is distinguished from the other two by yielding chlorine dioxide substantially free of chlorine. Mixtures of chlorine and chlorine dioxide are obtained in the other two processes.

The mechanism for the reaction of chloride with chlorate is illustrated below (eqs. 7–8) (35).

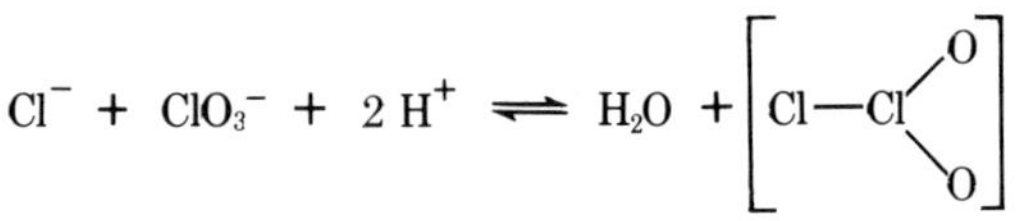

$$Cl^- + ClO_3^- + 2 H^+ \rightleftharpoons H_2O + \left[Cl-Cl\begin{matrix}O\\O\end{matrix} \right] \tag{7}$$

$$2\left[Cl-Cl\begin{matrix}O\\O\end{matrix} \right] \rightarrow Cl_2 + 2\ ClO_2 \text{ (rate determining)} \tag{8}$$

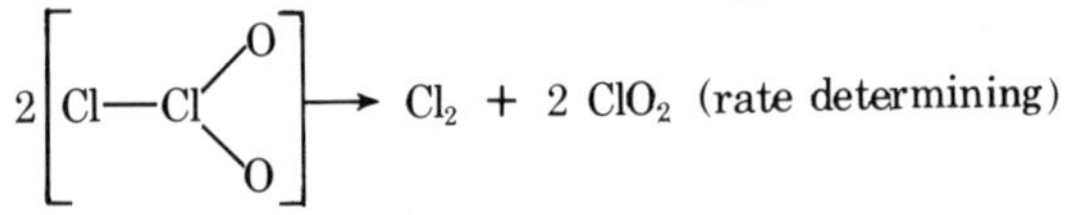

This and several other mechanisms of formation of chlorine dioxide from sodium chlorate have been reviewed (48). When the Cl^- to ClO_3^- ratio is high, equation 9 is responsible for a reduction in the yield of chlorine dioxide (69).

$$HClO_3 + 5\ HCl \rightarrow 3\ Cl_2 + 3\ H_2O \tag{9}$$

Present versions of the sulfuric acid process for chlorine dioxide are Erco's ER-3 process and Hooker's single vessel process (SVP). Both of these developed from the older ER-2 process that required large amounts of sulfuric acid (4.5–5.0 M) and produced a waste liquor containing much sulfuric acid and sodium bisulfate with small amounts of sodium chloride and chlorate. The ER-3 and SVP processes were developed in the late 1960s (70–74). A detailed description of the SVP has appeared in a patent assigned to Hooker (75). Metal ion catalysts and vanadium pentoxide are disclosed in the Hooker patents for improving the yield of chlorine dioxide.

A system using two or three chlorine dioxide generators in series, to provide high acidity in the first generator and low acidity in the last, is reported in Erco patents (76–77). Yields of chlorine dioxide, based on sodium chlorate ranging from 87.7–99.3%, are claimed. More than 33 plants are currently estimated to be using one version or another of the sulfuric acid-based chlorine dioxide process in the United States and Canada.

The hydrochloric acid process, also known as the Kesting process, was originally developed for making ClO_2 to be used in the manufacture of sodium chlorite. Sodium chloride, suitable for recycle to a chlorate cell, is formed instead of sodium sulfate which creates a disposal problem if it cannot be consumed in a pulping operation (78–83).

The sulfur dioxide process involves the reaction of gaseous sulfur dioxide with a sodium chlorate solution containing some sulfuric acid. This older process is often employed to produce chlorine dioxide intended for conversion to sodium chlorite since the gas produced contains little or no Cl_2. A single-vessel generator patented by Hooker (84) is used to precipitate sodium bisulfate *in situ* at high acidity and sodium sulfate at lower acidity.

An increase in chlorine dioxide yields from 68–73.3% in early processes to 86–97.6%

in more recent processes has been achieved through use of sulfuric acid and small amounts of sodium chloride (85–86).

Manufacture of chlorine dioxide from sodium chlorate in electrolytic cells divided by permselective membranes has been proposed (87–89). The principles of electrodialysis (qv) and electrolysis are utilized in the operation of this method. The formation of chlorine dioxide by this process is essentially similar to the chemical methods of manufacture previously described. A possible economic advantage of the electrolytic method is the production of salt-free caustic soda; however, the cost of expensive membranes used in the cells may offset this advantage.

For small-scale uses chlorine dioxide is produced by chemical oxidation of sodium chlorite through reaction with chlorine or by anodic oxidation in an electrolytic cell. Sodium chlorite may also be converted to chlorine dioxide by disproportionation in the presence of acids.

Economic Aspects

Production statistics for chlorine dioxide can only be estimated since it must be generated captively. A reasonably accurate estimate can be made based on consumption of sodium chlorate by the pulp and paper industry. Table 1 shows that production of chlorine dioxide in this use has increased by a factor of 10 from 1955 to 1975.

Table 1. U.S. Production of Chlorine Dioxide, Thousands of Metric Tons

1955	1960	1965	1970	1975
7.8	22.4	50.3	78.2	79.9

Production of chlorine dioxide is energy-intensive because it requires electrolysis of sodium chloride to sodium chlorate. An increase in the cost of electricity by 1¢/(kW·h) increases the cost of chlorine dioxide by ca 11¢/kg. The cost of technical sodium chlorite required to produce 1 kg of ClO_2 is approximately $3.24 based on published 1977 prices ($1.79/kg) and a 90% yield.

Analysis

Analysis of chlorine dioxide relies on the use of reagents normally used for chlorine because of its similar reactivity. Difficulties arise, however, in attempting to analyze solutions containing mixtures of chlorine dioxide, chlorite, hypochlorite, and chloramines (qv). The volatility of chlorine dioxide from aqueous solution can also cause problems. The iodometric method of analysis is useful for standardizing solutions of chlorine dioxide. Reaction of chlorine dioxide with potassium iodide in mildly alkaline solution (ca pH 9) liberates one equivalent of iodine (eq. 10) which may then be titrated with thiosulfate by standard procedures.

$$2 ClO_2 + 2 KI \rightarrow 2 KClO_2 + I_2 \tag{10}$$

Since chlorine dioxide disproportionates slowly at ca pH 9, it is important to carry out this determination by pipetting the ClO_2 solution into an iodide solution buffered at pH 9 with $Na_2CO_3/NaHCO_3$ or phosphates. This method is suitable for analysis of ClO_2 in the presence of chlorite since ClO_2^- does not react with iodide under these conditions. However, chlorine dioxide and chlorite are both reduced to chloride by KI in strong mineral acid (eqs. 11–12):

$$2\ ClO_2 + 10\ KI + 8\ HCl \rightarrow 10\ KCl + 5\ I_2 + 4\ H_2O \qquad (11)$$

$$ClO_2^- + 4\ KI + 4\ HCl \rightarrow 2\ I_2 + 4\ KCl + Cl^- + 2\ H_2O \qquad (12)$$

A determination of ClO_2^- in the presence of ClO_2 can be performed by first titrating ClO_2 at ca pH 9 followed by a titration of $ClO_2 + ClO_2^-$ in strong acid. There is an amperometric method for the determination of free chlorine, chloramines, chlorine dioxide, and chlorite in any combination (90). However, the method requires a time-consuming disproportionation reaction of chlorine dioxide to chlorite and chlorate at pH 12 prior to titration of hypochlorite. The spectrophotometric determination of ClO_2 at λ = 360 nm in solutions containing minor amounts of other chlorine oxidants such as Cl_2, OCl^-, $HOCl$, and chlorite has been widely employed in laboratory studies (21,23).

Water treatment applications require colorimetric methods that are applicable at concentrations of 0.5–10 mg ClO_2/L. These are summarized in Table 2.

Table 2. Colorimetric Determination of Chlorine Dioxide

Name of method	Reagent	pH	Color change	Reference
OTO	o-tolidine plus oxalic acid	1.0–2.0	yellow, 400–490 nm	91
ACVK	Acid Chrome Violet K	8.1–8.4	colorless, 550 nm	92
tyrosine	tyrosine (p-hydroxy phenylalanine)	4.5–4.6	red, 490 nm	93
H-acid	8-amino-1-naphthol-3,6-disulfonic acid	4.1–4.3	bluish pink, 525 nm	94
DPD	diethyl-p-phenylene diamine	6.0–6.5	red, 555 nm	95–97

The DPD method (97) appears to be the most dependable and convenient since the red color of the oxidized reagent develops rapidly and persists over a reasonable time period. It can be applied to mixtures of Cl_2, ClO_2, and ClO_2^- as follows: chlorine reacts with malonic acid before $\frac{1}{5}\ ClO_2$ is determined at pH 6.0–6.5; the sum of $Cl_2 + \frac{1}{5}\ ClO_2$ is found at pH 6.0–6.5 without prior addition of malonic acid; a third sample, acidified with dilute H_2SO_4 (pH 0–1), reacts with KI to liberate I_2; and the iodine equivalent to the sum of $Cl_2 + ClO_2 + ClO_2^-$ is determined with DPD after adjustment of the pH.

Health and Safety Factors

Chlorine dioxide has long been recognized as being biologically active and early studies (98–100) indicated that it possesses both bactericidal and viricidal properties as measured against *E. coli.*, common water pathogens, and mouse-adapted poliomyelitis viruses. The apparent mechanism of bacterial disinfection by chlorine dioxide is disruption of protein synthesis (101) (see Disinfectants and antiseptics).

Chlorine dioxide is effective against 11 strains of viruses (102). On an equimolar basis chlorine dioxide is approximately ten times as efficient as chlorine(I) (HOCl +

OCl⁻) in water with high chlorine demand, and twice as efficient in the absence of demand. Chlorine dioxide is effective for the detoxification of botulism toxin in water (103). It is a toxic chemical at relatively low levels of concentration in either air or water. Exposure of rats to 1 ppm ClO$_2$ in air for 5 h/d, 5 d/wk for 2 mo had no adverse effects; however, 10 ppm ClO$_2$ in air for 2 h/d for 30 d caused localized bronchopneumonia and elevated leukocyte counts (104–106). Short-term exposure (15 min, 2–4 times per day for 1 mo) to 5 ppm ClO$_2$ did not alter the blood composition or lung histology of rats. Since 1977, the NIOSH standard for maximum average concentration of chlorine dioxide in air is 0.1 ppm on a time-weighted average basis (107). Short term exposure to higher concentrations may cause irritation of the eyes and nose; however, the effect is only temporary after the source of irritation has been eliminated. The maximum working concentration is ca 5 g ClO$_2$/L H$_2$O at room temperature if explosive ClO$_2$ concentrations in the vapor space are to be avoided; however, the vapor pressure of chlorine dioxide even in dilute aqueous solutions is significant and certain precautions are required to prevent over exposure to the chlorine dioxide being released.

Toxic effects of chlorine dioxide in drinking water of rats have been evaluated in a two-year study (108). No toxic effect on rats was observed at 10 mg/L; however, at 100 mg/L the mortality rate increased. A more recent study indicates that 0.5 mg/L ClO$_2$ produces no abnormal effects on rats but at 5 mg/L there is a significant reduction in weight gain; 10 and 100 mg/L have a pronounced toxic effect. A maximum residual concentration of 0.4–0.5 mg/L ClO$_2$ in drinking water is suggested as a result of this study (109).

Concentrations of ClO$_2$ above ca 10% in air may cause low-order explosions or puffs. Good ventilation and effective hoods are necessary when working with chlorine dioxide to minimize or eliminate toxic inhalation hazards.

Uses

Uses for chlorine dioxide rely on its unique oxidizing properties and broad-spectrum biological activity. The bulk of chlorine dioxide is consumed for the bleaching of wood pulp (see Pulp). Chlorine dioxide is produced from sodium chlorate at the mill for captive use. Estimated United States volume was 96,600 metric tons in 1971. Chlorine dioxide has many small-scale applications in which the high costs and hazards of on-site production from sodium chlorate cannot be justified. A safe and efficient method for on-site preparation of chlorine dioxide is the reaction of dilute solutions of sodium chlorite and chlorine in a chlorine dioxide generator (Fig. 2).

The formation of carcinogenic chloroform and other trihalomethanes in drinking water can be prevented by treating with chlorine dioxide instead of chlorine (110–111). Chlorine dioxide is effective in controlling phenolic tastes, odors, manganese, and iron in potable water. Other water treatment uses include biological control of cooling-tower water, eg, in cases where chlorine demand is high, and in food processing flumes, paper mill white-water systems, and sewage plant effluents (112). It is also employed for bleaching textiles and in controlling odors in rendering and fish meal plants (113).

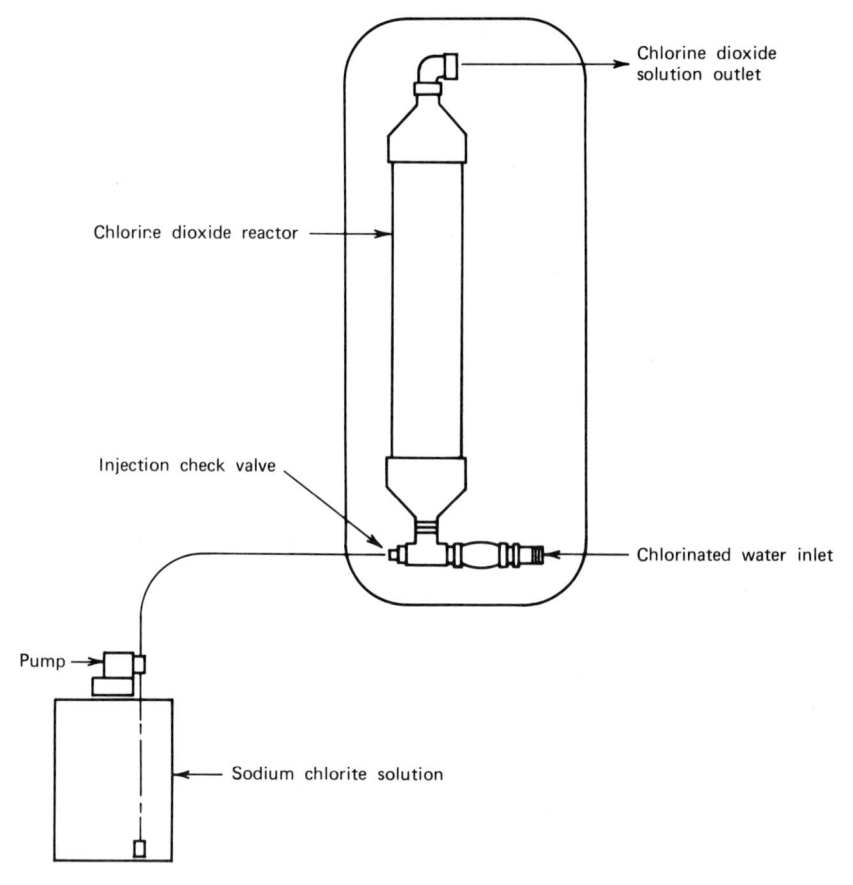

Figure 2. Apparatus for continuous on-site generation of dilute chlorine dioxide solution. From Olin Water Services (114); a similar apparatus can be obtained from Wallace and Tiernan, a division of Pennwalt Corporation.

CHLOROUS ACID AND CHLORITES

Physical Properties

Most of the physical and chemical properties of $HClO_2$ have been inferred from investigations of acidified solutions of alkali chlorites. The existence of $HClO_2$ is based primarily upon spectroscopic evidence. Absorption spectra of chlorous acid and chlorite ion (6,16–20) exhibit no transitions in the visible region. The corresponding salts appear white when pure, although decomposition yielding traces of chlorine dioxide may impart a faint greenish tinge. In chlorite ion the Cl–O bond length is ca 0.156 nm and the bond angle is 108–110° (115–118). Consistent with this structure, the three vibrational bands in the low-temperature ir spectrum of matrix-isolated sodium chlorite are ($\bar{\nu}$ in cm^{-1}) $\bar{\nu}_1$, 795 (^{35}Cl), 790 (^{37}Cl); $\bar{\nu}_2$, 401; and $\bar{\nu}_3$, 831.3 (^{35}Cl), 823.2 (^{37}Cl). The $\bar{\nu}_1$ and $\bar{\nu}_3$ transitions are broad and ill-defined bands in the ir spectrum of crystalline $NaClO_2$ at room temperature (119).

Crystallization of sodium chlorite from aqueous solution yields $NaClO_2 \cdot 3H_2O$

[49658-21-1] below 37.4°C and anhydrous NaClO₂ [7758-19-2] above 37.4°C (120). The triclinic trihydrate has two Na⁺ and two ClO₂⁻ ions in its unit cell (117) and there are four NaClO₂ in the unit cell of monoclinic NaClO₂ (118). Important macroscopic properties of sodium chlorite are summarized in Table 3.

Table 3. Physical Properties of Sodium Chlorite

Property	Analytical grade, NaClO₂ (mol wt = 90.44)	Technical grade
mp	180–200°C, decomp	218–219°C
appearance	colorless crystalline solid, deliquescent	white flakes or powder; deliquescent
density, g/cm³		
crystals	2.468	2.19
bulk, packed	1.176	1.10

The solubilities of pure and technical grade sodium chlorite are shown in Figure 3.

An aqueous solution containing 50 wt % technical sodium chlorite tends to supercool with no spontaneous crystallization occurring down to 0°C. Seeding with crystals of anhydrous NaClO₂ does not induce crystallization at this temperature, although crystallization can begin upon seeding with trihydrate (121).

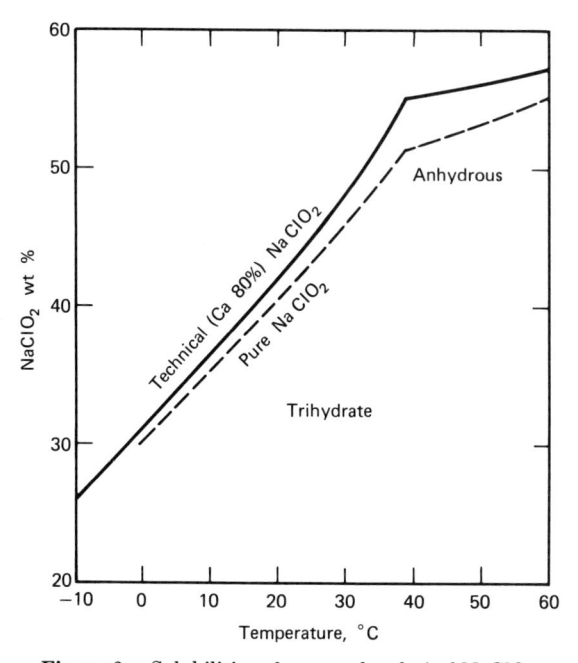

Figure 3. Solubilities of pure and technical NaClO₂.

Chemical Properties

The enthalpy and free energy of formation for $HClO_2$ at 298°K are $\Delta H° = -51.9$ kJ/mol (-12.4 kcal/mol) and $\Delta F° = 5.9$ kJ/mol (1.4 kcal/mol) (122–123). Corresponding data for aqueous chlorite ion are $\Delta H° = -66.6$ kJ/mol (-15.9 kcal/mol) and $\Delta F° = 17.2$ kJ/mol (4.1 kcal/mol) (15).

Dissociation constants of $HClO_2$ as determined by several investigators are summarized in Table 4.

Mechanistic interpretations of decomposition in acidic solutions offer four alternative nonviolent processes: (1) Slow disproportionation in concentrated neutral and alkaline solutions requiring temperatures near boiling (eq. 13) (128):

$$3\ ClO_2^- \rightarrow 2\ ClO_3^- + Cl^- \qquad (13)$$

(2) Disproportionation, accompanied by release of approximately 5% oxygen based on chlorite consumed, upon heating of dry $NaClO_2$ at 175–200°C (128); (3) Rapid photochemical decomposition yielding substantial quantities of oxygen and chloride in addition to chlorate. The stoichiometry follows equation 14 at 8.4 pH and equation 15 at 4.0 pH (129):

$$6\ ClO_2^- \xrightarrow{h\nu} 2\ ClO_3^- + 4\ Cl^- + 3\ O_2 \qquad (14)$$

$$10\ ClO_2^- \xrightarrow{h\nu} 2\ ClO_3^- + 6\ Cl^- + 3\ O_2 + 2\ ClO_4^- \qquad (15)$$

(4) Decomposition of acidified solutions of chlorite accompanied by formation of chloride, chlorine dioxide, and chlorate in variable quantities normally without formation of chlorine.

Several chlorites explode or detonate when struck or heated. These include the salts of the heavy metals ions Hg^+, Tl^+, Pb^{2+}, Cu^{2+}, and Ag^+, and ammonium and tetramethylammonium chlorite. However, $NaClO_2$ is not shock-sensitive unless contaminated with combustibles. The decomposition and disproportionation reactions of sodium chlorite in acidic media, ie, the reactions belonging to category (4) above, have much significance in industrial processes, including bleaching and generation of chlorine dioxide. The kinetics and stoichiometries of these reactions depend on the chlorine(III) ($HClO_2 + ClO_2^-$) and hydrogen ion concentrations, and on the presence of chloride ion (1,7,16,42,128a). Reactions carried out without chloride ion initially present tend to approximate the stoichiometry (eq. 16):

$$4\ HClO_2 \rightarrow 2\ ClO_2 + ClO_3^- + Cl^- + 2\ H^+ + H_2O \qquad (16)$$

Addition of chloride ion, eg, acidification with dilute HCl, increases the reaction rate and changes the stoichiometry (eq. 17):

Table 4. Acid Dissociation Constants of Chlorous Acid

K_a	Temperature, °C	Reference
1.07×10^{-2}	19–20	124[a]
1.01×10^{-2}	23	125
0.49×10^{-2}	25	126
1×10^{-2}	25	127
1.10×10^{-2}	25	124[b]

[a] By extrapolation of K_a to infinite dilution.
[b] Calculated from thermodynamic data.

$$5\ HClO_2 \xrightarrow{Cl^-} 4\ ClO_2 + H^+ + Cl^- + 2\ H_2O \tag{17}$$

This equation reflects the theoretical maximum of chlorine dioxide obtainable by disproportionation of chlorous acid. However, more chlorine dioxide can be obtained by the reaction of chlorites with certain oxidizing agents.

At least three mechanisms have been proposed for acid decomposition reactions of chlorites. Mechanism 1 (eq. 18–23) for reactions in the presence of 1.2–2.0 M HClO$_4$ and initial absence of chloride (21):

Mechanism 1

$$HClO_2 + HClO_2 \rightarrow HOCl + H^+ + ClO_3^- \tag{18}$$

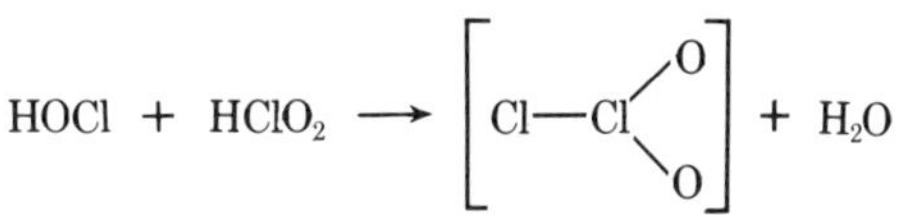

$$HOCl + HClO_2 \rightarrow [Cl-Cl(O)(O)] + H_2O \tag{19}$$

$$HOCl + H^+ + Cl^- \rightleftharpoons Cl_2 + H_2O \tag{20}$$

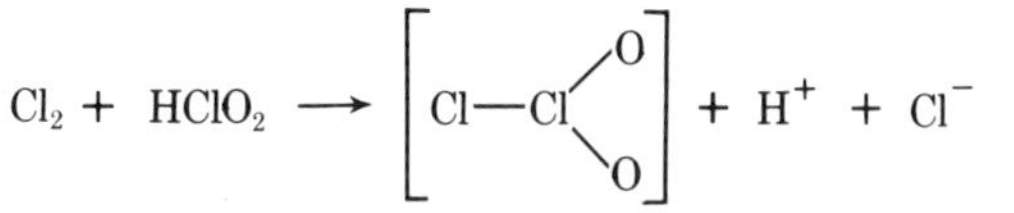

$$Cl_2 + HClO_2 \rightarrow [Cl-Cl(O)(O)] + H^+ + Cl^- \tag{21}$$

$$[Cl-Cl(O)(O)] + H_2O \rightleftharpoons Cl^- + ClO_3^- + 2\ H^+ \tag{22}$$

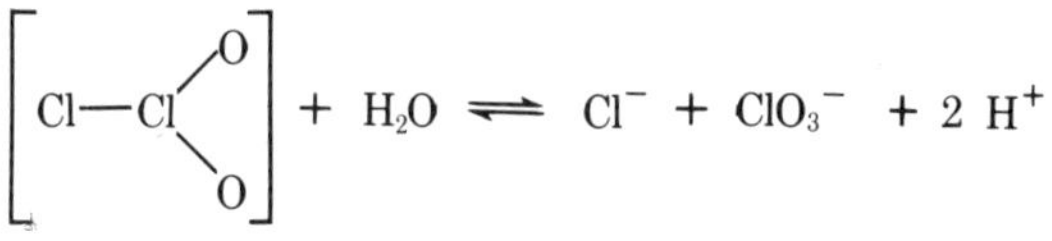

$$2\ [Cl-Cl(O)(O)] \rightarrow Cl_2 + 2\ ClO_2 \tag{23}$$

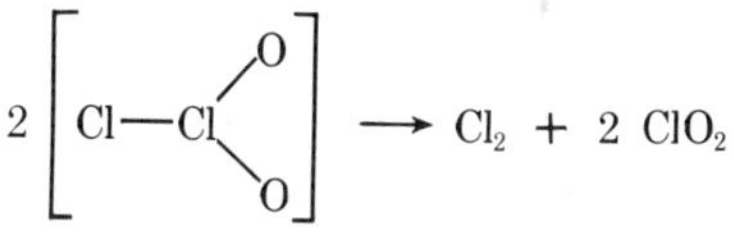

The intermediate [Cl–ClO$_2$] (31) is either hydrolyzed to Cl$^-$ and ClO$_3^-$ via the concentration-independent equilibrium (eq. 22) or metathesis via the concentration-dependent second-order reaction (eq. 23). This suggests that the yield of chlorine dioxide can be optimized by measures that increase the steady state concentration of [Cl–ClO$_2$].

The catalytic effect of chloride ion can be attributed to Mechanism 2 (eqs. 24–27), which is faster than Mechanism 1 and initiated by chloride ion to give the rate-controlling reaction shown below:

Mechanism 2

$$HClO_2 + Cl^- + H^+ \rightarrow 2\ HOCl \tag{24}$$

This step may be followed by:

$$HOCl + Cl^- + H^+ \rightleftharpoons Cl_2 + H_2O \tag{25}$$

and

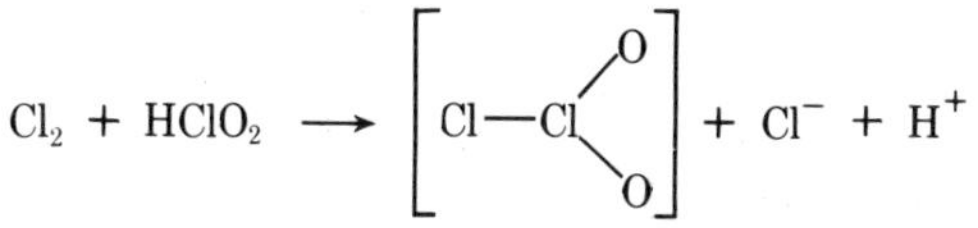

$$Cl_2 + HClO_2 \rightarrow [Cl-Cl(O)(O)] + Cl^- + H^+ \tag{26}$$

or by the slower oxidation of $HClO_2$ by hypochlorous acid:

$$HOCl + HClO_2 \rightarrow \left[Cl-Cl\begin{smallmatrix}O\\ \diagdown\\ \diagup\\ O\end{smallmatrix} \right] + H_2O \qquad (27)$$

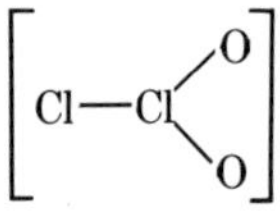

The final steps are the first and second order reactions of $[Cl-ClO_2]$ as noted in Mechanism 1. Mechanism 2 predicts correctly that the yield of chlorine dioxide should increase with increasing concentrations of hydrogen and chloride ions.

A third decomposition, Mechanism 3 (eq. 28), assumes significance when $[H^+]$ is lowered so that $[HClO_2] \approx [ClO_2^-]$, because $HClO_2$ reacts more rapidly with ClO_2^- than it does with itself, thus causing equation 28 to replace equation 18 as the rate-controlling step. Mechanism 3:

Mechanism 3

$$HClO_2 + ClO_2^- \rightarrow HOCl + ClO_3^- \qquad (28)$$

The decomposition of ClO_2^- in the absence of initial chloride, is determined by Mechanisms 1 and 3 at 0–3 pH (37). Chlorine is not a product in reactions with dilute acids because of the rapidity of reactions of Cl_2 and HOCl with chlorous acid and chlorite (eqs. 26 and 27). However, chlorine has been reported as a product in reactions with more concentrated HCl (128a,130,131) although this has not been confirmed in a subsequent study (132).

Oxidations with chlorine and hypochlorous acid in acidic and neutral media are in general agreement with Mechanisms 1–3 with the important exception that the slow initial steps in decompositions (eqs. 18, 24, and 28) are bypassed, so that the oxidations proceed far more rapidly.

Overall stoichiometries are consistent with the following reactions in acidified and neutral solutions (eqs. 29–34):

Acidified (pH $\ll$ pK_a of $HClO_2$):

$$Cl_2 + 2\,HClO_2 \rightarrow 2\,ClO_2 + 2\,Cl^- + 2\,H^+ \qquad (29)$$

$$Cl_2 + HClO_2 + H_2O \rightarrow ClO_3^- + 2\,Cl^- + 3\,H^+ \qquad (30)$$

$$HOCl + 2\,HClO_2 \rightarrow 2\,ClO_2 + Cl^- + H^+ + H_2O \qquad (31)$$

$$HOCl + HClO_2 \rightarrow ClO_3^- + Cl^- + 2\,H^+ \qquad (32)$$

Weakly acidic and neutral (pH $\gg$ pK_a of $HClO_2$):

$$HOCl + 2\,ClO_2^- \rightarrow 2\,ClO_2 + Cl^- + OH^- \qquad (33)$$

$$HOCl + ClO_2^- \rightarrow ClO_3^- + Cl^- + H^+ \qquad (34)$$

Chlorate is formed rather slowly in alkaline solutions. The fact that chlorine and HOCl are indeed the oxidants and ClO_2^- or $HClO_2$ the reducing agent has been demonstrated in radioactive tracer studies (31,40).

It is of technical interest to manipulate the reaction conditions to maximize the yield of chlorine dioxide, since chlorite converted to chlorate is lost for purposes of chlorine dioxide generation. Chlorine dioxide formation is maximized by increasing concentrations of chlorine and sodium chlorite. Addition of a strong mineral acid has the same effect because it drives both equilibria: $H_2O + [Cl_2O_2] \rightleftharpoons ClO_2^- + Cl^- + 2$

H^+; and $Cl_2 + H_2O \rightleftharpoons Cl^- + H^+ + HOCl$ to the left, thus inhibiting consumption of Cl_2O_2 via the undesired route (eq. 28) and accelerating formation of ClO_2. The same reasoning can be applied to chloride ion.

The majority of reported data are in good agreement with these fundamental considerations. For instance, only 7.8% of ClO_2^- at 0.024 M is oxidized to ClO_3^- by 0.02 M chlorine with perchloric acid at 0.4 M, as opposed to 28% at 7.5 pH (31). Chlorate ion is the main product at pH 8–10 in contrast to pH 4 where ClO_2 is the predominant product (128a). In reactions carried out in the presence of 0.2 M $HClO_4$ with the ratio of $NaClO_2$ to Cl_2 equal to 2.0, 90% of ClO_2^- is converted to ClO_2 when $[NaClO_2]$ is 10^{-2} M vs 30% conversion when $[NaClO_2]$ is 10^{-4} M (133). Additional data are reported in references 35, 48, 134–139. Aside from chlorine and hypochlorous acid, chlorite (and chlorous acid) can also be oxidized by bromine (140), nitrogen dioxide (141), and peroxydisulfate ion (142). Electrochemical oxidation on a Pt anode in presence of Na_2SO_4 yields chlorine dioxide (143).

Common reducing agents are oxidized by chlorous acid and chlorites, although ClO_2^- is a rather weak oxidant in neutral and more so in alkaline solutions. Iodide ion present in excess is thus oxidized to iodine in neutral and acidic solutions (144); an excess of chlorite yields IO_3^- via I_2 (145). Sulfide and sulfite ions are oxidized to sulfate in acidic solution (144,146). Cyanate and sulfate ion are formed by thiocyanate ion oxidation (147). Lower valent metal ions such as V^{3+}, Cr^{2+}, Fe^{2+}, Mn^{2+}, Ni^{2+}, Co^{2+}, and U^{4+} are oxidized to higher oxidation states and ClO_2^- is reduced to Cl^- (148–151).

Hydrogen peroxide reduces chlorite only slowly at 40–80°C and pH 3.7–9.0 (152). At room temperature, chlorous acid is stabilized by H_2O_2 (153), probably because its decomposition product, ClO_2, is readily reduced to $HClO_2$ in accordance with equation 1. Acidified solutions of sodium chlorite are corrosive to iron and its alloys; therefore, titanium has been proposed as a suitable material of construction. Sodium nitrate may be added as corrosion inhibitor, eg, as in the commercial product Textone 50 which contains approximately 50% $NaClO_2$ and 35% $NaNO_3$.

Organic Chemistry. Mixing solid sodium chlorite with combustibles may result in violent explosions occurring spontaneously or upon grinding, sparking or shock. However, many organic compounds are oxidized only partially and comparatively slowly by ClO_2^- in aqueous solution. Aldehydes (including certain carbohydrates with CHO functionality) are readily oxidized to the corresponding carboxylic acids in weakly acidic and neutral media (eq. 35) (17,154–158).

$$R-CHO + H^+ + ClO_2^- \rightarrow R-COOH + HOCl \qquad (35)$$

The HOCl produced interacts with chlorite yielding chlorine dioxide and some Cl^- and ClO_3^- (eqs. 25 and 26). The reaction has been used to activate sodium chlorite with formaldehyde for bleaching (42). However, ClO_2 is not the only species that accounts for the bleaching properties of sodium chlorite. In fact, much bleaching technology is aimed at suppressing the formation of chlorine dioxide.

Manufacture and Shipment

Dilute chlorous acid essentially free of other compounds can be obtained by the reaction of barium chlorite [14674-74-9] with an equivalent of dilute sulfuric acid and separating the insoluble $BaSO_4$ (eq. 36).

$$Ba(ClO_2)_2 + H_2SO_4 \rightarrow BaSO_4 \downarrow + 2\ HClO_2 \qquad (36)$$

The barium chlorite is readily prepared from a solution of Ba(OH)$_2$ and a predetermined amount of hydrogen peroxide with ClO$_2$. Certain metals, such as aluminum, magnesium, zinc, and cadmium react with ClO$_2$ in accordance with equation 37:

$$M + x\,ClO_2 \rightarrow M(ClO_2)_x \tag{37}$$

Manufacture of sodium chlorite by the reaction of chlorine dioxide with sodium amalgam has been proposed (159–161), but commercialization of this process has not been attempted since yields are unsatisfactory owing to reduction of ClO$_2^-$ to Cl$^-$.

The commercial manufacture of sodium chlorite depends entirely on chlorine dioxide made from sodium chlorate (see Chlorine dioxide, synthesis and manufacture). In general, chlorine dioxide is absorbed in caustic soda containing a reducing agent. Only hydrogen peroxide continues to be used for this purpose because other reducing agents tend to introduce impurities or coloration. Older processes based on reduction with carbonaceous materials have only historical significance (162–164). Raw material requirements for hydrogen peroxide exceed theoretical amounts (eq. 1) since H$_2$O$_2$ decomposes rapidly in NaOH. Also, an excess of H$_2$O$_2$ is important to prevent disproportionation yielding chlorate (87,165). A small excess of NaOH is also maintained to stabilize the product. Sodium chloride may be added to reduce the NaClO$_2$ assay to approximately 80 wt % of dry product which is desirable for product safety and reactivity reasons. Concentration and drying of the product may be carried out by spray, drum or other drying equipment.

Flaked commercial product is packed in polyethylene-lined drums; solutions containing 50 wt % of technical grade product may be shipped by tank truck or tank car. Shipment by air is prohibited. Carloads or truckloads require a "dangerous" placard, and lcl and ltl shipments a yellow label. The trade names Textone and C-2 apply to dry and solution products.

Economic Aspects

Commercially significant quantities of sodium chlorite are produced in three areas of the noncommunist world, ie, Japan, the European Economic Community (EEC), and the United States.

The Japanese Ministry of International Trade and Industry (MITI) reports sodium chlorite production in Japan at 6200 metric tons in 1973 and 5400 t in 1974. However, it is not clear if statistics are kept on a consistent NaClO$_2$ basis. Nippon Soda is the major Japanese producer.

EEC capacity has been estimated at ca 18,000 metric tons in 1977 with consumption ca 16,800 t. The French concern Produits Chimiques Ugine Kuhlmann, largest producer in the EEC, is also a manufacturer of hydrogen peroxide and sodium chlorate, two essential raw materials for sodium chlorite.

Olin Corporation is the sole producer in the United States. Mathieson Alkali Works, a predecessor company of Olin, pioneered the manufacture and many applications of sodium chlorite in the mid-1940s.

Free-world demand for sodium chlorite has a history of depending on expansionary and recessive phases of the general economy as some of the consuming industries, ie, textiles, pulp, and paper are rather sensitive to general economic conditions. United States demand has become less dependent on the fortunes of the textile industry as consumption of sodium chlorite by this industry has declined from approximately 40% of the United States total demand in 1972 to less than 30% in 1976.

Consumption for water treatment applications has grown since 1975 with continued growth in excess of GNP (gross national product) rates expected through the mid-1980s.

United States prices averaged over all types of products, ie, solution, dry, and bulk, as well as packaged, have risen from ca 88¢ in 1970 to ca $1.32 in 1976 per kg of technical grade (80% NaClO$_2$) product.

Trade publications for the chemical process industry listed solid NaClO$_2$, technical grade, in carload lots, at $1.28/kg in 1976 and at $1.74/kg in 1977. Costs are sensitive to the cost of sodium chlorate.

Health and Safety Factors

The toxic effects of sodium chlorite on warm-blooded animals, including man, are related to its character as oxidant. Hemoglobin is oxidized to methemoglobin which inactivates erythrocytes for purposes of oxygen transport. The LD$_{50}$ of NaClO$_2$ is ca 140 mg/kg in rats; a single dose of 30 g could be lethal to a human adult (166). Investigations of chlorite-induced methemoglobia in cats have indicated possible hemolytic effects, ie, single, comparatively large sublethal doses caused breakdown of erythrocyte membranes and release of heme into the serum (167). However, albino rats were found to tolerate 100 mg NaClO$_2$/L in drinking water for 2 yr without increased mortality (108). Dolphins and seals appear to tolerate 0.5–2.0 mg/L NaClO$_2$ in an aquarium with no ill effects (168). Toxicity data concerning several warm-blooded and aquatic species are summarized in Table 5.

Literature on chlorites as disinfectants is far less voluminous than that on the more effective chlorine dioxide, although sodium chlorite does have bactericidal and algicidal properties.

No wood or lumber, eg, wooden floors, should be present where the product is handled. Spillages of sodium chlorite solutions must not be cleaned with rags. Spilled solutions or solids should be flushed with copious quantities of water.

Contact with acids results in the release of toxic and explosive chlorine dioxide. This gaseous decomposition product may reach explosive concentrations when sodium chlorite comes into contact with concentrated acid. Fires involving NaClO$_2$ can be extinguished by application of large quantities of water. Unopened drums in the affected area should be cooled by spraying water onto them. The solid should be removed

Table 5. Animal Toxicity of Sodium Chlorite[a]

Species	Test	Toxicity
mallard duck	dietary LC$_{50}$[b]	10,000 ppm
bobwhite quail	dietary LC$_{50}$[b]	10,000 ppm
monkeys	eye irritation[c]	no irritation
blue gill, *Lepomis Macrodisus*	48 h TL$_{50}$[d]	208 mg/L
daphnia magna	48 h TL$_{50}$[d]	0.29 mg/L
	no effect level	0.10 mg/L
rainbow trout, *Salmo Gairdneri*	48 h TL$_{50}$[d]	50 mg/L

[a] Source: Olin Corp.
[b] LC$_{50}$ = concentration which kills 50% of the test organisms.
[c] Section 191.12, *Federal Hazardous Substances Act*.
[d] TL$_{50}$ = toxic level for 50% of the test organisms.

from its containers with a clean, dry metal scoop that is reserved for this chemical only. Protective equipment, eg, goggles, neoprene gloves, and apron, should be worn by handlers. Contaminated clothing should be washed quickly and thoroughly with water to avoid fire hazards. Drums containing the product must be stored upright in a cool, dry place. Covers should always be replaced immediately after removal of material to avoid absorption of moisture that can lead to caking. Containers should not be dropped, rolled, or skidded and should be washed thoroughly with water when empty.

Skin, eyes, or mucous membranes having been exposed to sodium chlorite should be washed for 15 min in gently flowing, clean water. A physician must be called immediately in case of eye contact or if skin irritation persists. Persons having ingested sodium chlorite should drink large quantities of water or milk. This should be followed with milk of magnesia, vegetable oil, or beaten eggs. A physician must be called immediately.

Uses

A significant fraction of the sodium chlorite consumed in the United States serves as source of chlorine dioxide in applications where required volumes are comparatively small. Applications of sodium chlorite include the bleaching and stripping of textiles, and employment as an industrial disinfectant and oxidant. Potential uses in pulping and pulp bleaching processes have been described (169–171). Sodium chlorite is used as an oxidant for removal of nitrogen oxide pollutants from industrial off-gases (172–174). Consumption of sodium chlorite by the U.S. textile industry has declined in recent years, although worldwide textile bleaching uses are by no means insignificant. Chlorite bleaching processes are broadly applicable to a variety of cellulosic and synthetic fibers, including cottons, acetates, rayons, and synthetics such as polyesters, acrylics, and nylons. Cotton is not degraded by sodium chlorite, since oxidation reactions are highly selective toward lignin and hemicellulose components of the fiber. It has also been claimed that chlorite processes are superior to other bleaching methods for synthetics but the need to bleach these fibers has apparently diminished as improvements in the manufacture of synthetics resulted in greater brightness before bleaching.

The specific reactivity of chlorite with certain malodorous and highly toxic compounds such as unsaturated aldehydes, mercaptans, thioethers, and hydrogen sulfide and cyanide can be employed to scrub off-gases from rendering plants and similar facilities engaged in the recovery of animal fats and production of animal-feed proteins (175).

Sodium chlorite has also been evaluated as disinfectant in the wet processing of hides although wide-spread use is doubtful since chlorite appears to weaken the leather (176–177). Other disinfectant applications include slime and microbial control in paper mills (178) and silicate formulations (179). However, chlorine dioxide is usually preferred to sodium chlorite in biocidal applications (see Industrial antimicrobial agents).

Patents on application of sodium chlorite in manufacturing industries have appeared in the areas of detoxification of cyanide-containing plant effluents (180), etching of metals (181), destruction of dyestuffs in textile mill effluents (182), and bleaching of wood veneers in lumber and furniture industries (183–185).

BIBLIOGRAPHY

"Chlorites and Chlorine Dioxide" under "Chlorine Compounds, Inorganic" in *ECT* 1st ed., Vol. 3, pp.

696–707, by J. F. White, Mathieson Chemical Corporation; "Chlorous Acid, Chlorites, and Chlorine Dioxide" in *ECT* 2nd ed., Vol. 5, pp. 27–50, by H. L. Robson, Olin Mathieson Chemical Corporation.

1. W. Bray, *Z. Physik. Chem.* **54,** 569 (1906); W. Bray, *Z. Anorg. Allg. Chem.* **48,** 217 (1906).
2. F. E. King and J. R. Partington, *J. Chem. Soc.*, 925 (1926).
3. N. W. Taylor, *J. Am. Chem. Soc.* **48,** 854 (1926).
4. A. H. Nielson and P. J. H. Woltz, *J. Chem. Phys.* **20,** 1878 (1952).
5. J. L. Pascal, A. C. Pavia, and J. Potier, *J. Mol. Struct.* **13,** 381 (1972).
6. D. Leonesi and G. Piantoni, *Ann. Chim. Rome* **55,** 668 (1965).
7. B. Barnett, *Ph.D. Thesis,* University of California, 1935.
8. J. D. Dunitz and K. Hedberg, *J. Am. Chem. Soc.* **72,** 3108 (1950).
9. J. B. Coon, *Phys. Rev.* **85,** 746 (1951).
10. K. Hedberg, *J. Chem. Phys.* **19,** 509 (1951).
11. T. G. Kujwmzelis, *Physik. Z.,* 665 (1938).
12. R. F. Curl, Jr. and co-workers, *Phys. Rev.* **121,** 1119 (1961).
13. D. Sundhoff and H. J. Schumacher, *Z. Physik. Chem.* **B28,** 17 (1935).
14. I. E. Flis, *J. Appl. Chem. USSR* **37,** 684 (1964).
15. D. D. Wagman and co-workers, *Natl. Bur. Stand. US Tech Note* **270-1,** 1/124,25 (1965).
16. W. Buser and H. Hänisch, *Helv. Chim. Acta* **35,** 2547 (1952).
17. F. Stitt and co-workers, *Anal. Chem.* **26,** 1478 (1954).
18. M. Konopik, J. Derkosch, and E. Berger, *Monatsh.* **84,** 214 (1953).
19. H. L. Friedman, *J. Chem. Phys.* **21,** 319 (1953).
20. T. Chen, *Anal. Chem.* **39,** 804 (1967).
21. R. G. Kieffer and G. Gordon, *Inorg. Chem.* **7,** 235 (1967).
22. K. Schaefer, *J. Physik. Chem.* **93,** 312 (1919).
23. F. Lenzi, *Ph. D. Thesis,* University of Toronto, 1965.
24. J. E. Bennett and D. J. E. Ingram, *Phil. Mag.* **1**(8), 109 (1956).
25. J. E. Bennett, D. J. E. Ingram, and D. Schonland, *Proc. Phys. Soc.* **A69,** 556 (1956).
26. D. J. E. Ingram, *Free Radicals as Studied by Electron Resonance,* Butterworth's Scientific Publications, London, 1958, pp. 225–226.
27. J. B. Coon, *J. Chem. Phys.* **14,** 665 (1946).
28. J. F. Haller and W. W. Northgraves, *Tappi* **38,** 199 (1955).
29. M. Bigorgne, *Compt. Rend.* **236,** 1966 (1953).
30. U.S. Pat. 2,683,651 (July 13, 1954), H. B. Williamson and C. A. Hample (to Cardox Corp.).
31. H. Taube and H. Dodgen, *J. Am. Chem. Soc.* **71,** 3330 (1949).
32. H. Fukutomi and G. Gordon, *J. Am. Chem. Soc.* **89,** 1362 (1967).
33. J. Halpern and H. Taube, *J. Am. Chem. Soc.* **74,** 375 (1952).
34. U. Glabisz, *Przem. Chem.* **13,** 508 (1957).
35. G. Holst, *Ind. Eng. Chem.* **42,** 2359 (1950).
36. C. Castellani-Bisi, *Annal. Chim. Rome* **49,** 2056 (1959).
37. C. C. Hong and W. H. Rapson, *Can. J. Chem.* **46,** 2053 (1968); C. C. Hong, *Thesis,* University of Toronto, 1966.
38. R. H. Rapson, *Tappi* **39,** 554 (1956).
39. C. F. Goodeve and F. V. Richardson, *J. Am. Chem. Soc.* **59,** 294 (1937).
40. Ref. 31, p. 2501.
41. A. Reychler, *Bull. Soc. Chim. Fr.* **25,** 663 (1901).
42. J. F. White, M. C. Taylor, and G. P. Vincent, *Ind. Eng. Chem.* **34,** 782 (1942).
43. G. Gordon and G. Kokoszka, *unpublished results,* 1967.
44. B. O. Lindgren and C. M. Svahn, *Acta Chem. Scand.* **20,** 211 (1966).
45. B. O. Lindgren, C. M. Svahn, and G. Widmark, *Acta Chem. Scand.* **19,** 7 (1965).
46. D. H. Rosenblatt and co-workers, *J. Org. Chem.* **28,** 2790 (1963).
47. D. H. Rosenblatt and co-workers, *J. Org. Chem.* **33,** 1649 (1968).
48. G. Gordon, R. G. Kieffer, and D. H. Rosenblatt, *Prog. Inorg. Chem.* **15,** 208 (1972).
49. F. S. Becker, J. K. Hamilton, and W. E. Lucke, *Tappi* **48,** 60 (1965).
50. K. Bhaduri, *Z. Anorg. Chem.* **84,** 113 (1913).
51. J. W. T. Spinks and J. M. Porter, *J. Am. Chem. Soc.* **56,** 264 (1934).
52. R. H. Zinius and C. B. Purves, *Tappi* **43,** 27 (1960).
53. C. Kawasaki and T. Horio, *J. Pharm. Soc. Jpn.* **74,** 904 (1954).

54. C. Kawasaki, T. Horio, and G. Hamada, *J. Pharm. Soc. Jpn.* **74,** 907 (1954).
55. E. Schmidt and K. Brunsdorf, *Chem. Ber.* **55B,** 1529 (1922).
56. E. Schmidt, W. Haag, and L. Sperling, *Chem. Ber.* **58B,** 1394 (1925).
57. G. Gianola and J. Meybeck, *Assoc. Tech. Ind. Papet.* **1,** 25 (1960).
58. U. Glabisz, *Chem. Stosow Ser. A.* **10,** 211 (1966).
59. R. S. E. Ingols and G. M. Ridenour, *Water Sewage Works* **95,** 187 (1948).
60. K. Ogawa and T. Naito, *Jpn. Anal.* **3,** 421 (1954).
61. K. Paluch, *Rocz. Chem.* **38,** 35 (1964).
62. K. Paluch, *Rocz. Chem.* **39,** 1539 (1965).
63. C. D. Logan, R. M. Husband, and C. B. Purves, *Can. J. Chem.* **33,** 82 (1955).
64. K. Paluch, *Rocz. Chem.* **38,** 43 (1964).
65. S. Skramovsky, Z. Tauer, and J. Novotny, *Coll. Czech. Chem. Commun.* **20,** 718 (1955).
66. K. Paluch and D. Dziewonska, *Rocz. Chem.* **41,** 1285 (1967).
67. R. A. Somsen, *Tappi* **43,** 154 (1960).
68. P. S. Fredricks, B. O. Lindgren, and O. Theander, *Acta Chem. Scand.* **21,** 2895 (1967).
69. F. Lenzi and W. H. Rapson, *Pulp Paper Mag. Can.* **63,** T-442 (1962).
70. U.S. Pat. 3,733,395 (May 15, 1973), W. A. Fuller (to Hooker Chemical).
71. U.S. Pat. 3,563,702 (Feb. 16, 1971), H. D. Partridge and co-workers (to Hooker Chemical).
72. Can. Pat. 825,084 (Oct. 14, 1969), W. H. Rapson (to Electric Reduction Co. of Canada).
73. U.S. Pat. 3,933,987 (Jan. 20, 1976), A. C. Schulz (to Hooker Chemicals and Plastics).
74. Can. Pat. 826,577 (Nov. 4, 1969), J. D. Winfield and co-workers (to Electric Reduction Co. of Canada).
75. U.S. Pat. 3,816,077 (June 11, 1974), W. A. Fuller and co-workers (to Hooker Chemical).
76. U.S. Pat. 3,793,439 (Feb. 19, 1974), W. H. Rapson (to Erco Industries).
77. U.S. Pat. 3,789,108 (Jan. 29, 1974), W. H. Rapson (to Erco Industries).
78. Can. Pat. 956,783 (Oct. 29, 1974), D. G. Hatherly (to Erco Industries).
79. Can. Pat. 922,661 (Mar. 13, 1973), G. D. Westerlund (to Chemetics International).
80. U.S. Pat. 3,404,952 (Oct. 8, 1968), G. O. Westerlund (to Chemech Engineering).
81. U.S. Pat. 3,524,728 (Aug. 18, 1970), G. O. Westerlund (to Chemech Engineering).
82. U.S. Pat. 3,607,027 (Sept. 21, 1971), G. O. Westerlund (to Chemech Engineering).
83. U.S. Pat. 3,929,974 (Dec. 30, 1975), J. D. Winfield (to Erco Industries).
84. U.S. Pat. 3,933,988 (Jan. 20, 1976), H. J. Rosen (to Hooker Chemicals and Plastics).
85. U.S. Pat. 2,481,240 (Sept. 6, 1949), W. H. Rapson and co-workers (to Canadian International Paper).
86. U.S. Pat. 2,598,087 (May 27, 1952), M. Wayman and co-workers (to Canadian International Paper).
87. U.S. Pat. 2,616,783 (Nov. 4, 1952), E. Wagner (to Degussa).
88. U.S. Pat. 3,904,496 (Sept. 9, 1975), C. J. Harke and co-workers (to Hooker Chemicals and Plastics).
89. U.S. Pat. 3,904,495 (Sept. 9, 1975), J. D. Eng and co-workers (to Hooker Chemicals and Plastics).
90. J. F. Haller and S. S. Listek, *Anal. Chem.* **20,** 639 (1948).
91. J. A. McCarthy, *J. N. Engl. Water Works Assoc.* **59,** 252 (1945).
92. J. A. Myhrstad and J. E. Samdal, *J. Am. Water Works Assoc.* **61,** 205 (1969).
93. A. W. Hodgden and R. S. Ingols, *Anal. Chem.* **26,** 1224 (1954).
94. M. A. Post and W. A. Moore, *Anal. Chem.* **31,** 1872 (1959).
95. A. T. Palin, *Water Sewage Works* **107,** 457 (1960).
96. D. B. Adams and co-workers, *Proc. Soc. Water Treat. Exam. Engl.* **15,** 117 (1966).
97. A. T. Palin, *J. Inst. Water Eng. Engl.* **21,** 587 (1967).
98. G. M. Ridenour and R. S. Ingols, *J. Am. Water Works Assoc.* **36,** 561 (1947).
99. G. M. Ridenour and E. H. Armbruster, *J. Am. Water Works Assoc.* **41,** 537 (1949).
100. G. M. Ridenour and R. S. Ingols, *Am. J. Publ. Health* **36,** 639 (1946).
101. M. A. Benarde and co-workers, *Appl. Microbiol.* **15,** 257 (1967).
102. J. E. Smith and J. L. McVey, *Prepri. Pap. Natl. Meet. Div. Environ. Chem., Am. Chem. Soc.* **13,** 177 (1973).
103. Y. M. Morozov, *Tr. Saratov Med. Inst.* **73,** 40 (1970).
104. J. Paulet and S. Desbrousses, *Arch. Mal. Prof. Med. Trav. Secur. Soc.* **31,** 97 (1970).
105. *Ibid.,* **33,** 56 (1972).
106. *Ibid.,* **35,** 797 (1974).

107. H. E. Christensen, T. T. Lugenbyhl, eds., *Registry of Toxic Effects of Chemical Substances,* U.S. Department of Health, Education and Welfare, Public Health Service, Center for Disease Control, National Institute for Occupational Safety and Health, Rockville, Md., 1975, p. 307.
108. Medical College of Virginia, unpublished report for Mathieson Alkali Works, Feb. 7, 1949.
109. S. A. Fridlyand and G. Z. Kagan, *Gig. Sanit.* **36,** 18 (1971).
110. *Preliminary Assessment of Suspected Carcinogens in Drinking Water,* EPA Report to Congress, Office of Toxic Substances, EPA Washington, D.C., 1975, p. 43.
111. J. M. Symons and co-workers, *Interim Treatment Guide for the Control of Chloroform and other Trihalomethanes,* Water Supply Research Div., Municipal Environ. Res. Lab., EPA, Cincinnati, Ohio, June 1976, pp. 18, 39.
112. W. J. Ward, *Annual Meeting 1976,* Cooling Tower Institute, Houston, Texas, 1976.
113. G. C. White, *Handbook of Chlorination,* Van Nostrand Reinhold Company, New York, 1972, pp. 596–627.
114. U.S. Pat. 4,013,761 (Mar. 26, 1977), W. J. Ward and co-workers (to Olin Corporation).
115. A. W. Searcy, *J. Chem. Phys.* **28,** 1237 (1958).
116. V. Tazzoli and co-workers, *Acta Cryst.* **31,** 1032, 2750 (1975).
117. C. Tarimci, E. Schempp, and S. C. Chang, *Acta Cryst.* **31,** 2146 (1975).
118. C. Tarimci, R. D. Rosenstein, and E. Schempp, *Acta Cryst.* **31,** 610 (1975).
119. D. E. Tevault, F. K. Chi, and L. Andrews, *J. Mol. Spectrosc.* **51,** 450 (1974).
120. G. Cunningham and T. S. Oey, *J. Am. Chem. Soc.* **77,** 799 (1955).
121. R. L. Doerr, *unpublished data,* Olin Corporation, 1977.
122. F. D. Rossini and co-workers, *Natl. Bur. Stand. US Circ.* **500,** 25 (1952).
123. D. D. Wagman and co-workers, *Natl. Bur. Stand. US Tech. Note* **270–3,** 1/264,28 (1968).
124. G. F. Davidson, *J. Chem. Soc.,* 1649 (1954).
125. M. W. Lister, *Can. J. Chem.* **30,** 879 (1952).
126. K. Tachiki, *J. Chem. Soc. Jpn.* **65,** 346 (1944); T. Naito, *J. Chem. Soc. Jpn. Ind. Chem. Sect.* **65,** 1016 (1962).
127. I. E. Flis, *J. Appl. Chem. USSR* **29,** 689 (1956).
128. (a) M. C. Taylor, *Ind. Eng. Chem.* **32,** 899 (1940); (b) G. R. Levi, *Atti Accad. Lincei* **31**(I), 370 (1922).
129. G. M. Nabar, C. R. Ramachandran, and V. A. Shenai, *Indian J. Technol.* **2,** 11 (1964).
130. U. Glabisz, *Rocz. Chem.* **39,** 141 (1965).
131. U. Glabisz and J. Minczewski, *Chem. Anal. Warsaw* **9,** 131 (1964).
132. J. Kepinski and G. Blaszkiewicz, *Talanta* **13,** 357 (1966).
133. F. Emmenegger and G. Gordon, *Inorg. Chem.* **6,** 633 (1967).
134. E. Keating, *Paper Mill News* **49,** (1953); E. Keating, *Papier,* **9,** 155 (1952).
135. K. Luther and F. McDougall, *Z. Physik. Chem.* **55,** 477 (1906).
136. F. Foerster and P. Dolch, *Z. Elektrochem.* **23,** 137 (1917).
137. D. G. H. Daniels and J. K. Whitehead, *Chem. Ind.,* 1214 (1957).
138. W. Masschelein, *Ind. Eng. Chem.* **6,** 137 (1967).
139. F. Böhmländer, *Wasser-Abwasser* **104,** 518 (1963).
140. G. R. Levi and M. Tabet, *Gazz. Chim. Ital.* **65,** 1138 (1935).
141. C. Bertoglio Riolo, *Gazz. Chim. Ital.* **85,** 1698 (1955).
142. M. L. Granstrom and G. F. Lee, *J. Am. Water Works Assoc.* **50,** 1453 (1958).
143. Brit. Pat. 692,763 (June 10, 1953), (to Farbenfabriken Bayer A.G.).
144. D. T. Jackson and J. L. Parsons, *Ind. Eng. Chem. Anal. Ed.* **9,** 14 (1937); J. DeMeeus and J. Sigalla, *J. Chim. Phys.* **63,** 453 (1966).
145. D. M. Kern and C.-H. Kim, *J. Am. Chem. Soc.* **87,** 5309 (1965).
146. G. Ishi, *Kogyo Kagaku Zasshi* **65,** 1013 (1962).
147. P. Spacu and co-workers, *Rev. Chim. Acad. Rep. Popul. Roumaine* **3,** 127 (1958).
148. R. D. Cornelius and G. Gordon, *Inorg. Chem.* **15,** 1002 (1976).
149. R. C. Thompson and G. Gordon, *Inorg. Chem.* **5,** 557 (1966).
150. G. R. Levi and E. ResGarrini, *Gazz. Chim. Ital.* **87,** 7 (1957).
151. D. M. H. Kern and G. Gordon, *Theory and Structure of Complex Compounds,* Pergamon Press Inc., Long Island City, N.Y., 1964, p. 655.
152. A. Prokopcikas and J. Valsiuniene, *Lietuvos TSR Mokstu Akad. Darbai B* **89,** (1963); **79,** (1964).
153. U.S. Pat. 2,358,866 (Sept. 26, 1944), J. D. MacMahon (to Mathieson Alkali Works).
154. H. F. Launer, W. K. Wilson, and J. H. Flynn, *J. Res. Natl. Bur. Stand.* **51,** 237 (1953).
155. H. F. Launer and Y. Tomimatsu, *J. Am. Chem. Soc.* **76,** 2591 (1954).

156. A. Jeanes and H. S. Isbell, *J. Res. Natl. Bur. Stand.* **27,** 125 (1941).
157. H. S. Isbell and L. T. Sniegoski, *J. Res. Natl. Bur. Stand.* **68A,** 301 (1964).
158. B. O. Lindgren and T. Nilsson, *Acta Chem. Scand.* **27,** 888 (1973).
159. U.S. Pat. 2,926,996 (Mar. 1, 1960), W. Koostra (to N.V. Koninklijke Nederlandsche Zoutindustrie).
160. Ger. Pat. 1,059,891 (June 25, 1959), R. Hirschberg and H. Hund (to Farbwerke Hoechst).
161. Sh. S. Shchegol, *Khim. Khim. Tekhnol.* **1,** 357 (1958).
162. U.S. Pat. 2,092,945 (Sept. 14, 1937), G. P. Vincent (to Mathieson Alkali Works Inc.); U.S. Pat. 2,565,209 (Aug. 21, 1951), F. H. Dole (to Mathieson Chemical Corp.).
163. Brit. Pat. 687,246 (Feb. 11, 1953), (to Tennants Consolidated, Ltd.).
164. Jpn. Pat. 29-7576 (Nov. 18, 1954), S. Yamamoto (to Asahi).
165. Ger. Pat. 1,567,465 (Sept. 5, 1974), R. Paetsch and co-workers (to Degussa).
166. J. Musil and co-workers, *Sci. Pap. Inst. Chem. Technol. Prague,* 327 (1964).
167. P. Heffernan, *private communication,* Municipal Environmental Research Laboratory, EPA Cincinnati, Ohio, 1977.
168. R. P. Dempster, *Steinhart Aquarium Publication,* Sept. 1970.
169. J. Janci and J. Farkas, *Vysk. Pr. Odboru Pap. Celul.* **18,** V21 (1973).
170. G. Wegener, *Papier* **29,** 429 (1975).
171. Th. N. Kleinert, *Holzforsch. Holzverwert.* **24,** 12 (1972).
172. Jpn. Kokai 74-05,875 (June 19, 1974), S. Takasaki and co-workers (to Fuji Kasui Kogyo Co., Ltd.).
173. Jpn. Kokai 74-130,361 (Dec. 13, 1973), T. Senjo and M. Kobayashi (to Fuji Kasui Kogyo Company, Ltd.; Sumitomo Metal Ind., Ltd.).
174. Jpn. Kokai 75-73,892 (June 18, 1975), T. Shibata and co-workers (to Matsushita Electric Industrial Co., Ltd.).
175. H. D. Brand, *Fette Seifen Anstrichm.* **77,** 354 (1975).
176. C. A. Money, *J. Am. Leather Chem. Assoc.* **69,** 112 (1974).
177. D. R. Cooper, *J. Soc. Leather Technol. Chem.* **57,** 19 (1973).
178. U.S. Pat. 3,046,185 (July 24, 1962), J. E. Buonanno and J. C. Shore (to Metro Atlantic, Inc.).
179. U.S. Pat. 3,336,236 (Aug. 15, 1967), R. J. Michalski (to Nalco Chemical Company).
180. Jpn. Pat. 74-32,270 (Aug. 29, 1974), T. Hoshizumi (to Sumitomo Chemical Company Ltd.).
181. Brit. Pat. 1,310,238 (Mar. 14, 1973), S. S. Tulsi (to Oxy Metal Finishing Ltd.).
182. Jpn. Kokai 73-27,076 (Apr. 10, 1973), K. Niijima and co-workers (to Saitama Prefecture).
183. Jpn. Kokai 75-148,507 (Nov. 28, 1975), K. Fuse and M. Tanimoto (to Japan Carlit Company, Ltd.).
184. Jpn. Kokai 74-13,305 (Feb. 5, 1974), T. Hosokawa and co-workers (to Hodogaya Chemical Company, Ltd.).
185. N. Levitin, *Forest Prod. J.* **25,** 28 (1975).

MANFRED G. NOACK
RICHARD L. DOERR
Olin Corporation

CHLORIC ACID AND CHLORATES

Chlorates are salts of chloric acid [7790-93-4], HClO₃, which is fairly stable in cold water solution in concentrations up to ca 30%. Upon heating, chlorine and chlorine dioxide may be evolved, depending upon the strength of the solution. Concentration of chloric acid by evaporation under reduced pressure may be carried to ≥40% accompanied by the evolution of chlorine and oxygen and the formation of perchloric acid in proportions approximating:

$$8\ HClO_3 \rightarrow 4\ HClO_4 + 2\ H_2O + 3\ O_2 + 2\ Cl_2 \tag{1}$$

Chloric acid is a strong oxidizing agent but its oxidizing properties vary somewhat with the pH and temperature of the solution. Its reaction with organic substances may be quite violent in strong solutions. Sulfur dioxide is oxidized to sulfuric acid by solutions of chloric acid. In dilute solutions chlorine is formed and may react further; in concentrated solutions chlorine dioxide is liberated.

$$2\ HClO_3 + 5\ H_2SO_3 \rightarrow 5\ H_2SO_4 + H_2O + Cl_2 \tag{2}$$

$$Cl_2 + H_2O + H_2SO_3 \rightarrow H_2SO_4 + 2\ HCl \tag{3}$$

$$2\ HClO_3 + H_2SO_3 \rightarrow H_2SO_4 + H_2O + 2\ ClO_2 \tag{4}$$

Several oxides of chlorine are known but the dioxide, ClO₂, is the only one of industrial importance. It is a yellow–orange gas above its boiling point of 11°C; the liquid sp gr at 0°C is 1.642 (1). Its freezing point is −59°C, and the solid below this temperature resembles potassium dichromate in appearance. Its preparation from sodium chlorate [7775-09-9], for use as a bleach, is described below.

Another oxide of chlorine is the hexoxide, Cl₂O₆, a dark red liquid below its boiling point of 203°C; specific gravity is 2.023 at its freezing point of 3.5°C (2). This oxide can exist in either the monomeric (ClO₃) or dimeric (Cl₂O₆) form, but in the gaseous state it is entirely monomeric. It may be prepared by mixing streams of inert gases containing chlorine dioxide and ozone.

$$ClO_2 + O_3 \rightarrow ClO_3 + O_2 \tag{5}$$

It is the least explosive of the oxides of chlorine, but is a powerful oxidizing agent, reacting violently with all forms of organic matter and explosively with water. Like the other oxides of chlorine, it is very dangerous at high concentrations and no safe procedures for handling such concentrations are known (2).

SODIUM AND POTASSIUM CHLORATE

Properties

Sodium chlorate, NaClO₃, forms cubic crystals, mp 248°C, decomposition point 265°C, sp gr 2.49, n_D^{20} 1.515. The crystals are slightly hygroscopic and this property has limited the use of sodium chlorate in industry for some applications. The crystals are optically active; either or both d and l forms may be crystallized from solutions, depending upon the seed crystals. An unstable trigonal form, of similar structure to sodium nitrate, may separate from solutions supersaturated with the cubic salt.

Potassium chlorate [3811-04-9], KClO₃, crystallizes in the monoclinic system, usually as short prisms, mp 368°C, decomposition point 400°C, sp gr 2.32, n_D^{20} 1.440. It is nonhygroscopic.

The solubility of sodium chlorate is much greater than that of potassium chlorate, as shown in Figure 1. Figure 2 represents the Lowenherz phase diagram for the reciprocal salt system $NaClO_3$–$NaCl$–KCl–$KClO_3$ in water at 30°C, and indicates the effects of mutual solubilities of the salts involved.

On thermal decomposition, both sodium and potassium chlorate may produce the corresponding perchlorate (5). Mixtures of potassium chlorate with metal oxide catalysts, principally with manganese dioxide, are employed as a laboratory source of oxygen. The evolution of oxygen starts at about 70°C and becomes strong at 100°C, below the fusion point (6). The molten salts are powerful oxidizing agents. Mixtures of chlorates with organic materials have been employed as explosives; however, because of their extreme sensitivity to shock and unpredictability, such mixtures are not classed as permissible explosives in the United States. A mixture of sodium and potassium chlorate with any combustible organic or inorganic matter should be regarded as dangerous (7). Chlorates in neutral and alkaline solutions at room temperature do not show oxidizing properties. In acid solution chlorates are a source of chloric acid.

Manufacture

Chlorates may be prepared by the chlorination of a hypochlorite solution. Most of the chlorate is manufactured by the electrolysis of sodium chloride solution in electrochemical cells without diaphragms. Potassium chloride can be used for chlorate electrolysis for the direct production of potassium chlorate, but since sodium chlorate is so much more soluble, the production of the sodium salt is generally preferred. Potassium chlorate may be obtained from the sodium chlorate by a metathesis with potassium chloride.

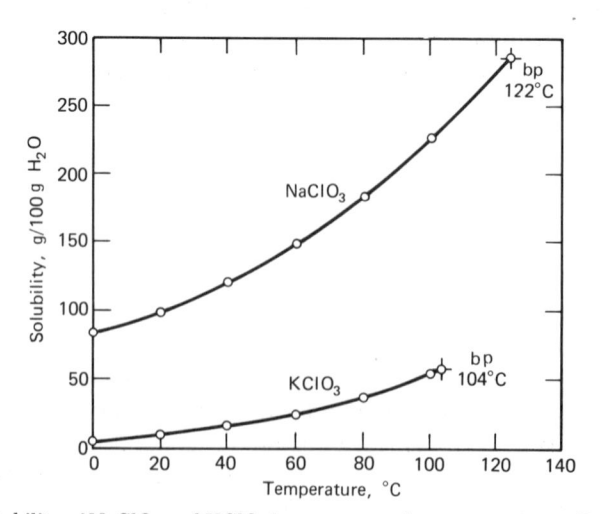

Figure 1. Solubility of $NaClO_3$ and $KClO_3$ in water at various temperatures. Data are mainly from Seidell (3).

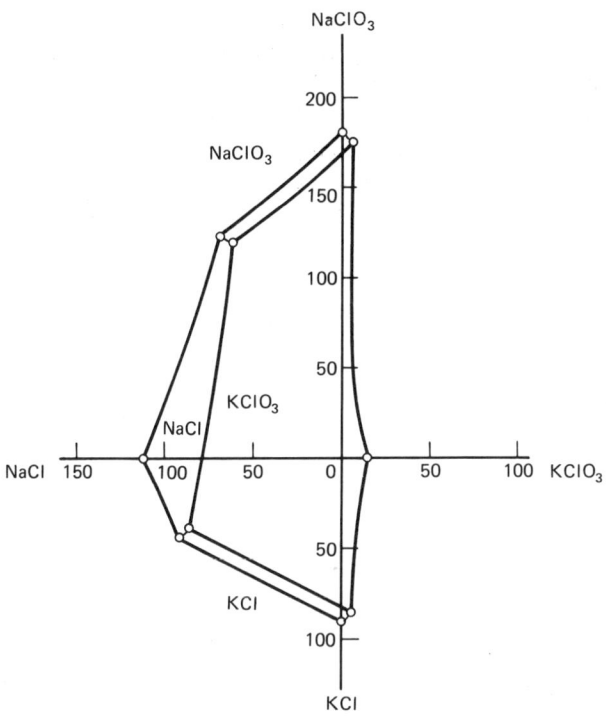

Figure 2. Löwenherz phase diagram for the system NaClO$_3$–NaCl–KCl–KClO$_3$–H$_2$O at 30°C (mol/1000 mol H$_2$O) (4).

Cell Design. Sodium chlorate cell designs vary widely between individual chlorate producers. In general, the sodium chlorate production system consists of cells and a holding volume usually in a closed loop. The cells generate active chlorine, and the holding volume acts as a reactor for further conversion to the product (8). The cells have ranged from small monopolar cells (graphite anodes–mild steel cathodes) with an annual capacity of 22–90 metric tons of sodium chlorate per year (9) to large multipolar or bipolar cell assemblies (10) (graphite sheet acting as anode one side and cathode on the other side) that are of ≤1800 t annual capacity.

In principle, the production of sodium chlorate in an electrolytic cell is a simple operation (see Electrochemical processing). Direct current is passed through a cell containing an anode and cathode without a diaphragm in an electrolyte that contains sodium chloride, sodium chlorate, sodium dichromate, and with possible trace impurities of calcium, magnesium, and sulfate at a pH of ca 6.9. The sodium chloride is converted to sodium chlorate with the evolution of hydrogen. The gas evolution at the cathode agitates the electrolyte in the cell.

At the present time, the electrolytic cells for sodium chlorate manufacture are undergoing a rapid change as the result of new technology. The changes affecting sodium chlorate cell design are: (1) development of noble metal-coated titanium anodes for the chlorine and chlorate industry; (2) new cell design to take advantage of coated titanium anodes; (3) OSHA requirements for improved working conditions in manufacturing plants; (4) EPA requirement of less air pollution; (5) increased cost and reduced availability of graphite; and (6) increased cost of electric power.

The use of noble metal-coated titanium anodes, first developed for the caustic–chlorine cell and later applied to the production of sodium chlorate, caused a major change in sodium chlorate cell design (11) (see Alkali and chlorine products; Metal anodes). The new anodes are generally referred to as dimensionally-stable anodes. They are electrodes having an electrically conductive film-forming base material supporting a surface coating or deposit of at least one member of the group of noble metals, noble metal alloys or noble metal oxides. Suitable film-forming substrates are titanium, tantalum, or niobium.

A comparison of operating conditions using graphite and noble metal-coated titanium anodes in a chlorate cell is shown in Table 1.

Table 1. Sodium Chlorate Cell

	Anodes	
Condition	Graphite	Coated titanium
temperature, °C	>40	ca 80
current efficiency, %	85	92
current density, A/cm^2	0.043	0.27
cathode–anode spacing, cm	0.6	0.3
sodium chloride concentration, g/L	>110	<100
cathode material	mild steel	mild steel
cooling water	large flow	low flow
voltage, drop, V	3.7	3.4

New sodium chlorate cells were designed to take advantage of the characteristics of the new anodes. The best operating conditions for a sodium chlorate cell using the new anodes are quite different from a cell using graphite anodes. Table 2 lists some of the new commercial cells using coated titanium anodes.

The new sodium chlorate cells are either monopolar or bipolar. It is difficult to find acceptable cathode material for the bipolar cells. Practically all the new cells have connections for the recovery of evolved hydrogen. In the older graphite cells the hydrogen and other gases from the cell were vented to the atmosphere. Air pollution controls and the increased costs of fuel have made hydrogen scrubbing and recovery a necessity (see Hydrogen).

Table 2. New Sodium Chlorate Cells

Company	Type of cell	Ref.
Chemetics International, Ltd.	monopolar	12
Diamond Shamrock Corp.	bipolar	13
Diamond Shamrock & Huron Chemical Ltd.	bipolar	14
Hooker Chemical Corp.	monopolar	15
Hooker Chemical Corp.	bipolar	16
Kema Nardis A.B.	bipolar	17
Kerr-McGee Chemical Corp.	monopolar	18
Krebs	monopolar	19
Pennwalt Corp.	?	20
Standard Chemical Ltd.	?	21
Solvay & Cie	monopolar	22

Table 3 is a comparison of the results and operating conditions of sodium chlorate cells with coated titanium anodes with cells using graphite anodes.

The anode cost per metric ton of sodium chlorate produced has not been determined for the noble metal-coated titanium anodes. This cost varies with the type of noble metal and the cell operating conditions. Laboratory and pilot plant data from new cells indicate that anode cost per metric ton of sodium chlorate produced will be less than for the graphite cells.

Other materials that have been used as anodes in sodium chlorate cells are magnetite (7) and lead dioxide (23).

The cost of power is the major expense in producing sodium chlorate. Production of one metric ton of sodium chlorate requires 4500–5800 kW·h of power and ca 565 kg of sodium chloride.

Cell Chemistry. In the sodium chlorate cell, free chlorine is formed at the anode, and hydrogen and hydroxyl ions are formed at the cathode. Chlorine reacts rapidly with water to form hypochlorous and hydrochloric acids (24).

$$Cl_2 + H_2O \rightleftharpoons HOCl + H^+ + Cl^- \tag{6}$$

Because the electrodes are close together and have no diaphragm separation, the chlorine may also diffuse through the solution and react with the hydroxyl ions formed at the cathode to produce hypochlorous acid and chloride ion.

$$Cl_2 + OH^- \rightleftharpoons HOCl + Cl^- \tag{7}$$

The solution around the anode becomes strongly acidic and the solution in contact with the cathode is alkaline. In addition to this pH gradient, the concentration of hypochlorous acid also establishes a gradient, relatively high around the anode and relatively low around the cathode.

It is usually assumed that the chlorate is formed by two separate simultaneous reactions (25): by the oxidation of hypochlorite ion by free hypochlorous acid (eq. 8), and by the electrochemical formation of chlorate through discharge of the hypochlorite ion at a potential approximately equal to that for discharge of the chloride ion (eq. 9).

$$2 HClO + ClO^- \rightleftharpoons ClO_3^- + 2 Cl^- + 2 H^+ \tag{8}$$

$$6 ClO^- + 3 H_2O \rightarrow 2 ClO_3^- + 6 H^+ + 4 Cl^- + 1.5 O_2 + 6 e^- \tag{9}$$

Table 3. Sodium Chlorate Cell

Condition	Graphite anode	Noble metal-coated titanium anode
voltage drop per cell, V	3.7	3.4
power consumption per metric ton NaClO$_3$, kW·h	5800	4535
current efficiency, %	85	93
hydrogen collection	no	yes
current density anode, A/cm^2	0.043	0.27
operating temperature, °C	40	60–80
dichromate concentration in electrolyte, g/L	5	1
sodium chlorate concentration in electrolyte, g/L	550	650
minimum NaCl concentration in electrolyte, g/L	110	80

Equilibrium is established in the cell between hypochlorous acid and hypochlorite ions in which as many hypochlorite ions are formed in unit time as are consumed by both reactions plus other possible side reactions. Equation 8 may be considered to occur with a current efficiency of 100%; the overall process is as follows:

$$NaCl + 3 H_2O \rightarrow NaClO_3 + 3 H_2 \qquad (10)$$

However, the current efficiency is only 66.7% in the case of equation 9. With appropriate choice of operating conditions, efficiencies of 85–95% can be obtained (26–27).

The most favorable conditions for equation 8 are a temperature range of 60–75°C and pH 6.5–7.0; the optimum pH is about 6.9. However, this reaction is quite slow since it takes place in the body of the liquor rather than at or near the anode surfaces. In designing chlorate cell installations, provision must be made for adequate liquor retention time by including additional storage space. In continuous-flow installations a separate tank or reactor may be provided to hold the cell effluent. It has been suggested that a chemical reaction does not take place but that chlorate is formed electrochemically at the anode (28–29).

In the chlorate cells a number of side reactions may take place that can seriously reduce the current efficiency: those causing reduction losses at the cathode, and those resulting in oxidation losses at the anode (30).

Both hypochlorite and chlorate formed in the cell may be reduced at the cathode. Since hypochlorite is much more easily reduced, only this reaction need be considered. It has been shown (31) that reduction losses: increase proportionally with hypochlorite concentration; are inversely proportional to current density; increase with temperature at about 3% per °C; decrease with increasing sodium chloride content at high hypochlorite concentrations, even when chlorate content is increasing; and are independent of pH between 6.5–10.5 provided the hypochlorite concentration is held constant. These observations can be explained on the assumption that the rate of reduction of hypochlorite is limited by the rate at which the hypochlorite is moved into contact with the electrode surface through migration, liquor circulation, convection, and diffusion (32). Factors affecting such movement, such as increased viscosity with increased sodium chlorate concentration, will influence the reduction losses accordingly.

The historical development of the electrochemistry and technology of chlorates is described in ref. 33. The first commercial use of electrolysis of chloride solutions to produce chlorate began in France in 1866 using diaphragm cells to avoid cathodic reduction (25). However, since 1898, following the discovery by Müller (34) that chromate ion would depress such reduction, sodium chromate has been used for this purpose in cells without diaphragms. Usually 2–3 g/L of sodium chromate is added. This forms a protective coating of a hydrated mixture of chromium oxides (Cr_2O_3 and CrO_3) which is permeable only to water and to certain ions, eg, H^+, but not ClO^- which can, therefore, no longer be reduced. Chromate also provides some chromate–dichromate buffering which tends to stabilize the pH of the solution.

Oxidation losses at the anode can vary within a wide range, and many of the older proposed mechanisms cannot explain the observed results. In general, the following oxidation losses can occur:

Loss of oxygen from hypochlorite:

$$HClO + H_2O \rightarrow O_2 + 3 H^+ + Cl^- + 2 e^- \qquad (11)$$

Loss of oxygen from water:

$$2 H_2O \rightarrow O_2 + 4 H^+ + 4 e^- \qquad (12)$$

Oxidation of chlorate to perchlorate:

$$ClO_3^- + H_2O \rightarrow ClO_4^- + 2\,H^+ + 2\,e^- \qquad (13)$$

Chlorine losses in cell gas:

$$Cl_2\,(aq) \rightarrow Cl_2\,(g) \qquad (14)$$

Breakdown of hypochlorite in solution:

$$2\,HClO \rightarrow 2\,HCl + O_2 \qquad (15)$$

Oxidation losses are independent of pH between 6.5–10.5 if the hypochlorite concentration is kept constant (31). Therefore equation 12 is of little significance. A certain amount of perchlorate formation takes place on magnetite anodes, but with graphite and coated titanium anodes equation 13 is negligible. The loss of chlorine in the cell gas is low as long as the pH is not allowed to decrease much below the neutral point. The pH can be controlled by addition of acid, but excessive additions of hydrochloric acid will result in increased chlorine losses. Since chlorine and hydrogen form explosive mixtures, a hazardous condition may result. Usually chlorine evolution is kept below 0.5%.

There is essentially no reaction of hypochlorite in solution according to equation 15 unless catalyzed by uv or certain heavy metal ions.

The major portion of the oxidation losses stem from anodic oxygen evolved from hypochlorite according to equation 11. The following relationships exist: (*1*) oxidation losses appear to be directly proportional to the hypochlorite concentration; (*2*) oxidation losses are an inverse function of the current density; and (*3*) oxidation losses increase with temperature at the rate of about 2% per °C (31).

There is also some discharge of hydroxyl ion at the anode to form oxygen (eq. 16), or with graphite to form carbon dioxide (eq. 17).

$$4\,OH^- \rightarrow O_2 + 2\,H_2O + 4\,e^- \qquad (16)$$

$$4\,OH^- + C \rightarrow CO_2 + 2\,H_2O + 4\,e^- \qquad (17)$$

Since equation 17 is favored by higher temperatures, chlorate cells with graphite anodes are usually held at 35–45°C to reduce graphite loss. Also, both reactions are markedly increased at low sodium chloride concentrations, particularly <100 g/L.

Graphite is also consumed by the direct attack of hypochlorous acid:

$$C + 2\,HClO \rightarrow CO_2 + 2\,H^+ + 2\,Cl^- \qquad (18)$$

The anode current density on graphite anodes is usually 0.032–0.043 A/cm^2 in order that the graphite loss is mainly from discharge of hydroxyl ion and thus varies with the useful work of the cell. Carbon dioxide formation accounts for about 80% of the total loss (35). Some graphite flakes and must be settled or filtered out of the cell effluent before further processing; a small amount may be oxidized to soluble compounds. In commercial installations, the graphite loss usually amounts to 10–25 kg/t of sodium chlorate produced. The use of coated titanium anodes is providing a satisfactory solution to this problem.

The operating parameters for an electrochemical sodium chlorate cell are summarized in refs. 36–43.

Chlorate Recovery. The sodium chlorate cell effluent is usually passed to holding tanks to permit the conversion of hypochlorite to chlorate to proceed as far as possible. Any residual hypochlorite, usually ca 1 g/L, is destroyed with some organic material such as a formate or urea. The chromate remaining in the liquor provides corrosion protection for the steel equipment used in subsequent steps. The liquor is usually filtered to remove insoluble particles, and may be concentrated by evaporation, yielding a crop of sodium chloride. The hot concentrated liquor is then cooled to yield sodium chlorate crystals that are separated by centrifuging. The filtrate is returned either to the evaporator or to the cell feed.

Alternatively, a high chlorate concentration of ca 600 g/L may be maintained in the cell effluent. The concentration step is omitted, and a crop of sodium chlorate is recovered directly by cooling the filtered cell effluent. The mother liquor is resaturated with sodium chloride and is returned to the cells. The sodium chlorate may be recrystallized if desired. Some of the sodium chlorate is redissolved and sent to perchlorate cells for conversion to perchlorate. The rest of the chlorate solution is allowed to react with a hot solution of potassium chloride and cooled to yield crystalline potassium chlorate (see Fig. 2).

Chlorate Analysis. Chlorate ion is determined by reaction with a reducing agent. Ferrous sulfate is preferred for quality control work (44) but other reagents, such as arsenous acid, stannous chloride, and potassium iodide, have also been used (45). When ferrous sulfate is used, a measured excess of the reagent is added to a strong hydrochloric acid solution of the chlorate for reduction, after which the excess ferrous sulfate is titrated with an oxidant, usually potassium permanganate or potassium dichromate.

Chlorate may also be determined as chloride after conversion with sulfur dioxide or ferrous sulfate solution or by adding zinc to an acetic acid solution. The chloride is determined volumetrically or gravimetrically as silver chloride (46). The gravimetric method is preferred for maximum accuracy.

Production and Shipment

To produce chlorine dioxide, sodium chlorate and sulfuric acid are fed into reactors (usually two in series) together with a reducing agent, ie, sulfur dioxide, methanol, hydrochloric acid, or sodium chloride (47). Compressed air is sparged into the reactors to agitate the charge and to remove the chlorine dioxide gas as it is formed. This is absorbed in water in a scrubbing tower to produce a solution of 0.6–1.0% ClO_2 for direct bleaching of pulp.

The first such chlorine dioxide generating unit was placed in operation in Canada in 1950 and the first United States unit was built in 1952 (48). At present, such units are in operation at all major bleached-chemical pulp mills in the United States and Canada.

Several variations of the chlorine dioxide generating process have been patented (49–52). In general, the formation of chlorine dioxide may be represented by the following (53–58):

$$HClO_3 + HCl \rightarrow HClO_2 + HClO \qquad (19)$$

$$HClO_3 + HClO_2 \rightarrow 2\ ClO_2 + H_2O \qquad (20)$$

A reaction that also occurs is:

$$HClO + HCl \rightarrow Cl_2 + H_2O \qquad (21)$$

The overall reaction is:

$$2 \text{ HClO}_3 + 2 \text{ HCl} \rightarrow 2 \text{ ClO}_2 + \text{Cl}_2 + 2 \text{ H}_2\text{O} \tag{22}$$

These reactions produce 2 moles of ClO_2 for each mole of Cl_2. When reducing agents other than chloride are used, much less chlorine is evolved because these reducing agents serve to reduce either HOCl or Cl_2 to chloride to maintain the above reactions. If insufficient chloride is present for equations 19 and 20, the reducing agent will reduce chlorate to chloride until the chloride concentration is adequate. Chlorine is thus produced along with chlorine dioxide to an extent determined by the competition between equation 21 and the formation of chloride by reduction. The addition of at least some chloride with the chlorate increases the efficiency of chlorine dioxide production, but it also increases the amount of free chlorine evolved. The presence of a small percentage of chlorine with the chlorine dioxide does not harm the resulting bleach. If a large amount of chlorine is present in the exit gas, only a portion is absorbed in the scrubbing tower used to prepare the bleach solution. Most of the Cl_2 passes to a secondary scrubber and is either absorbed by caustic solution, or used in other processes (see also Chlorites under Chlorine oxygen acids; Bleaching agents; Pulp).

A typical specification for sodium chlorate of technical grade is: $NaClO_3$, 99.5% min; NaCl, 0.12% max; moisture, 0.20% max; clear solution.

The sodium chlorate intended for shipment is usually dried in rotary driers to less than 0.2% moisture content, and is loaded into shipping containers, or is stored in moisture-free bins or silos prior to packaging. Some sodium chlorate is shipped as a solution containing: sodium chloride, ca 200 g/L (15 wt %, 3.4M); sodium chlorate, ca 350 g/L (26 wt %, 3.3M); and chromium, 130 ppm. This solution may be obtained directly from the sodium chlorate cell or it may be prepared from the cell solution. The mixture is sold in tank car amounts.

Dry crystalline sodium chlorate is shipped in steel drums of 45, 100, and 270 kg net, but for larger consumers it is also shipped in bulk in tank trucks and trailers (22 metric ton, min) and in 100 m^3 (3500 ft^3) tank hopper cars.

It may be unloaded very easily as a slurry or solution by recirculating hot water from a storage tank into the tank car or truck. Details for such handling are available from the principal producers.

Economic Aspects

Sodium chlorate and potassium chlorate are the most important salts of chloric acid.

The manufacture and use of chlorates has continued to be one of the fastest growing branches of the heavy chemical industry. The growth of sodium chlorate has been 5%/yr (59). The 1974 United States production was 182,700 metric tons (60). The estimated United States demand for sodium chlorate in 1975 was 181,400 t (59). Canada's sodium chlorate usage in 1975 was estimated at 127,000–136,000 t (61).

Sodium chlorate production in the United States increased from 72,560 metric tons in 1959 to 152,100 t in 1975, with an output capacity of over 226,750 t. U.S. capacity is expected to rise over 400,000 t/yr in 1979 (62). Major United States producers of sodium chlorate in 1975 are given in Table 4.

In 1977 the price of sodium chlorate was ca 33¢/kg in carload lots, and the price of potassium chlorate was ca 32¢/kg in 136-kg steel drums and carload lots. The match industry consumes ca 11,500 t $KClO_3$/yr (see Pyrotechnics).

Table 4. 1975 United States Sodium Chlorate Production Capacity[a]

Producer	Estimated capacity, t/yr
Brunswick Chemical, Brunswick, Ga.	7,250
Georgia-Pacific, Bellingham, Wash.	3,175
Hooker, Columbus, Miss.	59,850
Hooker, Niagara Falls, N.Y.	14,000
Hooker, Taft, La.	40,800
Huron Chemicals, Butler, Ala.	3,625
Huron Chemicals, Riegelwood, N.C.	6,325
Kerr-McGee Chemical Corp., Hamilton, Miss.	29,000
Kerr-McGee Chemical Corp., Henderson, Nev.	29,000
Pacific Engineering, Henderson, Nev.	5,450
Pennwalt, Calvert City, Ky.	28,100
Pennwalt, Portland, Ore.	15,500
Total	242,075

[a] New U.S. plants in 1978–1979 include IMC, Orrington, Me., 40,000 t/yr; Olin, McIntosh, Ala., 20,000 t/yr; and ERCO, Monroe, La., 25,000 t/yr (62).

OTHER CHLORATES

Barium chlorate [10294-38-9], $Ba(ClO_3)_2 \cdot H_2O$, colorless monoclinic crystals, mp 120°C ($-H_2O$), sp gr 3.18, n_D^{20} 1.562, is prepared by the reaction of barium chloride and sodium chlorate in solution; it precipitates on cooling and is purified by recrystallizing. It is used in pyrotechnics.

Lithium chlorate [13453-71-9], $LiClO_3$, rhombic needles, mp 124–129°C, decomposes on heating to 270°C, is one of the most soluble salts known; it is very hygroscopic. It is prepared by adding lithium chloride to sodium chlorate solution; sodium chloride precipitates, the liquor is concentrated, and the lithium chlorate is filtered and dried. It has limited use in pyrotechnics.

Safety

Chlorates are strong oxidizing agents. Dry materials, such as cloth, leather, or paper contaminated with chlorate may be ignited easily by heat or friction. Extreme care must be taken to ensure that chlorates do not come in contact with heat, organic materials, phosphorus, ammonium compounds, sulfur compounds, oils, greases or waxes, powdered metals, paint, metal salts (especially copper), and solvents (63). Chlorates should be stored separately from all flammable materials in a cool, dry, fireproof building.

Easily washable clothing should be worn when working with chlorates and should be changed and washed daily, or more often if contaminated by dust or solution. Clothing splashed with chlorate solution should be removed before it dries. Shoes and gloves should be rubberized. Leather shoes or other leather articles should not be worn. Goggles, face shields, and dust respirators should be worn where necessary to protect against dust, splashing, or spillage. Workers should bathe before leaving the working area.

All spillage or dusting should be avoided or removed immediately. Adequate ventilation should be provided to prevent dust from settling on structural members or walkways. Smoking, sparks, or open flames must be rigidly avoided. Any solid chlorate not in use must be kept tightly covered in the original metal containers. Skidding or sliding of such containers must also be avoided.

Ordinary handling of chlorates or their solutions presents no serious health hazard other than the extreme danger of flammability resulting when combustible materials impregnated with such solutions become dry. Ingestion or excessive inhalation of sodium and potassium chlorate dust should be avoided.

If clothing impregnated with chlorate catches fire, it should be deluged with water. Fire blankets should never be used; they are intended to smother a fire, whereas chlorate, by its very nature, supplies its own oxygen. Thus a fire blanket would only confine the heat to the body.

For more detailed information on precautionary and first aid procedures see references 64–65.

Uses

The major use (78%) of sodium chlorate is in the conversion to chlorine dioxide bleach by the pulp and paper industry. Chemical wood pulp bleached with chlorine dioxide has superior brightness over pulps bleached with other reagents. The strength of the cellulose fiber is not degraded; thus a whiter and stronger paper is obtained with chlorine dioxide. However, because of its hazardous properties, chlorine dioxide can not be shipped. It is therefore generated by the pulp producers at the bleaching plant (see Pulp).

The second important use (12%) of sodium chlorate is as an intermediate in the production of other chlorates and of perchlorates (see Perchloric acid and perchlorates under Chlorine oxygen acids and salts).

The use of sodium chlorate as a herbicide (qv) amounted to about 4550 metric tons in 1975.

Another large agricultural use of sodium chlorate is as a defoliant for cotton and as a desiccant for soybeans to remove the leaves prior to mechanical picking. Additives such as magnesium chloride, urea, or sodium carbonate may be used to reduce the fire hazard from the resulting dead leaves.

In ore processing, sodium chlorate is an important oxidizing agent, particularly for uranium ores. Minor uses of sodium chlorate include the preparation of certain dyes and the processing of textiles and furs.

Potassium chlorate is used mainly in the manufacture of matches. A typical specification for this use would be: $KClO_3$, 99.7% min; insoluble in water, 0.01% max; Cl^-, 0.03% max; moisture, 0.05% max; screen test, 99.5% at least 250 μm (60 mesh), not less than 92% at least 74 μm (200 mesh). In pyrotechnics, chlorates may be mixed with certain organic compounds such as lactose to give a relatively cool flame, so that certain dyes may be incorporated in the mixture to give colored flares. Potassium chlorate is also used in pharmaceutical preparations, for heating pads, and for other minor purposes; sodium chlorate cannot be used in place of potassium chlorate for these uses because of its hygroscopicity.

BIBLIOGRAPHY

"Chlorates" under "Chlorine Compounds, Inorganic" in *ECT* 1st ed., Vol. 3, pp. 707–716, by H. L. Robson,

Mathieson Chemical Corporation; "Chloric Acid and Chlorates" under "Chlorine Oxygen Acids and Salts" in *ECT* 2nd ed., Vol. 5, pp. 50–61, by T. W. Clapper and W. A. Gale, American Potash & Chemical Corporation.

1. F. E. King and J. R. Partington, *J. Chem. Soc.*, 925 (1926); P. L. Gilmont, *Tappi* **51,** 62A (1968).
2. C. F. Goodeve and F. A. Todd, *Nature (London)* **132,** 514 (1933).
3. A. Seidell, *Solubilities of Inorganic and Metal Organic Compounds,* 3rd ed., Van Nostrand Co., New York, 1940, pp. 784, 1250.
4. A. Nallet and R. A. Paris, *Bull. Soc. Chim. Fr.,* 488 (1956).
5. J. C. Schmacher, *ACS Monographs,* No. 146, 1960, pp. 77.
6. H. M. McLaughlin and F. E. Brown, *J. Am. Chem. Soc.* **50,** 782 (1928).
7. I. Kabik, *U.S. Bur. Mines Inf. Circ.* **7340,** (1945).
8. M. M. Jaksic and co-workers, *J. Electrochem. Soc.* **116,** 1316 (1969); **117,** 414 (1970).
9. U.S. Pat. 2,515,614 (July 18, 1950), J. C. Schumacher (to Western Electrochemical Co.).
10. U.S. Pat. 3,503,858 (Mar. 31, 1970), G. J. Crane (to Huron Nassau Ltd.).
11. D. Landolt and N. Ibl, *J. Appl. Electrochem.* **2,** 201 (1971).
12. Brochure, Chemetics International Ltd., 1827 West Fifth Ave., Vancouver, British Columbia, Canada.
13. U.S. Pat. 3,791,947 (Feb. 12, 1974), R. F. Loftfield (to Diamond Shamrock Corp.).
14. U.S. Pat. 3,819,503 (June 25, 1974), H. V. Casson, J. S. Bennett, and R. E. Loftfield (to Diamond Shamrock Corp. and Huron Chemicals Ltd.).
15. U.S. Pat. 3,732,153 (May 8, 1973), C. J. Harke, J. C. Parkinson, and J. E. Currey (to Hooker Chemical Corp.).
16. U.S. Pat. 3,518,180 (June 30, 1970), M. P. Grotheer (to Hooker Chemical Corp.).
17. *Electrochem. Prog.,* 4 (Dec. 1973).
18. U.S. Pat. 3,598,715 (Aug. 10, 1971), D. N. Goens and T. W. Clapper (to American Potash & Chemical Corp); U.S. Pat. 3,676,315 (July 11, 1972), D. N. Goens and T. W. Clapper (to Kerr-McGee Chemical Corp.).
19. *Sodium Chlorate—New Process,* technical bulletin, Krebs, 61 Rue Pouchet 75, Paris 17, Fr., June 1970.
20. *Chem. Process.* **39**(9), 60 (1976).
21. *Eur. Chem. News,* 14 (Dec. 13, 1974).
22. *Solvay (Belgium) Process for the Production of Sodium Chlorate, brochure,* Solvay et Cie, Societe Anonyme, rue du Prince Albert 33, B 1050 Brussels, Belg.
23. Yu V. Dobrov, L. M. Elina, and V. A. Grinevich, *Elektrokhimiya* **10,** 1116 (1974).
24. F. Foerster, *Trans. Am. Electrochem. Soc.* **46,** 23 (1925).
25. R. Bauer, *Chem. Ing. Tech.* **34,** 376 (1962).
26. M. M. Jaksic, *J. Electrochem. Soc.* **121,** 70 (1974).
27. A. R. Despic, M. M. Jaksic, and B. Z. Nikolic, *J. Appl. Electrochem.* **2,** 337 (1972).
28. V. A. Shlyapnikov, *Elektrokhimiya* **7,** 1080 (1971).
29. V. A. Shlyapnikov and T. S. Filippov, *Elektrokhimiya* **2,** 1165 (1966); **4,** 15 (1968); **5,** 806 (1969).
30. L. Hammar and G. Wranglen, *Electrochim. Acta* **9,** 1 (1964).
31. G. Wranglen, *Tek. Tidskr.* **92,** 197 (1962).
32. T. R. Beck, *J. Electrochem. Soc.* **116,** 1038 (1969).
33. J. C. Schumacher, *J. Electrochem. Soc.* **116,** 68C (1969).
34. E. Müller, *Z. Elektrochem.* **5,** 469 (1899); **7,** 398 (1901); **8,** 909 (1902).
35. M. Janes, *Trans. Electrochem. Soc.* **92,** 23 (1948).
36. M. M. Jaksic and co-workers, *J. Electrochem. Soc.* **116,** 394 (1969).
37. M. M. Jaksic, *J. Appl. Electrochem.* **3,** 219 (1973).
38. V. de Valera, *Trans. Faraday Soc.* **49,** 1338 (1953).
39. N. Ibl and D. Landolt, *J. Electrochem. Soc.* **115,** 713 (1968).
40. J. R. Newberry and co-workers, *J. Electrochem. Soc.* **116,** 114 (1969).
41. J. Newman, *Ind. Eng. Chem.* **60**(4), 12 (1968).
42. R. B. MacMullin, *Electrochem. Technol.* **1,** 5 (1963).
43. T. R. Beck, *J. Electrochem. Soc.* **119,** 320 (1972).
44. A. J. Boyle, V. V. Hughey, and C. C. Casto, *Ind. Eng. Chem. Anal. Ed.* **16,** 370 (1944).
45. I. M. Kolthoff and R. Belcher, *Volumetric Analyses,* 2nd ed., Vol. 3, Interscience Publishers, New York, 1957.

46. F. P. Treadwell and W. J. Hall, *Analytical Chemistry,* 9th ed., Vol. II, John Wiley & Sons, Inc., New York, 1948.
47. W. H. Rapson, *Can. J. Chem. Eng.,* 262 (Dec. 1958).
48. *Chem. Week* **93,** 111 (Sept. 7, 1963).
49. U.S. Pat. 2,280,938 (Apr. 28, 1942), G. P. Vincent (to Mathieson Alkali Works).
50. U.S. Pat. 2,863,722 (Dec. 9, 1958), W. H. Rapson (to Hooker Chemical Corp.).
51. U.S. Pat. 2,895,801 (July 21, 1959), W. W. North-Graves and B. H. Nicolaisen (to Olin Mathieson Chemical Corp.).
52. U.S. Pat. 2,936,219 (May 10, 1960), W. H. Rapson (to Hooker Chemical Corp.).
53. W. H. Rapson, *Tappi* **39,** 554 (1956).
54. H. Dodgen and H. Traube, *J. Am. Chem. Soc.* **71,** 2501 (1949).
55. H. Taube and H. Dodgen, *J. Am. Chem. Soc.* **71,** 3330 (1949).
56. F. A. Lenzi and W. H. Rapson, *Pulp Pap. Mag. Can.* **63,** T-442 (Sept. 1962).
57. C. C. Hong, F. Lenzi, and W. H. Rapson, *Can. J. Chem. Eng.* **45,** 349 (1967).
58. E. S. Atkinson, *Pulp Pap.,* 19 (Aug. 5, 1968).
59. *Chem. Mark. Rep.,* 9 (Oct. 27, 1975).
60. *Current Industrial Reports, Inorganic Chemicals,* Bureau of Census, U.S. Department of Commerce, Washington, D.C., Jan.–Dec. 1975.
61. *Can. Chem. Process.,* 10 (Jan. 1976).
62. *Chem. Week* 30 (Oct. 4, 1978).
63. *The Chlorate Manual, Technical Bulletin 1105,* Kerr-McGee Chemical Corporation, Oklahoma City, Okla., 1972.
64. *Data Sheet D-371,* National Safety Council, 425 N. Michigan Ave., Chicago, Ill. 60611.
65. *Safety Data Sheet SD-42,* Manufacturing Chemists Association, 1625 Eye St., N.W., Washington, D.C. 20006.

T. W. CLAPPER
Kerr-McGee Corporation

PERCHLORIC ACID AND PERCHLORATES

The compounds of chlorine are most stable when the chlorine atom is in either its lowest or highest oxidation state. The chlorine atom in perchloric acid [7601-90-3] and perchlorate compounds has valence +7, thus forming a large group of relatively stable compounds.

The most useful property of perchlorate compounds is their oxidizing capability. Safe, reproducible oxidations using perchlorates can be achieved under controlled conditions. These compounds provide a concentrated oxygen source; eg, one volume of ammonium perchlorate [7790-98-9], NH_4ClO_4, contains as much oxygen as 2000 volumes of air.

Aqueous perchlorate solutions exhibit very little oxidizing power when they are dilute and cold. However, hot concentrated perchloric acid is a powerful oxidizing agent and extreme caution is required in contact with oxidizable matter. Acidified concentrated solutions of perchlorate salts must also be handled with caution.

There are no naturally occurring deposits of commercially valuable perchlorates. Small amounts of potassium perchlorate [7778-74-7] occur in Chilean nitrate deposits (≤0.5%) (1).

The discovery and development of perchloric acid and perchlorates has been detailed by Schumacher (2).

Although potassium perchlorate was first prepared by von Stadion ca 1816 (3), it was not until the start of the next century that serious investigation of perchlorate compounds was undertaken. These investigations coincided with the rapid development of electrochemistry (see Electrochemical processing). Essentially all perchlorate compounds have been made since that time either directly or indirectly by electrochemical oxidation of chlorine compounds.

Commercial production began in Sweden for use in making explosives. Production in all countries probably did not exceed 1800 metric tons per year until 1940 when production increased during World War II to ca 18,000 t/yr to supply the demands of the missile and rocket industries. Known world production capacity was estimated to be ca 40,000 t/yr by 1963, slightly more than one-half was ammonium perchlorate. Production decreased after 1963. Present production is estimated to be ca 27,000 t/yr. Actual production figures are difficult to obtain because ammonium perchlorate is a strategic munitions material. Future production is expected to be largely dependent on the activity of space programs where ammonium perchlorate is used as the principal oxidizer in solid propellant rocket motors. Other uses, however, are being found that could increase the demand. For example, perchlorates are being used in slurry blasting agents and are being investigated as feed additives for fattening cattle (4–8) (see Explosives and propellants).

Properties

Chlorine Heptoxide (Dichlorine Heptoxide). The anhydride [10294-48-1] of perchloric acid, Cl_2O_7, is a colorless, volatile, oily liquid, d 1.82 g/cm^3 at 20°C, mp −91.5°C, and bp 84°C at 101.3 kPa (1 atm) and 0°C at 3.2 kPa (23.7 mm Hg). It explodes violently upon concussion or contact with flame or iodine. It does not react with sulfur, wood, or paper when cold. It may be prepared by dehydration of perchloric acid with phosphorus pentoxide (9):

$$2 \text{ HClO}_4 + \text{P}_2\text{O}_5 \rightarrow \text{Cl}_2\text{O}_7 + 2 \text{ HPO}_3$$

or by electrolysis of 55–73% HClO_4 at a cold anode (0 to $-55°\text{C}$). The Cl_2O_7 accumulates in the anode space as a separate phase (10). Other preparations include photochemically induced reaction of chlorine and ozone at 0 to $-10°\text{C}$ (11) and heating anhydrous $\text{Mg(ClO}_4)_2$ with P_2O_5 under vacuum and collecting the distillate in a receiver cooled to $-78°\text{C}$ (12).

The ir and Raman spectra of Cl_2O_7 have been recorded in the vapor, liquid, and solid phases (13), the uv and visible extinction coefficients have been measured (14), and the mass spectrum has been determined (15).

Chlorine heptoxide decomposes to chlorine and oxygen at low pressures 0.2–10.7 kPa (1.5–80 mm Hg) and at temperatures of 100–120°C (16).

Perchloric Acid. Commercial perchloric acid is an aqueous solution containing ≤72% HClO_4, usually 60–62% or 70–72% HClO_4. More concentrated solutions including anhydrous perchloric acid can be prepared but such solutions are hygroscopic and unstable.

Perchloric acid is more highly ionized in water than HCl, HNO_3, or H_2SO_4; thus, it is a strong mineral acid.

Perchloric acid is hygroscopic. The commercial 72% HClO_4 and 60–62% HClO_4 solutions approach the compositions $\text{HClO}_4\cdot 2\text{H}_2\text{O}$ and $2\text{HClO}_4\cdot 7\text{H}_2\text{O}$, respectively, two of the hydrates of perchloric acid. The 70–72% HClO_4 is an azeotrope of 28.4% H_2O–71.6% HClO_4 with a normal bp of 203°C (17–18).

A summary of the physical properties of perchloric acid and its hydrates is given in Table 1.

The density, viscosity, surface tension, and refractive index of aqueous perchloric acid solutions have been reported (2); perchloric acid and its hydrates are reviewed in reference 19.

The cost of perchloric acid is many times that of other common inorganic acids. For this reason, perchloric acid is seldom used in commercial applications for its strong acid properties alone and then only when a nonoxidizable acid is essential. As the

Table 1. Perchloric Acid and Its Hydrates

Formula	CAS Registry No.	Mol wt	wt %, HClO_4	mp, °C	bp, °C	Density, 20°C; appearance	$H_f°$, kJ/mol[a]
HClO_4	[7601-90-3]	100.46	100	−112	110 (expl)	1.7677; colorless, explosive, and shock sensitive	−40.6 −129.3, aq, m = 1
$\text{HClO}_4\cdot\text{H}_2\text{O}$	[60477-26-1]	118.47	84.8	+50	dec	1.7756; colorless, oily liquid	−382.2, crys
$\text{HClO}_4\cdot 2\text{H}_2\text{O}$	[13445-00-6]	136.49	73.6 (coml = 72%)	−17.5	203	1.65; colorless	−688, liq
$2\text{HClO}_4\cdot 5\text{H}_2\text{O}$	[34099-94-0]	291.00	69.1	−29.8			
$\text{HClO}_4\cdot 3\text{H}_2\text{O}$	[35468-32-7]	154.51	65.0	−37 (α) −43.2 (β)			
$2\text{HClO}_4\cdot 7\text{H}_2\text{O}$	[41371-23-1]	327.03	61.5	−41.4			−131.4

[a] To convert J to cal, divide by 4.184.

concentration and temperature are increased the reactivity becomes increasingly influenced by the oxidizing properties of the acid. Hot, concentrated perchloric acid will react vigorously, perhaps explosively, with oxidizable substances. Most organic compounds, under such conditions, are nearly quantitatively oxidized to carbon dioxide and water.

Hot concentrated perchloric acid is used for determining metallic elements in samples containing organic or other oxidizable matter ("liquid fire" reactions) (20–22). Perchloric acid in combination with other acids has been used for the destruction of organic matter prior to elemental analysis (23).

The value of wet ashing techniques for determination of trace elements in organic compounds has been well established (24–25). A recent study of wet oxidation by perchloric acid and its mixtures with other acids showed that compounds with N-methyl, S-methyl, C-methyl, and pyridyl moieties are the most resistant to oxidation (26). Only pyridine compounds survived wet oxidation with a combination of sulfuric, nitric, and perchloric acids. Vanadium (V), cerium(III), and copper(II) exert a catalytic influence on some wet oxidations.

Perchloric acid decomposes thermally into oxygen and Cl_2, HCl, ClO_2, and Cl_2O (27).

Chlorine monoxide, ClO_3 and ClO_4 have also been identified as intermediates in the thermal decomposition of $HClO_4$. The primary step in the homogeneous decomposition of $HClO_4$ is probably $HO-ClO_3 \rightarrow \cdot OH + \cdot ClO_3$ followed by further reaction to more stable species. Decomposition with cupric oxide (CuO) as catalyst presumably proceeds first by adsorption of H^+ and ClO_4^- on the catalyst followed by bimolecular decomposition of surface perchlorate ions and formation of water (eq. 1):

$$2\,H^+ + 2\,ClO_4^- \rightarrow H_2O_{(g)} + 2\,ClO_2 + \tfrac{3}{2}\,O_2 \qquad (1)$$

In a minor step, the decomposition of ClO_4^- on the surface yields ClO and Cl_2 (28).

Ammonium Perchlorate. The water solubility of ammonium perchlorate is given in Table 2. Ammonium perchlorate has very high solubility in ammonia (138 g/100 g NH_3 at 25°C) (29). Its dissociation constant in ammonia is 5.4×10^{-3} (30). Physical properties of ammonium perchlorate are given in Table 3.

The use of ammonium perchlorate as an oxidizer in rocket propellants has prompted extensive investigation of its thermal decomposition (16,31–33).

Table 2. The System $NH_4ClO_4 \cdot H_2O$

Temp, °C	g NH_4ClO_4/100 g soln (aq)[a]	g NH_4ClO_4/100 mL satd soln (aq)[b]
−2.7, eutectic	9.8	
0	10.74	11.56
20		20.85
25	20.02	
40		30.58
45	28.02	
60	33.64	39.05
75	39.45	
80		48.19
100		57.01

[a] Ref. 35.
[b] Ref. 36.

Table 3. Properties of Ammonium Perchlorate and the Group IA—Alkali Metal Perchlorates[a]

Perchlorate	CAS Registry No.	mp, °C	bp, °C	Crystalline form and phase transition temp[b]	Refractive index	Density[c], g/cm^3	Soly in 100 parts H$_2$O[c]	ΔH_f°[c], kJ/mol at 25°C
NH$_4$ClO$_4$	[7790-98-9]	(dec)		orthorhombic to cubic at 513 K	1.4824 1.4828[e] 1.4868[e]	1.95	20.0	−290.5
LiClO$_4$	[7791-03-9]	236–247[d]	470 dec	orthorhombic		2.43	37.8	−384.0
LiClO$_4$·3H$_2$O	[13453-78-6]	95; −3H$_2$O at 130	470 dec			1.84		
NaClO$_4$	[7601-89-0]	482 dec		orthorhombic to cubic at 577–586 K	1.4606[e] 1.4617	2.499	170[0]	−385.7
NaClO$_4$·H$_2$O	[7791-07-3]	130 dec			1.4731[e]	2.02	66[0]	
KClO$_4$	[7778-74-7]	580 −610[b]	653 dec	orthorhombic to cubic at 573–579 K	1.4717[e] 1.4724[e] 1.876[e]	2.5298	1.99	−433.5
RbClO$_4$	[13510-42-4]	281	606 dec	orthorhombic to cubic at 551–554 K	1.4701	2.9	1.26	−434.6
CsClO$_4$	[13454-84-7]	224 dec	575, to CsCl	orthorhombic to cubic at 492–497 K	1.4752 1.4788 1.4804	3.3274[d]	1.9	−434.6

[a] Ref. 36.
[b] Additional crystallographic properties are given in ref. 37.
[c] At 20°C, except where noted.
[d] To convert J to cal, divide by 4.184.
[e] Ref. 2.

Ammonium perchlorate undergoes a reversible crystallographic transition from low temperature orthorhombic to cubic structure at 240°C. The low temperature, bipyramidal, orthorhombic form has cell dimensions $a_o = 0.9202$ nm, $b_o = 0.5816$ nm, and $c_o = 0.7449$ nm. The polymorphic change is attributable to the onset of free rotation of the perchlorate ions.

The heat of transition from orthorhombic to cubic crystal structures has been computed to be 9.6 ± 0.8 kJ/mol (2.3 ± 0.2 kcal/mol) (34).

The cubic high temperature form contains four molecules per unit cell which has a cube edge of 0.763 nm and a density of 1.76 g/cm³ as compared with 1.95 g/cm³ for the low-temperature form.

The specific heat of ammonium perchlorate is 1.29 J/(g·K) [0.309 cal/(g·K)] at 15–240°C and 1.53 J/(g·K) [0.365 cal/(g·K)] above the transition temperature.

On heating ammonium perchlorate, slow weight loss starts at about 210°C. Loss is caused by a combination of sublimation and decomposition (eq. 2). The low temperature decomposition products have been identified as NH_3, $HClO_4$, N_2, O_2, H_2O, NO_2, Cl_2, and ClO_2 (38). The decomposition below 300°C is approximately represented by equation 3 (39–41):

$$NH_4ClO_4 \rightarrow NH_3 + HClO_4 \tag{2}$$

$$4\ NH_4ClO_4 \rightarrow 2\ Cl_2 + 3\ O_2 + 8\ H_2O + 2\ N_2O \tag{3}$$

Above 300°C the proportion of nitric oxide increases. Above 350°C the decomposition can be represented by equation 4:

$$10\ NH_4ClO_4 \rightarrow 2.5\ Cl_2 + 2\ N_2O + 2.5\ NOCl + HClO_4 + 1.5\ HCl + 18.75\ H_2O + 1.75\ N_2 + 6.375\ O_2 \tag{4}$$

The gas-phase ignition mechanism of ammonium perchlorate using skimming transient mass spectroscopy was studied (42). It was determined that decomposition of NH_4ClO_4, at heating rates representative of actual propellant systems, was predominantly to ammonia and perchloric acid and that final decomposition takes place in the gas phase. Product gas spectra were similar to the solid-phase decomposition spectra except that the former did not show the existence of Cl_2.

Decomposition and sublimation of ammonium perchlorate is suppressed under an atmosphere of ammonia. The decomposition can be inhibited until a temperature of 340°C is reached (43). At 345–350°C a fast reaction begins with a subsequent explosion owing to proton transfer and autocatalytic oxidation of NH_3 by perchloric acid or its decomposition products.

Decomposition can be retarded by addition of other substances besides NH_3 formed in decomposition of NH_4ClO_4, as well as substances that react with the NH_4ClO_4 or its decomposition products to form less reactive compounds, and by addition of decomposition inhibitors (44). Especially effective inhibitors are NH_4Cl, NH_4F, CdF_2, ZnF_2, and $PbCl_2$. Urea and dicyandiamide are effective inhibitors at elevated temperatures.

Prior mechanical and thermal treatment was found to affect the isothermal decomposition of NH_4ClO_4 at 215–235°C (45).

Ammonium perchlorate heated to 215–235°C slowly decomposes, losing up to 30% of the initial weight, then the decomposition ceases. The porous, solid residue is still pure ammonium perchlorate which regains its original properties by exposure to solvent vapor (39). Decomposition decreased from the usual 30% to approximately 20% when the NH_4ClO_4 was compressed before heating.

Differential thermal analysis of small samples of commercial ammonium perchlorate normally shows two exotherms on heating above the transition temperature. The first exotherm initiates in the 275–320°C range. The initiation and magnitude of the exotherm are influenced by impurities present in the ammonium perchlorate. The first exotherm is absent in pure NH_4ClO_4. The common impurities in commercial ammonium perchlorate influencing this first peak are chlorates and chromates (46). The second exothermic decomposition is initiated at ca 400–470°C. Initiation temperature of this exotherm also depends on purity of the NH_4ClO_4, pressure, and composition of the atmosphere over the sample. With larger samples of ammonium perchlorate the heat released during the first exothermic decomposition is sufficient to initiate the second one.

The effects of a large number of additives on the thermal decomposition of ammonium perchlorate have been studied. A partial list includes compounds of Ni, Cu, Zn, Fe, Cd, Al, Mn, and Cr (47).

Ammonium perchlorate is used primarily as the oxidizer in rocket and missile propellant systems. It has been used in the production of explosives (48).

Group 1A (Alkali) Perchlorates. Compounds have been identified where perchlorate has been combined with one or more elements of every group in the periodic table except Group O (the inert gases). Representative compounds will be mentioned but no attempt will be made to list or describe all known compounds.

Potassium perchlorate is of historical significance in that it was the first perchlorate discovered (2–3), and it is used in pyrotechnics (qv).

Lithium perchlorate is of special interest because it contains the highest percentage of oxygen of the metal perchlorates (60.1% O_2, even higher than NH_4ClO_4 which contains 54.5%). It would find more extensive use if not for its hygroscopic properties.

Some of the physical properties of alkali metal perchlorates are given in Table 3. All Group IA perchlorates are white or colorless. Solubility of perchlorate salts in water decreases in the order Na > Li > NH_4 > K > Rb > Cs. The higher solubility of $NaClO_4$ makes it useful as an intermediate for production of all other perchlorates by double metathesis reactions and controlled crystallization. The low solubility of K, Rb, and Cs perchlorates has made possible their determination by gravimetric analytical procedures. All except the lithium salt undergo crystalline-phase transformations from orthorhombic to cubic on heating. Only lithium perchlorate appears to have a definite melting point without decomposition in a narrow temperature range around the melting point.

Lithium perchlorate forms a simple eutectic with lithium nitrate at 172°C and a composition of 53.5 mol % $LiClO_4$ and 46.5 mol % $LiNO_3$ (49).

Lithium perchlorate has a greater solubility than one gram per gram of solvent in methyl alcohol, ethyl alcohol, n-butyl alcohol, acetone, and ethyl ether. Sodium perchlorate also has appreciable solubility in these solvents (50). The high solubility of $LiClO_4$ in organic solvents has been used in a gel-like explosive consisting of 59.4 wt % $LiClO_4$, 30.5 wt % methyl alcohol, and 10 wt % of 80% H_2O_2. It has a sp gr of 1.60 and an explosion velocity of 3600 m/s (51).

Cordes and Smith (52) found that the initial solid-phase decomposition rates of $NaClO_4$, $KClO_4$, $RbClO_4$, and $CsClO_4$ below 410°C are essentially the same. The major products are O_2 and ClO_3^-. The decomposition rates for both sodium and potassium perchlorates increase with temperature starting 30–40°C below their melting points (465°C for $NaClO_4$, 580°C for $KClO_4$).

The similarity of the specific rates of oxygen evolution suggests a mechanism dependent primarily on the nature of the perchlorate ion and indicates a simple homogeneous solid-phase reaction (eq. 5).

$$MClO_4 \rightarrow MClO_3 + \tfrac{1}{2} O_2 \ (M = Na, K, Rb, Cs) \tag{5}$$

At higher temperatures, approaching or above the melting points of the perchlorates, the decomposition reactions become more complex giving rise to simultaneous formation of several products.

Decomposition of lithium perchlorate was studied at lower temperatures (53). The initial reaction was the same as for other alkali metal perchlorates (eqs. 6–7)

$$2 ClO_4^- \rightarrow 2 ClO_3^- + O_2, \tag{6}$$

followed by ClO_3^- decomposition

$$ClO_3^- \rightarrow Cl^- + \tfrac{3}{2} O_2 \tag{7}$$

At higher temperatures and higher degrees of perchlorate conversion, the decomposition of $LiClO_4$ is autocatalytic because the decomposition is promoted by Cl^- (54).

Group IB Perchlorates. Both copper(I), $(CuClO_4 \cdot H_2O)$ [17031-33-3], and copper(II), $Cu(ClO_4)_2$ [13770-18-8], perchlorates are known. The divalent copper perchlorates form a series of hydrates $(Cu(ClO_4)_2 \cdot 2H_2O$ [17031-32-2]; $Cu(ClO_4)_2 \cdot 4H_2O$ [17031-32-2]; $Cu(ClO_4)_2 \cdot 6H_2O$ [10294-46-9]; and $Cu(ClO_4)_2 \cdot 7H_2O$ [67632-65-9]).

Copper perchlorate also forms a number of complexes with ammonia, ammonia and water, pyridine, and with organic derivatives of these compounds. A large number of copper–organic-perchlorate complexes have been reported.

Copper perchlorate is an effective burning rate accelerator for solid propellants (55).

Silver perchlorate [7783-93-9] has been used extensively for the preparation of other perchlorates by the following reaction (eq. 8):

$$AgClO_4 + MX \rightarrow AgX + MClO_4 \tag{8}$$

where MX is either an inorganic or organic halide.

The silver salt is very deliquescent and forms a monohydrate that can be dehydrated at 43°C. The anhydrous salt is light sensitive. Its solubility is 557 g/100 g H_2O at 25°C. It has high solubility in organic solvents including benzene, toluene, aniline, pyridine, nitrobenzene, chlorobenzene, glycerol, and glacial acetic acid. Its solubility in toluene is 101 g/100 g solvent. It is insoluble in chloroform, carbon tetrachloride, and ligroin. Explosions of silver perchlorate have been reported (56–58).

Silver perchlorate forms complexes with benzene, toluene, aniline, pyridine, and dioxane (2).

The dissociation of silver perchlorate in aqueous solutions of acetic, perchloric, and nitric acids has been studied (59).

Gold also forms organic perchlorate complexes, eg, $(C_6H_5)_3AsAu(C_6F_5)_2ClO_4$ [42774-61-8].

Group IIA (Alkaline Earth) Perchlorates. Physical properties of the alkaline earth metal perchlorates are given in Table 4. Group IIA perchlorates form many hydrates, ammoniates, and other solvate compounds.

The salts may be prepared in the anhydrous state by heating ammonium perchlorate with the corresponding hydroxide or carbonate. The more basic metals react more rapidly and at lower temperatures.

Table 4. Properties of Group IIA (Alkaline Earth) Perchlorates

Perchlorate	CAS Registry no.	Density, g/cm^3	mp, °C	Soly, g/100 g water
Be(ClO$_4$)$_2$	[56044-34-9]			
Be(ClO$_4$)$_2$.2H$_2$O	[39527-87-2]		<80	
Be(ClO$_4$)$_2$.4H$_2$O	[7787-48-6]			59.5
Mg(ClO$_4$)$_2$	[10034-81-8]	2.60	251 dec	99.6
Mg(ClO$_4$)$_2$.2H$_2$O	[18716-62-6]			
Mg(ClO$_4$)$_2$.4H$_2$O	[22465-13-0]			
Mg(ClO$_4$)$_2$.6H$_2$O	[13446-19-0]	1.970	185	v sol
Ca(ClO$_4$)$_2$	[13477-36-6]	2.651	270 dec	188.6
Ca(ClO$_4$)$_2$.4H$_2$O	[15627-86-8]			
Sr(ClO$_4$)$_2$	[13450-97-0]			309.7
Ba(ClO$_4$)$_2$	[13465-95-7]	3.2	505.	33
Ba(ClO$_4$)$_2$.3H$_2$O	[10294-39-0]	2.74	400 dec	198

The hydrates may be prepared readily by reaction of the metal oxide, carbonate, or hydroxide with perchloric acid.

Beryllium perchlorate dihydrate has been prepared by allowing BeCl$_2$ to react with HClO$_4$ or by heating BeCl$_2$ with HClO$_4$.H$_2$O at 60°C (60). An oxo- or hydroxy containing compound is formed at 150°C. At 190–265°C HClO$_4$ is lost and the solid residue is Be$_4$O(ClO$_4$)$_6$ [39455-86-2]. Complete decomposition to BeO occurs on heating at 290°C.

The strong affinity of anhydrous magnesium perchlorate for water gives it great value as a dehydrating agent (61). It also strongly absorbs methanol, ethanol, acetone, pyridine, acetonitrile, ammonia, and nitromethane vapors (62).

The other anhydrous alkaline earth perchlorates have absorption properties similar to those of MgClO$_4$.BaClO$_4$, which is used as a desiccant (see Drying agents).

Group IIB Perchlorates. The perchlorates of zinc [13637-61-1], cadmium [13760-37-7], and mercury have been reported. Mercury forms both monovalent [13932-02-0] and divalent [7616-83-3] compounds.

Group IIIA Perchlorates. Perchlorate compounds have been reported containing each of the Group IIIA elements.

Boron perchlorates occur as double salts with alkali perchlorates, eg, Cs[B(ClO$_4$)$_4$] [33152-95-3] (63).

Aluminum perchlorate [14452-95-3] forms a series of hydrates having 3, 6, 9, or 15 moles of water per mole of Al(ClO$_4$)$_3$. The anhydrous salt can be prepared from the trihydrate by drying at 145–155°C over phosphorus pentoxide under reduced pressure (64).

The compound may also be obtained by the reaction of anhydrous aluminum chloride with silver perchlorate in organic solvents such as methanol or benzene (65).

Upon heating the hydrates of this salt, dehydration occurs simultaneously with hydrolysis at 178°C to yield the basic salt, Al(OH)(ClO$_4$)$_2$ [67632-67-1], and upon heating to 264°C the salt decomposes to aluminum oxide (66). The salt is completely ionized in dilute aqueous solutions and is soluble in nitromethane, nitrobenzene, acetonitrile, and 2-ethoxyethanol.

Group IIIB Perchlorates and Inner Transition Metal Perchlorates. The perchlorates of yttrium and lanthanum are known. Similarly, the trivalent perchlorate compounds of the lanthanide series of inner transition metal compounds have nearly all been prepared (67–70) and the specific heats of their solutions reported (71).

Tetravalent cerium perchlorate [14338-93-3], Ce(ClO$_4$)$_4$, is also known.

Uranium perchlorate [14989-40-3] has been made (72) (see Actinides).

Group IVA Perchlorates. A large number of perchlorates containing organic carbon have been reported (73). Amine perchlorates (74), diazonium perchlorates (75), oxonium perchlorates (76–77), and perchlorate esters (78–80) have all been prepared. Extreme caution should be used in working with organic perchlorates. Many decompose violently when heated, contacted by other reagents, or subjected to mechanical shock.

The amine perchlorates are generally crystalline salts that are quite stable at ordinary temperatures. When heated they volatilize or decompose. Heat or mechanical shock will cause them to explode.

Diazonium perchlorates are examples of the most explosive compounds known.

Perchloric acid forms oxonium and carbenium salts with aldehydes, ketones, and ethers (76–77). The salts are shock sensitive and are decomposed by moisture.

The perchlorate esters are explosive, very shock sensitive liquids (78–79). Trichloromethyl perchlorate [67632-66-0] is formed by the reaction between carbon tetrachloride and silver perchlorate in the presence of a little hydrogen chloride. This colorless liquid detonates on contact with organic substance or on heating (81).

The arylmethyl perchlorates are formed by the reaction of perchloric acid and the corresponding alcohol or by methathesis of the chloride with silver perchlorate in nitrobenzene solvent. They are highly colored crystalline materials of relatively high melting points.

Triazonium perchlorate monohydrate [61017-17-2] has been prepared by the reaction of Ba(ClO$_4$)$_2$ with triazonium methane sulfonate and can be used as a thermally stable propellant oxidizer (82).

Lead perchlorate [13637-76-8] forms a series of hydrates. Anhydrous solutions of lead perchlorate in methyl alcohol are explosive.

Group IVB Perchlorates. Titanium tetraperchlorate [13498-15-2] has been prepared by reaction of Cl(ClO$_4$) [27218-16-2] and TiCl$_4$ at −45 to 20°C; chlorine is evolved in the reaction (83–85).

Group VA Perchlorates. The Group VA perchlorates are particularly important; ie, the nitrogen perchlorates have been used as oxidizers in rocket propellant systems.

Hydrazine perchlorate [13762-80-6] and hydrazine diperchlorate [13812-39-0] have been investigated as oxidizers for propellant systems. They are powerful and sensitive oxidants. Extreme care must be used in handling them to prevent explosions. The difficulty in obtaining the anhydrous compounds with acceptable crystal size and shape has discouraged their use. Hydrazine perchlorate is made by neutralizing an aqueous solution of hydrazine with aqueous perchloric acid (86).

Hydroxylammonium perchlorate [15588-62-2] has been made by the reaction of sodium perchlorate with hydroxylammonium sulfate or hydroxylammonium chloride. The salts Na$_2$SO$_4$ and NaCl are precipitated from the reaction mixture by addition of low boiling alcohols or ethers (87). Hydroxylammonium perchlorate has also been prepared by the following reaction (eq. 9) (88):

$$(NH_3OH)_2SO_4 + Ba(ClO_4)_2 \rightarrow 2\ (NH_3OH)ClO_4 + BaSO_4 \qquad (9)$$

Hydroxylammonium perchlorate is unstable to impact and to friction. Dta shows an exotherm at 178–220°C corresponding to its decomposition to ammonium perchlorate and oxygen. The resulting impure NH_4ClO_4 undergoes exothermic decomposition at 313–370°C.

Nitrosyl perchlorate [15605-28-4] is made by passing a mixture of nitric oxide and nitrogen dioxide into 72% perchloric acid. The anhydrous salt may be obtained by partially drying the hydrate over phosphorus pentoxide in an atmosphere of nitrogen oxides followed by final desiccation in vacuo. It decomposes at less than 100°C (eq. 10):

$$2\ NOClO_4 \rightarrow 2\ ClO_2 + N_2O_5 + \tfrac{1}{2} O_2 \qquad (10)$$

Nitrosyl perchlorate reacts with water and evolves nitrogen oxides. It reacts violently with dry ether and with primary amines.

Bismuth perchlorate pentahydrate [66172-92-7] can be prepared by dissolving bismuth oxide in 40% perchloric acid (89). The salt is converted to the bismuthyl compound (trihydrate) on addition of water followed by evaporation over calcium chloride. Further drying converts the $BiOClO_4 \cdot 3H_2O$ [67632-68-2] to $BiOClO_4 \cdot H_2O$ [66172-93-8].

Group VB Perchlorates. Vanadyl perchlorate [67632-69-3], $VO(ClO_4)_3$, has been prepared by the following reaction (eq. 11):

$$VOCl_3 + 3\ Cl_2O_4 \rightarrow VO(ClO_4)_3 + 3\ Cl_2 \qquad (11)$$

The reaction is carried out in the cold (−45 to ca 20°C). The vanadyl perchlorate has been suggested for use in gas generating compositions (84).

Group VIA Perchlorates. A perchlorate compound containing sulfur, diperchlorate sulfate [43059-05-8], $SO_4(ClO_4)_2$, was produced by low temperature electrolysis of 12 N H_2SO_4 and 3 N $HClO_4$ solution. The compound probably results from reaction of the radicals $ClO_4 \cdot$ and $S_2O_8 \cdot$ formed on discharge at the anode of the electrolytic cell (90). The compound is a strong oxidizing agent. Reaction with acetone, benzene, toluene, alcohols, and fats at 25°C is exothermic and explosive. The compound is soluble in CCl_4 and halide-substituted Freon without chemical reaction.

Group VIB Perchlorates. Both divalent and trivalent chromium perchlorate [13931-95-8, 13527-21-9] have been reported. The trivalent salt forms a large series of hydrates.

Anhydrous chromyl perchlorate [60499-74-3] has been prepared by the cold reaction (eq. 12) and has been suggested for use in gas generating systems (84).

$$CrO_2Cl_2 + 2\ Cl_2O_4 \xrightarrow{-45\ \text{to}\ 20°C} 2\ Cl_2 + CrO_2(ClO_4)_2 \qquad (12)$$

Group VIIA Perchlorates. Fluorine perchlorate [37366-48-6], $FClO_4$, is formed from the reaction of elemental fluorine with 60–70% aqueous perchloric acid (91). The compound is normally a gas (mp −167.5°C, bp −15.9°C). It is extremely reactive and is explosive in all states.

Interestingly, perchloryl fluoride [7616-94-6], fluorine chlorate, $FClO_3$, the acyl fluoride of perchloric acid, is a very stable compound. Also, it is normally a gas (mp −147.7°C, bp −46.7°C). It can be prepared by electrolysis of a saturated solution of sodium perchlorate in anhydrous hydrogen fluoride. Some of its uses are as an oxidant in rocket fuels and as a gaseous dielectric for transformers.

Perchloryl fluoride can be obtained in 85–90% yield from a warm reaction mixture of $KClO_4$, HF, and SbF_5 at 40–50°C (92).

The properties, reactions, and applications of perchloryl fluoride have been reviewed (93–94) (see Fluorine compounds, inorganic).

Group VIIB Perchlorates. Both divalent and trivalent manganese perchlorate, [13770-16-6, 13498-03-8] are known.

Group VIII Perchlorates, The Transition Elements. Perchlorate compounds of Fe, Co, Ni, Rh, and Pd have been described (95). These perchlorates are usually colored.

Double perchlorate salts have been reported. For example, anhydrous $HClO_4$ does not react with $CoCl_2$ at room temperature, but in the presence of $CsClO_4$ [13454-84-7], a $CsCo(ClO_4)_3$ hydrate [13478-33-6] is formed (95).

The perchlorate ion (ClO_4^-) has been considered as a noncoordinating or a poorly-coordinating anion. However, when water has been rigorously excluded anhydrous complexes can be prepared (96). For example, nickel complexes $NiL_6(ClO_4)_2$, $NiL_4(ClO_4)_2$, and $NiL_2(ClO_4)_2$, where L is CH_3CN have been shown to contain ionic, unidentate, and bidentate perchlorate by ir spectroscopy and other physical methods. Coordination is from the metal through the oxygen to the chlorine atom. Perchlorate complexes of nickel (97), cobalt (98), copper (99), sodium (100), and tin (101) have been reported. The metal in each case is also coordinated with an organic group (CH_3CN, CH_3, pyridyl, etc) (see Coordination compounds).

Johansson has also reviewed the role of the perchlorate as a coordinating ligand in solution (102).

Manufacture of Perchloric Acid

Commercial preparation of perchloric acid began in 1830 when Serullas prepared perchloric acid by the thermal decomposition of chloric acid (103). Since then it has been made by electrolysis of chloric acid solution; by irradiating aqueous solutions of chlorine dioxide (104); by electrolysis of hydrochloric acid solution (105); by passing ozonized air through a solution of hypochlorous acid or sodium hypochlorite (106); by ion exchange; and by electrodialysis of perchlorate salts using ion-selective membranes (107).

Commercial preparation has largely been by the chemical reactions of sodium perchlorate and hydrochloric acid (108). This process was used commercially by Hooker Electrochemical Company for some years. In this process sodium chloride was first oxidized electrolytically to sodium chlorate.

Perchloric acid has been produced by direct electrolytic oxidation of cold (18°C), very dilute (0.5 N) hydrochloric acid (109–110). The anode was platinum and the cathode copper or silver. The product after electrolysis contained 25–30 g $HClO_4$/L with small amounts of HCl and chloric acid. The perchloric acid was evaporated to produce the 60% acid of commerce. Current efficiency was in the 30–40% range.

Workers at Merck A.G. have developed a more attractive method for the direct electrochemical production of perchloric acid by electrolysis of chlorine in cold dilute perchloric acid (111). Perchloric acid (40%) is chilled to −5°C and saturated with chlorine. The cold Cl_2–$HClO_4$ solution is electrolyzed in the anolyte of the diaphragm electrolytic cell in which the chlorine at the platinum anode is converted to perchloric acid. The anolyte discharge is continuously recycled to the chiller and the chlorine absorber. Temperature rise through the cell is about 8°C. A portion of the perchloric acid passes through the permeable membrane into the catholyte. Hydrogen is generated at the silver cathode. The enriched perchloric acid catholyte is discharged into a collection tank where it is withdrawn at a controlled rate to feed a packed vacuum column. Water, the small amount of hydrochloric acid, and chlorine carried along with the perchloric acid are removed from the acid and condensed, accumulated, and intermittently returned to the anolyte recycle loop. The vacuum column is heated at the bottom to help drive the HCl, Cl_2, and water from the perchloric acid. The system is controlled so that high purity perchloric acid of the desired concentration is withdrawn from the bottom of the vacuum column, and cooled and accumulated in a product tank.

Anhydrous perchloric acid is prepared by vacuum distillation from a mixture of 72% $HClO_4$ and fuming sulfuric acid. The distillation is carried out in glass apparatus at pressures of ≤ 0.13 kPa (≤ 1 mm Hg). Yields of 75% have been reported (112). The freshly prepared anhydrous acid is a clear, colorless, hygroscopic liquid. It is unstable at ordinary temperatures; it slowly becomes yellow or brown and explosive. Decomposition is temperature dependent so that it can be stored for long periods at very low temperatures. At ordinary temperatures spontaneous explosion can be expected after 20–30 d.

Manufacture of Perchlorates

The history of the manufacture of perchlorates is outlined in references 2 and 113.

Of the several possible methods for producing perchlorates only the electrochemical oxidation of lower valence chlorine-containing compounds is commercially important.

Although several of the metal perchlorate compounds can be made by direct electrooxidation of the chloride or chlorate salts and most of them can be made by the reactions of perchloric acid with the oxide, hydroxide, or carbonate of the metal, essentially all of the commercial perchlorates are made by metathesis with sodium perchlorate. Most of the sodium perchlorate is made by electrooxidation of sodium chlorate.

Electrochemical Process. The reaction to form perchlorate at the anode of an electrolytic cell can be written as (eq. 13):

$$ClO_3^- + H_2O \rightarrow ClO_4^- + 2\,H^+ + 2\,e^- \qquad (13)$$

Some oxygen is also produced at the anode and is the main cause of the inefficiency of the process. Hydrogen is produced from reduction of water at the cathode.

Anode mechanisms for the formation of perchlorate ions have been proposed involving discharge of the chlorate ion at the anode to form a chlorate radical (114–116). The unstable chlorate radical undergoes further reaction with water to form perchlorate ion and hydrogen ion.

A mechanism was proposed that involved formation of active oxygen at the anode by electrolysis of water followed by either reaction of the active oxygen with chlorate to give perchlorate or with a second atom of active oxygen to form oxygen gas (117).

The mechanism for perchlorate formation has also been proposed to involve mass transport of chlorate ion which reacts with an adsorbed oxygen species at the anode surface. The reacting, adsorbed oxygen is then replaced by adsorbed oxygen generated by the oxidation of water (118–120).

Smooth platinum or lead dioxide anodes are usually used to make perchlorate because they can sustain high anodic potentials with a minimum of anodic dissolution. Current efficiencies are higher with smooth platinum (95% compared to 85% for lead dioxide), and platinum is less fragile. Lead dioxide anodes, however, are less expensive than platinum and can be used to electrolyze solutions low in chlorate ion concentration.

Sodium Perchlorate. With the exception of the direct anodic oxidation of chlorine to perchloric acid (111) previously discussed, all commercial production of perchlorates is by the electrolytic oxidation of sodium chlorate to sodium perchlorate. The sodium chlorate is produced electrochemically from sodium chloride (see Chloric acid and chlorates).

Table 5 shows the electrolytic cell operating data for various producers of sodium perchlorate.

Legendre (121) has outlined the pretreatment of electrolyte feed to cells for various processes and the aftertreatment of the electrolyzed cell product. Pretreatment involves removal of unwanted impurities or introduction of additives to improve cell performance. Aftertreatment includes removal of additives or impurities followed by either evaporation, crystallization, crystal recovery (centrifugation), and drying, if solid NaClO$_4$ is desired, or direct use of the clarified electrolyte in double exchange (metathesis) reactions to produce NH$_4$ClO$_4$, KClO$_4$, or other perchlorate products.

The cell tank is always constructed of the same material as the cathodes. The tank is negatively charged and is thus cathodically protected from corrosion. Iron or steel are commonly used as cathodic materials, Chedde Pechiney, however, uses bronze. The anodes are usually platinum or a platinum-plated substrate; Pacific Engineering uses lead dioxide-plated graphite substrate anodes.

The starting electrolyte contains 300–700 g NaClO$_3$/L depending on the process. It also contains 50–700 g NaClO$_4$/L if solid sodium perchlorate is recovered as a final product and the mother liquor from the crystallization and crystal recovery is recycled to the cells as in the Chedde Pechiney process.

In all processes heat must be removed from the electrolyte either by internal cell cooling or by circulation through an external heat exchanger.

Platinum loss increases with increasing temperature. However, operating voltage decreases with increasing temperature and the operating temperature selected for each process (30–60°C) is chosen to minimize operating costs.

Platinum loss increases and current efficiency decreases as the NaClO$_3$ concentration decreases. Platinum loss increases rapidly as the NaClO$_3$ concentration in the electrolyte decreases below 100 g NaClO$_3$/L and becomes excessive at NaClO$_3$ concentrations less than 50 g NaClO$_3$/L (119–120).

Sodium dichromate is added to the electrolyte in platinum anode cells to inhibit reduction of perchlorate at the cathode. Sodium fluoride is used in the lead dioxide anode cells to improve current efficiency.

The lead dioxide anodes used by Pacific Engineering and Production Company for perchlorate manufacture consist of β lead dioxide plated on graphite substrate. The graphite provides strength, support, and good electrical connection to the otherwise fragile lead dioxide (123–125). Graphite-substrate lead dioxide anodes are suitable for use in high amperage cells for the production of both chlorates and per-

Table 5. Operational Data for Electrooxidation of Sodium Chlorate to Sodium Perchlorate[a]

Property	Pacific Engineering Corp.	Kerr-McGee Chemical Corporation 1st Stage	Kerr-McGee Chemical Corporation 2nd Stage	Chedde Pechiney	Cardox Corp.	Elektrochemie Turgi
amperage, kA	5	5		1.7–3.5	0.5	1.5–3.0
voltage, V	4.75	6.0		5.8–6.2	6.2–6.8	
anode current density, kA/m^2	1.5	3.1		4.5	5.2	
anode	PbO$_2$–graphite	platinum		platinum	platinum on copper	platinum
cathode	316 SS	steel		bronze	steel	iron
anode–cathode spacing, cm	0.5			<0.5	3	0.2–0.5
electrolyte composition						
start						
NaClO$_3$, g/L	700	600	100	300	650	700
NaClO$_4$, g/L	0	0	500	700	0	
end						
NaClO$_3$, g/L	3	100	5	80	20	4–10
NaClO$_4$, g/L	1050	500	600	1100	800	900
Na$_2$Cr$_2$O$_7$ conc, g/L		5		<1	1	2
pH	6.5	6.5		9–10	10	8
temperature, °C	60	40–45		60	50	30
current efficiency, %	85	90	70	95	97	80–90
energy consumption, kW·h/kg	2.45	3.0		2.77	2.93	
operational mode	batch	continuous		continuous	batch	continuous

[a] Based, in part, on references 121 and 122.

chlorates (126–127). There is a process for using lead dioxide anodes in the production of sodium perchlorate directly from sodium chloride rather than starting with sodium chlorate (128). Current efficiencies of 50–60% and energy consumption of 12–13 kW·h/kg of $NaClO_4$ produced are reported. The process has the advantage that it is unnecessary to separate and redissolve the $NaClO_3$ from the chlorate plant electrolyte before preparation for perchlorate manufacture.

Ammonium Perchlorate. Extremely high purity ammonium perchlorate can be made by the direct reaction of ammonia with aqueous perchloric acid solutions (eq. 14):

$$NH_3 + HClO_4 \rightarrow NH_4ClO_4 \qquad (14)$$

The necessity for such extremely high purity ammonium perchlorate is limited to research applications, and only laboratory quantities are made in this way.

The commercial product is manufactured by the double exchange reaction of sodium perchlorate and ammonium chloride (129–130).

Properties of ammonium perchlorate and the alkali metal perchlorates are given in Table 3. Not only is the solubility of NH_4ClO_4 much less than the solubility of $NaClO_4$, it is also much less soluble than the other ion pairs (NaCl, NH_4Cl) in the double exchange reaction (eq. 15):

$$NaClO_4 + NH_4Cl \rightarrow NH_4ClO_4 + NaCl \qquad (15)$$

and the ammonium perchlorate separates in high purity as the crystalline anhydrous salt.

A flow diagram for a 45 metric ton per day ammonium perchlorate plant is shown in Figure 1 (130). The reactor product is crystallized in a vacuum-cooled crystallizer and a crop of ammonium perchlorate is removed in the NH_4ClO_4 centrifuge. The

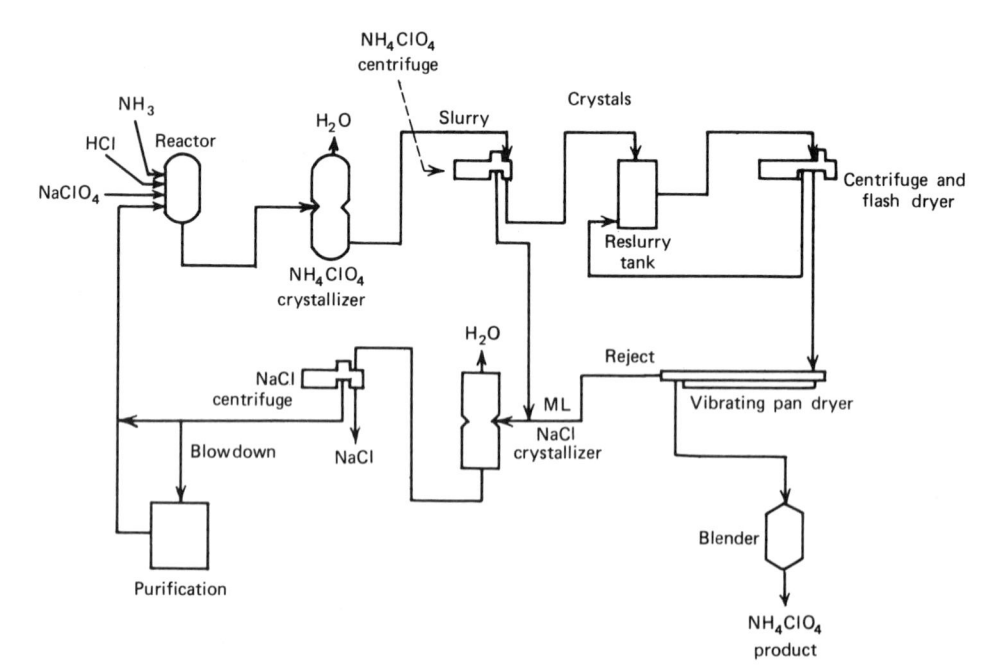

Figure 1. Flow diagram of 45 t/d ammonium perchlorate plant (130). ML = mother liquor.

crystalline product is reslurried, recentrifuged, dried, and blended for shipment. Mother liquor is concentrated in an evaporator to precipitate sodium chloride. The depleted mother liquor from the sodium chloride crystallization is then returned for further reaction with $NaClO_4$, HCl, and NH_3.

The product meets the following typical specification for a propellant-grade ammonium perchlorate:

Property	Value
moisture, %	0.10 max
water insolubles, %	0.04 max
chloride, as NH_4Cl, %	0.15 max
sulfated ash, as $NaClO_4$, %	0.30 max
oxidants, as NH_4ClO_3, %	0.05 max
perchlorates, as NH_4ClO_4, %	99.0 min
pH of water solution	4.3–5.8

Sodium chloride from the mother liquor, and potassium perchlorate from potassium impurities in NaCl or water, can be of concern for some applications. The sodium chloride can be removed by recrystallization and the potassium can be removed from the pregnant ammonium perchlorate before crystallization by ion exchange with solid ammonium perchlorate (slurry or in a column) (131).

A process has been patented for making ammonium perchlorate from sodium perchlorate, ammonia, carbon dioxide (rather than the usual HCl), and water (132).

Shipping Regulations

Perchloric acid and perchlorates are shipped in regulation glass or metal drum containers specified by the DOT hazardous materials regulations (133). Perchloric acid and the inorganic perchlorates are classified as oxidizers and require oxidizer shipping labels. Shipping is forbidden on passenger-carrying aircraft or railcars.

Shipment of perchloric acid of concentrations greater than 72% is forbidden.

Ammonium perchlorate may be shipped in steel drums with plastic liners. Large shipments are made in portable aluminum containers holding up to 2.27 metric tons. Lower side or hopper type product discharge openings are not permitted.

Sodium or magnesium perchlorate may be shipped in tank cars wet with ≥10% water which is equally distributed.

All perchlorates not specifically covered by DOT regulations should have special precautionary labels affixed recognizing the specific fire or explosion hazards expected from the individual perchlorate.

Economic Aspects

The annual use of 70% perchloric acid is estimated to be ca 450 metric tons. Anhydrous $HClO_4$ has no commercial uses because of its instability. The 70% $HClO_4$ is not sold in bulk. It is sold only in 3.63-kg bottles for $50 each.

Ammonium perchlorate is the workhorse oxidizer for the solid propellant industry and the principal commercial perchlorate. In bulk its prices are $0.65–1.10/kg; as a specialty item, its price can be as high $2.55/kg.

More $NaClO_4$ is manufactured than NH_4ClO_4; however, very little $NaClO_4$ is sold on the open market. It is used captively, especially as an intermediate in the manufacture of other perchlorates. About 725 t/yr, priced at 41¢/kg, is used for explosives.

Methods of Analysis

Methylene blue gives violet precipitates with perchlorates that are soluble in hot water and can be used for qualitative analysis. Precipitation of perchlorate with a standard excess of methylene blue has also been used as a quantitative method (134).

Potassium, rubidium, and cesium salts precipitate perchlorate from cold ethanol–water solutions and may be used to detect or determine perchlorate (135).

Tetraphenylarsonium chloride precipitates the perchlorate ion and may be used for its gravimetric or volumetric determination (136–137). Tetraphenylphosphonium chloride is also a good precipitation agent for small concentrations of perchlorate (138).

The classical method for determining appreciable amounts of perchloric acid or perchlorate salts involves conversion of the perchlorate to chloride by fusion with sodium carbonate in a platinum dish. The fusion breaks down all chlorates or perchlorates to NaCl. The residue after fusion is dissolved in water or dilute nitric acid and the total chlorides determined by the usual volumetric or gravimetric methods. The total chloride value is corrected by subtracting the chloride equivalent of $NaClO_3$ and NaCl initially present in the sample. The corrected chloride value is then used to determine the perchlorate salt initially present.

Ammonium chloride may be substituted for sodium carbonate in fusion if alkali metal perchlorates are being determined. The excess NH_4Cl volatilizes leaving only metal chlorides in the residue after fusion.

Perchlorate ion can be quantitatively determined in the $1-10^{-5}M$ range using ion-specific electrodes (qv) (139–141). Using direct dilution and careful control of ionic strength, pH, and temperature, standard samples containing known concentrations of ClO_4^- in phosphate buffer are used and electrode response has been found to be linear with perchlorate concentrations over small concentration ranges.

Ammonium perchlorate content of a sample can be determined by reaction with DMF in MEK with nitrogen purging to prevent CO_2 interference, followed by titration of the acid produced with standard sodium methylate solution to a pale-green thymol blue end point.

The ammonium perchlorate content of a crystalline sample can also be determined by the hydrogen ion liberated by reaction with formaldehyde to give hexamethylenetetraamine. The assumption is made that all ammonium ion is present as NH_4ClO_4.

Health and Safety Factors

All perchlorate compounds, owing to their high oxygen content, can undergo from vigorous to explosive reactions when oxidizable substances are present. Such reactions once initiated cannot be quenched by smothering because the oxygen to promote them originates in the perchlorate.

Perchlorate compounds vary widely in their associated hazards. It is recommended that small quantities be used under shielded conditions whenever new perchlorate compounds or new applications for known perchlorate compounds are under investigation. The following generalized considerations apply.

Perchloric Acid. Anhydrous perchloric acid should be prepared only as required for research purposes and in small quantities. It should not be stored for extended periods and should be kept at the lowest available temperature during storage. It must not be permitted to come into contact with oxidizable matter.

Perchloric acid in concentrations up to 72% is stable. Accidental contact with oxidizable materials must be avoided (cloth, paper, leather, etc), and purposeful contact with organic matter must be conducted under carefully controlled conditions. Reaction with metals can give hydrogen which can then be oxidized explosively by the perchlorate.

Perchlorates. Perchlorate compounds can be divided into three general classes of hazard: (1) Perchlorates in which all elements in the compounds are at their highest valence. The alkali and alkaline earth perchlorates are representative of this class of compounds. Unless contaminated, these compounds are very stable. They can be heated to temperature exceeding 400°C with only minor decomposition. Only if heated to temperatures causing perchlorate ion breakdown, will gases be generated that might build explosive pressures if confined. With these compounds the hazards are associated with mixtures containing oxidizable substances. (2) Inorganic perchlorates in which the cation contains elements having a variable valence. These perchlorates decompose at lower temperatures than the above. The decomposition can be strongly exothermic, generating gases at a sufficiently fast rate to become explosive. Perchlorates in this class are very useful but require proper handling; eg, ammonium perchlorate, if pure, is stable below 200°C and does not decompose rapidly until temperatures approaching 400°C are reached. (3) Organic perchlorates must always be handled with great caution because both oxidizing and reducing moieties are present within the same molecule. Initiation of reaction is relatively easy and once initiated can become violent.

Work clothing must be washed frequently but immediate washing is necessary if clothing is contaminated with perchlorate dusts or solutions; contaminated clothing must be kept wet until removed and washed.

Fires involving perchlorates can be extinguished by deluging with water but not by smothering. Manufacturing facilities provide jump tanks or deluge showers to quickly extinguish such fires.

Contamination of wood or other organic materials in work areas must be avoided. Contaminated organic materials must be kept wet until they can be burned under controlled conditions. Ordinary petroleum oils and greases must be avoided on handling equipment. Halocarbon lubricants can be used (142).

Ordinary handling of the three common perchlorates, NH_4ClO_4, $NaClO_4$, $KClO_4$ has not produced any evidence of serious health hazards other than the danger of flammability. Inhalation or ingestion of perchlorates should be avoided.

Perchlorates have been shown to influence the iodine balance of the normal human thyroid gland (143).

Perchloric acid given orally or subcutaneously to rats, mice, and dogs showed a specific antithyroid action and induced abnormalities in hepatic, renal, cardiovascular, and hemopoietic functions (144).

No known perchlorate-induced health effects have been observed in long-time workers in perchlorate plants.

From a 1907–1974 literature survey, Burrows and Dacre concluded that the acute toxicity of perchlorate to aquatic animals and microorganisms is very low, with toxic levels probably exceeding 1000 mg/L for periods of 24 hours or longer (145). Fish, leeches, and tadpoles all survive indefinitely in water with perchlorate levels above 500 mg/L. Chronic effects may appear at lower levels (36 mg/L for some species) owing to the antithyroid effect of perchlorate.

No aquatic organism has been found to be affected by perchlorate in the 1 mg/L range. Some bacteria are capable of metabolizing perchlorate. For example, there is a process for purification of wastewaters from perchlorates and chlorates that utilizes the microorganism *Vibrio dechloraticans* Cuznesove B-1168 for metabolism and removal of perchlorates from mixtures of industrial and domestic wastes.

Uses

The principal uses of the perchlorates involve the use of ammonium perchlorate as the oxidizer in the propellant of rockets and missiles (see Explosives and propellants); the use of perchloric acid in analytical chemistry or for research purposes; and the use of perchlorates in explosive and pyrotechnic formulations (see Pyrotechnics). Perchloric acid is also used to make high purity metal perchlorates. Sodium perchlorate is now used in slurry blasting formulations. Magnesium perchlorate is used as the electrolyte in dry cells (see Batteries).

Steel plates have been bonded using equal amounts of a two-part epoxy resin adhesive and NH_4ClO_4 (146). When separation of plates is desired, the assembly is heated to ca 300°C causing the adhesive layer to undergo self-supported burning until it vanishes, then the plates can be disassembled nondestructively.

Research in the USSR has been reported on the use of ammonium perchlorate as a feed supplement for cattle (4–6), sheep (7), and poultry (8). Controlled weight gains (depending on feed type and dosage) in test livestock were increased 3–31% by addition of NH_4ClO_4 to the feed. Feed expenditure was also reduced 7–18%. The optimum dose was 2–5 mg/kg. No negative effects were observed with doses five to ten times this level. The ammonium perchlorate behaves as a thyrostatic agent.

Perchlorates have been included in oxygen generating systems (147). Such devices are used for generating oxygen for life support in such systems as submarines, aircraft, spaceships, bomb shelters, and for use in breathing apparatus.

Processes have been developed for the recovery of potassium from Great Salt Lake and Dead Sea brines and bitterns using perchlorate to precipitate the potassium as $KClO_4$ (148–149) (see Chemicals from brine).

BIBLIOGRAPHY

"Perchloric Acid and Perchlorates" under "Chlorine Compounds, Inorganic" in *ECT* 1st ed., Vol. 3, pp. 716–729, by H. L. Robson, Mathieson Chemical Corporation and J. C. Schumacher, Western Electrochemical Company; "Perchloric Acid and Perchlorates" under "Chlorine Oxygen Acids and Salts" in *ECT* 2nd ed., Vol. 5, pp. 61–84, by Joseph C. Schumacher and Robert D. Stewart, American Potash & Chemical Corporation.

1. H. Beckhurts, *Arch. Pharm.* **224,** 333 (1886).
2. J. C. Schumacher, *Perchlorates, Their Manufacture & Uses, ACS Monograph 146,* Reinhold Publishing Corp. New York, 1960, 256 pp.
3. F. von Stadion, *Am. Chem. Phys.* **8,** 406 (1818).
4. P. N. Razumovskii and co-workers, *Khim Sel'sk. Khoz.* **14**(1), 71 (1976).

5. P. N. Razumovskii, G. S. Semanin, and G. I. Balk, *Kompleksn. Ispol'z Biol. Akt. Veshchestv. Korml. S-kh Zhivotn, Mater. Vsas. Soveshch* **1,** 370 (1973).
6. A. S. Solan and co-workers, *Zhivolnovodstvo* (11), 63 (1974).
7. V. I. Mikhailov, B. R. Gotsulenko, and V. P. Kardivari, *Zhivotnovodstvo* (5), 83 (1976).
8. H. G. Pena and co-workers, *Poult. Sci.* **55**(1), 188 (1976).
9. C. F. Goodeve and J. Powney, *J. Chem. Soc.*, 2078 (1932).
10. E. V. Kasalkin, A. A. Rakov, and V. I. Veselovski, *Elektrokhimiya* **3,** 1034 (1967).
11. R. W. Davidson and D. G. Williams *J. Phys. Chem.* **77,** 2515 (1973).
12. N. Kolarov and co-workers, *God. Vissh. Khimikotekhnol. Inst. Sofia* **19**(1), 245 (1974).
13. J. D. Witt and R. M. Hammaker, *J. Chem. Phys.* **58,** 303 (1973).
14. C. Lin, *J. Chem. Eng. Data* **21,** 411 (1976).
15. H. F. Cordes and S. R. Smith, *J. Chem. Eng. Data* **15,** 158 (1970).
16. P. Jacobs, W. McCarthy, and H. M. Whitehead, *Chem. Rev.* **69,** 551 (1969).
17. H. J. Van Wyk, *Z. Anorg. Chem.* **32,** 115 (1902).
18. *Ibid.*, **48,** 1 (1906).
19. S. Fugimoto, *Kagaku Kogo* **21,** 641 (1970).
20. G. F. Smith, *Mixed Perchloric, Sulfuric and Phosphoric Acids and Their Application in Analysis*, G. Frederick Smith Chemical Co., Columbus, Ohio, 1942.
21. G. F. Smith, *Anal. Chem. Acta* **8,** 397 (1953).
22. *Ibid.*, **17,** 175 (1957).
23. E. Kahane, *L'Action de l'Acide Perchlorique sur les Matières Organique*, Hermann et Cie, Paris, Fr., 1934.
24. H. Diehl and G. F. Smith, *Talanta* **2,** 209 (1959).
25. T. T. Gorsuch, *The Destruction of Organic Matter*, Pergamon Press, Oxford, Eng., 1970.
26. G. D. Martinie and A. A. Schilt, *Anal. Chem.* **48**(1), 70 (1976).
27. G. A. Vorob'eva and co-workers, *Zavod. Lab.* **42**(1), 25 (1976).
28. F. Solymosi and J. H. Block, *J. Catal.* **42,** 173 (1976).
29. H. Hunt and L. Boncyk, *J. Am. Chem. Soc.* **55,** 3528 (1933).
30. E. N. Ginyonova and V. A. Pleskov, *J. Phys. Chem. U.S.S.R.* **8,** 345 (1936).
31. P. Barret, *Cah. Therm.* **4,** 13 (1924).
32. C. N. R. Rao and B. Prakash, *Natl. Stand. Ref. Data Ser. Natl. Bur. Stand.* **56,** 28 (Nov. 1975).
33. F. Solymosi, *Acta Phys. Chem.* **19**(1–2), 67 (1973).
34. M. M. Markowitz and D. A. Boryta, *J. Am. Rocket Soc.* **32,** 1941 (1962).
35. F. A. Freeth, *Rec. Trav. Chim.* **43,** 475 (1924).
36. J. A. Dean, ed., *Lange's Handbook of Chemistry*, 11th ed., McGraw-Hill Book Company, New York, 1973.
37. Ref. 32, p. 38.
38. H. Osada and E. Sakamoto, *Kogyo Kayaku Kyokaiski* **24**(5), 236 (1963).
39. L. L. Bircumshaw and B. H. Newman, *Proc. R. Soc. London* **A227,** 115 (1954).
40. *Ibid.*, **A227,** 288 (1955).
41. L. L. Bircumshaw and T. R. Phillips, *J. Chem. Soc.*, 4741 (1957).
42. K. Jakus, *U.S. NTIS, A.D. Report 1975 AD-A017285*, 1976, p. 9.
43. D. Thomas and co-workers, *Prog. Vac. Microbalance Tech.* **3,** 281 (1975).
44. A. P. Glaskova, *AIAA J.* **13,** 438 (1975).
45. V. R. Pai Verneker and K. Rajishivar, *J. Solid State Chem.* **17**(1–2), 27 (1976).
46. J. C. Petricciani and co-workers, *J. Phys. Chem.* **64,** 1309 (1960).
47. T. L. Boggs, D. E. Zurn, and H. F. Cordes, *AIAA Publ.* **75,** 233 (1975).
48. Swed. Pat. 8487 (Nov. 27, 1897), O. F. Carlson.
49. M. M. Markowitz, *J. Phys. Chem.* **62,** 827 (1958).
50. H. H. Willard and G. F. Smith, *J. Am. Chem. Soc.* **45,** 286 (1923).
51. Jpn. Kokai 7492,310 (Sept. 3, 1974) and 73/1039 (Dec. 28, 1972), K. Shigematsu, Y. Ikeda, and S. Mitsui (to Technical Research and Development Inst., Japan Defense Agency).
52. H. F. Cordes and S. R. Smith, *J. Phys. Chem.* **72,** 2189 (1968).
53. H. F. Cordes and S. R. Smith, *J. Phys. Chem.* **78,** 776 (1974).
54. M. M. Markowitz and D. A. Boryta, *J. Phys. Chem.* **65,** 1419 (1961).
55. K. A. Hofmann and A. Zedtwitz, *Ber. Dtsch. Chem. Ges.* **42,** 2031 (1909).
56. R. Binkley, Jr., *J. Am. Chem. Soc.* **62,** 3524 (1940).
57. N. V. Sidgwick, *Chemical Elements and their Compounds*, Clarendon Press, Oxford, Eng., 1950.
58. F. Hein, *Chem. Tech. Berlin* **9,** 97 (1957).

59. A. R. Rodriquez and C. Pointrenaud, *Anal. Chim. Acta* **87**(1), 125 (1976).
60. L. B. Serezhkina and co-workers, *Dokl. Akad. Nauk SSR* **211**(1), 123 (1973).
61. G. F. Smith, *Dehydration Studies Using Anhydrous Magnesium Perchlorate*, G. Frederick Smith Chemical Company, Columbus, Ohio, 1951.
62. A. L. Bacerella, D. F. Dever, and E. Grunwald, *Anal. Chem.* **27**, 1833 (1955).
63. N. V. Krivtsov, V. P. Babaeva, and V. Ya Rosolovsku *Zh. Neorg. Khim.* **18**, 353 (1973).
64. J. Gonzales de Barcia and E. Moles, *Bol. Acad. Cienc. Exactas Fis. Nat. Madrid* **2**, 8 (1936).
65. J. G. Acerete and R. U. Lacal, *Rev. Acad. Cienc. Exactas Fis. Quim. Nat. Zaragoza* **9**, 11 (1954).
66. A. A. Zinov'ev and A. I. Chudinova, *Zh. Neorg. Khim.* **1**, 1722 (1956).
67. *Technical Data Bulletin R-4*, Lindsay Chemical Company, West Chicago, Ill.
68. F. H. Spedding and S. Jaffe, *J. Am. Chem. Soc.* **76**, 884 (1954).
69. Hiroshi Oguro, *Bunseki Kagaku* **25**, 468 (1976).
70. A. S. Karnaukhov, T. Ya. Ashikhima, and N. N. Runov, *Sb. Nauch Tr. Yaroslav Gos. Ped. Int.* (144), 48 (1975).
71. F. H. Spedding, J. L. Baker, and J. P. Watters, *J. Chem. Eng. Data* **20**(2), 189 (1975).
72. V. Jedinakova, *Sb. Vys. Sk. Chem. Technol. Praze Anorg. Chem. Technol.* **B18**, 113 (1974).
73. H. Burton and P. F. G. Praill, *Analyst* **80**, 4 (1955).
74. K. A. Hofmann and co-workers, *Ber. Dtsch. Chem. Ges.* **43**, 2624 (1910).
75. K. A. Hofmann and H. Arnoldi, *Ber. Dtsch. Chem. Ges.* **39**, 3146 (1906).
76. K. A. Hofmann, H. Kirmreuther, and A. Thal, *Ber. Dtsch. Chem. Ges.* **43**, 183 (1910).
77. K. A. Hofmann, A. Metzler, and H. Lecher, *Ber. Dtsch. Chem. Ges.* **43**, 178 (1910).
78. J. Meyer and W. Spormann, *Z. Anorg. Allgem. Chem.* **228**, 341 (1936).
79. K. A. Hofmann, A. Zedtivitz, and H. Wagner, *Ber. Dtsch. Chem. Ges.* **42**, 4390 (1910).
80. J. Radell and Connolly, *ASD Tech. Dept. Air Force Systems Command*, 61-109 Aeronautical Systems Division, Wright-Patterson Air Force Base, Ohio, May 1961.
81. L. Birkenbach and J. Goubeau, *Naturwissenschaften* **18**, 530 (1930).
82. U.S. Pat. 3,976,753 (Aug. 24, 1976), L. R. Grant (Rockwell International).
83. C. J. Schack, D. Pilipovich, and O. Christe, *Inorganic Nuclear Chemistry-Herbert H. Nyman Memoirs, Volume,* Rocketdyne Div., Rockwell Int'l., Canoga Park, Calif., 1976, pp. 207–208.
84. U.S. Pat. 4,012,492 (Mar. 15, 1977), C. J. Schack and D. Pilipovich (to U.S.A. Sect. of the Navy).
85. N. V. Krivtsov, V. P. Babaeva, and V. Ya Rosolovsku, *Izv. Akad. Nauk SSSR, Ser. Khim.* (8), 1692 (1976).
86. R. Salvadori, *Gazz. Chim, Ital.* **37**(II), 32 (1907).
87. U.S. Pat. 3,420,621 (Jan. 7, 1969), J. W. Watters, R. E. Farncomb, and M. J. Cziesla (to the United States of America).
88. A. A. Zinov'ev and I. A. Zakharova, *Zhurn. Neorg. Khim.* **5**, 775 (1960).
89. F. Fichter and E. Jenny, *Helv. Chim. Acta* **6**, 225 (1923).
90. G. F. Potapova, A. A. Rakov, and V. I. Veselovskii, *Elektrokhimiya* **9**, 1054 (1973).
91. W. Oechsli, *Z. Electrochem.* **9**, 807 (1903).
92. C. A. Wamser and co-workers, *Inorg. Syn.* **14**, 29 (1973).
93. W. M. Khutorelskii, L. V. Okhlobyslina, and A. A. Fainzil'berg, *Usp Khim.* **36**, 377 (1967).
94. F. X. Powell and E. R. Lippincott, *J. Chem. Phys.* **32**, 1883 (1960).
95. S. V. Loginov and co-workers, *Khim. Kompleksn. Soedin.* **3**(12), 389 (1975); R. E. Elson and J. E. Stucky, *J. Inorg. Nucl. Chem.* **35**, 1029 (1973); V. E. Kalinina and co-workers, *Khim. Takhnol.* **19**(8), 1287 (1976).
96. M. R. Rosenthal, *J. Chem. Educ.* **50**, 331 (1973).
97. A. E. Wickenden and R. A. Krause, *Inorg. Chem.* **4**, 404 (1965).
98. J. Lewis, R. S. Nyholm, and G. A. Rodley, *Nature* **207**, 73 (1965).
99. W. R. McWhinnie, *J. Inorg. Nucl. Chem.* **26**, 21 (1964).
100. G. H. W. Millburn and co-workers, *Chem. Commun.*, 1188 (1968).
101. H. C. Clark and R. J. O'Brien, *Inorg. Chem.* **2**, 740 (1963).
102. L. Johansson, *Coord. Chem. Rev.* **12**, 241 (1974).
103. G. S. Serullas, *Ann. Chem. Phys.* **45**, 270 (1830).
104. N. A. E. Millon, *Ann. Chim. Phys.* **7**, 298 (1843).
105. H. Kolbe, *Ann. Chem.* **64**, 237 (1847).
106. T. Fairley, *Br. Assoc. Adv. Sci. Rep.* **44**, 57 (1874).
107. S. Vaclav, S. Tichy, and A. Regner, *Chem. Prum.* **16**, 577 (1966).
108. U.S. Pat. 2,392,861 (Jan. 15, 1946), J. C. Pernert (to Oldbury Electrochemical Co.).

109. H. M. Goodwin and E. C. Walker, *Chem. Metall. Eng.* **25,** 1093 (1921); U.S. Pats. 913,944 (Apr. 23, 1917), 1,271,633 (July 9, 1918), (to Genesee Chemical Co.)
110. H. M. Goodwin and E. C. Walker, *Trans. Electrochem. Soc.* **40,** 157 (1921).
111. W. Müller and P. Jönck, *Chem. Ing. Tech.* **35,** 78 (1963).
112. G. F. Smith, *J. Am. Chem. Soc.* **75,** 184 (1953).
113. J. C. Schumacher, *J. Electrochem. Soc.* **116**(2), 68c (1969).
114. W. Oechsli, *Z. Electrochem.* **9,** 807 (1903).
115. N. V. S. Knibbs and H. Palfreeman, *Trans. Faraday Soc.* **16,** 402 (1920).
116. K. Sugino and S. Aoyagi, *J. Electrochem. Soc.* **103,** 166 (1956).
117. C. W. Bennett and E. L. Mack, *Trans. Electrochem. Soc.* **29,** 323 (1916).
118. M. P. Grother and E. H. Cook, *Electrochem. Technol.* **6,** 221 (1968).
119. E. H. Cook and M. P. Grother, *paper presented at the Los Angeles Meeting of the Electrochemical Society, May 1971.*
120. U.S. Pat. 3,475,301 (Oct. 28, 1969), E. H. Cook and M. P. Grother (to Hooker Chemical Corporation).
121. A. Legendre, *Chem. Ing. Tech.* **34,** 379 (1962).
122. E. Hausmann and E. Kramer, *Chem. Ing. Tech.* **43,** 170 (1971).
123. *Chem. Eng.* **72,** 82 (July 19, 1965).
124. U.S. Pat. 2,945,691 (July 19, 1960), F. D. Gibson, Jr., (to Pacific Engineering & Production Co. of Nevada).
125. U.S. Pat. 3,634,216 (Jan. 11, 1972), F. D. Gibson, Jr., R. L. Thayer, and B. B. Halker (to Pacific Engineering & Production Co. of Nevada).
126. K. C. Narasimhaw and H. V. K. Udupa, *J. Electrochem. Soc.* **123,** 1294 (1976).
127. M. Nagalingam and co-workers, *Chem. Ing. Tech.* **41,** 1301 (1969).
128. U.S. Pat. 3,493,478 (Feb. 3, 1970), H. V. K. Udupa and co-workers (to Central Electrochemical Research Institute, Karalkudi, S. Rly., India).
129. J. C. Schumacher and D. R. Stern, *Chem. Eng. Prog.* **53,** 428 (1957).
130. J. E. Reynolds and T. W. Clapper, *Chem. Eng. Prog.* **57**(11), 138 (1961).
131. U.S. Pat. 3,781,412 (Dec. 25, 1973), R. C. Rhees (to Kerr-McGee Chemical Corp.).
132. U.S. Pat. 3,218,121 (Nov. 16, 1965), L. E. Tufts (to Hooker Chemical Corporation).
133. R. M. Griziano, *Agent, Hazardous Materials Regulations, Tariff No. 31,* Department of Transportation, Washington, D.C., Mar. 1, 1977.
134. F. D. Snell and C. T. Snell, *Colorimetric Methods of Analysis,* Vol. II, D. Van Nostrand Company, Princeton, N. J., 1957, pp. 718–719.
135. H. H. Willard and H. Diehl, *Advanced Quantitative Analysis,* D. Van Nostrand Company, Princeton, N. J., 1944, pp. 254–257.
136. G. M. Smith, *Anal. Ed. Ind. Eng. Chem.* **11,** 186 (1939).
137. *Ibid.,* p. 269.
138. H. H. Willard, *private communication,* Oct. 24, 1958.
139. R. J. Bacquk and R. J. Dubois, *Anal. Chem.* **40,** 685 (1968).
140. T. M. Hseu and G. A. Rechnitz, *Anal. Lett.* **1,** 629 (1968).
141. M. J. Smith and S. E. Manahan, *Anal. Chem. Acta.* **48,** 315 (1969).
142. C. M. Olson, *J. Electrochem. Soc.* **116**(1), 33-C (1969).
143. H. Buergi and co-workers, *Eur. J. Clin. Invest.* **4**(1), 65 (1974).
144. L. N. Selivanova, I. G. Koltunova, and E. N. Vorob'eva, *Gig. Tr. Prof. Zabol.* (8), 33 (1973).
145. D. Burrows and J. C. Dacre, *U.S. NTIS, AD-A, Rep. No. 010660,* Army Med. Bioeng. Res. Dev. Lab. Fort Detrick, Md. 1975, pp. 95.
146. U.S. Pat. 3,993,524 (Nov. 23, 1976), Y. Okada and S. Kensho (to Nissan Motor Co. Ltd.).
147. U.S. Pat. 3,993,514 (Nov. 23, 1976), E. J. Pacanowsky and E. A. Martino (to Thiokol Corporation).
148. J. A. Epstein and co-workers, *Hydrometallurgy* **1,** 39 (1975).
149. D. R. George, J. M. Riley, and J. R. Ross, *paper presented at the 62nd National Meeting of the Institute of Chemical Engineers,* Salt Lake City, Utah, May 21–24, 1967.

R. C. RHEES
Pacific Engineering & Production Co.

CHLORITES. See Chlorine oxygen acids and salts.

CHLORITES (MINERALS). See Silica.

CHLOROCARBONS AND CHLOROHYDROCARBONS

Survey, 668
Methyl chloride, 677
Methylene chloride, 686
Chloroform, 693
Carbon tetrachloride, 704
Ethyl chloride, 714
Other chloroethanes, 722
Dichloroethylenes, 742
Trichloroethylene, 745
Tetrachloroethylene, 754
Allyl chloride, 763
Chloroprene, 773
Chlorinated paraffins, 786
Chlorinated derivatives of cyclopentadiene, 791
Chlorinated benzenes, 797
Benzene hexachloride, 808
Ring-chlorinated toluenes, 819
Benzyl chloride, benzal chloride, and benzotrichloride, 828
Chlorinated naphthalenes, 838
Chlorinated biphenyl and related compounds, 844

SURVEY

Chlorination of a variety of hydrocarbon feedstocks produces many valuable and useful chlorinated solvents, intermediates, and chemical products. The chlorinated derivatives provide an important means of upgrading the value of industrial chlorine. In 1974, a total of 9.9×10^9 metric tons of chlorine was produced in the United States of which 4.6×10^9 t was used in the production of chlorinated hydrocarbon derivatives.

The major industrial chlorinated hydrocarbon derivatives include vinyl chloride monomer, vinylidene chloride, 1,2-dichloroethane (ethylene dichloride), methyl chloride, methylene chloride, chloroform, carbon tetrachloride, 1,1,1-trichloroethane, trichloroethylene, tetrachloroethylene, mono- and dichlorobenzenes, and hexachlorocyclopentadiene. Eighty-eight percent of the 1,2-dichloroethane produced in the United States is converted to vinyl chloride used in the production of poly(vinyl chloride). Large quantities of chloroform and carbon tetrachloride are used as chemical intermediates in the manufacture of fluorohydrocarbons (see Fluorine compounds, organic). Methylene chloride, 1,1,1-trichloroethane, trichloroethylene, and tetrachloroethylene have wide and varied use as solvents (see Solvents, industrial). Chlorobenzenes are important chemical intermediates. Adducts of hexachlorocyclopentadiene [77-47-4], C_5Cl_6, with unsaturated monomers provide a versatile series of flame retardant additives (see Flame retardants, halogenated). Hexachlorocyclopentadiene is also a key intermediate in the manufacture of chlorinated insecticides (see Insect control technology). U.S. demand for some of the more important chlorinated hydrocarbon derivatives is listed in Table 1. Chlorinated solvents have many applications including use in adhesives (qv), aerosol products (see Aerosols), extraction solvents (see Extraction), industrial solvent blends, paint and coating solvents (see Paint; Coatings, industrial), and pharmaceuticals (qv). They are also used in drycleaning, vapor degreasing in the electronic industry, metal cleaning, textile processing, and as reaction media. Metal cleaning, the largest single application, consumes in excess of 400,000 t of chlorinated solvents annually (1) (see Metal surface treatments). The use of perchloroethylene in drycleaning (qv) is the second largest solvent application. Estimated U.S. demand (Bureau of Census) for the four solvents in 1976 was 891,309 metric tons (see Table 1). Individual solvent demands include 226,796 t methylene chloride, 233,600 t 1,1,1-trichloroethane, 127,006 t trichloroethylene, and 303,907 t (670 million lb) of perchloroethylene.

Vinyl chloride monomer showed an average annual growth of 8.2% from 1970–1976 (see Vinyl polymers). 1,2-Dichloroethane, which is used to produce vinyl chloride, also showed substantial growth (4.4%). The dramatic decline in production and use of trichloroethylene in 1970–1976 was offset in large part by increases in demand for 1,1,1-trichloroethane (6.8% growth) in the same period. The proposed bans on aerosol grades of chlorofluorocarbons will contribute to substantial decreases in chloroform and carbon tetrachloride demand in the future. However, methylene chloride has shown an annual growth of 4–4.4% in this period. Use of methylene chloride as a vapor-pressure depressant in nonfluorocarbon aerosol formulations will continue to expand. Chlorinated hydrocarbon derivatives are produced by several processes. Aliphatic hydrocarbon feedstocks for chlorination reactions include methane (natural gas), ethane, ethylene (qv), propylene (qv), and propane (see Gas, natural; Hydrocarbons).

In the United States, very little acetylene is used as a chlorination feedstock for production of trichloroethylene and tetrachloroethylene solvents (see Acetylene-derived chemicals). 1,2-Dichloroethane, an important intermediate for chlorinated hydrocarbon products, is produced by catalytic chlorination of ethylene in either the vapor or liquid phase or by oxychlorination of ethylene. Thermal dehydrochlorination of 1,2-dichloroethane produces vinyl chloride and coproduct hydrogen chloride. Hydrogen chloride is commonly recycled to an oxychlorination system to produce additional 1,2-dichloroethane.

Table 1. U.S. Demand for Chlorinated Hydrocarbons

Compound	CAS Registry no.	Thousands of metric tons 1970	1976	Change per year, %
vinyl chloride	[75-01-4]	1534	2289	+8.2
vinylidene chloride	[75-35-4]	54	70	+5.0
1,2-dichloroethane	[540-59-0]	3773	4773	+4.4
methyl chloride	[74-87-3]	192	168	−2.0
methylene chloride	[75-09-2]	182	227	+4.4
chloroform	[67-66-3]	109	125	+2.4
carbon tetrachloride	[56-23-5]	459	385	−2.7
1,1,1-trichloroethane	[71-55-6]	166	234	+6.8
trichloroethylene	[79-01-6]	274	127	−8.9
perchloroethylene	[127-18-4]	320	304	−2.7
chlorobenzene	[108-90-7]		163	
o-dichlorobenzene	[95-50-1]			
p-dichlorobenzene	[106-46-7]		68	

Vinylidene chloride (1,1-dichloroethylene) is produced by dehydrochlorination of 1,1,2-trichloroethane [25323-89-1]. Hydrogen chloride can be added at low temperature to vinylidene chloride in the presence of ferric chloride to make 1,1,1-trichloroethane. Thermal chlorination of 1,2-dichloroethane is one route to commercial production of trichloroethylene and tetrachloroethylene.

Chlorinated methanes are made by the direct thermal chlorination of methane or the methanol–hydrogen chloride reaction. The latter produces methyl chloride which is thermally chlorinated to yield more highly chlorinated methanes. Chlorination of benzene in the presence of a catalyst such as ferric chloride, or oxychlorination with hydrogen chloride, yields primarily monochlorobenzene. Additional chlorination gives the dichloro isomers and some higher analogues.

Hexachlorocyclopentadiene (hex), a clear liquid, is prepared by the complete chlorination of either pentanes or cyclopentadiene.

Typical manufacturing processes for C_1 and C_2 chlorohydrocarbons are shown in Figure 1.

General Properties of Chlorinated Hydrocarbons

Progressive chlorination of a hydrocarbon molecule yields a sequence of liquids, solids, or both, of increasing nonflammability, density, and viscosity, as well as improved solubility for a large number of inorganic and organic materials. Specific heat, dielectric constant, and water solubility of a solvent exhibit a progressive decrease with increasing chlorine content.

All chlorinated hydrocarbons are susceptible at elevated temperatures to pyrolysis breakdown which liberates hydrogen chloride. Olefinic chlorinated derivatives are oxidized in the presence of ultraviolet light to give hydrogen chloride, phosgene, and chlorinated acetyl chlorides (acid derivatives). Saturated aliphatic chlorine derivatives are usually quite stable to oxidation, although 1,1,2-trichloroethane exhibits appreciable oxidation as contrasted with the stable 1,1,1-trichloroethane isomer. Alcohols and amines are often added to oxidation-sensitive solvents to minimize this mode of degradation.

Although many chlorinated hydrocarbons attack aluminum, proprietary organic inhibitors permit commercial use of reactive solvents such as 1,1,1-trichloroethane and trichloroethylene in both cold and hot cleaning of aluminum. Commercial use of many chlorinated derivatives imposes stress on the stability of the solvent. Inhibitors that may be classified as antioxidants (qv), acid acceptors, and metal stabilizers are added to minimize these stresses. Hydrogen chloride, often a product of degradation, can be neutralized by an added epoxide acting as an acid acceptor. All of the chlorinated derivatives may hydrolyze at a slow but finite rate when dissolved in, or in contact with, a water phase. Again, hydrogen chloride is the product of concern because of potential corrosion of equipment and containers. Hydrolysis cannot normally be stopped by addition of an inhibitor, but is controlled by minimum water contact and removal of the metal, such as iron containers, in contact with the solvent.

All volatile organic solvents are toxic to some degree. Excessive vapor inhalation of the volatile, commercial chlorinated solvents and the central nervous system depression that results is the greatest hazard in industrial use of these solvents. Proper equipment and operating procedures permit safe use of solvents such as methylene chloride, 1,1,1-trichloroethane, trichloroethylene, and tetrachloroethylene in both

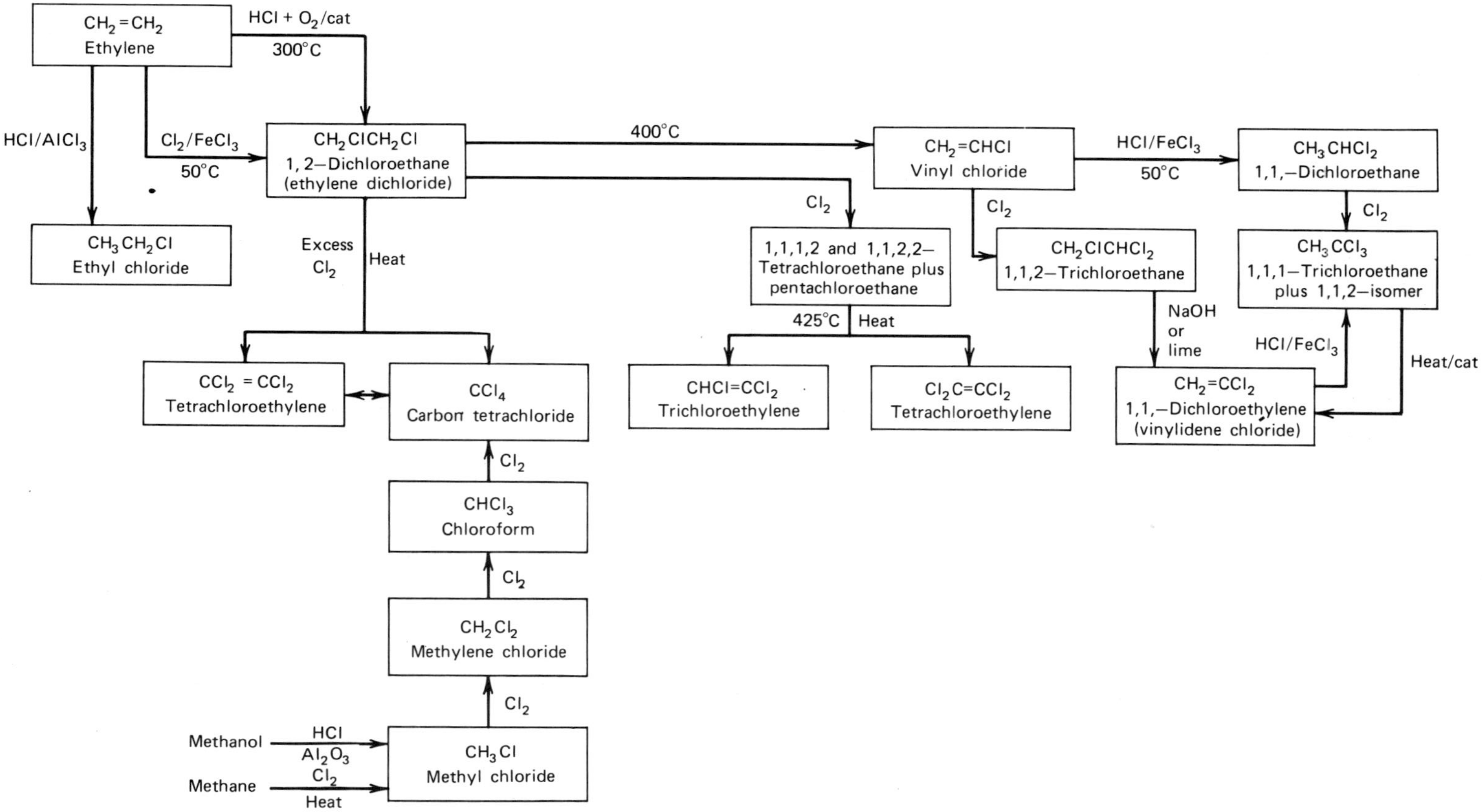

Figure 1. Manufacturing processes for C_1 and C_2 chlorohydrocarbons.

cold and hot metal-cleaning operations. The degree of toxicity of a solvent cannot be predicted from its chlorine content or the structure of chlorinated derivatives. For example, one of the least toxic metal-cleaning solvents is 1,1,1-trichloroethane which has a recommended threshold limit value (TLV) of 350 ppm. However, the 1,1,2-trichloroethane isomer is one of the most toxic chlorinated derivatives, with a TLV of only 10 ppm.

Types of Aliphatic Chlorination Reactions

Substitution Chlorination. The substitution of chlorine for hydrogen atoms in a hydrocarbon is an important commercial chlorination process. The chlorination of pentane by the Sharples Solvent Corporation in 1929 has been cited as the first commercial substitution chlorination process in the industry (2). The Chemische Fabrik Griesheim Electron in Germany began production of carbon tetrachloride from methane in the late twenties, carrying out the reaction over pumice at 500°C. They also pioneered the chlorinolysis reaction for making carbon tetrachloride and tetrachloroethylene. Since the addition of chlorine to an olefinic double bond is a much easier reaction than the chlorination of methane and its homologues, the chlorination of unsaturated and aromatic hydrocarbons had been an established industrial process long before the substitution process was developed.

Activated chlorine used for the substitution reaction is obtained by either thermal or photochemical means. The thermal method requires temperatures of at least 250°C to start the reaction. The large reaction exotherm demands close temperature control by cooling or dilution. Thermal chlorination is inexpensive and less sensitive to inhibition than the photochemical process. Mercury arc lamps are the normal source of light for the latter furnishing wavelengths from 300–500 nm.

Chlorination of methane yields all four possible chlorinated derivatives: methyl chloride, dichloromethane (methylene chloride), trichloromethane (chloroform), and tetrachloromethane (carbon tetrachloride). The reaction proceeds by a radical chain mechanism, as shown in equations 1 through 8.

Chain initiation
$$Cl_2 \xrightleftharpoons{heat/light} 2\,Cl\cdot \tag{1}$$

Chain propagation
$$CH_4 + Cl\cdot \rightarrow CH_3\cdot + HCl \tag{2}$$
$$CH_3\cdot + Cl_2 \rightarrow CH_3Cl + Cl\cdot \tag{3}$$
$$CH_3Cl + Cl\cdot \rightarrow CH_2Cl\cdot + HCl \tag{4}$$
$$CH_2Cl\cdot + Cl_2 \rightarrow ClCH_2Cl + Cl\cdot \tag{5}$$

Chain termination
$$CH_3\cdot + Cl\cdot \rightarrow CH_3Cl \tag{6}$$
$$Cl\cdot + Cl\cdot \rightarrow Cl_2 \tag{7}$$
$$CH_3\cdot + CH_3\cdot \rightarrow CH_3CH_3 \tag{8}$$

Chlorine atoms obtained from the dissociation of chlorine molecules by thermal or photochemical energy react with a methane molecule to form hydrogen chloride and a methyl free radical. The radical reacts with an undissociated chlorine molecule to give methyl chloride and a new chlorine atom necessary to continue the reaction. Other more highly chlorinated products are formed in a similar manner. Chain termination may proceed by way of several of the examples cited in equations 6, 7, and 8. The initial radical-producing catalytic process is inhibited by oxygen. In some commercial processes, small amounts of air are deliberately added to discourage chlorination beyond the monochloro stage.

Methane, chlorine, and recycled chloromethanes are fed to a tubular reactor at a reactor temperature of 490–530°C to yield all four chlorinated methane derivatives. Similarly, chlorination of ethane produces ethyl chloride and higher chlorinated ethanes. The process is employed commercially to produce 1,1,1-trichloroethane. 1,1,1-Trichloroethane is also produced via chlorination of 1,1-dichloroethane with 1,1,2-trichloroethane as a coproduct. Hexachlorocyclopentadiene is formed by a complex series of chlorination, cyclization, and dechlorination reactions. First, substitutive chlorination of pentanes is carried out by either photochemical or thermal methods to give a product with 6–7 atoms of chlorine per mole of pentane. The polychloropentane product mixed with excess chlorine is then passed through a porous bed of Fuller's earth or silica at 350–500°C to give hexachlorocyclopentadiene. Cyclopentadiene is another possible feedstock for the production of hexachlorocyclopentadiene.

Addition Chlorination. Chlorination of olefins, such as ethylene, by the addition of chlorine is commercially important and can be carried out either as a catalytic vapor-phase or as a liquid-phase process. The reaction is influenced by light, the walls of the reactor vessel, and inhibitors such as oxygen, and has a radical chain type of mechanism. Some ionic pathways may also be involved because the chlorine addition is accelerated by the presence of ferric chloride, aluminum chloride, antimony pentachloride, or cupric chloride. A typical commercial process for the preparation of 1,2-dichloroethane (ethylene dichloride) is the chlorination of ethylene at 40–50°C in the presence of ferric chloride. The introduction of 5% air to the chlorine feed prevents further substitution chlorination of the 1,2-dichloroethane to 1,1,2-trichloroethane. The addition of chlorine to tetrachloroethylene under photochemical chlorination conditions has been investigated by Leermakers and Dickenson (3). It is suggested that this chlorination, which is strongly inhibited by oxygen, proceeds by a radical chain mechanism as shown in equations 9–13.

$$Cl_2 \underset{}{\overset{light}{\rightleftharpoons}} 2\ Cl \cdot \tag{9}$$

$$Cl \cdot + Cl_2C{=}CCl_2 \rightarrow CCl_3CCl_2 \cdot \tag{10}$$

$$CCl_3CCl_2 \cdot + Cl_2 \rightarrow CCl_3CCl_3 + Cl \cdot \tag{11}$$

$$2\ CCl_3CCl_2 \cdot \rightarrow 2\ Cl_2C{=}CCl_2 + Cl_2 \tag{12}$$

or

$$2\ CCl_3CCl_2 \cdot \rightarrow CCl_3CCl_3 + Cl_2C{=}CCl_2 \tag{13}$$

The chlorination of ethylene with a ferric chloride catalyst may indeed be a polar reaction, as shown by equations 14–16 (4).

$$FeCl_3 + Cl_2 \rightarrow (FeCl_4^- Cl^+) \tag{14}$$

$$(FeCl_4^- Cl^+) + CH_2{=}CH_2 \rightarrow (CH_2ClCH_2^+ FeCl_4^-) \tag{15}$$

$$(CH_2ClCH_2^+ FeCl_4^-) \rightarrow CH_2ClCH_2Cl + FeCl_3 \tag{16}$$

Hydrochlorination. The addition of hydrogen chloride to simple olefins in the absence of peroxides takes place by an electrophilic mechanism; the orientation is in accord with Markovnikov's rule. The addition occurs in two steps with formation of an intermediate carbonium ion, as in equation 17. Addition of the chloride ion (eq. 18) results in the hydrochlorination derivative. Metal chloride catalysts are

$$RCH{=}CH_2 + HCl \rightarrow R{-}\overset{\vee}{C}H{-}CH_3 + Cl^- \tag{17}$$

$$R{-}\overset{\vee}{C}H{-}CH_3 + Cl^- \rightarrow RCHCl{-}CH_3 \tag{18}$$

normally used in commercial processes for the hydrochlorination of olefinic derivatives. For example, hydrochlorination of ethylene at temperatures lower than 100°C and in the presence of aluminum chloride yields ethyl chloride. A commercial process for the preparation of 1,1-dichloroethane employs the addition of hydrogen chloride to vinyl chloride in the liquid phase at 50°C in the presence of ferric chloride.

The hydrochlorination of olefins is a weakly exothermic reaction, with heats of reaction ranging from 4–21 kJ/mol (1–5 kcal/mol). The hydrochlorination of acetylene is more exothermic, 184 kJ/mol (44 kcal/mol).

Dehydrochlorination. The commercial production of vinyl chloride depends on the thermal dehydrochlorination of 1,2-dichloroethane. Barton and co-workers (5) have established a radical chain mechanism for the vapor-phase thermal dehydrochlorination of 1,2-dichloroethane at 350–515°C. The reaction is greatly accelerated by the presence of such reagents as chlorine and oxygen, and retarded or inhibited by olefins and alcohols. A typical cracking furnace for the production of vinyl chloride involves a residence time of 3–10 s. The reactor products, on a molar basis, contain an average of 39% vinyl chloride, 39% hydrogen chloride, 22% unconverted 1,2-dichloroethane, and trace amounts of various by-products. The use of activated carbon impregnated with barium chloride permits the cracking operation to proceed at temperatures in the range from 200–350°C. Most vinyl chloride produced by industry is made by thermal, noncatalytic cracking of 1,2-dichloroethane.

Refluxing 1,1,2-trichloroethane with calcium hydroxide and water gives a good yield of 1,1-dichloroethylene (6). Dehydrochlorination of chlorinated derivatives, such as 1,1,2-trichloroethane, may be carried out with a variety of catalytic materials, including Lewis acids such as aluminum chloride. Dehydrochlorination of the 1,1,1-trichloroethane isomer with catalytic amounts of a Lewis acid also yields 1,1-dichloroethylene. Dehydrochlorination of 1,1,2-trichloroethane at 500°C in the presence of a copper catalyst gives a different product, ie, 1,2-dichloroethylene. Addition of small amounts of a chlorinating agent, such as chlorine, promotes dehydrochlorination in the gaseous phase, probably by initiating a chain reaction involving formation of an organic radical that splits off chlorine with the formation of a double bond. The dehydrochlorination of 1,2-dichloroethane in the presence of chlorine, as shown in equations 19 and 20, is a typical example.

$$CH_2ClCH_2Cl + Cl\cdot \rightarrow CH_2Cl\dot{C}HCl + HCl \tag{19}$$

$$CH_2Cl\dot{C}HCl \rightarrow CH_2=CHCl + Cl\cdot \tag{20}$$

Chlorinolysis. Excess chlorine in high-temperature chlorination can cleave the C–C bonds of a hydrocarbon to give chlorinated derivatives of shorter chain length. A well known commercial process involving this technique is thermal chlorination of hydrocarbon feedstocks with chlorine to produce carbon tetrachloride and tetrachloroethylene with hydrogen chloride as by-product. The hydrocarbon feedstock may include hydrocarbons up to C_3 and any partially chlorinated derivatives. The yields can be varied widely by controlling recycled streams; eg, recycling carbon tetrachloride increases the tetrachloroethylene yield.

A typical reactor operates at 600–900°C with no catalyst and a residence time of 10–12 s. It produces a 92–93% yield of carbon tetrachloride and tetrachloroethylene, based on the chlorine input. The major steps in the process include: (*1*) chlorination of the hydrocarbon, (*2*) quenching of reactor effluents, (*3*) separation of hydrogen chloride/chlorine, (*4*) recycling of chlorine to the reactor, and (*5*) distillation to separate reaction products from the hydrogen chloride by-product. Advantages of this process include the use of cheap raw materials, flexibility in the ratios of carbon tetrachloride/tetrachloroethylene produced, and utilization of waste chlorinated residues that are used as a feedstock to the reactor. The large amount of hydrogen chloride by-product can be recycled to an oxychlorination unit or sold as anhydrous or aqueous hydrogen chloride, or both.

Typical reactions using either 1,2-dichloroethane or propane to produce carbon tetrachloride and tetrachloroethylene by the chlorinolysis reaction are shown in equations 21 through 24. Continued removal of tetrachloroethylene and recycling of carbon tetrachloride can result in a net zero production of carbon tetrachloride. From 1,2-dichloroethane:

$$CH_2ClCH_2Cl + 4\ Cl_2 \rightarrow CCl_4 + \tfrac{1}{2}\ Cl_2C\!=\!CCl_2 + 4\ HCl \tag{21}$$

$$2\ CCl_4 \rightleftharpoons Cl_2C\!=\!CCl_2 + 2\ Cl_2 \tag{22}$$

From propane:

$$C_3H_8 + 8\ Cl_2 \rightarrow CCl_4 + Cl_2C\!=\!CCl_2 + 8\ HCl \tag{23}$$

$$2\ CCl_4 \rightleftharpoons Cl_2C\!=\!CCl_2 + 2\ Cl_2 \tag{24}$$

Oxychlorination. Oxychlorination is an important process for the production of 1,2-dichloroethane. Hydrogen chloride, available as a by-product from other integrated chlorination processes, produces chlorine in the presence of oxygen (air) and a cupric chloride catalyst (eq. 25). Dehydrohalogenation of 1,2-dichloroethane to produce vinyl chloride is the ultimate aim of the process. The overall reaction involving ethylene, hydrogen chloride, and oxygen to produce vinyl chloride with water as the only by-product, is shown in equation 26. Ito has reviewed the oxychlorination process (7).

$$2\ HCl + \tfrac{1}{2}\ O_2 \xrightarrow[270°C]{CuCl_2} Cl_2 + H_2O \tag{25}$$

$$CH_2\!=\!CH_2 + HCl + 1/2\ O_2 \rightarrow CH_2\!=\!CHCl + H_2O \tag{26}$$

Raschig developed the first commercial oxychlorination process in 1928 to make chlorobenzene which was then hydrolyzed to phenol. The Durez plant in North Tonawanda, New York, put on-stream in 1937, used this process.

The first large-scale commercial oxychlorination process for vinyl chloride was put on-stream in 1958 by the The Dow Chemical Company. This plant, employing a fixed-tube reactor containing a catalyst of cupric chloride on an active carrier, produced 1,2-dichlorethane from ethylene. The high temperatures involved in the reaction were moderated by a suitable diluent. The average heat output from the reaction is 116 kJ/mol (50,000 Btu/lb mol). In a typical oxychlorination reaction, preheated gas streams at temperatures of 180–200°C are fed onto a fixed catalyst bed containing 2–4% copper impregnated on an activated alumina. The reaction occurs in a 15–22 s residence time period on the catalyst bed at a temperature of 230–315°C. Yields based on ethylene are about 93%.

Feeding 1,2-dichloroethane, hydrogen chloride, and oxygen onto a fluidized bed at 400°C produces trichloroethylene and tetrachloroethylene. The catalyst bed consists of cupric chloride and potassium chloride on graphite. A modified oxychlorination technique known as the Transcat process has been developed by the Lummus Company. The feedstock can be a saturated hydrocarbon or chlorohydrocarbon and the process is suited to the production of C_1 and C_2 chlorohydrocarbons.

Thermal Cracking. Thermal chlorination of ethylene yields the two isomers of tetrachloroethane, 1,1,1,2 and 1,1,2,2. Introduction of these tetrachloroethane derivatives into a tubular-type furnace at temperatures of 425–455°C gives good yields of trichloroethylene and tetrachloroethylene (8). In the cracking of the tetrachloroethane stream, introduction of ferric chloride into the 460°C vapor phase reaction zone improves the yield of trichloroethylene product.

Chlorinated Aromatic Derivatives

Aromatic compounds may be chlorinated with chlorine in the presence of a catalyst, such as iron, ferric chloride, or other Lewis acids. The halogenation reaction involves the electrophilic displacement of the aromatic hydrogen by halogen. Introduction of a second chlorine atom into the monochloro aromatic structure leads to ortho and para substitution. The presence of a Lewis acid favors polarization of the chlorine molecule, thereby increasing its electrophilic character. Because the polarization does not lead to complete ionization, the reaction should be represented as shown in equation 27.

$$\text{C}_6\text{H}_6 + \text{Cl}-\text{Cl}\cdots\text{AlCl}_3 \longrightarrow [\text{C}_6\text{H}_6\text{Cl}]^+ \text{AlCl}_4^- \longrightarrow \text{C}_6\text{H}_5-\text{Cl} + \text{HCl} + \text{AlCl}_3 \quad (27)$$

Continuous chlorination of benzene at 30–50°C in the presence of a Lewis acid typically yields 85% monochlorobenzene. Temperatures in the range of 150–190°C favor the dichlorobenzene products. The para isomer is produced in a ratio of 2–3 to 1 of the ortho isomer.

Other methods of aromatic ring chlorination include use of a mixture of hydrogen chloride and air in the presence of a copper–salt catalyst, or sulfuryl chloride in the presence of aluminum chloride at ambient temperatures.

Free-radical chlorination of toluene successively yields benzyl chloride, benzal chloride, and benzotrichloride. Related chlorination agents include sulfuryl chloride, *tert*-butyl hypochlorite, and *N*-chlorosuccinimide which yield benzyl chloride under the influence of light, heat, or radical initiators.

BIBLIOGRAPHY

"Survey" under "Chlorocarbons and Chlorohydrocarbons" in *ECT* 2nd ed., Vol. 5, pp. 85–92, by D. W. F. Hardie, Imperial Chemical Industries Ltd.

1. G. M. Rekstad, *Factory*, 27 (Jan. 1974).
2. W. Hirschkind, *Ind. Eng. Chem.* **41**(12), 2749 (1949).
3. J. A. Leermakers and R. G. Dickenson, *J. Am. Chem. Soc.* **54**, 3853, 4648 (1932).
4. H. P. Rothbaum, I. Ting, and P. W. Robertson, *J. Chem. Soc.*, 980 (1948); D. A. Evans, T. R. Watson, and P. W. Robertson, *J. Chem. Soc.*, 1624 (1950).
5. D. H. R. Barton, *J. Chem. Soc.*, 148, 165 (1949).
6. P. W. Sherwood, *Ind. Eng. Chem.* **54**, 29 (1962).
7. Y. Ito, *J. Soc. Org. Syn. Chem. Jpn.* **23**(1), 33 (1965).
8. S. Tsuda, *Chem. Eng.*, 74 (May 4, 1970).

WESLEY L. ARCHER
Dow Chemical U.S.A.

METHYL CHLORIDE

Methyl chloride [74-87-3] (chloromethane, monochloromethane), CH_3Cl, at ordinary temperatures and pressures is a colorless gas with an ethereal, nonirritant odor and sweet taste. Methyl chloride is handled commercially as a liquid. It is miscible with the principal organic solvents and only slightly soluble in water. The dry liquid is stable and noncorrosive; however, in the presence of moisture, the liquid slowly decomposes and becomes corrosive to metals, particularly aluminum, zinc, and magnesium. Gaseous methyl chloride is moderately flammable. Prolonged exposure to high concentrations of the vapor can produce severe toxic effects. Methyl chloride is used mainly in the manufacture of silicones, tetramethyllead (TML, antiknock agent), synthetic rubber, and methylcellulose, and as a general methylating agent; its refrigerant and extractant applications now have secondary importance.

Impure methyl chloride was produced in the laboratory as early as 1835 by Dumas and Peligot, who heated wood spirit (crude methyl alcohol) with a mixture of sulfuric acid and common salt. It was later made by Schiff and by Walker and Johnson by the reaction of phosphorus chlorides with methyl alcohol. One of the first preparations of pure methyl chloride was probably that of Groves in 1874. Groves passed hydrogen chloride into a boiling solution of zinc chloride in twice its weight of wood spirit. Berthelot obtained the compound by chlorinating methane.

During the last quarter of the nineteenth century, methyl chloride was manufactured on a small scale in Europe for use as a refrigerant and in the synthesis of dyes. Large-scale production began in the United States about 1920, chiefly to meet refrigerant requirements. Manufacture in the United Kingdom began in 1930. Production increased greatly after 1943 when methyl chloride was required as the starting material for methyl silicones and fluorinated refrigerants. After World War II, methyl chloride production in the United States increased more than tenfold.

Physical and Chemical Properties

The physical properties of methyl chloride are listed in Table 1. Values for vapor pressure, density of liquid and saturated vapor, and enthalpy of liquid methyl chloride are given in Table 2 (1). Table 3 lists the values for viscosity of methyl chloride (2).

Methyl chloride is the simplest chlorinated hydrocarbon. Dry methyl chloride in the absence of air does not decompose at an appreciable rate at temperatures approaching 400°C, even in contact with many metals. Thermal dissociation is virtually complete at 1400°C. Oxidative breakdown of the gas requires temperatures of several hundred °C. Methyl chloride is decomposed by an open flame to give hydrogen chloride and carbon dioxide, with possible formation of small amounts of carbon monoxide and phosgene. Henderson and Hill (3) demonstrated that the burning velocity of the simple chloroparaffins is inversely proportional to chlorine content. Methyl chloride has a burning velocity of 10.9 cm/s whereas monochlorobutane burns at 31.6 cm/s.

Table 1. Physical Properties of Methyl Chloride

Property	Value
mp, °C	−97.7
bp, °C	
101.3 kPa[a]	−23.73
13.3 kPa[a]	−61.7
1.3 kPa[a]	−92.1
sp gr	
liq, 20/4°C	0.920
gas, 0°C, 101.3 kPa[a] (air = 1)	1.74
dens, g/L, sea level, 45° latitude, 0°C, 101.3 kPa[a]	2.3045
n_D	
liq, −23.7°C	1.3712
gas, 25°C	1.0007
surface tension, mN/m (= dyn/cm)	
0°C	19.5
10°C	17.8
20°C	16.2
specific heat, liq J/g[b]	
20°C	1.599
−15 to 30°C (av)	1.574
C_p, 25°C, 103.4 kPa[a]	0.199
C_v, 25°C	0.155
critical temperature, °C	143.1
critical pressure, kPa[a]	6679.2
critical density, g/cm^3	0.353
critical volume, cm^3/g	2.833
flash point (open cup)	below 0°C
autoignition temperature, °C	632
flammability limits in air, vol %	10.7–17.4
combustion velocity, cm/s	10.9
diffusivity in air, 25°C, 101.3 kPa[a], cm^2/s	0.105
coefficient of cubical expansion, liquid, −30 to 30°C (av)	0.00209
dielectric constant	
liq, −25°C	12.93
gas, 21°C	1.0109
dipole moment, C·m[c]	6.20×10^{-30}
heat of formation, ideal gas, 25°C, kJ/mol[b]	−81.92
free energy of formation, ideal gas, 25°C, kJ/mol[b]	−58.41
latent heat of fusion, J/g[b]	129.7
latent heat of evaporation at bp, J/g[b]	428.4
solubility of methyl chloride in water, 25°C, g/100 g H$_2$O	0.48
solubility of water in methyl chloride, 25°C, g/100 g methyl chloride	0.0725
solubility of methyl chloride gas, 20°C, 101.3 kPa[a], mL/100 mL solvent	
benzene	4723
carbon tetrachloride	3756
glacial acetic acid	3679
absolute alcohol	3740

[a] To convert kPa to mm Hg, multiply by 7.5.
[b] To convert J to cal, divide by 4.184.
[c] To convert C·m to debye, divide by 3.336×10^{-30}.

Table 2. Vapor Pressure, Density, and Enthalpy of Methyl Chloride

Temperature, °C	Vapor pressure, kPa[a]	Density Liquid, g/mL	Density Satd vapor, m³/kg	Enthalpy, kJ/kg[b] Satd liquid	Enthalpy, kJ/kg[b] Vaporization	Enthalpy, kJ/kg[b] Satd vapor
−60	15.6	0.936	2.235	409.82	460.54	870.36
−50	28.0	0.953	1.295	424.42	425.18	876.60
−40	47.4	0.970	0.794	439.12	443.60	882.72
−30	76.7	0.986	0.508	454.12	434.54	888.66
−20	118.8	1.003	0.338	469.22	425.22	894.44
−10	177.2	1.022	0.233	484.42	415.59	900.01
0	255.7	1.042	0.1648	500.00	405.07	905.07
10	358.2	1.064	0.1198	516.12	393.64	909.76
20	489.3	1.086	0.0891	531.57	382.50	914.07
30	652.5	1.110	0.0675	547.65	370.88	917.93
40	851.6	1.135	0.0520	563.96	357.39	921.32
50	1092	1.164	0.0480	580.39	343.90	924.29
60	1375	1.196	0.0324	597.01	329.75	926.76

[a] To convert kPa to mmHg, multiply by 7.5.
[b] To convert J to cal, divide by 4.184.

Table 3. Viscosity of Methyl Chloride

Temperature, °C	Viscosity, mPa·s (= cP) Vapor	Viscosity, mPa·s (= cP) Liquid
−40	0.0086	0.349
−17.8	0.0094	0.298
4.4	0.0101	0.263
26.7	0.0108	0.237
48.9	0.0115	0.217
71.1	0.0122	0.200
93.3	0.0128	0.186
115.6	0.0134	0.175

Methyl chloride breaks down to hydrogen chloride, hydrogen, and carbon when in contact with reduced nickel in the presence of excess hydrogen at 210°C (4). It is reduced to methane by heating with calcium hydride at 180°C (5).

If the liquid is heated in the presence of moisture, slow hydrolysis to methanol and hydrogen chloride occurs below 100°C. Methyl chloride is readily hydrolyzed by boiling dilute sodium hydroxide solution. At 120°C and 618.5 kPa (90 psi) methyl chloride saturated with water decomposes at the rate of ca 1 g/100 mL H_2O per h. A crystalline hydrate [20604-36-8], $CH_3Cl \cdot 6H_2O$, is formed by subjecting an aqueous solution of methyl chloride to low temperatures. The hydrate decomposes at 7.5°C and 101.3 kPa (1 atm).

Dry methyl chloride is unreactive with all common metals except the alkali and alkaline earth metals, magnesium, zinc, and aluminum. In dry ether solution, methyl chloride reacts with sodium to yield ethane by the Wurtz synthesis:

$$2\ CH_3Cl + 2\ Na \rightarrow CH_3CH_3 + 2\ NaCl$$

By the same reaction, methyl chloride can be condensed with higher chloroparaffins to give propane, butane, etc. Methyl chloride reacts with magnesium to form the Grignard reagent methylmagnesium chloride, CH_3MgCl, which has been applied to the synthesis of alcohols and to the preparation of intermediates for the formation of silicone polymers. The reaction with zinc is similar to that with magnesium. Studies have shown that dry methyl chloride vapor does not attack aluminum or its alloys at temperatures up to 60°C. In a damp atmosphere, the alloys, particularly those containing magnesium, are attacked. Reactions with aluminum catalyzed by aluminum chloride can take place with explosive violence.

Methyl chloride can be converted into methyl iodide or bromide by refluxing in acetone solution in the presence of sodium iodide or bromide. The reactivity of methyl chloride and other aliphatic chlorides in substitution reactions can often be increased by using a small amount of sodium or potassium iodide as in the formation of methyl aryl ethers. Methyl chloride and potassium phthalimide do not readily react to give N-methylphthalimide unless potassium iodide is added. The reaction to form methyl cellulose and the Williamson synthesis to give methyl ethers are catalyzed by small quantities of sodium or potassium iodide.

Methyl chloride reacts with ammonia in alcoholic solution or in the vapor phase (Hofmann reaction) to form a mixture of the hydrochlorides of methylamine, dimethylamine, trimethylamine, and tetramethylammonium chloride. With tertiary amines, methyl chloride forms quaternary derivatives.

$$CH_3Cl + NR_3 \rightarrow [R_3NCH_3]^+ Cl^-$$

Methyl chloride, as a typical aliphatic chloride, may be used in the Friedel-Crafts reaction:

$$CH_3Cl + C_6H_6 \xrightarrow{AlCl_3} C_6H_5CH_3 + HCl$$

Methylation of aromatic nuclei can also be achieved in the vapor state over a solid catalyst.

Manufacture

There are two principal processes for industrial production of methyl chloride: chlorination of methane, and reaction of hydrogen chloride and methanol. Several variants of both processes are used. The methanol–hydrogen chloride reaction yields methyl chloride as the sole product. Chlorination of methane yields other chlorohydrocarbons in substantial amounts; indeed, under certain conditions, methyl chloride may not be the principal product. Because the coproducts, eg, methylene chloride, chloroform, and carbon tetrachloride, are as commercially important as methyl chloride, methane chlorination can be regarded as a multiple-product process rather than one with several by-products. Hydrogen chloride is often the determining factor in choosing a route to produce methyl chloride. If a cheap abundant supply exists, the economics favor hydrochlorination. If hydrogen chloride can be sold either as anhydrous material or hydrochloric acid, then chlorination is favored.

Thermal chlorination of methane was first put on an industrial scale by Hoechst in Germany in 1923. At that time, high pressure methanol synthesis from hydrogen and carbon monoxide provided a new source of methanol for production of methyl chloride by reaction with hydrogen chloride (see Oxo process; Methanol). Prior to 1914 attempts were made to establish an industrial process for methanol by hydrolysis of methyl chloride obtained by chlorinating methane.

Chlorination of Methane. Methane can be chlorinated thermally, photochemically, or catalytically. Thermal chlorination, the most difficult method, may be carried out in the absence of light or catalysts. It is a chain reaction limited by the presence of oxygen. The first step in the reaction is the thermal dissociation of the chlorine molecules for which the activation energy is about 84 kJ/mol (20 kcal/mol), some 33 kJ (8 kcal) higher than for catalytic chlorination. The chlorine atoms react with methane to form hydrogen chloride and a methyl radical. The methyl radical in turn reacts with a chlorine molecule to form methyl chloride and another chlorine atom which can continue the reaction. The methane raw material may be natural gas, coke oven gas, or gas from petroleum refining.

In a typical thermal chlorination process, the chlorine–methane mixture, with an excess of methane is fed to a reactor where it mixes with gas previously subjected to reaction. By regulating the gas rate, and consequently the heat produced by the chlorination reaction, the temperature is maintained at about 400°C. The gas from the reactor is cooled and the hydrogen chloride removed by water washing in a packed tower. Hydrochloric acid thus obtained is generally commercially salable and is, therefore, advantageous to the economics of the process. Finally, the gas is scrubbed with caustic liquor, dried by refrigeration, and further cooled to effect its liquefaction. Uncondensed methane is returned to the reactor. The liquefied gas is then fractionally distilled. A typical reaction product yields on fractionation 35 wt % methyl chloride, 45 wt % methylene chloride, and 20 wt % chloroform plus a small amount of carbon tetrachloride. The relative amounts of these components can be changed by varying the reaction conditions.

Variants of the methane chlorination route are the Hoechst method (6), in which a small quantity of oxygen is present; and the British Celanese method (7), which uses sulfur monochloride either as catalyst or as the solvent in which the reaction takes place (8). Chlorination of methane to methyl chloride has been effected in a ternary melt of potassium-, cuprous-, and cupric chlorides at 425–500°C and in a fluidized bed of alumina gel impregnated with a potassium chloride–copper chloride melt; both methods produce approximately the same product mixture of methyl chloride and higher chlorohydrocarbons (9). The McBee-Hass technique of controlled high temperature chlorination (10) can be used to vary the ratios of the chloromethanes in the product from almost 100% methyl chloride to carbon tetrachloride exclusively.

Methanol–Hydrogen Chloride Reaction. Liquid mixtures of methanol and hydrochloric acid slowly yield methyl chloride even at 0°C (11–12). A difference of over 42 kJ (10 kcal) between the activation energies of esterification of methanol with carboxylic acids and reaction with hydrogen chloride points to a fundamental difference between the two reactions (13). In the former the hydrogen atom not the hydroxyl group is replaced; in the latter the hydroxyl is replaced by chlorine.

The methanol–hydrogen chloride process is typically carried out as follows: vaporized methanol and hydrogen chloride, mixed in equimolar proportions, are preheated to 180–200°C. Reaction occurs on passage through a converter packed with 1.68–2.38 mm (8–12 mesh) alumina gel at ca 350°C. Other catalysts may be used, eg, cuprous or zinc chloride on active carbon or pumice. Phosphoric acid on activated carbon has also been used as the catalyst at temperatures of 280–320°C (14). Space velocities of up to 300 h^{-1} (volumes of gas at STP per hour per volume catalyst space) are employed. The product gas is cooled, water-scrubbed, and liquefied. Yields of up to 95% on the methanol are commonly obtained. Gamma alumina has been used as a catalyst at 295–340°C to obtain 97.8% yields of methyl chloride (15).

The methanol process may also be carried out in the liquid phase by refluxing the alcohol at 150°C (16) with hydrochloric acid in the presence of dissolved zinc chloride. An organic medium may be used where the methanol and hydrogen chloride react in the presence of an unspecified catalyst at ambient temperature (17).

Other Reactions Producing Methyl Chloride. Many other reactions lead to the formation of methyl chloride; some have been proposed as the bases of industrial processes. Good yields of methyl chloride are produced in the reaction of methyl ether with hydrogen chloride in the presence of water at about 80–240°C, and under sufficient pressure to maintain the water as a liquid. Dimethyl ether for this process is obtained as waste from methyl cellulose manufacture (18). Dimethyl sulfate reacts with aluminum chloride at ordinary temperature or with sodium chloride at elevated temperature to give methyl chloride (19–20). Some methyl chloride results when monochlorodimethyl ether is decomposed with zinc (21). Methane, heated with phosgene at 400°C, is chlorinated to methyl chloride (22) as follows:

$$CH_4 + COCl_2 \xrightarrow{\Delta} CH_3Cl + CO + HCl$$

Methanol, heated at 250°C with chloroform or carbon tetrachloride in contact with active carbon, is converted in part to methyl chloride (23). Methyl chloride has been produced from methoxymagnesium chloride, CH_3OMgCl, a by-product from the manufacture of certain organo–silicon compounds, by heating at over 200°C (24).

Handling

Methyl chloride is transported and stored as liquefied gas under pressure in cylinders, tank cars, and tank trucks. The usual cylinder capacities in the United States are 43.56 and 63.5 kg. Cylinders may be fitted with either a short curved or long straight dip-pipe for dispensing liquid methyl chloride. Valves are normally steel and are fitted with a lead or compressed asbestos-fiber outlet washer. To allow for liquid expansion in the cylinders, the weight of methyl chloride contained is limited to 0.83% of the weight of water which would completely fill the cylinder at 40°C, or 0.78% of that weight at 65°C. The latter limit is observed when the full cylinders will be subjected to tropical temperatures. Similar filling ratios are employed for other containers.

In the United States and the United Kingdom, methyl chloride is supplied in horizontal cylinders of 544–590 kg capacity with valves similar to those mentioned above. Methyl chloride tank cars in the United States have a capacity of 18.1 or 35.4 metric tons and tank cars are usually fitted with four valves grouped together on a manhole cover under a removable protective dome. Two of these valves connect with dip-pipes extending to the bottom of the tank; the remaining two are connected, respectively, to a pressure gage and to the pressure-balancing pipe for use during the transfer of methyl chloride to storage vessels.

When in use, cylinders of methyl chloride are stacked valve downwards on a suitable stand and connected to the consuming plant by an adapter. To obtain all the methyl chloride, a cylinder may be heated by steam jet or by wrapping with an electric-strip resistance heater controlled to a maximum temperature of 50°C. Compressed nitrogen or natural gas can be used to maintain the flow to the receiving vessel when transferring methyl chloride from a tank car or tank truck; compressed air must never be employed for this purpose. The pipe system carrying methyl chloride should be grounded as a precaution against ignition by static electrical discharge. Liquid trapped between closed valves after methyl chloride has been discharged could expand and burst a joint in the line. To prevent this, the valve at the discharging vessel should be closed first; after the pressure has fallen, the valve at the discharge end of the line can be closed. Fusible safety plugs fitted to tanks and other vessels containing methyl chloride under pressure are customary in the United States but are not used in the United Kingdom. Methyl chloride containers in the United States are tested in accordance with rules established by the ICC and the Bureau of Explosives. Leakage of methyl chloride from plant or storage vessels must not be investigated by means of a halide lamp; in most instances a leak can be readily located by application of soap solution as the escaping gas forms observable bubbles. Odoriferous or lacrimatory warning agents are sometimes added to methyl chloride used in refrigerators.

Economic Aspects

Production and sales data for methyl chloride, as reported by the U.S. Tariff Commission, are given in Table 4. Methanol hydrochlorination was used to produce about 64% of the methyl chloride in 1969, and about 98% by 1974. Methyl chloride is used primarily in the manufacture of silicones and tetramethyllead (see Silicon compounds; Lead compounds).

Standards

Methyl chloride is generally marketed in refrigeration and technical grades. The former should be very pure to prevent attack by impurities on refrigeration equipment. It is usually produced from the technical grade by fractional distillation; the use of added stabilizers has not been reported. The moisture content of the refrigeration grade is generally less than 75 ppm. A representative technical grade contains not more than the following indicated quantities (in ppm) of impurities: water, 100; acid (as HCl), 10; methyl ether, 20; methanol, 50; acetone, 50; vinyl chloride, 100; ethyl chloride, 100; residue, 100. No free chlorine should be detectable. Traces of higher chlorides are generally present in methyl chloride produced by chlorination of methane. The boiling

Table 4. United States Methyl Chloride Production and Price Statistics

Year	Production[a]	Sales Quantity[a]	Sales Total value[b]	Unit value, $/kg
1945	13.47	12.52	4.14	0.33
1955	16.47	11.84	3.13	0.26
1965	85.05	43.00	6.64	0.15
1970	191.74	79.83	10.56	0.13
1971	198.45	87.59	11.59	0.13
1972	205.71	94.35	10.40	0.11
1973	246.80	103.10	13.64	0.13
1974	223.67	97.39	19.32	0.20
1975	166.24	65.59	20.24	0.31
1976	168.88			

[a] Thousands of metric tons.
[b] Millions of dollars.

point should be between −24 and −23°C, and 5–95% should distill within a range of about 0.2°C. Both refrigeration and technical grades should be clear, colorless, and free from visible impurities.

Analysis

There is no specific color or other reaction by which methyl chloride can be detected or identified. Gas chromatography can be used to determine the amount of methyl chloride present in a mixture of chlorocarbons. Quality testing of methyl chloride for appearance, water content, acidity, nonvolatile residue, residual odor, methanol, and acetone is routinely done by production laboratories. Water content is determined with Karl Fischer reagent using the apparatus by Kieselbach (25). Acidity is determined by titration with alcoholic sodium hydroxide solution. The nonvolatile residue, consisting of oil or waxy material, is determined by evaporating a sample of the methyl chloride at room temperature. The residue is examined after evaporation for the presence of odor. Methanol and acetone content are determined by gas chromatography.

Toxicity

Methyl chloride is one of the more toxic of the chlorinated hydrocarbons and there is no adequate warning of the presence of harmful concentrations. The delay in the development of symptoms is characteristic of the toxic effect of methyl chloride. The signs and symptoms of intoxication may not develop for several hours after termination of exposure and may become progressively worse for several days before improvement begins or death occurs. Repeated exposure to low concentrations damages the central nervous system, and, less frequently, the liver, kidneys, bone marrow, and cardiovascular system (26). Methyl chloride intoxication causes headache, blurred vision, loss of coordination, and reversible personality change involving moroseness, depression, and anxiety. Routine urine and blood tests have no diagnostic value (27). Massive methyl chloride inhalation has produced myocardial damage (28). Daily exposure to concentrations of 500 ppm is extremely dangerous, even for a period of two weeks or less (29). The 1976 OSHA Standard places a limit of 100 ppm for an 8-h time-weighted average.

Uses

The principal uses of methyl chloride in recent years as reported by the U.S. Tariff Commission, are given in Table 5.

Table 5. Uses of Methyl Chloride

Use	1970	1972	1974
silicones intermediate	38%	43%	50%
tetramethyllead intermediate	38%	38%	30%
butyl rubber (catalyst solvent)	5%	4%	5%
miscellaneous	19%	15%	15%

BIBLIOGRAPHY

"Methyl Chloride" under "Chlorine Compounds, Organic" in *ECT* 1st ed., Vol. 3, pp. 738–746, by P. J. Ehman, W. O. Walker, and W. R. Rinelli, Ansul Chemical Company; "Methyl Chloride" under "Chlorocarbons and Chlorohydrocarbons" in *ECT* 2nd ed., Vol. 5, pp. 100–111, by D. W. F. Hardie, Imperial Chemical Industries, Ltd.

1. N. B. Vargaftik, *Tables on the Thermophysical Properties of Liquids and Gases,* John Wiley & Sons, Inc., New York, 1975, p. 362.
2. K. S. Wilson and co-workers, *Chem. Eng. News* **21,** 1254 (1943).
3. H. T. Henderson and G. R. Hill, *J. Phys. Chem.* **60,** 874 (1956).
4. F. K. Beilstein, *Handbuch Der Organischen Chemie,* 4th ed., Vol. 1, Springer-Verlag, Berlin, Ger., 1918, p. 59.
5. P. Lebeau, *Compt. Rend.* **140,** 1042, 1264 (1905).
6. Ger. Pat. 841,588 (July 30, 1943), H. Petri (to Farbwerke Hoechst).
7. Brit. Pat. 674,900 (July 2, 1952), E. B. Thomas and F. Hindley (to British Celanese, Ltd.).
8. Can. Pat. 524,469 (May 11, 1950), E. B. Thomas and F. Hindley (to British Celanese, Ltd.).
9. E. Gorin, C. M. Fontana, and G. A. Kidder, *Ind. Eng. Chem.* **40,** 2128 (1948).
10. E. T. McBee and co-workers, *Ind. Eng. Chem.* **34,** 296 (1942).
11. C. L. Hoffpauir and R. T. Demint, *Text. Res. J.* **21,** 81 (1951).
12. S. R. Carter and J. A. V. Butler, *J. Chem. Soc.* **125**(I), 963 (1924).
13. C. N. Hinshelwood, *J. Chem. Soc.* (I), 599 (1935).
14. Ger. Pat. 478,126 (Nov. 28, 1924), K. Dachlauer and E. Eggert (to I. G. Farbenindustrie A.G.).
15. Brit. Pat. 1,230,743 (May 5, 1971), (to The Dow Chemical Company).
16. Ger. Pat. 671,086 (Jan. 31, 1939), H. Klein and C. Pfaundler (to I. G. Farbenindustrie A.G.).
17. *Chem. Eng.* **69,** 72 (July 23, 1962).
18. U.S. Pat. 2,084,710 (June 22, 1937), H. M. Spurlin (to Hercules Powder Co.).
19. A. A. Schamschwun, *Br. Chem. Phys. Abstr. A* **11,** 147 (May 1940).
20. R. F. Weinland and K. Schmid, *Ber. Dtsch. Chem. Ges.* **38**(II), 2327; (III), 3696 (1905).
21. M. Fileti and A. deGaspari, *Gazz. Chim. Ital.* **27**(II), 293 (1897).
22. Ger. Pat. 292,089 (Nov. 20, 1914), A. Hochstetter.
23. Ger. Pat. 664,321 (Aug. 30, 1938), F. Rothweiler (to I. G. Farbenindustrie A.G.).
24. Brit. Pat. 646,620 (Sept. 13, 1948), H. A. Clark (to The Dow Corning Corp.).
25. R. Kieselbach, *Anal. Chem.* **21,** 1578 (1949).
26. N. I. Sax, *Dangerous Properties of Industrial Materials,* 4th ed., Van Nostrand Reinhold Company, New York, 1975, p. 915.
27. J. D. C. MacDonald, *J. Occup. Med.* **6**(2), 81 (1964).
28. M. Gummert, *Z. Ges. Inn. Med. U. Ihre Grenzgebiete* **16,** 677 (1961).
29. L. T. Fairhall, *Industrial Toxicology,* Williams and Wilkins, Baltimore, Md., 1957.

R. C. AHLSTROM, JR.
J. M. STEELE
Dow Chemical U.S.A.

METHYLENE CHLORIDE

Methylene chloride [75-09-2] (dichloromethane, methylene dichloride), CH_2Cl_2, is rapidly becoming one of the most important halogenated industrial solvents. It was first prepared in 1840 by Regnault who chlorinated methyl chloride in sunlight. During the past ten years annual growth has been 8–10%. Growth is expected to equal or exceed these figures during the next few years as methylene chloride replaces solvents that have been restricted because of toxicity or environmental reasons. The established markets for methylene chloride are also expected to remain strong because it is one of the least toxic of the common industrial solvents.

Physical and Chemical Properties

The physical properties of methylene chloride are listed in Table 1 and the binary azeotropes in Table 2.

Methylene chloride is a clear, colorless, volatile liquid with a mild ethereal odor. Though only slightly soluble in water, it is completely miscible with other grades of chlorinated solvents, diethyl ether, and ethyl alcohol in all proportions. It dissolves in most other common organic solvents. Methylene chloride is also an excellent solvent for many resins, waxes, and fats, and hence is well suited to a wide variety of industrial uses. Methylene chloride alone exhibits no flash or fire point. However, as little as 10 volume percent acetone or methyl alcohol is capable of producing a flash point.

Methylene chloride is one of the more stable of the chlorinated hydrocarbon solvents. Its initial thermal degradation temperature is 120°C in dry air (1). This temperature decreases as the moisture content increases. The reaction produces mainly HCl with trace amounts of phosgene. Decomposition under these conditions can be inhibited by addition of small quantities (0.0001–1.0%) of phenolic compounds, eg, phenol, hydroquinone, p-cresol, resorcinol, thymol, and 1-naphthol (2). Stabilization may also be effected by the addition of small amounts of amines (3), or a mixture of nitromethane and 1,4 dioxane. The latter diminishes attack on aluminum and inhibits iron-catalyzed reactions of methylene chloride (4). The addition of small amounts of epoxides can also inhibit aluminum reactions catalyzed by iron (5). On prolonged contact with water, methylene chloride hydrolyzes very slowly, forming HCl as the primary product. On prolonged heating with water in a sealed vessel at 140–170°C, methylene chloride yields formaldehyde and hydrochloric acid as shown by equation 1 (6).

$$CH_2Cl_2 + H_2O \rightarrow HCHO + 2\ HCl \tag{1}$$

Prolonged heating with water at 180°C results in formic acid, methyl chloride, methanol, hydrochloric acid, and some carbon monoxide.

Dry methylene chloride does not react with the common metals under normal conditions; however, a reaction with aluminum can be initiated by the addition of small amounts of other halogenated solvents or an aromatic solvent (7). Iron catalyzes the reaction and this can be significant in the handling and storage of methylene chloride and in the formulation of products, eg, in aluminum aerosol containers of pigmented paints, where the conditions necessary for the reaction are commonly found. A typical reaction in this process is shown in equation 2 (see Friedel-Crafts reactions).

$$CH_2Cl_2 + C_6H_5CH_3 \xrightarrow{AlCl_3} CH_3C_6H_4CH_2Cl + HCl \tag{2}$$

Table 1. Properties of Methylene Chloride

molecular weight	84.92
boiling point, at 101.3 kPa[a], °C	39.8
freezing point, °C	−96.7
specific gravity, at 20°C	1.320
density, at 20°C kg/m^3	1315.7
vapor density (air = 1.02)	2.93
diffusivity in air, m^2/s	9×10^{-5}
refractive index at 20°C	1.4244
coefficient of cubical expansion (20–35°C)	0.0014
viscosity at 20°C, mPa·s (= cP)	0.43
surface tension, N/m (= dyn/cm) at 20°C	0.02812
at 30°C	0.02654
heat of vaporization at 20°C, kJ/kg[b] (Btu/lb)	329.23 (141.7)
thermal capacity, liq, 15–45°C (kJ/kg·K)[b]	1.171
heat capacity at 25°C, J/mol[b]	54.09
heat of combustion, MJ/kg[b]	7.1175
critical density, kg/m^3	472
critical temperature, °C	245.0
critical pressure, MPa[c]	6.171
vapor pressure, kPa[a] at 0°C	19.6
at 20°C	46.5
at 30°C	68.1
solubility in water at 20°C, g/kg	13.2
water solubility in methylene chloride at 20°C, g/kg	1.4
kauri-butanol value	136
auto-ignition temperature, °C	640
flash point (ASTM) D1310-67	none
explosive limits at 25°C, vol % in air	14–25
electrical properties at 24°C	
dielectric strength, V/cm (V/100 mils)	94.488 (24.000)
specific resistivity at 24°C, Ω·cm	1.81×10^8
dielectric constant at 24°C, 100 kHz	10.7

[a] To convert kPa to mm Hg, multiply by 7.5.
[b] To convert J to cal, divide by 4.184.
[c] To convert MPa to atm, divide by 0.101.

Further dechlorination may occur with the formation of substituted diphenylmethanes. If enough aluminum metal is present, the Friedel-Crafts reactions involved may generate considerable heat and smoke and substantial amounts of hydrogen chloride which reacts with more aluminum metal, rapidly forming AlCl$_3$. The addition of an epoxide inhibits the initiation of this reaction by consuming HCl.

In the gas phase, methylene chloride reacts with nitrogen dioxide at 270°C to yield a gaseous mixture consisting mainly of carbon monoxide, nitric oxide, and hydrogen chloride (8). A similar reaction occurs with chloroform, where phosgene is the principal product.

Methylene chloride is easily reduced to methyl chloride (qv) and methane by alkali metal ammonium compounds in liquid ammonia. When the vapor is contacted with reduced nickel at 200°C in the presence of excess hydrogen, hydrogen chloride and elementary carbon are produced. Heating with alcoholic ammonia at 100–125°C results in hexamethylenetetramine, (CH$_2$)$_6$N$_4$, a heterocyclic compound; with aqueous ammonia at 200°C, hydrogen chloride, formic acid, and methylamine are produced.

CHLOROCARBONS, -HYDROCARBONS (CH$_2$Cl$_2$)

Table 2. Binary Azeotropes of Methylene Chloride

Second component	bp, °C	bp of azeotrope, °C	Methylene chloride, wt %
water	100.0	38.1	98.5
methanol	64.7	39.2	94.0
tert-butyl alcohol	82.8	57.1	94.0
2-propanol	82.3	56.6	92.0
ethanol	78.4	54.6	88.5
iodomethane	42.5	39.8	79.0
propylene oxide	35.0	40.6	77.0
cyclopentane	49.5	38.0	70.0
diethyl ether	34.6	40.8	70.0
acetone	56.5	57.6	70.0
carbon disulfide	46.3	37.0	61.0
diethylamine	55.5	52.0	45.0

In the presence of chlorination catalysts, methylene chloride may be chlorinated to chloroform (qv) and carbon tetrachloride (qv). Bromochloromethane is produced by reaction of an excess of a mixture of methylene chloride and bromine with aluminum at 26 to 30°C (9).

When heated for 8 h at 200°C and 91.2 MPa (900 atm) pressure in the presence of aluminum, methylene chloride reacts with carbon monoxide to yield chloroacetyl chloride, CH$_2$ClCOCl (10).

Methylene chloride vapor reacts at 300–400°C with a mixture of silicon and reduced copper under a nitrogen atmosphere to give a mixture of organo-silicon derivatives (11) (see Silicon compounds).

Manufacture

Methylene chloride is produced industrially in the United States by two methods. The oldest method involves a direct reaction of excess methane (natural gas) with chlorine at high temperature (approximately 485–510°C). This process produces methyl chloride, chloroform, and carbon tetrachloride as coproducts. The reactor effluent also contains unreacted methane and hydrogen chloride which are usually separated from the chloromethanes by scrubbing with a refrigerated mixture of higher chloromethanes in which they are only slightly soluble. The methane, freed from acid by water scrubbing, is recycled to the chlorinator. The chloromethane, which contains the methylene chloride and its coproducts, passes to a sequence of fractionating columns after washing, alkali scrubbing, and drying. The temperature and raw material flow rates to the reactor can be controlled to produce the particular chloromethane desired.

The predominant method of manufacturing methylene chloride employs as a first step the reaction of hydrogen chloride and methanol with the aid of a catalyst to give methyl chloride. The reaction is generally carried out in the vapor phase where methanol and hydrogen chloride are continuously mixed in approximately equimolecular ratios and passed through a preheater maintained at ca 180°C. The gas mixture is then passed through a converter at about atmospheric pressure and a temperature of 340 to 350°C. The externally heated converter is packed with previously ignited alumina gel of 1.7–2.4 mm (8–12 mesh) size or a similar catalyst such as zinc chloride on pumice, cuprous chloride, or activated carbon.

Space velocities at STP of about 275 m^3/(h·m^3) of gross catalyst volume are generally used (12). The primary product of this reaction is methyl chloride, which is then fed to reactors similar to those used in the methane process where it is combined with chlorine to produce methylene chloride, chloroform, and carbon tetrachloride. Condensers collect and purify the hot reaction gases leaving the converter in a manner similar to that described for the methane process (see also Methanol).

Methylene chloride can also be produced in the liquid phase at 100 to 150°C by refluxing and distilling an aqueous mixture containing methanol, hydrogen chloride, and zinc chloride. This method is not considered as efficient as the vapor-phase process, and is not as widely used. Methylene chloride can be produced by the reduction of higher chloromethanes, but the method has not achieved industrial significance (13). Chloroform is slowly reduced upon warming with trisilane, Si$_3$H$_8$, in the absence of air as shown in equation 3.

$$4 \text{ CHCl}_3 + \text{Si}_3\text{H}_8 \rightarrow 4 \text{ CH}_2\text{Cl}_2 + \text{Si}_3\text{H}_4\text{Cl}_4 \tag{3}$$

Ferrous hydroxide in the presence of alkaline hydroxides or carbonates reportedly reduces carbon tetrachloride to methylene chloride (14).

Economic Aspects

Table 3 Lists the U.S. producers of methylene chloride and their rated yearly capacities.

Since the product mix of chloromethanes is very flexible, production may be adjusted according to the demand for specific end products. In addition to the capacities listed in Table 3, Dow Chemical has announced an expansion of methylene chloride production scheduled for late 1980 that will boost capacity by 68,000 metric tons. Vulcan expanded capacity by 45,000 t/yr in late 1977. The U.S. demand for methylene chloride was 220,000 t in 1975 and 250,000 t in 1976. Annual consumption of 355,000 t is projected by 1980. Growth has remained remarkably strong in all major use areas, and could exceed projections owing to the withdrawal of fluorocarbons as aerosol propellants and foam blowing agents, and a rapidly increasing use of methylene chloride as a vapor degreasing solvent (see Aerosols; Air pollution). Government leg-

Table 3. U.S. Methylene Chloride Producers[a]

Producer	Capacity (thousands of metric tons per year)
Allied Chemical, Moundsville, W. Va.	22.7
Diamond Shamrock, Belle, W. Va.	45.4
Dow Chemical, Freeport, Tex.	90.7
Plaquemine, La.	81.6
Stauffer Chemical, Louisville, Ky.	27.2
Vulcan Materials, Geismar, La.	36.3
Wichita, Kans.	13.6
Total	*317.5*

[a] Source—*Chemical Marketing Reporter*, 1976.

islation that restricts the use of other solvents because of toxicity or flammability could result in widespread use of methylene chloride in other major markets, such as adhesives (qv), where use had previously been limited.

In recent years, exports of methylene chloride have substantially exceeded imports. However, with Dow Chemical's new plant of 450 t/yr rated capacity at Stade, Federal Republic of Germany, now on stream, exports may be reduced, making more material available for the anticipated fast growth of the U.S. market.

Standards and Analysis

Although methylene chloride is considered a very stable compound, small amounts of stabilizers are usually added at the time of manufacture. Additional stabilizers may be used to provide adequate protection against corrosion or solvent breakdown in specific applications. A representative commercial grade of methylene chloride has the following specifications:

Property	Value
distillation range at 101 kPa (1 atm), IBP-DP	39.4–40.4°C
specific gravity at 25/25°C	1.319–1.322
acidity (as HCl), maximum	5 ppm
nonvolatile matter, maximum	10 ppm
water, maximum	100 ppm
APHA color, maximum	10
free halogens	negative to test
residual odor	negative to test

Gas chromatographic or infrared techniques are commonly used to monitor the purity of methylene chloride shipments.

Handling

Because of its low boiling point, methylene chloride should be stored in a cool place away from direct sunlight. Storage containers may be constructed of mild or plain steel, galvanized or suitably lined. Aluminum is not recommended for bulk storage. All bulk storage tanks should be equipped with a vent dryer packed with calcium chloride or other appropriate dessicant to exclude moisture. Alternatively, the tank may utilize a dry inert gas pad with an appropriate pressure-vacuum relief valve. Methylene chloride is transported in drums, truck transports, rail cars, barges, and oceangoing ships.

Toxicity

Methylene chloride is one of the least toxic chlorinated methanes. The LD_{50} in rats is in the range of 1.6–3.0 g/kg body weight. The fatal dose for a 68–kg person ranges from 30 cm^3 (1 oz.) to 470 cm^3 (1 pint). Methylene chloride is painful and irritating if splashed directly into the eye. The ACGIH threshold limit value (TLV) for methylene chloride is 200 ppm (v/v) for an 8-h exposure.

Methylene chloride vapors have an anesthetic action. The odor threshold is approximately 300 ppm. At concentrations between 310 and 800 ppm, the odor is clearly identifiable but not unpleasant. In the range of 900 to 1200 ppm, the odor is pronounced and anesthetic effects with accompanying dizziness begin after 20 minutes exposure. Excessive exposure to levels greater than 2300 ppm causes lightheadedness, dizziness, nausea, headaches, and tingling or numbness of extremities. Mental alertness and physical coordination may also be impaired. Exposure to very high concentrations could result in unconsciousness or death.

Methylene chloride, applied to both intact and abraded skin of rabbits in doses as large as 0.5 g/kg body weight per day, five times per week, for a period of 90 days caused no apparent adverse effects. Absorption through the skin is not usually a hazard when good working practices are followed.

A recent determination that carbon monoxide might be a metabolite of methylene chloride in humans (15) suggests that unacceptable levels of carboxyhemoglobin would exist in the blood of persons exposed to methylene chloride vapors at concentrations greater than 500 ppm for extended periods of time. These conditions are rarely encountered in most industrial applications. However, as with any organic solvent, adequate ventilation should be provided to ensure compliance with all industrial and governmental regulations.

Studies in which pregnant rats and mice were exposed to 1250-ppm levels of methylene chloride for 7 hours a day on days 6–15 of gestation indicated no significant maternal embryonal fetal toxicity (16). Methylene chloride was shown to be nonteratogenetic to either animal at the concentration studied.

Continuous exposure of dogs (17), monkeys, rats, and mice to methylene chloride at 500 and 1000 ppm concentrations produced the following severe toxic effects on dogs, rats, and mice: dogs died after three weeks exposure to 1000 ppm and six weeks exposure to 500 ppm; 30% of the mice died after four weeks exposure to 500 ppm; rats survived 14 weeks exposure to 500 ppm but experienced subnormal weight gain. All three species exhibited significant growth and histopathological hepatic lesions after 14 weeks. In addition, the rats exhibited abnormal kidney histopathology. Fat stains disclosed mild fatty increases in monkey livers after 14 weeks exposure to 1000 ppm.

Continuous exposure to low-level methylene chloride concentrations of 100 ppm and 25 ppm for 14 weeks did not affect the spontaneous activity of mice (18).

Unlike many organic solvents, including hexane, heptane, benzene, xylene, toluene, gasoline, and particularly some of the other chlorinated and fluorinated solvents, methylene chloride does not cause cardiac arrhythmias in the presence of elevated epinephrine levels when inhaled at concentrations as high as 20,000 ppm (19).

Fatalities have occurred when unprotected workers have entered an unventilated tank or piece of equipment that contained high vapor concentrations of a chlorinated solvents such as methylene chloride. (See Trichloroethane for a description of proper tank or equipment entry).

Uses

For use in paint strippers, one of its first applications, methylene chloride is blended with other chemical components to maximize its effectiveness against specific coatings. Typical additives include alcohols, acids, amines or ammonium hydroxide, detergents, and paraffin wax. Other formulated products containing methylene chloride, such as adhesives, represent future growth markets.

Methylene chloride is used as an extraction solvent for the decaffeination of coffee (qv), spices, and beer hops because of its strong solvency power and stability (20–22) and is well suited for use in the manufacture of photographic film, and as a carrier solvent in the textile industry (see Dye carriers). Its use as a solvent for vapor degreasing of metal parts is increasing. Methylene chloride may also be blended with petroleum and other chlorinated hydrocarbons for use as a dip-type cleaner in the metal-working industry. There is a rapidly growing market for methylene chloride as a vapor-pressure depressant and solvent in aerosol mixtures (23). Aerosol applications are expected to become the largest consumer of methylene chloride within the next few years. Other applications include low-pressure refrigerants, air conditioning (qv) installations, and as a low-temperature heat-transfer medium (see Refrigeration; Heat transfer technology).

There are several uses for methylene chloride in chemical processing, including the manufacture of polycarbonate (qv) plastic from bisphenol and phosgene, the manufacture of photoresist coatings (see Photoreactive polymers), and as a solvent carrier for the manufacture of insecticide and herbicide chemicals. Methylene chloride is also used by the pharmaceutical industry as a process solvent in the manufacture of steroids, antibiotics, and vitamins, and to a lesser extent as a solvent in the coating of tablets (24) (see Solvents, industrial).

Methylene chloride is used extensively in the urethane foam industry as an auxiliary blowing agent for flexible foams (see Urethane polymers; Foams). Its use in this area is expected to increase significantly with further technological developments. The urethane foam industry also uses methylene chloride for cleaning foam heads and lines immediately after production runs.

Other uses include grain fumigation (25) and oil dewaxing (26).

BIBLIOGRAPHY

The "Chlorocarbons and Chlorohydrocarbons, Methylene Chloride" are treated in *ECT* 1st ed. under "Chlorine Compounds, Organic", Vol. 3, pp. 747–750, by Janet Searles and H. A. McPhail, E. I. du Pont de Nemours & Co., Inc. and in *ECT* 2nd ed. under "Chlorocarbons and Chlorohydrocarbons, Methylene Chloride", Vol. 5, pp. 111–118 by D. W. F. Hardie, Imperial Chemical Industries Ltd.

1. P. J. Carlisle and A. A. Levine, *Ind. Eng. Chem.* **24**(2), 146 (1932).
2. U.S. Pat. 2,008,680 (Mar. 3, 1931), P. J. Carlisle, C. R. Harris, and P. Johnson (to E. I. du Pont de Nemours & Company, Inc.).
3. U.S. Pat. 1,904,405 (Mar. 3, 1931), C. R. Harris (to Rossler and Hasslacher).
4. *Chem. Eng. News* **39**(44), 37 (1961).
5. U.S. Pat. 3,670,039 (June 13, 1972), T. A. Vivian (to The Dow Chemical Company).
6. Ger. Pat. 467,234 (Aug. 29, 1922) (to Holzverkohlungs Industrie A.G.).
7. W. L. Archer and T. Anthony, unpublished results, The Dow Chemical Company.
8. D. V. M. George and J. H. Thomas, *Trans. Faraday Soc.* **58**, 262 (1962).
9. U.S. Pat. 2,694,094 (Nov. 9, 1954), Z. J. Lobos (to Stop Fire, Inc.).
10. J. Lavaux, *Ann. Chim. Phys.* **20**, 436 (1910).
11. U.S. Pat. 1,190,659 (July 11, 1916), B. S. Lacy.
12. Faith, Keyes, and Clark, *Industrial Chemicals*, 4th ed., John Wiley & Sons, Inc., New York.
13. A. Stock and P. Stiebeler, *Ber. Deut. Chem. Ges* **56**, 1088 (1923).
14. Ger. Pat. 416,014 (Dec. 23, 1923) (to Société Chimique des Usines du Rhone).
15. R. D. Stewart and co-workers, *Science* **176**, 295 (1972).
16. B. A. Schwetz, B. K. L. Leong, and P. Z. Gehring, *Toxicol. and Appl. Pharmacol.* **32**(1), 84 (April, 1975).
17. J. D. MacEwen, E. H. Vernot, and C. C. Hajn, *Continuous Animal Exposure to Dichloromethane AMRL-TR-72-28*, Aerospace Medical Research Laboratory, Wright-Patterson Air Force Base, Ohio, pp. 1–33.

18. A. Thomas, M. K. Pinkerton, and J. A. Warden, *Effects of Low Level Dichloromethane Exposure on the Spontaneous Activity of Mice*, AMRL-TR-71-120 Aerospace Medical Research Laboratory, Wright-Patterson Air Force Base, Ohio, 1972, pp. 223–229.
19. C. F. Reinhardt, L. S. Mullin, M. E. Maxfield, *J. of Occup. Med.* **15**(12), (1973).
20. *Fed. Regist.* **32**, 12605 (Aug. 31, 1967).
21. U.S. Pat. 3,671,262 (June 20, 1972), A. B. Wolfson, J. M. Patel, B. Lawrence (to Procter & Gamble Co.).
22. G. Bornmann, and co-workers, *Deut. Lebensm. Rundsch* **64**(6), 167 (1968).
23. T. Anthony, *Drug Cosmet. Ind.* 46 (March, 1976).
24. D. Lefort des Ylouses and co-workers, *Ann. Pharm. Fr.* **31**(11), 647 (1973).
25. G. A. Polchaninova and N. I. Sosedov, *Vliyanie Mikroorganizmov Protravitelei Semena* 179 (1972); *Chem. Abstr.* **78**, 965d (1973).
26. S. Abramovich and S. A. Minichairova, *Tr. Bashk. Naucho. Issled, Inst. Pererab. Nefti.* **9**, 61 (1971).

T. ANTHONY
Dow Chemical, U.S.A.

CHLOROFORM

Chloroform [67-66-3] (trichloromethane, methenyl chloride), $CHCl_3$, at normal temperature and pressure is a heavy, water-white, volatile liquid with a pleasant, ethereal, nonirritant odor. Although chloroform is nonflammable, its hot vapor in admixture with vaporized alcohol burns with a green-tinged flame. Chloroform is miscible with the principal organic solvents and slightly soluble in water. It is less stable in storage than either methyl or methylene chloride. Chloroform decomposes at ordinary temperatures in sunlight in the absence of air, and in the dark in the presence of air. Phosgene is one of the oxidative decomposition products.

Chloroform was discovered in 1831 by Liebig and Soubeirain simultaneously. Liebig obtained chloroform by the action of alkali or chloral; and Soubeirain by reaction of bleaching powder with alcohol or acetone. Guthrie, in the U.S., is also alleged to have discovered chloroform in 1831. In 1839, Dumas produced chloroform by heating alkali with trichloroacetic acid; in the following year, Regnault obtained it by chlorinating methyl chloride. Chloroform was first used in medicine as a stimulant, taken internally, and as an inhalant in cases of asthma. In November, 1847, on the suggestion of Waldie, a Liverpool chemist, Simpson used chloroform as a total anesthetic in obstetrics.

Shortly after Simpson's successful use of chloroform in Edinburgh, Fraser began making pure anesthetic chloroform on a small scale in Nova Scotia.

In 1900, the Pennsylvania Salt Manufacturing Company initiated large scale production in the U.S. The Midland Chemical Company, a subsidiary of Dow Chemical, began to manufacture chloroform by reducing carbon tetrachloride in 1903. Chloroform was one of the first organic chemicals produced on a large scale in the U.S.

Chloroform was used chiefly as an anesthetic and in pharmaceutical preparations immediately prior to World War II. However, these uses have been banned (see Health and Safety). Annual output in both the U.S. and the United Kingdom was between 900 and 1350 metric tons. During the war, chloroform production in the U.S. was tripled, largely to meet the requirement for penicillin manufacture. Demand for chloroform continued to increase in the postwar period as its technical applications were extended. The estimated 1976 U.S. production capacity for chloroform was 136,000 t/yr. Chloroform production is less than that of the other three methane chlorination products, CH_3Cl, CH_2Cl_2, and CCl_4. Consumption continues to increase, however, at a comparatively rapid rate. Chloroform is now used primarily in the manufacture of monochlorodifluoromethane, a refrigerant, and as a raw material for polytetrafluoroethylene plastics (see Refrigeration; Fluorine compounds, organic).

Physical and Chemical Properties

The physical properties of chloroform are listed in Table 1.

Chloroform dissolves alkaloids, cellulose acetate and benzoate, ethyl cellulose, essential oils, fats, gutta-percha, halogens, methyl methacrylate, mineral oils, many resins, rubber, tars, vegetable oils, and a wide range of common organic compounds. A temperature increase occurs when chloroform is mixed with diethyl ether. Chloroform forms a series of binary azeotropes (1), the azeotrope with water boils at 56.1 °C and contains 97.2% chloroform. The ternary azeotrope with ethanol and water boils at 55.5°C and contains 4 mol % alcohol and 3.5 mol % water. At 25°C, chloroform dissolves 3.59 times its volume of carbon dioxide.

Chloroform slowly decomposes on prolonged exposure to sunlight in the presence or absence of air, and in the dark in the presence of air. The products of oxidative breakdown include phosgene, hydrogen chloride, chlorine, carbon dioxide, and water. At 290°C, chloroform vapor is not attacked by oxygen. In contact with iron and water, hydrogen peroxide is also produced, probably by the following reaction sequence (2):

$$CHCl_3 + O_2 \rightarrow (Cl_3COOH) \xrightarrow{[H_2]} Cl_3COH + H_2O_2$$

$$Cl_3COH \rightarrow COCl_2 + HCl$$

Oxidation with powerful oxidizing agents, eg, chromic acid, results in formation of phosgene and liberation of chlorine. Nitrogen dioxide at about 270°C oxidizes chloroform to a mixture of compounds including phosgene, hydrogen chloride, water, and carbon dioxide (3). Ozone forms a blue solution in chloroform and causes rapid decomposition.

Chloroform and water at 0°C form six-sided crystals of a hydrate, $CHCl_3.18\ H_2O$ [67922-19-4], which decomposes at 1.6°C. Chloroform does not decompose appreciably when in prolonged contact with water at ordinary temperature and in the absence of air. However, on prolonged heating with water at 225°C, decomposition to formic acid, carbon monoxide, and hydrogen chloride occurs. A similar hydrolysis takes place when chloroform is decomposed at elevated temperature by potassium hydroxide.

$$CHCl_3 + 4\ KOH \rightarrow HCOOK + 3\ KCl + 2\ H_2O$$

Chloroform resists thermal decomposition at temperatures up to about 290°C. Pyrolysis of chloroform vapor occurs at temperatures above 450°C, producing tetra-

Table 1. Physical Properties of Chloroform

Property	Value
molecular weight	119.38
refractive index at 20°C	1.4467
autoignition temperature, °C	above 1000
flash point, °C	none
melting point, °C	−63.2
101 MPa[a]	−43.4
507 MPa[a]	20.4
1216 MPa[a]	112.6
boiling point, 101 kPa,[b] °C	61.3
specific gravity	
0/4°C	1.52637
25/4°C	1.48069
60.9/4°C	1.4081
vapor density, 101 kPa,[b] 0°C, kg/m^3	4.36
surface tension, mN/m (=dyn/cm)	
air, 20°C	27.14
air, 60°C	21.73
water, 20°C	45.0
heat capacity, 20°C, kJ/(kg·K)[c]	0.979
critical temperature, °C	263.4
critical pressure, MPa[a]	5.45
critical density, kg/m^3	500
critical volume, m^3/kg	0.002
thermal conductivity, 20°C, W/(m·K)	0.130
coefficient of cubical expansion	0.001399
dielectric constant, 20°C	4.9
dipole moment, C·m[d]	3.84 × 10^{-30}
heat of combustion, MJ/(kg·mol)[a]	373
heat of formation, 25°C, MJ/(kg·mol)[c]	
gas	−89.66
liquid	−120.9
latent heat of evaporation, at bp, kJ/kg[c]	247
solubility of chloroform in water, g/kg H$_2$O	
0°C	10.62
10°C	8.95
20°C	8.22
30°C	7.76
solubility of water in chloroform, at 22°C, g/kg chloroform	0.806
viscosity, liquid, mPa·s (= cP)	
−13°C	0.855
0°C	0.700
15°C	0.596
20°C	0.563
30°C	0.510

vapor pressure	°C	kPa[b]	°C	kPa[b]
	−60	0.11	0	8.13
	−50	0.27	10	13.40
	−40	0.63	20	21.28
	−30	1.33	30	32.80
	−20	2.61	40	48.85
	−10	4.63	50	70.13

[a] To convert MPa to atm, multiply by 9.87.
[b] To convert kPa to mm Hg, multiply by 7.5.
[c] To convert J to cal, divide by 4.184.
[d] To convert C·m to debye, divide by 3.336 × 10^{-30}

chloroethylene, hydrogen chloride, and a number of chlorohydrocarbons in minor amounts (4–5). Pyrolysis in contact with hot pumice is catalyzed by vaporized iodine (1%), resulting in tetrachloroethylene, hexachloroethane, and carbon tetrachloride. Hexachlorobenzene, carbon monoxide, hydrogen chloride, and titanium tetrachloride are formed when chloroform vapor is decomposed by hot titanium oxide. In contact with potassium amalgam or red-hot copper, chloroform reacts to give acetylene.

$$2\ CHCl_3 + 6\ K[Hg] \rightarrow HC{\equiv}CH + 6\ KCl\ [Hg]$$

Small quantities of ethyl alcohol stabilize chloroform during storage. Various other stabilizers have been proposed, $CH_2{=}CHCH_2CH_2CN$, and methacrylonitrile (6).

Chloroform can be reduced to methane with zinc dust and aqueous alcohol. In the presence of ammonia, the reduction yields methylene chloride as well as methane.

Chloroform reacts readily with halogens or halogenating agents. Chlorination of the irradiated vapor is believed to occur by a chain reaction of the following type (7):

$$Cl_2 \xrightarrow{h\nu} 2\ Cl\cdot$$

$$Cl\cdot + CHCl_3 \rightarrow CCl_3\cdot + HCl$$

$$CCl_3\cdot + Cl_2 \rightarrow CCl_4 + Cl\cdot$$

$$CCl_3\cdot + CCl_3\cdot + Cl_2 \rightarrow 2\ CCl_4$$

At 225–275°C, bromination of the vapor yields bromochloromethanes: CCl_3Br, CCl_2Br_2, $CClBr_3$. Chloroform reacts with aluminum bromide to form bromoform, $CHBr_3$ (see Bromine compounds). Chloroform cannot be directly fluorinated with elementary fluorine; fluoroform, CHF_3, is produced from chloroform by reaction with hydrogen fluoride in the presence of a metallic fluoride catalyst (8). It is also a coproduct of monochlorodifluoromethane from the HF–$CHCl_3$ reaction over antimony chlorofluoride. Iodine gives a characteristic purple solution in chloroform but does not react even at the boiling point. Iodoform, CHI_3, may be produced from chloroform by reaction with ethyl iodide in the presence of aluminum chloride; however, this is not the route normally used for its preparation.

No decomposition occurs when boiling chloroform is in prolonged contact with anhydrous aluminum chloride; a double compound is formed from which unchanged chloroform is liberated by the action of water. With benzene in the presence of aluminum chloride, chloroform reacts to give triphenylmethane, $(C_6H_5)_3CH$, which is also formed from chloroform and phenylmagnesium bromide, C_6H_5MgBr (see Friedel-Crafts reaction; Grignard reaction).

In the presence of an alkali metal hydroxide at about 50°C, chloroform condenses with acetone to give 1,1,1-trichloro-2-methyl-2-propanol, chlorobutanol, chloretone, or acetone-chloroform) (9–10), as shown:

$$\begin{array}{c} CH_3 \\ {}\!\!\!\diagdown \\ {}CO + CHCl_3 \rightarrow \\ {}\!\!\!\diagup \\ CH_3 \end{array} \quad \begin{array}{c} CH_3OH \\ {}\!\!\!\diagdown\ \ \diagup \\ {}C \\ {}\!\!\!\diagup\ \ \diagdown \\ CH_3CCl_3 \end{array}$$

Chlorobutanol is a white crystalline substance with a camphor-like odor; its sedative, anesthetic, and antiseptic properties have given the compound some importance in the pharmaceutical industry (see Hypnotics, sedatives and anticonvulsants).

Chloroform reacts with aniline and other aromatic and aliphatic primary amines in alcoholic alkaline solution to form isonitriles (isocyanides, carbylamines), as shown:

$$CHCl_3 + C_6H_5NH_2 + 3\ KOH \rightarrow C_5H_5N\equiv C + 3\ KCl + 3\ H_2O$$

Phenyl isonitrile has a powerful characteristic odor; it is used as in qualitative testing for chloroform or primary aromatic amines (carbylamine test; see Analysis). Chloroform reacts with phenols in alkaline solution to give hydroxy-aromatic aldehydes (Reimer-Tiemann reaction); eg, phenol gives chiefly p-hydroxybenzaldehyde and some salicylaldehyde (11):

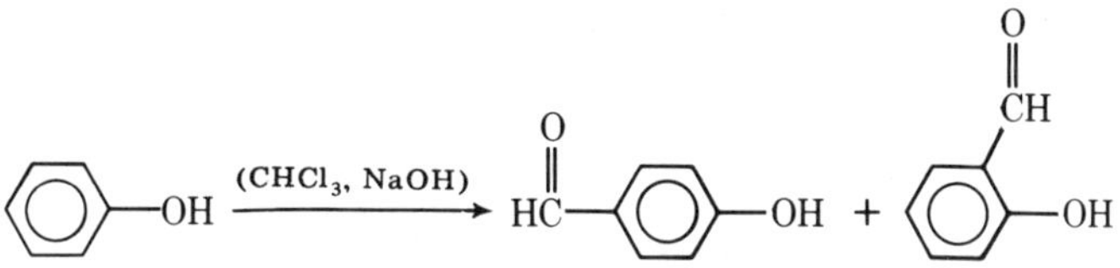

Chloroform combines with the inner anhydride of salicyclic acid to form a well-defined crystalline double compound (12):

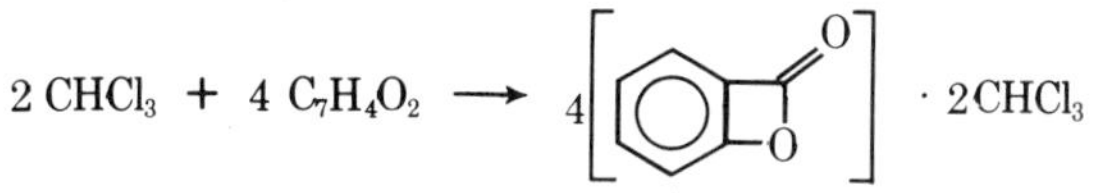

$$2\ CHCl_3 + 4\ C_7H_4O_2 \rightarrow 4[\text{phthalide-like}] \cdot 2\ CHCl_3$$

The above complex readily liberates chloroform when heated, this reaction has been used to produce very pure chloroform (see Salicylic acid).

Reactivities of several chlorinated solvents, including chloroform, with aluminum, iron, and zinc in both dry and wet systems have been determined, as have chemical reactivities in oxidation reactions and in reactions with amines (11). Unstabilized wet chloroform reacts completely with aluminum, attacks zinc at a rate of >250 μm/yr, and iron at <250 μm/yr. The dry, uninhibited solvent attacks aluminum and zinc at a rate of 250 μm/yr and iron at 25 μm/yr.

Manufacture

Many compounds containing either the CH_3CO— or $CH_3CH(OH)$— group yield chloroform on reaction with chlorine and alkali. Until recently, this method was used almost exclusively to produce chloroform from acetone or ethyl alcohol. Methane or methyl chloride chlorination is now the most common method of producing chloroform.

Methane Chlorination. Several variants of this process have been proposed or are in operation. Chlorination of methane to a mixture of chloromethanes has already been outlined (see Methylene chloride). For example, by controlling process conditions in some cases a single plant may produce either chloroform or carbon tetrachloride almost exclusively (12). In one patented variant (13), methane, chlorine, and carbon tetrachloride are fed to a fluidized bed at gas velocities of 6.1–27 cm/s and a temperature of 650–775°C. When the ratios of chlorine and carbon tetrachloride to methane in the feed mixture are between 1.75–2.60:1 and 0.75–1.25:1, respectively, chloroform is the predominant product.

In another patented variant (14), a mixture of 20 vol % methane and 60 vol % hydrogen is chlorinated in a fluidized bed at 325–450 °C to a mixture of chloroform and carbon tetrachloride. Again, the conditions are controlled to produce predominantly chloroform. The chlorine in this process is fed to the reactor in increments so that its concentration at any particular point in the fluidized bed does not exceed the ignition threshold concentration for a given bed temperature; the specified overall ratio of chlorine to methane hydrogen is 60–100% of the maximum stoichiometric ratio. The ignition threshold temperature decreases as the percentage of chlorine in the mixture increases.

Carbon Tetrachloride Reduction. Limited reduction of carbon tetrachloride to chloroform may be effected by reaction with hydrogen, methane (15), zinc dust (16), or ethyl alcohol (17). No commercial processes are based on these methods.

Hydrogen reduction is generally carried out using hydrogen that is generated by reaction of iron with hydrochloric acid. This reducing mixture is mechanically agitated with the carbon tetrachloride for three to four days at a constant temperature of 15°C until reaction is complete. The chloroform is removed from the iron oxychloride sludge by steam distillation.

Methane reduction of carbon tetrachloride occurs at 400–650°C in 0.1–20 s in a smooth-walled reaction chamber, added surface material is unnecessary.

Zinc-dust reduction is carried out on carbon tetrachloride vapor. To recover the zinc, the reaction products are condensed in ammonium chloride solution from which the complex $ZnCl_2 \cdot NH_4Cl$ crystallizes. The organic condensate typically contains 80% chloroform, 17% unchanged carbon tetrachloride, and 3% water.

Reduction with ethyl alcohol in a sealed vessel at 200°C for 25 hours yields only small quantities of chloroform and ethyl chloride. Ultraviolet irradiation of a carbon tetrachloride—alcohol mixture produces higher yields of chloroform, but the reaction proceeds very slowly.

Manufacture from Acetone. In this virtually obsolete process bleaching powder (see Bleaching agents) is slowly charged to water in an agitated mixing tank in a ratio of about 360 kg/m^3 (3 lb/gal). The resultant suspension is strained into the pot of a cast-iron still which is fitted with an agitator and heating and cooling coils. Acetone is introduced to the still in a ratio of 0.1 kg per kg of bleaching powder. The temperature is then increased to 62°C and the chloroform begins to distill. In this reaction the bleaching powder acts like a mixture of chlorine and calcium hydroxide:

$$2 \underset{\text{acetone}}{CH_3COCH_3} \xrightarrow{(Cl_2)} 2 \underset{\text{trichloroacetone}}{CCl_3COCH_3} \xrightarrow{(Ca(OH)_2)} 2 \underset{\text{chloroform}}{CHCl_3} + \underset{\text{calcium acetate}}{Ca(CH_3COO)_2}$$

The temperature is gradually raised until all the chloroform has distilled. The crude chloroform is purified by treatment with concentrated sulfuric acid and finally distilled from anhydrous quicklime. The yield is usually about 90% based on acetone charged, and the process may be operated continuously.

Chloroform may be produced from acetone by intermediate production of hexachloroacetone (18). Hexachloroacetone is mixed with a solution of an alkali metal hydroxide at a temperature below 60°C. The alkali metal salt of trichloroacetic acid is also formed.

$$CCl_3COCCl_3 + NaOH \rightarrow CHCl_3 + CCl_3COONa$$

Other Methods of Forming Chloroform. Chloroform can be produced by decomposition of pentachloroethane with aluminum chloride; perchloroethylene is the coproduct (19).

$$2\ CCl_3CHCl_2 \rightarrow 2\ CHCl_3 + CCl_2{=}CCl_2$$

Electrolysis of a solution of an alkali metal or alkaline earth metal chloride in aqueous alcohol produces chloroform by reaction of the liberated chlorine with the alcohol (20). Chloroform has also been prepared by reaction of carbon monoxide and hydrogen chloride under pressure at about 400°C in the presence of a catalytic oxide, and by reaction of water gas and hydrogen under pressure in contact with a catalyst containing chlorine. None of the above methods has become the basis of a commercial process.

Economic Aspects

Chloroform production figures, sales figures, and prices in the U.S. for the period 1960–1977 inclusive are presented in Table 2. Annual production of chloroform remained between 900 and 1350 metric tons during the 1930s when it was used primarily for anesthetic purposes and by the pharmaceutical and fine chemical industries. During and immediately after World War II, chloroform found a new application as an extractant in the production of penicillin. Chloroform production continued to expand for use in the manufacture of chlorofluoromethane refrigerants for domestic air conditioners (21), and as a starting material in the manufacture of polytetrafluoroethylene (PTFE). Over the same postwar period chloroform production in the United Kingdom increased at a similar rate although it was only about one tenth that in the U.S. Production in other European countries is estimated to be similar to that of the United Kingdom.

Production in 1975 was 119,000 metric tons in the U.S., a decline of 13% from the high 137,000 t produced in 1974. In 1976 third-quarter production reached an annual rate of almost 136,000 t (22).

Table 2. U.S. Chloroform Production and Price Statistics, Thousands of Metric Tons[a]

Year	Production	Sales	Unit Sales Value, ¢/kg	Purified	Technical
1960	34.67	25.40	22.1	55.1	37.5
1965	69.18	55.94	17.6	55.1	37.5
1970	108.81	79.33	13.2	56.2	38.6
1973	114.66	110.60	15.4	55.1	38.6
1974	136.89	114.44	24.3	55.1	37.5
1975	118.71	87.14	35.3	48.5	35.3
1976	132.39	120.39	35.3	48.5	44.1
1977	137.02			52.9	46.3

[a] According to U.S. International Trade Commission.

The strong demand for chloroform as a result of increased fluorocarbon-22 demand, coupled with a good economic recovery and equally strong demand for coproduced methylene chloride for use in solvent degreasing and paint removal, has buoyed prices for both products. The optimistic outlook has provided the incentive for producers to announce capacity expansions that should satisfy demand well into the next decade.

During the period from 1960 to 1973, list prices for chloroform remained substantially constant at 37.5–38.5 ¢/kg although the unit sales value of all chloroform sold actually declined from 22 ¢/kg in 1960 to a low of 13 ¢/kg in 1970–1971. The price decline was caused by overcapacity for production of chloromethanes, and by competition for merchant sales to fluorocarbon producers, who were also, despite good growth rates, generally in an overcapacity situation.

Increased costs of natural gas and electricity, methanol, chlorine, etc, caused prices to rise quickly; today's list prices are closer to actual selling prices.

Specifications and Standards

Technical-grade chloroform generally contains one or more stabilizers which vary according to specification requirements. Common stabilizers are industrial methylated spirit (0.2%), absolute alcohol (0.6–1%), thymol, t-butylphenol, or n-octylphenol (0.0005–0.01%). A representative technical-quality chloroform contains the following amounts of the indicated substances (maximum): water, 300 ppm; acid (as HCl), 2 ppm; methylene chloride, 200 ppm; bromochloromethane, 550 ppm; carbon tetrachloride, 1500 ppm; residue (on evaporation to dryness at 110°C), 20 ppm. Dissolved chlorine should not be detectable.

Analysis

In the known absence of bromoform, iodoform, chloral, and other halogenated methanes, the formation of phenyl isonitrile with aniline (see earlier) provides a simple and fairly sensitive, but nonspecific, test for the presence of chloroform (carbylamine test). Phenyl isonitrile formation is the identification test given in the *British Pharmacopoeia*. A small quantity of resorcinol and caustic soda solution (10% concn) added to chloroform results in the appearance of a yellowish-red color, fluorescing yellow-green. When 0.5 mL of a 5% thymol solution is boiled with a drop of chloroform and a small quantity of potassium hydroxide solution, a yellow color with a reddish sheen develops; the addition of sulfuric acid causes a change to brilliant violet, which diluted with water, finally changes to blue (23).

Chloroform may be estimated quantitatively by determining the amount of copper oxide produced when it is warmed with Fehling's solution (potassium cupritartrate) (24). An alternative procedure consists of heating the chloroform with concentrated alcoholic potassium hydroxide in a sealed tube at 100°C and determining the amount of potassium chloride produced (25).

Handling

Chloroform should be stored in sealed containers in a cool place. Glass containers should be dark green or amber. Technical-grade chloroform can be stored in lead-lined or mild steel containers of all-welded construction. When storage vessels are made of unlined steel, precautions are needed to prevent the entry of moisture. The technical grade is conveyed in galvanized steel drums; large quantities are shipped in tank trucks or tank cars. Chloroform is not subject to Department of Transportation shipment regulations.

Health and Safety

Chloroform does not present a serious industrial hazard if workers are adequately supervised and instructed in its proper handling. The principal hazard is damage to the liver and kidneys resulting from inhalation or ingestion. Inhalation of high concentrations may result in disturbances of equilibrium or loss of consciousness. Chloroform is mildly irritating to skin and mucous membranes upon contact, and to the alimentary tract upon ingestion. It is believed that medically significant quantities are not absorbed through intact skin.

The toxic effects of chloroform resemble those of carbon tetrachloride. The probable effects of exposure to various atmospheric concentrations of chloroform are summarized below (26).

Concentration (ppm)	(mg/L)	Response
205–310	1–1.5	smallest amount that can be detected by smell
390	1.9	endured for 30 minutes without complaint
1025	5	definite aftereffects; fatigue and headache still experienced hours after exposure
1025	5	dizziness, intracranial pressure, and nausea after 7 minutes exposure
1475	7.2	dizziness and salivation after a few minutes exposure
4100	20	vomiting, sensation of fainting
14,340–16,400	70–80	narcotic limiting concentration

In the past, chloroform was used extensively as a surgical anesthetic, but this use was abandoned because exposure to narcotic concentrations often resulted in sudden death from effects on the heart and circulation, or severe injury to the liver. In addition, chloroform for this and other consumer uses (see below) was banned by FDA in 1976 with the discovery that it is carcinogenic in mice (27). In 1978, the NRC reported that it is virtually impossible to establish a link between human bladder cancer and chloroform in drinking water (28). When splashed into the eye, chloroform causes local pain and irritation, but serious injury would not be expected. Skin contact for single, brief exposures ordinarily causes little or no local irritation.

Repeated or prolonged contact with the skin, especially under clothing, may result in local irritation and inflammation, and at elevated temperatures, such as in the presence of an open flame, chloroform decomposes to form by-products, including phosgene, chlorine, and hydrogen chloride, all of which are severe irritants to the respiratory tract.

Ingestion of chloroform is followed immediately by a severe burning in the mouth and throat, pain in the chest and abdomen, and vomiting. Loss of consciousness and liver injury may follow depending upon the amount swallowed. The tendency of chloroform to produce liver injury is significantly augmented in alcoholics and persons with nutritional deficiencies.

The most serious hazard of repeated exposure to chloroform inhalation is injury to the liver and kidneys. Evidence indicates that in man, repeated exposure to atmospheric concentrations well below the odor threshold may cause such injury. Industrial experience has shown that daily exposure to concentrations below 100 ppm may result in a variety of nervous system and alimentary tract symptoms, in the absence of demonstrable evidence of injury (29). Injury to the liver is similar to but somewhat less severe than that caused by carbon tetrachloride. Kidney injury is usually associated with, but less severe than, liver injury.

The recommended maximum time-weighted average concentration in the workroom atmosphere for 10-h daily exposure is 50 ppm but NIOSH recommends the exposure limit be reduced to 10 ppm per 10-h workday (27). It may be desirable to exclude alcoholics, persons with chronic disorders of the liver, kidneys, and central nervous system, and those with nutritional deficiencies from working with chloroform.

Treatment of chloroform poisoning is symptomatic; no specific antidote is known. Adrenalin should not be given to a person suffering from chloroform poisoning.

Uses

Although chloroform production continues to increase, it is used in smaller tonnages than the other three chlorinated methanes. Like most chlorinated solvents, it readily dissolves fats, greases, gums, oils, and waxes, and is frequently used as a dry cleaning spot remover and for removing fats from waste products (see Solvents, industrial). Miscellaneous uses include extraction and purification of penicillin, alkaloids, vitamins and flavors, and as an intermediate in the preparation of dyes and pesticides. Chloroform has also been used as a fumigant and insecticide. The anesthetic properties of chloroform were discovered in 1847 and as stated earlier this was its primary use. It has since been replaced as an anesthetic by safer and more versatile materials with fewer side effects (see Anesthetics). Chloroform has also been used in the formulation of cough syrups, tooth pastes, liniments, and toothache preparations. These latter uses were also banned by the FDA in 1976 (27).

Other minor uses in solvent adhesives and compounding of food-packaging plastics and resins will also be restricted for environmental reasons. By far the largest and fastest growing use for chloroform is in the manufacture of monochlorodifluoromethane or Fluorocarbon 22 (F-22), which has grown from 18,144 metric tons in 1960 to 59,874 t in 1975 when it accounted for approximately 95,254–97,522 t of chloroform consumption—about 80% of total chloroform output (22). However, future usage of chlorofluorocarbons may be limited by regulations to prevent ozone depletion in the upper atmosphere (see Air pollution; Aerosols; Ozone).

Approximately 70–75% of F-22 is sold for use as a refrigerant (see Refrigeration; Air conditioning), with minor amounts used as an aerosol propellant (see Aerosols). The rest is used to manufacture fluorocarbon resin intermediates such as tetrafluoroethylene.

BIBLIOGRAPHY

"Chloroform" in *ECT* 1st ed., Vol. 3, pp. 842–848, by Leonard Stievater, Jr., McKesson & Robbins, Inc., and R. J. Van Nostrand, Brown Company; "Chlorocarbons and Chlorohydrocarbons, Chloroform" in *ECT* 2nd ed., Vol. 5 pp. 119–127, by D. W. F. Hardie, Imperial Chemical Industries, Ltd.

1. I. Mellan, *Source Book of Industrial Solvents,* Reinhold Publishing Corp., New York, 1957, p. 126.
2. R. Neu, *Pharmazie* **3,** 251 (1948).
3. D. V. E. George and J. H. Thomas, *Trans. Faraday Soc.* **58,** (470), 262 (1962).
4. G. P. Semeluk and R. B. Bernstein, *J. Am. Chem. Soc.* **76,** (14), 373 (1954).
5. W. Ramsay and S. Young, *Jahresber Fort. Chem.* **628,** (1886).
6. U.S. Pat. 3,029,298 (April 10, 1962), F. S. Hirsekorn and J. H. Rains (to Frontier Chemical Co.).
7. I. H. Winning, *Trans. Faraday Soc.* **47,** 1084 (1951).
8. A. J. Rudge, *The Manufacture and Use of Fluorine and its Compounds,* Oxford University Press, Inc., New York, 1962, p. 68.
9. C. Willgerodt, *Ber. Deut. Chem. Ges.* **14,** 2451 (1881).
10. E. H. Huntress, *Organic Chlorine Compounds,* John Wiley & Sons, Inc., New York, 1948, pp. 285–288.
11. *Ind. Eng. Chem., Prod., Res., Dev.* **26**(2), 158 (1977).
12. *Chem. Eng.* **64**(3), 160, (1957).
13. U.S. Pat. 2,829,180 (April 1, 1958), R. N. Montgomery and J. J. Lukes (to Diamond Alkali Co.).
14. U.S. Pat. 2,585,469 (May 20, 1948), P. R. Johnson (to E. I. du Pont de Nemours & Co., Inc.).
15. U.S. Pat. 2,695,918 (Sept. 4, 1952), H. M. Pitt and H. Bender (to Stauffer Chemical Co.).
16. Span. Pats. 188,302 (May 20, 1949); 188,286 (May 19, 1949) (to Productos Riera, S.A.).
17. G. A. Razuvaev and Yu. A. Sorokin, *Zh. Obach. Khim.* **23,** 1519 (1953).
18. U.S. Pat, 2,695,918 (Nov. 30, 1954), E. E. Gilbert, D. H. Kelley, and C. Woolf (to Allied Chemical & Dye Corp.).
19. H. J. Prins, *J. Prakt. Chem.* **89,** 414 (1914).
20. Ger. Pat. 29,771 (March 7, 1884) (to Chemische Fabrik auf Aktiem (vorm. Schering)).
21. *Chem. Week* **12**(2), 83 (July 12, 1958).
22. *Chemical Profile,* Sept. 27, 1976; *Chemical Products, Synopsis,* Nov. 1976.
23. R. Dupouy, *Chem. News* **88,** 37 (1903).
24. E. Baudrimont, *J. Pharm. Chim.* **9,** 410 (1869).
25. L. deSaint-Martin, *Compt. Rend.* **106,** 494 (1888).
26. F. A. Patty, Editor, *Industrial Hygiene and Toxicology,* Vol. II, Interscience Publishers, Inc. New York, 1949.
27. *Chem. Week* **18**(14), 17 (April 7, 1976).
28. *Chem. Eng. News,* 8 (Oct. 16, 1978).
29. P. J. R. Challen and co-workers. *Br. J. Ind. Med.* **15,** 43, (1958).

H. D. DeShon
Dow Chemical U.S.A.

CARBON TETRACHLORIDE

Carbon tetrachloride [56-23-5] (tetrachloromethane), CCl_4, at ordinary temperature and pressure is a heavy, colorless liquid with a characteristic nonirritant odor; it is nonflammable. The vapor decomposes to give toxic products such as phosgene when in contact with a flame or very hot surface. Carbon tetrachloride contains 92 wt % chlorine. It is the most toxic of the chloromethanes and the most unstable upon thermal oxidation. The commercial product frequently contains added stabilizers. Carbon tetrachloride is miscible with many common organic liquids and is a powerful solvent for asphalt, benzyl resin (polymerized benzyl chloride), bitumens, chlorinated rubber, ethyl cellulose, fats, gums, rosin, and waxes.

Carbon tetrachloride is used principally as an intermediate in the manufacture of chlorofluoromethane for refrigeration, aerosol and blowing agent markets. In 1934 it was supplanted as the predominant dry cleaning agent in the United States by perchloroethylene (which is much less toxic and more stable). Carbon tetrachloride is also used in grain fumigation and a variety of solvent and chemical manufacturing applications.

Carbon tetrachloride was one of the first organic chemicals produced on a large scale. In the 1890s, commercial manufacturing processes were being investigated by the United Alkali Co. in England. At the same time, it was also produced in Germany, exported to the United States, and retailed as a spotting agent under the trade name Carbona. Large-scale production of carbon tetrachloride in the United States began about 1907. By 1914, annual production fell just short of 4500 metric tons and was used primarily for drycleaning and for charging fire extinguishers. During World War I, U.S. production of carbon tetrachloride expanded greatly; its use was extended to grain fumigation and the rubber industry. The demands of World War II also stimulated production and marked the beginning of its use as the starting material for chlorofluoromethanes, currently the most important application for carbon tetrachloride.

Physical and Chemical Properties

The physical properties of carbon tetrachloride are listed in Table 1.

Carbon tetrachloride readily dissolves stannic chloride, $SnCl_4$, but not ferric chloride, $FeCl_3$. Carbon tetrachloride forms a large number of binary and several ternary azeotropic mixtures; a partial list of the former is shown in Table 2.

Many polymer films (eg, polyethylene, polyacrylonitrile) are permeable to carbon tetrachloride vapor (1). Carbon tetrachloride vapor affects the explosion limits of several gaseous mixtures, eg, air–hydrogen and air–methane. The extinctive effect that carbon tetrachloride has on a flame, mainly because of its cooling action, is derived from its high thermal capacity (2).

As chlorination proceeds from methyl chloride to carbon tetrachloride, the length of the C–Cl bond is decreased from 0.1786 nm in the former to 0.1755 nm in the latter (3). At ca 400°C, thermal decomposition of carbon tetrachloride occurs very slowly, whereas at 900–1300°C dissociation is extensive, forming perchloroethylene and hexachloroethane, and liberating some chlorine. Subjecting the vapor to an electric arc also forms perchloroethylene and hexachloroethane, as well as hexachlorobenzene, elementary carbon, and chlorine.

Carbon tetrachloride is the chloromethane least resistant to oxidative breakdown.

Table 1. Physical Properties of Carbon Tetrachloride, CCl$_4$

Property	Value
mol wt	153.82
mp, °C	
101.3 kPa[a]	−22.92
21.3 MPa[b]	−19.5
62.8 MPa[b]	0
117.5 MPa[b]	19.5
bp, 101.3 kPa[a], °C	76.72
n_D, 15°C	1.46305
sp gr	
0/4°C	1.63195
20/4°C	1.59472
76/4°C	1.48020
autoignition temperature, °C	>1,000
flash point, °C	none
density of solid, g/cm^3	
−186°C	1.831
−80°C	1.809
vapor density, air = 1	5.32
surface tension, mN/m (= dyn/cm)	
0°C	29.38
20°C	26.77
60°C	18.16
specific heat, J/kg[c]	
20°C	866
30°C	837
critical temperature, °C	283.2
critical pressure, MPa[b]	4.6
critical density, kg/m^3	558
thermal conductivity, mW/(m·K)	
liquid, 20°C	118
vapor, bp	7.29
average coefficient of volume expansion,	
0–40°C	0.00124
dielectric constant, ϵ	
liquid, 20°C	2.205
liquid, 50°C	1.874
vapor, 87.6°C	1.00302
heat of formation, kJ/mol[c]	
liquid	−142
vapor	−108
heat of combustion, liquid, at constant volume,	
18.7°C, kJ/mol[c]	365
latent heat of fusion, kJ/mol[c]	2.535
latent heat of vaporization, kJ/kg[c]	194.7
viscosity, mPa·s(= cP)	
0°C	1.329
20°C	0.965
40°C	0.739
60°C	0.585
100°C	0.383
180°C	0.201
vapor pressure, kPa[a]	
−50°C	0.123
−20°C	1.323
0°C	4.410

Table 1. (continued)

Property	Value
20°C	11.94
40°C	28.12
60°C	58.53
150°C	607.3
200°C	1458
soly of CCl$_4$ in water, 25°C, g/100 g H$_2$O	0.08
soly of water in CCl$_4$, 25°C, g/100 g CCl$_4$	0.013

^{a} To convert kPa to mm Hg, multiply by 7.5.
^{b} To convert MPa to atm, divide by 0.101.
^{c} To convert J to cal, divide by 4.184.

Table 2. Azeotropic Mixtures of Carbon Tetrachloride

Second component	bp of azeotrope, °C	CCl$_4$, wt %
n-butyl alcohol	77	97.5
acetic acid	77	97
ethyl nitrate	75	84.5
ethyl alcohol	65	84
nitromethane	71	83
ethylene dichloride	76	79
acetone	56	11.5

One g of CCl$_4$ mixed with air and heated to 335°C in the presence of iron, produces 375 mg of phosgene. Only 2.4 mg of phosgene is produced from 1 g of chloroform under the same conditions (4). A cold mixture of carbon tetrachloride and water, seeded with crystals of chloroform hydrate, yields crystals of a hydrate that decomposes at 1.4–1.49°C [101.3 kPa (760 mm Hg)]. When mixed with excess water and heated to 250°C, carbon tetrachloride decomposes to carbon dioxide and hydrochloric acid; if the quantity of water is limited, phosgene is produced. This decomposition also occurs when wet carbon tetrachloride is exposed to uv irradiation (253.7 nm) at ordinary temperatures (5). Chloromethanes, hexachloroethane, and perchloroethylene are formed with steam at high temperatures. A similar decomposition occurs when carbon tetrachloride vapor is heated with some metallic oxides, eg, aluminum and magnesium oxides. An aqueous suspension of carbon tetrachloride droplets exposed to ultrasonic irradiation at ordinary temperature decomposes to carbon dioxide, chlorine, hydrogen chloride, perchloroethylene, and hexachloroethane. Dry carbon tetrachloride does not react with most commonly used construction metals, eg, iron and nickel; it reacts very slowly with copper and lead. Like the other chloromethanes, carbon tetrachloride is reactive (sometimes explosively) with aluminum and its alloys (6–8). The presence of moisture is probably a necessary requirement for the reaction with aluminum. When carbon tetrachloride is in contact with metallic sodium or potassium, or with a liquid alloy of both metals, shock may produce an explosion (9). On heating with sodium amalgam, decomposition takes place with formation of sodium chloride and liberation of carbon; at 400°C an analogous reaction takes place with mercury alone.

Carbon tetrachloride can be reduced to chloroform using zinc and acid. With potassium amalgam and water, carbon tetrachloride can be totally reduced to methane. It has been employed in the dehydrogenation of chloroethanes at 400–600°C in the presence of a catalyst (10):

$$2\ CHCl_2CHCl_2 + 2\ CCl_4 \rightarrow 3\ CCl_2{=}CCl_2 + 4\ HCl$$

When treated with aluminum bromide at 100°C, carbon tetrachloride is converted to carbon tetrabromide; reaction with calcium iodide, CaI_2, at 75°C gives carbon tetraiodide. With concentrated hydriodic acid at 130°C, iodoform, CHI_3, is produced. Carbon tetrachloride is unaffected by gaseous fluorine at ordinary temperatures. Replacement of its chlorine by fluorine is brought about by reaction with hydrogen fluoride at a temperature of 230–300°C and a pressure of 5.17–6.89 MPa (750–1000 psi), producing mainly dichlorodifluoromethane (11–12). Replacement of more than two chlorine atoms in carbon tetrachloride with fluorine from hydrogen fluoride requires other techniques (13).

Carbon tetrachloride forms telomers with ethylene and certain vinyl derivatives; those derived from ethylene have the general formula $CCl_3(CH_2CH_2)_nCl$ in which n is a small number. The products are liquid mixtures. Reaction of ethylene and carbon tetrachloride takes place under pressure and is induced by the presence of a peroxygen compound, eg, benzoyl peroxide (14–16). A similar telomerization takes place between carbon tetrachloride and vinyl acetate to yield a liquid mixture of saturated chlorohydrocarbons (17):

$$CCl_3{\text{(---}}CH_2CH{\text{---})}_n Cl \quad\quad n = 1{-}8$$
$$\underset{OOCCH_3}{|}$$

Benzene reacts with carbon tetrachloride in the presence of anhydrous aluminum chloride to give triphenylchloromethane; no tetraphenylmethane is formed (18). At elevated temperatures, carbon tetrachloride is attacked by silica gel forming a silicon oxychloride (19).

Manufacture

For many years chlorination of carbon disulfide was the only process used to manufacture carbon tetrachloride. In the 1950s, chlorination of hydrocarbons, particularly methane, became more popular in the United States.

Chlorination of Hydrocarbons. Chlorination of hydrocarbons at pyrolytic temperatures is often referred to as chlorinolysis because it involves a simultaneous breakdown of the hydrocarbons and chlorination of the molecular fragments. The quantity of carbon tetrachloride produced depends on the nature of the hydrocarbon starting material and conditions of chlorination. In the Hüls process, a 5:1 mixture (by vol) of chlorine and methane reacts at 650°C; the temperature is maintained by control of the gas flow rate. A heat exchanger cools the exit gas to 450°C and more methane is added to the gas stream in a second reactor. The principal by-product is perchloroethylene. When ethylene is substituted for methane in this process, perchloroethylene becomes the main product.

In another methane–chlorination process (20), the reactants are brought into contact with a fluid catalyst bed maintained at about 300°C by the heat of the chlorination reaction. The catalyst is fuller's earth of 0.15–0.25 mm (60–100 mesh) particle size, and bulk density of 450–480 kg/m^3 (28–30 lb/ft^3). The crude product contains approximately equal quantities of carbon tetrachloride and perchloroethylene. Conditions can be established so that carbon tetrachloride, rather than one of the other chloromethanes, is the principal product. Recycle streams sent to the reactor suppress the formation of unwanted coproducts by mass action (21).

The chlorination of chloroform has been studied (22), but has not yet provided a means of commercially producing carbon tetrachloride. A proposed process is chlorination of hot chloroform using 1% phosphorus trichloride catalyst in the presence of light (23).

A number of processes have been described for producing carbon tetrachloride by chlorination of various hydrocarbons other than methane, eg, acetylene (24), paraffinic hydrocarbons up to C_4 (25–28), eg, the Transcat process (29–30), propane (31), and naphthalene (32). These processes belong to the class of destructive chlorination (chlorinolysis) reactions. Substantial quantities of other chloro derivatives, such as perchloroethylene and hexachloroethane, are usually produced along with the carbon tetrachloride.

Other Processes. Pyrolysis of hexachloroethane at 300–420°C yields a mixture of carbon tetrachloride and perchloroethylene (33). Processes for production of carbon tetrachloride by chlorinating vaporized perchloroethylene at 700–800°C are chemically equivalent to pyrolysis of hexachloroethane, since that compound is presumably produced (34–35).

Purification and Stabilization. Crude carbon tetrachloride is customarily purified by neutralization and drying followed by distillation. A halogen acid contaminating carbon tetrachloride (eg, after use as a solvent in a chlorination process), may be readily removed by treatment with dry pellets of soda ash (36). Active fluorine compounds (eg, hydrogen fluoride) may be removed prior to distillation by means of silica, silicic acids, and silicates at 50–150°C (37). Some purification can be accomplished in the distillation stage by maintaining the carbon tetrachloride under total reflux for a prolonged period before starting the distillation (38).

Although in the dry state carbon tetrachloride may be stored indefinitely in contact with some metal surfaces, its decomposition upon contact with water or on heating in air makes it desirable, if not always necessary, to add a small quantity of stabilizer to the commercial product. A number of compounds have been claimed to be effective stabilizers for carbon tetrachloride, eg, alkyl cyanamides such as diethyl cyanamide (39), 0.34–1% diphenylamine (40), ethyl acetate (to protect copper) (41), up to 1% ethyl cyanide (42), fatty-acid derivatives (to protect aluminum) (43), hexamethylenetetramine (44), resins and amines (45), thiocarbamide (46), and a ureide, ie, guanidine (47).

Economic Aspects

During the years immediately preceding World War II, trichloroethylene had already begun to displace carbon tetrachloride from its then extensive market in the United States as a metal-degreasing and textile-drycleaning solvent. Carbon tetrachloride is more difficult to recover from degreasing operations, more readily hydrolyzed, and more toxic than trichloroethylene. By the 1940s, carbon tetrachloride was rapidly losing in competition not only with trichloroethylene, but with perchloroethylene as well. In 1948 only 33% of the solvent used by the drycleaning industry was carbon tetrachloride and 60% was perchloroethylene; two years later the ratio of perchloroethylene to carbon tetrachloride was three to one. This technological change is not reflected in past sales of carbon tetrachloride which exhibited a steady increase following World War II. It was at this time (ca 1950) that carbon tetrachloride found a new and rapidly expanding use as the starting material in the manufacture of fluorinated refrigerants, an application that by 1954 accounted for about half the total demand for carbon tetrachloride.

Chlorofluorocarbon gases were introduced as aerosol propellants in the late 1950s and early 1960s. From 1960 to 1970, carbon tetrachloride growth averaged 10.7% per year (see Table 3). From 1970 to 1974, as less expensive propellants became available, production of fluorocarbon propellants increased at a rate of only about 7.2%/yr even though overall aerosol consumption grew ca 10%/yr during the same period (see Aerosols). Currently, the decline of carbon tetrachloride production averages 4.9%/yr (48).

Below are United States producers and their capacities

Producer	Location	1978 Capacity, 10^3 t
Allied Chemical	Moundsville, W. Va.	3.6
Dow Chemical	Freeport, Tex.	61.2
Dow Chemical	Pittsburgh, Calif.	36.3
Dow Chemical	Plaquemine, La.	56.7
Du Pont	Corpus Christi, Tex.	186.0
FMC	South Charleston, W. Va.	136.1
Stauffer Chemical	Le Moyne, Ala.	90.7
Stauffer Chemical	Louisville, Ky.	15.9
Vulcan Materials	Geismar, La.	40.8
Vulcan Materials	Wichita, Kan.	27.2
	Total	654.5

List prices do not reflect actual market conditions because of overcapacity and the competitive nature of the propellant and aerosol markets. Sales to merchant fluorocarbon producers were negotiated at considerably lower prices than those shown in Table 3. When price controls ended in 1974, and raw-material and energy costs began to increase, carbon tetrachloride list prices also escalated, and reported unit sales values more nearly approximated published prices. Prices are expected to increase at ca 5%/yr as costs increase and production declines. Although present U.S. carbon tetrachloride capacity of 635,000 t is far in excess of demand, many installations are capable of producing other chlorinated hydrocarbons such as perchloroethylene and chloroform.

Standards and Analysis

A good technical grade of carbon tetrachloride contains not more than the following amounts of impurities: 1 ppm acidity, as HCl, 1 ppm carbon disulfide (if manufactured by carbon disulfide chlorination), 20 ppm bromine, 200 ppm water, and 150 ppm chloroform. The residue should not exceed 10 ppm on total evaporation. The product should give no acid reaction with bromophenol blue, and the starch–iodine test should indicate the absence of free chlorine. When heated with pyrocatechol, copper powder, and alcoholic sodium hydroxide, carbon tetrachloride gives a blue color

Table 3. Carbon Tetrachloride Production and Sales, 1960–1977

Year	Activity, thousands of metric tons Production	Sales	Imports	Total consumption, thousands of metric tons	Sales value Total, millions of $	Unit, $/kg	Price Drums, $/kg	Tanks, $/kg
1960	168.8	151.3	1.9	170.6	27.14	0.176	0.259	0.237
1965	269.3	231.1	4.5	278.9	37.49	0.154	0.259	0.237
1970	458.7	381.6	0.05	458.7	44.09	0.110	0.2701	0.248
1975	411.2	219.6	7.4	418.6	65.9	0.308	0.3527	0.2866
1976	388.6	208.2	3.3	391.8	60.3	0.287	0.3197	0.3197
1977	366.1[a]		4.1	370.2			0.3197	0.3197

[a] Preliminary data.

that changes to red on addition of hydrochloric acid. This color reaction is not produced by chloroform. Quantitative analysis of carbon tetrachloride may be done by first decomposing the sample free of organic and inorganic chlorides, heating in a sealed tube with alcoholic potash, and subsequently determining the potassium chloride formed as the silver halide. The Zeiss interference refractometer has been used to determine the concentration of carbon tetrachloride vapor in air (49).

Health and Safety Factors (Toxicology)

Carbon tetrachloride is the oldest and was the most extensively used chlorinated solvent in degreasing and drycleaning operations. Consequently, its narcotic and toxic properties have been the subject of much investigation: careful investigations have repeatedly shown carbon tetrachloride to be one of the most harmful of the common solvents (50).

Carbon tetrachloride is toxic by inhalation of its vapor and oral intake of the liquid. Inhalation of the vapor constitutes the principal hazard. Exposure to excessive levels of vapor is characterized by two types of response: (1) an anesthetic effect similar to that caused by compounds such as diethyl ether and chloroform, and (2) organic injury to the tissues of certain organs, in particular the liver and kidneys. This type of injury may not become evident until 1–10 days after exposure. The nature of the effect is determined largely by the vapor concentration but the extent or severity of the effect is determined principally by the duration of exposure (51).

Organic injury may result from single prolonged exposure to carbon tetrachloride vapor, or repeated short-duration exposures. Serious and fatal injuries are usually the result of a single prolonged exposure. Vapor concentrations of only a few hundred parts per million may be sufficient to cause injury. Symptoms of exposure include nausea and vomiting, headache, burning of eyes, throat, or both, drowsiness, abdominal pain or discomfort, weakness, and muscle stiffness and soreness. Prolonged or repeated exposure to carbon tetrachloride vapor or liquid may result in subacute or chronic poisoning. Consequently a threshold limit value of 10 ppm by volume of carbon tetrachloride in air has been established as a maximum safe concentration for daily 8-h exposure.

Occasional brief contacts of liquid carbon tetrachloride with unbroken skin do not produce irritation, although the skin may feel dry owing to removal of natural oils. Prolonged and repeated contacts may cause dermatitis, cracking of the skin, and danger of secondary infection. Carbon tetrachloride is apparently absorbed through the skin, but at such a slow rate that there is no significant hazard of systemic poisoning in normal industrial operations.

All persons who have occasion to use or handle carbon tetrachloride should be thoroughly instructed and adequately supervised in the proper methods of handling the substance to prevent or minimize exposure to the liquid or its vapors (52). In most situations adequate, usually forced, ventilation is necessary to prevent excessive exposure. Persons who use alcohol excessively or have liver, kidney, or heart diseases should be excluded from any exposure to carbon tetrachloride. All individuals regularly exposed to carbon tetrachloride should receive periodic examinations by a physician acquainted with the occupational hazard involved. These examinations should include special attention to the kidneys and the liver. There is no known specific antidote for carbon tetrachloride poisoning. Treatment is symptomatic and supportive. Alcohol, oils, fats, and epinephrine should not be given to any person who has been exposed to carbon tetrachloride. Following exposure, the individual should be kept under observation long enough to permit the physician to determine whether liver or kidney injury has occurred. Artificial dialysis (qv) may be necessary in cases of severe renal failure.

Handling and Storage

Carbon tetrachloride is not subject to ICC Regulations for transport but is, nonetheless, usually shipped in ICC-specification containers (52). Galvanized metal drums or barrels with 18.9-L (5-gal) or 208-L (55-gal) capacity, which conform to ICC Specification 17E STC, are normally used. Smaller quantities of carbon tetrachloride are shipped in nonreturnable tinned steel cans holding 3.79 L (1 gal). For bulk shipments, carbon tetrachloride may be transported in railroad tank cars with capacities of 30 m^3 (8,000 gal), 38 m^3 (10,000 gal), or 53 m^3 (14,000 gal), meeting ICC Specifications 103 and 103W, or tank trucks meeting ICC specifications MC 303 and MC 304. Any other safe and suitable containers conforming to the requirements of rail and motor classification tariffs (ie, lined tank cars and drums) can also be used.

Uses

Carbon tetrachloride was formerly used for metal degreasing and as drycleaning fluid, fabric-spotting fluid, fire-extinguisher fluid, grain fumigant, and reaction medium. However, as its toxicity became recognized, it was replaced by less-toxic chlorinated hydrocarbons in metal and fabric cleaning applications. During the 1950s the demand for carbon tetrachloride as a raw material in the manufacture of chlorofluorocarbons increased and the net result was continued growth for the product. In 1970, carbon tetrachloride was banned from all use in consumer goods in the United States. Its current principal applications include chlorofluorocarbon production, grain fumigation, and use as a reaction medium.

For more than 20 years, the principal use for carbon tetrachloride has been the production of the fluorocarbon gases: trichloromonofluoromethane (fluorocarbon 11, or F-11) and dichlorodifluoromethane (fluorocarbon 12, or F-12). Fluorocarbons 11 and 12 are made by the catalytic reaction of hydrogen fluoride with carbon tetrachloride. F-11 and F-12 are widely used as aerosol propellants in numerous personal-care and convenience products such as colognes and perfumes, deodorants, hair sprays, and shaving creams. F-12 is also used extensively as a refrigerant. Of the carbon tetrachloride currently produced, 95% is used in the manufacture of fluorocarbons 11 and 12. Aerosols (qv) which account for approximately 50% of the fluorocarbon 11 and 12 market, reached the peak of their popularity in 1973. The FDA and the EPA have ended (with a few exceptions) the manufacture of fluorocarbons for the aerosol market on October 15, 1978. Use of a fluorocarbon gas as a spray-can propellant was banned as of December 15, 1978, and spray cans containing these gases would be kept off the market after April 15, 1979. The two agencies have published the proposed regulations (53).

BIBLIOGRAPHY

"Carbon Tetrachloride" in *ECT* 1st ed., Vol. 3, pp. 191–200, by P. S. Brallier, Stauffer Chemical Company; "Carbon Tetrachloride" in *ECT* 2nd ed., Vol. 5, pp. 128–139, by D. W. F. Hardie, Imperial Chemical Industries Ltd.

1. V. L. Simril and A. Hershberger, *Mod. Plast.* **27**(10), 97 (1950).
2. H. F. Coward and G. W. Jones, *Ind. Eng. Chem.* **18,** 970 (1926).
3. J. Duchesne, *Trans. Faraday Soc.* **46,** 187 (1950).
4. W. B. Crammett and V. A. Stenger, *Ind. Eng. Chem.* **48,** 434 (1956).
5. E. H. Lyons and R. G. Dickens, *J. Am. Chem. Soc.* **57,** 445 (1935).
6. M. Stern and H. H. Uhlig, *J. Electrochem. Soc.* **100,** 543 (1953).
7. E. W. Lindeijer, *Chem. Weekblad* **46,** 571 (1950).
8. *Chem. Age* **63,** 6 (1950).
9. H. Staudinger, *Z. Angew. Chem.* **35,** 659 (1922); **38,** 578 (1925).
10. U.S. Pat. 2,525,589 (Sept. 25, 1947), O. W. Cass (to E. I. du Pont de Nemours & Co., Inc.).
11. Brit. Pat. 576,189 (Apr. 4, 1944), J. H. Brown and W. B. Whalley (to Imperial Chemical Industries, Ltd.).
12. U.S. Pat. 2,443,630 (June 22, 1948), E. T. McBee and Z. D. Welch (to Purdue Research Foundation).
13. A. J. Rudge, *The Manufacture and Use of Fluorine and Its Compounds*, Oxford University Press, Inc., New York, 1962, p. 67.
14. U.S. Pat. 2,770,661 (Aug. 11, 1953), T. Horlenko, F. Marcote, and O. V. Luke (to Celanese Corp. of America).
15. Brit. Pat. 627,993 (Sept. 25, 1946), (to United States Rubber Co.).
16. R. M. Joyce, W. E. Hanford, and J. Harmon, *J. Am. Chem. Soc.* **70,** 2529 (1948).
17. Can. Pat. 440,898 (Apr. 22, 1947), J. Harmon.
18. M. Gomberg, *Ber. Deut. Chem. Ges.* **33,** 3144 (1900).
19. R. K. Taylor, *J. Am. Chem. Soc.* **75,** 2521 (1953).
20. U.S. Pat. 2,676,998 (Apr. 27, 1954), T. F. Kunz (to Diamond Alkali Co.).
21. *Chem. Eng.* **64,** 260 (1957).
22. H. A. Taylor and W. E. Hanson, *J. Chem. Phys.* **7,** 418 (1939).
23. E. Ger. Pat. 10,545 (Oct. 7, 1952), B. Hennig and A. Pessel.
24. Brit. Pat. 513,235 (Apr. 4, 1938), (to I.G. Farbenindustrie A.G.).
25. Ger. Pat. 887,809 (Sept. 16, 1941), O. Fruhwirth (to Donau Chemie).
26. Belg. Pat. 541,591 (Sept. 26, 1955), (to Chempatents Inc.).
27. Belg. Pat. 541,961 (Oct. 10, 1955), (to Chempatents, Inc.).
28. Can. Pat. 505,152 (Aug. 17, 1954), R. G. Heitz and W. E. Brown (to The Dow Chemical Co.).
29. H. Riegel, H. D. Schindler, and M. C. Sze, *AICHE Meeting, New Orleans, La., Mar. 13, 1973.*
30. *Chem. Eng.* **81,** 114 (June 24, 1974).
31. E. T. McBee and L. W. Devaney, *Ind. Eng. Chem.* **41,** 803 (1949).
32. Fr. Pat. 1,031,823 (June 26, 1953), (to Deutsche Solvay-Werke).
33. F. S. Dainton and K. J. Ivin, *Trans. Faraday So.* **46,** 925 (1950).
34. Brit. Pat. 749,408 (Jan. 30, 1953), (to Société d'Electrochimie, d'Electro-Metallurgie et des Aciéries Electriques d'Uzine).
35. Ger. Pat. 680,659 (Oct. 4, 1937), M. Mugdan (to Consortium fur Elecktrochemische Industrie).
36. Can. Pat. 516,519 (Sept. 23, 2955), J. B. Lovell and M. C. Fuqua (to Esso Research and Engineering Corp.).
37. Can. Pat. 515,658 (Aug. 16, 1955), P. A. Florio and J. D. Calfee (to Allied Dye and Chemical Corp.).
38. E. H. Waters, *Chem. Ind. London,* 742 (1953).
39. U.S. Pat. 2,043,257 (Oct. 9, 1933), E. C. Missbach (to Stauffer Chemical Co.).
40. U.S. Pat. 2,094,368 (Jan. 23, 1935), E. C. Missbach (to Stauffer Chemical Co.).
41. U.S. Pat. 2,002,168 (Apr. 5, 1933), P. S. Brallier (to Niagara Smelting Corp.).
42. U.S. Pat. 2,043,260 (Oct. 9, 1933), E. C. Missbach (to Stauffer Chemical Co.).
43. O. F. A. Biginelli, *Chim. Ind.* **68,** 748 (1952).
44. U.S. Pat. 2,043,259 (Oct. 9, 1933), E. C. Missbach (to Stauffer Chemical Co.).
45. U.S. Pat. 2,387,784 (Oct. 23, 1945), E. O. Ohlmann (to The Dow Chemical Co.).
46. U.S. Pat. 2,043,258 (Oct. 9, 1933), E. C. Missbach (to Stauffer Chemical Co.).

47. U.S. Pat. 2,069,711 (Jan. 23, 1935), E. C. Missbach (to Stauffer Chemical Co.).
48. *Chemical Products Synopsis,* Mansville Chemical Products, Mansville, N.Y., Dec. 1976.
49. H. F. Smyth, *J. Ind. Hyg. Toxicol.* **18,** 277 (1936).
50. H. B. Elkins, *The Chemistry of Industrial Toxicology,* John Wiley & Sons, Inc., New York, 1950, p. 132.
51. *Health Hazards and Precautions for the Safe Handling and Use of Carbon Tetrachloride,* unpublished report, Biochemical Research Laboratory, The Dow Chemical Company, Midland, Mich., Sept. 1966.
52. *Manufacturing Chemists' Association Manual,* sheet SD-3, 1963.
53. *Fed. Reg.,* 113.1 (Mar. 17, 1978).

H. D. DeShon
Dow Chemical U.S.A.

ETHYL CHLORIDE

Ethyl chloride [75-00-3] (chloroethane), C_2H_5Cl, is a colorless, mobile liquid of bp 12.4°C, a nonirritant ethereal odor, and a pleasant taste. It is flammable and burns with a green-edged flame, producing hydrogen chloride fumes. Ethyl chloride is used principally in the manufacture of tetraethyllead (TEL), the antiknock additive to motor fuel, but also serves as an ethylating agent, solvent, refrigerant, and local and general anesthetic. It is less toxic than the chloromethanes.

Valentine, in the fifteenth century, produced ethyl chloride from hydrochloric acid and ethyl alcohol. Glauber prepared it in 1648 by digesting rectified spirit of wine with zinc chloride. Reynoso, in 1856, produced a mixture of ethyl chloride and diethyl ether by heating a mixture of alcohol and hydrochloric acid to 100°C in a sealed tube.

Ethyl chloride became an important large-volume chemical in 1922 when manufacture of tetraethyllead began in the United States. It had previously been manufactured primarily for use as an anesthetic and refrigerant; annual production had not exceeded several hundred metric tons in any of the producing countries. Use of ethyl chloride as a starting material for TEL makes it an automotive chemical, and its subsequent expansion is linked with the growth of the automobile industry. Prior to World War II, annual output of ethyl chloride in the United States exceeded 23,000 t, but only 230–275 t was used for purposes other than the manufacture of TEL. During the war, ethyl chloride production increased approximately five-fold. United States output in the postwar years has more than doubled. Since 1960, world production of ethyl chloride has remained fairly constant. However, production in the United States is expected to decrease until the mid 1980s owing to the requirement for unleaded fuel use in late model cars. Currently, 90% of the ethyl chloride produced is used for TEL manufacture. As there is no significant secondary market at this time, the future for ethyl chloride demand is dim.

Physical and Chemical Properties

The physical properties of ethyl chloride are listed in Table 1.

At 0°C, 100 g ethyl chloride dissolves 0.07 g water and 100 g water dissolves 0.447 g ethyl chloride. The solubility of water in the chloride increases sharply with temperature to 0.36 g/100 g at 50°C. Ethyl chloride dissolves many organic substances, such as fats, oils, resins, and waxes, and it is also a solvent for sulfur and phosphorus. It is miscible with methyl and ethyl alcohols, diethyl ether, ethyl acetate, methylene chloride, chloroform, carbon tetrachloride, and benzene.

Three binary azeotropes of ethyl chloride have been reported but the data are uncertain (3).

The C—Cl bond in ethyl chloride (0.176 nm) is slightly shorter than the corresponding bond in methyl chloride (0.1786 nm). Ethyl chloride displays thermal stability similar to that of methyl chloride. It is practically unchanged on heating to 400°C when decomposition to ethylene and hydrogen chloride begins (4). This decomposition as nearly complete at 500–600°C on pumice or about 300°C with the chlorides of nickel, cobalt, iron, and lead (but not with chlorides of sodium, potassium, and silver). Several inorganic salts (eg, lithium chloride, calcium sulfate), metals (eg, platinum, iridium), and oxides (eg, aluminum oxide and silica) also catalyze the cracking of ethyl chloride (5–6).

Ethyl chloride can be dehydrochlorinated to ethylene using alcoholic potash. Condensation of alcohol with ethyl chloride in this reaction also produces some diethyl ether. Heating to 625°C and subsequent contact with calcium oxide and water at 400–450°C gives ethyl alcohol as the chief product of decomposition. Ethyl chloride yields butane, ethylene, water, and a solid of unknown composition when heated with metallic magnesium for about 6 h in a sealed tube.

Ethyl chloride burns with a green-edged flame, producing hydrogen chloride fumes, carbon dioxide, and water.

It forms regular crystals of a hydrate with water at 0°C (7). Dry ethyl chloride can be used in contact with most common metals in the absence of air up to 200°C. Its oxidation and hydrolysis are slow at ordinary temperatures. Ethyl chloride yields ethyl alcohol, acetaldehyde, and some ethylene in the presence of steam with various catalysts, eg, titanium dioxide and barium chloride.

When ethyl chloride is chlorinated under light, both ethylidene and ethylene chlorides are formed; the latter in smaller quantity (8). Chlorination in the presence of antimony pentachloride at 100°C produces ethylene chloride almost exclusively. Photochemical bromination of ethyl chloride yields a series of bromochloroethanes, eg, $CH_3CHBrCl$, CH_3CBr_2Cl, $CHBr_2CBr_2Cl$ (9). In contact with iron wire at 100°C, ethyl chloride can be brominated to ethyl bromide and ethylene bromide. When a mixture of ethyl chloride and ethylene bromide is maintained for a prolonged period at 25°C in the presence of aluminum chloride, a redistribution of halogen atoms occurs forming ethyl bromide, ethylene chlorobromide, and ethylene chloride (10). Hydriodic acid reacts at 130°C with ethyl chloride to give ethyl iodide. Vapor-phase fluorination with nitrogen-diluted fluorine below 60°C and in contact with copper gauze results in formation of carbon tetrafluoride, monochlorotrifluoromethane, dichlorodifluoroethylene, and several chlorofluoroethanes (11). The chlorine in ethyl chloride can be replaced by fluorine by reaction with hydrogen fluoride in the presence of antimony fluoride (see Fluorine compounds, organic).

Table 1. Physical Properties of Ethyl Chloride

Property	Value
melting point, °C	−138.3
boiling point at 101 kPa[a], °C	12.4
surface tension, air, mN/m (= dyn/cm)	
5°C	21.20
10°C	20.64
specific gravity, vapor at 101 kPa[a] (air = 1)	2.23
specific gravity, liquid	
0/4°C	0.92390
20/4°C	0.8970
−20°C	1.3913
0°C	1.3798
refractive index of vapor, n_D^{25}	1.001
specific heat, liquid from −48.4 to 45°C[b], J/(kg·K)[c]	$1612 + 2.72t + 1.46 \times 10^{-2} t^2$
specific heat, vapor at 101 kPa[a], 40°C, J/mol[c]	1.017
critical temperature, °C	186.6
critical pressure, MPa[d]	5.27
thermal conductivity, W/(m·K)	
liquid	0.1467
vapor at bp	0.0095
coefficient of volume expansion, 0–15°C, av	0.00156
dielectric constant ϵ	
liquid, 170°C	6.29
vapor, 23.5°C	1.01285
dipole moment, C·m[e]	6.672×10^{-30}
heat of combustion, kJ/mol[c]	1327
heat of formation, kJ/mol[c]	
liquid	132.4
vapor	107.7
latent heat of evaporation at bp, J/g[c]	383.4
latent heat of fusion, J/g[c]	69.09
heat of adsorption, on homogeneous carbon, kJ/mol[c,f]	10.47
flash point, °C	
open cup	−43
closed cup	−50
ignition temperature, °C	519
explosive limits in air, vol %	3.16–15
explosive limits in oxygen, vol %	4.0–67.2
explosive limits with nitrous oxide (N_2O), vol %	2.1–32.8
viscosity, mPa·s (= cP)	
liquid, 5°C	0.292
20°C	0.260
vapor, 12.4°C	0.093×10^{-3}
35°C	0.0165×10^{-3}
vapor pressure, kPa[g]	
−30°C	15.2
−10°C	40.5
0°C	61.86
10°C	92.3
20°C	134.8
60°C	456.0
100°C	1165

[a] To convert kPa to mm Hg, multiply by 7.5.
[b] For example, specific heat at −30°C, 1542.5 J/kg; at 20°C, 1672 J/kg (1).
[c] To convert J to cal, divide by 4.184.
[d] To convert MPa to atm, divide by 0.101.
[e] To convert C·m to debye, divide by 23.336×10^{-30}.
[f] Ref. 2.
[g] To convert kPa to mm Hg, multiply by 7.5.

Reaction of ethyl chloride with an alcoholic solution of ammonia yields ethylamine, diethylamine, triethylamine, and tetraethylammonium chloride (12–13) (see Amines, lower aliphatic).

In the presence of Friedel-Crafts catalysts, gaseous ethyl chloride reacts with benzene at about 25°C to give ethylbenzene, three diethylbenzenes, and other more complex compounds (14) (see Xylenes and ethylbenzene). Aromatic compounds can generally be ethylated by ethyl chloride in presence of anhydrous aluminum chloride (see Friedel-Crafts reactions).

Ethyl chloride combines directly with sulfur trioxide to give ethyl chlorosulfonate, $C_2H_5OSO_2Cl$, and 2-chloroethylsulfonic acid, $CH_2ClCH_2SO_2OH$ (15).

Manufacture

There are three industrial processes for the production of ethyl chloride hydrochlorination of ethylene, reaction of hydrochloric acid with ethanol, and chlorination of ethane. About 90–95% of the ethyl chloride is produced in the United States by the hydrochlorination of ethylene. Ethanol has not been used in the United States since about 1972 because of its prohibitive cost. Thermal chlorination of ethane has the disadvantage of producing undesired by-products.

Hydrochlorination of Ethylene. The exothermic vapor-phase reaction between ethylene and hydrogen chloride can be carried out at 130–250°C under a variety of catalytic conditions.

$$CH_2=CH_2 + HCl \rightleftharpoons C_2H_5Cl \quad -56.1 \text{ kJ } (-13.4 \text{ kcal})$$

At 200–250°C equilibrium conversion falls off; nevertheless, the process is usually conducted at the higher temperature to achieve a practical rate of reaction. The higher temperature accelerates the reaction, but also causes the formation of polymerization products, which ultimately destroy the catalyst. In the United States, ethyl chloride is produced mainly by reaction of ethylene and hydrogen chloride under 0.1–0.3 MPa (1–3 atm) pressure at normal temperature in a 2% solution of aluminum chloride in ethyl chloride (16–18). Other variations are reaction at 175–400°C in contact with a thorium salt, eg, thorium oxychloride on silica gel (19); use of 1,1,2-trichloroethane as a solvent and aluminum chloride catalyst for reaction at −5 to 55°C at 0.1–0.9 MPa (1–9 atm) pressure (20); and reaction at high pressures in contact with a peroxygen catalyst (21). Use of ^{60}Co gamma radiation also produces ethyl chloride and n-butyl chloride from ethylene and hydrogen chloride (22).

The hydrogen chloride needed for hydrochlorination of ethylene may be a by-product of other chlorocarbon processes such as the cracking of 1,1,2,2-tetrachloroethane to trichloroethylene (23). In one important form of this tandem procedure (24–25), two C_2 gas streams are supplied to the process: one rich in ethane, the other rich in ethylene. Chlorination of the ethane-rich stream is carried out at high temperature. This stage is noncatalyzed. Hydrogen chloride from the ethane chlorination along with unreacted ethylene is passed to the reactor where the ethylene-rich stream is hydrochlorinated under pressure at a lower temperature in the presence of a granular catalyst. Ethylene dichloride is a by-product of the process.

Chlorination of Ethane. Ethane may be chlorinated thermally, catalytically, photochemically, or electrolytically. Monochlorination is favored because ethyl chloride chlorinates at about one quarter of the rate at which it is itself produced from ethane.

Thermal chlorination of ethane is generally carried out at 250–500°C. At ca 400°C, a free radical chain reaction takes place:

$$Cl_2 \rightarrow 2\ Cl\cdot$$

$$Cl\cdot + C_2H_6 \rightarrow HCl + C_2H_5\cdot$$

$$C_2H_5\cdot + Cl_2 \rightarrow C_2H_5Cl + Cl\cdot \text{ (chain carrier)}$$

The chlorine and ethane are brought together in a fluid bed of finely divided, inert, solid heat-transfer medium, eg, sand, at 380–440°C; the linear velocity of the gas is sufficient to maintain the finely divided solid in suspension within the reactor (26). The reaction may be carried out under pressure, eg, 207 kPa (30 psi). Cracking of the ethyl chloride results in production of some ethylene which is removed by subsequent chlorination to ethylene chloride at less than 55°C in a subsidiary reactor containing cupric chloride–alumina catalyst. Ethyl chloride is condensed by compression, and unreacted ethane and ethylene are returned to the main reactor. For example, when an ethane-to-chlorine ratio (by volume) of 5:1 was employed at an average temperature of 420°C, a yield of 78% ethyl chloride was obtained (26).

The chlorination of ethane may be catalyzed by bringing the reacting gases into contact with metal chlorides (27), or crystalline carbon (graphite) (28). Photochemical chlorination is not used industrially. Electrolytic chlorination, which involves passing a low voltage current through a catalytic mixture of $AlCl_3$–$NaAlCl_4$, has not been used on a large scale (29).

Reaction of Ethyl Alcohol and Hydrochloric Acid. For many years this reaction was the only established technical process for ethyl chloride but it was abandoned because of the high cost of alcohol when petrochemicals became available. Zinc chloride was used as the catalyst at 110–140°C and ethyl chloride and water were continuously distilled from the concentrated catalyst solution.

$$C_2H_5OH + HCl \rightarrow H_2O + C_2H_5Cl$$

In a variation (30), ethyl alcohol reacts with a 10% hydrochloric acid solution in the presence of a metallic chloride catalyst and a volatile liquid, eg, benzene, cyclohexane, trichloroethylene, or carbon tetrachloride. The mixture is heated to distill a ternary azeotrope. When condensed, the azeotrope separates into an aqueous layer containing unreacted alcohol, and a nonaqueous layer containing the azeotroping liquid. Ethyl chloride is recovered by fractional distillation; alcohol and the azeotroping liquid are recycled (see Azeotropic and extractive distillation).

In another patented variation (31), equimolecular parts of concentrated hydrochloric acid and ethyl alcohol are mixed with a 70% aqueous zinc chloride solution and passed into additional 70% zinc chloride solution.

From Diethyl Sulfate. Several processes have been proposed for manufacture of ethyl chloride based on the reaction,

$$(C_2H_5O)_2SO_2 + 2\ HCl \rightarrow 2\ C_2H_5Cl + H_2SO_4$$

In one (32), ethylene reacts with sulfuric acid in the presence of an antimony, tin, or bismuth catalyst to give diethyl sulfate, which subsequently reacts with hydrochloric acid at 40–110°C. Diethyl sulfate can react with sodium chloride to yield sodium sulfate and ethyl chloride, or alternatively, ethane may react with sulfur trioxide and sodium chloride (33). Ethyl chloride is produced when ethylene and hydrogen chloride are passed into an anhydrous liquid bath of ethyl hydrogen sulfate containing a bismuth catalyst (34). None of these processes has been commercialized.

Other Processes For Ethyl Chloride. 1,2-Dichloroethane and ethylene, after at least 3 min in the presence of anhydrous calcium sulfate at 250–350°C and pressures of 0.69–2.76 MPa (100–400 psi), yield a mixture of ethyl chloride and vinyl chloride (35).

$$CH_2ClCH_2Cl + CH_2{=}CH_2 \rightarrow C_2H_5Cl + CH_2{=}CHCl$$

Vinyl chloride can be reduced to ethyl chloride at elevated temperatures by reaction with excess hydrogen in contact with a hydrogenation catalyst (36).

In the presence of a heavy-metal chloride and water at 80–240°C, and under sufficient pressure to maintain the water in the liquid phase, diethyl ether reacts with hydrochloric acid to give ethyl chloride (37).

Economic Aspects

United States production, sale, and price statistics for ethyl chloride from 1956–1976, inclusive, are given in Table 2. The economic history of ethyl chloride is entirely dominated by the fact that its principal application is in the manufacture of tetraethyllead (TEL). Use of jet engines in aircraft, more widespread use of diesel engines in transportation, and the development of processes for increasing the octane content of gasoline have reduced the amount of TEL employed. Consequently, the steady increase in demand for ethyl chloride, which prevailed up to 1957, has stopped. Growth from 1965–1975 exhibited an average increase of 0.5%/yr. Projections of future growth indicate an average annual decline of 30–50% through 1980 (39).

The only important demand for ethyl chloride, other than its use in TEL manufacture, arises from the ethyl cellulose industry (see Cellulose derivatives). This demand, together with that generated by a few chemical, pharmaceutical, and solvent applications has caused a moderate increase during the postwar period.

Table 2. Ethyl Chloride Production and Price Statistics[a]

Year	Production	Sales quantity[b]	Total value[c]	Unit value, $/kg
1956	293.4	66.3	10,302	0.154
1960	247.9	86.7	14,606	0.176
1965	311.7	124.5	19,447	0.154
1970	308.2	124.0	17,998	0.132
1971	282.0	111.3	15,674	0.132
1972	261.6	88.2	13,192	0.132
1973	300.0	128.0	20,557	0.154
1974	301.1	127.0	25,137	0.198
1975	261.5	127.3	28,000	0.202
1976	304.0[d]			

[a] According to U.S. Tariff Commission Report.
[b] In thousands of metric tons.
[c] In thousands of dollars.
[d] Ref. 38.

Standards

Good technical-grade ethyl chloride should not contain more than the following quantities of the indicated impurities: water, 15 ppm; acid (as HCl), 120 ppm; residue on evaporation at 110°C, 50 ppm. Ethyl chloride does not require added stabilizers.

Health and Safety Factors (Toxicology)

Ethyl chloride is handled and transported in pressure containers under conditions similar to those applied to methyl chloride (qv). In the presence of moisture ethyl chloride can be moderately corrosive. Carbon steel is preferred for storage vessels and prolonged contact with copper should be avoided.

Ethyl chloride is readily absorbed into the body through mucous membranes, lungs, and skin. Although rapidly excreted by the lungs, its high solubility in blood prolongs total elimination from the body (40). Ethyl chloride is apparently not metabolized to any significant degree (41). Recovery of conciousness after exposure to ethyl chloride often entails an unpleasant hangover period (40). Experiments with animals provide evidence of kidney irritation and promotion of fat accumulation in the kidneys, cardiac muscle, and liver. Concentrations of 15–30 vol % in air are quickly fatal to animals; a concentration of 2% causes some unsteadiness; exposure to 1% concentration has no observable effect. A recent NIOSH report (38) suggests that exposure levels should be reduced to as low a level as possible, rather than the currently used 1000 ppm level, since it may be a human carcinogen on the basis of National Cancer Institute studies showing a statistically significant excess of cancer in laboratory animals.

Uses

For the manufacture of tetraethyllead by reaction of ethyl chloride with lead–sodium alloy (4 PbNa + 4 $C_2H_5Cl \rightarrow Pb(C_2H_5)_4$ + 3 Pb + 4 NaCl), see Lead compounds.

Ethyl cellulose, produced by the reaction of ethyl chloride with soda cellulose, is used mainly in the plastics and lacquer industries (42) (see Cellulose derivatives).

Ethyl chloride is used to some extent as an ethylating agent in the synthesis of dyestuffs and fine chemicals. Benzene can be ethylated by the reaction with ethyl chloride in the presence of a Friedel-Crafts catalyst (see Alkylation; Friedel-Crafts reactions). In one process (43), the hydrogen chloride liberated from the Friedel-Crafts reaction reacts with ethylene to produce more ethyl chloride, which is recycled to the main reactor. Ethylbenzene for production of styrene (qv), used in high tonnage in the manufacture of polystyrene (see Styrene plastics), is normally made by reaction of ethylene with benzene, or by reforming petroleum cycloparaffins (see BTX processes; Xylenes and ethylbenzene). Ethyl chloride is used as a solvent in the polymerization of olefins using Friedel-Crafts catalysts (44–45), and as a polymerization activator to produce polyquinoline from quinoline at high temperature (121–160°C) (46).

Ethyl chloride can also be used as a feedstock to produce 1,1,1-trichloroethane via thermal chlorination at temperatures of 375–475°C (47), or via a fluidized-bed reactor at similar temperatures (48).

Limited quantities of ethyl chloride are used in the production of aerosols (qv), as a refrigerant, and as a local and general anesthetic. When inhaled alone, ethyl chloride has an analgesic action similar to that of nitrous oxide (49); ethyl chloride may be used as an adjuvant to nitrous oxide in general anesthesia (50) (see Anesthetics; Refrigeration).

For a number of years, ethyl chloride has been used as the working substance at an electric power plant on the island of Ischia, in the Bay of Naples, where there is a temperature difference of over 30°C between the local thermal springs and the surrounding sea water (51) (see Heat exchange technology).

BIBLIOGRAPHY

"Ethyl Chloride" under "Chlorine Compounds, Organic" in *ECT* 1st ed., Vol. 3, pp. 751–760, by Robert Herzog, I. M. Skinner, G. W. Thomson, and Hymin Shapiro, Ethyl Corporation; "Ethyl Chloride" under "Chlorocarbons and Chlorohydrocarbons" in *ECT*, 2nd ed., Vol. 5, pp. 140–147, by D. W. F. Hardie, Imperial Chemical Industries, Ltd.

1. L. Riedel, *Z. Gesamte Kalte Ind.* **47,** 87 (1940).
2. C. Pierce and R. N. Smith, *J. Am. Chem. Soc.* **75,** 846 (1953).
3. L. H. Horsley and co-workers, *Azeotropic Data*, Vols. 1 and 2, Nos. 6 and 35 of *Advances in Chemistry Series,* American Chemical Society, Washington, D.C., 1952 (Vol. 1) and 1962 (Vol. 2).
4. D. H. R. Barton and K. E. Howlett, *J. Chem. Soc.,* 165 (1949).
5. G. M. Schwab and H. Noller, *Z. Electrochem.* **58,** 762 (1954).
6. A. Heinzelmann, R. Letterer, and H. Noller, *J. Monatsh. Chem.* **102,** 1750 (1971).
7. P. Villard, *Ann. Phys.* **11,** 384 (1897).
8. J. D'Ans and J. Kautzsch, *J. Prakt. Chem.* **80,** 310 (1909).
9. J. Denzel, *Ann. Chem. Liebigs* **195,** 189 (1879).
10. G. Calingaert and co-workers, *J. Am. Chem. Soc.* **62,** 1546 (1940).
11. L. A. Bigelow, *Chem. Revs.* **40,** 51 (1947).
12. C. E. Groves, *J. Chem. Soc.* **13,** 331 (1860).
13. A. W. Hofmann, *Ber. Dtsch. Chem. Ges.* **3,** 109 (1870).
14. M. Blau and J. E. Willard, *J. Am. Chem. Soc.* **75,** 330 (1953).
15. Th. Von Purgold, *Z. Chem. Ind. U.S.S.R.* **8,** 669 (1868).
16. Can. Pat. 448,020 (Apr. 20, 1948), E. V. Fasce (to Standard Oil Development Co.).
17. U.S. Pat. 3,345,421 (Nov. 24, 1961), M. D. Brown (to Halcon International, Inc.).
18. R. V. Chandhari and L. K. Doraiswamy, *Chem. Eng. Sci.* **29,** 349 (1974).
19. Can. Pat. 464,069 (Mar. 28, 1950), D. C. Bond and M. Savoy (to The Pure Oil Co.).
20. U.S. Pat. 2,140,927 (Sept. 29, 1936), J. E. Pierce (to The Dow Chemical Co.).
21. Can. Pat. 464,488 (Apr. 18, 1950), W. E. Hanford and J. Harman (to Canadian Industries Ltd.).
22. A. Terakawa, J. Nakanishi, and T. Kiruyama, *Bull. Chem. Soc. Jpn.* **39,** 892 (1966).
23. Brit. Pat. 505,196 (Nov. 5, 1937), A. A. Levine (to E. I. du Pont de Nemours & Co., Inc.).
24. U.S. Pat. 2,246,082 (Aug. 22, 1939), W. E. Vaughn and F. F. Rust (to Shell Development Co.).
25. *Ind. Eng. Chem.* **47,** 984 (1955); *Pet. Refiner* **34,** 149 (1955).
26. Brit. Pat. 667,185 (Mar. 9, 1949), P. A. Hawkins and R. T. Foster (to Imperial Chemical Industries, Ltd.).
27. U.S. Pat. 2,140,547 (Aug. 26, 1936), J. H. Reilly (to The Dow Chemical Co.).
28. Brit. Pat. 483,051 (Oct. 8, 1936), G. W. Johnson (to I. G. Farbenindustrie A.G.).
29. Belg. Pat. 654,985 (Apr. 28, 1965), (to Imperial Chemical Industries, Ltd.).
30. U.S. Pat. 2,516,638 (Mar. 3, 1947), J. L. McCrudy (to The Dow Chemical Co.).
31. Fr. Pat. 858,724 (Dec. 2, 1940), (to Société anon. des Matiéres colorantes et Produits chimiques de Saint-Denis).
32. Brit. Pat. 566,147 (Dec. 28, 1942), E. G. Galitzenstein and C. Woolf (to Distillers Co. Ltd.).
33. A. P. Giraitis, *Erdöl Kohle* **9,** 791 (1951).
34. U.S. Pat. 2,125,284 (Nov. 11, 1935), L. C. Chamberlain and J. L. Williams (to The Dow Chemical Co.).
35. U.S. Pat. 2,681,372 (Jan. 16, 1951), P. W. Trotter (to Ethyl Corp.).

36. Brit. Pat. 470,817 (Feb. 18, 1936), G. W. Johnson (to I. G. Farbenindustrie A.G.).
37. U.S. Pat. 2,084,710 (Aug. 21, 1935), H. N. Spurlin (to Hercules Powder Co.).
38. *Chem. Week,* 21 (Sept. 6, 1978).
39. *Chemical Profiles, Ethyl Chloride,* Schnell Publishing Co., New York, Apr. 1, 1977.
40. J. I. Murray Lawson, *Br. J. Anaesth.* **37,** 667 (1965).
41. F. A. Patty, P. Irish, and D. Fassett, eds., *Industrial Hygiene and Toxicology,* 2nd ed., Vol. 2, Interscience Publishers Inc., New York, 1963, p. 1275.
42. S. L. Bass, A. J. Barry, and A. E. Young in E. Ott, ed., *Cellulose and Cellulose Derivatives,* Interscience Publishers Inc., New York, 1946, p. 758 ff.
43. Brit. Pat. 581,145 (July 30, 1942), (to Standard Oil Development Co.).
44. U.S. Pat. 2,387,784 (Dec. 28, 1940), R. M. Thomas and H. C. Reynolds (to Standard Oil Development Co.).
45. G. P. Below and co-workers, *Kinet. Katal.* **8,** 265 (1967).
46. R. F. Smirnov, B. I. Tikhomirov, and A. I. Yakubehik, *Vysokomol. Soedin Ser. B* **13,** 395 (1971).
47. U.S. Pat. 3,706,816 (Sept. 22, 1969), A. Campbell and R. A. Carruthers (to Imperial Chemical Industries, Ltd.).
48. U.S. Pat. 3,012,081 (Sept. 29, 1960), F. Conrad and A. J. Haefner (to Ethyl Corp.).
49. J. D. Rochford and B. T. Broadbent, *Br. Med. J.* **7,** 664 (1943).
50. C. W. Lincoln, *Anesth. Analg.* **20,** 328 (1941).
51. *Chem. Age* **58,** 114 (1948).

THOMAS E. MORRIS
WILLIAM D. TASTO
Dow Chemical U.S.A.

OTHER CHLOROETHANES

1,1-DICHLOROETHANE

1,1-Dichloroethane [*75-34-3*], ethylidene chloride, ethylidene dichloride, $CH_3\text{-}CHCl_2$, is a colorless liquid with an ethereal odor, miscible with most organic solvents including other chlorinated solvents. It is employed as a solvent, but its largest industrial use is as an intermediate in the production of 1,1,1-trichloroethane.

Physical and Chemical Properties

The properties of 1,1-dichloroethane are listed in Table 1.

1,1-Dichloroethane decomposes at 356–453°C by a homogeneous first-order reaction to give vinyl chloride and hydrogen chloride (1–2). Dehydrochlorination on activated alumina (3–4), and magnesium sulfate or potassium carbonate (5) has been studied. Dehydrochlorination in the presence of anhydrous aluminum chloride (6) proceeds readily.

Table 1. Properties of 1,1-Dichloroethane

Property	Value
melting point, °C	−96.7
boiling point, °C	57.3
density at 20°C, g/L	1.1747
n_D^{20}	1.4166
viscosity at 20°C, mPa·s (= cP)	0.377
surface tension at 20°C, mN/m (= dyn/cm)	23.34
specific heat at 20°C, J/(g·K)[a]	
liq	1.087
gas	0.824
latent heat of vapor at 20°C, J/g[a]	280.3
critical temperature, °C	261.5
critical pressure, MPa[b]	5.06
flash point (closed cup), °C	−12.0
explosive limits in air at 25°C, % by vol	5.4–11.4
autoignition temperature, °C	458
heat of combustion, kJ/g[a]	12.57
dielectric constant, liq at 20°C	10.9
vapor pressure, kPa[c]	
10°C	15.37
20°C	24.28
30°C	36.96
solubility at 20°C, g	
dichloroethane in 100 g H$_2$O	0.55
H$_2$O in 100 g dichloroethane	0.097
binary azeotropes, bp, °C	
with 1.9% H$_2$O[c]	53.3$_{97\ kPa}$
with 11.5% ethanol	54.6

[a] To convert J to cal, multiply by 0.239.
[b] To convert MPa to atm, multiply by 9.87.
[c] To convert kPa to mm Hg, multiply by 7.5.

Refluxing with an aluminum coupon leads to complete metal consumption with formation of aluminum chloride and 2,3-dichlorobutane. Reaction with iron and zinc is minimal under dry conditions, whereas wet solvent (7% water phase) increases corrosion rates.

The 48-hour accelerated oxidation test with 1,1-dichloroethane at reflux temperatures gives a 0.025% yield of hydrogen chloride as compared to 0.4% HCl for trichloroethylene and 0.6% HCl for tetrachloroethylene.

Reaction with an amine gives low yields of chloride ion and the dimer 2,3-dichlorobutane, CH$_3$CHClCHClCH$_3$.

Manufacture

1,1-Dichloroethane is produced commercially from hydrogen chloride and vinyl chloride at 20–55°C in the presence of an aluminum, ferric, or zinc chloride catalyst (7–8). 1,1-Dichloroethane is usually an intermediate in the production of vinyl chloride and of 1,1,1-trichloroethane by photochlorination (9).

Toxicity

1,1-Dichloroethane, like all volatile chlorinated solvents, has an anesthetic effect and depresses the central nervous system at high vapor concentrations. The 1976 American Conference of Governmental Industrial Hygienists (ACGIH) recommends a time-weighted average (TWA) solvent vapor concentration of 200 ppm for worker exposure. The oral LD$_{50}$ of 1,1-dichloroethane in rats is 14.1 g/kg, classifying it as essentially nontoxic by oral ingestion.

1,2-DICHLOROETHANE

1,2-Dichloroethane [107-06-2] (ethylene chloride, ethylene dichloride, s-dichloroethane), CH$_2$ClCH$_2$Cl, is a colorless, volatile liquid with a pleasant odor, stable at ordinary temperatures. It is miscible with other chlorinated solvents and soluble in common organic solvents as well as having high solvency for fats, greases, and waxes.

Physical and Chemical Properties

The physical properties of 1,2-dichloroethane are listed in Table 2.

Pyrolysis. Pyrolysis of 1,2-dichloroethane in the temperature range of 340–515°C gives vinyl chloride, hydrogen chloride, and traces of acetylene (1,11). The decomposition is accelerated by chlorine, bromine, bromotrichloromethane, or carbon tetrachloride (12). Catalytic dehydrochlorination of 1,2-dichloroethane on activated alumina (3) and metal carbonate and sulfate salts (5) has been reported.

Hydrolysis. Heating 1,2-dichloroethane with excess water at 60°C in a nitrogen atmosphere produces hydrogen chloride at a rate of 1.32×10^{-6} meq/h per mL of aqueous phase (13). At 74°C the rate was 9.40×10^{-6} and at 100°C the rate increased to 3.94×10^{-3}. This pseudo first-order hydrolysis reaction occurs mainly in the aqueous phase where the first product, ethylene chlorohydrin, is further hydrolyzed to ethylene glycol.

Hydrolysis at 160–175°C and 1.5 MPa (15 atm) in the presence of an acid catalyst gives ethylene glycol, which is also obtained in the presence of aqueous alkali at 140–250°C and up to 4.0 MPa (40 atm) pressure (14).

Oxidation. Atmospheric oxidation of 1,2-dichloroethane at room or reflux temperatures generates some hydrogen chloride and results in solvent discoloration. A 48-hour accelerated oxidation test at reflux temperatures gives only 0.006% hydrogen chloride (15). Addition of 0.1–0.2 wt % of an amine (eg, diisopropylamine) protects the 1,2-dichloroethane against oxidative breakdown.

Corrosion. Corrosion of aluminum, iron, and zinc by boiling 1,2-dichloroethane has been studied (15). Dry and refluxing 1,2-dichloroethane completely consumed a 2024 aluminum coupon in a 7-day study, whereas iron and zinc were barely attacked. Aluminum was attacked less than iron or zinc by refluxing with 1,2-dichloroethane containing 7% water. Corrosion rates in µm/yr (mils penetration per year or mpy) in dry solvent are 0.254 (0.01) for iron and 3.05 (0.12) for zinc. In the wet solvent, the corrosion rate for iron increases to 145 µm/yr (5.7 mpy) and for zinc to 1.2 mm/yr (47 mpy). Corrosion rate for aluminum in the wet solvent is 2.36 mm/yr (93 mpy) as compared to complete dissolution in the dry solvent.

Table 2. Properties[a] of 1,2-Dichloroethane

Property	Value
melting point, °C	−35.3
boiling point, °C	83.7
density at 20°C, g/L	1.2529
n_D^{20}	1.4451
viscosity at 20°C, mPa·s (= cP)	0.84
surface tension at 20°C, mN/m (= dyn/cm)	31.38
specific heat at 20°C, J/(g·K)[b]	
liq	1.288
gas	1.066
latent heat of vapor 20°C, J/g[b]	323.42
latent heat of fusion, J/g[b]	88.36
critical temperature, °C	290
critical pressure, MPa[c]	5.36
critical density, g/L	0.44
flash point, °C	
closed cup	17
open cup	21
explosive limits in air at 25°C, % by vol	6.2–15.6
autoignition temperature in air, °C	413
thermal conductivity, liq at 20°C, W/(m·K)[d]	0.143
heat of combustion, kJ/g[b]	12.57
heat of formation, kJ/(g·mol)[b]	
liquid	157.3
vapor	122.6
dielectric constant	
liquid, 20°C	10.45
vapor, 120°C	1.0048
dipole moment, C·m (debye)	5.24×10^{-30} (1.57)
coefficient of cubical expansion, mL/g, 0–30°C	0.00116
vapor pressure, kPa[e]	
10°C	5.3
20°C	8.5
30°C	13.3
solubility at 20°C, g	
1,2-dichloroethane in 100 g H$_2$O	0.869
H$_2$O in 100 g 1,2-dichloroethane	0.160
azeotropes[f], bp, °C	
with 19.5% H$_2$O	72
with 5% H$_2$O and 17% ethanol	66.7

[a] See ref. 10b for additional property data.
[b] To convert J to cal, divide by 4.184.
[c] To convert MPa to atm, divide by 0.101.
[d] To convert W/(m·K) to (Btu·ft)/(h·ft^2·°F), divide by 1.73.
[e] To convert kPa to mm Hg, multiply by 7.5.
[f] See ref. 10a for additional binary azeotropes.

Nucleophilic Substitution. The kinetics of the bimolecular nucleophilic substitution of the chlorine atoms in 1,2-dichloroethane with NaOH, NaOC$_6$H$_5$, (CH$_3$)$_3$N, pyridine, and CH$_3$COONa in aqueous solutions at 110–120°C has been studied (16).

The reaction of sodium cyanide with 1,2-dichloroethane in methanol at 50°C to give 3-chloropropionitrile proceeds very slowly. Dimethyl sulfoxide as a solvent for the reaction greatly enhances nucleophilic substitution of the chlorine atom. Further reaction of sodium cyanide at room temperature gives acrylonitrile (qv), $CH_2{=}CHCN$ (17) within a few minutes.

$$ClCH_2CH_2Cl + CN^- \xrightarrow{k_1} ClCH_2CH_2CN + Cl^- \text{ (slow)} \qquad (1)$$

$$ClCH_2CH_2CN + CN^- \xrightarrow{k_2} CH_2{=}CHCN + Cl^+ \text{ (fast)} \qquad (2)$$

Miscellaneous. 1,2-Dichloroethane reacts with toluene in the presence of Friedel-Crafts catalysts such as $AlBr_3$, $AlCl_3$, $GaCl_3$, and $ZrCl_3$ (18). Polycondensation products are formed by disproportionation of the primary ditolylethane product.

Ammonolysis of 1,2-dichloroethane with 50% aqueous ammonia at 100°C is a commercial process for ethylenediamine (19) (see Diamines).

Manufacture

1,2-Dichloroethane is produced by the catalytic vapor- or liquid-phase chlorination of ethylene. Most liquid-phase processes use ferric chloride as the catalyst. Other catalysts claimed in the patent literature include aluminum chloride, antimony pentachloride, and cupric chloride. The chlorination is carried out at 40–50°C with 5% air added to prevent substitution chlorination of the product.

$$CH_2{=}CH_2 + Cl_2 \text{ (5\% air)} \xrightarrow[40-50°C]{FeCl_3} ClCH_2CH_2Cl \qquad (3)$$

Oxychlorination of ethylene has become the second important process for 1,2-dichloroethane. The process is usually incorporated into an integrated vinyl chloride plant in which hydrogen chloride, recovered from the cracking of 1,2-dichloroethane to vinyl chloride, is recycled to an oxychlorination unit. The hydrogen chloride by-product is used as the chlorine source in the chlorination of ethylene in the presence of oxygen and copper chloride catalyst:

$$2\ CH_2{=}CH_2 + 4\ HCl + O_2 \xrightarrow[270°C]{CuCl_2} 2\ ClCH_2CH_2Cl + H_2O \qquad (4)$$

A fluidized-bed oxychlorination reactor developed by Goodrich is claimed to provide very good temperature control (20). A large number of patents deal with the catalyst technology (21–24).

Economic Aspects and Uses

A major portion (83%) of U.S. 1,2-dichloroethane production is converted to vinyl chloride monomer (see Vinyl polymers). 1,2-Dichloroethane is also a starting material for chlorinated solvents such as 1,1,1-trichloroethane, trichloroethylene, and perchloroethylene. Other uses include as a reactant to prepare ethylenediamines, and as an additive in tetraethyllead (TEL) antiknock mixtures (see Lead compounds) (see Table 3). The latter use will, of course, decrease in future years because of increased marketing of unleaded gasolines.

Yearly growth rate of 1,2-dichloroethane production in the United States has been 9% from 1969 to 1974. Bulk transport prices of 20¢/kg have been steady over this period. The 1975 world production capacity of 1,2-dichloroethane and projected increases are listed in Table 4.

Table 3. U.S. Production and Uses of 1,2-Dichloroethane, Metric Tons

Use	1963	1974	Increase, %
vinyl chloride	560,000	3,900,000	600
other chlorohydrocarbons	150,000	550,000	250
ethylenediamine	80,000	130,000	65
TEL[a] antiknock mixtures	73,000	106,000	45
Total production		4.77×10^6 [b]	

[a] TEL = tetraethyllead.
[b] The 16th highest-volume product in the United States.

Table 4. World 1,2-Dichloroethane Production Capacity [a], Metric Tons

Country or Region	1975	Projected[b]
North America	6,158	6,785
United States and Puerto Rico	5,801	5,801
South America	254	1,150
Brazil	167	714
Europe	>8,258	>14,498
Belgium	914	1,202
France	>1,900	>2,259
Germany, Federal Republic of	1,680	>1,680
Italy	1,420	1,969
Netherlands	485	1,141
Spain	260	1,468
United Kingdom	965	965
Asia	>3,353	>5,468
Japan	3,051	3,676
Australia	60	60
Africa	>52	>119
Total	>18,135	>28,080

[a] Data from the *Chemical Economics Handbook* (27). With permission of Stanford Research Institute.
[b] As of 1975.

Toxicity

1,2-Dichloroethane, at high vapor concentrations (above 200 ppm), can cause central nervous system depression and gastrointestinal upset characterized by mental confusion, dizziness, nausea, and vomiting. Definite liver, kidney, and adrenal injuries may occur at the higher vapor levels. The recommended 1976 ACGIH vapor exposure TWA standard for 1,2-dichloroethane is 50 ppm. However, in 1978 NIOSH recommended a reduction in the current standard to 5 ppm (25a). The odor threshhold for 1,2-dichloroethane is 50–100 ppm and thus odor does not serve as a good warning against possible overexposure.

1,2-Dichloroethane is one of the more toxic chlorinated solvents by inhalation (25b). The highest nontoxic vapor concentrations in chronic exposure studies with various animals range from 100 to 200 ppm (25–26). 1,2-Dichloroethane exhibits a low single-dose oral toxicity in rats, LD_{50} is 680 mg/kg (25b). Repeated skin contact should be avoided since the solvent can cause defatting of the skin, severe irritation, and moderate edema. Eye contact may have slight to severe effects.

1,1,1-TRICHLOROETHANE

1,1,1-Trichloroethane [71-55-6] (methyl chloroform), CH_3CCl_3, is a colorless, nonflammable liquid with a characteristic ethereal odor. It is miscible with other chlorinated solvents and soluble in common organic solvents. The compound was first prepared by Regnault about 1840.

1,1,1-Trichloroethane is among the least toxic of the chlorinated solvents used in industry today. The commercial metal-cleaning grades contain added inhibitors that make usage acceptable for all common metals including aluminum. It has excellent solvency for various greases, oils, tars, and waxes and a wide range of organic materials (see Solvents, industrial).

Physical and Chemical Properties

The physical properties of 1,1,1-trichoroethane are given in Table 5.

Pyrolysis. The pyrolysis of 1,1,1-trichloroethane at 325–425°C proceeds by a simultaneous unimolecular and radical chain mechanism to yield 1,1-dichloroethylene and hydrogen chloride (28). 1,1,1-Trichloroethane vapors mixed with air and passed over hot metal surfaces form relatively little phosgene at temperatures up to 370°C (29).

Hydrolysis. 1,1,1-Trichloroethane heated with water at 75–160°C under pressure and in the presence of sulfuric acid or a metal chloride catalyst decomposes to acetyl chloride, acetic acid, or acetic anhydride (30). However, hydrolysis under normal use conditions proceeds slowly (31). The hydrolysis is 100–1000 times faster with trichloroethane dissolved in the water phase than vice versa.

Refluxing 1,1,1-trichloroethane with ferric and gallium chloride hydrate salts gives hydrogen chloride, acetic acid, 1,1-dichloroethylene, and the dehydrated salts (32).

Oxidation. 1,1,1-Trichloroethane is stable to oxidation when compared to olefinic chlorinated solvents like trichloroethylene and tetrachloroethylene. Use of a 48-h accelerated oxidation test gave no hydrogen chloride whereas trichloroethylene gave 0.4 wt % HCl and tetrachloroethylene gave 0.6 wt % HCl (15).

Corrosion. The corrosion rates of 1,1,1-trichloroethane with metals in dry and wet environments have been reported (15). Refluxing uninhibited 1,1,1-trichloroethane reacts vigorously with aluminum to give aluminum chloride, 2,2,3,3-tetrachlorobutane, 1,1-dichloroethylene, and hydrogen chloride. Adequate metal inhibitors, however, prevent this reactivity and allow the solvent to be used in aluminum metal-cleaning applications. Dry, uninhibited 1,1,1-trichloroethane is not very corrosive with iron or zinc; corrosion rate with iron is <2.54 μm/yr (<0.1 mpy) and with zinc <25.4 μm/yr (<1.0 mpy). Addition of 7% water increases corrosion rates to 254 μm/yr (<10.0 mpy) for iron and >254 μm/yr (>10.0 mpy) for zinc. The presence of both water and ethanol increases iron or tin attack at reflux. Highest metal loss with tin-plated iron coupons occurred at a solvent composition of 30% 1,1,1-trichloroethane, 55% ethanol, and 15% water (33). Addition of water and/or ethanol increases the solubility of the metal chloride products obtained from the reacting metal surface.

Inhibitions. Organic inhibitors for proprietary grades of 1,1,1-trichloroethane are (1) acid acceptors, namely epoxide compounds used to neutralize small amounts of hydrogen chloride normally formed during solvent use; and (2) metal stabilizers, namely organic compounds that deactivate metal surfaces and remove or complex

Table 5. Properties of 1,1,1-Trichloroethane

Property	Value
melting point, °C	−33.0
boiling point, °C	74.0
density at 20°C	1.3249
n_D^{20}	1.4377
viscosity at 20°C, mPa·s (= cP)	0.858
surface tension at 25°C, mN/m (= dyn/cm)	25.54
specific heat at 20°C, J/g[a]	
liq	1.004
gas	0.782
latent heat of vapor at 20°C, J/g[a]	248.11
critical temperature, °C	311.5
critical pressure, MPa[b]	4.48
flash point (closed cup), °C	none
explosive limits in air at 25°C, % by vol	8.0–10.5
autoignition temperature, °C	537
heat of combustion, kJ/g[a]	6.69
dielectric constant, liq at 20°C	7.5
vapor pressure[c], kPa[d]	
20°C	13.3
40°C	31.7
solubility at 20°C, g	
trichloroethane in 100 g H$_2$O	0.095
H$_2$O in 100 g trichloroethane	0.034
binary azeotropes, bp, °C	
with 4.3% H$_2$O	65.0
with 23.0% methanol	55.5
with 17.4% ethanol	64.4
with 17% isopropyl alcohol	68.2
with 17.2% t-butyl alcohol	70.2

[a] To convert J to cal, divide by 4.184.
[b] To convert MPa to atm, multiply by 9.87.
[c] Antoine constants for 1,1,1-trichloroethane
$A = 7.76632, B = 1204.66, C = 226.671$

where $\log_{10}$ pressure (kPa) $= A - \left(\dfrac{B}{T + C}\right)$

T = temperature in °C.
[d] To convert kPa to mm Hg, multiply by 7.5.

trace amounts of metal chloride salts that might form. Several hundred United States and foreign patents have been issued on 1,1,1-trichloroethane stabilization usually with the aid of a Lewis base, such as an amino or carbonyl compound (34) (see Corrosion and corrosion inhibitors).

Dehydrochlorination. 1,1,1-Trichloroethane over activated alumina (3) or anhydrous aluminum chloride at 0°C (6) gives rapid hydrogen chloride evolution and polymer formation (6). Aluminum fluoride is the most active metal fluoride catalyzing dehydrochlorination (35). Several chlorinated solvents, including 1,1,1-trichloroethane, are dehydrochlorinated on hopcalite catalyst (MnO$_2$–CuO–Cu) at 305–315°C (36). Other catalytic materials include, MgSO$_4$ and K$_2$CO$_3$ (5) and molecular sieves (qv) containing cations of H$^+$, Mg^{2+}, Li$^+$, Na$^+$, or K$^+$ (37).

Miscellaneous. 1,1,1-Trichloroethane reacts with olefins, $CH_2\!\!=\!\!CRR'$, in the presence of $P(O)(NMe_2)_3$ and $FeCl_2$ at 130°C, to give compounds of the type $CH_3CCl_2CH_2CRR'Cl$ (38).

Fluorination of 1,1,1-trichloroethane with anhydrous hydrogen fluoride at 144°C gives both 1,1-dichloro-1-fluoroethane [1717-00-6], CH_3CCl_2F, and 1-chloro-1,1-difluoroethane [75-68-3], CH_3CClF_2 (39).

Reactivity with amines (15) and mercaptan groups (40) is low.

Manufacture

In the most important process, vinyl chloride (obtained from 1,2-dichloroethane, see above) is hydrochlorinated to 1,1-dichloroethane which is then thermally or photochemically chlorinated:

$$CH_2\!\!=\!\!CHCl + HCl \rightarrow CH_3CHCl_2 \xrightarrow[\text{uv}]{\text{Cl}_2 \text{ vapor phase}} CH_3CCl_3 + HCl$$

In a second process, hydrogen chloride is added to 1,1-dichloroethylene in the presence of a $FeCl_3$ catalyst:

$$CH_2\!\!=\!\!CCl_2 + HCl \xrightarrow[30°C]{FeCl_3} CH_3CCl_3$$

A process of minor importance utilizes a continuous noncatalytic chlorination of ethane which produces 1,1,1-trichloroethane and a number of other products, depending on the reaction conditions.

Economic Aspects

1,1,1-Trichloroethane is produced in the United States by The Dow Chemical Company, PPG Industries Inc., and Vulcan Materials Company. Several European and Japanese companies also produce large amounts yearly. Over 70% of the production is based on the vinyl chloride-1,1-dichloroethane process, 20% on the 1,1-dichloroethylene process, and about 10% on the direct chlorination of ethane.

The estimated U.S. demand (Bureau of the Census) for 1,1,1-trichloroethane in 1976 was 230,000 metric tons. Total world demand for 1,1,1-trichloroethane in 1972 was 290,000 t with an estimated growth rate of 12% per year through 1975 (41). U.S. production capacity was estimated to be about 390,000 t by 1978. Production capacity outside the United States was estimated to be approximately 383,000 t in 1977. Bulk U.S. prices averaged $0.35/L in 1972 and $0.56/L in 1975 (41).

Toxicity, Safety

1,1,1-Trichloroethane is among the least toxic of the industrial chlorinated solvents (42).

The acute oral LD_{50} dosages for 1,1,1-trichloroethane for guinea pigs, mice, rabbits, and rats range from 5 to 12 g/kg. Injection of radioactive $CH_3{}^{14}CCl_3$ in doses of 700 mg/kg into rats demonstrated that 99% of the tracer solvent was eliminated by respiration.

Vapor inhalation causes depression of the central nervous system (dizziness, light-headedness). The LC_{50} for rats is 18,000 ppm/3 h and 14,000 ppm/7 h. The 1976

Threshold Limit Value (TLV) for 1,1,1-trichloroethane suggested by the ACGIH is 350 ppm.

Entry into a tank that has contained any chlorinated or any easily evaporated solvent requires special procedures to ensure worker safety. The heavier vapors tend to concentrate in unventilated spaces. The proper tank entry procedure requires positive ventilation, testing for residue solvent vapor and oxygen levels, and the use of respiratory equipment and rescue harness. Monitoring the tank from outside is also important.

The use of the appropriate gas mask is permissible in vapor concentrations of less than 2% and when there is no deficiency of atmospheric oxygen, but not for exposures exceeding one-half hour.

Skin exposure to 1,1,1-trichloroethane can cause irritation, pain, blisters, and even burning. Eye exposure may produce irritation, but should not cause serious injury.

Uses

Inhibited grades of 1,1,1-trichloroethane are used in hundreds of different industrial cleaning applications. 1,1,1-Trichloroethane is preferred over trichloroethylene or tetrachloroethylene because of its lower toxicity. Additional advantages of 1,1,1-trichloroethane include optimum solvency, good evaporation rate, and no fire or flash point as determined by standard test methods. Common uses include cleaning of electrical equipment, motors, electronic components and instruments, missile hardware, paint masks, photographic film, printed circuit boards and various metal and certain plastic components during manufacture. Approximately equal quantities of 1,1,1-trichloroethane are used in cold and hot cleaning applications (see Metal surface treatments).

1,1,1-Trichloroethane and other chlorinated solvents are used for vapor degreasing (43–49). 1,1,1-Trichloroethane is an excellent solvent for development of photoresist polymers used in printed circuit board manufacture (see Photoreactive polymers; Integrated circuits).

The use of solvents like 1,1,1-trichloroethane to supplement or replace present aqueous textile processing and finishing techniques is a new application (50–52) (see Dye carriers).

Use as a solvent in consumer nonflammable adhesive formulations is a growth area. Certain neoprene adhesives formulated with 1,1,1-trichloroethane show superior peel strength properties.

Addition of 20–30% 1,1,1-trichloroethane to metal cutting fluids significantly increases tool life and/or cutting speeds in difficult drilling and tapping operations.

1,1,2-TRICHLOROETHANE

1,1,2-Trichloroethane [79-00-5] (vinyl trichloride), $CH_2ClCHCl_2$, is a colorless, nonflammable liquid with a pleasant odor, miscible with chlorinated solvents, and (as is 1,1,1-trichloroethane) soluble in the other common organic solvents.

Physical and Chemical Properties

The physical properties are given in Table 6.

Hydrolysis and Oxidation. Hydrolysis of 1,1,2-trichloroethane is much slower than that of 1,1,1-trichloroethane (31).

1,1,2-Trichloroethane is also much more prone to oxidative degradation than 1,1,1-trichloroethane. A 48-h accelerated oxidation test at reflux temperatures gave 0.84% hydrogen chloride with 1,1,2-trichloroethane but none with 1,1,1-trichloroethane (15). An alcohol, such as sec-butyl alcohol, will protect the solvent against oxidation and reaction with metals. An epoxide compound such as 1,2-butylene oxide can also be added to the solvent as an acid acceptor.

Corrosion. Uninhibited 1,1,2-trichloroethane is corrosive to aluminum, iron, and zinc at reflux temperatures (15). An aluminum coupon dissolves completely in dry and refluxing 1,1,2-trichloroethane, whereas the corrosion rate with iron is <254 μm/yr (<10 mpy), and >254 mμ/yr (>10 mpy) with zinc.

Addition of a water phase results in complete dissolution of all three metals. The water phase increases the solubility of the metal chloride reaction products and accelerates their removal from the reacting metal surface. The overall corrosion of iron and zinc in 1,1,1-trichloroethane is less than in 1,1,2-trichloroethane. Wet 1,1,1-trichloroethane gives the same corrosion rates as dry 1,1,2-trichloroethane: <254 μm/yr (<10 mpy) with iron and >254 μm/yr (>10 mpy) with zinc.

Nucleophilic Substitution. Reaction of 1,1,2-trichloroethane with the sulfhydryl group (—SH) in cysteine gave 75 ppm chloride ion in a 3-d exposure at 35°C (40). The identical experiment with 1,1,1-trichloroethane gave less than 5 ppm chloride ion.

Table 6. Properties of 1,1,2-Trichloroethane

Property	Value
melting point, °C	−37.0
boiling point, °C	113.7
density at 20°C	1.4432
n_D^{20}	1.4711
viscosity at 20°C, mPa·s (= cP)	1.20
surface tension at 20°C, mN/m (= dyn/cm)	32.52
specific heat, liq, at 20°C, J/(g·°C)[a]	1.113
latent heat of vapor at bp, J/g[a]	284.5
autoignition temperature, °C	460.0
thermal conductivity, liq at 20°C, W/(m·K)[b]	0.134
heat of combustion, vapor, MJ/(mol)[a]	1.098
vapor pressure at 20°C, kPa[c]	2.51
solubility, g	
1,1,2-trichloroethane in 100 g H$_2$O at 20°C	0.45
H$_2$O in 100 g 1,1,2-trichloroethane at 25°C	0.12
binary azeotropes, bp, °C	
with 97% methanol	64.5
with 57% tetrachloroethylene	112.0
with 70% ethanol	77.8

[a] To convert J to cal, divide by 4.184.
[b] To convert W/(m·K) to (Btu·ft)/(h·ft^2·°F), divide by 1.73.
[c] To convert kPa to mm Hg, multiply by 7.5.

Dehydrochlorination. 1,1,2-Trichloroethane is easily dehydrochlorinated by a number of catalytic reagents to give 1,1-dichloroethylene and some 1,2-dichloroethylene. Refluxing with Ca(OH)$_2$ and water gives 1,1-dichloroethylene (53). The rate of reaction is faster with the 1,1,2-trichloroethane than with 1,1,1-trichloroethane. Anhydrous aluminum chloride also gives rapid dehydrochlorination (6,53). In this reaction the 1,1,1-isomer reacts faster. 1,1,2-Trichloroethane mixed with nitrogen and 0.1% chlorine is dehydrochlorinated at 450°C in the presence of NaCl particles to give a 1:1 mixture of 1,1-dichloroethylene and *trans*-1,2-dichloroethylene (54). Other catalytic materials include MgSO$_4$ and K$_2$CO$_3$ (5), molecular sieves containing metal cations (37), and zinc dust (55).

In thermal dehydrochlorination up to 500°C *cis*- and *trans*-1,2-dichloroethylene, 1,1-dichloroethylene, and hydrogen chloride were obtained (1).

Manufacture

1,1,2-Trichloroethane is produced in the United States directly or indirectly from ethylene. For example, by chlorination of 1,2-dichloroethane, a product from ethylene (56):

$$CH_2=CH_2 + Cl_2 \rightarrow ClCH_2CH_2Cl \xrightarrow[Cl_2]{\Delta} CH_2ClCHCl_2 + HCl \quad (5)$$

Oxychlorination of ethylene with hydrogen chloride and oxygen at 280–370°C on a fluidized CuCl$_2$-KCl (on attapulgite) catalyst bed yields 1,2-dichloroethane and 1,1,2-trichloroethane, along with some higher chlorinated ethanes (22).

1,1,2-Trichloroethane is also a coproduct in the thermal chlorination of 1,1-dichloroethane to produce 1,1,1-trichloroethane. Vapor chlorination favors the 1,1,1-isomer, whereas reaction in the liquid phase may give much higher ratios of 1,1,2-trichloroethane.

Toxicity

1,1,2-Trichloroethane is much more toxic than 1,1,1-trichloroethane in acute exposure studies (57). The 1976 ACGIH recommended TWA value for 1,1,2-trichloroethane is 10 ppm.

An acute lethal dose (LC$_{50}$) for vapor exposure to 1,1,2-trichloroethane in the rat is 2,000 ppm for a 4-h exposure. The same lethal effect occurs at 18,000 ppm vapor levels during 3 hours of exposure with the 1,1,1-trichloro isomer. The oral LD$_{50}$ for 1,1,2-trichloroethane in rats is 0.1–0.2 g/kg, classifying it as moderately toxic. Liver and kidney damage occurs at even lower dosages. Skin adsorption is a possible route of overexposure.

Uses

The principal use of 1,1,2-trichloroethane is as a feedstock intermediate in the production of 1,1-dichloroethylene. 1,1,2-Trichloroethane is also used in limited applications where its high solvency for chlorinated rubbers, etc, is needed.

1,1,1,2-TETRACHLOROETHANE

1,1,1,2-Tetrachloroethane [*630-20-6*], CCl$_3$CH$_2$Cl, is used as a feedstock for the production of solvents such as trichloroethylene and tetrachloroethylene.

Physical and Chemical Properties

Physical properties of 1,1,1,2-tetrachloroethane are listed in Table 7.

Pyrolysis. Thermal decomposition of 1,1,1,2-tetrachloroethane affords tetrachloroethylene (by disproportionation), hydrogen chloride and trichloroethylene (58). The yield of the latter is increased in the presence of ferric chloride (59). Other catalytic materials include: FeCl$_3$–KCl mixture (60), AlCl$_3$ (6), the complex of AlCl$_3$ with nitrobenzene (61), activated alumina (3), Ca(OH)$_2$ (62–63), and NaCl (54).

Oxidation. Oxidation of 1,1,1,2-tetrachloroethane in the presence of ionizing radiation gives dichloroacetyl chloride, Cl$_2$CHCOCl (64). The gas-phase photochlorination of 1,1,1,2-tetrachloroethane in the absence and presence of oxygen has been studied (62).

Manufacture

1,1,1,2-Tetrachloroethane is often an incidental by-product in the manufacture of chlorinated ethanes. It can be prepared by heating the 1,1,2,2-isomer with anhydrous aluminum chloride or chlorination of 1,1-dichloroethylene at 40°C (65).

Toxicity

Rats exposed to 1000 ppm of vapors for 4–7 h/d for 8 days showed ataxia, decreased body weight and growth rate, and minimal central fatty metamorphosis of the liver (66). (Trichloroethanol and trichloroacetic acid were urinary metabolites.) Tetrachloroethylene did not show any of these effects. The single-dose oral LD$_{50}$ in male rats is 1.87 g/kg body weight and 3.73 g/kg for female rats (67).

Table 7. Properties of 1,1,1,2-Tetrachloroethane

Property	Value
melting point, °C	−68.7
boiling point, °C	130.5
density, g/L at 20°C	1.5465
n_D^{20}	1.4822
viscosity at 20°C, mPa·s (= cP)	1.501
surface tension at 20°C, mN/m (= dyn/cm)	32.13
latent heat of vapor, at 20°C, J/g[a]	243.50
solubility at 20°C, g	
H$_2$O in 100 g 1,1,1,2-tetrachloroethane	0.056

[a] To convert J to cal, divide by 4.184.

1,1,2,2-TETRACHLOROETHANE

1,1,2,2-Tetrachloroethane [79-34-5], acetylene tetrachloride, $CHCl_2CHCl_2$, is a heavy, nonflammable liquid with a sweetish odor. It is miscible with the chlorinated solvents and shows high solvency for a number of natural organic materials. It is also a solvent for sulfur and a number of inorganic compounds, eg, sodium sulfite.

Physical and Chemical Properties

The physical properties of 1,1,2,2-tetrachloroethane are listed in Table 8.

Pyrolysis. 1,1,2,2-Tetrachloroethane, like the 1,1,1,2-isomer, is thermally degraded with or without a catalytic agent to give trichloroethylene, tetrachloroethylene, and hydrogen chloride (58–60). An effective catalyst may include ferric chloride (59), $FeCl_3$–KCl (60), activated alumina (3), aluminum chloride (6), aluminum chloride in nitrobenzene (61), calcium hydroxide (53,63), molecular sieves containing metal cations (37), or a Fe–Cr–K oxide catalyst (68).

Table 8. Properties[a] of 1,1,2,2-Tetrachloroethane

Property	Value
melting point, °C	−42.5
boiling point, °C	146.3
density at 20°C, g/L	1.593
n_D^{20}	1.4942
viscosity at 20°C, mPa·s (= cP)	1.77
surface tension at 20°C, mN/m (= dyn/cm)	34.72
specific heat, J/(g·°C)[b]	
liq at 20°C	1.13
gas at 146.3°C	0.920
latent heat of vapor at 20°C, J/g[b]	230.5
critical temperature, °C	388
critical pressure, MPa[c]	3.99
thermal conductivity, W/(m·K)[d]	0.134
heat of combustion, vapor, kJ/g[a]	5.786
dielectric constant, liq at 20°C	8.00
dipole moment, C·m(debye)	6.17×10^{-30} (1.85)
coefficient of cubical expansion, mL/g	0.00103
vapor pressure, kPa[e]	
0°C	0.176
20°C	0.647
100°C	25.2
solubility at 25°C, g	
1,1,2,2-tetrachloroethane in 100 g H_2O	0.32
H_2O in 100 g 1,1,2,2-tetrachloroethane	0.11
sulfur in 100 g 1,1,2,2-tetrachloroethane at 120°C	100
binary azeotrope[f], bp, °C	
with 31% H_2O	93.2

[a] For additional physical property data over large temperature ranges, see ref. 10c.
[b] To convert J to cal, divide by 4.184.
[c] To convert MPa to atm, divide by 0.101.
[d] To convert W/(m·K) to (Btu·ft)/(h·ft²·°F), divide by 1.73.
[e] To convert kPa to mm Hg, multiply by 7.5.
[f] See ref. 10a for additional binary azeotropes.

Thermal cracking of both 1,1,1,2- and 1,1,2,2-tetrachloroethane gives a 95% conversion to trichloroethylene and tetrachloroethylene (69).

Dehydrochlorination and Chlorination. The simultaneous chlorination and dehydrochlorination of 1,1,2,2-tetrachloroethane proceeds via formation of a labile intermediate, Cl_3CCHCl_2 (70). Chlorination of tetrachloroethane to hexachloroethane is accelerated by 315–354 nm light (71). Heating a mixture of tetrachlorethane vapors and chlorine over active charcoal at 400°C gives carbon tetrachloride and hydrogen chloride (72).

Miscellaneous. Air oxidation of 1,1,2,2-tetrachloroethane under ionizing radiation gives dichloroacetyl chloride (64).

Contact of 1,1,2,2-tetrachloroethane with strong alkali gives dichloroacetylene and an explosion may result.

Manufacture

1,1,2,2-Tetrachloroethane is produced by direct chlorination or oxychlorination utilizing ethylene as a feedstock. In most cases, 1,1,2,2-tetrachloroethane is not isolated, but immediately thermally cracked at 454°C to give the desired trichloroethylene and tetrachloroethylene products (69). A two-stage chlorination of 1,2-dichloroethane to give 1,1,2,2-tetrachloroethane has been patented (73). High-purity 1,1,2,2-tetrachloroethane is made by chlorinating acetylene.

Toxicity

1,1,2,2-Tetrachloroethane has a time weighted average (TWA) of 5 ppm as recommended by the ACGIH (1976). Skin adsorption may also pose an exposure hazard. 1,1,2,2-Tetrachloroethane is one of the most toxic chlorinated hydrocarbons (74–75). The liver is most affected.

The reported lethal oral dose for dogs is 0.3 mL/kg body weight. Rats survive a 4-h vapor exposure at 500 ppm but not 4 hours at 1000 ppm (76). Cats and rabbits exposed to 100–160 ppm 1,1,2,2-tetrachloroethane vapors for 8 to 9 hours daily for 4 weeks did not show any organ damage (77). Injuries to workers have been reported at much lower vapor concentrations (74).

Studies of 1,1,2,2-tetrachloroethane-^{14}C metabolism in the mouse showed that half of the dose was expired as carbon dioxide, and 30% as oxidized products in the urine within a 3-day period; 4% of the solvent was released through the lungs chemically unchanged and 16% remained in the body after three days (78).

Uses

The only major use of 1,1,2,2-tetrachloroethane is as a feedstock in the manufacture of trichloroethylene, tetrachloroethylene, and 1,2-dichloroethylene. Although it is an excellent solvent, its use should be discouraged in view of its high toxicity.

PENTACHLOROETHANE

Pentachloroethane [76-01-7], $CHCl_2CCl_3$, is a colorless, heavy, nonflammable liquid with a chloroform-like odor; it is miscible with the common organic solvents.

Physical and Chemical Properties

Physical properties of pentachloroethane are listed in Table 9.

The kinetics and mechanism of the pyrolysis of pentachloroethane in the temperature ranges of 407–430°C and 547–592°C have been studied (79–81). Tetrachloroethylene and hydrogen chloride are the two major pyrolysis products.

Various catalytic materials promote dehydrochlorination including $AlCl_3$ (6,53), $AlCl_3$–nitrobenzene complex (61), activated alumina (3), and $FeCl_3$ (59). Chlorination in the presence of anhydrous aluminum chloride gives hexachloroethane. Dry pentachloroethane does not corrode iron at temperatures up to 100°C. It is slowly hydrolyzed by water at normal temperatures and oxidized in the presence of light to give trichloroacetyl chloride.

Manufacture

Pentachloroethane can be made by chlorinating 1,1,2,2-tetrachloroethane under ultraviolet light (82), or trichloroethylene at 70°C in the presence of ferric chloride, sulfur, or ultraviolet light (83). Oxychlorination of ethylene gives pentachloroethane as well as lower chlorinated hydrocarbons (22).

Toxicity

The toxicity of pentachloroethane is similar to that of the tetrachloroethanes (84). The strong narcotic effect of pentachloroethane is even greater than that of chloroform. Significant pathological changes in the liver, lungs and kidneys of cats was observed at vapor concentrations of 121 ppm given 8 to 9 hours daily for 23 days (77). Metabo-

Table 9. Properties of Pentachloroethane

Property	Value
melting point, °C	−29
boiling point, °C	161.95
density at 20°C, g/L	1.678
n_D^{20}	1.5035
viscosity at 20°C, mPa·s (= cP)	2.45
surface tension at 20°C, mN/m (= dyn/cm)	33.77
specific heat, liq at 20°C, J/(g·°C)[a]	0.900
latent heat of vapor at bp, J/g[a]	182.4
thermal conductivity, liq at 20°C, W/(m·K)[b]	0.130
heat of combustion, kJ/g[a]	4.25
vapor pressure at 20°C, kPa[c]	0.444
solubility at 20°C, g	
pentachloroethane in 100 g H_2O	0.05
H_2O in 100 g pentachloroethane	0.03
binary azeotrope[d], bp, °C	
with 43.4% H_2O	95.1$_{97\ kPa}$

[a] To convert J to cal, divide by 4.184.
[b] To convert W/(m·K) to (Btu·ft)/(h·ft²·°F), divide by 1.73.
[c] To convert kPa to mm Hg, multiply by 7.5.
[d] See ref. 10a for additional binary azeotropes.

lism in the mouse is reported to give trichloroethanol and trichloroacetic acid in the urine. Expired air from the mouse contained trichloroethylene and tetrachloroethylene, indicating dechlorination as well as dehydrochlorination (85).

Uses

Pentachloroethane is a good solvent for cellulose acetate, certain cellulose ethers, and for natural gums and resins, but its high toxicity has discouraged these uses. Pentachloroethane is still used as an intermediate in some tetrachloroethylene processes.

HEXACHLOROETHANE

Hexachloroethane [67-72-1] (perchloroethane), CCl_3CCl_3, is a white crystalline solid with a camphorlike odor. Hexachloroethane is nonflammable and has a number of minor industrial uses which are, however, limited because of its toxic nature. Crystalline hexachloroethane is a minor product in many industrial chlorination processes of saturated and unsaturated C_2 hydrocarbons. Hexachloroethane is also obtained by refluxing carbon tetrachloride with metals such as aluminum (40).

Physical and Chemical Properties

Physical properties of hexachloroethane are listed in Table 10.

Hexachloroethane is thermally cracked in the gaseous phase at 400–500°C to give tetrachloroethylene, carbon tetrachloride, and chlorine (86). The thermal decompo-

Table 10. Properties of Hexachloroethane

Property	Value
melting point (sublimes), °C	185.0
boiling point, °C	186.0
density at 20°C	2.094
crystal structure	
rhombic	up to 46°C
triclinic	46–71°C
cubic	above 71°C
specific heat, liq at 25°C, J/(g·°C)[a]	0.728
latent heat of vapor at bp, J/g[a]	194.1
heat of combustion, kJ/g[a]	3.073
vapor pressure[c], kPa	
20°C	0.028
100°C	5.07
150°C	34.8
solubility at 22.3°C, g	
hexachloroethane in 100 g H_2O	0.005
binary azeotropes[b], bp, °C	
with 30% phenol	173.7

[a] To convert J to cal, divide by 4.184.
[b] See ref. 10a for additional binary azeotropes.
[c] To convert kPa to mm Hg, multiply by 7.5.

sition may occur by means of a radical chain mechanism involving $\cdot C_2Cl_5$, $\cdot Cl$, or $\cdot CCl_3$ radicals. The decomposition is inhibited by traces of nitric oxide.

Powdered zinc reacts violently with hexachloroethane in alcoholic solutions to give the metal chloride and tetrachloroethylene; aluminum gives a less violent reaction (87). Triethanolamine in the presence of copper powder gives a good yield of tetrachloroethylene. Photochemical reaction with quinuclidine in a benzene solution gives a 66% yield of 2-chloroquinuclidine (88).

$$\text{quinuclidine} \xrightarrow{C_2Cl_6} \text{2-chloroquinuclidine}$$

Hexachloroethane is unreactive with aqueous alkali and acid at moderate temperatures. However, when heated with solid caustic above 200°C or with alcoholic alkalies at 100°C, decomposition to oxalic acid takes place.

Trichloromethyl Free Radical. Degradation of carbon tetrachloride by photochemical, radiolytical, or ultrasonic energy produces the trichloromethyl free radical which on dimerization gives hexachloroethane. Chloroform under strong x-ray irradiation also gives the trichloromethyl radical intermediate and hexachloroethane as final product.

Manufacture

Hexachloroethane is formed in minor amounts in many industrial chlorination processes designed to produce lower chlorinated hydrocarbons. Chlorination of tetrachloroethylene, in the presence of ferric chloride, at 100–140°C is one convenient method of preparing hexachloroethane. Photochemical chlorination of tetrachloroethylene under pressure and below 60°C has been patented as a method of producing hexachloroethane (89).

Toxicity

Hexachloroethane is considered to be one of the more toxic chlorinated hydrocarbons. The 1976 ACGIH recommended time weighted average (TWA) for hexachloroethane is 1 ppm or 10 mg/m^3 of air. Skin adsorption is a route of possible exposure hazard. The primary effect of hexachloroethane is depression of the central nervous system (90). Pentachloroethane and tetrachloroethylene are major metabolites of hexachloroethane in sheep (91).

Uses

Hexachloroethane, like carbon tetrachloride and 1,1,1-trichloroethane, can be used to formulate extreme pressure lubricants (92–93). For example, lubricating oils containing 0.02–3.0 wt % (as halogen) of hexachloroethane reduce the abrasion of exhaust valve seats in internal combustion engines (94) (see Lubrication).

Hexachloroethane has been suggested as a degasifier in the manufacture of aluminum and magnesium metals. Hexachloroethane has been used as a chain transfer agent in the radiochemical emulsion preparation of propylene-tetrafluoroethylene copolymer (95).

740 CHLOROCARBONS, -HYDROCARBONS (OTHER)

Other uses of hexachloroethane are as moth repellent, plasticizer for cellulose esters, anthelmintic in veterinary medicine, rubber accelerator, and as a component in fungicidal and insecticidal formulations.

BIBLIOGRAPHY

The "Chlorocarbons and Chlorohydrocarbons—Other Chloroethanes" are treated under "Chlorine Compounds, Organic" in *ECT* 1st ed., Vol. 3, "Ethylidene Chloride, Ethylene Chloride," pp 760–764, "1,1,1-Trichloroethane," pp. 764–765, "1,1,2-Trichloroethane," pp. 765–767 by John Conway, Carbide and Carbon Chemicals Corp., "1,1,2,2-Tetrachloroethane," pp. 767–771, "Pentachloroethane," pp. 771–773, both by Janet Searles and H. A. McPhail, E. I. du Pont de Nemours & Company, Inc., "Hexachloroethane," pp. 773–774 by Jesse Werner, General Aniline & Film Corp., General Aniline Works Division; "Chlorocarbons and Chlorohydrocarbons—Other Chloroethanes," in *ECT* 2nd ed., Vol. 5, 149–170 by D. W. F. Hardie Imperial Chemical Industries, Ltd.

1. D. H. R. Barton, *J. Chem. Soc.* 148 (1949).
2. H. Hartmann, H. Heydtmann, and G. Rinck, *Z. Physik. Chem.* Frankfurt **28,** 71 (1961).
3. S. Kiyonori and A. Shuzo, *Nippon Kagaku Kaishi* (10), 1945 (1974).
4. S. Kiyonori, *Bull. Chem. Soc. Jpn.* **47**(10), 2406 (1974).
5. P. Andreu and co-workers, *Am. Quim.* **65**(11), 931 (1969).
6. N. K. Taikova, A. E. Kulikova, and E. N. Zil'berman, *Zh. Org. Khim.* 4(11), 1880 (1968).
7. U.S. Pat. 2,007,144 (July 2, 1934), H. S. Nutting, P. S. Petrie, and M. E. Huscher (to The Dow Chemical Co.).
8. U.S.S.R. Pat, 470,512 (May 15, 1975), Yu. A. Treger, I. Mokroisova, and S. M. Velichko.
9. R. Muradka, *Asahi Garasu Kenkyu Hokoku* **16,** 123 (1966).
10. (a) L. E. Horsley, *Azeotropic Data, Advances in Chemistry Series* No. 6, American Chemical Society, Washington, D.C., 1952; also see *Azeotropic Data—II,* No. 35 published in 1962. (b) R. W. Gallant, *Hydrocarbon Process.* **45**(7), 111 (1966). (c) R. W. Gallant, *Hydrocarbon Process.* **46**(12), 119 (1967).
11. K. A. Holbrook, R. W. Walker, and W. R. Watson, *J. Chem. Soc. B,* 577 (1971).
12. S. Inokawa and co-workers, *Kogyo Kagaku Zasshi* **67**(10), 1540 (1964).
13. W. L. Howard and T. L. Moore, unpublished results from The Dow Chemical Company.
14. U.S. Pat. 2,148,304 (Feb. 21, 1939), J. D. Ruys and H. R. McCombie (to Shell Development Co.).
15. W. L. Archer and E. L. Simpson, *I and EC Prod. R and D,* **16**(2), 158 (June 1977).
16. K. Okamoto and co-workers, *Bull. Chem. Soc. Jpn.* **40**(8), 1917 (1967).
17. G. E. Ham and J. Stevens, *J. Org. Chem.* **27,** 4638 (1962); U.S. Pat. 3,206,499 (Sept. 14, 1965), G. E. Ham (to The Dow Chemical Co.).
18. S. Kunichika, S. Oka, and T. Sugiyama, *Bull. Inst. Chem. Res. Kyoto Univ.* **48**(6), 276 (1970).
19. Z. Leszczynski, J. Strzelecki, and D. Zelazko, *Przemysl. Chem.* **44**(6), 330 (1965).
20. *Hydrocarbon Process. Petrol. Refiner* **44,** 289 (1965).
21. Fr. Pat. 1,577,105 (Aug. 1, 1965), H. Riegel (to Lummus Co.).
22. Fr. Pat. 1,555,518 (Jan. 31, 1969), A. Antonini, P. Joffre and F. Laine (to Produits Chimiques Pechiney Saint-Gobain).
23. Jpn. Pat. 71 33,010 (Sept. 27, 1971), K. Miyauchi, Y. Sato, and S. Okamoto (to Mitsui Toatsu Chemicals Co.).
24. Ger. Pat. 2,106,016 (Sept. 16, 1971), C. H. Cather (to PPG Industries Inc.).
25. (a) *Chem. Eng. News,* 6 (Sept. 25, 1978); (b) D. D. Irish in F. A. Patty, ed., *Industrial Hygiene and Toxicology,* Interscience Publishers, a division of John Wiley & Sons, Inc., New York, 1963, pp. 1280–1284.
26. H. C. Spencer and co-workers, *A.M.A. Arch. Ind. Hyg. Occupational Med.* **4,** 482 (1951).
27. J. L. Blackford, "Ethylene Dichloride," *Chemical Economics Handbook, 651.5031A,* Stanford Research Institute, Menlo Park, Calif., Nov. 1975.
28. D. H. R. Barton and P. F. Onyon, *J. Am. Chem. Soc.* **72,** 988 (1950).
29. W. B. Crummett and V. A. Stenger, *Ind. Eng. Chem.* **48,** 434 (1956).
30. U.S. Pat. 1,870,601 (Aug. 9, 1932), E. C. Britton and W. R. Reed (to The Dow Chemical Co.).
31. S. C. Stowe and C. F. Raley, unpublished work from The Dow Chemical Company, also W. L. Howard and J. D. Burger.
32. M. E. Hill, *J. Org. Chem.* **25,** 1115 (1960).

33. W. L. Archer, *Aerosol Age* **12**(8), 16 (1967).
34. U.S. Pat. 3,444,248 (May 13, 1969), W. L. Archer (to Dow Chemical Co.); U.S. Pat. 3,452,108 (June 24, 1969), W. L. Archer (to Dow Chemical Co.); U.S. Pat. 3,452,109 (June 24, 1969), W. L. Archer (to Dow Chemical Co.); U.S. Pat. 3,454,659 (July 8, 1969), W. L. Archer (to Dow Chemical Co.); U.S. Pat. 3,468,966 (Sept. 23, 1969), W. L. Archer (to Dow Chemical Co.); U.S. Pat. 3,472,903 (Oct. 14, 1969), W. L. Archer (to Dow Chemical Co.); U.S. Pat. 3,546,305 (Dec. 8, 1970), W. L. Archer (to Dow Chemical Co.); U.S. Pat. 3,681,469 (Aug. 1, 1972), W. L. Archer (to Dow Chemical Co.).
35. S. Okazaki and M. Komata, *Nippon Kagaku Kaishi* (3), 459 (1973).
36. J. K. Musick and F. W. Williams, *Am. Soc. Mech. Eng. Pap.* 75 ENAs 17 (1975).
37. I. Mochida and Y. Yoneda, *J. Org. Chem.* **33**(5), 2161 (1968).
38. T. Sato, M. Seno, and T. Asahara, *Seisan-Kenkyu* **24**(6), 230 (1972).
39. J. H. Brown and W. B. Whalley, *J. Soc. Chem. Ind.* **67**, 332 (1948).
40. W. L. Archer, unpublished work from The Dow Chemical Co.
41. J. L. Blackford, "1,1,1-Trichloroethane" *Chemical Economics Handbook 697,2031A*, Stanford Research Institute, Menlo Park, Calif., Oct. 1975.
42. D. M. Aviado and co-workers, *Methyl Chloroform and Trichloroethylene in the Environment*, CRC Press Inc., Cleveland, Ohio, 1976, pp. 5–44.
43. W. G. Rollo and A. O'Grady, *Can. Paint and Finish.*, 15 (Oct. 1973).
44. W. L. Archer, *Cleaning Stainless Steel, ASTM STP538*, American Society for Testing and Materials, Philadelphia, Pa., 1973, pp. 54–64.
45. W. L. Archer, *Met. Prog.*, 133 (Oct. 1974).
46. L. E. Musgrave, REP-2231 prepared under contract, AT (29-1)-1106 for the U.S. Atomic Energy Commission, Aug. 8, 1974.
47. P. Goerlich, *Ind.-Lackier-Betr.* **43**(11), 383 (1975).
48. J. C. Blanchet, *Surfaces* **14**(94), 51 (1975).
49. L. Skory, J. Fulkerson, and D. Ritzema, *Product Finishing* **38**(5), (1974).
50. H. A. Farber and G. P. Souther, *Am. Dyest. Rep.* **57**, 934 (1968).
51. J. J. Willard, *Text. Chem. Color.* **4**(3), 62 (1972).
52. G. P. Souther, *Am. Dyest. Rep.* **59**, 23 (1970).
53. A. Suzuki, H. Iwata, and J. Nakamura, *Kogyo Kagaku Zasski* **69**(10), 1903 (1966).
54. Ger. Pat. 1,928,199 (Dec. 4, 1969), S. Berkowitz (to FMC Corp.).
55. T. Alfrey, H. C. Haas, and C. W. Lewis, *J. Am. Chem. Soc.* **74**, 2097 (1952).
56. U.S. Pat. 3,919,337 (Nov. 11, 1975), S. C. Gordon and A. N. Theodore (to Diamond Shamrock Corp.).
57. Ref. 25, p. 1291.
58. D. H. R. Barton and K. E. Howlett, *J. Chem. Soc.* 2033 (1951).
59. Fr. Pat. 2,057,606 (June 25, 1971), (to Toa Gosei Chemical Industry Co. Ltd.); U.S. Pat. 3,732,322 (to Toa Gosei Chemical Industry Co., Ltd.).
60. Jpn. Pat. 75-34,003 (Nov. 5, 1975), T. Uchino, K. Sato, and M. Takeuchi (to Asahi Glass Co.).
61. U.S. Pat. 3,304,336 (Feb. 14, 1967), W. A. Callahan (to Detrex Chemical Ind.).
62. D. Gillotay and J. Olbregts, *Int. J. Chem. Kinet.* **8**(1), 11 (1976).
63. Z. Roh, *Chem Prum.* **25**(6), 294 (1975).
64. U.S.S.R. Pat. 195,445 (March 25, 1976), V. A. Poluektov and co-workers.
65. Ger. Pat. 530,649 (July 4, 1929), (to I. G. Farbenindustrie A.G.).
66. W. N. Piper and G. L. Sparschu, unpublished work from The Dow Chemical Company, 1969.
67. F. L. Dunn and J. M. Norris, unpublished work from The Dow Chemical Company, 1969.
68. R. B. Valitov and co-workers, *Neftekhimiya* **15**(6), 917 (1975).
69. S. Tsuda, *Chem. Eng.* 74 (May 4, 1970).
70. N. N. Lebedev, V. F. Shvets, and V. A. Averýanov, *Kinet. Katal.* **12**(3), 560 (1971).
71. J. A. Pearce, *Can. J. Res.* **24F**, 369 (1946).
72. Fr. Pat. 836,979 (Jan. 31, 1939), (to I. G. Farbenindustrie A. G.).
73. Bel. Pat. 602,840 (April 20, 1961), (to PPG Industries Inc.).
74. Ref. 25, p. 1292–1294.
75. J. B. Sherman, *J. Trop. Med. Hyg.* **56**, 139 (1953).
76. H. F. Smyth, Jr., *Am. Ind. Hyg. Assoc. Quart.* **17**, 129 (1956).
77. H. B. Lehmann and F. Flury, *Toxicology and Hygiene of Industrial Solvents*, Trans. by F. King and H. F. Smyth, William and Wilkins, Baltimore, 1943.
78. S. Yllner, *Acta Pharmacol. Toxicol.* **29**, 499 (1971).
79. T. J. Houser and R. B. Bernstein, *J. Am. Chem. Soc.* **80**, 4439 (1958).

80. T. J. Houser and T. Cuzcano, *Int. J. Chem. Kinet.* **7**(3), 331 (1975).
81. V. A. Aver'yanov and G. F. Lebedeva, *Kinet. Katal.* **16**(4), 1073 (1975).
82. Ger. Pat. 248,982 (May 7, 1911), (to Salzbergiverk Neustassfurt and Teilnehmer).
83. Ger. Pat. 843,843 (April 14, 1942), R. Decker and H. Holz (to Wacker Chemie, G.m.b.H.).
84. See ref. 25, pp. 1296–1297.
85. S. Yllner, *Acta Pharmacol. Toxicol.* **29**, 481 (1971).
86. J. Puyo and co-workers, *Bull. Soc. Lorraine. Sci.* **2**(2), 75 (1962).
87. A. Lamouroux and J. Meyer, *Mem. Poudres* **39**, 435 (1957).
88. B. H. Bakker and W. N. Speckamp, *Tetrahedron Lett.* (46), 4065 (1975).
89. U.S. Pat 2,440,731 (May 4, 1948), W. H. Vining and O. W. Cass (to E. 1. du Pont de Nemours & Co., Inc.).
90. Ref. 25, p. 1298.
91. J. S. L. Fowler, *Br. J. Pharmac.* **35**, 530 (1969).
92. Brit. Pat. 841,788 (July 20, 1960), J. S. Elliott and E. D. Edwards (to C. C. Wakefield and Co. Ltd.).
93. W. Davey, *J. Inst. Petrol.* **31**, 73 (1945).
94. Jpn. Pat. 74 17,803 (Feb. 16, 1974), K. Sugiura and T. Miyagawa (to Nippon Oil Co.).
95. Jpn. Pat. 75 22,083 (March 8, 1975), N. Suzuki, J. Okamoto, and O. Matsuda, (to Japan Atomic Energy Res. Institute).

WESLEY L. ARCHER
Dow Chemical U.S.A.

1,2-DICHLOROETHYLENE

1,2-Dichloroethylene [540-59-0] (1,2-dichloroethene) is also known as acetylenedichloride, dioform, α,β-dichloroethylene, and *sym*-dichloroethylene. It exists as a mixture of two geometric isomers:

trans-1,2-Dichloroethylene
[156-60-5]

cis-1,2-Dichloroethylene
[156-59-2]

The isomeric mixture is a colorless, mobile liquid with a sweet, slightly irritating odor resembling that of chloroform. The cis-trans proportions in a crude mixture depend upon the production conditions. The isomers have distinct physical and chemical properties and can be separated by fractional distillation.

1,2-Dichloroethylene can be produced by direct chlorination of acetylene at about 40°C. It is often produced as a by-product in the chlorination of chlorinated compounds (1) and recycled as an intermediate for the synthesis of more useful chlorinated ethylenes (2).

Physical and Chemical Properties

1,2-Dichloroethylene always consists of a mixture of the cis and trans isomers. The physical properties of both isomeric forms are listed in Table 1. Binary and ternary azeotrope data for the cis and trans isomers are given in Table 2.

Table 1. Physical Properties of the Isomeric Forms of 1,2-Dichloroethylene

Property	Trans	Cis
mol wt	96.95	96.95
mp, °C	−49.44	−81.47
bp, °C	47.7	60.2
density	1.2631[10]	1.2917[15]
n_D		
15°C	1.44903	1.45189
20°C	1.44620	1.44900
visc, mPa·s (=cP)		
−50°C	1.005	1.156
−25°C	0.682	0.791
0°C	0.498	0.577
10°C	0.447	0.516
20°C	0.404	0.467
surface tension, 20°C, mN/m (=dyn/cm)	25	28
latent ht of vaporization, kJ/kg[a]	297.9 (48°C)	311.7 (60°C)
heat capacity, 20°C, kJ/(kg·K)[a]	1.158	1.176
vapor pressure, kPa[b]		
20°C	5.3	2.7
−10°C	8.5	5.1
0°C	15.1	8.7
10°C	24.7	14.7
20°C	35.3	24.0
30°C	54.7	33.3
40°C	76.7	46.7
47.7°C	101	66.7
60.25°C		101
soly of the isomer in water, 25°C, g/100 g	0.63	0.35
soly of water in the isomer, 25°C, g/100 g	0.55	0.55
steam distillation point, 101 kPa,[a] °C	45.3	53.8
flash point, °C	4	6
explosion limit in air (vol%)	9.7–12.8	

[a] To convert J to cal, divide by 4.184.
[b] To convert kPa to mm Hg, multiply by 7.5.

The trans isomer is more reactive than the cis isomer in 1,2-addition reactions (3). The cis and trans isomers also undergo benzyne (C_6H_4) cycloaddition (4). The isomers dimerize to tetrachlorobutene in the presence of organic peroxides. Photolysis of each isomer produces a different excited-state (5–6). Oxidation of 1,2-dichloroethylene in the presence of a free-radical initiator or concentrated sulfuric acid produces the corresponding epoxide, which then rearranges to form chloroacetyl chloride (7).

The unstabilized grade of 1,2-dichloroethylene hydrolyzes slowly in the presence of water, producing HCl. Although unaffected by weak alkalies, boiling with aqueous

Table 2. Azeotropes of 1,2-Dichloroethylene Isomers

Binary azeotropes Second component	bp, °C	*trans*-Dichloroethylene bp of azeotrope, °C	Trans isomer in mixture, wt %	*cis*-Dichloroethylene bp of azeotrope, °C	Cis isomer in mixture, wt %
methanol	64.5			51.5	87
ethanol	78.2	46.5	94.0	57.7	90.2
water	100.0	45.3	98.1	55.3	96.65

Ternary azeotropes (ethanol, water, 1,2-dichloroethylene):

Ethanol in mixture, wt %	Water in mixture wt %	Trans isomer in mixture, wt %	Cis isomer in mixture, wt %	bp of azeotrope, °C
1.4	1.1	94.5		44.4
6.65	2.85		90.5	53.8

NaOH, may give rise to an explosive mixture due to monochloroacetylene formation.

Storage and Handling

Because 1,2-dichloroethylene is corrosive to metals, inhibitors are required for storage. The stabilized grades of the isomers can be used or stored in contact with most common construction materials. However, contact with copper or its alloys and with hot alkaline solutions should be avoided to preclude possible formation of explosive monochloroacetylene. The isomers do have explosive limits in air (see Table 1). However, the liquid, even when hot, burns with a very cool flame which self-extinguishes unless the temperature is well above the flash point. A red label is required for shipping 1,2-dichloroethylene.

Toxicity

1,2-Dichloroethylene is moderately toxic and can be absorbed by the skin. It has a TLV of 200 ppm (8). The sweet, chloroform-like odor of *cis*- and *trans*-dichloroethylene does not provide adequate warning of dangerously high vapor concentrations. Thorough ventilation is essential whenever the solvent is used. Symptoms of exposure include narcosis, dizziness, and drowsiness. No data are available on the chronic effects of exposure to low vapor concentrations over extended periods of time.

Uses

1,2-Dichloroethylene can be used as a low-temperature extraction solvent for organic materials such as dyes, perfumes, lacquers, and thermoplastics (9–11). It is also used as a chemical intermediate in the synthesis of other chlorinated solvents and compounds (1) (see Solvents, industrial).

BIBLIOGRAPHY

The "Chlorocarbons and Chlorohydrocarbons Dichloroethylenes" are treated in *ECT* 1st ed. under "Chlorine Compounds, Organic", Vol. 3, pp 786–787 by Jesse Werner, General Aniline & Film Corp., Aniline Works Division; "Chlorocarbons and Chlorohydrocarbons Dichloroethylenes" in *ECT* 2nd ed. Vol. 5, pp. 178–183 by D. W. F. Hardie, Imperial Chemical Industries, Ltd.

1. M. D. Rosenzweig, *Chem. Eng.* 105, (Oct. 18, 1971).
2. Jpn. Pat. 7,330,249, (Sept. 18, 1973) H. Takenobu and co-workers, (to Central Glass Co., Ltd.).
3. G. Berens and co-workers *J. Am. Chem. Soc.* **97**, 7076 (1975).
4. M. Jones, *Tetrahedron Lett.* (53), 5593 (1968).
5. R. Ausubel, *J. Photochem.* **4**, 2418 (1975).
6. R. Ausubel, *Int. J. Chem. Kinet.* **7**, 739 (1975).
7. U.S. Pat. 3,654,358 (April 4, 1977), J. Gaines (to The Dow Chemical Company).
8. 1977 AGGIH publication, American Conference of Governments Industrial Hygienists, p. 15.
9. C. Marsden, *Solvents Guide*, 2nd ed., Interscience Publishers, New York 19B, p. 181.
10. L. Scheflan, *The Handbook of Solvents*, D. Van Nostrand Co., New York, 1953, p. 266.
11. G. Hawley, *The Condensed Chemical Dictionary*, Van Nostrand Reinhold, New York 1977, p. 279.

VIOLETE L. STEVENS
Dow Chemical Company

TRICHLOROETHYLENE

Trichloroethylene [79-01-6], trichloroethene, $CHCl=CCl_2$, is a colorless, sweet smelling, volatile liquid, and a powerful solvent for a large number of natural and synthetic substances. It is nonflammable under conditions of normal use. In the absence of stabilizers, it is slowly decomposed (autoxidized) by air. The oxidation products are acidic, and corrosive. Stabilizers are added to all commercial grades. Trichloroethylene is moderately toxic and has narcotic properties.

Trichloroethylene was first prepared by Fischer in 1864. In the early 1900s, processes were developed in Austria for the manufacture of tetrachloroethane and trichloroethylene from acetylene. Trichloroethylene manufacture began in Germany in 1920 and in the United States in 1925. Demand was stimulated by improvements in metal degreasing techniques during the 1920s and by the growth of dry-cleaning businesses during the 1930s.

The market grew steadily until 1970. Since that time trichloroethylene has come under increasing attack as an atmospheric pollutant and emissions have been severely restricted (see Air pollution).

Physical and Chemical Properties

The physical properties of trichloroethylene are listed in Table 1. Trichloroethylene is immiscible with water but miscible with many organic liquids and it is a versatile solvent. It does not have a flash or fire point. However, it does exhibit a flammable range when high concentrations of vapor are mixed with air and exposed to high-energy ignition sources (see Table 1).

Table 1. Properties of Trichloroethylene

Property	Value		
molecular weight	131.39		
melting point, °C	−87.1		
boiling point, °C	86.7		
specific gravity			
liquid			
20/4°C	1.465		
100/4°C	1.325		
vapor[a] at bp	4.54		
vapor density at bp, kg/m^3	4.45		
n_D			
liquid, 20°C	1.4782		
vapor, 0°C	1.001784		
viscosity, mPa·s (= cP)			
liquid			
20°C	0.58		
60°C	0.42		
vapor at 60°	10,300		
surface tension at 20°C, mN/m (= dyn/cm)	26.4		
heat capacity at 20°C, J/(kg·K)[b]			
liquid	941		
vapor	653		
critical temp, °C	271.0		
critical pressure, MPa[c]	5.02		
thermal conductivity, W/(m·K)			
liquid	138.5		
vapor, at bp	8.34		
coeff cubical expansion, liq, 0–40°C	0.00119		
dielectric constant, liquid, at 16°C	3.42		
dipole moment, C·m[d]	3.0×10^{-30}		
heat of combustion, MJ/kg[b]	7.325		
heat of formation, MJ/(kg·mol)[b]			
liquid	4.18		
vapor	−29.3		
latent heat of evaporation at bp, kJ/kg[b]	240		
explosive limits, vol % in air			
25°C	8.0–10.5		
100°C	8.0–52		
vapor pressure[e], kPa[f]			
Antoine constants	A	B	C
	5.94606	1187.51	214.474
solubility, g			
H$_2$O in 100 g trichloroethylene			
0°C	0.010		
20°C	0.0225		
60°C	0.080		
trichloroethylene in 100 g H$_2$O			
20°C	0.107		
60°C	0.124		

[a] Air = 1.
[b] To convert J to cal, divide by 4.184.
[c] To convert MPa to atm, divide by 0.101.
[d] To convert C·m to debye, divide by 3.336×10^{-30}.
[e] $\log_{10} P = A - \left(\dfrac{B}{T+C}\right)$.
[f] To convert kPa to mm Hg, multiply by 7.5.

From the industrial point of view, the most important reactions of trichloroethylene are atmospheric oxidation and degradation catalyzed by aluminum chloride. The autoxidation is catalyzed by free radicals and is greatly accelerated by elevated temperature and exposure to light, especially ultraviolet radiation. The degradation is believed to proceed either by dimerization to form hexachlorobutene, or by addition of oxygen to give intermediates (1) and (2).

Compound (1) decomposes to form dichloroacetyl chloride, Cl$_2$HCCOCl, which in the presence of water, decomposes to dichloroacetic acid and hydrochloric acid (HCl), with consequent increases in the corrosive action of the solvent on metal surfaces. Compound (2) decomposes to yield phosgene, carbon monoxide, and hydrogen chloride with an increase in the corrosive action on metal surfaces (see also Tetrachloroethylene).

In the presence of aluminum, oxidative degradation or dimerization supply HCl for the formation of aluminum chloride, which catalyzes further dimerization to hexachlorobutene. The latter is decomposed by heat to give more HCl. The result is a self-sustaining pathway to solvent decomposition. Sufficient quantities of aluminum can cause violent decomposition (1).

All commercial grades of trichloroethylene are stabilized against autoxidation and AlCl$_3$-catalyzed degradation. Amine-stabilized products are still sold today, but most vapor-degreasing grades contain neutral inhibitor mixtures (2–4) including a free-radical scavenger, such as an amine or pyrrole, to prevent the initial oxidation reaction. Epoxides, such as butylene oxide and epichlorohydrin, are added to scavenge any free HCl and AlCl$_3$.

Trichloroethylene is not readily hydrolyzed by water. Under pressure at 150°C, it gives glycolic acid, CH$_2$OHCOOH, with alkaline hydroxides. Reaction with sulfuric acid (90%) yields monochloroacetic acid, CH$_2$ClCOOH. Hot nitric acid reacts with trichloroethylene violently, producing complete oxidative decomposition. Under carefully controlled conditions, nitric acid gives trichloronitromethane (chloropicrin) and dinitrochloromethane (5). Strong alkalies, dehydrochlorinate trichloroethylene with production of spontaneously explosive and flammable chloroacetylenes. Dichloroacetylene, C$_2$Cl$_2$, can also be formed from trichloroethylene in the presence of epoxides and ionic halides (6).

In the presence of catalysts, trichloroethylene is readily chlorinated to pentachloro- and hexachloroethane. Bromination yields 1,2-dibromo-1,1,2-trichloroethane [13749-38-7]. The analogous iodine derivative has not been reported. Fluorination with hydrogen fluoride in the presence of antimony trifluoride produces 2-chloro-1,1,1-trifluoroethane [75-88-7] (7). Elemental fluorine gives a mixture of chlorofluoro derivatives of ethane, ethylene, and butane.

Liquid trichloroethylene has been polymerized by irradiation with ^{60}Co γ rays or 20-keV x-rays (8).

Trichloroethylene has a chain-transfer constant of <1 when copolymerized with vinyl chloride (9), and is used extensively to control the molecular weight of poly(vinyl chloride) polymer (see Vinyl polymers).

A variety of trichloroethylene copolymers have been reported, none with commercial significance. The alternating copolymer with vinyl acetate has been patented as an adhesive (10) and as a flame retardant (11–12). Copolymerization with 1,3-butadiene and its homologues has been reported (13–15). Other comonomers include acrylonitrile (16), isobutyl vinyl ether (17), maleic anhydride (18), and styrene (19).

Terpolymers have been made with vinyl chloride–vinylidene chloride (20) and vinyl acetate–vinyl alcohol (21).

Manufacture

As late as 1968, 85% of the production capacity in the United States was based on acetylene, but rising acetylene costs reduced this figure to 8% by 1976 (22), and now most trichloroethylene is made from ethylene or dichloroethane.

From Acetylene. The acetylene-based process consists of two steps. First acetylene is chlorinated in tetrachloroethane. The reaction is exothermic (402 kJ/mol or 96 kcal/mol) but is maintained at 80–90°C by the vaporization of solvent and product. Catalysts include ferric chloride, and sometimes phosphorus chloride and antimony chloride (23).

The product is then dehydrohalogenated to trichloroethylene at 96–100°C in aqueous bases such as $Ca(OH)_2$ (24), or by thermal cracking, usually over a catalyst (23) such as barium chloride on activated carbon or silica or aluminum gels at 300–500°C. The yield of trichloroethylene (22) is about 94% based on acetylene (see Acetylene-derived chemicals). A major disadvantage of the alkaline process is the loss of chlorine as calcium chloride. In thermal cracking the chlorine can be recovered as hydrochloric acid, an important feedstock in many chemical processes. Since it poisons the catalysts during thermal cracking, all ferric chloride must be removed from the tetrachloroethane feed (23). Tetrachloroethane can also be cracked to trichloroethylene without catalysts at 330–770°C, but considerable amounts of tarry by-products are formed.

Chlorination of Ethylene. Dichloroethane, produced by chlorination of ethylene, can be further chlorinated to trichloroethylene and tetrachloroethylene. The exothermic reaction is carried out at 280–450°C. Temperature is controlled by a fluidized bed, a molten salt bath, or the addition of an inert material such as perchloroethylene. The residence time in the reactor varies from 2 to 30 s, depending on conditions (23). Catalysts include potassium chloride and aluminum chloride (25), fuller's earth (26), graphite (27), activated carbon (28), and activated charcoal (26).

Maximum conversion to trichloroethylene (75% of dichloroethane feed) is achieved at a chlorine to dichloroethane ratio of 1.7:1. Tetrachloroethylene conversion reaches a maximum (86% conversion of dichloroethane) at a feed ratio of 3.0:1 (23).

Oxychlorination of Ethylene or Dichloroethane. Ethylene or dichloroethane can be chlorinated to a mixture of tetrachloroethylene and trichloroethylene in the presence of oxygen and catalysts. The reaction is carried out in a fluidized-bed reactor at 425°C and 138–207 kPa (20–30 psi). The most common catalysts are mixtures of potassium and cupric chlorides. Conversion to chlorocarbons ranges from 85–90%, with 10–15% lost as carbon monoxide and carbon dioxide (23). Temperature control is critical. Below 425°C, tetrachloroethane becomes a major product (57.3 wt % of crude product at 330°C) (29). Above 480°C, excessive burning and decomposition reactions occur. Product ratios can be controlled, but less readily than in the chlorination process. Reaction vessels must be constructed of corrosion-resistant alloys.

Other Routes. A unique process that produces vinyl chloride, trichloroethylene, dichloroethane, and trichloroethane simultaneously has been developed by Produits Chemiques Pechiney-Saint-Gobain in France (30). Dichloroethylene is chlorinated directly at low temperature to tetrachloroethane, which is then thermally cracked to give trichloroethylene and hydrochloric acid. The dichloroethylene feed is coproduced with vinyl chloride in a hot chlorination reactor, using chlorine and ethylene as feedstocks.

A Japanese process, developed by Taogosei Chemical Company, chlorinates ethylene directly in the absence of oxygen, at about 811 kPa (8 atm) pressure and 100–130°C (31). The products are tetrachloroethanes and pentachloroethane, which are then thermally cracked at 912 kPa (9 atm) and 429–451°C to produce a mixture of trichloroethylene, perchloroethylene, and hydrochloric acid.

Shipping and Storage

Trichloroethylene is shipped in both compartmented and uncompartmented tank trucks and tank cars, and also in 208-L (55-gal) steel drums. It is stored in mild-steel tanks equipped with breathing vents and driers to prevent accumulation of moisture. The solvent can be transferred through seamless black-iron pipes with gasketing materials of compressed asbestos, asbestos reinforced with metal, or asbestos impregnated with Teflon or Viton using centrifugal or positive-displacement pumps of cast iron or steel. Small quantities of trichloroethylene may be stored safely in amber or green glass containers.

Trichloroethylene is toxic, therefore, all containers should bear warning labels against breathing vapors, ingesting the liquid, splashing solvent in eyes or on skin and clothing, and using it near an open flame.

Although the flammability hazard is very low, ignition sources should not be present when trichloroethylene is used in highly confined or unventilated areas. Tanks in which flammable concentrations could develop should be grounded to prevent buildup of static electric charges.

Economic Aspects

Worldwide capacities and production figures (in thousands of metric tons) are: Western Europe capacity 440, production 297.0; Japan capacity 155, production 84.6; and United States capacity 214, production 127.5. The sources for this data are the Japanese Ministry of International Trade and Industry and the United States International Trade Commission.

United States production and price statistics are presented in Table 2. The trichloroethylene market in the United States has been shrinking sharply since 1970 owing to pressures from environmental and safety legislation. Similar but less intense pressures have weakened the markets in Europe and Japan (32) (see under Health and safety factors).

In 1966, the Los Angeles Air Pollution Control Board designated trichloroethylene a photochemically reactive solvent that decomposes in the lower atmosphere, con-

Table 2. United States Trichloroethylene Production and Prices [a]

Year	Production, thousands of metric tons	Price [b], ¢/kg
1956	157.0	23.70
1960	160.0	28.11
1965	197.1	22.60
1970	277.1	19.29
1972	193.5	21.50
1974	176.0	28.11
1976	137.4	41.34
1977	118.2	46.20

[a] Source: U.S. International Trade Commission.
[b] Tank-truck quantities delivered.

tributing to air pollution. In 1970 all states were required to submit pollution control plans to EPA to meet national air quality standards (see Air pollution). Many followed the example of California and restricted the emission of trichloroethylene.

Concurrently, the Occupational Safety and Health Administration (OSHA) was established and given power to set exposure levels for industrial chemicals. For worker exposure to trichloroethylene vapor, OSHA set a maximum 8-h time-weighted average (TWA) concentration of 100 ppm. This severely restricted certain applications, and many organizations converted to other chlorinated solvents. As a result, the U.S. production of trichloroethylene declined by 50% from a peak of 277,100 metric tons in 1970 to 137,400 metric tons in 1976.

This decline was hastened by a series of plant shutdowns between 1971 and 1973 resulting primarily from the high costs of the acetylene-based process. No new production capacity is planned in the United States for the foreseeable future.

Shortages, together with rapidly escalating fuel and feedstock prices, have led to a dramatic increase in the price of trichloroethylene, which has more than doubled between 1972 and 1976. Unless there is a drastic change in the future worldwide energy situation, this upward price trend is expected to continue.

Specifications and Standards

Commercial grades of trichloroethylene, formulated to meet use requirements, differ in the amount and type of added inhibitor. The grades sold in the United States include: (1) a neutrally inhibited vapor-degreasing grade; (2) an alkaline- (amine) inhibited vapor-degreasing grade; (3) a technical grade for use in formulations; (4) an extraction grade; (5) a high-purity, low-residue grade; and (6) a paint application grade. *U.S. Federal Specification O-T-634b* lists specifications for a regular and a vapor-degreasing grade.

Apart from added stabilizers, commercial grades of trichloroethylene should not contain more than the following amounts of impurities: water, 100 ppm; acidity (as HCl), 5 ppm; insoluble residue, 10 ppm. Free chlorine should not be detectable. Test methods have been established by ASTM to determine the following characteristics of trichloroethylene: acid acceptance, acidity or alkalinity, color, corrosivity on metals, nonvolatile-matter content, pH of water extractions, relative evaporation rate, specific gravity, water content, water-soluble halide-ion content, and halogen content (33).

Health and Safety Factors (Toxicity)

Trichloroethylene is intrinsically toxic, primarily because of its anesthetic effect on the central nervous system. Overexposure can lead to unconsciousness and death. Because it is widely used—more than 200,000 workers are exposed to trichloroethylene in the United States annually (34)—its physiological effects have been extensively studied (see Industrial hygiene and toxicology).

Exposure occurs almost exclusively by vapor inhalation, followed by rapid absorption into the bloodstream. At concentrations of 150–186 ppm, 51–70% of the trichloroethylene inhaled is absorbed. Metabolic breakdown occurs by oxidation to chloral hydrate, followed by reduction to trichloroethanol, part of which is further oxidized to trichloroacetic acid (35–37). Absorbed trichloroethylene that is not metabolized is eventually eliminated through the lungs (38).

The OSHA maximum time-weighted average (TWA) concentration has been set at 100 ppm for 8-h exposure.

Trichloroethylene is a central nervous system depressant, and high vapor concentrations will cause headache, vertigo, tremors, nausea and vomiting, fatigue, intoxication, unconsciousness, and death.

It is estimated that concentrations of 3000 ppm cause unconsciousness in less than ten minutes (39). Anesthetic effects have been reported at concentrations of 400 ppm after 20 min exposure. Decrease in psychomotor performance at a trichloroethylene concentration of 110 ppm has been reported (38), whereas other studies find no decrease up to concentrations of 200 ppm (40–43).

Victims of overexposure to trichloroethylene should be removed to fresh air and medical attention should be obtained immediately. A self-contained breathing device should be used wherever high vapor concentrations are expected.

The distinctive odor of trichloroethylene may not necessarily provide adequate warning of exposure, because it quickly desensitizes olfactory responses.

Fatalities have occurred when unprotected workers have entered unventilated areas with high vapor concentrations of trichloroethylene or other chlorinated solvents. For a complete description of proper entry to vessels containing any chlorinated solvent vapor, see under Trichloroethane.

Ingestion of trichloroethylene may cause liver deterioration, kidney malfunction, cardiac arrythmia, and coma (38); vomiting should not be induced and medical attention should be obtained immediately.

Protective gloves and aprons should be used to prevent skin contact, which may cause dermatitis (44–46). Eyes should be washed immediately after contact or splashing with trichloroethylene.

The National Cancer Institute reported in 1975 that massive oral doses of trichloroethylene caused liver tumors in mice, but not in rats (47). However, the American Conference of Governmental Industrial Hygienists (ACGIH) does not consider any chemical "an occupational carcinogen of any practical significance" when mouse tumors are caused by daily oral dosages above 500 mg per kg of body weight. This specifically excludes trichloroethylene from consideration as a carcinogen (48). Additional studies are being conducted by the Manufacturing Chemists Association.

Teraterogenetic studies conducted by Dow Chemical showed no effect of trichloroethylene (49).

Uses

Approximately 80% of the trichloroethylene produced in the United States is consumed in the vapor degreasing of fabricated metal parts (see Metal surface treatment); the remaining 20% is divided equally between exports and miscellaneous applications (22). In 1970, trichloroethylene accounted for 82% of all the chlorinated solvents used in vapor degreasing. By 1976, that share had declined to 42%. (Estimates were done by The Dow Chemical Company). A variety of miscellaneous applications include use of trichloroethylene as a component in adhesive and paint-stripping formulations, a low-temperature heat-transfer medium, a nonflammable solvent carrier in industrial paint systems, and a solvent base for metal phosphatizing systems. Trichloroethylene is used in the textile industry as a carrier solvent for spotting fluids and as a solvent in waterless preparation, dying, and finishing operations (see Dye carriers).

Trichloroethylene is widely used as a chain-transfer agent in the production of poly(vinyl chloride). An estimated 4500–6800 metric tons are consumed annually in this application (see Vinyl polymers).

BIBLIOGRAPHY

"Chlorocarbons and Chlorohydrocarbons, Trichloroethylene" treated under "Chlorine Compounds, Organic" in *ECT* 1st ed., Vol. 3, pp. 788–794 by Janet Searles and H. A. McPhail, E. I. du Pont de Nemours & Co., Inc.; "Trichloroethylene" under "Chlorocarbons and Chlorohydrocarbons" in *ECT* 2nd ed. Vol. 5, pp. 183–195, by D. W. F. Hardie, Imperial Chemical Industries Ltd.

1. L. Metz and A. Roedig, *Chem. Ing. Technick* **21**, 191 (1949).
2. U.S. Pat. 2,795,623 (June 11, 1957), F. W. Starks (to E. I. du Pont de Nemours & Co., Inc.).
3. U.S. Pat. 2,818,446 (Dec. 31, 1957), F. W. Starks (to E. I. du Pont de Nemours & Co., Inc.).
4. Brit. Pat. 794,700 (May 7, 1958), H. B. Copelin (to E. I. du Pont de Nemours & Co., Inc.).
5. R. B. Burrows and L. Hunter, *J. Chem. Soc.*, 1357 (1932).
6. B. Dobinson and G. E. Green, *Chem. Ind.*, 214 (Mar. 4, 1972).
7. A. J. Rudge, *The Manufacture and Use of Fluorine and its Compounds*, Oxford University Press (for Imperial Chemical Industries Ltd.), Cambridge, Mass., 1962, p. 71.
8. H. L. Cornish, Jr., *U.S. At. Energy Comm. TID-21388*, 1964.
9. J. Pichler and J. Rybicky, *Chem. Prum.* **16**, 559 (1966).
10. Jpn. Pat. 72 45,415 (Nov. 16, 1972), Kimimura, Takayoshi, and S. Wataru (to Hoechst Gosel Co. Ltd.).
11. U.S. Pat. 3,846,508 (Nov. 5, 1974), D. H. Heinert (to The Dow Chemical Co.).
12. U.S. Pat. 3,907,872 (Sept. 23, 1975), D. H. Heinert (to The Dow Chemical Co.).
13. Ger. Pat. 719,194 (Mar. 26, 1942), H. Kopff and C. Rautenschauch (to I.G. Farbenindustrie, A.G.).
14. Z. Jedlinski and E. Grzywa, *Polimery* **11**, 560 (1966).
15. Pol. Pat. 53,152 (Feb. 28, 1967), E. Grzywa and Z. Jedlinski (to Zaklady Chemiczne "Oswiecim").
16. S. U. Mullik and M. A. Quddus, *Pak. J. Sci. Ind. Res.* **12**(3), 181 (1970).
17. T. A. DuPlessis and A. C. Thomas, *J. Polym. Sci. Polym. Chem. Ed.* **11**, 2681 (1973).
18. R. A. Siddiqui and M. A. Quddus, *Pak. J. Sci. Ind. Res.* **14**(3), 197 (1971).
19. H. Asai, *Nippon Kagaku Zasshi* **85**, 252 (1964).
20. E. Krotki and J. Mitus, *Polimery* **9**, 155 (1964).
21. Jpn. Pat. 71 01,719 (Jan. 16, 1971), Shimokawa and Wataru (to Hekisto Gosei Co. Ltd.).
22. J. L. Blackford, "Trichloroethylene," in *Chemical Economics Handbook*, Stanford Research Institute, Menlo Park, Calif., 697.301A-697.302Y, Nov. 1975.
23. L. M. Elkin, *Process Economics Program, Chlorinated Solvents, Report No. 48*, Stanford Research Institute, Menlo Park, Calif., Feb. 1969.
24. Ger. Pat. 901,774 (Nov. 3, 1940), (to Wacker Chemie, G.m.b.H.).
25. U.S. Pat. 2,140,548 (Dec. 30, 1938), J. H. Reilly (to The Dow Chemical Co.).
26. Brit. Pat. 673,565 (June 11, 1952), (to Diamond Alkali).

27. U.S. Pat. 2,725,412 (Nov. 29, 1955), F. Conrad (to Ethyl Chem. Co.).
28. Neth. Appl. 6,607,204 (Nov. 28, 1966), F. Sanhaber (to Donau Chemic).
29. Fr. Pat. 1,435,542 (March 7, 1966), A. C. Schulz (to Hooker Chemical).
30. M. D. Rosenzweig, *Chem. Eng.* **78**(24), 105 (Oct. 18, 1971).
31. S. Tsuda, *Chem. Eng.* **77**(10), 74 (May 4, 1970).
32. *Eur. Chem. News* **30**(769), 6 (Jan. 14, 1977).
33. *1976 Annual Book of ASTM Standards, Part 30,* ASTM, Easton, Md., 1976.
34. *Criteria Document: Recommendations for an Occupational Exposure Standard for Trichloroethylene,* NIOSH, Contract No. HSM 73-11025, Nov. 1973.
35. B. Soucek and D. Vlachove, *Br. J. Ind. Med.* **17,** 60 (1960).
36. V. Bartonicek, *Br. J. Ind. Med.* **19,** 134 (1962).
37. M. Ogata, Y. Takatsuka, and K. Tomokuni, *Br. J. Ind. Med.* **28,** 386 (1971).
38. D. M. Aviado and co-workers, *Methyl Chloroform and Trichloroethylene in the Environment,* CRC Press, Cleveland, Ohio, 1976.
39. E. O. Longley and R. Jones, *Arch. Environ. Health* **7,** 249 (1963).
40. R. D. Steward and co-workers, *Arch. Environ. Health* **20,** 64 (1970).
41. G. J. Stopps and W. McLaughlin, *Am. Ind. Hyg. Assoc. J.* **29,** 43 (1967).
42. R. J. Vernon and R. K. Ferguson, *Arch. Environ. Health* **18,** 894 (1964).
43. R. K. Ferguson and R. J. Vernon, *Arch. Environ. Health* **29,** 462 (1970).
44. K. Kadlec, *Cesk. Dermatol.* **38,** 395 (1963).
45. S. M. Peck, *J. Am. Med. Assoc.* **125,** 190 (1944).
46. J. M. Schirren, *Berufs-Dermatosen* **19,** 240 (1971).
47. *Carcinogenesis Bioassay of Trichloroethylene,* NCI-CG-TR-2, U.S. Dept. of HEW, Feb. 1976, 197 pp.
48. *TLVs® Threshold Limit Values for Chemical Substances and Physical Agents in the Workroom Environment with Intended Changes for 1976,* ACGIH, Cincinnati, Ohio, 1976.
49. B. A. Schwetz, B. K. Leong, and P. J. Gehring, *Toxicol. Appl. Pharmacol.* **32,** 84 (1975).

W. C. McNeill, Jr.
Dow Chemical U.S.A.

TETRACHLOROETHYLENE

Tetrachloroethylene[127-18-4], tetrachloroethene, perchloroethylene, $CCl_2=CCl_2$, is known as perc and sold under a variety of trade names. It is a nonflammable liquid with a pleasant, ethereal odor, and the most stable of the chlorinated ethanes and ethylenes, requiring only small amounts of stabilizers. It is a powerful solvent for many substances, and is used in drycleaning, metal degreasing, and textile processing (see Drycleaning; Metal surface treatments; Solvents, industrial; Textiles).

Tetrachloroethylene was first prepared in 1821 by Faraday by the thermal decomposition of hexachloroethane. Manufacture began before World War I in the United Kingdom and Germany, and in the United States in 1925. In the interwar period, development of the small drycleaning unit stimulated demand in the United States, but trichloroethylene remained the principal drycleaning agent in the United Kingdom and Europe. After World War II, tetrachloroethylene began to assume dominance in Europe as well.

Physical and Chemical Properties

The physical properties of tetrachloroethylene are presented in Table 1. Tetrachloroethylene dissolves sulfur, iodine, mercuric chloride, and appreciable amounts of aluminum chloride. It is a solvent for a variety of organic compounds, as well as a large number of substances such as fats, oils, tars, rubber, and resins. It is miscible with the chlorinated organic solvents and most other common solvents and forms about sixty binary azeotropes (1).

Commercially pure, stabilized tetrachloroethylene can be used in contact with the common construction metals at about 140°C, even in the presence of air, water, and light. It resists hydrolysis at temperatures up to 150°C (2). In the absence of catalysts, air, or moisture, tetrachloroethylene is stable to about 500°C. It affords various decomposition products depending on conditions (3), but mostly hydrogen chloride and phosgene (4). In the absence of light tetrachloroethylene is unaffected by oxygen (5). However, under uv radiation in the presence of air or oxygen, tetrachloroethylene undergoes autoxidation to trichloroacetyl chloride. An intermediate stage in this reaction is the formation of the peroxy compounds (1) and (2).

Compound (1) undergoes a rearrangement to give trichloroacetyl chloride and oxygen. Compound (2) breaks down to yield two molecules of phosgene. The rearrangement

Table 1. Properties of Tetrachloroethylene

Property	Value
molecular weight	165.83
melting point, °C	−22.7
boiling point at 101 kPa[a], °C	121.2
specific gravity, liquid	
10/4°C	1.63120
20/4°C	1.62260
30/4°C	1.60640
120/4°C	1.44865
vapor density at bp at 101 kPa[a], kg/m^3	5.8
viscosity mPa·s (= cP)	
liquid	
15°C	0.932
25°C	0.839
50°C	0.657
75°C	0.534
vapor at 60°C	9900
surface tension, mN/m (= dyn/cm)	
15°C	32.86
30°C	31.27
thermal capacity, kJ/(kg·K)[b]	
liquid of 20°C	0.858
vapor at 100°C	0.611
thermal conductivity, mW/(m·K)	
liquid	126.6
vapor at bp	8.73
heat of combustion	
constant pressure with formation of aq HCl, kJ/(mol)[b]	679.9
constant volume at 18.7°C, kJ/(mol)[b]	831.8
latent heat of vaporization at 121.2°C, kJ/(mol)[b]	34.7
critical temperature, °C	347.1
critical pressure, MPa[c]	9.74
latent heat of fusion, kJ/(mol)[b]	10.57
heat of formation, kJ/(mol)[b]	
vapor	−25
liquid	12.5
n_D at 20°C	1.50547
dielectric constant at 1 kHz, 20°C	2.20
electrical conductivity at 20°C, 10^{15} (Ω·m)$^{-1}$	55.8
coefficient cubical expansion at 15–90°C, av	0.001079
vapor pressure, kPa[a]	
−20.6°C	0.1333
13.8	1.333
40.0	5.466
60.0	13.87
80	30.13
100	58.46
121.2	101.3
solubility at 25°C, mg	
tetrachloroethylene in 100 g H$_2$O	15
H$_2$O in 120 g tetrachloroethylene	8

[a] To convert kPa to mm Hg, multiply by 7.5.
[b] To convert J to cal, divide by 4.184.
[c] To convert MPa to atm, divide by 0.101.

to trichloroacetyl chloride predominates. The slow decomposition of tetrachloroethylene that occurs on prolonged storage in light and the presence of air is the result of autoxidation, which is inhibited by amines or phenols. Generally, tetrachloroethylene requires less stabilizer than trichloroethylene.

Excess hydrogen, in the presence of reduced nickel catalyst at 220°C, results in total decomposition to hydrogen chloride and carbon.

Photochlorination of tetrachloroethylene yields hexachloroethane. This reaction was originally observed by Faraday. Bromination to a mixture of monobromotrichloroethane and dibromodichloroethane is readily effected by heating with aluminum bromide at 100°C (4). In the presence of zirconium fluoride catalyst at 225–400°C, a mixture of hydrogen fluoride and chlorine give 1,2,2-trichloro-1,1,2-trifluoroethane, $CClF_2CCl_2F$ (7).

Tetrachloroethylene reacts explosively with butyllithium in petroleum ether solution (8). An explosive reaction also occurs with metallic potassium at its melting point, but not with sodium (9).

When heated at 110–120°C with o-benzenedithiol, o-$C_6H_4(SH)_2$, in the presence of sodium ethoxide, tetrachloroethylene gives 2,2'-bis-1,3-benzdithiolene (3) (10).

(3)

Formaldehyde and concentrated sulfuric acid at 80°C give 2,2-dichloropropanoic acid (11). In the presence of dibenzoyl peroxide copolymers with styrene, vinyl acetate, methyl acrylate, and acrylonitrile (12–13) are formed.

Corrosion of aluminum, iron, and zinc by tetrachloroethylene, which is negligible unless water is present, can be inhibited by the addition of stabilizers (14).

Manufacture

For many years tetrachloroethylene was produced almost exclusively from acetylene and chlorine via trichloroethylene (see Acetylene-derived chemicals). However, because of the high cost of acetylene and the cost of recovery of chlorine values, other hydrocarbons are now employed as feedstocks.

From Hydrocarbons. Simultaneous chlorination and pyrolysis processes utilize various hydrocarbons such as methane, ethane, propane, or higher paraffins (or their chlorinated derivatives). Typical reactions involved, eg, based on propane, are as follows:

$$CH_3-CH_2-CH_3 + 8\,Cl_2 \longrightarrow CCl_2{=}CCl_2 + CCl_4 + 8\,HCl$$

$$2\,CCl_4 \longrightarrow CCl_2{=}CCl_2 + 2\,Cl_2$$

In this process (see Fig. 1), chlorine, a light hydrocarbon, and several recycle streams are mixed and fed to a chlorination furnace which is maintained at 550–700°C. The products are carbon tetrachloride and tetrachloroethylene; the latter is probably formed largely by pyrolysis of the former. The effluent gases from the chlorinator are quenched, after which the chlorinated hydrocarbons are separated from the quenching medium in a blow-back column. The mixture of chlorohydrocarbons is then frac-

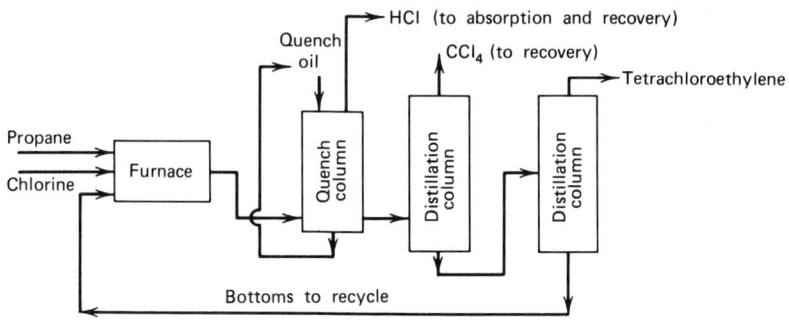

Figure 1. Flow diagram of the simultaneous chlorination and pyrolysis of propane.

tionated, and the more volatile carbon tetrachloride is recycled to the furnace. The crude tetrachloroethylene in the bottom fraction is purified by distillation, and the bottoms from this operation are also recycled to the chlorination furnace. The overall yield of tetrachloroethylene is higher than 95%, based on chlorine consumption. Ethane is preferred for this process in the United States since it is the least expensive raw material. About 40% of current United States tetrachloroethylene production is based on ethane and propane.

From Ethylene Dichloride. Tetrachloroethylene is a coproduct with trichloroethylene (qv) in the single-stage oxychlorination of ethylene dichloride (dichloroethane) with chlorine. The ratio of trichloroethylene to perchloroethylene can be varied to some degree by adjusting mole feed ratios of ethylene dichloride, chlorine, and oxygen. The reactions involved are as follows:

$$2\ ClH_2C-CH_2Cl + 5\ Cl_2 \rightarrow Cl_2HC-CHCl_2 + Cl_3C-CHCl_2 + 5\ HCl$$

$$Cl_2HC-CHCl_2 + Cl_3C-CHCl_2 \rightarrow Cl_2C=CHCl + 2\ HCl + Cl_2C=CCl_2$$

$$4\ HCl + O_2 \rightarrow 2\ H_2O + 2\ Cl_2$$

overall: $8\ ClH_2C-CH_2Cl + 6\ Cl_2 + 7\ O_2 \rightarrow 4\ ClHC=CCl_2 + 4\ Cl_2C=CCl_2 + 14\ H_2O$

As illustrated in Figure 2, ethylene dichloride, chlorine, oxygen, steam, and recycled chlorinated compounds are fed to a fluid-bed reactor employing an inexpensive oxychlorination catalyst such as potassium chloride and cupric chloride. The reactor is maintained at about 425°C and a pressure of 138–207 kPa (20–30 psi). In the PPG reactor (15) the feedstock can be either ethylene or chlorinated hydrocarbons (alone or in various combinations) to yield trichloroethylene, tetrachloroethylene, or a mixture of the two.

After vent scrubbing, the condensed crude product and the weak hydrochloric acid by-product are separated, and the crude product is dried by azeotropic distillation (qv). In the tetrachlor–trichlor column, the crude product is split into two streams, one rich in trichloroethylene and the other in tetrachloroethylene. The latter, containing trichloroethane, tetrachloroethylene, and components with boiling points higher than tetrachloroethylene is fed to the trichloroethane still. The overheads from this column are recycled to the reactor and the bottoms are fed to the tetrachloroethylene column. The overhead from this column is 99.9+ wt % purity tetrachloroethylene; it is neutralized with ammonia, washed, and dried. The bottoms from the tetrachloroethylene column are fed to a column that removes the heavier tars and carbon; the overheads are recycled to the reactor. About 35% of the tetrachloroethylene

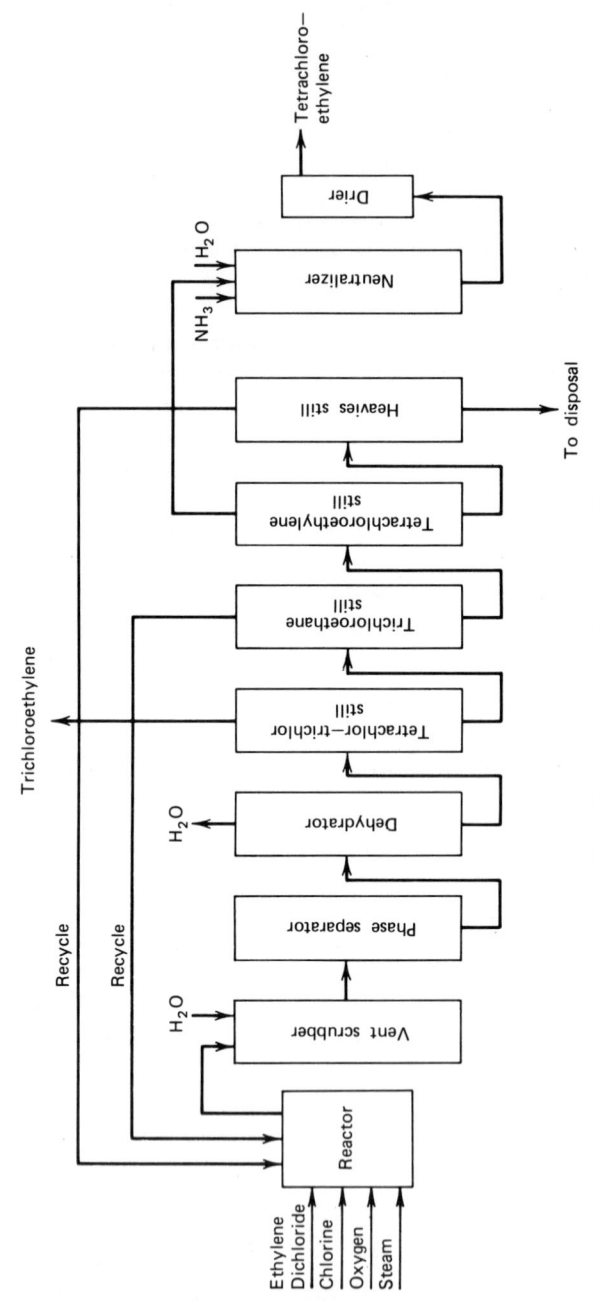

Figure 2. Flow diagram of the single-stage oxychlorination process.

produced in the United States is made from ethylene dichloride by this and other processes.

Shipping and Storage

Tetrachloroethylene is shipped by barge, tank car, tank truck, and in 208-L (55-gal) drums. It is stored in mild steel tanks equipped with breathing vents and chemical driers. It can be transferred through seamless black iron pipes, with gasketing materials of compressed asbestos, asbestos reinforced with metal, or asbestos impregnated with Teflon or Viton, employing centrifugal or positive displacement pumps of cast iron or steel construction.

Small quantities of tetrachloroethylene may be stored safely in green or amber glass containers. Tetrachloroethylene is toxic (see below) and all containers should bear warning labels.

Economic Aspects

The 1974 distribution of world capacity and demand is given in Table 2. Total United States capacity in 1976 was approximately 460,000 metric tons. Individual manufacturers and their capacities are listed in Table 3. United States production and sales data are given in Table 4.

Table 2. 1974 Tetrachloroethylene World Production Capacity and Demand, Thousands of Metric Tons

Area	Capacity	Demand
United States	474	331
Europe	517	431
Japan	83	57
Canada	26	15
Latin America	1	11
Total	1101	845

Specifications

Commercial grades of tetrachloroethylene differ in the amount and type of added inhibitor. Grades include (1) a neutrally inhibited vapor-degreasing grade; (2) a dry-cleaning grade; (3) a technical grade for use in formulations; and (4) a high-purity, low-residue grade. U.S. Federal Specification O-T-236A covers tetrachloroethylene.

Apart from stabilizers, commercial grades should not contain more than 0.01 wt % water, 0.0005 wt % acidity (as HCl), and 0.001 wt % insoluble residue. Free chlorine should not be detectable. ASTM has established standard test methods to determine acid acceptance, acidity or alkalinity, color, corrosivity on metals, nonvolatile-matter content, pH of water extractions, relative evaporation rate, specific gravity, water content, water-soluble halide ions, and halogens (16).

Table 3. United States Tetrachloroethylene Producers and Their Capacities

Producer	Location	1976 Capacity[a], thousands of metric tons
Diamond Shamrock	Deer Park, Tex.	74.8
Dow Chemical	Freeport, Tex.	54.4
Dow Chemical	Pittsburgh, Calif.	9.1
Dow Chemical	Plaquemine, La.	68.0
Ethyl	Baton Rouge, La.	22.7
Hooker Chemical	Taft, La.	18.1
PPG	Lake Charles, La.	90.7
Stauffer Chemical	Louisville, Ky.	31.7
Vulcan Materials	Geismar, La.	68.0
Vulcan Materials	Wichita, Kansas	22.7
Total		460.2

[a] Based on announced capacities and trade estimates, without allowing for possible shutdowns or mothballing. Capacities are flexible, since other chlorinated hydrocarbons can be made in the same equipment. (Not included in this table is DuPont's captive tetrachloroethylene production at Corpus Christi, Tex., which is used for the synthesis of fluorocarbons.)

Table 4. United States Tetrachloroethylene Production and Sales[a], Thousands of Metric Tons

Year	Production	Sales[b]	Price[c], $/t
1955	80.7	72.8	243
1960	95.0	84.9	259
1965	194.8	174.6	220
1969	288.1	277.4	215
1970	320.6	290.4	226
1971	319.6	296.6	226
1972	333.0	332.2	226
1973	320.1	333.1	226
1974	333.1	321.5	243
1975	308.0	267.3	353

[a] Source: U.S. International Trade Commission.
[b] Domestic and export.
[c] Tank prices. Basis: 1955–1956, delivered east; 1957–1970, delivered; 1971–1975, industrial grade, consumers, delivered.

Toxicity

The toxicity of tetrachloroethylene is due primarily to its anesthetic effect on the central nervous system. Overexposure can lead to unconsciousness and death. Because of its widespread usage, its effects on the body have been extensively studied.

Exposure to tetrachloroethylene occurs almost exclusively by vapor inhalation, followed by absorption into the bloodstream. Subsequently, 20% of the absorbed tetrachloroethylene is metabolized and eliminated through the kidneys (17–19). Metabolic breakdown occurs by oxidation to trichloroacetic acid and oxalic acid. The part that is not metabolized is eliminated through the lungs (20–21).

Tetrachloroethylene is a central nervous system depressant, causing headache, vertigo, tremors, nausea and vomiting, fatigue, unconsciousness, and death (20–26). It is estimated that concentrations of 1500 ppm cause unconsciousness in less than 30 min. Anesthetic effects have been reported at concentrations of 280 ppm after a 2-h exposure (23,27).

Victims of overexposure should be given fresh air and, if necessary, artificial respiration and medical attention. A self-contained breathing device should be used wherever high concentrations of tetrachloroethylene vapor are suspected (28).

The distinctive odor of tetrachloroethylene does not necessarily provide adequate warning. Because tetrachloroethylene quickly desensitizes olfactory responses, persons can suffer exposure to vapor concentrations in excess of TLV limits without smelling it. The TLV and OSHA's time-weighted average allowable vapor concentration for tetrachloroethylene is 100 ppm (29–31).

Fatalities have occurred when unprotected workers have entered unventilated tanks or pieces of equipment that contained high vapor concentrations of tetrachloroethylene (see Trichloroethane, toxicity, for a description of proper tank or equipment entry).

Ingestion of a small amount of tetrachloroethylene is unlikely to cause permanent injury. If solvent is swallowed, vomiting should be induced and qualified medical attention obtained immediately. Repeated exposure of skin to liquid tetrachloroethylene may result in dermatitis. Protective gloves and aprons should be used. Tetrachloroethylene causes temporary reddening and stinging of the eye. Discomfort can be minimized by washing the eyes immediately with water.

Uses

The major application for tetrachloroethylene is in drycleaning (qv). Approximately 80% of all drycleaners use it as their primary cleaning agent. It is also used in the vapor degreasing and cold cleaning of metals, in textile processing and finishing, and as a chemical intermediate in the manufacture of several fluorocarbons (see Fluorine compounds). The estimated use pattern during 1976 is presented below.

drycleaning	66%
textile processing	13%
metal degreasing	13%
fluorocarbon manufacture	3%
miscellaneous and exports	5%

BIBLIOGRAPHY

"Tetrachloroethylene" under "Chlorine Compounds, Organic" in *ECT* 1st ed., Vol. 3, pp. 794–798, by Janet Searles and H. A. McPhail, E. I. du Pont de Nemours & Co., Inc.; "Tetrachloroethylene" under "Chlorocarbons and Chlorohydrocarbons" in *ECT* 2nd ed., Vol. 5, pp. 195–203, by D. W. F. Hardie, Imperial Chemical Industries, Ltd.

1. L. H. Horsley and co-workers, *Adv. Chem. Ser.* (6), 32 (1952).
2. W. L. Howard and T. L. Moore, unpublished paper, The Dow Chemical Company, 1966.
3. M. G. Gonikberg, V. M. Zhulin, and V. P. Butuzov, *Bull. Acad. Sci. USSR Div. Chem. Sci.*, 739 (1956).
4. R. P. Marquardt, unpublished paper, The Dow Chemical Company.
5. E. Müller and K. Ehrmann, *Ber. Deut. Chem. Ges.* **69,** 2210 (1936).
6. P. Goldfinger, *J. Chim. Phys.* **55,** 234 (1958).

7. U.S. Pat. 2,850,543 (Sept. 2, 1958), C. Woolf (to Allied Chemical Corp.).
8. W. R. H. Hurtley and S. Smiles, *J. Chem. Soc.* 2269 (1926).
9. L. D. Rampino, *Chem. Eng. News* **36,** 62 (1958).
10. C. S. Marvel, F. D. Hager, and D. D. Coffman, *J. Am. Chem. Soc.* **49,** 2328 (1927).
11. H. J. Prins, *Rec. Trav. Chim.* **51,** 473 (1932).
12. K. W. Doak, *J. Am. Chem. Soc.* **70,** 1525 (1948).
13. F. R. Mayo, F. M. Lewis, and C. Walling, *J. Am. Chem. Soc.* **70,** 1529 (1948).
14. W. L. Archer and E. L. Simpson, *I&EC Product R&D* **16,** 158 (June 1977).
15. *Chem. Eng.,* 90 (Dec. 1, 1969).
16. *1976 Annual Book of ASTM Standards,* ASTM, Easton, Md., 1976.
17. S. Yllner, *Nature (London)* **191,** 82 (1961).
18. M. Ikeda and co-workers, *Br. J. Ind. Med.* **29,** 328 (1972).
19. M. Ogata and co-workers, *Br. J. Ind. Med.* **28,** 386 (1971).
20. R. D. Stewart and co-workers, *Arch. Environ. Health* **20,** 224 (1970).
21. R. D. Stewart and co-workers, *Report number NIOSH-MCOW-ENVM-PCE-74-6,* The Medical College of Wisconsin, Milwaukee, Wisc., 1974, 172 pp.
22. C. P. Carpenter, *J. Ind Hyg. Toxic* **19,** 323 (1937).
23. V. K. Rowe and co-workers, *Arch. Ind. Hyg. Occup. Med.* **5,** 556 (1952).
24. R. D. Stewart, *Arch. Environ. Health* **2,** 516 (1961).
25. P. D. Lamson, *Am. J. Hyg.* **9,** 430 (1929).
26. R. Patel, *J. Am. Med. Assoc.* **223,** 1510 (1973).
27. T. C. Tuttle, *Final Report for Contract HSM99-73-35,* Westinghouse Behavioral Services Center, Columbia, Md., 1976, 124 pp.
28. *Chlorinated Solvents-Toxicity, Handling Precautions, First Aid,* The Dow Chemical Company, Midland, Mich., 1976.
29. *American National Standard Z37.22-1967,* American National Standards Institute Inc., New York, 1967.
30. *Documentation of the Threshold Limit Values for Substances in Workroom Air,* 3rd ed., ACGIH, 1971, pp. 201–202.
31. *Fed. Reg.* Vol. **36**(105), Part II (May 29, 1971).
32. *Chem. Week,* 10 (Jan. 18, 1978).

S. L. KEIL
Dow Chemical U.S.A.

ALLYL CHLORIDE

Recently, no significant advances have been made in the basic technology used for the manufacture of allyl chloride [107-05-1], $CH_2=CH-CH_2Cl$. Uses and consumption of allyl chloride have continued to grow and changes in use patterns that have occurred recently are noted in this review. The current status of allyl chloride with regard to industrial exposure limits and toxicology is discussed in the section on Safety and Handling.

Allyl chloride is the most important of the allyl compounds. The development just before World War II of an economic and commercially feasible process for the manufacture of allyl chloride was a significant event in the chemical industry. For over eighty years, allyl chloride had remained a laboratory chemical until research in the field of hydrocarbon halogenation revealed a reaction that permitted synthesis from basic raw materials.

Application of the development was delayed until after the war; the first commercial-scale production of allyl chloride began in 1945. Within five years, almost one quarter of the glycerol marketed in the United States was derived from allyl chloride, and developments in plastics and resins involving other allyl chloride derivatives were exerting strong pressure on allyl chloride production capacity. Production has increased steadily, and in 1977 it was ca 180,000 metric tons at a price of 70.5¢/kg.

The interest that promoted the research effort resulting in the development of a means for direct synthesis of allyl chloride stemmed from consideration of its chemical properties. Allyl chloride is reactive as both an organic halide and an olefin. Additions can be made at the double bond, replacements at the chlorinated carbon, or both; hence, modifications can be directed to any point of the structure by proper selection of reagents and conditions. Furthermore, the allyl group is usually reactive in other molecules providing important starting points for further synthesis.

The first commercially produced derivative was allyl alcohol (see Allyl compounds), formed by dilute alkaline hydrolysis (1). The other major commercial derivatives of allyl chloride are epichlorohydrin (see Chlorohydrins) and glycerol (qv) (2); epichlorohydrin is formed by chlorohydrination of the double bond with subsequent elimination of hydrogen chloride, glycerol by chlorohydrination of the double bond and hydrolysis of the chlorine substituent.

Prior to the 1930s, chlorination of propylene had been studied extensively. The only significant reaction known under the conditions employed was the additive reaction producing 1,2-dichloropropane, $CH_2ClCHClCH_3$. Dehydrochlorination of the dichloride to allyl chloride by reaction with alkali was studied by several investigators. The principal product was 1-chloropropene, $CHCl=CHCH_3$, an allyl chloride isomer which is less interesting because it lacks a functional group on the third carbon. Subsequent work revealed that some allyl chloride could be formed by passing dichloropropane over calcium chloride at about 350°C but the reaction was slow; the principal unsaturated monochloride was again 1-chloropropene and tars and hydrocarbons were formed that rapidly poisoned the catalyst (3).

Direct pyrolysis of 1,2-dichloropropane at 600–700°C is fairly satisfactory (3–4). When properly controlled, 30% of the dichloride is decomposed per pass with yields of about 50% allyl chloride and 35% 1-chloropropene on the basis of consumed dichloride. Allyl chloride can be recovered by several techniques, but the overall process is not industrially attractive.

Research in the mid-1930s directed toward developing an understanding of hydrocarbon halogenation mechanisms led to the discovery that at considerably elevated temperatures substitution at saturated carbons occurred rather than the well-known addition at the double bond (5). In the case of propylene chlorination, favorable yields of allyl chloride made commercial production feasible (6) (see under Manufacture).

A later development is the discovery that allyl chloride can be obtained by reaction of propylene, hydrogen chloride, and oxygen in the presence of a supported lithium chloride catalyst (7).

Physical and Chemical Properties

Allyl chloride is a colorless liquid with a pungent odor. Some of its physical properties are given in Table 1. See also references 8 and 9.

Allyl chloride exhibits reactivity as an olefin and as an organic halide. Its activity as a chloride is enhanced by the presence of the double bond, but its activity as an olefin is somewhat less than that of propylene. Allyl chloride participates in most types of reactions characteristic of either functional group; reactions can be directed by control of conditions, selection of reagents, and provision of suitable catalysts. Allyl chloride does not polymerize well by free radical techniques (see Allyl monomers and polymers).

Typical Additions to the Double Bond. Chlorine, bromine, and iodine chloride at temperatures below the inception of the substitution reaction produce the 1,2,3-trihalides. High-temperature halogenation by a free radical mechanism leads to unsaturated dihalides $CH_2=CHCHClX$ (see under Manufacture). Hypochlorous and hypobromous acids add to form glycerol dihalohydrins, principally the 2,3-dihalo isomer. Hydrogen halides normally add to form 1,2-dihalides, although an abnormal addition of hydrogen bromide is known, leading to 3-bromo-1-chloropropane; the

Table 1. Properties of Allyl Chloride

Property	Value
sp gr of liquid, 20°C	0.9392
soly in water, 20°C, wt %	0.36
viscosity, 20°C, mPa·s (= cP)	0.336
freezing point, °C	−134.5
bp at 101 kPa[a], °C	44.96
vapor pressure, kPa[a]	$\log P = 19.1403 - 2098.0/T$ $-4.2114 \log T$
latent heat of vaporization at bp, kJ/mol[b]	29.04
specific heat, J/(g·°C)[b]	
vapor at 100°C	0.962
liquid	1.318
heat of combustion, vapor, kJ/mol[b]	1.844
flammability limits, air, vol %	
lower limit	3.28
upper limit	11.15
flash point, closed cup, °C	−31.7

[a] To convert kPa to mm Hg, multiply by 7.5.
[b] To convert J to cal, divide by 4.184.

reaction is believed to proceed by a free radical mechanism. 3-Bromo-1-chloropropane can be converted to cyclopropane by treatment with metallic zinc or sodium.

Water can be added by treatment with sulfuric acid at ambient or lower temperatures, followed by dilution with water. The product is 1-chloro-2-propanol.

Simple Replacements of Chlorine. The alkaline hydrolysis of allyl chloride to allyl alcohol is described under Allyl compounds. Passing allyl chloride and steam over calcium or potassium chloride is also reported to lead to the alcohol (8). Acidic hydrolysis in the presence of cuprous chloride is also known (10).

Other selected replacement reactions are summarized in Table 2.

Formation of Allyl Esters. Allyl esters are formed by reaction of allyl chloride with sodium salts of appropriate acids under conditions of controlled pH. Esters of the lower alkanoic, alkenoic, alkanedioic, cycloalkanoic, benzenecarboxylic, alkylbenzene carboxylic, and aromatic dicarboxylic acids may be prepared in this manner (11).

Formation of Nitrogen Compounds. Mono-, di-, and triallyl amines are prepared by reaction with ammonia, the ratio of reagents determines product distribution; with sufficient time and excess of allyl chloride, tetraallylammonium chloride and triallylamine predominate. Mixed amines are prepared in similar fashion by using a substituted amine in place of ammonia; they may also be prepared with allylamine and a suitable organic chloride.

Synthesis of Complex Molecules. Many syntheses of complex substances benefit from the reactivity of allyl chloride, which facilitates introduction of allyl groups into other structures, and from the subsequent activity of the allyl substituent.

Mixed allyl ethers may be prepared from allyl chloride and the appropriate alkoxide or alcohol–alkali mixture. Polyol ethers, especially those having more than one allyl group, form resinous polymers (see Allyl monomers and polymers). Allyl aryl ethers undergo the Claisen rearrangement to allyl-substituted phenols and are used as starting points for several syntheses, eg,

allyl phenyl ether → 2-allylphenol → 2-methylcoumaran

Allyl Grignard reagent is readily prepared from allyl chloride by the usual procedures (see Grignard reaction).

Several alkylation reactions are known; either the olefin or chloro- group may be involved. The reactions of allyl chloride with benzene are typical of reactions in-

Table 2. Some Replacement Reactions of Allyl Chloride

Reagent	Product	Remarks
CaI_2	C_3H_5I	
CuCN	C_3H_5CN	
NaNCS	C_3H_5SCN	cold
NaNCS	C_3H_5NCS	hot
$Na_2S_2O_3$	$C_3H_5SSO_2ONa$	
KSH	$C_3H_5SC_3H_5$	major product
	C_3H_5SH	minor product
Na_3AsO_3, then HCl	$C_3H_5AsO(OH)_2$	

volving the double bond. In the presence of ferric or zinc chloride, the products are 2-chloropropylbenzene and 1,2-diphenylpropane, as shown in the following equations:

$$C_6H_6 + ClCH_2CH=CH_2 \rightarrow [C_6H_5CH_2CH=CH_2] + HCl \rightarrow C_6H_5CH_2CHClCH_3$$
2-chloropropylbenzene

$$C_6H_5CH_2CHClCH_3 + C_6H_6 \rightarrow C_6H_5CH_2CH(C_6H_5)CH_3 + HCl$$
1,2-diphenylpropane

When aluminum chloride containing traces of water is used as a catalyst, further reaction occurs and products include 9,10-diethylanthracene and n-propylbenzene. The general ferric chloride-catalyzed reaction can be carried out with a number of benzene derivatives (see Friedel-Crafts reactions). The 1-aryl-2-chloropropanes may be converted to 2-amino derivatives, ie, amphetamine (Benzedrine), $C_6H_5CH_2CHNH_2CH_3$, and its analogues (see Psychopharmacological agents; Appetite-suppressing agents).

Several allylation reactions are known, frequently using a metallo–organic derivative of the compound being allylated, or a strongly electropositive metal in conjunction with the reactants; Grignard reactions are in this group. For example, allyl chloride reacts with sodamide in liquid ammonia to produce benzene; when sodamide is in excess, hexadiene dimer is the principal product, with some trimer and tetramer (C_{24}, six double bonds). Allylation at carbon atoms alpha to polar groups is used in preparation of α-allyl-substituted ketones and nitriles. Preparation of β-diketone derivatives, methionic acid derivatives, and malonic ester, cyanoacetic ester, and β-keto-ester derivatives, etc, involving substitution on a carbon atom between two polar groups, is particularly facile.

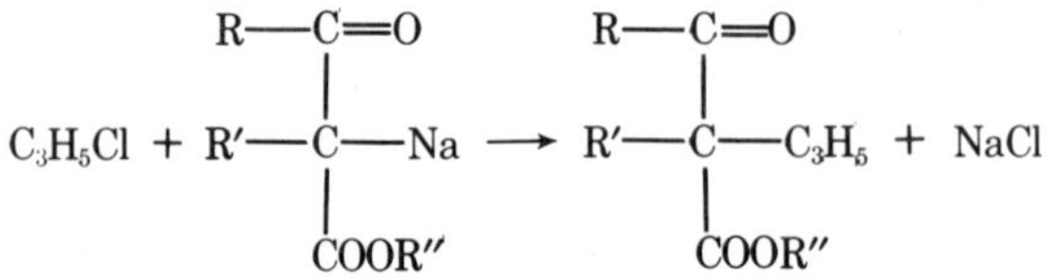

$$C_3H_5Cl + R'\!-\!\underset{\underset{COOR''}{|}}{\overset{\overset{R-C=O}{|}}{C}}\!-\!Na \rightarrow R'\!-\!\underset{\underset{COOR''}{|}}{\overset{\overset{R-C=O}{|}}{C}}\!-\!C_3H_5 + NaCl$$

Manufacture

Although other reactions leading to allyl chloride are known, the high-temperature substitutive chlorination of propylene is the route used commercially.

Mechanism. Allyl chloride is synthesized directly by chlorinating propylene under conditions favoring substitution of a chlorine atom for a hydrogen atom on the saturated carbon; the double bond is preserved, and hydrogen chloride is produced as a by-product, as shown below:

$$CH_2=CHCH_3 + Cl_2 \rightarrow CH_2=CHCH_2Cl + HCl$$

The reaction occurs only at temperatures above 300°C. The exotherm has been calculated as 111.7 kJ/mol (26.7 kcal/mol) at 355 K (1).

High-temperature substitutive halogenation of olefins proceeds by a free radical chain mechanism (12):

$$Cl_2 \rightarrow 2\ Cl\cdot$$
$$Cl\cdot + CH_2=CHCH_3 \rightarrow HCl + [CH_2\!\cdots\!CH\!\cdots\!CH_2]\cdot|$$

$$[CH_2\!=\!CH\!=\!CH_2]\cdot + Cl_2 \rightarrow CH_2\!=\!CH\!-\!CH_2Cl + Cl\cdot$$

The substitution is not a direct replacement since the chlorine may enter at a position different from that which the hydrogen is removed by allylic rearrangement of the radical intermediate (13).

Bromine substitution occurs more readily than chlorine substitution at a given temperature; this was attributed to the higher degree of thermal dissociation of bromine (14). It has been observed that the high temperature chlorination of either allyl chloride or 1-chloropropene leads to the same mixture of dichlorides, namely 90% 1,3-dichloropropene and 10% 3,3-dichloropropene (13). Some allylic rearrangement from the 1,3-form to the 3,3-form is conceivable, but the reverse rearrangement is considered unlikely. It is believed that this identical product distribution can best be explained by a free radical mechanism consistent with that postulated for the formation of allyl chloride, as shown in the following equations:

$$Cl_2 \rightarrow 2\,Cl\cdot$$

$$\left.\begin{array}{c} CH_2\!=\!CHCH_2Cl \\ \text{or} \\ CHCl\!=\!CHCH_3 \end{array}\right\} + Cl\cdot \rightarrow HCl + \left\{\begin{array}{c} CH_2\!=\!CH\dot{C}HCl \\ 10\%\uparrow\downarrow 90\% \\ CHCl\!=\!CH\dot{C}H_2 \end{array}\right\}$$

$$\left\{\begin{array}{c} CH_2\!=\!CH\dot{C}HCl \\ 10\%\uparrow\downarrow 90\% \\ CHCl\!=\!CH\dot{C}H_2 \end{array}\right\} + Cl_2 \rightarrow Cl\cdot + \left\{\begin{array}{c} CH_2\!=\!CHCHCl_2\ 10\% \\ \text{and} \\ CHCl\!=\!CHCH_2Cl\ 90\% \end{array}\right\}$$

It was finally observed that the distribution of unsaturated monochloride isomers obtained by high-temperature chlorination of propylene differs from the distribution resulting from pyrolysis of 1,2-dichloropropane. Compositions of unsaturated monochloride fractions given in reference 14 are the following:

Unsaturated monochlorides	High-temperature chlorination, %	Pyrolysis, %
$CH_2\!=\!CHCH_2Cl$	96	55–70
cis- and trans-$CHCl\!=\!CHCH_3$	1	30–40
$CH_2\!=\!CClCH_3$	3	5

The distribution of products of high-temperature chlorination is affected by mixing conditions, and there is some evidence that at least 2-chloropropene is formed by cracking 1,2-dichloropropane (15). The discrepancy in monochloride distribution is so marked, however, that it is concluded that 1,2-dichloropropane is not an intermediate in the principal high temperature allyl chloride reaction.

Reaction Variables. Temperature is the only variable that affects the monochloride substitution reaction directly, at least within the range of conditions usually encountered. Pressure, mole ratio, residence time, impurities, and degree of mixing affect side reactions (1).

Chlorine and propylene, in contact with each other at temperatures below approximately 200°C, react largely by addition. As the temperature is increased substitution may be observed; at 300°C the substitution reaction predominates. At extremely high temperatures, eg, 600°C and above, degradation reactions set in and the yield of allyl chloride decreases while production of heavier boiling products such as benzene increases.

The other unsaturated monochloride isomers are formed to some extent as co-products, as noted above. The chief secondary products are, however, the unsaturated dichlorides formed by further chlorination of allyl chloride. Some dichloropropane is always present and degradation products include benzene, tars, and carbon. More than twenty-four components are known or suspected in the reaction product.

The usual feed impurities encountered are water and propane. Water may react with chlorinated hydrocarbons to form undesirable by-products and may also provide a corrosive environment, and propane and other hydrocarbons are chlorinated to undesirable saturated chlorides.

Commercial Practice. Industrial-scale allyl chloride facilities perform three major processing operations–feed preparation, reaction, and product recovery (1,15). Variations in design and operation from the simplified scheme of Figure 1 are common and the following discussion emphasizes general principles rather than practice at any one location.

Feed Preparation. Both reactor feeds should be dry and reasonably pure to limit yield losses. Propylene purification measures vary with the feedstock; fractionation is employed if necessary, as are standard drying techniques (see Feedstocks).

The reaction liberates enough heat to permit adiabatic reactor operation if some preheat is supplied. To limit the addition reaction, which proceeds readily at low temperature, the feed is preheated before the reactants are mixed, but it is usually not possible to preheat to a level high enough to prevent any addition chlorination

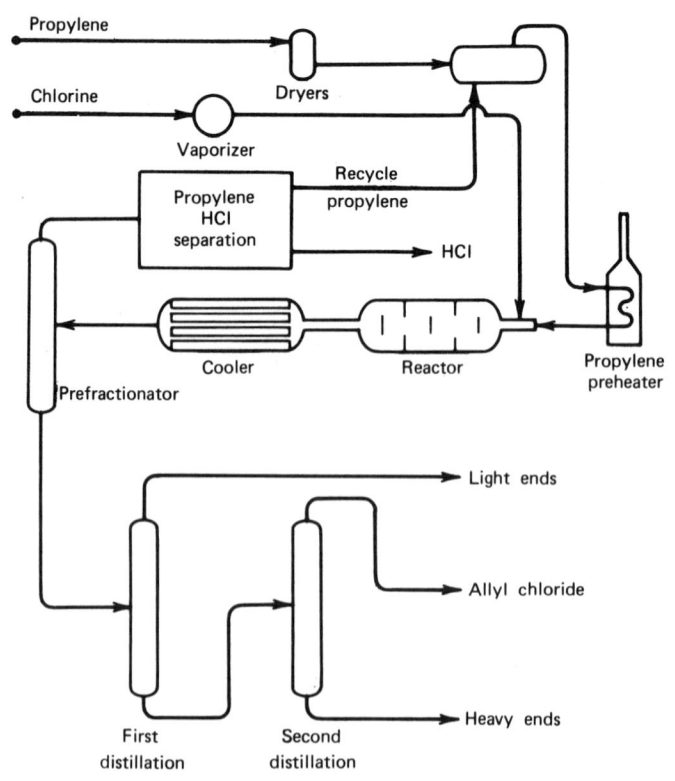

Figure 1. Simplified flow scheme for the production of allyl chloride.

even at the entrance to the reactor without exceeding the optimum reactor temperature near the exit from the reactor.

Chlorine is fed as a vapor. Exchange with hot water is adequate although any suitable vaporizing facilities may be used.

Reaction System. The reaction is carried out in an adiabatic reactor at 500–510°C and 205 kPa (15 psig). Feed is introduced under conventional instrument control through a suitable mixing nozzle. Reaction temperature is controlled by balancing feed mole ratio and propylene preheat temperature.

Commercial reactors utilize in excess of 99.9% of the chlorine. The chlorine that is not converted in the reactor is completely consumed by reaction with olefinic materials as the product is cooled.

The reactor product is cooled to about 50°C in conventional exchangers before further processing. Gradual fouling in the reactor system is experienced, and parallel reactor-cooler trains arranged for alternate operation are often provided. An average of two weeks between cleanouts has been reported (15). Steel construction may be used throughout the system except for chloride-resistant materials in valve trim, etc.

Product Recovery. Allyl chloride is separated from the product stream by a series of fractionations. After initial removal of hydrogen chloride and propylene, the organic chloride fraction is separated in a conventional two-step distillation. Other schemes are possible; for example, it is reported that, because of preferential absorption, kerosene can be used to extract allyl chloride from other reaction gases (16).

The initial removal of hydrogen chloride and propylene is accomplished in a fractionating column. The overhead stream is subsequently split; propylene is recycled and hydrogen chloride taken off to other use or to disposal. The choice of means of splitting depends to some extent upon ultimate utilization of hydrogen chloride. Fractional distillation may be employed if anhydrous acid is desired, or hydrogen chloride can be selectively absorbed into water to produce hydrochloric acid (17).

In the distillation of the organic chloride fraction, low-boiling constituents are taken overhead in the first column and allyl chloride in the second. The heavy-boiling fraction taken off as a bottom product in the second column is made up largely of unsaturated dichlorides. This stream, after further processing, has been found useful as a soil fumigant for nematode control (18) (see Poisons, economic).

Safety and Handling

This discussion represents a condensation of information believed to be reliable and is intended as a guide but not as a substitute for the large body of information available on safety and handling of allyl chloride. Before transportation or use, publications available from the manufacturers and suppliers of allyl chloride should be consulted for detailed information on its properties, toxicity, handling precautions, and emergency procedures (19). Serious consideration must be given not only to the hazards associated with allyl chloride but also to those of its derivatives and byproducts, some of which may be more noxious than the parent compound.

Health Hazards. Allyl chloride is a toxic, extremely flammable, severely irritating compound (20). The liquid can be fatal if swallowed. Contact with skin or eyes can cause severe burns and permanent injury. The vapors are severely irritating to the eyes, nose, throat, and lungs. Repeated or prolonged exposure can cause severe and lasting lung, liver, and kidney injury. High concentrations can cause death.

Eye irritation is the most common complaint of allyl chloride workers. In one study (21), some employees subjected to chronic exposure to allyl chloride concentrations ranging from a low of 1 ppm to a high of 113 ppm exhibited symptoms indicative of early stages of liver damage. However, these abnormal conditions returned to normal within 6 months on termination of exposure.

Although there are considerable toxicity data on allyl chloride, there has been no report of lifetime observations of animals exposed to repeated inhalation of allyl chloride vapors. Animal inhalation studies are being conducted by the Manufacturing Chemists Association to determine the potential hazard from long-term exposure to the vapors. The results of these studies are not yet available (see Industrial hygiene and toxicology).

Exposure Limits. A concentration of 1 ppm of allyl chloride in air has been established as the threshold limit value (TLV) for worker exposure by both the American Congress of Governmental Industrial Hygienists (ACGIH) and OSHA as the permissible level for employee exposure averaged over an 8-h work shift. The U.S. Department of Health, Education and Welfare (20) recommends that: "Exposure to allyl chloride vapor shall be controlled so that employees are not exposed at a concentration greater than 1.0 part per million parts of air (ppm) by volume (approximately 3.1 mg/m^3 of air) determined as a TWA concentration for up to a 10-h workday in a 40-h work week, or at a ceiling concentration of 3.0 ppm (9.4 mg/m^3) for any 15-min sampling period."

Precautionary Measures. Prevention of exposure is the best precautionary measure. Allyl chloride should be used in closed systems with good ventilation. Natural ventilation should be aided by mechanical ventilation in order to minimize the risk of exposure.

Allyl chloride has a pungent, disagreeable, garlic-like odor. The odor can be detected at concentrations of the order of 3–6 ppm. However, concentrations that produce no high degree of external irritation may be capable of causing chronic intoxication. It should be noted that these concentrations are higher than the OSHA permissible level. Therefore, workplace monitoring will be required to ensure low enough concentrations. Furthermore, olfactory fatigue may deaden the sense of smell. Therefore, it is incumbent that action be taken immediately upon detecting the odor of allyl chloride in order to avoid exposure.

Respiratory equipment approved for use with allyl chloride by NIOSH should be readily available and must be worn if it becomes necessary to remain in areas where concentrations exceed the prescribed limits. Impervious clothing, rubber gloves, and chemical goggles should be worn if there is danger of contact with the liquid. After any contact, protective gear should be washed before reuse. Thorough and immediate cleansing of rubber equipment is necessary to remove residual traces; rubber articles should be discarded if inspection shows signs of deterioration. Leather articles are penetrated easily and offer no protection; leather goods such as shoes cannot be decontaminated readily and should be destroyed and discarded.

Emergency Treatment. Because response to exposure may not be immediately apparent, every exposure should be treated promptly. Persons overcome by exposure to the vapors should be removed to a safe area and given artificial respiration if not breathing. If respiration is difficult, administer oxygen. In any case of exposure to significant vapor concentrations, medical attention should be obtained immediately at any sign of respiratory distress or eye irritation.

In case of contact with the liquid, contaminated clothing should be removed and the affected body parts washed immediately. The eyes should be flushed continuously with low pressure clear water for at least 15 min; medical attention should be obtained promptly. It must be remembered that the effects of exposure may be delayed several hours; early treatment can do much to alleviate the severity of the damage.

If swallowed, large quantities of a bland fluid such as water or milk should be given and vomiting induced. Medical attention should be obtained at once. Vomiting should not be induced nor should anything be given orally to an unconscious person.

Spills and Leaks. Besides the health hazards from exposure to the vapors or liquid, spills and leaks introduce a fire hazard typical of hydrocarbons. Its high vapor pressure, low flash point, and high vapor density can lead to the formation of flammable mixtures which can travel long distances along the ground to a source of ignition.

Spills should be confined and not permitted to enter any public water system. The proper authorities should be notified at once if such contamination occurs. Large spills should be diked and pumped into clean salvage tanks. The residue and small spills should be soaked up on an absorbent material, placed in a leak-proof container for disposal, and labeled flammable and toxic.

The combustion products from burning allyl chloride contain hydrogen chloride, and may, under limited oxygen conditions, form carbon monoxide and phosgene. Personnel trained for fire fighting must wear full protective clothing and self-contained breathing apparatus when in the fire area. Storage tanks exposed to fire should be cooled with water.

Reactivity Hazards. Because of its high reactivity, exploratory syntheses with allyl chloride should be undertaken first on a small scale, with due caution, to observe the speed of the reaction and the amount of heat evolved. In recovering allyl chloride and its reaction product from a reaction mixture, continuous distillation is preferable to large-scale batch distillation to avoid heating large volumes of material that may inadvertently contain catalytic or reactive substances.

Allyl chloride can react violently with chemically active metals such as sodium, zinc, magnesium, aluminum, and their alloys. Allyl chloride can react vigorously and explosively with strong sulfuric acid, with anhydrous metal halides, such as stannic, aluminum, and ferric chlorides, and with metal alkyls such as triethylaluminum and butyllithium, even when these reagents are slowly added at low temperature. Allyl chloride reacts readily with basic materials such as caustic soda and amines. Allyl chloride hydrolyzes in the presence of water to form hydrochloric acid and dangerously toxic allyl alcohol.

General Handling. Moist allyl chloride is corrosive; nickel, Monel, and other chloride-resistant materials are usually recommended for handling. Dry allyl chloride can be handled, if not excessively hot, in mild steel, cast iron, stoneware, and red brass; aluminum is not considered satisfactory.

Uses

In commercial application allyl chloride is the parent compound for a number of useful derivatives. At first the most important derivative was glycerol (qv). New synthetic routes to glycerol, competition for glycerol from other polyols, and developments in the resin field have changed the use pattern markedly. Currently, most allyl chloride produced is used in a wide variety of resins and polymers (see Allyl monomers and polymers).

The most important derivative of allyl chloride is epichlorohydrin (see Chlorohydrins), a basic building block for epoxy resins (qv). Epichlorohydrin is also used in the preparation of a variety of other polymeric materials such as specialty elastomers, cationic flocculants, and paper-treating resins.

Glycerol, allyl alcohol, and the allyl amines are other significant derivatives of allyl chloride. Although neither allyl chloride nor allyl alcohol polymerize readily, allyl esters, such as diallyl phthalate, can be polymerized to yield polymers with excellent electrical, physical, and optical properties (22). It was also discovered that allyl amines and, in particular, diallyl dialkyl quaternary ammonium compounds can be polymerized and copolymerized to yield high molecular weight cationic polymers which find considerable use in water treatment (see Water; Flocculants) (23).

Other uses for allyl chloride include the synthesis of medicinal derivatives such as barbiturates and diuretics and the anesthetic cyclopropane (see Anesthetics). Several specialty resins are derived from allyl esters and polyesters which may be made directly from allyl chloride (11). Resin uses also include a number of copolymers and interpolymers of allyl chloride with, eg, acrylonitrile (24), vinylidene cyanide (25), styrene, and diallyl esters (26); allyl chloride also serves as a catalyst (27) or modifier in production of other resins. Allyl amines are used in the synthesis of agricultural chemicals to improve the selectivity of herbicides (qv) (28). Sodium allyl sulfonate prepared by reaction with sodium sulfite is used as a component in metal plating solutions (29).

BIBLIOGRAPHY

"Allyl Chloride" under "Chlorine Compounds, Organic" in *ECT*, 1st ed., Vol. 3, pp. 800–806, by Harold G. Vesper, Shell Development Company; "Allyl Chloride" Under "Chlorocarbons and Chlorohydrocarbons" in *ECT* 2nd ed., Vol. 5, pp. 205–214, by B. H. Pilorz, Shell Chemical Company.

1. A. W. Fairbairn, H. A. Cheney, and A. J. Cherniavsky, *Chem. Eng. Progr.* **43**(6), 280 (1947).
2. D. L. Yabroff and J. Anderson, *Third World Petroleum Congress, Proc., Sect. V*, 22 (1951).
3. E. C. Williams, *Trans. Am. Inst. Chem. Eng.* **37**(1), 157 (1941); *Chem. Met. Eng.* **47**, 834 (1940).
4. U.S. Pat. 2,207,193 (July 9, 1940), H. P. A. Groll (to Shell Development Co.).
5. U.S. Pat. 2,130,084 (Sept. 13, 1938), H. P. A. Groll, G. Hearne, J. Burgin, and D. S. LaFrance (to Shell Development Co.).
6. E. C. Williams, "Modern Petroleum Research" in *Proc. Am. Petrol. Inst. Sect. II (Marketing)*, Nov. 16, 1938; *Ind. Eng. Chem., News Ed.* **16**, 630 (Dec. 10, 1938).
7. U.S. Pat. 2,966,525 (Dec. 27, 1960), D. E. Steen (to Monsanto Chemical Co.).
8. *Allyl Chloride,* Bulletin Code 164-178, The Dow Chemical Co., Midland, Mich., 1958.
9. *Allyl Chloride,* Technical Publication SC 49-8, Shell Chemical Co., San Francisco, 1949.
10. U.S. Pat. 2,475,364 (July 5, 1949), G. H. Van de Griendt and L. M. Peters (to Shell Development Co.).
11. U.S. Pat. 2,939,879 (June 7, 1960), A. De Benedictis (to Shell Oil Co.).
12. F. F. Rust and W. E. Vaughan, *J. Org. Chem.* **5**(5), 472 (1940).
13. G. Hearne and co-workers, *J. Am. Chem. Soc.* **75**(6), 1392 (1953).
14. H. P. A. Groll and G. Hearne, *Ind. Eng. Chem.* **31**(12), 1530 (1939).
15. J. B. Henderson and N. H. McKay, *Paper, Meet. Am. Chem. Soc.,* Southwest Region, Little Rock, Ark, Dec. 5, 1952.
16. A. Zielinski and S. Suknarowska, *Przemysl Chem.* **13**, 279 (1957).
17. C. F. Oldershaw and co-workers, *Chem. Eng. Progr.* **43**(7), 371 (1947).
18. W. Carter, *Science* **97**, 383 (Apr. 23, 1943); *J. Econ. Entomol.* **37**(1), 117 (1944).
19. *Allyl Chloride Toxicity and Safety Bulletin SC:195-76* and *Allyl Chloride Safety Notice SC:196-76,* Shell Chemical Company, 1976.
20. *Criteria for a Recommended Standard . . . Occupational Exposure to Allyl Chloride,* HEW Publication No. (NIOSH) 76-204, GPO, 1976.

21. M. Hausler and R. Lenich, *Arch. Toxicol. Berlin* **23,** 209 (1968).
22. C. E. Schildnecht, *Allyl Compounds and Their Polymers,* John Wiley & Sons, Inc., New York, 1973.
23. U.S. Pat. 2,656,337 (Oct. 20, 1953), J. R. Caldwell (to Eastman Kodak Co).
24. U.S. Pat. 2,650,911 (July 9, 1940), H. Gilbert, F. F. Miller, and L. Falt (to B. F. Goodrich Co.).
25. U.S. Pats. 2,569,959 (Aug. 2, 1951), 2,569,960 (Aug. 2, 1951), 2,592,211 (Apr. 8, 1952), 2,597,202 (May 20, 1952), P. O. Tawney (to U.S. Rubber Co.).
26. U.S. Pat. 2,721,887 (Oct. 25, 1955), H. Pines and V. N. Ipatieff (to Universal Oil Products Co.).
27. G. B. Butler and R. J. Angelo, *J. Am. Chem. Soc.* **79,** 3129 1957; U.S. Pat. 2,926,161 (Feb. 28, 1960), G. B. Butler, R. J. Angelo, and A. Cranshaw (to Peninsular Chem Research, Inc.).
28. U.S. Pat. 3,362,810 (Jan. 9, 1968), J. M. Deming (to Monsanto Company); Anon, *Chem. Week* **43,** (Apr. 25, 1973); U.S. Pat. 3,923,494 (Dec. 2, 1975), E. G. Teach (to Stauffer Chemical Co.).
29. U.S. Pat. 2,523,191 (Sept. 19, 1950), H. Brown (to Udylite Corporation).

ALDO DeBENEDICTIS
Shell Chemical Company

CHLOROPRENE

Chloroprene [*126-99-8*], 2-chloro-1,3-butadiene, was discovered in 1930 by Carothers and Collins (1–2) during research on the synthesis of divinylacetylene. The new compound was named in analogy to isoprene, 2-methyl-1,3-butadiene, the building block of natural rubber. Chloroprene reacts with oxygen to form peroxides. It enters into addition reactions with halogens, hydrogen halides, hypohalous acid, and mercaptans. It undergoes cycloaddition with itself to form cyclic dimers, and with other dienophiles to form a variety of Diels-Alder adducts. Its most useful reaction is that of free radical polymerization to form high molecular weight elastomeric polymers (see Elastomers, synthetic-neoprene). Most chloroprene is polymerized to make polychloroprene, a synthetic rubber used in wire and cable covers, gaskets, automotive parts, adhesives, caulks, flame-resistant cushioning, and other applications requiring chemical, oil, and weather resistance, or high gum strength (3).

Until the late 1960s all chloroprene was produced from acetylene by the following reactions:

$$2HC \equiv CH \xrightarrow{(CuCl)} HC \equiv C-CH=CH_2$$

monovinylacetylene (MVA)

$$HC \equiv C-CH=CH_2 + HCl \xrightarrow{(CuCl)} H_2C=CClCH=CH_2$$

An alternative route, known in the laboratory since 1930, uses 1,3-butadiene. This route became economical about 1965 as the price of acetylene increased and that of butadiene decreased. The butadiene route involves three steps which constitute the predominant commercial method of manufacturing (see Manufacture).

Physical Properties

The physical properties of chloroprene have been extensively investigated (Table 1). Chloroprene is a colorless, mobile, volatile liquid with an ethereal odor similar to that of ethyl bromide. It is very slightly soluble in water (<1%) and miscible with most organic solvents.

Table 1. Physical Properties of Chloroprene[a]

Property	Value
mol wt	88.54
melting point, °C	-130 ± 2
boiling point at 101 kPa[b], °C	59.4
critical temperature, °C	261.7
vapor pressure (T in K, p in kPa[b])	$\log_{10} p = 6.652 - 1545/T$
viscosity at 25°C, mPa·s(= cP)	0.394
density at 20°C, g/mL	0.9585
average coefficient of volumetric expansion (20–61°C), K^{-1}	0.001235
refractive index, n_D^{20}	1.4583
flash point (ASTM, open cup), °C	-20
latent heat of vaporization, kJ/g[c]	
0°C	0.332 8
60°C	0.302 7
specific heat, kJ/(kg·K)[c]	
liquid at 20°C	1.314
gas at 100°C	1.038 3
thermal conductivity (where t is °C)	
mW/(m·K),	$2.410 \times 10^{-5} + 0.160 \times 10^{-5} t$
dielectric constant at 27°C	4.9

[a] Ref. 4.
[b] To convert kPa to mm Hg, multiply by 7.5.
[c] To convert J to cal, divide by 4.184.

The absorption of chloroprene in the uv (5) and ir (6), the Raman spectrum (7), and the nuclear magnetic resonance spectrum (8) have been determined. The observed shifts of spectral bands, relative to butadiene, toward longer wave lengths reveal the strong resonance effect of the chlorine atom. The data also indicate that chloroprene is planar and transoid with respect to the C—C single bond (6–7,9).

Chemical Properties

General Chemical Properties. The chemical properties of chloroprene are mainly a product of the electronic interactions between the conjugated butadiene structure and the chlorine atom in the molecule. The chlorine atom strongly enhances the free-radical activity of the molecule but decreases its activity in ionic and Diels-Alder reactions. Thus chloroprene polymerizes much more readily but forms adducts with maleic anhydride less readily than does either butadiene or isoprene. Chloroprene autoxidizes easily, polymerizes spontaneously at room temperature, and forms cyclic dimers on prolonged standing in the presence of polymerization inhibitors. Many inorganic and organic compounds add to the double bonds, usually in the 1,4 manner, although several 1,2 and 3,4 additions are also known. Additions occur by homopolar, heteropolar, and coordination mechanisms to give straight-chain or cyclic products. The chlorine atom, like that of vinyl chloride, is very unreactive; only traces of chloride are split off on refluxing with concentrated alcoholic sodium hydroxide, alcoholic silver nitrate, or pyridine.

The general chemistry of chloroprene has been extensively studied (10). The discussion in this article is limited to dimers and important isomers and analogues of chloroprene.

Dimers. Although the tendency of chloroprene to undergo second-order addition to itself to form a mixture of dimers was recognized as early as 1931 (2), the chemistry of dimerization and the complex mixtures formed in the process were elucidated much later (11–16).

Dimerization is independent of free radical catalysts and inhibitors. The initial rate of dimerization in bulk monomer is about 0.35%/d at room temperature and about 1%/h at the boiling point of chloroprene. Hence small amounts of dimer are present in all polychloroprenes.

The initial dimer products include *trans*-1,2-dichloro-1,2-divinylcyclobutane, the corresponding cis isomer, 1,4-dichloro-4-vinylcyclohexene, 1-chloro-4-(1-chlorovinyl)cyclohexene, and 2-chloro-4-(1-chlorovinyl)cyclohexene.

After longer times or higher temperatures, other products are formed from rearrangements of initial products containing allylic chlorine. Especially interesting is 1,6-dichloro-1,5-cyclooctadiene from the rearrangement of the *cis*-cyclobutane isomer.

High pressure accelerates the dimerization reaction considerably, resulting in products rich in the chlorovinyl cyclohexene isomers. From this and from the absence of 2,4-dichloro-4-vinylcyclohexene in the reaction mixture, Stewart (15–16) has postulated that the chlorovinyl isomers are formed by a compact, four-centered transition state (presumably a normal Diels-Alder reaction), and the remaining products are formed by a less compact transition state, presumably diradical in nature.

Isomers and Analogues. The two position isomers of chloroprene and 2,3-dichloro-1,3-butadiene are significant in the chemistry of the polymer neoprene; the first two because they are impurities which arise in either of the chloroprene synthesis routes, and the latter because it is an important comonomer. A brief discussion of their preparation and properties, based largely on reference 10, follows.

1-Chloro-1,3-butadiene [627-22-5], bp 66–67°C, n_D^{20} 0.954, is frequently referred to as α-chloroprene. Its cis and trans isomers are by-products of most reactions leading to 2-chlorobutadiene. Small amounts are formed by hydrochlorination of vinylacetylene; varying amounts, depending on conditions, are formed by chlorination of butadiene, the isomerization of 1,4-dichloro-2-butene to 3,4-dichloro-1-butene, and the dehydrochlorination of 3,4-dichloro-1-butene. A mixture containing about 85% trans isomer is obtained by dehydrochlorination of *cis*-1,4-dichloro-2-butene (from *cis*-butenediol) with sodium amide in mineral oil (17). A mixture rich in *cis*-1-chlorobutadiene is similarly obtained from *trans*-1,4-dichloro-2-butene with sodium amide or other alkaline reagents. The cis diene can be obtained in high purity by treating this mixture with maleic anhydride to remove trans isomer and chloroprene (18). The cis and trans isomers can be equilibrated by the action of iodine in benzene at 100°C, giving a mixture of 70% cis and 30% trans (19), or thermally at about 550°C, giving about 40% cis and 60% trans (18). 1-Chlorobutadiene polymerizes and copolymerizes with chloroprene at a lower rate. Physical properties do not distinguish adequately between the cis and trans isomers, but spectral and chromatographic methods have been given by Viehe (19).

4-Chloro-1,2-butadiene [25790-55-0], isochloroprene, bp 88°C, n_D^{20} 1.4775, d_4^{20} 0.9891, is prepared by hydrochlorination of vinylacetylene in the absence of cuprous chloride or other isomerizing materials. In the presence of such agents, it is rapidly isomerized to chloroprene. Kinetics of the process have been discussed by Dolgopol'skii (20). Isomerization is irreversible but the reactivity of isochloroprene is otherwise that expected of an allylic halide (21).

2,3-Dichloro-1,3-butadiene [1653-19-6], bp 98°C, n_D^{20} 1.4890, d_4^{20} 1.1829, is significant because it is very reactive in copolymerization with chloroprene and it is available from by-products of both the acetylene and the butadiene routes to chloroprene (22). In the acetylene route, 1,3-dichloro-2-butene is allylically chlorinated and then dehydrochlorinated:

$$CH_3-CCl=CH-CH_2Cl \xrightarrow{Cl_2} CH_2=CClCHCl-CH_2Cl$$

$$CH_2=CCl-CHCl-CH_2Cl \xrightarrow{NaOH} CH_2=CCl-CCl=CH_2$$

In the butadiene route, chlorination of butadiene or dichlorobutenes gives the stereo isomers of 1,2,3,4 tetrachlorobutane. Dehydrochlorination of the stereoisomers gives a mixture of the three possible dichlorobutadienes:

$$C_4H_6, C_4H_6Cl_2 \xrightarrow{Cl_2} ClCH_2CHClCHClCH_2Cl \text{ (meso, } dl\text{)}$$

$$ClCH_2CHClCHClCH_2Cl \xrightarrow{NaOH} CH_2=CCl-CCl=CH_2 + CH_2=CCl-CH=CHCl$$

$$+ CH_2=CH-CCl=CHCl-$$

The *meso*-tetrachloride, arising from the chlorination of *trans*-1,4-dichlorobutene, produces a higher yield of the desired diene than the *dl*-tetrachloride. Complete purification of 2,3-dichlorobutadiene from its isomers is commercially undesirable because of its high reactivity, but the mixture of isomers can often be used to achieve many of the same results in polymerization (23).

The C_4-chloroacetylenes that are isomeric with chloroprene do not yet have any uses. However, the four linear possibilities and one of the cyclic structures have been reported (10).

Manufacture

Because most chloroprene is currently produced from butadiene (qv), only this route is described. A complete discussion of the acetylene route can be found in ref. 10 (see also Acetylene-derived chemicals).

Conversion of Butadiene to Chloroprene. The three essential steps in the production of chloroprene from butadiene are chlorination, isomerization, and caustic dehydrochlorination, as shown by the following equations:

Chlorination

$$CH_2=CH-CH=CH_2 + Cl_2 \rightarrow ClCH_2-CH=CH-CH_2Cl + ClCH-CHCl-CH=CH_2$$

Isomerization

$$ClCH_2-CH=CH-CH_2Cl \xrightleftharpoons{catalyst} CH_2=CH-CHCl-CH_2Cl$$

Dehydrohalogenation

$$CH_2=CH-CHCl-CH_2Cl + NaOH \rightarrow CH_2=CH-CCl=CH_2 + NaCl + H_2O$$

Chlorination of Butadiene. The need for nonacetylene routes to chloroprene fostered investigation of butadiene chlorination. Initially, liquid-phase chlorination was studied (24–27), but vapor-phase chlorination produced cleaner, more reliable yields (28–42). Vapor-phase chlorination leads to a near equilibrium mixture of 3,4-dichloro-1-butene and the cis and trans isomers of 1,4-dichloro-2-butene. The reaction mechanism is predominantly free radical.

Commercial production of dichlorobutenes from butadiene is based almost entirely on vapor phase chlorination. A wide variety of process conditions have been described (40–43). Critical features in commercial equipment are adequate mixing of the gas streams to prevent over-chlorination; avoidance of condensed phases in which the product can be chlorinated in preference to butadiene; and completion of the chlorination to avoid by-product formation in the refining operation. Reaction heat is absorbed either by using a large excess of butadiene (44) or by using a smaller excess along with an additional diluent such as HCl (40).

Isothermal laboratory studies (41) have demonstrated that at temperatures above 250°C the yield of dichlorobutenes attains a maximum of 90.5% as the ratio of 3,4 to 1,4 addition products increases. As the temperature rises, there is increased formation of monochlorobutadienes, trichlorobutene, and 4-vinylcyclohexene.

The preferred temperature range in commercial operation (45) is 290–330°C, and conversions of butadiene are generally 10–25%. Yields of mixed dichlorobutenes are in the range of 85–95%.

Figure 1 depicts a general flow diagram of an integrated plant for chlorination, refining, isomerization, and dehydrochlorination of butadiene to produce chloroprene.

The crude chlorination products are condensed from the excess butadiene which is then recycled to the reactor. Systems for refining the crude product vary considerably, depending on the degree of integration with subsequent isomerization and dehydrochlorination steps, and whether the 1,4-dichloride is desired separately for other uses or is to be completely isomerized to the 3,4-isomer. The streams to be separated are (*1*) low boiling impurities, mainly 1- and 2-chlorobutadiene, (*2*) purified 3,4-dichloro-1-butene, (*3*) 1,4-dichloro-2-butene, and (*4*) higher boiling by-products including trichlorobutenes, tetrachlorobutanes, telomers, and tars. The boiling ranges for these streams at atmospheric pressure are 60–70°C, 115–125°C, 150–160°C, and above 160°C, respectively. Commercial refining is carried out at reduced pressure to minimize decomposition of the high boiling fractions.

Process hazards in chlorination, in addition to the recognized hazards of handling butadiene, chlorine, and chlorinated hydrocarbons, include the potent vesicant properties of 1,4-dichloro-2-butene and some of the higher boiling by-products.

Isomerization of Dichlorobutenes. Where the sole use of dichlorobutenes is dehydrohalogenation to 2-chlorobutadiene, the 1,4-dichloro-2-butene must be isomerized to 3,4-dichloro-1-butene. In the presence of a catalyst (at 100°C), a liquid mixture of dichlorobutenes equilibrates to a composition of 21% 3,4-dichloro-1-butene, and 7% *cis*-1,4-, and 72% *trans*-1,4-dichloro-2-butene. The vapor phase above this liquid contains 52, 6, and 42%, respectively, of the isomers. Thus the lower boiling 3,4-dichlorobutene can be obtained as the sole product by distilling the vapor phase and recycling the undesired 1,4-dichloro-2-butene to the catalyst-containing liquid phase.

Many isomerization catalysts have been reported in the partial list of references (46–59). Copper metal or cuprous chloride, generally in the presence of a solubilizing agent, are preferred. The catalyst species actually present in the isomerization mixture

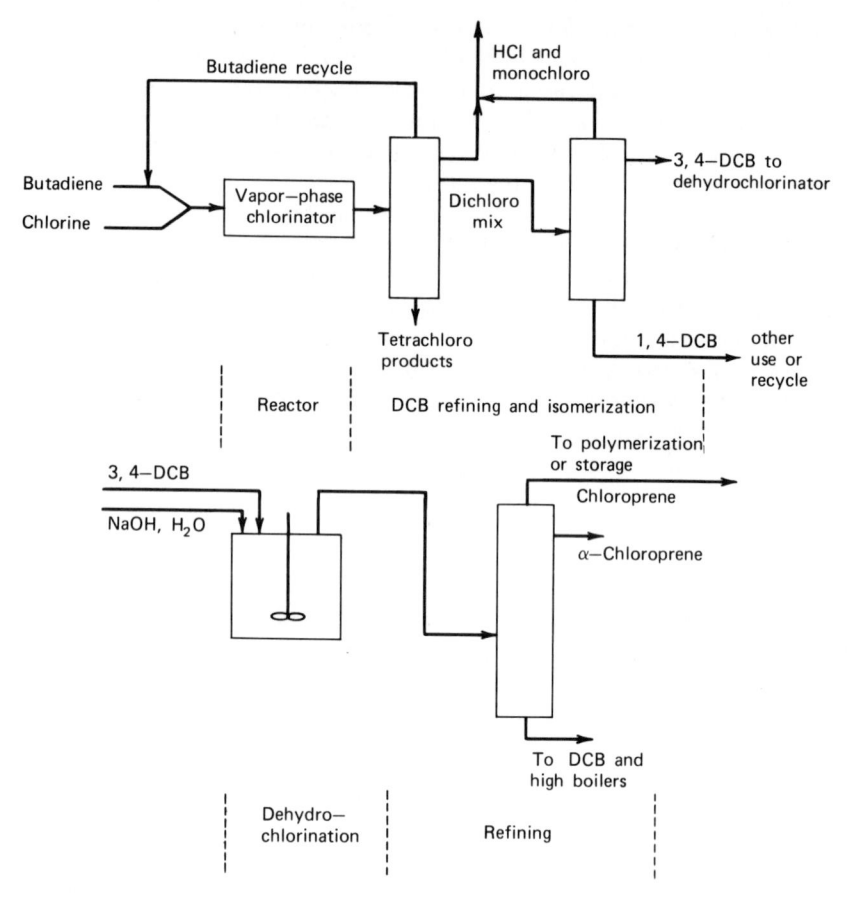

778 CHLOROCARBONS, -HYDROCARBONS (CHLOROPRENE)

Figure 1. Conversion of butadiene to chloroprene. 3,4-DCB is 3,4-dichloro-1-butene; 1,4-DCB is 1,4-dichloro-2-butene.

have not been defined because the system is complicated by the tendency of dichlorobutenes to oxidize cuprous salts, and by the presence of HCl from side reactions. The major yield loss in the isomerization step is from some dehydrochlorination to give 1-chlorobutadiene and HCl. Equipment must be constructed of glass or highly inert metals to resist the severe corrosiveness of the reaction mixture.

Dehydrohalogenation of 3,4-Dichloro-1-butene. Dehydrochlorination of 3,4-dichloro-1-butene with alkali produces chloroprene in excellent yields. This reaction, initially studied by Carothers (60) and more recently by Hearne and LaFrance (61) and Crocker and Turk (62), appears to occur via hydroxyl ion attack on the proton at the β carbon with subsequent concerted or simultaneous elimination of the terminal chlorine atom. Alternatively a two-step process mechanism has been suggested (63).

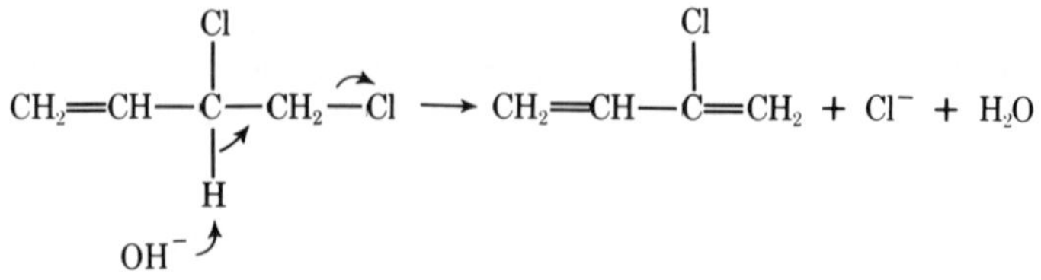

$$CH_2=CH-\underset{\underset{OH^-}{\overset{H}{|}}}{\overset{Cl}{\underset{|}{C}}}-CH_2-Cl \longrightarrow CH_2=CH-\overset{Cl}{\underset{|}{C}}=CH_2 + Cl^- + H_2O$$

Thermal dehydrochlorination of 3,4-dichloro-1-butene leads predominately to 1-chlorobutadiene (63–64). Other dehydrohalogenation conditions tested have included the use of ion-exchange resins (65) and liquid ammonia (66).

In the modern commercial processes, 3,4-dichloro-1-butene, relatively free of 1-chlorobutadiene and 1,4-dichlorobutenes, is dehydrochlorinated with aqueous caustic solution to produce chloroprene. In the simplest method dichlorobutene and a solution of 5–15% sodium hydroxide in water are added to a well-stirred reactor at 80–110°C. Chloroprene containing residual dichlorobutene vaporizes along with some water. The spent caustic–salt solution is discharged after removal of residual organic material. The product stream is given a crude distillation, either before or after drying, to recover unreacted dichlorobutene for recycling to the reactor. Final purification includes drying under reduced pressure, removing the last of the dichlorobutene and any other higher-boiling impurities, and reducing the 1-chlorobutadiene content to an acceptable level for polymerization.

Neither dichlorobutene nor chloroprene is appreciably soluble in water. Increasing the concentration of sodium hydroxide in the feed solution actually retards the dehydrochlorination, presumably by decreasing the already slight solubility. Several patents have been issued for catalysts that permit reaction at higher caustic concentration as well as at lower temperatures where more selective reaction occurs (67–72). Intense agitation, or the use of various surfactants, is required (73). Major yield loss occurs because of polymerization in the reactor unless the ingredients are adequately inhibited (74–75), otherwise the main yield loss in the reaction is the formation of 1-chlorobutadiene which is either introduced as an impurity in the feedstock or results from dehydrochlorination of the 1,4-dichlorobutene impurity:

$$CH_2=CH-CHCl-CH_2Cl \xrightarrow{-HCl} CH_2=CH-CCl=CH_2 + CH_2=CH-CH=CHCl$$

$$CH_2Cl-CH=CH-CH_2Cl \xrightarrow{-HCl} CH_2=CH-CH=CHCl$$

Depending on temperature, caustic concentration, and reagent purity, the normal process results in a 1–3% yield of this isomer which must be reduced in concentration to obtain polymerization-grade monomer. Since the isomers differ in boiling point by about 8°C, concentration of the impurity sufficient to permit purging adds significant expense.

Other impurities and yield losses include chlorination–dehydrochlorination products of butadiene impurities, dimerization products of chloroprene, and alcoholic hydrolysis products of the allylic chlorides present. Overall chemical yield of chloroprene is generally over 95%. Process yield may be somewhat less, depending on dichlorobutene and chloroprene losses in purging impurities from the system.

Other Routes to Chloroprene. It is apparent that the chemistry of chloroprene synthesis involves the conversion of a linear C_4 compound to 2-chloro-1,3-butadiene. The successful routes from acetylene and butadiene involve highly unsaturated carbon chains and multiple-step processes. Efforts to reduce the number of steps have not been successful. The single-step chlorination of butadiene, which gives mixtures of chloroprene and α-chloroprene (76–77), is not economical.

The routes to chloroprene from most other C_4 hydrocarbons available from petroleum cracking have been reviewed by the Distillers group (78). Their studies have included production of chloroprene by chlorination of mixtures of butenes and butadiene to chlorobutenes and dichlorobutanes with appropriate separation and recycling. The complex processes were not developed commercially (79–80).

Chloroprene is formed by thermal or catalytic dehydrohalogenation of 1,2,3- or 2,2,3-trichlorobutanes, 2,3-dichloro-1-butene, or 1,2- or 1,3-dichloro-2-butene (81–83).

Production of chloroprene by oxychlorination of 2-chloro-2-butene and by pyrolysis of 1-methyl-2,2-dichlorocyclopropane, or propylene and dichlorocarbene has been described (84–89). Other methods described for the preparation of chloroprene include pyrolysis of chlorobutenyl quaternary ammonium compounds (90–91), and dehydrochlorination and decomposition of 3,4-dichlorosulfolane obtained from the reaction of butadiene, chlorine, and sulfur dioxide (92).

Waste Disposal. High boilers and still heels from chloroprene refining, mainly 1,3-dichloro-2-butene, are best disposed of by special incineration in a waste organic chlorides burner equipped with a hydrochloric acid recovery unit. A significant amount of 1,3-dichloro-2-butene is purified and converted to 2,3-dichlorobutadiene for use as a comonomer in chloroprene polymerization.

The major by-product of the butadiene route to chloroprene is the spent caustic–salt solution discharged from the dehydrochlorination reactor. This stream is neutralized with HCl, freed of volatile organics, and clarified in large setting tanks to comply with rigid specifications. It is then either pumped into deep wells or added to natural brines such as sea water.

The still heels and high boilers from chlorobutene and chloroprene refining are disposed of in an organic chloride incinerator.

Storage, Handling, and Shipment

Purified chloroprene may be used directly, stored uninhibited under nitrogen at low temperature for short times, or inhibited and cooled for long term storage. At room temperature, even with adequate inhibitor and in small quantity, pure chloroprene rapidly becomes unfit for most uses because of dimerization (2,93–95). Dimerization, bulk polymerization, and the formation of autocatalytic "popcorn" polymer, which is insoluble in the monomer, are avoided in commercial production by refrigeration to <0°C, or addition of inhibitors where higher temperature or prolonged storage are required. Phenothiazine, *tert*-butyl catechol, picric acid, and the ammonium salt of N-nitroso-N-phenyl-hydroxylamine have been used to prevent bulk polymerization (2,93). Nitric oxide (96) and several nitroso compounds (97–99) have been recommended to prevent growth of popcorn polymers, particularly during distillation, where many of these inhibitors are believed to act by decomposition to nitric oxide and perhaps other inhibiting species.

Chloroprene should be handled with due regard to polymerization hazards, its flammability and toxicity. Volume concentrations of 4–20% in air are explosive, and a flash point of −20°C (ASTM, open cup) has been reported (10). Toxic symptoms have been reported from inhalation at 83 ppm and a TLV in air of 25 ppm has been recommended (100). The odor threshold is considerably lower than this, particularly with aged samples containing dimers.

Aged samples of chloroprene that have been exposed to air may contain or deposit dangerous amounts of peroxides.

Historically, chloroprene has been consumed by polymerization at a plant adjacent to its manufacturing site, and transported via pipeline. Today, significant quantities of chloroprene are transported by tank car and tank trucks. The inhibited liquid is cooled to −10°C before loading into insulated tanks and shipment times are closely monitored.

Economic and Energy Factors

Chloroprene production is controlled by the demand for polychloroprene synthetic rubbers which represent about 5% of the total demand for elastomers (see Elastomers, synthetic). Table 2 lists world annual consumption of dry polychloroprene, excluding the Union of Soviet Socialist Republics. Chloroprene production was probably 10% higher than these figures in view of latex manufacture and polymerization yield losses.

Chloroprene is manufactured by the DuPont Co. at two locations in the United States and one in Northern Ireland. It is also produced by Denki Kagaku in a plant formerly operated by Petrotex Chemicals Company in the United States, by Bayer A.G. and Knapsach-Griesheim A.G. in the Federal Republic of Germany, by Distugil in France, by Denki Kagaku, Showa, and Toyo Soda in Japan, and in several plants in the Union of Soviet Socialist Republics.

The relative costs of acetylene and butadiene have been a major factor in the economics of chloroprene manufacture. Acetylene was favored until the early 1960s. The growth of the polyethylene industry led to relatively inexpensive by-product butadiene from the manufacture of ethylene (qv). Lower investment chloroprene processes using by-product butadiene were developed in the mid 1960s. About 80% of the current world capacity for chloroprene uses butadiene as raw material.

Estimates have been made of the energy cost in MJ/kg for the processes by which chloroprene is manufactured. As shown in Table 3, the energy consumed by the

Table 2. World Consumption of Dry Polychloroprene [a]

Year	Annual consumption, thousands of metric tons
1935	0.14
1940	2.6
1945	46
1950	51
1955	95
1960	135
1965	199
1970	254
1975	280 (est)

[a] Not including the Union of Soviet Socialist Republics.

Table 3. Energy Cost of Chloroprene Manufacture, MJ/kg [a]

	Acetylene route	Butadiene route
processing energy	27.7	22
total energy [b]	230.0	141

[a] To convert MJ/kg to Btu/lb, divide by 0.002324.
[b] Including energy cost of acetylene or butadiene.

acetylene route, including purification of chloroprene, is somewhat higher than the energy consumed by the butadiene route. The energy cost difference is more pronounced if the energy content of the starting materials is included, as shown by the last line in the table (see Acetylene).

Specification, Standards, and Quality Control

Polymerization-grade chloroprene is at least 98% pure. It is a clear, water-white to pale green liquid, depending on the type and concentration of inhibitors present. It should be substantially free of peroxides and dissolved polymer.

Impurities in butadiene-based chloroprene depend largely on butadiene and dichlorobutene purification. Butenes in the butadiene feed stream produce dichlorobutanes which are inseparable from 3,4-dichloro-1-butene. The dichlorobutanes lose HCl in the dehydrochlorination step to form several vinylic chlorobutenes that are difficult to separate from chloroprene (101). 1-Chlorobutadiene arising from the dehydrohalogenation of 1,4-dichloro-1-butene requires careful control of the butene purification process to maintain a uniformly low level of 1-chlorobutadiene in the chloroprene.

Analytical methods used to control the quality of chloroprene are now almost entirely based on instrumental techniques. Gas chromatography and ir spectroscopy are used almost exclusively for organic impurities in the process (see Analytical methods). Peroxides have been measured by reaction with excess ferrous sulfate and back titration with titanous chloride. Inhibitor analysis is specific for the inhibitor used. Total solids by evaporation must be used with care because of the high boiling dimers present. Dissolved polymer can be determined by alcohol precipitation, but inclusion of dimers may preclude accurate results.

Health and Safety Factors (Toxicology)

The tendency of chloroprene to peroxidize, polymerize, and burn constitutes an acute hazard. Its odor threshold is about 1 ppm in air. It can be detected in water by odor at levels from 0.1–0.5 mg/L at room temperature. However, the odor level is stronger if there is significant dimer present.

Chloroprene, like most volatile organic chlorides, is physiologically active. It is an anesthetic. Inhalation, ingestion, or absorption through the skin can result in reduced blood pressure, loss of appetite, indigestion, and the appearance of albumen, reducing substances, and bile pigments in the urine. It can cause dermatitis, conjunctivitis, and hepatic or renal damage. A threshold limit value of 25 ppm in the workroom atmosphere was established by the American Conference of Governmental Industrial Hygienists in 1965 (100).

The effects of long term exposure of workers to low levels of chloroprene are currently under investigation. There are reports of mutagenic, embryotoxic and teratogenetic effects, as well as interference with fertility in test animals. Epidemiological studies of DuPont employees have not supported suggestions that chloroprene is a carcinogen and animal carcinogenicity tests in the Union of Soviet Socialist Republics have also been negative. Several independent tests of the effects of chronic exposures of animals to chloroprene are now in progress.

Up-to-date references and a detailed review of chloroprene toxicity are listed in

Criteria Document on Chloroprene issued by the United States National Institute for Occupational Safety and Health in the latter part of 1977.

BIBLIOGRAPHY

"Chloroprene" under "Chlorocarbons and Chlorohydrocarbons" in *ECT* 2nd ed., Vol. 5, pp. 215–231, by P. S. Bauchwitz, E. I. du Pont de Nemours & Co., Inc.

1. U.S. Pat. 1,950,431 (Mar. 13, 1934), W. H. Carothers and A. M. Collins (to E. I. du Pont de Nemours & Co., Inc.).
2. W. H. Carothers and co-workers, *J. Am. Chem. Soc.* **53,** 4203 (1931).
3. P. R. Johnson, *Rubber Chem. Technol.* **49,** 650 (1976).
4. P. S. Bauchwitz, J. B. Finlay, and C. A. Stewart, Jr., in E. C. Leonard, ed., *Vinyl and Diene Monomers,* Part II, John Wiley & Sons, Inc., New York, 1971, p. 1160.
5. W. C. Price and A. D. Walsh, *Proc. Roy. Soc. (London) Ser. A* **174,** 220 (1940).
6. G. J. Szasz and N. Sheppard, *Trans. Faraday Soc.* **49** 358 (1953).
7. M. I. Batuev and co-workers, *Proc. Acad. Sci. USSR Chem. Sect. (Eng. Transl.)* **132,** 543 (1960); V. N. Nikitin and co-workers, *Optika i Spektroskopiya Akad. Nauk SSSR Otd. Fiz.-Mat. Nauk, Sb. Statei* **2,** 330 (1963); *Chem. Abstr.* **59,** 14768 (1968); M. E. Movsesyan, Zh. O. Ninoyan, and L. T. Badalyan, *Dokl. Akad. Nauk Arm. SSR,* **40,** 205 (1965); *Chem. Abstr.* **63,** 12531 (1965).
8. A. A. Bothner-By and R. K. Harris, *J. Am. Chem. Soc.* **87,** 3445 (1965); A. A. Bothner-By and D. Jung, *J. Am. Chem. Soc.* **90,** 2342 (1968).
9. P. A. Akishin, L. V. Vilkov, and M. Tatevskii, *Proc. Acad. Sci. USSR Phys. Chem. Sect. (Eng. Transl.)* **118,** 1 (1958).
10. Ref. 4, pp. 1149–1183.
11. A. C. Coke and W. R. Schmitz, *J. Am. Chem. Soc.* **77,** 3056 (1950).
12. I. N. Nazarov and A. J. Kuznetsova, *Zh. Obsch. Khim.* **30,** 139 (1960).
13. F. Hrubak and J. Webr, *Makromol. Chem.* **104,** 275 (1967).
14. N. C. Billingham and co-workers, *Nature (London)* **213,** 494 (1967).
15. C. A. Stewart, Jr., *J. Am. Chem. Soc.* **93,** 4815 (1971).
16. *Ibid.* **94,** 635 (1972).
17. V. L. Heasley and B. R. Lais, *J. Org. Chem.* **33,** 2571 (1968).
18. U.S. Pat. 3,149,172 (Sept. 15, 1964), L. J. Hughes (to Monsanto Chemical Company).
19. H. G. Viehe, *Chem. Ber.* **97,** 598 (1964); *Angew. Chem.* **75,** 793 (1963).
20. I. M. Dolgopol'skii and Yu. V. Trenke, *Zh. Obshch. Khim.* **33,** 773 (1963); I. M. Dolgopol'skii, Yu V. Trenke, and M. Kh. Blyumental, *Zh. Obshch. Khim.* **33,** 1071 (1963).
21. S. A. Vartanyan and Sh. O. Badanyan, *Izv. Akad. Nauk Arm. SSR Khim. Nauk.* **15,** 231 (1962); *Chem. Abstr.* **59,** 3755 (1963).
22. G. J. Berchet and W. H. Carothers, *J. Am. Chem. Soc.* **55,** 2004 (1933).
23. U.S. Pat. 3,833,545 (Jan. 19, 1972), S. G. Fogg (to British Petroleum Chemicals International, Ltd.).
24. I. E. Muskat and H. E. Northrup, *J. Am. Chem. Soc.* **52,** 4043 (1930).
25. A. A. Petrov and N. P. Sopov, *Zh. Obsch. Chim.* **15,** 981 (1945).
26. A. N. Pudovic, *Zh. Obsch. Khim.* **19,** 1179 (1945).
27. K. Mislow and H. M. Hellman, *J. Am. Chem. Soc.* **73,** 244 (1951).
28. U.S. Pat. 2,299,477 (Oct. 20, 1943), G. W. Hearne and D. S. LaFrance (to Shell Development Company).
29. U.S. Pat. 2,453,089 (Nov. 2, 1948), G. H. Morey and co-workers (to Commercial Solvents).
30. U.S. Pat. 2,484,042 (Oct. 11, 1949), P. Mamler (to Publicker Industries, Inc.).
31. U.S. Pat. 2,581,929 (Jan. 8, 1952), K. C. Eberly and R. J. Reid (to Firestone Tire & Rubber Company).
32. U.S. Pat. 3,050,568 (Aug. 21, 1962), R. P. Arganbright (to Monsanto Chemical Company).
33. Brit. Pat. 669,338 (Apr. 2, 1952), R. C. Chuffart (to Imperial Chemical Industries).
34. Brit. Pat. 676,691 (July 30, 1952), R. C. Chuffart (to Imperial Chemical Industries).
35. Brit. Pat. 798,027 (July 16, 1958), H. P. Crocker and co-workers (to Distillers Company, Ltd.).
36. Brit. Pat. 798,393 (July 23, 1958), C. W. Capp and co-workers (to Distillers Company, Ltd.).
37. Brit. Pat. 800,787 (Sept. 2, 1958), F. J. Bellringer and H. P. Crocker (to Distillers Company, Ltd.).
38. Ger. Pat. 1,115,236 (Oct. 19, 1961), M. Minsinger (to BASF).

39. Ger. Pat. 1,118,189 (Nov. 30, 1961), N. W. Luft and co-workers (to Hans J. Zimmer Verfahrenstechnik).
40. R. E. Taylor and G. H. Morcy, *Ind. Eng. Chem.* **40,** 432 (1948).
41. P. M. Colling, "The Vapor Phase Chlorination of 1,3-Butadiene," a dissertation, University of Texas, Austin, Tex., June 1963.
42. Brit. Pat. 798,028 (July 16, 1958), F. J. Bellringer and H. P. Crocker (to Distillers Company, Ltd.).
43. Brit. Pat. 661,806 (Nov. 28, 1951), (to E. I. du Pont de Nemours & Co., Inc.).
44. Brit. Pat. 914,920 (Jan. 9, 1963), P. G. Caudle and H. P. Crocker (to Distillers Company, Ltd.).
45. A. J. Besozzi, W. H. Taylor, and C. W. Capp, *Preprint, Division of Petroleum Chemistry, American Chemical Society, New York Meeting, Aug. 27, 1972.*
46. U.S. Pat. 2,242,084 (May 13, 1941), O. Nicodemus and W. Schmidt (to I. G. Farben A.G.).
47. U.S. Pat. 2,422,252 (June 17, 1947), J. A. Otto (to General Chemical Company).
48. Brit. Pat. 569,719 (June 6, 1945), A. S. Carter (to E. I. du Pont de Nemours & Co., Inc.).
49. U.S. Pat. 2,446,475 (Aug. 3, 1948), G. W. Hearne and D. S. LaFrance (to Shell Development Company).
50. Fr. Pat. 1,326,286 (Mar. 25, 1963), C. W. Capp (to Distillers Company, Ltd.).
51. U.S. Pat. 2,911,450 (Nov. 3, 1959), D. E. Welton (to E. I. du Pont de Nemours & Co., Inc.).
52. U.S. Pats. 3,049,573; 3,049,572 (Aug. 14, 1962), R. F. Stahl and C. Woolf (to Allied Chemical Corp.).
53. Fr. Pat. 1,332,045 (June 12, 1963); 1,327,067 (Apr. 8, 1963), J. M. Guilhamou, M. Prilleux, and P. Verrier (to Esso Corp.); H. Olive and S. J. Olive, *Organomet. Chem.* **29,** 307 (1971).
54. Ger. Offen. 1,802,385 (Nov. 6, 1969), T. Yamara and co-workers (to Sumitomo Chemical Company).
55. Jpn. Pat. 688,451; 688,453 (Apr. 2, 1968), M. Kitabatake and co-workers (to Showa Denko).
56. Brit. Pat. 1,171,948 (Nov. 26, 1968), A. Oshima (to Kanegafuchi Chemical Industry Company).
57. Belg. Pat. 763,115 (Aug. 17, 1971), P. J. N. Brown and C. W. Capp (to BP Chemicals International, Ltd.); 769,238 (Dec. 29, 1971); 770,899 (Feb. 8, 1972); 772,534 (Mar. 13, 1972), P. J. N. Brown (to BP Chemicals International, Ltd.).
58. U.S. Pat. 3,515,760 (June 2, 1970), D. D. Wilde (to E. I. du Pont de Nemours & Co., Inc.).
59. U.S. Pat. 3,819,730 (June 6, 1975), B. T. Nakata (to E. I. du Pont de Nemours & Co., Inc.).
60. U.S. Pat. 2,038,538 (Apr. 28, 1936), W. H. Carothers (to E. I. du Pont de Nemours & Co., Inc.).
61. U.S. Pat. 2,430,016 (Nov. 4, (1947), G. W. Hearne and D. S. LaFrance (to Shell Development Company).
62. U.S. Pat. 2,999,888 (Sept. 12, 1961), H. P. Crocker and K. H. W. Turck (to Distillers Company, Ltd.).
63. H. Tominaga and co-workers, *Kogyo Kagaku Zasshi* **74**(2), 199 (1971).
64. U.S. Pat. 2,391,827 (Dec. 25, 1945), M. L. Adans and G. W. Hearne (to Shell Development Company).
65. Jap. Pat. Appl. 3470/67, Fukuoka and co-workers (to Denka-Kagaku Kogyo Kabushiki Kaisha).
66. Brit. Pat. 1,147,258 (Apr. 2, 1969), A. B. Root and H. O. Wolf (to E. I. du Pont de Nemours & Co., Inc.).
67. U.S. Pat. 3,876,716 (Apr. 8, 1975), J. B. Campbell (to E. I. du Pont de Nemours & Co., Inc.).
68. U.S. Pat. 3,755,476 (Aug. 28, 1973), J. W. Crary and R. E. Tarney (to E. I. du Pont de Nemours & Co., Inc.).
69. U.S. Pat. 3,754,044 (Aug. 21, 1973), C. A. Hargreaves, II, and S. J. Piaseczynki (to E. I. du Pont de Nemours & Co., Inc.).
70. U.S. Pat. 3,639,493 (Feb. 1, 1972), J. B. Campbell (to E. I. du Pont de Nemours & Co., Inc.).
71. U.S. Pat. 3,639,492 (Feb. 1, 1972), J. B. Campbell (to E. I. du Pont de Nemours & Co., Inc.).
72. U.S. Pat. 3,622,641 (Nov. 23, 1971), J. W. Crary (to E. I. du Pont de Nemours & Co., Inc.).
73. Brit. Pat. 1,055,064 (Jan. 11, 1967), R. Lauterbach and H. Schwarz (to Farbenfabriken Bayer A.G.).
74. U.S. Pats. 2,942,037; 2,942,028 (June 21, 1960), P. A. Jenkins (to Distillers Company, Ltd.).
75. U.S. Pats. 2,926,205 (Feb. 23, 1960), F. J. Bellringer (to Distillers Company, Ltd.); 2,948,662 (Aug. 9, 1960), (to Distillers Company, Ltd.).
76. Ger. Pat. 1,115,236 (Oct. 19, 1961), M. Minsinger (to Badische Anilin und Soda-Fabrik A.G.).
77. U.S. Pat. 2,581,929 (Jan. 8, 1952), K. C. Eberley and R. J. Reid (to Firestone Tire and Rubber Company).
78. C. E. Hollis, *Chem. Ind.,* 1030 (Aug. 2, 1969).
79. F. J. Bellringer and C. E. Hollis, *Hydrocarbon Process.* **47**(11), 127 (1968).

80. U.S. Pat. 2,948,760 (Aug. 9, 1960), C. W. Capp, H. P. Crocker, and F. E. Salt (to Distillers Company, Ltd.).
81. U.S. Pat. 2,524,383 (Oct. 3, 1950), G. W. Hearne and M. L. Adams (to Shell Development Company).
82. Ger. Pat. 1,219,472 (June 23, 1966), K. Seenewald, W. Vogt, and H. Baader (to Knapsack Griesheim A.G.).
83. Ger. Pat. 1,244,158 (July 13, 1967), H. Baader and co-workers (to Knapsack Griesheim A.G.).
84. Brit. Pat. 961,856 (June 24, 1964), R. P. Arganbright (to Monsanto Chemical Company).
85. U.S. Pat. 3,079,445 (Feb. 26, 1963), R. P. Arganbright (to Monsanto Chemical Company).
86. U.S. Pat. 3,149,171 (June 27, 1964), R. P. Arganbright (to Monsanto Chemical Company).
87. Ger. Pat. 1,144,261 (Feb. 28, 1963), J. W. Engelsma and R. V. Holden (to Shell Internationale Research Maatschappi N.V.).
88. Ger. Pat. 1,235,293 (Mar. 3, 1967), C. Finger and co-workers (to Esso A.G.).
89. Brit. Pat. 986,060 (Mar. 17, 1965), F. Strain and P. D. Barlett (to Pittsburgh Plate Glass Company).
90. A. T. Babayan, N. G. Vartanyan, and I. Y. Zurabov, *Zh. Obsch. Khim. (Eng. Transl.)* **25,** 1567 (1955).
91. A. T. Babayan, *Izv. Uchv. Zardennii. Khim. Khim. Tekhnol.* **2,** 594 (1959); *Chem. Abstr.* **54,** 8592 (1960).
92. U.S. Pat. 2,922,826 (June 26, 1960), H. L. Johnson and A. P. Stuart (to Sun Oil Company).
93. A. C. Cope and W. R. Schmitz, *J. Am. Chem. Soc.* **72,** 3056 (1950).
94. K. Bouchal, J. Coupek, S. Porkorny, and F. Hrabak, *Makromol. Chem.* **137,** 95 (1970).
95. C. A. Stewart, Jr., *J. Am. Chem. Soc.* **93,** 4815 (1971).
96. U.S. Pat. 2,926,205 (Feb. 23, 1960), F. J. Bellringer (to Distillers Company, Ltd.).
97. G. E. Ham, ed., *High Polymers,* Vol. XVIII, Interscience Publishers, New York, 1964, p. 720.
98. U.S. Pat. 2,395,649 (Feb. 26, 1946), F. C. Wagner (to E. I. du Pont de Nemours & Co., Inc.).
99. Brit. Pat. 858,444 (Jan. 11, 1961), (to Farbenfabriken Bayer A.G.).
100. *Threshold Limit Values for 1973, American Conference of Governmental and Industrial Hygienists Bulletin,* ACGIH, 1973.
101. U.S. Pat. 3,396,089 (Aug. 6, 1968), K. Sennewalk and co-workers (to Knapsack A.G.).

PAUL R. JOHNSON
E. I. du Pont de Nemours & Co., Inc.

CHLORINATED PARAFFINS

Chlorinated paraffins are classified as chlorinated hydrocarbons that have the general formula $C_x H_{(2x-y+2)} Cl_y$. They were first prepared in 1858 by P. A. Bolley (1). Significant commercial uses did not develop until the early 1930s when they were first used for fire-retardant and waterproof canvas material, and in the metal working industry as extreme pressure additives for lubricating oils. The raw materials used for chlorination consist of petroleum fractions such as normal paraffins, at least 98% linear, and wax fractions averaging as many as twenty-four carbon atoms. There are a number of raw materials available; however, those used for the production of chlorinated paraffins fall into three categories: (1) a C_{12} fraction that normally includes C_9–C_{14} hydrocarbons, (2) a C_{15} fraction that normally includes C_{13}–C_{17} hydrocarbons, and (3) a C_{24} fraction that normally includes C_{20}–C_{30} hydrocarbons. The selection of a particular raw material is dependent on the desired properties of the finished chlorinated paraffin. Isoparaffins (usually <1%), aromatics (usually <100 ppm), and metal contamination are kept as low as economically feasible since their presence results in products with undesirable properties.

Physical and Chemical Properties

Commercial chlorinated paraffins have 20–70% chlorine content. The bulk of the manufactured products fall within the 40–70% Cl range. Table 1 contains a list of commercial chlorinated paraffins by chlorine content. The important physical properties of the chlorinated paraffins include viscosity, solubility, color, and thermal instability. For a given paraffin, increasing chlorine content increases viscosity and specific gravity. With chlorinated paraffins of the same chlorine content, lower viscosities are observed for the lower molecular weight paraffins.

Chlorinated paraffins are miscible with many organic solvents, eg, aliphatic, aromatic, and terpene hydrocarbons; chlorinated aliphatic and aromatic hydrocarbons; hydrogenated naphthas; and ketones, esters, and drying oils. They are insoluble in water, glycerol, and the glycols. Water emulsions can be made with the use of proper emulsifying agents.

Table 2 lists several physical properties of typical commercial chlorinated paraffins (2).

Chlorinated paraffins are relatively inert materials. Prolonged exposure to heat or light can cause dehydrochlorination. Depending upon conditions of exposure, this can occur rapidly causing a darkening of the material. The presence of aluminum, zinc, and iron will catalyze the dehydrochlorination.

Stabilizers are normally added for storage purposes. Special or higher amounts of stabilizer may be used if conditions warrant. Common stabilizers used include epoxidized soybean oils, pentaerythritol, organometallic tin compounds, or certain lead or cadmium compounds.

Table 1. Commercial Chlorinated Paraffins[a]

Chlorine content, %	Average molecular formula	Diamond Shamrock Corp.	Keil Chemical	Dover Chemical	Plastifax	ICI	Neville	Pearsall
40–42	$C_{24}H_{44}Cl_6$	Chlorowax 40	CW-170	Paroil 140	Plastichlor 42-170	Cereclor 42	Unichlor 40-170	CPF-0004
48–54	$C_{24}H_{42}Cl_8$	Chlorowax 50	CW-220-50	Paroil 150S	Plastichlor 50-220	Cereclor 48	Unichlor 50-450	CPF-0020
70	$C_{24}H_{29}Cl_{21}$	Chlorowax 70		Chlorez 700		Cereclor 70	Unichlor 70AX	FLX-70
50–52	$C_{15}H_{26}Cl_6$			Paroil 1048		Cereclor S52		FLX-0008
60–65	$C_{12}H_{19}Cl_7$	Chlorowax 500C	CW-85-60	Paroil 160	Plastichlor P-59 P-65	Cereclor 60L	Unichlor 60L-60	FLX-0012
70	$C_{12}H_{15}Cl_{11}$	Chlorowax 70L	CW-200-70	Paroil 170HV	Plastichlor P-70	Cereclor 70L		

[a] United States products. European producers include BASF, Hoechst, Bayer, and Ugine, among others.

Table 2. Physical Properties of Chlorinated Paraffins

Paraffin feedstock	Wax				C_{13}–C_{17}	C_{10}–C_{13}
chlorine content, %	39	42	48	70	52	60
density at 25°C, g/mL	1.12	1.17	1.23	1.65	1.25	1.36
viscosity at 25°C, Pa·s[a]	0.7	3.0	12.5	solid	1.6	3.5
color (Gardner)	2	2	2	white	1	1
refractive index	1.501	1.505	1.516		1.510	1.516
pour point, °C	−20	0	10		−10	−10
heat stability, % HCl after 4 h at 175°C	0.2	0.2	0.25	0.15	0.10	0.10

[a] To convert Pa·s to poise, multiply by 10.

Manufacture

Chlorinated paraffins are produced by passing chlorine gas into a liquid paraffin. Because of the corrosive nature of hydrogen chloride and chlorine, special care must be given to materials of construction. Ultraviolet light is often used to promote chlorination, especially at higher chlorine contents. Metal chlorides can cause darkening by decomposition.

Chlorine feed rates and reaction temperatures differ slightly among the different producers. The reaction is exothermic and temperatures are usually kept at 90–100°C.

Manufacture of resinous chlorinated paraffins (70% chlorine content) requires the use of a solvent during the chlorination step. Additional steps of solvent stripping and grinding the product are necessary.

Economic Aspects

The annual United States production of chlorinated paraffins in 1965–1975 is given in Table 3.

The growth in usage of chlorinated paraffins is dependent on two conditions: (1) that stringent fire retardancy regulations are passed by government agencies and (2) that prices of primary plasticizers (qv) become high enough to make secondary plasticizers attractive.

Table 3. United States Chlorinated Paraffin Production and Statistics[a]

Year	Production, metric tons	Sales Quantity, metric tons	Sales Total value,[b]	Sales Unit value, $/kg
1965	19,845	19,973	5,698	0.29
1970	26,477	25,816	7,560	0.29
1973	33,822	34,475	11,581	0.33
1974	35,386	34,002	17,024	0.51
1975	32,894	32,273	18,934	0.60
1976	34,522	31,153	20,040	0.64
1977		37,046	23,127	0.62

[a] According to U.S. International Trade Commission Reports.
[b] In thousands of dollars.

Handling and Storage

Liquid chlorinated paraffins are available in standard 208-L (55-gal) drums, tank trucks, and tank cars. Special points to consider in handling and storage are the product's viscosity, specific gravity and temperature limitations.

The viscosity of a liquid chlorinated paraffin increases substantially as its temperature is reduced. Table 4 shows typical viscosities of liquid chlorinated paraffins at various temperatures.

If shipments of these products are exposed to extremely cold weather, it may be necessary to heat shipping containers before they are unloaded. Warm water, not steam, should be used for tank truck or tank car shipments. Chlorinated paraffins should be heated to no higher than 65°C. Even at this temperature, prolonged exposure can cause decomposition.

Storage tanks can be 6.35 mm or 9.53 mm welded mild steel, but extreme caution must be exercised to exclude moisture. Internal coatings (polyester or other plastics) should be used to prevent discoloration from iron contamination. Tanks with stainless steel, nickel, glass or lead linings can also be used. Glass-reinforced polyester tanks are especially recommended for chlorinated paraffin storage.

Solid chlorinated paraffins are shipped in 22.7-kg polyethylene lined bags or 113.4-kg fiber drums.

Toxicity

Chlorinated paraffins are classed as nontoxic compounds by ingestion or by dermal application and are not eye irritants in accordance with the procedures specified in the regulations of the Federal Hazardous Substance Labeling Act (3).

A 60% chlorinated paraffin (ie, Chlorowax 500C) tested under the above procedures resulted in an acute oral LD_{50} for male albino rats greater than 21.5 mL/kg of body weight. Acute dermal LD_{50} for albino rabbits is greater than 10.0 mL/kg of body weight. Test patches containing Chlorowax 500C were applied to the skin of 200 human male and female panelists. No clinically significant adverse effects were observed on any subject. Single application to the eyes of rabbits produced only mild erythema in four of six rabbits (4). Similar results have been obtained on other chlorinated paraffins such as Chlorowax 40 (5).

Table 4. Viscosity (in Pa·s[a]) vs Temperature

Chlorine, %	Temperature, °C				
	0	25	38	60	72
60	27.2	2.0	0.65	0.15	0.070
43	28.0	3.0	1.1	0.35	0.066
48	120.0	10.0	3.3	0.9	0.25

[a] To convert Pa·s to P, multiply by 10.

Uses

In the United States, approximately 50% of chlorinated paraffins are used as extreme pressure lubricant additives in the metal working industry. Twenty-five percent are used in plastics including fire retardant and water repellent coated fabrics. The remainder is used in rubber (qv), caulks, and sealants. In the United Kingdom, 50% of chlorinated paraffins are used as secondary plasticizers in the vinyl industry. This difference in use is the result of economic considerations.

Liquids. *Paints* (6–8). Liquid chlorinated paraffins are chemically inert plasticizers. In paint formulations employing hard resins, they are used to make films more flexible and eliminate film embrittlement. Lower viscosity products are used where high vehicle solids are required at relatively low viscosities. These products are also easier to emulsify. Intermediate viscosity, 40% chlorine products are widely used to plasticize chemical-resistant coatings which are based on chlorinated rubber and styrene–butadiene resins.

Plastics (9–12). Chlorinated paraffins are useful in plastic compounds as secondary plasticizers because of high fire retardant efficiency, good heat and light stability, high resin compatibility, lack of odor, and low toxicity.

Adhesives, Mastics, and Caulks (13). Chlorinated paraffins are used as general purpose plasticizers in these applications because of good permanence (low volatility); nondrying and nonpolymerizing properties; chemical resistance; and good moisture resistance (see Sealants).

Lubricants. Chlorinated paraffins have been used for years in the metal working industry as an extreme-pressure additive in various fluids. They are compatible with a wide variety of cutting oils to provide both extreme-pressure activity and as a boundary lubricant. If only extreme-pressure lubrication is desirable, a lower viscosity material can be used. For boundary lubrication a more viscous material must be used (see Lubrication).

Resinous Materials. *Paints.* Resinous chlorinated paraffins are used extensively as binder components for various solvent-base paints (6–8) to obtain improved chemical and water resistance; increased hardness and improved adhesion; and higher vehicle solids at working viscosity. In addition, chlorinated paraffins impart improved fire retardancy to water-based fire retardant coatings of the intumescent type and better color retention on exposure in exterior acrylic resin emulsion paints.

Plastics. Resinous chlorinated paraffins, combined with a synergist such as antimony oxide, form a highly efficient flame retardant system for many types of plastic (9–12) (see Flame retardants).

Polyolefins, polystyrene, polyesters, epoxides, polyurethane, and other flammable polymers can be made flame retardant by adding various levels of resinous chlorinated paraffin and antimony oxide. Polymers that are processed over 177°C require special stabilized grades. A processing temperature of 232°C is usually used as an upper limit for chlorinated paraffins.

BIBLIOGRAPHY

"Chlorinated Paraffins" under "Chlorine Compounds, Organic" in *ECT* 1st ed., Vol. 3, pp. 781–786, by H. M. Roberts, Imperial Chemical Industries Ltd.; "Chlorinated Paraffins" under "Chlorocarbons and Chlorohydrocarbons" in *ECT* 2nd ed., Vol. 5, pp. 231–240, by D. W. F. Hardie, Imperial Chemical Industries Ltd.

1. P. A. Bolley, *Justus Liebigs Ann. Chem.* **106,** 230 (1858).
2. Diamond Shamrock Corp., *Technical Data—Chlorowax & Delvet*, 1977.
3. *Fed. Reg.*, (Aug. 12, 1961), et seq.
4. Diamond Shamrock Corp., *Industrial Health Studies on Chlorowax 500C*, 1963.
5. Diamond Shamrock Corp., *Industrial Health Studies on Chlorowax 40*, 1963.
6. W. E. Allsebrook, *Paint Manuf.* **42,** 40 (1972).
7. H. Eckhart and G. Grimm, *Farbe Lack* **73,** 36 (1971).
8. K. S. Ford, *J. Oil Colour Chem. Assoc.* **55,** 584 (1972).
9. K. M. Bell, B. W. McAdam, and H. J. Caesar, *Kunststoffe* **59,** 344 (1969).
10. A. Hofmann, *Kunststoffe* **61,** 811 (1971).
11. U.S. Pat. 3,668,155 (June 6, 1972), C. Raley, Jr. (to The Dow Chemical Co.).
12. U.S. Pat. 3,755,227 (Aug. 28, 1973), R. A. Gray and D. G. Brady (to Lippon Zeib Co., Ltd.).
13. U.S. Pat. 3,694,305 (Sept. 26, 1972), S. M. Munawaar (to Compac Corp.).

B. A. SCHENKER
Diamond Shamrock Corp.

CHLORINATED DERIVATIVES OF CYCLOPENTADIENE

The two compounds hexachlorocyclopentadiene [77-47-4] (**1**) (HCCP) and octachlorocyclopentene [706-78-5] (**2**) (OCCP) are the only known commercially significant cyclic C_5 chlorocarbons produced from cyclopentadiene. They are conveniently discussed together as they are interconvertible under appropriate conditions, and they can be prepared from the same process (1). Hexachlorocyclopentadiene is commercially important as an intermediate for many insecticides, polyester resins, and flame retardants.

Hexachlorocyclopentadiene. Hexachlorocyclopentadiene (**1**) is a nonflammable liquid with a characteristically pungent, musty odor. The pure compound is light lemon-yellow. Impurities may produce a greenish tinge.

Octachlorocyclopentene. Octachlorocyclopentene (**2**) has a very faint camphoric odor. It freezes to a white solid at room temperature and is quite stable on prolonged storage. At its atmospheric reflux temperature (283°C) it dissociates slightly into hexachlorocyclopentadiene and chlorine.

Physical Properties

The physical properties of hexachlorocyclopentadiene and octachlorocyclopentene are listed in Table 1 (2).

Chemical Properties

The commercial use of hexachlorocyclopentadiene is based on its reactivity as a conjugated unsaturated material. It has a high order of reactivity with a variety of dienophilic materials in the Diels-Alder reaction as well as a variety of addition and substitution reactions. The products of this reaction are generally 1:1 adducts containing a hexachlorobicyclo[2.2.1] heptene structure. The products invariably contain the constituent derived from the olefin in the *endo* position although *exo* isomers are possible but are present only in small amounts, if at all.

Reference 3 is a comprehensive summary of the chemistry of perchlorocyclopentenes and cyclopentadienes; it describes the character and conditions for a large number of these reactions.

Manufacture

A number of manufacturing processes for hexachlorocyclopentadiene have been described and commercialized. Three alternative chemical routes are described below:

Process Employing Cyclic Hydrocarbons with Alkaline Hypochlorite. The first report describing the preparation and identification of hexachlorocyclopentadiene was made in 1930 (4). In this process, freshly prepared cyclopentadiene is mixed with ca 6–10 moles of alkaline hypochlorite solution per mole of cyclopentadiene using vigorous

Table 1. Physical Properties

	HCCP (1)	OCCP (2)
mol wt	272.79	343.71
sp gr, 20°C	1.710	1.816
bp, °C	239	285
mp, °C	11.34	38.2
viscosity, mPa·s (= cP)	6.97, 25°C	30.87, 45°C
100°C	1.72	7.51
latent heat of vap, J/g[a]	176.6	163
latent heat of fusion, J/g[a]	41.8	
specific heat	0.21	0.22
refractive index, n_D^t	1.5649[20]	1.5683[45]
surface tension, mN/m (= dyn/cm)	47	
vapor pressure[b]	°C kPa	°C kPa
	62 0.13	99.5 0.13
	106 1.3	147 1.3
	139.5 5.3	181 5.3
	164 13.3	207 13.3

[a] To convert J to cal, divide by 4.184.
[b] To convert kPa to mm Hg, multiply by 7.5.

agitation at ca 40°C. The hexachlorocyclopentadiene is recovered by fractional distillation. The reaction product always contains appreciable quantities of lower chlorinated cyclopentadienes. A thorough study of this reaction was reported in 1952; the maximum yield of C_5Cl_6 was 75% (5). Separation of pure product from the crude material is a major problem. Conversion of impurities to high-boiling condensation products (6) and recrystallization from solvents (7) have been proposed. The crude form has been used to prepare certain derivatives such as chlordane. However, for most applications isolation of a high purity compound is required. Modifications of this process have been operated on a commercial scale by Velsicol Chemical Corp. for the purpose of producing chlordane.

Processes Employing Aliphatic Hydrocarbons. Preparation of a polychlorinated straight- or branched-chain hydrocarbon followed by cyclization to give octachlorocyclopentene or hexachlorocyclopentadiene has been described. In 1932 Prins reported the preparation of monochloropentene by condensation of hexachloropropene with trichloroethylene (8). Alkaline dehydrochlorination followed by cyclization in the presence of aluminum chloride (9) gave octachlorocyclopentene.

Preparation of hexachlorocyclopentadiene by the Prins technique followed by thermal dechlorination of octachlorocyclopentene at 470–480°C was reported in 1947 (10).

A process in which the reactions of substitution-chlorination, dehydrochlorination, cyclization, and dehydrochlorination are carried out consecutively and continuously was developed (11–12). The first stage involves preparation of polychloropentanes by chlorination of pentane in the liquid phase and the second stage involves vapor-phase chlorination and ring closure using a porous surface-active inorganic catalyst. Yields of hexachlorocyclopentadiene of greater than 90% based on the hydrocarbon consumption are reported. This process has been operated on a commercial scale by Hooker Chemicals & Plastics Corp. By suitable modifications the process can also be adapted to the preparation of octachlorocyclopentene (1). A similar process was patented in the USSR in 1957 using tetrachloropentene as the feed (13). Detailed kinetic studies of this process have been reported (14–19).

Processes Employing Chlorination of Cyclic Hydrocarbons. Shell Development Company in the United States and N.V. de Bataasfshe Petroleum Maatschappij in other countries have disclosed in various patents the essential features of a chlorination process which they developed. The first step in the process involves liquid phase chlorination of cyclopentadiene at a temperature of about 50°C to produce tetrachlorocyclopentane. Some substitution takes place giving a product of nominal $C_5H_5Cl_5$ composition. In the Lidov process (20) catalytic chlorination over phosphorus pentachloride or arsenous oxide at 175–275°C gives a reported yield of about 95% octachlorocyclopentene which is then thermally dechlorinated to hexachlorocyclopentadiene. A procedure has been patented for obtaining hexachlorinated cyclopentane in one step by injecting the cyclopentadiene into a circulating photochlorinator (21). The hexachlorinated cyclopentane is then dehydrochlorinated over an active surface catalyst at elevated temperatures to produce hexachlorcyclopentadiene. Thermal substitution chlorination of pentachlorinated cyclopentane has also been reported. The polychlorinated cyclopentane is then vaporized in the presence of chlorine and is chlorinated to the octachlorocyclopentene and thermally dechlorinated to form hexachlorocyclopentadiene. Recently, Lummus has patented a process for converting the chlorinated cyclopentane to hexachlorocyclopentadiene using a molten salt mixture (22). Fluidized-bed reactors for this purpose have also been patented (23–24).

Table 2. Insecticides Produced From Hexachlorocyclopentadiene

Compound	CAS Reg. No.	Common name	Structure no.	Uses
endo,exo-1,2,3,4,10,10-hexachloro-1,4,4a,5,8,8-hexahydro-1,4:5,8-dimethanonaphthalene	[309-00-2]	Aldrin HHDN		no longer sold in the U.S.
endo,exo-3,4,5,6,9,9-hexachloro-1a,2,2a,3,6,6a,7,7a-octahydro-2,7:3,6-dimethenaphth[2,3-b]oxirene	[60-57-1]	Dieldrin		no longer used in the U.S.
endo,endo-5,6,7,8,9,9-hexachloro-1,2,3,4,4a,5,8,8a-octahydro-1,4:5,8-dimethanonaphth[2,3-b]oxirene	[72-20-8]	Endrin	(3)	registered for some grains and mouse control on orchard floors
1,2,4,5,6,7,8,8-octachloro-3a,4,7,7a-tetrahydro-4,7-methanoindane	[57-74-9]	Chlordane Kypchlor Octachlor Synklor	(4)	use in U.S. limited to termite control
1,4,5,6,7,8,8-heptachloro-3a,4,7,7a-tetrahydro-4,7-methano-1H-indene	[76-44-8]	Heptachlor		no longer sold in the U.S.
6,7,8,9,10,10-hexachloro-1,5,5a,6,9,9a-hexahydro-6,9-methano-2,4,3-benzodioxathiepin 3-oxide	[115-29-7]	Endosulfan Beosit Chlortiepin Cyclodan Insectophene Malix Thifor Thimiel Thiodan Thionex Thiosulfan	(5)	used widely on a number of crops
bis(pentachlorocyclopentadienyl)	[2227-17-0]	Dienochlor Pentac	(6)	greenhouse ornamentals
dodecachloroctahydro-1,3,4-metheno-1H-cyclobuta[cd]pentalene	[2385-85-5]	Mirex Dechlorane Fire Ant Bait		no longer used in U.S.

Octachlorocyclopentene and hexachlorocyclopentadiene are mutually interconvertible. The equilibrium conditions and reaction rates for the reaction have been described (10–12,20). The dechlorination reaction is accelerated in the presence of metals or metal salts of the transition elements.

$$C_5Cl_8(g) \rightleftharpoons C_5Cl_6(g) + Cl_2(g)$$

Shipment and Storage

Hexachlorocyclopentadiene is classified under DOT regulations as a Class B poison. Shipping labels must list it as a poisonous liquid, NOS. It may be shipped in DOT 103 tank cars or 208-L (55-gal) lined steel drums.

If moisture is rigorously excluded, HCCP can be stored without harming product or container. However, to avoid any possibility of iron contamination, storage tanks

Table 3. Flame Retardant Chemicals From Hexachlorocyclopentadiene

Compound name	CAS Reg. No.	Common name	Structure no.
1,4,5,6,7,7-hexachlorobicyclo-[2.2.1]hept-5-ene-2,3-dicarboxylic acid	[115-28-6]	Het Acid Chlorendic Acid	(see 7)
1,4,5,6,7,7-hexachlorobicyclo-[2.2.1]hept-5-ene-2,3-dicarboxylic anhydride	[115-27-5]	Het Anhydride Chlorendic Anydride	(7)
1,2,3,4,7,8,9,10,13,13,14,14-dodecachloro-1,4,4a,5,6,6a,7,10,10a,11,12,12a-dodecahydro-1,4:7,10-dimethanodibenzo[a,e]-cyclooctene	[13560-89-9]	Dechlorane Plus	(8)

Table 4. Producers and Prices of HCCP Products

Compound	Producer	Price, $/kg 1972	1977
Endrin	Shell International Chem. Corp.; Velsicol Chem. Corp.	5.40	5.40
Chlordane	Velsicol Chem. Corp.	1.70	1.59
Endosulfan	F.M.C. Corp. Hoeshst A.G. I. Pi. Ci. Makteshim-Agan Velsicol Chem. Corp.	3.30–4.40	7.72
Dienochlor	Hooker Chemical & Plastics Corp.	17.09	19.29
Het Acid	Hooker Chemical & Plastics Corp.	0.82	2.31
Het Anhydride (Chlorendic Anhydride)	Velsicol Chem. Corp.	0.77	1.81
Dechlorane Plus	Hooker Chemical & Plastics Corp.	1.54	2.87

should be nickel-clad, or lined with baked phenolic resin, or glass lined. Steam coils and pumps in contact with HCCP should be of nickel. Piping should be nickel, nickel-plated, glass-lined steel or phenolic lined. Valves should be of nickel, porcelain, or glass lined. Diaphragms in diaphragm valves should be of Kel-F or Teflon.

Health and Safety Factors (2,25)

Hexachlorocyclopentadiene is toxic and has been included on the EPA's Toxic Pollutant list (26). It can cause skin burns. It can be readily absorbed through the skin; therefore, it is necessary to avoid prolonged or repeated contact. Persons handling HCCP should wear elbow length neoprene gloves, protective goggles or face shields, and full clothing. Adequate ventilation should be provided when HCCP is handled in a closed area. Self-contained air masks or full face canister gas masks of the acid gases and organic vapors type should be available at all times. Toxicity tests on albino rats established on acute oral LD_{50} of 300–630 mg/kg of body weight. Acute inhalation tests in rats of <2 mg/L caused death in rats. It is described as a nonflammable lachrimator that activates tear ducts and inflames the eyes. HCCP is highly toxic to a variety of fish species.

Uses

Tables 2 and 3 list commercial products derived from HCCP. Table 4 lists producers and prices.

BIBLIOGRAPHY

"Chlorinated Derivatives of Cyclopentadiene," under "Chlorocarbons and Chlorohydrocarbons," in *ECT* 2nd ed., Vol. 5, pp. 240–252, by R. R. Whetson, Shell Development Company.

1. U.S. Pat. 2,899,370 (August 11, 1959), D. S. Rosenberg (to Hooker Chemical & Plastics Corp.).
2. *Hexachlorocyclopentadiene*, Hooker Chemical & Plastics Corp., Data Sheet No. 815A, April, 1969.
3. H. E. Ungnade and E. T. McBee, *Chem. Rev.* **58**, 249 (1958).
4. F. Strauss, L. Kollek, and W. Heyn, *Ber.* **63B**, 1868 (1930).
5. R. Riemschneider, *La Chemica e l'Industria* **34**, 266 (1952).
6. U.S. Pat. 2,658,085 (Nov. 3, 1953), M. Kleinman (to Arvey Corp.).
7. U.S. Pat. 2,927,947 (March 8, 1960), G. Liedtke (to Schering Akt-Ges).
8. H. J. Prins, *Rec. Trav. Chim.* **57**, 1065 (1938).
9. *Ibid.*, 659 (1938).
10. J. A. Krynitsky and R. W. Bost, *J. Am. Chem. Soc.* **69**, 1918 (1947).
11. U.S. Pat. 2,650,942 (Sept. 1, 1953), A. H. Maude and D. S. Rosenberg (to Hooker Chemical & Plastics Corp.).
12. U.S. Pat. 2,714,124 (July 26, 1955), A. H. Maude and D. S. Rosenberg (to Hooker Chemical & Plastics Corp.).
13. U.S.S.R. Pat. 108,590, N. N. Melnikov, S. D. Volodkovich, and L. M. Kogan (Nov. 25, 1957).
14. L. M. Kogan, N. M. Burmakin, and N. V. Cherniak, *J. Gen. Chem.* (USSR) **28**, 27 (1957).
15. L. M. Kogan and N. M. Burmakin, *Zh. Prkl Khim* **37**, 869 (1964).

16. L. M. Kogan, *Pron.* **5,** 448 (1959).
17. L. M. Kogan and N. M. Burmakin, *Zh. Prikl Khim.* **31,** 1585 (1958).
18. L. M. Kogan and N. M. Burmakin, *Zh. Org. Khim.* **28,** 27 (1958).
19. L. M. Kogan and N. M. Burmakin, *Zh. Prikl Khim* **33,** 1653 (1960).
20. Brit. Pat. 703,202 (January 27, 1954), R. E. Lidov (to N.V. deBataafsche Petroleum Maatschippij); U.S. Pat. 2,900,420 (August 18, 1959) (to Shell Chemical).
21. U.S. Pat. 3,637,479 (Jan. 25, 1972), D. S. Rosenberg and D. U. Spinney (to Hooker Chemical & Plastics Corp.).
22. U.S. Pat. 4,036,897 (July 19, 1977), V. A. Strangio, H. Reigel, and M. Sze (to Lummus Corp.).
23. U.S. Pat. 3,364,269 (Jan. 16, 1968), M. Minsinger (to Badische Anilin & Soda Fabrik).
24. U.S. Pat. 3,649,699 (March 14, 1972), K. K. Aoki and A. L. McMaster; U.S. Pat. 3,748,281 (July 24, 1973) (to BASF Wyandotte Corp.).
25. *PCl Hexachlorocyclopentadiene,* Velsicol Chemical Corp., Product Bulletin No. 50101-3, April 8, 1976.
26. *Fed. Reg.* **43**(21), 4108 (1978).

JAMES E. STEVENS
Hooker Chemical & Plastics Corp.

CHLORINATED BENZENES

Twelve chlorinated benzenes can be formed by replacing some or all of the hydrogen atoms of benzene with chlorine atoms. With the exceptions of 1,3-dichlorobenzene, 1,3,5-trichlorobenzene, and 1,2,3,5-tetrachlorobenzene, they are produced readily by chlorinating benzene in the presence of a Friedel-Crafts catalyst (see Friedel-Crafts reactions). The usual catalyst is ferric chloride, either as such or generated *in situ* by exposing a large surface of iron to the liquid being chlorinated. Each compound, except hexachlorobenzene, can be further chlorinated, hence the product is always a mixture of chlorinated benzenes. Pure compounds are obtained by distillation and crystallization.

Chlorobenzenes were first synthesized in the middle of the nineteenth century. The first direct chlorination of benzene was reported in 1905 (1). Commercial production was initiated in 1909 by the former United Alkali Company in England (2). In 1915, the Hooker Electrochemical Company began operation of its first chlorobenzenes plant in the U.S. with a capacity of ca 8200 metric tons per year at Niagara Falls.

The Dow Chemical Company also started their U.S. production of chlorobenzenes in 1915 (3). Chlorobenzene was the first and has been the dominant commercial product for 50 years. Large quantities of chlorobenzene were used during World War I to produce the military explosive picric acid.

In the mid 1920s, the Dow Chemical Company developed two processes which also consumed large amounts of chlorobenzene. In one, chlorobenzene was hydrolyzed with ammonium hydroxide in the presence of a dissolved copper catalyst to produce aniline. This process was used for more than 30 years. The other process produced phenol by hydrolyzing chlorobenzene with aqueous sodium hydroxide under high temperature and pressure conditions (4–5). Independently, the I. G. Farbenwerke in Germany developed an equivalent process. Plants were built in several European countries after World War II. The I.C.I. plant in England operated until 1965.

In the 1930s, the Raschig Company in Germany developed a different chlorobenzene–phenol process in which chlorobenzene vapor was hydrolyzed with steam using a calcium phosphate catalyst to produce phenol and HCl (6). Recovered HCl reacted with air and benzene over a copper oxide catalyst (Deacon Catalyst) to produce chlorobenzene and water (7–8). A similar process was developed in the United States by the Bakelite Division of Union Carbide, which operated a plant for many years. The Durez Company licensed the Raschig process and built a plant in the United States. The company was later taken over by the Hooker Chemical Corp. which made significant process improvements.

In recent years, production of phenol from cumene has become the dominant process and most of the chlorobenzene hydrolysis processes have been discontinued (see Cumene). A plant in Argentina is now the only vapor-phase hydrolysis phenol plant operating in the world.

Physical and Chemical Properties

The important physical properties for the chlorobenzenes are listed in Table 1. Vapor pressure as a function of temperature is correlated by the Antoine equation:
$$\log_{10} P \text{ (kPa)} = A - B/(T + C) - 0.875097$$
$$(1 \text{ kPa} = 7.5 \text{ mm Hg})$$
$$\log_{10} P \text{ (mm Hg)} = A - B/(T + C)$$
T is the temperature in °C, and A, B, C are the Antoine constants (listed in Table 1 for each compound).

The chlorine atom in chlorobenzene is sufficiently labile to be hydrolyzed. Hydrolysis with steam at 500°C over a calcium phosphate catalyst was used in the Raschig phenol process. Hydrolysis with aqueous sodium hydroxide at 400°C under high pressure was used in the Dow phenol process (4–5). The mechanism of this reaction has been extensively studied. The transient moiety benzyne has been shown to dominate the mechanism (9–10). In the Dow aniline process, with aqueous ammonia in the presence of a soluble copper catalyst, the chlorine is replaced by the —NH$_2$ group.

The hydrolysis of one of the chlorine atoms in 1,2,4,5-tetrachlorobenzene with sodium hydroxide dissolved in a suitable alcohol solvent produces 2,4,5-trichlorophenol. The reaction must be carried out under extremely restricted conditions to prevent the formation of the very toxic compound, 2,3,7,8-tetrachlorodibenzo-p-dioxin (TCDD) (see Industrial hygene and toxicology).

TCDD

Nitration of chlorobenzenes with nitric acid has wide industrial applications.

$$\text{C}_6\text{H}_{5-n}\text{Cl}_n + \text{HNO}_3 \longrightarrow \text{C}_6\text{H}_{4-n}(\text{NO}_2)\text{Cl}_n + \text{H}_2\text{O} \quad n = 1, 2, 5$$

Table 1. Physical Properties of Chlorobenzenes

	Chloro-benzene [108-90-7]	1,2-Di-chloro-benzene [95-50-1]	1,3-Di-chloro-benzene [541-73-1]	1,4-Di-chloro-benzene [106-46-7]	1,2,3-Trichloro-benzene [87-61-6]	1,2,4-Trichloro-benzene [120-82-1]	1,3,5-Tri-chloro-benzene [108-70-3]	1,2,3,4-Tetra-chloro-benzene [634-66-2]	1,2,3,5-Tetra-chloro-benzene [634-90-2]	1,2,4,5-Tetra-chloro-benzene [95-94-2]	Penta-chloro-benzene [608-93-5]	Hexa-chloro-benzene [1118-74-1]
mol wt	112.56	147.005	147.005	147.005	181.45	181.45	181.45	215.90	215.9	215.9	250.35	284.80
mp	−45.34	−16.97	−24.76	53.04	53.5	17.15	63.5	46.0	51	139.5	85	228.7
bp, 101.3 kPa[a]	131.7	180.4	173.0	174.1	218.5	213.8	208.5	254.9	246	248.0	276	319.3
critical temperature, °C	359.2	417.2	415.3	407.5		453.3		450		489.8		551
critical pressure, kPa[a]	4519	4031	4864	4109		3718		3380		3380		2847
critical density, kg/L	0.3655	0.411	0.458	0.411		0.447		0.40		0.475		0.518
Antoine constants												
A	7.046324	7.143024	7.072644	7.002424		7.136684		7.159274		7.284164		6.66747
B	1482.156	1703.916	1629.811	1578.149		1790.267		1930.023		2003.495		1654.17
C	224.115	219.352	215.821	208.84		206.283		196.213		207.038		117.536
liquid density, kg/L	1.10118	1.3022	1.2828	1.2475		1.44829		1.70		1.833(s)		1.596
viscosity, mPa·s (= cP)	0.756	1.3018	1.0254					3.37				
heat capacity for liquid, J/g[b]	1.339	1.159		1.188		1.008		1.259		1.142		
heat of fusion, J/g[b]	90.33	86.11	85.98	123.8		85.78		64.52		112.2		89.62
heat of vaporization, J/g[b]	331.1	311.0	296.8	297.4		280.0		268.9		221.8		190.8
explosive limits of vapor in air, 101.3 kPa[a], vol %	1.3–7.1					None		none to 205°C				
flash point, °C (ASTM method D56-70, closed cup)	28	71		67		99				none		
standard heat of formation of liquid, J/g[c]	−95.90	−125.23	−145.73	−284.6(s)		−263.1						−460(s)
thermal conductivity of liquid, W/(m·K)	0.127	0.121		0.105		0.108						
refractive index of liquid, n_D^{25}	1.5219	1.5492	1.54337	1.52849 (55°C)		1.56933						
dielectric constant of liquid	5.621	9.93	5.04	2.41		2.24						
surface tension, mN/m (= dyn/cm)	32.65	36.61	36.20	31.4		38.54		21.6				

[a] To convert kPa to mm Hg, multiply by 7.5.
[b] To convert J to cal, divide by 4.184.
[c] Ref 11.

Monochloronitrobenzenes and 1,2-dichloro-4-nitrobenzene are intermediates for herbicides and insecticides. Pentachloronitrobenzene is used as a soil fungicide. Chlorobenzene undergoes acid-catalyzed condensation with trichloroacetaldehyde to produce DDT. This application has been severely limited owing to ecological complications.

$$C_6H_5Cl + Cl_3CCHO \xrightarrow{H_2SO_4} (Cl\text{-}C_6H_4)_2CHCl_3 + H_2O$$

Manufacture

The production of any chlorinated benzene is a multiple product operation. A plant for any chlorobenzene must produce HCl and some other chlorinated benzenes. Only limited control can be exercised over the product ratios. Chlorinated benzenes can be produced by the vapor phase chlorination of benzene using air and HCl as chlorinating agents; this is the first stage of the Raschig phenol process (7–8). The energy costs for this process are so high that it is no longer used commercially and it could never have been operated economically to produce chlorobenzenes as main products. Chlorine and benzene react in the vapor phase at 400–500°C to give a different distribution of products (12), but such a process would be much more costly than conventional liquid-phase operations.

All the chlorobenzenes are now produced by chlorination of benzene in the liquid phase. Ferric chloride is the most common catalyst. A recent study indicates that the $FeCl_3 \cdot H_2O$ complex is probably the most effective catalyst (13). Although precautions are taken to keep water out of the system, it is possible that this complex catalyst is present in most operations owing to traces of water in benzene entering the reactor.

The liquid-phase chlorination of benzene is an ideal example of a set of sequential reactions. The possible reactions are shown in Figure 1, with the main rate constants indicated. For convenience, one of the rate constants, K_5, is given the value of unity, and the others shown as relative rates. Chlorine concentration does not enter the calculation procedure as it is the same for all reactions. Two classical papers have been published modeling the chlorination of benzene through the dichlorobenzenes (14–15). A reactor system may be simulated with the relative rate equations and flow equation. The batch reactor gives the minimum ratio of $-Cl_{n+1}/-Cl_n$. This can be approximated by a plug-flow reactor or a multistage stirred reactor. A single-stage stirred reactor will produce the highest $-Cl_{n+1}/-Cl_n$ ratio. When chlorobenzene is the desired product, control over the dichlorobenzenes to monochlorobenzene ratio is effected primarily by controlling the extent of chlorination. The low di/mono ratio is obtained at the expense of energy used in recycling the unreacted benzene.

In the liquid-phase chlorination, 1,3-dichlorobenzene is found only in small quantity, and 1,3,5-trichlorobenzene and 1,2,3,5-tetrachlorobenzene are undetectable. The ratios of 1,4- to 1,2-dichlorobenzene with various catalysts are shown in Table 2.

Iodine plus antimony trichloride is effective in selectively chlorinating 1,2,4-trichlorobenzene to 1,2,4,5-tetrachlorobenzene (22).

The chlorination reaction is exothermic. The heat liberated is ca 1.83 kJ/g Cl_2 (437 cal/g Cl_2). Heat is removed in some cases by circulating the reaction liquid through a suitable cooler (see Heat exchange technology). In other cases, chlorination occurs

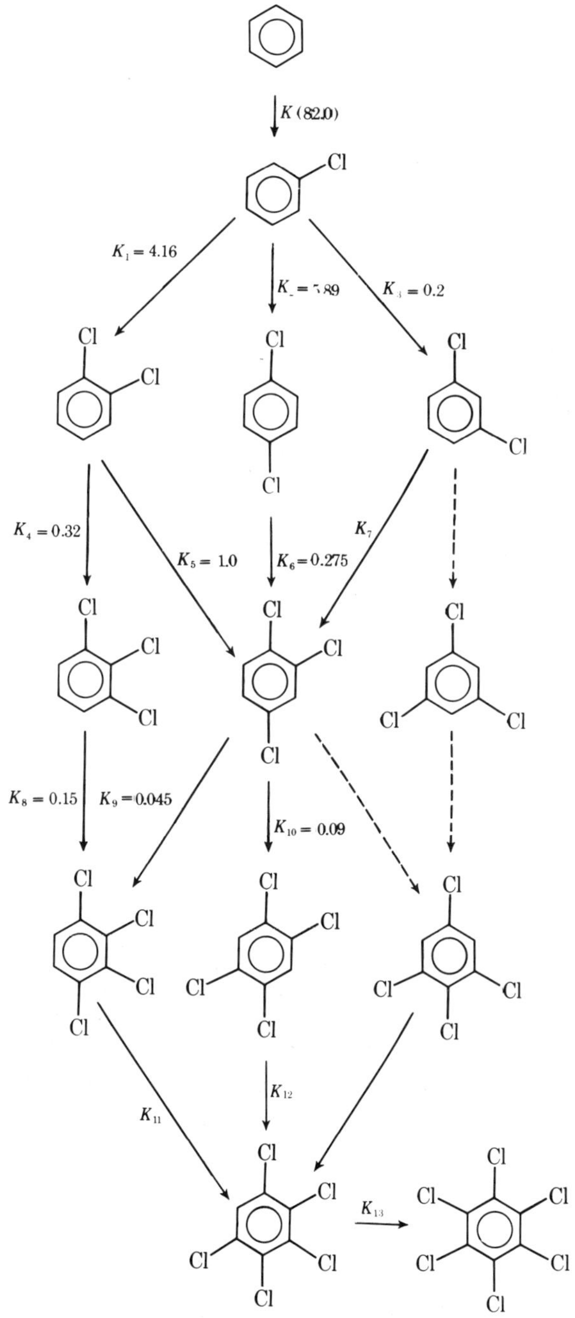

Figure 1. Sequential reactions showing relative rates in the liquid chlorination of benzene.

Table 2. Ratio of 1,4- to 1,2-Dichlorobenzene with Various Catalysts

Ratio	Catalyst	Ref.
1.4	FeCl$_3$	
1.2–1.7	several	16
2.2–3.3	FeCl$_3$, AlCl$_3$, or SbCl$_3$ and organic sulfur cmpd	17
2.0–2.65	FeS and organic sulfur cmpd	18
2–4	FeCl$_3$ or SbCl$_3$ and sulfur	19
3–5	SbCl$_3$ and sulfur	20
2.3	SnCl$_4$ and/or TiCl$_4$ and AlCl$_3$	21

at the boiling point: the heat of the reaction is removed from the reactor by the vaporizing liquid. The latter procedure has the disadvantage of operating at a higher temperature, but has the advantage of allowing a low-inventory reactor system which saves equipment costs, reduces operating hazards, and makes heat recovery possible.

Benzene chlorination reactors are subject to design and operating hazards. Stagnant areas must be avoided in reactor design as they allow chlorination to the tetra- and pentachlorobenzenes. These compounds have low solubility in the liquid and can cause plugging. Another hazard is the equivalent of spontaneous combustion. The temperature can rise locally to a point where the reaction $C_6H_6 + 3\ Cl_2 \rightarrow 6\ C + 6\ HCl$ can occur, principally in the vapor phase. Large amounts of HCl gas are released in this exothermic reaction which proceeds out of control. This phenomenon can also occur when the chlorine concentration builds up in the reactor if the normal chlorination catalyst is inactivated by any cause, such as an operation error which allows a sudden input of some water.

Since HCl is present in most parts of the equipment, corrosion is always a potential problem. Chlorine and benzene, or any recycled material, must be free of water to prevent corrosion and deactivation of catalyst. The reactor product contains HCl and iron. In some plants the product is neutralized with aqueous NaOH before distillation. In others, it is handled in a suitably designed distillation train, which includes a final residue still from which FeCl$_3$ can be removed with the high boiling tars.

Chlorobenzene mixtures behave in distillation as ideal solutions. In a continuous distillation train heat may be conserved by using the condensers from some units as the reboilers for others.

The separation capability of the stills for the polychlorobenzenes is limited. The distillates are always mixtures of close boiling isomers. Further separation is done by crystallizations, with the mother liquors being recycled to the stills. The two solid chlorobenzenes produced, 1,4-dichlorobenzene and 1,2,4,5-tetrachlorobenzene, require multiple crystallizations to obtain both a high quality product and good recovery. HCl is a constant by-product in the manufacture of chlorobenzenes. It is usually recovered by passing the gas stream through a scrubber tower over which a reactor mixture containing chlorination catalyst is circulated. This removes any unreacted chlorine that may have passed through the reactors. The HCl is then passed through one or more scrubbing towers in which high boiling chlorobenzenes are used as the solvent to remove the organic content. The absorbent in the final tower is refrigerated to the lowest possible temperature.

The HCl gas is absorbed in water to produce a 30–40% HCl solution. If the HCl must meet a very low organic content specification, a charcoal bed is used ahead of the HCl absorber, or the aqueous HCl solution product is treated with charcoal. Alternatively, the reactor gas can be compressed and passed to a distillation column with anhydrous 100% liquid HCl as the distillate; the organic materials are the bottoms and are recirculated to the process. Any noncondensible gas present in the HCl feed stream is vented from the distillation system and scrubbed with water. Any plant at times produces unwanted isomers. This requires an incinerator, capable of burning chlorinated hydrocarbons to HCl, H_2O, and CO_2, equipped with an efficient absorber for HCl (see Incinerators). An alternative to burning is the dechlorination using hydrogen over a suitable catalyst.

$$\text{C}_6\text{H}_{6-n}\text{Cl}_n + H_2 \longrightarrow \text{C}_6\text{H}_{7-n}\text{Cl}_{n-1} + HCl \qquad n = 1\text{--}6$$

The ultimate product could be benzene. Dechlorination can be done in the vapor phase with palladium, platinum, copper, or nickel catalysts (23–26) or in the liquid phase with palladium catalysts (27). The vapor phase dechlorination of 1,2,4-trichlorobenzene is reported to give good yields of 1,3-dichlorobenzene (24,26).

Another alternative to burning is rearrangement of the undesired isomer. This technique is practiced extensively in the petroleum industry, for example, in the production of xylene isomers using a $HF\text{--}BF_3$ catalyst–extractor system (28) (see BTX processing). Polychlorinated benzenes are considerably more resistant to rearrangement than are the isomeric hydrocarbon mixtures. Some patents have been issued to cover rearrangements using aluminum chloride catalyst (29–31). A $HF\text{--}SbCl_5$ catalyst system is also reported to be effective in converting dichlorobenzenes to 1,3-dichlorobenzene (32). There have been no reported commercial operations using these technologies to date.

Storage, Shipment, and Handling

Chlorobenzenes are stored in manufacturing plants in liquid form in black iron containers. Aluminum and aluminum alloys are not suitable. Mono-, 1,2-di, and 1,2,4-trichlorobenzenes are liquids at room temperature and are shipped in bulk in steel or stainless steel tank trucks or tank cars. 1,4-Dichlorobenzene and 1,2,4,5-tetrachlorobenzene are shipped either in molten form in insulated steel tank cars with heater coils, or as flake or granular solid in suitably sealed containers (paper bags, fiber packs, or drums).

Eye and skin contact as well as inhalation of vapors or dusts should be avoided when handling chlorobenzenes. They are generally considered nonflammable materials, with the exception of chlorobenzene which has a flash point of 34.5°C and is a flammable solvent based on DOT standards.

Chlorobenzenes are stable compounds; only under excessive heating at high temperatures will they decompose slowly and release some HCl gas and traces of phosgene. Chlorobenzenes are relatively toxic to fish. Spills should be contained and burned.

Economic Aspects

The chlorobenzene operations in the United States were developed primarily from the uses of chlorobenzene in the manufacture of phenol, aniline, and DDT. With the large production of monochlorobenzene, the other chlorobenzenes were produced in small quantities as by-products and their production rates were controlled with little difficulty to meet the market demands. As a result of the shift to other aniline and phenol processes and the restricted production of DDT, chlorobenzene production today is small.

Chlorination has to be carried out to higher conversions to produce more of the other chlorobenzenes. This poses a significant process problem. Production must match the market, or the unwanted material must be destroyed. Uses must also be found for the HCl by-product. Any producer must have chlorine available at a reasonable price, and the relative market price of HCl to chlorine cost becomes an important factor. A few of the typical prices (Nov. 1978) are shown below:

Chlorobenzenes	Price, ¢/kg
monochlorobenzene	57
1,2 dichlorobenzene	75
1,4 dichlorobenzene	55
1,2,4 dichlorobenzene	66

Specifications, Analyses and Quality Control

No widely established trade specifications exist for any of the chlorobenzenes. Each producer's standards are subject to modification by agreement with the customer.

All of the chlorobenzenes show readily separated and identifiable peaks by glc. This method is used almost exclusively for plant and quality control, and for sales specifications. Typical analyses in units of wt % are:

Chlorobenzene: benzene <0.05, dichlorobenzenes <0.1.

1,4-Dichlorobenzene: chlorobenzene and trichlorobenzenes <0.1, 1,2- and 1,3-dichlorobenzene each <0.5.

1,2-Dichlorobenzene is sold as two grades: (1) Technical: chlorobenzene <0.05, trichlorobenzenes <1.0, 1,2-dichlorobenzene 80.0, other isomers <19.0; (2) Purified 1,2-dichlorobenzene, produced by redistilling the technical product in a very efficient still: chlorobenzene <0.05, 1,2,4-trichlorobenzene <0.2, 1,2 dichlorobenzene 98.0.

1,2,4-Trichlorobenzene: chlorobenzene <0.1, dichlorobenzenes <0.5, tetrachlorobenzenes <0.5, 1,2,4-trichlorobenzene ca 97.0.

1,2,4,5-tetrachlorobenzene: 97.0.

Health and Safety Aspects (Toxicology)

In general, all of the chlorobenzenes are less toxic than benzene. Liquid chlorobenzenes produce mild to moderate irritation upon skin contact. Continued contact may cause roughness or a mild burn. Solids are capable of causing only mild irritation. Absorption through the skin is slow. Consequently, with short-time exposure over a limited area, no significant quantities will enter the body.

Contact with eye tissue at normal temperature causes pain, mild to moderate irritation, and possibly some transient corneal injury. Prompt washing with large quantities of water is helpful in minimizing the adverse effects of eye exposure.

The data from some single dosage oral toxicity tests, expressed as LD_{50}, are reported in Table 3. The values reported in the order of 1 g/kg or greater indicate a low acute oral toxicity. In animals, continued ingestion of chlorobenzenes over a long time can cause kidney and liver damage.

The Threshold Limit Value (TLV), the vapor concentration in ppm by volume, to which humans may be exposed for an 8-h working day for many years without adverse effects is also reported in Table 3. The saturated vapor concentration of the chlorobenzenes at 20°C listed in Table 3 are well above the TLV values; well-designed ventilation is required for working areas. A few reported cases of kidney and liver damage in humans may have been caused by repeated exposure to some of the chlorobenzenes.

Fires involving chlorobenzenes liberate HCl and possibly some phosgene; thus inhalation of the fumes must be avoided.

Also included in Table 3 are data on toxicity to fish. Spills into streams or lakes are likely to cause damage to fish life. The chlorobenzenes, because they are denser than water, tend to sink to the bottom and may persist in the area for a very long time. However, recent data indicate that dissolved 1,2,4-trichlorobenzene can be biodegraded by microorganisms from waste water treatment plants and also has a tendency to slowly dissipate from the water by volatilization (34).

Uses

Since the 1940s, large quantities of chlorobenzene have been used in the production of DDT, a widely used insecticide. However, since DDT causes serious ecological problems, its use has been severely restricted.

The production of monochlorobenzene in the United States has declined from a peak of 275,000 metric tons in 1960 to 146,000 t in 1977 (35), well below the estimated capacity of 298,000 t/yr. Chlorobenzene is now used mainly as a solvent, and to produce chloronitrobenzenes as intermediates for dyes, herbicides, and parathion insecticides. Production of 1,2-dichlorobenzene has grown from 11,200 t in 1960 to 36,000 t in 1975 (35). Much of the increase resulted from increased use as a process solvent for the manufacture of toluene diisocyanates (see Isocyanates). 1,2-Dichlorobenzene is also used to make 3,4-dichloroaniline, an intermediate for dyes and some agricultural chemicals, and has found limited use as a heat transfer fluid. For the latter application, some trace impurities must be removed to increase thermal stability. This is done by refluxing the 1,2-dichlorobenzene with $FeCl_3$ or $TiCl_4$ (36), and redistilling the material after removing the catalyst. 1,4-Dichlorobenzene is a solid at room temperature. Its main use is as mothballs and room deodorant blocks. The production of 1,4-dichlorobenzene increased slightly from 29,000 t in 1960 to 35,000 t in 1972, and then decreased to 31,800 t in 1975 (35). Growth will probably continue to be slow. 1,3-Dichlorobenzene has many potential uses and a number of patents cover its production (24,26,29–32), but only very limited commercial production has been reported to date. It was produced in Germany by chlorination of 1,3-dinitrobenzene at 200°C (37). Production of dichlorobenzenes in 1975 was only about 40% of the reported production capacities. 1,2,4-Trichlorobenzene has limited uses as a solvent and as a dye carrier in the textile industry (see Dye carriers). Mixtures of trichlorobenzenes have been

Table 3. Toxicity of Chlorinated Benzenes

Compound	Fish toxicity, no observed adverse effect level, mg/L in H_2O	LD_{50}, g/kg	TLV (inhalation), ppm[a]	Saturated concentration, ppm by vol at 20°C
chlorobenzene	<3 (Rainbow Trout)[b]	2.9 (rat)	75	11,900
	16 (Fathead)[c]	2.8 (rabbit)		
1,2-dichlorobenzene	3 (Fathead)[c]	0.8–2.0 (guinea pig)	50	1,125
1,4-dichlorobenzene	0.7 (Bluegill)[b]	1 (rat)	75	1,570
	5 (Fathead)[c]	4 (guinea pig)		
1,2,4-trichlorobenzene	2 (Fathead)[c]	1 (rat)	[d]	
1,2,4,5-tetrachlorobenzene	<1 (Fathead)[c]	<1 (rat)	[d]	260

[a] Volume per volume of air.
[b] 96-h dynamic test (33).
[c] 72-h static test (33).
[d] No TLV suggested.

produced by decomposition of the unwanted isomers in the production of the gamma isomer of hexachlorocyclohexane, an insecticide (see Benzene hexachloride).

1,2,4,5-tetrachlorobenzene is used exclusively as the raw material for 2,4,5-tri-acid and its esters), and bactericides (eg, hexachlorophene) (see Disinfectants). All other chlorinated benzenes have essentially no significant industrial applications.

BIBLIOGRAPHY

The "Chlorinated Benzenes" are treated in *ECT* 1st ed. under "Chlorine Compounds, Organic," Vol. 3; "Monochlorobenzene," pp. 812–817, by L. A. Kolker, Kolker Chemical Works, Inc., and Noland Poffenberger, The Dow Chemical Co.; "o-Dichlorobenzene," pp. 817–818, by Noland Poffenberger, The Dow Chemical Co.; p-Dichlorobenzene," pp. 819–822, by Axel Heilborn, Niagara Alkali Co.; in *ECT* 2nd ed. under "Chlorocarbons and Chlorohydrocarbons, Chlorinated Benzene," Vol. 5, pp. 253–267 by D. W. F. Hardie, Imperial Chemical Industries Ltd.

1. J. B. Cohen and P. Hartley, *J. Chem. Soc.* **87,** 1360 (1905).
2. D. W. F. Hardie, *A History of the Chemical Industry in Widnes,* Imperial Chemical Industries, Ltd., 1950, p. 155.
3. M. Campbell and H. Hatton, *Herbert H. Dow: Pioneer in Creative Chemistry,* Appleton-Century-Crofts Inc., New York, 1951, p. 114.
4. U.S. Pat. 1,607,618 (Nov. 23, 1926), W. J. Hale and E. C. Britton (to The Dow Chemical Co.).
5. W. J. Hale and E. C. Britton, *Ind. Eng. Chem.* **20,** 114 (1928).
6. R. M. Crawford, *Chem. Eng. News* **25**(1), 235 (1947).
7. Ger. Pat. 539,176 (Nov. 12, 1931), W. Prahl (to F. Raschig G.m.b.H.).
8. Ger. Pat. 575,765 (April 13, 1933), W. Prahl and W. Mathes (to F. Raschig G.m.b.H.).
9. L. Luttrighaus and D. Ambrose, *Chem. Ber.* **89,** 463 (1956).
10. J. D. Roberts and A. T. Bottini, *J. Am. Chem. Soc.* **79,** 1458 (1957).
11. D. R. Stull, E. F. Westrum, and G. C. Sinke, *The Chemical Thermodynamics of Organic Compounds,* John Wiley & Sons, Inc., New York, 1969.
12. Brit. Pat. 388,818 (Mar. 6, 1933), T. S. Wheeler (to ICI).
13. H. van den Berg and R. M. Westerink, *Ind. Eng. Chem. Fundamental* **15**(3), 164 (1976).
14. M. F. Bourion, *Ann. Chim. (Paris)* **14**(9), 215 (1920).
15. R. B. MacMullin, *Chem. Eng. Progress* **44**(3), 183 (1948).
16. H. F. Wiegandt and P. R. Lantos, *Ind. Eng. Chem.* **43,** 2167 (1951).
17. U.S. Pat. 3,226,447 (Dec. 28, 1965), G. H. Bing and R. A. Krieger (to Union Carbide Australia Ltd.).
18. Ned. Pat. 7413614 (Oct. 16, 1974), S. Robota, R. Paolieri, and J. G. McHugh (to Hooker Chemicals).
19. U.S. Pat. 1.946,040 (Feb. 6, 1934), W. C. Stoesser and F. B. Smith (to The Dow Chemical Company).
20. U.S. Pat. 2,976,330 (Mar. 21, 1961), J. Guerin (to Société Anonyme).
21. U.S. Pat. 3,636,171 (Jan. 18, 1972), K. L. Krumel and J. R. Dewald (to The Dow Chemical Company).
22. U.S. Pat. 3,557,227 (Jan. 19, 1971), M. M. Fooladi (to Sanford Chem. Co.).
23. U.S. Pat. 2,826,617 (Mar. 11, 1958), H. E. Redman and P. E. Weimer (to Ethyl Corp.).
24. U.S. Pat. 2,943,114 (June 28, 1960), H. E. Redman and P. E. Weimer (to Ethyl Corp.).
25. U.S. Pat. 2,886,605 (May 12, 1959), H. H. McClure, J. S. Melbert, and L. D. Hoblit (to The Dow Chemical Company).
26. U.S. Pat. 2,866,828 (Dec. 30, 1958), J. A. Crowder and E. E. Gilbert (to Allied Chemical Corp.).
27. U.S. Pat. 2,949,491 (Aug. 16, 1960), J. J. Rucker (to Hooker Chemical Corp.).
28. S. Ariki and A. Ohira, *Chem. Econ. & Eng. Rev.* **5**(7), 39 (1973).
29. U.S. Pat. 2,666,085 (Jan. 12, 1954), J. T. Fitzpatrick (to Union Carbide).
30. U.S. Pat. 2,819,321 (Jan. 7, 1958), B. O. Pray (to Columbia-Southern Chemical Corp.).
31. U.S. Pat. 2,920,109 (Jan. 5, 1960), J. W. Angelkorte (to Union Carbide Corp.).
32. Yu G. Erykalov and co-workers, *Zh. Org. Khim.* **9,** 348 (1973).
33. *Standard Methods for Examination of Water and Waste Water,* 14th ed., American Public Health Association, Washington, D.C., 1975, p. 800.

34. P. Simmons, D. Branson, and R. Bailey, *Biodegradability of 1,2,4-trichlorobenzene,* paper presented at the 1976 American Association of Textile Chemicals and Colorists International Technical Conference.
35. *Synthetic Organic Chemicals, United States Production and Sales,* United States International Trade Commission, Washington, D.C., 1960–1978.
36. U.S. Pat. 2,856,438 (Feb. 1, 1956), A. Procko (to Columbia-Southern Chemical Corp.).
37. *British Intelligence Objectives Subcommittee,* **986,** 151.

CHE-I KAO
NOLAND POFFENBERGER
Dow Chemical U.S.A.

BENZENE HEXACHLORIDE

Benzene hexachloride [608-73-1], (1,2,3,4,5,6-Hexachlorocyclohexane, Agrocide, Ambiocide, Benzanex, Benzex, BHC, Gammacide, Gammacoid, Gamaspra, Gamtox, Gyben, Hexachlorane, HCH, Hexdow, Isatox, Lintox, Lexone, Trives-T), a mixture having a high level of insecticidal activity, is the fully saturated product formed by light catalyzed addition of chlorine to benzene. This reaction produces a number of stereoisomeric compounds of the composition $C_6H_6Cl_6$, with varying amounts of both underchlorinated and overchlorinated compositions. This mixture, commonly called BHC, is one of the major families of chlorohydrocarbon insecticides (see Insect control technology).

The benzene hexachloride mixture is no longer used commercially in the United States. It is processed primarily either to provide a product with increased content of the gamma isomer or to provide the pure gamma isomer, commonly called lindane. It is the only isomer with significant insecticidal activity. In warm blooded animals important differences in pharmacological effects exist between isomers, eg, the alpha and gamma isomers are central nervous system depressants (1). Excellent sources of activity information are reported in refs. 2 and 3.

The benzene hexachloride mixture containing about 14% gamma isomer is a brown-to-white amorphous powder with a musty odor reminiscent of the new-mown-hay odor of phosgene. This mixture is processed by selective crystallization to a crystalline solid containing 40–48% gamma isomer. This product, called Fortified Benzene Hexachloride (FBHC), is no longer used in the United States but is still produced and sold internationally as an insecticide, or as a raw material for production of lindane (Gamma BHC, Gamaphex, Gammalin, Gammex, Gammexane, Jacutin, Lindafor, Lindagam, Lintox, Novigam, Silvanol, and Gamma HCH). The latter compound is a crystalline free-flowing solid containing 99+% gamma isomer showing high activity in a wide variety of insects.

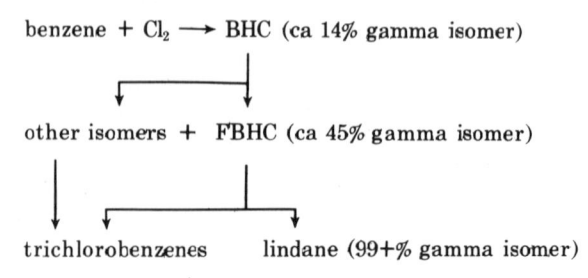

benzene + Cl₂ ⟶ BHC (ca 14% gamma isomer)

other isomers + FBHC (ca 45% gamma isomer)

trichlorobenzenes   lindane (99+% gamma isomer)

The undesired isomers remaining are often thermally treated to produce trichlorobenzenes.

Benzene hexachloride was first prepared by Faraday in 1825 by the addition of chlorine and benzene in sunlight. In 1833 Mitscherlich demonstrated important structural information by the dehydrochlorination of crude BHC with alkali to form trichlorobenzenes. In 1884 Meunier isolated the first two isomers alpha and beta, but gave no indication of their structures. Van der Linden in 1912 isolated the former as well as the gamma (lindane) and delta isomers. He concluded that the four isomers then known resulted from distribution of hydrogen and chlorine atoms either above or below the plane six-membered ring. The epsilon isomer was detected in 1947 (4). The eta and theta isomers were detected from the photochlorination of benzene tetrachloride (5). Identification of the final member of the series, the iota isomer, was claimed in 1969 (6), but as yet no definitive structure proof has been completed.

The insecticidal activity of BHC against the Colorado beetle was first noted in 1940 (7). In a search for an alternative to the expensive rotenone, ICI in 1942 discovered that the gamma isomer of BHC was active against grain weevils, the turnip flea beetle and other insects (8). Large quantities of BHC were used during World War II under the name Aphtiria.

Physical Properties

Structure of the Benzene Hexachloride Isomers. The misnomer benzene hexachloride has been associated with the stereoisomers of 1,2,3,4,5,6-hexachlorocyclohexane (9).

Although slightly imprecise, the conformational isomers are called chair and boat forms in this article. Terms such as half chair, skew boat, distorted torsion angles, etc, are more precise but add little to this level of understanding. Crystallographic structure determinations and nmr studies have given thorough structural details (10–13).

The stereoisomer least likely to interconvert will be the one with three or more chlorines in equatorial positions. The following shows chair–chair interconversion:

where a = axial; e = equatorial.

There are sixteen theoretical isomers. Chair–chair interconversions reduce this number to eight. The favorable conformations (bold-face) have three or more equatorial chlorines.

e e e e e e ⇌ a a a a a a	beta isomer
a e e e e e ⇌ e a a a a a	delta isomer
a a e e e e ⇌ e e a a a a	alpha isomer
a e a e e e ⇌ e a e a a a	theta isomer
a e e a e e ⇌ e a a e a a	epsilon isomer
a a a e e e ⇌ e e e a a a	gamma isomer
a a e a e e ⇌ e e a e a a	eta isomer
a e a e a e ⇌ e a e a e a	iota isomer

Of the eight isomers only alpha and eta are enantiomorphs. The eta isomer, on chair-chair interconversion, becomes superimposable on its mirror image. Therefore, only the alpha isomer is optically active. The *levo* enantiomer of the alpha isomer has been isolated by reaction of racemic modification of the alpha isomer with the optically active alkaloid brucine (14). The subsequent dehydrochlorination to trichlorobenzene only occurred with one enantiomer and not the other.

Analytical and structural determinations have validated the structures of five of the isomers (10–13).

The accepted structures for the eight isomers of benzene hexachloride are shown below.

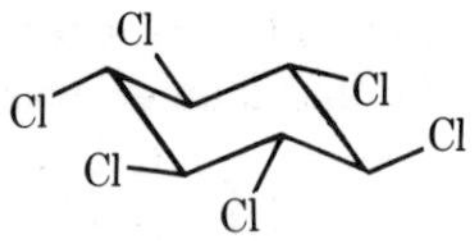

beta[319-85-7] 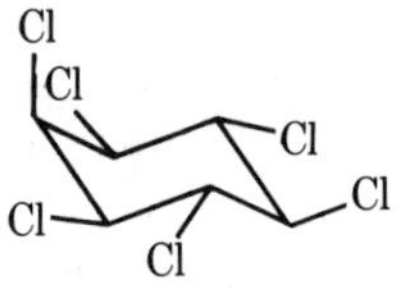   delta[319-86-8] 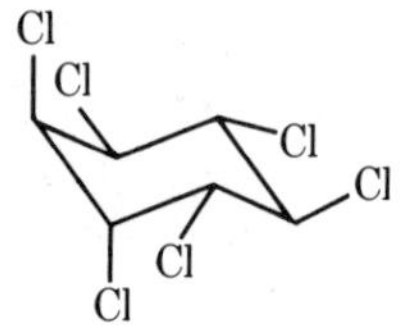   (+−)alpha[319-84-6] 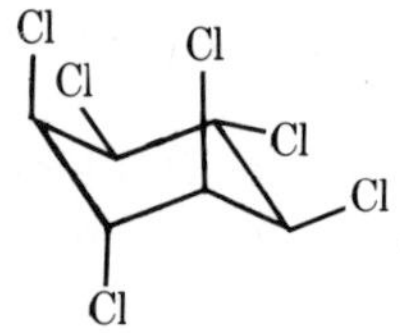   gamma[58-89-9](lindane)

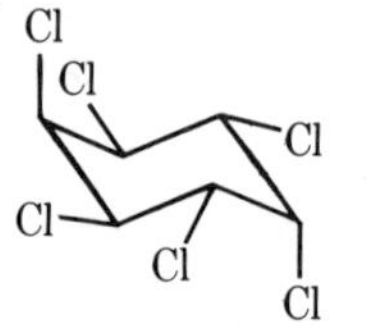

epsilon[6108-10-7] 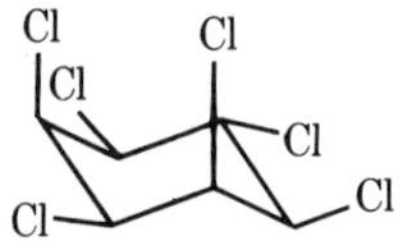   theta[6108-13-0] 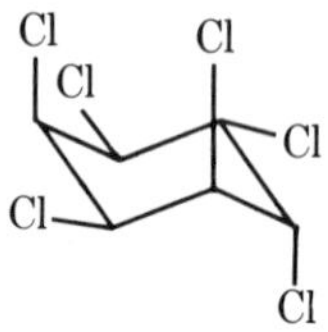   eta[6108-12-9] 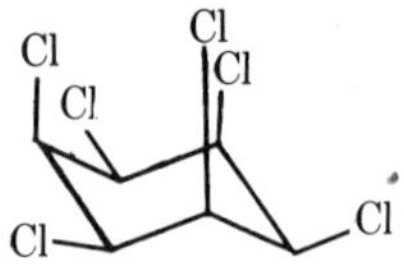   iota[6108-11-8]

The structure–activity relationship for the insecticidal activity of the gamma isomer has yet to be explained. An interesting observation is that the gamma isomer is the only BHC stereoisomer which is known to sustain chair–chair interconversion at ambient temperature (the nmr spectrum is a single sharp line showing equivalence of all hydrogens at room temperature), and has the largest spherical volume of any of the common isomers. A volume–activity correlation has been suggested (14). Others have tried to relate the toxicity of the gamma isomer to DDT (15) since the effect on insects is similar and the dimensions of both are similar. Many isosteres of the gamma isomer have been prepared with alkyl, methoxy, or halo substituents. These showed activity less than lindane but did relate to hydrophobic–lipophilic properties of the substituents. All isosteres provoked central nervous system changes.

Technical BHC. Technical BHC is a brownish-to-white solid having a musty odor which melts around 63°C. It can be distilled with no decomposition. The composition of the various isomers in benzene hexachloride depends on conditions of manufacture but is commonly as follows:

Isomer	%
α	65
β	7
γ	14
ε	4
others	10
Total	100

The principal reported physical constants of the five major isomers are shown in Tables 1 and 2.

Solubilities of the benzene hexachloride isomers in several organic solvents are shown in Table 2.

Table 1. Comparison of the Physical Constants of Lindane and Some of the Other BHC Isomers[a]

Isomer	Isomer in technical BHC, %	mp, °C	Dipole moment C·m × 10^{-30}[b]	Refractive index, n_D^{20}	λ_{max}, nm	Crystal
α	55–70	159.2	7.41	1.626	125.8	monoclinic prisms
β	5–14	311.7	0	1.633	134.6	cubic (octahedral)
γ	10–18	112.9	6.0:12.0	1.644	132.2	monoclinic plates, prisms
δ	6–10	140.8	7.34	1.576–1.674	118.1	fine plates
ε	3–4	218.2	0	1.635	139.6	monoclinic needles or hexagonal monoclinic crystals

[a] Ref. 16.
[b] To convert C·m to debye, divide by 3.336×10^{-30}.

Table 2. Solubility of the Various Isomers at 20°C, in g/100 g Solvent[a]

Solvent	Alpha	Beta	Lindane (ca 100% gamma)	Delta	Epsilon
ethyl acetate	12.7	6.9	35.7	58.5	
acetone	13.9	10.3	43.5	71.1	33.2
acetic acid	4.2	1	12.8	25.6	
ethyl alcohol	1.8	1.1	6.4	24.2	4.2
benzene	9.9	1.9	28.9	41.1	
chloroform	6.3	0.3	24.0	13.7	2
dioxane	33.6	7.8	31.4	58.9	
ether	6.2	1.8	20.8	35.4	3
carbon tetrachloride	1.8	0.3	6.7	3.6	0.5
toluene	9	2.1	27.6	41.6	

[a] Ref. 17.

The odor of BHC has been attributed to both under- and over-chlorinated cyclohexanes as well as chlorocyclohexenes. Many schemes have been proposed for odor removal but a simple carbon treatment (18) appears easiest. The pure gamma isomer retains only a very faint odor.

Fortified BHC. Fortified Benzene Hexachloride (FBHC) is a brown-to-white solid which melts near 78°C. The composition of the various isomers in FBHC depends on manufacture but is commonly as follows:

Isomer	%
α	19
β	10
δ	21
γ	44
ε	1
others	15

Lindane. Lindane is a crystalline white free flowing solid, 99+% gamma isomer which melts at 112°C. The pure gamma isomer has three modifications, one rhombic, and two monoclinic which are enantiotropic to the rhombic form. The melting points of the two stable forms are very close together. These isomers are relatively stable chemically and do not lose hydrogen chloride when heated with traces of metals and their salts. They are also stable to light and oxidative conditions.

Chemical Properties

With the exception of the β isomer, the isomers of BHC are dehydrochlorinated by alkali at 60°C to give primarily 1,2,4-trichlorobenzene. This is formed largely from the alpha isomer (60–80%). The γ and δ isomers, in addition, yield lesser quantities of 1,2,3- and 1,3,5-trichlorobenzenes. Cristol (19) showed that in the case of the α and γ isomers the dehydrochlorination takes place by a simple second-order reaction, whereas the reaction of the δ isomer is kinetically more complex. Less stringent reaction conditions give chlorinated cyclohexenes.

When BHC is chlorinated in carbon tetrachloride, various substitution derivatives are formed. By use of liquid chlorine, higher substituted derivatives can be obtained (20–21). BHC vapor in admixture with oxygen (air) and chlorine at 470°C in the presence of a catalyst containing alumina, copper chloride, and potassium chloride yields a mixture of hexachlorobenzene and pentachlorobenzene (22). High temperature chlorination in a chlorobenzene solvent in the presence of ferric chloride also results in the quantitative conversion of BHC to hexachlorobenzene (23).

BHC reacts readily with sulfur at 240–290°C, with elimination of two chlorine atoms and formation of a sulfur link between two pentachloro residues (24):

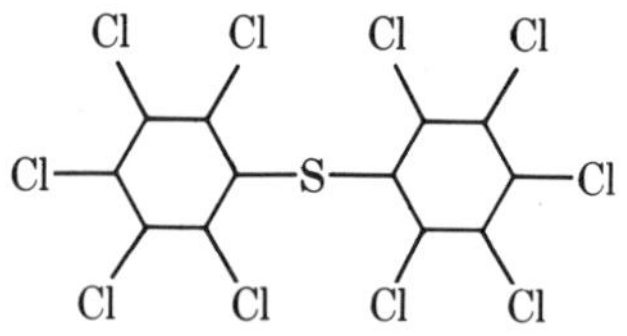

This derivative, known as SPC (Sulphure de Polychlorocyclane), has found some application in Europe as a pest control agent.

With elementary sulfur at 240–290°C, p-dichlorobenzene and 1,2,4,5-tetrachlorobenzene are formed with evolution of hydrochloric acid, sulfur dichloride, and hydrogen sulfide. Small quantities of 1,2,3- and 1,3,5-trichlorobenzenes are also produced (25).

Analysis of products of tritiated gamma isomer showed that tritiation occurs with dechlorination, and retention or inversion of the steric conformation, and the resulting pentachlorocyclohexane is labeled at the dechlorinated position (26).

Boiling water or steam treatment of BHC liberates only traces of hydrochloric acid, but at 200°C in sealed tubes 1,2,4-trichlorobenzene is formed. To inhibit decomposition, BHC is stored for long times at temperatures above ambient with added 1% sodium thiosulfate.

Treatment of BHC with zinc dust in acids dechlorinates the material to benzene. Although generally inert to strong acids, BHC is decomposed to benzene by chlorosulfonic acid or sulfur trioxide.

Manufacture

Technical benzene hexachloride is produced by the addition-chlorination of benzene in the presence of free radical initiators such as visible or ultraviolet light, x-rays, or gamma rays.

The only reported kinetic studies of the photochemical chlorination of benzene were carried on in the vapor phase (27).

There is probably a secondary substitution process occurring since chlorobenzene and both hepta- and octachlorocyclohexane appear in small amounts under most experimental and production conditions.

Technical benzene hexachloride is produced by either batch or continuous methods in either stirred tank type reactors or tubular reactors. In a typical process, benzene is chlorinated at 15–25°C in a glass reactor at 101 kPa (1 atm) with excess benzene. It is necessary to exclude oxygen and substitution catalysts, eg, iron. When ca 5–8% benzene is chlorinated (measured commonly by specific gravity), the beta isomer begins to precipitate. Benzene and excess chlorine are removed by evaporation at 85–88°C. If the desired product is to be what is commonly called technical benzene hexachloride, part of the benzene is removed first by evaporation at ambient pressure. Residual benzene is stripped at reduced pressure. Recovered benzene is condensed, scrubbed and returned to the reactor. The molten product is steam-stripped to remove traces of benzene and then cast into shallow pans. The gamma content of a typical product is 12–14%.

The gamma-isomer content can be increased significantly by the reaction of benzene and chlorine in a nonreactive solvent of high dielectric constant such as methylene chloride (27–28). In these solvents, gamma isomer contents as high as 26% have been reported. There are many reported variants of the above process for obtaining a higher gamma-isomer content (29–30).

Since the gamma isomer is more soluble than the alpha or beta isomers in most common solvents, including benzene, enrichment of gamma isomer is obtained by crystallization of the less soluble isomers in the region of gamma saturation. The solid phase of alpha and beta isomers can then be separated by filtration or centrifugation. The concentration of the gamma isomer in the solution phase will depend on the solvent used. For most producers the solvent used is benzene, giving a concentrate containing 40–45% of the gamma isomer. With other solvents, such as chlorinated hydrocarbons, a gamma content as high at 60% may be obtained through these extraction methods. The concentrate, known commercially as fortified BHC (FBHC), contains all the gamma, delta, and epsilon isomers present in the original crude benzene hexachloride, as well as minor amounts of hepta- and octachlorocyclohexanes and residual alpha and beta isomers. This mixture has a low and indistinct crystallization temperature. The melt can be supercooled to room temperature without inducing crystallization. Once solidified, the last crystal point is in the region of 70–80°C.

No commercial process has yet been found which permits recovery of all of the gamma isomer present in crude benzene hexachloride. Because of these inevitable losses and the attendant processing costs, some of the gamma isomer is marketed as technical BHC. However, there are many applications for which the undesirable properties of this crude material, such as the musty, acrid odor, and phytotoxicity of some isomers, are sufficiently objectionable to warrant the premium price of the purified material. Household and home garden formulations, seed treatment, etc, are made with lindane for these reasons. Lindane is the only BHC composition sold in the United States.

Numerous procedures have been described and patented for isolation of the highly desired gamma isomer (31–32). Only two of these procedures are known to be used commerically for lindane preparation: the supersaturation process, and the fluid classification process. Intermediate steps in concentrating the gamma isomer may employ one or more of the other procedures, however, before carrying out the primary stage of isolating the gamma isomer.

The usual solvent for the supersaturation process is a lower primary alcohol, generally methanol. Alcohol–water systems are also described in several patents. An appreciable portion of the gamma isomer is left in the concentrate when the limit of supersaturation of alpha isomer is reached. Further recovery of gamma isomer can be obtained by removal of the alcohol solvent and crystallization of a gamma-rich concentrate from a solvent having very different solubility characteristics for the various isomers such as carbon tetrachloride or a paraffinic hydrocarbon.

Preparation of lindane by the supersaturation process requires recovery of the gamma isomer from a solution containing a high concentration of impurities. Final purification of the resultant lindane is troublesome, and trace impurities are difficult to eliminate. If mixed crystals are obtained by crystallization from solution of alpha, beta, and gamma isomers, the very small alpha and beta crystals (33) can be separated from the gamma crystals by decanting a suspension of the smaller particles.

The delta isomer and minor soluble impurities are first separated from the alpha, beta, and gamma isomers by preparation of a saturated solution of the gamma isomer in an alcoholic solvent. Alcohols, eg, methanol or isopropanol, are very effective for this separation and also yield excellent crystal forms for subsequent crystal classification. An extremely pure lindane is obtained by this process, which accounts for virtually all of the production in the United States.

Although interest in new BHC–FBHC–lindane production methods has been limited in recent years, it has by no means disappeared. A scheme has been patented that provides for integrated production of lindane with full recovery of HCl and trichlorobenzenes (34). In addition, a mathematical model of stream structure has been proposed for a methanol extraction apparatus for production of enriched hexachlorocyclohexane (35).

Economic Aspects

Commercial production was begun by Hooker Chemical Corporation in the United States in early 1946. New producers were announced almost monthly until, by 1951, sixteen United States manufacturers released a flood of 53,000 metric tons of benzene hexachloride in various forms, equivalent to ca 8200 t of gamma isomer. The market was able to absorb only 60% of this output and production in subsequent years declined steadily to reach a fairly stable level of ca 2000 t/yr of the gamma isomer in all forms by 1960. This output was supplied by five domestic producers with a reported productive capacity about twice that of current market requirements. This oversupply became apparent and slowly producers dropped out of production.

During the 1960s and 1970s the toxic effects of the undesired isomers became known. As a result, by 1977 there was only one United States producer of lindane. The 1977 sales price was $7.72/kg.

World consumption of BHC and lindane (in metric tons) was:

Year	BHC	Lindane
1970	42,000	16,000
1971	28,000	13,000
1972	25,000	12,000
1973	30,000	8,000
1974	25,000	4,000

There is still activity in this field, eg, there is a new lindane–BHC facility capable of producing 11,000 t/yr in Spain (36).

The general prognosis for BHC is a fast decline in developed countries with a slow decline in developing countries where the lower prices are attractive. The situation is similar for FBHC.

Analytical Methods

There are no definitive chemical tests for the presence of any of the isomers of benzene hexachloride since they are all isomeric. Except for the beta isomer, they are similar in reactivity.

Although many analytical methods have been used for analysis of these isomers, there are only two that are significant in terms of time of analysis and accuracy. The ir method (37) is nondestructive and repeatable but, because of interferences from overchlorinated materials, its accuracy is suspect. The gas chromatography methods (38–39) are of similar repeatability (±1%). Degradation of the delta isomer occurs when conditions are not carefully maintained. Accurate determination of the minor isomers can be obtained by use of internal standards.

Thin layer chromatography has been used for the qualitative analysis of benzenehexachloride components (7). The analysis is quick and semiquantitative. By use of micro-thin layer chromatography detection levels of 0.025 µg of the individual isomers were observed.

Lindane itself can be estimated either by reaction with o-toluidine followed by measurement of the absorbance in a water/2-butanol mixture containing 0.22% ammonium molybdate (40) or by a similar sequence using ethanolamine in the presence of 0.1% p-nitrobenzenediazonium fluoborate (41).

Health and Safety Factors (Toxicology)

Many of the reports of the toxicology of BHC–FBHC–lindane are contradictory or plagued by experimental protocols using contaminated materials and controls. In the case of BHC–FBHC, the isomer content is extremely variable or not reported at all.

At the liver, hydroxylating enzymes act to make the isomer more polar (as an alcohol or phenol) for secretion as a sulfate or glucosiduronide. All BHC isomers can undergo this fate either directly or after HCl elimination, except the beta isomer which is difficult to eliminate. The more easily an isomer can assume an aromatic nature by HCl evolution, the more easily it can be hydroxylated and subsequently eliminated.

Much of the recent work has been attempted on lindane (2). Most of the following refers to this isomer since it is the active ingredient, and has none of the objectionable beta isomer.

Lindane is used as an insecticide in crop protection, and also as a therapeutic agent in human and veterinary medicine. Its manufacture and its practical application for a period of more than 20 years yield a wealth of experience which together with results of experiments in animals provide the basis for this information on toxicology.

After intake of relatively high quantities either in single or repeated dosage, lindane is fully excreted from the animal body within a short time, ie, ca one day. With repeated application an equilibrium is quickly established between intake and excretion. After cessation of medication the residues which prevail, mainly in body fat, are completely eliminated within days or a few weeks. When present at high levels in the tissues, lindane like a great number of other therapeutic agents stimulates a greater activity of the microsomal enzymes of the liver, and thereby its own metabolism and excretion are enhanced. Residues found in foods reaching the consumer, however, lie well below the threshold value required to produce this effect in experimental animals.

As apparent from the toxicity data given in Table 3, lindane can be classified as a chemical of intermediate acute toxicity. In long-term feeding experiments it shows a much higher no-effect dose value than other chlorinated hydrocarbons, which places lindane in a special position within this group. Even the reversible liver alterations known as chlorinated hydrocarbon liver are not considered to be of pathological character but of an adaptive nature. They need not be taken into consideration when fixing the no-effect level for lindane. The no-effect dosage for rats and dogs in two year feeding studies with lindane was 1.25 mg/kg. The acceptable daily intake (ADI) of 0.0125 mg/kg for man, as tentatively proposed by FAO/WHO has, in comparison with market basket analysis data, a safety factor of ca 100 over the no-effect level (actually, a factor of as much as 2000–4000 over the no-effect dose from animal experiments). Lindane residues are practically of no hygienic importance for the consumer. The real problem in connection with lindane is, however, the past and continued use of technical BHC, from which lindane is produced. Technical BHC used in plant protection and other fields of pest control not only confers taint and off-flavor to foods of plant and animal origin, but produces considerable residues, especially of the alpha and beta isomers, which create hygienic problems. These residues of alpha- and beta-BHC are carried on in the food chain and therefore use of technical BHC should be discouraged.

Other tests with mice, rats, and rabbits have shown that lindane has no teratogenetic effect. A three-generation test with rats proved that fertility as well as survival and development of the young were in no way impaired. In the dominant lethal test and in the host-mediated assay no indications of a mutagenic effect in warm-blooded

Table 3. Acute Toxicity (LD$_{50}$) and No-Effect Levels of Some Chlorinated Hydrocarbon Pesticides for Rats[a]

	Lindane (γ-BHC)	Technical BHC	α-BHC	β-BHC	Technical DDT	Chlordane	Heptachlor	Aldrin	Dieldrin	Endrin
oral LD$_{50}$, mg/kg of body weight	125	600	500	>6000	250	450	90	67	87	7–43
no-effect level from 2-yr feeding tests, ppm in the food	25–50	10	10	<10	1	<2.5	<1	<1	<1	1

[a] Refs. 42–44.

animals were apparent for lindane. This was also verified for man by showing no chromosomal changes even after many years of production experience.

Investigations on the oncogenicity of lindane in mice and rats have shown that neoplasms—typical rodent hepatoma—have occurred only in mice and that there is no effect for a 100–300-ppm level of lindane in the daily diet. This effect-level is much higher than that for alpha-BHC or the technical material.

During manufacture and also in the case of pest control operators, strict adherence to industrial hygiene requirements must be observed. It is particularly necessary to ensure the MAC (minimum acceptable concentration) value is not exceeded (0.05 mg/m^3). Increased concentrations in the air are most likely to occur under conditions of poor ventilation and high ambient temperatures which will increase the vaporization of γ-BHC. If factory hygiene principles are properly observed, there is no danger to production operators even with long periods of employment and exposure (see Industrial hygiene and toxicology). Lindane does not have a dangerous accumulative effect and therefore falls outside the classification of persistent chlorinated carbon pesticides (see Trace and residue analysis).

Residues of lindane in warmblooded animals and birds and other vertebrates are in general low and do not constitute a hazard to terrestrial wildlife. Levels of lindane in surface waters are in general in the order μg/L (ppb) and thus below the critical level for fish and other water organisms. Under practical field conditions lindane is generally of low phytotoxicity.

Uses

Lindane as one of the oldest chlorinated insecticides has generated many uses. A list covering 200 pages is available in the monograph *Lindane* (2). The fields of use discussed are: field crops, vegetable crops, fruit crops, viticulture, ornamentals, pasture and forage crops, forestry, soil treatment, seed treatment (seed dressing), locust control, timber protection (termite and ant control), stored materials and products, agricultural and industrial stores, public health, and veterinary hygiene).

No significant use apart from those due to its biological properties have been observed.

BIBLIOGRAPHY

"Benzene Hexachloride" under "Chlorine Compounds, Organic" in *ECT* 1st ed., Vol. 3, pp. 808–812, by J. J. Jacobs, Consulting Chemical Engineer; "Benzene Hexachloride" under "Chlorocarbons and Chlorohydrocarbons" in *ECT* 2nd ed., Vol. 5, pp. 267–281, by D. W. F. Hardie, Imperial Chemical Industries Ltd.

1. R. I. Metcalf, *Organic Insecticides*, Interscience Publishers, New York, 1955.
2. E. Ulmann, *Lindane*, Verlag K. Schillinger, Freiburg im Breisgau, Berlin, Ger., 1972.
3. G. T. Brooks, *Chlorinated Insecticides*, CRC Press, Cleveland, Ohio, 1974.
4. A. J. Kolka, H. D. Orloff, and M. E. Griffing, *J. Am. Chem. Soc.* **76,** 3940 (1954).
5. K. C. Kauer, R. B. DuVall, and R. L. Alquist, *Ind. Eng. Chem.* **39,** 1335 (1947).
6. K. Visweswariab and S. K. Majumder, *Chem. Ind.* **64,** 379 (1969).
7. A. Dupire and M. Raucourt, *Compt. Rend. Acad. Agric. Fr.* **29,** 470 (1943).
8. R. Slade, *Chem. Ind.* **40,** 314 (1945).
9. H. D. Orloff, *Chem. Rev.* **54,** 347 (1954).
10. K. Hayamizu and co-workers, *Tetrahedron* **28,** 779 (1972).
11. G. Smith, C. Kennard, and A. White, *Cryst. Struct. Commun.* **5,** 683 (1976).
12. *Ibid.*, p. 687.

13. G. Smith, C. Kennard, and A. White, *J. Chem. Soc. Perkin II*, 614 (1976).
14. J. L. Cohen, W. Lee, and E. J. Lien. *J. Pharma. Sci.* **63**, 1069 (1974).
15. G. Holan, *Nature (London)* **232**, 644 (1971).
16. Ref. 3, p. 184.
17. Ref. 2, p. 34.
18. M. V. Strongin, V. Kulikova, and N. N. Mansurova, *Z. Khim.* **39**, 833 (1966).
19. S. J. Cristol, *J. Am. Chem. Soc.* **71**, 1894 (1949).
20. T. Oiva and co-workers, *Botyu-Kagaku* **13**, 23 (1949); **14**, 42 (1949).
21. T. Van der Linden, *Rec. Trav. Chim.* **57**, 217 (1938).
22. Brit. Pat. 737,563 (Sept. 28, 1955), R. T. Foster (to Imperial Chemical Industires Ltd.).
23. Ger. Pat. 879,836 (June 17, 1953), E. Himmen (to Farbenfabriken Bayer A.G.); F. Becke and L. Wurtele, *Chem. Ber.* **91**, 1011 (1958); Ger. Pat. 955,231 (Jan. 17, 1957), F. Becke and H. Sperber (to Badische Anilin und Soda-Fabrik A.G.).
24. M. J. Guilhan, *Compt. Rend. Acad. Agric. Fr.* **33**, 101 (1947).
25. F. Becke, *Angew. Chem.* **72**, 867 (1960).
26. M. Hamada and E. Kawano, *Agric. Biol. Chem.* **324**, 1272 (1970).
27. H. P. Smith, W. A. Noyes, Jr., and E. H. Hart, *J. Am. Chem. Soc.* **55**, 4444 (1933).
28. W. Bissenger and co-workers, *Ind. Eng. Chem.*, **51**, 523 (1959).
29. C. Woo, *Huu Hsuch* **3**, 99 (1966).
30. G. Rendko, *Chem. Zvesti* **15**, 741 (1961), *Chem. Abstr.* **55**, 25813d (1961).
31. G. Strongin, V. Kulikova, and M. Mansurova, *Z. Khim.* **39**, 833 (1966).
32. G. Strongin and co-workers, *Tr. Khim. Tekknol.* **2**, 315 (1966); *Chem. Abstr.* **64**, 18490c (1966).
33. D. S. Rosenberg, *Chlorine*, R. E. Kruger Publishing Co., Hungtington, N.Y., 1962.
34. U.S. Pat. 2,773,103 (Dec. 4, 1956), G. Calingaert (to Ethyl Corporation).
35. R. P. Bulankin and co-workers, *Khim. Sredstva Zachchity Rost.* **6**, 5 (1976).
36. *Eur. Chem. News*, (Sept. 3, 1976).
37. W. Fisher and W. Ebin, *Z. Anal. Chem.* **188**, 176 (1962).
38. A. Davis and H. Joseph, *Anal. Chem.* **39**, 1016 (1967).
39. W. Cochrane and R. Maybury, *J. Assoc. Off. Anal. Chem.* **56**, 1324 (1973).
40. G. Baluja Mau Guim and A. Barba, *Rev. Agroquim. Tecnol. Aliment.* **5**, 489 (1965).
41. K. Visweswariah, S. Majumder, and M. Hayaram, *Murochem. J.* **17**, 26 (1972).
42. A. J. Lehman, *Assoc. Food & Drug Off. U.S. Q. Bull.* **15**, 122 (1951).
43. *Ibid.*, **16**, 47 (1952).
44. K. W. Jager, *Aldrin, Dieldrin, and Telodrin,* Amsterdam, London, and New York, 1970.

JAMES G. COLSON
Hooker Chemical & Plastics Corp.

RING-CHLORINATED TOLUENES

The ring-chlorinated derivatives of toluene form a group of stable, colorless compounds. Not all chlorotoluene isomers can be prepared by direct chlorination. Indirect routes to chlorotoluenes include reactions involving the replacement of the amino, chlorosulfonyl, hydroxyl, and nitro groups by chlorine and the use of the sulfonic acid and amino groups to orient substitution followed by their removal from the ring.

The first systematic study of chlorine reaction on toluene was carried out in 1866, by Beilstein and Geitner. Over the next forty years many studies were performed to isolate and identify the various chlorination products (1). Manufacture of chlorotoluenes was initiated in the early 1930s by the Hooker Electrochemical Company (Hooker Chemicals & Plastics Corp.) and the Heyden Chemical Corporation (Tenneco) in the United States; at present, Hooker remains the major U.S. producer.

Mono- and dichlorotoluenes are used primarily as chemical intermediates in the manufacture of pesticides, dyestuffs, pharmaceuticals, and peroxides and as solvents. Total annual production was limited prior to 1960 but has expanded greatly since that time.

MONOCHLOROTOLUENES

Physical Properties

o-Chlorotoluene (1-chloro-2-methylbenzene, OCT) is a mobile, colorless liquid with a penetrating odor similar to chlorobenzene. *o*-Chlorotoluene is miscible in all proportions with many organic liquids such as aliphatic and aromatic hydrocarbons, chlorinated solvents, lower alcohols, ketones, glacial acetic acid, and di-*n*-butylamine; it is insoluble in water, ethylene and diethylene glycols, and triethanolamine.

p-Chlorotoluene (1-chloro-4-methylbenzene, PCT) and *m*-chlorotoluene (1-chloro-3-methylbenzene, MCT) are mobile, colorless liquids with solvent properties similar to those of the ortho isomer.

o- And *p*-chlorotoluenes form binary azeotropes with various organic compounds including alcohols, acids, and esters (2). Oxygen indexes (minimum percentage of oxygen in an oxygen–nitrogen atmosphere required to sustain combustion after ignition) for the chlorotoluene isomers are ortho 19.2, meta 19.7, and para 19.1 (3) (see Flame retardants). *o*- And *p*-chlorotoluenes form stable ionic complexes with antimony pentachloride (4). Physical properties of the monochlorotoluene isomers are listed in Table 1 (5–9).

Chemical Properties

At moderate temperatures and pressures the monochlorotoluenes are stable to the action of steam, alkalies, amines, and hydrochloric and phosphoric acids. Reactions can be divided into three classes: reactions of the aromatic ring, reactions of the methyl group, and reactions involving the chlorine substituent.

820   CHLOROCARBONS, -HYDROCARBONS (TOLUENES)

Table 1. Physical Properties of the Monochlorotoluenes

Property		Ortho [95-49-8]	Meta [108-41-8]	Para [106-43-4]
mol wt		126.59		
mp, °C		−35.6	−47.8	7.5
bp, °C		159.2	161.7	162.4
density, kg/m³	20°C	1082.5	1072.2	1069.7
	25°C	1077.6		1065.1[24.4]
	30°C	1072.7		
refractive index, n_D^t	20°C	1.52680	1.5214[19]	1.5211
	25°C	1.52221		1.5193[24.4]
surface tension, N/m[a]		0.03344[20]		0.03224[25]
		0.03233[30]		0.02922[30]
dielectric constant at 20°C		4.73	5.55	6.20
viscosity (dynamic), Pa·s[b]				9.0×10^{-5}
dipole moment, C·m[c]		4.80×10^{-30}	5.97×10^{-30}	
heat of vaporization, kJ/mol[d]		43.01	42.18	42.475
vapor density (air = 1)				4.37
vapor pressure, °C at kPa[e]				
0.13		5.4	4.8	5.5
1.3		43.2	43.2	43.8
5.3		72.0	73.0	73.5
13.3		94.7	96.3	96.6
53.3		137.1	139.7	139.8

[a] To convert N/m to dyn/cm, multiply by 1000.
[b] To convert Pa·s to P, multiply by 10.
[c] To convert C·m to debye, divide by 3.336×10^{-30}.
[d] To convert kJ to kcal, divide by 4.184.
[e] To convert kPa to mm Hg, multiply by 7.5.

Reactions of the Aromatic Ring. Ring chlorination of o-chlorotoluene yields a mixture containing all four possible dichlorotoluene isomers. 2,5-Dichlorotoluene is the isomer formed in greatest quantity, representing up to ca 60% of the isomer mixture (10–11). The four isomers are also formed in the mononitration of o-chlorotoluene. Nitration of p-chlorotoluene produces a mixture of 66% 4-chloro-2-nitrotoluene and 34% 4-chloro-3-nitrotoluene (12). Chlorosulfonation of o-chlorotoluene yields 2-chloro-5-chlorosulfonyltoluene as the major product. Sulfonation of p-chlorotoluene with 20% oleum gives the 2-sulfonated product in 68% yield (13).

Chloromethylation of o-chlorotoluene reportedly gives α,3-dichloro-p-xylene as the sole product (14). With p-chlorotoluene, a mixture containing 63% 4-chloro-2-chloromethyltoluene and 37% 4-chloro-3-chloromethyltoluene is formed (15).

Reactions of the Methyl Group. Most monochlorotoluene uses relate to products derived from reactions of the methyl group. The polychlorinated toluenes usually undergo the same reactions, but frequently differ in reaction rates and reported yields.

Chlorination under free-radical conditions leads successively to the chlorinated benzyl, benzal, and benzotrichlorides (qv). Chlorinated benzaldehydes and benzoic acids can be obtained under both liquid- and vapor-phase oxidation conditions (16). Vapor-phase catalytic ammoxidation with oxygen and ammonia produces chlorobenzonitriles (17). p-Chlorophenylacetonitrile is produced in high yield from the reaction of p-chlorotoluene and cyanogen chloride at 650–700°C (18).

Halogen Reactions. Both displacement and benzyne mechanisms are involved in the hydrolysis of chlorotoluenes to cresols with aqueous sodium hydroxide (19). Displacement with retention of configuration is favored at lower temperatures. At 340°C, o-chlorotoluene yields a near equimolar mixture of o- and m-cresols. The hydrolysis of chlorotoluenes over rare earth phosphates at 400–500°C produces cresols with greater than 90% retention of isomer position (20).

o-Chlorotoluene with sodium in liquid ammonia gives a mixture of 67% o-toluidine and 33% m-toluidine (21). Thiophenols are produced from the reaction of chlorotoluenes with hydrogen sulfide at 500–600°C (22). Aluminum chloride and boron trifluoride–hydrogen fluoride catalyze isomerization of the chlorotoluenes to the equilibrium mixture (31% ortho, 44% meta, 25% para) (23). Transalkylation is a significant side reaction. Dehalogenation of the aromatic nucleus occurs readily with hydrogen and noble metal hydrogenation catalysts (24).

Preparation

The chlorination of toluene has been carried out with a wide variety of chlorinating agents, catalysts, and reaction conditions. The ratio of formed ortho and para isomers can vary over a wide range. Most recent studies have concentrated on developing means for increasing the para isomer content owing to its greater commercial significance. The meta isomer must be prepared by indirect means since only a small amount, <1%, is formed by direct chlorination.

Chlorinations with Elemental Chlorine. Monochlorotoluene mixtures containing more than 70% ortho isomer are obtained by chlorination with certain Lewis acid catalysts including the chlorides of aluminum, tin, titanium, and zirconium (24(a)–25). Toluene chlorination in an emulsion with an equal volume of water produces chlorotoluene containing 75% ortho isomer (26). Chlorination with ferric chloride as catalyst gives a product with 63% ortho isomer content. Para isomer contents of 45–55% are obtained through the use of certain specific metal sulfides or cocatalyst systems consisting of specific metal salts and sulfur, inorganic sulfides, or divalent organic sulfur compounds (24a,27–33).

Noncatalytic nuclear chlorination of toluene in a variety of solvents has been reported. Isomer distributions vary from approximately 60% ortho in hydroxylic solvents, eg, acetic acid, to 60% para in solvents, eg, nitromethane, acetonitrile, and ethylene dichloride (34–35). Reaction rates are relatively slow and these systems are particularly appropriate for kinetic studies.

Good yields of chlorotoluenes can be obtained by the reaction of toluene (qv) with molecular proportions of certain Lewis acid halides as the chlorinating agent (36). With ferric chloride at 50–60°C, the isomer mixture contains 13% ortho, 87% para, and less than 1% meta isomers. The reaction is accompanied by the formation of polymeric by-products and several modifications have been proposed to improve product yields (37–38).

Hydrogen chloride has been used as the chlorinating agent in both liquid- and vapor-phase systems. Nitric acid catalysis in aqueous hydrochloric acid at 80°C and 170 psig oxygen pressure gives a high conversion and yield of monochlorotoluenes with 61% ortho isomer content (39). Oxychlorination of toluene with oxygen and hydrogen chloride in the vapor phase over supported copper and palladium catalysts leads to chlorotoluene mixtures of up to 60% para content along with varying amounts of side chain chlorinated products (40–41).

p-Chlorotoluene is produced via the chlorine (42) or chlorotris(triphenylphosphine)rhodium (43) catalyzed desulfonylation of p-toluenesulfonyl chloride. Pure chlorotoluene isomers are available from diazotization of the toluidine isomers followed by reaction with cuprous chloride (Sandmeyer Reaction). This is the preferred method of preparing m-chlorotoluene.

The rate of chlorination of toluene relative to that of benzene is about 345 (44). Chlorination is usually carried out below 70°C with the reaction proceeding at a profitable rate at temperatures as low as 0°C. Chlorine efficiency is high and toluene conversion to a monochlorotoluene mixture can be carried to about 90% with the formation of only a few percent of dichlorotoluenes. In most catalyst systems, the percent of para isomer formed increases with a decrease in chlorination temperature. The reaction is exothermic with ca 139 kJ/mol (33 kcal/mol) produced per mole of monochlorotoluene. Required catalyst concentrations are low, generally on the order of several tenths of a percent or less.

Only trace amounts of side chain chlorinated products are formed with suitably active catalysts. It is usually desirable to remove reactive chlorides prior to fractionation in order to minimize the risk of equipment corrosion. The separation of o- and p-chlorotoluenes by fractionation requires a high efficiency isomer-separation column. The small amounts of m-chlorotoluene formed in the chlorination cannot be separated by fractionation and remains in the p-isomer fraction.

The toluene feed should be essentially free of paraffinic impurities which may produce high boiling residues that foul heat transfer surfaces. Trace water contamination has no effect on product composition. Steel can be used as construction material for catalyst systems containing iron. However, glass-lined equipment is usually preferred and must be used with other catalyst systems.

Both batch and continuous processes are suitable for commercial chlorination. The progress of the chlorination is conveniently followed by specific gravity measurements.

Handling and Shipment

Monochlorotoluenes can be shipped in bulk in steel tank cars and steel or aluminum tank trucks. Lined or unlined steel drums are suitable for drum shipment. Storage vessels are vented to a safe atmosphere and should be protected with suitable diking; the contents should be protected against static charge when transferring. Freight classification under DOT regulations for the monochlorotoluenes is combustible liquid NOS and, for truck transport, chemical NOI. The vapor forms flammable mixtures with air, thus, suitable ventilation should be provided and sources of ignition avoided.

Economic Aspects

The principal nations in which chlorotoluenes are produced are the United States, Germany, and Japan. Since the number of manufacturers is small, and much of the production is consumed captively, statistics covering the amounts produced are not available. World production of o- and p-chlorotoluenes is estimated at several tens of thousands of metric tons yearly. For commercially produced polychlorotoluenes volumes are 100–1000 t/yr.

Identification and Analysis

Several procedures for the analysis of chlorotoluene mixtures by gas chromatography have been described (45–46). The use of liquid crystal stationary phases is the preferred method for separating m- and p-chlorotoluene, a separation not easily obtained with conventional columns (47). Prepacked columns are commercially available (48). The presence of benzylic impurities in chlorotoluenes is determined by standard methods for hydrolyzable chlorine. Proton (49) and ^{13}C nmr shifts, characteristic ir absorption bands, and principal mass spectral peaks have been summarized (49a). Sources of reference spectra are listed. A procedure for using macroreticular resins for analysis of trace chlorotoluenes in water has been described (50).

Health and Safety Factors (Toxicology)

Chlorotoluene administered orally as a 70% emulsion in turkey red oil has LD_{50} values of 5.4 and 3.4 mL/kg in chickens and mice, respectively (51). In an acute, oral toxicity study on Halso 99, a solvent consisting primarily of o-chlorotoluene, the estimated median lethal dose (LD_{50}) value was determined as 3410 mg/kg (slightly toxic) by stomach tube administration to Sprague-Dawley albino rats (52). In similar tests, p-chlorotoluene and 2,4-dichlorotoluene had respective LD_{50}s of 1920 mg/kg and >4640 mg/kg.

In an acute inhalation toxicity study, the LC_{50} value was determined to be 7119 ppm for o-chlorotoluene as compared to 4168 ppm for technical o-dichlorobenzene (male albino rats, 4 h exposure time) (52). For o-chlorotoluene, the TLV (threshold limit value), as established by the American Conference of Governmental Industrial Hygienists, is 50 ppm (250 mg/m^3) time-weighted average (53).

Based on primary skin irritation studies, Halso 99 solvent has been classified as a mild or moderate skin irritant (52). Moderate conjunctival irritation was produced by a single installation of undiluted Halso 99 into the eyes of albino rats. The irritation gradually subsided and disappeared by the fifth day. Fluorescein staining on the seventh day revealed no corneal damage (52).

Two products were isolated from a study of the metabolism of p-chlorotoluene by *Pseudomonas putida:* (+) *cis*-4-chloro-2,3-dihydroxy-1-methylcyclohex-4,6-diene and 4-chloro-2,3-dihydroxy-1-methylbenzene (54). The enzymatic dehydrogenation of the former compound to the latter was demonstrated.

HIGHER CHLOROTOLUENES

Dichlorotoluenes

Of the six dichlorotoluenes, only the 2,4, 2,5, and 3,4 isomers are available from direct chlorination of monochlorotoluenes. Physical properties of the dichloro and other higher chlorotoluenes are given in Table 2.

2,3-Dichlorotoluene (1,2-dichloro-3-methylbenzene) is present in about 10% concentration in o-chlorotoluene chlorination mixtures and is best prepared by the Sandmeyer reaction on 3-amino-2-chlorotoluene. 2,4-Dichlorotoluene (2,4-di-

Table 2. Physical Properties of the Higher Chlorotoluenes

Toluene	CAS Registry No.	mp, °C	bp, °C	n_D^t	Density at 20°C, kg/m³	mol wt
2,3-dichloro	[32768-54-0]	5	208.3	1.5511^{20}		
2,4-dichloro	[95-73-8]	−13.5	201.1	1.5480^{22}	1249.8	
2,5-dichloro	[19398-61-9]	5	201.8	1.5449^{20}	1253.5	161.03
2,6-dichloro	[118-69-4]		200.6	1.5507^{20}	1268.6	
3,4-dichloro	[95-75-0]	−15.3	208.9	1.5471^{20}	1256.4	
3,5-dichloro	[25186-47-4]	26	201.2	1.5438^{20}		
2,3,4-trichloro	[7359-72-0]	43–44	244			
2,3,5-trichloro	[56961-86-5]	45–46	229–231			
2,3,6-trichloro	[2077-46-5]	45–46	118[a]			195.48
2,4,5-trichloro	[6639-30-1]	82.4	229–230[b]			
2,4,6-trichloro	[23749-65-7]	38				
3,4,5-trichloro	[21472-86-6]	45–45.5	246–247[c]			
2,3,4,5-tetrachloro	[1006-32-2]	98.1				
2,3,4,6-tetrachloro	[875-40-1]	92	266–276			229.93
2,3,5,6-tetrachloro	[1006-31-1]	93–94				264.37
pentachloro	[877-11-2]	224.5–225.5	301			

[a] At 2.4 kPa (18 mm Hg).
[b] At 95.4 kPa (716 mmHg).
[c] At 102.4 kPa (768 mm Hg).

chloro-1-methylbenzene) represents 80–85% of the dichlorotoluene fraction in chlorination of p-chlorotoluene with antimony trichloride (55) or zirconium tetrachloride (56) catalysts. It is separated from 3,4-dichlorotoluene by fractional distillation. 2,5-Dichlorotoluene (1,4-dichloro-2-methylbenzene) is formed in up to 60% yield in the sulfide cocatalyzed chlorination of o-chlorotoluene (10–11) and can be isolated in 99% purity by crystallization. Pure 2,5-dichlorotoluene is also available from the Sandmeyer reaction on 2-amino-5-chlorotoluene.

2,6-Dichlorotoluene (1,3-dichloro-2-methylbenzene) is obtained from the Sandmeyer reaction on 2-amino-6-chlorotoluene (57), ring chlorination and desulfonylation of p-toluene sulfonyl chloride (58), and from the chlorination and dealkylation of 4-tert-butyltoluene (59) or 3,5-di-tert-butyltoluene (60). Side chain chlorination of 2,6-dichlorotoluene does not proceed beyond 2,6-dichlorobenzal chloride. 3,4-Dichlorotoluene (1,2-dichloro-4-methylbenzene) is formed in up to 40% yield in the chlorination of p-chlorotoluene catalyzed by metal sulfides or metal halide–sulfur compound cocatalyst systems (61). 3,5-Dichlorotoluene (1,3-dichloro-5-methylbenzene) is prepared by diazotization of 2- or 4-amino-3,5-dichlorotoluene followed by reaction with alcohol.

Trichlorotoluenes

The chlorination of toluene and o- and p-chlorotoluenes produces mixtures of 2,3,6- and 2,4,5-trichlorotoluenes containing small amounts of the 2,3,4- and 2,4,6-isomers. Chlorination of toluene with iron or ferric chloride catalysts gives mixtures containing nearly equal amounts of 2,4,5- and 2,3,6-trichlorotoluene isomers (25,62). Starting with o-chlorotoluene, the product contains greater than 60% 2,3,6-isomer

(63). Chlorination of p-chlorotoluene with metal sulfides as catalyst gives a trichlorotoluene fraction of greater than 75% 2,4,5-isomer content (64). The other four trichlorotoluenes are available from the Sandmeyer reaction on suitable amines.

Tetra- and Pentachlorotoluenes

2,3,4,6-Tetrachlorotoluene (1,2,3,5-tetrachloro-4-methylbenzene) is prepared from the Sandmeyer reaction on 3-amino-2,4,6-trichlorotoluene. 2,3,4,5-Tetrachlorotoluene (1,2,3,4-tetrachloro-5-methylbenzene) is the major isomer in the further chlorination of 2,4,5-trichlorotoluene. Exhaustive chlorination of p-toluenesulfonyl chloride, followed by hydrolysis to remove the sulfonic acid group, yields 2,3,5,6-tetrachlorotoluene (1,2,4,5-tetrachloro-3-methylbenzene) in good yield (65). Pentachlorotoluene (pentachloromethylbenzene) is formed in 90% yield by the ferric chloride catalyzed chlorination of toluene in carbon tetrachloride or hexachlorobutadiene solution (66). Oxidation of pentachlorotoluene with excess sulfur trioxide, followed by hydrolysis of the intermediate pentachlorobenzyl disulfooxonium hydroxide inner salt produces pentachlorobenzyl alcohol in 91% yield (67).

Uses

Chlorotoluenes are used as intermediates in the pesticide, pharmaceutical, peroxide, dye, and other industries. First generation derivatives, principally side chain chlorinated products, are converted further to end products. The major use for p-chlorotoluene is in the manufacture of p-chlorobenzotrifluoride, a key intermediate in dinitroaniline and diphenyl ether herbicides (68) (see Herbicides). Lesser quantities are consumed in the manufacture of p-chlorobenzyl chloride, p-chlorobenzaldehyde, p-chlorobenzoyl chloride, p-chlorobenzoic acid, and 2,4- and 3,4-dichlorotoluenes. p-Chlorotoluene is an intermediate for a class of novel polyketone polymers (69).

Toluene chlorination isomer mixtures and mixtures with high o-chlorotoluene content are used widely as solvents (70). Solvent applications include reaction solvent, dye carrier (qv) formulations, sludge solvent, and paint (qv) and rubber stripping formulations (see Dyes and dye intermediates). Lesser quantities are used as intermediates through first generation derivatives similar to those of the para isomer.

2,4-Dichlorotoluene is used as a herbicide intermediate, and for manufacture of 2,4-dichlorobenzyl chloride and 2,4-dichlorobenzoyl chloride. 2,6-Dichlorotoluene is applied as a herbicide and dyestuff intermediate. 2,3,6-Trichlorotoluene is used in small quantities as a herbicide intermediate. The other polychlorotoluenes have limited industrial application.

BIBLIOGRAPHY

1. J. B. Cohen and H. D. Dakin, *J. Chem. Soc.* **79,** 1111 (1901).
2. L. H. Horsley and co-workers, *Azeotropic Data III, Advances in Chemistry Series,* No. 116, American Chemical Society, Washington, D.C., 1973.
3. G. L. Nelson and J. L. Webb, *J. Fire Flammability* **4,** 325 (1973).
4. R. G. Makitra, Ya. M. Tsikanchuk, and D. K. Tolopko, *Zh. Obshch. Khim.* **45,** 1917 (1975).
5. J. Timmermans, *Physico-Chemical Constants of Pure Organic Compounds,* Elsevier Publishing Co., Inc., New York, 1950, pp. 297–298.
6. V. Sĕdivec and J. Flek, *Handbook of Analysis of Organic Solvents,* John Wiley & Sons, Inc., New York, 1976, pp. 164–168, 398.

7. K. Raznjevic, *Handbook of Thermodynamic Tables and Charts*, McGraw-Hill Book Co., New York, 1976, tables 27-1 and 30-2.
8. R. R. Dreisbach, *Physical Properties of Chemical Compounds I, Advances in Chemistry Series*, American Chemical Society, Washington, D.C., 1955, No. 15, p. 139.
9. A. L. McClellan, *Tables of Experimental Dipole Moments*, W. H. Freeman and Company, San Francisco, Calif., 1963, p. 243.
10. Ger. Offen. 2,523,104 (Nov. 25, 1976), H. Rathjen (to Bayer A.G.).
11. U.S. Pat. 4,031,146 (June 21, 1977), E. P. DiBella (to Tenneco Chemicals Inc.).
12. Jpn. Kokai 75 151,828 (Dec. 6, 1975), M. Matsui, T. Kitsukawa, K. Sato, and T. Ogawa (to Mitsubishi Chem. Ind. Co., Ltd.).
13. Y. Muramoto and H. Asakura, *Nippon Kagaku Kaishi* **6**, 1070 (1975).
14. H. Stephen, W. F. Short, and G. Gladding, *J. Chem. Soc.* **117**, 510 (1920).
15. E. Kuimova and B. M. Mikhailov, *Zh. Org. Khim.* **7**, 1436 (1971).
16. B. Chopra and V. Ramakrishnan, *Indian Chem. J. Ann.* **38**, (1972).
17. Jpn. Kokai 68 10,623 (May 4, 1968), T. Yoshino and co-workers (Nitto Chemical Industry Co., Ltd.).
18. R. A. Grimm and J. E. Menting, *Ind. Eng. Chem. Prod. Res. Dev.* **14**, 158 (1975).
19. A. L. Bottini and J. D. Roberts, *J. Am. Chem. Soc.* **79**, 1458 (1957).
20. Ger. Offen. 2,162,756 (July 27, 1972), W. L. Kehl and R. J. Rennard (to Gulf Research & Development Co.).
21. R. Levine and E. R. Biehl, *J. Org. Chem.* **40**, 1835 (1975).
22. M. G. Voronkov and co-workers, *Zh. Org. Khim.* **11**, 1132 (1975).
23. G. A. Olah and M. W. Meyer, *J. Org. Chem.* **27**, 3464 (1962).
24. M. Kraus and V. Bazant in J. W. Hightower, ed., *Proceedings of the Fifth International Conference on Catalysis, Palm Beach, Fla.*, North-Holland Publishing Co., Amsterdam, 1972.
24a. I. G. Farben, *New Observations in the Chlorination of Toluene, PB-17658*, National Technical Information Service, Springfield, Va., Jan. 23, 1936, frames 2247–2256.
25. U.S. Pat. 3,000,975 (Sept. 19, 1961), E. P. DiBella (to Heyden Newport Chemical Co.).
26. Brit. Pat. 691,504 (May 13, 1953), (to Farbenfabriken Bayer).
27. Neth. Appl. 6,511,484 (Mar. 3, 1966), (to Hooker Chemicals & Plastics Corp.).
28. U.S. Pat. 3,226,447 (Dec. 28, 1965), G. H. Bing and R. A. Krieger (to Union Carbide Australia Ltd.).
29. Brit. Pat. 1,153,746 (May 29, 1969), (to Tenneco Chemicals Inc.).
30. Brit. Pat. 1,163,927 (Sept. 10, 1969), (to Tenneco Chemicals Inc.).
31. U.S. Pat. 3,317,617 (May 2, 1967), E. P. DiBella (to Tenneco Chemicals, Inc.).
32. U.S. Pats. 4,031,142, 4,031,147 (June 21, 1977), J. C. Graham (to Hooker Chemicals & Plastics Corp.).
33. U.S. Pat. 4,031,144 (June 21, 1977), E. P. DiBella (to Tenneco Chemicals Inc.).
34. L. M. Stock and A. Himoe, *Tetrahedron Lett.* (13), 9 (1960).
35. L. M. Stock and A. Himoe, *J. Am. Chem. Soc.* **83**, 4605 (1961).
36. P. Kovacic in G. A. Olah, ed., *Friedel-Crafts & Related Reactions*, Vol. IV, Interscience Publishers, Inc., a division of John Wiley & Sons, Inc., New York, 1965, Chapt. XLVIII, pp. 111–127.
37. Jpn. Pat. 49 76,828 (Nov. 1, 1973), J. T. Traxler (to International Minerals & Chemical Corp.).
38. Ger. Offen. 2,230,369 (Jan. 18, 1973), K. Sawazaki, H. Fujii, and M. Dehura (to Nikkei Kako Co., Ltd. and Sugai Chem. Ind. Ltd.).
39. C. M. Selwitz and V. A. Notaro, *Prepr. Div. Pet. Chem. ACS*, **17**, (4), E37-46 (1972).
40. Jpn. Kokai 73 81,822 (Nov. 1, 1973), R. Fuse, T. Inoue, and T. Kato (to Ajinomoto Co., Inc.).
41. A. B. Salomonov, P. P. Gertsen, and A. N. Ketov, *Zh. Prikl. Khim.* **43**, 1612 (1970).
42. B. Miller, *J. Org. Chem.* **38**, 1243 (1973); U.S. Pat. 3,844,917 (Oct. 29, 1974).
43. J. Blum, *Tetrahedron Lett.* (26), 3041 (1966).
44. P. B. D. DeLaMare and P. W. Robertson, *J. Chem. Soc.*, 279 (1943).
45. V. S. Kozlova and co-workers, *Zh. Anal. Khim.* **27**, 826 (1972).
46. A. E. Habboush and A. H. Tamush, *J. Chromatogr.* **53**, 151 (1970).
47. H. Kelker and E. Von Schivizhoffen in J. C. Giddings and R. A. Keller, eds., *Advances in Chromatography*, Vol. 6, Marcel Dekker, Inc., New York, 1968, pp. 247–297.
48. G. C. Reporter, Vol. 1, No. 1, Supelco, Inc., Bellefonte, Pa., Mar., 1976.
49. J. G. Lindberg, G. Y. Sugiyama, and R. L. Mellgren, *J. Magn. Reson.* **17**, 112 (1975).
49a. J. G. Grasselli and W. M. Richey, eds., *Atlas of Spectral Data and Physical Constants For Organic Compounds*, 2nd ed., Vol. IV, CRC Press Inc., Cleveland, Ohio, 1975.
50. G. A. Junk and co-workers, *J. Chromatogr.* **99**, 745 (1974).

51. Jpn. Kokai 73 36,320 (May 29, 1973), K. Tsunoda and J. Hirakoso (to Chuo Kagaku and Co., Ltd.).
52. *Bulletin #767; Halso 99,* Toxicity, Spec. Chem. Div., Hooker Chemicals & Plastics Corp., 1973.
53. *Documentation of the Threshold Limit Values For Substances In Workroom Air with Supplements For Those Substances Added or Changed Since 1971,* American Conference of Governmental Industrial Hygienists, 3rd ed., 1971, Second Printing 1974, pp. 302–303.
54. D. T. Gibson and co-workers, *Biochemistry* **7,** 3795 (1968).
55. U.S. Pat. 4,006,195 (Feb. 1, 1977), S. Gelfand (to Hooker Chemicals & Plastics Corp.).
56. U.S. Pat. 3,366,698 (Jan. 30, 1968), E. P. DiBella (to Tenneco Chemicals Inc.).
57. Ger. Pat. 1,237,552 (Mar. 30, 1967), J. T. Hackmann and co-workers (to Shell Research Ltd.).
58. Fr. Pat. 1,343,178 (Nov. 15, 1963), (to Shell Int'l. Res. Maat. N.V.).
59. Brit. Pat. 1,110,030 (Apr. 18, 1968), C. F. Kohll, H. D. Scharf, and R. Van Helden (to Shell Int'l. Res. Maat. N.V.).
60. Neth. Appl. 6,907,390 (Nov. 17, 1970), D. A. Was (to Shell Intl. Res. Maat. N.V.).
61. U.S. Pat. 4,031,145 (June 21, 1977), E. P. DiBella (to Tenneco Chemicals Inc.).
62. U.S. Pat. 3,219,688 (Nov. 23, 1965), E. D. Weil and co-workers (to Hooker Chemicals & Plastics Corp.).
63. H. C. Brimelow, R. L. Jones, and T. P. Metcalfe, *J. Chem. Soc.,* 1208 (1951).
64. U.S. Pat. 3,692,850 (Sept. 19, 1972), E. P. DiBella (to Tenneco Chemicals, Inc.).
65. R. Nishiyama and co-workers, *Yuki Gosei Kagku Kyokai Shi* **23,** 515, 521 (1965).
66. Jpn. Kokai 70 28,367 (Sept. 16, 1970), M. Ishida (to Kureha Chemical Ind. Co., Ltd.).
67. V. Mark and co-workers, *J. Am. Chem. Soc.* **93,** 3538 (1971).
68. F. M. Ashton and A. S. Crafts, *Mode of Action of Herbicides,* John Wiley & Sons, Inc., New York, 1973, pp. 10–24, 438–448.
69. U.S. Pat. 3,914,298 (Oct. 21, 1975), K. J. Dahl (to Raychem Corp.).
70. *Halso 99 Chlorinated Solvent,* Specialty Chemicals Div., Hooker Chemicals & Plastics Corp.

SAMUEL GELFAND
Hooker Chemical & Plastics Corp.

BENZYL CHLORIDE, BENZAL CHLORIDE, AND BENZOTRICHLORIDE

The chlorination of toluene in the absence of catalysts that promote nuclear substitution occurs preferentially in the side chain. The reaction is promoted by heat or free-radical initiators such as ultraviolet light or peroxides. Chlorination takes place in a stepwise manner and can be controlled to give good yields of the intermediate chlorination products.

Nearly all of the benzyl chloride [100-44-7], benzal chloride [98-87-3], and benzotrichloride [98-07-7] manufactured is converted to other chemical intermediates or products by reactions involving the chlorine substituents of the side chain. Each of the compounds has a single major use that consumes a large portion of the compound produced. Benzyl chloride is utilized in the manufacture of benzyl butyl phthalate, a vinyl resin plasticizer; benzal chloride is hydrolyzed to benzaldehyde; benzotrichloride is converted to benzoyl chloride.

Several related compounds, primarily ring-chlorinated derivatives, are also commercially significant. p-Chlorobenzotrichloride is converted to p-chlorobenzotrifluoride, an important intermediate in the manufacture of dinitroaniline herbicides (qv).

The side-chain chlorinated toluenes were first obtained from starting materials other than toluene; benzyl chloride was obtained from benzyl alcohol and hydrochloric acid by Cannizzaro in 1853; benzal chloride from benzaldehyde and phosphorus pentachloride by Cahours in 1848; and benzotrichloride from benzoyl chloride and phosphorus pentachloride by Schischkoff and Rosing in 1858. Early producers of benzyl chloride in the United States were Heyden Chemical Company and Allied Chemical in the middle 1920s.

Physical Properties

Benzyl chloride [(chloromethyl)benzene, α-chlorotoluene], $C_6H_5CH_2Cl$, formula weight 126.58, is a colorless liquid with a very pungent odor. Its vapors are irritating to the eyes and mucous membranes, and it is classified as a powerful lachrimator. The physical properties of pure benzyl chloride are given in Table 1 (1–6).

Thirty six binary azeotropic systems and nine ternary azeotropic systems that contain benzyl chloride have been reported (7). Benzyl chloride is insoluble in cold water, but decomposes slowly in hot water to give benzyl alcohol. It is miscible in all proportions, at room temperature, with most organic solvents.

The flash point of benzyl chloride is 67°C (closed cup); 74°C (open cup); autoignition temperature is 585°C; lower flammability limit: 1.1% by volume in air.

Benzal chloride [(dichloromethyl)benzene, α,α-dichlorotoluene, benzylidene chloride], $C_6H_5CHCl_2$, formula weight 161.03, is a colorless liquid with a pungent, aromatic odor. Eight binary azeotropic systems containing benzal chloride have been reported (7). Benzal chloride is insoluble in water at room temperature, but is miscible with most organic solvents.

Benzotrichloride [(trichloromethyl)benzene, α,α,α-trichlorotoluene, phenylchloroform], $C_6H_5CCl_3$, formula weight 195.48, is a colorless, oily liquid with a pungent odor. It is soluble in most organic solvents but it reacts with water and alcohol. Three binary azeotropes containing benzotrichloride have been reported (7). For benzotrichloride the flash point is above 99°C (Fisher tag closed cup); 127°C (Cleveland open cup); and the fire point (Cleveland open cup) above 247°C (8).

Table 1. Physical Properties of Benzyl Chloride, Benzal Chloride, and Benzotrichloride

Property	Benzyl chloride	Benzal chloride	Benzotrichloride
freezing point, °C	−39.2	−16.4	−4.75
boiling point, °C	179.4	205.2	220.6
density, t, kg/m^3	1113.5$_4^4$	1256$_{14}^{14}$	1374$_4^{20}$
	1104$_{15}^{15}$		
	1100$_{20}^{15}$		
refractive index, n_D^t	1.54124^{15}		
	1.5392^{20}	1.5502^{20}	1.55789^{20}
surface tension, mN/m (= dyn/cm)	19.50$^{179.5}$	20.20$^{203.5}$	38.03^{20}
	0.03765^{20}		
dipole moment (dil benzene soln) C·m[a]	6.24 × 10^{-30}	6.9 × 10^{-30}	7.24 × 10^{-30}
diffusion of vapor in air, D_0, cm^2/s	0.066		
vapor density (air = 1)	4.34		6.77
heat of combustion, kJ/mol[b]	3708.7		
specific heat (101 kPa[c]), J/(kg·K)[b]	1351.4		
volume coefficient of expansion	0.000972		
heat of vaporization, kJ/mol[b]	461.24		
vapor pressure, °C at kPa[c]			
0.13	22.0	35.4	45.8
0.67	47.8	64.0	73.7
1.33	60.8	78.7	87.6
5.33	90.7	112.1	119.8
8.00	100.5	123.6	130.0
13.3	114.2	138.3	144.3
26.7	134.0	160.7	165.6
53.3	155.8	187.0	189.2

[a] To convert C·m to debye, divide by 3.336 × 10^{-30}.
[b] To convert J to cal, divide by 4.184.
[c] To convert kPa to mm Hg, multiply by 7.50.

The solubilities of benzyl chloride, benzal chloride, and benzotrichloride in water have been calculated by a method devised for compounds with a significant hydrolysis rate (9).

Chemical Properties

The reactions of benzyl chloride, benzal chloride, and benzotrichloride may be divided into two classes: (*a*) the reaction of the side chain containing the halogen, and (*b*) the reactions of the aromatic ring.

Reactions of the Side Chain. Benzyl chloride is hydrolyzed slowly by boiling water and more rapidly at elevated temperature and pressure in the presence of alkalies (10). Reaction with aqueous sodium cyanide, preferably in the presence of a quaternary ammonium chloride, produces phenylacetonitrile in high yield (11). In the presence of suitable catalysts benzyl chloride reacts with carbon monoxide to produce phenylacetic acid (12–14). Benzyl esters are formed by heating benzyl chloride with the sodium salts of acids; benzyl ethers by reaction with sodium alkoxides. The ease of ether formation is improved by the use of phase-transfer catalysts (qv) (15).

The benzylation of a wide variety of aliphatic, aromatic, and heterocyclic amines has been reported. Benzyl chloride is converted into mono-, di-, and tribenzylamines by reaction with ammonia. Benzylaniline results from the reaction of benzyl chloride with aniline. Reaction with tertiary amines yields quaternary ammonium salts; with trialkylphosphines, quaternary phosphonium salts are formed.

Benzyl chloride readily forms a Grignard compound by reaction with magnesium in ether. Benzyl chloride is oxidized first to benzaldehyde, and then to benzoic acid. Reaction with ethylene oxide produces the benzyl chlorohydrin ether ($C_6H_5CH_2OCH_2CH_2Cl$) (16). Benzylphosphonic acid is formed from the reaction of benzyl chloride and triethyl phosphite, followed by hydrolysis (17).

Benzyl chloride reacts with alkali hydrogen sulfides, sulfides, and polysulfides to yield benzenethiol, dibenzyl sulfide, and dibenzyl polysulfide, respectively. With sodium cyanate it forms benzyl isocyanate (18).

Benzyl chloride reacts with benzene in the presence of a Lewis acid catalyst, to give diphenylmethane. It undergoes self condensation to form polymeric oils and solids (19). With phenol, benzyl chloride produces a mixture of o- and p-benzylphenol.

Benzal chloride is hydrolyzed to benzaldehyde under both acid and alkaline conditions. Typical conditions include reaction with steam in the presence of ferric chloride or a zinc phosphate catalyst (20) and reaction at 100°C with water containing an organic amine (21). Cinnamic acid is formed by heating benzal chloride and potassium acetate with an amine as catalyst (22).

Benzotrichloride is hydrolyzed to benzoic acid by hot water, concentrated sulfuric acid, or dilute aqueous alkali. Benzoyl chloride is produced by the reaction of benzotrichloride with an equimolar amount of water or an equivalent of benzoic acid. The reaction is catalyzed by Lewis acids such as ferric chloride and zinc chloride. Reaction of benzotrichloride with other organic acids or with anhydrides yields mixtures of benzoyl chloride and the acid chloride derived from the acid or anhydride. Benzotrifluoride is formed by the reaction of benzotrichloride with anhydrous hydrogen fluoride under both liquid- and vapor-phase reaction conditions.

Aromatic Ring Reactions. In the presence of an iodine catalyst chlorination of benzyl chloride yields a mixture consisting mostly of the ortho and para compounds. With strong Lewis acid catalysts such as ferric chloride, chlorination is accompanied by self condensation. Nitration of benzyl chloride with nitric acid in acetic anhydride gives an isomeric mixture containing about 33% ortho, 15% meta, and 52% para isomers (23); with benzal chloride, a mixture containing 23% *ortho*, 34% *meta*, and 43% *para* nitrobenzal chlorides is obtained.

Chlorosulfonation of benzotrichloride with chlorosulfonic acid (qv) (24) or with sulfur trioxide (25) gives m-chlorosulfonyl benzoyl chloride in high yield. Nitration with nitronium salts in sulfolane gives 68% m-nitrobenzotrichloride along with 13% of the ortho and 19% of the para isomers (26). Nitrobenzotrichloride is also obtained in high yield with no significant hydrolysis when nitration with a mixture of nitric and sulfuric acids is carried out below 30°C (27). 2,4-Dihydroxybenzophenone is formed by the uncatalyzed reaction of benzotrichloride with resorcinol in hydroxylic solvents (28) or in benzene containing methanol or ethanol (29). Benzophenone derivatives are formed from a variety of aromatic compounds by reaction with benzotrichloride in aqueous or alcoholic hydrofluoric acid (30).

Benzotrichloride with zinc chloride as catalyst reacts with ethylene glycol to form 2-chloroethyl benzoate (31). Perchlorotoluene is formed by chlorination with a solution of sulfur monochloride and aluminum chloride in sulfuryl chloride (32).

Manufacture

Benzyl chloride is manufactured by the thermal or photochemical chlorination of toluene at 65–100°C (33). At lower temperatures the amount of ring chlorinated by-products is increased. The chlorination is usually carried to no more than about 50% toluene conversion in order to minimize the amount of benzal chloride formed. Overall yield based on toluene is more than 90%. Various materials, including phosphorus pentachloride, have been reported to catalyze the side-chain chlorination. These compounds and others such as amides also reduce ring chlorination by complexing metallic impurities (34).

Under typical liquid-phase chlorination conditions the maximum conversion to benzyl chloride of about 70% is reached after reaction of about 1.1 moles of chlorine per mole of toluene (35). Higher yields of benzyl chloride have been claimed: 80% for low temperature chlorination (36); 80–85% for light-catalyzed chlorination in the vapor phase (37) and 93.6% for continuous chlorination above 125°C in a column packed with glass rings (38).

An 80% yield of benzyl chloride is obtained with sulfuryl chloride as chlorinating agent. Yields of >70% of benzyl chloride are obtained by the zinc chloride-catalyzed chloromethylation of benzene.

In commercial practice chlorination may be carried out either in batches or continuously. Glass-lined or nickel reactors may be used. Because certain metallic impurities such as iron catalyze ring chlorination and self-condensation, their presence must be avoided. The crude product is purged of dissolved hydrogen chloride, neutralized with alkali, and distilled. Chlorine efficiency is high; muriatic acid made by absorbing the by-product hydrogen chloride in water is usually free of significant amounts of dissolved chlorine.

Benzyl chloride undergoes self condensation relatively easily at high temperatures or in the presence of trace metallic impurities. The risk of decomposition during distillation is reduced by the use of various additives including lactams (39) and amines (40,41). Lime, sodium carbonate, and triethylamine are used as stabilizers during storage and shipment. Other soluble organic compounds that are reported to function as stabilizers in low concentration include DMF (42), arylamines (43), and triphenylphosphine (44).

Benzal chloride can be manufactured in 70% yield by chlorination with 2.0–2.2 moles of chlorine per mole of toluene. The benzal chloride is purified by distillation. Benzal chloride is also formed by the reaction of dichlorocarbene (CCl_2:) with benzene (44a).

Further chlorination at a temperature of 100–140°C with ultraviolet light yields benzotrichloride. The chlorination is normally carried to a benzotrichloride content of greater than 95% with a low benzal chloride content. After purging with inert gas to remove hydrogen chloride, the crude product is utilized directly or purified by distillation. Under batch conditions chlorine efficiency during the latter stages of the chlorination is low. Product quality and chlorine efficiency can be improved by carrying out the chlorination continuously in a multistage system (45). Additives such as phosphorus trichloride are used to complex metallic impurities. Contaminants or reaction conditions that cause darkening and thereby reduce light penetration must be avoided if the chlorination is to be efficient (46).

An extensive kinetic study of the photochlorination of toluene in a continuous annular reactor has been carried out (35). The radiation-initiated chlorination of toluene has also been investigated (47–49).

Handling and Shipment

As is the case during manufacture, contact with those metallic impurities that catalyze Friedel-Crafts condensation reactions must be avoided. The self-condensation reaction is exothermic and the reaction can accelerate producing a rapid buildup of hydrogen chloride pressure in closed systems.

Benzyl chloride is available in both anhydrous and stabilized forms. Both forms can be shipped in glass carboys, nickel and lined-steel drums, and nickel tank trucks and tank cars. Stabilized benzyl chloride can be shipped in unlined and lacquer-lined drums, and tank trucks or cars of construction other than nickel. Glass-lined tanks are the first choice for bulk storage of anhydrous benzyl chloride; lead-lined, nickel, or ceramic tanks can also be used.

Benzyl chloride is classified by DOT as chemicals NOIBN, Corrosive Liquid (white label required). A detailed Chemical Safety Data Sheet is available (50).

Benzotrichloride is classified under DOT regulation as a corrosive liquid NOS; Freight Classification Chemical NOI. It is shipped in lacquer-lined steel drums and nickel-lined tank trailers (8). Benzal chloride is handled and shipped in a manner generally similar to benzotrichloride.

Economic Aspects

Table 2 shows the production and sales of benzyl chloride through 1972 as reported by the U.S. Tariff Commission. Statistics are not available for benzal chloride or benzotrichloride, or for benzyl chloride after 1972.

U.S. production of benzyl chloride in 1976 is estimated to be about 45,000 metric tons at a selling price of ca $0.66/kg. There is a wide discrepancy between the production and sales volumes in Table 2 indicating that the bulk of the benzyl chloride manufactured was used in captive processing.

Production of benzal chloride is small, with U.S. production probably amounting to no more than several thousand metric tons annually. U.S. production of benzotrichloride is probably greater than ten thousand metric tons annually, most of which is consumed captively in the producing plants.

Table 2. Production and Sales of Benzyl Chloride

Year	Total production, metric tons	Sales, metric tons	Unit value, $/kg
1956	5,700	1,050	0.44
1960	9,700	2,800	0.44
1965	28,200	4,350	0.40
1970	34,100	8,950	0.31
1971	34,000	9,550	0.29
1972	36,500	8,550	0.29

Identification and Analysis

The side chain chlorine contents of benzyl chloride, benzal chloride, and benzotrichlorides are determined by hydrolysis with methanolic sodium hydroxide followed by titration with silver nitrate. Total chlorine determination, including ring chlorine, is made by standard combustion methods (51). Several procedures for the gas chromatographic analysis of chlorotoluene mixtures have been described (52–53). Proton and ^{13}C nuclear magnetic resonance shifts, characteristic infrared absorption bands, and principal mass spectral peaks have been summarized including sources of reference spectra (54). Procedures for measuring trace benzyl chloride in air (55) and in water (56) have been described.

Health and Safety Factors (Toxicology)

In rats the LD_{50} of benzyl chloride administered subcutaneously in oil solution is 1000 mg/kg (slightly toxic) (57). The LC_{50}s in mice and rats are 80 and 150 ppm (2-h inhalation exposure), respectively. Exposure to 20–200 ppm for 2 h produces mucous membrane irritation and conjunctivitis.

Benzyl chloride is carcinogenic in rats, producing local sarcomas when administered by subcutaneous injection. No increase in the incidence of lung tumors was observed in mice injected intraperitoneally. Benzyl chloride was weakly mutagenic in tests with strain TA 100 of *Salmonella typhimurim* (58).

Benzyl chloride is absorbed through the lungs and gastrointestinal tract. In man, air concentrations of 32 ppm cause severe irritation of the eyes and respiratory tract. In the U.S. the permissable level of benzyl chloride in the work environment has been established at 1 ppm (5 mg/m^3). Odor threshold is given as 0.047 ppm (59).

Vapors of both benzal chloride and benzotrichloride are strongly irritating and lacrimatory. Reported toxicities for benzal chloride are: LD_{50} (mice) 467 mg/kg (moderately toxic) (60) LC_{50} (inhalation, rats) 82 ppm (61). For benzotrichloride: LD_{50} (rats) 6000 mg/kg (slightly toxic) (62); lowest published lethal dose (frog) 2150 mg/kg (63), LC_{50} (inhalation, rats) 30 ppm (61); toxic dose level (inhalation rats) 125 ppm/4 h (63).

Uses

Nearly all uses and applications of benzyl chloride are related to reactions of the active halide substituent. More than two thirds of benzyl chloride produced is used in the manufacture of benzyl butyl phthalate, a plasticizer used extensively in vinyl flooring and other flexible poly(vinyl chloride) uses such as food packaging. Other major uses are the manufacture of benzyl alcohol and of benzyl chloride-derived quaternary ammonium compounds, each of which consumes more than 10% of the benzyl chloride produced. Smaller volume uses include the manufacture of benzyl cyanide, benzyl esters such as benzyl acetate, butyrate, cinnamate, and salicylate, benzylamine, and benzyldimethylamine, and *p*-benzylphenol. In the dye industry benzyl chloride is used as an intermediate in the manufacture of triphenylmethane dyes. First generation derivatives of benzyl chloride are processed further to pharmaceutical, perfume, and flavor products.

Nearly all of the benzal chloride produced is consumed in the manufacture of benzaldehyde (qv). The major part of benzotrichloride production is used in the manufacture of benzoyl chloride (see Benzoic acid). Lesser amounts are consumed in the manufacture of benzotrifluoride, as a dyestuff intermediate, and in producing hydroxybenzophenone ultraviolet light stabilizers (see Uv stabilizers).

Benzyl-derived quaternary ammonium compounds are used widely as cationic surface active agents and as germicides, fungicides and sanitizers. Benzyl alcohol is used in a wide spectrum of applications including pharmaceuticals and perfumes, as a solvent, and as a textile dye assistant.

Derivatives

Ring-Substituted Derivatives. The ring-chlorinated derivatives of benzyl chloride, benzal chloride, and benzotrichloride are produced by the direct side-chain chlorination of the corresponding chlorinated toluenes or by one of several indirect routes if the required chlorotoluene is not readily available.

Physical constants of the main ring chlorinated derivatives of benzyl chloride benzal chloride, and benzotrichloride are given in Table 3.

o-, *p*-, 2,4 And 3,4-dichlorobenzyl chloride, benzal chloride, and benzotrichlorides are manufactured by side-chain chlorination of the appropriate chlorotoluene.

p-Chlorobenzotrichloride (1-chloro-4-trichloromethylbenzene) can be prepared by peroxide catalyzed chlorination of *p*-toluenesulfonyl chloride or di-*p*-tolysulfone (65). 2,4-Dichlorobenzotrichloride (1,3-dichloro-4-trichlormethylbenzene) is obtained by the chlorination of 2-chloro-4-chlorosulfonyltoluene (64).

3,4-Dichlorobenzyl chloride (1,2-dichloro-4-chloromethylbenzene) containing some 2,3-dichlorobenzyl chloride is produced by the chloromethylation of *o*-dichlorobenzene in oleum solution (66). Chlorination of 2-chloro-6-nitrotoluene at 160–185°C gives a mixture of 2,6-dichlorobenzal chloride and 2,6-dichlorobenzyl chloride (67).

The ring chlorinated benzyl chlorides are used in the preparation of quaternary ammonium salts, and as intermediates for pharmaceuticals and pesticides. *p*-Chlorobenzyl chloride is an intermediate in the manufacture of the rice herbicide, Saturn ((*S*-4-chlorobenzyl-*N*,*N*-diethylthiolcarbamate) (68). *o*- and *p*-Chlorobenzal chlorides (1-chloro-2- and 4-dichloromethylbenzenes) are starting materials for the manufacture of *o*- and *p*-chlorobenzaldehydes.

o- And *p*-chloro- and 2,4- and 3,4-dichlorobenzotrichlorides are intermediates in the manufacture of the corresponding chlorinated benzoic acids and benzoyl chlorides. Fluorination of the chlorinated benzotrichlorides produces the chlorinated benzotrifluorides, intermediates in the manufacture of dinitroaniline and diphenyl ether herbicides (69) (qv).

2,6-Dichlorobenzal chloride is used in the manufacture of 2,6-dichlorobenzaldehyde and 2,6-dichlorobenzonitrile (70). With the exception of certain products used in the manufacture of herbicides, the volume of individual compounds produced is small, amounting to no more than several hundred tons annually for any individual compound.

Side-Chain Chlorinated Xylene Derivatives. Only a few of the nine side-chain chlorinated derivatives of each of the xylenes are available from direct chlorination. All three of the monochlorinated compounds, α-chloro-*o*-xylene (1-(chloromethyl)-2-methylbenzene), α-chloro-*m*-xylene ((1-(chloromethyl)-3-methylbenzene)), α-chloro-*p*-xylene ((1-(chloromethyl)-4-methylbenzene)) are obtained in high yield from

Table 3. Physical Constants of the Main Ring-Chlorinated Derivatives of Benzyl Chloride, Benzal Chloride, and Benzotrichloride

Benzene derivative	CAS Reg. No.	Common name	mp, °C	bp, °C	n_D^{20}	Density, d_4^t, kg/m³
1-chloro-2-(chloromethyl)	[611-19-8]	o-chlorobenzyl chloride	−17	217	1.5330^{20}	1270_4^0
1-chloro-3-(chloromethyl)	[621-20-2]	m-chlorobenzyl chloride		215–16[a]		1269.5_4^{15}
1-chloro-4-(chloromethyl)	[104-83-6]	p-chlorobenzyl chloride	31	222	1.5554^{20}	
1-chloro-2-(dichloromethyl)	[88-66-4]	o-chlorobenzal chloride		228.5	1.5670^{16}	1399_{15}^{15}
1-chloro-3-(dichloromethyl)	[15145-69-4]	m-chlorobenzal chloride		235–7		
1-chloro-4-(dichloromethyl)	[13940-94-8]	p-chlorobenzal chloride		236[b]		
2,4-dichloro-1-(chloromethyl)	[94-99-5]	2,4-dichlorobenzyl chloride	−2.6	248	1.5761^{20}	1407_4^{20}
1,3-dichloro-2-(chloromethyl)	[2014-83-7]	2,6-dichlorobenzyl chloride	39–40	117–119[c]		$1412_{15.6}^{25}$
1,2-dichloro-4-(chloromethyl)	[102-47-6]	3,4-dichlorobenzyl chloride	37–37.5	241		1519_4^{20}
1-chloro-2-(trichloromethyl)	[2136-89-2]	o-chlorobenzotrichloride	29.4	264.3	1.5836^{20}	1495^{14}
1-chloro-3-(trichloromethyl)	[2136-81-4]	m-chlorobenzotrichloride		255	1.4461^{20}	1495_4^{30}
1-chloro-4-(trichloromethyl)	[5216-25-1]	p-chlorobenzotrichloride		245	1.4463^{20}	
1,3-dichloro-2-(dichloromethyl)	[81-19-6]	2,6-dichlorobenzal chloride		250		
1,2-dichloro-4-(dichloromethyl)	[56961-84-3]	3,4-dichlorobenzal chloride		257		1518_{22}^{22}
2,4-dichloro-1-(dichloromethyl)	[134-25-8]	2,4-dichlorobenzal chloride	47–48	155–159[d]		
1,2-dichloro-4-(trichloromethyl)	[13014-24-9]	3,4-dichlorobenzotrichloride	25.8	283.1	1.5886^{20}	1591_4^{20}

[a] At 100.4 kPa (753 mm Hg).
[b] At 100.7 kPa (755 mm Hg).
[c] At 1.87 kPa (14 mm Hg).
[d] At 2.67 kPa (20 mm Hg).

partial chlorination of the xylenes. 1,3-Bis(chloromethyl)benzene can be isolated in moderate yield from chlorination mixtures (71–72).

The fully chlorinated products, 1,3-bis(trichloromethyl)benzene [881-99-2], and 1,4-bis(trichloromethyl)benzene [68-36-0], are manufactured by exhaustive chlorination of *meta* and *para* xylenes. For the meta compound, ring chlorination cannot be completely eliminated in the early stages of the reaction. The xylene hexachlorides are intermediates in the manufacture of the xylene hexafluorides and of iso- and terephthaloyl chlorides (see Phthalic acids).

1-(Dichloromethyl)-2-(trichloromethyl)benzene, the end product of exhaustive side-chain chlorination of *o*-xylene (73) is an intermediate in the manufacture of phthalaldehydic acid.

BIBLIOGRAPHY

"Benzyl Chloride, Benzal Chloride, and Benzotrichloride" are treated under "Chlorine Compounds Organic" in *ECT*, 1st ed., Vol. 3, pp. 822–826, by R. L. Clark and C. P. Neidig, Heyden Chemical Corporation; and "Benzyl Chloride, Benzal Chloride, and Benzotrichloride" under "Chlorocarbons and Chlorohydrocarbons" in *ECT* 2nd ed., Vol. 5, pp. 281–289, by H. Sidi, Heyden Newport Chemical Corporation.

1. *Handbook of Chemistry and Physics,* 58th ed., 1977–1978, CRC Press Inc., Cleveland, Ohio, pp. C-522, 523, 527, 528, 738, D-198.
2. *International Critical Tables,* Vol. 5, McGraw-Hill Book Co., New York, pp. 62, 111, 169.
3. R. R. Dreisbach, *Physical Properties of Chemical Compounds* 1, no. 15 in *Advances in Chemistry Series,* American Chemical Society, Washington, D.C., 1955, pp. 141–143.
4. A. L. McClellan, *Tables of Experimental Dipole Moments,* W. H. Freeman and Co., San Francisco, Calif., 1963, pp. 232, 237–238, 243.
5. J. Timmermans and Mme. Hennant-Roland, *J. Chim. Phys.* **32,** 501 (1935).
6. D. R. Stull, *Ind. Engr. Chem.* **39,** 525 (1947).
7. L. H. Horsley and co-workers, *Azeotropic Data III,* no. 116 in *Advances in Chemistry Series,* American Chemical Society, Washington, D.C., 1973.
8. *Spec. Chem. Div.,* Hooker Chemicals & Plastics Corp., Data Sheet No. 728-D.
9. K. Ohnishi and K. Tanabe, *Bull. Chem. Soc. Japan,* **44,** 2647 (1971).
10. U.S. Pat. 3,557,222 (Jan. 19, 1971), H. W. Withers and J. L. Rose (to Velsicol Chemical Corp.).
11. Brit. Pat. 1,336,883 (Nov. 14, 1973), H. Coates, R. L. Barker, R. Guest, and A. Kent (to Albright & Wilson, Ltd.).
12. J. K. Stille and P. K. Wong, *J. Org. Chem.* **40,** 532 (1975).
13. Ger. Offen. 2,259,072 (June 20, 1974), M. E. Chahawi and H. Richtzenhain (to Dynamit Nobel A.G.).
14. Ger. Offen. 2,035,902 (Feb. 4, 1971), M. Foa, L. Cassar, and G. P. Chiusoli (to Montecatini Edison S.P.A.).
15. H. H. Freedman and R. A. DuBois, *Tetrahedron Lett.* **38,** 3251 (1975).
16. Jpn. Kokai 75 67,942 (May 29, 1975), S. Komori.
17. Brit. Pat. 1,366,600 (Sept. 11, 1974), F. J. Harris and H. L. Brown (to Scottish Agric. Ind. Ltd.).
18. Ger. Offen. 2,449,607 (April 30, 1975), Y. Inamoto, H. Kitano, Y. Tanaka, F. Tanimoto and co-workers (to Kao Soap Co., Ltd.).
19. H. C. Haas, D. I. Livingston, and M. Saunders, *J. Polym. Sci.* **15,** 503 (1955).
20. U.S. Pat. 3,524,885 (Aug. 18, 1970), A. J. Deinet (to Tenneco Chemicals Inc.).
21. Jpn. Pat. 69 12,132 (June 2, 1969), H. Funamoto (to Kureha Chem. Ind. Co. Ltd.).
22. Jpn. Kokai 73 81,830 (Nov. 30, 1973), K. Shinoda and K. Kobayashi (to Kureha Chem. Ind. Co. Ltd.).
23. F. DeSarlo and co-workers, *J. Chem. Soc. B* 719 (1971).
24. U.S. Pat. 3,290,370 (Dec. 12, 1966), E. D. Weil and R. J. Lisanke (to Hooker Chemical Corp.).
25. U.S. Pat. 3,322,822 (May 30, 1967), S. Gelfand (to Hooker Chemical Corp.).
26. G. Grynkiewicz and J. H. Ridd, *J. Chem. Soc. B* 716 (1971).
27. U.S. Pat. 3,182,091 (May 4, 1965), O. Scherer, H. Hahn, and N. Münch (to Farb. Hoechst Akt.).

28. U.S. Pat. 3,769,349 (Oct. 30, 1973), M. Yukutomi, Y. Tanaka, S. Genda and M. Kitauri (to Kyodo Chemical Co. Ltd.).
29. Ger. Offen. 2,208,197 (Aug. 30, 1973), B. Lachmann and H. J. Rosenkrantz (to Bayer A.G.).
30. Ger. Offen. 2,451,037 (April 29, 1976), K. Eiglmeier (to Hoechst A.G.).
31. U.S. Pat. 3,050,549 (Aug. 21, 1962), S. Gelfand (to Hooker Chemical Corp.).
32. M. Ballester, C. Molinet, and J. Castañer, *J. Am. Chem. Soc.* **82,** 4254 (1960).
33. *Faith, Keyes, and Clark's Industrial Chemicals* 4th ed., John Wiley & Sons, Inc., New York, 1975, pp. 145–148.
34. U.S. Pat. 2,695,873 (Nov. 30, 1954), A. J. Loverde (to Hooker Electrochemical Co.).
35. H. G. Haring and H. W. Knol, *Chem. Process Eng.* **45,** 540, 619, 690 (1964); **46,** 38 (1965).
36. G. Benoy and L. DeMayer, *Compt Rend. 27th Congr. Intern Chim. Ind.*, Brussels, 1954 2; *Industrie Chim. Belg. 20* Spec. No. 160-2 (1955).
37. G. V. Asolkar and P. C. Guha, *J. Indian Chem. Soc.* **23,** 47 (1946).
38. A. Scipioni, *Ann. Chim. (Rome)* **41,** 491 (1951).
39. U.S. Pat. 3,715,283 (Feb. 6, 1973), W. Bockmann (to Bayer Akt.).
40. Czeck. Pat. 159,100 (June 15, 1975), J. Besta and M. Soolek.
41. Brit. Pat. 1,410,474 (Oct. 15, 1975), C. H. G. Hands (to Albright and Wilson Ltd.).
42. Jpn. Kokai 73 05,726 (Jan. 24, 1972), N. Kato and Y. Sato (to Mitsui Toatsu Chemicals Inc.).
43. Jpn Kokai 73 05,725 (Jan. 24, 1972), N. Kato and Y. Sato (to Mitsui Toatsu Chemicals Inc.).
44. U.S. Pat. 3,535,391 (Oct. 20, 1970), G. D. Kyker (to Velsicol Chemical Co.).
44a. Brit. Pat. 1,390,394 (April 9, 1975), A. D. Forbes, R. C. Pitkethly, and J. Wood (to Brit. Petrol. Co. Ltd.).
45. Ger. Offen. 2,152,608 (April 26, 1973), W. Böckmann and R. Hornung; D.T. 2,227,337 (Aug. 28, 1975), (to Bayer A.G.).
46. Jpn. Kokai 76 08,223 (Jan. 23, 1976), M. Fuseda and K. Ezaki (to Hodogaya Chemical Co. Ltd.).
47. J. Y. Yang, C. C. Thomas, Jr., and H. T. Cullinan, *Ind. Eng. Chem. Process Res. Develop.* **9,** 214 (1970).
48. H. T. Cullinan, Jr. and co-workers, in ref. 47, p. 222.
49. B. F. Ives, H. T. Cullinan, Jr., and J. Y. Yang, *Nucl. Technol* **18,** 29 (1973).
50. *Benzyl Chloride-Chemical Safety Data Sheet SD-69,* Revised 1974; Manufacturing Chemists Assoc., Washington, D.C.
51. W. Kirsten, *Anal. Chem.* **25,** 74 (1953).
52. D. A. Solomons and J. S. Ratcliffe, *J. Chromatog.* **76,** 101 (1973).
53. R. Ramakrishnan and N. Subramanian, *J. Chromatog.* **114,** 247 (1975).
54. J. G. Grasselli and W. M. Richey, eds., *Atlas of Spectral Data and Physical Constants for Organic Compounds,* 2nd ed., Vol. IV, CRC Press Inc., Cleveland, 1975.
55. B. B. Baker, Jr., *J. Am. Ind. Hyg. Assoc.* **35,** 735 (1974).
56. G. A. Junk and co-workers, *J. Chromatog.* **99,** 745 (1974).
57. *IARC Monogr. Eval. Carcinog. Risk Chem. Man* **11,** 217–23 (1976), Toxbib., 77, 50224.
58. J. McCann and co-workers, *Proc. Nat. Acad. Sci.* **72,** 5135 (1975).
59. A. C. Stern, ed., *Air Pollution,* Vol. II, Academic Press, New York, 1968, p. 325.
60. V. V. Stankevich and V. I. Osetrov, *Gigiena i Fisiol. Truda, Proizv. Toksikol., Klinika Prof. Zabolevanii,* 96 (1963).
61. T. V. Mikhailova, *Gig. Tr. Prof. Zabol.* **8,** 14 (1964).
62. N. I. Sax, *Dangerous Properties of Industrial Materials,* 4th ed., Van Nostrand Reinhold Company, New York 1975.
63. H. E. Christensen, ed., *Registry of Toxic Effects of Chemical Substances,* 1976 ed., U.S. Dept. of Health, Education, and Welfare, Rockville, Md.
64. U.S. Pat. 3,230,268 (Jan. 18, 1966), K. Kobayashi and N. Ishino (to Fuso Chemical Co. Ltd.).
65. Jpn. Kokai 75 25,534 (Mar. 18, 1975), K. Kobayashi, N. Ishimo, and T. Nobeoka (to Fuso Chemical Co. Ltd.).
66. Brit. Pat. 951,302 (March 4, 1964) (to Monsanto Canada Ltd.).
67. Ger. Pat. 1,237,552 (March 30, 1967), J. T. Hackmann, J. Yates, T. J. Willcox, P. T. Haken, and D. A. Wood (to Shell Research Ltd.).
68. U.S. Pat. 3,914,270 (Oct. 21, 1975), K. Makoto, H. Kamata, and K. Masuro (to Kumiai Chem. Ind. Co. Ltd.).
69. F. M. Ashton and A. S. Crafts, *Mode of Action of Herbicides,* John Wiley & Sons, New York, 1973, pp. 10–24, 438–448.
70. U.S. Pat. 3,458,560 (July 29, 1969), R. A. Carboni (to E.I. du Pont de Nemours & Co.).

71. U.S. Pat. 2,994,653 (April 27, 1959), G. A. Miller (to Diamond Alkali Co.).
72. E. Clippinger, *ACS Petrol. Div. Prep.* **15**(1), B 37 (1970).
73. Ger. Offen. 2,535,696 (Feb. 17, 1977), P. Riegger, H. Richtzenhain, and G. Zoche (to Dyanmit Nobel A.G.).

SAMUEL GELFAND
Hooker Chemicals & Plastics Corp.

CHLORINATED NAPHTHALENES

In 1833 Laurent discovered that naphthalene could be chlorinated to obtain waxlike materials (1). Aylsworth first described the technological potential of these materials as impregnants for paper, wood, and fibers in 1909–1913 (2). Industrial applications for the products gradually increased, and their significance to industrialized countries grew with the demand created by the two world wars. However, inexperience, unconcern, and carelessness resulted in incidents of disease among people and animals exposed to chloronaphthalenes. Use of chloronaphthalenes in the United States began to decline in the 1960s owing largely to growing concern over health hazards.

There are 77 possible chloronaphthalenes, not all are known or rigidly identified materials. In the absence of substitution–chlorination catalysts, additive chlorination of naphthalene may be promoted by uv irradiation or free radical catalysts, or may occur in the presence of the metal halide catalysts in polychlorination to yield chlorohydronaphthalenes. Conformational as well as positional isomers are possible in additive chlorination. Identification of such compounds in the pure state can be very complicated.

Nomenclature

The numbering or lettering of positions on the naphthalene ring that might be encountered in the chemical literature is shown in the following structures.

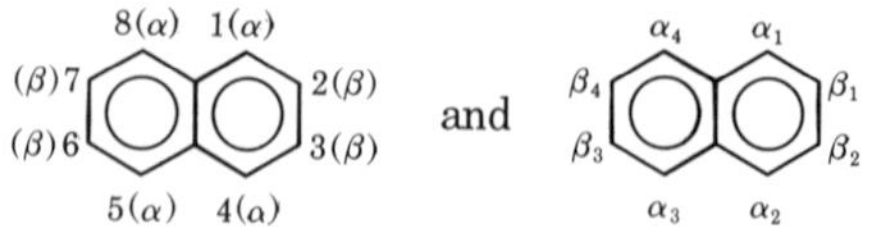

In the older literature, certain disubstitution products of naphthalene have been designated by Greek prefixes: 1,5 = *ana,* 1,8 = *peri,* 2,6 = *amphi,* and 2,7 = *pros.*

Physical Properties

Monochloronaphthalenes. A selection of physical properties of monochloronaphthalenes is given in Table 1.

Table 1. Some Physical Properties of Monochloronaphthalenes

Property	1-Chloronaphthalene [90-13-1]	2-Chloronaphthalene [91-58-7]
mp, °C	−2.3	59.5–60
bp, °C	260	259
refractive index, n_D^{20}	1.6326	
viscosity at 25°C, mPa·s (= cP)	2.94	
crystal form		platelets
density, g/cm³	1.1890^{25}	1.2656^{16}

Di- and Polychloronaphthalenes. The available melting and boiling point data and theoretical chlorine contents for purified materials are given in Table 2.

Industrially Important Chloronaphthalenes. The industrially significant chloronaphthalenes are generally not pure materials, but mixtures of isomers of mono- and/or polychloronaphthalenes. The reason for this is a combination of cost, performance, and property factors.

The commercial products range from thin liquids to hard waxes to a high melting solid, with melting points from ca −40 to 180°C. Their specific gravities are 1.2–2.0 at 25°C. The solids give melts of low viscosity.

A manufacturer does not necessarily offer the complete array of products. In 1976 the materials offered included 96% monochloronaphthalene (largely the α-isomer), and mixtures of mono- and dichloronaphthalenes; tri- and tetrachloronaphthalenes; tetra-, penta-, and hexachloronaphthalenes; octachloronaphthalene; and blends of chloronaphthalenes with polymers or bitumens. Liquid chloronaphthalenes are soluble in almost all organic solvents. Waxy or solid chloronaphthalenes are soluble in chlorinated solvents, aromatic solvents, and petroleum naphthas, and are compatible with petroleum waxes, chlorinated paraffins, polyisobutylenes, and plasticizers such as dioctyl phthalate and tricresyl phosphate. Chloronaphthalenes have very good chemical and thermal stability, and low flammability. Their volatilities range from modest to extremely low (octachloronaphthalene); flash points range from 135 to >430°C.

Chloronaphthalene mixtures containing tri- to hexachloronaphthalenes are excellent electrical insulating materials. They have d-c resistivities of ca 100 TΩ·cm at 25°C, low loss factors, and high dielectric constants (important in automotive and electronic component applications) (see Insulation, electric).

Chemical Properties

1-Chloronaphthalene can be made by the ferric chloride-catalyzed chlorination of molten naphthalene at 100–110°C. The neutralized crude product is distilled and the fraction boiling at 259–260°C contains ca 91% 1-chloronaphthalene and 9% 2-chloronaphthalene. 1-Chloronaphthalene undergoes the usual electrophilic substitution reactions such as nitration, sulfonation, and chloromethylation. Its conversion to 1-naphthol, eg, by aqueous caustic at 260–300°C in the presence of Cu catalysts (3) or with potassium *tert*-butoxide in dimethyl sulfoxide (4) has been studied but has not yet yielded a sufficiently pure product for its principal use as the starting material for the insecticide, carbaryl.

Table 2. Some Physical Properties of Di- and Selected Polychloronaphthalenes

Chloronaphthalene isomer	CAS Registry No.	mp, °C	bp, °C
di-(36.0% Cl)			
1,2-	[2050-69-3]	37	295–298
1,3-	[2198-75-6]	61.5–62	291
1,4-	[1825-31-6]	71–72	287
1,5-	[1825-30-5]	106.5–107	
1,6-	[2050-72-8]	48.5–49	
1,7-	[2050-73-9]	63.5	
1,8-	[2050-74-0]	89–89.5	
2,3-	[2050-75-1]	119.5–120.5	
2,6-	[2065-70-5]	137–138	285
2,7-	[2198-77-8]	115–116	
tri-(45.9% Cl)			
1,2,3-	[50402-52-3]	84	
1,2,4-	[50402-51-2]	92	
1,2,5-	[55720-33-7]	79	
1,2,6-	[51570-44-6]	92.5	
1,2,7-	[55720-34-8]	88	
1,2,8-	[55720-35-9]	83	
1,3,5-	[51570-43-5]	103	
1,3,6-	[55720-36-0]	81	
1,3,7-	[55720-37-1]	113	
1,3,8-	[55720-38-2]	85	
1,4,5-	[2437-55-0]	133	
1,4,6-	[2737-54-9]	68	
1,6,7-	[55720-39-3]	109	
2,3,6-	[55720-40-6]	91	
tetra-(53.3% Cl)			
1,2,3,4-	[20020-02-4]	198	
1,2,3,5-	[53555-63-8]	141	
1,2,3,7-	[55720-41-7]	115	
1,2,4,6-	[51570-45-7]	111	
1,2,4,7-	[67922-21-8]	144	
1,2,5,6-	[67922-22-9]	164	
1,2,5,7-	[67922-23-0]	114	
1,2,6,8-	[67922-24-1]	125–127	
1,3,5,7-	[53555-64-9]	179	
1,3,5,8-	[31604-28-1]	131	
1,3,6,7-	[55720-42-8]	120	
1,4,5,8-	[3432-57-3]	183	
1,4,6,7-	[55720-43-9]	139	
penta-(59.0% Cl)			
1,2,3,4,5-	[67922-25-2]	168.5	
1,2,3,4,6-	[67922-26-2]	147	
1,2,3,5,7-	[53555-65-0]	171	
hexa-(63.5% Cl)			
1,2,3,4,5,7-	[67922-27-4]	194	
hepta-(67.2% Cl)			
1,2,3,4,5,6,8-	[58863-15-3]	194	
octa-(70.3% Cl)			
1,2,3,4,5,6,7,8-	[2234-13-1]	197.5–198	

2-Chloronaphthalene cannot be isolated easily from the product of the direct chlorination of naphthalene but can be made readily from 2-naphthylamine (a carcinogen) via the diazonium salt by the Sandmeyer reaction (5). Its vapor-phase hydrolysis over a copper-on-silica catalyst gives 2-naphthol (6).

Pure isomers can be prepared by displacing amine substituents with chloride (via the Sandmeyer reaction), or by treating hydroxyl-, sulfonyl chloride-, or nitro-substituted naphthalenes with phosphorus pentachloride. A good yield of fairly pure octachloronaphthalene can be made by chlorination of naphthalene in the presence of an antimony pentachloride catalyst to a fp of the melt of 160°C, followed by recrystallization from benzene.

Manufacture

Commercial chloronaphthalenes are manufactured by the metal halide-catalyzed chlorination of molten naphthalene to the desired chlorination stage at a temperature slightly above the melting point of the desired product. Because of the possible presence of unreacted chlorine and entrained or vaporized organics, the hydrogen chloride evolved must be treated appropriately when used for hydrochloric acid manufacture to avoid environmental problems. Crude chloronaphthalenes are treated with soda ash or caustic soda, fractionated under reduced pressure, and purified with activated clay.

Economic Aspects

Manufacturers of certain chloronaphthalenes and the trade names of their products are given in Table 3.

The production of chloronaphthalenes in the Federal Republic of Germany in 1972 was reportedly 75,000 metric tons (7). In the United States, production in 1956 was reported as 3500 t, but in 1976 had declined to a fraction of that amount; sales prices were in the $1.38–1.94/kg range. The decline in the United States was caused by rising costs, competing products, shifting markets, and increasingly stringent industrial health and safety regulations (see Industrial hygiene and toxicology).

Table 3. Manufacturers of Certain Chloronaphthalenes

Nation	Manufacturer	Trade name
United States	Koppers Company, Inc.[a]	Halowax materials
FRG	Bayer A.G. (Mobay Chemical in U.S.)	Nibren waxes
United Kingdom	Imperial Chemical Industries, Ltd. (ICI in U.S.)	Seekay waxes
France	Prodelec	Clonocire products
Italy	Caffaro	Cerifal materials

[a] Stopped manufacture in 1977.

Analysis and Shipment

Wet analysis of the chlorinated naphthalenes for chlorine content is described in ref. 8. Gas chromatographic methods are applicable (9); the use of high-performance liquid chromatography has been described (10).

Carbon steel drums are employed for shipment of liquid chloronaphthalenes; fiber pack drums or cardboard boxes are used for solids in flake, slab, or powder form.

Health and Safety Aspects (Toxicology)

Skin contact with the vapor, fumes, or dust of the chlorinated naphthalenes, especially the highly chlorinated products, may cause dermatitis and acneform lesions, usually around the face and neck. Inhalation of the vapor or fumes of chloronaphthalenes may result in liver disease. Several chloronaphthalene-related deaths due to acute atrophy or necrosis of the liver were reported in the 1930s and 1940s. The acne symptoms have been described variously as chloronaphthalene-related acne, cable rash, chloracne (this term is not specific for the chloronaphthalene-related acne), and Perna-disease. X-disease of cattle, a type of hyperkeratosis, has been ascribed to ingestion of feed contaminated with machinery lubricants containing chloronaphthalenes. The following chloronaphthalenes appear on a list of guidelines issued by the OSHA (11). Note that the mono- and dichloronaphthalenes are absent from this list. A warning note on skin exposure hazards has been added.

Compound	TLV, in ca mg/m^3 (of air)
trichloronaphthalene	5
tetrachloronaphthalene	2
pentachloronaphthalene	0.5
hexachloronaphthalene	0.2
octachloronaphthalene	0.1

Precautions in the industrial handling of chlorinated naphthalenes include strict personal cleanliness during and after work, the use of fat-free barrier creams, protective clothing, periodic changes of clothing, efficient exhaust systems for the work place, handling in closed systems, good housekeeping, and informed employee and supervisory practices. Preemployment and periodic physical examinations of personnel have also been recommended.

Uses

There are no commercial uses for purified di- or polychloronaphthalene isomers with the exception of octachloronaphthalene. Purified dichloronaphthalene is potentially useful as a starting material for poly(arylene sulfides) by reaction with sulfur and soda ash (12) (see Polymers containing sulfur). Octachloronaphthalene reacts with sodium hydroxide in dimethyl sulfoxide to give heptachloro-1-naphthol, allegedly useful as a fungicide (qv), algicide, and bactericide (13). Fusion of octachloronaphthalene with sulfur at ca 315–400°C yields a tetrachloronaphtho-bis-dithiole which can be used as a yellow colorant in plastics (14) (see Colorants for plastics).

Monochloronaphthalenes and mixtures of mono- and dichloronaphthalenes have been used or recommended for chemical-resistant gage fluids and instrument seals, as heat exchange fluids, high boiling specialty solvents, for color dispersions, as engine crank case additives to dissolve sludges and gums, and as ingredients in motor tuneup compounds. Monochloronaphthalene has also been used as a raw material for dyes and as a wood preservative with fungicidal and insecticidal properties, although the latter application is no longer practiced in the United States.

The tri- and higher-chlorinated naphthalene products have been used as impregnants for condensers and capacitors and dipping–encapsulating compounds in electronic and automotive applications; as temporary binders in the manufacture of ceramic components (see Ceramics); in paper coating and impregnation; in precision casting of alloys; in electroplating stop-off compounds; as additives in gear oils and cutting compounds; in flameproofing (see Flame retardants) and insulation (qv) of electrical cable and conductors (eg, as bitumen blends); as moisture-proof sealants; and as separators in batteries. Technical octachloronaphthalene has been recommended as an additive in grinding-wheel media and cutting-oil coolants. Blends of polychloronaphthalenes with polyisobutylenes are used as masking compounds in electroplating (qv).

BIBLIOGRAPHY

"Chlorinated Naphthalenes" under "Chlorine Compounds, Organic" in *ECT* 1st ed., Vol. 3, pp. 832–837, by Jesse Werner, General Aniline & Film Corp., General Aniline Works Division; "Chlorinated Naphthalenes" under "Chlorocarbons and Chlorohydrocarbons" in *ECT* 2nd ed., Vol. 5, pp. 297–303, by D. W. F. Hardie, Imperial Chemical Industries Ltd.

1. A. Laurent, *Ann. Chim. Phys.* **52,** 275 (1833).
2. U.S. Pats. 914,222; 914,223 (Mar. 2, 1909); 1,111,289 (Sept. 22, 1913), J. W. Aylsworth.
3. U.S. Pat. 3,413,357 (Nov. 26, 1968), K. F. Bursack, E. L. Johnston, and H. J. Moltzan (to Frontier Chemical Co., Division of Vulcan Materials Co.).
4. R. H. Hayles, J. S. Bradshaw, and D. R. Pratt, *J. Org. Chem.* **36,** 314 (1971).
5. E. Pfeil, *Angew. Chem.* **65,** 155 (1953).
6. Brit. Pat. 1,178,836 (Jan. 21, 1970), A. H. Gilbert (to Imperial Chemical Industries Ltd.).
7. *Ullmann's Encyklopadie der Technischen Chemie,* 4th ed., Vol. 9, Verlag Chemie, Weinheim/Bergstr., FRG, 1975.
8. H. G. Treibl in F. D. Snell and L. S. Ettre, eds., *Encyclopedia of Industrial Chemical Analysis,* Vol. 16, John Wiley & Sons, Inc., New York, 1972, pp. 220–222.
9. F. A. Beland and R. D. Greer, *J. Chromatogr.* **84,** 59 (1973).
10. U. A. T. Brinkman and A. DeKok, *J. Chromatogr.* **129,** 451 (1976).
11. *Code of Federal Regulations,* Title 29, Subtitle B, Chapt. XVII, "Occupational Safety and Health Administration, Department of Labor," Part 1910, "Occupational Safety and Health Standards," Subpart Z, "Toxic and Hazardous Substances," 1910.1000 (a), July 1, 1976.
12. H. A. Smith in H. Bikales, ed., *Encyclopedia of Polymer Science and Technology,* Vol. 10, John Wiley & Sons, Inc., New York, 1969, p. 653 ff.
13. U.S. Pat. 3,651,154 (Mar. 21, 1972), E. Klingsberg (to American Cyanamid Co.).
14. U.S. Pat. 3,636,048 (Jan. 18, 1972), E. Klingsberg (to American Cyanamid Co.); *Tetrahedron* **28,** 963 (1972).

HANS DRESSLER
Koppers Company, Inc.

CHLORINATED BIPHENYLS AND RELATED COMPOUNDS

Biphenyl (diphenyl), terphenyls, higher polyphenyls, or mixtures of these compounds, can be chlorinated to give a wide range of products that have outstanding chemical and thermal stabilities. Individual isomers which range from liquids to high melting crystalline solids have been prepared in the laboratory by various synthetic routes. However, for normal commercial applications direct chlorination of these hydrocarbons results in mixtures of isomers that have quite different physical properties. Some of these mixtures were widely used in the United States in a range of commercial applications until the early 1970s. Although the production and sale of the polychlorinated biphenyls was discontinued in late 1977, they are still present in many of the transformers and capacitors now in use. The term PCB is commonly used as an abbreviation for polychlorinated biphenyl. The majority of PCBs are mixtures of isomers of trichlorobiphenyl [25323-68-6], tetrachlorobiphenyl [26914-33-0], pentachlorobiphenyl [25429-29-2], and small amounts of dichlorobiphenyl [25512-42-9] and hexachlorobiphenyl [26601-64-9].

Domestic U.S. production of polychlorinated biphenyls was stopped in October 1977 because of the tendency of these products to accumulate and persist in the environment owing to low degradation rates, and because of toxic effects. Governmental actions have resulted in the control of the use, disposal, and production of the PCBs in nearly all world areas including the U.S., and the complete ban of PCBs in Japan and Sweden. Polychlorinated biphenyls were also produced in the United Kingdom and Japan and are still being produced in Germany, Italy, France and the U.S.S.R. Registered trademarks used commercially by the producers of chlorinated biphenyls or polyphenyls are Aroclor (Monsanto Company, U.S.), Clophen (Farbenfabriken Bayer A.G., FRG), Fenclor (Caffaro, Italy), Kanechlor (Kanefaguchi Chemical Co., Japan), Pyralene (Prodelec, France) and Sovtol (U.S.S.R.).

The chemical structure of the polychlorinated biphenyls has been known for nearly 100 years. Commercial production was initiated in the United States in 1929 in response to the electrical industry's need for an improved dielectric insulating fluid (nonconductor of direct current) for use in transformers and capacitors which would also provide increased fire resistant benefits.

Mineral oil, the product formerly used for these applications, had stability and flammability problems that made it potentially hazardous. Power surges in electrical equipment can cause arcs, and a sustained high energy arc can ignite the oil. The unique properties of the PCBs, inertness and fire resistance as well as dielectric insulation, made them ideally suited for those applications where high voltage arcing could occur and result in fires and/or explosions, damage to equipment and hazards to people.

The fire-resistant nature of the polychlorinated biphenyls, terphenyls, and higher polyphenyls combined with outstanding thermal stability made them excellent choices as hydraulic and heat transfer fluids alone or in formulations (see Hydraulic fluids; Heat exchange technology). They were also used to improve the waterproofing characteristics of surface coatings and offered many advantages to the manufacturer of carbonless copy paper, printing inks, plasticizers, special adhesives, lubricating additives, and vacuum pump fluids.

In the late 1960s the first signs of potential environmental problems appeared. A Swedish biologist, using a new and totally unique analytical technique, identified

PCBs for the first time as interference peaks in DDT determinations in the bodies of fish (1). Analytical methods had to be developed to separate the PCBs from other chlorinated hydrocarbons, like DDT, that were known to be present in the environment, and then techniques had to be developed to enable researchers to identify the chemicals in very small amounts (ppb) (see Trace and residue analysis).

In 1968, about 1000 people in Japan became ill from eating rice oil heavily contaminated with several thousand ppm of PCBs as a result of an undetected leak in the heat transfer equipment used to heat the rice oil (2).

When scientific investigations first confirmed the presence of PCBs in the environment in the United States in 1970, and long before information regarding the potential impact of this contamination was determined, Monsanto as the sole U.S. manufacturer voluntarily began a program to terminate sales of PCBs for those applications that were likely to result in environmental contamination. Major applications affected by this withdrawal were in carbonless paper, fire resistant hydraulic fluids, heat transfer fluids and plasticizers. These restrictions resulted in the reduction of the use of PCBs to approximately half its previous level. At the same time, because of the lack of replacement products in the electrical power distribution industry, arrangements were made to continue to supply the PCBs for use in transformers and capacitors. It was felt that this would restrict entry into the environment until satisfactory replacements could be developed for these critical electrical applications.

The steps taken did reduce the amount of potential environmental contamination by the PCBs. However, because of the extremely slow biodegradation rates and the development of improved analytical techniques, PCBs were found in many places as researchers in industry, universities, research institutes, and government laboratories intensified their efforts to identify, remove and determine the long term effects of these materials.

By late 1976, it was apparent that replacement products were becoming available to the electrical industry and Monsanto made arrangements to completely withdraw from the manufacture of the PCBs. Concurrently governmental actions had resulted in restrictions on the use and application of the PCBs and the Toxic Substances Control Act (3) contained provisions for discontinuance of their use and eventual disposal (4–5).

Physical and Chemical Properties

The polychloro polyphenyls can no longer be used commercially. The reader is therefore referred to a number of reviews on their preparation, chemistry, and properties (6–7).

The individual isomers of the chlorobiphenyls vary from liquids to waxes to crystalline solids. However, in the commercial process, mixed isomers are produced which give products having properties quite different, particularly in crystallinity and liquid range, compared to the individual isomers.

Chlorinated biphenyls are considered to be generally chemically inert. However, they react with certain materials under high temperature conditions; eg, with sodium hydroxide under extreme conditions they yield phenolic materials.

The chlorinated biphenyls are insoluble in water, glycerol, and glycols but are soluble in most of the common organic solvents. The chlorinated biphenyl materials are quite resistant to oxidation. They are permanently thermoplastic in the higher chlorination levels and are considered extremely fire resistant. Generally they are compatible with metals making them extremely useful in energy transfer applications. Similar phenomena made them useful as extreme pressure additives to lubricants. However, under elevated temperatures, the chlorine can react with metal to give corrosion.

The polychlorinated biphenyls have reasonably high dielectric constants, high volume resistivities and dielectric strengths, and low power factors.

Health and Safety Factors (Toxicology)

During the long years of use of the PCBs, much work was done to determine their gross toxicity. Exposures to large concentrations were examined extensively in animal studies as well as skin testing. In general, they were found to be relatively innocuous materials with some cases of irritation to human skin. However, these studies did not indicate that the materials were of such a nature that continued use should be stopped.

Prolonged exposure to PCB vapor at high temperature was known to lead to systemic toxic effects. Inhalation studies on animals indicated that the maximum safe concentration of vapor was in the range of 0.5–1.0 mg of the lower chlorinated biphenyl mixtures per m^3 of air. Therefore, limits were set as to the maximum allowable concentration per 8-h working day of 1.0 mg of the lower chlorinated biphenyl compounds per m^3 of air and 0.5 mg of the more highly chlorinated biphenyl compounds per m^3 of air. Although the chlorinated biphenyls were not normally skin irritants, their solvent action could remove natural protective oils and fats which led to drying and cracking of the skin.

All of this information indicated that normal care should be taken in the use of the PCBs, but until the incidents cited above there were no indications that additional precautions were necessary. As noted, with the development of improved analytical techniques, it was found that small amounts of the PCBs did accumulate in the food chain. Since the PCBs' ability to persist in the environment for long times was not known, they were introduced into the environment by open burning or incomplete incineration, by vaporization from paints, coatings and plastics, by direct entry or leakage into sewage and streams, by dumping of waste materials and by various disposal techniques that did not destroy the material.

The occurrence of PCBs in water in extremely low levels (now measurable) has undoubtedly occurred for long periods of time. PCBs are fat soluble and are stored in the lipids of animals. They resist metabolic changes and tend to be concentrated in animals high in the food chain. It was found that the higher the chlorine content of PCBs, the more stable they are to biodegradation.

Action was taken first to restrict the PCBs to closed applications followed by complete removal from the market and cessation of domestic manufacture. These steps will not completely remove the chlorinated biphenyls from the environment since they will persist because of low degradation rates. Therefore, it is important that all materials containing PCBs be properly disposed of to prevent increasing amounts in the environment; however, some biodegradation does occur, therefore, it can be expected that the concentration will eventually decrease.

The PCBs are destroyed by incineration at high temperatures (>1100°C) with long residence time in properly designed incinerators (qv). This service is best provided by firms specializing in such work. Ordinary incinerators used to dispose of organic matter may not be satisfactory for the disposal of PCBs since they may tend to vaporize the product instead of converting them to carbon dioxide, water and hydrogen chloride.

On April 18, 1978, new regulations became effective in the United States concerning the storage and disposal of PCBs. These regulations specify incineration as the only acceptable method of PCB disposal unless, by reason of the inability to dispose of the waste or contaminated materials in this manner, clearance is obtained from the EPA to dispose of the materials in another way (2,8–9).

The major products still in wide use where disposal is important are capacitors and transformers or other electrical equipment containing PCBs. These units can last for periods up to 20 years or more and, therefore, correct disposal practices should be observed for such equipment. The amount of PCBs present in environmental materials can be determined using ASTM D3304-74, "Standard Method for Analysis of Environmental Materials for Chlorinated Biphenyls."

Uses

An extremely wide range of uses were developed for the polychlorinated polyphenyls before they were removed from the market. The largest application was in the electrical industry for which they were originally developed. Since this use continued after most other applications were stopped, and because of the long life of capacitors and transformers, the ultimate disposal of PCBs from such uses must be controlled. Guidelines for such controls and ultimate disposal have been developed by both the electrical industry and the U.S. government (2,8–9).

Some of the most important uses of PCBs were as follows: As a dielectric medium in transformers, either alone or in blends with other materials such as trichlorobenzene; as the dielectric impregnating medium in capacitors; as plasticizers; as ingredients in lacquers, paints and varnishes and adhesives; as water proofing compounds in various types of coatings; as lubricants or lubricant additives under extreme conditions; as heat transfer fluids; as fire resistant hydraulic fluids; as vacuum pump fluids; and as air compressor lubricants. Their success in such diverse applications was due to their unique blend of fire resistance, thermal and oxidative stability, electrical characteristics, solvency, inertness and liquid range.

BIBLIOGRAPHY

The "Chlorocarbons and Chlorohydrocarbons, Chlorinated Biphenyl and Related Compounds," are treated under "Chlorinated Diphenyls" under "Chlorine Compounds, Organic" in *ECT* 1st ed., Vol. 3, pp. 826–832, by C. F. Booth, Monsanto Chemical Company; for *ECT* 2nd ed., see ref. 6.

1. S. Jensen, *PCB Conference,* National Swedish Environmental Protection Board, Research Secretariat, Solna, Sweden, December 1970.
2. Interdepartmental Task Force on PCBs, *Polychlorinated Biphenyls and the Environment, COM-72-10419,* Washington, D.C., 1972.
3. *Public Law 94-469,* Oct. 11, 1976.
4. *Fed. Regist.* **43**(34), 7150 (1978).
5. *Fed. Regist.* **43**(110), 24802 (1978).
6. H. L. Hubbard, *Chlorinated Biphenyl and Related Compounds, ECT* 2nd ed., Vol. 5, 1964, pp. 289–297.
7. O. Hutzinger, S. Safe, and V. Zitko, *The Chemistry of PCBs,* CRC Press, Cleveland, Ohio, 1974.

8. Versar, Inc., *Final Report, PCBs in the United States: Industrial Use and Environmental Distribution*, Report to U.S. EPA, Task I: Contract No. 68-01-3259, 1976.
9. *Proceedings of the National Conference on Polychlorinated Biphenyls, Chicago, Ill., 19–21, Nov. 1975*, EPA-560/6-75-004, 1976.

ROGER E. HATTON
Monsanto Company

CHLOROCARBONS AND CHLOROHYDROCARBONS—VINYL CHLORIDE. See Vinyl polymers.

CHLOROCARBONS AND CHLOROHYDROCARBONS—VINYLIDENE POLYMERS. See Vinylidene chloride.

CHLOROHYDRINS

The chlorohydrins are organic compounds containing one or more chlorine atoms and one or more hydroxyl groups. Chlorohydrins, with the possible exception of Jaconine [480-75-1], are unknown in natural products and are prepared by synthetic methods.

In the systematic nomenclature of chlorohydrins, the alcohol-carbon receives the lowest possible number; the chlorine group, or groups, are then numbered. Examples are given below, together with common names that are in frequent use.

ClCH$_2$CH$_2$OH	CH$_3$CHCH$_2$OH | Cl	CH$_3$CHCH$_2$Cl | OH	ClCH$_2$CH$_2$CH$_2$OH	ClCH$_2$CHCH$_2$O
2-chloro-1-ethanol [107-07-3] (ethylene chlorohydrin)	2-chloro-1-propanol [78-89-7] (propylene β-chlorohydrin)	1-chloro-2-propanol [127-00-4] (propylene α-chlorohydrin)	3-chloro-1-propanol [627-30-5] (trimethylene chlorohydrin)	1-chloro-2,3-epoxypropane [106-89-8] (epichlorohydrin)

Epichlorohydrin, a chloro ether, is included in this article because of its ease of preparation from glycerol 2,3-dichlorohydrin [616-23-9] and because its reactions are similar to those of the glycerol dichlorohydrins (in many cases epichlorohydrin is actually an intermediate). Epichlorohydrin is the only chlorohydrin sold on a large scale.

The 1,2-chlorohydrins are prepared commercially by the addition of hypochlorous acid to olefins (chlorohydrination). Propylene and glycerol chlorohydrins are industrially important.

Chlorohydrins in which the hydroxyl and chlorine groups are not on adjacent carbons (such as trimethylene chlorohydrin), are unimportant commercially since they usually cannot be prepared directly from the olefin. They are synthesized by reaction of dihydric alcohols or cyclic ethers with hydrochloric acid or by partial hydrolysis of dichlorides.

The chlorohydrins are usually liquids of varying viscosity. With the exception of ethylene chlorohydrin, they are not miscible with water and are less soluble than the corresponding glycols. The chlorohydrins are soluble in ethyl alcohol, acetone, and ethyl ether, and moderately soluble in hydrocarbon solvents. Lower molecular weight chlorohydrins can be distilled at atmospheric pressure, and higher molecular weight chlorohydrins *in vacuo*. Their boiling points are between those of the corresponding glycols and dichlorides.

The chlorohydrins are slightly unstable and yellow on standing. Chemically, they behave either as alcohols or chlorides and show the typical reactions of these compounds. The chlorine in 1,2-chlorohydrins is very labile and reacts metathetically with bases through intermediate formation of an epoxide. Dehydrohalogenation of 1,2-chlorohydrins to epoxides, and hydrolysis to glycols, occur readily (see Epoxidation).

Bromohydrins and iodohydrins can be prepared similarly to chlorohydrins. For example, an olefin reacts with iodine in the presence of water and a suitable oxidizing agent, such as oxygen catalyzed by nitrous acid (1). In addition, iodohydrins can be obtained readily from the parent chlorohydrin by reaction with alkali metal iodides. Fluorohydrins are prepared by partial hydrolysis of chloro- or bromofluorides since the fluorine group hydrolyzes more slowly. The halogen atoms of the bromo- and iodohydrins are more reactive than those of the corresponding chlorohydrins. Thus, the bromohydrins and iodohydrins are often used as intermediates in organic synthesis.

Chlorohydrins are generally prepared by addition of hypochlorous acid to olefins (chlorohydrination) or by reaction of hydrochloric acid with epoxides or glycols. The addition of the chlorine occurs preferentially at the least substituted carbon.

The hypochlorous acid is usually prepared *in situ;* the olefin and the chlorine are added to the water simultaneously. Major side reactions that occur during chlorohydrination are chlorination (formation of dichlorides) and formation of chloro ethers. The chlorination can be inhibited by addition of ferric or cupric chlorides, by carrying out the chlorohydrination in the presence of air, and by minimizing contact between olefin and chlorine. Ether formation can be minimized by avoiding high chlorohydrin concentrations. By-product formation becomes pronounced when liquid olefins are used. In this case yields are improved by highly dispersing the olefin; styrene chlorohydrin [1674-30-2] is prepared in 78% yield if styrene is emulsified with a suitable detergent. The use of solvents such as acetone or pyridine during chlorohydrination has been described (2–3).

By-product formation can be minimized in the laboratory by using "bound" hypochlorous acid such as monochlorourea or *tert*-butyl hypochlorite. Industrially, these synthetic methods have not been used, since they are usually too costly except for specialty items. However, a patent claiming the preparation of propylene chlorohydrin by chlorohydrination of propylene with *tert*-butyl hypochlorite was issued in 1976 (4).

Analytical methods for a variety of chlorohydrins have been reviewed (5).

ETHYLENE CHLOROHYDRIN

Ethylene chlorohydrin, ClCH$_2$CH$_2$OH, formerly an important intermediate in the manufacture of ethylene oxide, is no longer produced commercially in the United States. Its preparation and reactions apply, in most cases, to chlorohydrins generally.

Ethylene chlorohydrin is a colorless liquid possessing an ether-like odor. It is miscible with water and many organic liquids and dissolves compounds ranging from inorganic salts to cellulose ethers.

Physical Properties

Physical properties of ethylene chlorohydrin are shown in Table 1. The vapor pressure at 17.5–127.5°C is defined by the following Antoinone equation:

$$\log P(\text{kPa}) = 6.96431 - 2727.4/T + 1.75 \log T - 0.0067805\, T$$

(To convert kPa to mm Hg, multiply by 7.5.)

Chemical Properties

The hydroxyl group undergoes most reactions characteristic of alcohols, such as esterification (qv) or ether formation. The chlorine function undergoes most of the reactions characteristic of alkyl chlorides. The reaction with ammonia to give a mixture of mono, di-, and triethanolamines has been reviewed (6) and studied (7) (see Alkanolamines).

Metathesis reactions between ethylene chlorohydrin and alkali metal salts yield a variety of products as shown in the following equation:

$$\text{ClCH}_2\text{CH}_2\text{OH} + \text{MX} \rightarrow \text{XCH}_2\text{CH}_2\text{OH} + \text{MCl}$$

where M = Na, K, etc, and X = F (8), Br (9), CN (10), or S.

Table 1. Physical Properties of Ethylene and Propylene Chlorohydrins

Property	2-Chloroethanol	2-Chloro-1-propanol	1-Chloro-2-propanol
mp, °C	−62.6		
bp, °, 101.3 kPa[a]	128.7	133–134	127.4
10.7 kPa[a]			78–81
1.33 kPa[a]	29		
flash point, °C (CC)	57.2		51.7
density, g/mL	1.2045 (d$_{20}^{20}$)	1.1092 (d$_4^{20}$)	
heat of vaporization, J/g[b], at 126.5°C	556		
surface tension, mN/m[c]	38.9		32.1
n_D^{20}	1.44197	1.437, 1.4390	1.4387, 1.4394
dipole moment, C·m[d]	5.84 × 10^{-30}		
viscosity, mPa·s[e]	3.43		4.67

[a] To convert kPa to mm Hg, multiply by 7.5.
[b] To convert J to cal, divide by 4.184.
[c] mN/m = dyn/cm.
[d] To convert C·m to D, divide by 3.336 × 10^{-30}.
[e] mPa·s = cP.

Replacement of the chlorine group by fluorine or bromine is enhanced by employing ethylene glycol solvent at elevated temperatures. The yield of 2-hydroxypropionitrile obtained from the reaction involving NaCN is improved by addition of a catalytic amount of CuCN (11).

In a number of reactions both the chlorine and hydroxyl groups are involved. Hydrolysis is characteristic of these reactions, as shown below, although this is not readily apparent:

$$ClCH_2CH_2OH + H_2O \xrightarrow{NaHCO_3} OHCH_2CH_2OH$$

At 105°C in sodium bicarbonate solution, hydrolysis to ethylene glycol is complete in 1 h. In water alone, the reaction is 15% completed at 97°C in 12 h (12–13). The high rate of hydrolysis of ethylene chlorohydrin as compared to alkyl chlorides, eg, ethylene dichloride, is readily explained by the intermediate formation of ethylene oxide by internal displacement of the chlorine atom by the alkoxide ion. Ethylene oxide may be removed continuously as formed (14–19).

Preparation and Manufacture

Preparation of ethylene chlorohydrin can be accomplished by one of several methods. Addition of hydrochloric acid to ethylene oxide (20), prepared by direct oxidation, is well suited for a laboratory procedure.

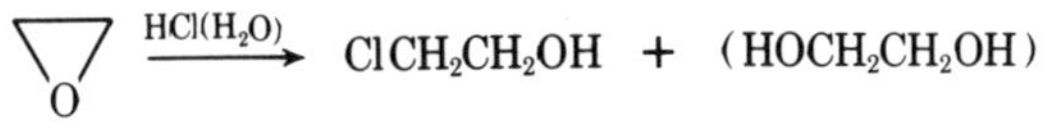

$$\underset{O}{\triangle} \xrightarrow{HCl(H_2O)} ClCH_2CH_2OH + (HOCH_2CH_2OH)$$

Formation of ethylene glycol by reaction with water is minimized by the addition of a soluble chloride (21) or by the reaction of anhydrous hydrogen chloride with ethylene oxide in the presence of ferric chloride and NaH_2PO_4 (22).

In the manufacture of ethylene chlorohydrin, dilute hypochlorous acid reacts with ethylene gas (23).

$$CH_2{=}CH_2 + HOCl \rightarrow ClCH_2CH_2OH$$

As established by Gomberg and others (24–26), hypochlorous acid is best produced by reaction of chlorine with water (see Bleaching agents; Chlorine oxygen acids). The reaction is an equilibrium and may be shifted to the right by using a large amount of

$$Cl_2 + H_2O \rightleftarrows HOCl + HCl$$

water or neutralization of HCl. This displacement of the equilibrium position also serves to minimize the concentration of chlorine, which reacts with ethylene to give ethylene chloride as a by-product.

Many types of process equipment have been described for carrying out the chlorohydrination of ethylene. In one procedure a dispersion of chlorine and ethylene is bubbled through a tower containing water (27). The exotherm is controlled by means of a cooling coil. Chlorine is added at the proper rate to ensure its complete reaction, minimizing formation of ethylene chloride and the need for chlorine recycling. The ethylene chlorohydrin is recovered by distillation as the water azeotrope. Suitable acid-resistant materials of construction that are used in the process to guard against corrosion include brick-lined towers with rubber backing, and piping constructed of

852   CHLOROHYDRINS

glass, ceramics, high-silica iron, tantalum, or Hastelloy alloys. A flow diagram is given in reference 28.

Anhydrous or 98% ethylene chlorohydrin is no longer commercially available in the United States.

Health and Safety Factors

Inhalation or contact with the skin should be carefully avoided because of the toxic properties of ethylene chlorohydrin (29–30). Concentrations in the air should be held below 5 ppm. Aspiration of higher concentrations results in severe, sometimes fatal, damage to the internal organs and nervous system (30–31). Rubber gloves afford little protection because harmful amounts can pass through rubber and be absorbed by the skin (32). In case of accidental contact with ethylene chlorohydrin, wet clothing should be removed at once, and the skin areas should be washed with soap and water.

Uses

Within the chemical industry a revolution has occurred that has altered the role of ethylene chlorohydrin as the principal source of ethylene oxide (qv). The latter is used extensively for the production of ethylene glycol and glycol ethers (see Glycols; Ethers) and as an intermediate in a variety of applications. Most, if not all, of the ethylene oxide used today is produced by the direct oxidation of ethylene over silver catalysts. Facilities for ethylene chlorohydrin have, in many cases, been converted to the production of propylene chlorohydrin, the dehydrochlorination of which yields propylene oxide (qv), a useful chemical that is not readily obtained by direct air oxidation of propylene.

PROPYLENE AND POLYMETHYLENE CHLOROHYDRINS

Propylene Chlorohydrins

1-Chloro-2-propanol (propylene α-chlorohydrin) is a colorless liquid of a faint ethereal odor. It forms an azeotrope with water boiling at 95.4°C at 101 kPa (1 atm) and contains 54.2 wt % chlorohydrin. 2-Chloro-1-propanol (propylene β-chlorohydrin), forms an azeotrope with water boiling at 96°C and containing 15.15 mol % propylene chlorohydrin. Other physical properties of the propylene chlorohydrins are listed in Table 1.

A number of methods are available for the preparation of propylene chlorohydrin mixtures; the acid-catalyzed reaction of allyl chloride (see Chlorocarbons) with water (33–34) gives 1-chloro-2-propanol free of 2-chloro-1-propanol

$$CH_2{=}CHCH_2Cl \ + \ H_2O \ \xrightarrow{H_2SO_4} \ CH_3\underset{\underset{OH}{|}}{C}HCH_2Cl$$

2-Chloro-1-propanol (propylene β-chlorohydrin) is prepared by the addition of hydrochloric acid to allyl alcohol (allyl chloride is obtained as a by-product) (35) (see Allyl compounds).

$$CH_2{=}CHCH_2OH \ + \ HCl \ \longrightarrow \ CH_3\underset{\underset{Cl}{|}}{C}HCH_2OH$$

The propylene chlorohydrin isomers may be separated by distillation at ca 6.7 kPa (50 mm Hg) (36). Mixed chlorohydrin isomers are obtained by reactions involving propylene glycol or propylene oxide with hydrogen chloride (36–37).

Industrially, propylene chlorohydrin (mixed isomers) is prepared by chlorohydrination of propylene (37) using facilities similar to those described for ethylene chlorohydrin.

Propylene chlorohydrin undergoes many reactions characteristic of alcohols or alkyl halides (38). Dehydrochlorination gives propylene oxide which was formerly entirely manufactured by this route. However, most newer propylene oxide plants are based on the epoxidation of propylene with a hydroperoxide, such as *tert*-butyl hydroperoxide. Propylene oxide, an intermediate of major commercial significance, is used in the production of polyurethane foams and other applications; hydration of propylene oxide gives propylene glycol, which is used in unsaturated polyesters, cosmetics, drugs, and food applications (see Glycols).

Trimethylene Chlorohydrin

Trimethylene chlorohydrin (3-chloro-1-propanol) is a colorless oil of agreeable odor which decomposes with discoloration and liberation of hydrogen chloride on distillation at ordinary pressures (see Table 2).

Trimethylene chlorohydrin is best prepared in the laboratory by the action of hydrogen chloride on trimethylene glycol (39). A process involving the interaction of ethylene, formaldehyde, and hydrogen chloride using zinc chloride catalyst has been described (40) for the synthesis of trimethylene chlorohydrin.

A variety of reactions has been reported (38) for trimethylene chlorohydrin. The most important ones involve the formation of cyclopropane and trimethylene oxide (41) derived from the chlorohydrin via the dichloride and chloroacetate, respectively.

Tetramethylene Chlorohydrin

Tetramethylene chlorohydrin [928-51-8] (4-chloro-1-butanol) is a colorless, oily liquid (see Table 2). Attempted distillation at higher temperatures results in decomposition to hydrogen chloride and tetrahydrofuran, $\underline{C}H_2CH_2CH_2CH_2\underline{O}$ (42).

The chlorohydrin is most conveniently prepared in 96% yield by the reaction of hydrogen chloride with tetrahydrofuran in the presence of zinc chloride catalyst (43). A process which gives tetramethylene chlorohydrin as one of the major products involves the reaction of ethyl chloride with ethylene oxide using aluminum chloride catalyst (44). Frequently, tetramethylene chlorohydrin is generated from tetrahy-

Table 2. Physical Properties of Trimethylene and Tetramethylene Chlorohydrins

	3-Chloro-1-propanol	4-Chloro-1-butanol
bp, °C, at kPa[a]		
101.3	165	
16		84–85
1.93	64–66	
density, d_4^{20}, g/mL	1.1318	1.0083
n_D^{20}	1.4459, 1.4485	1.4518

[a] To convert kPa to mm Hg, multiply by 7.5.

drofuran and allowed to react *in situ* with a derivative-forming reagent as, for instance, in the formation of esters (45–46).

MISCELLANEOUS CHLOROHYDRINS

Detoeuf (47) reports the chlorohydrination of several olefins such as 2-octene, cyclohexene, and others.

Chlorohydrination of unsaturated ethers occurs readily; thus diallyl ether yields di(3-chloro-2-hydroxypropyl) ether, bp (266 Pa or 2 mm Hg) 138–139°C.

Unsaturated aldehydes react with hypochlorous acid; acrolein yields 2-chloro-3-hydroxypropionaldehyde [28598-66-5] (48) and crotonaldehyde (49) yields 2-chloro-3-hydroxybutyraldehyde.

Unsaturated fatty acids react with *tert*-butyl hypochlorite to give the chlorohydrins of the corresponding acids (50).

Chlorohydrin ethers can be prepared by reaction of chloroalkoxy derivatives with olefins. The olefin reacts with chlorine in an alcohol solution; thus, ethylene in ethanol gives an 85% yield of the ethyl ether of ethylene chlorohydrin (chloroethyl ethyl ether) (51).

Styrene chlorohydrin [1674-30-2] $C_6H_5CHOHCH_2Cl$, (bp at 0.8 kPa or 6 mm Hg 110–111°C), is prepared in 85% yield (52) from styrene and *tert*-butyl hypochlorite, or by reaction of styrene with chlorine and water in the presence of an emulsifying agent (53).

Excellent reviews of the various methods of preparing chlorohydrins are references 3 and 54.

GLYCEROL CHLOROHYDRINS

The glycerol chlorohydrins are esters of hydrochloric acid and glycerol (qv). Both the mono- and dichlorohydrins occur in two isomeric forms (55).

$ClCH_2CHCH_2OH$	$HOCH_2CHCH_2OH$	$ClCH_2CHCH_2Cl$	$ClCH_2CHCH_2OH$
OH	Cl	OH	Cl
α-monochlorohydrin	β-monochlorohydrin	α,γ-dichlorohydrin	α,β-dichlorohydrin

The chlorohydrins of glycerol are important intermediates, because of their reactivity and their many functional groups. The monochlorohydrins find applications in the synthesis of glycerol derivatives (esters, amines, etc). Almost all of the dichlorohydrins produced from allyl chloride find a captive use in the production of epichlorohydrin.

GLYCEROL MONOCHLOROHYDRINS

α-Monochlorohydrin [96-24-2] (3-chloro-1,2-propanediol) is a colorless, slightly viscous liquid of faint, pleasant odor which freezes to a glass at liquid-air temperatures. It partially decomposes on heating to 140°C. Other physical properties are shown in Table 3.

Table 3. Physical Properties of Glycerol Mono- and Dichlorohydrins

	3-Chloro-1,2-propanediol	2-Chloro-1,3-propanediol	1,3-Dichloro-2-propanol	2,3-Dichloro-1-propanol
bp °C at kPa[a]				
101.3	213		175	182
4.3			86	
2.4		146		
1.9	119–119.5	124.5–125		
1.8				81–81.5
1.6			75	
0.13	83			
0.07	80.9			
flash point, °C (OC)	138			91
density, d_4^{20}, g/mL	1.3204	1.4831	1.3645	1.3607, 1.3616
heat of combustion, kJ/g[b]	1680 ± 0.5	1700.7	13.2	13.33
n_D^{20}	1.4810		1.48375	1.4819, 1.48491
viscosity, mPa·s[c]	159	300		
soly in water, 25°C		moderate	15.6 g/100 g	12.7 g/100 g

[a] To convert kPa to mm Hg, multiply by 7.5.
[b] To convert kJ to kcal, divide by 4.184.
[c] mPa·s = cP.

β-Monochlorohydrin [497-04-1] (2-chloro-1,3-propanediol) is a colorless viscous liquid. It is miscible with acetone, lower alcohols, and ethers, and is moderately soluble in benzene (see Table 3).

Reactions

Most reactions of the glycerol monochlorohydrins have been reportedly carried out with the α-monochlorohydrin, although a mixture of the two isomers may have been present in many cases.

The hydroxyl groups of glycerol monochlorohydrin undergo the typical reactions of alcohols. They may be esterified with organic and inorganic acids. Borates, arsenates, and phosphates are readily formed from boric acid, arsenious acid, and phosphorus oxychloride, respectively (28).

As a glycol, glycerol monochlorohydrin reacts with ketones and aldehydes to form cyclic acetals (1,3-dioxolanes).

ClCH$_2$CHCH$_2$OH + RR'CHO → [dioxolane with CH$_2$Cl, R, R' substituents] + H$_2$O
       |
       OH

The reaction with formaldehyde can be used to extract the chlorohydrin from an aqueous solution (56). Addition of hydrogen chloride to a suspension of trioxymethylene (trimer of formaldehyde) in α-monochlorohydrin results in α-monochlorohydrin-β,γ-bis(chloromethyl) ether.

ClCH$_2$CHCH$_2$OH + (CH$_2$O)$_3$ $\xrightarrow{HCl}$ ClCH$_2$CHCH$_2$OCH$_2$Cl
       | |
       OH   OCH$_2$Cl

856   CHLOROHYDRINS

When the chlorine group is replaced with an organic group under alkaline conditions, glycidol is the intermediate (transitory in many cases). Since the epoxide linkage is preferentially opened to form α-substituted glycol derivatives, α-substituted glycerols are the main or exclusive products regardless of the structure of the monochlorohydrin. Originally, this mechanism was not known, and the often reported syntheses of β substituted glycerols from β-monochlorohydrins should be viewed with suspicion since the α-derivative was probably obtained in all cases.

Glycidol (2,3-epoxy-1-propanol) may be isolated from the products of mild saponification of either isomeric chlorohydrin with alkaline catalysts; α-chlorohydrin hydrolyzes faster than the β isomer. Glycidol is commercially available in development quantities. Further hydrolysis of glycidol yields glycerol.

According to the above scheme, reactions of glycerol chlorohydrin with phenols yield α-phenyl ethers; with alcoholates, α-alkyl ethers; with acid salts, the α esters; with ammonium hydroxide, 3-amino-1,2-propanediol; and with amines, the corresponding 3-amino derivatives.

GLYCEROL DICHLOROHYDRINS

α,γ-Dichlorohydrin [96-23-1] (1,3-dichloro-2-propanol) is a colorless, slightly viscous liquid with a rather sweet odor; the dielectric constant is ca 11.7 (γ = 60 cm) and ca 15.5 (γ = 600 cm) at 20°C, and 11.95 (γ = 60 cm) at 19.6°C. It is miscible with alcohols, acetone, benzene, and ether, and immiscible with petroleum ether. Other physical properties are shown in Table 3. It forms many azeotropes (57).

α,β-Dichlorohydrin [616-23-9] (2,3-dichloro-1-propanol) is a colorless slightly viscous liquid. Its dielectric constant is ca 10.2 (γ = 60 cm) and about 16.1 (γ = 600 cm) at 20°C, and solubility in water, 25°C, 12.7 g/100 g of solvent. Its miscibility with organic solvents is the same as that of α,γ-dichlorohydrin (see Table 3). For azeotropes, see reference 57.

Reactions

Reactions of the glycerol dichlorohydrins, as with those of the monochlorohydrins, can be divided into three categories—reactions involving both the chlorine and hydroxyl groups, and reactions involving either the hydroxyl or the chlorine.

Most reactions involving both hydroxyl and chlorine are base catalyzed and epichlorohydrin is formed as the (generally transitory) intermediate. In fact, most of the reactions can be more conveniently carried out with epichlorohydrin, which is readily available. In this equation, B can be OH, RO, C$_6$H$_5$O, etc:

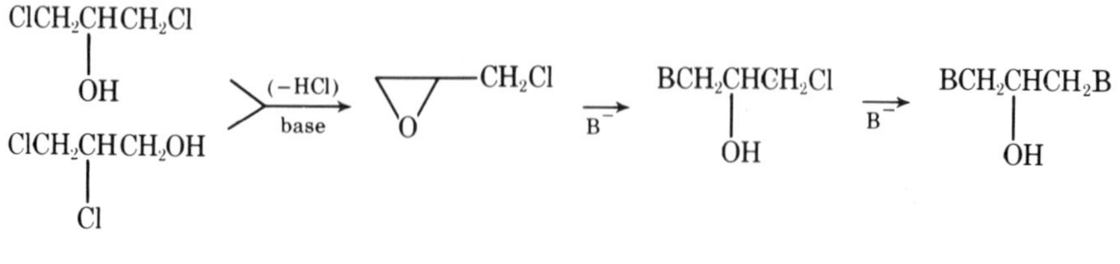

The most important synthetic application of the dichlorohydrins, developed by Shell, is the formation of epichlorohydrin which, by hydrolysis, yields glycerol monochlorohydrin and glycerol (B = OH). With ammonia (B = NH$_2$), 1,3-diamino-2-

propanol is formed; amines yield 1,3-diamino derivatives, and caustic in excess methanol (B = CH$_3$O$^-$) yields glycerol α,γ-dimethyl ethers. With acid salts (B = RCOO$^-$), α,γ-diesters are formed. In addition to these examples, numerous other derivatives can be made by this reaction scheme.

The hydroxyl group of the dichlorohydrin can undergo the typical reactions of alcohols, such as esterification with organic acids or acid chlorides.

The chlorine groups react under proper conditions with sodium azide to form mono- and diazido derivatives. They react with sodium tetrasulfide to form sulfur-containing polymers, which have found no commercial applications.

The p-nitrobenzoate of glycerol α,γ-dichlorohydrin melts at 55–60°C; the derivative of glycerol α,β-dichlorohydrin melts at 35.5–37°C.

Manufacture

Glycerol was the main source for the preparation of the chlorohydrins until the process for direct substitutive chlorination of propylene (qv) to allyl chloride paved the way for the synthesis of chlorohydrins by chlorohydrination of allyl chloride.

From Propylene. As mentioned previously, the chlorohydrination of allyl chloride yields a mixture of dichlorohydrins which can be used in the manufacture of epichlorohydrin and glycerol.

The chlorohydrination yields a mixture of approximately 70% α,β- and 30% α,γ-glycerol dichlorohydrins. The crude mixture is used in the manufacture of epichlorohydrin.

Since the solubility of allyl chloride in water is small, side reactions that can occur in the nonaqueous phase (such as chlorination) have to be minimized by choosing suitable conditions. In the continuous process, the amount of water insoluble allyl chloride in the reaction mixture is minimized.

An aqueous hypochlorous acid solution is brought into intimate contact with allyl chloride which is in fine dispersion in water. After the first reaction the mixture is taken to the separator where two phases form. The aqueous layer is recycled and more chlorine is added to generate hypochlorous acid.

The presence of chloride ions in the chlorohydrination mixture also leads to increased formation of by-products such as 1,2,3-trichloropropane. This can be avoided by distilling the hypochlorous acid solution, or, in the laboratory, by the use of a chloroamide such as chlorourea, or *tert*-butylhypochlorite.

By-product chloroethers result mainly from high chlorohydrin concentrations.

Many improvement patents have been published in the last decade, and a review article on chlorohydrination was published in France in 1964 (3).

Other Syntheses. Pure glycerol α-monochlorohydrin may be obtained from the sulfuric acid-catalyzed hydrolysis of epichlorohydrin. An elegant specific synthesis of the α-monochlorohydrin consists of hydroxylating allyl chloride with hydrogen peroxide in the presence of osmium tetroxide catalyst (58). The α-monochlorohydrin may be obtained in good purity by the addition of hydrochloric acid to glycidol. By a similar reaction, addition of hydrochloric acid to epichlorohydrin yields 99.6% α,γ- and 0.4% α,β-glycerol dichlorohydrins at -10°C. The mechanism of this reaction and the glycidol-hydrochloric acid reaction is discussed in detail in reference 59.

Glycerol α,β-dichlorohydrin may be obtained in ca 90% yield by chlorination of acrolein, followed by reduction, or by the chlorination of allyl alcohol (60–61).

858 CHLOROHYDRINS

Health and Safety Factors

The glycerol mono- and dichlorohydrins have low vapor pressures at moderate temperatures, but care should be taken to avoid exposure by inhalation. No maximum permissible concentrations have been set. However, since dichlorohydrins readily form epichlorohydrin, workers using glycerol dichlorohydrins (and monochlorohydrins) should take the same precautions that are recommended for epichlorohydrin. The chlorohydrins are harmful when taken internally and are absorbed through the skin. Both mono- and dichlorohydrins have been shown by animal experiments to have a narcotic and depressant effect on the heart, circulation, and respiration.

Uses

The glycerol monochlorohydrins are available as specialty products. One manufacturer (Dixie Chemicals, Houston, Texas) quoted ca $4.50/kg (1977) for development quantities. No current use for the monochlorohydrin is known.

The glycerol dichlorohydrins are produced on a large scale for captive use; they are hydrolyzed to epichlorohydrin. Manufacturers could probably offer the dichlorohydrin in large quantities, if desired. Since most reactions of glycerol dichlorohydrins can be carried out with epichlorohydrin, the latter compound, which is available in 99% purity, has replaced the dichlorohydrins in these applications. Both the glycerol mono- and dichlorohydrins have good solvent properties; however, their toxicities preclude their use in most cases.

EPICHLOROHYDRIN

Physical Properties

$$\underset{O}{\triangle}\!\!-CH_2Cl$$

Epichlorohydrin (62) (chloromethyl oxirane) is a colorless, mobile liquid with an irritating, chloroform-like odor; mp, −57.2°C; bp: 30–32°C (at 1.3 kPa or 10 mm Hg), 116.11 (at 101.3 kPa or 760 mm Hg); dt/dp at 116.11°C, 0.33°C/kPa; coefficient of expansion at 20°C, 0.00104 per °C; d_4^{20} 1.18066; n_D^{20} 1.43805; latent heat of vaporization (calcd), 37.9 kJ/mol (9.06 kcal/mol) at the bp; heat of combustion 18.943 kJ/g (4.50 kcal/g), flash point (Tag open cup), 40.6°C; specific conductance, 34 × 10^{-9} S/cm at 25°C; dielectric constant (λ = 60 cm), 20.8 at 21.5°C; viscosity, 1.03 mPa·s (= cP) at 28°C; vapor density (air = 1, at boiling point of epichlorohydrin) 3.19; and autoignition temperature 415.6°C. The flammable limits in air (vol %) are: lower 3.8, upper 21.0, and the evaporation rate (butyl acetate = 1) is 1.35. Epichlorohydrin is soluble in most organic solvents, the solubility in water is 6.6 wt % at 20°C. It forms azeotropes with a wide variety of organic liquids (5,63). The azeotrope with water boils at 88°C and contains 75 wt % epichlorohydrin.

By virtue of the asymmetric carbon, epichlorohydrin exists as a racemic mixture containing equal amounts of dextro- and levorotatory forms. The latter has been reported to racemize upon distillation at atmospheric pressure, but may be distilled unchanged under reduced pressure, bp 92–93°C at 48.0 kPa (360 mm Hg).

Reactions

The epoxide and chlorine groups of epichlorohydrin are both potentially reactive sites. In general, however, transformations of epichlorohydrin occur initially by ring opening of the more reactive epoxide group. In reactions with compounds containing active hydrogens such as alcohols, acids, and amines, epichlorohydrin serves as an excellent starting material for preparation of glycerol α-monochlorohydrins. In the presence of basic reagents, secondary reactions can occur by the elimination of hydrogen chloride to form glycidyl-substituted products. These in turn can undergo further additive reactions. By suitable adjustment of reaction conditions, epichlorohydrin can be an intermediate in the synthesis of a wide variety of products including derivatives of α-monochlorohydrin, dichlorohydrins, glycidol, and glycerol.

Transformations such as these are conveniently illustrated by the reactions of epichlorohydrin with alcohols. In the presence of acidic substances such as stannic chloride (64), glycerol α-chlorohydrin-γ-ethers are formed in high yields.

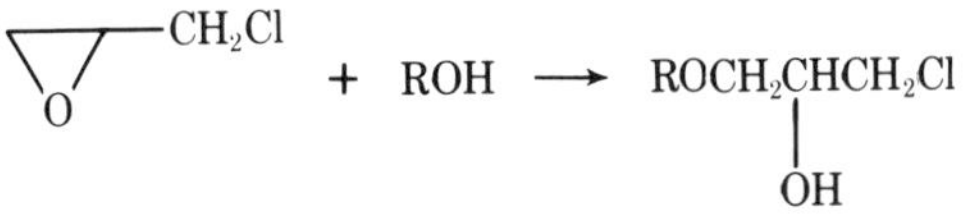

$$\text{epichlorohydrin} + \text{ROH} \longrightarrow \text{ROCH}_2\text{CHCH}_2\text{Cl}$$
$$|$$
$$\text{OH}$$

The monochlorohydrin ether can be converted to the glycidyl ether (65) by reaction with sodium hydroxide or to the glycerol α-monoether (66) by hydrolysis using acid catalysts. The glycidyl ether can in turn be converted to the glycerol α,γ-diether by alcoholysis (64).

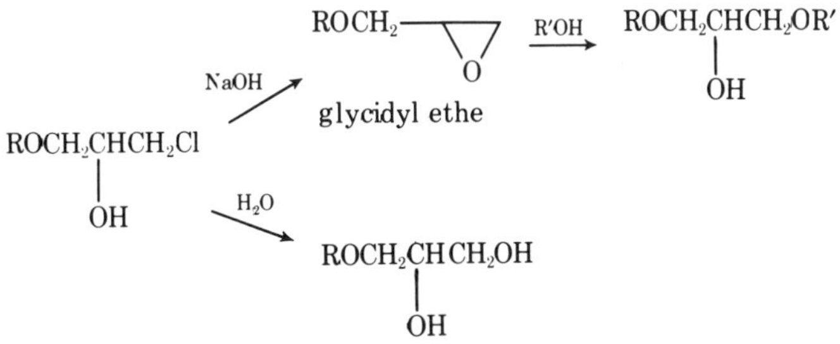

$$\text{ROCH}_2\text{CHCH}_2\text{Cl} \xrightarrow{\text{NaOH}} \text{ROCH}_2\text{–glycidyl ether} \xrightarrow{\text{R'OH}} \text{ROCH}_2\text{CHCH}_2\text{OR'}$$
$$\text{OH} \xrightarrow{\text{H}_2\text{O}} \text{ROCH}_2\text{CHCH}_2\text{OH}$$

Phenols (64) react with epichlorohydrin in a manner analogous to alcohols although under milder conditions. The full range of products from α-chlorohydrin-γ-ethers to glycidyl ethers and glycerol α,γ-diethers can be obtained.

Ammonia and primary and secondary amines react with epichlorohydrin by initial ring opening of the epoxide group. Although the reactions can in some cases be stopped at the chlorohydrin (67) or glycidyl amine stages (68) frequently an excess of amine (69,70) or ammonia (71) is employed to yield the glycerol α,γ-diamino derivatives.

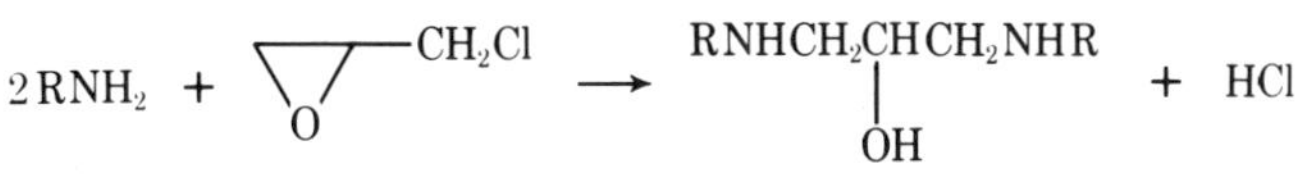

$$2\,\text{RNH}_2 + \text{epichlorohydrin} \longrightarrow \text{RNHCH}_2\text{CHCH}_2\text{NHR} + \text{HCl}$$
$$|$$
$$\text{OH}$$

Epichlorohydrin reacts with carboxylic acids to yield a mixture of α- and β-monoesters and with carboxylic acid anhydrides to form diesters, as shown below:

860 CHLOROHYDRINS

$$RCO_2H + \underset{O}{\triangle}\!\!-CH_2Cl \longrightarrow \underset{\underset{OH}{|}}{RCO_2CH_2CHCH_2Cl} + \underset{\underset{CH_2OH}{|}}{RCO_2CHCH_2Cl}$$
$$\qquad\qquad\qquad\qquad\qquad\qquad \alpha\text{-ester} \qquad\qquad \beta\text{-ester}$$

$$(RCO)_2O + \underset{O}{\triangle}\!\!-CH_2Cl \longrightarrow \underset{\underset{RCO_2CHCH_2Cl}{|}}{RCO_2CH_2}$$

Glycidyl esters are formed directly in high yields by reaction of sodium salts of carboxylic acids with epichlorohydrin (72).

β-Esters of α,γ-dichlorohydrin are obtained by reaction of epichlorohydrin with acid chlorides (73).

Reaction of epichlorohydrin with organometallic compounds leads to the formation of substituted chlorohydrins, as shown in the following equation:

$$RM + \underset{O}{\triangle}\!\!-CH_2Cl \xrightarrow{H_2O} \underset{\underset{OH}{|}}{RCH_2CHCH_2Cl}$$

Products of this general structure were obtained using Grignard reagents (74) and from aryl lithium compounds (75). 2-Penten-4-yn-1-ol was isolated from sodium acetylide in liquid ammonia (76).

Hydration of epichlorohydrin to give glycerol α-monochlorohydrin proceeds slowly at room temperature but is accelerated by heat or traces of acids. Hydrochloric acid opens the epoxide ring of epichlorohydrin to give glycerol α,γ-dichlorohydrin. Hydrobromic and hydroiodic acids behave similarly; hydrofluoric acid converts epichlorohydrin to a polymer. Phosphoric, sulfuric, nitric, and perchloric acids form esters with epichlorohydrin.

Epichlorohydrin is converted to the iodohydrin by alkali metal iodides, and by sodium sulfite, sodium bisulfite, sodium thiosulfate (77), and sodium cyanide (78) to addition compounds.

The diethyl ether of glycerol α-monochlorohydrin and polymeric materials were obtained from the reaction of epichlorohydrin with diethyl ether in the presence of boron trifluoride (79). The allyl ether of glycerol α,γ-dichlorohydrin was isolated (70% yield) by reaction of epichlorohydrin with allyl chloride (80) in the presence of cuprous ions. Addition compounds of epichlorohydrin and trimethylchlorosilane have been described (81,82).

Reactions of epichlorohydrin with aldehydes or ketones using various Lewis acid catalysts give 1,3-dioxolanes.

$$\underset{O}{\triangle}\!\!-CH_2Cl + R\overset{O}{\overset{\|}{C}}R' \xrightarrow{\text{Lewis acid}} \text{1,3-dioxolane with } CH_2Cl, R, R'$$

$$R' = H \text{ or alkyl}$$

Preparation and Manufacture

In the most commonly used process for manufacturing epichlorohydrin, allyl chloride from the high-temperature chlorination of propylene is chlorohydrinated with chlorine water to give a mixture of isomeric glycerol chlorohydrins. This mixture in turn is dehydrochlorinated with an alkali and epichlorohydrin is separated by steam stripping and purified by subsequent distillation. Contact times in all steps must be carefully controlled to minimize hydrolysis of epichlorohydrin. Relatively few improvement patents have been issued in the last two decades (83–89).

Another method for producing epichlorohydrin (90) starts with the chlorination of acrolein to give 2,3-dichloropropionaldehyde. Reduction of the aldehyde with *sec*-butyl alcohol using aluminum *sec*-butoxide catalyst (Meerwein-Ponndorf) gives glycerol β,γ-dichlorohydrin. Dehydrochlorination with lime forms epichlorohydrin.

Epichlorohydrin can also be prepared by the epoxidation of allyl chloride with peracids (91,92,93), perborates (94), or by epoxidation with *tert*-butyl hydroperoxide over vanadium (95), tungsten (95), or molybdenum (95–96) compounds or by oxidation with air over a cobalt catalyst (97).

July 1977 list prices were $0.93/kg for car lots, delivered, and $1.00/kg in car lot drums, delivered. U.S. production in 1973 was estimated at 157,000 metric tons.

Health and Safety Factors (Toxicology)

Epichlorohydrin (62) is a toxic, severely irritating compound with a chloroform-like odor. OSHA does not permit employee exposure to concentrations exceeding 5 ppm. NIOSH recommends lowering the limit to 0.5 ppm. A 5 ppm concentration is not detectable by odor; at 25 ppm the odor is recognized by most people. If contacted with the skin or eyes, the liquid can cause severe burns and permanent injury and may cause lasting liver, lung, and kidney injury. Some people become sensitized on exposure to very small quantities. In addition, a moderate carcinogenic activity has been demonstrated in rats and epichlorohydrin is thus a suspect human carcinogen according to the American Conference of Governmental Industrial Hygienists (ACGIH) (98).

Epichlorohydrin is flammable and can form explosive mixtures with air at elevated temperatures within certain limits. Epichlorohydrin may polymerize and burst its container when heated in a fire. Toxic gases such as hydrogen chloride and, in certain circumstances, carbon monoxide and phosgene are released in a fire involving epichlorohydrin. It is classified as a 1C flammable liquid.

Because of the tendency of epichlorohydrin to react violently, exploratory syntheses using this compound should be conducted on a small scale, with adequate shielding and with provision for rapid cooling. High concentrations of epichlorohydrin in the reaction mixture can be avoided by adding it slowly as the reaction progresses or by the use of inert solvents. Epichlorohydrin is unusually sensitive to inorganic bases, mineral acids, and metal halides, such as stannic or ferric chlorides, and may react violently in their presence.

Uses

Manufacture of glycerol (qv) and epoxy resins(qv) consume most of the epichlorohydrin currently produced. Smaller volumes are used in a variety of applications including synthesis of glycerol and glycidol derivatives which serve as intermediates for plasticizers, dyestuffs, surfactants, and pharmaceuticals. In addition, epichlorohydrin is used in the manufacture of castings, adhesives, stabilizers for chlorine-containing materials, anion-exchange resins, polymers and paper-sizing agents.

BIBLIOGRAPHY

"Epichlorohydrin" in *ECT* 1st ed., Vol. 3, pp. 865–869, P. H. Williams, Shell Development Company; "Glycerol Chlorohydrins" in *ECT* 1st ed., Vol. 3, pp. 857–865, P. H. Williams, Shell Development Company; "Propylene, Trimethylene, and Tetramethylene Chlorohydrins" in *ECT* 1st ed., Vol. 3; pp. 856–857, M. G. Gergel and Max Revelise, Columbia Organic Chemicals Company; "Chlorohydrins" in *ECT* 2nd ed., Vol. 5, pp. 304–324, G. D. Lichtenwalter and G. H. Riesser, Shell Chemical Company.

1. J. W. Conforth, D. T. Green, *J. Chem. Soc. C* (6) 846–9 (1970).
2. A. Guver, A. Bieler, and E. Pedrazetti, *Helv. Chim. Acta* **39,** 423 (1956).
3. J. Myszkovski and co-workers, *Chimie et Industrie* **91,** 654 (1964).
4. OLS (Germ. Pat. Discl. 2,541,526 (April 1, 1976), A. Gelbein, J. T. Know (to Lummus Co.).
5. G. H. Riesser and J. G. Riesser, "Halohydrins," *Encyclopedia of Chemical Analysis,* John Wiley & Sons, Inc., Vol. 14, 1971, pp. 153–178.
6. F. L. Resen, *Oil Gas J.* **51**(16), 102 (1952).
7. V. A. Krishnamurthy and M. R. A. Rao, *J. Indian Inst. Sci.* **40,** 145 (1958).
8. H. Kitano, *J. Chem. Soc. Japan* **58,** 119 (1955).
9. K. Fukui and co-workers, *J. Chem. Soc. Japan* **58,** 600 (1955).
10. E. C. Kendall and B. McKenzie, *Organic Syntheses,* Vol. 1, John Wiley & Sons, Inc., New York, 1941, p. 256.
11. Jpn. Pat. 153,924 (Nov. 28, 1942), (to Kanegabuchi Textile Co.).
12. U.S. Pat 1,442,386 (Jan. 16, 1923), G. O. Curme and C. O. Young (to Union Carbide Co.).
13. U.S. Pat. 1,695,250 (Dec. 11, 1928), G. O. Curme (to Carbide and Carbon Chemicals Corp.).
14. Brit. Pat. 286,850 (March 15, 1928), K. H. Saunders and H. Signall (to British Dyestuff Corp.).
15. Ger. Pat. 299,682 (March 3, 1920), (to Badische Anilin- und Soda-Fabrik A.G.).
16. U.S. Pats. 1,446,872–1,446,874 (Feb. 27, 1923), B. T. Brooks (to Chandeloid Chemical Co.).
17. U.S. Pat. 1,589,358 (June 22, 1926), J. N. Burdick (to Carbide and Carbon Chemicals Corp.).
18. U.S. Pat. 1,792,668 (Feb. 17, 1931), J. Weber, H. Schrader, and E. Wiedbrauek (to T. Goldschmidt A.G.).
19. U.S. Pat. 1,986,082 (Jan. 1, 1935), F. B. Thole, S. F. Birch, and W. D. Scott (to Anglo-Persian Oil Co.).
20. G. Forsberg and L. Smith, *Acta Chem. Scand.* **1,** 577 (1947).
21. Brit. Pat. 660,835 (Nov. 14, 1951), H. S. Davies (to American Cyanamid).
22. U.S.S.R. Pat. 130, 502 (Aug. 5, 1960), V. S. Etlis and coworkers.
23. T. Jen, *Union Ind. Research Inst. Rept. (Taiwan),* **35,** (1958).
24. C. Ellis, *The Chemistry of Petroleum Derivatives,* Vol. 2, Reinhold Publishing Corp., New York, 1937, pp. 515–518.
25. H. Tropsch and R. Kassler, *Mitt. Kohlenforsch. Inst. Prag.* **1,** 16 (1931).
26. U.S. Pat. 1,295,339 (Feb. 25, 1919), K. P. McElroy.
27. U.S. Pats. 1,456,916; 1,456,959 (May 29, 1923), G. O. Curme and C. O. Young (to Carbide and Carbon Chemicals Corp.).
28. A. Dupire, *Compt. Rend.* **202,** 2086 (1930); **214,** S2 (1942).
29. E. L. Middleton, *J. Ind. Hyg. Toxicol.* **12,** 265 (1930).
30. H. F. Smyth, J. Seaton, and L. Fischer, *J. Ind. Hyg. Toxicol.* **23,** 259 (1941).
31. A. F. Bush and co-workers, *J. Ind. Hyg. Toxicol.* **31,** 352 (1949).
32. F. Ballota et al., *Brit. J. Ind. Med.* **10,** 161 (1953).
33. A. Devial, *Bull. Soc. Chim. Belges* **39,** S9 (1930).
34. O. U. Magidson and V. M. Fedosova, *Med. Prom. USSR 11* (3), 25 (1957).
35. A. Dewael, *Bull Soc. Chim. Belges* **33,** 504 (1924).
36. G. Forsberg and L. Smith, *Acta. Chem. Scand.* **1,** 57S (1947).
37. N. Koddoo, *Chem. Eng.* **9**(9), 149 (1952).
38. E. H. Huntress, *Organic Chlorine Compounds,* John Wiley & Sons, Inc., New York, 1948.
39. C. S. Marvel and H. O. Calvary, *Organic Syntheses,* Coll. Vol. 1, John Wiley & Sons, Inc., New York, 1948, p. 535.
40. Brit. Pat. 465,467 (Jan. 3, 1937), (to I. G. Farbenindustrie, A.G.).
41. C. R. Noller, *Organic Syntheses,* Vol. 29, John Wiley & Sons, Inc., New York, 1949, p. 92.

42. W. R. Kirner and G. H. Richter, *J. Am. Chem. Soc.* **51,** 2503 (1939).
43. D. Starr and R. M. Hixon, *Organic Syntheses,* Vol. 2, John Wiley & Sons, Inc., New York, 1949, p. 571.
44. Brit. Pat. 354,992 (July 25, 1930), (to I. G. Farbenindustrie, A.G.).
45. Brit. Pat. 642,489 (Sept. 6, 1950), (to British Celanese Ltd.).
46. M. E. Synerholm, *Organic Syntheses,* Coll. Vol. 3, John Wiley & Sons, New York, 1949, pp. 29, 30.
47. A. Detoeuf, *Bull. Soc. Chim. France* **31,** 169 (1922).
48. H. Schulz and H. Wagner, *Angew. Chem.* **62,** 105 (1950); Brit. Pat. 573,720 (Dec. 4, 1945), A. Staudinger and K. W. H. Tuerck (to Distillers Co., Ltd.).
49. G. A. Ropp, W. E. Craig, and U. Raaen, *Organic Syntheses,* Vol. 33, John Wiley & Sons, Inc., New York, 1953, p. 15.
50. Belgian Pat. 538,330 (1955), (to Rohm & Haas.).
51. Ger. Pat. 537,696 Aug. 18, 1928), P. Ernst (to A. Wacker, Gesellschaft für Elektrochemische Industrie); A. K. Selezner, *Zh. Prikl. Khim.* **27,** 650 (1954).
52. W. E. Hanby and N. H. Rvdon, *J. Chem. Soc.,* 114 (1946).
53. Fr. Pat. 1,104,348 (Nov. 18, 1955), (to Degussa G.m.b.H.).
54. Houben-Weyl, *Methodern der organischen Chemie,* Band V/3, Georg Thieme Verlag, p. 768ff (1962).
55. P. B. D. Dela Mave and J. G. Pritchard, *J. Chem. Soc.,* 3990 (1954). L. Bjellerup and L. Smith, *Kgl. Fysiograf. Sällskap Lund. Förk.* **24,** 21 (1954).
56. U.S. Pat. 2,406,713 (Aug. 27, 1946), M. Senkus (to Commercial Solvents).
57. M. Lecat, *Tables Azeotropiques,* Vol. 1, 3rd. suppl., Brussels, 1940; F. K. Beilstein, *Handbuch der organischen Chemie,* Vol. 1, 3rd suppl., 4th ed., Springer-Verlag, Berlin, 1918–1943, pp. 1427.
58. Can. Pat. 440,807 (April 15, 1947), L. Rosenstein (to Shell Development Co.).
59. L. Smith, *Acta Chem. Scand.* **4,** 1375 (1950).
60. OLS (Ger. Pat. Discl.) 2,007,867 (Aug. 21, 1971), D. Freudenberger and H. Fernholz (to Farbwerke Hoechst).
61. U.S. Pat. 2,860,146 (Nov. 11, 1958), K. Fuhrman, H. V. Finch, and G. W. Hearne (to Shell Development Co.); G. P. Gibson, *Chem. Ind.* **50,** 949,970 (1931).
62. *Epichlorohydrin,* Shell Chemical Corp., New York, 1959; *Toxicity and Safety Bulletin,* Shell Chemical, Houston, 1976.
63. L. H. Horsley, *Ind. Eng. Chem., Anal. Ed.* **19,** 50S (1947).
64. U.S. Pats. 2,327,053 (Aug. 17, 1943) and 2,380,185 (July 10, 1945), K. E. Marple, E. C. Shokal, and T. W. Evans (to Shell Development Co.).
65. U.S. Pat. 2,314,039 (March 16, 1943), T. W. Evans, K. E. Marple, and E. C. Shokal (to Shell Development Co.).
66. A. Fairborne, G. P. Gibson, and D. W. Stephens, *Chem. Ind.* **49,** 1021 (1930).
67. F. B. Dains and co-workers, *J. Am. Chem. Soc.* **44,** 2637 (1922).
68. F. Zetzsche and F. Aeschlimann, *Helv. Chim. Acta* **9,** 70S (1926).
69. I. T. Strukov, *Khim. Farm. Prom.,* **11**(2), (1934).
70. C. K. Ingold and E. Rothstein, *J. Chem. Soc.,* 1666 (1931).
71. U.S. Pats. 1,985,885 (Jan. 1, 1935); 2,065,113 (Dec. 22, 1936), R. R. Bottoms (to Girdler Corp.).
72. E. B. Kester, C. J. Gaiser, and M. E. Lazar, *J. Org. Chem.* **8,** 550 (1943).
73. G. S. Whitby, *J. Chem. Soc.* 1458 (1926).
74. H. Normant, *Compt. Rend.* **219,** 163 (1944).
75. H. Gilman, B. Hofferth, and J. B. Honeycutt, *J. Am. Chem. Soc.* **74,** 1594 (1952).
76. L. J. Haynes and co-workers, *J. Chem. Soc.* 1583 (1947).
77. Ger. Pat. 865,597 (Feb. 2, 1953), P. Schlack (to Kunstseidefabrik Bobingen).
78. C. C. J. Culvenior and co-workers, *J. Chem. Soc.* 3123 (1950).
79. H. Meerwein and co-workers, *Ann. Chem. Liebiyz* **566,** 50 (1950).
80. U.S. Pat. 2,608,586 (Aug. 26, 1952), S. A. Ballard, R. C. Morris, and J. L. Van Winkle (to Shell Development Co.).
81. K. A. Adrianov and co-workers, *Bull. Acad. Sci. USSR, Div. Chem. Sci.* 469 (1955).
82. M. S. Malinovskii and M. K. Romantsevich, *Zhur. Obshch. Khim.* **27,** 1680 (1957).
83. Jpn. Pat. 75 16,341 (June 12, 1975), S. Takakuwa, A. Goto, and S. Ogura (to Osaka Soda Co., Ltd.).
84. U.S. Pat. 2,714,122 (July 26, 1955), W. C. Smith, J. Anderson, and J. C. Bloom (to Shell Development Co.); U.S. Pat. 2,714,123 (July 26, 1955), G. F. Johnson (to Shell Development Co.).

85. Ger. Pat. 1,285,993 (Jan. 2, 1968), H. Berthold and K. Funke (to Veb Leuna-Werke "Walter Ulbricht").
86. U.S. Pat. 2,993,077 (July 18, 1961), F. C. Trager (to Pittsburgh Plate and Glass Co.).
87. Fr. Pat. 1,412,886 (Oct. 1, 1965), (to Chemische Werke Huels A.G.).
88. Fr. Pat. 1,328,311 (May 31, 1960), D. Brown (to Scientific Design Co.).
89. S. H. Kazimov, A. S. Rzaeva, and G. Z. Ponomareva, *Khim. Prom.* **49,** 824 (1973).
90. U.S. Pat. 2,860,140 (Nov. 11, 1958), K. E. Furman, H. Fiach, and G. W. Hearne (to Shell Development Co.).
91. Brit. Pat. 784,620 (Oct. 9, 1957), B. Phillips and P. S. Starcher (to Union Carbide Corp.).
92. U.S. Pat. 3,799,949 (March 26, 1974), R. Keller and co-workers (to Degussa).
93. OLS (Ger. Pat. Discl.) 1,942,557 (March 18, 1971), A. Kleemann and co-workers (to Degussa).
94. Fr. Pat. 1,447,267 (June 1, 1965), J. C. Bruenie and N. Crenne (to Société Chimique des Usines Rhône-Poulenc).
95. Jpn. Pat. 70 17,645 (June 18, 1970), S. Sakan, M. Sano, and K. Hattori (to Japanese Chemical Industries).
96. L. A. Oshin, G. A. Shakhovtseva and B. E. Krasotkina, *Neftekhimija* (USSR), **15,** 281 (1975).
97. I. Ya. Mokrousva, L. A. Oshin, and Y. A. Tregar, *Kinet. Katal.* **17**(2), 515 (1976).
98. *Chem. Week* **17** (Aug. 16, 1978).

GREGOR H. RIESSER
Shell Development Company

CHLOROPHENOLS

Physical Properties

All of the nineteen possible chlorinated phenols are commercially available. Table 1 lists a few of the pertinent physical properties.

Commercially, 2,4-dichlorophenol and pentachlorophenol are the most important. Also important are 2,4,5-trichlorophenol, and *ortho-* and *para-*chlorophenol. The first three will be dealt with in considerably more detail than the others.

Monochlorophenols

Three isomers exist for monochlorophenol [*25167-80-0*]. Currently the 2- and 4-monochlorinated phenols are utilized in modest volumes. Main uses are: as intermediates in dyestuffs and in the manufacture of higher chlorinated phenols (such as 2,4-dichlorophenol), as preservatives; and 4-chlorophenol is used as an intermediate in the manufacture of the disinfectant Clorophene [*120-32-1*] (2-benzyl-4-chlorophenol) (see Disinfectants), and quinizarin 1,4-dihydroxyanthraquinone.

Some work has been carried out in an attempt to get isomer selectivity between chlorination at the 2 (ortho) and 4 (para) positions of phenol. Treatment of phenol with SO_2Cl_2 in the presence of Fe, $FeCl_3$, or $ZnCl_2$ gave primarily the para isomer (1). Another study consisted of chlorinating phenol with *tert*-butyl hypochlorite and

Table 1. Physical Properties of the Chlorophenols

Compound	CAS Registry Number	bp, °C	mp, °C	Dissociation constant at 25°C, K_a
2-chlorophenol	[95-57-8]	175–176	8.7	3.2×10^{-9}
4-chlorophenol	[106-48-9]	219	40–41	6.6×10^{-10}
3-chlorophenol	[108-43-0]	215–217	32.8	1.4×10^{-9}
2,4-dichlorophenol	[120-83-2]	210–211	43–44	2.1×10^{-8}
2,6-dichlorophenol	[87-65-0]	219–220	67	1.6×10^{-7}
2,3-dichlorophenol	[576-24-9]	206	58	3.6×10^{-7}
2,5-dichlorophenol	[583-78-8]	212–213	58	4.5×10^{-7}
3,4-dichlorophenol	[95-77-2]	253	65	4.1×10^{-8}
3,5-dichlorophenol	[591-35-5]	233	68	1.2×10^{-7}
2,4,6-trichlorophenol	[88-06-2]	246	68	3.8×10^{-8}
2,4,5-trichlorophenol	[95-95-4]	245–246	68	3.7×10^{-8}
2,3,4-trichlorophenol	[15950-66-0]		83.5	2.2×10^{-8}
2,3,5-trichlorophenol	[933-78-8]	255	62	4.3×10^{-8}
2,3,6-trichlorophenol	[933-75-5]	272	101	7.4×10^{-8}
3,4,5-trichlorophenol	[609-19-8]	275	101	1.8×10^{-8}
2,3,4,6-tetrachlorophenol	[58-90-2]	64 (3.0 kPa)[a]	69–70	4.2×10^{-6}
2,3,4,5-tetrachlorophenol	[4901-51-3]		116–117	1.1×10^{-7}
2,3,5,6-tetrachlorophenol	[935-95-5]		115	3.3×10^{-6}
pentachlorophenol	[87-86-5]	309–310	190	1.2×10^{-5}

[a] To convert kPa to mm Hg, multiply by 7.5.

chlorine. The results indicated that higher concentrations and higher temperatures during the reaction tended to favor the formation of the ortho isomer (2). Also the effect of solvents during the chlorination of phenol (and p-chlorophenol) have been studied (3). Solvents such as acetonitrile, nitromethane, and bis-2-chloroethyl ether diminish ortho substitution because of steric hindrance and the formation of stable six-membered hydrogen-bonded complexes with phenol.

Sodium p-phenol sulfonate and phenyl phosphate (prepared from phenol) have both been subjected to chlorination. The former was subsequently desulfonated to give 2-chlorophenol, and the latter was hydrolyzed in acid or base to give p-chlorophenol in 90% yields (4–5). Monochlorinated phenols have also been obtained via the hydrolysis of di- and trichlorinated benzenes. Hydrolysis of 1,2-dichlorobenzene with KOH or NaOH (6) as well as 1,3-dichlorobenzene and 1,4-dichlorobenzene with cupric salts and hydroxylamine (7) gives the *ortho, meta,* and *para* chlorophenols, respectively. The vapor phase hydrolysis of 1,2,4-trichlorobenzene over a phosphate catalyst at temperatures of 370–500°C yields m-chlorophenol (8). One chlorine atom appears to be hydrolyzed, and the other is replaced with hydrogen. Dehalogenations on polysubstituted chlorophenols have been carried out to yield mixtures of monochlorinated phenols (9).

o-And p-chlorocumenes have been oxidized to form the corresponding peroxides which in turn have been converted to the corresponding chlorophenols (10). In addition, m-chlorophenol has been prepared by the oxidation of p-chlorobenzoic acid with stoichiometric quantities of copper(II) oxide as the oxidizing agent (11).

Dichlorophenols

The main dichlorophenol of commercial interest is 2,4-dichlorophenol. It is used in large volumes in the manufacture of 2,4-dichlorophenoxyacetic acid (2,4-D), and is conveniently manufactured via the chlorination of phenol. In a patented process phenol is dissolved in liquid SO_2 and treated with cold gaseous chlorine to give 98% pure 2,4-dichlorophenol (12). If the above procedure is applied to 2-chlorophenol, the product contains 10% 2,6-dichlorophenol. 1,2,3-Trichlorobenzene when sulfonated yields a mixture of 2,3,4-trichlorobenzenesulfonic acid and 2,3,4-trichlorobenzene-1,5-disulfonic acid. Treatment of this mixture with alkali hydroxide followed by acid hydrolysis yields 2,3-dichlorophenol (13). 3,4-Dichlorophenol can also be obtained via preparation of the corresponding 3,4-dichlorocumene (propylene plus o-dichlorobenzene) and subsequent oxidation followed by treatment with sulfuric acid (14).

Several investigations have been carried out on the hydrolysis of 1,2,4-trichlorobenzene. Treatment with copper, iron, or zinc halides as catalysts has resulted in the formation of 3,5-dichlorophenol (15,16). The hydrogenation of polychlorinated phenols over catalysts consisting of heavy metals, with the addition of $FeSO_4$, $CuSO_4$, Na_2S, or sulfur yielded 3-chlorophenol and 3,5-dichlorophenol (17). 3,5-Dichlorophenol has also been prepared via the partial dehalogenation of polychlorophenol using catalysts prepared from group VIII metals and sulfur or sulfides (18). A wide variety of dichlorophenols have been separated by the use of countercurrent dissociation extraction (19).

Trichlorophenols

The most important analogue in the trichlorinated phenol series is 2,4,5-trichlorophenol. It is used as an intermediate in the manufacture of the herbicides 2,4,5-trichlorophenoxyacetic acid (2,4,5-T) and 2-(2,4,5-trichlorophenoxy)propionic acid (silvex), the germicide hexachlorophene and the insecticides trichloronate and Fenchlorophos. It has also been used in the past as an intermediate in the synthesis of the herbicide erbon (2-(2,4,5-trichlorophenoxy)ethyl 2,2-dichloropropionate). 2,4,5-Trichlorophenol is currently marketed by The Dow Chemical Company and is used as an antifungal agent in several applications (20): in adhesives as a preservative in polyvinyl acetate emulsions; in the automotive industry to preserve rubber gaskets; and in textiles to preserve emulsions used in the rayon industry. 2,4,5-Trichlorophenol is generally applied as a direct additive (ie, directly into molten rubber), or dissolved in the oil phase of emulsions.

The sodium salt of 2,4,5-trichlorophenol is also used as a fungicide and bactericide. Uses include (21): adhesives (qv), as with 2,4,5-trichlorophenol; in cooling water as an inhibitor of microbial growth in recirculating water; in foundry core wash, to prevent breakdown of oils and scum formation; in leather dressing and finishes, to prevent the decomposition of nitrogenous compounds; in metal working fluids, to prevent breakdown of oils, emulsifying agents, and other components. An aqueous solution of the sodium salt is normally used in these applications.

2,4,5-Trichlorophenol is manufactured via the chlorination of benzene with 4 moles of chlorine to form 1,2,4,5-tetrachlorobenzene. The tetrachlorobenzene is then hydrolyzed in base to form the desired product. Care needs to be taken to minimize the formation of 2,3,7,8-tetrachlorodibenzo-p-dioxin in this reaction. 2,4,6-Trichlorophenol, produced via the chlorination of phenol, has been manufactured and sold by Dow as a preservative. Chlorination of 2,4,5-trichlorophenol in acidic solution gives 2,4,4,6-tetrachloro-2,5-cyclohexadienone and 2,2,4,5,6,6-hexachloro-3-cyclohexenone (22). In a similar experiment the chlorination of 2,4,6-trichlorophenol also yielded 2,2,4,5,6,6-hexachloro-3-cyclohexen-1-one (23). 2,4,5-Trichlorophenol is obtained as the main product in the chlorination of 3,4-dichlorophenol (24). Mixtures of mono-, di-, and tri-substituted phenols have been obtained via the reduction of 2,3,4,6-tetrachlorophenol (25). 2,4,6-Trichlorophenol has been obtained in yields of 97.5% by spraying chlorine into molten phenol (26).

Tetrachlorophenols

2,3,4,6-Tetrachlorophenol is available commercially and is used as a preservative. Little mention is made in the chemical literature of other tetrachlorophenols. They are present in small quantities in pentachlorophenol (27) and are formed during the breakdown of pentachlorophenol in the soil (28).

The chlorination of phenol using specifically potassium tellurate, K_2TeO_3, as catalyst has been found to give exclusively tetrachlorinated phenols (29). Isomers of tetrachlorophenol and pentachlorophenol have been separated via gradient elution with ion-exchange resins (30). The mixture is adsorbed onto Dowex 2-X8 resin, which is washed with sodium acetate, acetic acid, and methanol.

A variety of chlorophenols in aqueous solutions can be degraded with γ radiation to yield inorganic chloride and oxalic acid (31).

Pentachlorophenol

Pentachlorophenol (Penta, PCP) and its sodium salt are used extensively as antimicrobial agents. Pentachlorophenol is used as an antifungal agent in the following applications (32): in the wood industry as a perservative to control termites and fungus growth in building poles, posts, lumber, etc (most extensive use); in the construction industry, to control molds on inert building surfaces such as, tile roofs, and concrete blocks; in the leather industry, to impart mold resistance for upper leather in shoes; in the paint industry, for self protection of protein-based latex paints. Pentachlorophenol is added directly to the formulations or in an appropriate organic solvent (see Fungicide; Industrial antimicrobial agents).

The sodium salt of pentachlorophenol is used as an antifungal and antibacterial, and has applications in the following areas (33): adhesives—antimicrobial protection of adhesives based on starch, vegetable, and animal proteins during manufacture, storage and service life; construction materials—control of mold growth on inert surfaces as with pentachlorophenol; leather—prevention of hide deterioration and in the treatment of solutions during tanning; paint—aid in the shelf preservation of protein-based latex paints; petroleum—prevent the growth of bacteria in drilling muds; photographic solutions—control of fungus and slime; pulp and paper—protection and preservation of processing materials, stored pulp and fiberboard against mildew rot, and termites; textiles—protection of finished yarns and cloth against molding during storage; water treatment—control of algal, fungal, and bacterial induced slimes in industrial recirculating water. The sodium salt of pentachlorophenol is utilized primarily in an aqueous solution.

Pentachlorophenol is manufactured via the chlorination of phenol at 100–180°C with various catalysts (34–36). Some of the catalysts that have been used include AlCl$_3$, FeCl$_3$, activated carbon and quinoline, tellurium and some tellurium salts (29). Only AlCl$_3$, however, is used on a practical basis in this chlorination. The chemical reaction and production scheme are given in equation 1 and Figure 1, respectively (37).

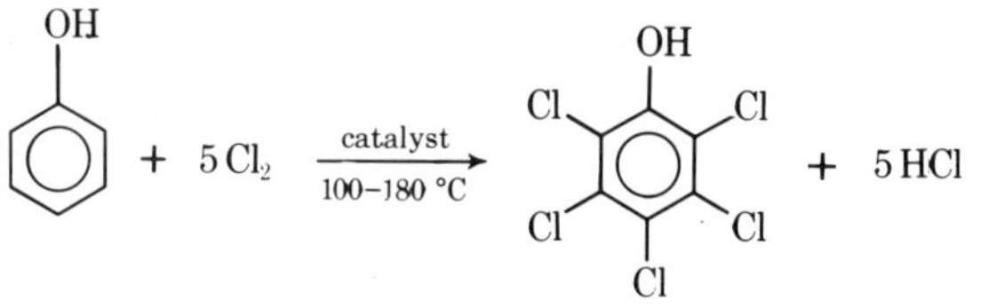

$$\text{C}_6\text{H}_5\text{OH} + 5\,\text{Cl}_2 \xrightarrow[100-180\,°C]{\text{catalyst}} \text{C}_6\text{Cl}_5\text{OH} + 5\,\text{HCl} \qquad (1)$$

The chlorination is done neat with a starting temperature between 65–130°C (preferably 105°C). After three to four atoms of chlorine have been substituted onto phenol, the temperature of the reaction mixture is increased to maintain a temperature approximately 10°C over the melting point of the chlorinated mixture. The reaction is complete in 5–15 h. Catalyst concentration is critical and is about 0.0075 mol/mol of phenol. The HCl and chlorine gases from the primary reactor are treated further with phenol to form pure HCl gas and a mixture of chlorinated phenols. The HCl gas can be oxidized back to chlorine gas and recycled. The partially chlorinated phenols can be separated or recycled to the primary reactor. Crude pentachlorophenol has been purified by distillation under reduced pressure in the presence of 0.05–2 wt % amine or alkanolamine (38).

Pentachlorophenol has also been prepared in 92% yields by the hydrolysis of hexachlorobenzene in an alkaline solution at 240°C and 3.62 MPa (525 psi) (39). Industry sources, however, agree that this method has never been used in any commercial preparations (37).

Chlorination of pentachlorophenol in acetic acid results in the formation of 2,3,4,5,6,6-hexachloro-2,4-cyclohexadien-1-one and small amounts of hexachlorophenol (2,3,4,4,5,6-hexachloro-2,5-cyclohexadien-1-one) and "δ-octachlorophenol" (2,2,3,4,5,5,6,6-octachloro-3-cyclohexen-1-one) (40). Hexachlorophenol [599-52-0] has been obtained in 70% yields by gradually adding 3–5% iodine to pentachlorophenol in hot carbon tetrachloride saturated with chlorine (41). Hexachlorophenol can be converted to pentachlorophenol by reduction with SO$_2$ (42).

Dehalogenations are usually carried out on pentachlorophenols using alumina granules impregnated with CuCl$_2$ (43).

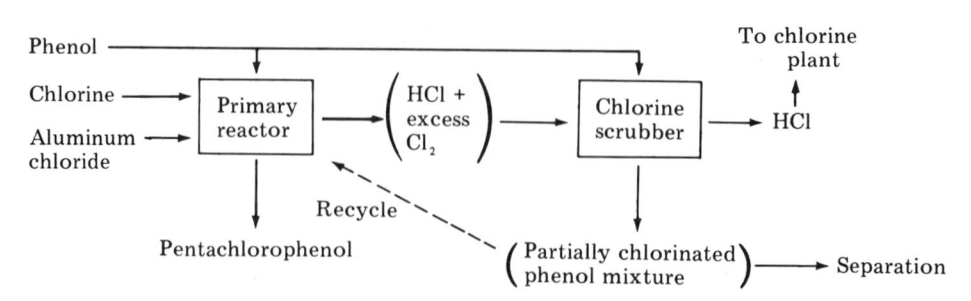

Figure 1. Production scheme for pentachlorophenol.

Analytical Methods

A wide variety of methods exist for the analysis of chlorinated phenols. These include acid–base titrations, quantitative reactions of the phenolic hydroxyl group such as acetylation or phthalation, and identification via ir spectroscopy. The analytical method that is gaining the most acceptance is gas chromatography (glc). This method furnishes suitable methods of assay as well as a means to separate and/or quantify the concentrations of chlorophenol isomer impurities that are present.

Economic Aspects

Information regarding the volumes of 2,4-dichlorophenol and pentachlorophenol used in the United States can be approximated from the amounts of phenol used in the manufacture of pentachlorophenol and 2,4-D (Table 2) (44).

Currently there are four U.S. producers of pentachlorophenol. These are (annual plant capacity (1977) is given in parentheses): Dow (7,735 t), Monsanto (11,830 t), Reichhold (9,100 t), and Vulcan (7,280 t) (37).

Health and Safety Factors

Toxicity. Some acute oral LD_{50} rat toxicities are given for the more common chlorinated phenols in Table 3.

Many of the chlorinated phenols are readily absorbed through the skin in toxic amounts. All chlorophenols are irritating to both the skin and eyes, also the dusts are very irritating to the respiratory tract. A review of the toxicities of chlorinated phenols is given in reference 46.

Polychlorinated phenols are the starting materials for the synthesis of the chlorinated dibenzo-p-dioxins (see eq. 2). The degree of toxicity displayed by the dioxins can vary from very extreme to nontoxic and is dependent upon the exact positions that are occupied by chlorine atoms on the aromatic rings.

chlorinated dibenzo-p-dioxin (2)

Table 2. Phenol Usage and Prices

Year	Amount of 2,4-DCP[a], t/yr	Phenol needed for 2,4-D[b], t/yr	Bulk prices 2,4-D, ¢/kg	Phenol needed for penta[c], t/yr	¢/kg penta[d]	Amount of penta, t/yr
1962	17,380	10,010	88	6,370	46	17,199
1966	28,645	15,470	73	7,280	44	19,656
1970	17,381	10,010	68–78	8,190	37	22,113
1974	23,706	13,650	79	7,280	42	19,656

[a] 2,4-DCP is 2,4-dichlorophenol.
[b] Approx 0.57 metric ton of phenol is required to produce 1.0 t 2,4-D. (2,4-dichlorophenoxyacetic acid).
[c] The Chemical Marketing Reporter.
[d] Approx 0.37 t phenol is required to produce 1.0 t Penta (pentachlorophenol).

Table 3. Rat Toxicities for Common Chlorinated Phenols

Compound	LD$_{50}$ (rats), g/kg[a]
2-chlorophenol	0.67
3-chlorophenol	0.57
4-chlorophenol	0.67
2,4-dichlorophenol	0.58
2,4,5-trichlorophenol	2.83
2,4,6-trichlorophenol	0.82
tetrachlorophenol	0.14
pentachlorophenol	0.18

[a] Ref. 45.

As an example, if one starts with 2,4,5-trichlorophenol, the corresponding dioxin formed is the very toxic 2,3,7,8-tetrachlorodibenzo-p-dioxin. For further discussions of the toxicities and synthesis of this class of compounds, see references 47 and 48.

Environmental Aspects. A considerable amount of information is available regarding stability and toxicity of chlorinated phenols in the environment (49). Three generalizations have been reached regarding chlorophenols in the environment: (a) chlorophenols are much more environmentally stable than the parent unsubstituted phenol (50), (b) as the number of chlorine atoms increase the rate of decomposition decreases (51), and (c) compounds containing a meta chlorine (eg, 3-chlorophenol and 2,4,5-trichlorophenol) are more persistent than compounds lacking a chlorine atom in positions meta to the hydroxyl group (52). The three conclusions above are quite general and do not always agree with specific experimental data.

Table 4 gives three properties that relate to environmental stability and indicates the environmental properties of some of the chlorinated phenols.

Table 4. Environmental Data on Some Chlorinated Phenols

Compound	Soly, g/100 g H$_2$O	Partition coefficient, water-octanol	Microbial decomposition in soil (days to disappear)[a] Dunkirk soil	Mardin soil
phenol	6.6	1.46[b]	2	1
2-chlorophenol	<0.1	2.15[b]	14	47
3-chlorophenol	0.26	2.50[b]	>72	>47
4-chlorophenol	2.71	2.39[b]	9	3
2,4-dichlorophenol	0.45	3.06[c]	9	5
2,5-dichlorophenol	sparingly soluble	3.20[c]	>72	
2,4,5-trichlorophenol	0.12	3.72[d]	>72	>47
2,4,6-trichlorophenol	0.09	3.62[c]	5	13
2,3,4,6-tetrachlorophenol	0.10	4.10[c]	>72	
pentachlorophenol	14–19 ppm		>72	

[a] Ref. 52.
[b] Ref. 53.
[c] Ref. 54.
[d] Ref. 55.

It is evident that the water solubility tends to decrease and the water-octanol partition coefficient increases as more chlorines are added to the phenol ring. This is an indication that higher chlorinated phenols will be more persistent in the environment than the lower chlorinated analogues. The time required for complete disappearance of phenols in Dunkirk and Mardin soils is given in Table 4. This was determined by monitoring spectophotometrically the rate of disappearance of the substrate after being added to the soils. The experimental data are in fair agreement with points one and three of the generalizations given above.

Photodegradation studies have been carried out on 2,4-dichlorophenol and pentachlorophenol. These indicate that a variety of intermediates are formed (56–57) (see Trace and residue analysis).

BIBLIOGRAPHY

The "Chlorophenols" are treated in *ECT* 1st ed. under "Phenol and Phenols", Vol. 10, pp. 317–320, by Calvin Golumbic, Bureau of Mines, U.S. Department of the Interior, and under "Chlorophenols" in *ECT* 2nd ed., Vol. 5 pp. 325–339, by J. D. Doedens, The Dow Chemical Company.

1. Jpn. Pat. 17,808 (Sept. 18, 1967), N. Shindo, Y. Ura, F. Suzuki, and T. Hosono (to Nissan Chemical Industries Lts.).
2. W. D. Watson, *J. Org. Chem.* **39,** 1160 (1974).
3. V. I. Igoshev and A. D. Sobolev, *Ufim. Khim. Zavod, Ufa, USSR Zh. Org. Khim.* **4,** 1832 (1968).
4. U. D. Simonov and co-workers, *Prom. Obraztsy, Tovarnye Znaki* **47** (20), 29 (1970); *Chem. Abstr.* **74,** 12823M (1971).
5. Neth. Appl. 6,609,527 (Jan. 9, 1967) (to Sumitomo Chemical Company Ltd.).
6. Jpn. Pat. 34,327 (Nov. 21, 1972), F. Ryozo and O. Eiji (to Mitsui Toatsu Chemicals Company Ltd.).
7. Jpn. Pat. 17,372 (July 31, 1969), T. Yosei, O. Noboru, K. Toshio, Y. Akira, and M. Takashi (to Asahi Chemical Industry Company Ltd.).
8. A. I. Naumov and Z. G. Lapteva, *Zhur. Obshchei. Khim.* **26,** 1647 (1956).
9. Brit. Pat. 793,426 (April 16, 1958), (to Dow Chemical Company).
10. J. Slosar and V. Sterba, *Chem. Prumysl* **15,** 206 (1965); *Chem. Abstr.* **63,** 4195E (1965).
11. U.S. Pat. 2,852,567 (Sept. 16, 1958), R. D. Barnard and R. H. Meyer (to The Dow Chemical Company).
12. U.S. Pat. 2,759,981 (Aug. 21, 1956), B. O. Pray and D. N. Sukow (to Columbia-Southern Chemical Corporation).
13. Czech. Pat. 105,282, 105,281 (Oct. 15, 1962), O. Schiessl and Z. Stota.
14. U.S. Pat. 2,854,448 (Sept. 30, 1958), A. H. Widiger, Jr. (to The Dow Chemical Company).
15. W. Langenbeck, H. Furst, and G. Reinisch, *J. Prakt. Chem.* **2,** 308 (1955).
16. Jpn. Pat. 17,807 (Sept. 18, 1967), N. Shindo, Y. Ura, F. Suzuki, and T. Hosono (to Nissan Chemical Industries).
17. Ger. Offen. 2,259,433 (June 6, 1974); 2,344,926 (April 3, 1975), K. Wedemeyer, E. Koppelmann, and W. Evertz (to Bayer A.G.).
18. Ger. Offen. 2,344,925 (April 3, 1975), W. Kiel, K. Wedemeyer, and W. Eventz (to Bayer A.G.).
19. M. H. Milnes, *Proc. Int. Solvent Extr. Cont.* **1,** 983 (1974).
20. Dow Product Literature. Section I-2 Dowicide 2 Antimicrobial, (1976), Form No. 192-477-76, The Dow Chemical Company, Midland, Mich.
21. Dow Product Literature. Section I-10, Dowicide B Antimicrobial (1976), Form No. 192-478-76, The Dow Chemical Company, Midland, Mich.
22. P. Svec, A. M. Sorensen, M. Zbirovsky, *Org. Prep. Proced. Int.* **5,** 209 (1973).
23. Czech. Pat. 157,460 (April 15, 1975), P. Svec and M. Zbirovsky.
24. U.S. Pat. 2,756,260 (July 24, 1956), A. H. Widiger, Jr. (to The Dow Chemical Company).
25. Ger. Pat. 1,109,701 (June 29, 1961), A. E. Brainerd, Jr. and N. Poffenberger (to The Dow Chemical Company).
26. V. A. Nekrosova and V. I. Zetkin, *Zh. Prikl, Khim.* **38,** 1407 (1965).

27. U.S. Pat. 2,947,790 (Aug. 2, 1960), F. J. Shelton, T. S. Hodgins, and C. L. Allyn (to Reichold Chemicals).
28. A. Ide and co-workers, *Agr. Biol. Chem.* **36,** 1937 (1972).
29. W. Rodziewicz, J. Dobrowolski, and W. Wojnowski, *Przemysl Chem.* **37,** 645 (1958); *Chem. Abst.* **53,** 17946G (1959).
30. N. E. Skelly, *Anal. Chem.* **33,** 271 (1961).
31. E. Gilbert and H. Gueston, *Vom Wasser* **41,** 359 (1973).
32. Dow Product Literature, Section I-6, Dowicide 7 Antimicrobial, 1976, Form No. 192-89-72, The Dow Chemical Company, Midland, Mich.
33. Dow Product Literature, Section I-11, Dowicide G-ST Antimicrobial, Form No. 192-475-76, The Dow Chemical Company, Midland, Mich.
34. Jpn. Pat. Appl. 7017 ['61] (June 9, 1958), I. Saikawa (to Toyama Chemical Industry Company).
35. Ger. Pat. 838,701 (Sept. 3, 1958), C. Thönnessen and L. Seelmann (to Dr. F. Raschig G.m.b.H.).
36. U.S. Pat. 2,938,059 (May 24, 1960), W. Rodziewicz, J. Dobrowolski, and W. Wojnowski.
37. *Pentachlorophenol—A Wood Preservative,* Memorandum for the Office of Pesticide Programs, March 1, 1977, E.P.A. Subcommittee No. 6, American Wood Preserves Institute, 1651 Old Meadow Road, McLean, VA.
38. U.S. Pat. 3,816,268 (June 11, 1974), W. D. Watson, E. H. Kobel, and H. Erwin (to The Dow Chemical Company).
39. U.S. Pat. 3,051,761 (Aug. 28, 1962), G. MacBeth and R. G. Heitz (to The Dow Chemical Company).
40. L. Denivelle and R. Fort, *Bull. Soc. Chim. France* 459 (1958).
41. *Ibid.,* 1834 (1956).
42. Czech. Pat. 137,655 (July 15, 1970), L. Zatloukaland and M. Kosina.
43. Fr. Pat. 2,161,861 (Aug. 17, 1973), George S. Rivier (to S. A. Progil).
44. S. A. Cogswell, *Phenol-Salient Statistics,* Chemical Economics Handbook, Stanford Research Institute, Menlo Park, Calif., April, 1975, p. 686.5022 M.
45. M. Windholz, ed., *The Merck Index,* 9th ed., Merck and Company, Rahway, New Jersey, 1976.
46. F. A. Patty, ed., *Industrial Hygiene and Toxicology,* Vol. II 2nd ed, Wiley-Interscience, New York, 1963, pp. 1396–1405.
47. *Environmental Health Perspectives, Experimental Issue No. 5,* Sept. 1973, U.S. Dept. of Health Education and Welfare, N.I.H.
48. R. D. Arsenault, *Pentachlorophenol and Contained Chlorinated Dibenzodioxins in the Environment,* American Wood Preserves Association, Washington, D.C., 1976.
49. P. H. Howard and P. R. Durkin, *Preliminary Environmental Hazard Assessments of Chlorinated Naphthalene Silicones Fluorocarbons, Benzene Polycarbonates, and Chlorophenols,* p. 204–263. Phillip H. Howard and Patrick R. Durkin, Syracuse University Research Corporation, Syracuse, NY (prepared for E.P.A. N.T.I.S. Report #PB-238074), Nov. 1973, pp. 204–263.
50. R. S. Ingols, P. E. Gaffney, and P. C. Stevenson, *J. Water Pollut. Contr. Fedr.* **38,** 629 (1966).
51. C. W. Chamber, H. H. Tabak, and P. W. Kabler, *J. Water Poll. Contr. Fedr.* **35,** 1517 (1963).
52. M. Alexander and M. J. H. Aleem, *J. Agr. Food Chem.* **9,** 44 (1961).
53. T. Fujita, J. Iwasa, and C. Hantsch, *J. Am. Chem. Soc.* **86,** 5175 (1964).
54. M. Stockdale and M. Selwyn, *Eur. J. Biochem.* **21,** 565 (1971).
55. A. Leo, C. Hantsch, and D. Elkins, *Chem. Rev.* **71,** 525 (1971).
56. D. G. Crosby and H. O. Tutass, *J. Agr. Food Chem.* **14,** 596 (1966).
57. K. Manakata and M. Kuwahara, *Residue Rev.* **25,** 13 (1969).

E. R. FREITER
Dow Chemical U.S.A.

CHLOROPHYLL. See Dyes, natural.

CHLOROPRENE, CH_2=CHCCl=CH_2. See Chlorocarbons and chlorohydrocarbons.

CHLOROSULFURIC ACID

A. W. Williamson described chlorosulfuric acid [7790-94-5] preparation and properties in 1854 (1). It is a highly reactive compound containing equimolar quantities of HCl and SO_3. The clear, colorless liquid is actually an equilibrium mixture of chlorosulfuric acid with minor amounts of hydrogen chloride, sulfur trioxide, and some related compounds. The acid reacts violently with water, producing heat and dense white fumes of hydrochloric acid and sulfuric acid. It reacts with almost all organic materials, in some cases with charring. Its uses are principally in organic synthesis as a sulfating, sulfonating, or chlorosulfonating agent. It is preferred in many applications because it yields the desired isomers.

The main application for chlorosulfuric acid is as an intermediate in the production of synthetic detergents, drugs and dyestuffs. It has also been used as a smoke-forming agent in warfare (see Chemicals in war).

The structure of the compound was debated for many years until Dharmatti (2) in 1941 showed by magnetic susceptibility measurements that the chlorine was bonded directly to the sulfur atom. Raman spectra studies by Gerding (3) and Gillespie and Robinson (4) have provided additional confirmation of the structure. Consequently chlorosulfuric acid, $ClSO_2OH$, can be considered the monoacid chloride of sulfuric acid since one chlorine atom has replaced one hydroxyl group.

Following the nomenclature of the old theory of types, chlorosulfuric acid was named sulfuric chlorohydrin. The commercial designation is chlorosulfonic acid and it has been also been called chlorohydrated sulfuric acid and chlorohydrosulfurous acid.

Physical Properties

Heating chlorosulfuric acid results in the equilibrium formation of sulfuryl chloride, sulfuric acid, pyrosulfuryl mono- and dichlorides, and pyrosulfuric acid. There is evidence of the formation of higher polyacids such as $HS_4O_{12}Cl$. Heating beyond the boiling point results in decomposition into sulfur dioxide, chlorine, and water. Distillation tends to degrade the acid rather than purify it; therefore, the physical constants reported reflect the influence of varying amounts of these impurities. The values given in Table 1 are considered to be the most reliable available. In addition, values have been reported for heats of formation (5), vapor pressure data (6), infrared spectra (7) and thermal constants of mixtures with SO_3 (8).

Chlorosulfuric acid is miscible with sulfur trioxide, 100% sulfuric acid and pyrosulfuryl chloride in all proportions; mixtures with sulfuric trioxide are used as smoke-forming agents. The properties of such mixtures have been described (3,9,13) (see Chemicals in war). Mixtures of chlorosulfuric acid and pyrosulfuryl chloride form an azeotrope when distilled (14).

Chlorosulfuric acid is soluble in *sym*-tetrachloroethane ($C_2H_2Cl_4$), chloroform, and dichloromethane, but essentially insoluble in carbon tetrachloride and carbon disulfide. It is soluble in acetic acid and acetic anhydride, and in trifluoroacetic acid and its anhydride as well as in sulfuryl chloride. It reacts with alcohols, ketones, diethyl ether, and dimethyl sulfoxide, although some literature references report the use of the latter two as solvents at low temperatures. Caution should be used working with

Table 1. Physical Properties of Chlorosulfuric Acid

Property	Value	References
molecular weight	116.531	
freezing point, °C	−81 to −80	
boiling point (with decomp), °C	151–152	
vapor pressure, in Pa^a and K	$\log P = 11.496 - \dfrac{2752}{T}$	
vapor density		
(measured) at 216°C, kg/m^3	2.4	
(calculated)	4.04	
density, kg/m^3,		
d_4^{20}	1753	
d_4^t (0–100°C)	$1784.7 - 1.616t + 1.21t^2 \times 10^{-3} - 4.1t^2 \times 10^{-6}$	
$d^{-70°C}$	1900	16
$d^{-10°C}$	1800	
viscosity, mPa·s (= cP),		
−31.6°C	10.0	
−17.8	6.4	
15.6	3.0	
49	1.7	
specific heat, J/(kg·K)^b	1.18×10^3	
heat of formation from elements at 25°C, J/mol^b	-597.1×10^3	
heat of vaporization, J/g^b	452–460	
heat of solution in water at 18°C, J/mol^b	168.6×10^3	
index of refraction, n_D^{14}	1.437	
dielectric constant at 15°C	60 ± 10	17
electrical conductivity, at 25°C (Ω·cm)$^{-1}$	0.2–0.3×10^{-3}	18
in sulfuric acid		19
in liquid HCl		20

^{a} To convert Pa to mm Hg, multiply by 0.0075.
^{b} To convert J to cal, divide by 4.184.

any solvent, since reaction may occur when the temperature is increased or in the presence of catalysts. The solubility of hydrogen chloride in chlorosulfuric acid is approximately 2.4 mol/(L·MPa) [0.24 mol/(L·atm)] at 20–25°C (15) but decreases rapidly with increasing temperature.

Chemical Properties

Chlorosulfuric acid is a strong acid containing a relatively weak sulfur–chlorine bond. It is a powerful sulfating and sulfonating agent, a fairly strong dehydrating agent, and a specialized chloridating agent. In most of its applications it is used to form sulfates, sulfonates, sulfonyl chlorides, and occasionally other chlorine derivatives with such organic compounds as hydrocarbons, alcohols, phenols, and amines. The reactions of chlorosulfuric acid are the result of attachment of an —SO$_3$H group to give a sulfate or sulfonate. The general reactions are:

$$R-OH + ClSO_3H \rightarrow R-O-SO_3H + HCl$$

$$R-H + ClSO_3H \rightarrow R-SO_3H + HCl$$

In the presence of excess acid, sulfonyl chlorides are formed.

$$C_6H_6 + 2\,ClSO_3H \rightarrow C_6H_5SO_2Cl + HCl + H_2SO_4$$

Sulfamation forms a —C—N—S— bond as in $C_6H_{11}NHSO_3Na$, sodium cyclohexyl sulfamate. A sulfonyl chloride group (—SO_2Cl) is attached to an aromatic group (chlorosulfonation) or to an alkoxy group (chlorosulfation) as in $CH_3CH(CH_3)$—O—SO_2Cl, isopropyl chlorosulfate. In sulfone formation two aromatic rings are attached to an —SO_2— group as in C_6H_5—SO_2—C_6H_5, diphenyl sulfone (see Sulfur compounds).

The fluorine analogue of chlorosulfuric acid, fluosulfuric acid, is considerably more stable than chlorosulfuric acid because of the strong fluorine–sulfur bond. In the other direction bromosulfuric acid decomposes in air at $-30°C$, and it has not been possible to synthesize iodosulfuric acid (21). Many salts and esters of chlorosulfuric acid are known, most of them are relatively unstable or hydrolyze readily in moist air.

Strong dehydrating agents such as phosphorus pentoxide or sulfur trioxide convert chlorosulfuric acid to its anhydride, pyrosulfuryl chloride (disulfuryl dichloride), $S_2O_5Cl_2$. Analogous trisulfuryl compounds have been identified in mixtures with sulfur trioxide by the use of Raman spectroscopy (3,10). When boiled in the presence of mercury salts or other catalysts, chlorosulfuric acid decomposes quantitatively to sulfuryl chloride, SO_2Cl_2, and sulfuric acid.

$$2 \, ClSO_2OH \xrightarrow[HgX_2]{\Delta} SO_2Cl_2 + H_2SO_4$$

The reverse reaction has been claimed as a preparative method (22), but it appears to proceed only under special conditions. Noncatalytic decomposition at elevated temperatures also generates sulfuryl chloride, chlorine, sulfur dioxide, and other compounds.

The acid is rather slow to react with aliphatic hydrocarbons unless a double bond or some other reactive group is present. This fact permits straight-chain fatty alcohols such as lauryl to be converted to the corresponding sulfate without the degradation or discoloration experienced with the more vigorous reagent sulfur trioxide. A number of literature articles refer to "chlorosulfonation" of paraffins but this is generally misleading, mixtures of chlorine and sulfur dioxide (or sulfuryl chloride) are used rather than chlorosulfuric acid. A review of the reactions of chlorosulfuric acid is given by Jackson (23).

Manufacture

Modern plants manufacture chlorosulfuric acid by the direct union of sulfur trioxide with dry hydrogen chloride gas. Reaction is continuous with heat removal to maintain the temperature at 50–70°C. The sulfur trioxide may be in the form of 100% liquid or gas, as obtained from boiling oleum (fuming sulfuric acid), or may be present as a dilute gaseous mixture obtained directly from a contact sulfuric acid plant. In some older, discontinued processes, hydrogen chloride reacts directly with oleum, followed by distillation, or an alkali or alkaline earth chloride reacts with oleum.

The reaction of sulfur trioxide and hydrogen chloride takes place spontaneously with evolution of a large quantity of heat. If the reaction is allowed to proceed adiabatically, the high temperatures reached cause partial decomposition of the product and formation of unwanted contaminants. However, if the product is to be used only as a smoke-forming agent and not as a chemical intermediate, quality is not critical

876 CHLOROSULFURIC ACID

and adiabatic or high-temperature operation can be used. The design and operation of this type of plant is described in several *U.S. Publication Board Reports* (24–25).

Two processes for the manufacture of high-quality acid are described in a number of patents (26–36). The most common features of these processes are continuous flow operation, two or more vessels in series for gas-liquid contacting, heat exchangers for controlling temperatures in the range 60–120°C, and use of excess chlorosulfuric acid as a solvent during at least part of the reaction.

Older manufacturing plants were typically fabricated of cast iron or steel, in some cases with aluminum-alloy condensers. In recent years there has been a demand, especially from the synthetic detergent industry, for acid containing only traces of iron. To satisfy this demand, plants now in use are fabricated of glass, glass-lined steel, or other noncorroding materials.

Economic Aspects

Although specific figures are not available, annual production of chlorosulfuric acid prior to the late 1930s was apparently relatively small. World War II brought about a substantial increase, both for use in the production of military smoke screens and as an intermediate for sulfa drugs. When the war ended, the growth of the synthetic detergent industry replaced military consumption, and production continued to increase.

Although production figures are not available, there are 23 listed manufacturers in Europe, plus units in Japan and the USSR (37). Major producers in the United States are the Monsanto Co. and E. I. du Pont de Nemours & Co. which have an estimated combined capacity exceeding 90,000 metric tons per year.

The price for chlorosulfuric acid was stable at $91/t over the period 1957 through 1962. It then slowly escalated to $103/t where it remained stable from 1969 through 1974. In 1975 and 1976 the price averaged $115/t and in 1977 listed prices range from $187 to $209/t (38).

Specifications and Standards

No formal industry-wide specifications for chlorosulfuric acid exist, each producer or user establishing individual requirements as needed. Typical commercial chlorosulfuric acid meets the following specifications:

Property	Value
appearance	mobile liquid
color	200 APHA max
turbidity	16 APHA max
assay, ClSO$_3$H	98.5% min
iron as Fe	5 ppm max
free SO$_3$	0.7% max

Average analyses of commercial acid are:

total ClSO$_3$H	99.4%
total chlorides as HCl	31.2%
sulfuric acid	0.2%

free SO$_3$	0.4%
iron as Fe	1.0 ppm
Al	<1.0 ppm

For many uses iron content and color are not critical. In these cases the acid is generally stored and shipped in steel equipment, with changes in these properties resulting. After shipping or storing in steel the acid typically contains <0.0025–0.0050% iron (Fe), and ranges in color from pale yellow to amber. U.S. military specification MILC 379A applies to a mixture of chlorosulfuric acid and sulfur trioxide.

The acid may be shipped in tank car or smaller quantities via common carrier, but may not be shipped in the same car or truck with foodstuffs or explosives. Freight classification is Corrosive Material. Detailed shipping specifications may be found in reference 39.

Analytical and Test Methods

Total acidity and total chlorides can be determined by conventional techniques after hydrolyzing a sample. Satisfactory procedures for determining hydrogen chloride and free sulfur trioxide have been described in references 15 and 40. Small amounts of both hydrogen chloride and free sulfur trioxide can be found in the same sample because of the equilibrium nature of the liquid. Procedures for the direct determination of pyrosulfuryl chloride have been described in references 41–42, but are not generally required for routine analysis. Small concentrations of sulfuric acid can be measured by electrical conductivity.

Spot tests for detecting chlorosulfuric acid are based on the use of powdered tellurium, which gives a cherry-red color, and powdered selenium, which gives a moss-green color.

Health and Safety Factors

Personal Safety and First Aid. All personnel handling chlorosulfuric acid should be thoroughly familiar with Chemical Safety Data Sheet SD-33 (Chlorosulfonic Acid) published by the Manufacturing Chemists' Association (39).

Chlorosulfuric acid is a strong acid which reacts vigorously with water. Thus, the principal hazard is contact with the liquid acid which will severely burn body tissue. The vapor is also hazardous and is extremely irritating to the skin, eyes, nose, and throat. In the case of contact, the affected area should be immediately and thoroughly flushed with water and a physician consulted.

When exposed to the atmosphere, chlorosulfuric acid fumes react with the moisture of the air and release highly irritating and corrosive hydrochloric acid fumes and sulfuric acid mist which may cause delayed lung damage. The vapor has such a sharp and penetrating odor that the inhalation of toxic quantities is unlikely unless it is impossible to escape the fumes. The U.S. Department of Labor (OSHA) regulation limits exposure to hydrogen chloride to 5 ppm in air (7 mg/m^3), and exposure to sulfuric acid mist in any 8-h period to a weighted average of 1 mg/m^3 (Title 29, Part 1910.1000 Air Contaminants). In the case of inhalation the affected individual should be removed to fresh air immediately, artificial respiration or oxygen administered if necessary, and a physician called.

Case histories of some industrial personnel exposures have been described (43–44).

878 CHLOROSULFURIC ACID

Personal protective equipment should be used to protect a worker whenever contact with the acid could be encountered. The following acid-proof protective equipment for all operating and maintenance personnel should be available: chemical safety goggles, hard hat with brim, safety shoes with boots, jacket, and gauntlet gloves. Also, shirt and trousers of wool or acrylic fiber should be worn. For emergencies or where there is a possibility of considerable exposure, the protective equipment should include a complete acid suit with hood, gloves and boots and respiratory protective equipment such as self-contained breathing apparatus, positive-pressure hose mask, air-line mask or an industrial canister-type gas mask.

Safety showers, water hydrant and hose and eye wash fountains should be easily accessible in all areas where chlorosulfuric acid is handled, including unloading stations and storage areas.

Fire Hazard. Although chlorosulfuric acid itself is not flammable, it may cause ignition by contact with combustible material. Open fires, open lights and matches should not be used in or around tanks or containers because hydrogen gas released by the action of chlorosulfuric acid on metal forms explosive mixtures with air. Water, carbon dioxide and dry-chemical fire extinguishers should be kept readily available.

Storage. The acid should never stand in a line completely sealed between two closed valves or check valves since excessive pressures caused by liquid thermal expansion can create hazardous acid sprays or pipe rupture. All lubricants and packing materials in contact with chlorosulfuric acid must be chemically resistant to this acid. Use of flange guards on all flange joints is recommended.

Spills and Waste Disposal. Accidentally spilled chlorosulfuric acid should be washed off immediately with large volumes of water. Care should be taken as water and chlorosulfuric acid react violently during flushing. Since hydrochloric acid gas is evolved, the water should be applied from a distance, preferably upwind. No one should be allowed downwind of the contaminated area while the flushing is taking place. If water flushing is impossible, the contaminated area should be covered with dry sand, ashes or gravel. Remaining traces of acid should be neutralized with soda ash or lime. Again caution should be exercised because of the potentially violent reaction of any concentrated residual acid. Provide ventilation for spills indoors.

Washings from large spills must be neutralized with an alkaline material before they are discharged into a sewer system.

Small quantities of chlorosulfuric acid can be disposed of by dilution with large volumes of water and neutralization, if necessary, with soda ash or a mixture of soda ash and sand.

Information on waste disposal is contained in reference 39. Any disposal method must comply with local, state and federal pollution-control regulations.

Uses

Since its reactivity and versatility in organic synthesis are high, the list of uses for chlorosulfuric acid is very large. Some of the major uses are as follows:

Detergents, including sulfates, particularly lauryl sulfate, sulfates of olefins or unsaturated oils, sulfates of polyoxypropylene glycol, sulfonates of alkylated diphenyl ether, ethoxylates, etc (see Surfactants).

Pharmaceuticals such as sulfa drugs, synthetic sweeteners, anticoagulants, phenolphthalein, substituted sulfuric acids and salts, diuretics, and active chlorine agents for disinfection (see Disinfectants).

Dyes and pigments including acid dyes, vat dyes, monoazo dyes, phthalocyanine dyes, and surface treatment of polyethylene or polyester fibers and films.

Ion-exchange resins. Thermosetting resins are produced by the reaction of indoles and formaldehyde with chlorosulfuric acid as a catalyst. Sulfonation of glycolphthalate or benzylchloride-naphthalene resins with chlorosulfuric acid produces water soluble resins (see Resins). Chlorosulfuric acid is used as a sulfonating agent in the preparation of resin-based ion-exchange materials.

Catalysis. Chlorosulfuric acid exhibits catalytic properties in reactions of the following types: esterification of aliphatic acids in both liquid and vapor phase, alkylation of olefinic hydrocarbons, preparation of alkyl halides from olefinic halides and isoparaffins with tertiary hydrogen, and preparation of unsaturated ketones from olefins and anhydrides of fatty acids.

Miscellaneous uses include preparation of pesticides, plasticizers, tanning agents, textile and paper specialties, fluorocarbons, rubber and plastic release agents, as vulcanization aid for isoolefin copolymers, stabilizing agent, condensing agents, source of anhydrous hydrogen chloride, and separating agent for mixtures of sulfur dioxide and chlorine.

BIBLIOGRAPHY

The "Chlorosulfuric Acids" are treated under "Chlorosulfonic Acid" in *ECT* 1st ed., Vol. 3, pp. 885–889, by D. P. Shedd, Monsanto Chemical Company; "Chlorosulfonic Acid" in *ECT* 2nd ed., Vol. 5. pp. 357–363, by J. R. Donovan, Monsanto Company.

1. A. W. Williamson, *Proc. Royal Soc. (London)* **7**, 11 (1854).
2. S. S. Dharmatti, *Proc. Indian Acad. Sci. Sect. A* **13**, 359 (1941).
3. H. Gerding, *J. Chem. Phys.* **46**, 118 (1948).
4. R. J. Gillespie and E. A. Robinson, *Can. J. Chem.* **40**, 644 (1962).
5. G. W. Richards and A. A. Woolf, *J. Chem. Soc. A* (7), 1118 (1967).
6. L. P. Ryadneva and A. S. Lenskii, *Zh. Prikl. Khim.* **36**, 2413 (1963).
7. R. Savoie and P. A. Giguere, *Can. J. Chem.* **42**, 277 (1964).
8. A. S. Lenskii and co-workers, *Zh. Neorgan. Khim.* **9**, 1147 (1964).
9. E. W. Balson and N. K. Adam, *Trans. Faraday Soc.* **44**, 412 (1948).
10. R. J. Gillespie and E. H. Robinson, *Can. J. Chem.* **39**, 2179 (1961); **40**, 675 (1962).
11. S. A. Kudryavtsev and co-workers, *Zh. Prikl. Khim.* **14**, 478 (1941).
12. R. Macy, *Chem. Corps J.* **1**, 36 (1947).
13. R. J. McCallum and E. L. Tollefson, *Can. J. Res. Sect. F.* **26**, 241 (1948).
14. C. R. Sanger and E. R. Riegel, *Proc. Am. Acad. Arts Sci.* **47**, 673 (1912); *Z. Anorg. Allgem. Chem.* **76**, 79 (1912).
15. E. Korinth, *Agnew. Chem.* **72**, 108 (1960).
16. M. Schmidt and G. Talsky, *Chem. Ber.* **92**, 1539 (1959).
17. R. J. Gillespie and R. F. M. White, *Trans. Faraday Soc.* **54**, 1846 (1958).
18. P. Walden, *Z. Anorg. Allgem. Chem.* **29**, 371 (1902).
19. J. Barr and co-workers, *Can. J. Chem.* **39**, 1266 (1961).
20. M. E. Peach and T. C. Waddington, *J. Chem. Soc.* 2680 (1962).
21. M. Schmidt and G. Talsky, *Z. Anorg. Allgem. Chem.* **303**, 210 (1960).
22. U.S. Pat. 1,554,870 (Sept. 22, 1925), R. H. McKee and C. M. Salls.
23. K. E. Jackson, *Chem. Revs.* **25**, 81 (1939).
24. R. E. Richardson and co-workers, *U.S. Dept. Comm. Office Tech. Serv. PB Rept. 218*, 1945.
25. W. A. M. Edwards and co-workers, *U.S. Dept. Comm. Office Tech. Serv. PB Rept. 34005* (*BIOS Final Report 243, Item 22*, May 1946; *U.S. Dept. Comm. Office Tech. Serv. PB Rept. L34005-S* (*FIAT Tech. Bull. T12*, March 1947).

26. U.S. Pat. 1,013,181 (Jan. 2, 1912), A. Klages and H. Vollberg.
27. U.S. Pat. 1,422,335 (July 11, 1922), T. L. Briggs (to General Chemical Co.).
28. U.S. Pat. 2,311,619 (Feb. 16, 1943), N. A. Laury (to American Cyanamid Co.).
29. U.S. Pat. 2,377,642 (June 5, 1945), R. B. Mooney and G. E. Wentworth (to Imperial Chemical Industries Ltd.); Brit. Pat. 561,841 (1945), (to Imperial Chemical Industries Ltd.).
30. Ger. Pat. 543,758 (May 25, 1929), K. Dachlauer (to I. G. Farbenindustrie A.G.).
31. Ger. Pat. 914,733 (July 8, 1954), H. Beyer.
32. Jpn. Pat. 2,665 (May 1, 1957), M. Kawamoto and H. Ejiri (to Mitsubishi Chemical Industries Co.).
33. U.S.S.R. Pat. 113,664 (Aug. 20, 1958), A. S. Lenskii and co-workers.
34. Neth. Appl. 6,154,410 (May 9, 1966), (to BASF A.G.).
35. Jpn. Pat. 70 24,648 (Aug. 17, 1970), (to Mitsubishi Chemical Industries Co.).
36. Jpn. Pat. 76 10,840 (April 7, 1976), (to Mitsubishi Chemical Industries Co.).
37. *Chem. Sources Europe*, 1976–1977.
38. *Chemical Marketing Reporter*.
39. *Chlorosulfonic Acid, Chemical Safety Data Sheet SD-33*, Manufacturing Chemists' Assoc., Inc., Washington, D.C., 1949.
40. W. Seaman and co-workers, *Anal. Chem.* **22,** 549 (1950).
41. J. H. Payne, Jr., *Ph.D. Thesis*, Purdue University, Lafayette, Indiana, June, 1947.
42. G. V. Zavorov, *Zavodsk. Lab.* **27,** 1208 (1961).
43. L. Arzt, *Dermatol. Wochschr.* **97,** 995 (1933).
44. F. Roulet and O. Straub, *Arch. Gewerbepathol. Gewerbehyg.* **10,** 451 (1941).

General References

J. W. Mellor, *A. Comprehensive Treatise on Inorganic and Theoretical Chemistry*, Vol. 10, Longmans, Green & Co., London, 1920, pp. 684–692.
Gmelins Handbuch der Anorganischen Chemie, System 9, Part A, Verlag Chemie GmbH, Weinhein/Bergstrasse, 1953, pp. 484–485.
R. Pointeau in Paul Pascal, ed., *Nouveau Traite de Chimie Minerale,* Vol. 13, part 2, Masson and co-workers, Paris, 1961, pp. 1557–1562.
Technical Bulletins published by E. I. du Pont de Nemours & Co., Inc., and Monsanto Company.

H. O. BURRUS
E. I. du Pont de Nemours & Co., Inc.